Metal Forming 2024

20th Metal Forming International Conference held in Kraków, Poland on September 15-18, 2024

Editors

Danuta Szeliga
Krzysztof Muszka

AGH University of Krakow, Poland

Peer review statement

All papers published in this volume of "Materials Research Proceedings" have been peer reviewed. The process of peer review was initiated and overseen by the above proceedings editors. All reviews were conducted by expert referees in accordance to Materials Research Forum LLC high standards.

Published under License by **Materials Research Forum LLC**
Millersville, PA 17551, USA

Published as part of the proceedings series
Materials Research Proceedings
Volume 44 (2024)

ISSN 2474-3941 (Print)
ISSN 2474-395X (Online)

ISBN 978-1-64490-324-7 (Print)
ISBN 978-1-64490-325-4 (eBook)

Distributed worldwide by

Materials Research Forum LLC
105 Springdale Lane
Millersville, PA 17551
USA
https://www.mrforum.com

Manufactured in the United State of America
10 9 8 7 6 5 4 3 2 1

Table of Contents

MANUFACTURING

MATERIALS

MODELLING

PROCESSES

Preface

This issue of MRP contains a collection of papers presented at the 20th Metal Forming International Conference held in Kraków, Poland on September 15-18, 2024. Metal Forming 2024 is the 20th in a series of International Conferences organized since 1974 by AGH University of Krakow, Poland. From 1994 to 2010 the Conference was organized biannually, jointly with the University of Birmingham, UK. The latter was replaced by the University of Toyohashi in 2010, when the Conference went, for the first time, to Japan. Metal Forming 2012 was again organized in Krakow, Poland by AGH University of Science and Technology, together with the University of Toyohashi and, for the first time, by the University of Palermo, Italy, which has joined the Organization team, and hosted the following Conference in 2014. Due to the COVID-19 pandemic, Metal Forming 2020 was organized online by AGH and Metal Forming 2022 was organized, for the first time, in a hybrid format by the China Society for Technology of Plasticity, CMES and Taiyuan University of Technology.

This time, during the 20th International Conference on Metal Forming, approximately 200 regular presentations and 4 specially invited keynote lectures were presented.

The papers in this volume showcase the latest advancements in metal forming science and technology. With over 100 contributions submitted by authors from universities, research institutes, and industry worldwide, 82 have been published following a rigorous review process. The contents cover a broad range of research topics, from incremental forming and metal forming of 3D printed materials to the numerical modeling of processes and systems. The volume also highlights the latest achievements in material characterization techniques for predicting microstructure evolution and mechanical properties during or after thermomechanical processing. A significant portion of the contributions focuses on constitutive and numerical modeling, as well as process and system optimization. The manuscripts encompass a diverse array of materials, including metal powders, lightweight systems, advanced high-strength steels, multiphase materials, and high-entropy alloys.

We extend our gratitude to the reviewers, especially the members of the Scientific and Steering Committees of the Conference, for their hard work and critical yet constructive feedback, which has been instrumental in maintaining the high scientific standards of the Conference. We hope these proceedings serve as a valuable source of information and inspiration for academics, researchers, engineers, and students.

Danuta Szeliga
Krzysztof Muszka
AGH University of Krakow, Poland

Committees

Steering Committee

Scientific Committee

Marion Merklein, University of Erlangen-Nuremberg, Germany
Wojciech Misiołek, Lehigh University, USA
Pierre Montmittonet, Cemef – MINES ParisTech, France
Sebastian Mróz, Czestochowa University of Technology, Poland
Masaaki Otsu, University of Fukui, Japan
Eric Palmiere, University of Sheffield, UK
Henryk Paul, Institute of Metallurgy and Materials Science, Polish Academy of Science, Poland
Elena Pereloma, University of Wollongong, Australia
Cecilia Poletti, Graz University of Technology, Austria
Ulrich Prahl, TU Bergakademie Freiberg, Germany
Łukasz Rauch, AGH University of Krakow, Poland
Konstantin Redkin, Whemco Inc., USA
Siegfreied Schmauder, University of Stuttgart, Germany
Christof Sommitsch, Graz University of Technology, Austria
A.H. Ton van Den Boogaard, University of Twente, Netherlands
Hiroshi Utsunomiya, Osaka University, Japan
Xinyun Wang, Huazhong University of Science and Technology, China
Bradley Wynne, University of Strathclyde, UK
Ming Yang, Tokyo Metropolitan University, Japan
Guoqun Zhao, Shandong University, China

Conference Chairs

Danuta Szeliga, AGH University of Krakow, Poland
Krzysztof Muszka, AGH University of Krakow, Poland
Livan Fratini, University of Palermo, Italy
Fabrizio Micari, University of Palermo, Italy
Yohei Abe, Toyohashi University of Technology, Japan
Tomoyoshi Maeno, Yokohama National University, Japan

Local Organizing Committee

Anna Smyk, AGH University of Krakow, Poland
Monika Stefańska-Kądziela, AGH University of Krakow, Poland
Krzysztof Bzowski, AGH University of Krakow, Poland

Sponsors

AGH University of Krakow, Poland

University of Palermo, Italy

Toyohashi University of Technology, Japan

Yokohama National University, Japan

DEFORMATION

Microstructure and hot deformation behavior of twin roll cast ZAX210 magnesium wire

KADEN Christoph[1,a *], GHORPADE Vishal[1,b], ULLMANN Madlen[1,c] and PRAHL Ulrich[1,d]

[1]Institute of Metal Forming, Technische Universität Bergakademie Freiberg, Bernhard-von-Cotta-Straße 4, 09599 Freiberg, Germany

[a]christoph.kaden@imf.tu-freiberg.de, [b]vishal.ghorpade@student.tu-freiberg.de, [c]madlen.ullmann@imf.tu-freiberg.de, [d]ulrich.prahl@imf.tu-freiberg.de

Keywords: Twin-Roll Casting, Mg-2Zn-1Al-0.3Ca, Flow Curve, Processing Map

Abstract. The hot deformation behavior of a Mg-2Zn-1Al-0.3Ca (ZAX210) wire, manufactured by twin roll casting was investigated by hot compression tests conducted at different temperatures (250 – 400°C) and strain rates (0.01 - 10 s^{-1}). The analyses of the microstructure and the texture are determined by an optical microscope and scanning electron microscope. A processing map is developed to identify optimal process parameters. The twin roll cast state shows the presence of $\{10\bar{1}2\}$ extension twins, $\{10\bar{1}1\}$ compression twins, double twins, and a weak prismatic texture. Furthermore, regions with small recrystallized grains as well as a small segregation line extend across the center of the wire. The heat treatment at 420°C for 2 h results in a grain structure, consisting of equiaxed grains with an average grain size of approximately 32 µm in the center and 17 µm at the edges of the wire. The heat treatment further reduces the segregation and weakens the texture by spreading the basal plane along the transverse direction. In contrast to the initial state, only $\{10\bar{1}2\}$ extension twins can be observed. The experimental results of the compression tests show that the flow stress increases with higher strain rates and decreasing temperatures. The course of the flow curves indicate that dynamic recrystallization occurs during hot deformation. The activation energy for plastic deformation of the twin roll cast and heat-treated wire is calculated to 150 kJ/mol. The process map identifies optimal process parameters, while deviations from these parameters result in crack formation, primarily at grain boundaries.

Introduction

As the lightest engineered metal, Magnesium and its alloys have garnered significant attention in the last decades. The low density, high strength-to-weight ratio and good castability make it an appealing choice across various industries, ranging from automotive [1] and aerospace [2] to biomedical applications [1,3]. However, the use of magnesium alloys is limited by their poor formability at room temperature due to inadequate slip systems. There are several methods to improve formability by achieve grain refinement and texture modification. One approach is to add rare earth elements as alloying elements [4,5]. A more cost-effective alternative is the use of calcium as an alloying element [6]. In addition to improving formability at room temperature, calcium improves corrosion resistance [7] and strength at elevated temperatures [8]. In the field of additive manufacturing, magnesium wires are particularly important as they are also used to weld components [9]. The conventional method of manufacturing magnesium wires involves casting followed by multiple extrusions with intermediate heat treatments [10]. However, this approach has drawbacks such as multiple process steps and length limitations. A cost and energy saving alternative is the combination of twin-roll casting and wire rolling [11]. By combining the casting and rolling processes, semi-finished products with near-net-shape geometry can be produced in a single step using the twin-roll casting process. The chosen process parameters for the subsequent wire rolling process are critical, making the study of the hot forming behavior of magnesium alloys of great interest. Kittner et al. [12] determined an activation energy of 180.5 kJ/mol for twin-roll cast strips of the alloy ZAX210. Temperatures above 370 °C and strain rates between 3 s^{-1} and 10 s^{-1} led to instabilities. For twin-roll cast wire, Arndt et al. [13] determined an activation energy of 159.0 kJ/mol and an instability range for temperatures between 250 °C and 375 °C and strain rates between 0.01 s^{-1} and 10 s^{-1} for the alloy AZ31. The present study focuses on the microstructural investigation and hot forming behavior of twin-roll cast and heat treated ZAX210 alloy wires. Based on the obtained stress-strain curves and microstructural observations, process maps were developed to derive optimum parameters for hot forming of the ZAX210 alloy.

Material and Methods

The investigated wire was produced at the twin-roll casting plant of the Institute for Metal Forming at the TU Bergakademie Freiberg [14]. In this process, magnesium ingots were melted in a furnace and fed through a heated nozzle at a temperature of 690°C into the rolling gap between two rotating, water-cooled rolls. Upon contact with the rolls, the melt begins to solidify from the outside inward and undergoes initial deformation. The chemical composition of the magnesium ingots used is shown in Table 1. The casting speed is 4.0 m/min, the oval cross-section of the twin-roll cast wire is 20 mm x 9.2 mm.

Table 1: Chemical composition of the ZAX210 alloy (wt.%).

Zn	Al	Ca	Mn	Fe	Ni	Cu	Mg
2.000	0.741	>0.180	0.037	0.006	0.003	0.001	Bal.

Heat treatment was carried out according to previous studies at 420°C for 2 hours [12]. The samples were placed in a preheated furnace and quenched in water (room temperature) after the holding time. Compression test specimens with a diameter of 5 mm and a height of 10 mm were then machined from the heat-treated specimens. Compression tests were performed using a BÄHR DIL 805 D quenching and deformation dilatometer at temperatures of 250, 300, 350, and 400°C and strain rates of 0.01, 0.1, 1, and 10 s^{-1} up to an equivalent logarithmic strain of 1. First, the samples were heated to the deformation temperature at a heating rate of 5 K/s. To ensure a uniform temperature throughout the sample volume, the temperature was held for 5 minutes. To preserve the deformed microstructure, the samples were quenched at a cooling rate of 100 K/s after forming. After data acquisition, stress-strain curves were determined and corrected for temperature and friction. For microstructural studies, the samples were sectioned, ground with silicon carbide paper, polished with OPS, and finally etched with a solution of ethanol, acetic acid, picric acid, and distilled water. The microstructure of the longitudinal section of the sample was examined using a Keyence VHX-6000 optical microscope and a ZEISS GeminiSEM 450 scanning electron microscope. Texture measurements were performed using an electron backscatter diffraction (EBSD) detector.

The developing of the processing map is based on the dynamic materials model (DMM) [15]. The processing map is a superimposition of a power dissipation and an instability map and represents the response of a material to the given process parameters. During a forming process, the total power P dissipates in the generation of heat (G content) and in microstructural changes (J co-content). The relationship can be expressed as follows:

$$P = G + J = \int_0^{\dot{\varepsilon}} \sigma d\dot{\varepsilon} + \int_0^{\sigma} \dot{\varepsilon} d\sigma \ . \quad (1)$$

The efficiency of power dissipation (η) represents the power dissipation used for microstructural changes and is described as follows:

$$\eta = \frac{2 \cdot m}{m + 1} \ , \tag{2}$$

where m is the strain rate sensitivity. To obtain the power dissipation map, the efficiency parameter is plotted as a function of temperature and strain rate. The instability map can be obtained with the instability parameter ($\xi(\dot{\varepsilon})$), given by:

$$\xi(\dot{\varepsilon}) = \frac{\partial \ln\left(\frac{m}{m+1}\right)}{\partial \ln \dot{\varepsilon}} + m \leq 0 \ . \tag{3}$$

When the value of $\xi(\dot{\varepsilon})$ is negative, metallurgical instability during plastic flow occurs.

Results and Discussion

Microstructure of Twin-Roll Cast and Heat-Treated Alloy. The initial state for the subsequent investigations is the twin-roll cast and heat-treated condition. Fig. 1a shows the corresponding microstructure. The already minimal centerline segregation after twin-roll casting was completely reduced by the heat treatment. In addition, globular grains have formed. At the upper edge, the average grain size is 17 µm (19 µm at the lower edge). In the center, there is a larger grain size distribution with an average grain size of 32 µm (Fig. 2). The chemical composition of individual regions can be determined by EDX measurements (Fig. 1b, c). After heat treatment, the precipitates are relatively finely distributed within the Mg solid solution, especially at the edges. The precipitates contain high amounts of Zn and almost equal amounts of Al and Ca. According to Katsarou et al [16], these are most likely the $Mg_6Ca_2Zn_3$ phase. Furthermore, twins can be observed in the microstructure. The misorientation distribution (Figure 1d) shows a distinct peak between 85 and 90 degrees, caused by $\{10\bar{1}2\}$ extension twins. There is no obvious preferred orientation from the pole figures. The intensity maxima are distributed along the transverse direction and are weakly pronounced (Fig. 3).

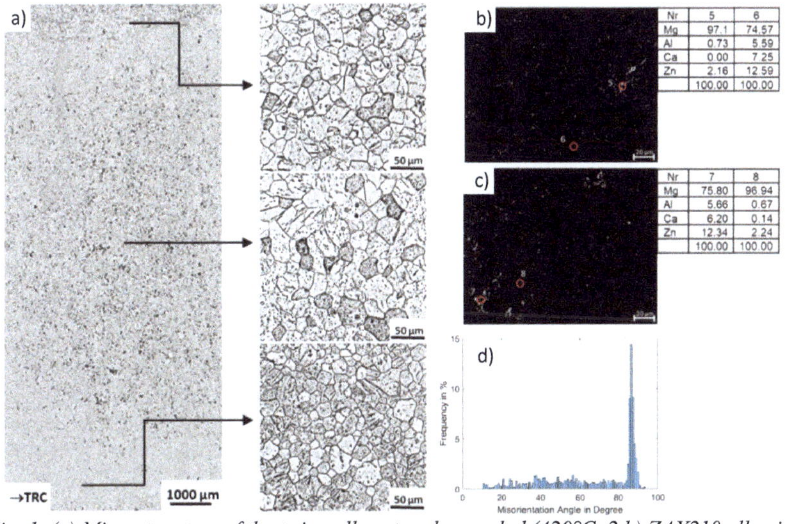

Nr	5	6
Mg	97.1	74.57
Al	0.73	5.59
Ca	0.00	7.25
Zn	2.16	12.59
	100.00	100.00

Nr	7	8
Mg	75.80	96.94
Al	5.66	0.67
Ca	6.20	0.14
Zn	12.34	2.24
	100.00	100.00

Fig. 1. (a) Microstructure of the twin-roll cast and annealed (420°C, 2 h) ZAX210 alloy in longitudinal view, EDS measurement of the (b) edge and (c) middle of the wire and (d) misorientation angle distribution.

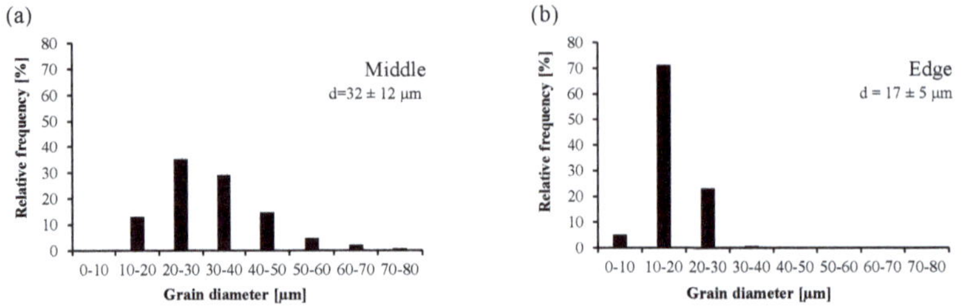

Fig. 2. Grain size distribution of the twin-roll cast and heat-treated (420°C, 2 h) ZAX210 alloy, (a) wire center and (b) upper edge area.

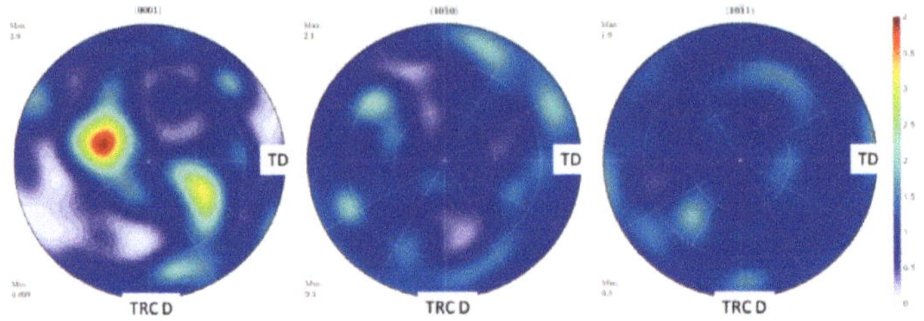

Fig. 3. (0001)-basal, (10$\bar{1}$0)-prismatic and (10$\bar{1}$1)-pyramidal pole figures of the twin-roll cast and annealed (420°C, 2 h) ZAX210 alloy of wire center calculated from electron backscatter diffraction (EBSD).

Flow Behavior. Fig. 4 shows the stress-strain curves of the twin-roll cast and heat-treated ZAX210 alloy at temperatures ranging from 250°C to 400°C and strain rates of 0.01 s^{-1}, 0.1 s^{-1}, 1 s^{-1}, and 10 s^{-1}. The curves demonstrate that the flow behavior, i.e., strain hardening and softening, is influenced by both temperature and strain rate. At the onset of deformation, strain hardening takes place, which is marked by an increase in dislocation density. This leads to a rapid rise in the stress-strain curve at low degrees of deformation. The flow stress increases as the strain rate increases and the temperature decreases, indicating strain hardening. Furthermore, the peak flow stress increases at higher equivalent strain values due to the greater dislocation multiplication at higher strain rates. This leads to dislocations obstructing each other, resulting in an increase in flow stress. Additionally, there is insufficient time for diffusion-controlled softening processes to occur. The flow stress decreases with increasing temperature or decreasing strain rate due to slower dislocation multiplication, longer duration for dynamic recrystallization, and enhanced diffusion at higher temperatures. Furthermore, the critical resolved shear stress (CRSS) for non-basal slip systems decreases with increasing temperature. At high temperatures and low strain rates (400°C, 0.01 s^{-1}), there is only a slight increase in the stress-strain curve, resulting in minimal strain hardening. Once the maximum flow stress is reached, the curve decreases due to softening mechanisms such as dynamic recrystallization. However, a significant decrease in flow stress is only observed at 350°C and 400°C and 10 s^{-1}. At other deformation parameters, the stress-strain curve remains nearly constant. Additionally, the effect of strain rate on flow stress appears to be dependent on temperature. At 250°C, the strain rate (excluding 0.01 s^{-1}) has minimal impact on flow stress, but its influence becomes more significant as the temperature increases. This observation aligns with Rao's findings for the calcium-containing alloy ZX11 [17].

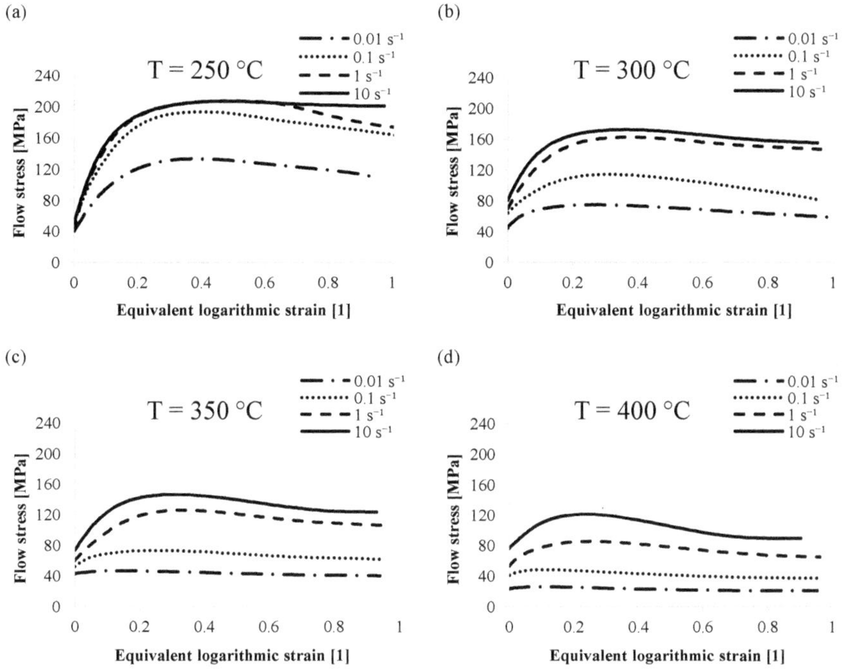

Fig. 4. Flow curves of the twin-roll cast and heat-treated ZAX210 alloy at different forming speeds and temperatures: (a) 250°C, (b) 300°C, (c) 350°C and (d) 400°C.

Analysis of Hot Deformation Behavior of Twin-Roll Cast and Heat-Treated ZAX210 Alloy. As shown in the stress-strain curves, the flow behavior depends on temperature, strain rate, and degree of deformation. The relationships between flow stress and deformation parameters can be described using the Arrhenius equations, developed by Sellars and Tegart [18]. The relationship, valid for all stress conditions, is expressed as:

$$\dot{\varepsilon}=A\,[\sinh(\alpha\sigma)]^{n}\exp\left(-\frac{Q}{RT}\right)\ , \tag{4}$$

Where $\dot{\varepsilon}$ is the strain rate (s^{-1}), A is a material constant, n is the stress exponent, α is a fitting parameter (MPa^{-1}), Q represents the activation energy for plastic flow (kJ/mol), σ is the peak flow stress (MPa), R is the gas constant (8.314 J/(mol·K)), and T signifies the deformation temperature (K). Utilizing the mean values obtained from the slopes of the graphs depicting the linear dependency among flow stress, strain rate, and temperature, the parameters can be determined. Accordingly, the following values are obtained: $A = 6.62585 \times 10^{12}\ s^{-1}$, n = 7.335, $\alpha = 0.0092\ MPa^{-1}$, and Q = 149.68 kJ/mol. These determined parameters are valid for temperatures ranging from 250°C to 400°C and strain rates from 0.01 s^{-1} to 10 s^{-1}. The activation energy Q serves as an indicator of the difficulty in initiating deformations during hot forming. The calculated value exceeds the lattice diffusion of magnesium (135 kJ/mol). The high value of Q can be attributed to the fine microstructure and the presence of precipitates. Small grains lead to more grain boundaries, which hinder the movement of dislocations. Likewise, precipitates lead to an obstruction of dislocation movement, making self-diffusion more difficult [12,19].

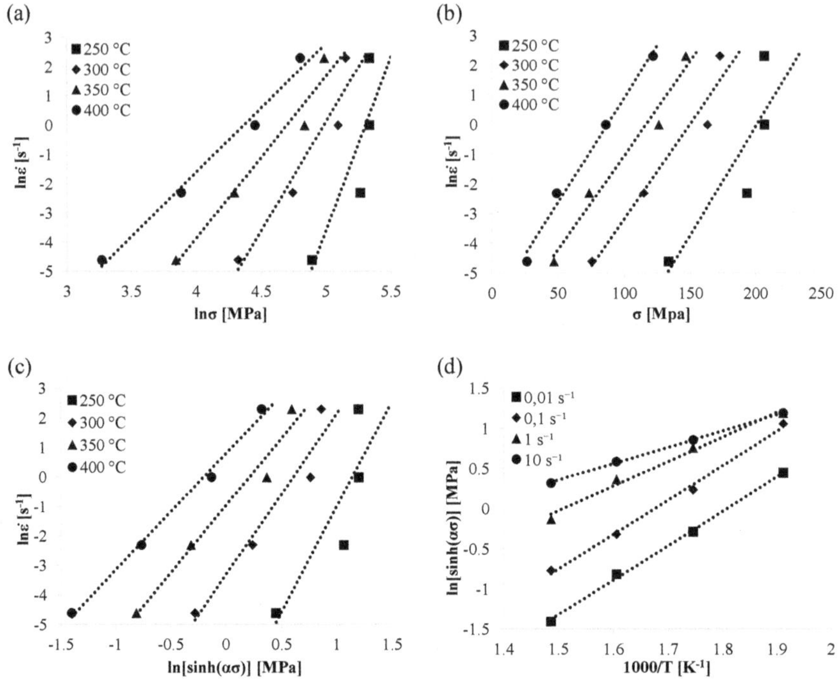

Fig. 5. Relationship between ln(ε̇) and (a) lnσ and (b) σ as a function of the forming temperature as well as (c) ln(ε̇) and ln[sinh(ασ)] at different forming temperatures and (d) ln[sinh(ασ)]$^{-1}$/T as a function of the strain rate.

To describe the the influence of temperature and strain rate on the material flow behavior, the Zener-Hollomon parameter Z is employed. Taking Eq. 4 into account, the following formula is derived for the calculation of Z:

$$Z=\dot{\varepsilon}\exp\left(\frac{Q}{RT}\right)=A\,[\sinh(\alpha\sigma)]^{n}. \tag{5}$$

The linear relationship between ln Z and ln[sinh(ασ)] is depicted in Fig. 6. The linear regression was determined with an accuracy of (R^2) = 0.94. According to Eq. 5, the stress exponent n can be calculated to 5.01. This value can be utilized to estimate the dominant mechanism of plastic flow. Dislocation climb creep is cited in the literature as the deformation mechanism for a stress exponent ranging between 4 and 6 [20].

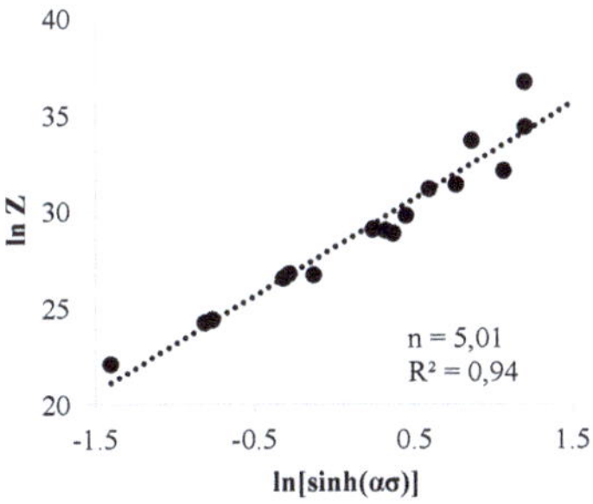

Fig. 6. Relationship between ln Z and ln[sinh(ασ)].

Processing Map. The process map for the twin-roll cast and heat-treated ZAX210 wire is shown in Fig. 7a. It is valid for an equivalent logarithmic strain of 0.4, as it reflects a typical rolling pass. The numbers along the contour lines correspond to the power dissipation efficiency η. The gray shaded area indicates the instability range. This range is observed at temperatures between 250°C and 350°C, with strain rates ranging from 0.14 s^{-1} to 1.8 s^{-1}. For temperatures above 350°C, the range shifts to strain rates of 1 s^{-1} to 10 s^{-1}. Under these conditions, crack formation occurs, preferably along grain boundaries (Fig. 7b). In contrast, Fig. 7c displays a crack-free microstructure with dynamic recrystallization at grain boundaries and twin boundaries. Under these parameters, hot forming is feasible without issues. The instability at 400°C and high strain rates also aligns with the findings of Kittner et al. [12] for strip material.

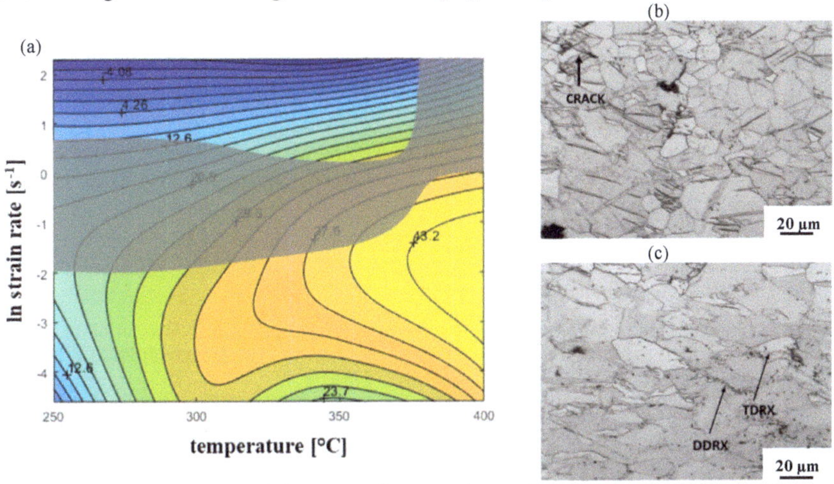

Fig. 7. (a) Processing Map of the twin-roll cast and annealed (420°C, 2 h) ZAX210 alloy at an equivalent strain of 0.4, (b) microstructure within the instability domain (400°C; 10 s^{-1}) and (c) resulting microstructure in stable domain (350°C; 0.1 s^{-1}).

Conclusions

In this study, to hot forming behavior of a twin-roll cast and heat-treated wire of the ZAX210 alloy was investigated. The following conclusions can be drawn:

(1) The determined stress-strain curves exhibit a typical behavior, with the flow stress

increasing with increasing strain rate or decreasing temperature.

(2) Using the Arrhenius equation, an activation energy for plastic deformation of 150 kJ/mol was determined.

(3) The stress exponent of 5.01 suggests dislocation climb creep as the dominant deformation mechanism.

(4) The process map for an equivalent logarithmic strain of 0.4 shows an instability range from 0.14 s^{-1} to 1.8 s^{-1} for temperatures ranging from 250 °C to 350 °C. Above 350 °C, the range shifts from 1 s^{-1} to 10 s^{-1}. Particularly at high temperatures and high strain rates, cracks propagate along grain boundaries.

Funding

The authors would like to thank the Federal Ministry for Economic Affairs and Climate Action for supporting this research work through the project "CLEAN-Mag: CO2-neutral production of lightweight magnesium components", project no. 03LB3080A, part of the Technologietransfer-Programm Leichtbau (TTP LB).

References

[1] B. Liu, J. Yang, X. Zhang, Q. Yang, J. Zhang, X. Li, Development and application of magnesium alloy parts for automotive OEMs: A review, J. Magnes. Alloy. 11 (2023) 15–47. https://doi.org/10.1016/j.jma.2022.12.015

[2] J. Bai, Y. Yang, C. Wen, J. Chen, G. Zhou, B. Jiang, X. Peng, F. Pan, Applications of magnesium alloys for aerospace: A review, J. Magnes. Alloy. 11 (2023) 3609–3619. https://doi.org/10.1016/j.jma.2023.09.015

[3] T. Zhang, W. Wang, J. Liu, L. Wang, Y. Tang, A review on magnesium alloys for biomedical applications, Frontiers in Bioengineering and Biotechnology (2022). https://doi.org/10.3389/fbioe.2022.953344

[4] L. M. Calado, M. J. Carmezim, F. Montemor, Rare Earth Based Magnesium Alloys - A Review on WE Series, Frontiers in Materials (2022). https://doi.org/10.3389/fmats.2021.804906

[5] H. Cai, Z. Wang, L. Liu, J. Su, Y. Li, F. Guo, Grain Refinement Mechanism of Rare Earth Elements (Ce, Y and Gd) on AZ91 Magnesium Alloy at Different Cooling Rates, Inter Metalcast (2023). https://doi.org/10.1007/s40962-023-01204-5

[6] Y. Li, B. Liu, J. Wang, Y. Sun, H. Li, Effect of Rare Earth and Ca Micro-Alloying on the Microstructure and Anisotropy of AZ31 Magnesium Alloy Sheets, Metals 12 (2022) 1724. https://doi.org/10.3390/met12101724

[7] D. Wan, Y. Wang, S. Dong, Y. Xue, G. Han, F. Yang, H. Tang, J. Kang, G. Zeng, J. Xu, Improving corrosion resistance of high strength Mg-Zn-Y alloy through Ca addition, Corrosion Engineering, Science and Technology 57 (2022) 789–795. https://doi.org/10.1080/1478422X.2022.2127637

[8] C. Hou, F. Qi, L. Wang, L. Lu, N. Zhao, J. She, Y. Zhou, X. Ouyang, Effects of Ca addition on the microstructure and mechanical properties of Mg–Zn–Sn–Mn alloys, J. Mater. Res. Technol. 25 (2023) 3884–3900. https://doi.org/10.1016/j.jmrt.2023.06.194

[9] S.K. Manjhi, P. Sekar, S. Bontha, A. Balan, Additive manufacturing of magnesium alloys: Characterization and post-processing, Int. J. Lightweight Mater. Manuf. 7 (2024) 184–213. https://doi.org/10.1016/j.ijlmm.2023.06.004

[10] M. Nienaber, M. Braatz, N. Ben Khalifa, J. Bohlen, Property profile development during wire extrusion and wire drawing of magnesium alloys AZ31 and ZX10, Materials & Design 224 (2022) 111355. https://doi.org/10.1016/j.matdes.2022.111355

[11] M. Moses, M. Ullmann, R. Kawalla, U. Prahl, Improving Mechanical Properties of Twin-Roll Cast AZ31 by Wire Rolling, MSF 1016 (2021) 957–963. https://doi.org/10.4028/www.scientific.net/MSF.1016.957

[12] K. Kittner, M. Ullmann, T. Henseler, R. Kawalla, U. Prahl, Microstructure and Hot Deformation Behavior of Twin Roll Cast Mg-2Zn-1Al-0.3Ca Alloy, Materials (Basel, Switzerland) 12 (2019). https://doi.org/10.3390/ma12071020

[13] F. Arndt, S. Berndorf, M. Moses, M. Ullmann, U. Prahl, Microstructure and Hot Deformation Behaviour of Twin-Roll Cast AZ31 Magnesium Wire, Crystals 12 (2022) 173. https://doi.org/10.3390/cryst12020173

[14] Marie Moses, Twin Roll Casting of Magnesium Wire: An Innovative Continuous Production Route, Frontiers in Bioengineering and Biotechnology. https://doi.org/10.37904/metal.2019.812

[15] Y.V.R.K. Prasad, K.P. Rao, S. Sasidhara, Hot Working Guide. A Compendium of Processing Maps, ASM International (2015).

[16] L. Katsarou, K. Suresh, K.P. Rao, N. Hort, C. Blawert, C.L. Mendis, H. Dieringa, Microstructure and Properties of Magnesium Alloy Mg-1Zn-1Ca (ZX11), Magnesium Technol. (2015). https://doi.org/10.1007/978-3-319-48185-2_78

[17] K. Rao, K. Suresh, Y. Prasad, C. Dharmendra, N. Hort, H. Dieringa, High Temperature Strength and Hot Working Technology for As-Cast Mg–1Zn–1Ca (ZX11) Alloy, Metals 7 (2017) 405. https://doi.org/10.3390/met7100405

[18] C.M. Sellars, W.J. McTegart, On the mechanism of hot deformation, Acta Metall. 14 (1966) 1136–1138. https://doi.org/10.1016/0001-6160(66)90207-0

[19] J. Qi, Y. Du, B. Jiang, M. Shen, Hot deformation behavior and microstructural evolution of Mg–Zn–Ca–La alloys, J. Mater. Res. 33 (2018) 2817–2826. https://doi.org/10.1557/jmr.2018.265

[20] C. Zhaoyun, T. Rui, D. Zichao, Plastic Flow Characteristics of an Extruded Mg-Li-Zn-RE Alloy, Rare Metal Mater. Eng. 42 (2013) 1779–1784. https://doi.org/10.1016/S1875-5372(14)60004-7

Materials Research Proceedings 44 (2024) 11-22 https://doi.org/10.21741/9781644903254-2

Effect of ribs on torque in bending of rebars

HIGAKI Satoshi[1,a *], NAITO Haruto[2,b], GO Tomoki[2,c], SASADA Masahiro[2,d] and TANAKA Tatsuya[2,e]

[1]Toyo Kensetsu Kohki CO., LTD. 4-15, 2-chome, Sangenya Higashi, Taisho-ku, Osaka-city, Osaka, Japan

[2]Faculty of Science and Engineering, Department of Mechanical Engineering, Doshisha University, Miyakodani1-3, Tatara, Kyotanabe-City, Kyoto, Japan

[a]s_higaki@toyokensetsukohki.co.jp, [b]cguf4065@mail4.doshisha.ac.jp, [c]ctwh0505@mail4.doshisha.ac.jp, [d]msasada@mail.doshisha.ac.jp, [e]tatanaka@mail.doshisha.ac.jp

Keywords: Bending, Rebar, Torque, Finite Element Method, Deformation

Abstract. Deformed steel bars (Rebar) are bent into various shapes as construction materials and used with concrete. Bending machines are used for processing, and torque is required to bend rebars. Understanding the torque required for rebars with variable cross-sectional shape is crucial. The purpose of this study is to clarify the relationship between torque and material deformation. Rebars and round steel bars were bent using a simple bending machine. A high-speed camera was used to observe the material deformation during the bending process. Elastic-plastic finite element analysis was also conducted to evaluate the relationship between the material deformation and torque of the rebars during bending. The following conclusions were obtained. When the circular roller contacts the round steel bar and starts bending the bar, the torque increases rapidly. Subsequently, the torque does not change easily and remains constant. On the other hand, the torque of the rebar also increases rapidly when the circular roller comes into contact. Thereafter, the torque repeats increasing and decreasing. It was confirmed that the torque required to bend a rebar exceeds that required to bend a round steel bar with an outside diameter similar to the minimum outside diameter of the rebar. The fluctuation in torque during rebar bending occurs due to the bending of the smaller outside diameter section of the rebar, which is further away from the fulcrum roller toward the circular roller. This contrasts with the bending process of round steel bars, where bending primarily occurs in the area in contact with the fulcrum roller. As the bending of a rebar progress, part of the rebar may move away from the fulcrum roller. When a section located on the rebar receiver side from the fulcrum roller moves away from the fulcrum roller, the torque required to progress the rotation angle of the circular roller was reduced. The torque required for bending rebars is found to be between the torque required for bending round steel bars of the smallest and largest diameters of rebars.

Introduction

Rebars are steel bars characterized by their surface ribs and nodes, commonly used as construction materials. To shape rebars, a bending machine is used. The bending machine used in this study consists of a fulcrum roller, a circular roller, and a rebar receiver. The rebar is bent by the rotation of the circular roller around the fulcrum roller. The circular roller pushes the rebar and bends the rebar along the fulcrum roller. It is important to clarify the torque required to rotate the circular roller to design and develop such a bending machine. Although there is no problem in designing and manufacturing bending machines based on past experience and knowledge, it is useful to determine the change in torque during the bending process in order to further advance the machine in the future.

Higaki, S [1,2] reported on the shape of deformed rebars that have undergone bending and on the displacement of deformed rebars during the bending process. In addition, there are reports [3,4] on research related to the bending of round steel bars, such as flattening control of circular rings and shape bending of steel, which are applied to circular tubes. There are reports [5] on bending moments of square and circular cross-sections, such as bending yielding of steel. For rebars, there is a study on the performance evaluation of rebars at cryogenic temperatures [6], a study on bending and tempering [7], and a study on stress distribution in the bending section of rebars [8].

The purpose of this study is to clarify the relationship between material deformation and torque variation in rebars with axially varying cross-sectional shapes. Material deformation of round steel bars and rebars was experimentally observed and elastoplastic finite element analysis was used to determine the material deformation, based on the results of the elasto-plastic analysis, the following is reported. The experimental results demonstrate a rapid increase in rebar torque upon contact with the circular roller. Subsequently, the torque fluctuates, increasing and decreasing repeatedly, without a further significant increase. The minimum value of the increase or decrease in torque of the rebar was close to the torque required to bend a round steel bar with the smallest outside diameter of the rebar. The increase from the minimum value to the maximum value in the increase and decrease of the torque of the rebar is considered to be related to the rib of the rebar making contact with the fulcrum roller. Therefore, bending of the smaller outside diameter section of the rebar, which is slightly further away from the fulcrum roller toward the circular roller, was occurred.

Finite Element Analysis conditions

Fig. 1 illustrates a schematic diagram of the bending machine used. First, the material is placed so that it contacts the fulcrum roller and rebar receiver. Subsequently, the circular roller is rotated with the center of the fulcrum roller as the rotation center to the position of θ_F=180°. During this time, the rebar is bent along the fulcrum roller. Elastic-plastic finite element analysis was performed using the analysis software Simufact Forming2022. Fig. 2 shows the analytical model. The analysis was performed for the case where the circular roller was rotated around the fulcrum roller up to θ_F=180°. Fig. 3 illustrates an enlarged view of the rebar model. A hexahedral element with a side length of 0.5 mm was used. The material was assumed to be elastoplastic. The mechanical properties of the materials considered in the analysis are listed in Table 1. The stress-strain relationship after yielding was considered as illustrated in Fig. 4, based on results obtained from tensile tests. The circular roller, fulcrum roller, and rebar receiver were assumed to be rigid bodies. Coulomb friction was considered in the contact between the rigid and elastoplastic bodies.

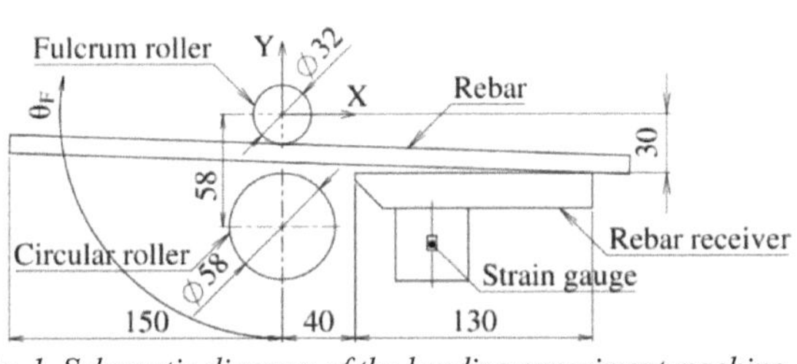

Fig. 1. Schematic diagram of the bending experiment machine.

The circular roller of the bending machine has a built-in bearing and can rotate on its own axis. FEM was also performed on a round steel bar made of the same material as rebar. The diameters of round steel bars are φ10.1 mm and φ8.8 mm. In the FEM analysis, the circular roller was set so that it could not rotate, and the coefficient of friction between the circular roller and the material was set to 0.01. In the experiment, it was observed that the nodes of the rebars were crushed and a new surface appeared. Therefore, the coefficient of friction was set to 0.3 for the contact between the fulcrum roller and the rebar. On the other hand, the coefficient of friction between the round steel bar and the fulcrum roller was set to 0.15. The coefficient of friction between the rebar receiver and the material was also set to 0.15. The FEM analysis was performed with an increment of θ_F= 2.37°.

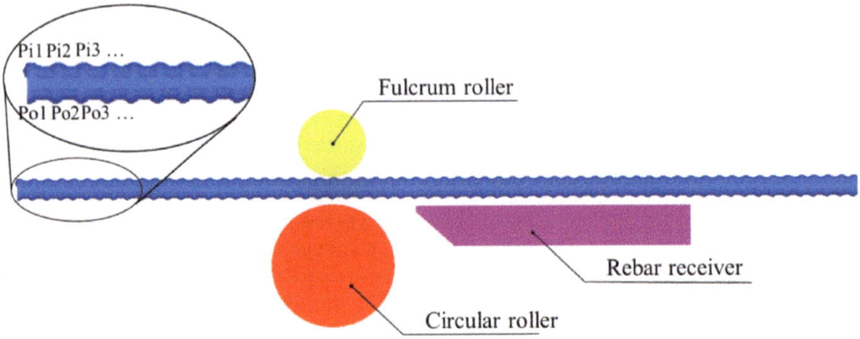

Fig. 2. Overall view of the analysis model.

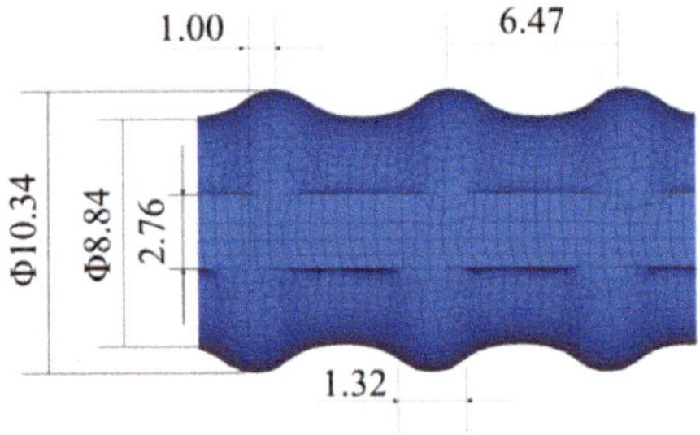

Fig. 3. Part of an analysis model of rebar.

Table 1. Mechanical properties.

Yield stress,σ(MPa)	435.9
Young's modulus,E(GPa)	205
Poisson's ratio,ν	0.3
Friction coefficient,μ	0.01,0.15,0.3

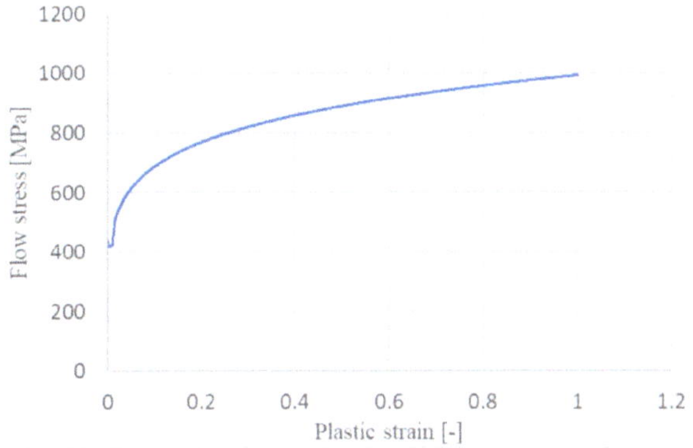

Fig. 4. True stress-plastic strain diagram.

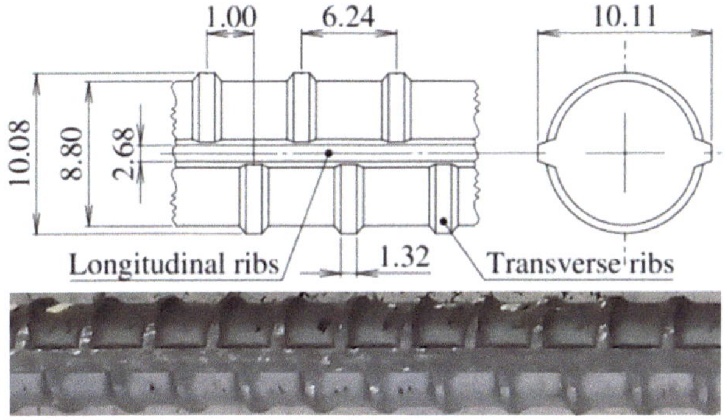

(a) Schematic of rebar (b) Photograph of rebar
Fig. 5. Detail of rebar.

Experimental materials and conditions

Experiments were conducted to compare the results with the analytical results. Fig. 5 illustrates a schematic diagram of a rebar. The rebar is nominal D10, material SD345 (JIS G 3112), and 400 mm long. For comparison, φ8.8 and φ10.1 round steel bars machined from rebars of nominal size D13 and material SD345 were also prepared for testing and measurement. Note that the geometry of the round steel bars is the same as in the elasto-plastic finite element analysis. However, the geometry of the rebars differs from that of the elasto-plastic finite element analysis model in several areas. The maximum outside diameter of the rebar is φ10.1 mm, which is slightly different from the FEM model of rebar. The minimum outside diameter is φ8.8 mm, which is the same value as the FEM model of rebar. Therefore, the experimental results for the rebars were treated as a reference when considering the results of the elasto-plastic finite element analysis. The rotation speed was 9.8 rpm. During the bending process, a high-speed camera was placed at a distance of 290 mm vertically upward from the surface where the material was placed to capture video. The resolution was 2592 × 2048 pixels and the frame rate was 60 fps. Simultaneously, the torque was measured by a torque meter installed on the motor shaft of the bending machine.

Materials Research Proceedings 44 (2024) 11-22 https://doi.org/10.21741/9781644903254-2

Results and discussion

Material deformation during processing. Fig. 6 shows the deformed shape of a round steel bar. Fig. 7 shows the FEM results about the deformed shape of the round steel bar. At a rotation angle of θ_F=60°, the rebar between the fulcrum roller and rebar receiver is deflected. In the FEM result and experimental results at θ_F=180°, a gap was observed between the fulcrum roller and the round bar. Based on the moving image of material deformation taken during the experiment, tracking points were set on the round steel bar and trajectories were obtained using motion analysis software. Fig.8 shows the trajectory. In section P_{i20}, it can be confirmed that the displacement is in the direction away from the fulcrum roller.Thus, the analytical and experimental results are in qualitatively good agreement.

Fig. 9 shows the deformation of a rebar. Fig. 10 shows the deformation of the rebar obtained by FEM. At θ_F=60°, the rebar between the fulcrum roller and the rebar receiver is deformed. At θ_F=180°, there is a gap between the fulcrum roller and the rebar. The deformation shape at θ_F=180° shows that the rebar, which has already been bent is separated from the fulcrum toller. As the rotation angle increases from θ_F=60° to θ_F=180°, the portion bent along the fulcrum roller becomes longer. And a slight axial displacement of the rebar occurs.

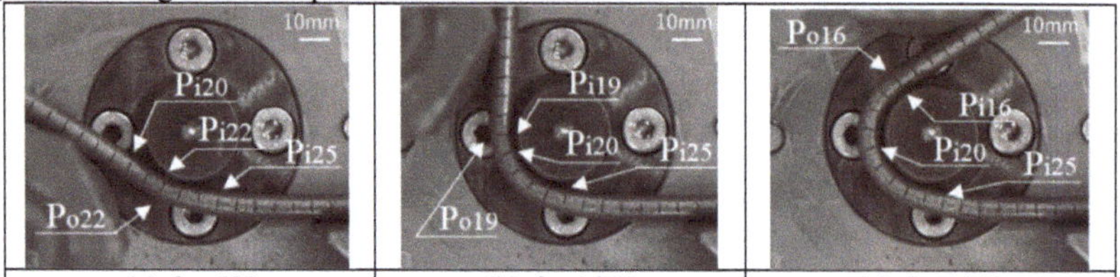

(a) θ_F=60°	(b) θ_F=120°	(c) θ_F=180°

Fig. 6. Deformed shape for round steel bars with an outer diameter of 8.8 mm by EXP.

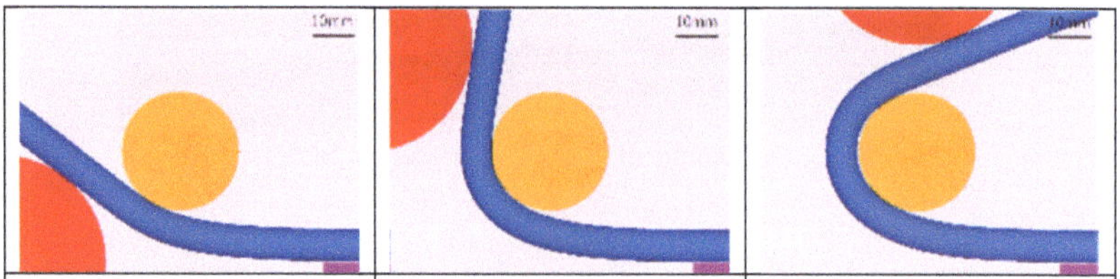

(a) θ_F=60°	(b) θ_F=120°	(c) θ_F=180°

Fig. 7. Deformed shape for round steel bars with an outer diameter of 8.8 mm by FEM.

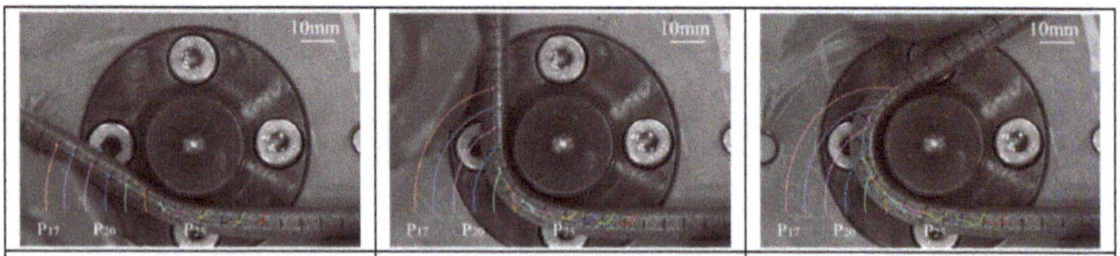

(a) θ_F=60°	(b) θ_F=120°	(c) θ_F=180°

Fig. 8. Tracking point trajectory for round steel bars with an outer diameter of 8.8 mm.

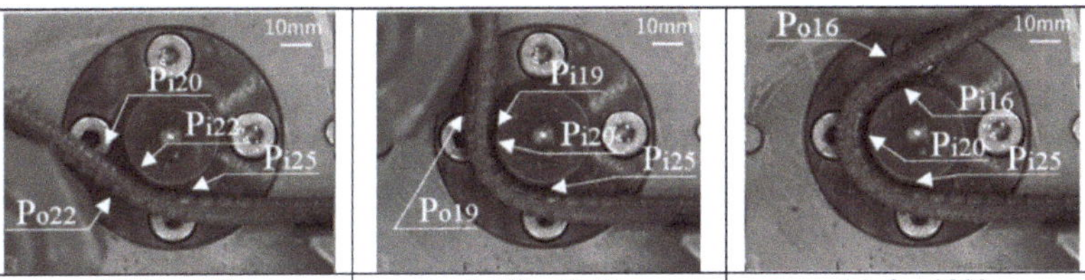

(a) θ_F=60°	(b) θ_F=120°	(c) θ_F=180°

Fig. 9. Deformed shape for rebars by EXP.

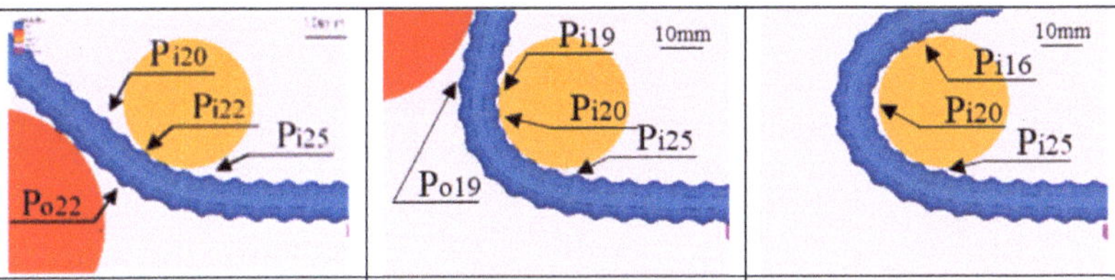

(a) θ_F=60°	(b) θ_F=120°	(c) θ_F=180°

Fig. 10. Deformed shape for rebars by FEM.

Relationship between torque and rotation angle θ_F. Fig. 11 shows the change in torque measured in the experiment. Torque for rebars and round steel bars is constant after increasing. It can be seen that the processing of rebars progresses with repeated increases and decreases in torque. The magnitude of torque associated with bending rebars is less than that required to bend φ10.1 round steel bars. The minimum torque for rebars is close to that obtained by bending round steel bars with an outer diameter of φ8.8. The position of the figure center position of the cross-section of the rebar and that of the φ8.8 round steel bar are different, and since these results were obtained when the same diameter fulcrum roller was used for bending, further consideration of the handling of the figure center position is considered necessary.

Fig. 12 shows the torque variation obtained from the elastoplastic finite element analysis. The experimental and analytical results for round steel bar are in good agreement. The experimental and analytical results for rebar are also in good agreement. Therefore, we proceeded to consider torque in bending round steel bars and rebars using the results of elastoplastic finite element analysis. The torque for bending rebar can be considered with reference to the torque when bending a round steel bar whose outer diameter is the same as the minimum outer diameter of the rebar.

Increase and decrease in torque change for rebars. The material deformation was confirmed to clarify the relationship between the fluctuated torque and material deformation during the bending of rebar. In Fig. 12, one of the smallest values of torque that repeatedly increases and decreases is at θ_F=110.6°. Fig. 13 shows the results obtained by FEM at θ_F=110.6°. The equivalent strain, the plastic strain increment of the equivalent strain, and the region in contact with the rigid body are shown. The section numbers shown in Fig. 2 should be included in these figures to indicate the movement of sections. Rib P_{i20} is in contact with the fulcrum roller. As illustrated in Fig. 13(c), and the bending process is considered to be performed with the rib P_{i20} as the fulcrum. As can be seen from Fig. 13(b), the plastic strain increment between ribs P_{i19} and P_{i21} is large, suggesting that the bending process is progressing in this region. To compare rebars and round steel bars, Fig. 14 shows the equivalent strain, the plastic strain increment of equivalent strain, and the contact state for bending of φ8.8 round steel bar. In the case of a round steel bar, the round bar contacts the fulcrum roller over a wide area. Although not exactly the same due to the effects of the ribs, a similar trend extends over the region of greater strain for round steel bars and rebars. In other words, at θ_F=110.6°, the rebar is considered to be bent along the fulcrum roller in the region where the outer diameter of the bar is small; this can be inferred from the fact that torque values similar to those of φ8.8 round steel bars were observed for rebars.

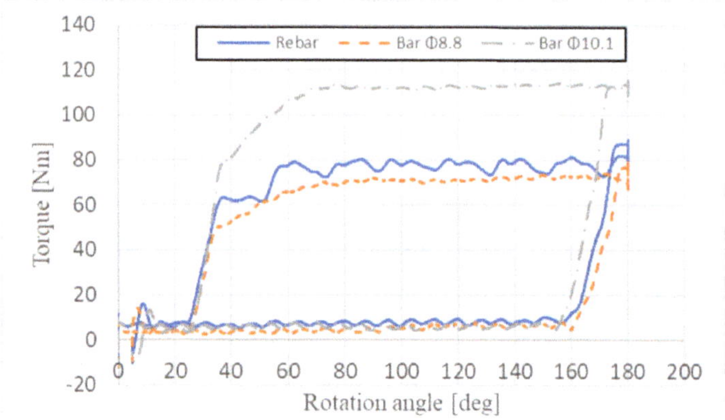

Fig. 11. Torque comparison between rebar and round steel bar in EXP.

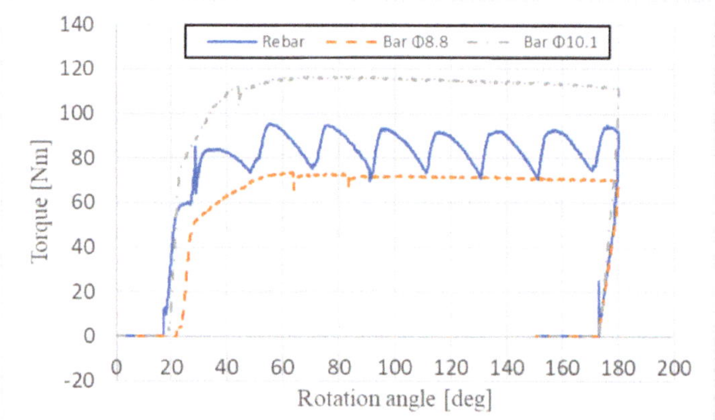

Fig. 12. Torque comparison between rebar and round steel bar in FEM.

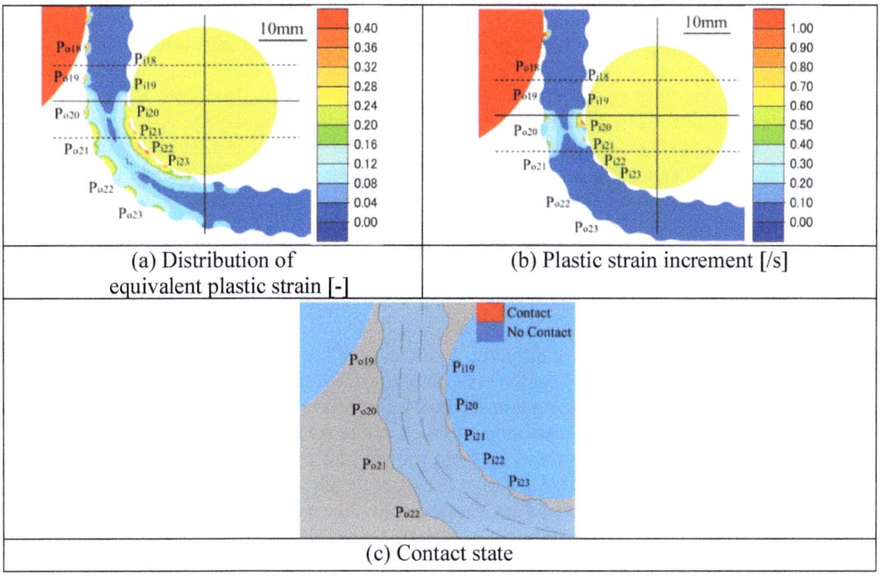

(a) Distribution of equivalent plastic strain [-]

(b) Plastic strain increment [/s]

(c) Contact state

Fig. 13. Analytical results for rebar at θ_F=110.6°.

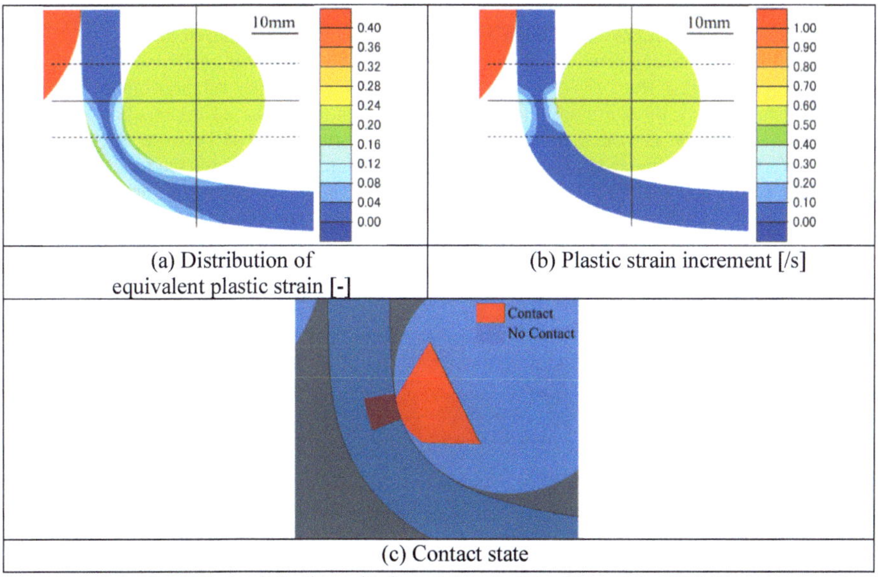

(a) Distribution of equivalent plastic strain [-]

(b) Plastic strain increment [/s]

(c) Contact state

Fig. 14. Analytical results for φ8.8 round steel bar at θ_F=110.6°.

 Materials Research Forum LLC
https://doi.org/10.21741/9781644903254-2

Consider the case where the circular roller is rotated from θ_F when the torque is minimum value, Fig. 15 shows the equivalent strain distribution at θ_F=112.9° when the torque changes from minimum to maximum, at θ_F=117.6° when the torque reaches its maximum value, and then at θ_F=131.7° when the torque reaches its minimum value again. Fig. 16 shows the plastic strain increment at these times. Fig.17 shows the area in contact with the rigid body. As can be seen from Fig. 16(a), for θ_F=112.9°, the plastic strain increment at rib P_{i19} is higher. Furthermore, as illustrated in Fig. 17(a), rib P_{i19} is in contact with the fulcrum roller. As illustrated in Fig. 16(a), at θ_F=112.9°, the plastic strain increment in the area with the smallest diameter between rib P_{i19} and P_{i18} begins to increase. It is thought that this area begins to bend anew as θ_F increases. As bending progresses from θ_F=112.9° to θ_F=117.6°, the plastic strain increment between ribs P_{i19} and P_{i18} is higher. It is thought that the torque will increases as the bending process in this area progresses.

Fig. 18(b) shows the plastic strain incremental at θ_F=117.6° from the FEM results for round steel bar. Comparing Fig. 16(b) and Fig. 18(b), θ_F=117.6° is the same, but the position where the plastic strain increment is higher is different. In the case of round steel bars, an area with a large plastic strain increment is located near the horizontal line passing through the center of the fulcrum roller, which is represented by a solid line. In the case of rebars, on the other hand, an area with a high plastic strain increment is located closer to the circular roller than to the solid horizontal line. Thus, the rebar is bent at a location away from the circular roller due to the presence of a rib. Therefore, the external force exerted on the rebar from the circular roller to advance the bending process is greater than for round steel bars. Consequently, the torque required to rotate the circular roller is expected to increase. This difference in bending position between round steel bar and rebar is thought to be one of the reasons for the increased torque.

The torque reaches its minimum value when the circular roller is further rotated from θ_F=117.6° to θ_F=131.7°. Fig.17(c) shows that P_{i19} is the contact position at θ_F=131.7°. From Fig. 16(c), it can be observed that the plastic strain increment between the ribs P_{i19} and P_{i18} is smaller, and the plastic strain increment of the rebar on the circular roller side is smaller than that of the rib P_{i18}. The fact that the torque decreases after reaching its maximum value can be attributed in part to the separation of rib P_{i20} from the fulcrum roller. Fig. 19 shows the relationship between the displacement in the x direction and the rotation angle θ_F with respect to the position of rib P_{i20}. The angle of rotation at which rib P_{i20} contacts the fulcrum roller is between θ_F=110.6° and θ_F=117.6°. The dotted line in the figure indicates the rotation angle at which the torque reaches its minimum value, and the solid line indicates the rotation angle at which the torque reaches its maximum value. As θ_F increases from θ_F=112.9°, the displacement begins to move in the negative direction of the X axis. In other words, during the bending process with rib P_{i19} as the fulcrum, rib P_{i20} moves away from the fulcrum roller. At this time torque required to advance the rotation of the circular roller is reduced. In addition, other effects on the increase or decrease in torque need to be studied in the future.

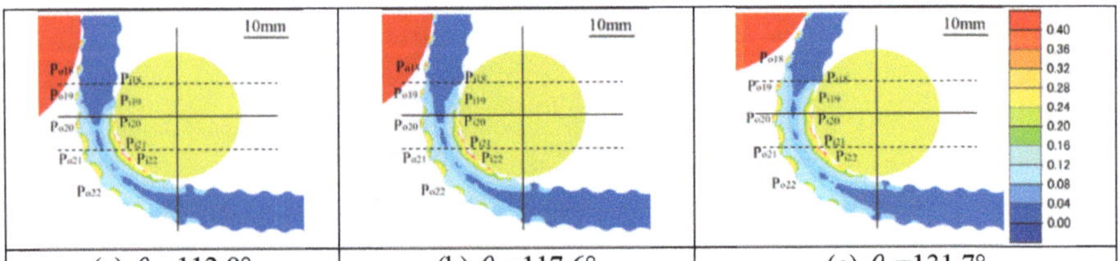

(a) θ_F=112.9°	(b) θ_F=117.6°	(c) θ_F=131.7°

Fig. 15. Distribution of equivalent plastic strain for rebar [-].

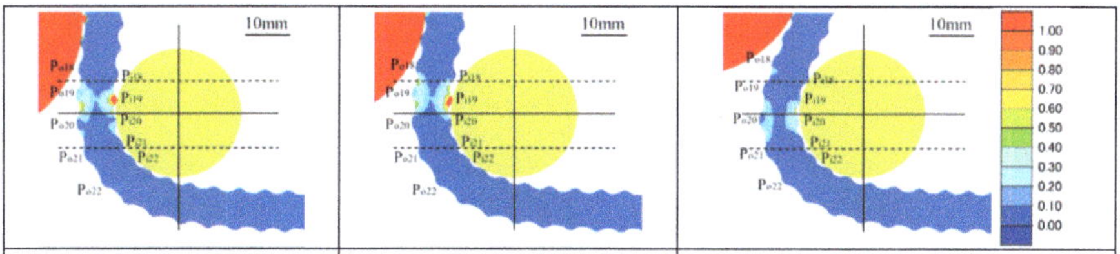

(a) θ_F=112.9° (b) θ_F=117.6° (c) θ_F=131.7°

Fig. 16. plastic strain increment for rebar [/s].

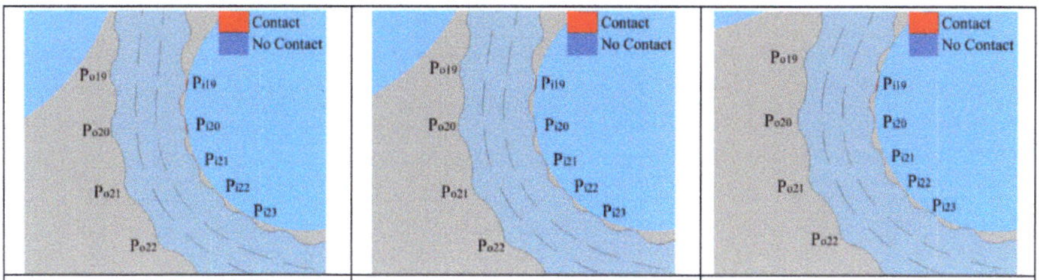

(a) θ_F=112.9° (b) θ_F=117.6° (c) θ_F=131.7°

Fig. 17. Contact state between rebar and fulcrum roller.

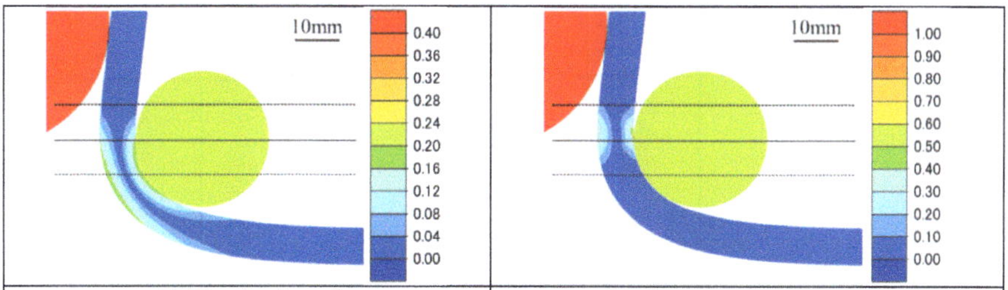

(a) Distribution of equivalent plastic strain [-] (b) Plastic strain increment [/s]

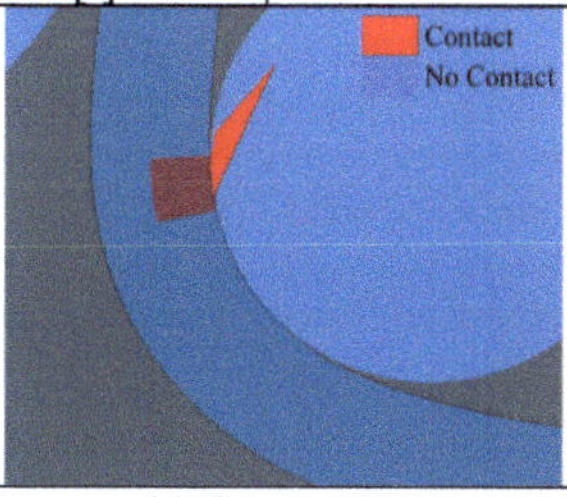

(c) Contact state

Fig. 18. FEM results for round steel bar with an outer diameter of 8.8 at θ_F=117.6°.

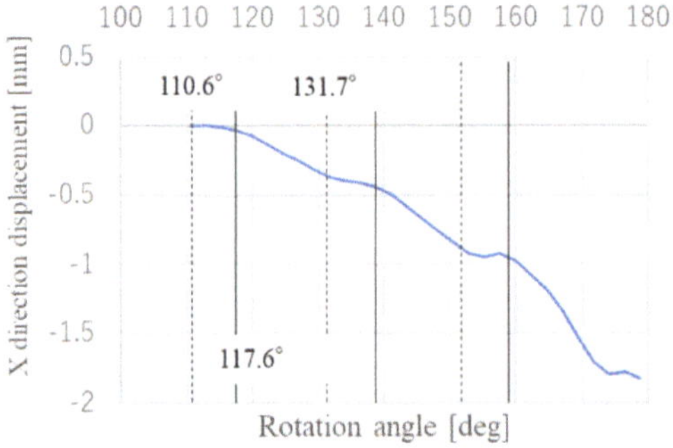

Fig. 19. X direction displacement of P_{i20} for rebar.

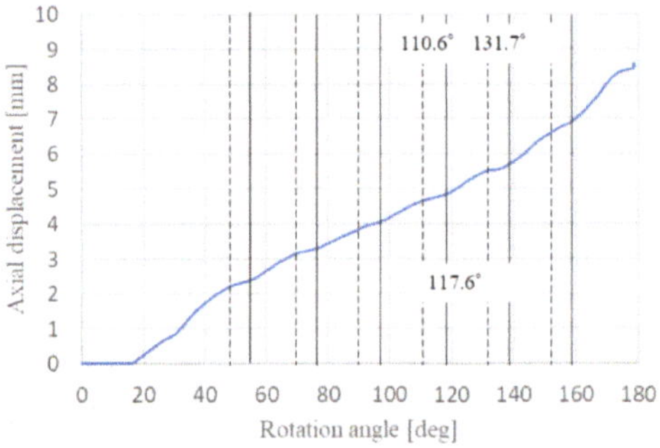

Fig. 20. Axial displacement of rebar.

Axial displacement. It was mentioned earlier that a part of the rebar separates from the fulcrum roller during the bending process. This may be related to the axial displacement of the material at the point where it contacts the rebar receiver. Fig. 20 shows the relationships between the rotation angle and axial displacement. The dotted line in the figure shows the θ_F at minimum torque and the solid line shows the θ_F at maximum torque. As the rotation angle θ_F increases, the axial displacement also increases. When the torque changes from the minimum to the maximum value, in other words, from the dotted line to the solid line, the increase in axial displacement tends to decrease. Therefore, it can be observed that this phenomenon is related to the separation of the rib from the fulcrum roller reported in section Increase and decrease in torque change for rebars.

Conclusion

To clarify the relationship between material deformation and torque during the bending of rebars, experiments and finite element analysis were conducted on D10 rebars, φ8.8 round steel bars, and φ10.1 round steel bars. As a result, the following conclusions were reached.

(1) In the case of rebars, the torque to rotate the circular roller increases as the rotation angle increases. Furthermore, the torque does not increase significantly when the rotation angle increase. In addition, torque is fluctuating. On the other hand, for the round steel bars, the torque increases at the beginning of the process and then becomes constant. It is thought

that the bending was successively performed by rotating the circular roller for both rebars and round steel bars.

(2) In the case of rebars, a fluctuation of torque was observed in the relationship between torque and the rotation angle. It is thought that fluctuation of torque occurred due to the effect of transverse ribs. The results of the FEM showed that the material deformation of rebars differ from that of round steel bars due to the transverse ribs. The results of FEM confirming the plastic strain increment suggest that the round steel bar is bent near the region in contact with the fulcrum roller. Compared to round steel bars, in the case of rebars, areas with large plastic strain increments are slightly deviated in the direction of the circular rollers. These differences are expected to result in a higher value of torque required for bending rebars than for round steel bars with an outer diameter equal to the minimum value of the outside diameter of the rebar.

(3) The torque for the rebar decreases when the section closer to the rebar receiver than the section serving as the fulcrum moves away from the fulcrum roller.

(4) It was confirmed that axial displacement occurred in the rebars as the circular roller rotated. The results of the FEM analysis demonstrated that the trends of increasing axial displacement of the rebar were related to torque fluctuation.

References

[1] S. Higaki, T. Go, K. Mizuno, M. Sasada, T. Tanaka, Effect of Diameter of Fulcrum Rolleron Shape of Rebar in Bending, Proceedings of the 14th International Conference on the Technology of Plasticity - Current Trends in the Technology of Plasticity ICTP 2023, Volume 4, 404-413.

[2] S. Higaki, H. Nishida, Y. Koike, M. Sasada, T. Tanaka, Effect of transfer ribs on longitudinal displacement of rebars in bending, Procedia Manuf. 50 (2020) 253-256. https://doi.org/10.1016/j.promfg.2020.08.047

[3] K. Takahashi, T. Watanabe, T. Kuboki, M. Murata, K. Ono, K. Yano, Reduction in Flatness Using Axial Compressive Force in Rotary Draw Bending of Circular Tube, J. Japan Soc. Tech. Plast. 49 (2008) 896-900. https://doi.org/10.9773/sosei.49.896

[4] T. Murota, J. Endow, Experimental Analysis on Plastic Deformation of Tubes in Uniform Bending, J. Japan Soc. Tech. Plast. 23 (1982) 343-349.

[5] T. Nishihara, S. Taira, On the Yielding of Steel under Bending Moment (Report 2), Trans. Japan Soc. Mech. Eng. 18 (1952) 90-94.

[6] T. Miura, K. Funamoto, K. Seto, Impact Resistance Ability in Bended Point of Reinforcing Bars at Very Low Temperature, Proceedings of JSCE 557 (1997) 15-22.

[7] H. Okimoto, K. Turuoka, Investigation of the fracture properties of deformed bars by the bend-and-return process, J. Struct. Constr. Eng. AIJ 442 (1992) 33-41.

[8] K. Sasaki, Y. Hisari, H. Igarashi, T. Miyagawa, Analysis of the strain and stress on bent section of reinforcing bar by finite element method, Proceedings of the Japan Concrete Institute 30 (2008) 987-992.

Analysis of the influence of spinning temperature on the coordinated deformation of Mg/Al composite tube

ZHENG Jinbin[1], SHU Xuedao[1*], LI Zixuan[1] and XU Haijie[1]

[1]Faculty of Mechanical Engineering and Mechanics, Ningbo University Ningbo, 315211, China

shuxuedao@nbu.edu.cn

Keywords: Composite Tubes, Spinning Process, Simulation Analysis, Temperature, Coordinated Deformation

Abstract. Currently, the focus of research in the realm of Mg/Al layered metal composites predominantly lies within the domain of plate structures, while the development of Mg/Al bimetallic composite tubes (BCTs), which hold significant utility, remains in its nascent exploratory phase. Among the various forming methodologies applied to BCTs, the spinning process has garnered considerable attention due to its advantageous attributes encompassing minimal forming loads, uncomplicated tooling requirements, cost-effectiveness, and high dimensional precision. This study delves into the outcomes of forming Mg/Al BCTs at diverse spinning temperatures via a synergy of theoretical frameworks and simulation experiments facilitated by the Simufact Forming software platform. A novel approach is presented for assessing the degree of coordinated deformation at the interface by scrutinizing the interface strain gradient, elucidating the intrinsic correlation between the interface strain gradient and the bonding strength at the interface. Through meticulous analysis, it is discerned that the BCTs exhibit the least interface strain gradient and axial extension disparity when subjected to a spinning temperature of 350°C. Consequently, in the context of spinning preparation for Mg/Al BCTs, a spinning temperature of approximately 350°C is recommended to engender Mg/Al BCTs characterized by enhanced harmonized deformation properties.

Introduction

In response to the escalating demands for materials with enhanced comprehensive performance in modern production, single-metal materials have encountered limitations in meeting the requisite utilization prerequisites. Consequently, over the past few decades, researchers have increasingly turned their attention towards composite materials, leading to a proliferation of diverse composite materials and corresponding fabrication methodologies. One notable avenue within this landscape is Laminated Metal Composites (LMCs) [1]. LMCs represent a class of composite materials that are synthesized via various composite technologies, facilitating the establishment of robust metallurgical bonding at the interface between two or more metals possessing disparate physical, chemical, and mechanical properties [2]. By harnessing the potential of LMCs, it becomes feasible to amalgamate the exceptional attributes inherent to each constituent material, while simultaneously leveraging a "phase compensation effect" that enables the preservation of the distinctive characteristics of individual metal or alloy components[3]. Through judicious combinations, LMCs possess the capacity to compensate for inherent weaknesses and engender superior comprehensive performance.

Magnesium alloys have garnered significant utilization in a diverse range of industries such as aerospace, electronics, automotive, and high-speed rail due to their notable attributes including low density [4], high specific strength [5], and elevated stiffness [6]. However, owing to the inherent chemical reactivity of magnesium and its alloys, the formation of loose and porous oxide films when exposed to atmospheric conditions [7] restricts their application in corrosive environments. On the other hand, aluminum alloys possess good corrosion resistance attributed to the formation of a thin and compact oxide layer on their surface [8]. Consequently, the development of Mg/Al bimetallic composites presents an opportunity to expand the potential applications of Mg alloys while capitalizing on the excellent properties offered by both lightweight metals to compensate for their respective limitations, thereby offering a wide-ranging scope for practical implementation.

Due to the hexagonal close-packed (HCP) crystal structure of Mg and its alloys, only two slip systems can be activated at room temperature, leading to distinct plastic deformation characteristics compared to Al alloys [9]. This discrepancy poses challenges in achieving metallurgical bonding at the bimetallic interface through solid-state bonding methods. Therefore, thermal processing techniques, such as hot rolling, casting, annealing, and hot spinning, are commonly employed in the solid-state bonding of Mg/Al bimetallic composites. Currently, research on Mg/Al layered metal composites primarily focuses on sheet configurations (Al/Mg clad sheets), with the exploration of Mg/Al BCTs for broader applications still in the preliminary stages [10]. Numerous traditional material forming process can be employed in the fabrication of BCTs. For instance, Zhan et al. [11] investigated the bulging formation of square Mg/Al BCTs and observed that the bulging process only achieved mechanical bonding at the interface, lacking metallurgical bonding. Zhang et al. [12] explored the effect of interlayer chromium (Cr) on the bonding interface of composite cast Al/steel BCTs, revealing that the incorporation of interlayer Cr effectively impeded the formation of intermetallic compounds (IMCs) and enhanced the interface bonding strength. Notably, the spinning process offers distinct advantages such as low forming loads, simplicity in tooling requirements, cost-effectiveness, and excellent dimensional accuracy, thus finding extensive application in composite pipe fabrication [13,14]. As an example, Samandari et al. [15] successfully fabricated Al/St BCTs utilizing the spinning process. In the study of spinning processes for Mg/Al BCTs, Zheng et al. [16] identified that multi-pass spinning process significantly improved the surface quality of the composite tubes.

In the fabrication process of bimetallic composite materials, if the deformation of the two constituent materials is not synchronized, shear stress and tensile stress may arise at the interface. When these stresses exceed the shear strength of the interface, separation of the composite parts occurs. Hence, in order to mitigate excessive stresses and ensure proper coordination of deformations, appropriate processing parameters must be carefully chosen to harmonize the deformation of the two materials comprising the bimetallic composite pipe. Bambach et al.[17] suggested that similar flow stress between the two materials is essential during bimetallic rolling to facilitate coordinated deformation of the composite plate. However, the concept of coordinated deformation observed in rolling forming of composite plates does not directly apply to spinning process due to the asymmetrical distribution of rolls in the spinning process. In the context of bimetallic composite plates, where the rolling forces on the two plates are equal, achieving similar flow stress aids in coordinated material deformation. Nevertheless, this principle is not universally applicable beyond sheet rolling processes [18].In the spinning process, differing pressures are exerted on the inner and outer tubes, resulting in a lower radial stress on the inner tube. Consequently, when both tubes exhibit similar flow stress, their deformations become uncoordinated. Thus, the principle of similar flow stress, although conducive to coordinated material deformation in theory, is not suitable for spinning forming. Typically, the outer tube necessitates higher flow stress or the inner tube requires lower flow stress to ensure harmonized deformation of both tubes.

Drawing from the aforementioned concepts, this study employs finite element simulation experiments. The spinning simulation experiment is conducted on a composite tube comprising 7075 Al alloy and AZ31b Mg alloy within a temperature range of 300-400°C. Throughout the spinning process, variations in stress and strain at the interface of the inner and outer tubes are observed. These discrepancies evolve with temperature fluctuations, resulting in non-synchronized

Materials Research Forum LLC
https://doi.org/10.21741/9781644903254-3

deformation at the interface, as evidenced by axial displacement of the two tubes. Analysis of the results indicates that maintaining an optimal spinning temperature of approximately 350°C for Mg/Al BCTs is crucial. This research serves as a foundational framework for the development of Mg/Al BCTs with tailored structural characteristics and performance attributes.

Finite element simulation setting and material parameter analysis

The spinning forming method is shown in Fig. 1, and the Mg/Al BCTs is assembled on a fixture equipped with a triangular chuck. During spinning, the triangular chuck drives the Mg/Al BCTs to rotate, and the mandrel also rotates. At the same time of rotation, the roller starts to feed radially from one end of the triangular chuck. When the preset thinning rate is reached, it begins to feed axially.

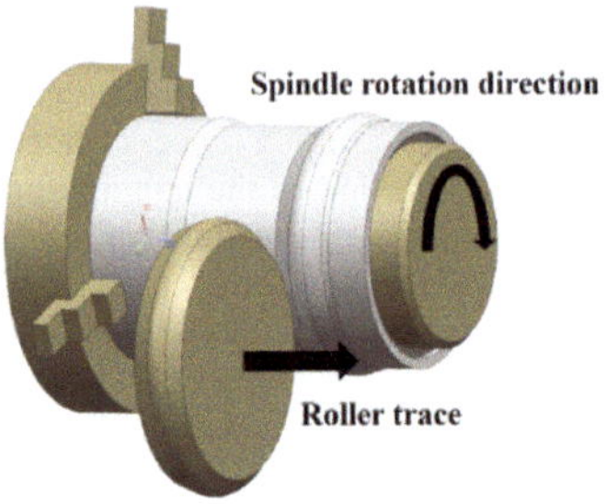

Fig. 1. Composite tubes spinning process diagram.

The mold model are imported into Simufact Forming finite element analysis software after modeling in NX10.0. Among them, 7075 Al alloy material uses the material model provided in the software, and the material parameters of AZ31b Mg alloy use the stress-strain curve studied by Zhang et al. [19], which is imported into the finite element software and simulated. The specific finite element simulation parameters are shown in Table 1.

Table 1. Finite element simulation setting.

The parameter name	The numerical
Spindle speed/ (r/min)	300
Roller feed ratio (mm/r)	3
Spinning temperature (°C)	300/325/350/375/400
Thickness of Al tube (mm)	6
Thickness of Mg tube (mm)	6
Tube length (mm)	150
Thinning rate (%)	30

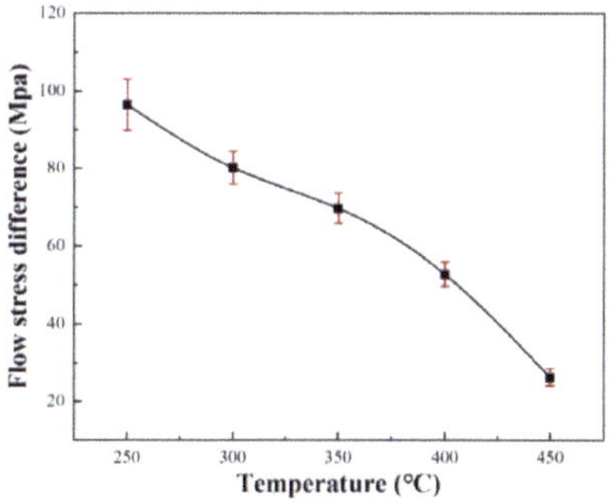

Fig. 2. The difference of flow stress at different temperatures.

In the spinning process employed in this study, the high-speed rotation of the spindle maintains the strain rate of the deformation zone of the composite tube material at approximately 20/s. Consequently, the primary factor influencing the material's flow stress during spinning is the forming temperature. An analysis of the average flow stress difference between Mg alloy and Al alloy within the temperature range of 250°C-450°C was conducted, with the results depicted in Fig. 2. The analysis reveals that the flow stress disparity between Mg alloy and Al alloy decreases as the spinning temperature increases. Specifically, the maximum difference is recorded at 96.5MPa at 250°C, while the minimum difference is observed at 26.45 MPa at 450°C. Contrary to composite rolling processes, in spinning process, the proximity of the flow stress values of the two materials does not necessarily yield optimal results. It is imperative to regulate the spinning temperature to maintain a specific stress difference between the inner and outer tubes, ensuring coordinated deformation. Therefore, a detailed analysis of the simulation results within the temperature range of 300°C-400°C is essential to ascertain the optimal spinning conditions for preserving deformation coordination.

Analysis of results

Axial displacement analysis of Mg/Al BCTs.

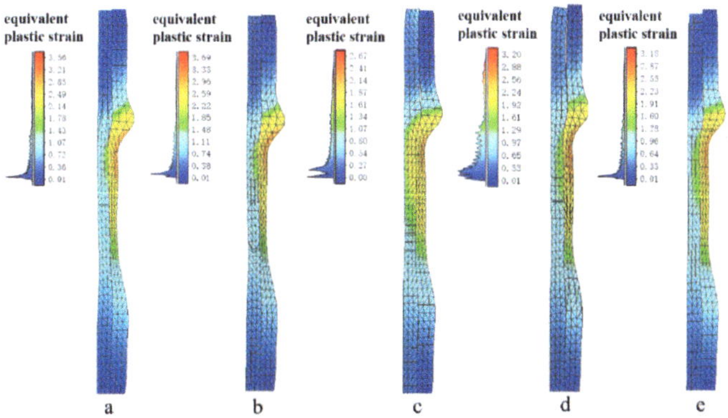

Fig. 3. Finite element simulation results at different temperatures (a) 300°C, (b) 325°C, (c) 350°C, (d) 375°C, (e) 400°C.

The finite element simulation outcomes are depicted in Fig. 3, which illustrates the equivalent strain cloud diagram of the spinning temperature range between 300°C and 400°C. The results indicate that the maximum equivalent strain value of the Mg/Al BCTs is 2.67 at a spinning temperature of 350°C, while it reaches 3.69 at 300°C.To better assess the coordinated deformation degree of the Mg/Al BCTs, this study adopts the axial displacement of the BCTs as a measure of the deformation level. In particular, the axial displacement difference between the Mg tube and the Al tube is utilized to characterize the coordinated deformation degree of the BCTs. The resulting outcome is displayed in Fig. 4.

Figure 4 illustrates that as the spinning temperature increases, the overall axial displacement of the Mg tube and Al tube demonstrates a decreasing trend. However, the axial displacement difference between the two tubes exhibits an initial decrease followed by an increase, resulting in an overall upward trend in displacement difference. Notably, at 350°C, the displacement difference becomes negative, reaching -0.51. This signifies that at 350°C, the axial elongation of the Mg tube surpasses that of the Al tube. Furthermore, it is evident from the figure that for spinning temperatures exceeding 350°C, the axial displacement difference between the two tubes exhibits a notable increasing trend. This can be attributed to the decreased deformation resistance of the Al tube and an associated increase in its plastic deformation ability as the processing temperature rises. Consequently, the axial displacement of the Al tube increases, leading to a larger axial displacement difference between the two tubes. Thus, the elevation of the spinning temperature does not promote improved deformation coordination in the Mg/Al BCTs. To gain further insights into this phenomenon, the radial equivalent stress distribution of the spinning simulation results at 350°C is analyzed.

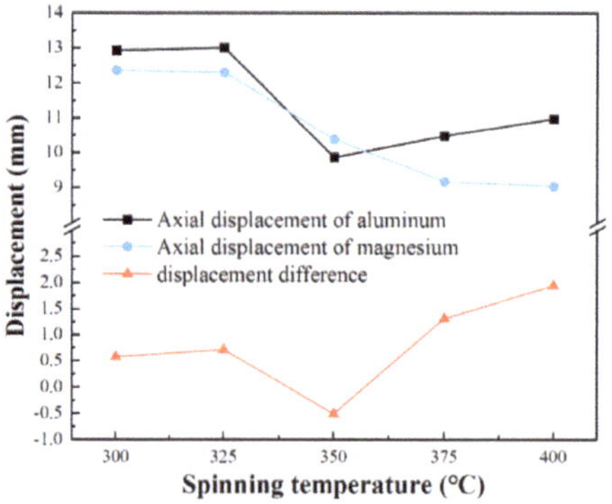

Fig. 4. Axial displacement extension of Mg/Al BCTs at different temperatures.

Analysis of equivalent stress and strain of Mg/Al BCTs. Fig. 5 presents the radial equivalent stress distribution of the wall of Mg/Al BCTs during the spinning process, the radial equivalent stress distribution of the BCTs wall shows a stepped distribution, which also corresponds to what has been said before: the inner and outer tubes need a 'flow stress difference' to show a better coordinated deformation state. When the theoretical flow stress difference of Mg/Al BCTs is greater than or less than the actual stress difference of the inner and outer tubes in the spinning forming process, the coordinated deformation state of the inner and outer tubes becomes worse, which also just explains that when the temperature is too high or too low, the axial displacement difference of the inner and outer tubes is at a higher value. Although this stress difference is reflected in the axial displacement, it has a small amplitude change. But the performance is not very obvious. On the bonding interface of the BCTs, although the interface of the BCTs begins to bond with the application of temperature and pressure, the interface will produce tangential and axial stress due to the uncoordinated deformation, thus affecting the bonding strength of the BCTs interface. when the tangential and axial stress is greater than the shear strength of the composite interface, it will directly lead to the separation of the already composite interface.

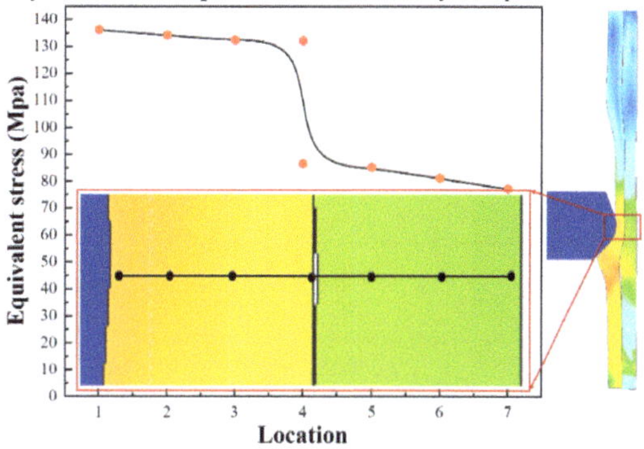

Fig. 5. Radial equivalent stress distribution of BCTs.

The findings from the equivalent strain analysis of the interface, specifically in the radial direction of the Mg/Al BCTs, are depicted in Fig. 6. The figure indicates that at 350°C and 400°C, the radial distribution of equivalent strain in the Mg/Al BCTs exhibits a step-like pattern similar to the equivalent stress distribution. However, a notable distinction is observed at 350°C, where the step at this interface nearly diminishes. This observation aligns with the equivalent strain cloud diagram for Mg/Al BCTs at different temperatures, as illustrated in Fig. 3. In the 350°C cloud diagram, the distribution of strain change appears more uniform, consistent with the axial position difference of the BCTs presented in Fig. 4. Notably, the axial displacement difference at 350°C demonstrates a lesser magnitude compared to other temperatures. Consequently, it can be anticipated that a reduced ladder-like distribution of radial equivalent strain in the Mg/Al BCTs corresponds to an improved coordinated deformation degree. This, in turn, leads to reduced shear stress (impacting the bonding interface) and heightened bonding strength of the interface.

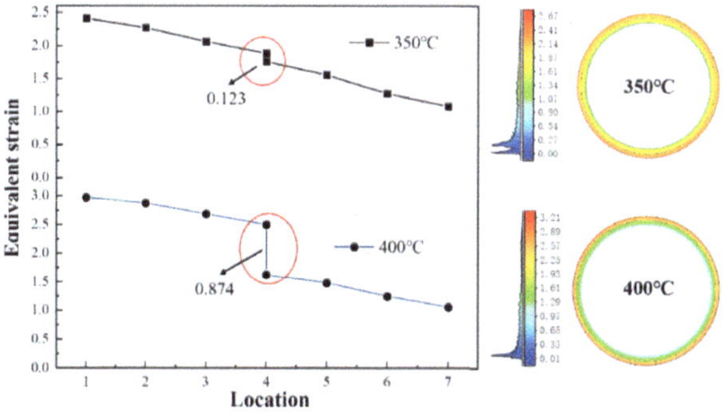

Fig. 6. The radial equivalent strain distribution of the Mg/Al BCTs at 350°C and 400°C.

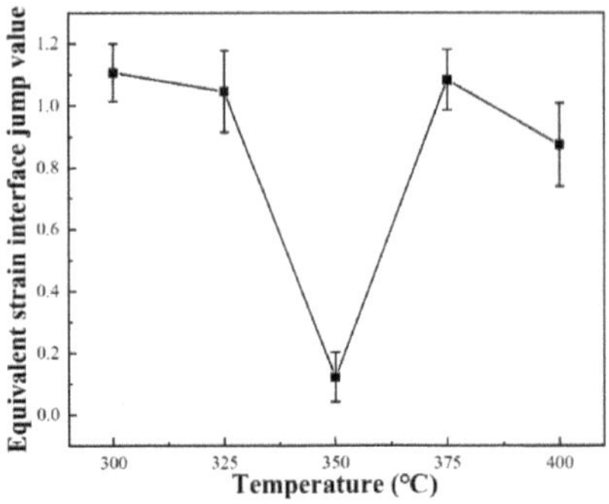

Fig. 7. The change of equivalent strain jump value at the interface of Mg/Al BCTs at different temperatures.

The equivalent strain jump values at the interface of the BCTs mentioned above are analyzed. The results are shown in Fig. 7. The results show that when the spinning temperature is 350 °C, the equivalent strain jump value at the interface of the BCTs is at the lowest level, which is 0.123. This indicates that when the spinning temperature of the BCTs is 350°C, the relative strain generated at the interface of the BCTs is small, and the shear stress generated at the interface due to the relative displacement is also at a low level. When the temperature changes to 300°C-325°C and 375°C-400°C, the jump value of the interface strain of the BCTs is at a high level, and the maximum value reaches 1.108. At the same time, in the two temperature ranges, the difference of the axial displacement of the BCTs also shows a high level.

Conclusion

Different from the process of rolling Mg/Al composite plate, in the spinning process of Mg/Al BCTs, due to the different contact stress between the inner and outer tubes, the conclusion that 'similar flow stress is more conducive to the deformation coordination of composite materials' in the previous rolling research is not applicable to the spinning process of BCTs. Therefore, this paper studies the finite element simulation results of the BCTs in the temperature range of 300°C-400°C, and deeply analyzes its deformation and behavior. The following conclusions are obtained:

1. When the spinning temperature rises, the axial displacement of Mg tube and Al tube shows a downward trend as a whole. However, the axial displacement difference between the Mg tube and the Al tube shows a trend of decreasing first and then increasing, and the overall displacement difference shows an upward trend. The minimum value is -0.51 mm, which is located at the spinning temperature of 350°C.

2. There is a stress-strain jump value in the bonding interface of Mg/Al BCTs. It is precisely because of the jump value of stress that leads to the phenomenon of strain jump value. The larger the jump value, the more unfavorable the interface bonding.

3. At a temperature of 350°C, the minimum strain jump value of 0.123 is observed. This finding signifies that when the spinning temperature of the BCTs is maintained at 350°C, the relative strain generated at the BCTs' interface remains minimal. Consequently, the shear stress arising from the relative displacement at the interface also remains at a low level. Simultaneously, this observation corresponds to the phenomenon of the smallest axial displacement difference among the BCTs occurring at 350°C.These findings provide evidence supporting the recommendation to maintain the spinning temperature within the range of 350°C during the spinning process of Mg/Al BCTs. This temperature range ensures the attainment of coordinated deformation in the BCTs, consequently enhancing the bonding strength of the interface.

Acknowledgement
This study was funded by the National Natural Science Foundation of China (U23A20629, No.52375346, No.52205403), the Natural Science Foundation of Zhejiang, China (No. LZ22E050002), Priority funding for post-doctoral research project in Zhejiang Province (No. 348842), Ningbo Science and Technology Major Project (No. 2022Z064, No.2023Z011).

References

[1] T. Zhang, H. Xu, Z. J. Li, Development and Present Situation of Layered Metal Composites, J. Eng. Sci. 43 (2021) 6775.

[2] X.T. Liu, T.A. Zhang, J.Z. Cui, Technology of clad metal production and its latest progress. Materials Reports 16 (2002) 41.

[3] G.M. Tian, X.M. Li, Y.Q Zhao, Research status of processing technology of laminated metal composite, Mater. China 11 (2013) 696.

[4] L. Liu, X. Chen, F. Pan, A review on electromagnetic shielding magnesium alloys, J. Magnes. Alloy. 9 (2021)16. https://doi.org/10.1016/j.jma.2021.10.001

[5] A. Meher, M. Mahapatra, P. Samal, P.R. Vundavilli, Study on effect of TiB2 reinforcement on the microstructural and mechanical properties of magnesium RZ5 alloy based metal matrix composites, J. Magnes. Alloy. 8 (2020) 780-792. https://doi.org/10.1016/j.jma.2020.04.003

[6] Y. Yang, X. Chen, J. Nie, K. Wei, Q. Mao, F. Lu, Y. Zhao, Achieving ultrastrong Magnesium-lithium alloys by low-strain rotary swaging, Mater. Res. Letter 9 (2021) 255. https://doi.org/10.1080/21663831.2021.1891150

[7] Y. Song, E.H. Han, K. Dong, D. Shan, C.D. Yim, B.S. You, Effect of hydrogen on the corrosion behavior of the Mg-xZn alloys, J. Magnes. Alloy. 2 (2014) 208-213. https://doi.org/10.1016/j.jma.2014.10.002

[8] C.C. Zhi, Z.Y. Wu, L.F. Ma, Z. Huang, Z. Zheng, H. Xv, J. Weitao, J. Lei, Effect of thickness ratio on interfacial structure and mechanical properties of Mg/Al composite plates in differential temperature asymmetrical rolling, J. Mater. Res. Technol. 24 (2023) 8332-8347. https://doi.org/10.1016/j.jmrt.2023.05.031

[9] D.C. Foley, M. Maharbi, K.T. Hartwig, I. Karaman, L.J. Kecskes, S.N. Mathaudhu, Grain refinement vs. crystallographic texture: mechanical anisotropy in a magnesium alloy, Scripta Mater. 64 (2011)193-196. https://doi.org/10.1016/j.scriptamat.2010.09.042

[10] N. Shi, WX. Wang, T.T. Zhang, Finite element simulation and analysis of staggered spinning forming of Magnesium/Aluminum composite pipe, Journal of Taiyuan University of Technology 52 (2021) 269-273.

[11] L.Q. Wang, G. Wang, J.L. Yang, AZ31 / 7475 square bimetal composite tube gas bulging-cold contraction bonding process, J. Plast. Eng. 28 (2021) 69-76.

[12] D. Zhang, G. Zhang, H. Yu, W. Lv, K. Wen, H. Xu, Controlling interfacial composition and improvement in bonding strength of compound casted Al/steel bimetal via Cr interlayer, J. Mater. Res. Technol. 23 (2023) 4385-4395. https://doi.org/10.1016/j.jmrt.2023.02.053

[13] N. Shi, Microstructure evolution and interface bonding mechanism of 6061Al/AZ3lBMg composite pipe by stagger spinning, Taiyuan University of Technology, 2020.

[14] Z.X. Li, S. Rezaei, T. Wang, Recent advances and trends in roll bonding process and bonding model: A review, Chinese Journal of Aeronautics 36 (2023) 36-74.

[15] M. Samandari, K. Abrinia, A. Akbarzadeh, Properties and Mechanism of Al/St Bimetal Tube

Bonding Produced by Cold Spin-Bonding (CSB) Process, Transactions of the Indian Institute of Metals 70 (2017).

[16] J.B. Zheng, X.D. Shu, S.Y. Chen, Forming Quality Analysis of Mg-Al Composite Pipes of Multi-Pass Power Spinning, ASME International Mechanical Engineering Congress and Exposition, Proceedings (IMECE), October 30- November 3, 2022, Columbus, Ohio, USA.

[17] M. Bambach, M. Pietryga, A. Mikloweit, A finite element framework for the evolution of bond strength in joining-by-forming processes, J. Mater. Process. Technol. 214 (2014) 2156-2168. https://doi.org/10.1016/j.jmatprotec.2014.03.015

[18] S. Guenther, G. Schwich, G. Hirt, Investigation of bond formation behaviour in composite ring rolling, J. Mater. Process. Technol. 275 (2018) 116364. https://doi.org/10.1016/j.jmatprotec.2019.116364

[19] Z. Zhang, Basic research on plastic deformation behavior of AZ31B magnesium alloy, Central South University, 2011.

Hot local compression and die quench ausforming of quenchable steel sheet

MAENO Tomoyoshi[1,a *], IKEDA Yuto[2,b], HANADA Akihiro[2,c] and MORI Ken-ichiro[3,d]

[1]Division of Systems Research, Faculty of Engineering, Yokohama National University, Yokohama, 240-8501, Kanagawa, Japan

[2]Department of Mechanical Engineering, Materials Science, and Ocean Engineering, Graduate School of Engineering Science, Yokohama National University, Yokohama, 240-8501, Kanagawa, Japan

[3]Department of Mechanical Engineering, Toyohashi University of Technology, Toyohashi, 441-8580, Aichi, Japan

[a]maeno-tomoyoshi-yf@ynu.ac.jp, [b]ikeda-yuto-cm@ynu.jp, [c]hanada-akihiro-xb@ynu.jp, [d]mori.kenichiro.rx@tut.jp

Keywords: Hot Stamping, Compression, Thermomechanical Treatment, Ausforming, Die Quenching

Abstract. Characteristics of hot local compression were investigated to apply hot stamping to a plate forging process. A heated quenchable steel sheet was locally compressed in a thickness direction and held with a punch at the bottom dead centre for die quenching. The effect of a forming temperature on reduction ratio and hardness was examined. Holding at the bottom dead centre improved the compression ratio due to plastic deformation during martensite transformation. Additionally, the quenched hardness of the locally compressed portion increased due to ausformig.

Introduction

The cold forging from plates produces functional and small parts with complex shapes. It is called plate forging [1] or sheet-bulk forming [2]. The plate forging produces small functional parts such as gear, cam, plate with pin, stepped plate, etc. Although drawing, bending and shearing are the main deformations in the stamping of the thin sheets, local compression in a thickness direction, backward extrusion and semi-shearing are often used in plate forging [1]. The productivity and material yield of plate forging is higher than that of machining and bulk forging. However, the compression in the thickness direction for local thinning significantly increases the forming load due to friction [3, 4]. When the required hardness of the product is high, batch or continuous heat treatment is performed. Therefore, there is no flexibility in production plans.

On the other hand, hot stamping produces high-strength panel parts for automobiles. The forming load is remarkably reduced by heating the steel sheets, and springback is prevented. Moreover, the stamped parts are hardened by quenching with dies; thus, ultra-high strength steel parts with a tensile strength of approximately 1.5 GPa are obtained [5]. Furthermore, the production control is flexible because the die quenching as heat treatment in hot stamping is processed individually.

The combination of plate forging and hot stamping is attractive for reducing the forming load and omitting the heat treatment process after forming [6]. In plate forging, the degree of material deformation is greater than that of sheet metal forming. In addition, the required shape accuracy in plate forging is also severe. It is necessary to understand the deformation and die quench characteristics under compression in the thickness direction, which are the main factors in plate forging.

In the present study, hot local compression and die quenching of a quenchable thick steel sheet were performed to investigate the feasibility of hot plate forging. The compression and die-quenching behaviour were investigated. The effect of a forming temperature on reduction ratio and hardness was examined. Furthermore, a mechanism of improvement in hardness at the compressed portion was discussed.

Hot local compression of quenchable thick steel sheet

Fig. 1 shows the concept of hot plate forging. For hot plate forging with die quenching, Complex shape forming by progressive stamping in plate forging and individual heat treatment in hot stamping are combined to achieve one-piece flow manufacturing. In addition, since the material is heated and softened, the high-forming load in compression in the thickness direction will be reduced.

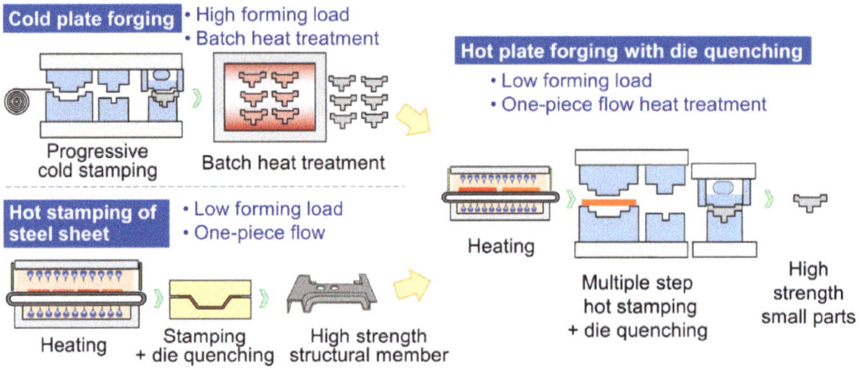

Fig. 1. Flexible production control by hot plate forging with die quenching.

Hot local compression with die-quenching of a quenchable thick steel sheet is illustrated in Fig. 2. A boron steel sheet of N22CB with a thickness of 4 mm is cut into 100×120 and used for local compression. The chemical composition of the N22CB quenchable steel sheet is shown in Table 1. This sheet is similar to a 22MnB5 commonly used in hot stamping. The blank sheet is heated in an electric furnace at 950°C for 7 minutes and then transferred to the die. The centre of the heated sheet is compressed with a cylindrical punch after natural cooling to a forming start temperature T_f. The punch is held at the bottom dead centre for t_h seconds for die quenching. The diameter of the punch is 15 mm, and a gap of 2.5 mm is provided around the punch between it and the sheet holder. Compression is performed without lubrication. An 800 kN Mechanical servo press is used, the average press speed during compression is 30 mm/s, and the time from contact between the sheet and punch to the bottom dead centre is 0.1 s at a nominal compression ratio r = 75%. The temperature of the heated sheet is measured by a radiation thermometer calibrated with a thermocouple.

Table 2 shows the conditions used in the experiment. The penetration ratio of the punch to the initial plate thickness under no load was assumed as the nominal compression rate r. The sheet material used transforms into a ferrite phase at 670°C during air cooling, so under the condition of T_f = 650°C, transformation occurs before compression and is not hardened. For comparison, cold compression without heating was also performed.

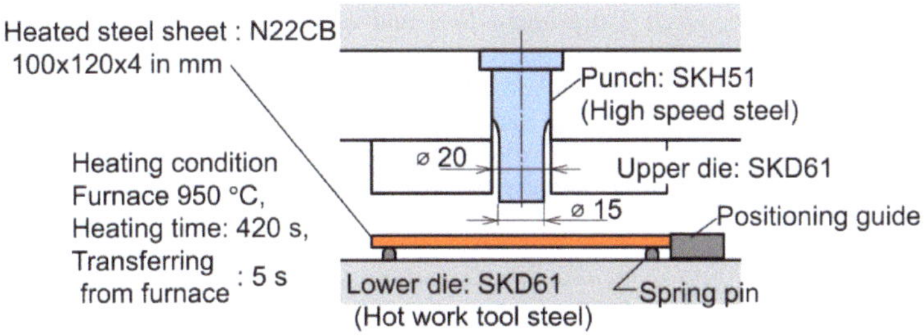

Fig. 2. Hot local compression and die-quenching of quenchable thick steel sheet.

Table 1. Chemical composition of N22CB quenchable steel sheet

C	Si	Mn	B
0.22	0.15	0.45	35 ppm

Table 2. Used condition for hot local compression and die-quenching.

Nominal compression ratio r [%]	Temperature at compression T_f [°C]	Holding time for die-quenching t_h [s]
25,50,75	room, 650, 750, 840	0, 1.5, 5.5, 20

Results of Hot local compression in thickness direction of quenchable thick steel sheet

The cross sections of the compressed portion obtained at different compression temperatures for r = 50%, t_h = 20 s are shown in Fig. 3. A protrusion is observed around the compressed portion for heated conditions. On the other hand, the underside of the compressed portion is curved for the cold and at T_f = 650°C. Because the punch and sheet holder descend together, the sheet is bent and then returned to a flat by sandwiching the sheet holder. However, for cold and at 650°C as under transformation temperature, springback occurred, and the sheet was not flat.

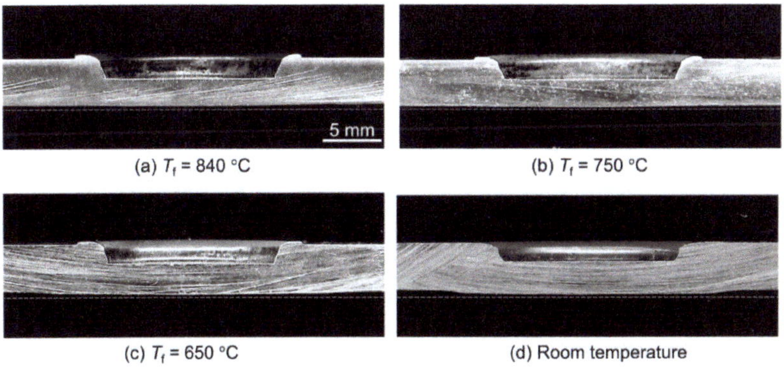

Fig. 3. Cross sections of compressed portion obtained at different compression temperature for r = 50%, t_h = 20s.

Materials Research Forum LLC
https://doi.org/10.21741/9781644903254-4

The relationships between a maximum load and obtained compression ratio in hot local compression for T_f = 840°C without sheet holder are shown in Fig. 4. The sheet holder was removed to evaluate the forming load because the load increased rapidly when the sheet holder interacted with it. When obtaining the same reduction, the maximum load at T_f = 840°C was approximately 40% lower than that of cold compression. In addition, the obtained compression ratio was close to the nominal one due to a decrease in the elastic deformation of the punch.

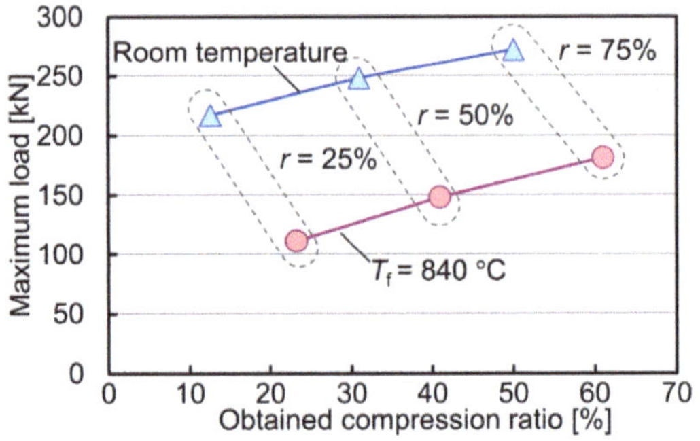

Fig. 4. Relationship between maximum load and obtained compression ratio in hot local compression for T_f = 840°C without sheet holder.

The effect of holding time on the obtained compression ratio for hot local compression for T_f = 650°C and 840°C are shown in Fig. 5. For T_f = 840°C, an increase in compression is observed in the early stages of holding time at the bottom dead centre. In contrast, it does not occur for T_f = 650°C.

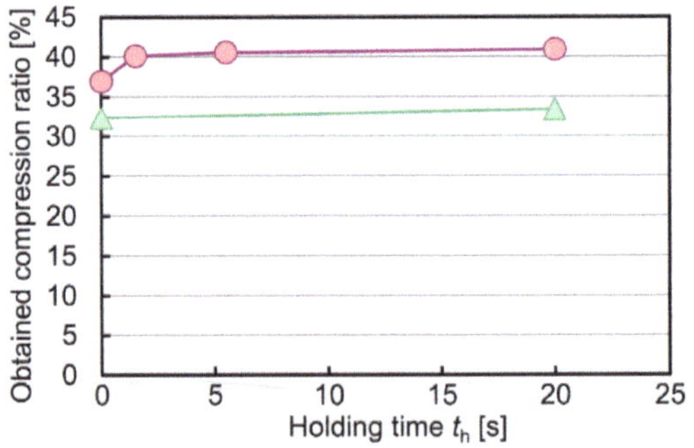

Fig. 5. Effect of holding time on obtained compression ratio for hot local compression for T_f = 650°C and 840°C.

Mechanism of increasing in compression ratio by holding at bottom dead centre

The mechanism of increase in compression ratio by holding at the bottom dead centre for T_f = 840°C was investigated. The relationship between sheet temperature at the compressed portion after 0.5 s from compression and holding time for T_f = 840°C r = 50 % is shown in Fig.6. For reference, the Ms point and Mf point of a 22MnB5 steel sheet with a similar composition are also shown. The Martensite start (M_s) and finish (M_f) temperature of a 22MnB5 steel sheet with a similar composition are also referenced in Fig 6. The Martensitic transformation has not started because the temperature is maintained at 700°C at $t_h = 0$ s. At $t_h = 1.5$ s, the temperature is below 400°C, and martensitic transformation begins, and the transformation appears to be ongoing until $t_h = 5\text{-}10$ s.

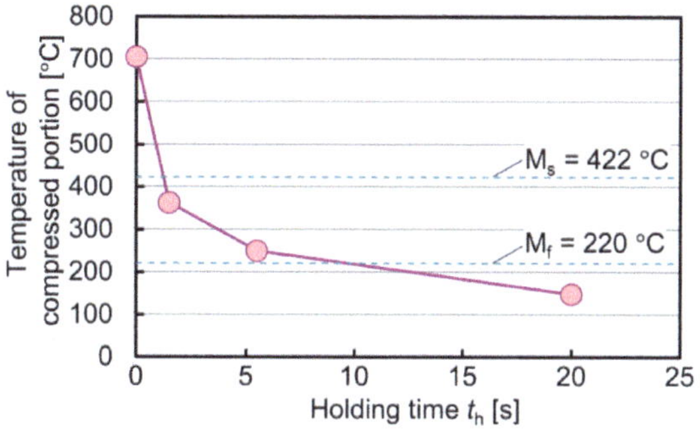

Fig. 6. Relationship between temperature at compressed portion after 0.5 s from compression and holding time for T_f = 840 °C r = 50 %.

The compression load-time curves during holding at the bottom dead centre for r = 50% are given in Fig. 7. The step noise at 10 s is due to the specifications of the servo press and can be ignored. The load decreases monotonically for the cold an T_f = 650°C as under transformation temperature. It is thought that the holding load decreased monotonically due to creep deformation and thermal shrinking of the sheet. On the other hand, for T_f = 750°C and 840°C, the load decreases until 1.5 s and then increases from 1.5 to 5 s. After that, the load shows a gradual decrease again.

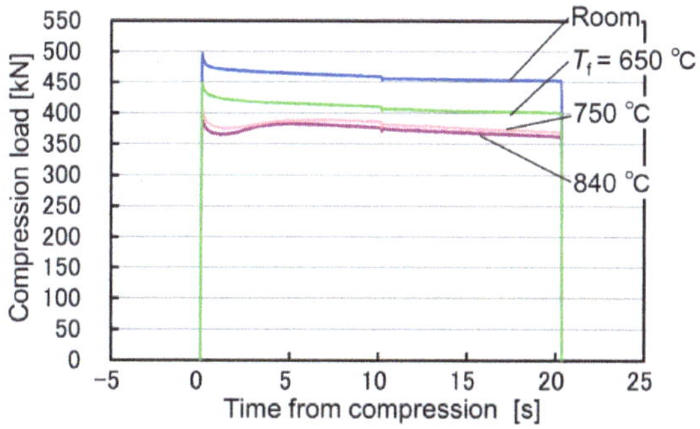

Fig. 7. Compression load-time curves during holding at bottom dead centre for r = 50%.

The mechanism of improvement of compression ratio by holding at a bottom dead centre in local hot compression is illustrated in Fig. 8. When without holding at the bottom dead centre, volume expansion by face transition from fcc to bcc or bct structures occurs without tool restraint. Hence, volume expansion occurs in isotropic, and naturally, the thickness of the compressed part increases. On the other hand, for holding at the bottom dead centre, the die fixes the expansion in a thickness direction during a martensitic transformation of the compressed portion. As a result,

Materials Research Forum LLC
https://doi.org/10.21741/9781644903254-4

the compressed portion is plastically deformed in the radial direction, and the reduction ratio is improved during holding at the bottom dead centre.

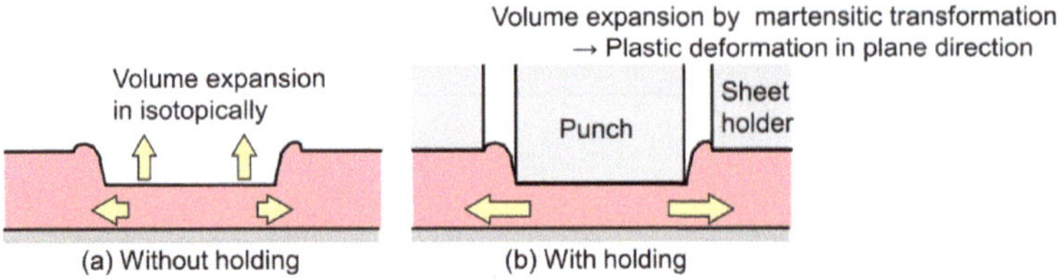

Fig. 8. Mechanism of improvement of compression ratio by holding at bottom dead centre in local hot compression.

Effect of local compression on die-quenched hardness

The hardness distributions of the compressed portion for T_f = 840°C are shown in Fig. 9. The hardness of water-quenching without compression is also shown for reference. The hardness peripheral to the compressed portion, i.e., the non-compressed portion, is slightly lower than water-quenching due to auto-tempering with the low cooling rate. Meanwhile, for the compressed part, the hardness increases as the compression ratio increases, exceeding 600HV at r = 75%.

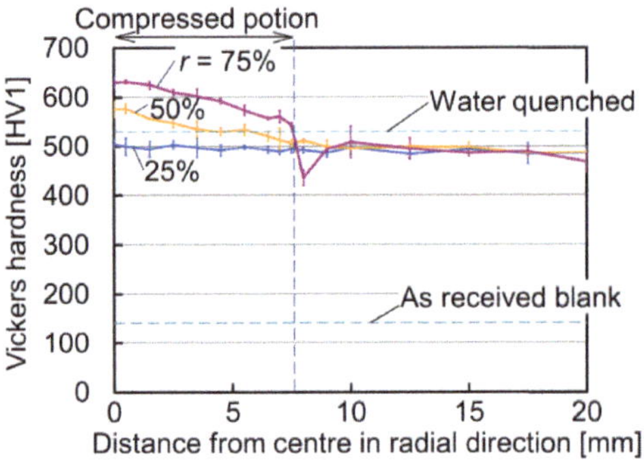

Fig. 9. Hardness distributions of compressed portion for T_f = 840°C.

The microstructures of a cross-section at the centre of a hot local compressed portion after die quenching for T_f = 840°C and t_h = 20 s are shown in Fig. 10. The microstructure was obtained by scanning electron microscopy after etching with Nital. The microstructure of water quenched without compression is also shown for comparison. All microstructures show the martensite. In addition, the appearance becomes finer as the compression ratio increases, and the structure at r = 75% is noticeably finer than that of r = 25% and water-cooling. The increase in hardness in the compressed portion shown in Fig. 9 is thought to be due to the refinement of the martensitic structure. This hardness increase by compression is similar to ausforming, which is composed of rapid cooling, plastic deformation, and quenching [7]. As shown in Fig. 6, the compression was performed at 840°C-700°C. Hence, it can be said to be modified ausforming [8]. In this process, a compressed portion is quenched by punch holding immediately after local compression without transferring to the next stage. Since the plastically deformed austenite is die-quenched without recrystallization, modified ausforming has been achieved. Ausforming has little practical use because it is difficult to control the forming temperature and timing of quenching it before recrystallization. In plate forging hot stamping, which combines large plastic deformation and die-quenching, the application of ausforming is highly feasible. It is expected to increase not only hardness but also toughness by Ausforming [9].

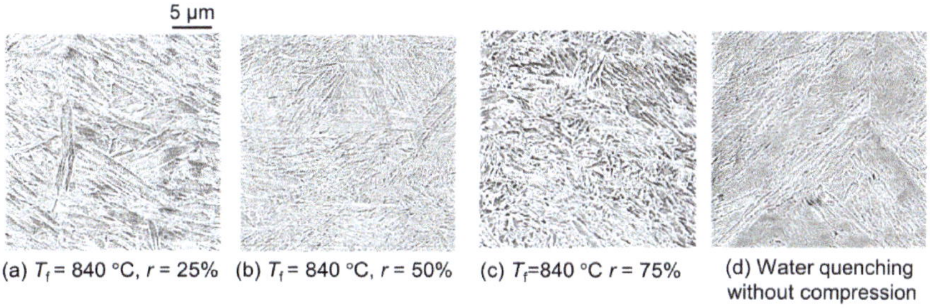

Fig. 10. Microstructure of cross-section at centre of hot local compressed portion after die quenching for T_f = 840°C and t_h = 20 s.

A required holding time for die quenching tends to be extended in hot plate forging due to the high heat capacity of a thick steel sheet. This is a bottleneck in the production line. Direct water cooling is applied to hot stamping to reduce the holding time [10]. However, although applying only water quenching omits a holding, the improvement in compression ratio by holding will be lost. Furthermore, ausforming may not be achieved due to recrystallisation caused by a cooling delay in the transfer for water cooling. Therefore, direct water cooling was applied after short holding at the bottom dead centre. The transferring time between holding and water cooling is 8 s.

The effect of a holding time and water-cooling on a hardness distribution for r = 50% and T_f = 840°C is shown in Fig. 11. At T_h = 1.5, the sheet was transferred to a water bath from the tool during martensitic transformation, resulting in insufficient quenching. For water cooling after holding T_h = 5.5 s as that martensite transformation finished, a hardness distribution similar to that of T_h = 20 s was obtained.

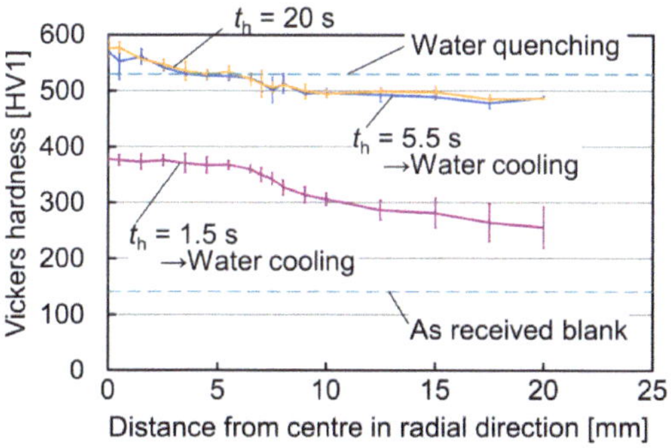

Fig. 11. Effect of holding time and water-cooling on hardness distribution for r = 50% and T_f = 840°C.

Conclusion

In this study, hot local compression of a quenchable thick steel sheet was performed to apply hot stamping having die quenching to plate forging. The following results were obtained. The forming load for hot local compression is significantly low compared to cold forming, and it is advantageous in plate forging, which often uses compression in the thickness direction.

- Holding at the bottom dead centre provides a die quenching and improves the compression ratio. The mechanism is plastic deformation induced by volume expansion during martensitic transformation under holding with the punch tool.
- The die-quenched hardness after local compression is increased by ausforming. The microstructure of that became finer.
- The holding time at the bottom dead centre was reduced using water cooling and die quenching together. Provided the holding was kept until the completion of martensitic transformation, the improvement in compression ratio by holding and hardening by ausforming was obtained even for water cooling.

Hot plate forging is a multi-step forming process similar to progressive stamping. Since there is a temperature drop at each forming stage, an appropriate order of forming stages, such as bending, shearing, etc., should be discussed. In addition, the required cooling rate and the relationship between the strain and the hardness after quenching should be examined in ausforming using die quenching.

References

[1] K. Mori, T. Nakano, State-of-the-art of plate forging in Japan, Prod. Eng. Res. Devel. 10 (2016) 81–91. https://doi.org/10.1007/s11740-015-0648-1

[2] D. Gröbel, R. Schulte, P. Hildenbrand, M. Lechner, U. Engel, P. Sieczkarek, S. Wernicke, S. Gies, A.E. Tekkaya, B.A. Behrens, S. Hübner, M. Vucetic, S. Koch, M. Merklein, Manufacturing of functional elements by sheet-bulk metal forming processes, Prod. Eng. Res. Devel. 10 (2016) 63–80. https://doi.org/10.1007/s11740-016-0662-y

[3] T. Maeno, K. Osakada, K. Mori, Reduction of friction in compression of plates by load pulsation, Int. J. Mach. Tool. Manuf. 51 (2011) 612–617. http://dx.doi.org/10.1016/j.ijmachtools.2011.03.007

[4] T. Maeno, K. Mori, A. Hori, Application of load pulsation using servo press to plate forging of stainless steel parts, J. Mater. Proc. Tech. 214 (2014) 1379–1387. https://doi.org/10.1016/j.jmatprotec.2014.01.018

[5] K. Mori, P.F. Bariani, B.-A. Behrens, A. Brosius, S. Bruschi, T. Maeno, M. Merklein, J. Yanagimoto, Hot stamping of ultra-high strength steel parts, CIRP Annals 66 (2017) 755–777. https://doi.org/10.1016/j.cirp.2017.05.007

[6] K Mori, T Maeno, M. Tsuchiya, T. Nanya, Inclusion of hot stamping operations in progressive-die plate forging of tailored high strength gear part, Int. J. Adv. Manuf. Technol., 90 (2017) 3585–3594. DOI: 10.1007/s00170-016-9705-y

[7] A. Tekkaya, J. Allwood, P. Bariani, S. Bruschi, J. Cao, S. Gramlich, P. Groche, G. Hirt, T. Ishikawa, C. Löbbe, J. Lueg-Althoff, M. Merklein, W.Z. Misiolek, M. Pietrzyk, R. Shivpuri, J. Yanagimoto, Metal forming beyond shaping: Predicting and setting product properties, CIRP Ann. Manuf. Technol. 64 (2015) 629–653. https://doi.org/10.1016/j.cirp.2015.05.001

[8] C.M. Yen, C.A. Stickels, Lamellate fracture in 5150 steel processed by modified ausforming, Metall. Trans. 1 (1970) 3037–3047. https://doi.org/10.1007/BF03038417

[9] M.W. Vaughan, P. Samimi, S.L. Gibbons, R.A. Abrahams, R.C. Harris, R.E. Barber, I. Karaman, Exploring performance limits of a new martensitic high strength steel by ausforming via equal channel angular pressing, Scr. Mater. 184 (2020) 63–69. https://doi.org/10.1016/j.scriptamat.2020.03.011

[10] T. Maeno, K. Mori, M. Fujimoto, Improvements in productivity and formability by water and die quenching in hot stamping of ultra-high strength steel parts, CIRP Ann. Manuf. Technol. 64 (2015) 281–284. https://doi.org/10.1016/j.cirp.2015.04.128

The influence of dynamic characteristics of six-bar linkage mechanical press transmission system on the accuracy of slider

ZHAI Hua[1,a*], ZHANG Yingjie[1,b], SONG Zhaoai[1,c], ZANG Congyun[2,d], ZHANG QingLin[2,e], LV Pin[3,f] and LI Jiajie[1,g]

[1]Anhui Province Key Lab of Aerospace Structural Parts Forming Technology and Equipment, Hefei University of Technology

[2]Jiangsu CPTEK Servo Technology Co.,Ltd., Liyang, 213300, China

[3]Tai'an Hualu Forging Machine Tool Co., Ltd., Taian, 271000, China1 Heidelberg, Germany

[a]jxzhaihuajx@hfut.edu.cn, [b]zyjcq@foxmail.com, [c]2304655692@qq.com, [d]zangchongyun@cptek.com.cn, [e]zhangqinglin@cptek.com.cn, [f]lupin@hfut.edu.cn, [g]2022170266@mail.hfut.edu.cn

Keywords: Six Link Mechanical Press, Dynamic Model, Drive System, Dynamic Characteristis, Orthogonal Experimental Method

Abstract. The six-bar linkage mechanical press is a crucial equipment in the preparation of metal bipolar plates for fuel cells. The dynamic characteristics of the transmission system have a significant impact on the precision of the slide operation, which directly affects the quality of the bipolar plate forming. In this article, A dynamic model of the six-bar linkage transmission system under complex forming workloads was established to research dynamic characteristics. The dynamic characteristics of each bar was analyzed under load conditions. Combining with the orthogonal experimental method, we investigate the various parameters influence on the press slide precision. Through an six-factor orthogonal experiment, the results show that the down-push rod dynamic characteristics are the most important factor affecting the slide stroke precision under the transmission system's workload conditions. Therefore, it is necessary to optimize the down-push rod design and manufacturing process. The results of this study have a certain reference value for the design and optimization of the six-bar linkage mechanical press, and the improvement of the quality of metal bipolar plate preparation for fuel cells.

Introduction

As a critical component within the fuel cell framework, the metal bipolar plate facilitates pivotal functions such as the direction of reactive gases, current guidance, provision of structural support, drainage, and acting as a cooling component[1]. The six-bar mechanical press is a significant apparatus in the fabrication of fuel cell metal bipolar plates, with the dynamical characteristics of its transmission system having a profound impact on the precision of the slider operation, thereby directly influencing the forming quality of the bipolar plates[2].In the analysis of mechanisms, the dynamic force and positioning accuracy of mechanical presses are crucial [3]. Some scholars have pointed out that factors such as clearance, friction, stiffness, and geometric deviation have significant impacts on them [4-6]. Numerous simulations and experiments [7-11] have also confirmed that compared to the rigid model, the rigid-flexible coupling model considering clearance and lubrication fits better with experimental data. The existence of clearance and friction as well as the stress deformation of the connecting rods can significantly affect the dynamic characteristics and slider accuracy of the press. Some scholars [12-14] have considered the reliability of dynamic accuracy for multi-clearance multi-connecting rod, and improved the dynamic accuracy of mechanical presses through comprehensive optimization of mathematical models, closed-loop control, and other methods.

This manuscript focuses on the six-bar mechanical press, establishes a simulated dynamic model for the six-bar mechanism. It explores the deformation and strain of each bar post load application, and with the orthogonal experimental method, the impact of parameters post deformation under force on the operational precision of the press slider has been improved.

Construction and analysis of the simulated dynamic model for the six-bar mechanism

Construction of the kinematic model for the six-bar mechanism. The kinematic relationship of the six-bar mechanism is illustrated in Fig. 1. The primary design parameters of the six-bar mechanism include the crank length L1, the pull rod length L2, the lengths of the two sides of the triangle frame L3 and L4, the push rod length L5, the fixed point coordinates x and y connecting the triangle frame and the machine frame , the slider offset e, and the angle between the two sides of the triangle frame θ. When these parameters are fixed, the kinematic trajectory of the entire mechanism is fully determined. The kinematic expression of the six-bar mechanism is shown in Eq. 1. By substituting the parameters in Table 1 for calculation, the slider stroke curve shown in Fig. 2 can be obtained.

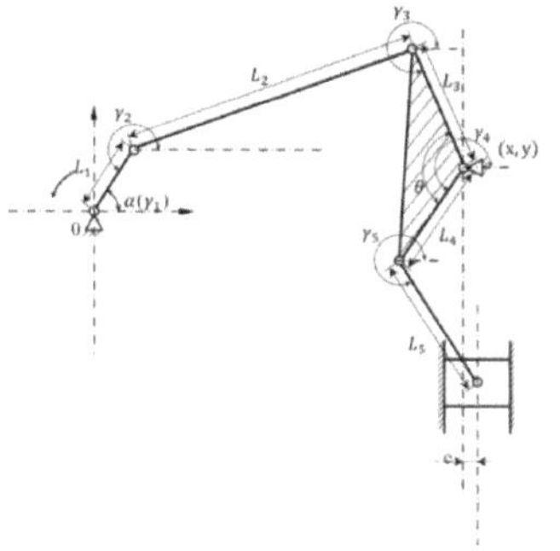

Fig. 1. Schematic diagram of the kinematic relationship of the six-bar mechanism.

The variable γ_i represents the angle between each rod and the horizontal direction, while S corresponds to the y-component of the slider's coordinates.

Table 1. Parameter specific values.

Parameter	Value
L_1/mm	113
L_2/mm	1000
L_3/mm	330
L_4/mm	270
L_5/mm	500
x/mm	1050
y/mm	100
e/mm	0
θ/°	134

$$\begin{cases} L_1 \cos\gamma_1 + L_2 \cos\gamma_2 + L_3 \cos\gamma_3 - x = 0 \\ L_1 \sin\gamma_1 + L_2 \sin\gamma_2 + L_3 \sin\gamma_3 - y = 0 \\ x + L_4 \cos\gamma_4 + L_5 \cos\gamma_5 - e = 0 \\ y + L_4 \sin\gamma_4 + L_5 \sin\gamma_5 - S = 0 \end{cases} \quad (1)$$

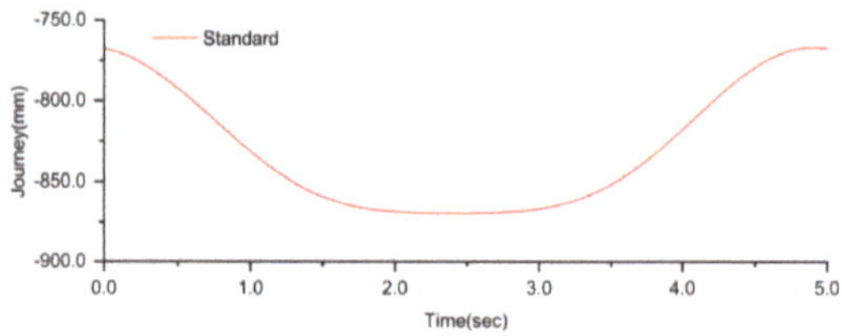

Fig. 2. Standard stroke curve.

Under the assumption that only the kinematic constraints between the links are considered, neglecting factors such as the deformation of the links and clearances at the hinge points, it can be assumed that the motion of the slider is perfectly accurate. Therefore, the trajectory of the slider under this ideal condition can be taken as the reference.

Establishment of dynamic model of six-link multi-flexible body simulation. Using dynamic analysis software software, a simulation model of a six-bar linkage can be established. Rotational joint constraints are set at each hinge point, while a translational joint constraint is set at the slider. All the linkages except the slider of the six-link mechanism are flexibly processed. The crank rotates at a speed of 72 degrees per second. When the slider reaches the loading point, an upward load of 80 tons is applied. When the slider reaches the unloading point, the load force is removed. The loading point is located 20 mm above the bottom dead center, while the unloading point is located 1mm above the bottom dead center. The cross-section of each link is a rectangular shape with dimensions of 40 mm × 40 mm.

Based on the provided conditions and procedures, the simulation model is configured, and solving the model yields the slider's displacement curve under working load as well as the deformation and force results for each link. As shown in Fig. 3 below, it presents a comparison between the slider's displacement curve considering the deformation of the six-bar linkage mechanism and the standard displacement curve.

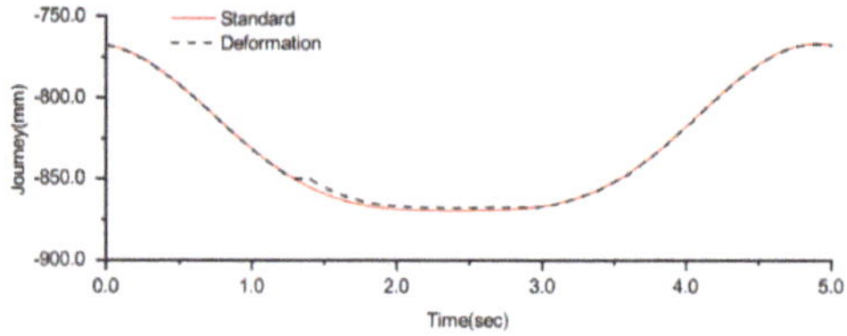

Fig. 3. Comparison diagram of load slide block stroke and standard stroke.

From Fig. 3, there is a certain difference between the slider's displacement under load and the standard displacement curve. This indicates that the dynamic characteristics of the transmission system after being loaded have a significant influence on the operational precision of the slider. Furthermore, it can be observed that after reaching the loading point where the load force is applied, the slider's displacement remains unchanged for a period of time. This implies that during this period, the six-bar linkage begins to undergo elastic deformation due to the applied load force, which is the main reason for the change in slider precision. Due to the applied load force, the six-bar linkage mechanism remains in a state of compression deformation as it operates between the load point and the unloading point. If the overall stiffness of the six-bar linkage mechanism is insufficient, it may lead to instances where the slider starts its return stroke before reaching the unloading point and the bottom dead center. This significantly reduces the formed product quality of the six-bar linkage mechanical press.

Analysis of the influence of the deformation of each connecting rod on the running accuracy of the slider

Deformation and condition of each bar under load. Based on the meticulous analysis of the multi-body soft-body dynamics simulation, the deformation and force distribution of each link under the load conditions are illustrated in Fig. 4 and Fig. 5, respectively.

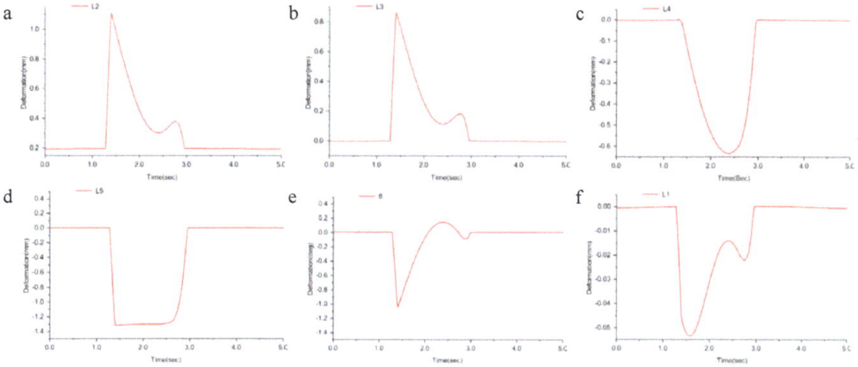

Fig. 4. Deformation diagram of each connecting rod under load. (a) L1 bar load deformation diagram; (b) L2 bar load deformation diagram; (c) L3 bar load deformation diagram; (d) L4 bar load deformation diagram; (e) L5 bar load deformation diagram; (f) θ angle load deformation diagram.

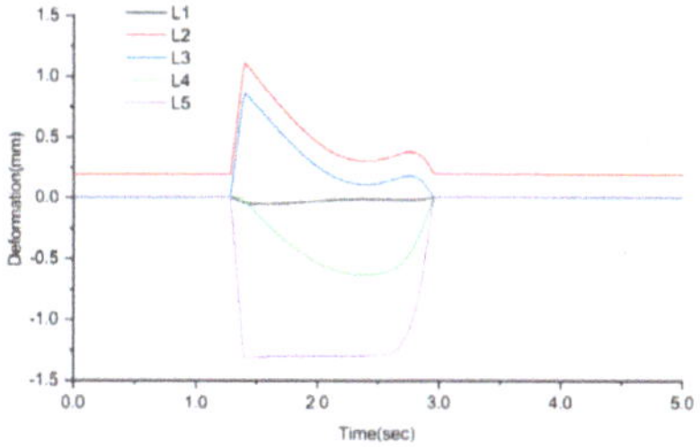

Fig. 5. Comparison diagram of load deformation of each connecting rod.

From Fig. 4 and Fig. 5, The slider causes the maximum deformation in link L5, and the triangular frame also undergoes deformation in terms of its angle. Additionally, the maximum deformation of each link in the six-bar mechanism at the point of force application and the deformation of each link at the bottom dead center position can be determined. By utilizing the formula for strain calculation, the strain values for each link can be obtained and are presented in Table 2.

Table 2. Results of deformation and strain of each bar under loading.

NO.	Maximum deformation at the load point /mm	Maximum strain at the load point	Deformation at the lower dead point /mm	Strain at the lower dead point
L1	-0.0532	-0.4708×10^{-3}	-0.0142	-0.1257×10^{-3}
L2	1.1091	1.1091×10^{-3}	0.3038	0.3083×10^{-3}
L3	0.8617	2.6112×10^{-3}	0.1139	0.3452×10^{-3}
L4	-0.6344	-2.3496×10^{-3}	-0.6298	-2.3326×10^{-3}
L5	-1.3154	-2.6308×10^{-3}	-1.293	-2.586×10^{-3}

Likewise, the maximum deformation at the load point for the angle θ is -1.0429°, with a corresponding maximum strain of -7.78×10^{-3}. At the bottom dead center position, the deformation is 0.1431°, and the strain is 1.0679×10^{-3}. Based on the analysis presented in Table 2, it can be concluded that under the loaded condition, link L5 exhibits the highest deformation and strain at both the point of force application and the bottom dead center position.

Analysis of the Influence of rod deformation on the accuracy of slider load point. Six-bar linkage mechanism undergoes deformation of its members when subjected to loading at the position of the load point. In comparison to the standard stroke curve, the slider exhibits a phenomenon of lagging behind as it crosses the load point. This lagging phenomenon results in the crank reaching a specific phase while the slider fails to reach its ideal position during the interval between the slider's movement from the load point to the unloading point, thereby reducing the precision of slider motion.

The numerical values of the maximum deformation of each member at the load point can be obtained from Table 2. The results have shown that when considering only the tensile and compressive stiffness of each member, increasing the stiffness of the members of the six-bar linkage mechanism in various ways will result in a decrease in the maximum deformation of each member at the load point. Additionally, different combinations of stiffness values for each member correspond to a set of maximum deformations for the members, as shown in Eq. 2.

$$f(E_1, E_2, E_3, E_4, E_5) = [\Delta L_1, \Delta L_2, \Delta L_3, \Delta L_4, \Delta L_5] \tag{2}$$

Where, E_i is the stiffness of each bar, while ΔL_i denotes the maximum deformation of each bar at the location of load. When only the stiffness of each bar of the given six-bar mechanism is increased, the maximum deformation of each bar at the load point exhibits a relationship as depicted in Eq. 3.

$$|\Delta L_i| < |\Delta L_{i0}| \tag{3}$$

Where, ΔL_{i0} represents the maximum deformation of each bar at the point of load under the state given in Section Establishment of dynamic model of six-link multi-flexible body simulation, as shown in Table 2.

Using the phase angle φ_0 at the point of load when the slider runs to the standard stroke as the reference value, we take the intermediate value ΔL_i between the static state and the maximum deformation value in Table 2 in the six-bar linkage mechanism, and add it to the initial parameters in Table 1 as the design value of the bar length of the six-bar linkage mechanism. At this design value, we calculate the phase angle φ of the crank when the slider runs to the load position under the rigid state. By comparing the difference Δφ between this phase angle φ and the phase angle φ_0 obtained at the standard stroke, the accuracy of the load point under the stiffness E_i of the slider can be compared with the selected intermediate value ΔL_i as the corresponding maximum deformation at the load point.

An analysis of the influence of the accuracy of the slider's load point on various link deformations can be conducted using the orthogonal experimental method. The orthogonal experimental method is a statistical experimental design method used to study the influence of multiple factors on the response variable. Compared with the method of full factorial full level experiments, orthogonal experimental method can determine the main factors that affect the experimental indicators through fewer experiments [15-17]. By employing a special design matrix (orthogonal table), this method can effectively evaluate the main factors and interaction effects within a limited number of experimental trials.

First, the assessment index is established as the difference $\Delta\varphi$ between the phase angle φ of the load point position under different factors and the phase angle φ_0 of the load point under the standard stroke.

Secondly, only the influence of link deformations on the accuracy of the slider's load point is considered, and the main factors are determined to be: L1, L2, L3, L4, and L5. The reference parameters are as shown in Table 1, with each parameter having an offset range of $[0, \Delta L_{i0}]$. As obtained from the aforementioned simulation, the phase angle $\varphi_0 = 122.16°$ of the load point under the standard stroke is determined.

Considering that the included angle θ of the triangle rod is one of the design values and also varies under load, θ is likewise considered as an auxiliary factor. Therefore, there are a total of 6 factors, and to ensure greater accuracy in the analysis results, each factor is divided into 5 levels. It is finally decided to use the orthogonal table $L_{25}(5^6)$, as shown in Table 3.

Table 3. $L_{25}(5^6)$ Orthogonal table.

NO.	Factor A	Factor B	Factor C	Factor D	Factor E	Factor F
1	1	1	1	1	1	1
2	1	2	3	4	5	2
3	1	3	5	2	4	3
4	1	4	2	5	3	4
5	1	5	4	3	2	5
6	2	1	5	4	3	5
7	2	2	2	2	2	1
8	2	3	4	5	1	2
9	2	4	1	3	5	3
10	2	5	3	1	4	4
11	3	1	4	2	5	4
12	3	2	1	5	4	5
13	3	3	3	3	3	1
14	3	4	5	1	2	2
15	3	5	2	4	1	3
16	4	1	3	5	2	3
17	4	2	5	3	1	4
18	4	3	2	1	5	5
19	4	4	4	4	4	1
20	4	5	1	2	3	2
21	5	1	2	3	4	2
22	5	2	4	1	3	3
23	5	3	1	4	2	4
24	5	4	3	2	1	5
25	5	5	5	5	5	1

The levels corresponding to each factor are shown in Table 4.

Table 4. Factor and level comparison table.

Level	Factor					
	L1	L2	L3	L4	L5	θ
	A	B	C	D	E	F
1	-0.0106	0.2218	0.1725	-0.1269	-0.2631	-0.2086
2	-0.0213	0.4436	0.3447	-0.2538	-0.5262	-0.4172
3	-0.0319	0.6655	0.5170	-0.3806	-0.7892	-0.6257
4	-0.0426	0.8873	0.6894	-0.5075	-1.0523	-0.8343
5	-0.0532	1.1091	0.8617	-0.6344	-1.3154	-1.0429

Based on Table 3 and Table 4, conduct the experimental design, calculate the phase angle differences for each experimental combination, and enter the results into the right-hand column of the table, as shown in Table 5.

Table 5. Experimental data.

NO.	Factor						$\Delta\varphi$
	A	B	C	D	E	F	
1	1	1	1	1	1	1	1.296
2	1	2	3	4	5	2	4.032
3	1	3	5	2	4	3	4.032
4	1	4	2	5	3	4	4.896
5	1	5	4	3	2	5	5.04
6	2	1	5	4	3	5	5.04
7	2	2	2	2	2	1	1.872
8	2	3	4	5	1	2	2.736
9	2	4	1	3	5	3	4.608
10	2	5	3	1	4	4	4.752
11	3	1	4	2	5	4	4.896
12	3	2	1	5	4	5	5.616
13	3	3	3	3	3	1	2.592
14	3	4	5	1	2	2	5.216
15	3	5	2	4	1	3	3.456
16	4	1	3	5	2	3	3.6
17	4	2	5	3	1	4	3.6
18	4	3	2	1	5	5	5.472
19	4	4	4	4	4	1	3.312
20	4	5	1	2	3	2	3.168
21	5	1	2	3	4	2	3.312
22	5	2	4	1	3	3	3.456
23	5	3	1	4	2	4	4.176
24	5	4	3	2	1	5	4.32
25	5	5	5	5	5	1	3.888
K_{1j}^{avr}	3.8592	3.6288	3.7728	4.0384	3.0816	2.592	
K_{2j}^{avr}	3.8016	3.7152	3.8016	3.4848	3.7504	3.6928	
K_{3j}^{avr}	4.3552	3.8016	3.8592	3.8304	3.8304	3.8304	
K_{4j}^{avr}	3.8304	4.4704	3.888	4.0032	4.2048	4.464	
K_{5j}^{avr}	3.8304	4.0661	4.3552	4.5216	4.5792	5.0976	
R_j	0.5536	0.8416	0.5824	1.0368	1.4976	2.5056	

In order to assess the impact of various factor levels on performance indicators, the phase angle differences at all levels of factor j under level i are summed and divided by the number of levels, yielding the average of the experimental indicators at level i of factor j. The calculation process at level *1* of factor A is illustrated in Eq. (4).

$$K_{1A}^{avr} = \frac{1}{5}(1.296 + 4.023 + 4.023 + 4.896 + 5.04) = 3.8592 \quad (4)$$

The calculated results K_{ij}^{avr} and R_j should be filled into the corresponding positions in Table 5.
We conducted a sorting of the obtained range R_j, and thereby established the primary and secondary relationships of the factors' impact on the indicators.

$$F \rightarrow E \rightarrow D \rightarrow B \rightarrow C \rightarrow A \quad (5)$$

Analysis of the primary and secondary relationships reveals that, apart from the auxiliary factor of the angle φ of the tripod, under the load condition, the push rod L5 has the most significant impact on the accuracy of the point of load.
We divided the range R_j by the interval ε_i at the horizontal of the response factor (as shown in Table 6) to obtain the unit deformation range R_{i0} (as depicted in Table 6).

Table 6. Unit deformation range table.

Factor	Range R_j	Level interval ε_j	Unit deformation range R_{j0}
A	0.5536	0.01064	52.0331
B	0.8416	0.22182	3.7941
C	0.5824	0.17254	3.3754
D	1.0368	0.12688	8.1715
E	1.4976	0.26308	5.6927
F	2.5056	0.20858	12.0127

The analysis of Table 6 reveals that, apart from the auxiliary factor F, the deformation of the crank under unit deformation has a significantly higher impact on the accuracy of the slider's capacity to occur than the deformation of other rods. However, under general loading conditions, the crank is relatively short and experiences much smaller forces compared to other components, resulting in deformation far smaller than that of the other rods under the same tensile and compressive stiffness. Therefore, the impact on the accuracy of the slider's load point is not as significantly as that of the push rod L5.
Analysis of the impact of increased rod stiffness on the accuracy of slider movement. The deformation of each rod has a significant impact on the accuracy of slider movement. We investigated the influence of increased rod stiffness on the accuracy of slider movement by separately enhancing the stiffness of each rod in the simulation model. According to the stiffness calculation formula, the stiffness of a rod can be altered by modifying its cross-sectional area. Therefore, we increased the stiffness of each rod by sequentially changing the cross-sectional area from 30 mm×30 mm to 100 mm×100 mm in the simulation model. Subsequently, we conducted simulation calculations to obtain the travel curves of the slider after the stiffness of each rod was increased, as well as the travel curves of the slider after the stiffness of all rods was increased, as depicted in Fig. 6.

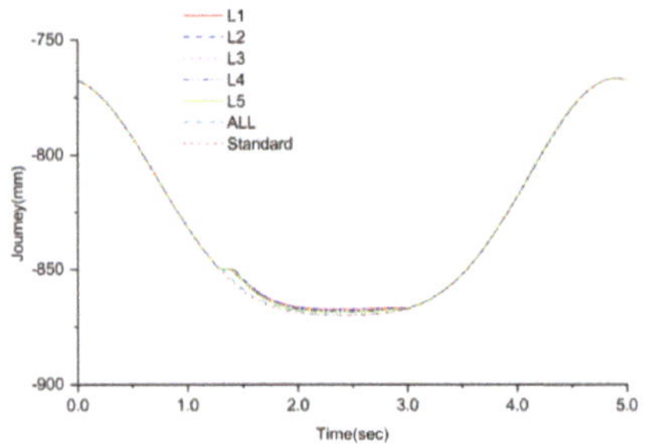

Fig. 6. Comparison diagram of stroke after change of stiffness of different connecting rod.

In Fig. 6, L1-L5 respectively represent the travel curves of the slider after the corresponding connecting rod stiffness is increased, while ALL represents the travel curve of the slider after all rod stiffness are increased, and standard represents the standard travel curve of the slider. Analysis of Fig. 6 reveals that after altering the stiffness of each slider, the travel curves in the unloaded travel segment are generally consistent, whereas in the loaded travel segment, significant differences are observed among the respective curves. The travel curves of each slider in the loaded travel segment are illustrated in Fig. 7.

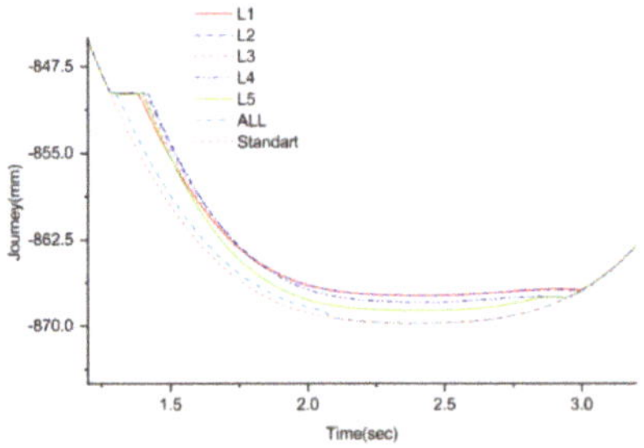

Fig. 7. Comparative diagram of loading stroke after the change of different connecting rod stiffness.

From Fig. 7, apart from the travel curve of the slider after increasing the stiffness of all connecting rods, the travel curve of the slider after increasing the stiffness of the push rod L5 closely approximates the standard curve during the loaded travel segment. Hence, it can be inferred that increasing the stiffness of the push rod L5 effectively enhances the operational precision of the slider. Compared to enhancing the stiffness of the push rod L5, increasing the stiffness of the crank can effectively ameliorate the lag phenomenon in the slider's movement when reaching the load point. Similarly, it has been observed that after enhancing the stiffness of all rods, there is a notable improvement in the operational precision of the slider during the loaded travel segment, and the slider has gained the capability to traverse the unloading point and reach the bottom dead center. Enhancing the overall stiffness of the six-bar linkage mechanism is an effective approach to address the inability of the slider to traverse the unloading point and reach the bottom dead center.

Conclusion

1) The dynamic characteristics of the six-bar linkage mechanical press transmission system significantly impact the precision of the slide motion.

2) Among the members of the six-bar linkage mechanism, the lower connecting rod L5 has the most significant influence on the accuracy of the capacity location after being subjected to loading.
3) The crank rod deformation in the six-bar linkage mechanism has the greatest impact on the precision of the capacity location of the slide.
4) Enhancing the stiffness of the lower connecting rod L5 in the six-bar linkage mechanism can effectively improve the precision of the slide motion.
5) Increasing the stiffness of the crank L1 can effectively ameliorate the lag phenomenon occurring when the slide reaches the capacity location.
6) Enhancing the overall stiffness of the six-bar linkage mechanism is an effective approach to address the phenomenon where the slide fails to pass the unloading point and reach the bottom dead center.

Acknowledgements
The experimental part of this paper is supported by Jiangsu CPTEK Servo Technology Co.,Ltd. This work was supported by Anhui Province Key Project No. 202103a07020010, and No. 202304a05020074, flexible molding equipment for carbon fiber composite materials, JZ2023GQBK0070.

References

[1] X. Li, I. Sabir, Review of bipolar plates in PEM fuel cells:flow-field designs, Int. J. Hydrogen Energy 30 (2005) 359-371. https://doi.org/10.1016/j.ijhydene.2004.09.019

[2] M. Xiao, G. Geng, G. Li, H. Li, R. Ma, Analysis on dynamic precision reliability of high-speedprecision press based on Monte Carlo method, Nonlinear Dynamics 90 (2007) 2979-2988. . https://doi.org/10.1007/s11071-017-3857-7

[3] V.V. Telegin, A.M. Kozlov, V.I. Sakalo, Solid modeling and dynamic analysis of mechanisms of press-forging machines, Procedia Eng. 206 (2017) 1258-1263. https://doi.org/10.1016/j.proeng.2017.10.628

[4] Z. Chval, M. Cechura, Optimization of power transmission on mechanical forging presses, Procedia Eng. 69 (2014) 890-896. https://doi.org/10.1016/j.proeng.2014.03.067

[5] N. Wagner, T. Päßle, Screening-based evidence of the impact of geometrical deviations on inner stresses on mechanical press, Procedia Manuf. 27 (2019) 185-191. https://doi.org/10.1016/j.promfg.2018.12.063

[6] Q. Tiana, P. Flores, H.M. Lankaranic, A comprehensive survey of the analytical, numerical and experimental methodologies for dynamics of multibody check for updates mechanical systems with clearance or imperfect joints, Mechanism and Machine Theory 122 (2018) 1-57. https://doi.org/10.1016/j.mechmachtheory.2017.12.002

[7] O. Muvengei, J. Kihiu, B. Ikua, Numerical study of parametric effects on the dynamic response of planar multi-body systems with differently locted frictionless revolute clearance joints, Mechanism and Machine Theory 53 (2012) 30-46.

[8] Zheng Feng Bai, Yang Zhao, Dynamics modeling and quantitative analysis of multibody systems including revolute clearance joint, Precision Engineering 36 (2012) 554-576.

[9] P. Flores, R. Leine, C. Glocker, Modeling and analysis of rigid multibody systems with translational clearance joints based on the nonsmooth dynamics approach, Multibody Dynamics: Computational Methods and Applications 107-130 (2011).

[10] Xinyu Dong, Yu Sun, Xuze Wu, et al. Dynamic modeling and performance analysis of

toggle-linkage presses considering mixed clearances and flexibility, International Journal of Non-Linear Mechanics, 147 (2022) 104243.

[11] Y. Chen, Y. Sun, C. Chen, Dynamic analysis of a planar slider-crank mechanism with clearance for a high speed and heavy load press system, Mechanism and Machine Theory 98 (2016) 81-100.

[12] X. Chen, S. Gao, Dynamic accuracy reliability modeling and analysis of planar multi-link mechanism with revolute clearance, European J. Mech. A Solids 90 (2021) 104317. https://doi.org/10.1016/j.euromechsol.2021.104317

[13] F. Hoppe, C. Pihan, P. Groche, Closed-loop control of eccentric presses based on inverse kinematic models, Procedia Manuf. 29 (2019) 240-247. https://doi.org/10.1016/j.promfg.2019.02.132

[14] Y. Li, Y. Su, S. Wang, Dimensional synthesis for multi-linkage of high-speed mechanical press, Procedia Eng. 81 (2014) 1639-1644. https://doi.org/10.1016/j.proeng.2014.10.204

[15] Z. Yuan, X. Chen, H. Zeng, K. Wang, J. Qiu, Identification of the elastic constant values for numerical simulation of high velocity impact on dyneema® woven fabrics using orthogonal experiments, Compos. Struct. 204 (2018) 178-191. https://doi.org/10.1016/j.compstruct.2018.07.024

[16] S. Zhang, Y. Mao, F. Liu, H. Xu, Z. Qu, X. Liao, Multi-objective optimization and evaluation of PEMFC performance based on orthogonal experiment and entropy weight method, Energ. Convers. Manage. 291 (2023) 117310. https://doi.org/10.1016/j.enconman.2023.117310

[17] U. Karaguzel, M. Bakkal, E. Budak, Mechanical and thermal modeling of orthogonal turn-milling operatio, Procedia CIRP 58 (2017) 287-292.

Research on circumferential tension-axial shear loading of thin-walled aluminum alloy tubes

ZHANG Shuning[1], WANG Xiaosong[2,3,a]*, YU Bingzhai[1] and LIU Gang[2,3]

[1]School of Materials Science and Engineering, Harbin Institute of Technology, No.92, West Dazhi Street, Harbin, 150001, China

[2]National Key Laboratory of Precision Hot Processing of Metals, Harbin Institute of Technology, No.92, West Dazhi Street, Harbin, 150001, China

[3]Institute of High Pressure Fluid Forming, Harbin Institute of Technology, No.92, West Dazhi Street, Harbin, 150001, China

[a]hitxswang@hit.edu.cn

Keywords: Thin-Walled Tube, Tensile and Shear Stress State, Circumferential Tension-Axial Shear, Strain Path, Aluminum Alloy

Abstract. The deformation of parts with complex shapes is under a tensile and shear stress state. The deformation characteristics of materials under the tensile and shear stress state are beneficial for improving the prediction accuracy of the deformation of parts with complex shapes. A loading method of circumferential tension-axial shear of thin-walled tubes using two semi-circle mandrels and one specially designed tubular specimen is proposed in this paper. An experimental device was developed according to this method using servo electric cylinders. A thin-walled 5052 aluminum tube was tested and the strain path of the center point of the deformation zone was analyzed. This loading method can be used to investigate the characteristics of tubular material under the tensile and shear stress state.

Introduction

Parts with complex shapes deform under a tensile and shear stress state [1, 2]. Understanding the deformation characteristics of materials under a tensile and shear stress state is beneficial for improving the simulation accuracy of the deformation of parts with complex shapes [3]. Tension and shear loading tests of thin-walled sheets are achieved by applying shear force in one direction and tensile force in the orthogonal direction. Mohr conducted the tension and shear loading experiment of TRIP700 steel sheets. It is indicated that the prediction accuracy of the Mises criterion is poor for the condition when the ratio of tensile stress to shear stress is smaller than 0.5 [4]. Pijlman utilized the tension and shear loading results of AA5182 aluminum alloy sheets to establish an anisotropic constitutive relation based on the Vegter model. The strain distribution of parts with irregularly shaped sections during the deep-drawing process is accurately predicted [5]. Zheng investigated the influence of pre-deformation on the formability of 6K21-T4 aluminum alloy sheets. It is indicated that the ultimate strain of the sheet is reduced by the equi-biaxial tensile and shear pre-deformation when the ratio of tensile stress to shear stress is smaller than 1, while the ultimate strain is increased when the ratio is larger than 1 [6].

Due to the limitations of tube shape, research on the tensile and shear deformation of tubes has not been reported so far. Wang used a pair of semi-circle mandrels to exert force along the axial direction on a specially designed tubular specimen. The pure shear deformation along the axial direction is achieved at the interface plane between the two half tubes of the specimen [7]. Based on the principle of axial pure shear loading of tubes, the circumferential tension and axial shear test is proposed by applying a tensile force perpendicular to the interface plane. An experimental device was developed and a thin-walled 5052 aluminum tube was tested. The strain path of the tube during the tensile and shear deformation process was analyzed in this paper. This loading method can be used to investigate the characteristics of materials under a tensile and shear stress state.

Principle of circumferential tensile-axial shear deformation

Based on the axial pure shear principle of thin-walled tubes, the circumferential tensile-axial shear deformation is achieved, as shown in Fig. 1. The interface plane of the two half tubes of the specimen is the pre-set deformation zone. A tensile force F_θ which is perpendicular to the interface plane of the two half tubes is applied to the mandrel, which causes circumferential stress σ_θ in the deformation zone. Under the combined loading of axial force F_z and tensile force F_θ, the deformation zone is under a tensile and shear stress state.

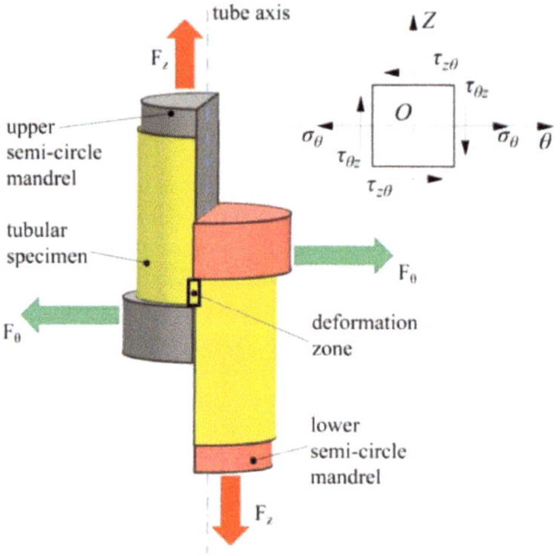

Fig. 1. Principle of tensile and shear deformation of tubes.

The interface plane is subjected to shear stress and tensile stress. The shear stress is calculated according to Eq. 1:

$$\tau_{\theta z} = \frac{F_z}{2*L*t} \tag{1}$$

and tensile stress according to Eq. 2:

$$\sigma_\theta = \frac{F_\theta}{2*L*t} \tag{2}$$

where L is the real-time axial length of the deformation zone, and t is the real-time wall thickness of the tube.

Development of the circumferential tension-axial shear experimental device

As shown in Fig. 2, the experimental device for circumferential tension-axial shear loading is composed of a frame, servo electric cylinders, and a Digital Image Correlation (DIC) measuring system. The specimen and the mandrels are placed along the horizontal direction. A servo electric cylinder placed along the horizontal direction is connected to the lower mandrel, applying a movement along the axial direction of the tube to the mandrel. Two servo electric cylinders are placed along the vertical direction. The two vertical servo electric cylinders are connected to the upper and the lower mandrels respectively, applying vertical tensile force to the mandrels. There are force sensors installed on each servo electric cylinder to measure the force. In the deformation zone, the shear stress is along the axial direction of the tube, while the tensile stress is along the circumferential direction. To ensure that the tensile force and axial force are orthogonal, a pair of linear guide rails are installed on the two vertical sides of the frame. The vertical servo electric cylinders are installed on the guide rails. The frame and the cylinders are shown in Fig. 3, wherein the maximum load of each servo electric cylinder is 37 kN.

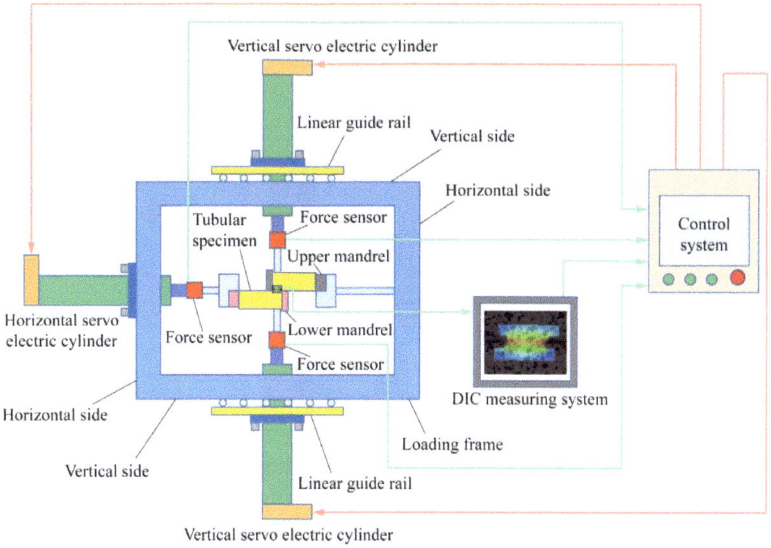

Fig. 2. Principal of circumferential tension-axial shear loading device.

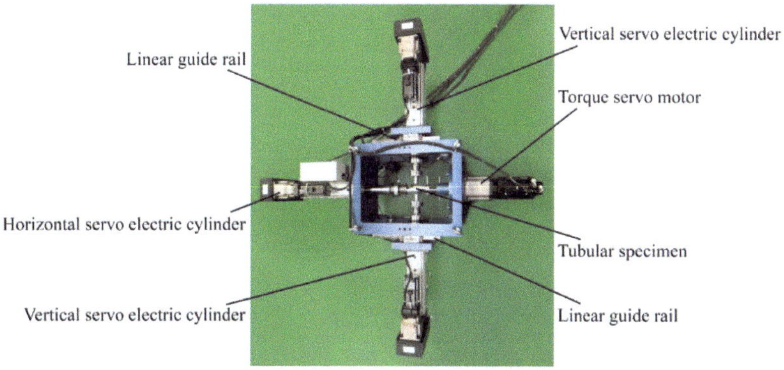

Fig. 3. Frame and cylinders of the experimental device.

Circumferential tension-axial shear loading experiments

A thin-walled 5052 aluminum alloy tube with a nominal dimension of 50×1.2mm was used in the experiment. The tubular specimen is shown in Fig. 4. The length of the deformation zone L is set to L/t = 10. Long grooves with the half-circle notch are machined at both ends of the deformation zone to minimize the influence of the force from the mandrel on the stress state. The axial length of the slots Ls is set to Ls/t = 2.5, the width d to d/t = 1, and the notch radius R to R/t = 0.5. The outer diameter D of the semi-circle mandrel is 47.5 mm, leaving a gap of 0.05 mm between the mandrel and the inner wall of the tube.

(a) Specimen

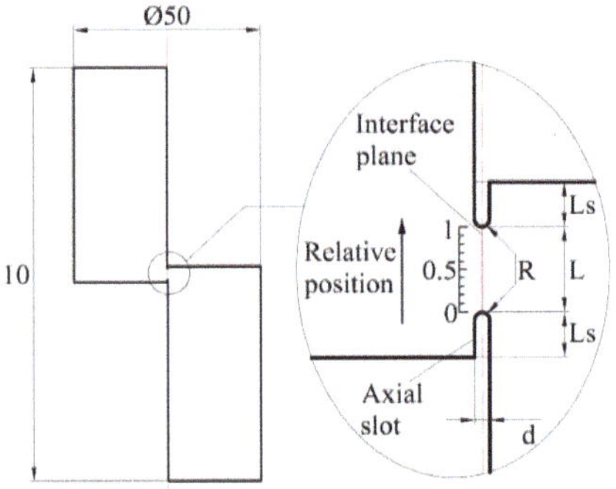

(b) Dimension(unit:mm)

Fig. 4. Tubular specimen.

Experiments were carried out using the loading plans in Table 1. The cracked specimens are shown in Fig. 5. For Loading plan 1, the crack surface in the deformation zone is flat, while the crack surface is inclined relative to the tube axis for plan 2. For both plans, no significant necking is exhibited on the specimen before cracked.

Materials Research Forum LLC
https://doi.org/10.21741/9781644903254-6

Table 1. Loading plan.

No.	Loading parameters
1	The ratio of tensile stress to shear stress is 0.5 until cracked
2	The ratio of tensile stress to shear stress is 2 until cracked

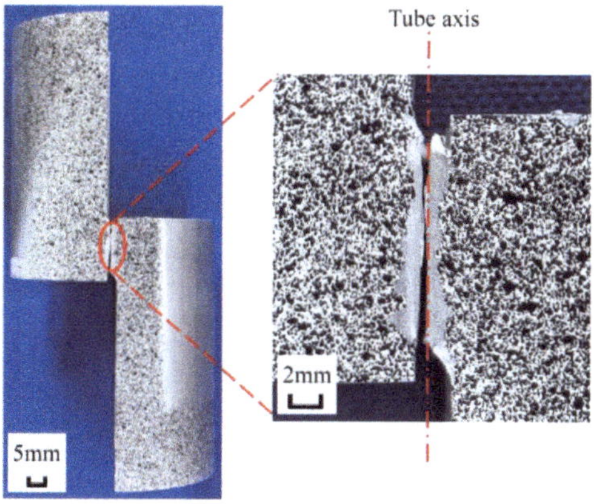

(a) Plan 1

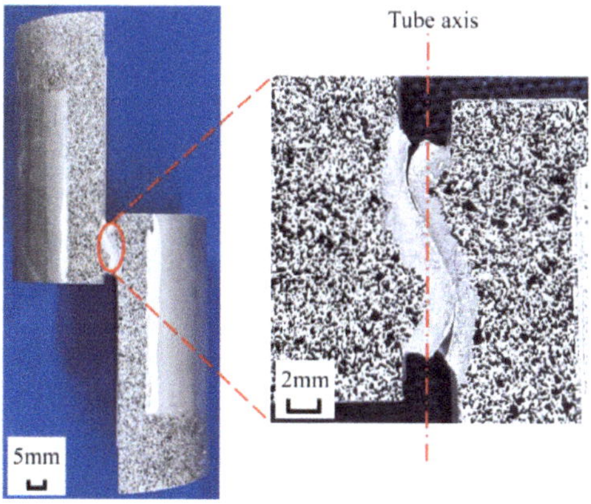

(b) Plan 2

Fig. 5. Cracked specimen.

The axial load-displacement curve and the circumferential load-displacement curve are shown in Fig. 6(a) and (b), respectively. The endpoint of each curve corresponds to its maximum value.

Materials Research Forum LLC
https://doi.org/10.21741/9781644903254-6

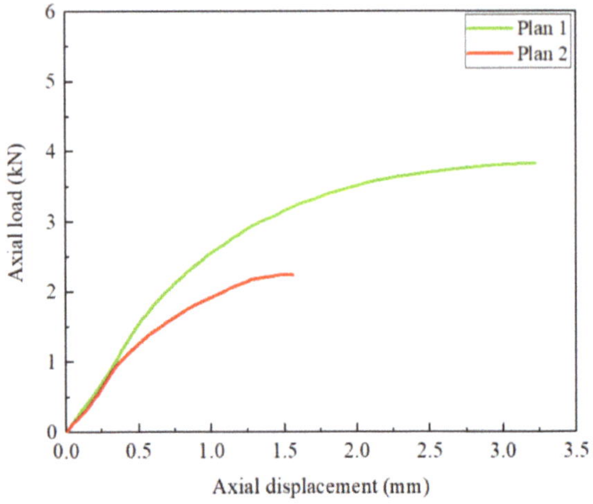

(a) Axial load-axial displacement

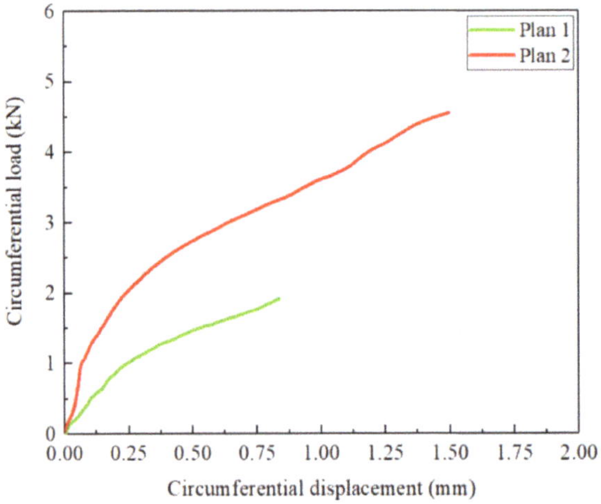

(b) Circumferential load-circumferential displacement

Fig. 6. Load-displacement curve.

The strain path of the center point of the deformation zone is shown in Fig. 7. The strain path of the tube under the circumferential plane strain tensile deformation is also shown. For plan 1, the strain path is in the range of pure shear to uniaxial tension, and the strain path for plan 2 is between plane strain tension and uniaxial tension. It is shown that the strain path between pure shear and plane strain can be achieved using this method by changing the ratio of shear stress to tensile stress.

 Materials Research Forum LLC
https://doi.org/10.21741/9781644903254-6

The next work is to investigate the influence of the shear stress on the forming limit of the thin-walled tubes under tensile-shear stress state.

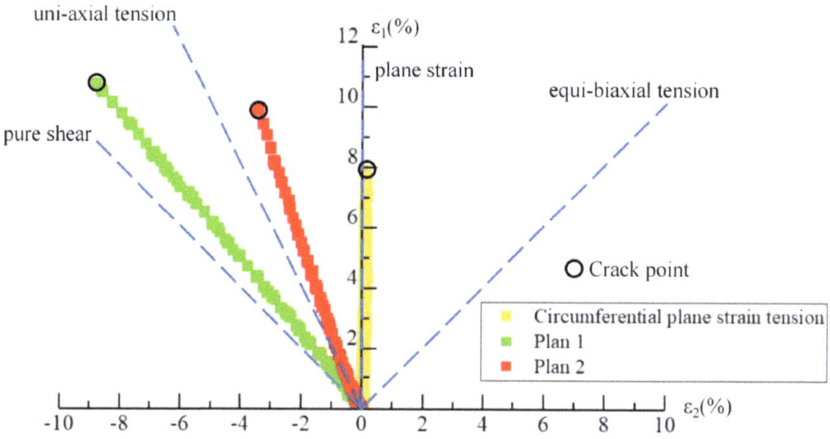

Fig. 7. Strain path.

Conclusions

A loading method for circumferential tension-axial shear of thin-walled tubes is proposed in this paper, an experimental device is developed and an aluminum tube is tested. The strain path of the tubular specimen between pure shear to plane strain tension can be achieved using this experimental device. The deformation characteristics and forming limit under a tension-shear stress state will be investigated in future work.

Acknowledgments

Financial assistance of the National Natural Science Foundation of China, Project 51975147, is acknowledged.

References

[1] C.P. Dick, Y.P. Korkolis, Mechanics and full-field deformation study of the ring hoop tension test, Int. J. Solids. Struct. 51 (2014) 3042–3057. https://doi.org/10.1016/j.ijsolstr.2014.04.023

[2] F.J. Gomez, M.A. Martin Rengel, J. Ruiz-Hervias, M.A. Puerta, Study of the hoop fracture behaviour of nuclear fuel cladding from ring compression tests by means of non-linear optimization techniques, J. Nucl. Mater. 489 (2017) 150-157. https://doi.org/10.1016/j.jnucmat.2017.03.043

[3] S.N. Zhang, Y.B. Pei, X.S. Wang, G. Liu, In-situ EBSD observation of texture evolution and lattice rotation behavior in shear deformation of thin-walled aluminum alloy tubes, J. Mater. Res. Technol. 29 (2024) 449–460. https://doi.org/10.1016/j.jmrt.2024.01.150

[4] D. Mohr, M. Oswald, A new experimental technique for the multi-axial testing of advanced high strength steel sheets, Exp. Mech. 48 (2008) 65–77. http://dx.doi.org/10.1007/s11340-007-9053-9

[5] H.H. Pijlman, Sheet material characterisation by multi-axial experiments (Doctoral thesis), University of Twente, The Netherlands, 2001.

[6] L.H. Zheng, Z.J. Wang, H. Song, Experimental Method for Multistage Loading Tests with Various Prestrain Paths, Exp. Mech. 59 (2019) 51–63.

[7] X.S. Wang, S.N. Zhang, M.C. Fu, W.L. Hu, G. Liu, Investigation on influence of mandrel shape on shear stress in pure shearing test of thin-walled aluminum alloy tubes, Procedia Manuf. 50 (2020) 609-612. http://dx.doi.org/10.1016/j.promfg.2020.08.109

Materials Research Proceedings 44 (2024) 60-69 https://doi.org/10.21741/9781644903254-7

Numerical investigation of buckling behaviour of grid stiffened panel with opening during creep age forming

ZHOU Qingheng[1], XIAO Xuelian[1], KANG Shaoming[1], HUANG Xia[1,2], LI Yong[1], LI Dongsheng[1] and ZHOU Wenbin[1,*]

[1]School of Mechanical Engineering and Automation, Beihang University Beijing 100191, China

[2]AVIC Manufacturing Technology Institute, Beijing 100024, China

* zhouwb@buaa.edu.cn

Keywords: Formability, Buckling, Stiffened Panel with Opening, Simplified Model, Finite Element Method

Abstract. This study focuses on the prediction of the formability of the integrally stiffened panels with opening due to buckling during the loading stage and heating stage of the creep age forming (CAF) process. The simplified model of integrally stiffened panels with opening was established to investigate the buckling behavior of the stiffened panel subjected to pure bending moment at room temperature and ageing temperature. The influence of the main geometrical parameters (stiffener thickness and stiffener height) on the critical buckling stress and critical buckling strain of the stiffened panel at the two temperatures was studied by finite element simulation. With the increase of the stiffener thickness and the decrease of the stiffener height, the critical buckling stress will increase to about 570 MPa at room temperature and 475 MPa at ageing temperature, which makes the formability limit increase correspondingly. When the stiffened panel is in the elastic region, the non-dimensional critical buckling coefficient is reduced by 20%, compared with the traditional stiffened panel. When the stiffened panel enters the plastic region, the critical buckling strain increases and the forming limit enhances accordingly. The research results offer a significant reference for the structural optimization design of the stiffened panel, presenting a simplified approach for studying other special shapes of stiffened panels, which is helpful to ensure that the workpiece will not be buckled during the forming process.

Introduction

Integrally stiffened panels with opening have a highly potential to replace conventional riveting stiffened panel in aerospace and marine structures [1], such as the aircraft fuselage, to achieve lightweight [2], provide ventilation channels, and accommodate features such as pipes or cables. Creep ageing forming (CAF), as a stiffened panel forming process, plays an important role in meeting the industrial demand [3,4]. The workpiece is loaded by bending moment at room temperature and heated to the ageing temperature for creep process, and then unloaded when the deformation is finished [5]. In the forming process, the framed part of the stiffened panel is under high pressure stress and will result in buckling [6]. Therefore, it is necessary to simplify the integrally stiffened panels with opening and investigate the buckling behavior of the key region at room temperature and ageing temperature in order to predict the forming limit.

The buckling behavior of stiffened panels subjected to bending moment or shear load has been studied extensively. Zhou et al. [7] successfully predicted the forming limit of the traditional stiffened panel by using the buckling modes simulated by linear finite element method to obtain the initial defects, and carried out nonlinear finite element simulation based on the Riks method.

Lian et al. [8] studied the buckling behavior of conventional stiffened panels under shear load, and the results showed that the increasing of the equivalent skin thickness leads to the increasing of the critical buckling stress. Based on theoretical calculation, Liu et al. [9] established the

Materials Research Forum LLC
https://doi.org/10.21741/9781644903254-7

enhanced elastoplastic damage coupled model of stiffened panel under uniaxial compression. Based on buckling test, Pevzner et al. [10] studied buckling behavior of stiffened panels with different cross sections, such as J-form stiffeners and T-form stiffeners. However, few researchers have studied the buckling behavior of special shapes of stiffened panels, such as panels with opening.

In this study, the buckling behavior of integrally stiffened panels with opening subjected to pure bending at room temperature and ageing temperature was studied by finite element simulation, and the forming limits during the loading stage and heating stage of CAF were predicted successfully. Firstly, the overall buckling simulation of the integrally stiffened panel with opening was carried out, and it is found that the buckling occurs mainly in the position between two transverse bars. Based on the simulation results, the key region of the integrally stiffened panel with opening model was simplified, and the corresponding simulation simplified model was established. The influence of different geometrical parameters, such as stiffener thickness and stiffener height, on the buckling critical stress and buckling critical strain of stiffened panel was studied by Abaqus software. At the same time, non-dimensional critical buckling coefficient k_{cr} is introduced to compare the forming ability of the stiffened panel with opening with the traditional stiffened panel.

Problem definition

In this paper, the buckling behavior of the integrally stiffened panel with opening under pure bending moment M_x is studied. The integrally stiffened panel with opening is shown in Fig. 1. The overall simulation with whole stiffened panel was carried out by linear finite element method, and it is found that the buckling mainly occurs in the position between the two transverse bars, as shown in Fig. 2. The top of the stiffeners experienced high pressure stress during the forming process of the integral stiffened panel, leading to the potential buckling and instability deformation of the longitudinal stiffeners at both ends of the center frame, as illustrated in Fig. 1, resulting in the forming failure of the integral panel. In the simulation process of this paper, the key region of the integrally stiffened panel model was simplified. The simplified model is shown in Fig. 3, where the geometric parameters include the distance between the two transverse stiffeners a, the length of the transverse stiffeners b, the thickness of the skin t_{sk}, the height of the longitudinal stiffeners h and the thickness of the stiffened panel t_{st}. The simplified model can represent the buckling behavior of the whole stiffened panel. In this study, some geometric parameters of stiffened panels were fixed as a = 140 mm, b = 90 mm, t_{st}/t_{sk} = 1.

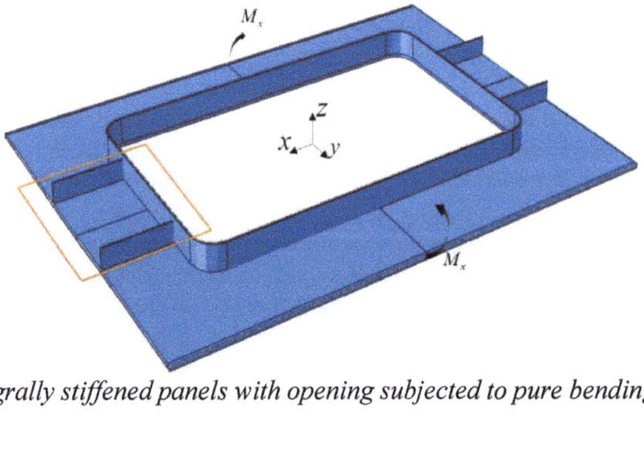

***Fig. 1**: Integrally stiffened panels with opening subjected to pure bending moment*

***Fig. 2**: The overall simulation of integrally stiffened panel*

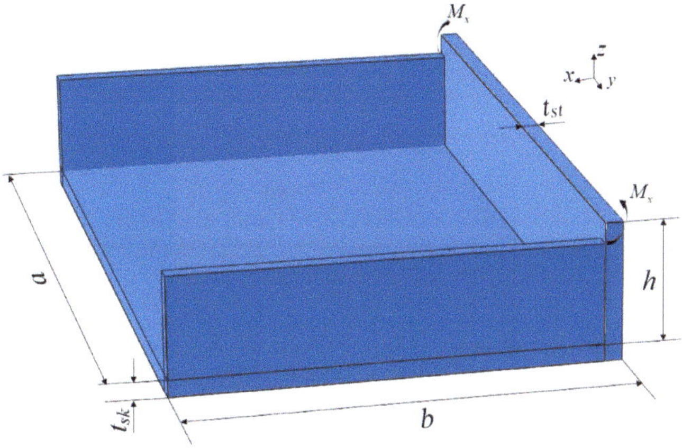

***Fig. 3**: Simplified FE model of integrally stiffened panels with opening*

Finite element simulation

Material and properties

The integrally stiffened panel material used in this paper is aluminum alloy 7050. In industry, in order to achieve the T7451 state of aluminum alloy, it is often aged at 120 °C for 4 to 6 hours after solid solution heat treatment, and then aged at 160 °C for about 20 h [11]. Thus, in this study, after solution heat treatment (475, 1h) and quenching, the material was aged at 120 °C for 5 h. The tensile test was carried out at room temperature and 160 °C respectively. The material properties of AA7050 are shown in Table 1. The engineering stress-strain curves at room temperature (RT) and ageing temperature (160 °C) are shown in Fig. 4. It can be seen from Fig. 4 that the elastic modulus of the alloy at two temperatures is basically the same, and the yield strength decreases under high temperature conditions.

Table 1. *Mechanical properties for AA7050-T7451 at the room temperature and 160 °C*

Temperature	E(GPa)	σ_y(MPa)
Room temperature	73.07	537.4
160 °C	73.07	457.5

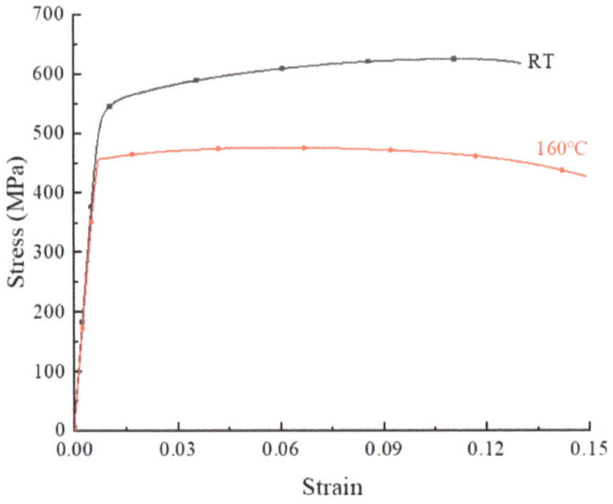

Fig. 4: *Engineering stress-strain behaviour of AA7050 at the room temperature (RT) and the ageing temperature (160 °C)*

FE method

The simplified finite element model is shown in Fig. 5. The R4S shell element is used for modeling, which can obtain higher solution accuracy [12]. The stiffened panel is subjected to pure bending moments at two transverse edges. The two loading edges of the model are constrained to the simply supported condition: the displacement of the two lateral edges of the fixed skin in the z-direction, the middle line of the skin in the y-direction, and the two transverse edges of the longitudinal stiffener in the x-direction. At room temperature and ageing temperature, different elastic-plastic parameters are assigned to the material. The room temperature simulation applied the room temperature material data, and the high temperature simulation applied the corresponding data. In the simulation process, the elastic modulus $E = 73071$, Poisson's ratio $\nu = 0.33$ [13].

In the process of finite element simulation in this paper, the model is simplified based on the result of linear finite element simulation, which has been mentioned in Fig. 2. In the simplified model, the pure bending is applied on the stiffened panel, and the longitudinal stiffened part between the two transverse stiffeners is the main region of buckling instability. Therefore, in this paper, the stiffener thickness and stiffener height are selected as variables to carry out research. The stress at the midpoint of the longitudinal stiffener is taken as the buckling stress of the simplified model, and the ratio of the displacement at both ends of the longitudinal stiffener to the length of the longitudinal stiffener is defined as the average strain.

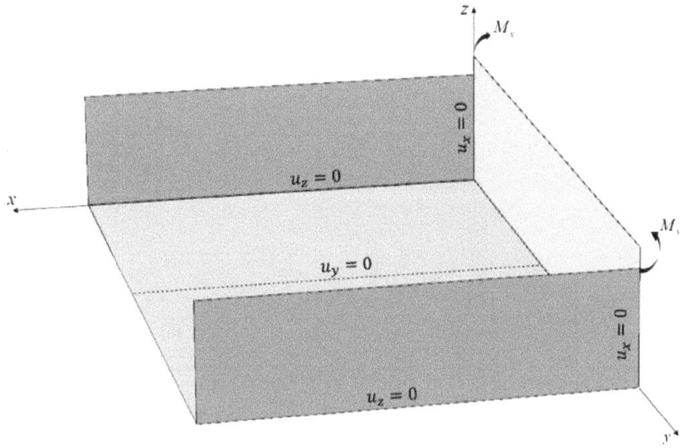

Fig. 5*: Boundary conditions of Simplified FE model of integrally stiffened panels with opening*

Results and discussion
Two sets of finite element simulation at room temperature and 160 °C, the commercial software Abaqus was used to investigate the buckling behavior of simplified model of the integrally stiffened panel under different geometric parameters (thickness and height of the stiffened panel), and to predict the buckling critical stress and strain of the stiffened panel during loading and heating.

Effect of stiffener thickness
In this section, based on the results of finite element simulation, the effects of the stiffener thickness t_{st} at RT and 160 °C on the buckling behavior of a simplified model under bending are discussed. In the simplified model, the ratio of bottom skin length a, bottom skin width b, stiffener height h, skin thickness to stiffener thickness were taken as 140 mm, 90 mm, 20 mm, and 1.0, respectively. The test range of stiffener thickness is from 1 mm to 2.8 mm. The average compressive stress and average strain at the top midpoint of the stiffener are obtained by finite element simulation. Fig. 6 shows the relationship between the average normalized stress $\sigma / \sigma_y$and the average strain on the left and right sides of the top midpoint of the stiffener under RT with different thickness of the stiffener.

When t_{st}= 1, 1.5 mm, the average stress increases linearly with the increase of the average strain in the initial stage, and remains stable when buckling occurs. When the stiffened panel is stressed in the elastic zone, the bifurcation point is easy to observe and calculate. As shown in Fig. 6, when the average strain reaches 0.3% and 0.6% respectively, the buckling stresses under corresponding conditions are obtained, which are 196.7 MPa and 388.1 MPa respectively. However, when t_{st}= 2, 2.5, 2.8 mm, the stress of the stiffened panel first lies in the linear zone, and when it enters the plastic zone, the stress increases slightly with the increase of the average strain. When the average strain reaches 0.9%, 1.5% and 2.2%, respectively, the corresponding buckling stresses are 533.2 MPa, 573.4 MPa and 583.5 MPa. At the same time, due to the increase of the stiffener thickness, the buckling phenomenon is gradually less obvious.

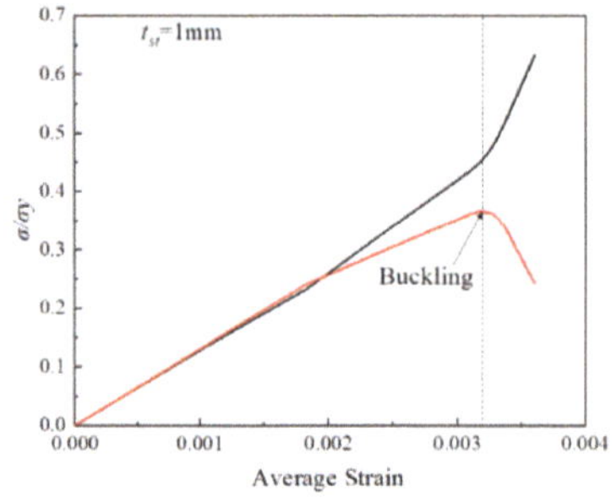

(a) t_{st} = 1 mm

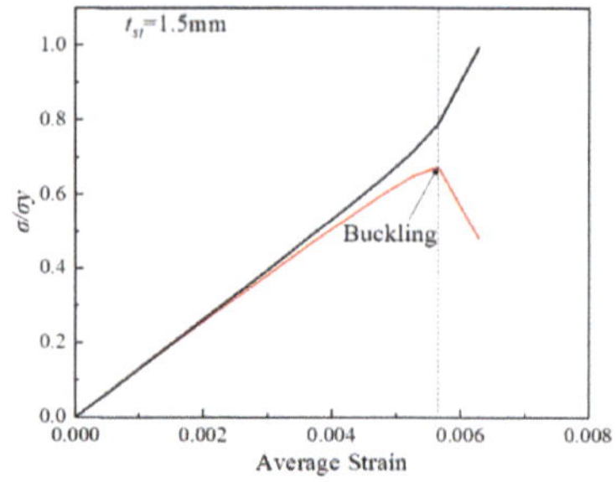

(b) t_{st} = 1.5 mm

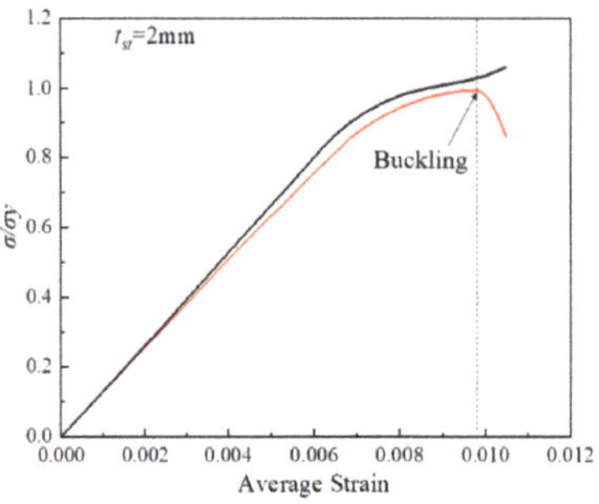

(c) t_{st} = 2 mm

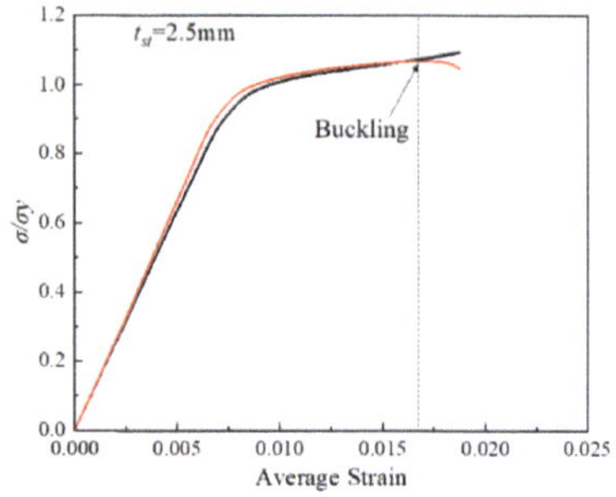

(d) t_{st} = 2.5 mm

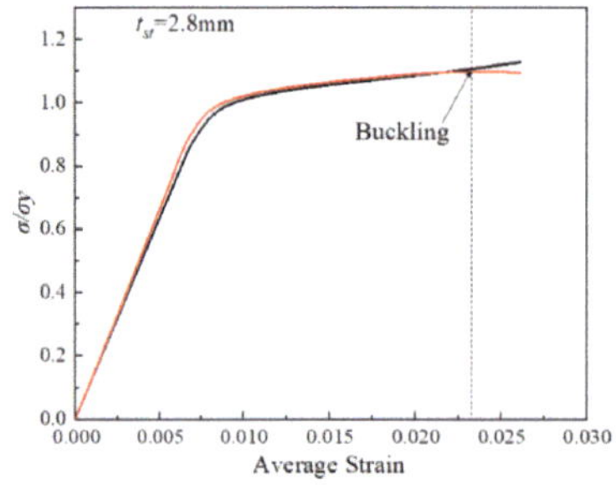

(e) t_{st} = 2.8 mm

***Fig. 6**: Average normalised stress of left side (red line) and right side (black line) of top stiffener versus average strain obtained from non-linear FE simulations at the room temperature with different stiffener thickness.*

Fig. 7 shows the relationship between the average normalized stress and the average strain at the left and right sides of the midpoint at the top of the stiffener at 160 °C. When t_{st} = 1, 1.5 mm, the buckling of the stiffened panel occurs in the elastic zone, and the buckling stresses are 196.1 MPa and 368.7 MPa, respectively. When t_{st} = 2, 2.5 and 2.8 mm, the buckling of the stiffened panel occurs in the plastic zone, and the buckling stresses are 462.4 MPa, 471.8 MPa and 486.7 MPa, respectively. The variation trend of average normalized stress at ageing temperature is similar to that at room temperature.

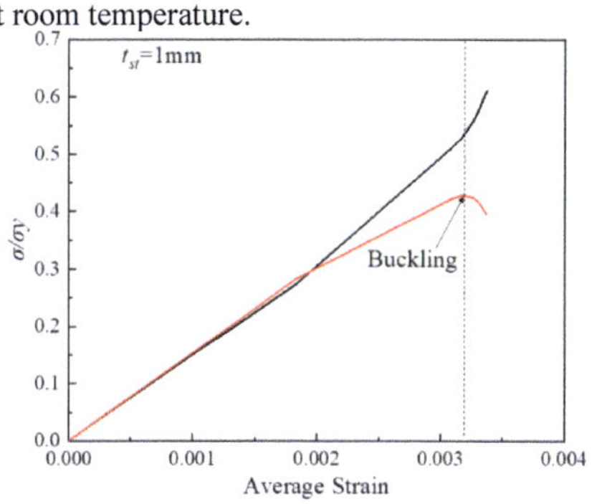

(a) t_{st} = 1 mm

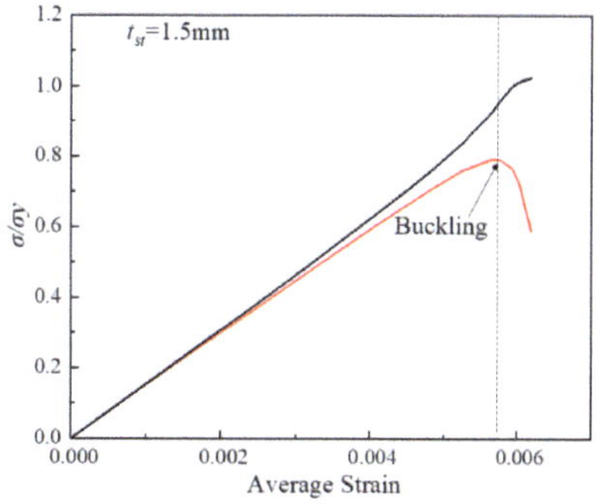

(b) t_{st} = 1.5 mm

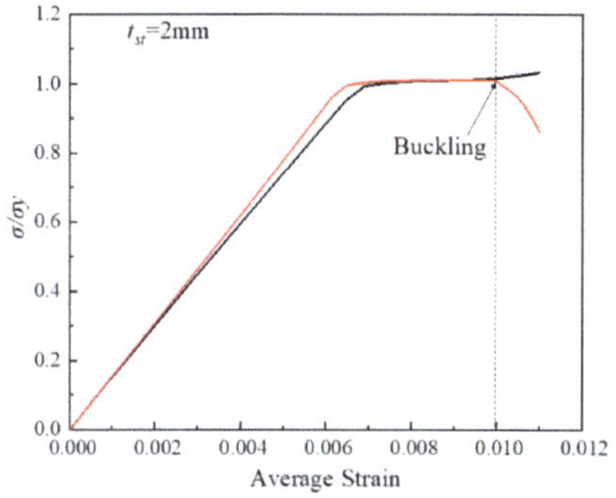

(c) t_{st} = 2mm

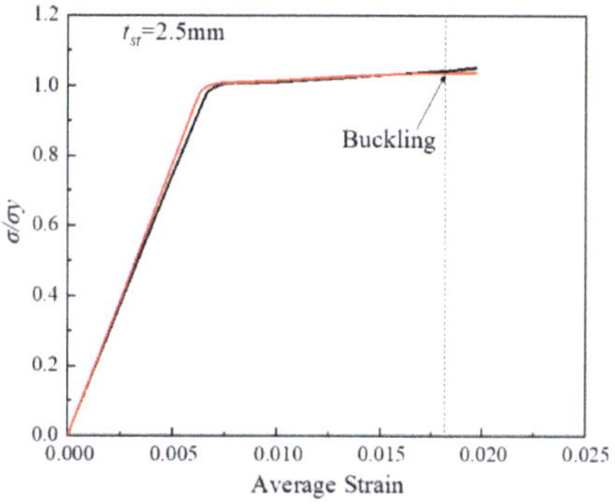

(d) t_{st} = 2.5 mm

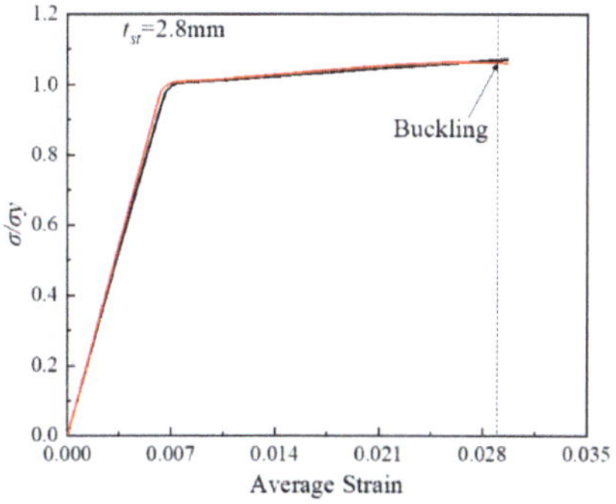

(e) t_{st} = 2.8mm

Fig. 7: *Average normalised stress of left side (red line) and right side (black line) of top stiffener versus average strain obtained from non-linear FE simulations at 160 °C with different stiffener thickness.*

The critical buckling stress and critical buckling strain under different stiffener thickness at RT and 160 °C are shown in Fig. 8. Due to the same elastic material parameters, the buckling critical stress and critical strain increase similarly with thickness at two different temperatures. When the structure enters the plastic zone, the critical buckling stress at ageing temperature is less than that at room temperature, and the average reduction is about 15%. However, greater buckling strain is produced at ageing temperature, which is about 40% higher than that at room temperature. This increases the forming limit. Therefore, in the case of plasticity, the stiffened panel can be heated first and then deformed by loading.

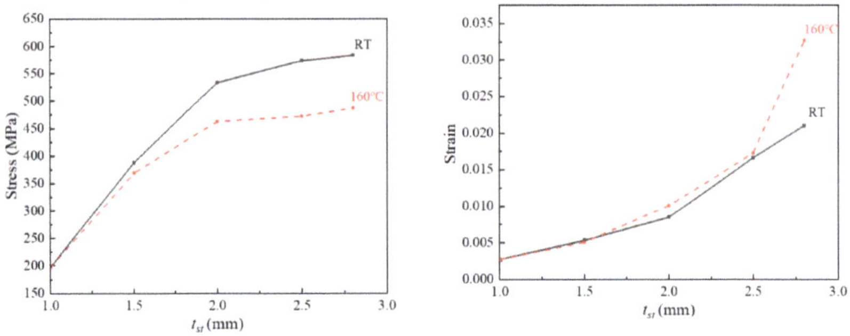

(a) Evolution of buckling critical stress with increased stiffener thickness

(b) Evolution of buckling critical strain with increased stiffener thickness

***Fig. 8**: Critical buckling stress and critical strain versus stiffener thickness t_{st} at the room temperature (solid line) and 160 °C (dashed line)*

For the elastic region, the non-dimensional critical buckling coefficient k_{cr} is introduced to study the influence of geometric size and temperature on the buckling response of the stiffened panel. The expression is as follows [14,15]:

$$k_{cr} = \frac{12(1-v^2)\sigma h^2}{E\pi^2 {t_{st}}^2} \tag{1}$$

Where v is Poison's ratio, E is elastic modulus, and σ is the corresponding buckling stress under t_{st}, 0.33, 73071, and 196.7 respectively.

The results show that when t_{st} = 1 mm, k_{cr} = 1.17 for stiffened panel with opening. For traditional stiffened panels, when t_{st} = 1 mm, k_{cr} = 1.5 [14]. This shows that for stiffened panels with empty frames, although the weight of the stiffened panel is reduced, the buckling stress is inevitably reduced. Therefore, it is necessary to choose the right combination of stiffened thickness and height to ensure the forming limit under the premise of meeting the lightweight.

Effect of stiffener height

In this section, based on the results of finite element simulation, the effects of the stiffener height h at RT and 160 °C on the buckling behavior of a simplified model subject to bending are discussed. In the simplified model, the length of the bottom skin a, the width of the bottom skin b, the stiffener thickness t_{st}, the ratio of the skin thickness to the stiffener thickness were taken as 140 mm, 90 mm, 2 mm and 1.0, respectively. Stiffener height test range is from 17 mm to 35 mm. The critical buckling stress and critical buckling strain decrease with the increase of stiffener height. At both temperatures, when the stiffener height is less than 25 mm, the buckling occurs in the elastic region. When the stiffener height is greater than 25 mm, the buckling occurs in the plastic region, and at the same stiffener height, the critical buckling stress decreases by about 16% and the critical buckling strain increases by about 9%.

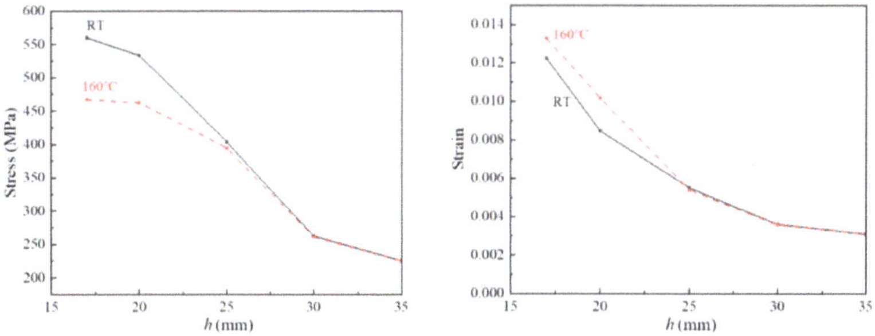

(a) Evolution of buckling critical stress with stiffener height increased (b) Evolution of buckling critical strain with stiffener height increased

***Fig. 9**: Critical buckling stress and critical strain versus stiffener height at the room temperature (solid line) and 160 °C (dashed line)*

Conclusions

In this study, the key region in the forming process of the integrally stiffened panel was simplified, and the forming limits in the loading and heating stages of the CAF process were predicted by finite element simulation. The elastoplastic buckling behavior of AA7050 under the pure bending moment at room temperature and ageing temperature (160 °C) was studied. The reduction of the stiffener height or the increase of stiffener thickness can help to avoid buckling instability in the forming process. In the elastic region, the non-dimensional critical buckling coefficient k_{cr} is 1.16 for the stiffened panel with opening. It reduced by 20%, compared with the traditional stiffened panel, which is 1,5. However, this kind of stiffened panel more meets the demand of lightweight. When the stiffened panel enters the plastic zone, the increase of temperature will enhance the forming limit. Therefore, if the buckling does not occur during the loading process, then the buckling will not occur during the heating process, and the forming limit of the workpiece can be improved by heating before loading

References

[1] J.K. Paik, S. van der Veen, A. Duran, M. Collette, Ultimate compressive strength design methods of aluminum welded stiffened panel structures for aerospace, marine and land-based applications: A benchmark study, Thin-Walled Structures 43(10) (2005) 1550-1566. https://doi.org/10.1016/j.tws.2005.06.003

[2] W. Zhou, Y. Li, Z. Shi, J. Lin, An analytical solution for elastic buckling analysis of stiffened panel subjected to pure bending, International Journal of Mechanical Sciences 161-162 (2019) 105024. https://doi.org/10.1016/j.ijmecsci.2019.105024

[3] F. Lyu, Y. Li, X. Huang, Z. Shi, Y. Zeng, J. Lin, An investigation of creep age forming of AA7B04 stiffened plates: Experiment and FE modelling, Journal of Manufacturing Processes 37 (2019) 232-241. https://doi.org/10.1016/j.jmapro.2018.11.031

[4] L. Zhan, J. Lin, T.A. Dean, M. Huang, Experimental studies and constitutive modelling of the hardening of aluminium alloy 7055 under creep age forming conditions, International Journal of Mechanical Sciences 53(8) (2011) 595-605. https://doi.org/10.1016/j.ijmecsci.2011.05.006

[5] L. Zhan, J. Lin, T.A. Dean, A review of the development of creep age forming: Experimentation, modelling and applications, International Journal of Machine Tools and Manufacture 51(1) (2011) 1-17. https://doi.org/10.1016/j.ijmachtools.2010.08.007

[6] Y. Yan, H.-b. Wang, M. Wan, Prediction of stiffener buckling in press bend forming of integral panels, Transactions of Nonferrous Metals Society of China 21(11) (2011) 2459-2465. https://doi.org/10.1016/S1003-6326(11)61037-6

[7] W. Zhou, Q. Rong, Z. Shi, Y. Zeng, Non-linear finite element investigation of formability limit by buckling in creep age forming of stiffened panels, Procedia Manufacturing 50 (2020) 625-629. https://doi.org/10.1016/j.promfg.2020.08.112

[8] C. Lian, P. Wang, X. Chen, F. Liu, K. Yuan, J. Zheng, Z. Yue, Experimental and numerical research on the analysis methods for buckling and post-buckling of inclined stiffened panel under shear load, Thin-Walled Structures 195 (2024) 111374. https://doi.org/10.1016/j.tws.2023.111374

[9] M. Liu, Y. Gao, D. Zhang, Z. Wang, D. Wu, B. Yu, Y. Lei, An enhanced elastoplastic damage coupled model for compression characteristics analysis of continuous fiber-reinforced thermoplastic composite stiffened panel, Composite Structures 297 (2022) 115924. https://doi.org/10.1016/j.compstruct.2022.115924

[10] P. Pevzner, H. Abramovich, T. Weller, Calculation of the collapse load of an axially compressed laminated composite stringer-stiffened curved panel–An engineering approach, Composite Structures 83(4) (2008) 341-353. https://doi.org/10.1016/j.compstruct.2007.05.001

[11] J.-H. Zheng, C.M. Davies, J. Lin, R. Pan, C. Li, Constitutive modelling of a T74 multi-step creep ageing behaviour of AA7050 and its application to stress relaxation ageing in age formed aluminium components, Procedia Engineering 207 (2017) 281-286.https://doi.org/10.1016/j.proeng.2017.10.775

[12] Abaqus User's Manual (15.6). Providence, RI, USA: 2010.

[13] Y. Li, Z. Shi, J. Lin, Experimental investigation and modelling of yield strength and work hardening behaviour of artificially aged Al-Cu-Li alloy, Materials & Design 183 (2019) 108121. https://doi.org/10.1016/j.matdes.2019.108121

[14] W. Zhou, Z. Shi, J. Lin, Semi-analytical study of buckling response for grid-stiffened panels during creep age forming, Thin-Walled Structures 182 (2023) 110231. https://doi.org/10.1016/j.tws.2022.110231

[15] W. Zhou, Z. Shi, Y. Li, Q. Rong, Y. Zeng, J. Lin, Elastic-plastic buckling analysis of stiffened panel subjected to global bending in forming process, Aerospace Science and Technology 115 (2021) 106781. https://doi.org/10.1016/j.ast.2021.106781

Advanced elastic-plastic constitutive models for medium-thickness high-strength steel plates: Microstructure-based parameter identification and verification

HU Xiao[1,a *], ZHOU Yu[2], GAO Chenxin[3], SHI Liya[4], LIAO Yong[4], ZHANG Haiming[3,b *], WANG Feilong[1] and WANG Liangyun[1]

[1]Pangang Group Research Institute Co., Ltd., Panzhihua 617000, China

[2]Pangang Group Panzhihua Steel and Vanadium Co., Ltd., Panzhihua 617000, China

[3]School of Materials Science and Engineering, Shanghai Jiao Tong University, Shanghai, 200240, China

[4]Dongfeng Liuzhou Motor Co., Ltd., Liuzhou 545000, China

[a]yjyhuxiao@pzhsteel.com.cn, [b]hm.zhang@sjtu.edu.cn

Keywords: Constitutive Models, Medium-Thickness Plate, Crystal Plasticity, Yield Criteria, Kinematic Hardening

Abstract. Replacing castings with medium-thickness, high-strength steel sheet (with thickness exceeding 3 mm) stampings can save raw materials and energy, and many medium-thickness plate parts are used in commercial vehicle frame assemblies and cab mounts. However, there are obvious differences between the sheet metal forming mechanisms of medium-thickness plate parts and thin sheet parts. There is a lack of systematic research on the issues of anisotropy, springback, non-uniform material flow, and fracture that exist in the sheet metal forming of medium-thickness plate parts. In this study, we coupled the three-dimensional anisotropic yield criteria Hill 48 and the kinematic hardening models (Armstrong-Frederick) to develop elastic-plastic constitutive models for the forming simulation of medium-thickness high-strength steel plates. Taking the 5-mm-thick 750L high-strength steel as the research object, for the problem that the anisotropic yield criterion parameters (especially in the thickness direction) are difficult to calibrate, we employed high-resolution full-field crystal plasticity simulations to obtain the parameters of the yield criterion, and determined the material parameters of the kinematic hardening models by the inverse analysis method. Through this multiscale modeling technique, we obtained the complete constitutive parameters of the material, which were verified by mechanical experiments with nonlinear strain paths. The effectiveness of different model combinations in predicting the plastic behavior and mechanical responses on medium-thickness high-strength steel plates is further evaluated.

Introduction

In the evolving landscape of material processing engineering, advanced stamped components offer a promising avenue for enhancing material efficiency and energy conservation. Particularly, the use of medium-thickness, high-strength steel plates, defined here as those exceeding 3 mm in thickness, represents a significant step forward in this domain [1]. These materials find extensive applications across various sectors, notably in the automotive industry, where they are integral to the structural integrity and functionality of commercial vehicle frames. Components such as rear mount lower brackets, longitudinal beams, cross beams, and filter mounts are pivotal examples showcasing the material's versatility and reliability.

Despite their apparent advantages and growing adoption, the metal forming processes of medium-thickness, high-strength steel plates encapsulate a series of complex phenomena that distinctly set them apart from their thin sheet counterparts. The inherent challenges—plastic anisotropy, springback, non-uniform material flow, and fracture mechanisms—underscore a critical gap in our current understanding and predictive capabilities. The complicated behavior of these materials under forming conditions necessitates a comprehensive analytical framework that transcends conventional methodologies, thereby facilitating a deeper insight into their mechanical response and formability characteristics.

In addressing these challenges, the present study underscores a concerted effort to bridge the knowledge gap through the development of advanced elastic-plastic constitutive models, which are pivotal in simulating the plastic behavior of the medium-thickness steel with greater precision and fidelity. The integration of sophisticated while easily yield criterion—namely Hill 48 [2]— alongside the Armstrong-Frederick (A-F) kinematic hardening model [3], lays the groundwork for an innovative approach that marries theoretical rigor with practical applicability.

The choice of 5-mm-thick 750L high-strength steel as the focal point of this investigation is motivated by its widespread usage and representative nature within the context of medium-thickness applications. The intricacies associated with calibrating anisotropic yield criterion parameters, particularly along the thickness direction, pose a significant challenge that this study aims to surmount. By leveraging high-resolution full-field crystal plasticity (CP) simulations [4, 5], this study endeavors to extract material parameters that are otherwise elusive through conventional means. The subsequent calibration of kinematic hardening models via the inverse analysis method [6] further exemplifies the study's methodical approach to constitutive modeling.

This research culminates in a set of comprehensive constitutive parameters that embody the steel's intricate behavior under a spectrum of loading conditions. The validation of these models through a series of mechanical experiments, characterized by nonlinear strain paths, serves as a testament to their robustness and predictive accuracy.

The ensuing discourse is structured to provide an in-depth exploration of the aforementioned themes, beginning with a detailed exposition of the experimental details, multiscale constitutive, and CP modeling technique employed in this study. Subsequent sections will delve into the results and discussions, elucidating the empirical findings and theoretical implications that emerge from this research. Collectively, these insights aim to advance the field of material engineering, offering a detailed understanding of medium-thickness high-strength steel plate and paving the way for their optimized utilization in industrial applications.

Experiments and consititutive models

A series of experiments were meticulously designed to characterize the mechanical behavior and anisotropic properties of the medium-thickness high-strength steel plate. The material's stress-strain response and its plastic anisotropy are the key focal points of this investigation.

Uniaxial tensile testing. The uniaxial tensile tests were conducted using an Instron universal testing machine, complemented by the ARAMIS Digital Image Correlation (DIC) system for enhanced strain measurement accuracy, as shown in Fig. 1a. The tests targeted various material orientations to ascertain the anisotropy in mechanical properties. Specifically, samples were tested along three principal directions: transverse direction (TD), rolling direction (RD), and diagonal direction (DD, or 45° to the rolling direction). To ensure the robustness of the results, each direction was subjected to five replicate tests. Additionally, testing was expanded to include orientations of 22.5° and 62.5° to validate the yield criterion model's accuracy.

Sample dimensions for the tensile tests adhered to the GB/T 228.1-2021 (ISO6892-1:2019) standard for metallic material tensile testing at room temperature. The gauge length and width of the samples were set to 50mm and 12.5mm, respectively. Prior to testing, the thickness of each specimen was measured at three equidistant points along the gauge length, and the average value was adopted as the sample's thickness.

Compression testing. To complement the tensile data and capture the steel's mechanical behavior in the thickness direction, compression tests were performed on cubic specimens extracted from each plate, as shown in Fig. 1b. The dimensions for these specimens were 4 mm×4 mm×5 mm. Ensuring uniform contact and reducing friction between the specimens and the compression platens were crucial. This was achieved by grinding the sample's top and bottom surfaces and applying a thin layer of lubricant on the platens. The compression rate was standardized at 0.001/s, equating to 0.3mm/min, with the test concluding at a true strain of 0.2. The ARAMIS DIC system provided real-time strain readings throughout the compression tests.

Tensile-compression testing. To calibrate the kinematic hardening model for the steel, a series of tensile-compression experiments were conducted. The experimental campaign was structured in two sequential stages: tensile loading followed by unloading, and subsequent compression testing. This approach was employed for each material variant along both RD and TD. For each direction, the sample was subjected to three distinct pre-strain levels during the tensile loading stage, and each pre-strain level test was repeated three times. Subsequent to the tensile test, small rectangular specimens were extracted from the pre-strained sample for compression testing, as shown in Fig. 1c. The dimension of these samples was tailored to 8 mm×5 mm×5 mm. The compression was conducted at a rate of 0.5 mm/min, targeting a total deformation of ~20%.

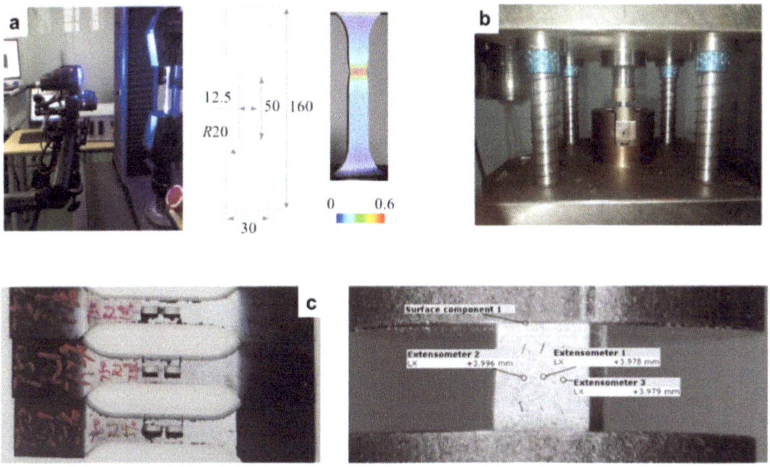

Fig. 1. The tensile test setup integrated with the DIC system, the geometric dimensions of the tensile specimens and the strain distribution contour of the specimen after necking. (b) The specialized apparatus for the compression test of the thick steel plate. (c) The pre-strained specimens for cyclic loading experiments, the sampling images for the sequential compressive test, and the schematic diagram of the reverse compression experiment.

Microstructure characterization. To understand the initial microstructural characteristic of the steel as well as get the data for building the microstructural representative volume element (RVE) for CP modeling, Electron backscatter diffraction (EBSD) experiments were used to characterize the steel's microstructure and texture. Samples measuring 6 mm×6 mm×5 mm were extracted from the as-received steel plate, with clear demarcations for the RD, TD, and ND. These samples were bisected to isolate surfaces for EBSD characterization, specifically choosing the RD-TD plane from both the surface and intermediate section of the plate.

Materials Research Forum LLC
https://doi.org/10.21741/9781644903254-8

The EBSD characterization was conducted using a NOVA NanoSEM 230 scanning electron microscope, with scanning voltage of 20 kV, beam size of 6, and working distance of 13 mm. The electron microscope operated at a magnification of 2000×, targeting an EBSD characterization area of 100 μm × 100 μm. The scan step size was determined to be 0.2 μm, resulting in a grid of 500 × 500. Fig. 2 shows the characterized microstructure of the as-received steel. The phases detected included ferrite (and martensite)—BCC phase and very little austenite and cementite, and the austenite and cementite constituents were ignored in the building of RVE.

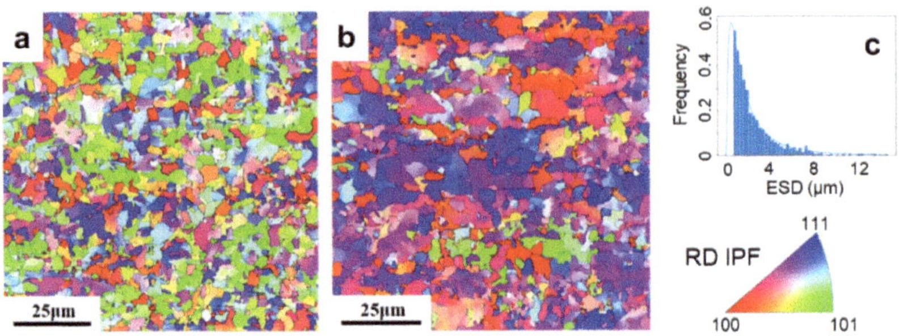

Fig. 2. The initial microstructure of the steel characterized by EBSD on the RD-TD surface (a) and RD-TD section at the half thickness (b). (c) The frequency distribution of grain size. The EBSD maps are colored with the RD inverset pole figure (IPF).

Multiscale material models

Hill 48 yield criterion. The Hill 48 yield criterion is foundational in the field of material science and engineering, particularly for characterizing the anisotropic behavior of metals during plastic deformation. The criterion extends the von Mises yield condition to accommodate anisotropic materials. The criterion is expressed as a quadratic function of the stress components, tailored to capture the anisotropy in different material directions. The general form of the Hill 48 criterion for orthotropic materials (common in rolled metal sheets) is:

$$\sqrt{F(\sigma_{22}-\sigma_{33})^2+G(\sigma_{33}-\sigma_{11})^2+H(\sigma_{11}-\sigma_{33})^2+2L\sigma_{23}^2+2M\sigma_{31}^2+2N\sigma_{12}^2} \quad (1)$$

Here, F, G, H, L, M, N are material constants that reflect the anisotropic properties of the material. These constants are typically determined experimentally, often through tensile tests in different directions relative to the material's microstructure.

The Armstrong-Frederick kinematic hardening model. The A-F kinematic hardening model is a well-regarded constitutive framework used to describe the behavior of materials undergoing cyclic loading and the Bauschinger effect. The fundamental idea behind the A-F model is to enhance the classical kinematic hardening rule by incorporating a nonlinear term that describes the evolution of the backstress α_{ij}, which is given by:

$$d\alpha_{ij}=\frac{2}{3}Cd\varepsilon_{ij}^{p}-\gamma\,\alpha_{ij}d\bar{\varepsilon}^{p} \quad (2)$$

Here, $d\alpha_{ij}$ represents the increment in backstress, $d\varepsilon_{ij}^{p}$ is the increment in plastic strain, $d\bar{\varepsilon}^{p}$ is the equivalent plastic strain increment. C and γ are material constants; C controls the rate at which the backstress evolves with plastic deformation, and γ is a dynamic recovery parameter that dictates the rate at which the backstress is reduced as the material undergoes continued plastic deformation.

Crystal plasticity modelling. Determining material parameters for three-dimensional anisotropic yield criterion presents a significant challenge, particularly for the parameters related to the thickness direction anisotropy of metal sheet/plate. In recent years, virtual laboratory techniques based on CP have emerged as a powerful tool for identifying parameters of advanced yield criteria. The core approach involves using a calibrated CP model to simulate and predict the mechanical response of materials under arbitrary stress states. By generating a multitude of such stochastic stress points and employing optimization algorithms, material parameters of advanced yield criteria can be conveniently ascertained. It is imperative, however, that the CP and the microstructural RVE models are sufficiently accurate.

A finite strain rate-dependent CP model is employed in this study. The CP model adopts a simple power-law expression for the flow equation, which delineates the relationship between shear strain rate and resolved shear stress as follows,

$$\dot{\gamma}^{\alpha} = \dot{\gamma}_0 \left(\frac{|\tau^{\alpha}|}{g^{\alpha}}\right)^{\frac{1}{m}} \mathrm{sign}(\tau^{\alpha}) \quad (3)$$

where $\dot{\gamma}^{\alpha}$ is the reference slip rate on the specific slip system, g^{α} represents the slip resistance of slip system α, and m is the strain rate sensitivity coefficient. g^{α} signifies the resistance against dislocation movement encountered during plastic deformation. A modified phenomenological model, incorporating the is adopted to describe evolution g^{α} as follows,

$$\dot{g}^{\alpha} = \begin{cases} 0, & \int \dot{\gamma}^{\alpha}\, dt < \gamma_L \\ \sum_{\beta=1}^{n} h_{\alpha\beta} |\dot{\gamma}^{\beta}|, & \int \dot{\gamma}^{\alpha}\, dt \geq \gamma_L \end{cases} \quad (4)$$

$$h_{\alpha\beta} = h_0[q + (1-q)\delta^{\alpha\beta}] \left|1 - \frac{g^{\beta}}{g_s^{\alpha}}\right|^{a} \mathrm{sign}\left(1 - \frac{g^{\beta}}{g_s^{\alpha}}\right) \quad (5)$$

where γ_L, h_0, g_s, and a are material constants representing the maximum shear strain before the vanish of Lüders straining, reference self-hardening modulus, saturation slip resistance, and hardening exponent, respectively; the initial CRSS is denoted as g_0. The parameter q denotes the ratio of interaction strengths among various slip systems, characterizing the distinction between self-hardening and latent hardening of slip system. More details about the adopted CP model can be refered to [4].

A pivotal element in conducting the CP simulation is the deployment of high-resolution RVE that capture the microstructural intricacies of the studied steel. In this work, the open-source modeling software DREAM.3D [7] was employed to construct a polycrystalline RVE model for the steel, leveraging the steel's EBSD data as a foundational input. The reconstructed RVE embodies the heterogeneity and anisotropy inherent in the steel's microstructure (grain size, morphology, and orientation obtained from EBSD data), serving as a critical foundation for subsequent CP simulations.

Results and Discussion

Fig. 3 presents the experimental results of the mechanical tests of the steel. The results indicate that the steel plate exhibits strong anisotropic mechanical behavior, as evidenced by the varying stress-strain curves and r values along different directions. The RD curve shows a continuous stress increase with strain, featuring a steep initial elastic region followed by a gradual transition into plastic deformation, indicating strain hardening. The TD curve is similar to RD, but with a slightly higher yield point and ultimate strength, which could be due to effect of strong crystallographic texture of the as-received steel, as shown in Fig. 2. The DD curve, represented by a dashed line, lies between RD and TD, suggesting that the mechanical properties in this direction are intermediate.

Moreover, the r value, a measure of the material's ability to resist thinning or thickening when stretched, varies with direction. The r value is highest along the RD, intermediate along DD, and lowest along TD. This difference in r values suggests that the steel plate is most ductile along the RD and least along TD, which can significantly affect sheet metal forming operation.

As previously mentioned, the EBSD data illustrated in Fig. 2 was used to construct the RVE necessary for CP simulations. The reconstructed high-resolution RVE presented in Fig 4a encompasses approximately 10,000 equiaxed grains, with grain size and orientation distributions derived from the EBSD data. Utilizing the established RVE and the stress-strain curves from Fig. 3 (RD and TD), a inverse analysis method was employed to determine the material parameters of the CP model, as detailed in Table 1. Fig. 4b displays a comparison between the flow stress curves predicted by the CP simulation and from the experiments. The simulated results correlated with the experimental data satisfactorily, affirming the calibrated CP model's adeptness in describing the steel's mechanical behavior along various orientations, inclusive of the Lüders stress plateau and work-hardening.

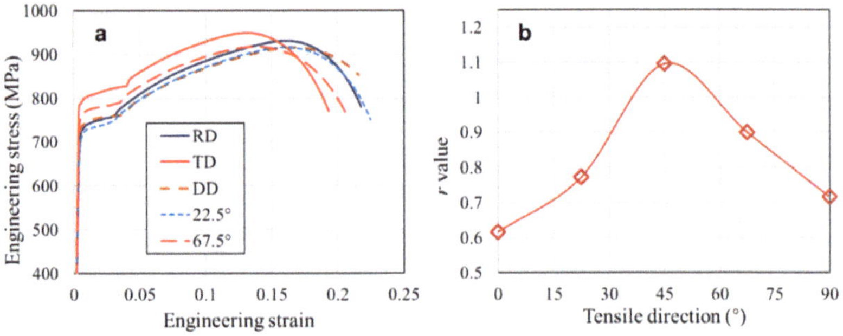

Fig. 3. Engineering stress-strain curves and r-values of the steel tensile tested along different directions.

Table 1. Materials parameters of the CP for the medium-thickness steel. The parameters were identified by the inverse analysis method.

Slip system	g_0^α/MPa	g_s^α/MPa	h_0/MPa	a	γ_L	$\dot{\gamma}_0$/s^{-1}	q	n
$\{\bar{1}10\}\langle 111\rangle$	300	435	825	1.04	0.0595	0.001	1.4	20
$\{\bar{2}11\}\langle 111\rangle$	319	449	2555	1.34				

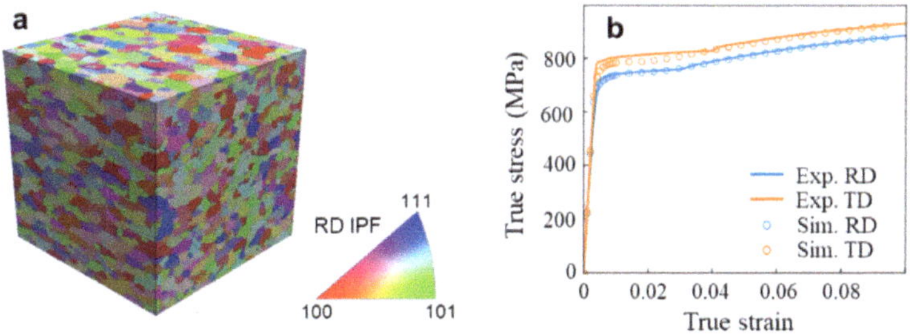

Fig. 4. (a) The RVE constructed from the EBSD data for CP simulations. (b) The comparison between the flow stress curves in RD and TD predicted by the calibrated CP simulation and from the experiments.

Building upon the previously calibrated CP model, we employed it as a virtual laboratory to conduct simulations under 100 random stress states. This approach yielded mechanical response data of the polycrystalline RVE under arbitrary stress conditions. Subsequently, we used these 100 stress data points and applied the Levenberg–Marquardt algorithm to fit the six material parameters of the Hill 48 yield criterion, as depicted in Table 2. Fig. 5 illustrates the yield loci predicted by the Hill 48 criterion along with the r values for different tensile directions. The results conspicuously demonstrate that the identified Hill 48 yield criterion adeptly characterizes the steel's anisotropic mechanical behavior and plastic flow.

Table 2. Materials parameters of Hill 48 yield criterion calibrated by the CP based virtual laboratory.

F	G	H	L	M	N
0.5329	0.6177	0.3823	1.9511	1.5949	1.8354

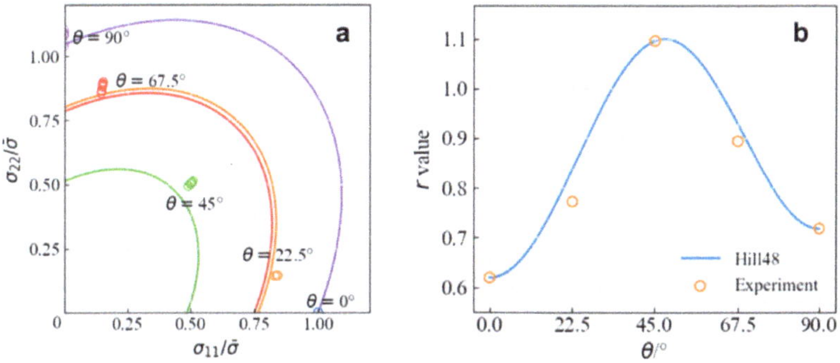

Fig. 5. (a) The comparison between the experimental yield loci (open symbols) and those predicted by the calibrated Hill 48 yield criterion. (b) The comparison between the experimental r values (open symbols) and those predicted by the calibrated Hill 48 yield criterion

We further integrated the calibrated Hill 48 yield criterion with the A-F kinematic hardening model and the Swift strain hardening model to develop a comprehensive three-dimensional elastic-plastic constitutive model tailored for the medium-thickness steel plate. Specifically, to address the Lüders yield plateau phenomenon exhibited by the steel, we modified the Swift model to accurately capture the Lüders plateau occurring during tensile deformation, as shown in Eq. 6.

$$\bar{\sigma} = \begin{cases} A(\bar{\varepsilon}^p + \varepsilon_0)^n + a(\bar{\varepsilon}^p - \varepsilon_b)^2 - a(\varepsilon_L - \varepsilon_b)^2 & \bar{\varepsilon}^p < \varepsilon_L \\ A(\bar{\varepsilon}^p + \varepsilon_0)^n & \bar{\varepsilon}^p > \varepsilon_L \end{cases} \quad (6)$$

The inverse analysis method combined with the particle swarm optimization algorithm were employed to identify the material parameters of this constitutive model, as shown in the Table 3. Fig. 6 presents the stress-strain curves of the steel under reverse loading following three different pre-strain levels in tension, including both experimental and simulated predictions. The results indicate a pronounced Bauschinger effect of the steel, and the established constitutive model adeptly describes the steel's work hardening, the Bauschinger effect upon reverse loading, and kinematic hardening behavior.

In summary, the elastic-plastic constitutive model established in this study, along with the material parameters calibrated using CP simulation and inverse analysis method, effectively characterizes the anisotropic behavior, mechanical behavior (including work hardening and the Lüders yield plateau), and kinematic hardening behavior of the medium-thickness steel plate.

Table 3. Materials parameters of the A-F kinematic hardening model and the Swift model for the steel. The parameters were identified by the inverse analysis method.

a/GPa	ε_b	ε_L	A/MPa	ε_0	n	C/GPa	γ
577	0.0216	0.0209	925	0.0118	0.212	64.2	219

Conclusions

This study focuses on medium-thickness high-strength steel plate widely applied in the automotive industry, examining their anisotropic mechanical behavior and Bauschinger effect through systematic mechanical testing. The steel exhibits pronounced anisotropic mechanical properties and plastic flow, the Luders yield plateau, and significant Bauschinger effect. We integrated the Hill 48 yield criterion, Armstrong-Frederick kinematic hardening model, and the Swift strain hardening model to develop a three-dimensional elastoplastic constitutive model tailored for the forming operation of medium-thickness steel plate. Utilizing high-resolution crystal plasticity simulations as a virtual laboratory, we employed inverse analysis to calibrate material parameters of the anisotropic yield criterion, and further determined the complete parameters for the constitutive model. The results demonstrate that the established constitutive model and its parameters accurately describes the complex mechanical behaviors and deformation characteristics of the medium-thickness steel plate.

The models and the multiscale modelling technique developed in this paper provide an essential reference for constitutive model development and numerical simulation of medium-thickness plates.

Materials Research Forum LLC
https://doi.org/10.21741/9781644903254-8

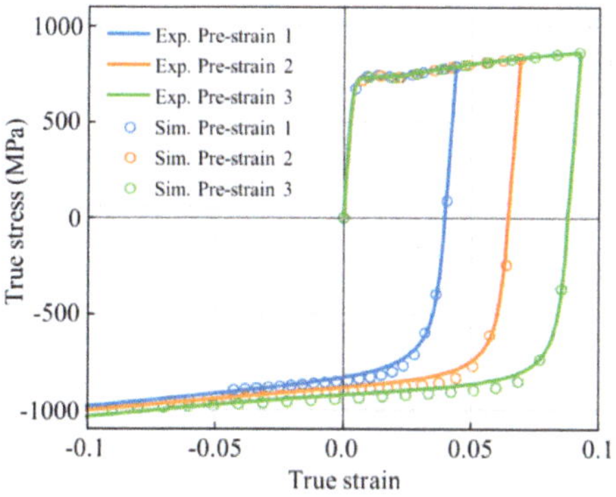

Fig. 6. Stress-strain curves for reverse loading following different pre-strain levels. The solid lines represent experimental data. Corresponding simulation results are depicted with open circles, aligning with their respective experimental curves. This comparison highlights the constitutive model's accuracy in capturing the mechanical response of the steel after varying degrees of pre-straining.

Acknowledge
Financial assistance of the National Natural Science Foundation of China (52075329) is acknowledged.

References

[1] C. Zhu, P. Gu, Y. Yu, Z. Tao, H. Zhang, Layered fracture prediction of QStE700 medium-thickness steel plate under tensile loading, Proceedings of the Institution of Mechanical Engineers, Part B: J. Eng. Manuf. 235 (2021)1144-1153. https://doi.org/10.1177/0954405420978035

[2] R. Hill, E. Orowan, A theory of the yielding and plastic flow of anisotropic metals, Proceedings of the Royal Society of London Series A Math. Phys. Sci. 193 (1948) 281-297. https://doi.org/doi:10.1098/rspa.1948.0045

[3] C.O. Frederick, P.J. Armstrong, A mathematical representation of the multiaxial Bauschinger effect, Mater. High Temp. 24 (2007) 1-26. http://dx.doi.org/10.3184/096034007X207589

[4] H. Zhang, M. Diehl, F. Roters, D. Raabe, A virtual laboratory using high resolution crystal plasticity simulations to determine the initial yield surface for sheet metal forming operations, Int. J. Plast. 80 (2016) 111-138. https://doi.org/10.1016/j.ijplas.2016.01.002

[5] A. Nascimento, S. Roongta, M. Diehl, I.J. Beyerlein, A machine learning model to predict yield surfaces from crystal plasticity simulations, Int. J. Plast. 161 (2023) 103507. https://doi.org/10.1016/j.ijplas.2022.103507

[6] H. Ma, Y. Li, H. Zhang, Q. Li, F. Chen, Z. Cui, A virtual laboratory based on full-field crystal plasticity simulation to characterize the multiscale mechanical properties of AHSS, Scientific Reports 12 (2022). https://doi.org/10.1038/s41598-022-09045-8

[7] M.A. Groeber, M.A. Jackson, DREAM.3D: A Digital Representation Environment for the Analysis of Microstructure in 3D, Integr. Mater. Manuf. Innov. 3 (2014) 56-72. https://doi.org/10.1186/2193-9772-3-5

Superplastic microstructure characteristics and unified constitutive model of TA32 titanium alloy

LIU Yang[1,2,a], WANG Chunhui[1,2,b], LI Zhiqiang[3,c], ZHANG Ning[3,d], ZHAO Bing[3,e] and SUN Chaoyang[1,2,f *]

[1]School of Mechanical Engineering, University of Science and Technology Beijing, Beijing 100083, China

[2]Beijing Key Laboratory of Lightweight Metal Forming, Beijing 100083, China

[3]AVIC Manufacturing Technology Institute, Beijing 100024, China

[a]yliu_ustb@126.com, [b]wangch@ustb.edu.cn, [c]zqlee98@126.com, [d]zn64112@163.com, [e]zhao6833@163.com, [f]suncy@ustb.edu.cn

Keywords: TA32 Titanium Alloy, Superplasticity, Deformation Characteristics, Constitutive Model

Abstract. The superplastic deformation behavior and microstructure characteristics of TA32 titanium alloy were studied at different deformation temperatures (920-960°C) and initial strain rates (1×10^{-3}-3×10^{-3} s^{-1}). A unified constitutive model was developed and coupled with internal state variables. Results show that TA32 alloy featured steady flow stress and uniform plastic deformability under the dynamic equilibrium between softening and hardening effects. Increasing temperature and strain induced $\alpha\rightarrow\beta$ phase transformation and α grain growth. Dislocation accumulation and grain boundary migration formed the dynamic recovery (DRV) and dynamic recrystallization (DRX) mechanisms. Using dislocation density, the proposed unified constitutive model built a quantitative relationship between superplastic flow stress and internal state variables such as grain size, with the correlation coefficient *R*, average absolute relative error (AARE), and root mean square error (RMSE) between the predicted and experimental stress values being 0.98, 6.91%, and 1.94, respectively. The unified constitutive model could predict well the flow stress and grain size evolution of TA32 titanium alloy during superplastic deformation.

Introduction

The recently developed near-α TA32 titanium alloy exhibits a proper match of thermal strength, fracture toughness, high-temperature creep resistance, and gas corrosion resistance at 550°C. It can satisfy the demands of microstructure uniformity and performance durability at high service temperatures, making it valuable for the aerospace and automobile lightweight fields [1, 2]. In addition, the superplastic forming (SPF) process utilizes the characteristics of titanium alloy, such as low deformation resistance above 900°C and no necking, to form components with complex structure. Therefore, it is needed to explore the mechanical behaviors, microstructure evolution, and physical-based constitutive model of TA32 titanium alloy under superplastic deformation conditions.

Generally, dislocation motion is the primary driver of alloy's microstructure evolution during viscoplastic deformation. Stored energy driven DRV/DRX and grain growth controlled by grain boundary migration are also closely associated with dislocation multiplication or annihilation. Several studies have investigated the microstructure characteristics of titanium alloys during high-temperature deformation and established the microstructure-based constitutive model to describe the multiscale deformation behavior. Babu et al. [3] investigated the deformation mechanisms of TC4 titanium alloy at thermal processing conditions and established a constitutive model based on immobile dislocation density and excess vacancy concentration. Xiao et al. [4] examined the spheroidization of α phase and DRV/DRX effects of β phase in Ti-55511 titanium alloy and established a dislocation density-based constitutive model expressing softening and work-hardening mechanisms. Bai et al. [5] developed a unified constitutive model for two-phase titanium alloys coupled with the α phase spheroidization phenomenon, taking into account the influence of adiabatic temperature change and the evolution of phase volume fraction on the plastic strain rates. Yasmeen et al. [6] investigated the evolution of grain size, dislocation density, and phase volume fraction of TA15 titanium alloy under superplastic deformation conditions and established a set of mechanism-based unified viscoplastic constitutive equations.

This paper utilizes microstructure morphology to analyze the phase distribution, grain boundary properties, and grain size evolution of TA32 alloy under superplastic deformation. Additionally, based on multiscale experimental data, a unified constitutive model of superplasticity is developed that describes macroscopic and microscopic behaviors. Ultimately, the tens of material constants coupled with each other are solved using the multi-objective optimization function. This is expected to provide theory-based guidance for SPF process for manufacturing complex components of TA32 alloy.

Experiments

Materials and methods. The experimental material is hot rolled TA32 sheet. The actual chemical composition is Ti-4.81Al-3.31Sn-2.29Zr-0.88Mo-0.50Nb-0.19Ta-0.17Si (in wt.%). The initial microstructure observed by MV6000 optical microscope (OM) is shown in Fig. 1(a). The black β phase is clustered between white α phase in the form of strips or clumps, while the equiaxed α grains have a non-uniform distribution in grain size. The tests were carried out by a LETRYDL-20T electronic universal testing machine equipped with a specialized clamp for superplastic deformation, as shown in Fig. 1(b). The sample shape and size are shown in Fig. 1(c). The tensile deformation temperatures were 920°C, 940°C, and 960°C, and the initial strain rates are 1×10^{-3} s^{-1} and 3×10^{-3} s^{-1}. The testing procedure is shown in Fig. 1(d). The EBSD observation was conducted using a JSM-7900F scanning electron microscope with a step size of 0.35 μm.

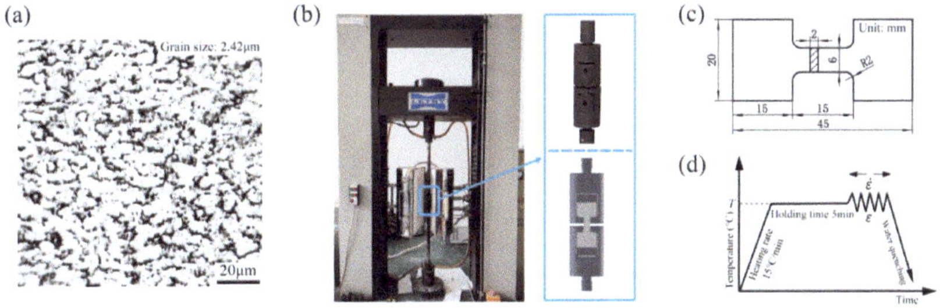

Fig. 1. As-received material and experimental methods: (a) OM image; (b) tensile testing system ; (c) sample size; (d) testing procedure.

Results. Fig. 2(a) exhibits the true stress-strain curves of TA32 alloy. At the initial stage of deformation process, dislocation multiplication and entanglement lead to a rapid increase in stress. Inevitably, the DRV and DRX accompany the deformation under high-temperature and load conditions, causing the dislocation annihilation, the plastic deformation resistance of TA32 alloy to decrease, and the degree of stress growth to decrease sharply. Even after reaching peak stress, the stress continues to decrease as strain increases. In addition, TA32 alloy's peak stress is less than 50 MPa during superplastic deformation. Notably, at 920°C and 1×10^{-3}-3×10^{-3} s^{-1}, the flow stress changes less as the strain increases. This enables TA32 alloy to form components with uniform microstructure and mechanical properties under complex stress state.

Materials Research Forum LLC
https://doi.org/10.21741/9781644903254-9

Fig. 2(b) shows the strain hardening exponent n of TA32 alloy at different conditions. The data shows that at 960°C and 1×10^{-3} s^{-1}, n values remain around 0.2. This indicates that the strain hardening effect always dominates, and the large plastic deformation process of TA32 alloy is relatively stable and may have an optimal uniform deformability (the critical strain value of geometry instability increases). The n values under the remaining deformation conditions decrease as strain increases, indicating the hardening effect is weakened while the softening effect is relatively enhanced and the uniform deformability decreases (the critical strain value of geometry instability decreases). In addition, for the same strain, the n value increases as temperature increases and strain rate decreases, causing hardening and a decrease in plastic deformability. This is due to the long high temperature, which causes excessive grain growth, grain boundary sliding to gradually reduce, and need for larger load to overcome the internal critical shear stress [7]. In particular, the curves at 920-940°C and 1×10^{-3} s^{-1} flatten out when the strain exceeds 0.7. This may be caused by the dislocation annihilation rate gradually approaching the dislocation accumulation rate, resulting in a dynamic equilibrium between the conflicting softening and hardening mechanisms.

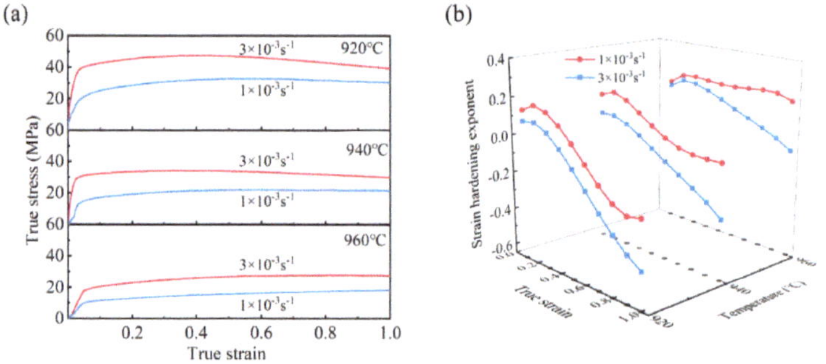

Fig. 2. (a) The true strain-stress curves and (b) strain hardening exponent.

Fig. 3 gives the α/β phase characteristics. Fig. 3(a), (b), (c), (e) and (f) demonstrate that increasing temperature and strain homogenize the two-phase microstructure and transform α to β phase. As a result, the two-phase grains become more equiaxed, and the phase boundaries are more clearly defined at 960°C/strain 1. Fig. 3(a) and (d) show that the phase volume fraction and microstructure morphology between α and β phases are similar at different strain rates. Apparently, heat and load cause small β grains to merge and grow together, while β phase gradually precipitates at α phase boundary. Eventually, the initially fragmented β phase changes to continuous or spherical. The α→β phase dynamic transformation is essentially driven by deformation storage energy and stress. The α→β diffusionless phase transformation occurs when the accumulated energy during deformation exceeds the energy barrier required for shear mode and lattice expansion. Furthermore, on one hand, high temperatures can reduce the difference in Gibbs free energy between the α and β phases and the energy required for lattice changes. On the other hand, strain concentration at the phase boundaries contributes to the change in the total free energy of the system, which promotes the phase transition through atomic diffusion. Thus, an appropriate increase in strain or temperature must result in an increase in β phase volume fraction [8].

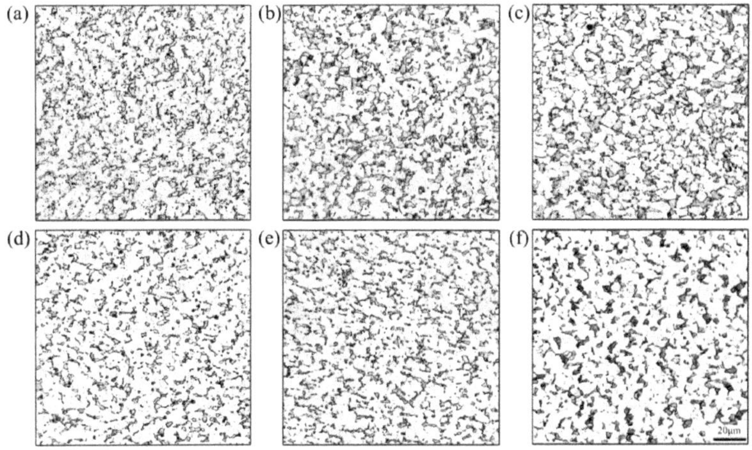

Fig. 3. The OM images of deformed microstructure: (a) 920°C/1×10^{-3} s^{-1}/strain 0.5; (b) 940°C/1×10^{-3} s^{-1}/strain 0.5; (c) 960°C/1×10^{-3} s^{-1}/strain 0.5; (d) 920°C/3×10^{-3} s^{-1}/strain 0.5; (e) 920°C/1×10^{-3} s^{-1}/strain 0.25; (f) 920°C/1×10^{-3} s^{-1}/strain 1.

Fig. 4 gives information on α grain morphology and grain boundary. A rise in temperature and strain causes the larger α_p and β grains and a more uniform α_p grain size distribution, as seen in Fig. 4(a), (b) and (c). High temperature activates grain boundary diffusion and atomic motion, causing small α grains to coalesce and grow into larger ones [9]. Additionally, LAGBs decrease with increasing temperature and strain. Generally, near LAGBs tend to have high geometrically necessary dislocation density and form channel-like structures. Dispersed intragranular dislocations gradually accumulate into dislocation walls or cells and cling more to channel-like structures, forming subgrain boundaries with higher misorientation angle, which convert LAGBs to middle-angle grain boundaries (MAGBs) (DRV effect form substructures). Moreover, under active grain boundary migration, MAGBs continue to absorb nearby free dislocations while subgrains rotate to coordinate plastic deformation. Eventually MAGBs transform into HAGBs, and the lattice distortion, non-uniform strain and high residual stress disappear, producing new equiaxed grains with uniform crystal structure (DRX effect form undistorted grains) [9, 10]. It should be noted that the increasing temperature and strain lead to the generation of finer and lamellar α_s grain (more β phase).

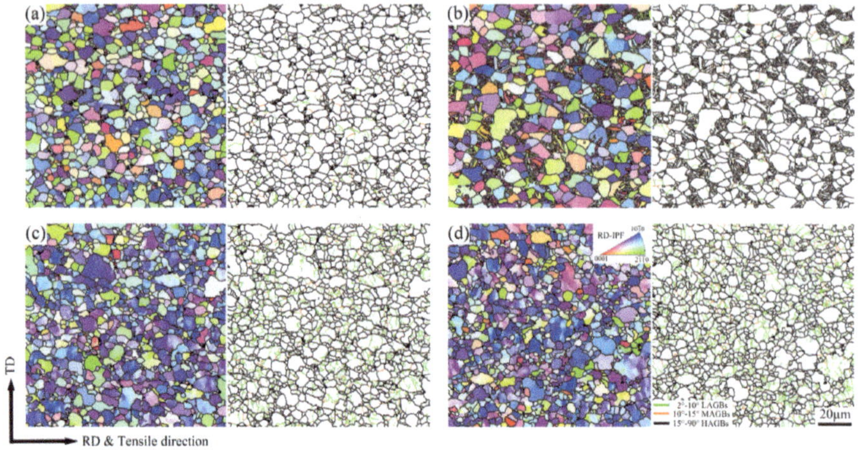

Fig. 4. The α grain morphology and grain boundary maps of deformed microstructure: (a) 920°C/1×10⁻³ s⁻¹/strain 1; (b) 960°C/1×10⁻³ s⁻¹/strain 1; (c) 920°C/1×10⁻³ s⁻¹/strain 0.25; (d) 920°C/3×10⁻³ s⁻¹/strain 0.25.

Construction of unified constitutive model

Flow rule. High rate dependency and viscoplasticity, which might be considered creep behavior, are characteristics of the superplastic deformation of titanium alloys. The basic form of the plastic stress equation combines the Rosserd equation, which describes strain hardening and strain rate hardening, and Norton's low, which describes steady state creep behavior, as follows [11]:

$$\dot{\varepsilon}_p = \left(\frac{\sigma_v}{K}\right)^n \tag{1}$$

where $\dot{\varepsilon}_p$ is plastic strain rate, σ_v is the viscoplastic stress, K is the coefficient of resistance, n is the viscosity exponent.

The true stress σ of titanium alloys during viscoplastic deformation is a linear combination of initial yield stress k, isotropic hardening stress H, and σ_v. The flow rule can be expressed as follows [12]:

$$\dot{\varepsilon}_p = \left\langle\frac{\sigma - H - k}{K}\right\rangle^n \tag{2}$$

where ⟨ ⟩ is the McCauley bracket, indicating that the formula in the bracket is valid only when the alloy enters the plastic deformation stage.

The H can be expressed as a function of dislocation density ρ, as follows:

$$\dot{H} = 0.5B\bar{\rho}^{-0.5}\dot{\bar{\rho}} \tag{3}$$

where B is the material constant.

The relationship between elastic strain and flow stress can be determined using Hook's law:

$$\sigma = E\left(\varepsilon_T - \varepsilon_p\right) \tag{4}$$

where E is the elastic modulus, ε_T is the total strain, ε_p is the plastic strain.

Modeling of dislocation density. The Kocks-Mecking (KM) dislocation evolution model, based on Roberts model, is applied to light alloys with an average grain size within 10 μm. The KM model considers the linear relationship between the dislocation accumulation during strain hardening and the dislocation annihilation facilitated by DRV. This can be expressed as follows:

$$\frac{d\rho}{d\varepsilon_p} = K_1\sqrt{\rho} - K_2\rho \tag{5}$$

where K_1 and K_2 are the material constants.

The normalized dislocation density evolution is expressed by considering the high-temperature deformation mechanism together as follows [13]:

$$\dot{\bar{\rho}} = A\left(1-\bar{\rho}\right)\left|\dot{\varepsilon}_p\right| - C\bar{\rho}^{n_2} \tag{6}$$

where $\bar{\rho}$ is the normalized dislocation density formulated as $\bar{\rho} = 1 - \rho_0/\rho$, ρ_0 is the initial dislocation density, ρ is the dislocation density at deformation, A, n_2, and C are the material constants. The normalized dislocation density is utilized to limit the dislocation density between 0 and 1, in order to accurately describe the dislocation density evolution in the constitutive model [14, 15].

Modeling of dynamic recrystallization. The equation for the volume fraction of dynamic recrystallization controlled by dislocation and strain is as follows [16]:

$$\dot{S} = \frac{q_1\left[x\bar{\rho} - \bar{\rho}_c\left(1-S\right)\right]\left(1-S\right)^{q_2}\dot{\varepsilon}_p^{\,q_3}}{d} \tag{7}$$

where $\bar{\rho}_c$ is the critial dislocation density for dynamic recrystallization to occur and can be formulated as $\bar{\rho}_c = q_4\dot{\varepsilon}_p^{\,q_5}$ [17], d is the average grain size, q_1, q_2, q_3, q_4 and q_5 are the material constants.

The evolution rate of dynamic recrystallized onset parameter x is expressed as follows:

$$\dot{x} = q_6\left(1-x\right)\bar{\rho} \tag{8}$$

where q_6 is the material constant.

The impact of dynamic recrystallization on dislocation density must be considered as follows [11]:

$$\dot{\bar{\rho}} = A\left(1-\bar{\rho}\right)\left|\dot{\varepsilon}_p\right| - C_1\bar{\rho}^{n_2} - \left(\frac{C_2\bar{\rho}}{1-S}\right)\dot{S} \tag{9}$$

where C_2 is the material constant.

Modeling of grain size. During the high-tempertaure deformation of alloys, the grains undergo a dynamic growth and refinement process. Considering the dynamic and static grain growth as well as recrystallization assist grain refinement, the rate of grain size evolution can be expressed as follows [18, 19]:

$$\dot{d} = w_1 d^{-\gamma_1} + w_2 \dot{\varepsilon}_p d^{-\gamma_2} - w_3 \dot{S}^{\gamma_3} d^{\gamma_4} \tag{10}$$

where w_1, w_2, w_3, γ_1, γ_2, γ_3 and γ_4 are the material constants.

When considering how grain size affects flow stress, it is important to modify the plastic strain rate as follows:

$$\dot{\varepsilon}_p = \left\langle \frac{\sigma - H - k}{K} \right\rangle^{n_1} \left(\bar{d}\right)^{-u} \tag{11}$$

where u is the material constant, $\bar{d}$ is the normalized grain size formulated as $\bar{d} = d/d_i$, d and d_i are the average grain size of initial and deformed material, respectively.

Determination of material constants. In the unified constitutive model presented above, A, u, n_2, γ_2, γ_3 and γ_4 are the temperature-independent material constants. The temperature-dependent material constants are shown in Table 1. k_0, K_0, n_0, B_0, C_{10}, C_{20}, w_{10}, w_{20}, w_{30}, γ_0, q_{10}, q_{20}, q_{30}, q_{40}, q_{50}, q_{60} and E_0 are material constants. Q_k, Q_K, Q_n, Q_B, Q_{C1}, Q_{C2}, Q_{w1}, Q_{w2}, Q_{w3}, Q_γ, Q_{q1}, Q_{q2}, Q_{q3}, Q_{q4}, Q_{q5}, Q_{q6} and Q_E are activation energies linked to the temperature-dependent parameters. R is the gas constant (8.314 J mol^{-1} K^{-1}).

Table 1. The temperature-dependent parameters and their corresponding formulation.

$k = k_0 \exp(Q_k/RT)$	$w_1 = w_{10} \exp(Q_{w1}/RT)$	$q_3 = q_{30} \exp(Q_{q3}/RT)$
$K = K_0 \exp(Q_K/RT)$	$w_2 = w_{20} \exp(-Q_{w2}/RT)$	$q_4 = q_{40} \exp(Q_{q4}/RT)$
$n_1 = n_0 \exp(Q_n/RT)$	$w_3 = w_{30} \exp(-Q_{w3}/RT)$	$q_5 = q_{50} \exp(-Q_{q5}/RT)$
$B = B_0 \exp(Q_B/RT)$	$\gamma_1 = \gamma_0 \exp(Q_\gamma/RT)$	$q_6 = q_{60} \exp(-Q_{q6}/RT)$
$C_1 = C_{10} \exp(-Q_{C1}/RT)$	$q_1 = q_{10} \exp(-Q_{q1}/RT)$	$E = E_0 \exp(Q_E/RT)$
$C_2 = C_{20} \exp(-Q_{C2}/RT)$	$q_2 = q_{20} \exp(Q_{q2}/RT)$	

The material constants are searched and optimized using the genetic algorithm (GA) toolbox in Matlab software. Firstly, the computational program for unified constitutive model was programmed using the forward Euler method and iterative approach. Where Eq. 12 shows the computational and iterative procedure for each physical variable formulated by forward Euler method. The computational program was then used to program the custom function required for the GA toolbox to function properly. Finally, the GA's objective function was defined as the residual error between the predicted and experimental stress and grain size values. The global objective function is the sum of various objective functions.

$$\begin{cases} \varepsilon_{i+1}^{p} = \varepsilon_{i}^{p} + \dot{\varepsilon}_{i}^{p}\Delta t \\ \varepsilon_{i+1}^{t} = \varepsilon_{i}^{t} + \dot{\varepsilon}_{i}^{t}\Delta t \\ K_{1,i+1} = K_{1,i} + \dot{K}_{1,i}\Delta t \\ K_{2,i+1} = K_{2,i} + \dot{K}_{2,i}\Delta t \\ \cdots \\ K_{i+1} = K_{1,i+1} + K_{2,i+1} + \ldots \\ \sigma_{i+1} = E\left(\varepsilon_{i+1}^{t} - \varepsilon_{i+1}^{p}\right) \\ \sigma_{i+1} = \sigma_{i} + \Delta t \end{cases} \tag{12}$$

where K is the internal state variable, i is the iteration counter, $i = 1, 2, 3, ..., n$.

To further simplify the GA's calculation process, the custom function is modified to adapt to different deformation conditions. Then, the temperature-independent material constants are fixed, and the remaining material constants are determined using linear fitting to account for the effect of temperature. Ultimately, the optimal material constants are computed, searched for, and verified repeatedly using the GA toolbox in Matlab.

Verification of unified constitutive model

Fig. 5(a), (b) and c show the comparison between predicted stress of unified constitutive model and experimental results. The predicted values generally fit experimental values well. However, the prediction accuracy decreases in the low strains (0.05~0.4) as the temperature increases. Generally, the alloy's constitutive model often has limitations in the predicted curves under various deformation conditions that follow similar trends. Additionally, material constants solved under one condition may limit the model's accuracy under other deformation conditions. For example, the predicted curves in Fig. 5(c) resemble those in Fig. 5(b). Fig. 5(d) shows the correlation between predicted and experimental stress values. Table 2 presents the errors between predicted and experimental stress values. The correlation coefficient R, AARE and RMSE between the predicted and experimental stress values are 0.98, 6.91% and 1.94, respectively. The correlation between predicted and experimental values was strong. Under different deformation conditions, the maximum values of AARE and RMSE are 13.38% and 2.25, respectively, while the minimum values of AARE and RMSE are 2.90% and 1.55, respectively. These results suggest that the unified constitutive model has high prediction accuracy and can effectively describe the mechanic behaviors of TA32 alloy under superplastic deformation conditions.

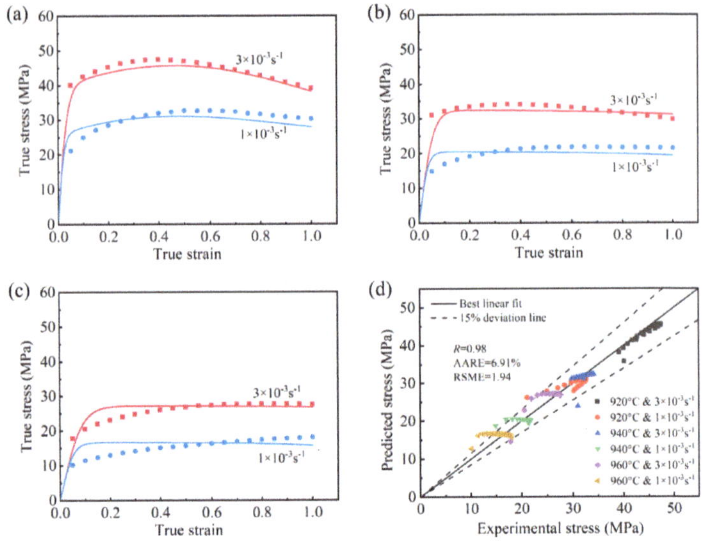

Fig. 5. Comparison between predicted stress values (curves) of constitutive model and experimental results (points): (a) 920°C, (b) 940°C and (c) 960°C; (d) correlation between predicted and experimental values

Table 2. Errors between predicted stress of constitutive model and experimental results at different deformation conditions.

	920°C and 1×10^{-3} s^{-1}	920°C and 3×10^{-3} s^{-1}	940°C and 1×10^{-3} s^{-1}	940°C and 3×10^{-3} s^{-1}	960°C and 1×10^{-3} s^{-1}	960°C and 3×10^{-3} s^{-1}
AARE	6.23%	2.90%	8.64%	4.21%	13.38%	6.07%
RMSE	2.05	1.55	1.90	1.96	2.25	1.86

Fig. 6(a) shows the comparison between predicted grain sizes of unified constitutive model and experimental results. The thermal effect causes an increase in grain size, while the dynamic recrystallization leads to a smaller grain size. So, the grain size changes dynamically during superplastic deformation. The average grain size of TA32 alloy decreases with increasing strain. On one hand, the more significant DRX effect generates finer grains. On the other hand, the water quenching following superplastic deformation of near-α titanium alloy induces the transformation of large β grains into lamellar and fine α_s grains. Fig. 6(b) illustrates the correlation between predicted and experimental normalized grain sizes, reflecting model's ability to predict internal state variable. The correlation coefficient *R*, AARE and RMSE between the predicted and experimental normalized grain size are 0.92, 4.44% and 0.051, respectively. The degree of dispersion is slightly high compared to stress. In sum, the unified constitutive model presented in this paper also has excellent accuracy in predicting internal state variable.

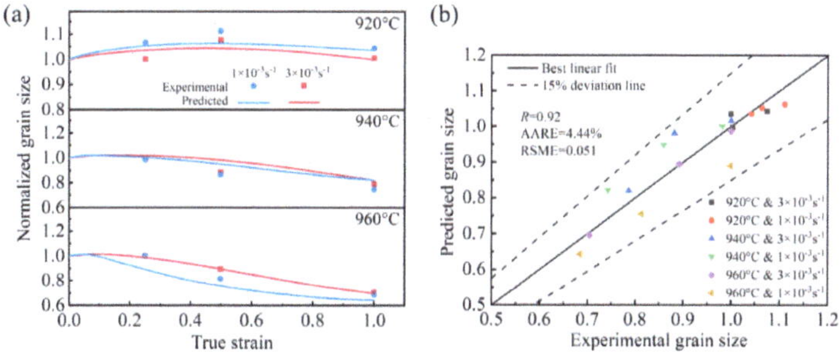

Fig. 6. (a) Comparison and (b) correlation between predicted normalized grain sizes (curves) of constitutive model and experimental results (points).

Conclusions

(1) The flow stress of TA32 alloy during superplastic deformation decreases as the temperature increases and strain rate decreases. Increasing temperature and strain rate promote the transformation of LAGBs to HAGBs, the disappearance of intragranular defects, α→β phase transformation, and the formation of more equiaxed grains. The microstructure of TA32 alloy varies little within the range of 1×10^{-3}-3×10^{-3} s^{-1}.

(2) Building upon the macroscopic and microscopic deformation characteristics of TA32 titanium alloy during superplastic deformation, a unified constitutive model is constructed, coupled with the grain size evolution. The model describes the interactions among physical variables such as flow stress, plastic strain rate, dislocation density, and average grain size and reflects the influence of microstructure on flow stress.

(3) The statistical results reveal the correlation coefficient R, AARE and RMSE between predicted and experimental stress values are 0.98, 6.91% and 1.94, respectively. Similarly, the correlation coefficient R, AARE and RMSE between predicted and experimental normalized grain size values are 0.92, 4.44% and 0.051, respectively. The unified constitutive model has a good prediction effect on the superplastic mechanic behavior and grain size evolution characteristics of TA32 alloy.

Acknowledgments

Authors acknowledge the funding supported by National Natural Science Foundation of China (No. U22A20186, 52211530441, 52175285) and Aeronautical Science Foundation of China (No. 2022Z047025001).

References

[1] S. Li, T.S. Deng, Y.H. Zhang, Y.Q. Liang, R.X. Li, T.H. Dong, Review on the creep resistance of high-temperature titanium alloy, Trans. Indian Inst. Met. 74 (2021) 215-222. https://doi.org/10.1007/s12666-020-02137-x

[2] C. Cheng, Y. Feng, Z.Y. Chen, H.E. Li, Q.J. Wang, X. Wang, Effect of annealing temperature on microstructure, texture and tensile properties of TA32 sheet, Mater. Sci. Eng. A 826 (2021) 141971. https://doi.org/10.1016/j.msea.2021.141971

[3] B. Babu, L.E. Lindgren, Dislocation density based model for plastic deformation and globularization of Ti-6Al-4V, Int. J. Plast. 50 (2013) 94-108. https://doi.org/10.1016/j.ijplas.2013.04.003

[4] Y.W. Xiao, Y.C. Lin, Y.Q. Jiang, X.Y. Zhang, G.D. Pang, D. Wang, K.C. Zhou, A dislocation density-based model and processing maps of Ti-55511 alloy with bimodal

microstructures during hot compression in α+β region, Mater. Sci. Eng., A 790 (2020) 139692. https://doi.org/10.1016/j.msea.2020.139692

[5] Q. Bai, J. Lin, T.A. Dean, D.S. Balint, T. Gao, Z. Zhang, Modelling of dominant softening mechanisms for Ti-6Al-4V in steady state hot forming conditions, Mater. Sci. Eng. A 559 (2013) 352-358. https://doi.org/10.1016/j.msea.2012.08.110

[6] T. Yasmeen, Z.T. Shao, L. Zhao, P. Gao, J. Lin, J. Jiang, Constitutive modeling for the simulation of the superplastic forming of TA15 titanium alloy, Int. J. Mech. Sci. 164 (2019) 105178. https://doi.org/10.1016/j.ijmecsci.2019.105178

[7] Y. Liu, Z.Q. Li, B. Zhao, T. Yang, C.Y. Sun, Superplastic deformation behavior and constitutive model of TA32 titanium alloy, Rare Metal Mat. Eng. 51 (2022) 3752-3761.

[8] J.W. Xu, W.D. Zeng, Q.Y. Zhao, D.D. Zhou, S.T. He, R.C. Jia, X.F. Zhou, Analysis of deformation behavior and microstructure of a near-alpha titanium alloy in single phase region, Mater. Sci. Eng. A 803 (2021) 140723. https://doi.org/10.1016/j.msea.2020.140723

[9] Y. Liu, Z.Q. Li, B. Zhao, C.y. Sun, Y.h. Feng, Microstructure evolution characteristics of near-α TA32 titanium alloy during superplastic tensile deformation, Mater. Sci. Eng., A 879 (2023) 145264. https://doi.org/10.1016/j.msea.2023.145264

[10] L. Despax, V. Vidal, D. Delagnes, M. Dehmas, H. Matsumoto, V. Velay, Influence of strain rate and temperature on the deformation mechanisms of a fine-grained Ti-6Al-4V alloy, Mater. Sci. Eng. A 790 (2020) 139718. https://doi.org/10.1016/j.msea.2020.139718

[11] J. Lin, T.A. Dean, Modelling of microstructure evolution in hot forming using unified constitutive equations, J. Mater. Process. Technol. 167 (2005) 354-362. https://doi.org/10.1016/j.jmatprotec.2005.06.026

[12] L. Yang, B.y. Wang, G. Liu, H.j. Zhao, J. Zhou, Hot tensile behavior and self-consistent constitutive modeling of TA15 titanium alloy sheets, J. Mater. Eng. Perform. 24 (2015) 4647-4655. https://doi.org/10.1007/s11665-015-1784-7

[13] J. Lin, M. Mohamed, D. Balint, T.A. Dean, The development of continuum damage mechanics-based theories for predicting forming limit diagrams for hot stamping applications, Int. J. Damage Mech. 23 (2013) 684-701. https://doi.org/10.1177/1056789513507731

[14] J. Lin, Y. Liu, D.C.J. Farrugia, M. Zhou, Development of dislocation-based unified material model for simulating microstructure evolution in multipass hot rolling, Philos. Mag. 85 (2005) 1967-1987. https://doi.org/10.1080/14786430412331305285

[15] J. Lin, Y. Liu, T.A. Dean, A review on damage mechanisms, models and calibration methods under various deformation conditions, Int. J. Damage Mech. 14 (2016) 299-319. https://doi.org/10.1177/1056789505050357

[16] E. Alabort, P. Kontis, D. Barba, K. Dragnevski, R.C. Reed, On the mechanisms of superplasticity in Ti-6Al-4V, Acta Mater. 105 (2016) 449-463. https://doi.org/10.1016/j.actamat.2015.12.003

[17] W. Roberts, B. Ahlblom, A nucleation criterion for dynamic recrystallization during hot working, Acta Mater. 26 (1978) 801-813. https://doi.org/10.1016/0001-6160(78)90030-5

[18] R. Sandström, R. Lagneborg, A model for hot working occurring by recrystallization, Acta Mater. 23 (1975) 387-398. https://doi.org/10.1016/0001-6160(75)90132-7

[19] J. Lin, A set of unified constitutive equations for modelling microstructure evolution in hot deformation, J. Mater. Process. Technol. 143-144 (2003) 281-285. https://doi.org/10.1016/s0924-0136(03)00472-2

Metal Forming 2024 — Materials Research Forum LLC
Materials Research Proceedings 44 (2024) 90-99 — https://doi.org/10.21741/9781644903254-10

Experimental formability evaluation for aluminium alloy sheets under hot stamping conditions

ZHANG Ruiqiang[1,a *], LI Jiaqi[1,b], SHI Zhusheng[1,c] and LIN Jianguo[1,d]

[1]Department of Mechanical Engineering, Imperial College London, London SW7 2AZ, UK

[a]r.zhang17@imperial.ac.uk, [b]jiaqi.li120@imperial.ac.uk, [c]zhusheng.shi@imperial.ac.uk, [d]jianguo.lin@imperial.ac.uk

Keywords: Aluminium Alloy, Hot Stamping, Formability, Necking, Fracture, Forming Limit Curves

Abstract. The process of Hot Form and Quench for aluminium alloys, known as HFQ®, has been developed and applied to manufacture lightweight, high-strength engineering panel components in the automotive industry. However, formability evaluation for the alloys under hot stamping conditions is challenging. In this study, a recently developed biaxial testing method has been applied to aluminium alloy AA6082 for formability evaluation at temperatures ranging from 440–510 °C and at a strain rate of 0.1 s^{-1}. This method involves heating cruciform specimens via the resistance heating system in the Gleeble, and deforming them until fracture via a customised biaxial tensile rig which transfers a uniaxial force into biaxial forces. The location for welding thermocouples on cruciform specimen surface for temperature feedback control in the Gleeble is investigated. Furthermore, temperature nonuniformity within the gauge area of cruciform specimens is quantified, and biaxial tensile tests on the specimens are carried out under different conditions. Both the limit strains at the onset of necking and at fracture are determined, and their dependency on the deformation conditions is analysed. It is found that the biaxial testing method is applicable to AA6082 for formability evaluation under hot stamping conditions. In addition, the limit major strains vary with the strain state, but exhibit a minor dependency on the temperature in the range investigated.

Introduction

Hot Form and Quench (HFQ®) for aluminium alloys is a hot stamping process for manufacturing lightweight, high-strength panel components for a wider range of industrial applications [1]. During the processes, the alloy is heated for solution heat treatment, and then quickly transferred into cold process dies for simultaneous forming and in-die quenching, followed by artificial aging to obtain the highest strength because of the precipitate hardening [2, 3]. Numerical simulations of hot stamping processes are essential to the design and optimisation of these processes [4, 5]. For accurate numerical simulations, it is essential to determine the thermo-mechanical properties of the alloy under hot stamping conditions for the development and calibration of materials constitutive models [6-8].

Formability refers to the capability of an alloy to undergo plastic deformation without experiencing failure [9]. Forming limit curves (FLCs) are the most widely used tools to quantitively evaluate the formability for sheet metals, which are constructed using the limit strains at the onset of necking in various strain states from uniaxial over plane-strain to equi-biaxial tensions [10-12]. Fracture forming limit curves (FFLCs) are more recently developed tools, which are constructed using the limit strains at fracture [13]. The Nakajima tests have been widely carried out to determine FLCs of sheet metals at room temperature [14-16], and even at high temperatures [17-19]. In the Nakajima tests, sheet metals are deformed by a hemispherical punch through thickness direction [20]. In order to reduce the friction between the punch and the blank, a suitable lubricant system should be applied, so that the fracture of the blank occurs within a certain region

[20]. For high temperature tests, the blank is heated in furnaces. However, due to the challenge to replicate the intricate thermal cycle in hot stamping, and the failure of lubricant system at high temperatures, it is extremely difficult to apply the Nakajima tests to determine FLCs and/or FFLCs for sheet metals under hot stamping conditions.

Recently, a biaxial testing method has been developed and successfully applied to boron steel for the determination of both FLCs and FFLCs under hot stamping conditions [21]. In this testing method, a type of cruciform specimen was designed, and heated using the resistance heating method. By leveraging the feedback control heating system in the Gleeble thermo-mechanical simulator, it has been found that the intricate thermal cycle for hot stamping can be replicated accurately. In addition, since the specimens are stretched in plane-stress state, lubricant systems are not required anymore as in the Nakajima tests. However, the applicability of the biaxial testing method to aluminium alloys, such as the cruciform specimen design and the heating method, is still unknown.

The present study aims to apply the newly developed biaxial testing method to aluminium alloy AA6082 for the determination of both FLCs and FFLCs under hot stamping conditions. Cruciform specimens of AA6082 are heated using the resistance heating method, and stretched under different loading, temperature and strain rate conditions. The digital image correlation (DIC) is applied to measure the full-field strain within the gauge area of the specimens. The strain data are consequently post-processed to determine the limit strains at necking and fracture for constructing both FLCs and FFLCs.

Methodology

Material and thermal cycle for hot stamping. Aluminium alloy AA6082 sheets with an initial thickness of 1.5 mm are used in the present study. The composition of the alloy is Al-0.87Si-0.33Fe-0.026Cu-0.51Mn-0.97Mg-0.044Cr0.025Zn-0.016Ti (wt. %). Fig. 1 shows the temperature cycle of AA6082 which mimics the conditions during industrial hot stamping processes. The alloy is heated to 535 °C and soaked at this temperature for 60 s to generate supersaturated solid solution. Subsequently, the alloy is quenched at 60°C/s to the target forming temperature for thermo-mechanical deformation.

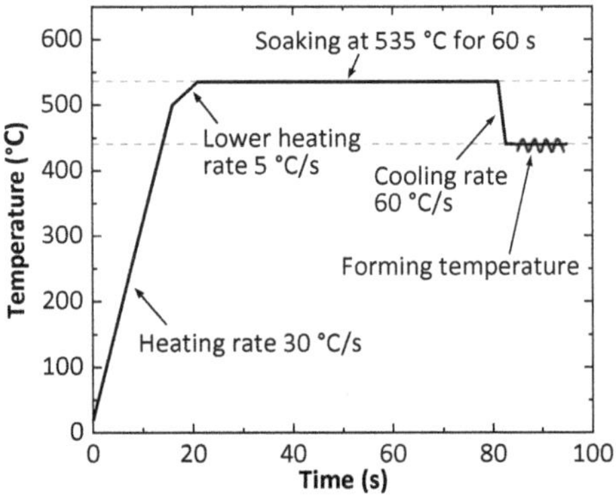

Fig. 1. Temperature cycle for AA6082 for hot stamping.

Biaxial testing method. The biaxial testing method, originally developed in Ref. [21], is applied to AA6082 for formability evaluation under hot stamping conditions in the present study. Fig. 2(a) shows the schematic diagram of utilising this biaxial testing method to determine forming limit curves (FLCs) and fracture forming limit curves (FFLCs) for the alloy. In this method, a biaxial

Materials Research Proceedings 44 (2024) 90-99 https://doi.org/10.21741/9781644903254-10

tensile system is used to heat and stretch cruciform specimens under various loading conditions. The biaxial tensile system consists of a Gleeble thermo-mechanical simulator and a customised biaxial tensile rig. During the testing, the biaxial tensile rig is installed inside the chamber of the Gleeble, as shown in Fig. 2(b), and transfers, through slider-crank mechanisms, uniaxial forces provided by the Gleeble into biaxial forces for stretching cruciform specimens. The Gleeble also provides electric currents to heat the specimens using the resistance heating method to precisely achieve the complex temperature cycle including rapid heating and quenching as shown in Fig. 1. During the deformation of the specimens, the strain fields within the gauge area are measured using the digital image correlation (DIC), and consequently post-processed using the spatio-temporal method, originally developed in Ref. [22], to determine the limit strains at the onset of necking and at fracture. Finally, those limit strains are used to construct both FLCs and FFLCs. Fig. 1(c) shows the experimental setup for formability evaluation, including the Gleeble, the high-speed camera (Photron FASTCAM Mini delivering 1,280 by 1,024 pixel resolution) for DIC, and the biaxial tensile rig together with the cruciform specimen, etc. Fig. 2(d) presents the geometry and dimensions of the cruciform specimen, originally developed in Ref. [21, 23], and its heating scheme. The gauge area of each specimen is located in the centre, and is intentionally thinned into a domed shape to concentrate and localise deformation within this region. Regarding the heating scheme, the upper arm of each specimen is connected to the positive electrode, and the lower arm is connected to the negative electrode. During the testing, electrical currents provided by the Gleeble, flow through the specimen and heat the specimen based on the phenomena of joule heating.

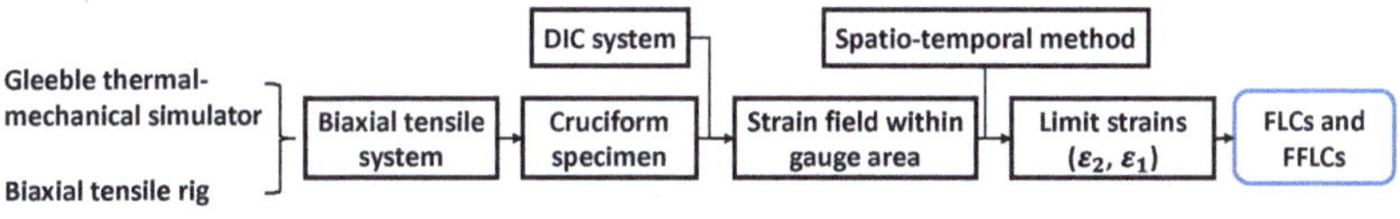

(a)

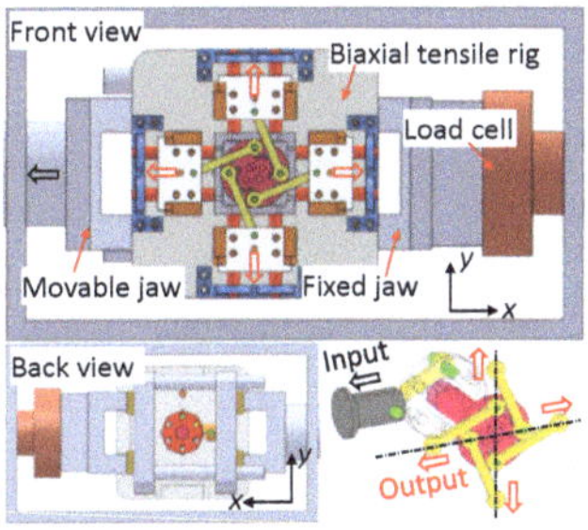

(b)

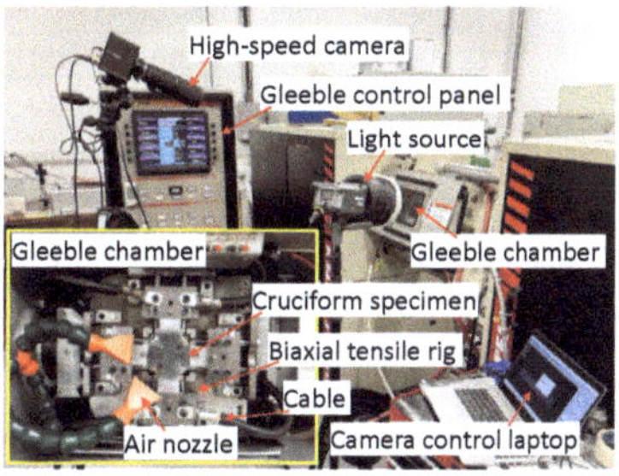

(c)

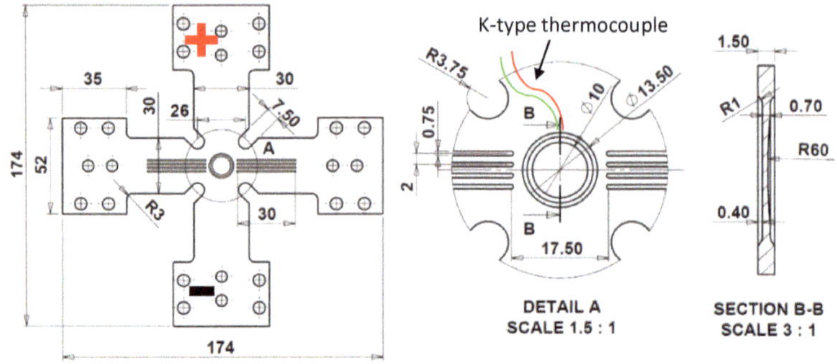

(d)

Fig. 2. Biaxial testing method for formability evaluation under hot stamping conditions, (a) Schematic diagram of utilising the biaxial testing method to determine forming limit curves (FLCs) and fracture forming limit curves (FFLCs), (b) Experimental setup, (c) Biaxial tensile rig and its setup in the Gleeble, and (d) Geometry and dimensions (unit: mm) of the cruciform specimen. Symbols '+' and '–' mean connecting positive and negative electrode, respectively.

Regarding the strain measurement using DIC, in the present study, a region of 35 mm by 28 mm on specimen surface which covers the gauge area throughout the deformation was recorded using the high-speed camera. The pictures were post-processed using the commercial software GOM Correlate 2018, with setting facet size to 19 pixels and point distance to 10 pixels.

Table 1 lists all testing conditions for formability evaluation in the present study, including loading state, temperature and strain rate. Each test was repeated twice to enhance the accuracy and reliability of the data obtained.

Table 1. Testing conditions for biaxial tensile tests for formability evaluation.

Test	Loading state	Temperature (°C)	Equivalent strain rate (s^{-1})
1	Equi-biaxial	440	0.1
2	Plane-strain	440	0.1
3	Uniaxial	440	0.1
4	Equi-biaxial	510	0.1
5	Plane-strain	510	0.1
6	Uniaxial	510	0.1

Results

Temperature distributions. It is well known that the heating system in the Gleeble results in a nonuniform temperature distribution within the gauge area of the specimen [24]. The temperatures at the locations A, B, and C, as indicated in Fig. 3, were therefore measured and quantified by using K-type thermocouples welded at these locations. Fig. 3 shows the results when controlling the temperature at A to reproduce the target temperature cycle specified in Fig. 1. As can be seen, the target temperature cycle at A was reproduced precisely. Furthermore, after the temperature stabilises, the location A has the highest temperature, followed by B and then C. As indicated in Fig. 2(d), in the present study, the measured temperature cycle at B was used to indirectly control the temperature at A. Additionally, the stabilised temperatures at B and C were quantified when the stabilised temperature at A reached different values. Table 2 lists the results. As can be seen, the difference between the temperatures at A and C, which represent the maximum temperature difference within the gauge area [21], is less than 20°C.

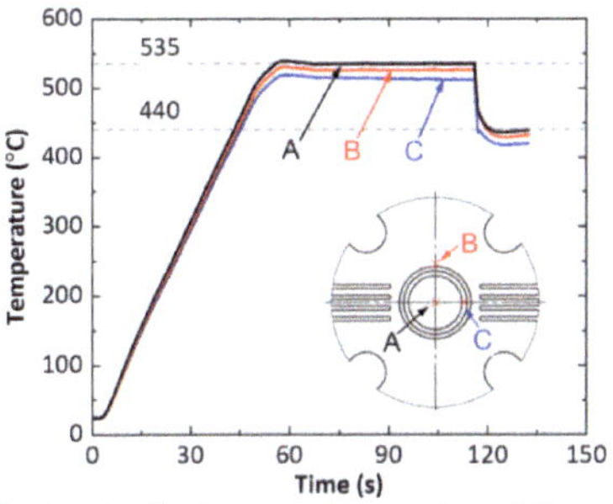

Fig. 3. Temperature distribution in the intersection region of the cruciform specimen, with an inset showing the locations of A, B and C.

Table 2. Stabilised temperature (°C) at the locations A, B and C of cruciform specimen during soaking and deformation. Different cases mean the stabilised temperature at A reaching different values.

Location	Case I	Case II	Case III	Case IV
A	535	510	440	370
B	525	498	432	361
C	513	490	420	357

Fracture cruciform specimens. Fig. 4 shows the fractured cruciform specimens of AA6082 under the different conditions listed in Table 1. As can be seen, the fracture occurred near the specimen centre in all the tests. This indicates the effectiveness of the specimen design for evaluating the formability for AA6082 [25], similarly to the applications to boron steel [21].

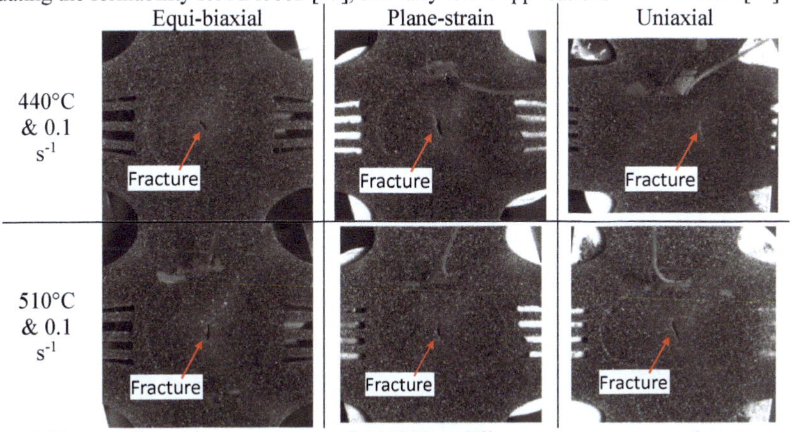

Fig. 4. Fractured cruciform specimens of AA6082 at different temperatures and strain rates.

Fig. 5 shows the total force applied to the biaxial tensile rig for stretching these cruciform specimens until the fracture; the force was measured by the load cell as indicated in Fig. 2(b). As can be seen, the force required to induce fracture in the specimens in plane-strain tension is smaller than that in plane-strain tension but higher than that in uniaxial tension. Additionally, a higher temperature results in lower forces needed to fracture the specimens. Similar phenomenon was observed in the applications to boron steel [21]. However, the fracture force required for AA6082 is significantly less than that for boron steel.

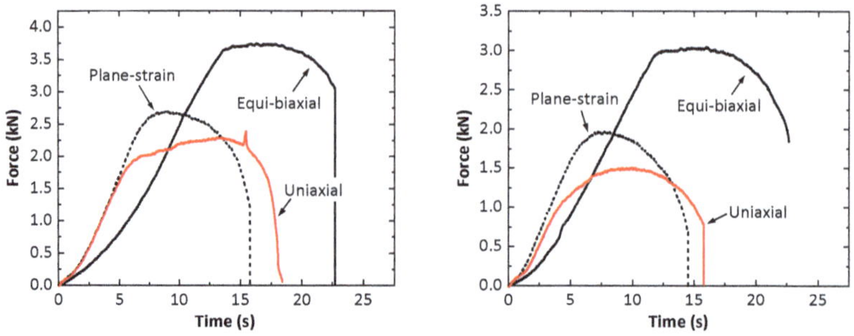

(a) (b)

Fig. 5 Total force applied to the biaxial tensile rig for stretching the cruciform specimens at (a) 440°C and 0.1 s^{-1} and (b) 510°C and 0.1 s^{-1}.

Fig. 6 shows the strain fields within the gauge area of the specimens, measured using DIC, in different loading states at 510°C and 0.1 s^{-1}. The percent values 0%, 50%, 80% and 98% represent the time normalised by the time at fracture. As can be seen, the full field strains within the gauge area before 98% were successfully measured in all the cases. Furthermore, as the time increases, the deformation within the gauge area increases, especially in the vicinity of the specimen centre. This is attributed to the nonuniform thickness (Fig. 2(d)) and the temperature distribution (Fig. 3) within the gauge area.

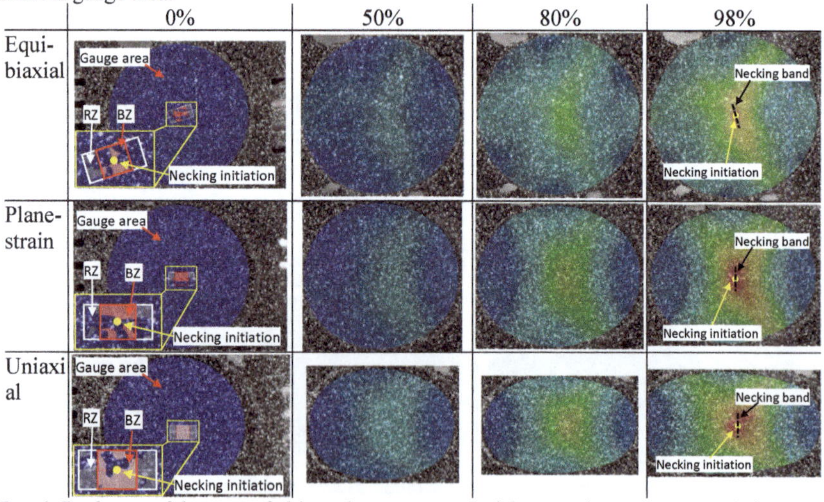

Fig. 6. Evolution of the strain fields in the gauge area of the cruciform specimens deformed in different loading states at 510°C and 0.1 s^{-1}. The percent values 0%, 50%, 80% and 98% represent the times normalised by the time at fracture.

Materials Research Proceedings 44 (2024) 90-99 https://doi.org/10.21741/9781644903254-10

FLCs and FFLCs for AA6082. Based on the strain fields measured using DIC, the limit strains at the onset of necking and at fracture were determined by using the spatio-temporal method. To apply this method, a base zone (BZ) and a reference zone (RZ) were selected around the location where necking occurred, as shown in Fig. 6. Both zones have the same size along the localised necking band, but differ in the dimension perpendicular to the band. Specifically, the initial size of BZ is 1.1 mm by 1.1 mm, while the size of RZ is 2.2 mm by 1.1 mm [21]. Fig. 7(a1), (b1) and (c1) shows the thinning within BZ against the thinning within RZ in different loading states at 510°C and 0.1 s^{-1}. Bilinear fitting was applied to the thinning data, both during the initial stages and the later stages of deformation. Subsequently, the onset of localised necking was identified at the intersection of the fitted lines. Fig. 7(a2), (b2) and (c2) shows the limit strains determined using the spatio-temporal method, together with the strain paths until fracture. Among the three strain states presented, the strain path obtained in equi-biaxial tension demonstrates the best linearity, with the associated strain ratio of approximately 0.6.

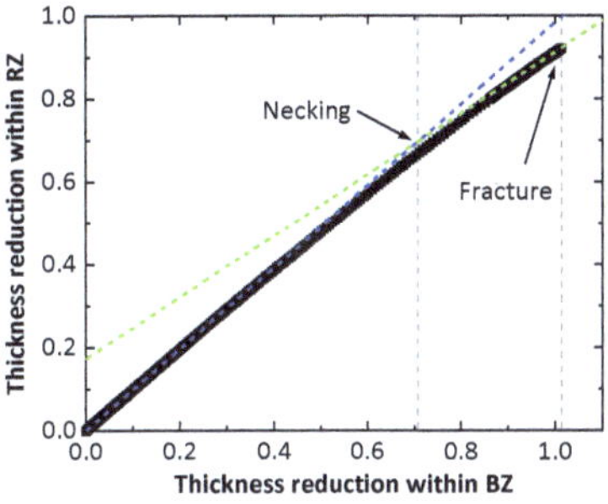

(a1)

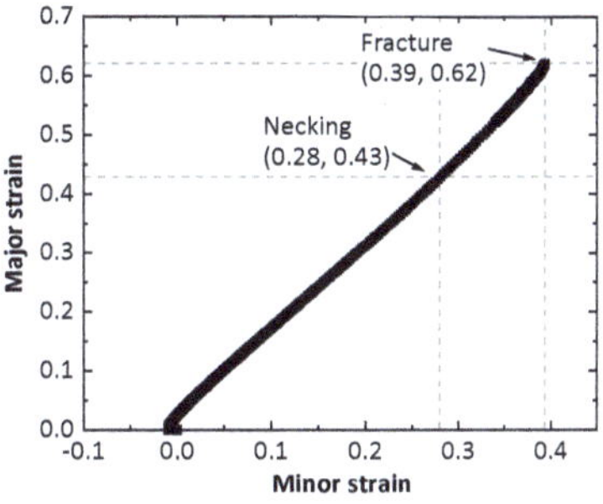

(a2)

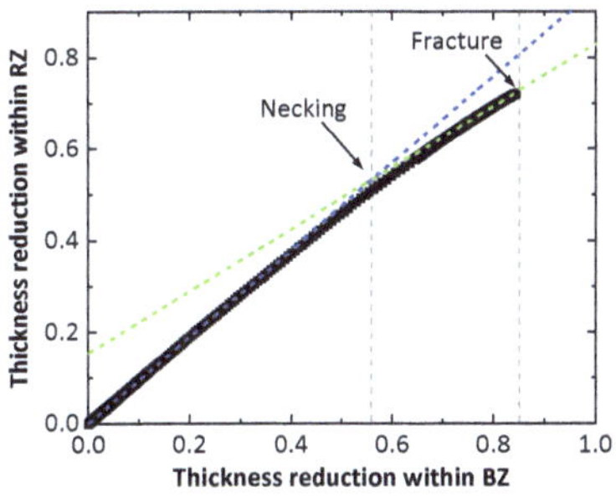

(b1)

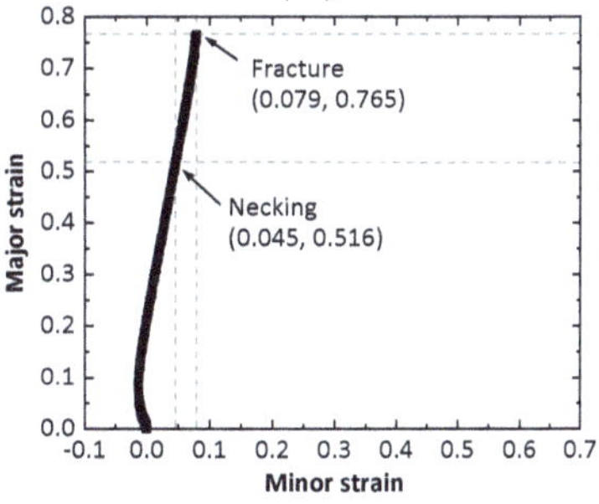

(b2)

Materials Research Forum LLC
https://doi.org/10.21741/9781644903254-10

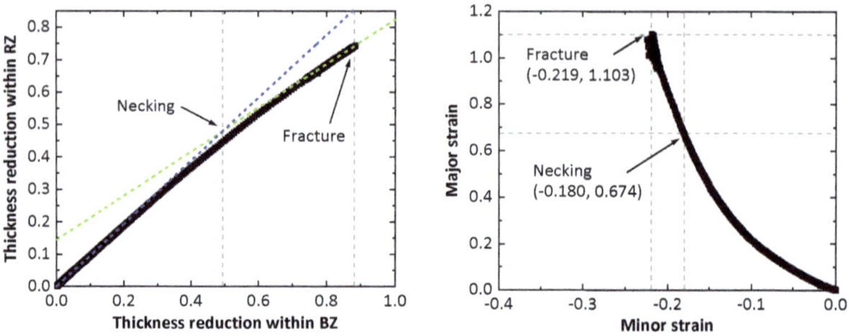

(c1) (c2)

Fig. 7. Determination of the limit strains at the onset of necking and at fracture using the spatio-temporal method for the specimens deformed at 510°C and 0.1 s^{-1}: the thinning within the base zone (BZ) against the reference zone (RZ) in (a1) equi-biaxial, (b1) plane-strain and (c1) uniaxial tensions, and the strain paths with the determined limit strains in (a2) equi-biaxial, (b2) plane-strain and (c2) uniaxial tensions.

Fig. 8 shows the collated limit strains at necking for FLCs and at fracture for FFLCs for AA6082 at 440°C and 0.1 s^{-1} and at 510°C and 0.1 s^{-1}, determined using the biaxial testing method in the present study. As can be seen, the limit major strain values at both necking and fracture are dependent on strain state; they are the highest in uniaxial tension, and decrease at a decreasing rate as the strain ratio increases. Additionally, the limit strain values are slightly dependent on the deformation temperature in the range investigated (i.e. 440–510°C) in the present study.

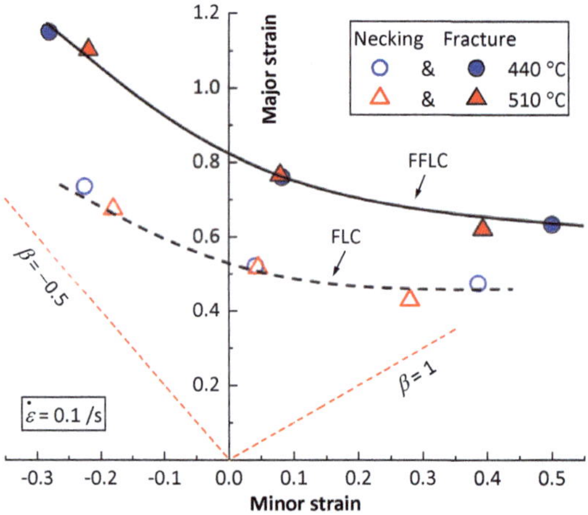

Fig. 8. Determined limit strains at necking for constructing forming limit curves (FLCs) and limit strains at fracture for constructing forming limit curves (FFLCs) for AA6082 under hot stamping conditions.

Conclusion

The formability of AA6082 sheets under hot stamping conditions at 440-510°C and 0.1 s^{-1} has been successfully evaluated using a recently developed biaxial testing method. The maximum temperature difference within the gauge area of the cruciform specimen is approximately 20°C. Furthermore, this testing method ensures the fracture initiation near the specimen centre, contributing to achieving almost linear strain paths with target strain ratios. The determined limit strains at both necking and fracture depend on the strain states, with a minor dependency on the deformation temperature investigated in this study.

Acknowledgement
This work was supported by the Engineering and Physical Sciences Research Council (EPSRC) [Grant number EP/R001715/1] on “Lightform: Embedding Materials Engineering in Manufacturing with Light Alloys”.

References

[1] J. Lin, T.A. Dean, R.P. Garrett, A.D. Foster, Process for forming metal alloy sheet components, International 'Patent' WO 2008059242 A2, (2008), May 22, 2008.

[2] K. Zheng, Y. Dong, J.-H. Zheng, A. Foster, J. Lin, H. Dong, T.A. Dean, The effect of hot form quench (HFQ®) conditions on precipitation and mechanical properties of aluminium alloys, Mater. Sci. Eng. A 761 (2019). https://doi.org/10.1016/j.msea.2019.06.027

[3] ASM Handbook, Heat Treating of Aluminum Alloys, 1991.

[4] Z.W. Xing, J. Bao, Y.Y. Yang, Numerical simulation of hot stamping of quenchable boron steel, Mater. Sci. Eng. A 499 (2009) 28-31. https://doi.org/10.1016/j.msea.2007.09.102

[5] H.S. Liu, Z.W. Xing, J. Bao, B.Y. Song, Investigation of the Hot-Stamping Process for Advanced High-Strength Steel Sheet by Numerical Simulation, J. Mater. Eng. Perform. 19 (2009) 325-334. https://doi.org/10.1007/s11665-009-9510-y

[6] M. Merklein, J. Lechler, Investigation of the thermo-mechanical properties of hot stamping steels, J. Mater. Process. Tech. 177 (2006) 452-455. https://doi.org/10.1016/j.jmatprotec.2006.03.233

[7] B.L. Ma, M. Wan, X.D. Wu, Z.Y. Cai, K.S. Diao, J.Q. Han, Investigation on forming limit of advanced high strength steels (AHSS) under hot stamping conditions, J. Manuf. Process. 30 (2017) 320-327. https://doi.org/10.1016/j.jmapro.2017.10.001

[8] M.S. Mohamed, A.D. Foster, J. Lin, D.S. Balint, T.A. Dean, Investigation of deformation and failure features in hot stamping of AA6082: experimentation and modelling, Int. J. Mach. Tool. Manuf. 53 (2012) 27-38. https://doi.org/10.1016/j.ijmachtools.2011.07.005

[9] S. Bruschi, T. Altan, D. Banabic, P. Bariani, A. Brosius, J. Cao, A. Ghiotti, M. Khraisheh, M. Merklein, A. Tekkaya, Testing and modelling of material behaviour and formability in sheet metal forming, CIRP Annals-Manuf. Tech. 63 (2014) 727-749

[10] S.P. Keeler, Determination of forming limits in automotive stampings, Sheet Met. Ind. 42 (1965) 683-691. https://doi.org/10.4271/650535

[11] G.M. Goodwin, Application of strain analysis to sheet metal forming problems in the press shop, SAE Tech. Pap., 1968, pp. 380-387. https://doi.org/10.4271/680093

[12] S.K. Paul, Controlling factors of forming limit curve: A review, Adv. Industr. Manuf. Eng. 2 (2021). https://doi.org/10.1016/j.aime.2021.100033

[13] A. Atkins, Fracture in forming, J. Mater. Process. Tech. 56 (1996) 609-618. https://doi.org/10.1016/0924-0136(95)01875-1

[14] M. Dilmec, H.S. Halkaci, F. Ozturk, M. Turkoz, Detailed Investigation of Forming Limit Determination Standards for Aluminum Alloys, J. Testing Evaluat. 41 (2013). https://doi.org/10.1520/jte104356

[15] Z. Chen, G. Fang, J.-Q. Zhao, Formability evaluation of aluminum alloy 6061-T6 sheet at room and elevated temperatures, J. Mater. Eng. Perform. 26 (2017) 4626-4637. https://doi.org/10.1007/s11665-017-2895-0

[16] E. Affronti, C. Jaremenko, M. Merklein, A. Maier, Analysis of Forming Limits in Sheet Metal Forming with Pattern Recognition Methods. Part 1: Characterization of Onset of Necking and Expert Evaluation, Materials 11 (2018). https://doi.org/10.3390/ma11091495

[17] P. Bariani, S. Bruschi, A. Ghiotti, A. Turetta, Testing formability in the hot stamping of HSS, CIRP Annals - Manuf. Tech. 57 (2008) 265-268. https://doi.org/10.1016/j.cirp.2008.03.049

[18] L. Ying, T. Gao, H. Rong, X. Han, P. Hu, W. Hou, On the thermal forming limit diagram (TFLD) with GTN mesoscopic damage model for AA7075 aluminum alloy: Numerical and experimental investigation, J. Alloy. Compd. 802 (2019) 675-693. https://doi.org/10.1016/j.jallcom.2019.05.342

[19] H. Rong, P. Hu, L. Ying, W. Hou, J. Zhang, Thermal forming limit diagram (TFLD) of AA7075 aluminum alloy based on a modified continuum damage model: Experimental and theoretical investigations, Int. J. Mech. Sci. 156 (2019) 59-73. https://doi.org/10.1016/j.ijmecsci.2019.03.027

[20] BS EN ISO 12004, Metallic materials — Sheet and strip — Determination of forming-limit curves Part 2: Determination of forminglimit curves in the laboratory, BSI British Standards, 2008. https://doi.org/10.3403/bseniso12004

[21] R. Zhang, Z. Shi, V.A. Yardley, J. Lin, Experimental studies of necking and fracture limits of boron steel sheet under hot stamping conditions, J. Mater. Process. Tech. 302 (2022). https://doi.org/10.1016/j.jmatprotec.2021.117481

[22] R. Zhang, Z. Shi, Z. Shao, T.A. Dean, J. Lin, A novel spatio-temporal method for determining necking and fracture strains of sheet metals, Int. J. Mech. Sci. 189 (2021). https://doi.org/10.1016/j.ijmecsci.2020.105977

[23] R. Zhang, Z. Shi, Z. Shao, V.A. Yardley, J. Lin, An effective method for determining necking and fracture strains of sheet metals, MethodsX 8 (2021). https://doi.org/10.1016/j.mex.2021.101234

[24] R. Zhang, J. Jiang, J. Lin, V.A. Yardley, Investigation of variability in apparent values of materials properties in thermo-mechanical uniaxial tensile tests on sheet metals, J. Manuf. Process. 101 (2023) 737-754. https://doi.org/10.1016/j.jmapro.2023.06.005

[25] R. Zhang, Z. Shao, Z. Shi, T.A. Dean, J. Lin, Effect of cruciform specimen design on strain paths and fracture location in equi-biaxial tension, J. Mater. Process. Tech. 289 (2021). https://doi.org/10.1016/j.jmatprotec.2020.116932

Plastic flow modelling under abrupt strain-path change and parameter identification method

ZHANG Tianyin[1,2,a] and HAN Xianhong[1,2,b *]

[1]National Engineering Research Center of Die & Mold CAD and State Key Laboratory of Metal Matrix Composites, Shanghai Jiao Tong University, 1954 Huashan Road, 200030, Shanghai, China

[2]School of Materials Science and Engineering, Shanghai Jiao Tong University, 1954 Huashan Road, 200030, Shanghai, China

[a]zhangtianyin@sjtu.edu.cn, [b]hanxh@sjtu.edu.cn

Keywords: Plastic Flow Rule, Corner Effects, Strain-Path Change, Parameter Identification

Abstract. Rational plastic flow modelling under abrupt strain-path change is crucial for accurate metal forming simulations. Previous experimental studies have revealed that many metals exhibit “corner effects” after abrupt strain-path change in the plastic stage, including a sharp vertex produced on the subsequent yield surface and apparent non-normality of the plastic flow direction. Some advanced models can give reasonable predictions on the above phenomenon, but they have not been widely used in metal forming simulations due to their complexity. A simple non-associated flow rule is developed here to describe corner effects. A stable stress update algorithm is developed for this rule based on the fully implicit backward Euler method, and its validation is verified by FE simulations of the nonlinear tension-torsion loading experiments. In order to determine the material parameters of these non-associated flow rules representing corner effects, a general identification method is proposed based on the equivalent tangential shear modulus reduction. The results show that, compared with the previous theories, the present rule performs better in predicting corner effects. In addition, although the apparent tangential shear modulus reduction is observed during the nonlinear experiments, the modulus does not seem to decrease to near zero.

Introduction

Reliable metal forming simulations put forward a higher request for the plastic flow modelling of complex mechanical responses, such as corner effects. Previous experimental studies [1-4] have revealed that many metals exhibit corner effects after abrupt strain-path change in the plastic stage, including a sharp vertex produced on the subsequent yield surface and apparent non-normality of the plastic flow direction. Corner effects are concomitant with the apparent decrease of shear modulus, thus playing an important role in the problems of strain localization, e.g., plastic buckling [5,6], shear band formation [7,8], necking forming limit [9,10].

The existence of vertexes on the static yield surface is still controversial because it may be caused by the visco-plasticity of metals [4,11]. Then the smooth yield surface is adopted by mainstream research and some advanced non-associated flow rules [2,7,10,12] have been developed to represent corner effects. These rules are constructed to describe the rotation behavior of the plastic strain rate towards the strain rate deviator. However, they have not been widely used in metal forming simulations due to their complexity in expression and implementation [6]. Furthermore, the parameter identification of these rules is also challenging due to the involvement of abrupt strain-path change.

This paper is structured as follows. Firstly, we review several typical non-associated flow rules representing corner effects, and construct a concise flow rule through mathematical simplification of the general framework. Then a parameter identification method based on the equivalent tangential shear modulus reduction is used to determine the parameters of A6063-T5. Finally, model validation is performed through comparison with the nonlinear tension-torsion experiment.

Non-associated flow rules representing corner effects

A specific type of non-associated flow rules has been proposed to describe corner effects. Assuming the smooth yield surface, the plastic flow rules are constructed to reflect the rotation of the plastic flow direction towards the strain rate direction, and can be expressed as a general framework:

$$\dot{\boldsymbol{\varepsilon}}^{\mathrm{p}} = \dot{\lambda}\left((1 - \tan\theta^{\mathrm{p}}\cot\theta)\mathbf{n} + \frac{\tan\theta^{\mathrm{p}}}{\sin\theta}\mathbf{l}\right) \tag{1}$$

where $\dot{\lambda}$ is the plastic multiplier. $\mathbf{n}$ is the unit normal of the yield surface, defined by $\mathbf{n} = \mathbf{N}/||\mathbf{N}||$ and $\mathbf{N} = \partial f/\partial\boldsymbol{\sigma}$ (f is the yield function). $\mathbf{l}$ is the unit tensor in the direction of the strain rate deviator, which is defined by $\mathbf{l} = \dot{\boldsymbol{\varepsilon}}^{\mathrm{dev}}/\|\dot{\boldsymbol{\varepsilon}}^{\mathrm{dev}}\|$. θ^{p} denotes the angle between the plastic strain rate and the normal $\mathbf{n}$; θ denotes the angle between the strain rate deviator and the normal $\mathbf{n}$, as shown in Fig. 1.

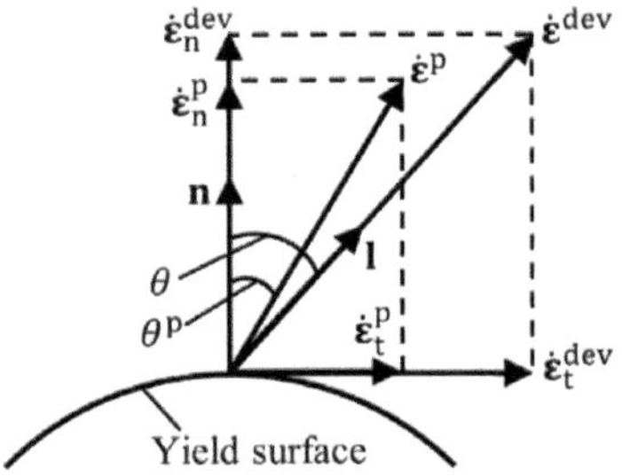

Fig. 1. Schematic of variables and tensors.

Several typical "corner models" can be included in this framework (Eq. 1) as specific forms. For the rule proposed by Simo [12]:

$$\theta^{\mathrm{p}} = \begin{cases} \theta & \text{for } 0 \le \theta < \theta_0 \\ \theta_0 & \text{for } \theta_0 \le \theta \le \pi/2 \end{cases} \tag{2}$$

where θ_0 is a material parameter representing the saturation value of θ^{p}.

The rule proposed by Yoshida [7] approximately satisfies:

$$\tan\theta^{\mathrm{p}} \approx \begin{cases} \tan\theta & \text{for } 0 \le \theta < \theta_0 \\ \frac{\pi/2-\theta}{\pi/2-\theta_0}\tan\theta & \text{for } \theta_0 \le \theta \le \pi/2 \end{cases} \tag{3}$$

where θ_0 is the material parameter.

The rule proposed by Yoshida and Tsuchimoto [2] approximately satisfies:

$$\theta^{\mathrm{p}} \approx \begin{cases} \theta & \text{for } 0 \le \theta < \theta_0 \\ c_3(\theta - \theta_0) + \theta_0 & \text{for } \theta_0 \le \theta \le \pi/2 \end{cases} \tag{4}$$

where c_3 and θ_0 are material parameters.

Here we construct a concise flow rule through simplifying Eq. 1 mathematically. Assuming that $K_c = \frac{\tan\theta^p}{\sin\theta}$ is a material constant, the plastic flow rule can be simplified as

$$\dot{\boldsymbol{\varepsilon}}^p = \dot{\lambda}\left(\left(1 - K_c(\mathbf{n}:\mathbf{l})\right)\mathbf{n} + K_c\mathbf{l}\right). \tag{5}$$

Eq. 5 does not contain the variable θ or the branch structure, so the gradient tensor of $\left(1 - K_c(\mathbf{n}:\mathbf{l})\right)\mathbf{n} + K_c\mathbf{l}$ with respect to stress can be easily given. Then a stable stress update algorithm is developed for this rule based on the fully implicit backward Euler method, and more information can be seen in the study [13].

Parameter identification

This section presents the parameter identification method of the plastic flow rules representing corner effects. Simultaneously considering the consistency condition, the isotropic Hooke's law and Eq. 1, the stress rate deviator $\dot{\boldsymbol{\sigma}}^{dev}$ can be decomposed into two terms, i.e., one depends on $\dot{\boldsymbol{\varepsilon}}_n^{dev}$ and the other depends on $\dot{\boldsymbol{\varepsilon}}_t^{dev}$:

$$\dot{\boldsymbol{\sigma}}^{dev} = 2G_n\dot{\boldsymbol{\varepsilon}}_n^{dev} + 2G_t\dot{\boldsymbol{\varepsilon}}_t^{dev} \quad \text{with} \quad \begin{cases} G_n = \left(1 - \frac{1}{\xi}\right)G \\ G_t = \left(1 - \frac{\tan\theta^p}{\xi\tan\theta}\right)G \end{cases} \tag{6}$$

where $\xi = 1 + \frac{H}{2G\|\mathbf{N}\|^2}$ (H is the strain hardening modulus and G is the elastic shear modulus), and its value is close to 1 for common metals since $H \ll G$. G_n and G_t are respectively the coefficients before $\dot{\boldsymbol{\varepsilon}}_n^{dev}$ and $\dot{\boldsymbol{\varepsilon}}_t^{dev}$.

As θ^p is a function of θ, $\tan\theta^p$ can be expressed as $F(\theta, k_1, k_2, \dots, k_j)$, where $k_1, k_2, \dots, k_j$ represent the parameters in a flow rule. Then the equivalent tangential shear modulus G_t can be written as

$$\frac{G_t}{G} \approx 1 - \frac{F(\theta, k_1, k_2, \dots, k_j)}{\tan\theta}. \tag{7}$$

The nonlinear tension-torsion loading experiments with abrupt strain-path change are considered for parameter identification. As shown in Fig. 2(a), in Step 1, the tube specimen is stretched until the axial logarithmic strain reaches e_1; in Step 2, the deformation mode is abruptly transformed to combined tension–torsion.

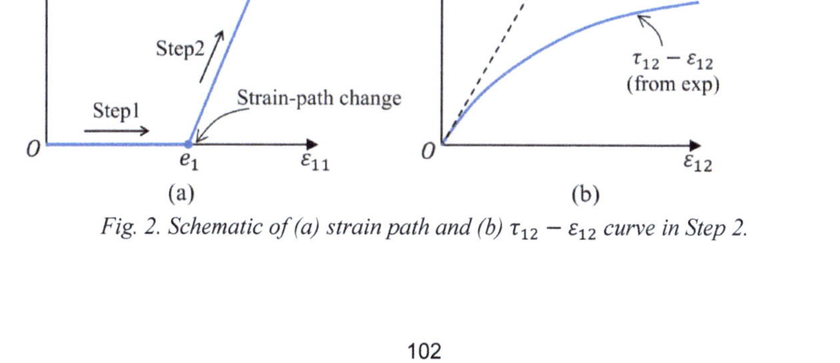

Fig. 2. Schematic of (a) strain path and (b) $\tau_{12} - \varepsilon_{12}$ curve in Step 2.

In the transient immediately after the strain-path change, the ratio of $\dot{\tau}_{12}$ and $\dot{\varepsilon}_{12}$ equals to $2G_t$, as illustrated in Fig. 2(b). Then the values of G_t can be measured and used for parameter identification in Eq. 7. Fig. 3 shows the G_t-data from several nonlinear tension-torsion loading experiments of the aluminum alloy A6063-T5. The parameters of different flow rules are determined through the least square fitting method. The present rule and the rule of Yoshida and Tsuchimoto [2] perform better compared with the associated flow rule and the previous theories proposed by Simo [12], Yoshida [7]. Besides, G_t does not reduce to near zero at least for A6063-T5.

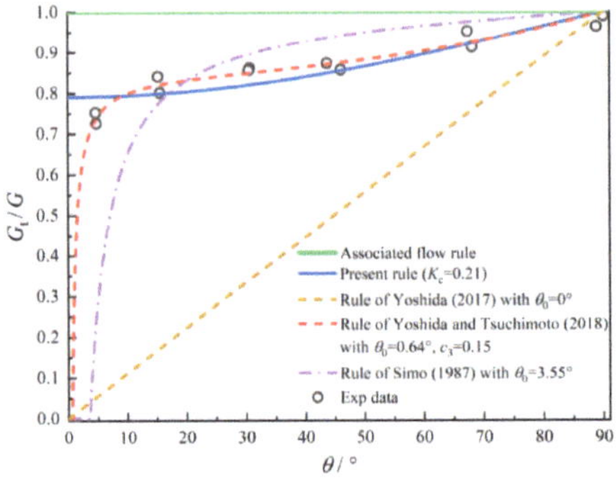

Fig. 3. Experimental data of G_t and predictions of different plastic flow rules.

The above parameter identification method can also be extended to other types of experiments with abrupt strain-path change, and more information can be seen in the study [4].

Model validation

The nonlinear tension-torsion experiment with $\theta \approx 67.5°$ is used for model validation. Here we quantify the plastic flow behavior of A6063-T5 after abrupt strain-path change through the angle β between the plastic strain rate and the loading path trace in the stress space of $\sigma_{12} - \sigma_{11}$, as shown in Fig. 4(a). $\alpha = \arctan\frac{\sigma_{12}}{\sigma_{11}}$ is defined to indicate the position of the current stress point.

The FEM simulations in this section are based on the commercial software ABAQUS. The boundary conditions are applied to a four-node shell element with fully integration. Such unit test simulation is sufficient to completely reproduce the above experimental condition ($\theta \approx 67.5°$). The present rule is implemented by the fully implicit stress update algorithm, while the other flow rules are implemented by the semi-implicit stress update algorithm. And the rule parameters of A6063-T5 have been identified in Fig. 3. The material is assumed to follow anisotropic yielding and isotropic hardening, respectively described by the Poly4 yield function and the Hollomon model.

The results ($\beta - \alpha$ curves) of different plastic flow rules are presented in Fig. 4(b). As the stress path in the early stage of Step 2 is almost the current yield surface, β is predicted to be about 90° by the associated flow rule, and such prediction apparently violates experimental observations. The non-associated flow rules representing corner effects improve the prediction performance of the plastic flow direction after abrupt strain-path change. Among them, the present rule and the rule proposed by Yoshida and Tsuchimoto [2] exhibit the best accuracy.

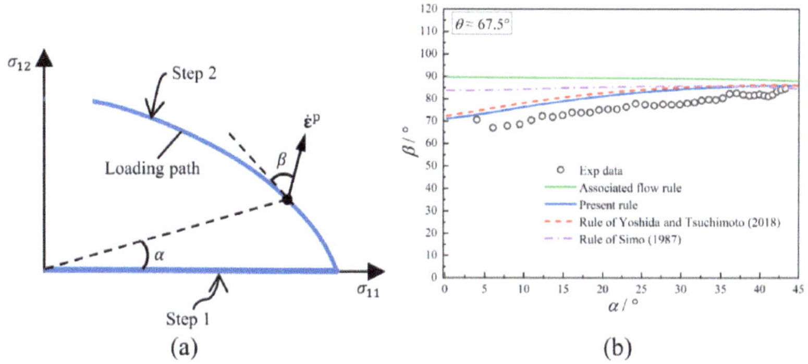

Fig. 4. (a) Schematic of β and α; (b) $\beta - \alpha$ curves predicted by different plastic flow rules.

Conclusions

Several typical non-associated flow rules representing corner effects have been implemented to predict the apparent non-normality of plastic flow in corner effects. The parameters of these plastic flow rules are identified through the reduction of tangential shear modulus. The results suggest that the present rule and the rule proposed by Yoshida and Tsuchimoto [2] exhibit the best accuracy in predicting the flow direction under abrupt strain-path change. Moreover, the tangential shear modulus does not reduce to near zero as reported before.

Acknowledgments

Support by the National Key Research and Development Programs of China (Grant No. 2022YFE0196600 and No. 2023YFB2504604) and the National Natural Science Foundation of China (Grant No. 52175349) are acknowledged.

References

[1] T. Kuwabara, M. Kuroda, V. Tvergaard, K. Nomura, Use of abrupt strain path change for determining subsequent yield surface: Experimental study with metal sheets, Acta Mater. 48 (2000) 2071-2079. https://doi.org/10.1016/S1359-6454(00)00048-3

[2] K. Yoshida, T. Tsuchimoto, Plastic flow of thin-walled tubes under nonlinear tension-torsion loading paths and an improved pseudo-corner model, Int. J. Plast. 104 (2018) 214-229. https://doi.org/10.1016/j.ijplas.2018.02.013

[3] K. Yoshida, N. Okada, Plastic flow behavior of fcc polycrystal subjected to nonlinear loadings over large strain range, Int. J. Plast. 127 (2020) 102639.

[4] T.Y. Zhang, X.H. Han, Parameter identification for the non-associated flow rules representing corner effects through the equivalent tangential shear modulus reduction after abrupt strain-path change, Int. J. Plast. 169 (2023) 103726.

[5] L. Ronning, O.S. Hopperstad, P.K. Larsen, Numerical study of the effects of constitutive models on plastic buckling of plate elements, Eur. J. Mech. A Solids 29 (2010) 508-522. https://doi.org/10.1016/j.euromechsol.2010.02.001

[6] A. Nasikas, S.A. Karamanos, S.A. Papanicolopulos, A framework for formulating and implementing non-associative plasticity models for shell buckling computations, Int. J. Solids Struct. 257 (2022) 111508. https://doi.org/10.1016/j.ijsolstr.2022.111508

[7] K. Yoshida, A plastic flow rule representing corner effects predicted by rate-independent crystal plasticity, Int. J. Solids Struct. 120 (2017) 213-225. https://doi.org/10.1016/j.ijsolstr.2017.05.004

[8] M. Kuroda, V. Tvergaard, Shear band development in anisotropic bent specimens, Eur. J. Mech. A Solids 23 (2004) 811-821. https://doi.org/10.1016/j.euromechsol.2004.05.006

[9] K. Ito, K. Satoh, M. Goya, T. Yoshida, Prediction of limit strain in sheet metal-forming processes by 3D analysis of localized necking, Int. J. Mech. Sci. 42 (2000) 2233-2248. https://doi.org/10.1016/S0020-7403(00)00004-7

[10] M. Kuroda, V. Tvergaard, A phenomenological plasticity model with non-normality effects representing observations in crystal plasticity, J. Mech. Phys. Solids 49 (2001) 1239-1263. https://doi.org/10.1016/S0022-5096(00)00080-6

[11] Y.F. Yang, T. Balan, Prediction of the yield surface evolution and some apparent non-normality effects after abrupt strain-path change using classical plasticity, Int. J. Plast. 119 (2019) 331-343. https://doi.org/10.1016/j.ijplas.2019.04.006

[12] J.C. Simo, A J2-flow theory exhibiting a corner-like effect and suitable for large-scale computation, Comput. Method Appl. Mech. Eng. 62 (1987) 169-194. https://doi.org/10.1016/0045-7825(87)90022-3

[13] T.Y. Zhang, M.S. Liu, X.H. Han, A non-associated flow rule with simple non-branching form representing the apparent non-normality effects after abrupt strain-path change, Int. J. Plast. 159 (2022) 103452. https://doi.org/10.1016/j.ijplas.2022.103452

Simulation of texture-related micro-indentation behavior of typical HCP alloy plates

ZHOU Xiangyu[1,a], ZHOU Kecheng[1,b], WU Peidong[2,c], XU Zhutian[1,d], PENG Linfa[1,e] *, LI Shuhui[1,f] and WANG Huamiao[1,3,g] *

[1]State Key Laboratory of Mechanical System and Vibration, Shanghai Jiao Tong University, Shanghai, 200240, China

[2]Department of Mechanical Engineering, McMaster University, Hamilton, ON, Canada, L8S 4L7

[3]School of Mechanics and Engineering Science, Shanghai University, Shanghai, 200072, China

[a]sjtuzhouxy@sjtu.edu.cn, [b]kecheng@sjtu.edu.cn, [c]peidong@mcmaster.ca, [d]zhutianxu@sjtu.edu.cn, [e]penglinfa@sjtu.edu.cn, [f]lishuhui@sjtu.edu.cn, [g]wanghm@shu.edu.cn

Keywords: Magnesium Alloy, Texture-Dependency, Crystal Plasticity, Twinning

Abstract: The mechanical behaviors of magnesium and its alloys, regulated by the slip and twinning mechanisms, are remarkably influenced by the crystalline orientation. A better understanding and quantification of the mechanisms in deformation is necessary. To this end, the crystal plasticity finite element (CPFE) simulations of magnesium alloy plates have been investigated in this study. The process of impression of single crystal samples with three different declination angles with 0°, 45°, and 90° is simulated. The corresponding detailed CPFE simulations reveal a texture-dependent spatial distribution of extension twins, and the twin morphologies and the indentation topography of single crystals have been discussed, highlighting the influence of the initial texture. Furthermore, the simulations uncover the accumulated shear strain of the deformation system varying with spatial position inside the sample, which is in favor of the analysis of the twin-favor zone.

Introduction

In the context of the pursuit of the low-carbon economy, magnesium and its alloys have triggered worldwide attention because of their lightweight, high specific strength, and excellent recyclability [1–4]. The strong anisotropic mechanical behavior of Mg alloys, influenced by the hexagonal close-packed (HCP) structure, the intrinsic polarity of the twin, and the crystal orientation [5–8], is currently a hindrance to their potential prevalent applications. To address this issue, a better understanding and quantification of the competition among the predominant microscopic mechanisms that govern the mechanical behavior of Mg alloys is crucial, particularly in regard to the role of the initial texture in shaping these mechanisms. At room temperature, slip and extension twin [9–13] are the major mechanisms behind the plastic deformation. Numerous studies have investigated the mechanical behaviors and underlying microscopic mechanisms of Mg alloys under homogeneous loading conditions [14–17] such as tension, compression, and simple shear, as well as inhomogeneous loading conditions like torsion [18,19] and bending [20,21].

The indentation test is a widely used method for determining the mechanical properties of materials, where the complexity of the stress state around an indenter may sophisticate the interactions between the various deformation modes. But there is much less fundamental understanding regarding the nature of plastic deformation activity especially the extension twin under such localized loading conditions. Gollapudi et al. [22] experimentally concurred that the extension twin and basal slip are responsible for the crystal-orientation dependent indentation morphologies of Mg alloy. Shin et al. [23] conducted nanoindentation on the basal and prismatic planes of single-crystalline Mg and found that extension twin is essential for strain accommodation. Bočan et al. [24] found a strong correlation between the nanomechanical behaviors of pure Mg and AZ31 alloy and the crystal orientation, which can be interpreted by the anisotropic activities of slip systems and extension twin.

Moreover, the means of the crystal plastic finite element (CPFE) method was employed for the interpretation of the indentation behaviors of magnesium, which provides easy access to factors that are difficult to obtain experimentally. Kitahara et al. [25] reported the twin morphologies beneath indent on low index planes in pure Mg single crystals and they concluded that the anisotropic development of the twin region is induced by the basal slip through CPFE simulations. Sánchez-Martín et al. [26] studied the activation and propagation of twinning in pure Mg grains with different orientations. The crystal-orientation dependent indentation topographies driven by the activity of extension twin have been observed in experiments and confirmed by CPFE simulations. Zambaldi et al. [27] presented the crystal-orientation dependent response of pure Mg during single-grain indentation. The importance of the twinning behaviors is emphasized in their research, but hardly a visually apparent twin distribution inside the sample can be found. Besides, the indentation behaviors and underly microscopic mechanisms of polycrystalline Mg alloys should be further investigated, especially the twin pattern. In the following, twin nucleation, propagation, and growth (TNPG) model [28] has been employed in a CPFE method to describe the twinning mechanism and analyze the indentation behavior of single crystal samples and randomly oriented sample. The texture-dependent indentation behaviors will be discussed.

Numerical procedures

Description of Crystal Plasticity Model. The CPFEM for indentation of polycrystalline Mg alloy in this paper can be found else wise [29]. It is briefly described for completeness. For both slip and twinning systems, the shear rate ($\dot{\gamma}^\alpha$) is formulated as

$$\dot{\gamma}^\alpha = \dot{\gamma}_0 \left|\frac{\tau^\alpha}{\tau_c^\alpha}\right|^{\frac{1}{m}} sgn(\tau^\alpha) \tag{1}$$

where $\dot{\gamma}_0$, m, τ^α, and τ_c^α refer to reference shear rate, rate sensitivity component, resolved shear stress (RSS), and critical resolved shear stress (CRSS), respectively. The evolution of CRSS in the deformation process is given by

$$\dot{\tau_c^\alpha} = \frac{d\hat{\tau}^\alpha}{d\Gamma} \Sigma_\beta h^{\alpha\beta} \left|\dot{\gamma}^\beta\right| \tag{2}$$

where Γ, $h^{\alpha\beta}$, $\hat{\tau}^\alpha$ are the accumulated shear strain in the grain ($\Gamma = \sum_\alpha \int \dot{\gamma}^\alpha dt$), the hardening coupling coefficients, which used to describe the resistance on system α associated with system β, and the threshold stress, respectively.

The threshold stress for a slip system α is related to the initial CRSS (τ_0^α), the back-extrapolated CRSS ($\tau_0^\alpha + \tau_1^\alpha$), the initial hardening rate (h_0^α), and the asymptotic hardening rate (h_1^α) in an extended Voce law.

$$\hat{\tau}^\alpha = \tau_0^\alpha + (\tau_1^\alpha + h_1^\alpha\Gamma)\left(1 - \exp\left(-\frac{h_0^\alpha}{\tau_1^\alpha}\Gamma\right)\right) \tag{3}$$

The stress relaxation associated with twinning is accounted for by employing the model proposed by Wu et al. [30], where the threshold stress for twinning system α is

$$\hat{\tau}^\alpha = \begin{cases} \tau_0^\alpha - \frac{\tau_0^\alpha - \tau_g^\alpha}{f_g^\alpha} f^\alpha & f^\alpha \le f_g^\alpha \\ \tau_0^\alpha + (\tau_1^\alpha + h_1^\alpha\Gamma)\left(1 - \exp\left(-\frac{h_0^\alpha}{\tau_1^\alpha}\Gamma\right)\right) & f^\alpha > f_g^\alpha \end{cases} \tag{4}$$

where f_g^α is the critical twin volume fraction, τ_g^α is the minimum twin resistance and f^α is the current twin volume fraction (TVF). Twinning is terminated within a grain if the TVF reaches a threshold value V^{th}.

Finite Element Model. The crystal plasticity finite element (FE) simulation was performed using ABAQUS v2022, a commercial software package. Fig. 1 shows the schematic diagram of models with three different declination angles: (I) 0°, (II) 45° and (III) 90°. To facilitate the following discussion, each model is divided into four parts based on the symmetry of the Vickers indenter: northern west (NW), northern east (NE), southern west (SW), and southern east (SE). The single crystal constitutive model described earlier was implemented in the ABAQUS UMAT subroutine to describe the constitutive response at every integration point of all elements.

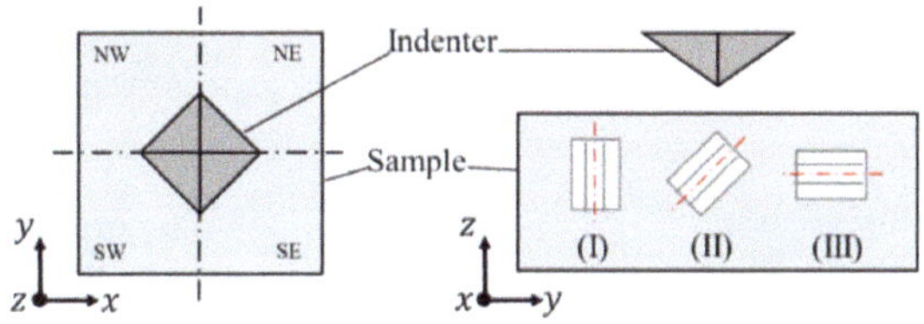

Fig. 1. The schematic diagram of indentation test model.

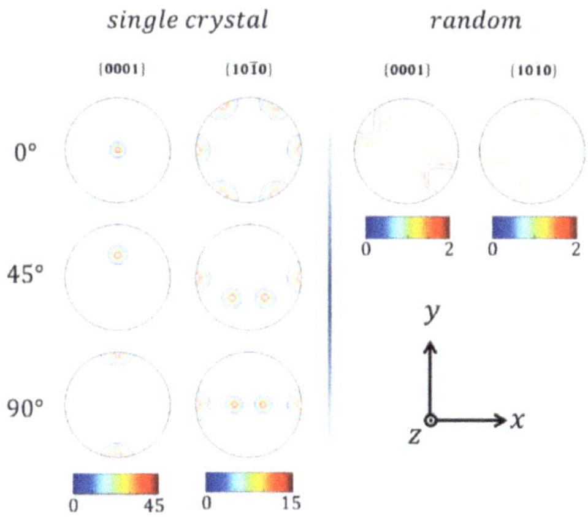

Fig. 2. The pole figures of indentation samples in terms of $\{\mathbf{0001}\}$ *and* $\{\mathbf{10\bar{1}}0\}$.

The pole figures of single crystal and random samples are presented in Fig. 2. To simulate the plastic deformation behavior of different samples, the crystal orientation information was incorporated into the crystal plasticity subroutine. Each element, which represents a single grain, was assigned a set of Euler angles to define its crystal orientation. Basal slip, prismatic slip, and pyramidal slip, and extension twin were considered in the modeling to accommodate the plastic deformation.

Results and discussion

Twin Morphologies. Fig. 3(a-c) show the simulated twin morphologies of the 0°, 45°, and 90° indentation samples. These morphologies correspond to the set of nodes where the TVF (Twinning Volume Fraction) is greater than 0.1. In the 0° sample, the twins are mainly concentrated in the

Materials Research Proceedings 44 (2024) 106-112 https://doi.org/10.21741/9781644903254-12

region that interfaces directly with the indenter and are confined to shallow layers. Conversely, the extension twin in the 45° and 90° samples is more extensive and penetrates deeper beneath the indenter-contact area. The convex hull formed by the twins has a distinct angular morphology and exhibits strong symmetry. In the case of single crystal-0, the indentation results show a hexagram-like pattern, which is associated with the six variants of the extension twin [30]. In the case of the single crystal-45 sample, the twin distribution mainly penetrates into the northern part of the sample, and the distribution of twins in this sample displays better symmetry with respect to the x=0 plane, as is also observed in the single crystal-90 sample.

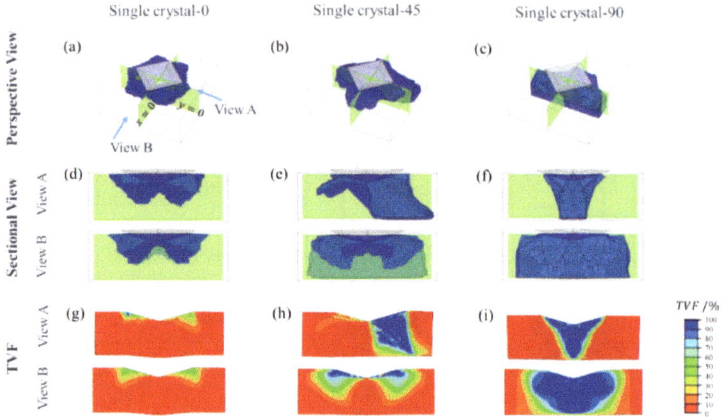

Fig. 3. The simulated 3D-twin morphology of (a) single cyrstal-0, (b) single crystal-45, and (c) single cyrstal-90 sample, (d–f) The 2D-twin morphology at x=0 section and y=0 section, (g–i) The TVF distribution at $\mathbf{x} = \mathbf{0}$ *section and* $\mathbf{y} = \mathbf{0}$ *section.*

Accumulated Shear Strain. Fig. 4 presents a comparison of the accumulated shear strain distribution between the simulated results of single crystal samples with different initial orientations and random sample. In single crystal 0 sample, basal slip is found to be the dominant deformation mechanism. Specifically, the accumulated shear strain resulting from basal slip increases to a maximum at approximately 20 microns and then decreases gradually. The accumulated shear strain of extension twin monotonically decreases from the surface. In contrast, a significant difference is evident between the northern part (NW, NE) and the southern part (SW, SE) in the accumulated shear strain of basal slip and extension twin in single crystal-45 sample, as illustrated in Fig. 4(b). The shear strain of the extension twin is conspicuously higher in the northern part compared to the southern part, resulting in the anisotropic distribution of the twin region (Fig. 3(e, h)). The shear strain of basal slip in the northern part is also higher than that in the southern part. Due to the symmetry of twin regions, a similar distribution pattern of shear strain also observed in the single crystal-90 sample, as shown in Fig. 3(f, i).

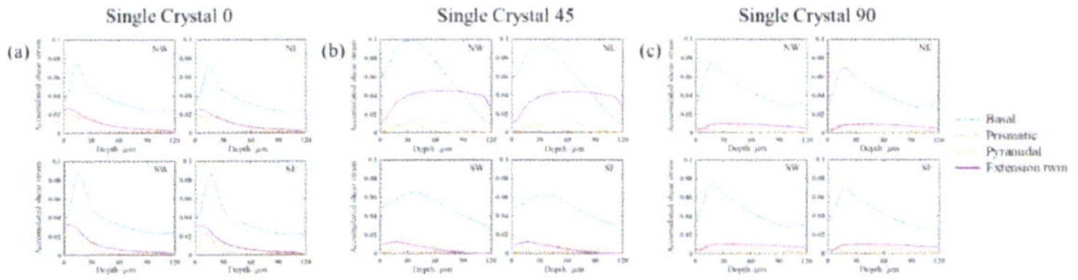

Fig. 4. The accumulated shear strain distribution along depth of basal slip, prismatic slip, pyramidal slip, and extension twin: (a) single crystal-0 sample, (b) single crystal-45 sample and (c) single crystal-90 sample.

The Indentation Morphology. Fig. 5 depicts the simulated indentation morphology of four samples. As shown in Fig. 5(d), no pile-up or sink-in effects are observed in random sample. In contrast, the texture-dependent indentation morphologies of single crystal samples (Fig. 5(a-c)) are highlighted. The single crystal samples exhibit smooth raised features on their surface. In the single crystal-0 samples, the pile-up profile occurs adjacent to the indenter. In the single crystal-45 samples, the pile-up profile is observed in the northern part, and the sink-in profile occurs around the indenter and expands along the x-axis. In the single crystal-90 sample, the pile-up profile is marginally apparent and the expansion of sink-in profile is more prominent. Notably, the extension twin in the samples can cause apparent extension along the c-axis and shrinkage perpendicular to the c-axis, resulting in pile-up and sink-in effects observed in the indentation morphology. Therefore, the twin morphology can be inferred from the indentation morphology, and vice versa.

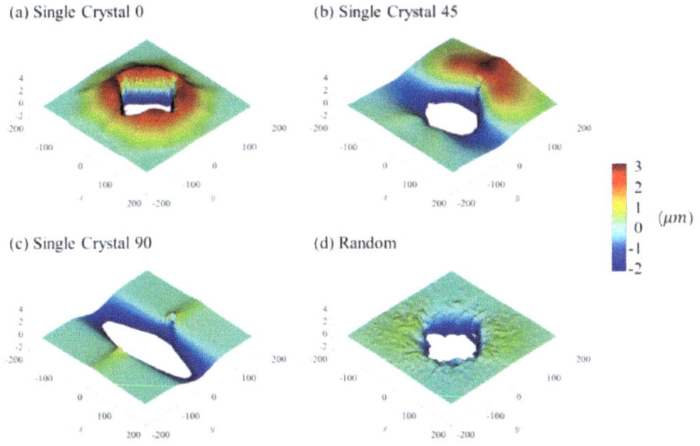

Fig. 5. The simulated indentation morphology of (a–c) single crystal texture samples, and (d) random texture sample.

Summary and Conclusions

In this work, three-dimensional crystal plasticity finite element (CPFE) simulations were performed using several initial textures, including three single crystal textures, and a random texture. The 3-D and 2-D twin morphologies, surface morphologies and accumulated shear strain of deformation mode are discussed, and they all show texture-dependent characteristics. The surface morphologies are strongly correlated with the internal twin distribution. Additionally, the shear strain distribution of the extension twin corresponds to its three-dimensional twin distribution. In the single crystal-0 sample, twin is confined to the shallow layer, but it can penetrate deep beneath the region directly under the indenter in the single crystal-90 sample. As for the single crystal-45 sample, the twin only extends into the northern parts.

References

[1] S.R. Agnew, J.F. Nie, Preface to the viewpoint set on: The current state of magnesium alloy science and technology, Scr. Mater. 63 (2010) 671–673. https://doi.org/10.1016/j.scriptamat.2010.06.029

[2] D. Eliezer, E. Aghion, F.H. (Sam) Froes, Magnesium Science, Technology and Applications, Adv. Perform. Mater. 5 (1998) 201–212. https://doi.org/10.1023/A:1008682415141.

[3] T.M. Pollock, Weight Loss with Magnesium Alloys, Science. 328 (2010) 986–987. https://doi.org/10.1126/science.1182848

[4] Y. Yang, X. Xiong, J. Chen, X. Peng, D. Chen, F. Pan, Research advances in magnesium and magnesium alloys worldwide in 2020, J. Magnes. Alloy. 9 (2021) 705–747. https://doi.org/10.1016/j.jma.2021.04.001

[5] B. Shi, C. Yang, Y. Peng, F. Zhang, F. Pan, Anisotropy of wrought magnesium alloys: A focused overview, J. Magnes. Alloy. 10 (2022) 1476–1510. https://doi.org/10.1016/j.jma.2022.03.006

[6] C.M. Cepeda-Jiménez, M.T. Pérez-Prado, Microplasticity-based rationalization of the room temperature yield asymmetry in conventional polycrystalline Mg alloys, Acta Mater. 108 (2016) 304–316. https://doi.org/10.1016/j.actamat.2016.02.023

[7] Y.B. Chun, C.H.J. Davies, Twinning-induced anomaly in the yield surface of highly textured Mg–3Al–1Zn plate, Scr. Mater. 64 (2011) 958–961. https://doi.org/10.1016/j.scriptamat.2011.01.044

[8] H. Qiao, X.Q. Guo, A.L. Oppedal, H. El Kadiri, P.D. Wu, S.R. Agnew, Twin-induced hardening in extruded Mg alloy AM30, Mater. Sci. Eng., A. 687 (2017) 17–27. https://doi.org/10.1016/j.msea.2016.12.123

[9] C.S. Roberts, Magnesium and its Alloys, John Wiley & Sons, Inc., New York, 1960. https://doi.org/10.1002/9781118985960.meh107

[10] F.E. Hauser, P.R. Landon, J.E. Dorn, Deformation and fracture mechanisms of polycrystalline magnesium at low temperatures, Transactions of American Society for Metals. 48 (1956) 986–1002.

[11] E.C. Burke, W.R. Hibbard, Plastic Deformation of Magnesium Single Crystals, JOM. 4 (1952) 295–303. https://doi.org/10.1007/BF03397694

[12] R.E. Reed-Hill, W.D. Robertson, Pyramidal slip in magnesium, Trans. TMS-AIME. 212 (1958) 256–259.

[13] E.W. Kelley, W.F. Hosford, Plane-strain compression of magnesium and magnesium alloy crystals, Trans Met Soc AIME. 242 (1968) 5–13.

[14] X. Lou, M. Li, R. Boger, S. Agnew, R. Wagoner, Hardening evolution of AZ31B Mg sheet, Int. J. Plast. 23 (2007) 44–86. https://doi.org/10.1016/j.ijplas.2006.03.005

[15] H. Wang, Y. Wu, P.D. Wu, K.W. Neale, Numerical analysis of large strain simple shear and fixed-end torsion of HCP polycrystals, Comput. Mater. Contin. 19 (2010) 255. https://doi.org/10.3970/cmc.2010.019.255

[16] H. Wang, B. Raeisinia, P.D. Wu, S.R. Agnew, C.N. Tomé, Evaluation of self-consistent polycrystal plasticity models for magnesium alloy AZ31B sheet, Int. J. Solids Struct. 47 (2010) 2905–2917. https://doi.org/10.1016/j.ijsolstr.2010.06.016

[17] M. Sabbaghian, N. Fakhar, P. Nagy, K. Fekete, J. Gubicza, Investigation of shear and tensile mechanical properties of ZK60 Mg alloy sheet processed by rolling and sheet extrusion, Mater. Sci. Eng., A. 828 (2021) 142098. https://doi.org/10.1016/j.msea.2021.142098

[18] P.D. Wu, H. Wang, K.W. Neale, On the large strain torsion of HCP polycrystals, Int. J. Appl. Mechanics. 04 (2012) 1250024. https://doi.org/10.1142/S175882511250024X

[19] H. Wang, X. Zhang, W. Wu, P.K. Liaw, K. An, Q. Yu, P. Wu, On the torsional and coupled torsion-tension/compression behavior of magnesium alloy solid rod: A crystal plasticity evaluation, Int. J. Plast. 151 (2022) 103213. https://doi.org/10.1016/j.ijplas.2022.103213

[20] D. Tang, K. Zhou, W. Tang, P. Wu, H. Wang, On the inhomogeneous deformation behavior of magnesium alloy beam subjected to bending, Int. J. Plast. 150 (2022) 103180. https://doi.org/10.1016/j.ijplas.2021.103180

[21] K. Zhou, X. Sun, H. Wang, X. Zhang, D. Tang, W. Tang, Y. Jiang, P. Wu, H. Wang, Texture-dependent bending behaviors of extruded AZ31 magnesium alloy plates, J. Magnes. Alloy (2023) S2213956723000294. https://doi.org/10.1016/j.jma.2023.02.003

[22] S. Gollapudi, M.A. Azeem, A. Tewari, U. Ramamurty, Orientation dependence of the indentation impression morphology in a Mg alloy, Scr. Mater. 64 (2011) 189–192. https://doi.org/10.1016/j.scriptamat.2010.09.041

[23] J.-H. Shin, S.-H. Kim, T.K. Ha, K.H. Oh, I.-S. Choi, H.N. Han, Nanoindentation study for deformation twinning of magnesium single crystal, Scr. Mater. 68 (2013) 483–486. https://doi.org/10.1016/j.scriptamat.2012.11.030

[24] J. Bočan, J. Maňák, A. Jäger, Nanomechanical analysis of AZ31 magnesium alloy and pure magnesium correlated with crystallographic orientation, Mater. Sci. Eng. A 644 (2015) 121–128. https://doi.org/10.1016/j.msea.2015.07.055

[25] H. Kitahara, T. Mayama, K. Okumura, Y. Tadano, M. Tsushida, S. Ando, Anisotropic deformation induced by spherical indentation of pure Mg single crystals, Acta Mater. 78 (2014) 290–300. https://doi.org/10.1016/j.actamat.2014.06.039

[26] R. Sánchez-Martín, M.T. Pérez-Prado, J. Segurado, J.M. Molina-Aldareguia, Effect of indentation size on the nucleation and propagation of tensile twinning in pure magnesium, Acta Mater. 93 (2015) 114–128. https://doi.org/10.1016/j.actamat.2015.04.005

[27] C. Zambaldi, C. Zehnder, D. Raabe, Orientation dependent deformation by slip and twinning in magnesium during single crystal indentation, Acta Mater. 91 (2015) 267–288. https://doi.org/10.1016/j.actamat.2015.01.046

[28] P.D. Wu, X.Q. Guo, H. Qiao, D.J. Lloyd, A constitutive model of twin nucleation, propagation and growth in magnesium crystals, Mater. Sci. Eng. A 625 (2015) 140–145. https://doi.org/10.1016/j.msea.2014.11.096

[29] H. Wang, P.D. Wu, J. Wang, C.N. Tomé, A crystal plasticity model for hexagonal close packed (HCP) crystals including twinning and de-twinning mechanisms, Int. J. Plast. 49 (2013) 36–52. https://doi.org/10.1016/j.ijplas.2013.02.016

[30] B. Selvarajou, J.-H. Shin, T.K. Ha, I. Choi, S.P. Joshi, H.N. Han, Orientation-dependent indentation response of magnesium single crystals: Modeling and experiments, Acta Mater. 81 (2014) 358–376. https://doi.org/10.1016/j.actamat.2014.08.042

Forming of lightweight helical gears by means of sheet-bulk metal forming

RECK Manuel[1,a *] and MERKLEIN Marion[1,b]

[1]Institute of Manufacturing Technology, Friedrich-Alexander-Universität Erlangen-Nürnberg, Egerlandstraße 13, 91058 Erlangen, Germany

[a]manuel.reck@fau.de, [b]marion.merklein@fau.de

Keywords: Gear, Sheet Metal, Extrusion, Cold Forming

Abstract. A rising demand on resource efficient and highly functional products causes conventional metal forming processes to the reach their limits. For this reason, sheet-bulk metal forming (SBMF) for the production of thin-walled components with various filigree functional elements has been widely investigated in recent years. However, there is a research gap in the understanding of approaches within this process class for producing helical gears. These are in high demand in industrial environments due to their smooth running capabilities. In this study, a process combination of deep drawing and lateral extrusion for the production of helical lightweight gears is therefore examined. In the reference variant, the mild deep-drawing steel DC04 with a sheet thickness of t_0 = 2.0 mm is formed with a 20° helical die. General process limits are derived depending on the die filling of the tool-side cavities. In addition, the influence of the gear angle on process and component target variables is analyzed and discussed.

Introduction

According to UN forecasts, the number of people living on earth will increase from around 7.8 billion at present to about 9.7 billion in 2050 [1]. This growth is closely linked to the economic market integration of developing and emerging countries, which will rise international resource consumption [1]. In view of the limited global resources, it is essential to fundamentally rethink conventional production techniques and develop sustainable production methods that are still economically viable [2]. The major challenge for companies is the increasing complexity and the growing number of product variants [2]. The mobility sector in particular is affected by significant structural change. With increasingly shorter product life cycles, customers expect more models and product variations [3]. At the beginning of the 1990s, the product ranges of car manufacturers consisted of around eight models. At present, the product ranges comprise three times as many vehicle models [3].

Due to the legal requirements to reduce CO2 consumption, the innovation efforts of car manufacturers are growing. The objective is no longer primarily the comfort of the customer, but the efficiency and environmental friendliness of the vehicle [4]. Research and development is focusing in particular on innovative, more efficient powertrain concepts and lightweight construction [4]. There is significant potential for improvement to be explored in the area of the combustion engine, with electric motors also increasingly gaining market share [5]. However, the weight savings that can be achieved by implementing lightweight design concepts on the entire vehicle ought not to be underestimated for both power train concepts [6]. With the use of modern steel materials and forming technologies, a weight reduction of up to 25.5 kg can be achieved on a standard mid-range vehicle [6].

Downsizing engines is one approach that can be taken at system level [7]. This means reducing technical parameters such as engine cubic capacity or weight while maintaining the same driving performance and dynamics required by the customer. From this point of view, this is only feasible with higher specific engine outputs [7]. As a result, individual key components are permanently used at their load limits. In manual transmissions, the synchronizer ring is an example of this. Due to the integration of different functional elements such as driver and toothing elements, this component has a high level of geometric complexity. Helical gears are in this context well established in the industry for their low noise development and smoother running. Therefore, they are also used in lightweight designs of drive trains for electric car motors [8].

Sheet-bulk metal forming (BMF), which is characterized by the application of bulk forming operations to sheet metal blanks, makes an important contribution to the development of resource-efficient manufacturing processes for metallic components [9]. In this process class, the material flow is not just limited to the sheet metal plane, but it is also possible to extrude functional elements from the sheet metal plane [9]. This enables the production of complicated sheet metal components with high component functionality. While an in-depth understanding of the process class has already been developed for components with spur gear elements and carrier structures, there is a research gap for helical-toothed lightweight components. The geometry of helical gears results in higher flow paths and tooth volumes than with comparable sheet metal forming of spur gears. The achievable form filling therefore poses a particular challenge for the established processes such as upsetting or deep drawing combined with extrusion.

Methodology and Objective

The main goal of the investigation is the identification of process specific challenges and mechanisms that arise from the forming of helical gear elements in a combined approach of sheet-bulk metal forming. To achieve this objective, a gearing geometry with industrial relevance is first designed. The focus is on the running capability and a light-weight approach. However, the limits of die filling of filigree gearing elements in extrusion processes on sheet metal must be taken into account. The fundamental understanding of processes for straight-toothed functional components is used for this purpose [10]. A single-stage tool system for the efficient forming is subsequently designed based on the selected component geometry. The tool system consists of a unit for deep drawing and the following lateral extrusion.

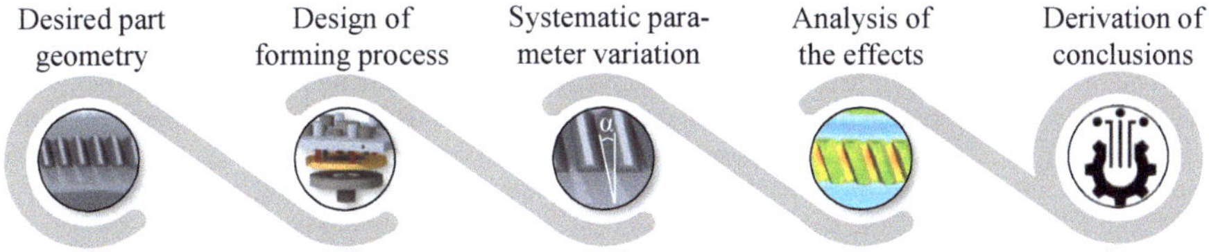

Fig. 1. Methodical approach of the research.

A systematic parameter variation moreover provides information about the effects that occur during the lateral extrusion of helical gear teeth. This knowledge is crucial for further research in measures of process improvement. Simulations of the finite element method are used to determine the process result parameters. This approach enables the elaboration of profound comprehension of the cause-effect relationships. The accuracy of the prediction of the applied numerical model was validated in [10] using process- and part-sided parameters. The flank angle is varied in accordance with industrial applications between 0 and 40 degrees. The analysis is based on the process target variables of forming force and tool load as well as on component-sided parameters like the material flow and the degree of deformation. Based on the research results, conclusions on the challenges and process limitations for the production of helical lightweight gears by sheet-bulk metal forming are drawn.

Tool setup and numerical model

In the process setup, the mild deep drawing steel DC04 is selected as workpiece material with an initial sheet thickness of 2.0 mm. Previous studies of this material have proven its suitability for SBMF extrusion processes with a high degree of deformation [10]. The tool setup for the forming of the lightweight helical geared part consist of a deep drawing stage for the component body and a lateral extrusion stage for obtaining functional elements. The associated process kinematic is shown in Fig. 2.

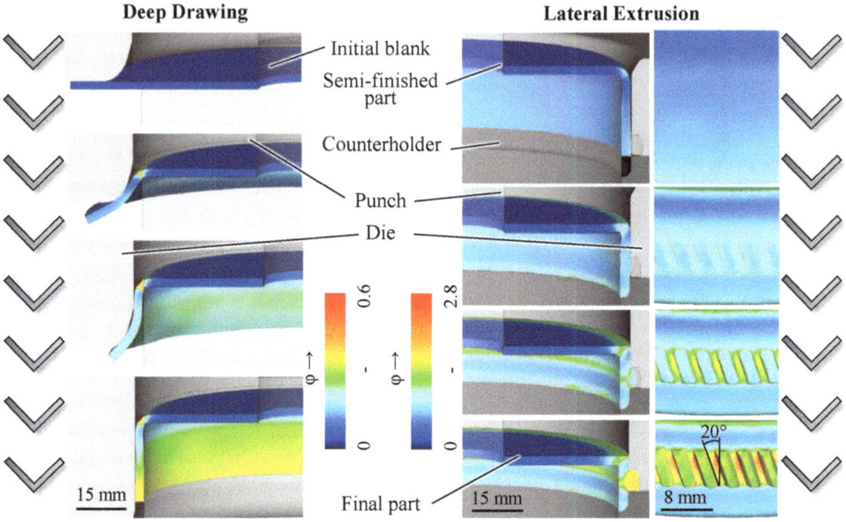

Fig. 2. Process kinematics of deep drawing and lateral extrusion.

The main body is manufactured using a conventional deep drawing process. At the beginning of the process, the sheet metal blank is clamped between a punch and a counterholder. The upper drawing ring is then moved downwards and forms the component frame. At the start of the subsequent lateral extrusion process, the sheet thickness is already increased from 2.0 mm to 2.25 mm. This results as the deep drawing process involves a drawing gap to prevent the ironing of the frame. In lateral extrusion, the synchronous downward movement of the punch and counterholder causes the workpiece to be displaced until contact is made with the lower die. The radius at the cup opening is then formed by the plastification of the component frame. The further increase in the process force introduced by the punch causes the plasticification of the material in the frame, which leads to the filling of the cavities between the workpiece and the tool. The cavities in the center of the die are designed as a negative form of the component-side teeth. This enables the cavities for the helical gearing elements to be filled as the process progresses.

Results and discussion

For the initial process investigation, the forming forces are analyzed for the reference setup with a gearing angle of 20°. The process forces for the deep drawing and the lateral extrusion are shown in Fig. 3(a). The linear ascent of the deep drawing force can be explained by the elastic strain shares at the beginning of forming. Due to the increase of friction force at the drawing ring radius, the reverse bending force and the ideal forming force caused by work hardening , the process force reaches its maximum of 61 kN at a stroke of 13.4 mm. As soon as the frame has finished bending at the outer radius of the counterholder, the process force decreases. The force-displacement curve is typical for deep drawing and has been verified in numerous publications [11]. With a maximum process force of 1.310 kN, the lateral extrusion forming of the helical-toothed functional elements far exceeds the force required for deep drawing.

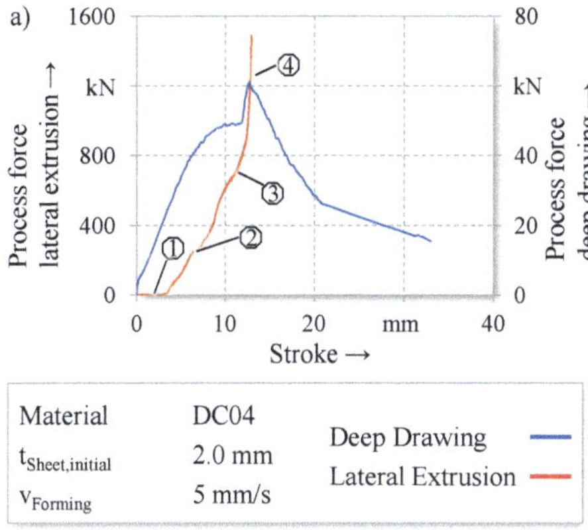

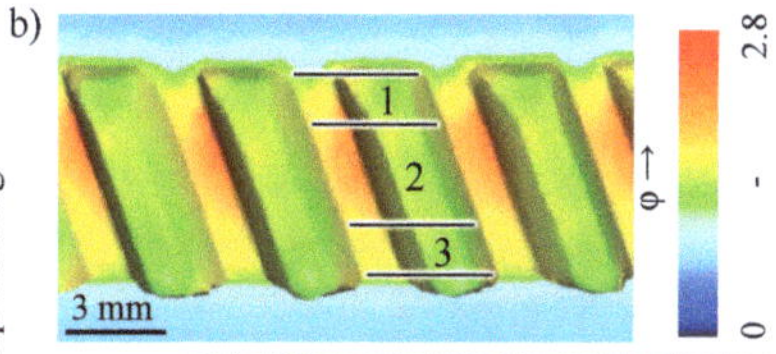

φ	1	2	3
Bottom	1.2 - 2.1	2.2 – 2.6	1.5 – 2.0
Left tooth flank	1.5 – 2.3	1.7 – 2.4	1.6 – 1.9
Tooth head	1.3 – 1.4	1.3 - 1.5	1.1 – 1.3
Right tooth flank	1.4 – 1.6	1.5 – 1.8	1.5 – 1.7

Fig. 3. Force-stroke curve of deep drawing and lateral extrusion (a) and the distribution of the degree of deformation of the finished part (b).

At the start of lateral extrusion, the process force is around 8 kN (1). In this phase of the process, the component is pushed down into the lower die. Consequently, only frictional forces between the component and the tool have to be overcome. The material is not yet plasticized. From a stroke of 3.0 mm, there is a steep increase in process force. This is due to the contact of the lower cup wall with the die-side radius and the resulting material deformation. The slight drop in the gradient from a stroke of 6.5 mm (2) can be explained by the formation of a fold at the corner between the component base and the cup wall. The upsetting of the frame causes an increase in the degree of deformation and consequently work hardening as the process progresses. The forming of the helical-toothed cavities begins at a stroke of approximately 9.0 mm. As the cavity filling increases (3), the process force also rises, resulting in a linear increase due to the almost complete contact between component and tool (4). In the comparable process for the forming of straight teeth elements, a maximum process force of about 1.050 kN is reached [10]. However, direct comparability is not given due to the significantly lower die filling in the straight tooth process. While a die filling of 95 % is used in the investigated process for more precise identification of the effects, the die filling in the compared process is only 63 %. At a die filling of 63%, the process of the helical gear parts is with 1.060 kN at a comparable level. The variation of the helix angle therefore appears to have only a minor influence on the force required for forming. To understand this behavior, the distribution of the degree of deformation for the helical gear elements is shown in Fig. 3(b) and the underlying material flow is discussed using Fig. 4.

In the tooth root, the degree of deformation is generally highest. This is the case for all differentiated areas 1-3. The localization of the highest degrees of deformation in this area is comparable with the results of the investigations of the lateral extrusion of spur gear components. The reason for this is the strong material deflection caused by the geometric flow restriction of the tooth heads on the tool side.

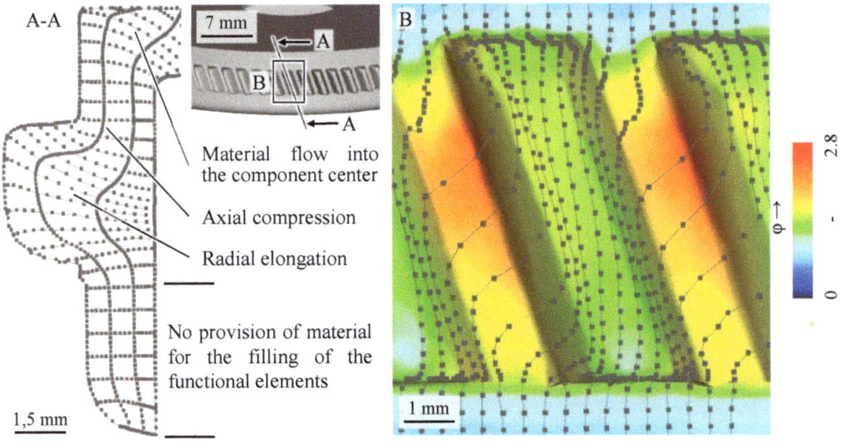

Fig. 4. Material flow analysis of the helical gear elements.

Detail B of Fig. 4 visualizes the strong elongation of the material in the tooth root due to the material deflection. The flow lines are tracked using a grid with connected points at a fixed distance during forming. Based on the change compared to the initial grid, the material flow can be estimated. In addition, a gradient of the degree of deformation results while higher values result in the left tooth root and decreasing values towards the left bottom area. According to the material flow visualization of section A-A, the upper component wall solely provides the material volume for the cavity filling. In the area below the toothing elements, the grid shows no distortion, which indicates little material flow. This circumstance therefore leads to a downward displacement of the material, which is subsequently deflected by the inclined die teeth. This axial displacement causes the gradient of the degree of deformation due to the changed angle of impact of the material between the left and right tooth flank. This results in an asymmetrical material flow between the right and left tooth flank. In addition, cross-section A-A shows the basic mechanism of filling the functional elements in radial direction. It is characterized by material displacement due to axial compression as well as radial expansion because of the hollow volume between the workpiece and the die cavities.

The asymmetrical material flow is moreover verified by analyzing the material flow velocity in the gearing elements. In this context, Fig. 5(a) shows the material flow velocity in the radial direction at different phases of form filling. There is a strong gradient in material flow velocity between the left and right side of the tooth. While the values for the left half of the tooth average 4.2 mm/s, the material flow velocity in the right half of the tooth reaches only an average of 3.1 mm/s. The main reason for the differences in material flow speed is the superposition of axial and radial material displacement (Fig. 4 section A-A) with the different deflection angles on the tool-side tooth flanks (Fig. 4 detail B). As a result of the asymmetrical material flow rates, the tooth cavity is filled unevenly, with the material of the left half of the tooth running ahead of the material of the right half. The tooth profile is therefore only symmetrical when the cavity is entirely filled.

Materials Research Forum LLC
https://doi.org/10.21741/9781644903254-13

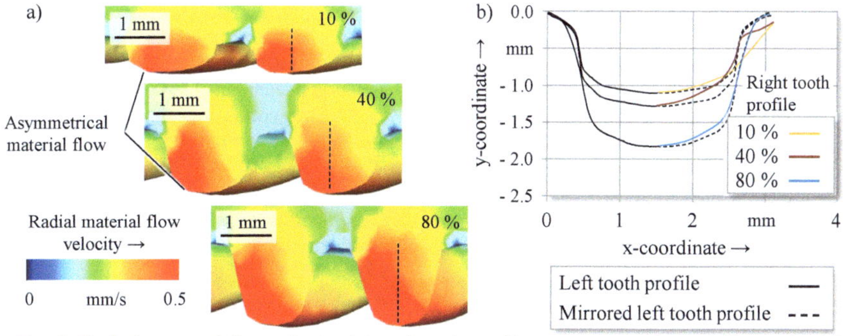

Fig. 5. Radial material flow velocity (a) and tooth profile comparison (b) at different die filling stages.

Fig. 5(b) shows the deviation between the left and right tooth profile, taking the left tooth flank as reference. In all three stages of die filling from 10-80 %, the right tooth is less filled at the tooth face and topland. As the form filling increases, the difference between the left and right tooth filling is equalized due to the smaller remaining cavity. Only a slight deviation remains, which is eliminated with complete die filling due to the geometric limitation of the die cavities. Since, depending on the strength of the blank material, the tool load often exceeds the permissible range during sheet metal forming of filigree gearing elements [12], it is often not possible to achieve complete die filling. It can therefore be concluded that the accuracy of the tooth flank profiles represents a decisive process limit due to the uneven form filling.

The tool life time is a decisive factor with regard to the economic efficiency of the lateral extrusion process at higher quantities. In this context, the tool lifetime is considerably affected by the external forces acting on the tools [13]. As the tool loads are expected to be highest in the die due to the fine cavities, this area is examined for the reference variant at a helix angle of 20°. Failure-critical stress components are evaluated in order to identify areas with a high risk of fracture. Stress concentrations are localized using the $\sigma_{v.Mises}$ (Fig. 6(a)), while the principal stresses σ_{Max} (Fig. 6(b)) and σ_{Min} (Fig. 6(c)) are used to determine the maximum tensile respectively compressive stresses.

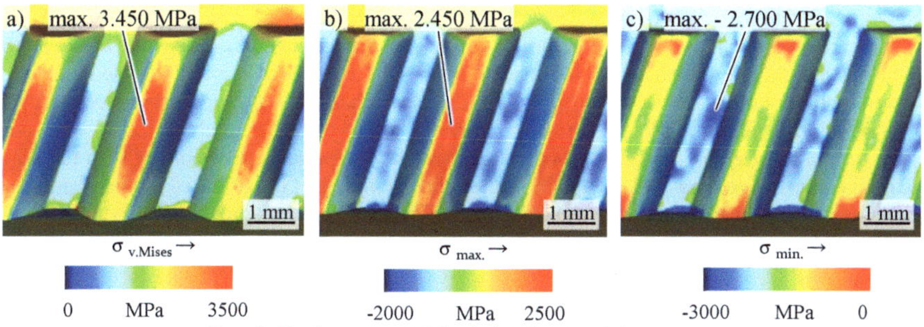

Fig. 6. Tool stresses of the helical geared forming die.

The evaluation is carried out at a form filling of 95 %. Stress concentrations are detected on the basis of $\sigma_{v.Mises}$ as well as σ_{Max} in the tooth base of the die cavities. Due to the superposition of the axial and radial material flow, a lateral bending stress on the tooth elements on the tool side is induced, which leads to a tensile load in the tooth foot area. The teeth themselves are primarily subject to pressure. These are in direct contact with the material during forming and are exposed to high normal pressures due to the deflection of the material. This behavior is also demonstrated in the investigation of spur gear dies with filigree functional elements [10]. Furthermore, no asymmetry of the tool loads can be detected. This leads to the conclusion that the axial force components acting when the material impacts the tooth flanks on the tool side are still low at a helix angle of 20°.

The transferability of the process and component-side results at a helix angle of 20° is examined in the following section with regard to their comparability with other helix angles. For this purpose, the angle is varied between 0° and 40° with a step width of 10° and the effects on the target variables of process force, work hardening and tool load are analyzed and discussed. Fig. 7 shows the influence of the helix angle variation on the tool loads.

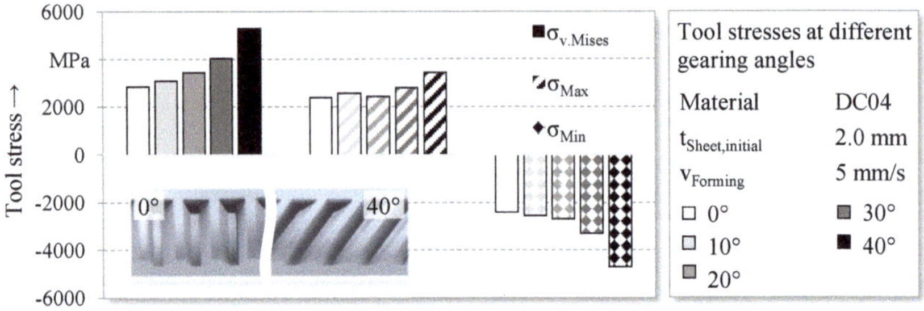

Fig. 7. Influence of the helix angle on the tool loads.

The maximum $\sigma_{v.Mises}$ detected with a spur die is 2.850 MPa and increases to up to 5.300 MPa (+85 %) when the toothing angle is increased. The maximum tensile stresses show no clear trend for the 0°-30° variants. Only with a helix angle of 40° do the maximum tensile stresses increase noticeably to up to 3.450 MPa. The bending load due to the axial material flow components therefore has an influence on the resulting tensile stresses at large helix angles. This this is due to the increasing lateral force components that act on the gearing as the angle increases. However, these only exceed the prevailing tensile stresses in the base area resulting from the compressive load on the teeth from an angle of 40°. In the investigated process, the compressive stress state shows a continuous increase in stresses with increasing flank angle. This increase is also due to the rising lateral forces.

An increase in the helix angle therefore has to be considered as having a negative effect on the tool loads and therefore also on the lifetime of the tools. The loads also exceed the load limits of conventional tool materials at a mold filling of 95 % at higher flank angles. Carbide tools are one possible solution to this challenge. These have great potential to withstand the forces in sheet bulk metal forming processes for helical-toothed lightweight components, particularly with regard to the high compressive loads [14]. To gain further understanding of the effect of the helix angle on process-sided parameters, Fig. 8(a) shows the reached maximum process force values at the different helix angles. Starting with a spur gear, the angle is increased in steps of 10° to 40°.

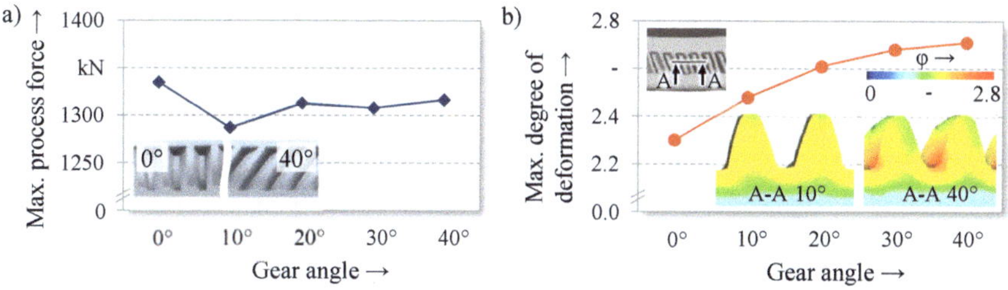

Fig. 8. Maximum process force (a) and maximum degree of deformation (b) at helix angles from 0° to 40°.

The helix angle does not unambiguously influence the maximum process force reached in lateral extrusion. This can be explained with the material flow behavior presented in Fig. 4. The material is displaced by the axial compression of the frame in a radial direction. During the cavity filling, axial material flow is rather superpositioned by the radial elongation. With regard to the relative movement between the workpiece and the tool-side gear teeth, it is primarily the axial displacement that leads to increased frictional forces as the gear tooth angle increases. The displacement of the material in the radial direction can be described as the material clinging to the tool wall and does not result in an increase in the process force due to the merely lower frictional forces at higher helix angles. From a volumetric point of view, the geometric flow restriction caused by the teeth on the tool side is independent of the tooth angle. Therefore, the required process force remains at a comparable level with varying helix angles. Despite higher axial forces, this effect is also demonstrated in the full forward extrusion of helical-toothed components [15]. In addition, the internal mechanical component properties are compared on the basis of the maximum degrees of deformation. Fig. 8(b) presents the corresponding maximum values as well as an exemplary cross-section of the teeth at a helix angle of 0° and 40°. The maximum dregree of deformation φ_{max} increases from 2.3 at 0° in a logarithmic saturation curve to $\varphi_{max} = 2.7$ at a helix angle of 40°. The increasing strain hardening results from the axial components of the material flow, which cause a deflection of the material on the left tooth flank. As the analysis is based on a die filling of 95%, the values on the right tooth flank are also increased in the 40° variant compared to the spur gear. This is due to the fact that the material is also deflected on the right tooth flank at the end of the forming process. For the gear application, the rise in the toothing angle can therefore be classified as positive, as the increased hardness of the teeth results in a reduction in wear [16].

To put the different effects into relation, Fig. 9 summarizes the mechanism of the toothing angle based on the model concept. The deflection of the axial material flow component (Fig. 4) increases with larger flank angle. This results in an increase in the lateral force F_l introduced, which acts on the toothing on the tool side. For this reason, the tensile loads in the base area between the teeth increase, as do the compressive loads in the tooth itself (Fig. 7). Furthermore, the axial force F_a and the radial force F_r are independent of the angle. The deflection angle also increases the hardening, particularly in the tooth area near the surface (Fig. 9).

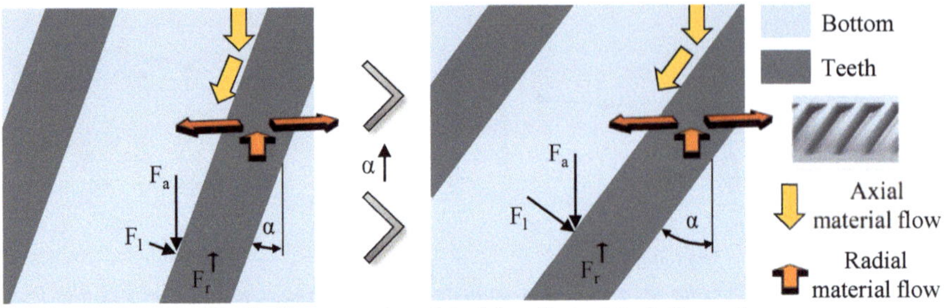

Fig. 9. Conceptual model of the effect of the helix angle.

Conclusions and Outlook

The investigation presents the general demonstration of the suitability of the process combination of deep drawing and lateral extrusion for the production of helical-toothed thin-walled lightweight components by means of sheet metal forming. It has been shown, that an asymmetric hardness distribution in the teeth results from the deflection of the axial material flow component. The asymmetric hardness distribution can be exploited for defining the working direction of the gear to gain a higher wear resistance. Moreover, two main process limits are identified. At low form filling, the primary process limit is the precision of the tooth flank profile due to asymmetric material flow. With high die filling, the flank profile is equalized, but the process restriction becomes the high tool loads, which can be addressed by the use of carbid forming tools to increase the compressive strength. Similar to forward extrusion process for helical gears, higher tooth angles in lateral extrusion improve the application properties due to the increased tooth hardness. This positive effect is contrasted by increased tool loads due to the higher lateral force components acting on the tooth flanks.

Further research should focus on the development of process-adapted ejector concepts. Due to the necessary rotation of the component for demolding and the associated contact pressures, there is a risk of deterioration in gear quality. Possible approaches in this context are rotating dies or inverse pre-correction of the gear geometry. In addition, further contributing factors such as the semi-finished product material, the sheet thickness and the gear geometry need to be investigated with regard to process and component-related results.

Acknowledgment

This work was supported by the German Research Foundation (DFG) within the scope of the Project 501091735 "Manufacturing of helical-toothed functional components from sheet metal by the development and analysis of a sheet-bulk metal forming process".

References

[1] World Population Prospects 2022: Summary of Results. (2022).

[2] R. Stark, G. Seliger, J. Bonvoisin, Sustainable Manufacturing, Cham: Springer, (2017).

[3] Wyman, O.: Massiver Wandel in der automobilen Wertschöpfungsstruktur. (2012).

[4] M. Kleiner, M. Geiger, A. Klaus, Manufacturing of Lightweight Components by Metal Forming, CIRP Annals 52 (2003) 521-542. https://doi.org/10.1016/S0007-8506(07)60202-9

[5] D.V. Pelegov, J.-J. Chanaron, Electric Car Market Analysis Using Open Data: Sales, Volatility Assessment, and Forecasting, Sustainability 15 (2022) 1. https://doi.org/10.3390/su15010399

[6] B. Klein, T. Gänsicke, Leichtbau-Konstruktion: Springer, (2019).

[7] R. Golloch, Downsizing bei Verbrennungsmotoren: Springer, (2005).

[8] T. Biermann, T. Smetana, Schaeffler lightweight differentials. Schaeffler Symposium, (2010).

[9] M. Merklein, J.M. Allwood, B.A. Behrens, A. Brosius, H. Hagenah, K. Kuzman, K. Mori, A.E. Tekkaya, A. Weckenmann, Bulk forming of sheet metal, CIRP Annals, 61 (2012) 725-745. https://doi.org/10.1016/j.cirp.2012.05.007

[10] F. Pilz, M. Merklein, Comparison of extrusion processes in sheet-bulk metal forming for production of filigree functional elements, CIRP J. Manuf. Sci. Tech. 26 (2019) 41-49. https://doi.org/10.1016/j.cirpj.2019.05.004

[11] H. Zein, O.M. Irfan, Optimization and Mapping of the Deep Drawing Force Considering Friction Combination, Applied Sci. 11 (2021) 19. https://doi.org/10.3390/app11199235

[12] F. Pilz, D. Gröbel, M. Merklein, Investigation of fatigue strength of tool steels in sheet-bulk metal forming: AIP Conference Proceedings, 2018.

[13] T. Trzepieciński, Approaches for Preventing Tool Wear in Sheet Metal Forming Processes, Machines 11 (2023) 6. https://doi.org/10.3390/machines11060616

[14] A. Nilsson, L. Kirkhorn, M. Andersson, J.E. Ståhl, Improved tool wear properties in sheet metal forming using Carbide Steel, a novel abrasion resistant cast material, Wear 271 (2011) 1280-1287. https://doi.org/10.1016/j.wear.2011.01.083

[15] C. Kiener, Kaltfließpressen von gerad- und schrägverzahnten Zahnrädern. Dissertation, Friedrich-Alexander-Universität Erlangen-Nürnberg, 2020.

[16] J. Lai, J. Liu, H. Hu, J. Yu, A. Zhang, W. Yuan, Z. Gao, Contribution of hardness and work hardening to the wear resistance of Pt-based metallic surfaces revealed by nanoscale reciprocating sliding, Applied Surf. Sci. 614 (2023) 156178. https://doi.org/10.1016/j.apsusc.2022.156178

MANUFACTURING

Metal Forming 2024 Materials Research Forum LLC
Materials Research Proceedings 44 (2024) 124-133 https://doi.org/10.21741/9781644903254-14

Design and simulation of cross wedge rolling process for hollow motor shaft

HAN M.[1,2,a], LIU J.X.[1,2], SHI M.J.[1,2], SONG H.W.[1], ZHANG S.H.[1] and CHENG M.[1,b*]

[1] Institute of Metal Research, Chinese Academy of Sciences, Shi-changxu Innovation Center for Advanced Materials, Shenyang 110016, China

[2] School of Materials Science and Engineering, University of Science and Technology of China, Shenyang 110016, China

[a]mhan22s@imr.ac.cn, [b]mcheng@imr.ac.cn

Keywords: Hollow Motor Shaft, Cross Wedge Rolling, Hollow Billet, Numerical Simulation

Abstract. Nowadays, electric cars are experiencing rapid development, in which the production of motor shaft is one of the important parts. The cross wedge rolling (CWR) process is expected to be an efficient and material-saving production method for motor shaft production. Traditional motor shafts are manufactured by forging and machining, which inevitably causes material waste. While using the CWR method to produce the motor shaft, the forming accuracy can be improved, keeping the material performance. It's possible to manufacture the hollow motor shaft billet directly and reducing the production cost greatly. Because of the asymmetric geometry profile of the motor shaft, the force is unbalanced during the forming process. The thermo-mechanical coupling finite element simulation is used to design and optimize the flat CWR tools. Then, the distribution of stress, strain, damage and temperature during the CWR process are analyzed. The traditional CWR process will easily cause central defects of the shaft such as micropores and voids. The hollow billet is used in the study. It helps to avoid such defects, also can be a new method to deal with the hollow motor shafts by CWR. Some disadvantages of traditional manufacture can be overcome.

Introduction

Motor shaft is a classic transmission shaft, and its demand is increasing with the development of electric cars. The hollow motor shaft has three advantages: low moment of inertia, lightweight structure and convenient defect detection, which is the mainstream trend of future development. The motor shaft has the characteristic of multiple steps, and the traditional manufacturing process is mechanical processing. The inner hole needs to be processed by deep hole processing, which leads to low material utilization and high processing difficulty. The cross wedge rolling (CWR) technology is a near-net shape metal working process with high forming precision and small cutting dimension allowance for forming the shaft parts [1,2]. Therefore, the use of CWR technology is expected to solve the above problems. However, there are two difficulties in producing hollow shafts for CWR. One is how to ensure the dimensional accuracy of the surface and inner hole of hollow shaft. The other one is how to rolling asymmetric structures [3,4].

The research on the hollow shaft manufactured by CWR is divided into two types: with mandrel and without mandrel [5-7]. Among them, using a mandrel can control the accuracy of the inner hole size, but the selection of mandrel size can only rely on experience, making tool design inconvenient. When the inner hole diameter is small, rolling without a mandrel can also obtain a rolled piece with roughly accurate inner hole size, and then simple processing of the inner hole can meet the requirements. Hollow motor shafts mostly use smaller inner hole sizes, so rolling without mandrel has the advantage of improving production efficiency [8].

At the same time, the motor shaft is asymmetric, and the traditional CWR mostly adopts double pieces symmetrical rolling method. However, due to the long length of the motor shaft, limited by the size of the CWR rolling mill, single piece forming can only be used. It leads to the asymmetric step characteristics at both ends of the motor shaft, which are prone to axial force imbalance during the forming process, and the rolled piece is flattened after being skewed, resulting in rolling failure. Therefore, in the process of tool design, attention should be paid to the issue of axial force balance [9,10].

In this paper the changes in temperature, stress, and strain of hollow shaft parts during cross wedge rolling were analyzed through tool design and finite element simulation of motor shafts, providing guidance for actual production.

Tool design and simulation of CWR process

Tool Design for target motor shaft. Fig. 1 shows the target workpiece (rough motor shaft). It can be observed that the shaft has asymmetric features: the steps at both ends are asymmetric and fewer steps on the left and more steps (which have significant difference in diameters) on the right. At the same time, there is a narrow and high step in the middle of shaft. Therefore, axial force balance and forming method of narrow step had to be considered before the rolling tools are designed.

Fig. 2 shows the CWR tool with special design. For the forming of narrow step, using the traditional CWR tool design method will result in excessive processing volume, so special tool design is needed to squeeze the narrow steps into shape. The metal flow during the extrusion process should not be too large, which can be solved through tool design: the billet is slowly divided in the initial stage of wedging, and then formed with a smaller spreading angle. The position of dividing the billet needs to be calculated through the volume of narrow steps. In order to maintain axial force stability at the positions where asymmetric steps are formed at both ends, the tool was designed for the left step to be formed in two stages.

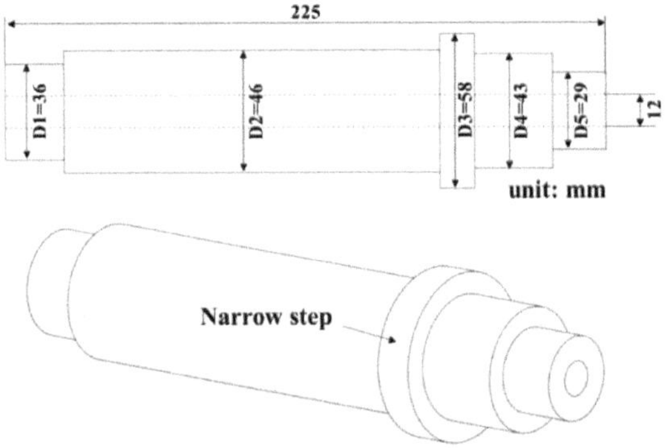

Fig. 1. Target rolled workpiece.

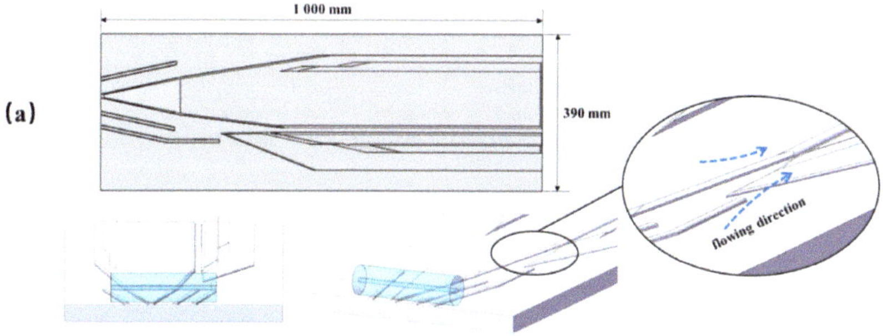

Fig. 2. Top view and special design of tools.

Finite element modeling. Finite element software DEFORM-3D was used for the simulation. The billet and tools were established, and their assembly relationship was adjusted, as shown in Fig. 3. Finally, the STL. format file was imported into the finite element software and simulation parameters were set further.

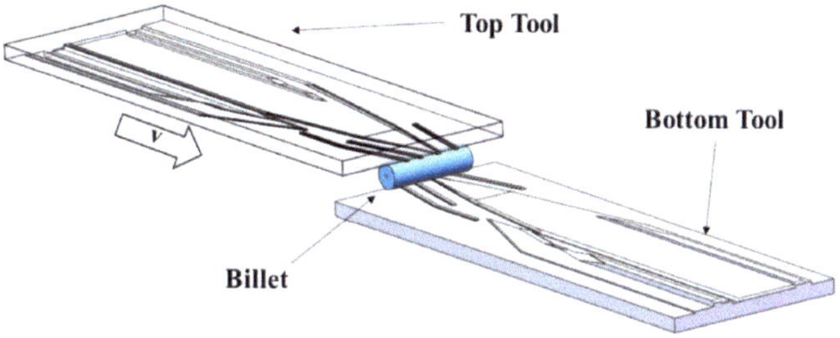

Fig. 3. Schematic diagram of assembly relationship between tools and billet.

During the CWR process of forming asymmetric parts, there are both radial compression and axial extension, as well as lateral extension. This results in non-linear stress-strain behavior of the material during the deformation process, as well as non-linear strain and displacement. The boundary conditions are very complex. Taking into account the above factors, establish a finite element model that conforms to the real forming process and make the following assumptions:

1. The plastic deformation of the billet is large, and its elastic deformation is ignored. It is set as a rigid plastic body, and the material model is selected from the 45 steel in the software material library.
2. Compared to the plastic deformation of the billet at high temperatures, the elastic deformation of the tool is too small to be ignored, and both sides of the tool can be regarded as rigid bodies.
3. In actual production, the friction groove can effectively avoid the phenomenon of slipping between the rolled piece and the tool. The friction type is selected as shear friction, and a higher friction coefficient (f=1.5) will be set.
4. The rolling process has a short duration, and the heat transfer coefficient between the billet and the environment, as well as the tool, can be considered as a constant value.

And the main parameters used in the simulation are shown in Table 1.

Table 1. Parameters used in the simulation.

Parameters	Number
Environment and tool temperature	20 °C
Environmental thermal conductivity coefficient	0.02 N/(mm·s·°C)
Thermal conductivity coefficient of tools and billet	12 N/(mm·s·°C)
Rolling speed	300 mm/s
Billet temperature	1 000 °C
Damage model	Normalized Cockroft & Latham: $f = \int_0^{\varepsilon} \frac{\sigma_1}{\sigma_i} d\varepsilon$

Results and discussion

Analysis of dimensional accuracy of rolled piece. During the rolling process, the surface of the billet comes into direct contact with the tool, generating a fast temperature drop. The surface metal is sensitive to temperature changes, and the plasticity and flowability of the material are affected by temperature, leading to rolling failure. Therefore, temperature of the billet and the rolling speed are the key factors affecting the dimensional accuracy of the rolled piece. The rolling speed should not be too low, and the selected rolling speed for this study is 300 mm/s. The simulation results of the rolled piece at 1000°C are shown in Fig. 4.

Observing the morphology of the rolled piece, the size of the inner hole varies in a wavy shape, with a diameter of 3.4 mm at the narrowest point and 14.6 mm at the widest point. The initial diameter of the inner hole of billet was 7.5 mm (Calculate the approximate value based on the principle of volume invariance). The rolling force caused the billet to elongate towards both ends, and the diameter of the inner hole in the wedged position reached its maximum. The diameter of the inner hole decreases due to radial compression at both ends. There was no collapse in the inner hole, so the subsequent precision machining steps can be simplified. The dimensions of each step from left to right on the outer diameter of the rolled piece are set to D1-D5, as shown in Fig. 1. The comparison between the target size and the simulated size is shown in Table 2. The outer diameter deviation values are all less than 0.8 %, which proves that the external dimensions meet the requirements. Both ends of the material head have concave centers, but none of them affect the required part of the rolled piece. Therefore, the size of the rolled piece simulated by finite element method meets the target requirements, providing guidance for the experiment.

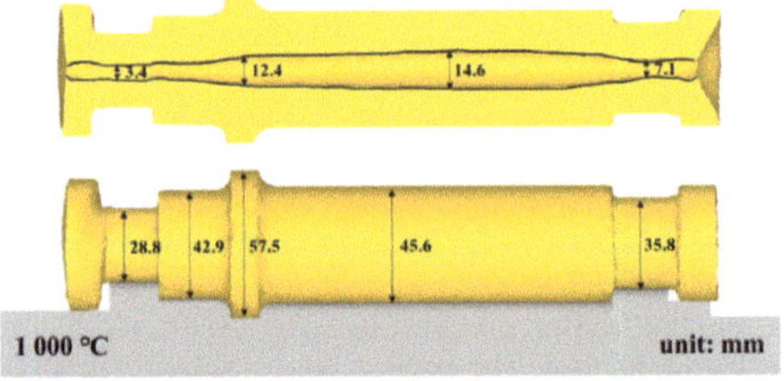

Fig. 4. The rolled workpiece at 1000°C.

Table 2. Target and simulation results of the selected cross-sections.

Position	Target (mm)	Simulation (mm)	Deviation (%)
D1	28.8	29	0.7
D2	42.9	43	0.2
D3	57.5	58	0.8
D4	45.6	46	0.8
D5	35.8	36	0.6

Analysis of temperature and stress. The CWR process lasts about 5.3 seconds, so it is of great significance to investigate the temperature changes of the billet. Fig. 5, 6 and 7 show the simulation results at a rolling temperature of 1000°C and a rolling speed of 300 mm/s, when the rolling speed is high, the collapse of hollow shaft may occur [11]. Fig. 5(a) shows the temperature changes on the surface and cross-section during the wedging process of the rolled piece; Fig. 5(b) shows the stress changes on the surface and cross-section during the wedging process of the rolled piece. It can be clearly observed that there is a significant temperature difference between the surface and the center of the rolled piece. The surface of the billet directly contacts the tool, and the effect of heat transfer reduces the surface temperature. However, plastic deformation occurs inside due to compression, and the deformation energy is converted into heat energy, leading to an increase in temperature. The most significant decrease in temperature occurs when the rolled piece comes into contact with the wedge first, and it is also the part with the largest deformation.

Fig. 6(a) and 6(b) respectively show the temperature and stress changes on the surface and cross-section of the rolled piece during the intermediate stage of rolling. It can be seen that the surface temperature has significantly decreased and the core temperature has increased. The stress is mainly concentrated at the position where the surface contacts the tool.

Fig. 7(a) and 7(b) respectively show the temperature and stress changes on the surface and cross-section of the rolled piece during the end of rolling stage. The temperature from the surface to the core is distributed in a stepped pattern, but there is still a certain amount of high temperature on the right side of the narrow step. This is because the rolled piece is subjected to a left axial force, and the left side is in close contact with the tool, resulting in a rapid temperature decrease. At this point, the strain is mainly concentrated in the step deformation zone with high shrinkage rate on the left side.

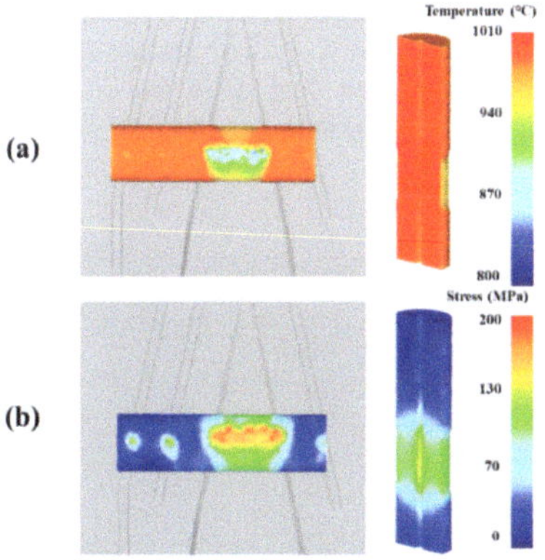

Fig. 5. Distribution of (a) temperature and (b) stress during CWR process (The beginning stage).

(a)

(b)

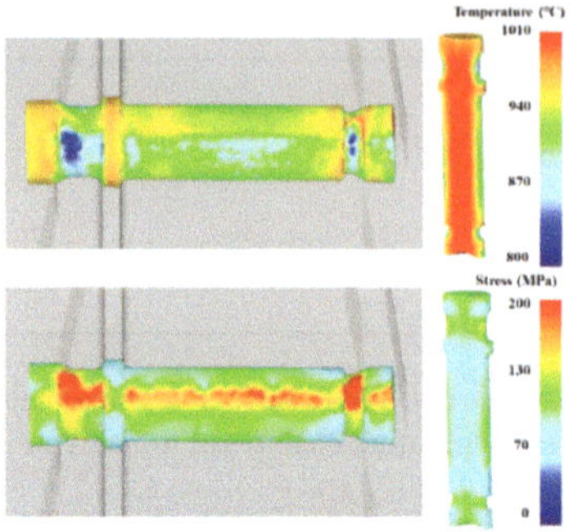

Fig. 6. Distribution of (a) temperature and (b) stress during CWR process (The intermediate stage).

(a)

(b)

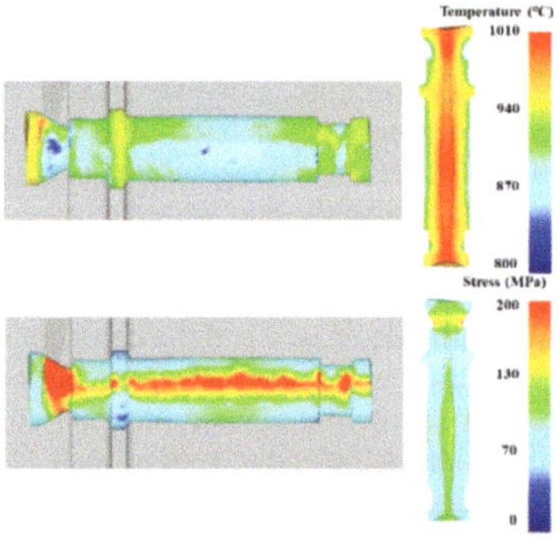

Fig. 7. Distribution of (a) temperature and (b) stress during CWR process (The end stage).

Fig. 8 shows the temperature and stress distribution on the cross-section of the rolled piece at different rolling stages. As the rolling progresses, the temperature gradually decreases from the core to the surface. And the stress is mainly concentrated in the area in contact with the tool. The position with the lowest temperature corresponds to the position with the highest stress, which makes the surface prone to damage. In the final stage, the deformation and stress are relatively large, which leads to the inner hole being compressed into an elliptical shape.

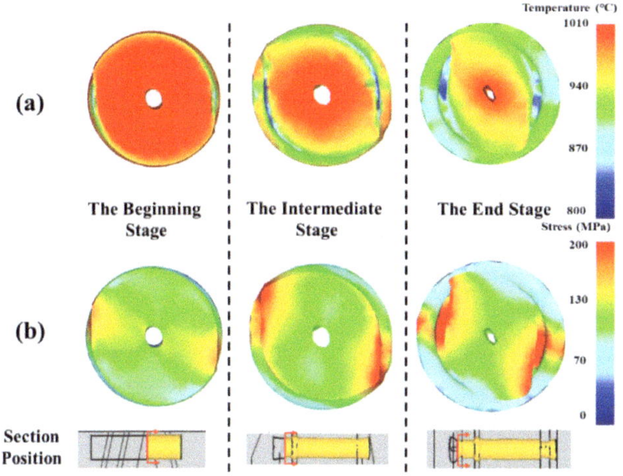

Fig. 8. Distribution of (a) temperature and (b) stress at cross section rolling piece during CWR process.

Analysis on the cross-section of the billet. Set six tracking points P1-P6 on the cross-section of the billet (Fig. 9), and then extract the temperature, stress, and strain of each point for comparative analysis to observe the trend of the forming process. The position selected for the cross-section is the position of wedging at the beginning of rolling, and the selection of this section is because it is the initial material segmentation position.

The temperature variation curves of each tracking point over time are shown in Fig. 10(a). It is observed that the temperature decreases more significantly as the distance from the center of the rolled piece increases, and the fluctuation of surface temperature becomes more pronounced as it approaches the center. This is due to the cyclic contact between the rolled piece and the colder tool. Due to the conversion of deformation work into thermal energy, the center temperature of the rolled piece shows a trend of first increasing and then decreasing, with a maximum temperature of 1040°C. The temperature curve at points P1-P4 shows a similar trend, indicating that the changes occurring during the deformation process are similar. Higher temperatures may lead to coarsening of the grain structure in the center of the rolled piece; And point P6 exhibits a very obvious temperature drop feature, and its temperature state exhibits significant fluctuations due to the rotation of the rolled piece.

Fig. 10(b) shows the variation curve of effective strain over time for each tracking point. Analyzing the curve, it can be observed that although the distance between the six tracking points is uniformly distributed, the strain is not uniform. The closer to the surface, the greater the increase in strain. This is because most of the deformation is carried out on the outer layer of the rolled piece. The strain at all six points shows a stage of drastic increase, and then the rolling proceeds to the finishing section. The curves of P1 and P2 are basically similar, belonging to the area with small deformation in the center, P2, P3, and P4 belong to the intermediate transition area, and P6 belongs to the area with the highest surface deformation. Compared with Fig. 11(a), it is found that the area of temperature cycling is also the area with the highest deformation, which may lead to defects such as cracks on the surface of the rolled piece.

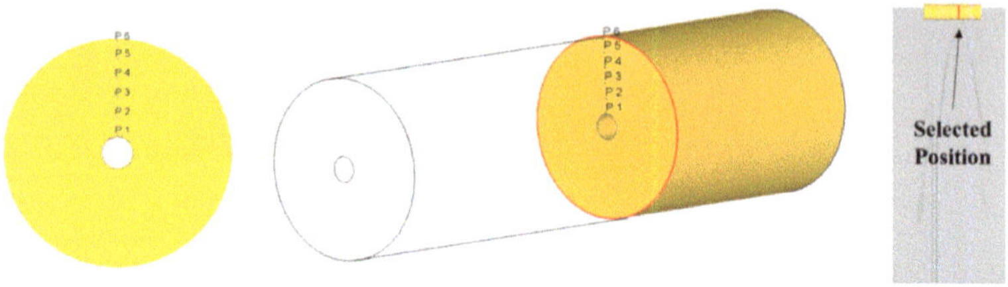

Fig. 9. Position of tracking points before rolling.

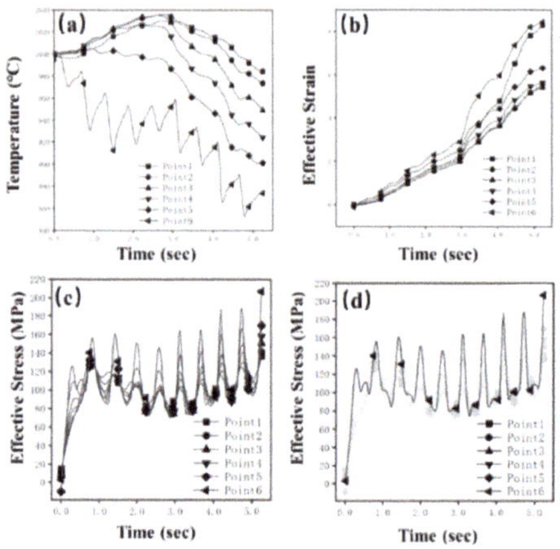

Fig. 10. The curves of temperature, effective strain and effective stress of tracking points over time.

Fig. 10(c) and (d) show the variation curves of effective stress at each tracking point over time. Analyzing the curve, it can be found that the stress at each point during the rolling process is roughly distributed in a wave shape, and the closer it is to the surface, the greater the amplitude of the wave. The fluctuation of stress is exactly opposite to the fluctuation of temperature. When the temperature is low, the stress is high, which may lead to damage.

Damage analysis. In the spreading section of the tool, the surface temperature of the rolled piece decreases and significant plastic deformation occurs, which may lead to cracking of the billet. The damage situation during the rolling process of the rolled piece is shown in Fig. 11. When the rolled piece contacts the wedge tip, damage begins to occur, and the damage value rapidly increases as the rolled piece enters the spreading section. To avoid the occurrence of cracks, it is necessary to optimize the widening section of the tool. While the widening angle should not be too large, the influence of the forming angle should also be considered.

For the center of the rolled piece, the damage increases with the elongation of the rolled piece, and the maximum damage value is located at the initial wedging position. This also corresponds to the position with the largest inner hole size.

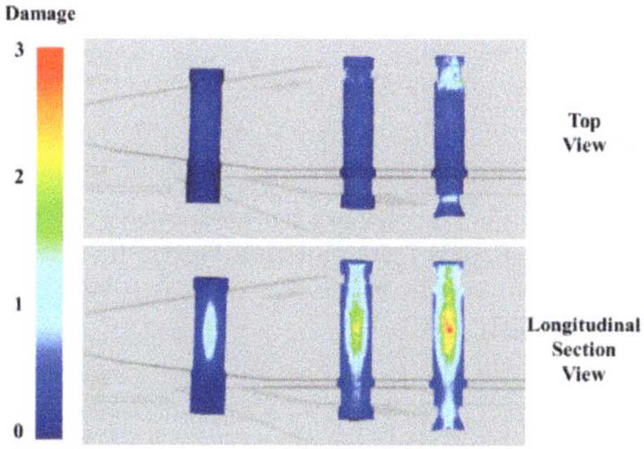

Fig. 11. Damage distribution during CWR process.

Summary

In this paper, the CWR preforming process of hollow motor shaft was studied, The main conclusions are as follows:

1. Hollow motor shaft can be produced through flat CWR technology at a rolling speed of 300 mm/s and a rolling temperature of 1000 °C.
2. The surface dimension deviation not exceeding 0.8%, the maximum and minimum size of the inner hole are 14.6 mm and 3.4 mm, respectively;
3. Steps with high cross-sectional shrinkage are prone to damage, and the corresponding inner holes are prone to elliptical shapes;
4. The damage in the inner hole area is mainly concentrated at the initial wedged position, which is related to the elongation of the rolled piece.

Acknowledgment

This work was financially supported by the International Science and Technology Cooperation Project in Liaoning Province of China (Grant No. 2022JH2/10700006, 2023JH2/10700014), the National Key R&D Plan of China (Grant No. SQ2018YFE0119600) and the Sino-Belarus Inter-Governmental S&T Cooperation project (Grant No. CB02-09).

References

[1] Z. Pater, J. Tomczak, T. Bulzak, J. Bartnicki, A. Tofil, Prediction of Crack Formation for Cross Wedge Rolling of Harrow Tooth Preform, Materials 12 (2019) 2287-2287. https://doi.org/10.3390/ma12142287

[2] P.A. Li, B.Y. Wang, C. Yang, H. Zhang, W. Li, J. Zhou, Study on necking defects, microstructure and mechanical properties of TC4 alloy by cross wedge rolling, Int. J. Mater. Forming 15 (2022). https://doi.org/10.1007/s12289-022-01718-z

[3] Z.M. Yu, W.F. Peng, X. Zhang, M. Oleksandr, V. Titov, Evolution of microstructure of aluminum alloy hollow shaft in cross wedge rolling without mandrel, J. Central South University 29 (2022) 807-820. https://doi.org/10.1007/s11771-022-4950-8

[4] X. Ren, Y.M. Huo, S.R.E. Hosseini, T. He, Z. Yan, F.A.O. Fernandes, A.B. Pereira, H. Ji, J. Bai, Z. Bian, X. Du, A multi-scale modelling by coupling cellular automata with finite element method and its application on cross-wedge rolling, Materials Today Commun. 37 (2023) 106976. https://doi.org/10.1016/j.mtcomm.2023.106976

[5] T. Bulzak, Assessment of the Susceptibility to Material Fracture in the Cross-Wedge Rolling Process with Concave Tools, Materials 15 (2022) 6605-6605. https://doi.org/10.3390/ma15196605

[6] A.Q. Tian, X.H. Xu, L. Sun, Y.L. Deng, Effects of interrupted ageing and asymmetric rolling on microstructures, mechanical properties, and intergranular corrosion behavior of Al-Mg-Si-Zn alloy, J. Central South University 29 (2002) 821-835.

[7] Z. Pater, T. Bulzak, J. Tomczak, Cross-Wedge Rolling of Driving Shaft from Titanium Alloy Ti6Al4V, Key Eng. Mater. 3939 (2016) 687-687. https://doi.org/10.4028/www.scientific.net/KEM.687.125

[8] Z. Pater, J. Tomczak, Experimental Tests for Cross Wedge Rolling of Forgings Made from Non-Ferrous Metal Alloys, Arch. Metall. Mater. 57 (2012) 919-928. https://doi.org/10.2478/v10172-012-0101-9

[9] Z. Pater, J. Tomczak, T. Bulzak, Rotary compression as a new calibration test for prediction of a critical damage value, J. Mater. Res. Tech. 9 (2020) 5487-5498. https://doi.org/10.1016/j.jmrt.2020.03.074

[10] P.A. Li, B.Y. Wang, P.N. Feng, J. Shen, J. Wang, Numerical and experimental study on the hot cross wedge rolling of Ti-6Al-4V vehicle lower arm preform, Int. J. Adv. Manuf. Tech. 118 (2021) 3283-3301. https://doi.org/10.1007/s00170-021-07979-3

[11] L.F. Lin, W.F. Peng, S. Zhu, Z. Wu, J. Zhu, Y. Shao, H. Li, M. Oleksandr, V. Titov, Cross wedge roll bonding process for laminated shafts forming: Interface microstructure, bonding mechanism, and parameter influence, J. Mater. Process. Tech. 317 (2023). https://doi.org/10.1016/j.jmatprotec.2023.117971

Effects of forming conditions on limit of drawing ratio during cold drawing and ironing process of Ti-6Al-4V alloy sheets

OKUDE Yusuke[1,a *], IWAOKA Taku[1,b], NAKAMURA Isao[1,c], MURAOKA Tsuyoshi[1,d] and KATAGIRI Takashi[1,e]

[1]Tokyo Metropolitan Industrial Technology Research Institute, 2-4-10, Aomi, Koto-ku, Tokyo, Japan

[a]okude.yusuke@iri-tokyo.jp, [b]iwaoka.taku@iri-tokyo.jp, [c]nakamura.isao@iri-tokyo.jp, [d]muraoka.tsuyoshi@iri-tokyo.jp, [e]katagiri.takashi@iri-tokyo.jp

Keywords: Sheet Metal Working, Drawing and Ironing, Warm Forming, Cold Forming, Press Motion

Abstract. The low ductility of Ti–6Al–4V alloy at forming temperatures below 780°C is still a problem. Ti–6Al–4V is usually hot press formed at 780 to 900°C. A hot press formed Ti–6Al–4V alloy has been utilized in various devices such as medical and aerospace equipment. However, hot press forming increases cost, especially because it uses a furnace. To reduce cost, hot stamping by electric heating has been applied to titanium sheets or continuous resistance heating has been applied to the bar forming of Ti–6Al–4V alloy. Both electric heating and continuous resistance heating reduce the heating time in the forming temperature range from 780 to 900°C. Hence, it is necessary to develop a new press forming for Ti–6Al–4V alloy sheets at forming temperatures below 780°C to improve productivity. We have developed a method to achieve of cold press forming of Ti–6Al–4V alloy sheets by combining press motion control and clearance between a die and a blank. In this study, we tried to form various forming conditions for developed press forming method. In this method, the following two processes were separated: punch motion process without BHF and BHF motion process without punch motion. These two processes were carried out alternately during deep drawing or drawing and ironing. Three forming parameters, namely, the punch speed S_{pi}, punch stroke P_{si}, and blank holding force BHF_i, were controlled separately and the effect of forming conditions on deformation behavior was investigated by using experiment and finite element analysis(FEA). As a result, the effects of forming conditions on limit of drawing ratio of Ti–6Al–4V alloy sheets were investigated.

Introduction

Ti-6Al-4V titanium alloy has been actively used in medical equipment and aero-space due to its characteristics of light weight, corrosion resistance, and high specific strength [1–3]. A hot press formed Ti–6Al–4V alloy has been utilized in various devices such as medical and aerospace equipment [4]. However, hot press forming increases cost, especially because it uses a furnace. To reduce cost, hot stamping by electric heating has been applied to Ti sheets or continuous resistance heating has been applied to the bar forming of Ti–6Al–4V alloy. Both electric heating and continuous resistance heating reduce the heating time in the forming temperature range from 780 to 900°C [5,6]. Hence, it is necessary to develop a new press forming for Ti–6Al–4V alloy sheets at forming temperatures below 780°C to improve productivity.

We developed a method to achieve of warm and cold press forming of Ti–6Al–4V alloy sheets by combining press motion control and clearance between a die and a blank [7,8]. In this study, we tried tube forming by developed press forming method. In this method, the following two processes were separated: punch motion process without BHF and BHF motion process without punch motion. These two processes were carried out alternately during deep drawing or drawing and ironing. Three forming parameters, namely, the punch speed S_{pi}, punch stroke P_{si}, and blank

Materials Research Forum LLC
https://doi.org/10.21741/9781644903254-15

holding force BHF_i, were controlled separately and the effect of forming conditions on deformation behavior was investigated by using experiment and finite element analysis(FEA). As a result, the LDR of Ti–6Al–4V alloy sheets in drawing and ironing was investigated by applying the developed incremental press forming method.

Experimental apparatus and formability of Ti-6Al-4V alloy sheet at warm temperature of deep drawing test

In order to understand the basic formability of cylindrical deep drawing of Ti-6Al-4V alloy sheet, cylindrical deep drawing with a general blank holding force (BHF) was first performed. Fig. 1 shows a schematic of deep drawing process, and Table 1 the dimensions of the dies of deep drawing process. The die material used was SKH51. Ti-6Al-4V alloy sheet and a blank with thickness of $t_0 = 0.5$mm were used as test materials. The test was performed with a blank diameter of ϕ 60 mm (Drawing ratio, $DR = 1.85$). Cylindrical deep drawing was carried out at the condition of 300°C forming temperature. Test was performed at the same temperature as the blank for the drawing die and the wrinkle control plate. The punch was kept at room temperature under all conditions. The punch speed S_p was set to three conditions of 9, 90 and 900 mm/min, and the constant wrinkle suppressing force BHF_{const} was applied at 4 kN. Fig. 2 shows the punch load—punch stroke curve and photographs of cups drawn under each set of conditions. In the each condition, fracture occurred at punch shoulder. Therefore, the low formability of Ti-6Al-4V alloy sheet was confirmed.

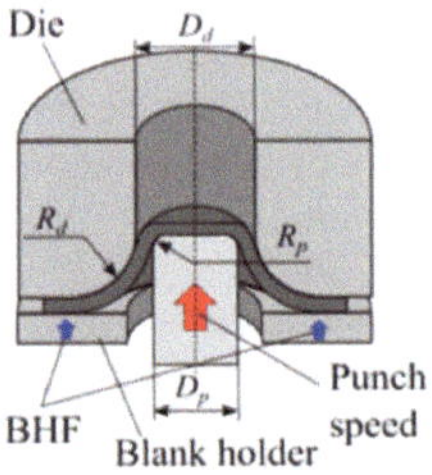

Fig. 1. Schematic of deep drawing process .

Table 1. Dimension of dies of deep drawing process.

Die	Diameter : D_d [mm]	35.0, 33.8, 31.6
	Shoulder radius : R_d [mm]	3.0
Punch	Diameter : D_p [mm]	32.5, 33.0, 30.8
	Shoulder radius : R_p [mm]	5.0

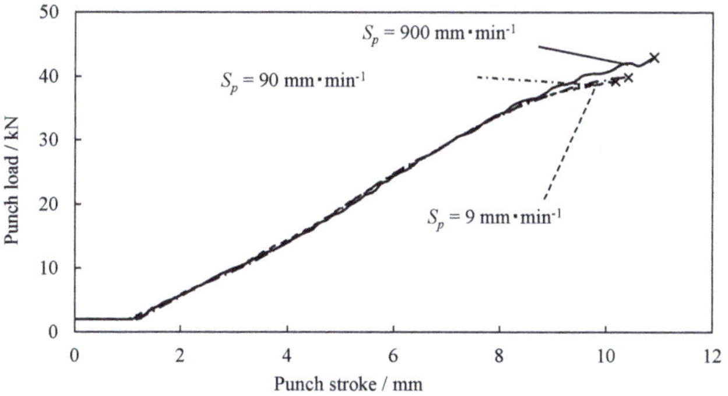

(a) Punch load — punch stroke curve

Conditions	Punch speed S_p [mm/min]		
	9	90	900
Photo. of drawn cups	 Occurrence of fracture at punch shoulder	 Occurrence of fracture at punch shoulder	 Occurrence of fracture at punch shoulder

(b) Photographs of cups drawn under each set of conditions

Fig. 2. Effects of punch speed on deformation behaviour of Ti-6Al-4V alloy sheet at 300°C.

Incremental press forming process method and formability of Ti-6Al-4V alloy sheet

In cylindrical deep drawing with a constant blank holding force, cracks occurred at the punch shoulder at any punching speed at a forming temperature of 300°C. In this study, we focused on the fact that wall thinning progressed due to tension due to wrinkle suppressing force from the start of the punch until the maximum punch load was reached during press forming. When the material is pressed into the mold with a punch under the condition of applying wrinkle suppressing force, the thickness of the punch shoulder is significantly reduced. Therefore, in order to perform press forming in the sequential forming developed in this study, two processes were separated: a punch motion process and BHF without punch motion wrinkle force loading process. Fig.3 shows the schematic of incremental press forming.

To confirm the effect of developed incremental press forming on deformation behavior, the warm and cold drawing and ironing of Ti-6Al-4V alloy sheet were carried out. In the incremental press forming, warm and cold drawing and ironing experiment was performed under the conditions of S_{pi} =90 mm/min, P_{si} =0.4 mm, and BHF_i =20 kN. The forming temperature 300°C and room temperature was utilized. Die dimensions D_d = 33.8 mm and D_p = 33.0 mm were selected for drawing and ironing. For lubrication, the surface oxide film of Ti-6Al-4V alloy was focused on. Ti-6Al-4V alloy sheet was treated with atmospheric oxide film at the conditions of 500°C for 2 hours before experiment. And for comparison, a cylindrical drawing was performed under the same lubricating conditions with a punch speed of 90 mm/min and blank holding force of 4 kN. Fig. 4 shows the effect of incremental press forming on formability of Ti-6Al-4V alloy sheet in warm drawing and ironing, and Fig. 5 the effect of incremental press forming on formability of Ti-6Al-4V alloy sheet in cold drawing and ironing. By the application incremental press forming method, the warm and cold drawing and ironing of Ti-6Al-4V alloy sheets was achieved.

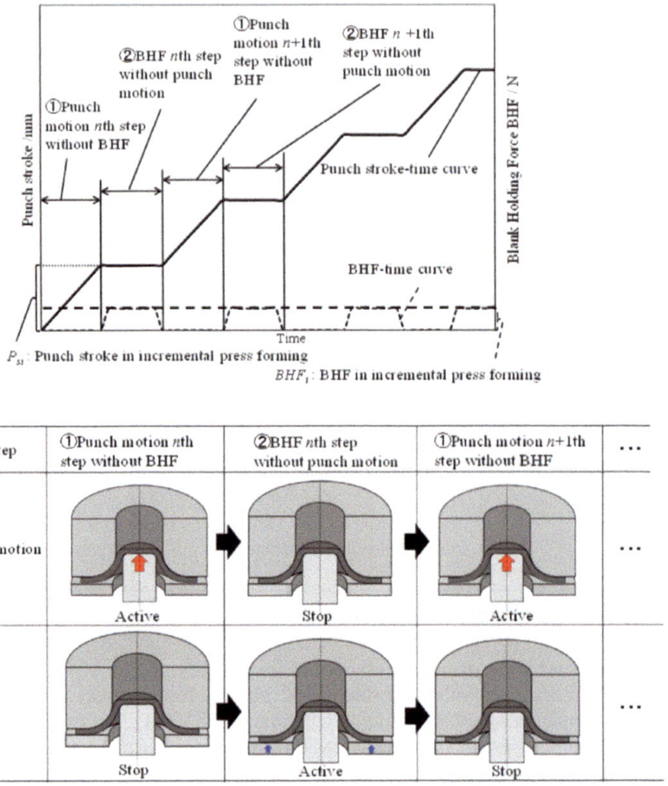

Fig. 3. Schematic of incremental press forming.

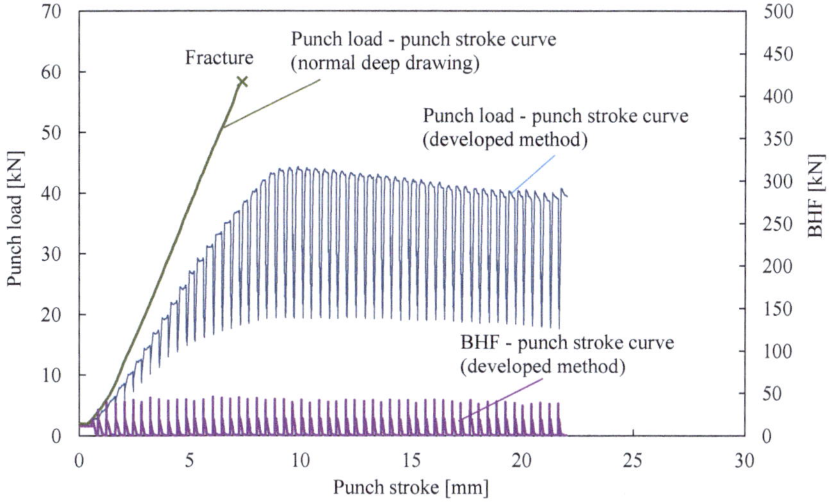

(a) Punch load－punch stroke curve

Normal deep drawing Incremental press forming

(b) Photographs of cups drawn under each set of conditions

Fig. 4. Effect of incremental press forming on deformation behavior at 300°C.

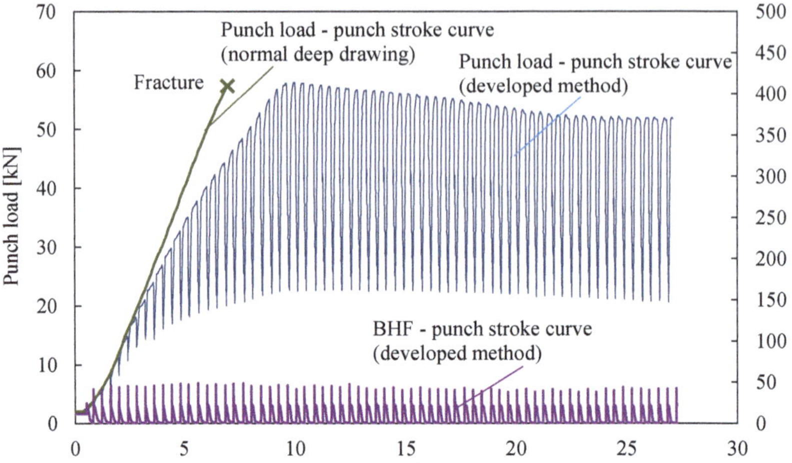

(a) Punch load—punch stroke curve

Normal deep drawing Incremental press forming

(b) Photographs of cups drawn under each set of conditions

Fig. 5. Effect of incremental press forming on deformation behavior at room temperature.

Drawing and ironing by incremental press forming and formability of Ti–6Al–4V alloy sheets to clarify the effect of the incremental press forming, FEA of drawing and ironing of Ti–6Al–4V alloy sheets was calculated. The finite element analysis code LS-DYNA3D was used for drawing and ironing simulation. Representative finite element models of workpieces (Ti–6Al–4V,t_0 = 0.5 mm, blank diameter = ϕ 60 mm) were constructed. Table 2 indicates the properties of the sheet materials of Ti–6Al–4V alloy sheets (room temperature). An elastoplastic material was assumed, satisfying an n power strain hardening law with the constitutive equation as follows,

$$\sigma = C(\varepsilon_n + \overline{\varepsilon_p})^n . \qquad (1)$$

where, C is plastic modulus, ε_n elastic strain, $\overline{\varepsilon_p}$ equivalent plastic strain, n work hardening exponential.

Fig. 6 shows the schematic and FEA model of deep drawing process and Table 3 the dimensions of dies. In FEA, the friction coefficients between dies and blank are set to 0.1 without lublication(punch and blank, blank holder and blank), and 0.05 with lubrication (die and blank). In the incremental forming, press forming was performed under the conditions of P_{si} = 0.4 mm, and BHF_i = 20 kN.

Fig.7 shows results of drawing and ironing process of Ti–6Al–4V alloy sheets (shear strain distribution, P_{si} = 0.4 mm, BHF_i = 20kN). At the point I, the workpiece was formed for the bottom of drawn cup by drawing process only. Hence, the punch load gradually increased until starting the drawing and ironing (point II). When the starting point of drawing and ironing is reached, the punch load increased sharply. And shear strain caused by ironing process occurred at the thickness of workpiece near the die radius end. Then, shear strain increased with the progress of ironing process as shown in Fig. 7 (point III). As the results, the drawing and ironing process was calculated by FEA, and ironing due to shear strain could be estimated by simulation as shown in Fig. 8 [8].

Table 2. Material property of blank at room temperature.

Material	Ti–6Al–4V
Density, ρ [kg/mm^3]	4.61×10^3
Modulus of elasticity E [GPa]	106.3
Poisson's ratio	0.3
C value [MPa] *	1724
n value*	0.19

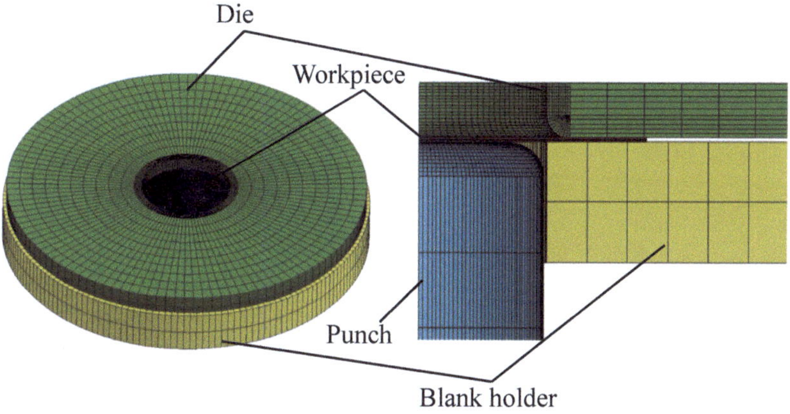

Fig. 6. Effect of incremental press forming on deformation behavior at room temperature.

Materials Research Forum LLC
https://doi.org/10.21741/9781644903254-15

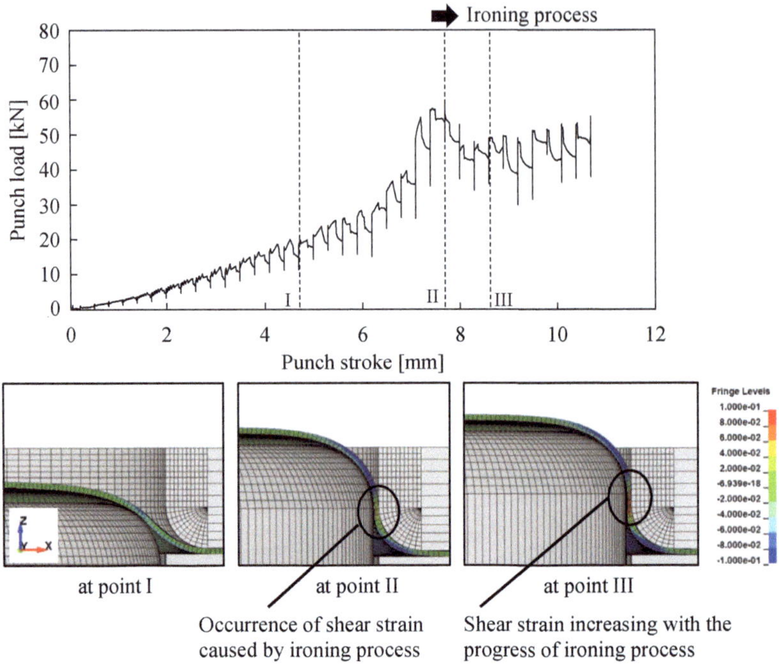

Fig. 7. FEA results of drawing and ironing process of Ti-6Al-4V alloy sheets (Shear strain distribution, $t_0 = 0.5$ mm, $P_{si} = 0.4$ mm, $BHF_i = 20$ kN).

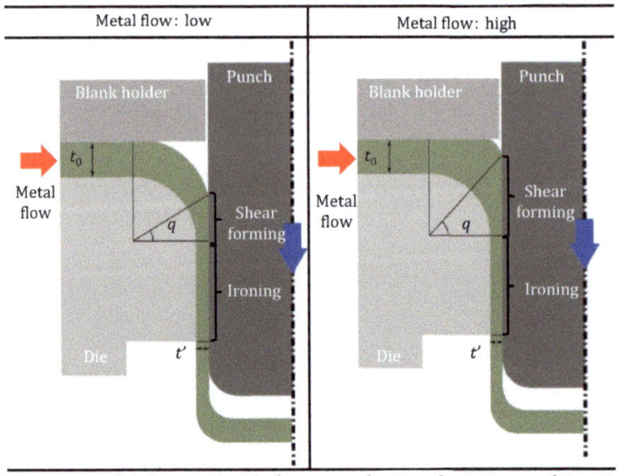

Fig. 8. Schematic of deformation behaviour during drawing and ironing [9].

Limit of Drawing Ratio During Cold Drawing and Ironing Process of Ti-6Al-4V Alloy Sheets
Finally, the Limit of Drawing Ratio (LDR) during cold drawing and ironing process were investigated. Ti-6Al-4V alloy sheet with thickness of t_0 = 0.5 mm were used. The eperiment was performed with a blank diameter of ϕ43, 50, 55 and 60 mm (*DR* = 1.4, 1.67, 1.79,). Die dimensions D_d = 31.6 mm and D_p = 30.8 mm were selected. The conditions of incremental press forming S_{pi} =90 mm/min, P_{si} =0.4 mm, and BHF_i =10, 20, 30, 40 kN were utilized. Fig. 9 shows the punch load – punch stroke curves and photographs of formed cups during drawing and ironing at the conditions ϕ43 (*DR* = 1.4) and BHF_i =30 kN. And Fig. 10 LDR of Ti-6Al-4V alloy sheets in cold drawing and ironing(D_d = 31.6 mm, D_p = 30.8 mm).

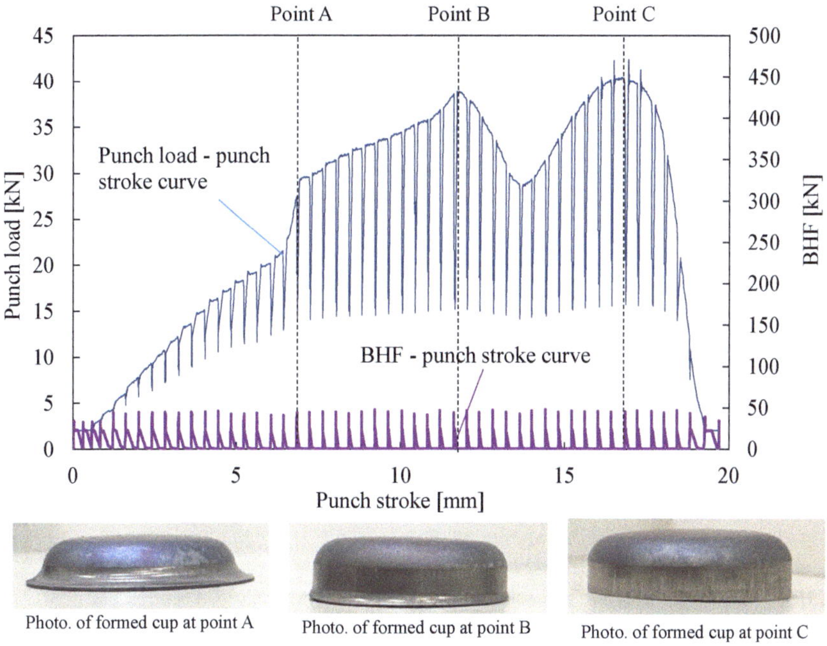

Fig. 9. punch load – punch stroke curves and photographs of formed cups during drawing and ironing at the conditions at room temperature (Blank diameter ϕ43, DR = 1.4 and BHF_i =30 kN).

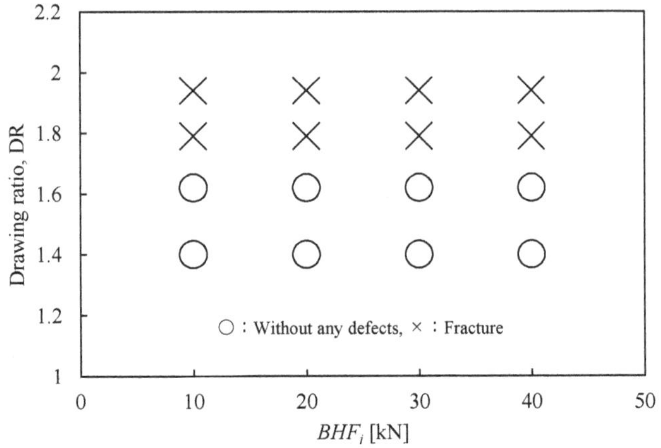

Fig. 10. LDR of Ti-6Al-4V alloy sheet in drawing and ironing at room temperature (D_d = 31.6 mm, D_p = 30.8 mm).

Summary

1) By the application incremental press forming method, the warm and cold drawing and ironing of Ti-6Al-4V alloy sheets was achieved.
2) The drawing and ironing process was calculated by FEA, and ironing due to shear strain could be estimated by simulation.
3) This study indicates the LDR during of Ti-6Al-4V alloy sheets in cold drawing and ironing (*Dd* = 31.6mm, D_p = 30.8mm).

References

[1] T. Maeno, K. Mori, Z. Hamedon, Proc. Manuf., 14th International Conference on Metal Forming 2012, 287-290.

[2] T. Maeno, K. Mori, Y. Yamashita, M. Nakayama, Z. Hamedon, Proc. 2011 Jpn. Spring Conf. Technol. Plast. 2011, 345-346.

[3] T. Maeno, K. Mori, Y. Yamashita, Proc. 2012 Jpn. Spring Conf. Technol. Plast. 2012, 67-68.

[4] D.J. Beal, R. Boyer, D. Sanders, ASM Handbook, 14B, Metalworking: Sheet Forming, 2006, 656-669.

[5] K. Yamazaki, T. Yamaguchi, Y. Murayama, K. Mori, T. Maeno, Jpn. Soc. Technol. Plast. 55 (2014) 1097-1101.

[6] J. Yanagimoto, R. Izumi, Continuous electric resistance heating—Hot forming system for high-alloy metals with poor workability, J. Mater. Process. Technol. 209 (2009) 3060-3068. https://doi.org/10.1016/j.jmatprotec.2008.07.010

[7] Y. Okude, T. Iwaoka, I. Nakamura, T. Katagiri, Development of Warm Forming Method for Ti-6Al-4V Alloy Sheets with Press Motion Control, J. Jpn.Soc. Technol. Plast. 61 (2020) 159-164. https://doi.org/10.9773/sosei.61.159

[8] Y. Okude, T. Iwaoka, I. Nakamura, T. Katagiri, T. Muraoka, J. Jpn.Soc. Technol. Plast. 62 (2021) 133-138.

[9] Meth. Eng. 37 (1994) 3323-3341.

Metal Forming 2024 Materials Research Forum LLC
Materials Research Proceedings 44 (2024) 144-150 https://doi.org/10.21741/9781644903254-16

Study on microstructural behavior of aluminum alloy thin-walled tube necking formed by three-roll skew rolling

CHEN Y.S.[1,a], SHU X.D.[1,b*], XIA Y.X.[1,c], HU J.[1,d], LI Z.X.[1,e], XU H.J.[1,f], ZHANG S.[2,g]

[1] Zhejiang Provincial Key Laboratory of Part Rolling Technology, Faculty of Mechanical Engineering and Mechanics, Ningbo University No.818, Fenghua Road, Jiangbei District, Ningbo, Zhejiang Provinc, China

[2] College of Mechanical and Automotive Engineering, Ningbo University of Technology No.201, Fenghua Road, Jiangbei District, Ningbo, Zhejiang Provinc, China

[a] 2211090084@nbu.edu.cn, [b] shuxuedao@nbu.edu.cn, [c] 2001090016@nbu.edu.cn, [d] 1569502208@qq.com, [e] lizixuan@nbu.edu.cn, [f] xuhaijie@nbu.edu.cn, [g] zhangsong@nbut.edu.cn

Keywords: Three-Roll Skew Rolling, Tube, Necking, Secondary Development, Dynamic Recrystallisation

Abstract. To address the issue of incomplete formation of low necking coefficient pipe fittings in hot extrusion processes, this study adopts a novel three-roll skew rolling (TRSR) technique to investigate the microstructural recrystallization distribution patterns during necking forming of aerospace thin-walled pull rods using TRSR. The principles of TRSR necking forming of pipe fittings are initially elucidated, followed by the establishment of a finite element model using simufact forming software and secondary development of the software to embed dynamic recrystallization volume and grain size models. The study analyzes the temperature distribution patterns and dynamic recrystallization behavior during necking forming of aluminum alloy thin-walled pipes via TRSR, aiming to contribute to the research on high-quality necking pipe fittings produced through TRSR.

Introduction

Rod-like fittings are important load-bearing connecting components for aircraft. Traditional processes for manufacturing such necking fittings mainly involve hot extrusion [1], which can lead to defects such as wrinkling, instability, and fracture in the manufactured fittings [2]. Guo et al. [3] analyzed the thickness distribution law by bidirectional extrusion and achieved a maximum deformation necking coefficient of 0.68 by utilizing the temperature gradient coefficient. Yi et al. [4] utilized electric-assisted extrusion to increase the thickness of fittings by up to 300%. However, research on necking forming of these fittings has mainly focused on low necking coefficient fittings obtained through two-stage hot extrusion, resulting in low manufacturing efficiency and the need for improved forming quality. In recent years, researchers such as Pater have employed three-roll skew rolling technology to manufacture axisymmetric components like hollow shafts for trains [5]. Shu [6] conducted a microstructure analysis of the forming process of a 30CrMoA hollow shaft using three-roll skew rolling, revealing the microstructural evolution of the formed part in various deformation stages. Wang et al. [7] discussed the wall thickness quality of three-roll skew rolling (TRSR) hollow shafts from the perspectives of forming mechanism and experimental design. Ye et al. [8] innovatively applied the three-roll skew rolling process to achieve thin-walled fitting necking and thickness increase, analyzing the influence of initial wall thickness on the quality of the formed part through finite element simulation. Xia et al. [9] verified the feasibility of forming aluminum alloy thin-walled fittings using three-roll skew rolling through roller profile design and finite element simulation. However, no one has yet analyzed the microstructural changes in necking forming of thin-walled fittings using three-roll skew rolling. In the rolling process, the more pronounced the refinement of grain size inside the rolled part, the better the quality of the formed part. This paper provides further guidance for high-quality necking fittings formed by three-roll skew rolling through analysis of temperature distribution laws during rolling and dynamic recrystallization grain size and volume fraction.

Principle of three-roll skew rolling forming and necking coefficient

The principle of necking forming for fittings using three-roll skew rolling is shown in Figure 1. It consists of three rolls and a traction clamp arranged in a clover-shaped pattern and offset by 8° from the center axis. The rolls rotate synchronously with the same direction, causing the rolled material to rotate in the opposite direction and giving it a certain axial movement. However, to ensure forming accuracy, the rolled material is also axially moved by the clamp while being synchronously rolled by the upper and lower rolls. In this way, the rolling of the fittings is completed.

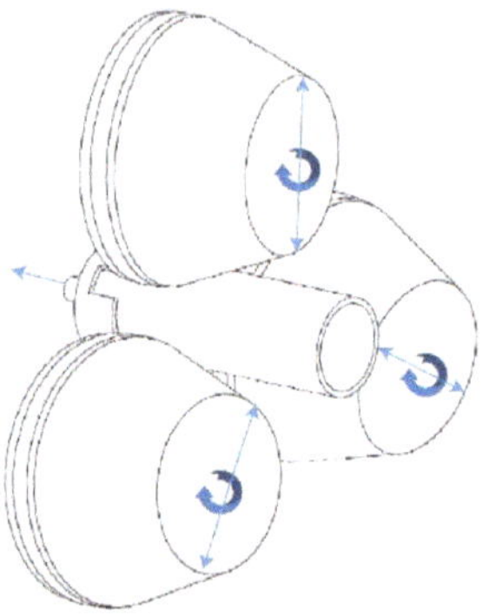

***Figure 1**. The principle of three-roll skew rolling forming*

In order to ensure the reasonable necking coefficient β of the pipe fitting and to guarantee the stable rolling process, there are certain limitations on the vertical movement of the rolls during the rolling process. The distance of roll movement x is related to the roll diameter D and the inner diameter d of the pipe fitting.

The geometric analysis of Figure 1 leads to the following deduction:

$$x < \frac{(\sqrt{3}-2)D}{2} - d \tag{1}$$

$$\beta = \frac{2x}{d} \tag{2}$$

Finite element model for necking forming of three-roll skew rolled pipe

Model establishment.

Firstly, a three-roll skew rolled pipe necking finite element model can be established by creating a 3D model using Solidworks software and saving it in STL format, then importing the model into Simufact Forming finite element software. The model consists of three parts: rolls, billet, and fixture, where the billet and fixture are set to be bonded while the rest are in contact. The billet has dimensions of 32mm × 100mm with a wall thickness of 2mm, and the angle between the roll axis and the billet axis is set to 8°. The roll speed n is 60 rpm, and the friction between the roll and billet is assumed to be shear friction with a coefficient of 0.9. The billet is made of 6061 aluminum alloy and is meshed using sheetmesh. Considering the thin wall thickness of the pipe, four layers of mesh are generated at the wall thickness location, resulting in a total of 8035 initial mesh elements. The empirical constitutive equation for the billet is [10]:

$$\dot{\varepsilon} = 1.1148 \times 10^{21}[\sinh(0.01877\sigma)]^{6.2813} \times exp\left[\frac{-307.528 \times 10^3}{RT}\right]$$

In this formula,
$\dot{\varepsilon}$ is the strain rate (s^{-1});
σ is the stress (Pa);
T is the temperature (°C);
the ideal gas law constant is R = 8.31J/(mol·K).

During the three-roll skew rolling process, the workpiece mainly undergoes plastic deformation. Due to the small amount of elastic deformation, the workpiece is defined as a plastic body, while the roll assembly and fixture are defined as rigid bodies. The initial temperature of the workpiece is set to 300°C, the roll temperature is set to 150°C, the fixture is set to 20°C, and the initial environmental temperature is set to 20°C.

3.2 Secondary development of materials based on empirical ontological modelling

By investigating the relationship between dynamic recrystallization grain size and recrystallization volume fraction model [11] and the deformation temperature, strain rate, as well as strain magnitude in the material forming process, a microstructure evolution model is established for the necking forming process of aluminum alloy thin-walled tube by three-roll skew rolling. Additionally, through secondary development, the model is embedded into simulation for the accurate simulation of dynamic recrystallization in aluminum alloy hot rolling forming process.

Simufact software conducts secondary development by compiling subroutines using the Fortran language. This development integrates dynamic recrystallization grain size model and volume fraction model for 6061 aluminum alloy. It also customizes user variables to facilitate post-processing result analysis.

***Table 1.** Custom variables and post-processing results*

Unit variable number	Variable symbol	Reprocessing results
1	DGR	Dynamic recrystallisation grain size
2	DRX	Dynamic recrystallisation volume fraction

Call ueloop and plotv subroutine programming, the dynamic recrystallisation grain size model and recrystallisation volume fraction model will be written into the ueloop subroutine, the parameters are read in a continuous loop, and finally the results of the relevant variables are output through the post-processing of the plotv subroutine, which is partially compiled subroutine as follows:

```
subroutine ueloop(m1,m2,iflag)
implicit logical (a-z)
ccccCalling public module parameters
#include "hards.cmn" !nstats
#include "creeps.cmn" !timinc
#include "arrays.cmn" !iintel
..........
cccDefine initial grain size
dimension grndat(1)
grndat(1)=50
.........
```

```
cccCalculation of dynamic recrystallisation grain size
drg=119.28*eplas**0.632*erate**(-0.107)*exp((-16540.8)/((incret-273)*8.314))
.........
cccVariable Result Output
subroutine plotv(v,s,sp,etot,eplas,ecreep,t,m,nn,kcus,ndi,nshear,jpltcd)
.........
if(jpltcd.eq.4) v=t(4) !1st user state variable
.........
```

Finite element result analysis

Temperature field distribution of rolled parts.

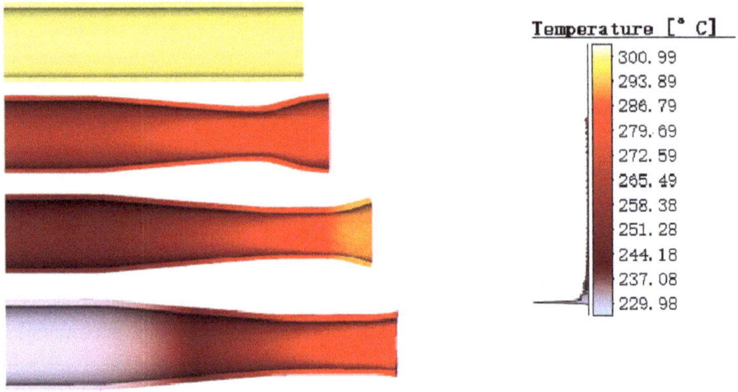

Figure 1*. Temperature of longitudinal section of the rolled part at each stage*

Figure 2 illustrates the temperature field distribution across the longitudinal cross-sections of the rolled component at different stages. During the rolling of the billet, the temperature decrease is most rapid in the contact area with the fixture, followed by the undeformed end regions, due to the heat conduction from the fixture and the rolling environment. During the necking forming rolling, the temperature of the component gradually decreases as the rolling proceeds, with a decrease of up to 30°C. Once the necking forming is completed, the temperature of the neck region stabilizes at 270°C. As the end of the component is formed, there is a slight increase in temperature within a range of 20°C due to the reduction in necking coefficient. Upon completion of the rolling process, the left end of the component, being in contact with air and the fixture, experiences significant heat dissipation.

Microstructure Distribution in the Necking Forming Section of the Rolled Component.

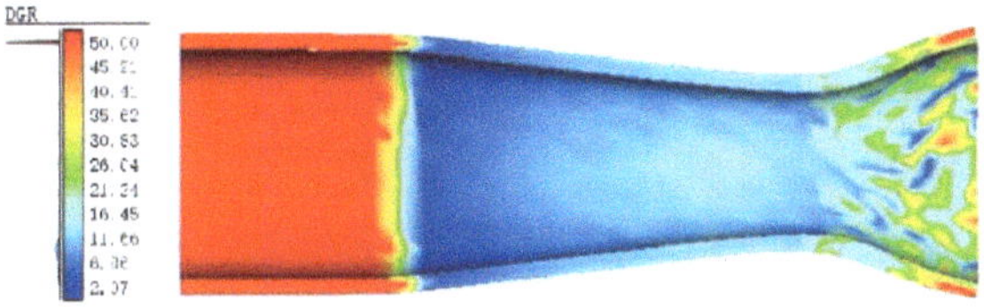

(a) Distribution of Recrystallized Grain Sizes across the Longitudinal Cross-Section of the Rolled Component

(b) Distribution of Recrystallization Volume Fraction across the Longitudinal Cross-Section of the Rolled Component

Figure 3*. Distribution of Dynamic Recrystallization Grain Sizes and Volume Fraction during the Necking Forming Stage*

Figure 3 illustrates the distribution of dynamic recrystallization grain sizes and dynamic recrystallization volume fraction across the longitudinal cross-section of the necking forming stage of the rolled component. During this stage, the dynamic recrystallization volume fraction is relatively low at the beginning of the rolling process and gradually increases. A significant grain refinement phenomenon occurs in the necking section of the component, where the grain size decreases from 50μm to as low as 8μm. Dynamic recrystallization mainly occurs in the necking forming area of the component, leading to noticeable grain refinement in the forming region, which gradually permeates from the surface to the core and inner wall of the component. On the right side of the component and in the undeformed end region, the grain size remains at 50μm. The boundary between the deformed and undeformed regions shows a distinct change in grain size, indicating conformity with dynamic recrystallization conditions.

During this stage, the deformation of the component gradually increases, accompanied by an increase in strain rate. The high distortion energy of grain boundaries activates dynamic recrystallization to a greater extent. As a result, the volume fraction of dynamic recrystallization continuously increases during the necking forming stage. By the completion of necking rolling, the recrystallization volume fraction has reached 60%, with even higher values observed on the surface of the component. This is consistent with the conditions conducive to the growth of recrystallized grain sizes.

Microstructure Distribution in the End Forming Section of the Rolled Component.

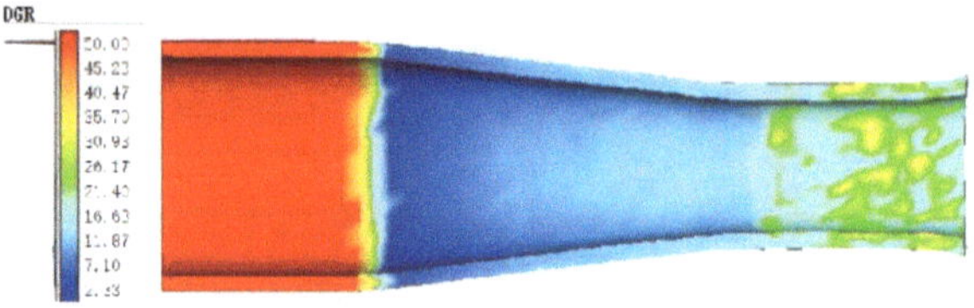

(a) Distribution of Recrystallized Grain Sizes across the Longitudinal Cross-Section of the Rolled Component

(b) Distribution of Recrystallization Volume Fraction across the Longitudinal Cross-Section of the Rolled Component

***Figure 4**. Microstructural Distribution of Dynamic Recrystallization Grain Sizes and Volume Fraction during the End Forming Stage*

Figure 4 illustrates the distribution of dynamic recrystallization grain sizes and volume fraction across the longitudinal cross-section of the end forming stage of the rolled component. During this stage, as the necking coefficient of the component gradually decreases, the strain increment at the end of the component increases, leading to higher temperatures and intense plastic deformation. Consequently, there is a more pronounced increase in recrystallization grain sizes, with the maximum grain size reaching 34μm at the surface in contact with the rolling mill rolls, and the maximum recrystallization volume fraction reaching 0.97. Upon completion of the rolling process, the dynamic recrystallization grain size refines from an initial 50μm to 10μm, achieving a grain refinement level of 80%. The distribution pattern of dynamic recrystallization grain sizes and volume fraction remains largely consistent throughout this stage.

Conclusion

(1) In the necking forming process of aluminum alloy thin-walled tubes using three-roll skew rolling, the temperature variation is relatively small. However, due to heat conduction from fixtures and air, there is rapid heat dissipation in both the formed and unformed stages of the rolled component. During the rolling stage, the temperature variation of the component is minor, remaining close to the initial temperature.

(2) Three-roll skew rolling for necking forming not only achieves low necking coefficients in a single pass but also results in significant grain size refinement of the component. The degree of grain refinement reaches 80%, thereby enhancing the mechanical properties of the thin-walled tubes.

Acknowledgments

This study was funded by the National Natural Science Foundation of China (grant number: 52375346) and the Ningbo Major Special Project (grant number: 2022Z064).

References

[1] Qiu Z Y, Xu X F, Fu C L, Xiao J and Xiong G L 2019 Experimental study and optimisation of shrinkage and thickening forming of 5A02 aluminium alloy thin-walled tubes and pipes by hot extrusion. Journal of Plasticity Engineering. 26 7-14.

[2] Li Z Z, Er Q Q, Wang B S, Su C, Yang L and Dong X J 2018 Influence and control of hot extrusion process on the organisation of 690 alloy. Rare Metal Materials and Engineering. 47 3776-3783.

[3] Guo Y Q, Xu C G, Han J T and Wang Z Y 2020 New technique of precision necking for long tubes with variable wall thickness. Frontiers of Mechanical Engineering. 15 622-630. https://doi.org/10.1007/s11465-019-0565-7

[4] Yi H, Xue F X, Li X D and Zhong G H 2022 Numerical simulation and experimental study of electrically assisted 5A02 thin-walled tube shrinkage thickening forming. Journal of Plasticity Engineering. 29 27-32.

[5] Pater Z 2017 FEM Analysis of loads and torque in a skew rolling process for producing axisymmetric parts Arch. Archives Of Metallurgy And Materials. 62 85–90. https://doi.org/10.1515/amm-2017-0011

[6] Shu C, Zhang S, Bidare P, Essa K, Abdel-Wahab, Shu X D et al 2022 Microstructure evolution of three-roll skew-rolling formed hollow axles with uniform wall thickness. International Journal Of Advanced Manufacturing Technology.121 4069-4085. https://doi.org/10.1007/s00170-022-09583-5

[7] Wang J T, Shu X D, Ye C Q, Li Z X, Li S Xe et al 2023 Study on forming quality of three-roll skew rolling hollow axle. International Journal Of Advanced Manufacturing Technology. 128 1089-1100. https://doi.org/10.1007/s00170-023-11893-1

[8] Ye B H, Shu X D, Xia Y X 2022 Wall thickness variation rule of three-roll slant rolling reduction section of aviation thin-walled pipe fittings. Journal of Ningbo University (Science and Technology Edition). 35 57-62.

[9] Xia Y X, Shu X D, Zhang S, Wang J T, Ye C Q 2022 Research On Integrated Forming Process Of Three-roll Reducing And Thickening Of Necking Section For Thin-walled Tube Of Aluminum Alloy 2A12.Metalurgija. 61 733-736.

[10] Qi Y J, Yu X P, Lv H Y, Pang G Y 2018 Dynamic Recrystallization Behavior of 6061 Aluminum Alloy. Special Casting & Nonferrous Alloys. 38 1185-1188.

[11] Mou Q J and Chen B J 2016 Analysis of dynamic recrystallisation during forging and forming of 6061 aluminium alloy wheels. Foundry technology. 37 2711-2713.

Materials Research Proceedings 44 (2024) 151-157 https://doi.org/10.21741/9781644903254-17

Development and industrial implementation of a biodegradable surgical wire production technology from ZnMg0.004 alloy: *In vitro* assessment of degradation rate and mechanical property changes

MILENIN Andrij[1,a *], KUSTRA Piotr[1,b], PAĆKO Marek[1,c], BYRSKA-WÓJCIK Dorota[1,d], SULEJ-CHOJNACKA Joanna[2,e], PÓŁROLNICZAK Joanna[2,f], BARWIŃSKI Roman[3,g] and KOPAŁA Marcin[3,h]

[1]AGH University of Kraków, al. Mickiewicza 30, 30-059 Kraków, Poland

[2]Łukasiewicz Research Network – Poznań Institute of Technology, Estkowskiego 6, 61-755 Poznań, Poland

[3]INTERMECH Sp. z o.o., Przemysłowa 9B, 37-450 Stalowa Wola, Poland

[a]milenin@agh.edu.pl, [b]pkustra@agh.edu.pl, [c]packo@agh.edu.pl, [d]byrska@agh.edu.pl, [e]joanna.sulej-chojnacka@pit.lukasiewicz.gov.pl, [f]joanna.polrolniczak@pit.lukasiewicz.gov.pl, [g]r.barwinski@intermech.com.pl, [h]m.kopala@intermech.com.pl

Keywords: Wire Drawing, Biodegradable Material, Surgical Wire, ZnMg Alloy, In Vitro Biocorrosion

Abstract. The paper discusses the technology for the production of the new biodegradable surgical wire, developed at AGH, along with the experience of its integration into INTERMECH's production processes. This wire's basic mechanical and corrosion properties are thoroughly examined. The developed technology is based on the use of a combination of hot and cold (in the last passes) drawing of a workpiece obtained by hot extrusion. Hot drawing allows to achieve the necessary ductility of the wire, while the final cold drawing improves the surface quality, which is important in the context of the biological properties of the wire. To assess biodegradation rates, *in vitro* testing was conducted within a natural corrosive environment (using bovine serum from Biowest). Mechanical tensile tests were performed on wires at various stages of corrosion, ranging from 3 to 30 days, in addition to the original wire. These tests were carried out using a Zwick 250 testing machine. The study revealed that the tensile strength of all produced wires decreases by less than 50% after 30 days of corrosion. Further experimental investigations were conducted to monitor changes in wire mass over time. The tests spanned 120 days, during which the corrosion rate declined from 2 mm/year in the initial days to approximately 0.05 mm/year in the later stages. This study indicates that when planning the use of this wire in bone surgery, a corrosion time of 1-2 years should be anticipated. In cases where the wire is exposed to intense bodily fluid flushing, this timeline may be shorter.

Introduction

The first use of metallic biodegradable surgical sutures dates back to the mid-19th century [1], employing pure magnesium and later its alloys [2]. The expectation was that these threads, with mechanical properties anticipated to surpass those of polymer threads, would address issues like reoperation (the subsequent removal of threads) and provide mechanical unloading to healing tissues during biodegradation. However, challenges such as excessive corrosion rates, hydrogen release during corrosion, and inadequate ductility of wires made from magnesium alloys when tying different surgical knots prompted a quest for alternative biodegradable metals and alloys.

A potential solution emerged with the usage of zinc alloys containing a small amount (up to 0.1% by weight) of magnesium [3]. The objective of this alloying was to enhance the mechanical properties of zinc to a level suitable for manufacturing biodegradable implants.

Biodegradable polymer-based sutures for soft tissue joining are prevalent in the market. A comparison of the mechanical properties between these threads and those made from magnesium alloys and zinc is detailed in [4]. Specifically, the study reveals that, under the influence of a surgical knot, both magnesium alloy wires and conventional polymer sutures experience approximately a 50% reduction in tensile strength. Contrastingly, zinc wire, subject to the same knot, exhibits only a 3% decrease in strength.

Surgical steel wire is the current standard for connecting hard tissues, such as bone fragments post-fracture. However, there is a lack of biodegradable alternatives in the market. Addressing this gap, [5] focuses on studying the production of biodegradable surgical wire for hard tissue connection. In [6], a laboratory-developed technology for manufacturing such wires from Zn-Mg alloys is presented. This process involves ingot production, extrusion, and hot and cold drawing, resulting in wire diameters ranging from 0.3 to 1 mm.

Paper [7] delves into the biological properties of the produced wire, determining the optimal magnesium content in zinc, leading to high cell viability, particularly with osteoblasts. This finding highlights the suitability of the ZnMg0.004 alloy as the primary material for biodegradable surgical wire production.

The technology was subsequently implemented at INTERMECH Sp. z o.o., focusing on producing biodegradable surgical wire for veterinary medicine.

This study aims to assess changes in the mechanical properties of wires under industrial production conditions, considering time and the degree of biocorrosion *in vitro*. These results hold significance both in the scientific realm and as practical insights for surgeons

Material and Methods

The work examines the ZnMg0.004 alloy, which has the following chemical composition (in wt. %): Mg 0.0034wt.%; Pb < 0.003 wt.%; Cd < 0.003 wt.%; Fe < 0.002 wt.%; Sn < 0.001 wt.%; Cu < 0.001 wt.%; Al < 0.001wt.%; Zn – balance.

A detailed description of the production process for ZnMg0.004 alloy is present below.

- Alloy formation: The process begins with the creation of the zinc-magnesium alloy by introducing magnesium into molten zinc using an induction furnace. This fusion results in the formation of the alloy.
- Ingot casting. The alloy is then cast into ingots with a diameter of 120 mm. These ingots serve as the initial material for further processing.
- Ingot cutting. Following casting, the ingots are precisely cut into billets, each with a height of 160 mm. This step prepares the alloy for subsequent processing stages.
- Chemical composition control: To ensure the alloy's chemical composition, the atomic absorption spectroscopy (AAS) method is employed for meticulous quality control.
- Extrusion. This production phase involves extruding the billet through a die with 30 channels, each having a diameter of 1.8 mm.
- Hot drawing (Fig 1, position 1): The extruded material is subjected to hot drawing, which commences at a temperature within the deformation zone ranging from 150 to 200°C. This process reduces the wire's diameter to 1.3 mm. The drawing speed is set at 10 mm/s.
- Cold drawing (Fig. 1, position 2): Subsequent drawing continues at room temperature, further reducing the wire's diameter from 1.3 mm to its final size of 0.8-1.0 mm.
- Cleaning process. During the drawing process, meticulous cleaning is carried out to maintain the wire's surface purity. The wire is carefully wiped with cotton wool and ethanol to eliminate any foreign substances that could potentially affect cytotoxicity.
- Ultrasonic cleaning (Fig. 1, position 4). At the penultimate stage, the wire undergoes an additional cleaning step, where it is immersed in an ultrasonic bath filled with water and a small quantity of liquid soap. This ensures the removal of any residual impurities.

- Final rinse: The wire is finally rinsed for 15 minutes in a solution of 70% alcohol, further enhancing its cleanliness and suitability for biomedical applications.
- Mechanical testing of the final wires in the manufacturing process can be performed in dedicated equipment (Fig. 1, position 5).

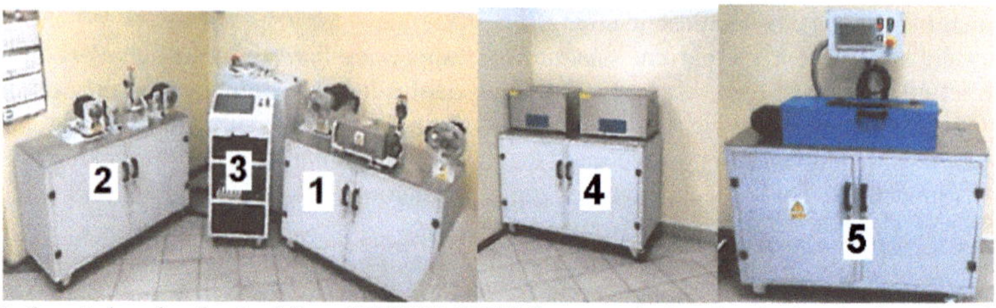

Fig. 1. Industrial technological line (INTERMECH Sp. z o.o.): 1 – the device for hot drawing; 2 – the device for cold drawing; 3 - the control electronic system; 4 – devices for ultrasonic cleaning; 5 - mechanical testing machine of the final wires.

Wire samples, post-production, underwent *in vitro* corrosion under laboratory conditions for varying durations: 3, 7, 14, 21, and 28 days. The wire ends were insulated against corrosion (Fig. 2) to facilitate mounting in a Zwick250 machine, used for mechanical tensile tests. These tests were conducted at room temperature with a deformation velocity of 20 mm/min. Fig. 3 illustrates samples before and after 28 days of corrosion. Three samples were used for each corrosion period to assess result accuracy.

The *in vitro* corrosion tests were performed in a mixture of bovine serum supplied by Biowest (catalog No. S0250, https://www.biowest.net. Accessed 2 Dec 2019). The protein content in the mixture was equal to 30 g/l. Sodium azide (0.3 wt%) and 20 mM of ethylenediaminetetraacetic acid were added to the solution to inhibit bacteria growth and bind calcium ions. The mixture was filtered through a sterile filter (pore size 20 μm). During corrosion tests, the pH value and the temperature of the corrosion medium were kept equal to 7.2–7.4 and 37 ± 0.1 °C, respectively. The protein concentration was monitored by a Genesys™20 spectrophotometer (Spectronic Instruments, USA). During the *in vitro* corrosion test all recommendations given by the ASTM F732-00 standard were satisfied.

Additionally, studies of changes in the mass of the samples over a period of up to 120 days were carried out on individual samples approximately 100 mm long. After each test period, the samples were removed from the corrosion medium and a change in their weight was determined using a laboratory balance with an accuracy of 0.00001 g.

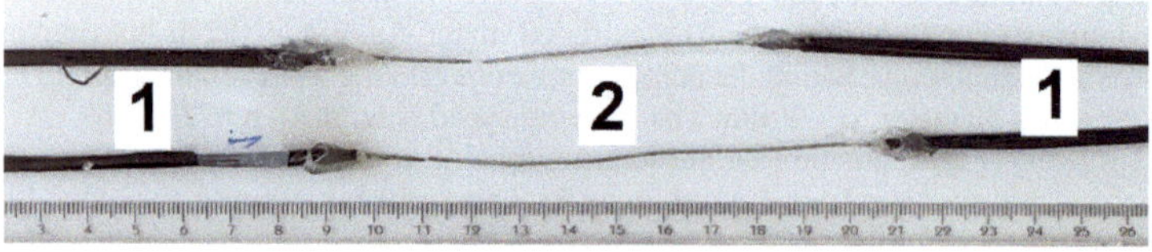

Fig. 2. Photos of samples after corrosion and strength tests: 1 – fragments of the sample, isolated from the corrosive medium; 2 – working part of the sample.

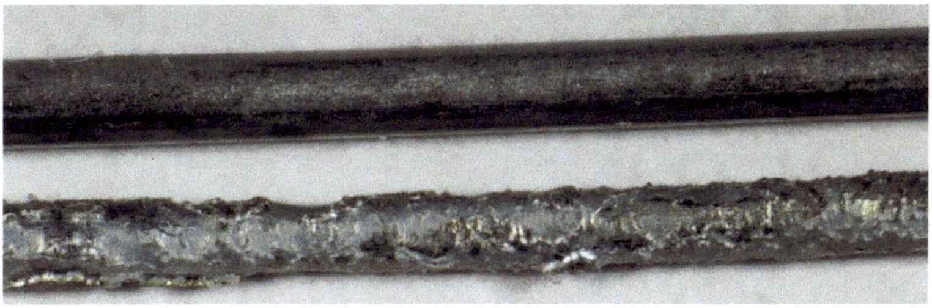

Fig. 3. Wire with an initial diameter of 1 mm before (top image) and after 28 days of corrosion (bottom image).

Results

The main research results are presented in Fig. 4. The change in the maximum force corresponding to wire rupture depending on the corrosion time is shown in Fig. 4,a. It follows from it that the decrease in wire strength over a month of corrosion was 33% and 47% for wire with an initial diameter of 1 and 0.8 mm, respectively. No statistically significant changes in the mechanical properties of the wire material (Young's modulus and tensile strength, Fig. 4,b and 4,c, respectively) were recorded. For these parameters, an increase in the measurement error is observed, which is associated with an increase in the measurement error of the wire diameter after corrosion, which becomes more uneven along the length than for the original wire (Fig. 3).

For the period corresponding to the time of corrosion of the samples before mechanical tests, the change in the mass of the samples is shown in Fig. 4,d. It follows from this that corrosion occurs most intensely in the first two weeks of the process. The total time of such tests was 120 days. During this period, the corrosion rate for wires with an initial diameter of 1.0 mm ranged from 2.3 mm/year in the initial 3 days to 0.19 mm/year after 28 days, decreasing to approximately 0.05 mm/year in the later stages. For wires with an initial diameter of 0.8 mm, these values were 2.17, 0.30, and 0.05 mm/year, respectively. Examples of SEM images depicting wires at different stages of corrosion are presented in Fig. 5. Ultimately, this study suggests that a full degradation time of 1-2 years should be expected. However, it's important to exercise caution in interpreting this assessment, as the rate of corrosion *in vitro* may significantly differ from that *in vivo*. A more accurate estimation of the time required for complete wire dissolution will be attainable upon completion of the entire cycle of clinical trials.

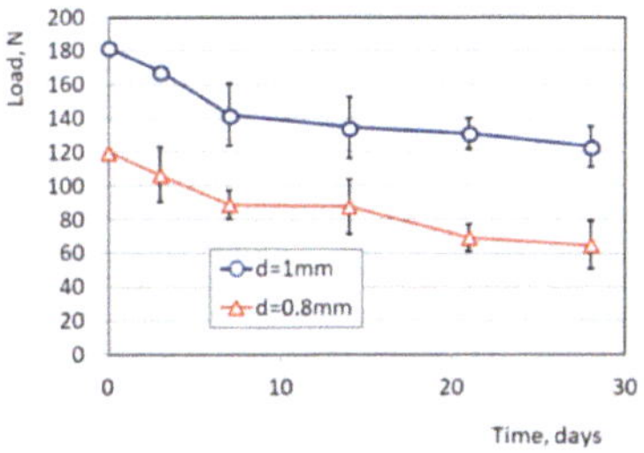

a)

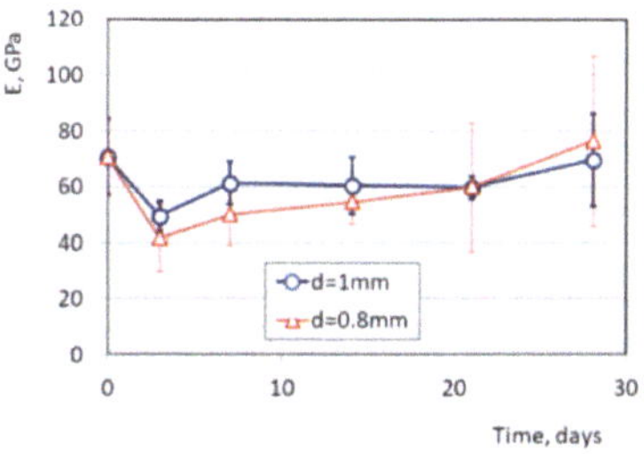

b)

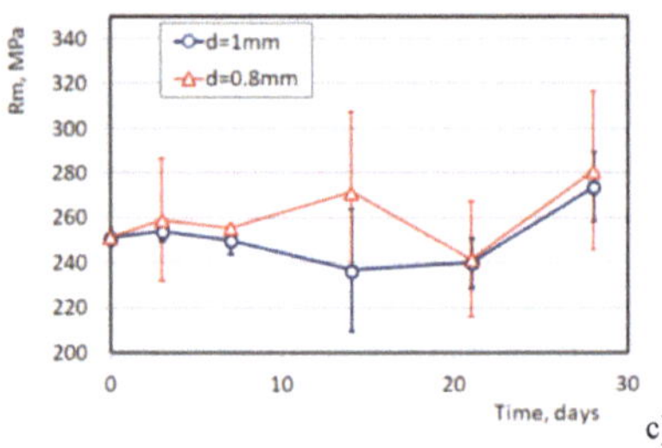

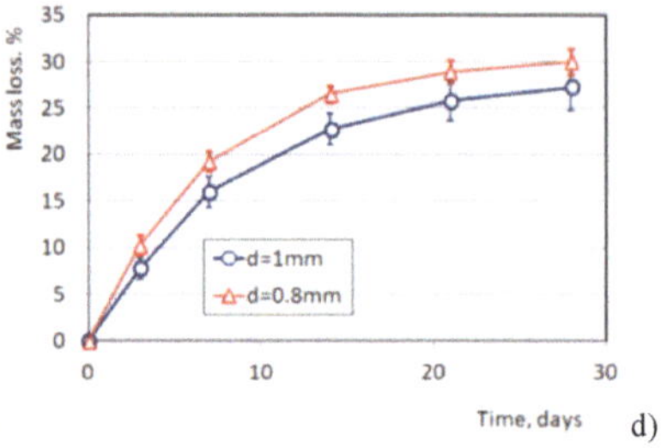

Fig. 4. Result of the mechanical tests of the ZnMg0.004 wire after corrosion (a, b, c) and mass loss of the wire during corrosion (d).

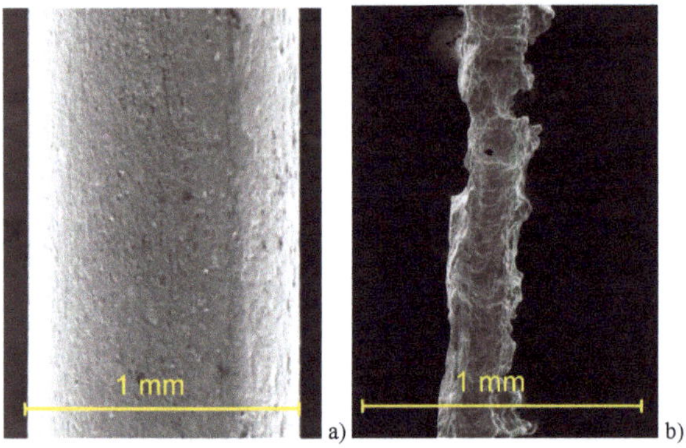

Fig. 5. SEM images of wire with an initial diameter of 1 mm made from ZnMg0.004 alloy after initial (a) and final stage (b) of corrosion in vitro.

Discussion

Of significant practical interest is the relationship between the maximum wire-breaking force and the amount of wire mass loss during corrosion. A representation of the experimental data in such coordinates is shown in Fig. 6a. Approximation of the obtained dependence using linear functions showed a high R^2 coefficient (R-squared coefficient of determination). This fact is a consequence of fairly uniform corrosion of the wire over its surface. As follows from the images in Fig. 5, even at the last stages of corrosion, wire made of ZnMg0.004 alloy does not lose continuity. In practice, the equations shown in Fig. 6a make it possible to predict changes in wire strength depending on mass loss due to corrosion. The obtained conclusion makes it possible to evaluate the loss of wire strength if there is only a curve for the change in wire mass during corrosion and to avoid repeating long and expensive mechanical tests of the material after various stages of corrosion.

It is also of interest to answer the question of whether this pattern holds for all materials. To find the answer, we will use a similar study of the strength of wires with an initial diameter of 1.8 mm from the MgCa0.9 and MgCa1.2 alloys after corrosion in the same corrosive environment [8]. Similar to Fig. 6a, we present the change in wire strength as a result of *in vitro* corrosion as a dependence of the breaking force on the loss of wire mass. The obtained dependences (Fig. 6b) are characterized by a much lower value of the R-squared coefficient of determination than in the case of the ZnMg0.004 alloy considered in this work. The reason for this difference is that the corrosion of magnesium alloys is more uneven over the surface of the wire than of zinc and its alloys.

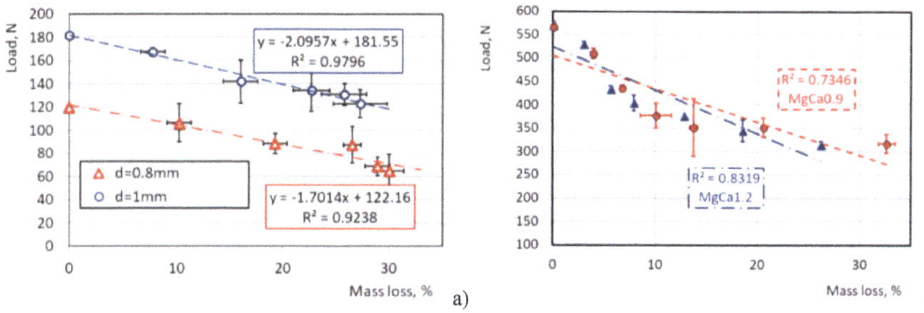

Fig. 6. Relationship between the ultimate load of wire and mass loss during corrosion: a – for ZnMg0.004 alloy; b – for MgCa alloys.

Summary

Biodegradable surgical wires with diameters of 0.8 and 1.0 mm, manufactured at INTERMECH from the ZnMg0.004 alloy, were investigated.

Based on strength tests after different stages of corrosion, it was found that the decrease in strength (breaking force) of the tested wires after 28 days of *in vitro* corrosion is less than 50% of their initial strength.

For the *in vitro* corrosion conditions tested, it was found that the decrease in wire strength is linearly proportional to the wire mass loss during corrosion, which is the result of uniform corrosion on the wire surface. Analogous tests for Mg-Ca alloys showed the lack of such proportionality due to the different nature of corrosion of Mg-Ca and Zn-Mg alloys.

Acknowledgments

The investigation is conducted within project no. POIR.04.01.04-00-0074/17 named: "Comprehensive development and preparation for the implementation of innovative implant solutions in the treatment of animals, surgical instruments for their implantology and biodegradable surgical thread for veterinary medicine" Action 4.1 "Research and Development", Subaction 4.1.4 "Application projects" Operational Program Smart Growth 2014-2020 co-financed from the European Regional Development Fund.

European Funds Smart Growth
Republic of Poland
The National Centre for Research and Development
European Union European Regional Development Fund

References

[1] F. Witte, The history of biodegradable magnesium implants: A review, Acta Biomater. 6 (2010) 1680–1692. https://doi.org/https://doi.org/10.1016/j.actbio.2010.02.028

[2] J.-M. Seitz, M. Durisin, J. Goldman, J.W. Drelich, Recent Advances in Biodegradable Metals for Medical Sutures: A Critical Review, Adv. Healthc. Mater. 4 (2015) 1915–1936. https://doi.org/https://doi.org/10.1002/adhm.201500189

[3] H. Jin, S. Zhao, R. Guillory, P.K. Bowen, Z. Yin, A. Griebel, J. Schaffer, E.J. Earley, J. Goldman, J.W. Drelich, Novel high-strength, low-alloys Zn-Mg (<0.1wt% Mg) and their arterial biodegradation, Mater. Sci. Eng. C 84 (2018) 67–79. https://doi.org/https://doi.org/10.1016/j.msec.2017.11.021

[4] A. Milenin, P. Kustra, M. Wróbel, M. Paćko, D. Byrska-Wójcik, Comparison of the stress relaxation of biodegradable surgical threads made of Mg and Zn alloys and some commercial synthetic materials, Arch. Metall. Mater. 64 (2019). https://doi.org/10.24425/amm.2019.129506

[5] A. Milenin, M. Wróbel, P. Kustra, M. Packo, D. Byrska-Wójcik, J. Sulej-Chojnacka, B. Płonka, Mechanical properties, crystallographic texture, and *in vitro* bio-corrosion of low-alloyed Zn–Mg, produced by hot and cold drawing for biodegradable surgical wires, Arch. Civ. Mech. Eng. 21 (2021) 163. https://doi.org/10.1007/s43452-021-00311-6

[6] P. Kustra, M. Wróbel, D. Byrska-Wójcik, M. Paćko, B. Płonka, M. Wróbel, J. Sulej-Chojnacka, A. Milenin, Manufacture technology, mechanical and biocorrosion properties of the Zn and ZnMg0.008 alloy wires designed for biodegradable surgical threads, J. Manuf. Process. 67 (2021) 513–520. https://doi.org/https://doi.org/10.1016/j.jmapro.2021.05.024

[7] A. Milenin, K. Łukowicz, K. Truchan, A.M. Osyczka, *In vitro* cytotoxicity of biodegradable Zn-Mg surgical wires in tumor and healthy cells, Acta Bioeng. Biomech. 23 (2021) 85–93. https://doi.org/10.37190/ABB-01927-2021-01

[8] A. Milenin, P. Kustra, D. Byrska-Wójcik, M. Wróbel, M. Paćko, J. Sulej-Chojnacka, S. Matuszyńska, B. Płonka, The effect of *in vitro* corrosion on the mechanical properties of metallic high strength biodegradable surgical threads, Arch. Civ. Mech. Eng. 20 (2020) 60. https://doi.org/10.1007/s43452-020-00062-w

Material flow analysis in friction stir consolidation during recycling aluminum alloy chips

LATIF Abdul[1,a *], PULEO Riccardo[1,b], INGARAO Giuseppe[1,c], MICARI Fabrizio[1,d] and FRATINI Livan[1,e]

[1]Department of Engineering, University of Palermo, Viale delle Scienze, Palermo, 90128, Italy

[a]abdul.latif@unipa.it, [b]riccardo.puleo01@unipa.it, [c]Giuseppe.ingarao@unipa.it, [d]fabrizio.micari@unipa.it, [e]livan.fratini@unipa.it

Keywords: Friction Stir Consolidation, Recycling, Material Flow, Aluminum Alloys

Abstract. Friction stir consolidation (FSC) is a solid-state process primarily employed to recycle aluminum machining scrap with the aim to meet the increasing demand of aluminum within the framework of sustainability goals. This technology has recently drawn the attention of many researchers due to its potential beyond the recycling approach to offer plausible new routes for alloying and upcycling metal scraps. Concerning the FSC process, a rotating tool with a certain force is applied to a given batch of chips mass enclosed inside a die chamber turning it into a consolidated billet by friction and stirring action of the tool. The obtained samples are characterized by non-homogenous properties as the non-symmetric nature of the process and therefore different strain levels occur while producing the recycled billet. This study was focused on developing a proper experimental setup to visualize material flow supported with preliminary numerical simulation. The adopted approach led to reveal the deformation during the FSC process in different billet regions by visualizing the shape change of the pre-embedded copper marker in the billet.

Introduction

Exploring sustainable recycling processes is the hot trend of the modern world in order to meet the increasing demand for aluminum for various industrial applications in a more sustainable way [1]. Conventionally recycling is performed by re-melting the scraps and transforming them into new products. However, the process itself has several limitations especially when dealing with scraps that have a higher surface area to volume ratio such as machining chips that are produced during manufacturing processes. Such kind of scraps have a higher affinity toward oxidation and therefore lead to higher material loss [2]. Thus, recycling such kinds of scraps by conventional remelting route is neither energy nor resource efficient.

Researchers are continuously exploring solid state techniques for recycling machining chips. These methods outperformed conventional remelting by turning input scraps into semi-finished or finished products by mechanical means or plastic deformation without undergoing the remelting phase. These methods are further categorized into powder metallurgy-based processes such as plasma sintering and plastic deformation-based processes like equal angle channel angle pressing, hot-pressing, high-pressure torsion, and variants of friction stir processing [3].

Recently friction-based solid-state processes such as friction stir welding and its variants like friction stir extrusion, friction stir deposition, and friction consolidation have drawn the attention of many researchers [4-6]. They offer flexibility ranging from welding metals parts to the recycling of the machining scraps. Concerning the field variables of these friction stir processes, it is important to study strain and strain rate, so the microstructure and material properties of the product can be understood and manipulated by varying the process parameters.

The most popular technique adopted by the researchers is inserting a marker in the form of copper or aluminium wire/foil inside the workpiece during the friction stir process. The actual deformation was then obtained by analyzing the deformation of the marker. In this regard, Fratini et al [7] investigated the material flow by analyzing the deformation state of a thin foil embedded at the adjoining edges of AA7075-T6 butt joint during the friction stir welding process. They observed that the material flow on the back of the tool, from the retreating side towards the advancing, is favoring bonding occurrence at the advancing side. Moreover, it worth mentioning that the process parameters, as well as the tool shape, affect the material flow. A qualitative approach was also applied by Seidel and Reynolds using the marker insert technique in FSW to visualize the material flow path after the tool had passed through the workpiece [8].

Friction stir deposition is a variant of the friction stir welding process which involves a layer-by-layer tool deposition to generate multiple metal substrates. Stubblefield et al. [9] developed a combined computational and experimental method for analyzing material flow during the FSD process by embedding a copper wire as a marker inside a AA6061-T6 feedstock. They observed copper particles on the advancing side that were found similar to what Fratini at al [7] observed during FSW.

Friction extrusion is also one of the process categories of FSW that generates large plastic strains and deformation through friction between a rotating die and the material to be extruded. Li et al. [10] developed an experimental approach to get an insight view on strain and strain rate in the FE process and their correlation with process parameters. In specific, they studied the material flow to determine the distribution of steady-state strain and strain rate in friction extrusion by analyzing the shape changes of the deforming markers.

Like FSE, the friction stir consolidation process belongs to the same friction stir welding process category. The process involves a rotating tool, which also applies a vertical load, for compacting and consolidating input chip-mass enclosed inside the die chamber [6]. The heat is generated due to the combined effect of the vertical load and the friction between the tool and the chips, allowing the chips to bond together [11]. The evolution of different process variables such as strain and strain rate are still unknown and are important to predict the properties of final consolidated products.

Therefore, this research is focused on analyzing material flow during the FSC process by using a cooper marker. The strain distribution is quantified considering the geometry change of the marker at a different position in the FSC billet. A full map of material flow in the friction stir consolidation process is determined and presented. In addition, a numerical simulation is performed, and a qualitative comparison is established between the numerical results and the experimental results. In fact, in shaping and forming processes like FSC, it is vital to understand the evolution of microstructure and other material properties with varying process parameters by obtaining the strain and strain rate.

Methodology

A 3 mm thick rolled sheet of AA7075 with average values of 150 HV and 155 μm for hardness and grain size, respectively, was turned into machining scraps through milling operations with tool rotational speed of 1250 rpm, feed rate of 280 mm/min, and cutting depth of 1 mm. FSC process was performed using ESAB LEGIO, a dedicated friction stir welding machine (Fig.1a). The chips were loaded inside the die chamber of H13 steel with a nominal diameter of 25 mm, and an H13 steel tool was employed to compact the chips. A preliminary consolidation step was applied by activating the tool rotation at 1500 rpm with a pressing force starting from 5 kN to 20 kN with 0.5 kN/s increment. Then a hole with a nominal diameter of 1 mm was drilled inside the partially consolidated billet and a copper wire of 1 mm diameter with height equal to the height of the billet (10 mm) was inserted inside the hole as a marker (Fig.1b,c and d). Copper has good ductility with good sensitivity to external deformation and therefore it has been utilized for studying the material flow in friction stir processing [7]. In order to ensure the precise location of the hole, the drill bit was guided by an auxiliary billet with a hole at a distance of 6.5 mm (1/2r) as shown in Fig. 1d. Then the billet was fully consolidated by applying a rotational speed of 1500 rpm and 20 kN pressing load to the tool for a processing time of 60 s.

After cooling the equipment, the billet was detached from the die and its surface was grind stepwise at a regular pitch distance of 0.5mm to reveal the marker across the billet height (Fig.1d). The geometry change of the markers was used as an index to quantify the strain distribution. The area lengths and average thicknesses of the markers in different layers were measured using ImageJ image processing software. Based on the geometry change, the true strain of the markers in the remnant billet was calculated according to the equations 1 to 5 specified by Li et al. [10]. By using the following equations, longitudinal strain, circumferential strain, radial strain, total strain, and effective strain were calculated at different depths of the billet.

$$\varepsilon_l = \ln\left(\frac{A_o}{A_n}\right) \quad (1)$$

$$\varepsilon_\theta = \ln\left(\frac{l_n}{d_o}\right) \quad (2)$$

$$\varepsilon_r = \ln\left(\frac{t}{d_o}\right) \quad (3)$$

$$\varepsilon_t = \varepsilon_l + \varepsilon_\theta + \varepsilon_r \quad (4)$$

$$\varepsilon_{eff} = \sqrt{(\varepsilon_l{}^2 + \varepsilon_\theta{}^2 + \varepsilon_r{}^2)} \quad (5)$$

Here ε_l, ε_θ, ε_r, ε_t and ε_{eff} are the longitudinal strain, circumferential strain, radial strain, total, and effective strain, respectively. A_0 is the original cross-sectional area of the marker, A_n is the area of the marker on a transverse cross-section. ln is the length of the marker in the circumferential (theta) direction, and d_0 is the original marker-wire diameter; t is the average thickness of the marker in the radial direction, and d_0 is the average chord of the original marker-wire diameter. By the definition of constant volume in continuum mechanics, the total strain ε_t, which is the sum of three mutually orthogonal strain components, the longitudinal strain, radial strain, and circumferential strain, should be zero. By plotting the strain values from the bottom to the top of the deformation zone, the strain evolution is revealed.

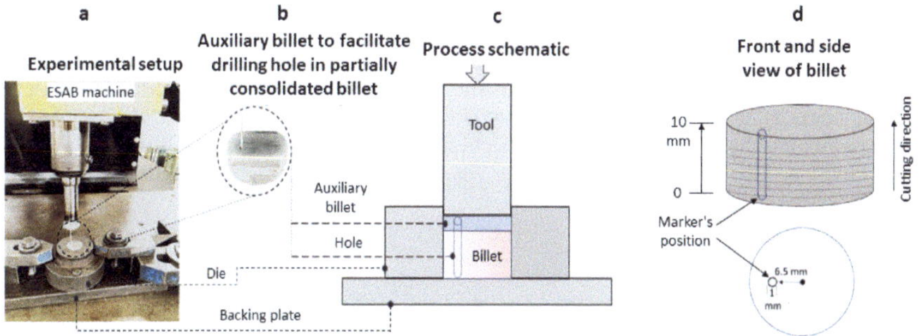

Fig. 1. Experimental setup and its schematic to analyze material flow during FSC process.

In order to further support the experimental results and get more insights into material flow during the FSC process, the numerical simulation was performed by using FEA commercial software DEFORM 3D. The tool, die and backing plate were modelled rigid bodies with a mesh size of 30000 elements while the billet of compacted chips was considered a porous object (25000 elements) with an embedded copper wire by implementing the Shima and Oyane [12] formulation on the mechanical properties definition of the material as depicted in Fig. 2. The inverse approach was adopted by setting suitable boundary conditions, in particular, constant friction equals to 0.4 and the heat transfer coefficient of 11 N/sec/mm/°C were considered.

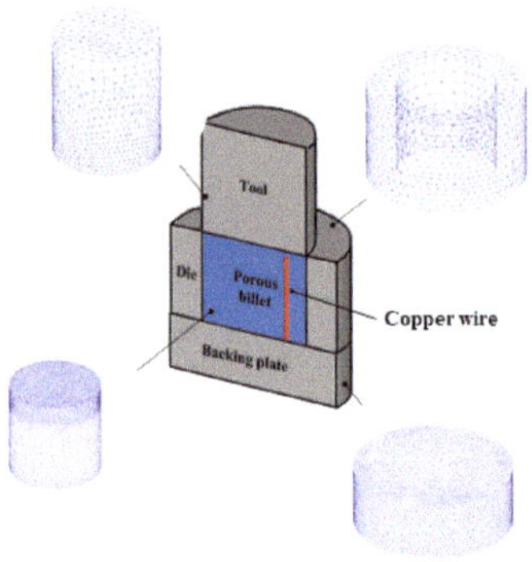

Fig. 2. Model for numerical simulation for material flow analysis.

Results

To visualize the deformation evolution and calculate the strain distribution during the deforming volume of FSC, the cross-sectional images of the AA7075 billet at different layers are presented in Fig. 3. Here, the bottom of the billet (contacting surface of billet and backing plate) was considered zero position Fig. 3. Each picture was labelled with the distance from the bottom of the billet. For example, 1 mm means the picture is taken from a cross section that is 1 mm above the bottom of the billet, in other words, 1 mm from the backing plate.

From the series of images, it is evident that no deformation was noticed at the bottom of the billet. However, the deformation occurred toward the top zone almost 3 mm below the top surface, or in other words the deforming zone is situated at a height between 7 mm and 10 mm from the bottom surface. As moving from the bottom surface towards the top, the marker starts to deform along the circumferential direction at a 7 mm distance till to the top surface. At the very top from 9 mm to 10 mm, the marker exhibited the highest deformation splitting and dispersing across the surface of the billet. It represents that this zone undergoes the highest strain during the FSC process.

Materials Research Proceedings 44 (2024) 158-166 https://doi.org/10.21741/9781644903254-18

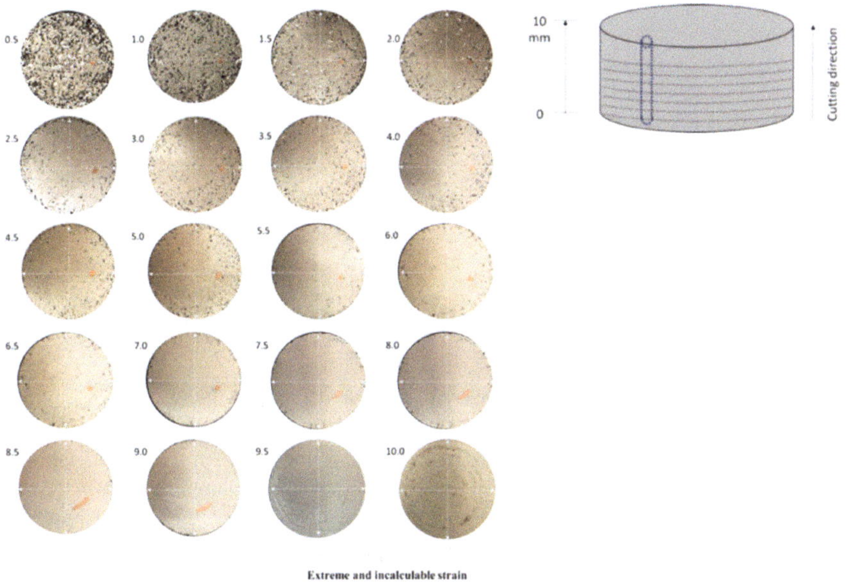

Fig. 3.Cross section of different layer of the FSC billet with respect its height.

The longitudinal strain (ε_l), circumferential strain (ε_θ), radial strain (ε_r), total strain (ε_t), and effective strain (ε_{eff}), at different depths in the deformation zone, are presented in Fig. 4. The circumferential strain increases from the position of 7 to 10 mm from the top of FSC billet and then decreases. The longitudinal strain is negative, and the absolute value continues to increase from the bottom to the top. It is important to note that the total strain is always close to zero, which indicates the obtained results are reliable. Based on the definition of the Von Mises effective strain in plasticity theory, see Eq. 5, effective strain (ε_{eff}) was calculated from three orthogonal strain components. That is, three mutually orthogonal strain components are not the principal strains. Plots of the calculated strain values were plotted with respect to their depths in the deformation zone shown in Fig. 4. As the top, most part (9-10 mm) underwent the highest strain, causing the marker to disperse across the billet section, the strain in this zone was incalculable.

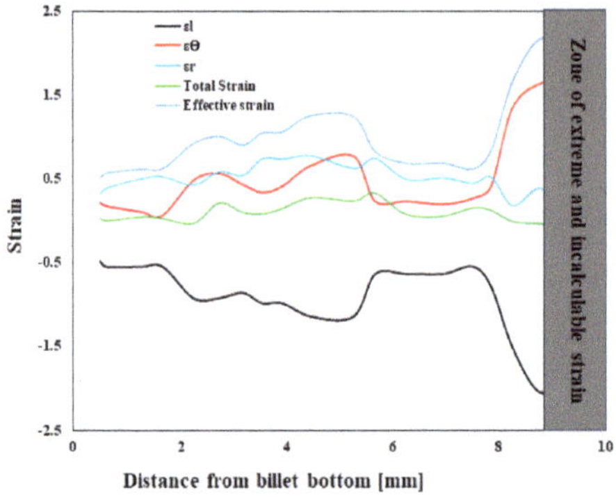

Fig. 4. Evolution of different strain across the FSC billet.

The simulation provides a qualitative overview of the material flow that characterizes the FSC process. At the end of the process, a point tracking and a flow net analysis were performed to emphasize the flow inside the porous billet. The flownet analysis has shown the flow in the entire cross section of the billet, while the horizontal point tracking was mainly added on the top surface of the billet to highlight how this surface moves during the FSC process (Fig. 5).

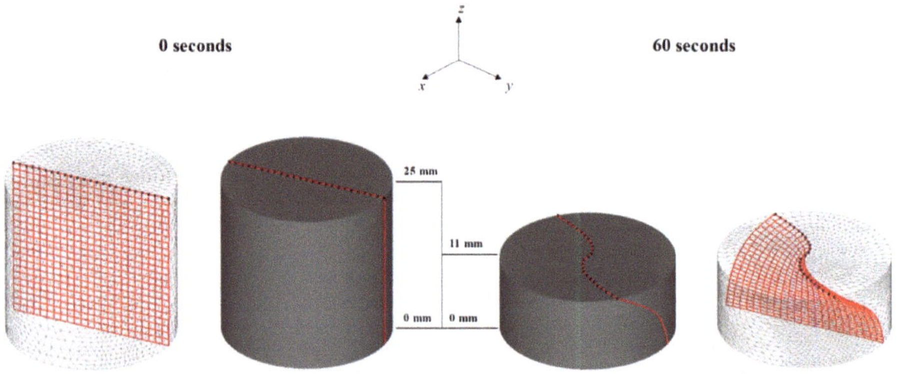

Fig. 5. Graphical visualization of the flownet (red grid) and the point tracking (black points) at the start (0 s) and the end of the process (60 s).

Additional results of the first flow analysis are presented in Fig. 6 where the rotation of the top surface is described by means of the displacement of a straight line of points. It is clearly visible that starting from the initial height of the billet the surface rotated by an angle of about 45°. Fig. 6 also shows that, due to the process configuration, the material close to the rotational center of the tool is characterized by a simple upsetting condition whilst the material close to the die surface is characterized by the highest rotation. In the same way, it is possible to distinguish and divide the 3D consolidated billet into two main zones: high strain level zone and low strain level zone (Fig. 7 and Fig. 8).

Materials Research Forum LLC
https://doi.org/10.21741/9781644903254-18

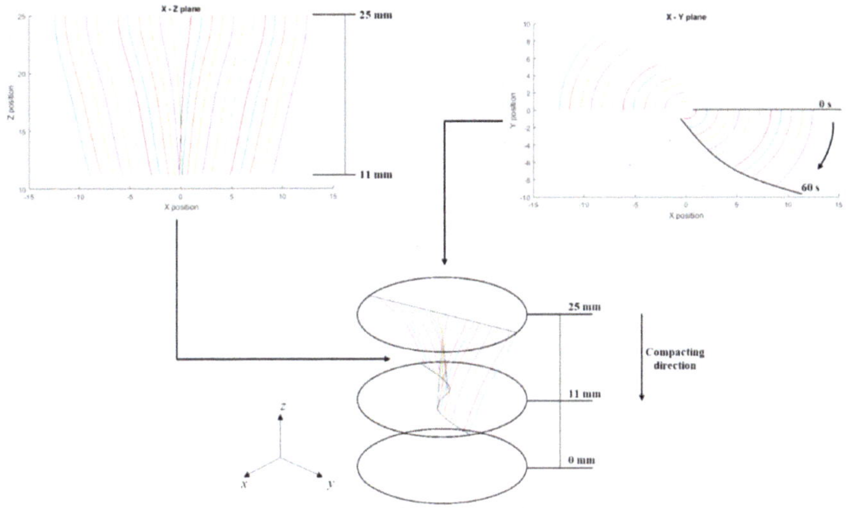

Fig 6. Point tacking visualization on MATLAB environment.

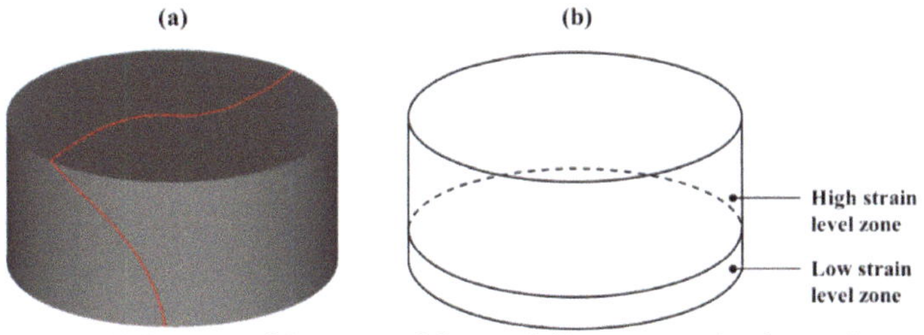

Fig. 7. (a) 3D view of the material flow and (b) two strain level zone division.

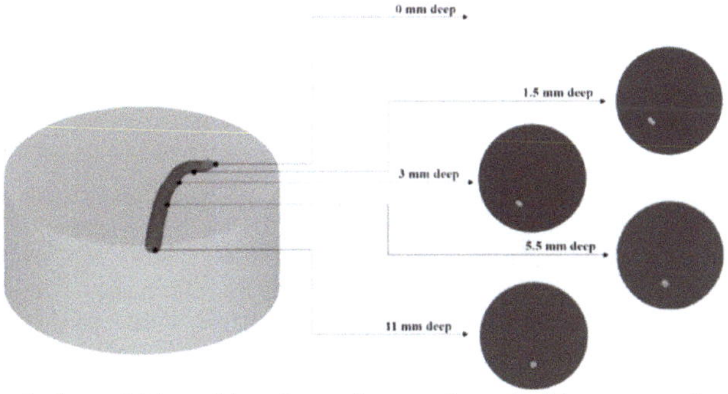

Fig. 8. Numerical acquisition of the wire surfaces at the end of the process, for different layers.

In order to further strengthen the arguments of the material flow occurrence at the top part and the absence of material flow at the bottom, the microstructure of the AA7075 consolidated billet was analyzed along its height at a radial distance of 6.5 mm as shown in Fig. 9. At the bottom of FSC billet, no significant change in the microstructure of the chips and bottom of the billet occurred. On the other hand, the formation of the new grain can be clearly seen at the top of the billet. These results depict that the bottom region is not affected by the strain and strain rate while it is expected that the top region is influenced by the combined effect of temperature, strain, and strain rate.

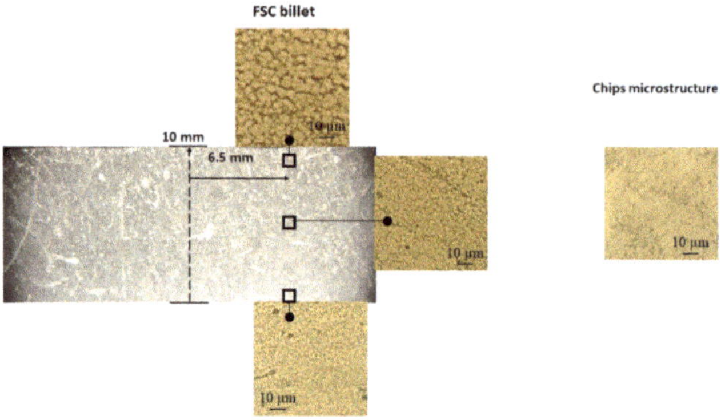

Fig. 9. Microstructure changes with respect to height of the FSC billet.

Conclusion

Although this study provided just a preliminary overview of the material flow during the FSC process. However, based on the previous discussion, the following conclusion can be drawn.
This study set up an experimental approach that depicted the material flow during the FSC of the AA7075 billet.

This research also will provide an effective avenue for obtaining the necessary strain values for establishing theoretical and numerical models for the FSC process.

In particular, the material flow occurred especially at the top zone of the billet, particularly in the zone between the top surface of the billet and 3 mm below.

The future work can be focused on improving the experimental setup for analyzing the material flow analysis, understanding the influence of the process parameters, and providing an improved numerical model to support material flow analysis results from the experimental data.

Acknowledgements

This study was carried out within the MICS (Made in Italy – Circular and Sustainable) Extended Partnership and received funding from the European Union Next-Generation EU (PIANO NAZIONALE DI RIPRESA E RESILIENZA (PNRR) – MISSIONE 4 COMPONENTE 2, INVESTIMENTO 1.3 – D.D. 1551.11-10-2022, PE00000004). This manuscript reflects only the authors' views and opinions, neither the European Union nor the European Commission can be considered responsible for them.

References

[1] D. Raabe, D. Ponge, P.J. Uggowitzer, M. Roscher, M. Paolantonio, C. Liu, H. Antrekowitsch, E. Kozeschnik, D. Seidmann, B. Gault, F. De Geuser, A. Deschamps, C.

Hutchinson, C. Liu, Z. Li, P. Prangnell, J. Robson, P. Shanthraj, S. Vakili, C. Sinclair, S. Pogatscher, Making sustainable aluminum by recycling scrap: The science of "dirty" alloys, Progress Mater. Sci. (2022) 100947. https://doi.org/10.1016/j.pmatsci.2022.100947

[2] J.R. Dufou, A.E. Tekkaya, M. Haase, T. Welo, K. Vanmeensel, K. Kellens, W. Dewulf, D. Paraskevas, Environmental assessment of solid state recycling routes for aluminium alloys: can solid state processes significantly reduce the environmental impact of aluminium recycling?, CIRP Ann. 64 (2015) 37–40. https://doi.org/10.1016/j.cirp.2015.04.051

[3] S. Shamsudin, M.A. Lajis , Z.W. Zhong, Evolutionary in solid state recycling techniques of aluminium: a review, Procedia CIRP 40 (2016) 256-261. https://doi.org/10.1016/j.procir.2016.01.117

[4] D. Baffari, A.P. Reynolds, A. Masnata, L. Fratini, G. Ingarao, Friction stir extrusion to recycle aluminum alloys scraps: Energy efficiency characterization, J. Manuf. Process. 43 (2019) 63-69. http://dx.doi.org/10.1016/j.jmapro.2019.03.049

[5] J.B. Jordon, P.G. Allison, B.J. Phillips, D.Z. Avery, R.P. Kinser, L.N. Brewer, K. Doherty, Direct recycling of machine chips through a novel solid-state additive manufacturing process, Mater. Design 193 (2020) 108850. https://doi.org/10.1016/j.matdes.2020.108850

[6] A. Latif, G. Ingarao, M. Gucciardi, L. Fratini, A novel approach to enhance mechanical properties during recycling of aluminum alloy scrap through friction stir consolidation, Int. J. Adv. Manuf. Tech. 119 (2022) 1989-2005. https://link.springer.com/article/10.1007%2Fs00170-021-08346-y

[7] L. Fratini, G. Buffa, D. Palmeri, J. Hua, R. Shivpuri, Material flow in FSW of AA7075–T6 butt joints: numerical simulations and experimental verifications, Sci. Tech. Weld. Join. 11 (2006) 412-421. https://doi.org/10.1179/174329306X113271

[8] T.U. Seidel, A.P. Reynolds, Visualization of the material flow in AA2195 friction-stir welds using a marker insert technique, Metall. Mater. Trans. A 32 (2001) 2879-2884. https://doi.org/10.1007/s11661-001-1038-1

[9] G.G. Stubblefield, K.A. Fraser, T.W. Robinson, N. Zhu, R.P. Kinser, J.Z. Tew, B.T. Cordle, J.B. Jordon, P.G. Allison, A computational and experimental approach to understanding material flow behavior during additive friction stir deposition (AFSD), Computat. Part. Mech. 1-15 (2023). https://ui.adsabs.harvard.edu/link_gateway/2023CPM....10.1629S/doi:10.1007/s40571-023-00578-x

[10] X. Li, M.R.E. Rabby, M. Ryan, G. Grant, A.P. Reynolds. Evaluation of orthogonal strain components in friction extrusion, J. Mater. Res. Tech. 15 (2021) 3357-3364. https://doi.org/10.1016/j.jmrt.2021.10.001

[11] T. Wei, A.P. Reynolds, Friction consolidation of aluminum chips, Friction Stir Weld. Process. VI (2011) 289-298. http://dx.doi.org/10.1002/9781118062302.ch34

[12] S. Shima, M. Oyane, Plasticity theory for porous metals, Int. J. Mech. Sci. 18 (1976) 285–291. https://doi.org/10.1016/0020-7403(76)90030-8

Surface integrity analysis of 304 stainless steel after slurry blasting

CHAI Ze-lin[1], ZHOU Cun-long[1*], GUO Rui[1] and JIANG Zhengyi[2]

[1]Shanxi Provincial Key Laboratory of Metallurgical Equipment Design and Technology, College of Mechanical Engineering, Taiyuan University of Science and Technology, Taiyuan 030024, China

[2]TYUT-UOW Joint Research Center, University of Wollongong, Wollongong, NSW 2522, Australia

zcunlong@tyust.edu.cn

Keywords: Slurry Blasting, Surface Integrity, Surface Roughness, Residual Stress, Corrosion Resistance

Abstract. The Eco Pickled Surface technology has attracted increasing attention due to its green and environmentally friendly characteristics. However, due to the complex internal structure of the oxide skin and the diverse bonding methods with the substrate surface, as well as the failure to establish a relationship between the substrate surface integrity and process parameters after descaling, the widespread application of this technology is restricted. The 304 stainless steel hot-rolled plates and strips had been researched in this article, and the finite element ANSYS/AUTODYN modules had been used to simulate the descaling process; An analysis was conducted on the surface integrity of 304 stainless steel after descaling, including roughness, residual stress, and corrosion resistance. The results showed that: The maximum temperature on the substrate surface under different abrasive particle sizes, blasting speeds, and blasting angles is 95°C;the impact range and depth of the substrate surface after slurry blast will increase with the increase of abrasive particle size and injection speed, The maximum hardened layer after slurry impact is 140 μm; The minimum corrosion current density on the substrate surface after slurry impact can reach 1.29*10-6A/cm2, the corrosion resistance of the substrate surface has been significantly improved, and the EBSD results show that the grains after slurry impact are concentrated in 4-19 μm, maximum 22 μm. The average grain size is 10.79 μm.

Introduction

The slurry descaling technology represented by EPS technology has attracted increasing interest and attention due to its green and environmentally friendly characteristics [1]. However, due to the diverse types and complex structures of oxide scales on metal surfaces of different materials, as well as the lack of sufficient critical strain data for oxide scale rupture, it is difficult to establish a comprehensive slurry impact process model, resulting in difficulty in controlling the surface integrity after descaling. With the advancement of technology, people have higher requirements for the surface roughness, residual stress, and corrosion resistance of plates and strips after leaving the factory. However, due to the lack of relationship between surface integrity and process parameters, the promotion of this technology is greatly affected. Among them, the impact energy of the slurry determined by the coupling of various process parameters directly affects the cracking and peeling of the oxide scale and the elastic-plastic deformation of the substrate surface. Therefore, analyzing the relationship between the impact energy of the slurry during the slurry throwing process and the integrity of the substrate surface after descaling is of great significance for the slurry throwing descaling process.

Finnie [2] assumes that when the abrasive particles of a rigid body impact elastic-plastic metal materials, the deformation follows the three-dimensional grinding theory, which involves three stages: sliding, plowing, and cutting. A material removal volume expression related to abrasive particle size and impact angle was established based on the chip volume generated by the abrasive particles within the operating distance. When the deformation is small, it is based on the theory of elastic-plastic deformation. Bitter [3] assumes that the abrasive particles of an elastic body impact elastic-plastic materials in two directions of motion, namely the direction perpendicular to the surface of the workpiece and the direction parallel to the workpiece. According to Hertz theory, the vertical direction is used to obtain the impact force, the diameter and depth of the impact crater, and thus establish a material deformation model; The horizontal direction is calculated based on the energy required to convert abrasive kinetic energy into shear volume. On this basis, Zhang Chengguang et al. used Hertz's law to incorporate the shape of abrasives in actual production and the elastic deformation of pits after impact into the material deformation model, making the predicted results closer to reality [4]. The Duan team conducted an experiment on the impact of abrasive impact on the microstructure of hot-dip galvanized steel layers. By measuring the morphology, roughness, residual stress, phase composition, and hardness of the samples, the impact of abrasive impact on the steel matrix was evaluated. The experimental results showed that after abrasive impact, the rigid matrix exhibited severe plastic deformation, and the tensile stress on the surface transformed into compressive residual stress. The surface roughness and hardness were greatly improved, The compression residual stress suppresses the dissolution of Fe atoms, resulting in a thinner coating. The refinement of grains and the generation of grain defects promote the diffusion of Fe and Zn in the early stage of the reaction, shorten the reaction time, and thus increase the coating thickness. The thickness of the entire coating of the sample decreases with the increase of surface roughness [5]. Some studies consider the deformation of metal surface contours, which is caused by the elastic-plastic deformation of the contour peaks being flattened, resulting in changes in the roughness of the metal surface [6]. Reference [7] indicates that when the abrasive particle size is optimal, the surface roughness Ra of the substrate after descaling can reach the quality level of high-quality acid washed boards (Ra1.6). Due to the fact that the deformation heat of the workpiece is rapidly carried away by the fluid during the impact process of abrasive slurry, and the surface roughness can reach very small, this technology has been widely applied in the polishing and precision grinding of ultra precision optical components, superhard materials Si3N4 and Al2O3 in recent years. The surface roughness of the processed workpiece reaches the nanometer level [8]. Vedansh Chaturvedi [9] conducted experimental research on the impact of abrasive slurry on 304 stainless steel plates, established an optimization model with lateral velocity, target distance, abrasive flow rate, and water pressure as optimization parameters, and surface roughness as the objective function, and optimized the parameters using sound to noise ratio technology. Sudhakar R studied the effects of abrasive particle size and velocity, nozzle lateral velocity, and target distance on surface roughness, and fitted the corresponding relationship between abrasive particle motion parameters and roughness values [10]. Scholars such as Benjamin Levy, Mohammed EI Mansori, and Mourad El Hadrouz have studied the impact of slurry abrasives on surface substrates through the concept of digital twins. They have captured abrasive trajectories using high-speed cameras and measured abrasive flow rates using flow sensors to fit the effects of abrasive particle size, density, and environmental pressure on substrate surface roughness. They have also proposed the concept of EASR, which only analyzes representative surfaces and no longer requires complete three-dimensional shape measurements of the entire surface, The use of EASR can effectively describe the characteristics of the substrate surface to create numerical models and optimize econometric characterization [11]. Scholars such as Navdeep Minhas, Ankit Thakur, Sumit Mehlwal, and others conducted experimental research on the surface treatment of AlSi10Mg alloy by slurry treatment. Three input parameters (slurry time, abrasive velocity, and abrasive particle size) were selected at three levels to set the matrix, and a quadratic model was established using response surface methodology. The experimental results showed that abrasive velocity had the greatest impact on surface roughness, with a positive effect (41.89%), followed by slurry time (29.29%) and abrasive particle size (23.71%) The negative impact is significant, and the predicted surface roughness values (maximum and minimum values) obtained under the optimal process parameters are consistent with the experimental measurement values (± 6%) [12]. The grain structure and mechanical properties will undergo changes after abrasive impact on the metal surface. The Yang Fan team from Tongji University conducted a random shot blasting experiment, and the results showed that residual compressive stress is a beneficial side effect of abrasive removal of metal surface oxide skin. It can effectively delay crack propagation, significantly extend the fatigue life of working parts, and establish the relationship between maximum residual stress and compressive stress depth with abrasive speed, impact angle, and abrasive particle size by developing ABAQUS/Python scripts, The impact angle is the largest influencing factor [13]. Moonchaleanporn, Pongporn, Sukrit Songkuea and other scholars conducted corrosion resistance experiments on 304L stainless steel abrasive under different sintering gas conditions. 304L stainless steel was sintered at two different heating rates under nitrogen and argon gas, and then subjected to abrasive impact. Subsequently, salt spray tests were conducted on the samples to determine their corrosion resistance; The experimental results show that the surface porosity and roughness of the shot material have significantly changed after impact, and the surface hardness has significantly increased. The corrosion resistance of 304L sintered under argon gas conditions is significantly higher than that under nitrogen gas conditions, and there is no strong theoretical support for the reasons for the differences [14]. Scholars such as Gurmider Singh and Gurmider Singh conducted abrasive impact tests on the surface of AZ31 alloy. After analyzing the factors affecting surface roughness, material removal rate, and corrosion resistance, they first proposed that abrasive impact on the surface of AZ31 alloy can improve metal biological activity. After abrasive impact, the rough surface provides strong mechanical locking for the initial stage of cell adhesion, resulting in good cell growth characteristics, Therefore, abrasive impact surfaces can also be applied in the medical industry [15]. The grains on the surface of the metal are refined and the microstructure becomes compact. The residual stress introduced after impact strengthens the metal surface, increases hardness, and at the same time, the yield limit and strength limit of the metal surface are also improved. The corrosion resistance of the rolled piece is correspondingly increased, and the service performance is also improved [16]. However, the relationship between corrosion resistance and the kinetic parameters of abrasive particles is still in the qualitative stage.

In recent years, our research group has studied the influence of jet process parameters on material surface pressure and roughness [17-18], but the impact of abrasive particle impact on substrate surface roughness and corrosion resistance is still in the experimental stage.

In summary, research on the mechanism of action of slurry on the substrate surface and its impact on corrosion resistance is still in the process of regular exploration and has not yet formed a systematic theoretical system, lacking theoretical guidance for regulating surface quality and performance. Therefore, this article intends to take 304 stainless steel as the research object, and reveal the synergistic mechanism of different process parameters on the surface roughness and corrosion resistance of the substrate by studying the deformation law and characteristics of the microstructure of the substrate material and its influence on the mechanical properties such as work hardening and residual stress, in order to achieve flexible control of the surface quality of the substrate; Ultimately, it provides theoretical and technical basis for the systematic study of the deformation mechanism of abrasive impact on metal surfaces, solving the optimization problem of various process parameters in acid free descaling process, and developing new technologies, devices, and processes for removing oxide scales.

Analysis of the impact mechanism of slurry blasting

T In the slurry blasting process, the slurry enters the blade of the blast machine from the blasting chamber and is accelerated under centrifugal force. Afterward, the slurry leaves the blade at a certain angle and generates a total impact force (F) on the oxide scale of the metal matrix surface. The vertical component is the impact force F1, and the horizontal component is the micro-cutting force F2, as shown in Fig. 1. Some studies have shown that the slurry blasting impact is a high-angle projectile greater than 50° with circular abrasive particles. Therefore, the effect of slurry blasting impact on the abrasive particles is dominant, while the horizontal micro-cutting effect is not significant[19-21]. After the oxide skin ruptures and peels off under external loads, the impact energy of abrasive particles will act on the surface of the substrate, causing elastic-plastic deformation of the substrate, thereby affecting the integrity of the substrate surface.

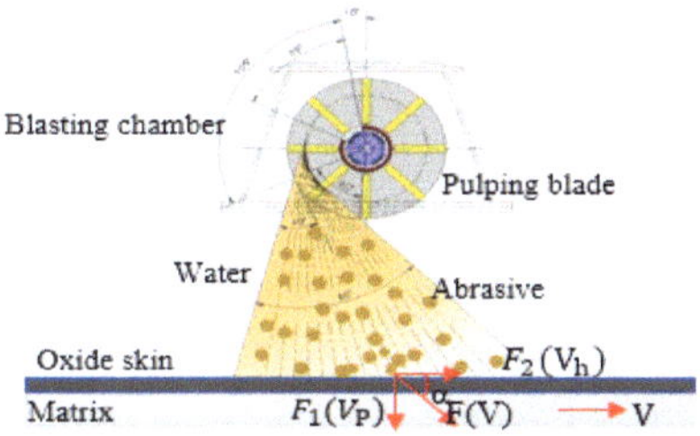

Fig. 1. Schematic diagram of slurry impact force.

Finite element analysis of slurry blasting
In order to analyze the surface quality of the substrate after slurry impact from an energy perspective, the finite element simulation analysis of the slurry process will be conducted below.

Establishment of finite element model. The finite element analysis was conducted using the ANSYS/AUTODYN module, while the AUTODYN modeling was carried out using the SPH method. The SPH method discretizes a fluid or solid into a series of particles that move without any connecting action, each particle being an independent interpolation point. By using specific calculation methods to combine the mechanical properties of each particle, it achieves an overall approximation of the actual mechanical properties of the modeled object [22].

In the slurry impact, the slurry is selected as a mixture of water and abrasive particles. The slurry in the slurry is a mixture model of water and abrasive particles, with a total of 20024 SPH particle units. The model parameters are shown in Table 1.

Using Lagrange to establish a workpiece model, 304 was selected as the target material in the material library, and the performance parameters were not changed; The modeling size of the target is 100 mm x 100 mm x 3 mm, with a total of 515121 Lagrange particle units. The edge length of the unit at the impact point is 0.1 mm, and the rest is 0.5 mm. The finite element model is shown in Fig. 2.

Table1. Finite element model parameters.

	bulasting medium	number of particles	Angle/°	cast/mm
Slurry blasting	steel shot、water	20024	90	340

Materials Research Proceedings 44 (2024) 167-181 https://doi.org/10.21741/9781644903254-19

The independent variables of finite element calculation are the initial velocity of the slurry and the particle size of the abrasive. In order to reduce computational complexity, finite element analysis assumes that: (1) the materials involved in the slurry impact process include water, abrasive, and 304 workpiece. (2) Both water particles and abrasive particles are spherical. (3) Water particles are randomly distributed and their velocity is the same as the initial velocity of the abrasive. (4) The time for abrasive particles to impact the surface of the substrate is 10-5 seconds.

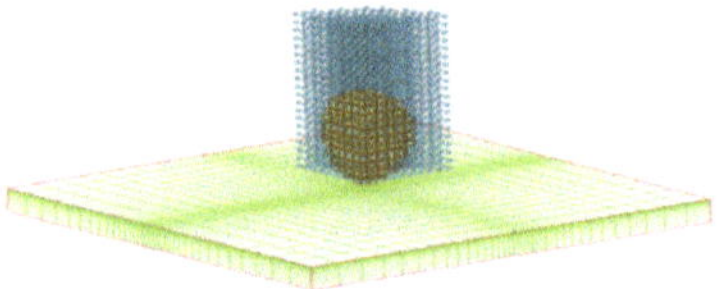

Fig. 2. Schematic diagrams of finite element model slurry blasting.

Instantaneous velocity of slurry impact on the surface of the substrate. The instantaneous velocity when the slurry impacts the surface of the substrate is shown in Fig. 3, and the abrasive particle size When the initial velocity is 80 m · s^{-1} at 1 mm, the instantaneous velocity of the abrasive particle impacting the substrate surface is 40-46 m · s^{-1}, and the velocity attenuation rate during the entire motion process is 43%~50%.

Fig. 3. Schematic diagrams of instantaneous velocity of during the impact of matrix surface by slurry blasting.

The instantaneous velocity of the slurry impacting the surface of the substrate decreases significantly. The reason for this phenomenon is that the water flow in the slurry abrasive slurry also has a higher initial velocity. When the water flow rebounds upon the substrate, a certain static pressure area is formed due to the continuity of the fluid, as shown in Figure 4. The abrasive particles are dispersed solid particles, so they can pass through the static pressure zone and impact the substrate, while the water flow will not enter this static pressure zone due to its continuity. The static pressure direction of the static pressure zone formed by the fluid is from the inside out, and the maximum static pressure value is at the center. When the initial velocity is 80 m · s^{-1}, the maximum static pressure is 69.1 kPa. When the abrasive particles pass through this area, due to the influence of static pressure resistance, the abrasive particle velocity will decrease by 43% to 50%.

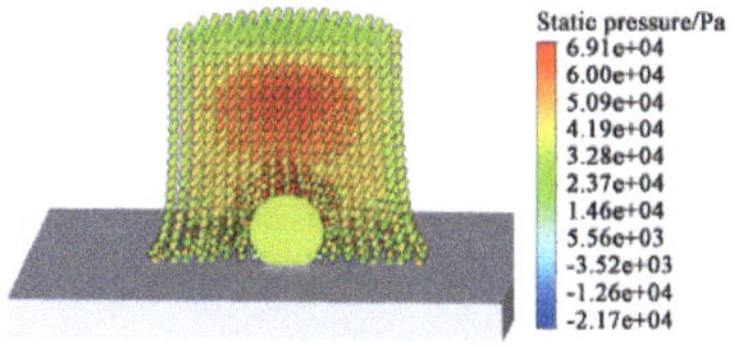

Fig. 4. Schematic diagram of static pressure zone finite element calculation.

Instantaneous velocity of slurry impact on the surface of the substrate. After slurry impact, plastic deformation occurs on the surface of the substrate, forming pits, as shown in Fig. 5. The impact depth and impact diameter of abrasive particles on the substrate have a significant impact on the surface quality of the substrate after impact. The finite element calculation results of the depth and diameter of slurry impact under different process parameters are shown in Fig. 6. The particle size of the slurry abrasive is determined by 0.6 mm increased to 1.0 mm, increase impact depth by 65%, increase impact diameter by 17%; The ejection velocity increased from 60 m · s^{-1} to 80 m · s^{-1}, the impact depth increased by 36%, and the impact diameter increased by 10%.

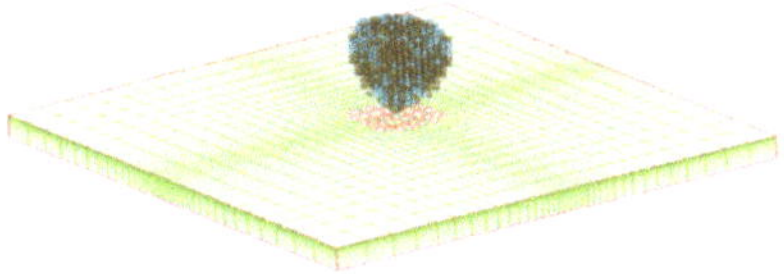

Fig. 5. Plastic deformation of matrix surface after impac.

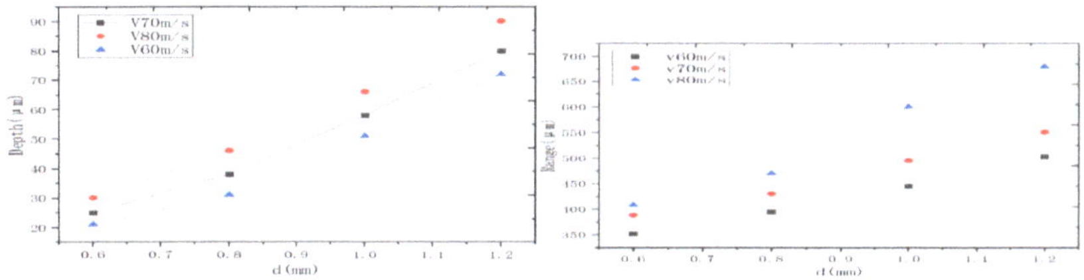

Fig. 6. impact depth (a) and impact diameter (b) of slurry blasting.

Instantaneous temperature of the substrate surface during slurry impact. The kinetic energy of abrasive particles is not only converted into impact energy on the surface of the substrate, but also a portion of the energy is converted into thermal energy due to the collision between the abrasive particles and the substrate. The collision of the two will cause an increase in the surface temperature of the substrate, while the mechanical properties of metal materials will undergo significant changes due to the increase in surface temperature. Therefore, it is necessary to analyze the thermal energy generated by the collision.

ROWE et al. [23] established a heat distribution ratio model between workpiece and abrasive particles from a microscopic perspective by studying the interaction between abrasive particles and matrix:

$$R = \left\{1 + \frac{0.97k_g}{\sqrt{r_g v_s k_w \rho_w c_w}}\right\}^{-1} \quad (1)$$

Where, k_g- the thermal conductivity of the abrasive material, r_g- the contact radius between the abrasive and the workpiece(mm), v_s- the abrasive impact velocity(m/s), k_w- the matrix thermal conductivity, ρ_w- the matrix density, c_w- the specific heat capacity of the matrix material.

Yu Zhipeng [24] established a three-dimensional temperature field equation for the collision of a single abrasive particle with a workpiece based on Eq. 1. The research results show that the collision temperature generated when the abrasive particle just contacts the surface is the highest, and the temperature generated by the collision of spherical abrasive particles is higher than that of other shapes. Therefore, when studying the instantaneous maximum temperature during shot blasting and slurry impact, the initial time of collision between the abrasive particles and the substrate should be selected. Assuming that the contact point on the surface of the workpiece is spherical, its diameter will be much smaller than the diameter of the pit, and its temperature rise will be greater than the calculated value of Eq. 1. Moreover, the energy conversion rate during instantaneous collision cannot be accurately calculated. In order to further calculate the instantaneous temperature at the beginning of the collision, finite element analysis was conducted on the temperature change during the collision process, as shown in Fig. 7.

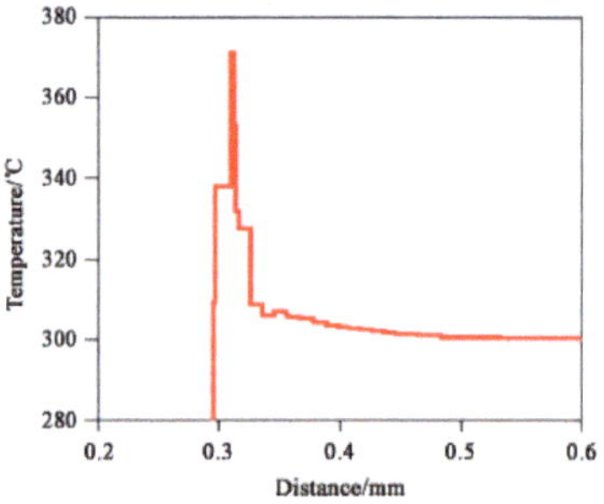

Fig .7. Finite element calculation results of temperature during abrasive impact process.

The results in Fig. 7 indicate that the center point of the collision contact has the highest temperature, and the edge gradually decreases; The center point of the slurry impact is offset due to the action of free fluid. When extracting the instantaneous temperature calculation results of finite element method under different process parameters for slurry impact, the instantaneous temperature change on the substrate surface is not significant with the change of process parameters, and the range of change is 60-97°C.

The results show that the entire process of slurry impact is significantly lower than that of shot blasting due to the instantaneous velocity during collision, and the presence of continuous fluid results in a large amount of heat generated during the entire process being carried away by high-speed water flow; Although the static pressure zone formed by continuous fluid may hinder the loss of heat to a certain extent, when discrete abrasive particles pass through the static pressure zone, they will also carry water into it. At the same time, the range of the static pressure zone formed is small, and the high-speed movement of the boundary fluid will still carry away the heat generated by the impact. Therefore, the surface temperature of the substrate will not change significantly during the entire process of slurry impact.

Experimental study on the surface integrity of 304 stainless steel substrate after slurry impact

The previous text used finite element analysis to study the mechanism of surface deformation of the substrate caused by slurry impact. In order to verify the finite element results and quantitatively analyze the integrity of the surface of 304 stainless steel after slurry impact, this paper conducted slurry impact tests on 304 stainless steel.

Test materials and methods. The slurry impact test was conducted on the self-developed slurry and slurry combined descaling experimental platform by our research group; The sample is a hot-rolled 304 stainless steel plate with surface oxide skin not removed, with dimensions of 100 mm * 100 mm * 3 mm; The abrasive grains are selected as cast steel pellets (with a Vickers hardness of 5 GPa and a density of 7.5g/cm3). The experimental parameters are shown in the Table 2.

Table 2. Experimental parameters for slurry throwing.

d	D/steel shoot	water	v(r/min)	cast	time
0.6mm	3L	4L	1600~2900	340mm	10s
0.8mm	3L	4L	1600~2900	340mm	10s
1.0mm	3L	4l	1600~2900	340mm	10s
1.2mm	3L	4L	1600~2900	340mm	10s

Analysis of Surface Roughness. The dense oxide layer on the matrix surface disappeared after the slurry impact. In addition, the plastic deformation caused by slurry impact leads to pits of different depths on the matrix surface (Fig. 8).

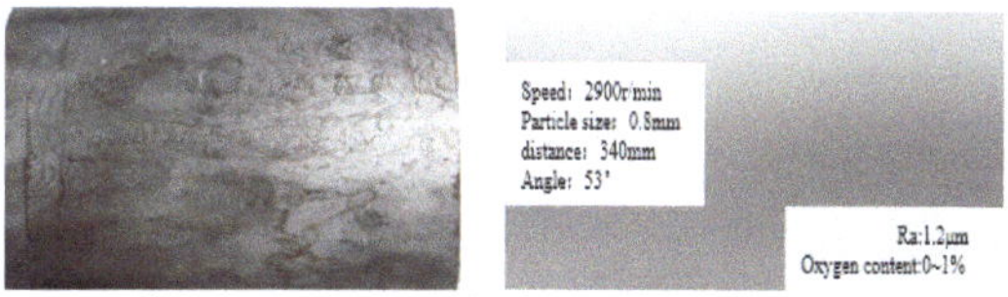

Fig. 8. Schematic diagram of 304 stainless steel after descaling for (a) 304 stainless steel sample before descaling, (b) 304 stainless steel sample after descaling.

The surface roughness after slurry impact was measured using the Mitutoyo SJ410 roughness measuring instrument, and the results are shown in Fig. 9.

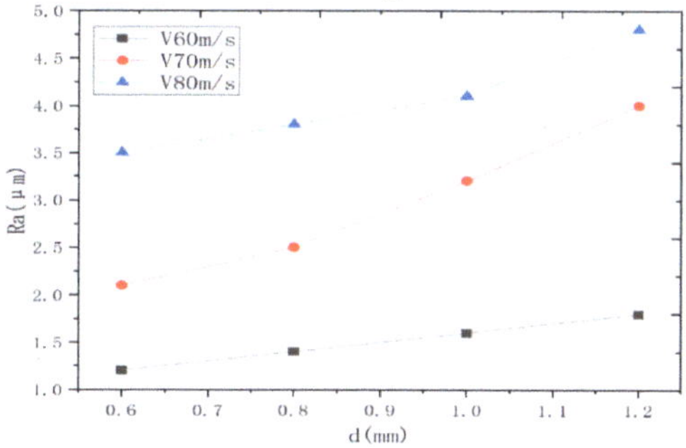

Fig. 9. Surface roughness of the substrate after slurry impact.

As shown in Fig. 9, with the increase of impact speed and abrasive particle size, the surface roughness of the matrix after impact increases. This indicates that as the impact energy of the slurry during the slurry throwing process increases, the degree of plastic deformation on the surface of the matrix increases, resulting in an increase in the depth of the wave valley. Therefore, the surface roughness increases. At the same time, it also indicates that when the abrasive particle size is less than 1mm, the impact speed has a greater impact on surface roughness than the abrasive particle size, When the abrasive particle size is greater than or equal to 1mm, the impact of abrasive particle size on surface roughness is higher than that of impact velocity.

Analysis of residual stress on the surface after slurry impact. Measure the residual stress on the surface of the substrate after impact using the IXRD residual stress tester. In order to accurately analyze the impact of slurry impact on residual stress on the substrate surface, this paper adopts a layered measurement of residual stress on the substrate surface. Firstly, the residual stress value on the 304 surface before the slurry experiment is tested. After the slurry impact, the sample is subjected to electrochemical polishing method, with a thickness of 10% μ Measure layer by layer in units of m until the test values are consistent with before the experiment, and this depth is the maximum depth. The experimental results are shown in Fig. 10.

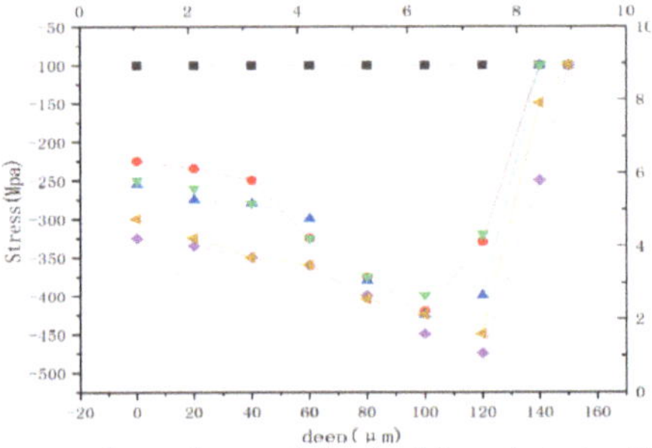

Fig. 10. Residual stress on the surface and depth of the substrate after slurry impact.

Fig. 10 shows that the maximum depth of residual stress after slurry impact is 120 μ M~140 μ m. And it shows a trend of first increasing and then decreasing, with the slope of the increasing process much lower than the decreasing slope, and the depth of the maximum residual stress is directly proportional to the impact energy. After slurry impact, a hardened layer is formed on the surface of the substrate due to residual stress, which improves surface strength and corrosion resistance, and improves service life to a certain extent. At the same time, the maximum depth of the hardened layer is 140 μ m. The impact on subsequent processes such as cold rolling and coating is minimal.

Analysis of surface corrosion resistance after slurry impact. Self corrosion current is a physical parameter that reflects electrochemical kinetics. The lower the corrosion current density, the stronger the corrosion resistance of the material; The self corrosion potential can indicate the tendency of a material to be corroded in a certain medium, and a larger or more positive value indicates that the material is less susceptible to corrosion [25]. In the anodic polarization zone, as the potential increases, the current density first increases exponentially. This curve represents the active dissolution zone, and the main reaction occurring in this stage is metal dissolution; Then the curve reaches a stage of gradual change. When the potential increases to a certain critical value, it reaches the passivation current density and passivation potential of the material. At this time, the current density will decrease or increase at an extremely slow rate with the increase of potential.

Materials Research Proceedings 44 (2024) 167-181 https://doi.org/10.21741/9781644903254-19

This is because as the voltage continues to scan towards the anode, a dense oxide film is formed on the surface of the substrate to protect it from corrosion or slow corrosion, This leads to passivation phenomenon. When the applied potential increases to a certain extent, the current density suddenly increases and continues to increase, and this potential is called the pitting potential. The oxide film of the sample is broken down by the applied voltage, resulting in pitting corrosion. The larger the pitting potential, the better the corrosion resistance of the material [26].

The electrochemical potentiodynamic polarization curves of unprocessed 304 stainless steel and samples after slurry impact are shown in Fig. 11.

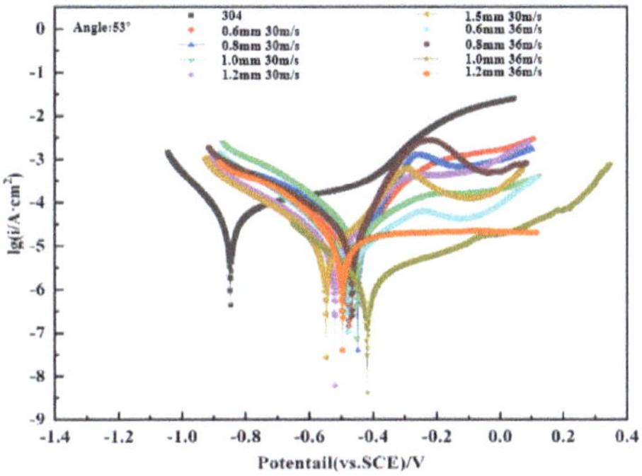

Fig. 11. Polarization curve after slurry impact.

From Fig.11, it can be seen that in the cathodic polarization zone, the corrosion current density on the surface of untreated 304 stainless steel substrate decreases sharply with the increase of potential. When the potential increases to -0.827 V, the current density increases again with the increase of potential. At this point, it enters the anodic polarization zone, and the intersection of the two curves at the junction of the anodic and cathodic polarization zones is the self corrosion potential value, which is approximately -0.827 V for the original sample. After slurry impact, the samples showed a corrected self corrosion potential compared to the original 304 stainless steel, indicating that the original 304 stainless steel had a higher corrosion tendency and faster corrosion rate during the activation stage. After slurry impact, the corrosion current density of the sample significantly decreased, with higher pitting potential and significantly enhanced passivation ability. The results indicate that the corrosion resistance of the substrate surface has significantly changed under the action of slurry impact energy after slurry impact.

Microscopic organization analysis. The metallographic diagram of the cross-section of the unprocessed sample and the matrix after slurry impact is shown in Fig. 12. From Fig. 12(a), it can be seen that the microstructure of 304 stainless steel is single-phase austenite, with clear and complete grain boundaries. The grains are basically equiaxed, and the grain size is relatively uniform, with a small amount of twinning structure present. Fig. 12(b) shows that after slurry impact, there is a clearly uncorroded black area at the boundary of the cross-section, which has not completely corroded within the same time, indicating a significant improvement in the corrosion resistance of the substrate surface after slurry impact.

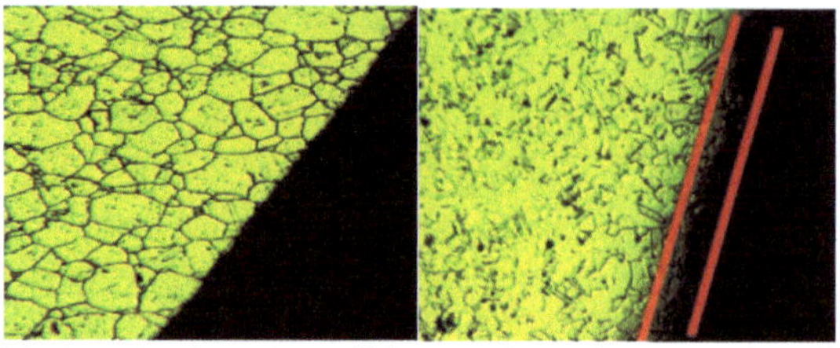

(a) Original section boundary （b）Boundary of cross-section after slurry throwing

Fig. 12. Crystal image before and after slurry throwing.

The IPF diagram of the EBSD test results is shown in Fig. 13(a), where it can be observed that the color representing the orientation within each grain of the original sample is relatively uniform. After slurry treatment, the results of the sample are shown in Fig. 13(b), where it can be seen that there is a clear preferred orientation of the grains on the surface of the matrix towards the<001>crystal plane.

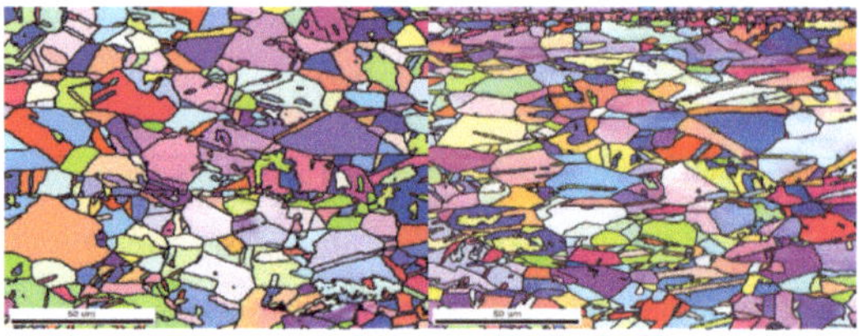

Fig. 13. IPF diagram of the EBSD.

The grain distribution of EBSD is shown in Fig. 14, and it can be seen from Fig. 14 that the grains on the surface layer of the matrix have undergone significant refinement after slurry treatment. Fig. 14(a) shows the original grain size, which can be seen to be mainly concentrated in the range of 10-30 μ Approximately m, with a maximum size of 41 μ m. Its average grain size is 24.03 μ M; The grain size diagram after slurry impact is shown in Fig. 14(b), with grains concentrated in 4-19 μ About m, maximum 22 μ m. The average grain size is 10.79 μ M. After the slurry impact, a large amount of twin structures were generated at the top of the matrix, which is due to the elastic-plastic deformation of the matrix and the formation of coherent twin boundaries. This twin boundary is a completely undistorted coherent crystal plane with low interfacial energy (about 1/10 of that of ordinary grain boundaries) and is very stable. They divide the original austenite grains into a large number of submicron small pieces, which have higher grain boundary density. This can promote the diffusion of Cr to the material surface and form a uniform and dense passivation film rich in Cr, reducing the contact area with the corrosive medium, preventing further erosion of the corrosive medium, and thus improving the corrosion resistance of the material.

Materials Research Proceedings 44 (2024) 167-181 https://doi.org/10.21741/9781644903254-19

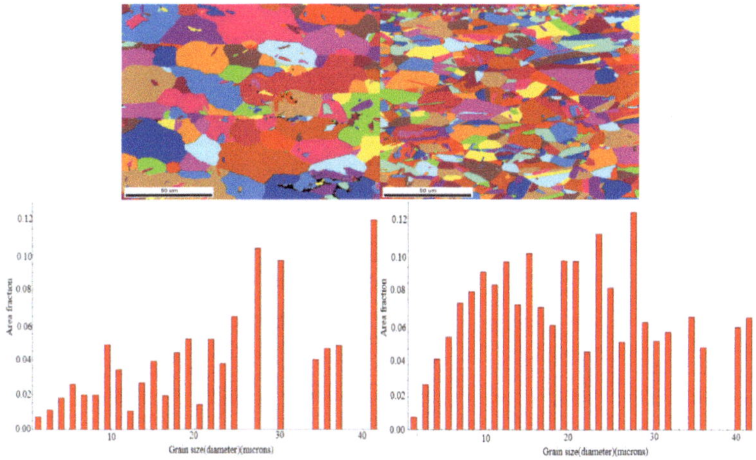

Fig. 14. Grain size of the EBSD.

The EBSD phase transition diagram is shown in Fig. 15, where the red area represents martensite and the green area represents austenite. Fig. 15(a) shows that there is only a single austenite phase in the matrix of the unprocessed 304 stainless steel sample. Fig. 15(b) shows that after slurry impact, a large amount of austenite transforms into martensite in the grains, with a martensite content of 9.8%. During the process of slurry impact, 304 stainless steel undergoes elastic-plastic deformation on the surface layer of the matrix under continuous impact of the slurry. The deformation energy provides a certain amount of energy for the phase transformation, achieving the minimum driving force required for martensitic transformation. Austenitic transformation induces martensitic transformation, which is mainly distributed at the austenite grain boundaries. The generation of martensite reduces the distance between austenite grain boundaries and improves the density of crystals. Therefore, the difficulty of ions entering the grain boundaries increases, thereby improving the corrosion resistance of the matrix surface.

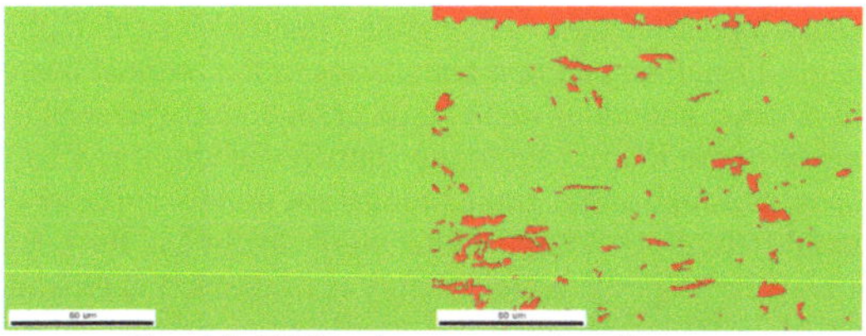

Fig. 15. EBSD phase transition diagram.

After slurry impact, the surface of the matrix undergoes elastic-plastic deformation, and the deformation reflected in the microstructure is dominated by dislocation slip. As shown in Fig. 16, the dislocation density significantly increases in the refined area of the substrate surface, and the intersection between dislocations will increase. This will lead to an increase in dislocation slip resistance, which to some extent prevents the slip of the crystal structure, ensures the stability of the crystal, delays the corrosion rate, and improves the corrosion resistance of the substrate.

Materials Research Proceedings 44 (2024) 167-181 https://doi.org/10.21741/9781644903254-19

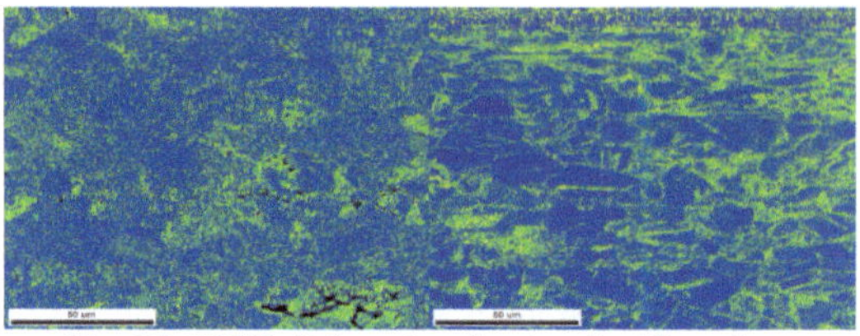

Fig. 16. EBSD dislocation density significantly.

The unprocessed grain boundary diagram of 304 stainless steel is shown in Fig. 17(a), and the grain boundary diagram after slurry treatment is shown in Fig. 17(b).

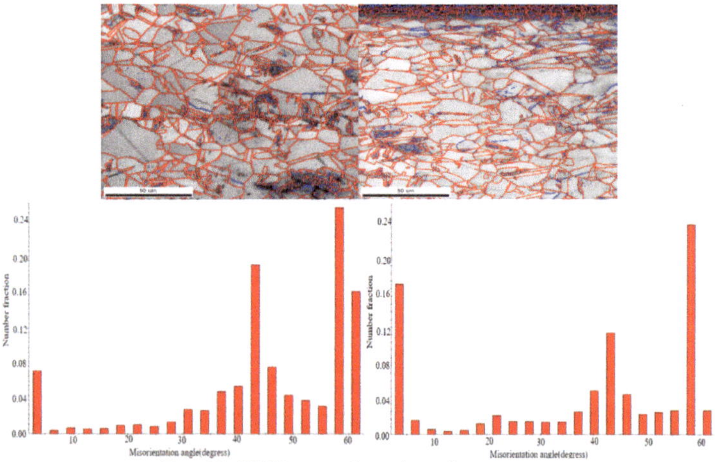

Fig. 17. EBSD grain boundary diagram.

The blue boundary represents small angle grain boundaries (with a phase difference of 2 °~10 °), and the red boundary represents large angle grain boundaries (with a phase difference greater than 10 °). The content of small angle grain boundaries in the original board is only 7.3%, while after slurry treatment, a large amount of small angle grain boundaries with a content of 21.9% are generated on the surface of the matrix. From the statistical chart, it can be seen that the original sheet is mainly composed of high angle grain boundaries, mainly concentrated at 40 °~60 °, with an average orientation difference of 45.52 °; After slurry treatment, there is a significant increase in small angle grain boundaries in the board, mainly concentrated at around 5 °, with an average orientation difference of 37.89 °. The energy of small angle grain boundaries is lower and more stable, resulting in better corrosion resistance.

Conclusions

This article analyzes the mechanism of elastic-plastic deformation on the surface of 304 stainless steel substrate under slurry impact, simulates the slurry process using finite element software, and finally conducts slurry impact tests to analyze the surface integrity of the 304 stainless steel substrate after impact. The conclusions are as follows:

(1) The maximum temperature on the substrate surface under different abrasive particle sizes, blasting speeds, and blasting angles is 95 °C, When other process parameters

are the same, the impact range and depth of the substrate surface after slurry blast will increase with the increase of abrasive particle size and injection speed. As the blasting angle increases, the impact depth will show a trend of increasing, and the blast range will show a trend of first increasing and then decreasing with the increase of ejection angle. When the ejection angle is 75 °,the blast range is the largest. When other process parameters are the same and the abrasive particle size increases.

（2） The surface roughness value of the metal substrate after slurry impact increases with the increase of abrasive particle size, and the difference between the surface roughness after impact gradually increases with the increase of abrasive particle size; The residual stress on the substrate surface after slurry impact will increase with the increase of abrasive particle size, blast speed, and blast angle; The maximum hardened layer after slurry impact is 140 μm. The residual stress shows a trend of first increasing and then decreasing with the depth of the layer. The minimum corrosion current density on the substrate surface after slurry impact can reach $1.29*10^{-6}A/cm^2$, the corrosion resistance of the substrate surface has been significantly improved, and the EBSD results show that the grains after slurry impact are concentrated in 4-19 μm About m, maximum 22 μm. The average grain size is 10.79 μm.

Reference

[1] K. Voges, A. Mueth, Eco-pickled surface：an environmentally advantageous alternative to conventional acid pickling, Iron Steel Technol. 4 (2008) 81-96.

[2] J.G.A. Bitter, A study of erosion phenomena: Part I, Wear 6 (1963) 5-21. https://doi.org/10.1016/0043-1648(63)90003-6

[3] J.G.A. Bitter, A study of erosion phenomena: Part II, Wear 6 (1963) 169-190. https://doi.org/10.1016/0043-1648(63)90073-5

[4] C. Zhang, Y. Zhang, F. Zhang, D. Luan, Study on Removal Model of Abrasive Waterjet Machining, J. Mech. Eng. 51 (2015) 188-196. https://doi.org/10.3901/JME.2015.07.188

[5] J. Li, A. Du, Y. Fan, X. Zhao, R. Ma, J. Wu, Effect of shot-blasting pretreatment on microstructures of hot-dip galvanized coating, Surf. Coat. Tech. 4 (20190 218-224. https://doi.org/10.1016/j.surfcoat.2019.02.075

[6] Ramprasad, G. Upadhyay, K. Hassan, Optimization MRR Of stainless steel 403 in abrasive water jet machining using Anova and Taguchi method, Int. J. Eng. Res. Appl. 5 (2015) 86-91.

[7] K. Voges, A. Mueth, Eco-pickled surface：an environmentally advantageous alternative to conventional acid pickling, Iron Steel Technol. 4 (2008) 81-96.

[8] C. Fengjun, A review of research on abrasive jet surface polishing, Surf. Technol. 44 (2015) 119-123. https://doi.org/10.16490/j.cnki.issn.1001-3660.2015.11.019

[9] V. Chaturvedi, D. Singh, Mutli Response optimization of process parameters of abrasive water jet machining for stainless steel AISI 304 using VIKOR approach coupled with signal to noise ratio mechnology, J. Adv. Manuf. Syst. 14 (2015) 107-121. https://doi.org/10.1142/S0219686715500080

[10] S.R. Lohar, P.R. Kubade. Current research and development in abrasive water jet machining (AWJM): A Review, Int. J. Sci. Res. 5 (2016) 996-999.

[11] B. Levy, M. El Mansori, M. El Hadrouz, S. Mezghani, A.-L. Beaudonnet, J. Cabrero, Smart tribo-peening process for surface functionalization through digital twin concept, Adv. Manuf. Technol. 4 (2021) 1-23. https://doi.org/10.1007/s00170-021-07143-x

[12] N. Minhas, A. Thakur, S. Mehlwal, Multi-variable Optimization of the Shot Blasting of Additively Manufactured AlSi10Mg Plates for Surface Roughness Using Response Surface Methodology, Arabian J. Sci. Eng. 4 (2021) 1-15.

[13] Z. Li, F. Yang, Y. Liu, Y. Gao, Numerical Simulation of Derusting Treatment of Steel Parts By Shot Blast, CMES 120 (2019) 486-500. https://doi.org/10.32604/cmes.2019.05187

[14] P. Moonchaleanporn, S. Songkuea, J. Klanpolrang, P. Payaksak, K. Wongpinkaew, N. Bunchoo, E. Viyanit, A. Manonukul, Improving corrosion resistance in sintered 304L stainless steel using shot blasting, Int. J. Mater. Product Technol. 59 (2019) 172-191. https://doi.org/10.1504/IJMPT.2019.102643

[15] G. Singh, S. Singh, C. Prakash, S. Ramakrishna, On investigating the soda-lime shot blasting of AZ31 alloy: Effects on surface roughness, material removal rate, corrosion resistance, and bioactivity, J. Magnesium Alloy. 9 (2021) 1272-1284. https://doi.org/10.1016/j.jma.2020.11.017

[16] U. Oberste-Lehn, C. Beamer, P. Gümpel, A. Karl, Corrosion resistance of low-temperature, surface-hardened stainless steel, Mater. Perform. 58 (2019) 42-46.

[17] X. Zhang, Mechanism and Experimental Study on Surface Oxidation Skin Rupture of Q235 Strip Steel, Taiyuan University of science and technology, 2016.

[18] D. Bian, Research on the critical fracture stress of Q235 oxide scale, Taiyuan University of science and technology, 2013.

[19] M. Parsi, K. Najmi, F. Najafifard, S. Hassani, B.S. McLaury, S.A. Shirazi, A comprehensive review of solid particle erosion modeling for oil and gas wells and pipelines applications, J. Natural Gas Sci. Eng. 21 (2014) 850-873. https://doi.org/10.1016/j.jngse.2014.10.001

[20] S. Wang, Simulation and Experimental Study on Shot Blasting Impact Process for Acid Free Descaling of Rolled Strip Steel, Beijing: University of Science and Technology Beijing, 2018.

[21] L. Zhai, Mechanism, Evaluation and Application of Shot Blasting Strengthening, Heat Treat. Tech. Equip. 29 (2008) 53-66.

[22] F. Chen, W. Ge, Review of smooth particle hydrodynamics methods for multiphase flow, J. Mech 53 (2021) 1-17.

[23] W.B. Rowe, M.N. Morgan, S.C.E. Black, B. Mills, A simplified approach to control of thermal damage in grinding, CIRP annals 45 (1996) 299-302. https://doi.org/10.1016/S0007-8506(07)63067-4

[24] Z. Yu, Study on the influence of shape and position characteristics of abrasive particles and their proportions on the thermodynamic coupling process of grinding, Liaoning: Northeastern University, 2019.

[25] H. Yang, Study on Preparation and Properties of Sol gel Anticorrosive Coating on Aluminum Alloy, Xi'an University of technology, 2021.

[26] Y. Wang, Overview of research on pitting corrosion of stainless steel, Corros. Sci. Protect. Tech. 27 (2015) 387-392.

Metal Forming 2024 Materials Research Forum LLC
Materials Research Proceedings 44 (2024) 182-192 https://doi.org/10.21741/9781644903254-20

New evaluation method for cold-forge solid phase bonding of dissimilar metals

OKAMOTO Haruki[1,a*], ABE Eiji[1,b], HARADA Hiroshi[1,c] and YUKAWA Nobuki[1,d]

[1]Department of Material Science and Engineering, Nagoya University, Furo-cho, Chikusa-ku, Nagoya, Aichi 464-8601, Japan

[a]okamoto@mp.material.nagoya-u.ac.jp, [b]abe.eiji@material.nagoya-u.ac.jp, [c]harada.hiroshi@material.nagoya-u.ac.jp, [d]yukawa@nagoya-u.jp

Keywords: Solid Phase Bonding, Forging, Steel, Aluminum

Abstract. Solid phase bonding is an effective method for joining dissimilar metals. Solid phase bonding is the bonding method without fusing metal matrix. Using this method, clad metal which is joined with sheet metals is in practice. However, the parameters to achieve stable cold-forge solid phase bonding in bulk are not well known. To investigate the parameters that affect the solid phase bonding state, a new test method to evaluate the strength of the bonding interface obtained by cold forging was developed. Using this method, test pieces bonded with S10C/A1070 and SUS304/A1070 were made in fluctuating interface bonding parameters (the surface area expansion ratio, the surface pressure). Although tensile test results showed large variation in joining strength at low value of bonding parameters, bonding was strong enough to exceed the A1070 base metal strength at the region of high value of bonding parameters. As it was found that the bonding parameters were able to measure the effect on joining strength in this experiment, it is considered that bonding parameters for solid phase bonding can be evaluated by the newly developed test method.

Introduction

To reduce the weight of automobiles, demand for technology to join dissimilar metals increases [1]. For instance, high-strength, high-rigidity steel materials and low-density aluminum alloy materials are tried to join. However, it is difficult to bond dissimilar metals (particularly Fe/Al) strongly by fusion welding because of the occurrence of thick intermetallic compounds (IMCs), such as Al_3Fe, Al_5Fe_2, and Al_2Fe [2]. These IMCs reduce the joining strength of the Fe/Al interface when the IMC thickness is in the range of 0.5–2μm or over [3, 4]. Solid phase bonding is an effective method for bonding dissimilar metals as thick IMCs are difficult to form. Bonding methods that utilize plastic deformation [4, 5] are classified into two types. Mechanical bonding such as lock seam processing [6] and mechanical clinching, a type of caulking joining method [1, 7]. Metallurgical methods that utilize metallic bonding through exposure and close contact with nascent surfaces caused by plastic deformation [8]. For instance, FSW [9], explosive welding [10], diffusion bonding [11, 12] and cold-forge solid-phase bonding. In this study, the latter, especially cold-forge solid-phase bonding was focused on. To bond steel and aluminum, both metals are deformed at the same time by forging, and the contact surface area of each metal is expanded. Surface area expansion breaks the brittle [13, 14] oxide layer on the metallic surface and exposes nascent surface. Pressure is applied to the nascent surface, and bonding is made by bringing it into contact (Fig. 1 [8]). Since this bonding method does not require external heat input, no or only a small amount of brittle IMC layer at the bonding interface is generated [12, 15]. Since no heat input is required, it is environmentally friendly and inexpensive. Clad metal bonded by the same mechanism is in practice. However, the parameters to achieve stable cold-forge solid phase bonding in bulk are not well known. Currently, the factors that are considered to affect bonding (bonding parameters) are the surface area expansion ratio of the base metal [16, 17], the surface

 Materials Research Forum LLC
https://doi.org/10.21741/9781644903254-20

pressure on the interface [16], and the relative slip amount [16, 18]. The goal is to quantitatively evaluate the influence of these bonding parameters on forge solid phase bonding. Previous research has shown that this method can provide a certain degree of bonding strength and relationship between bonding parameters and joining strength can be suggested to some extent too. However, it was not possible to make test pieces with a wide range of joining parameters. In addition, it can not measure joining strength accurately because mechanical clinching was found on the interface and the bonding interface was not plane. Due to the inadequacy of this methods, it was difficult to evaluate bondability with a wide range of bonding parameters and it was not possible to evaluate the strength of the solid phase bonding itself. In this study, to improve the problems, a new evaluate method to measure joining strength accurately was developed. With this new test method, it is possible to make test pieces bonded with a wide range of bonding parameters compared with previous one. Moreover, they are applied tensile load perpendicular to the plane bonding interface.

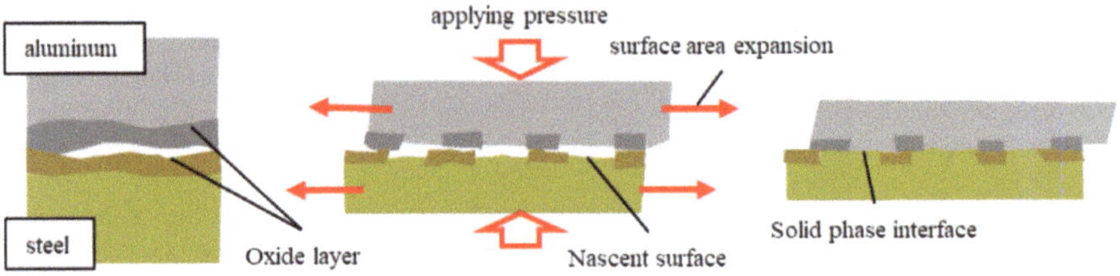

Fig.1. Image of Cold-Forged Solid Phase Bonding [8].

New Evaluation / Experimental Condition
A newly developed method to evaluate solid phase bonding is shown in Fig. 2. First, the center of the S10C or the SUS304 cylinders was hollowed out, and the A1070 cylinder was inserted into it. They were plugged into a rigid restraint die with a hole in the center and a height of 10mm. They were forged restraining the lower part with a concentric grooved restraining upper die. Solid phase bonding is expected to be achieved at the red line shown in the Fig. 2. With this forging method, it is possible to forge aluminum while applying strong back pressure, so to expand the surface area of steel and aluminum at the same time, which have significantly different yield stresses. As the lower part of the cylinder was restrained by a die, a sufficient grip section was secured on the aluminum part in performing the tensile test. Using this method, the affection of bonding parameters could be investigated. Tensile test pieces can be made with large wide bonding parameters by changing the thickness of steel (t_{Fe}), the diameter of aluminum (φd_{Al}), and reduction. In this study, to evaluate the joining strength, the tensile test was performed on test pieces prepared material combination of S10C/A1070 and SUS304/A1070. As shown in Fig. 3, tensile test pieces were cut from the center of forged metal by wire electric discharge machining (WEDM). To remove the heat effect area caused by WEDM, 0.25 mm of the outer diameter of the test piece was removed by turning [19]. Notches were added to the steel to hook the jig. Using this processed test piece, a tensile test was performed.

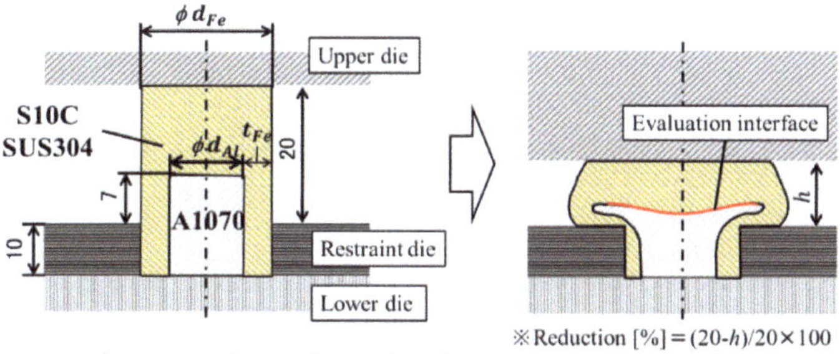

Fig. 2. Experimental overview for new evaluation method.

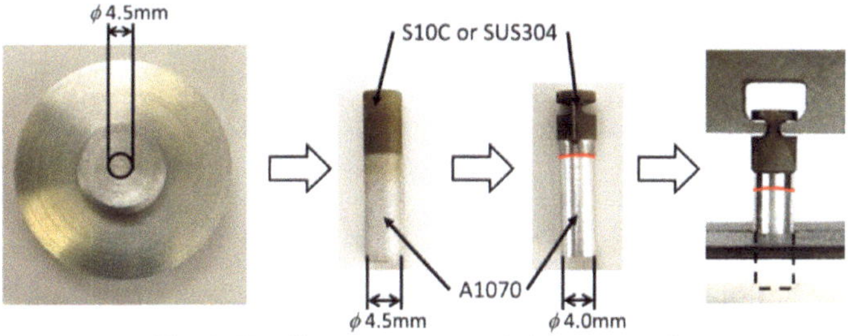

Fig. 3. Tensile test to measure joining strength.

In hollowing out the center of steel (S10C, SUS304), the end mill was thrust to the bottom of the steel. The profile curve was measured by using stylus surface roughness measurement (probe radius is 0.2 μm, tip angle is 60°, measurement speed is 0.25 mm/sec) to know the effect of roughness within a range of ±2mm in the radial direction from the center of S10C, SUS304 and A1070. The profile curve of S10C and SUS304 were shown in Fig. 4. In this result, the center of steel is about 70 μm higher than the location of 2 mm away from the center because the end mill can not cut the central area. The arithmetic average roughness (S10C, SUS304, A1070) was calculated. Results are shown in Table 1.

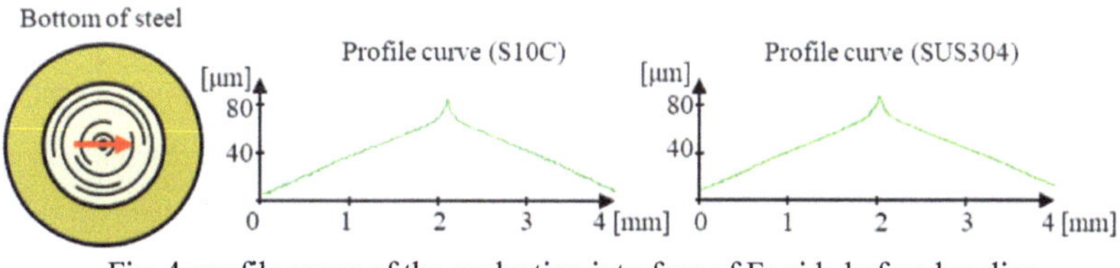

Fig. 4. profile curve of the evaluation interface of Fe side before bonding.

Table 1. Arithmetic average roughness (Ra) of S10C, SUS304, A1070.

Material	Ra [μm]
S10C	0.153
SUS304	0.096
A1070	0.112

New Evaluation / FEM Condition

To know the interface condition, bonding parameters (the surface area expansion ratio, the surface pressure, and the relative slip amount) were calculated using FEM. Finite element analysis software, DEFORM-2D was used. Conditions and assumptions for analysis were defined as shown in Table 2. FEM model was Axisymmetric, S10C, SUS304 and A1070 were regarded as elasto-plastic, die and restraint die were regarded as rigid.The compression speed was defined to match the one of the actual press machine. Flow stress for each material shown in Fig. 5 was determined by compression test with repeated lubrication. The friction conditions shown in Fig. 6 were determined so that the cross section diagram of the contact interface matched with experimental results. The coulomb friction factor was 0.15 when the surface area expansion ratio (= Π) was less than 2.0, and 0.4 when Π exceeded 2.0 on the interface of steel/aluminum. As upper die is a concentric grooved restraining and lower part of cylinder was completely restrained, other contact conditions were regarded as sticking friction. Thermal effect was neglected.

Table 2. FEM conditions.

Model	Axisymmetric
Deformation condition	S10C・SUS304・A1070 : Elasto-plastic Die・Restraint Die : Rigid
Materials	S10C-A1070, SUS304-A1070
Young’s modulus	S10C : 206[GPa] SUS304 : 197 [GPa] A1070 : 69[GPa]
Poisson’s ratio	S10C : 0.30 SUS304 : 0.30 A1070 : 0.33
Compression speed	0.607 [mm/sec]

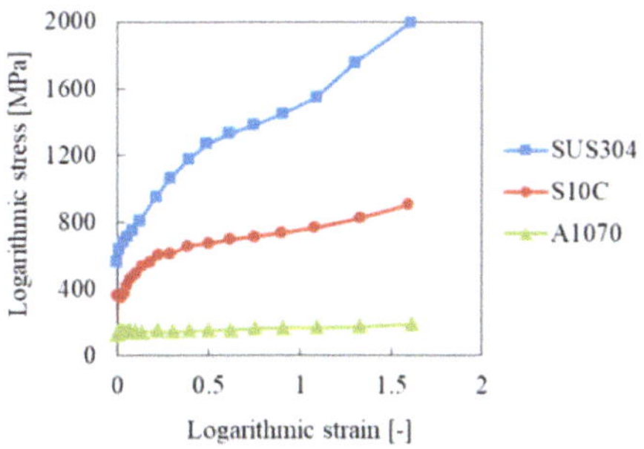

Fig.5 Flow stress

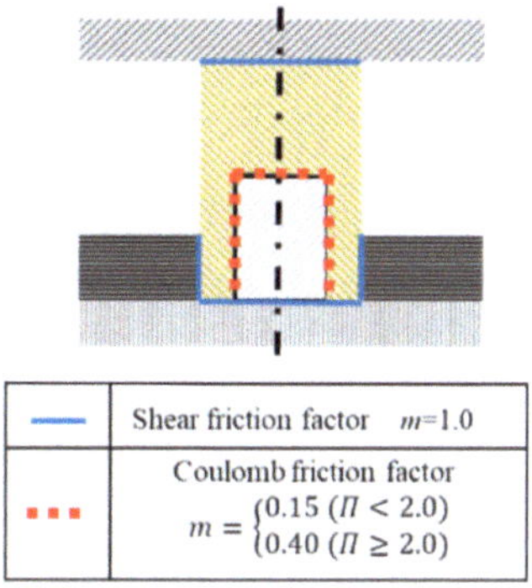

Fig.6 Friction condition

Result of FEM

Bonding parameters (the surface area expansion ratio, the surface pressure, and relative slip amount) were varied by changing the reduction, the diameter of aluminium (φd_{Al}), and the thickness of steel (t_{Fe}). Fig. 7 shows the calculated bonding parameter distribution along the S10C/A1070 and SUS304/A1070 interface obtained from the results of FEM under the conditions of t_{Fe}=3mm, φd_{Al}=10mm. The relationship between reduction and the surface pressure was shown in Fig. 7(a), (b). The black dashed line in Fig. 7 indicates the location where the test piece was taken. As reduction increased, the surface pressure increased. In SUS304/A1070, nearly twice the surface pressure compared with S10C/A1070 was obtained. The relationship between reduction and the surface area expansion ratio was shown in Fig. 7(c), (d). The surface area expansion ratio of SUS304/A1070 was small compared to S10C/A1070. In both metal combinations, the surface pressure and the surface area expansion ratio were almost constant in the location where the test piece was taken. The relationship between reduction and the relative slip amount was shown in Fig. 7(e), (f). Although the relative slip amount increased as the reduction increased, it was considered to be very small in the radial location of 0-2 mm compared to the relative slip amount in previous research [16, 18]. The cold-forge solid phase bonding of high tensile strength steel and aluminium alloy sheets by miwada [16], it is shown that approximately 0.1-0.4 mm of the relative slip amount was observed on the bonding interface. In this study, approximately 0-0.002 mm of the relative slip amount was observed. Therefore, only the surface area expansion ratio and the surface pressure were focused on.

Materials Research Forum LLC
https://doi.org/10.21741/9781644903254-20

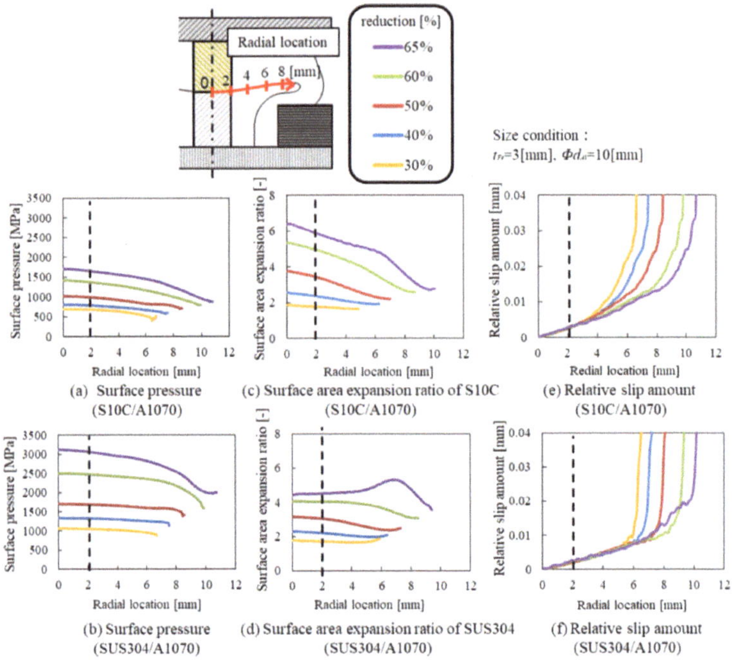

Fig. 7. Distribution of bonding parameters.

The effect of t_{Fe} and φd_{Al} changes on bonding parameters was investigated under the condition of a reduction of 60%. The relationship between t_{Fe} , φd_{Al}, and the surface area expansion ratio, the surface pressure were plotted shown in Fig. 8. The surface area expansion ratio and the surface pressure were average values in the radial location of 0-2 mm. Fig. 8(a) showed the distribution for S10C/A1070, and Fig. 8(b) showed the distribution for SUS304/A1070. The correspondence of shape on the graph was shown in Table 3. In both metal combinations, as t_{Fe} was increased, the surface pressure increased, and the surface area expansion ratio decreased. A similar tendency was observed when φd_{Al} was increased. It was also found that SUS304/A1070 is difficult to vary the surface area expansion ratio over a wide range compared to S10C/A1070. From the results of Fig. 8, it was suggested that it was possible to adjust the bonding parameters on the interface to some extent by changing the size of S10C, SUS304 and, A1070. As an example of the way to use, in S10C/A1070, by comparing a test piece made of t_{Fe}=4 mm, φd_{Al}=8 mm with t_{Fe}=3 mm, φd_{Al}=10 mm under the condition of a reduction of 60%, only the effect of surface area expansion ratio on solid phase bonding could be investigated while keeping the surface pressure at the same level. It is considered to be valid as a test method to investigate bonding conditions.

Materials Research Forum LLC
https://doi.org/10.21741/9781644903254-20

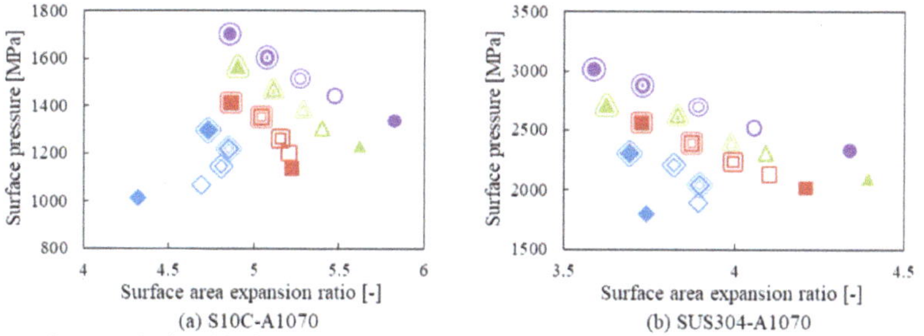

Fig. 8. Relationship between surface area expansion ratio and surface pressure.

Table 3. Correspondence of shapes on the graph.

		t_{Fe} [mm]				
		4	3.5	3	2.5	2
$\phi\, d_{Al}$ [mm]	12	◉	◎	◎	○	●
	10	▲	△	△	△	▲
	8	▣	▣	▣	□	■
	6	◈	◇	◇	◇	◆

Tensile Test Result

Tensile test pieces bonded with S10C/A1070 and SUS304/A1070 were made under the conditions of t_{Fe}=3mm, φd_{Al}=10mm. Fig. 9(a), (b) shows the results of tensile tests by the relationship between reduction and joining strength. The result of the metal combination of S10C/A1070 is shown in Fig. 9(a) and SUS304/A1070 is shown in Fig. 9(b). Fig. 10(a), (b) shows the relationship between the surface area expansion ratio (average value in 0-2mm radial distance) and joining strength. The result of the metal combination of S10C/A1070 is shown in Fig. 10(a) and SUS304/A1070 is shown in Fig. 10(b). And Fig. 11(a), (b) shows the relationship between the surface pressure (average value in 0-2 mm radial distance) and joining strength. The result of the metal combination of S10C/A1070 is shown in Fig. 11(a) and SUS304/A1070 is shown in Fig. 11(b). In Figs. 9, 10, 11, red plots show the results of tensile test pieces that could be measured joining strength, blue cross marks show them broken in cutting by WEDM. The surface area expansion ratio and the surface pressure are the average values on the interface of test pieces. In both metal combinations, as reduction increased, as the values of bonding parameters were increased, joining strength was increased. Some of the test pieces with low reduction were scarcely bonded, they were broken by cutting by WEDM due to residual stress [20]. On the other hand, test pieces forged with high reduction showed their joining strength about 130 MPa. To investigate the effect of work hardening in forging, micro Vickers hardness test (load was 0.245, time was 15s) was held as shown in Fig. 12. In range of 1.5mm×4.0 mm, about 100 points, Vickers hardness was measured and drew the distribution in three condition (S10C-A1070 reduction 50%, S10C-A1070 reduction 62.5%, SUS304-A1070 reduction 50%). From the test results in Fig. 13, it was recognized that work hardening was clearly observed in approximately 0-1 mm downward from the interface considering A1070 base metal Vickers hardness was 40HV0.025 in all experimental

Materials Research Proceedings 44 (2024) 182-192 https://doi.org/10.21741/9781644903254-20

conditions. It is thought that the joining strength which was the same level as the tensile strength of A1070 base metal was obtained under the condition of high reduction.

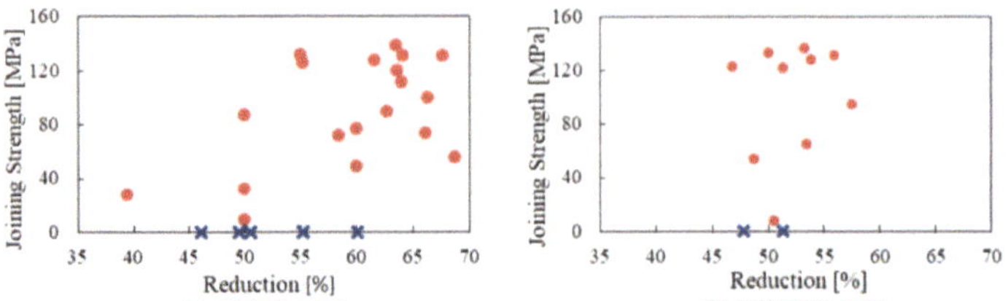

(a) S10C/A1070 (b) SUS304/A1070

Fig.9 Reduction vs joining strength

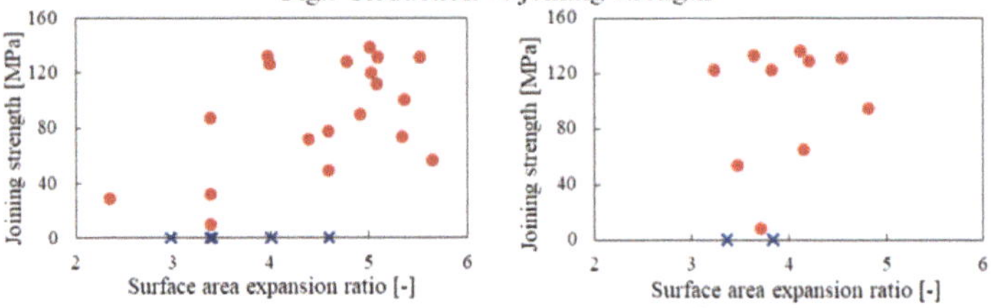

(a) S10C/A1070 (b) SUS304/A1070

Fig.10 Surface area expansion ratio vs joining strength

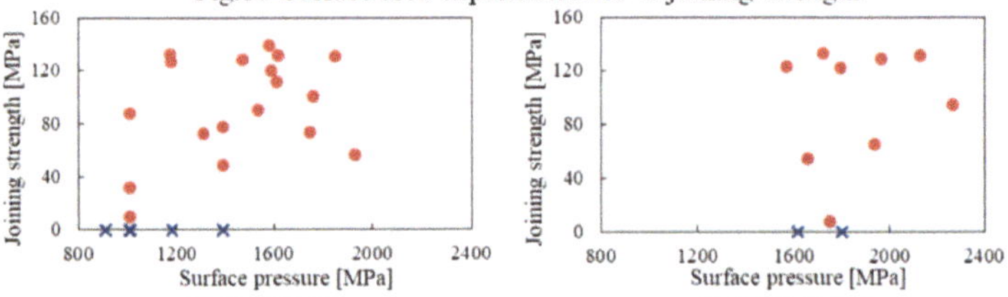

(a) S10C/A1070 (b) SUS304/A1070

Fig.11 Surface pressure vs joining strength

● : Test piece that tensile strength could be measured

✕ : Test piece broken by cutting with WEDM

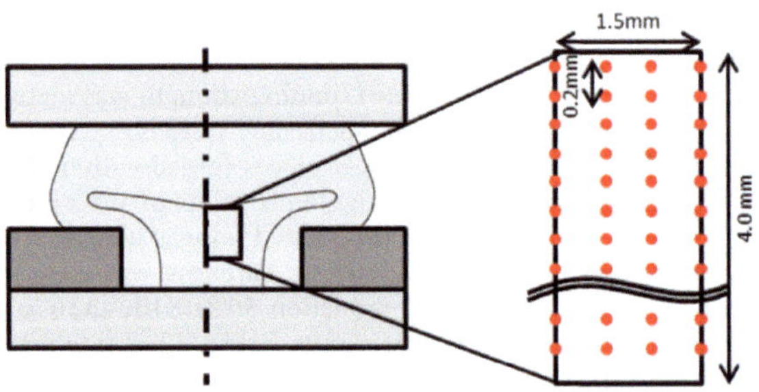

Fig. 12. Vickers hardness test measurement position.

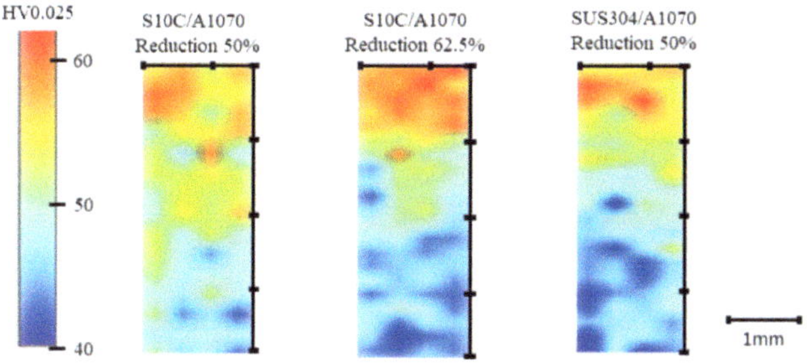

Fig. 13. Vickers hardness distribution.

Fig. 14(a), Fig. 15(a) show load-stroke diagram of the tensile test results of S10C/A1070 and SUS304/A1070 respectively which showed highest joining strength, and Fig. 14(b), Fig. 15(b) show backscattered electron image (BEI) of steel side fracture surface after the tensile test in the combination of S10C/A1070 and SUS304/A1070 respectively. S10C/A1070 was a macroscopically brittle failure, while SUS304/A1070 showed elongation from Fig. 14(a) and Fig. 15(a). The black part is the adhesion of the aluminum in BEI. The amount of adhesive rate of aluminum on the steel side fracture surface was obtained from these BEIs. SUS304/A1070 gained high adhesive rate compared to S10C/A1070 from Fig. 14(a) and Fig. 15(b). One of the reasons for the large variation of tensile test results is thought to be the low adhesive rate. When there are large defects in size or in number on the interface, it is easier to be in brittle fracture. Fig. 16(a), (b) showed relationship between adhesive rate and joining strength in the combination of S10C/A1070 and SUS304/A1070, respectively. It was shown that the higher the adhesive rate it was, the less variation there was and the more stable joining strength was showed. It seems that it shows the importance of increasing the adhesion rate in order to achieve stable and strong bonding.

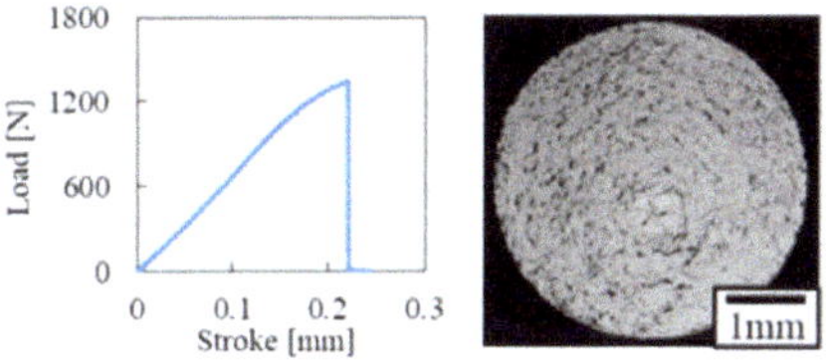

(a) Load-stroke diagram (b) BEI of steel fracture surface

Fig.14 S10C/A1070

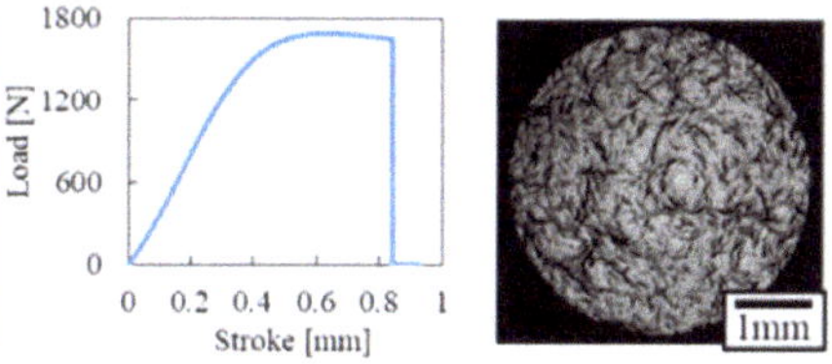

(a) Load-stroke diagram (b) BEI of steel fracture surface

Fig.15 SUS304/A1070

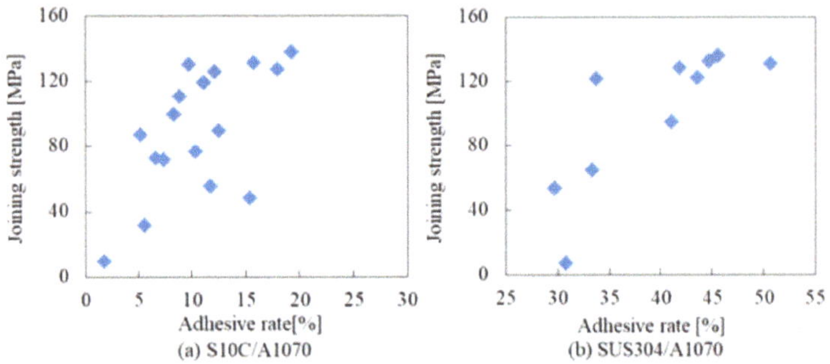

Fig. 16. Relationship between adhesive rate and joining strength.

Summary

The parameters required to achieve forge solid phase bonding was investigated by doing the experiment and FEM.

・The test method that can accurately measure the joining strength was developed.

・In this bonding method, Fe/Al could be bonded under a wide range of the surface area expansion rate and surface pressure. The constant distribution of the surface pressure and the surface area expansion ratio was obtained in the region of the interface of tensile test pieces.

・Using a newly developed test method, joining strength was measured in combinations of S10C/A1070 and SUS304/A1070 by changing reduction. In both combinations, as reduction increased, joining strength was increased.

・Although variation was observed, the effect of reduction was shown in the results of the tensile test. Fe/Al could be bonded strongly by cold-forge solid phase bonding, some test pieces showed joining strength was about 130MPa.

・Some test pieces in SUS304/A1070 were broken in ductility. Observing the BEI of the steel side fracture surface after the tensile test, nearly twice the adhesive rate of SUS304/A1070 was obtained compared with S10C/A1070. It is thought difference in adhesion rate affects fracture form.

References

[1] Fujimoto yuichiro, Urushiyama Yuta, Automotive multi-material strategy (2017) 115-127, in Japanese.

[2] ASM HANDBOOK vol.3 (2016) 133.

[3] H. Yamagisi, High-productivity and high-strength Fe/Al dissimilar metal joining by spot forge welding, Mater. Lett. 278 (2020) 128412. https://doi.org/10.1016/j.matlet.2020.128412

[4] K. Martinsen, Joining of dissimilar materials, CIRP Annals-Manuf. Tech. 64 (2015) 679-699. https://doi.org/10.1016/j.cirp.2015.05.006

[5] R.F. Tylecote, The Solid Phase Welding of Metals, Edward Arnold Ltd., (1968) 226-318.

[6] Y. Fang, W. Gai, H. Yang, B. Lu, Modelling and lock-seam effects on compressive behaviour of concrete-filled helical corrugated steel tubes, Structure 58 (2023) 105321. https://doi.org/10.1016/j.istruc.2023.105321

[7] Y. Abe, K. Mori, T. Kato, Joining of high strength steel and aluminium alloy sheets by mechanical clinching with dies for control of metal flow, J. Mater. Process. Tech. 212 (2012) 884-889. https://doi.org/10.1016/j.jmatprotec.2011.11.015

[8] N. Bay, Trans. ASME, J. Eng. Ind. 101 (1979) 121-127.

[9] X. Liu, Analysis of process parameters effects on friction stir welding of dissimilar aluminum alloy to advanced high strength steel, Metal. Des. 59 (2014) 50-62. https://doi.org/10.1016/j.matdes.2014.02.003

[10] S.H. Carpenter, R.H. Wittman, Annual Review of Mater. Sci. (1975) 177-199.

[11] M.G. Nicholas, R.M. Crispin, Diffusion bonding stainless steel to alumina using aluminium interlayers, J. Mater. Sci. 17 (1982) 3347-3360. https://doi.org/10.1007/BF01203505

[12] G. Chen, X. Chang, G. Liu, Q. Chen, F. Han, S. Zhang, Z. Zhao, Formation of metallurgical bonding interface in aluminum-steel bimetal parts by thixotropic-core compound forging, J. Mater. Process. Tech. 283 (2020) 116710. https://doi.org/10.1016/j.jmatprotec.2020.116710

[13] D. Chicot, Mechanical properties of magnetite (Fe3O4), hematite (α-Fe2O3) and goethite (α-FeO·OH) by instrumented indentation and molecular dynamics analysis, Mater. Chem. Phys. 129 (2011) 862-870. https://doi.org/10.1016/j.matchemphys.2011.05.056

[14] M.K. Tripp, The mechanical properties of atomic layer deposited alumina for use in micro- and nano-electromechanical systems, Sensors and Actuators A: Physical 130-131 (2006) 419-429. https://doi.org/10.1016/j.sna.2006.01.029

[15] H. Yamagishi, Spot Forge-Welding for Rapid Dissimilar Joining of Fe to Al to Produce an Intermetallic Compound-Free Interface, Mater. Trans. 62 (2021) 1576-1582. https://doi.org/10.2320/matertrans.MT-M2021080

[16] Y. Miwada, T. Ishiguro, E. Abe, N. Yukawa, T. Ishikawa, T. Suganuma, Cold Forge Spot-bonding of High Tensile Strength Steel and Aluminum Alloy Sheets, Procedia Eng. 81 (2014) 2006-2011. https://doi.org/10.1016/j.proeng.2014.10.272

[17] P. Groche, S. Wohletz, A. Erb, A. Altin, Effect of the primary heat treatment on the bond formation in cold welding of aluminum and steel by cold forging, J. Mater. Process. Tech. 214 (2014) 2040-2048. https://doi.org/10.1016/j.jmatprotec.2013.12.021

[18] R. Matsumoto, Improvement in bonding strength by applying circumferential sliding in cold copper/aluminum forge-bonding, J. Mater. Process. Tech. 307 (2022) 117685. https://doi.org/10.1016/j.jmatprotec.2022.117685

[19] L. Straka, I. Corny, J. Borzikova, Strojarstvo 51 (2009) 633-640.

[20] S. Ossenkemper, C. Dahnke, Analytical and experimental bond strength investigation of cold forged composite shafts, J. Mater. Process. Tech. 264 (2019) 190-199. https://doi.org/10.1016/j.jmatprotec.2018.09.008

Magnetic pulse-assisted semi-solid brazing of Cu/Al tube joints and oxide film removal mechanism

RUI Zhenglei[1,a*] and HUANG Shangyu[1,b*]

[1]School of Materials Science and Engineering, Wuhan University of Technology No.122, Luoshi Road, Hongshan District, Wuhan City, Hubei Province, China

[a]ruizhenglei@whut.edu.cn, [b]huangshy@whut.edu.cn

Keywords: MPASSB, Oxide Film Removal, Shear Rheology, Element Diffusion

Abstract. In this paper, the advantages of magnetic pulse welding, brazing and semi-solid forming are combined to propose magnetic pulse-assisted semi-solid brazing (MPASSB) to join copper and aluminum tubes. Based on 3D electromagnetic-structural-fluid finite element simulations, the high-speed impact of the aluminum tube on the filler metal will drive the high-speed shear rheology of the semi-solid filler metal. Based on 2D FEM-SPH finite element analysis, the aluminum tube will have a strong interaction with the filler metal during the high-speed impact on the filler metal, which is conducive to the removal of the oxide film on the surface of the copper-aluminum tube and the promotion of element diffusion. Microscopic characterization of the joints revealed that the interfacial oxide film had been completely removed, achieving interfacial metallurgical bonding, and no brittle copper-aluminum intermetallic compounds were found at the interface, with a shear strength of 40 MPa.

Introduction

Copper-aluminum dissimilar joints save energy, can be lightweight, and are widely used in vehicles, ships and other industries. However, when copper-aluminum joints are made, the dense aluminum oxide film on the aluminum surface prevents the surface wetting of the base material and metallurgical bonding of the joint. In addition, layers of brittle intermetallic compounds (IMCs), such as Cu_9Al_4 and $CuAl_2$, are easily formed at the joint interface, which seriously affect the strength and fatigue performance of the joint [1].

In order to better realize the effective metallurgical bonding at the interface of Cu/Al joints, fusion welding, ultrasonic welding, magnetic pulse welding, and brazing methods have been widely used. The traditional fusion welding technology has a large welding heat input, and it is difficult to control the welding heat input, so the brittle IMCs cannot be effectively suppressed and the mechanical properties are poor [2]. To further reduce the welding heat input, solid-phase welding (ultrasonic welding [3], friction stir welding [4], explosive welding [5], magnetic pulse welding [6], and resistance welding [7]) is widely used for copper-aluminum connections. However, the research and application of the above methods for the connection of small-diameter tubes are rarely reported.

Brazing can avoid the direct contact reaction of heterogeneous base materials through the transition buffer of filler metal, which can inhibit the generation of brittle IMCs to a certain extent. At the same time, it can also effectively alleviate the differences in the physical properties of base materials, and improve the strength of the joints and fatigue load resistance. Flame brazing is mostly used in the industrial production of Cu/Al tubes, but too high brazing temperatures are prone to the dramatic growth of brittle phases, which seriously affect the joint properties.

In order to further reduce the brazing temperature, this paper proposes a new method of MPASSB by combining the composite advantages of electromagnetic forming, semi-solid forming and brazing. This method uses electromagnetic force to promote high-speed plastic deformation of the aluminum tube and, at the same time, the aluminum tube impact and extrusion of semi-solid filler metal to promote high-speed shear rheology of the filler metal. Oxide film will be removed and element diffusion will be promoted. Moreover, the use of semi-solid forming characteristics will reduce the brazing temperature and inhibit the growth of brittle IMCs. In the paper, the effect of the assembly method on the removal of interfacial oxide layer and the joint microstructure will be analyzed in combination with finite element simulation and experimental results.

Experiment

The outer tube was made of 1060-O aluminum tube with an outer diameter of 19 mm and a thickness of 1 mm, and the inner tube was made of T2 copper tube with an outer diameter of 15 mm and a thickness of 1.5 mm. The filler metal is Zn-15Al, with a thickness of 400μm and an axial length of 10mm. Before welding, the outer wall of the copper tube, the inner wall of the aluminum tube, and the surface of the filler metal were polished and cleaned with alcohol to remove the oxide film and oil on the surface of the tubes and filler metal.

When clamping forming, the copper and aluminum tubes, filler metal, coils and field shaper are assembled according to Fig. 1(a), and the height between the top of the aluminum tube and the top of the inner wall of the field shaper is 12 mm. After the assembly is completed, the capacitor is charged and the clamping forming discharged, causing the Cu/Al tubes joint to form a stop, as shown in Fig. 1(a). After the clamping forming is completed, the capacitor is recharged after the area of the tube to be brazed is moved to the working area of the field shaper, and discharged again after the brazing temperature is reached, as shown in Fig. 1(b).

The outer tube impinges on the semi-solid filler metal at high speed under the action of electromagnetic force, causing it to flow at high speed, and the oxide film on the tube wall is crushed and sheared by the solid phase particles inside the filler metal, thus breaking and removing it, as shown in Fig. 2(c-d).

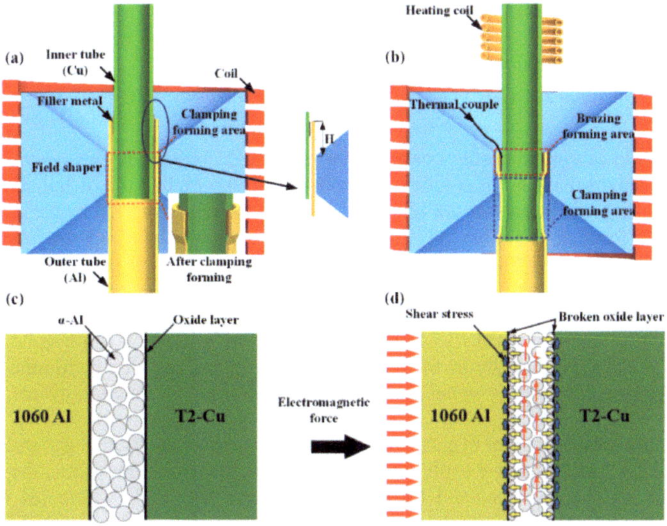

Fig. 1. Principle of MPASSB process: (a) clamping forming; (b) brazing forming; (c) state of tubes and filler metal before brazing forming; (d) oxide film removal and shear rheology of filler metal during brazing forming.

Simulation
For the 3D simulation, the air cell is added during brazing forming to provide a shear rheological region for the filler metal, as shown in Fig. 2(a-b). The finite element analysis region is shown in Fig. 2(b). Meanwhile, the Johnson-Cooker intrinsic model was used to characterize the high strain rate plastic deformation behavior during the forming process [8], and the filler metal shear rheology model was measured to be 1.79 Pa·s. The 2D simulation was performed by loading the electromagnetic force obtained from the 3D simulation into the 2D simulation.

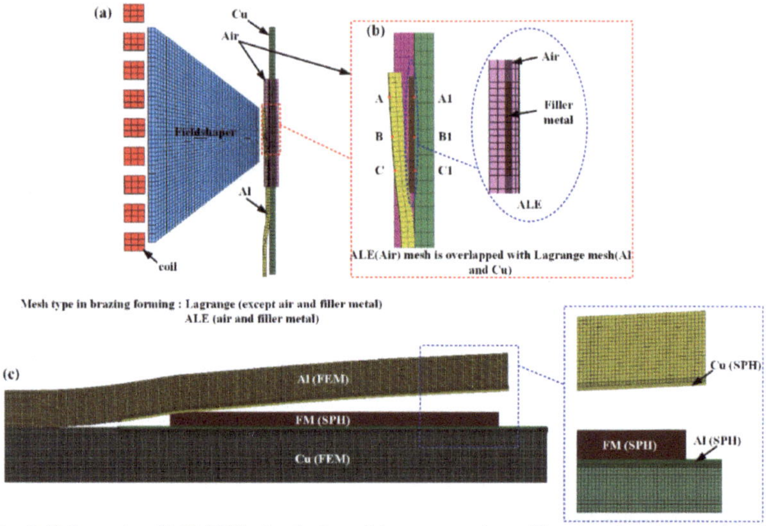

Fig. 2. Schematic of MPASSB simulation: (a) cross section of brazing forming model; (b) enlarged view of brazing forming area; (c) FEM-SPH model.

Results
Fig. 3 shows the compressive stress on the filler metal at different moments of brazing forming. Since the bottom of the aluminum tube is already in contact with the filler metal at the end of the clamping forming, the bottom of the filler metal is the first to be extruded by the aluminum tube during brazing forming, as shown in Fig.3(a). As time progresses, the aluminum tube continues to deform under the action of the electromagnetic force, and the contact area with the filler metal gradually increases, so that the top of the aluminum tube contacts the top of the filler metal at 0.027ms, and continues to extrude the top of the filler metal, as shown in Fig. 3(b). The middle area of the braze material is finally squeezed as shown in Fig. 3 (c). It can also be seen that the filler metal is ejected outwardly by the continuous squeezing of the aluminum tube. At the same time, the pressure on the filler metal decreases as the deformation time progresses, as shown in Fig. 3(d-f).

Fig. 4 shows the distribution of compressive and shear stresses in different areas of the interface between the Cu/Al tubes and the filler metal. As can be seen from the figure, the compressive stress at the interface between the Cu/Al tubes and the filler metal decreases from the top downward. Among them, the compressive stress in the top region of the interface is the largest, respectively 551.5 MPa and 494.1 MPa; the compressive stress in the middle region of the interface is the second largest, respectively 364.8 MPa and 410.6 MPa; and the compressive stress at the bottom of the interface is the smallest, respectively 113.4 MPa and 73.3 MPa. The shear stress at the interface between the Cu/Al tubes and the filler metal are slightly different compared to the compressive stress. Among them, the top region of the interface is the largest, respectively 56.2 MPa and 39.5 MPa. However, the values of shear stress in the middle and bottom regions of the interface between the aluminum/filler metal (Al/FM) or the copper/filler metal (Cu/FM) are essentially the same separately. The values in the middle and bottom region of the Al/FM interface are 33.1 MPa and 31.5 MPa, respectively, while the middle and bottom region of the Cu/FM interface are 14.9 MPa and 15.2 MPa.

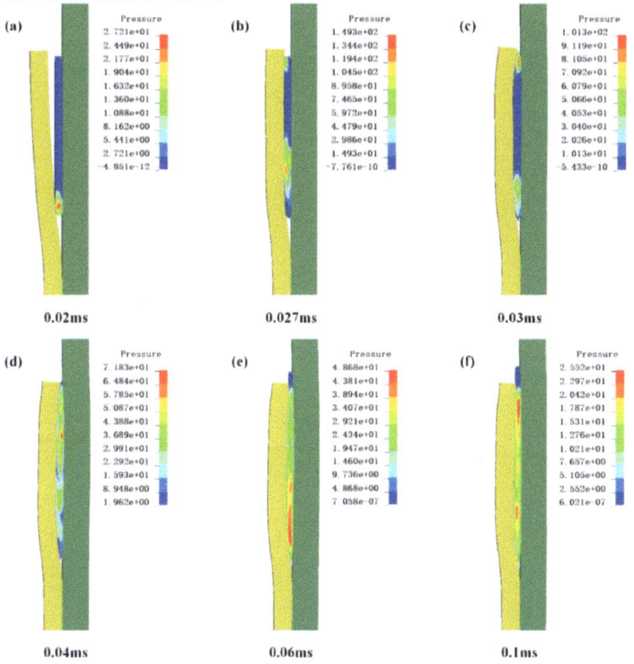

Fig. 3. Compressive stress of filler metal at different moments of MPASSB.

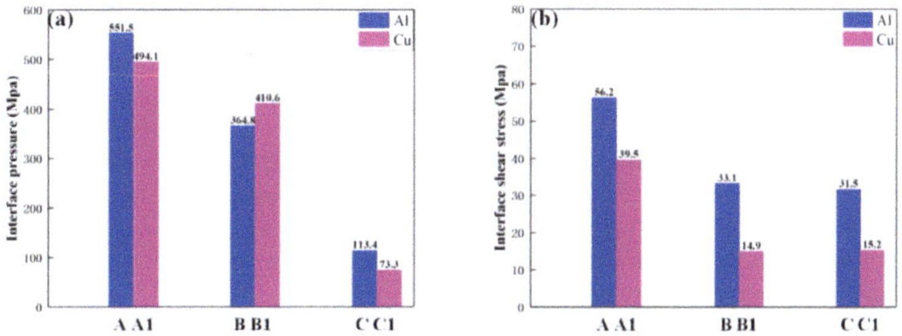

Fig. 4. Interface stress in MPASSB simulation: (a) compressive stress; (b) shear stress.

Fig. 5 shows the results of the Cu/Al tubes interface interaction obtained from the 2D FEM-SPH simulation. From Fig. 5(a-d), it can be seen that the aluminum tube will have certain interactions with the filler metal when it hits the filler metal at a high speed under the action of electromagnetic force, but there will be no waveform interface. After the aluminum tube and the filler metal are completely in contact, the surface of the aluminum tube will be peeled off by the filler metal to a certain extent under the combined effect of interfacial compressive and shear stresses, as shown in Fig. 5(e-f).

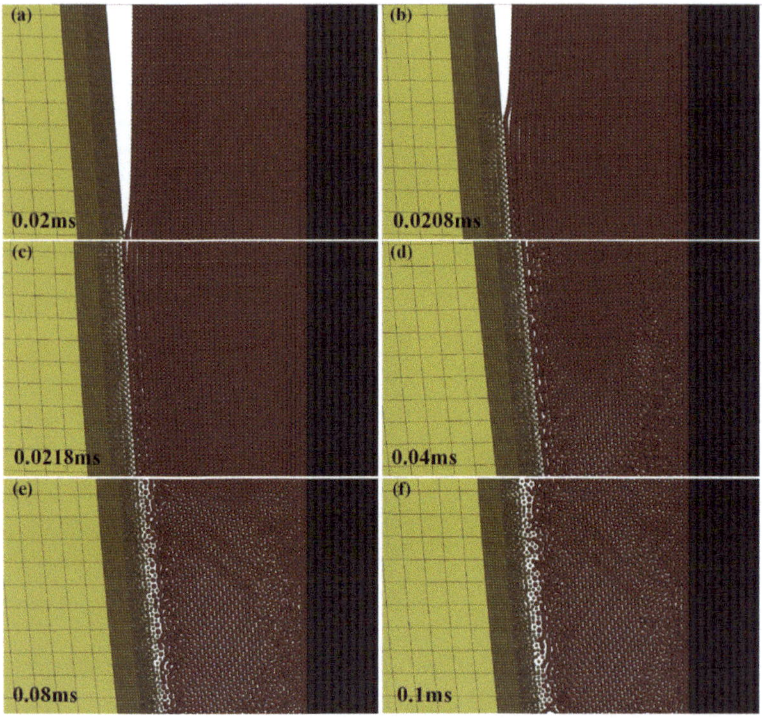

Fig. 5. Cu/Al tubes interface interactions of MPASSB.

The following Fig. 6 shows the organization and morphology of the top, middle and bottom regions of the specimen, and the analyzed area is shown in Fig. 2(b). As can be seen from the figure, the oxides at the interface of the Al/FM are completely removed, and α-Al solid solution perpendicular to the aluminum matrix is formed, which is conducive to improving the mechanical properties of the joint. Less α-Al and a large number of Cu/Zn phases were formed at the brazing joints. At the Cu/FM interface, effective element diffusion was happened in the top, middle and bottom regions, with the organization distribution of $Al_{4.2}Cu_{3.2}Zn_{0.7}$ and Cu/Zn phases, and no Cu/Al brittle IMCs were found at the interface.

Materials Research Proceedings 44 (2024) 193-200 https://doi.org/10.21741/9781644903254-21

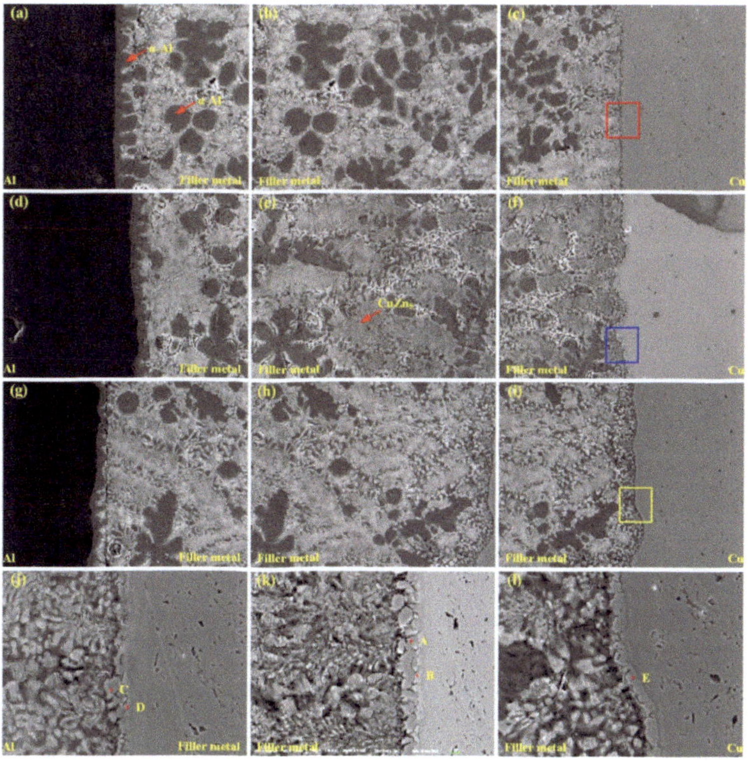

Fig. 6. Microstructure of the MPASSB joints: (a) top interface of Al/FM (b) top of filler metal; (c) top interface of Cu/FM; (d) middle interface of Al/FM; (e) middle of filler metal; (f) middle interface of Cu/FM; (g) bottom interface of Al/FM (h) bottom of filler metal; (i) bottom interface of Cu/FM; (j) enlargement of red box area in Fig. 6(c); (k) enlargement of blue box area in Fig. 6(f); (j) enlargement of yellow box area in Fig. 6(i).

Table 1. EDS analysis results of the points marked in Fig. 6.

Position	O	Al	Cu	Zn	Possible phase
A	1.5	38.1	47.8	12.7	$Al_{4.2}Cu_{3.2}Zn_{0.7}$
B	0.7	3.3	27.4	68.5	$CnZn_5$
C	0.8	32.6	49.7	17.0	$Al_{4.2}Cu_{3.2}Zn_{0.7}$
D	0.6	4.5	32.4	62.5	Cn_5Zn_8
E	0.6	3.6	28.4	67.4	$CnZn_5$

Fig. 7 shows the surface analysis and line analysis of the interface elements on the Al/FM and Cu/FM. As can be seen from the Fig. (a-f) and Fig. (a1-f1), at 440°C, Cu, Al and Zn atoms diffuse to a certain extent, but the ability of Cu atoms to diffuse to the Al/FM interface is weaker, which is mainly limited by the shorter brazing time and brazing temperature. Al atoms are relatively more on the Cu/FM interface, and the Al atoms in the filler metal diffuse to the interface to a greater degree, forming a ternary Al-Cu-Zn alloy phase. Finally, the mechanical properties were also tested in this paper with a shear strength of 40 MPa.

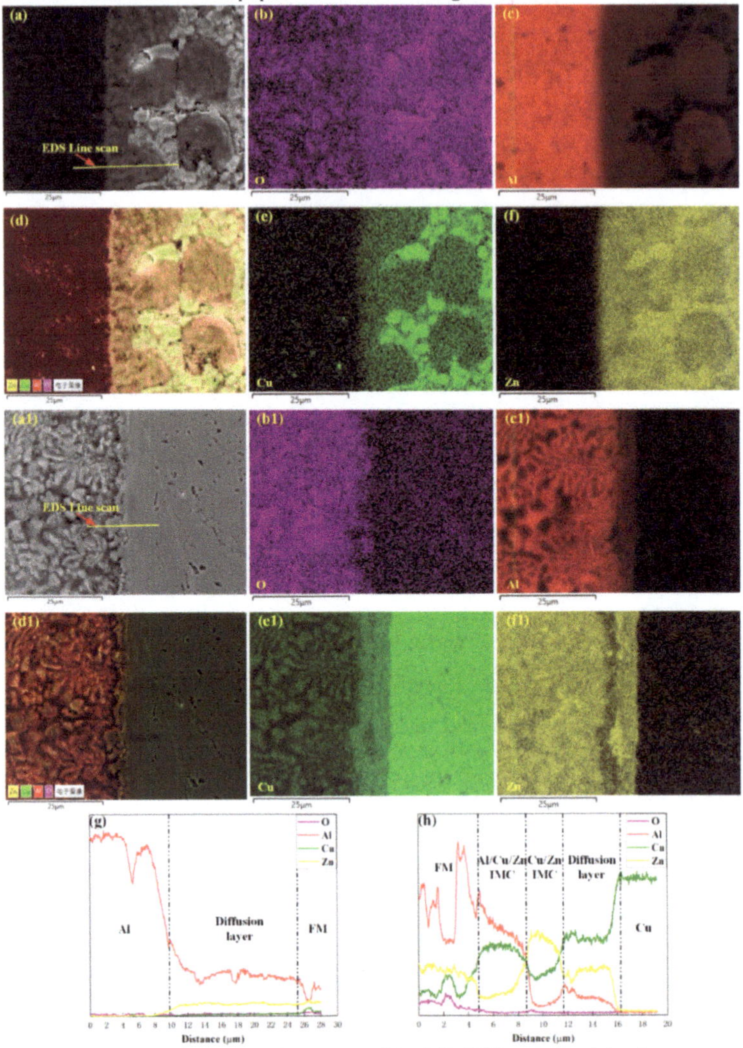

Fig. 7. EDS analysis at the top analysis area of the MPASSB sample: (a) enlargement of red green area in Fig. 6(a); (b) O; (c) Al; (d) combined EDS elemental mapping of Fig. 7(a); (e) Cu; (f) Zn; (a1) enlargement of red green area in Fig. 6(j); (b1) O; (c1) Al; (d1) combined EDS elemental mapping of Fig. 7(a); (e1) Cu; (f1) Zn; (g) elemental line scan analysis at the Cu/FM interface shown in Fig. 7(a); (h) elemental line scan analysis at the Cu/FM interface shown in Fig. 7(a1).

Conclusion

(1) MPASSB can be used for Cu/Al tubes joining;

(2) The interfacial compressive and shear stresses generated when the aluminum tube impacts the filler metal at high speed are the key to remove the oxide film and achieve interfacial metallurgical bonding;

(3) No visible Cu/Al brittle IMCs was found in the joints, proving that MPASSB can largely inhibit the growth of brittle IMCs.

References

[1] Y. Zhou, C. Li, T. Shen, D. Chen, X. Wang, Y. Ma,Interfacial microstructure of multi-layered Al-Cu joint by electromagnetic pulse welding, J. Mater. Res. Technol. 25 (2023) 2446-2454. https://doi.org/10.1016/j.jmrt.2023.06.034

[2] T. Ma, S. Panda, Characterisation of Cu-Al alloy lap joint using TIG Welding, Cirp J. Manuf. Sci. Technol. 35 (2021) 454-459. https://doi.org/10.1016/j.cirpj.2021.07.009

[3] H.T. Fujii, H. Endo, Y.S. Sato, H. Kokawa, Interfacial microstructure evolution and weld formation during ultrasonic welding of Al alloy to Cu, Mater. Charact. 139 (2018) 233-240. https://doi.org/10.1016/j.matchar.2018.03.010

[4] Y.Y. Zuo, P. Gong, S.D. Ji, Q.H. Li, Z.W. Ma, Z. Lv, Ultrasound-assisted friction stir transient liquid phase spot welded dissimilar copper-aluminum joint, J. Manuf. Process. 62 (2021) 58-66. https://doi.org/10.1016/j.jmapro.2020.11.019

[5] G. Carvalho, I. Galvao, R. Mendes, R. Leal, A. Loureiro, Weldability of aluminium-copper in explosive welding, Int. J. Adv. Manuf. Technol. 103 (2019) 1-11. https://doi.org/10.1007/s00170-019-03841-9

[6] J.S. Li, R.N. Raoelison, T. Sapanathan, Y.L. Hou, M. Rachik, Interface evolution during magnetic pulse welding under extremely high strain rate collision: mechanisms, thermomechanical kinetics and consequences, Acta Mater. 195 (2020) 404-415. https://doi.org/10.1016/j.actamat.2020.05.028

[7] Y. Zhang, L. Han, T.L. Sun, et al. Automatic Control System of Copper Pipe and Aluminum Pipe Butt Resistance Welding Machine; proceedings of the International Conference on Advanced in Control Engineering and Information Science (CEIS), Dali, PEOPLES R CHINA, F Aug 18-19, 2011. Elsevier Science Bv: AMSTERDAM, 2011.

[8] C.G. Wang, Q.X. Liu, G.Y. Li, et al. Study on mechanical properties and microstructural feature of magnetic pulse welding joint between Cu and Al sheets, Int. J. Adv. Manuf. Technol. 113 (2021) 1-13. https://doi.org/10.1007/s00170-021-06737-9

Solid state recycling operations for AA7075

HAN Chenyang[1,a], SIMONETTO Enrico[1], GHIOTTI Andrea[1], BRUSCHI Stefania[1], LATIF Abdul[2,b*], INGARAO Giuseppe[2], MICARI Fabrizio[2], FRATINI Livan[2]

[1]Department of Industrial Engineering, University of Padova, al. Via Venezia 1, Padova 35131, Italy

[2]Department of Engineering, University of Palermo, Viale delle Scienze E, 90128 Palermo, Italy

[a]chenyang.han@studenti.unipd.it, [b]abdul.latif@unipa.it

Keywords: Solid-State Recycling, Forming, AA Alloy

Abstract. Solid-state recycling offers a sustainable solution characterized by low energy consumption and heightened product efficiency compared to conventional melting-based methods. Among these, Friction Stir Extrusion processes emerge as an innovative and promising category within solid-state recycling, showcasing the ability to directly repurpose machining chips into solid products. However, as these processes primarily yield semi-finished products -typically in the form of wires, rods, or small billets- subsequent post-recycling forming processes become imperative for achieving net-shape or near-net-shape components. Consequently, a thorough evaluation of the mechanical properties of materials derived from solid-state processes is essential for ensuring the dependable design of process chains. This paper focuses on the calibration of constitutive parameters for raw materials obtained from aluminium chips within a novel process chain rooted in solid-state recycling operations. Cylindrical billets, resulting from friction stir consolidation (FSC), are tested in compressive conditions and investigated with respected of processing conditions to obtain ultra-fined grained materials through sever plastic deformation processes.

Introduction

The global aluminium industry faces the challenge of reducing carbon dioxide emissions and the greenhouse effect these emissions can have on global warming and climate change[1]. These emissions arise both directly from the production process and from the recycling process[2]. For aluminium alloy recycling, the traditional remelting recycling method usually results in re-oxidation of the metal, resulting in permanent loss of the metal. The remelting recycling process also generates higher energy consumption and recycling costs[3]. Solid State Recycling (SSR) process can avoid the remelting step and thus improve the recycling efficiency. Compared with conventional remelting recycling, direct conversion of aluminium and its alloy chips into compact specimens by extrusion leads to savings of 40% in materials, 26–31% in energy, and 16–60% in labour [4].

Friction Stir consolidation (FSC) is one of the SSR processes that has effectively recycled the metal chips into solid blocks, which can effectively reduce the impact on the environment during the recycling process. At the same time, compared with other SSR processes, FSC can provide more adequate oxide breaking and solidification effects[5].

The present study focuses on the calibration of constitutive parameters for raw materials obtained from aluminium chips within a novel process chain rooted in solid-state recycling operations. AA7075 alloy chips were friction stir consolidated into the shape of a billet, and subsequently respective billets were made into cylinders. The cylinders were subjected to compression testing, and the cut billets were subjected to microhardness testing. Finally, the thermal conditions in backward extrusion processes were investigated.

Application case

The application case refers to the cold impact backward extrusion of cans used to manufacture pressurized dispensers for cosmetic products. The typical manufacturing chain which counts several steps: (i) lubricant deposition on aluminium disks, by means a tumbling barrel that rotates at 16 RPM for 20 minutes in order to lubricate the external surface of the disks, (ii) impact extrusion of aluminium disks, (iii) trimming of the extruded cylinder because the extrusion process leaves the can slightly wavy at the top, washing and drying, (iv) internal and external spray coating, which is then cured in an oven in order to extend product life and preserve the integrity of the contents, (v) outside painting and printing and ending with (vi) edging and necking. The most severe step of the process chain is represented by the impact backward extrusion, since it exhibits the highest levels of pressure, temperature and surface expansion, which are particularly severe for the lubrication and the dies service life.

Materials

Specimens.

The starting material (as-received material) was AA7075 sheet with chemical composition listed in Table 1. The as -received material (see Fig. 1) is characterized by average values of 150 HV and 155 μm for hardness and grain size, respectively. The sheet was reduced into chips through milling operations with tool rotational speed of 1250 rpm, feed rate of 280 mm/min, and cutting depth of 1 mm. The selection of aluminum alloy AA 7xxx in the current studies was driven by its extensive use in the aerospace manufacturing industry where almost 90% of the input material turns into scrap [6]. In addition, AA7075 represents a class of wrought aluminum alloys with higher hardness value and relatively lower ductility. Thus, assessing the formability of AA7075 recycled billet can provide a better understanding to forecast quality of recycled billets with relatively good ductility such as AA6xxx, AA2xxx, and AA5xxx. This assessment can pave the way for transforming scrap materials into a value-added component for potential industrial applications.

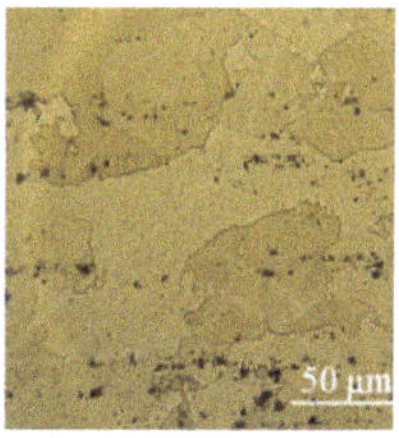

Fig. 1. *AA7075 as received.*

Table 1. *Nominal chemical composition of the AA7075.*

Al	Zn	O	Mg
88.07	5.07	2.60	42.25
Cu	**Cr**	**Fe**	**Mn**
1.76	0.2	0.03	0.02

Punch and dies.

The tool steel used for the punches is the EN HS 6-5-2 alloyed steel, commercially available with the name of Böhler S600 (see the chemical composition in **Table 2**).

The steel was thermally treated, in order to obtain a surface hardness of 62 (±1.5) HRC. The punches were machined to obtain a final surface roughness S_a of 1.735 (±0.001) μm. In the case of the die, the tool steel is the EN X37CrMoV51KU, commercially available with the name of Böhler W300 (see the chemical composition in **Table 2**).

The die steel was thermally treated, to obtain a surface hardness of 54 (±1.5) HRC. The final surface roughness Sa is equal to 0.563 (±0.001) μm. **Figure. 2** shows the tools surface and metal disc non coated in the as-delivered conditions.

Materials Research Proceedings 44 (2024) 201-208 https://doi.org/10.21741/9781644903254-22

Table 1. *Chemical composition of the tools grade.*

	C	Mn	Si	Cr	Mo	V	W
S600	0.9	-	-	4.1	5.0	1.8	6.2
W300	0.4	0.4	1.1	5	1.3	0.4	-

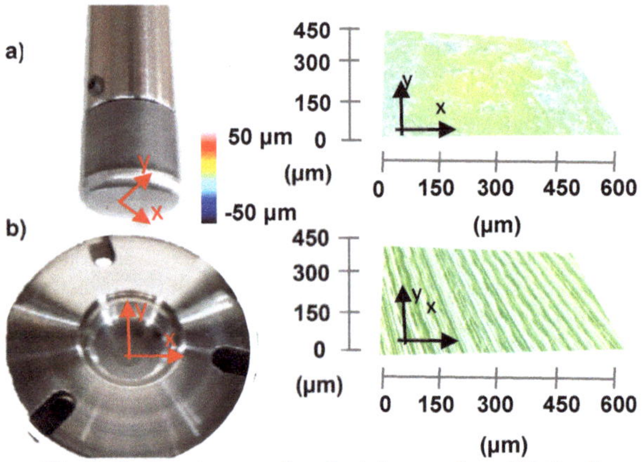

Fig. 2. *Surface topography of: a) the punch and b) the die.*

Experimental
Test specimens production.
The cleaning operation of the chips was performed by submerging them in acetone for 30 minutes. Then 15 g of input chip mass was loaded in a cylindrical die with a nominal diameter of 25.4 mm and then compacted at 5 kN force by a H13 steel cylindrical tool with a 25 mm diameter. The die and pressing tool system was integrated with ESAB-LEGIO (Fig. 3), a dedicated friction stir welding machine. Finally, the chips were consolidated at 20 kN with a high tool rotational speed of 1500 rpm, a processing time of 60 s and by adding the material in a single step, which means that the chips are consolidated in a single run [7].

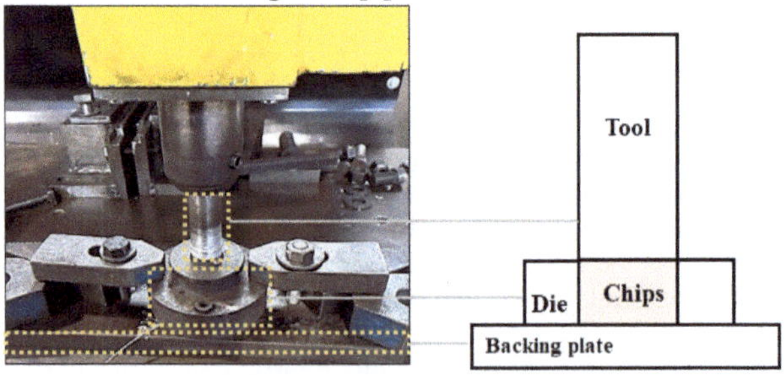

Fig. 3. *Experimental setup and process schematic of FSC process*

 Materials Research Forum LLC
https://doi.org/10.21741/9781644903254-22

Compression tests for material characterization.
Cylindrical samples for compression tests, having a diameter of 6 mm and a height of 8 mm, were obtained from stirred billets. Compression tests were performed on a Gleeble™ 3800 machine. The tests were carried out at 100°C, 200°C, and 300°C and a constant strain rate 1 s^{-1}. The result as is shown at Fig. 4.

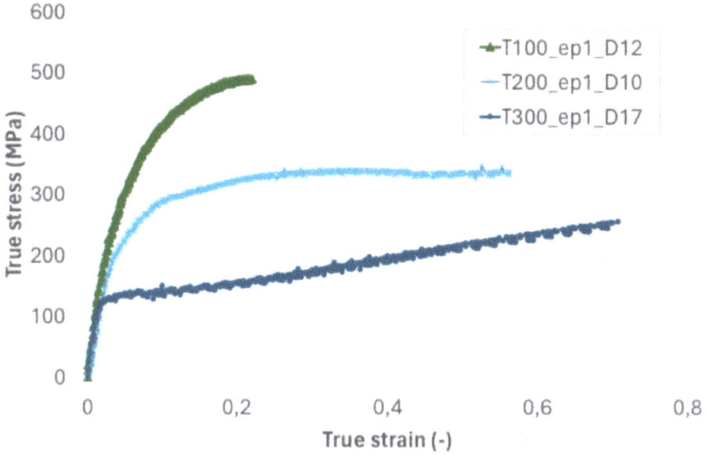

Fig. 4. *Compression test at different temperatures.*

Hardness tests.
Hardness (HV) measurements were carried out to assess uniformity of the mechanical properties in the manufactured sample. The hardness was evaluated by the Vickers hardness test. A load of 49 N (5 kg) was applied for a dwell time of 15 s along four loci along the cylindrical direction, i.e. at radius, r = 0 mm (L0), 6.5 mm, 9 mm, and 12.25 mm (L1, L2, and L3, respectively) as illustrated in Fig. 5. A constant pitch of 0.5 mm was set for the load points on each line. The highest hardness was shown by the central and top parts of the billet. The hardness values decrease as moving from centre to the edge (external surface). The bottom part exhibited the lowest hardness was values. The hardness of the whole section of the billet has been represented in Fig. 5. This gradation in hardness is caused by the process mechanics of the FSC process [8].

Axial distribution of hardness: The rotating tool is considered the hub of the heat. Maximum heat is gained by the top surface, and sound bonding is developed between the layers of the chips, and therefore, the top surface has high hardness. The intensity of the heat reaching the bottom surface is quite low and is not sufficient to cause a full solid bonding of the layers of the chip at the bottom, and thereby, an unconsolidated zone was observed at the bottom of the billet with a relatively low hardness value [7].

For radial hardness distribution: The hardness also follows a decreasing trend along the radial direction. It decreased from the central line to the line near the external surface [9]. L1 and L2 measuring loci, show a similar trend to L0 one but a slightly late drop occurred at L1 due to occurring material flow [10]. In turn, L3 shows an earlier drop in hardness value due to the upward movement of the bottom material at that radius.

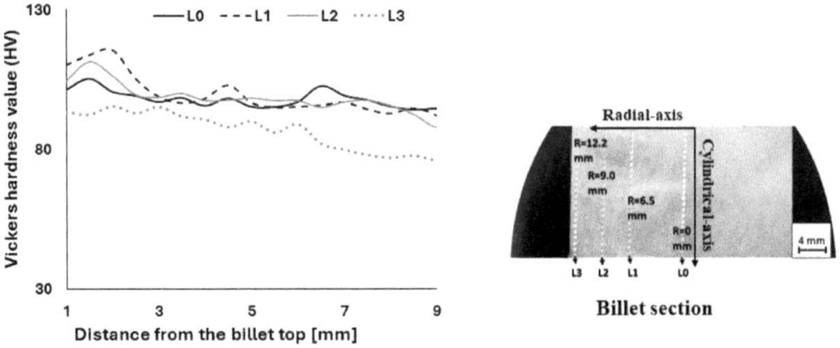

Fig 5. Material hardness along the FSC billet section

FE model

The FE software Deform 3D was used to simulate the impact extrusion process, using the Optimizer Tool embedded in the software suite for the inverse analysis. By means of the algorithm, the user can submit the forming problem directly to the software application, by choosing the criteria to be optimized and the variables to be modified. On this basis, the difference between the values of extrusion forces, recorded in the experimental tests and obtained from FE simulations respectively, was selected as objective function to be minimized, choosing the friction factor *m* according to the Tresca model as the parameter governing the convergence. The flow stress model used in the simulations was the Hansel-Spittel (H-S) equation, to take into account the effects of the temperature due to the adiabatic heating during the deformations:

$$\sigma = A \cdot e^{m_1 T} \cdot \varepsilon^{m_2} \cdot \dot{\varepsilon}^{m_3} e^{m_4/\varepsilon} \quad (1)$$

where σ is the material trues stress, ε the material equivalent true strain, $\dot{\varepsilon}$ the equivalent material strain rate, T the temperature and A, m_1, m_2, m_3, m_4 the constitutive parameters. The flow stress was investigated in the same range of temperatures and strain rates of the industrial process. Through a nonlinear regression analysis, the constants of the H-S equation were calculated to be implemented in the FE code. The simulations were reproduced under the conditions of **Table 4.**

***Table 4:** The mail process parameters used for the simulation and constitutive parameters for the material.*

Temperature [°C]			20, 100, 200	
Strain [-]			1	
Strain rate [s^{-1}]			1	
A	**m_1**	**m_2**	**m_3**	**m_4**
1309,03998	-0,00545	0,26205	0,00280	-0,00091

Punch, die and backing plate were modelled rigid bodies with a mesh size of 15000 elements while the FSC recycled billet was considered a plastic object of 35000 elements as shown in Fig. 6. Three different numerical simulations at punch speeds of 50 mm/s, 100 mm/s, and 200 mm/s, each with a total punch displacement of 7.5 mm, were performed under a constant friction coefficient of 0.1 and a heat transfer coefficient of 11 N/sec/mm/°C. A clearance of 2.5 mm was maintained between the punch and die for the billet extrusion.

To compare the influence of punch speed on the final state of the extruded billet, temperature profiles for all three simulations were compared specifically along the trajectory comprising 20 points from "A" to "B" as illustrated in Fig. 7. The trajectory was taken along the section of the billet, starting from the centre of the external wall of the billet "A" and completed at centre of the billet bottom surface. The results showed that higher punch speed led to impart higher temperature (Fig.7)

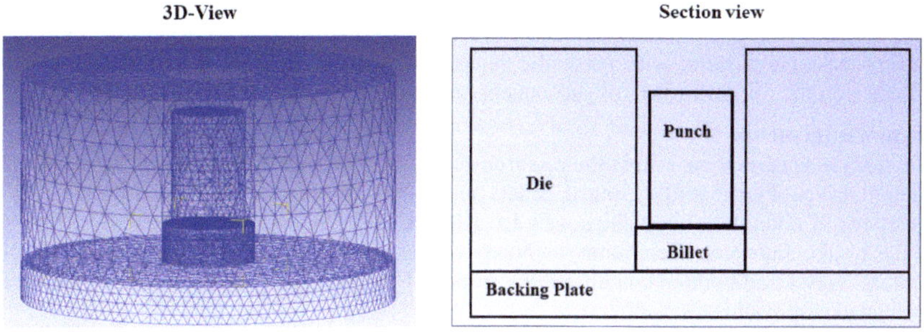

Fig. 6. *Numerical Model for extrusion process*

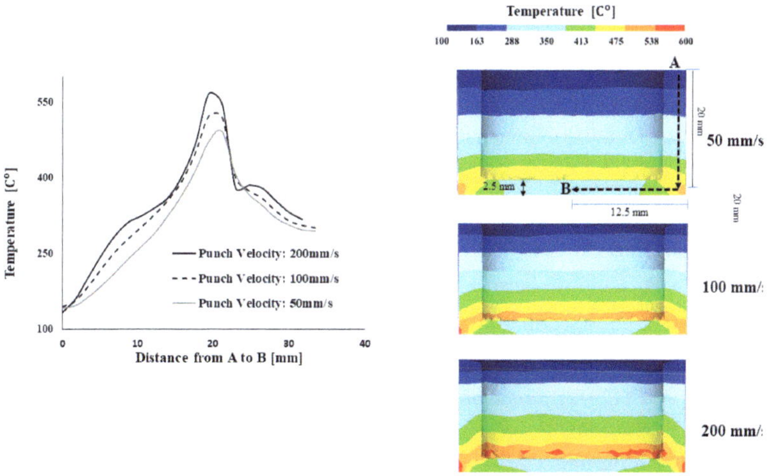

Fig. *7 Influence of punch speed on temperature of extruded profile*

Conclusions

The paper focuses on the calibration of constitutive parameters for raw materials obtained from aluminium chips within a novel process chain rooted in solid-state recycling operations. Cylindrical billets, resulting from friction stir consolidation, have been tested in compressive conditions and investigated with respected of processing conditions to obtain ultra-fined grained materials through sever plastic deformation processes. The description of the microstructural distribution has been described in previous studies [7]. The microstructure of the bottom part remained unaltered during the FSC process. Moving toward the top part of the billet, changes in microstructure (grain growth) and the improvement in billet quality (disappearance of defects) were noticed. Numerical models for the extrusion operation have been developed to assess the main process parameters, with particular regard to temperature since it might have a relevant influence on the microstructure of subsequent operations.

Acknowledgements

This study was carried out within the research activities of project "Finalizing processes for multi-material based Functionally Graded billets and wires obtained through solid state recycling operations of aluminum alloy chips – FULL RECYCLE" (2022X5RPSM PRIN 2022), financed by the by the European Community – Next Generation EU. This manuscript reflects only the Authors' views and opinions, neither the European Union nor the European Commission can be considered responsible for them.

References

[1] Saevarsdottir, G., Kvande, H., & Welch, B. J. (2020). Aluminum production in the times of climate change: The global challenge to reduce the carbon footprint and prevent carbon leakage. *Jom*, *72*, 296-308.

[2] Latif, A., Ingarao, G., & Fratini, L. (2022). Multi-material based functionally graded billets manufacturing through friction stir consolidation of aluminium alloys chips. *CIRP Annals*, *71*(1), 261-264.

[3] Duflou, J. R., Tekkaya, A. E., Haase, M., Welo, T., Vanmeensel, K., Kellens, K., ... & Paraskevas, D. (2015). Environmental assessment of solid state recycling routes for aluminium alloys: can solid state processes significantly reduce the environmental impact of aluminium recycling?. *Cirp Annals*, *64*(1), 37-40.

[4] Gronostajski, J., Marciniak, H., & Matuszak, A. (2000). New methods of aluminium and aluminium-alloy chips recycling. *Journal of materials processing technology*, *106*(1-3), 34-39.

[5] Puleo, R., Latif, A., Ingarao, G., Di Lorenzo, R., & Fratini, L. (2023). Solid bonding criteria design for aluminum chips recycling through Friction Stir Consolidation. *Journal of Materials Processing Technology*, *319*, 118080.

[6] Allwood, J. M., Cullen, J. M., Carruth, M. A., Cooper, D. R., McBrien, M., Milford, R. L., ... & Patel, A. C. (2012). *Sustainable materials: with both eyes open* (Vol. 2012). Cambridge, UK: UIT Cambridge Limited.

[7] Latif, A., Ingarao, G., Gucciardi, M., & Fratini, L. (2022). A novel approach to enhance mechanical properties during recycling of aluminum alloy scrap through friction stir consolidation. *The International Journal of Advanced Manufacturing Technology*, *119*(3), 1989-2005.

[8] Latif, A., Ingarao, G., & Fratini, L. (2022). Multi-material based functionally graded billets manufacturing through friction stir consolidation of aluminium alloys chips. *CIRP Annals, 71(1)*, 261-264.

[9] Latif, A., Gucciardi, M., Ingarao, G., & Fratini, L. (2021). Outlining the Limits of Friction Stir Consolidation as Used as an Aluminum Alloys Recycling Approach. In Proceedings of the International Conference on Sustainable Design and Manufacturing (pp. 169-180). Springer, Singapore

[10] Buffa, G., Baffari, D., Ingarao, G., & Fratini, L. (2020). Uncovering technological and environmental potentials of aluminum alloy scraps recycling through friction stir consolidation. International Journal of Precision Engineering and Manufacturing-Green Technology, 7, 955-964.

FEM study of process parameters in a novel superplastic forming of titanium alloy Ti-6Al-4V

LIU Jun[1,a], PARAMESWARAN Elangovan[1,b], YAKUSHINA Evgenia[1,c], ZUELLI Nicola[2,d], MILLIKEN David M.[2,e], HUMPHREYS Nick[3,f], GILL Les[4,g] and GREENOUGH Christopher[4,h]

[1]Advanced Forming Research Centre, University of Strathclyde, 85 Inchinnan Drive, Renfrew, PA4 9LJ, United Kingdom

[2]Metallics Research Centre with Boeing, Westway Industrial Business Park, Porterfield Road, Renfrew, PA4 8DJ, United Kingdom

[3]TIMET UK, 120 Holford Drive, Witton, Birmingham, B6 7UG, United Kingdom

[4]SDE Technology, Brixton Way, Harlescott, Shrewbury, Shropshire, SY1 3LB, United Kingdom

[a]j.liu@strath.ac.uk, [b]elangovan.parameswaran@strath.ac.uk, [c]evgenia.yakushina@strath.ac.uk, [d]nicola.zuelli2@boeing.com, [e]david.m.milliken2@boeing.com, [f]nick.humphreys@timet.com, [g]lesg@sde.technology, [h]chrisg@sde.technology

Keywords: Titanium, Superplastic Forming, Hot Drawing, Strain Rate

Abstract. A novel hybrid forming process involving the use of hot drawing along with superplastic forming (SPF) is studied here. The hot drawing stage helps in enhancing the formability and in fast deforming the sheet metal into a hollow shape with desired amount of material draw-in. During the subsequent SPF stage, gas pressure was applied onto the pre-formed part to complete the forming process at a targeted strain rate. With the hybrid process, titanium alloy Ti-6Al-4V sheets have been successfully formed in lab-scaled conditions at 800°C in 16 min. In this paper, finite element modelling (FEM) was used to demonstrate the effects of each stage (hot drawing and SPF) during the process. A plasticity model based on tensile test data was adopted as a material model for simulation. The pressure cycle which was predicted from the simulation has been used in the process to maintain the sheet forming at an average strain rate (e.g. 10^{-3} and 5×10^{-4} s^{-1}). Material draw-in and thickness distribution were used to compare and optimise the process parameters. The simulations have shown the capability of the model to be used for the hybrid superplastic forming process. The influences of varying process parameters, such as punch geometry, blank size, blank-holder force, friction coefficient and pressure cycle, were investigated by the simulations. It was found that the punch geometry and blank size played significant roles on the thickness uniformity of the final part, from which an optimised hot-drawing system that could lead to minimum thinning has been designed by FEM method.

Introduction

Titanium Ti-6Al-4V is a well-known lightweight alloy that exhibits characteristics, such as high strength-to-weight ratios and good superplastic behaviour for aerospace applications [1]. Superplasticity in titanium alloys has been known for over 50 years [2]. Superplastic forming (SPF) [3] is used by the relevant industry to form large and complex components in one operation at the targeted temperature and strain rate. The superplastic behaviour of Ti-6Al-4V alloy is achieved at high temperatures (typically above 900°C) and at low strain rates (usually lower than 10^{-3} s^{-1}) [4].

The formed product has excellent precision and a good surface finish. It also does not suffer from springback or residual stress. The benefits of SPF are its high formability, complex (and intricate) shape possibilities, lower tooling cost, labour and material saving through elimination of parts joining and secondary machining. However, the disadvantages of SPF are mainly the slow forming rate, high forming temperature and poor thickness control. In order to address those limitations, a hybrid SPF process which combines hot mechanical drawing and blowing has been developed to address the problems aforementioned [5].

Finite element modelling (FEM) is a numerical method commonly used to investigate the process feasibility and facilitate the SPF process design. Prediction of the process variables, e.g. pressure cycles, strain rate control, thickness evolution and tooling design, can be derived from these analyses. Most of the experiments and the SPF simulations are carried out by using commercial FEM codes, such as ABAQUS, ANSYS, MSC.MARC and PAM-STAMP. In superplastic forming, the ability of the simulation code to predict the deformation behaviour of the material is a key factor in developing optimum gas pressure cycles that can form the part in the least time and obtain the best results [6].

In this paper, a validated power-law material model has been implemented in PAM-STAMP to study the process feasibility and predict the process parameters. The influence of process parameters has been studied, including the level of material draw-in and thickness distribution. An optimised forming process window has been proposed by designing the tooling.

Material data

The flow stress behaviour at strain rates ranging from 10^{-4} to 10^{-1} s^{-1} during tensile testing can be characterised by the power law model as seen below:

$$\sigma = K \cdot \dot{\varepsilon}^m \cdot \varepsilon^n \tag{1}$$

where σ is the flow stress, $\dot{\varepsilon}$ is the strain rate, ε is the plastic strain, K is the stress index, m is the strain rate sensitivity index, and n is the strain hardening index. High-temperature tensile tests following ASTM E2448 [7] have been carried out to determine the superplastic behaviour (elongation, flow stress and strain rate sensitivity) at the targeted conditions. It is noted that the mechanism of strain hardening in superplastic flow is essentially due to grain growth, and therefore the n value is determined as a function of strain due to grain growth during tension. In this study, the strain rate sensitivity and strain hardening index were calibrated to be 0.505 and 0.05 at 900°C, respectively. The Young's modulus of 20 GPa was estimated for the material. The Poison's ratio of 0.3 was used in the simulation.

Simulation model

The model used for simulation is shown in Fig. 1(a). It comprises of the die, punch, blank holder, and blank. The tools (i.e. die, punch, and blank holder) were modelled using rigid elements and the blank was defined as standard deformable elements in PAM-STAMP. The simulation follows the experimental sequence and is divided into two stages. Firstly, the punch, which is modelled as a rigid tool, moves towards to the blank at a constant speed (16 mm/s) until it reaches a pre-defined displacement, as illustrated in Fig. 1(b). This phase is also called hot drawing or mechanical pre-forming. For the second phase of the process (Fig. 1(c)), the punch is inactive, and the sheet is deformed due to the applied pressure only. This phase is also called superplastic forming or blowing.

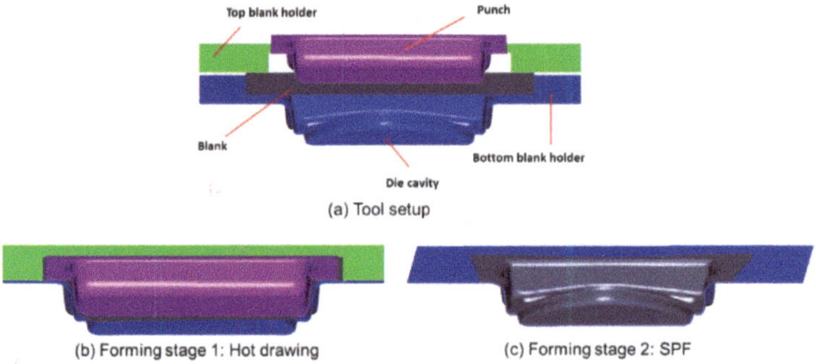

Fig. 1. Cross-sectional view of the simulation model.

Results and discussion

Material draw-in. The material draw-in length was obtained from the variation in width between the undeformed and the deformed sheets, as illustrated in Fig. 2. It can be seen that the maximum material draw-in level for the deformed sheet is found in the middle of each side, while the minimum draw-in is located at the blank corners. In this study, thickness distribution at the end of hot drawing and blowing stages will be compared to understand the effects of different parameters.

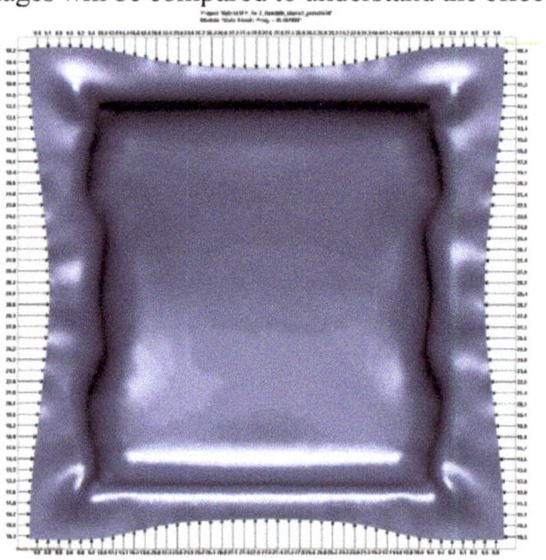

Fig. 2. Illustration of the material draw-in for a pre-deformed (hot-drawn) blank.

Effect of punch geometry.

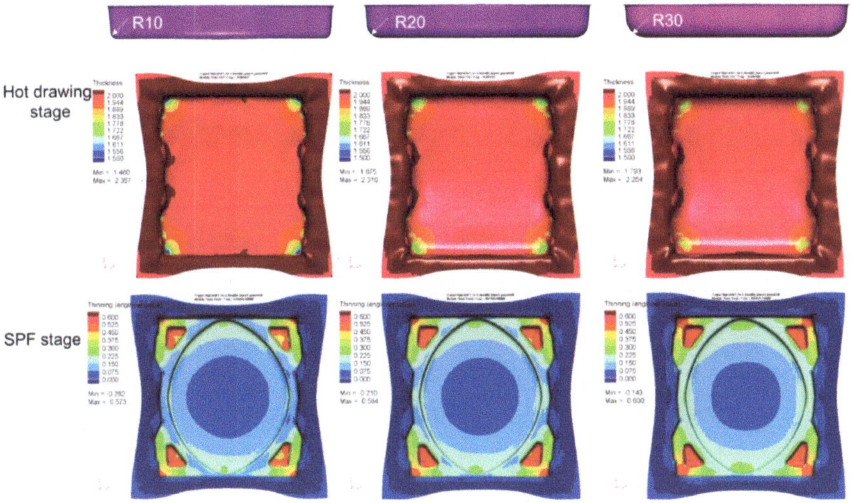

Fig. 3. Material thinning at hot-drawing and SPF stages with different punch radius (thickness unit: mm).

In the mechanical pre-forming (hot drawing) phase, punch is used to deform the material into the die cavity and control the material flow on the blank flanges. The effect of punch geometry has been studied. As shown in Fig. 3, a series of punch shapes with different radius (R10, R20 and R30) are compared.

According to the simulation results, with increase of the punch radius, there is a decrease of material draw-in. The predicted material thinning values after hot drawing (pre-forming) and superplastic forming (blowing) are also compared in Table 1. The formed parts show less material thinning with the punch radii of R10, as a result of more material draw-in at the pre-forming stage. Thus, a more uniform thickness can be achieved after SPF as more material is pre-deformed into the die cavity during the hot-drawing phase.

Table 1.The effect of punch radius on simulation results.

Punch radii	**R10**	**R20**	**R30**
Max material draw-in length at hot drawing stage	36.9 mm	31.9 mm	27.2 mm
Max material thinning at hot-drawing stage	0.270	0.176	0.103
Max material thinning at SPF stage	0.573	0.581	0.600

Effect of blank size. The FE simulation is used to optimise the blank size with which an optimal forming results are achieved. In the case, the punch radii of R20 is selected in simulation to study the effects of blank size/shape.

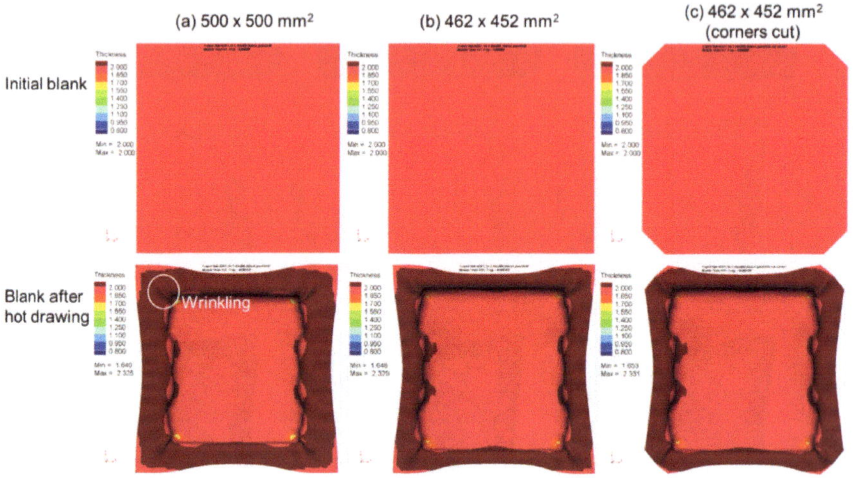

Fig. 4. The effect of blank size on the thickness distribution at the end of hot drawing (legend unit: mm).

Table 2. The effect of blank size on simulation results.

Blank size/shape	500x500 mm^2	462x452 mm^2	462x452 mm^2 (corners cut)
Max material draw-in length at hot drawing stage	28.2 mm	31.9 mm	32.3 mm
Max material thinning at hot-drawing stage	0.181	0.176	0.174
Max material thinning at SPF stage	0.582	0.581	0.575

As shown in Fig. 4, three blanks with different shapes are compared, in terms of the material draw-in and the thickness distribution. With decrease of the blank size, there is an increase of material draw-in, which leads to a more even thickness distribution with less material thinning on the formed part. It is also noted that the larger blank is more prone to wrinkling. As seen in Fig. 4(a), wrinkling is easily found on the corners of the hot-drawn parts as the blank holder is not designed in the region. Simulation results also indicate that more material draw-in can be available when the blank corners are trimmed off before hot drawing, as shown in Fig. 4(c), which will lead to a more even thickness distribution in the final part (Table 2).

Effect of blank-holder force during hot drawing. During hot drawing, blank holder force should be optimised to avoid wrinkling (as seen in Fig. 5) on the blank and minimise the frictional force between the blank and tool. The effect of blank holder force has been studied by comparing different levels of blank holder force, as shown in Table 3. With the increase of blank holder force, there is a decrease of material draw-in, which will eventually affect the thickness distribution on the formed part. Simulation has confirmed that an optimal blank holder force of 20 kN for the material deformed at the hybrid SPF condition.

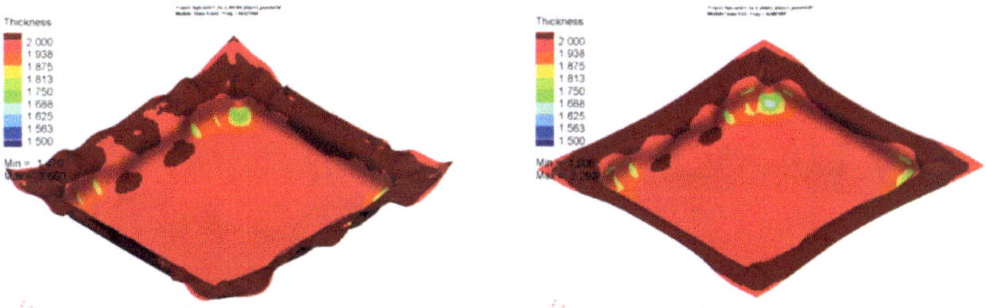

Fig. 5. Thickness distribution at end of hot drawing under different blank holder forces (legend unit: mm).

Table 3. The effect of blank holder force on simulation results.

Blank holder force (kN)	0	20	40	80
Max draw-in length	N/A (wrinkling)	31.9 mm	30.7 mm	28.7 mm
Max material thinning at hot-drawing stage	N/A (wrinkling)	0.176	0.187	0.197

Effect of friction. As in the superplastic forming process, it has been reported that material can be locked against the tool by friction and forming pressure, once the material contacts the surface of the die [5]. The friction effect was investigated in the simulation so as to develop the process, as shown in Table 4.

In comparison with different levels of friction, three friction coefficients were used to simulate the hybrid SPF process. It is observed that the material draw-in is affected significantly, showing less material draw-in under higher friction. In the superplastic forming stage, the thickness is affected by friction, as shown in Fig. 6, when comparing the cavity area of the formed parts. The maximum thinning occurred in the four 'pocket' regions. Ideally, a lower friction can deliver a more even thickness distribution in the formed part. However, it is well accepted that the friction coefficient is assumed to be 0.3 in real forming conditions.

Table 4.The effect of friction on simulation results.

Coefficient of friction	$\mu = 0.1$	$\mu = 0.3$	$\mu = 0.5$
Max material draw-in length	33.0 mm	31.9 mm	30 mm
Max material thinning at hot drawing	0.177	0.176	0.177
Max material thinning at SPF	0.533	0.581	0.595

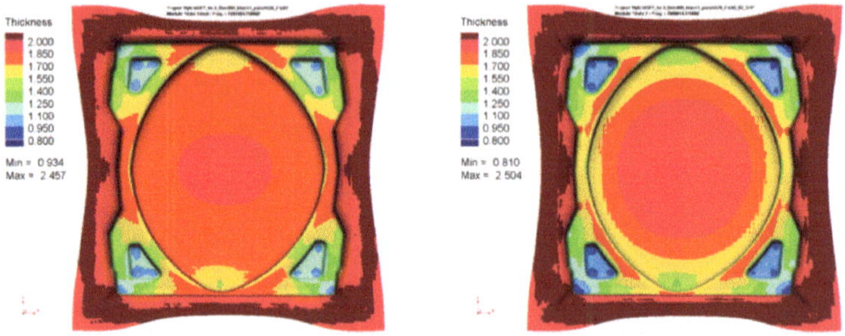

(a) μ = 0.1 (b) μ = 0.5

Fig. 6. Thickness distribution under different friction levels (legend unit: mm).

Effect of pressure cycle during blow forming. The material exhibits superplastic behaviour at a specified temperature and strain rate. In this study, the forming temperature is 900°C. The optimal strain rate during the superplastic forming stage is set to be between 5×10^{-4} and 10^{-3} s^{-1}.

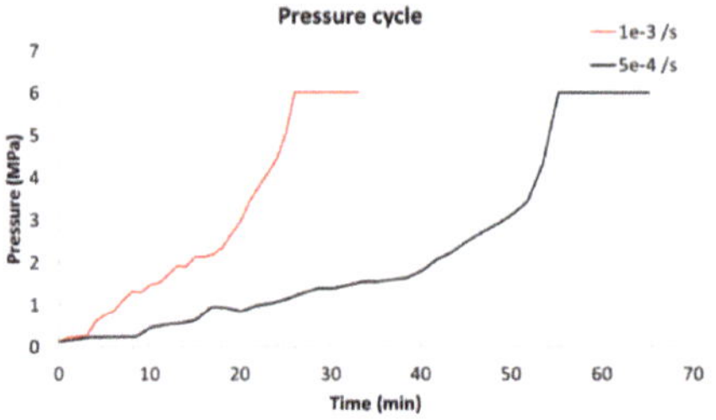

Fig. 7. Pressure cycles predicted by FE simulation at the targeting strain rates.

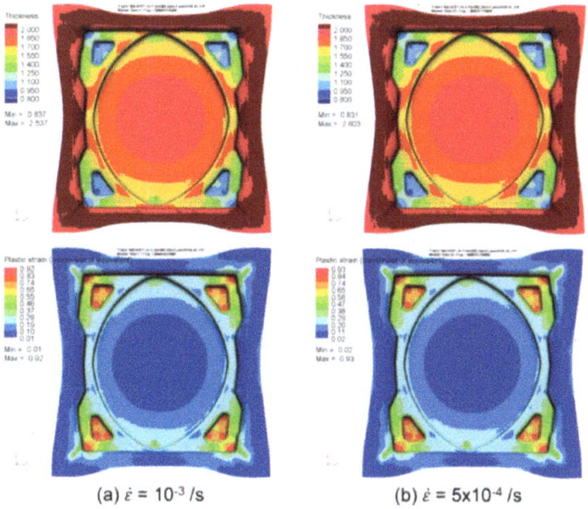

Fig. 8. Distribution of thickness and equivalent strain at different forming strain rates (thickness unit: mm).

Accordingly, the pressure cycles were predicted by FE simulation, as plotted in Fig. 7. It is predicted that the superplastic forming process is completed within 65 min and 33 min at a targeting strain rate of 5×10^{-4} and 10^{-3} s^{-1}, respectively. The thickness distribution at the corresponding strain rate is also shown in Fig. 8. Little difference in the distribution of thickness and equivalent strain is found between the two strain rates.

Conclusions

Finite element modelling was used to simulate the superplastic forming combined with a mechanical pre-forming process. Simulations showed the material localised thinning area, where the punch is in contact with the blank during hot drawing. The material draw-in and thickness distribution were predicted by introducing a power law material model based on the tensile test results at 900°C into PAM-STAMP. The influence of punch geometry, blank size, blank holder force, friction and strain rate was investigated in the simulations. It was observed that the level of material draw-in decreased, as the blank holder force or friction increased. In the comparison of the thickness distributions, a more uniform formed part can be achieved with the optimised tooling design (punch geometry and blank holder force) and process parameters (blank size and strain rate).

Acknowledgement

This work was supported by the National Aerospace Technology Exploitation Programme (NATEP), under the project HYFOTIC (reference no. 10063986). The authors also appreciated for the material donated to the project by TIMET UK.

References

[1] S.P. Akula, M. Ojha, K.L. Rao, A.K. Gupta, A review on superplastic forming of Ti-6Al-4V and other titanium alloys, Mater. Today Commun. 34 (2023) 105343.

[2] D. Lee, W.A. Backofen, Superplasticity in Some Titanium and Zirconium Alloys, Trans. Metall. Soc. AIME 239 (1967) 1034-1040.

[3] A.J. Barnes, Superplastic Forming 40 Years and Still Growing, J. Mater. Eng. Perform. 16 (2007) 440-454. https://doi.org/10.1007/s11665-007-9076-5

[4] T.G. Nieh, J. Wadsworth, O.D. Sherby, Superplasticity in Metals and Ceramics, Cambridge University Press, Cambridge, 1997.

[5] J. Liu, M.-J. Tan, Y. Aue-u-lan, M. Guo, S. Castagne, B.-W. Chua, Superplastic-like forming of Ti-6Al-4V alloy, Int. J. Adv. Manuf. Technol. 69 (2013) 1097-1104. https://doi.org/10.1007/s00170-013-5101-z

[6] J. Liu, J. Edberg, M.J. Tan, L.E. Lindgren, S. Castagne, A.E.W. Jarfors, Finite element modelling of superplastic-like forming using a dislocation density-based model for AA5083, Modell. Simul. Mater. Sci. Eng. 21 (2013) 025006. http://dx.doi.org/10.1088/0965-0393/21/2/025006

[7] ASTM E2448 Standard Test Method for Determining the Superplastic Properties of Metallic Sheet Materials, ASTM International (2022).

Response of vacuum vessel inner shell under low-speed impact

ZHAI Hua[1,a *], WANG Xuncheng[1,b], LI Cheng[1,c], MA Jianguo[2,d], LIU Zhihong[2,e *] and WU Yucheng[1,f]

[1]Anhui Province Key Lab of Aerospace Structural Parts Forming Technology and Equipment, Hefei University of Technology, Hefei, China

[2]Institute of Plasma Physics, Chinese Academy of Sciences, Hefei, China

[a]jxzhaihuajx@hfut.edu.cn, [b]1542365788@qq.com, [c]963734268@qq.com, [d]mjg@ipp.ac.cn, [e]zhliu@ipp.ac.cn, [f]ycwu@hfut.edu.cn

Keywords: Low Speed Impact, Vacuum Vessel Inner Shell, Impact Response, Simulation Rib Plate

Abstract. Nuclear fusion energy is a promising new energy, This vacuum environment in nuclear fusion reactor is provided by the precision vacuum vessel, which also provides a support structure for the installation of internal components of the fusion device. In this paper, the impact response of double layer vacuum chamber inner wall under transverse impact load is experimentally studied to manufacture the high quality vacuum chamber. From the drop weight experiments, the vacuum chamber wall under five groups of different impact energy is analyzed, and the impact response displacement, and energy absorption rate of the sample are compared and discussed. The experimental results show that the impact force and impact response displacement increase with the impact energy, and the error also increases. The influence of rib plate in the vacuum chamber inner wall on impact response displacement is analyzed. With the increase of the damage deformation of the vacuum chamber wall, the rebound amount of the impact block cannot increase linearly with the impact energy. The actual situation is simulated with software, and the results of experimental and simulated energy absorption are compared and explained.

Introduction

Nuclear fusion energy is a promising new energy [1]. In order to realize controllable nuclear fusion reaction, it is necessary to provide a very high vacuum environment to ensure the operation of plasma [2,3]. This vacuum environment is provided by the vacuum vessel with D-shaped ring structure through full penetration welding, which also provides a support structure for the installation of internal components of the fusion device. The design of the double shell structure can be filled with boron containing screen cooling water with shielding function. The whole annular vacuum vessel structure is divided into 8 sectors, and 1/8 sector is divided into 1/16 sector. 1/16 sector consist of multiple PS1-PS4 sections [4]. In the welding process, the internal stress caused by thermal expansion and cold contraction will lead to plastic shrinkage deformation. Reasonable welding sequence can reduce large shrinkage residual stress, and achieve the effect of reducing welding deformation. Keimasaki et al. [5] proposed the assembly sequence for the JT-60sa vacuum vessel. Xu et al. [6] further predicted the deformation of vacuum vessels based on inherent strain theory, and proposed six different welding sequences of PS3 section, and obtained the optimal results. They also obtained the conclusion that applying external constraints will significantly reduce welding deformation, but the welding residual stress will increase, and the annular stiffener has the function of restraining welding deformation. Chen et al. [7] proposed a thermo-elasto-plastic finite element method using shell element with section integration to study the influence of welding sequence on the structural deformation and residual stress of large stiffened plates. Ordieres J. [8] investigated the influence of welding fixture on welding deformation of ITER vacuum vessel PS1 by finite element method, the control effect of welding fixture on welding deformation is better than that of welding sequence.

The heat input of various welding methods is quite different, and under the condition of using the same welding method, different joint forms have different effects on the longitudinal, transverse and angular deformation. Ji et al. [9] studied the influence of narrow gap tungsten inert gas welding (ng-tig) and electronic restraint welding (EBW) on the welding results in the manufacturing process of CFETR by using nondestructive testing technology and mechanical tests. The impact absorption energy of the two welded joints is lower than that of the base metal, but the strength of ng-tig weld is higher than that of the base metal. P. Y.lee et al. [10] proposed all parts into a whole through bolts and shear keys. The finite element analysis shows that the maximum static stress intensity of the structure without welding is within the stress limit, and there will be no large deformation and stress concentration. At the same time, the fatigue life of the new structure is much higher than the design standard of ITER. Zhai et al. [11] used electron beam welding technology to prepare welded joints with different weld spacing, and studied the microstructure and mechanical properties under different weld spacing. The results showed that the grain size in the heat affected zone increased with the increase of weld spacing, and the joint with small weld spacing after welding had better performance.

The welding tooling of large structural parts mainly ensures the accuracy through fixtures and adjusting devices. Li et al. [12] calculated the mathematical optimization model of the position deviation of the clamping part according to the geometric relationship between the resultant force of all external forces on the clamping part in the spatial coordinate system and the total contact stiffness. Based on inherent strain theory, Ji et al. [13] analyzed the welding deformation of 1/32 vacuum vessel, the welding fixture and the reserved allowance for welding shrinkage were used to reduce the welding deformation. It is often necessary to detect the welding quality of the vacuum vessel sector according to the problems of weld surface cracks, incomplete penetration, weld leakage and workpiece surface cracks. Wang et al. [14] designed a dual matrix array probe to solve the limitations of the traditional ultrasonic testing, and used ultrasonic testing technology to carry out non-destructive testing of CFETR welding [15]. Yutonoguchi et al. [16] Evaluated the applicability of eddy current testing method to detect the welding defects of hydraulic pipes of ITER cladding module. G D'Amico et al. [17] predicted the mechanical behavior to control the fixture forces generated in the welding process. Caixas et al. [18] established a welding vacuum vessel thermal load model, compared the actual and accurate deformation with the simulation results. Zhang et al. [19] concluded that the largest deformation in the welding process occurred on the shell around the transition structure, and the main difference between welding stress and deformation occurred at the shell around the window.

Some relevant work have been performed to reduce the welding deformation of the vacuum vessel , but there is relatively little research on how to straighten the deformation after welding. In this paper, a low speed impact experiment device for the vacuum inner shell is designed. Through the impact response analysis of the vacuum inner shell after impact, the deformation caused by the inner wall after welding can be reduced to improve inconvenient manual operation and low efficiency, and provides a design basis for the further research and development of the vacuum vessel welding deformation straightening correction device.

Experimental research

Experimental specimen and low speed impact device. Due to the huge volume of the vacuum vessel, it needs a lot of materials to manufacture the equal scale model, and the process degree becomes correspondingly complex with the increase of the size in the processing and manufacturing process. Therefore, the samples in this experiment are scaled to produce the vacuum inner shell model. The samples size were reduced by 5 times. The vacuum inner shell becomes 10mm, and the total length and width will be reduced accordingly. The material is 316L stainless steel, whose performance parameters are shown in Table 1, which is convenient for processing and manufacturing and correction experiments.

Table 1. 316L stainless steel performance parameters.

Material	E(GPa)	μ	σ_s (MPa)	G (MPa)
316L	210	0.288	235	6100

As shown in Fig. 1(a), the vacuum inner shell used in the experiment is mainly composed of inner wall, outer wall, upper rib plate, middle rib plate and lower rib plate. The inner wall arc length is 695 mm, the outer wall arc length is 795 mm, the inner and outer wall thickness is 10 mm, and the distance is 50 mm, and the thickness of the three ribbed plates is 7 mm.

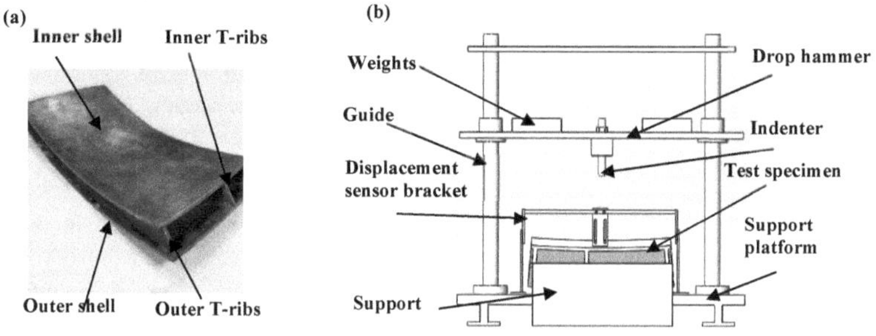

Fig. 1. Experimental specimen and low speed impact falling weight experimental device. (a) Experimental specimen. (b) Low speed impact falling weight experimental device.

Fig. 1(b) shows the experimental device and the components of the sketch. The drop hammer is guided vertically by two guide rails. These two parts are connected by ball bearings, and two standard weight can be fixed above the plate to change the mass of the falling weight. The vacuum inner shell is placed in the center of the test bench through two supports, and rigid constraints are set at both ends. The impact head uses a pressure sensor with hemispherical body, which is installed in the center of the slider. The displacement sensor is fixed above the vacuum inner shell with a bracket. The signal acquisition includes pressure signal acquisition, displacement signal acquisition and image acquisition. The image acquisition is realized by the Thousand Eye Wolf 2F01M high-speed camera. The pressure signal is transmitted through the BLSM sensor, with a testing range of 20000N. The small displacement sensor in the installation space of the impact head is the Miran KTR11 linear displacement sensor, with a range of 0-15 mm. The signal acquisition is realized by the Siemens S7 1200 programmable logic controller (PLC). The whole test process includes three steps.

1) Place the displacement sensor near the impact position, so that the displacement sensor is as close to the impact position as possible without interference with the indenter. Then connect the PLC with the displacement sensor and pressure sensor. Then adjust the high-speed camera, open the high-speed camera program, and adjust the position and focal length of the high-speed camera so that it can clearly capture the process of low-speed impact.

2) Place the corresponding standard weight above the falling weight and mark the height of the falling weight. In this experiment, set the heights of 300 mm, 400 mm, 500 mm and 600 mm respectively, and record them as four energy levels E1, E2, E3 and E4. Each energy level is subject to five groups of tests.

3) Release the falling weight from the specified height along the guide rail. In the process of energy transmission of the sample, the falling hammer will rebound several times. Finally, when the falling hammer is completely stationary, lift it to a certain height to release the remaining elastic deformation energy of the inner shell.

Experimental data discussion. 1) Impact energy. The height of the free falling body is divided into four grades. Therefore, without changing the mass of the falling hammer, the impact energy is also divided into four grades. First, two standard weights are placed, and then the mass of the impact block is added, the total is 28.4 kg, and the impact energy is 83.496 J, 111.328 J, 139.16 J, 166.992 J respectively.

2) Energy absorption rate of impact specimen. During the impact process, because the impact specimen is completely fixed, the kinetic energy generated during the falling process of the impact head will be transferred to the impact specimen, and then rebound itself. In the case of ignoring friction, at this time, the energy is only transferred between the impact head and the vacuum inner shell. The energy absorbed by the impact specimen can be obtained by subtracting the potential energy at the highest rebound point from the potential energy at the starting highest point of the impact head, This data is used to quantify the energy absorption capacity of the impact specimen and the impact response when bearing the impact of different energy levels.

3) Impact force and impact response displacement. Through the obtained data, we can analyze the relationship between the impact force and the vacuum inner shell deformation with the increase of energy level under different impact energy, and whether there is a functional relationship between them.

4) Other data. It mainly includes the impact speed, rebound speed and rebound displacement of the impact head in the falling stage. These data are mainly used to describe the impact process and calculate the previous data. The data acquisition method is to capture the impact and rebound process of the impact head through the high-speed camera, and then use the high-speed video target tracking measurement software to show the whole impact process by setting the trajectory of the tracking impact head, and draw the speed change curve through the software's own function.

There are four energy levels in total, and five groups of experiments are conducted for each level. The impact forces of the four energy levels are respectively expressed by F1,F2,F3,F4. The impact forces obtained are shown in Table 2. the impact response displacement of the four energy levels shall be represented by S1, S2, S3 and S4 respectively.

The average impact force of the four energy levels is 669.9219 N, 825.1736 N, 1054.36 N and 1253.42 N respectively. It can be found that as the impact height increases, the impact force will also increase. Under the same conditions of E1 energy level, the impact force on the segment changes gently, and under the same conditions of E2 and E3 energy levels, the impact force on the segment changes gradually. Under the same conditions of E4 energy level, the impact force changes obviously.

The impact response displacement under each group of energy levels is shown in the table above. The average value of impact response displacement is 0.026552 mm, 0.042558 mm, 0.056368 mm and 0.077706 mm respectively. With the gradual increase of impact energy, the impact response displacement increases, and the variation fluctuation of impact response displacement also increases. From the distribution relationship between the average value of impact force and impact response displacement of each group, the impact force is positively correlated with the impact displacement response, which indicates that the concave structure of the specimen has good impact resistance and excellent energy absorption performance.

From the above two sets of data, the impact force generated by the slider and the impact response displacement of the vacuum inner shell are roughly distributed in a relatively closed range. There are two main reasons for the same test error. First, because the range of the impact head is large, and the measured data is relatively small, which will magnify the test error. The second is the error caused by the flatness of the specimen itself and the flatness of the installation.

Table 2. Impact force results.

Group number	F_1(N)	S_1(mm)	F_2(N)	S_2(mm)	F_3(N)	S_3(mm)	F_4(N)	S_4(mm)
1	640.1476	0.02645	831.5538	0.04024	1027.922	0.05435	1310.07	0.08343
2	684.1002	0.02452	797.5261	0.04364	1025.798	0.05435	1186.72	0.07823
3	671.3397	0.02848	798.2350	0.03971	1119.464	0.05208	1209.641	0.07235
4	674.8843	0.02566	895.3559	0.04356	1086.054	0.06123	1209.557	0.07217
5	679.1377	0.02765	803.1973	0.04564	1012.562	0.05983	1351.112	0.08235

The low speed impact process can be recorded with a high-speed camera. After recording the speed of each frame rate, take the average value of each group of impact speed under the same energy level, and draw a speed frame rate diagram, as shown in Fig. 2. According to the height of free fall, the collision velocities of the four energy levels are 2.42 m/s, 2.8 m/s, 3.13 m/s and 3.43 m/s respectively. The collision velocities obtained by the high-speed camera are 2.42 m/s, 2.78 m/s, 3.21 m/s and 3.40 m/s respectively, and the error is within 5%. The results of the high-speed camera are available.

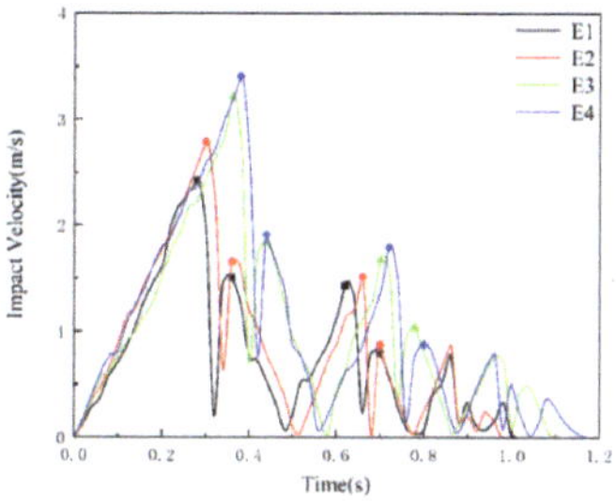

Fig. 2. Experimental speed frame rate diagram from high-speed camera.

With the increase of energy levels, the time of wave peaks successively moves backward, but the phase difference of wave length of each energy level curve is small. The height of the second wave peak and the third wave peak are basically the same, and the two represent the first rebound speed and the second collision speed respectively, indicating that the energy loss in the impact process can be basically ignored. Through the first impact speed and the first rebound speed, the energy absorption rate of the vacuum chamber wall can be obtained, as shown in the Table 3.

Table 3. Energy absorption rate of inner wall under different energy levels.

Energy level	First impact velocity(m/s)	First bounce velocity(m/s)	Energy absorption rate
E1	2.4256	1.6993	50.92%
E2	2.7865	1.854	55.73%
E3	3.2135	2.0581	58.98%
E4	3.4025	2.1449	60.26%

Materials Research Proceedings 44 (2024) 218-226 https://doi.org/10.21741/9781644903254-24

The change trend of the energy absorption rate increases with the increase of the energy level, the damage level will increase correspondingly, and the performance degradation of the sample will occur. Therefore, the rebound amount cannot cause a linear increase in the value with the linear increase of the energy level, so the result shows a positive correlation between the energy absorption rate and the impact energy level.

Simulation Research

The energy absorption rate under four energy levels is analyzed through finite element software, energy absorption rate simulating results under four energy levels were listed in Table 4.

Table 4. Energy absorption rate of vacuum chamber wall through simulation.

Energy level	Total energy(J)	Residual energy of impact block(J)	Energy absorption rate
E1	82.925	50.573	39.02%
E2	11.101	61.047	45.01%
E3	138.72	68.766	50.43%
E4	166.59	74.8	55.10%

The comparison of the energy absorption rate of the vacuum chamber wall between the experiment and the simulation is shown in Fig. 3. With the rise of the energy level, the energy absorption rate of the vacuum chamber wall gradually increases, and the change speed of the E1, E2, E3 energy absorption rate is consistent with the simulation results, but the experimental energy absorption rate is larger than the simulation results because the various collision speed is used to calculate the energy absorption rate. It does not mean that the vacuum chamber wall completely absorbs the remaining energy.

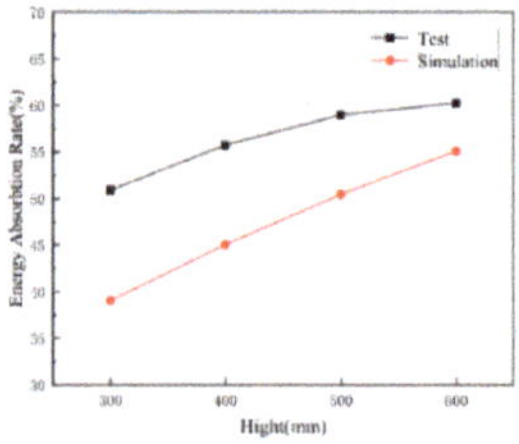

Fig. 3. Comparison of energy absorption rate of vacuum chamber wall between experiment and simulation.

In the experiment, the overall collision rebound caused by the unstable fixation of the vacuum chamber wall and the force of the support on the ground will also cause the loss of energy, resulting in the large value of the energy absorption rate in the experimental results.

The rib plate is an indispensable part of the vacuum chamber welding process, which plays a role in supporting the whole and strengthening the stability. Due to the limitations of the experimental device, it is temporarily impossible to analyze the influence of the rib plate on the whole vacuum chamber in mechanical straightening. The simulation software is used to act the impact energy on the positions close to the lower rib plate and the upper rib plate respectively to analyze the role of the rib plate in the correction of deformation.

When the impact position is close to the vertical ribs, the center and away from the outer T-ribs plate, the different deformation diagrams of the vacuum chamber wall were shown in Fig. 4.

respectively. It can be seen that the closer the collision center is to the position of the outer T-ribs in Fig. 4(a), the smaller the deformation under the same impact energy. The reason is that although the impact energy acts on the middle rib plate and the vertical ribs as shown in Fig. 4(c), the two rib plates are shorter than outer T-ribs, so the outer T-ribs plays a major role in resisting deformation in the inner wall of this vacuum chamber, and the deformation is proportional to the distance from the outer T-ribs.

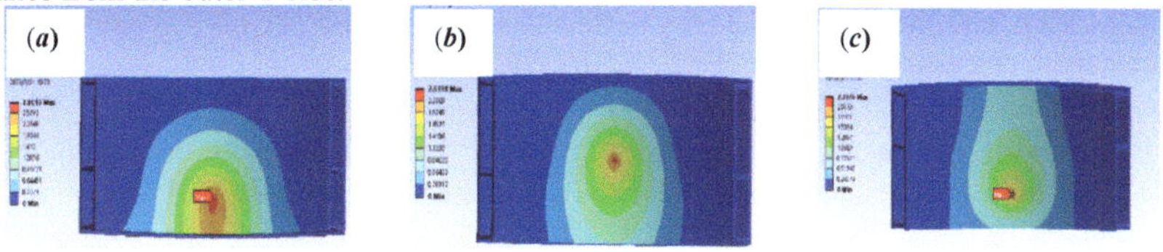

Fig. 4. The deformation diagram of the vacuum chamber wall when the impact position is close to the vertical ribs. (a) close to outer T-ribs (b) center (c) away from outer T-ribs.

Conclusion

In this paper, the impact response of vacuum chamber wall under low-speed impact is studied through experiments and simulation analysis. The following conclusions are drawn,

1) When the yield does not occur, the impact response displacement is positively correlated with the impact energy, and between different energy levels, with the increase of the energy level, the fluctuation of the impact force curve also gradually increases.

2) The change trend of the energy absorption rate of the wall in the vacuum chamber increases with the increase of the energy level, indicating that when the impact energy level increases, the damage level will increase correspondingly, and the performance of the sample will deteriorate. Therefore, the rebound amount cannot cause a linear increase in the value with the linear increase of the energy level, and the result shows a positive correlation between the energy absorption rate and the impact energy level.

3) According to the simulation results, the energy absorption rate of the vacuum chamber wall is relatively small, but the change trend is consistent with the experiment. The principle is that part of the energy makes the vacuum chamber wall rebound and act on the support as a whole during the experiment.

4) The rib plate plays a key role in the mechanical correction of vacuum chamber. The deformation under impact energy is proportional to the distance from the rib plate.

Declaration of competing interest

The authors declare that they have no known competing financial interests or personal relationships that could have appeared to influence the work reported in this paper.

Acknowledgments

This work was supported by Anhui Province Key Project No. 202304a05020074. Flexible molding equipment for carbon fiber composite materials, JZ2023GQBK0070.

References

[1] J. Zheng, J. Qin, K. Lu, M. Xu, X. Duan, G. Xu, J. Hu, X. Gong, Q. Zang, Z. Liu, L. Wang, R. Ding, J. Chen, P. Li, L. Xue, L. Cai, Y. Song, Recent progress in Chinese fusion research based on superconducting tokamak configuration, The Innovation 3 (2022) 100269. https://doi.org/10.1016/j.xinn.2022.100269

[2] P.V.E. McClintock, Controlled thermonuclear fusion, by Jean Louis Bobin, Contemporary Physics 58 (2017) 1-100.

[3] Y.T. Song, S.T. Wu, J.G. Li, B.N. Wan, P. Fu, M.Y. Ye, J. Zheng, K. Lu, X. Gao, S.M. Liu, X.F. Liu, M. Lei, X.B. Peng, Y. Cheng, Concept design of CFETR tokamak machine, IEEE T. Plasma Sci. 42 (2014) 503-509. https://doi.org/10.1109/TPS.2014.2299277

[4] M. Jianguo, W. Jiefeng, L. Zhihong, R&D of full-scale partial vacuum vessel mockup for future fusion engineering test reactor in China, J. Fusion Energy 34 (2015) 666-670.

[5] K. Masaki, Y.K. Shibama, S. Sakurai, K. Shibanuma, A. Sakasai, JT-60SA vacuum Vessel manufacturing and assembly, Fusion Eng. Des. 87 (2012) 742-746. https://doi.org/10.1016/j.fusengdes.2012.02.014

[6] L. Xiu, J. Wu, Z. Liu, J. Ma, X. Fan, H. Ji, X. Xia, Y. Li, Weld distortion prediction of the CFETR vacuum vessel by inherent strain theory, Fusion Eng. Des. 121 (2017) 43-49. https://doi.org/10.1016/j.fusengdes.2017.03.175

[7] Z. Chen, Z. Chen, R.A. Shenoi, Influence of welding sequence on welding deformation and residual stress of a stiffened plate structure, J. Ocean Eng. 106 (2015) 271-280. http://dx.doi.org/10.1016/j.oceaneng.2015.07.013

[8] J. Ordieres, E. Rodríguez, A. Bayón, J. Caixas, A. Barbensi, P. Guglielmi, Improvement of manufacturing jigs design for reduction of welding distortion in Vacuum Vessel PS1 through finite element analysis, Fusion Eng. Des. 146 (2019) 2168–2171. https://doi.org/10.1016/j.fusengdes.2019.03.141

[9] H. Ji, X. Xia, J. Wu, H. Wu, Z. Liu, J. Ma, Z. Liu, X. Shen, Welding technology development for the fabrication of vacuum vessel for CFETR, Fusion Eng. Des. 147 (2019) 111272. https://doi.org/10.1016/j.fusengdes.2019.111272

[10] P.Y. Lee, B.L Hou, G.D. Jian, et al. Redesign of gravity support system for ITER construction. Fusion Eng. Des. 85 (2010) 33-38.

[11] Y. Zhai, B. Huang, J. Zhang, B. Zhang, S. Liu, Q. Huang, Effect of weld spacing on microstructure and mechanical properties of CLAM electron beam welding joints, Fusion Eng. Des. 112 (2016) 440-449. https://doi.org/10.1016/j.fusengdes.2016.06.027

[12] B. Li, S.N. Melkote, Fixture clamping force optimisation and its impact on workpiece location accuracy, Int. J. Adv. Manuf. Technol. 17 (2001) 104-113. https://doi.org/10.1007/s001700170198

[13] H. Ji, J. Wu, H. Wu, Z. Liu, J. Ma, X. Fan, Analysis of welding deformation on CFETR 1/32 vacuum vessel mockup, Fusion Eng. Des. 144 (2019) 160-163. http://dx.doi.org/10.1016/j.fusengdes.2019.05.012

[14] R. Wang, Z. Liu, J. Wu, B. Jiang, B. Li, Design of DMA probe for the ultrasonic testing of CFETR vacuum vessel weld, Fusion Eng. Des. 146 (2019) 987-990. https://doi.org/10.1016/j.fusengdes.2019.01.136

[15] R. Wang, Z. Liu, J. Wu, B. Jiang, B. Li, Research on phased array ultrasonic testing on CFETR vacuum vessel welding, Fusion Eng. Des. 139 (2019) 124-127. https://doi.org/10.1016/j.fusengdes.2019.01.050

[16] Y. Noguchi, M. Tsunokai, K. Nakata, N. Takeda, Applicability of eddy current technique in in-bore NDT tool for ITER hydraulic pipe welds, Fusion Eng. Des. 146 (2019) 2571-2576. https://doi.org/10.1016/j.fusengdes.2019.04.044

[17] G. D'Amico, J. Caixas, A. LoBue, A. Bayon, In-process control of ITER vacuum vessel PS1 welding jig by means of FEM simulation and welding distortion analysis, Fusion Eng. Des. 146 (2019) 1843-1847. https://doi.org/10.1016/j.fusengdes.2019.03.048

[18] J. Caixas, J. Guirao, A. Bayon, L. Jones, J.F. Arbogast, A. Barbensi, A. Dans, A. Facca, E. Fernandez, J. Fernández, S. Iglesias, M. Jimenez, P. Jucker, G. Micó, J. Ordieres, J.M. Pacheco, R. Paoletti, G.P. Sanguinetti, V. Stamos, M. Tacconelli, Weld distortion prediction of the ITER vacuum vessel using finite element simulations, Fusion Eng. Des. 88 (2013) 2011-2014. https://doi.org/10.1016/j.fusengdes.2013.02.101

[19] J. Zhang, L. Yu, Y. Liu, H. Li, C. Liu, J. Wu, J. Ma, Z. Li, Effect of welding sequences on the welding stress and distortion in the CFETR vacuum vessel assembly using finite element simulation, Int. J. Pres. Ves. Pip. 175 (2019) 103930. https://doi.org/10.1016/j.ijpvp.2019.103930

Concurrent Improvement of strength, formability and SCC resistance of Al-Zn-Mg-Cu alloy by hot stamping after rapid heating and re-aging on paint baking treatment

HIROSAWA Shoichi[1,a*], KIM Junmo[1,b], MAENO Tomoyoshi[1,c], SUZUKI Yasushi[2,d] and YABUKI Yuuji[2,e]

[1]Graduate School of Engineering Science, Yokohama National University, 79-5 Tokiwadai, Hodogaya-ku, Yokohama 240-8501, Kanagawa, Japan

[2]R&D Division, G-TEKT Corporation, 4-7-28 Shinmeidai, Hamura 205-0023, Tokyo, Japan

[a]hirosawa@ynu.ac.jp, [b]kim-tunmo-dr@ynu.jp, [c]maeno-tomoyoshi-yf@ynu.ac.jp, [d]yasushi_suzuki.nv@g-tekt.jp, [e]yuuji_yabuki@g-tekt.jp

Keywords: Extra Super Duralumin, T6 Temper, High Strength, Formability, Corrosion Resistance, Medium Temperature

Abstract. In this study, the combined application of rapid heating for hot stamping and the subsequent re-aging during a paint baking treatment was newly developed for T6-tempered A7075 Al-Zn-Mg-Cu alloy sheets. From a systematic investigation, rapid heating to 255-260 °C and short-time holding for ~120 s was found to give the best combination of formability, hardness and stress corrosion cracking (SCC) resistance of the final products. In particular, re-aging treatment at 170 °C for 1.2 ks was well exploited not only for further improvement of SCC resistance but also for rebound of the decreased hardness by rapid heating. Therefore, the developed three-step heat treatment is regarded as a promising approach to producing high-strength, easily formable and corrosion-resistant Al-Zn-Mg-Cu alloy parts for automobiles.

Introduction

Recently, a replacement of steel-made parts by aluminum alloys is progressing in automobile industries, but most of those are 6xxx aluminum alloys with good formability, stress corrosion cracking (SCC) resistance, but medium strength. The application of high-strength 7xxx aluminum alloy parts to automobiles is partly attempted for further reduction of the weight and fuel consumption rate of the vehicles, but at present only limited number of automobile parts has been made from 7xxx aluminum alloys. This is because 7xxx aluminum alloys have problems concerning lower ductility at room temperature (RT) and less resistance to SCC. Therefore, a production of high-strength, easily formable and corrosion-resistant 7xxx aluminum alloy parts for automobiles is highly expected.

In aeronautics industries, O-tempered 7xxx aluminum alloy sheets are cold-stamped, solution-treated and then artificially aged, but this procedure is not good for automobile industries because of the prolonged aging treatment. Heat form quenching (HFQ) consisting of rapid heating for solution treatment, hot stamping and die quenching [1] still needs a prolonged aging treatment, although some parts for specialty cars have been made through this procedure. Therefore, three-step heat treatment developed by the authors [2] becomes useful because productivity can be improved by omitting both the solution and aging treatments in parts manufacturers. Fig. 1 shows a schematic illustration of three-step heat treatment utilized in this study. This heat treatment contains rapid heating of T6-tempered sheets to a medium temperature, holding at the temperature for 150 s maximum before hot stamping and die quenching, and re-aging by exploiting paint baking treatment in automobile companies. The rapid heating was intended to improve the formability of the T6-temperted high-strength sheets while minimizing the decrease in strength, whereas the final re-aging treatment not only further improves the SCC resistance but also rebounds the decreased strength by rapid heating.

Therefore, the objective of this study is to elucidate the effectiveness of combined application of rapid heating for hot stamping and the subsequent re-aging during paint baking treatment for producing high-strength, easily formable and corrosion-resistant Al-Zn-Mg-Cu alloy parts for automobiles.

Experiment

In this study, 2 mm-thick sheets of T6-tempered A7075 Al-Zn-Mg-Cu alloy were utilized as a starting material. As 1st step of three-step heat treatment in Fig. 1, the sheets were rapidly heated to 150-280°C at a heating rate of ~10°Cs^{-1} by a contact heating equipment, and then held at the temperature for 150 s maximum before hot stamping and die quenching (2nd step). The formability of the rapidly heated sheets was evaluated by high-temperature tensile test, deep drawing and 90° V-bending test. Some pieces of sheets were just heated and held for 150 s maximum in oil baths, followed by water-quenching without hot stamping. The heating and cooling rates of the two experiments were assured to be almost comparable by monitoring their temperature changes with thermocouples.

After stored at 20°C for 86.4 ks, re-aging treatment at 170°C for 1.2 ks was applied to the sheets with/without hot stamping, as the 3rd step of three-step heat treatment in Fig. 1. The resultant Vickers hardness and electrical conductivity were measured as indexes of mechanical strength and SCC resistance and utilized for establishing a process window with optimized heating conditions at 1st and 2nd step heat treatment for the best combination of formability, hardness and SCC resistance of the final products.

Furthermore, to directly prove the improved SCC resistance, corrosion test by immersing into a solution of NaCl was conducted for three-step heat-treated specimen (255°C120 s+170°C1.2 ks) together with T6-tempered and T73-tempered specimens. As illustrated in Fig. 2, four-point bending along ST direction under a constant stress of 330MPa was applied to the three specimens, and then imemersed into a solution of 3.5±0.1% NaCl (PH=6.4-7.2) at 25±3°C under a humidity of 40-75%. After repeating immersion (0.6 ks) and drying (3.0 ks) for 90 days, cracks and pitting on their RD-ND planes were observed by dark-field optical microscopy. Transgranular and intergranular precipitates in the three specimens were also observed by transmission electron microscopy.

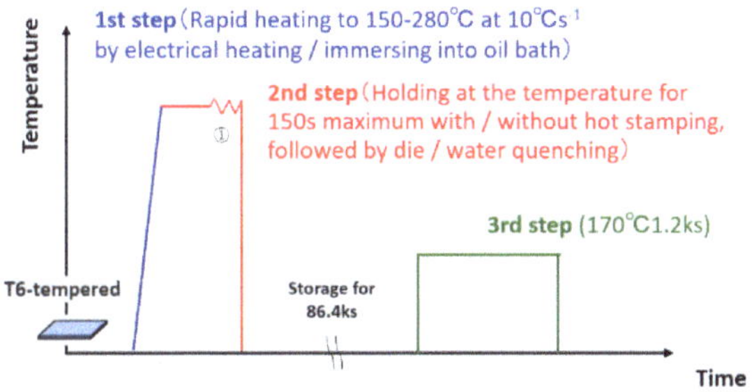

Fig. 1. Three-step heat treatment utilized in this study [2].

Materials Research Proceedings 44 (2024) 227-232 https://doi.org/10.21741/9781644903254-25

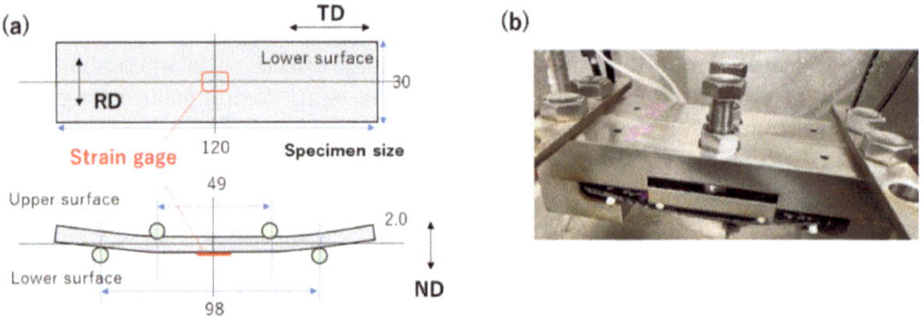

Fig. 2. The dimensions of corrosion test specimen under four-point bending along ND direction (a). The utilized V-bending equipment (b) was repeatedly immersed into a solution of NaCl and dried for 90 days.

Results and discussion

Fig. 3 shows stress-strain curves of T6-tempered sheets obtained from high-temperature tensile test at 150, 200 and 250 °C. As a result of rapid heating by 10°Cs^{-1} and holding for 10s, lower deformation resistance and larger elongation to fracture was obtained than those at RT; i.e. 540MPa of ultimate tensile stress (UTS) and 7% of elongation to fracture. Such improved formability at higher temperatures was also confirmed from the result of deep drawing in Fig. 4, in which rapid heating to 200 °C improved deep drawability from <1.46 (RT) to 1.88 in limiting drawing ratio (LDR). Fig. 5 compares springback and residual stress of V-bent sheets at RT to 250 °C. By rapid heating to 200 or 250°C, springback and residual stress after bending was reduced, resulting in the increase in bending radius / thickness ratio from <0.2 (RT) to >0.6 (200 °C). It should be noted that those are all positive evidence for the improved formability by rapid heating.

However, the higher the temperature or more prolonged the holding time, more decreased strength of the T6-temperted sheets was obtained due to overaging of the precipitation-hardened alloy. Fig. 6 shows changes in Vickers hardness and electrical conductivity after rapid heating to

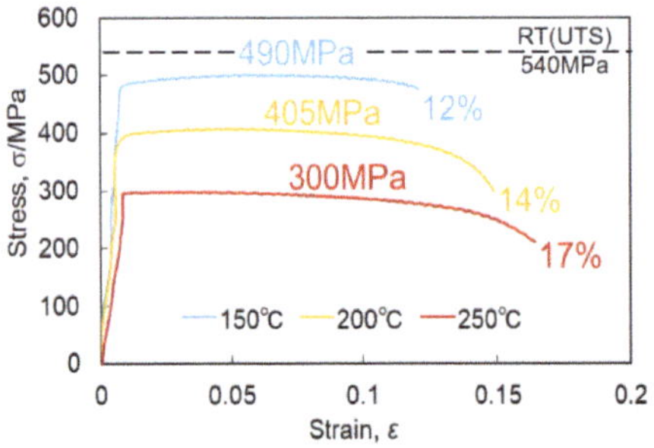

Fig. 3. Stress-strain curves of T6-tempered A7075 alloy sheets after rapid heating to 150, 200 and 250°C (Heating rate: 10°Cs^{-1}, holding time: 10 s, initial strain rate: 1s^{-1}).

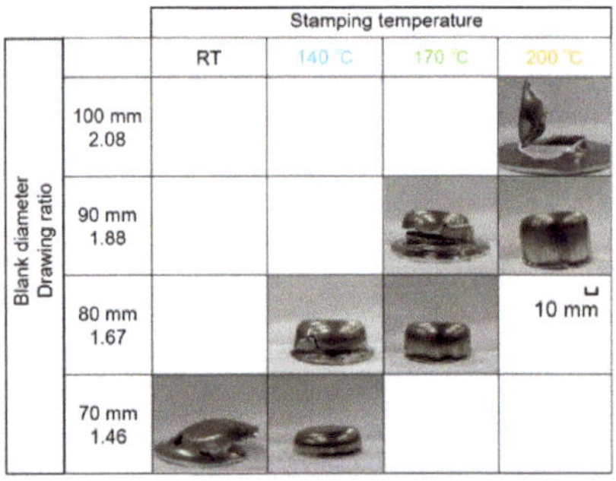

Fig. 4. Deep drawability of T6-tempered A7075 alloy sheets after rapid heating to 140, 170 and 200°C (Heating rate: 10°Cs^{-1}, holding time: 10 s).

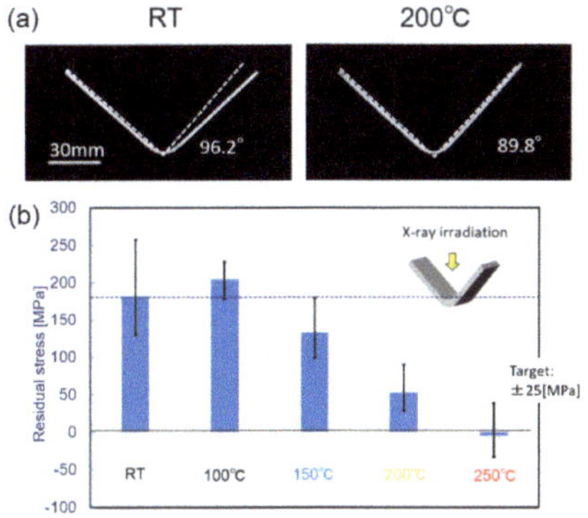

Fig. 5. Results of 90° V-bending test of T6-tempered A7075 alloy sheets at RT to 250°C (Punch radius: 9 mm). Residual stress after bending was also mesured by a X-ray analyzer.

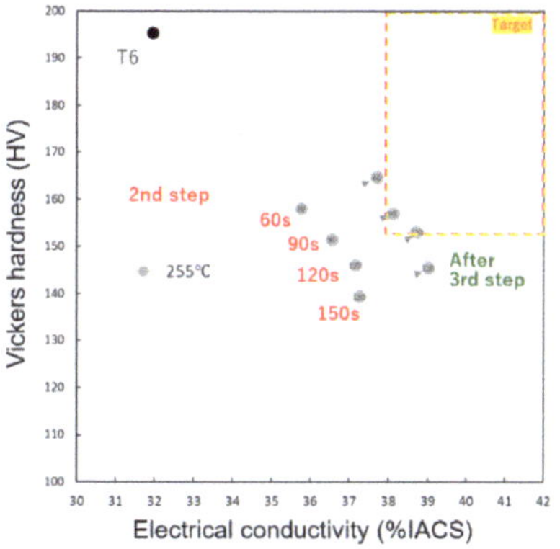

Fig. 6. Changes in Vickers hardness and electrical conductivity of T6-tempered A7075 alloy sheets after rapid heating to 255°C for 60-150 s and re-aging at 170 °C for 1.2 ks without hot stamping.

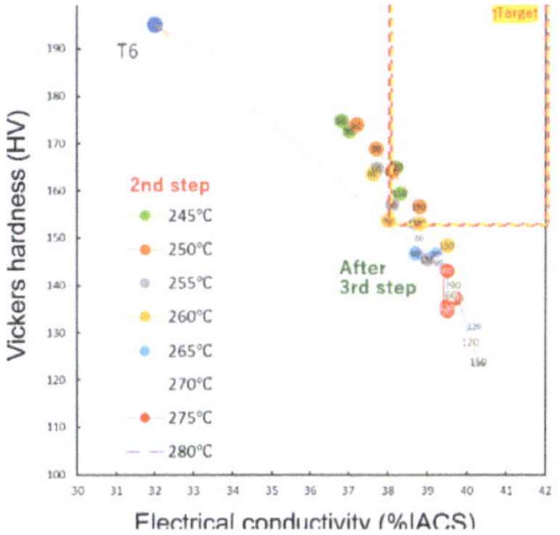

Fig. 7. Vickers hardness and electrical conductivity of T6-tempered A7075 alloy sheets after rapid heating to various temperatures and re-aging at 170°C for 1.2 ks without hot stamping. The numbers within plots indicate holding times in second at a given rapid heating temperature.

Table 1. Constructed process window of optimized heating conditions at 1st and 2nd step heat treatment for fulfilling target properties of the final products (>HV153 and >38.0 %IACS).

	60s	90s	120s	150s
245˚C	-	-	○	○
250˚C	-	-	○	○
255˚C	-	○	○	-
260˚C	-	△	○	-
265˚C	-	-	-	-
270˚C	-	-	-	-
275˚C	-	-	-	-
280˚C	-	-	-	-

255°C for 60-150 s (1st and 2nd step), and re-aging at 170°C for 1.2 ks (3rd step) without hot stamping. Monotonous decrease in hardness was observed with increasing holding time at 2nd step; i.e. from HV196 (T6-tempered) to HV138 (after 150 s holding), but the subsequent 3rd step heat treatment could successfully rebound the decreased hardness with further increase in electrical conductivity (See arrows in Fig. 6). Therefore, from the plots of all the resultant Vickers hardness and electrical conductivity after 3rd step heat treatment in Fig. 7, optimum heating conditions at 1st and 2nd step can be found. Table 1 is a constructed process window of optimized heating conditions at 1st and 2nd step heat treatment for fulfilling target properties of the final products (i.e. >HV153 and >38.0%IACS). It should be noted that these values were preset as compatible ones of A7075-T76 and -T73 alloys, where SCC resistance is regulated by their electrical conductivity [3].

Fig. 8 shows dark-field optical micrographs on RD-ND planes of T6-tempered and three-step heat treated (255°C120 s+170°C1.2 ks) specimens after four-point bending SCC test. Cracks were generated in the lower surface in the former, but only pitting corrosion with 0.1 mm depth was observed in the latter, suggesting that SCC resistance is significantly improved by the authors' developed three-step heat treatment [2].

In this study, to clarify the mechanism of such improved SCC resistance, TEM microstructures of transgranular and intergranular precipitates were observed. Fig. 9 shows TEM images of transgranular and intergranular precipitates in T6-tempered, two-step heat treated (255°C120 s)

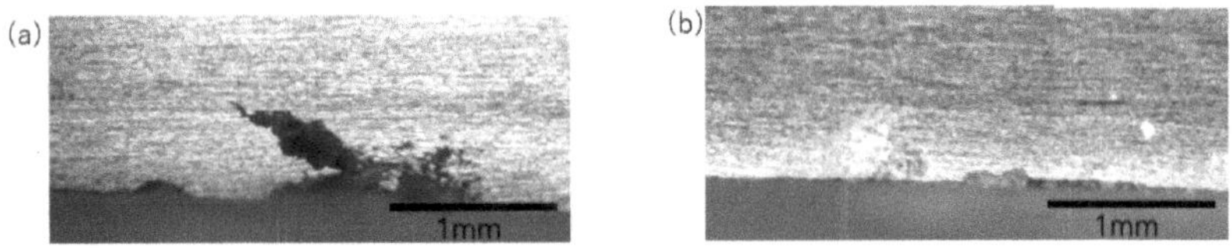

Fig. 8. Dark-field optical micrographs on RD-ND planes of (a) T6-tempered and (b) three-step heat treated (255°C120 s+170°C1.2 ks) specimens after four-point bending SCC test.

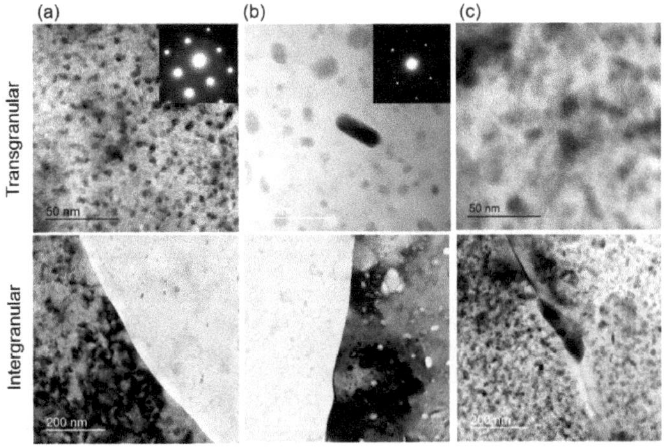

Fig. 9. TEM images of transgranular and intergranular precipitates in (a)T6-tempered, (b)two-step heat treated (255 °C120 s) and (c)three-step heat treated (255°C120 s+170°C1.2 ks) specimens.

and three-step heat treated (255°C120 s+170°C1.2 ks) specimens. By applying rapid heating and holding (1st and 2nd step), refined GP zones in the α-Al matrix were reverted, but coarser η' phase further grew as indicated in upper images of Fig. 9(a, b). However, 3rd step heat treatment resulted in the reprecipitation of GP zones (upper image of Fig. 9(c)), similar to the microstructural changes of precipitates during retrogression and re-aging (RRA) treatment [4]. As for intergranular precipitates along grain boundaries, on the other hand, sparsely formed η phase was observed only in the specimen after 3rd step heat treatment (lower image of Fig. 9(c)). Therefore, the same mechanism to the RRA treatment could be applicable to the improved SCC resistance accomplished by three-step heat treatment proposed in this study [2].

Summary

1. From a systematic investigation of the effects of 2nd step heat treatment, moderate decrease in hardness after rapid heating to ~250°C was found to improve not only the formability (i.e. lower deformation resistance and larger elongation to fracture) but also SCC resistance (i.e. increased electrical conductivity).
2. The subsequent re-aging treatment at 170°C for 1.2 ks (3rd step) could rebound the decreased hardness with further improvement in SCC resistance (i.e. increased electrical conductivity).
3. The optimized conditions of 1st and 2nd step heat treatment include 255°C for 120 s or 260°C for 120 s, and thus compatible hardness and electrical conductivity to those of A7075-T76 and T73 could be obtained by the newly developed three-step heat treatment in this study.

Acknowledgement

The authors would like to acknowledge the Light Metal Educational Foundation, Inc. and the Japan Aluminium Association for their generous support to this study.

References

[1] R.P. Garret, J. Lin, T.A. Dean, Solution heat treatment and cold die quenching in forming AA6xxx sheet components: Feasibility study, Adv. Mater. Res. 6-8 (2005) 673-680. https://doi.org/ 0.4028/ ww.scientific.net/AMR.6-8.673

[2] S. Hirosawa, T. Maeno, J. Kim, K. Nariya, Y. Suzuki, International Patent WO2023/053845 A1. Forming method of aluminum alloy sheets, (2023).

[3] Japanese standard association: JIS Handbook, Non-ferrous metals & metallurgy III, (2018), 886.

[4] B.M. Cina, U. S. Patent 3856584. Reducing the susceptibility of alloys, Particularly aluminum alloys, to stress corrosion cracking, (1974).

Effect of grain size on blank holder force of fine micro-blanking for Inconel 718 superalloy foils

JI Kaisheng[1,a], LIU Yanxiong[1,b *], ZHOU Wangbiao[1,c], WU Mengfei[1,d], ZHANG Han[1,e] and GONG Yucheng[1,f]

[1]Wuhan University of Technology,122 Luoshi Road, Wuhan, Hubei Province, China

[a]Jiks@whut.edu.cn, [b]liuyx@whut.edu.cn, [c]286637@whut.edu.cn, [d]347850@whut.edu.cn, [e]319998@whut.edu.cn, [f]273045@whut.edu.cn

Keywords: Fine Micro-Blanking, Grain Size, Blank Holder Force, Revised Empirical Formula

Abstract. Regarding product miniaturization, there is a noticeable difference between micro-blanking and macro-blanking in terms of the quality of the parts and the metal deformation behavior. In this paper, fine micro-blanking simulations were conducted on Inconel 718 foils with a thickness of 100 μm and different grain size to investigate the effects of different blank holder force on the fracture length of blanking surface. The results showed that the empirical (macro) formula of blank holder cannot apply to fine micro-blanking. The fracture length gradually decreases with the increase of blank holder force. And as the grain growth, the blank holder force to obtain a micro-blanking surface with full burnish length and no fracture zone was smaller. Based on the effect of grain size and blank holder force, a revised empirical formula of blank holder force was established for fine micro-blanking. Then, the fine micro-blanking experiments of Inconel 718 superalloy foils were conducted to evaluate the feasibility for the revised empirical formula of blank holder force. This work may contribute to the selection of forming force for fine micro-blanking applications.

Introduction

There are strict requirements for micro-hole components in many industrial domains, such as micro-electromechanical systems (MEMS), electrical/electronic devices, and medical systems [1, 2]. Various manufacturing technologies have emerged to satisfy these requirements. There are innovative techniques for manufacturing micro components including micro-hole like electrical discharge machining (EDM), exciter laser cutting, continuous force drilling, electromechanical drilling, and micro-blanking [3-7]. Micro-blanking is a straightforward, low-cost, highly productive approach that remits excellent mechanical properties and good surface quality [8, 9].

Micro-blanking is a plastic deformation technology that has been extensively researched for several decades [10-13]. Regarding product miniaturization, there is a noticeable difference between micro-blanking and macro-blanking in terms of the quality of the parts and the metal deformation behavior. Xu [14] et al. conducted forming limit experiments of metal sheets under different conditions based on miniaturized Holmberg and Marciniak tests. They found that the forming limit curve shifted downward with the decrease of the thickness-to-grain-size ratio. Geiger [15] et al. carried out theoretical analysis, numerical simulation, and experiments to find that the miniaturization of the workpiece dimensions can lead to a change in the boundary conditions in metal forming. Wang [16] et al. conducted micro bulk forming, and found that the billet size has a direct influence on the deformation behavior. This phenomenon is known as the “size effect”. The micro-forming process is accompanied by significant size effects, which reduce the formability of materials and the accuracy of parts [17]. Xu et al. [18]used brass foils with different grain size to conduct a serious of micro-blanking experiment and found the blanking force curves are repeatable for fine-grained specimens, but the curves of coarse-grained specimens has a tremendous difference for maximum blanking force and curve profile.

The fracture zone on the blanking surface will impact the service behavior and useful life of micro components and industrial equipment [19-21]. But the study results indicted many micro-blanking method can not to obtain the micro-blanking surface without the fractue zone, such as the rigid micro-blanking, laser dynamic micro-blanking, electromagnetism-assisted flexible micro-blanking, underwater shock wave-assisted flexible micro-blanking and ultrasound-assisted flexible micro-blanking [22-27]. In order to reduce the affect of fracture zone for service behavior and useful life, we proposed the fine micro-blanking method to fracture the micro components based on the fine blanking process and conducted micro-blanking experiments for Inconel 718 foils with different grain size [28, 29]. The results indicted the size effect during plastic deformation for submillimeter metal plates renders the empirical formula for (macro) blanking force inapplicable to fine micro-blanking. However, it is not possible to determine the exact blank holder force and counter force for fine micro-blanking.

Therefore, considering the problems existing in current fine micro-blanking research, this study conducted the finite element simulation to determine the exact process force. Through theoretical analysis, it is proved that increasing the blanking holder force is an effective method to improve the fine micro-blanking quality. The fine micro-blanking simulations were conducted on Inconel 718 foils with a thickness of 100 μm and different grain size to investigate the effects of different blank holder force on the fracture length of blanking surface. This study is helpful for further understanding the grain size effect, and provides technical guidance for the plastic forming of miniaturized key components.

Experimental materials and method

Inconel 718 sheets with thickness of 0.1 mm and were selected as experimental materials. Their chemical composition is shown in Table 1. Fig. 1 presents the original microstructure of the Inconel 718 foils. As shown in Fig. 1(a), the original grains were fine equiaxed crystals. The XRD pattern demonstrates that the matrix γ phase was dominant in the alloy, and there were no obvious γ', γ'', or δ phases. Therefore, it was assumed that there were no precipitated particles in the original foils.

Table 1. Chemical composition of Inconel 718 (wt. /%).

C	Cr	Ni	Co	Mo	Al	Ti	Fe	Nb	B	Mn
0.008	18.5	51.6	1.0	3.3	0.8	1.15	Bal.	5.5	0.006	0.35

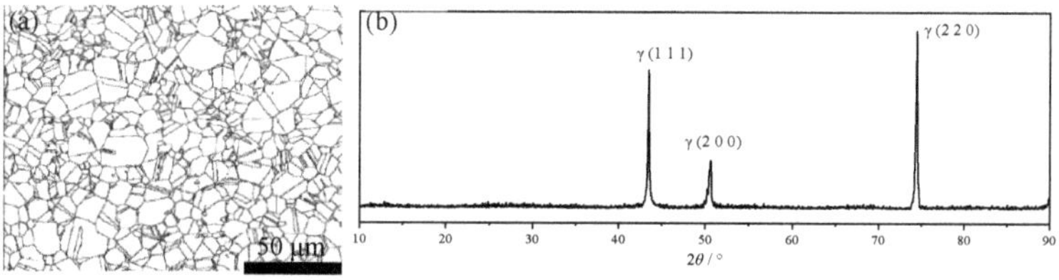

Fig. 1. Original microstructure of Inconel 718 foils: (a) microstructure; (b) XRD diffraction pattern.

The material is in the solid solution state as-received. To avoid the influence of the second phase and change the grain size on the on the mechanical properties of Inconel 718 foil, the time was 1h and temperatures of solid solutions were respectively 1050°C, 1100°C, and 1150°C, as all second phases dissolved by 1037°C. The grain size of Inconel 718 foils was dispalyed in Table 2.

Table 2. The average grain size after solid solution treatment for the Inconel 718 alloy.

Number	Average grain size / μm
O	9.3
A1	12.69
A2	52.04
A3	93.45

With the die structure and Deform-3D software, we applied FE simulation to analyze the deformation behavior of four different states of foils during Inconel 718 fine micro-blanking. The fine micro-blanking part is a round hole with thickness of 0.1 mm and radius of 0.25 mm. The FE model of the fine micro-blanking process including a punch, holder blank, metal foil, female die, and counter punch was established as shown in Fig. 2(a). We also built a quarter-size FE model based on the axisymmetry of the target part for further analysis at a reduced calculation burden and enhanced computation speed. The mesh type of the metal sheet in the FE model was a tetrahedral mesh, and the mesh was finely divided around the shear band to improve the accuracy of FE simulation. In Fig. 2(b), A and B indicate the symmetry surface of the FE model, and C is the finely divided mesh. The friction condition was the cold forming of steel dies, and its coulomb friction coefficient was 0.12. The speed of the punch movement was 0.2 mm/s, and the stroke of the punch was 0.12 mm. The Modified-Voce model and normalized Freudenthal fracture criteria were used as material models in the simulation.

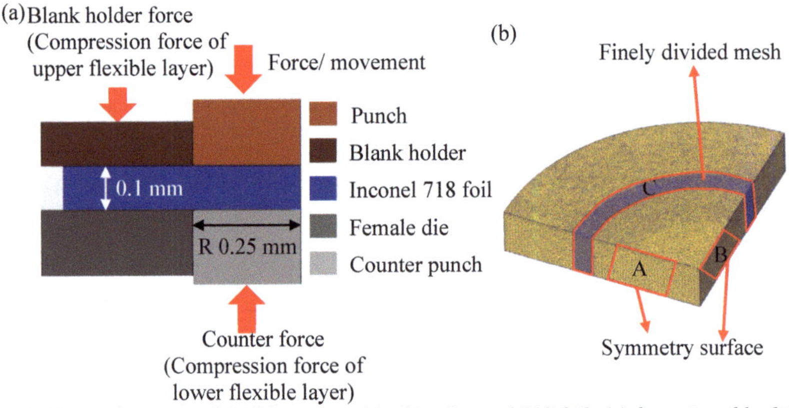

Fig. 2. Finite element model of fine micro-blanking Inconel 718 foil: (a) fine micro-blanking model, (b) workpiece mesh.

The materials constitutive model was applied to conducted the finite element simulation of fine micro-blanking, as shown in Eq. 1. The model includes the surface flow and internal flow of the material and considers the dependent variables and grain size.

$$\sigma=\eta m_1 a_1 \varepsilon^{n_1} + (1-\eta)\left(m_2 a_1 \varepsilon^{n_1} + \frac{a_2 \varepsilon^{n_2}}{\sqrt{d}} \right) \quad (1)$$

where η is the grain size factor, representing the ratio of the number of surface grains to the total number of grains in the sample. m_1 is the orientation factor of the grain on the surface, and m_2 is the orientation factor of the internal grain, a_1, a_2, q_1, q_2, n_1, and n_2 are material constants. These parameters are shown in Table 3 and the critical damage values of Inconel 718 foils with different grain sizes was displayed in Table 4.

Table 3. The parameters of material constitutive model.

Parameters	m_1	m_2	a_1/MPa	a_2/MPa	n_1	n_2
Value	2	3.06	263.78	8872.61	0.12	2.02

Table 4. The critical damage values of metal foils with the different grain size.

Number	Critical damage values
O	5.13
A1	3.48
A2	2.68
A3	2.22

Results and discussion

When the blank holder force is within a reasonable range, the blanking quality of micro-blanking pieces will be improved by the increase of blank holder force. Fine macro-blanking uses Eq. 2 to calculate the blank holder force.

$$F_{hb} = k_{bh} F_{max} \tag{2}$$

where F_{bh} is the blank holder force, F_{max} is the maximum blanking force and k_{bh} is the process coefficient of blank holder force. When k_{bh} is 0.5, the blanking quality of fine macro-blanking is good. Therefore, this study slected 0.5 as the k_{bh} to donducted the fintie element simulation for fine micro-blanking of Inconel 718. However, there are the fracture zone on the blanking surface, as shown in Fig. 3. And the fracture length is increase with grain size increase. It indicates the stress state in the shear deformation zone is changed from mean compressive stress to tensile stress, and cannot restrains the ductile fracture. In the other words, the empirical (macro) formula does not exactly to describe the fine micro-blanking when k_{bh} is 0.5. The small blank holder force caused the facture zone on the blanking surface. Therefore, the k_{bh} need to be improved.

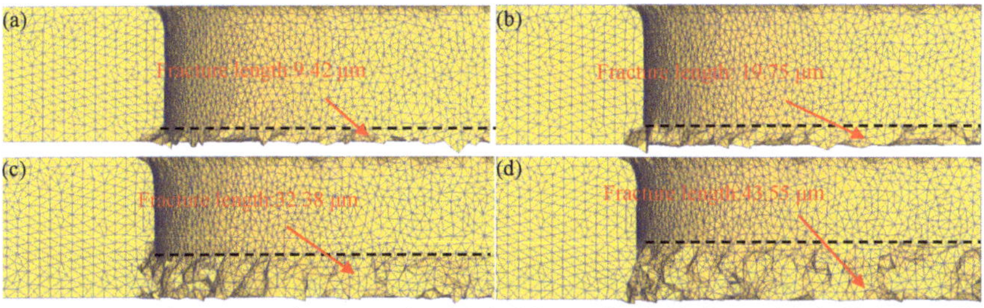

Fig. 3. Finite element simulation results: (a) d=9.3 μm, (b) d=12.69 μm, (c) d=52.04 μm, (d) d=93.45 μm.

The simulation results which using the different k_{bh} were dispalyed in Fig. 4. With the increase of process coefficient for blanking force, the fracture length of Inconel 718 foils with different grain sizes decreases gradually, until there is nofracture zone. It means that the process coefficient of blanking force no longer affect the quality of blanking surface when the blanking surface is fully brunish.

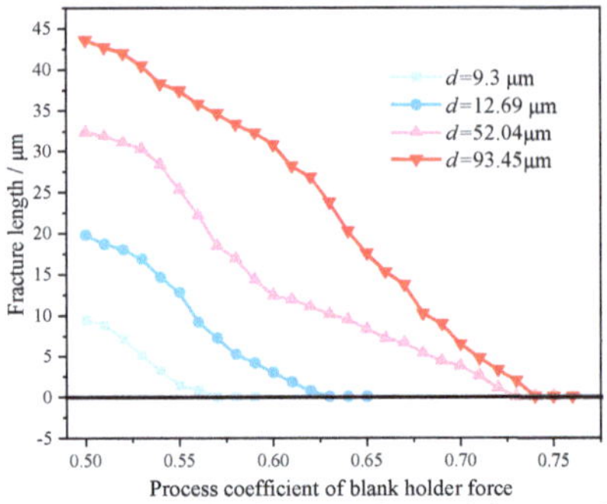

Fig. 4. The relationship between k_{bh} and fracture length.

The correlation analysis was conducted to explore the relationship between the k_{bh} and its influencing factors during fine micro-blanking for Inconel 718 foil, and the results are shown in Fig. 5. The results indicated the applicable k_{bh} increase with the grain size and yield-tensile strength ratio, decrease with the maximum tensile strength. Compared with the absolute value of correlation coefficient, it is more closely related to grain size and maximum tensile strength. And the maximum tensile strength is impacted by grain size.

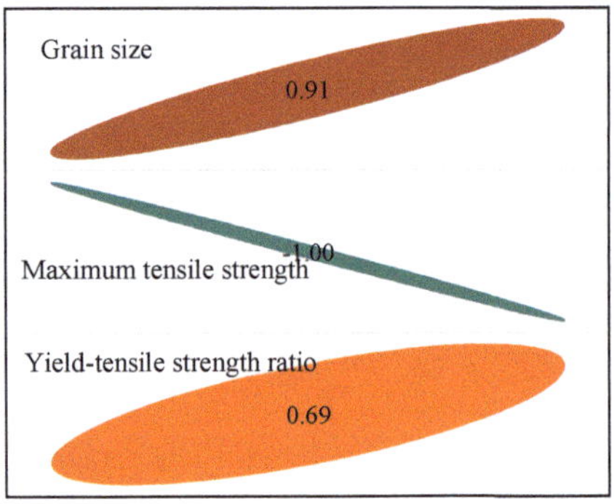

Fig. 5. The result of correlation analysis.

Therefore, we proposed that using the relationship between grain size and k_{bh} to modify empirical formula of blank holder force for fine micro-blanking, as shown in Fig. 6. To quantitatively evaluate the modified result, the relative error is introduced to assess the fitting accuracy. Through comparison and observation, it was found that the absolute relative errors of Inconel 718 thin plates with different grain sizes were all less than 2%. The modify empirical formula of blank holder force for fine micro-blanking can be described as follows:

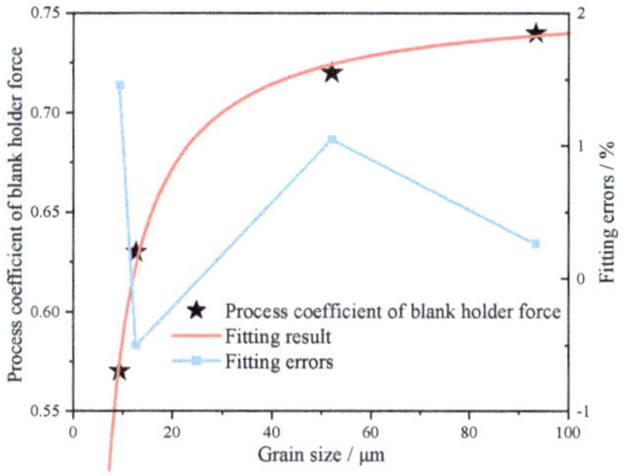

Fig. 6. The relationship between k_{bh} and grain size.

$$k_{bh-g} = 0.76 + \frac{1.69}{d} \quad (3)$$

The fine micro-blanking and modified empirical formula of blank holder force were used to fabricate the components of Inconel 718 foils with different grain size, as shown in Fig. 7. These size of components is similar to component model. And it is found that the fully burnish zone without fracture were obtained on by observing the blanking surfaces in the black frame.These results indicate that the modified empirical formula of blank holder force can be applied to manufacture micro components for fine micro-blanking.

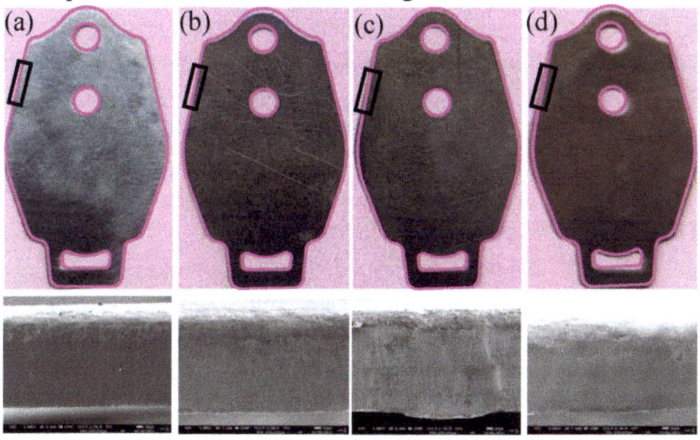

Fig. 7. Fine micro-blanking results: (a) d=9.3 μm, (b) d=12.69 μm, (c) d=52.04 μm, (d) d=93.45 μm.

Conclusion

In this study, we used fine micro-blanking technology and finite element simulation to determine the exact blank holder force and counter force for fine micro-blanking. When k_{bh} is 0.5, the empirical (macro) formula does not exactly to describe the fine micro-blanking and the fracture length is increase with grain size increase. With the increase of k_{bh}, the fracture length of Inconel 718 foils with different grain sizes decreases gradually, until there is nofracture zone. Due to k_{bh} is impacted by grain size, we established a modified empirical formula of blank holder force that is suitable for the fine micro-blanking based on the relationship between grain size and k_{bh}. This study lays a theoretical foundation for the miniaturized plastic forming of key components, and meets the needs of current development.

Acknowledgments
This work is grateful for the financial supports from the National Natural Science Foundation of China (52275371) and the National Key R&D Program of China (No. 2020YFA0714900).

References

[1] C.P. Raja, T.J.E.S. Ramesh, Influence of size effects and its key issues during microforming and its associated processes – A review, Eng. Sci. Technol. 24 (2021) 556-570. https://doi.org/10.1016/j.jestch.2020.08.007

[2] J.Q. Ran, M.W. Fu, W.L.J.I.J.o.P. Chan, The influence of size effect on the ductile fracture in micro-scaled plastic deformation, Int. J. Plast. 41 (2013) 65-81. https://doi.org/10.1016/j.ijplas.2012.09.002

[3] T. Masuzawa, J. Tsukamoto, M. Fujino, Drilling of Deep Microholes by EDM, CIRP Annals 38 (1989) 195-198. https://doi.org/10.1016/S0007-8506(07)62683-3

[4] H.K. Tönshoff, D. Hesse, J. Mommsen, Micromachining Using Excimer Lasers, CIRP Annals 42 (1993) 247-251. https://doi.org/10.1016/S0007-8506(07)62436-6

[5] M.S. Cheong, D.-W. Cho, K.F. Ehmann, Identification and control for micro-drilling productivity enhancement, Int. J. Mach. Tools Manuf. 39 (1999) 1539-1561. https://doi.org/10.1016/S0890-6955(99)00019-X

[6] S. Sharma, V.K. Jain, R. Shekhar, Electrochemical Drilling of Inconel Superalloy with Acidified Sodium Chloride Electrolyte, Int. J. Adv. Manuf. Technol. 19 (2002) 492-500. 10.1007/s001700200052. https://doi.org/10.1007/s001700200052

[7] Z. Shen, J. Zhang, H. Liu, X. Wang, Y. Ma, Reducing the rebound effect in micro-scale laser dynamic flexible forming through using plasticine as pressure-carrying medium, Int. J. Mach. Tools Manuf. 141 (2019) 1-18. https://doi.org/10.1016/j.ijmachtools.2019.03.004

[8] B.-Y. Joo, S.-H. Rhim, S.-I. Oh, Micro-hole fabrication by mechanical punching process, J. Mater. Process. Technol. 170 (2005) 593-601. https://doi.org/10.1016/j.jmatprotec.2005.06.038

[9] C. Zheng, C. Pan, Z. Tian, X. Zhao, L.J.O. Song, Laser shock induced incremental forming of pure copper foil and its deformation behavior, Opt. Laser Technol. 121 (2020) 105785. https://doi.org/10.1016/j.optlastec.2019.105785

[10] K. Ji, Y. Liu, S. Liu, Y. Zhang, A review on the microblanking technology of metallic foils, Int. J. Adv. Manuf. Technol. 114 (2021) 3185-3200. https://doi.org/10.1007/s00170-021-07096-1

[11] J. Cui, H. Liu, Y. Ma, M. Li, J. Gong, Y. Zhang, X. Wang, Flexible microblanking of amorphous alloys under laser dynamic loading, J. Manuf. Process. 56 (2020) 718-725. https://doi.org/10.1016/j.jmapro.2020.05.041

[12] F. Luo, F. Sun, K. Li, F. Gong, X. Liang, X. Wu, J. Ma, Ultrasonic assisted micro-shear punching of amorphous alloy, Mater. Res. Lett. 6 (2018) 545-551. https://doi.org/10.1007/s00170-017-0684-4

[13] Q.J. Zhao, J. Xu, C.J. Wang, D.B. Shan, B.J.A.M.R. Guo, Electromagnetic Micro-Punching Process of T2 Copper Foil, Adv. Mater. Res. 1120-1121 (2015) 1220-1225. https://doi.org/10.4028/www.scientific.net/AMR.1120-1121.1220

[14] Z.T. Xu, L.F. Peng, M.W. Fu, X.M. Lai, Size effect affected formability of sheet metals in micro/meso scale plastic deformation: Experiment and modeling, Int. J. Plast. 68 (2015) 34-54. https://doi.org/10.1016/j.ijplas.2014.11.002

[15] M. Geiger, F. Vollertsen, R. Kals, Fundamentals on the Manufacturing of Sheet Metal Microparts, CIRP Annals 45 (1996) 277-282. https://doi.org/10.1016/S0007-8506(07)63063-7

Materials Research Forum LLC
https://doi.org/10.21741/9781644903254-26

[16] G.C. Wang, W. Zheng, T. Wu, H. Jiang, G.Q. Zhao, D.B. Wei, Z.Y. Jiang, A multi-region model for numerical simulation of micro bulk forming, J. Mater. Process. Technol. 212 (2012) 678-684. https://doi.org/10.1016/j.jmatprotec.2011.05.023

[17] Y. Liu, C. Wang, H. Han, D. Shan, B. Guo, Investigation on effect of ultrasonic vibration on micro-blanking process of copper foil, Int. J. Adv. Manuf. Technol. 93 (2017) 2243-2249. 10.1007/s00170-017-0684-4.https://doi.org/10.1007/s00170-017-0684-4

[18] J. Xu, B. Guo, C. Wang, D. Shan, Blanking clearance and grain size effects on micro deformation behavior and fracture in micro-blanking of brass foil, Int. J. Mach. Tools Manuf. 60 (2012) 27-34. https://doi.org/10.1016/j.ijmachtools.2012.04.001

[19] Y. Liu, K. Ji, Y. Zhang, C. Yang, F. Yin, Y. Song, L. Hua, Energy consumption analysis for the fine blanking process, Int. J. Adv. Manuf. Technol. 130 (2024) 5743-5761. https://doi.org/10.1007/s00170-023-12917-6

[20] H. Mao, H. Chen, Y. Liu, K. Ji, A Novel Force Variation Fine Blanking Process for the High-strength and Low-plasticity Material, Metals 12 (2022) 458. https://doi.org/10.3390/met12030458

[21] S.N. Babaei, F. Schweinshaupt, S. Wipp, T. Herrig, U. Krupp, T. Bergs, Mechanism-driven fine blanking of high manganese steels, J. Mater. Process. 319 (2023) 118083. https://doi.org/10.1016/j.jmatprotec.2023.118083

[22] Y. Liu, C. Wang, B. Guo, D. Shan, M.J.M. Zhang, N. Letters, Softening effect of ultrasonic vibration on micro-blanking deformation behaviour of titanium foil, Micro Nano Lett. 12 (2017) 808-812. https://doi.org/10.1049/mnl.2017.0227

[23] X. Li, X. Li, R.D.K. Misra, Z. Chen, Grain size effect on shearing performance of copper foil: A polycrystal plasticity investigation, Mech. Mater. 166 (2022) 104212. https://doi.org/10.1016/j.mechmat.2022.104212

[24] H. Liu, M. Lu, X. Wang, Z. Shen, C. Gu, Y. Gu, Micro-punching of aluminum foil by laser dynamic flexible punching process, Int. J. Mater. Form. 8 (2015) 183-196. 10.1007/s12289-013-1159-2 https://doi.org/10.1007/s12289-013-1159-2

[25] D. Jiang-Hua, W. Wei, J. Xing-Ying, Z.J.S.G.X.J.o.P.E. Yan-Ran, Experimental investigation on electromagnetic assisted micro-piecing of brass foil, J. Plast Eng, 21 (2014) 58-62. https://doi.org/10.3969/j.issn.1007-2012.2014.03.012

[26] Y. Xiao, F. Sun, J.-q. Ran, B. Wang, J.-m. Zhong, J. Ma, F. Luo, Ultrasonic micro punching with flexible punch for thin stainless sheet metal, Int. J. Adv. Manuf. Technol, 108 (2020) 2763-2773. https://doi.org/10.1007/s00170-020-05544-y

[27] S. Tanaka, I. Bataev, M. Nishi, I. Balagansky, K. Hokamoto, Micropunching large-area metal sheets using underwater shock wave: Experimental study and numerical simulation, Int. J. Mach. Tools Manuf. 147 (2019) 103457. https://doi.org/10.1016/j.ijmachtools.2019.103457

[28] Y.X. Liu, K.S. Ji, Y.J. Zhang, Y.L. Song, F. Yin, Effect of grain size on fine micro-blanking for Inconel 718 foil, J. Manuf. Process. 107 (2023) 472-484. https://doi.org/10.1016/j.jmapro.2023.10.022

[29] K.S. Ji, Y.X. Liu, Y.J. Zhang, H. Zhang, Die design for fine micro-blanking process, IOP Conf. Ser.: Mater. Sci. Eng. 1270 (2022) 012044. https://doi.org/10.1088/1757-899X/1270/1/012044

Contour precision controlling of welding distortion in fabrication for stainless steel hyperbolic shells adopting mechanical constraints

ZHAI Hua[1,a,*], LI Cheng[1], LIU Zhihong[2], MA Jianguo[2], JI Haibiao[2], YAN Jianwen[3], LIU Qiong[3] and ZHU Xiaofeng[4]

[1]Anhui Province Key Lab of Aerospace Structural Parts Forming Technology and Equipment, Hefei University of Technology, Hefei 230009, China

[2]Institute of Plasma Physics, Chinese Academy of Sciences, Hefei 230031, China

[3]Hefei Metal Forming Intelligent Manufacturing Co., LTD, Hefei 230601, China

[4]China Institute of Atomic Energy, Beijing 102413, China

[a]jxzhaihuajx@hfut.edu.cn

Keywords: Double Shell Structure, Welding Distortion, Mechanical Constraint, Contour Precision Controlling, Inherent Strain Method

Abstract. Double-layer shell structures are widely used in the carapaces of vacuum vessels, cyclotrons, submarines and ships. The design of this double-layer structure considers the structural strength, sealing and special conditions. In the shell processing, the crisscross ribs are welded with the inner and outer shells to become parts. The welding deformation caused by multiple joints exceeds the tolerance, and external constraints are needed to ensure the forming accuracy of the contour. In this paper, the suppression effect of fixture and rigid constraint method on welding deformation of butt joint and T-joint was analyzed. The welding specimen of the double-layer shell structure was carried out, and the influence of different welding sequences on the final welding deformation was analyzed. Through external mechanical constraints, the deformation is reduced from 7.27 mm to 2.89 mm, which meets the tolerance requirements. The measured deformation is compared with the simulation results based on the inherent strain theory, which verifies the reliability of the numerical model and provides technical guidance for engineering manufacturing.

Introduction

The shell components of vacuum vessels, cyclotrons, submarines and ships are usually designed as double-layer shell structures, mainly considering strength, tightness, pressurization and special work conditions [1-2]. Due to the harsh service environment conditions, the processing accuracy, structural strength, and long-term service reliability of the double-layer shell components have high requirements. Cyclotrons and vacuum vessels in fusion reactor have a thick-walled double shell structure with intricate ribs and high-density, full-penetration welded joints [3]. The shell component is a large-scale structure with a complex hyperbolic curved shell, which usually adopts a multi-block segmented welding method. The assemble and welding of multi-block components forms the final shell components [4]. It is inevitable to cause independent block deformation and accumulate deformation in the process of continuous manufacturing. To improve the manufacturing accuracy of the shell, it is necessary to calculate and suppress the welding deformation of each manufacturing stage to ensure that the final contour precision of the shell can meet the tolerance requirements [5]. The welding deformation is positively correlated with the accumulated heat input during the welding process. Choosing reasonable welding method, welding sequence, welding parameters, and joint design can effectively reduce welding deformation. Usually, it is necessary to predict the amount of deformation through numerical simulation and then design a suitable fixture structure to suppress the generation of deformation. The finite element metho based on inherent strain has made significant achievements in predicting welding deformation of large structures, greatly improving computational efficiency while ensuring calculation accuracy [6]. Murakawa et al. proposed the concepts of inherent strain and inherent deformation to simulate welding deformation and residual stresses [7]. Deng et al. established a finite element method to accurately simulate the welding deformation of large-scale structures considering the initial root gap [8]. Wang et al. predicted and reduced the out-of-plane welding deformation of different types of welded joints in ship panel based on inherent deformation theory [9]. Lu et al. simulated the welding deformation of T-joint using the thermo-elastic-plastic method (TEPM) and the inherent strain method (ISM), and predicted the deformation successfully of side beam of bogie using the inherent strain [10]. Zhong et al. predicted the assembly welding deformation of the 1/8 vacuum vessel using equivalent transverse and longitudinal inherent strains, and reduced the welding deformation by optimizing the welding sequence and setting constraints [11].

In addition, the design of fixture is necessary, because the external constraints can effectively reduce the welding deformation. Ma et al. investigated the welding deformation of butt joint under fixture constrained conditions through experiments. The effects of different constraint positions on longitudinal shrinkage, transverse shrinkage and angular deformation are discussed in detail [12]. Zhang et al. compared the effects of preset stiffeners and anti-deformation method on suppressing the welding deformation of 30 mm T-joints [13]. Gharib et al. studied the effects of clamping time, clamping position and heat/cold release mode on the welding deformation of butt joints [14]. Xia et al. designed a special welding fixture for the 1/16 vacuum vessel, and carried out numerical simulation and experimental measurement to verify that the welding deformation is effectively reduced [15]. Ma et al. applied the inherent deformation calculated by constrained joints to large structures to predict welding deformation controlled by jigs. The jigs near the weld are more helpful in reducing the welding deformation [16]. At present, welding deformation control of large structures with multiple welds is still challenging. The inherent strain method can be used in numerical simulation, which also means that the inherent strain of the joint needs to be calculated first. The accuracy of inherent strain directly affects the prediction results of large components. In addition, the fixture is designed according to the results of deformation prediction, which requires higher simulation accuracy, otherwise it will lead to the increase of research and development cost in the manufacturing process. The difference of different welding structure characteristics makes the welding deformation suppression scheme different. There are few research literatures on welding deformation of double-layer shell structure. In this paper, according to the characteristics of double-layer structure, the optimal welding sequence, welding deformation mechanism and welding deformation suppression are studied, which brings data support to engineering manufacturing.

In this paper, the deformation suppression of typical welded joints of double-layer shell structure was studied. The clamping constraint was used to reduce the angular deformation of the butt joint, and the stiffener was used to reduce the deformation of the T-joint. The distribution of welding deformation before and after constraint was analyzed in detail. The optimal welding sequence of the double-layer shell was studied, and the welding deformation trend and mechanism under unconstrained condition were analyzed. The constraint system was used to suppress the deformation of the double-layer shell. Finally, the deformation is controlled within the tolerance range and the results are compared with the experimental measurement.

Numerical simulation method of welding deformation

In the calculation process of the welding temperature field, a nonlinear heat transfer equation is used to describe the heat transfer process generated by the welding arc inside the welded workpiece.

$$\rho c\frac{\partial T}{\partial t}(x,y,z,t) = -\nabla \cdot q(x,y,z,t) + Q(x,y,z,t) \quad (1)$$

where, ρ is the density of the materials, c is the specific heat capacity, T is the current temperature, q is the heat flux vector, Q is the internal heat generation rate, x, y and z are the coordinates in the reference system, t is the time and ∇ is the spatial gradient operator.

The double ellipsoidal volume heat source is used to represent the form of welding heat input [17]. The four parameters determine the shape of the heat source, a_1 is the length of the front semi-axes in the welding direction, a_2 is the length of the rear semi-axes in the welding direction, b is the semi-axes length in the direction of weld width, c is the semi-axes length in the depth direction of the weld seam. The heat flux distribution of the front and rear parts can be expressed as follows:

$$Q_1(x,y,z,t) = \frac{6\sqrt{3}f_1\eta UI}{a_1 bc\pi^{3/2}} \cdot \exp\left[-3\left(\frac{x-x_0-vt}{a_1}\right)^2 - 3\left(\frac{y-y_0}{b}\right)^2 - 3\left(\frac{z-z_0}{c}\right)^2\right] \quad (2)$$

$$Q_2(x,y,z,t) = \frac{6\sqrt{3}f_2\eta UI}{a_2 bc\pi^{3/2}} \cdot \exp\left[-3\left(\frac{x-x_0-vt}{a_2}\right)^2 - 3\left(\frac{y-y_0}{b}\right)^2 - 3\left(\frac{z-z_0}{c}\right)^2\right] \quad (3)$$

where, Q_1 and Q_2 are the effective power of the heat source in the front and rear hemispheres, respectively. f_1 and f_2 are the distribution coefficients of heat flux density in the front and rear hemispheres, and $f_1+f_2=2$. η is heat source efficiency. x_0, y_0, and z_0 are the starting point coordinates of the welding heat source, respectively. v is the welding speed, U is the welding voltage, and I is the welding current.

In the simulation model, the convection and radiation effects between the model and the external environment are considered. The convection heat loss q_c and the radiation heat loss q_r are as follows [18]:

$$q_c = -h_c(T_s - T_0) \quad (4)$$

$$q_r = -\varepsilon\sigma\left[(T_s + 273.15)^4 - (T_0 + 273.15)^4\right] \quad (5)$$

The mechanical behavior during welding is very complex. The inherent strain theory believes that the summation of thermal strain $\varepsilon_{thermal}$, plastic strain $\varepsilon_{plastic}$, creep strain ε_{creep}, and phase transition strain ε_{phase} after the welding process is the inherent strain, which is the source of the residual stress and deformation. The inherent strain $\varepsilon_{inherent}$ can be expressed as follows [19]:

$$\varepsilon_{inherent} = \varepsilon_{plastic} + \varepsilon_{thermal} + \varepsilon_{creep} + \varepsilon_{phase} \quad (6)$$

The magnitude and distribution of the inherent strain are mainly determined by the welding parameters, such as the types of the welded joint, welding heat source, material types, and geometric parameters. In Eq. 6, creep strain and phase transition strain are usually ignored because they are relatively small. Plastic strain is the main source of deformation. Welding distortion of a weld joint mainly including four components, longitudinal shrinkage (δ_x), transverse shrinkage (δ_y), longitudinal bending (θ_x) and transverse bending (angular distortion θ_y). The four components can be calculated using the following equation [19]:

$$\begin{cases} \delta_x = \dfrac{1}{h}\iint \varepsilon_x \mathrm{d}y\mathrm{d}z \\ \delta_y = \dfrac{1}{h}\iint \varepsilon_y \mathrm{d}y\mathrm{d}z \\ \theta_x = \dfrac{12}{h^3}\iint \varepsilon_x (z-h/2)\mathrm{d}y\mathrm{d}z \\ \theta_y = \dfrac{12}{h^3}\iint \varepsilon_y (z-h/2)\mathrm{d}y\mathrm{d}z \end{cases} \tag{7}$$

where, ε_x is the plastic strain in the welding direction (longitudinal direction), ε_y is the plastic strain in the transverse direction, h is the thickness.

The inherent deformation is obtained by integrating the plastic strain with Eq. 7, and the elastic analysis can be carried out by introducing the inherent deformation into the finite element model. The determination of the magnitude of the inherent strain depends on the prior thermo-elastic-plastic analysis. Reference [20-21] proposed an empirical method for calculating inherited strains based on extensive experimental research. Assumptions K and ξ are the longitudinal and transverse inherent strain coefficients respectively. After a lot of experiments and numerical simulations, K and ξ depend on the value of Q/h^2, as shown in Fig. 1.

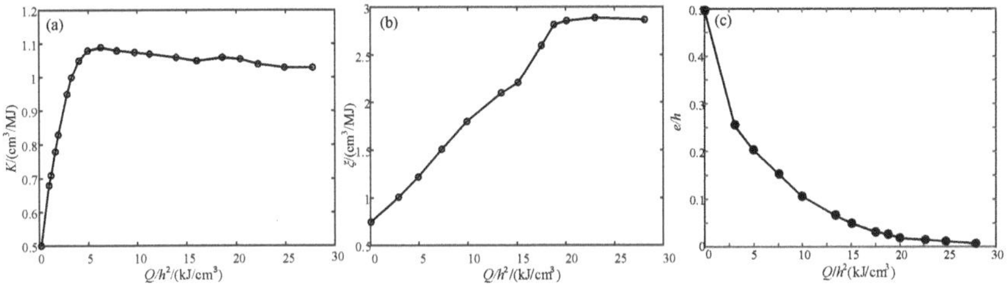

Fig. 1. The coefficient of empirical inherent strain method. (a) Longitudinal inherent strain coefficient. (b) Transverse inherent strain coefficient. (c) Eccentricity.

The magnitude range of Q/h^2 in engineering welding is 0~20 kJ/cm^3, and the magnitude range of medium and thick plates is generally 0~5 kJ/cm^3. Through linear curve fitting using the MATLAB, the approximate calculation formula can be obtained as follows:

$$K = \begin{cases} 0.08\times\dfrac{Q}{h^2}+0.7 & \left(\dfrac{Q}{h^2}<5\right) \\ 1.1 & \left(\dfrac{Q}{h^2}..5\right) \end{cases} \tag{8}$$

$$\xi = \begin{cases} 0.1025\times\dfrac{Q}{h^2}+0.75 & \left(\dfrac{Q}{h^2}<20\right) \\ 2.8 & \left(\dfrac{Q}{h^2}..20\right) \end{cases} \tag{9}$$

For multi-pass welding, the total volume of longitudinal inherent strain W_x and transverse inherent strain W_y can be expressed as [21]:

$$W_x = W_{xm} + 2S\varepsilon \tag{10}$$

$$W_y = \Sigma W_{yi} = \Sigma \xi_i Q_i \tag{11}$$

where, W_{xm} is the sum of longitudinal inherent strains for single pass welding; S is the area of the weld seam. ε is the yield strain. Q_i is the line energy of the i-layer weld, ξ_i is the inherent strain coefficient corresponding to the i-layer weld.

In the finite element model, the inherent strain application area can be set in the calculation model, and the anisotropy thermal expansion coefficient of the material in the inherent strain area can be defined. Unit temperature load can be applied to achieve the loading of the longitudinal and transverse inherent strains of the weld. The coefficient of thermal expansion α_x and α_y can be obtained using the following equations:

$$\alpha_x = \frac{W_x}{A\Delta T} \tag{12}$$

$$\alpha_y = \frac{W_y}{A\Delta T} \tag{13}$$

where, A is the sectional area of the inherent strain application unit; α_x corresponds to the coefficient of thermal expansion in the direction of longitudinal inherent strain, α_y corresponds to the coefficient of thermal expansion in the direction of transverse inherent strain; ΔT is the unit temperature load.

For different joint types, the applied area of inherent strain is different, as shown in Fig. 2. For butt joints, the thermal expansion coefficient can be applied to the rectangular area nearby of weld [21]:

$$A = 2a \times h' \tag{14}$$

where, a is the half-width of the natural inherent loading area, h' is the thickness of the inherent strain loading area, $h'=h-2e$, e is the eccentric value, also depending on the value of Q/h^2, as shown in Fig. 1(c).

For T-joints, the heat will be applied to the flange and web at the same time during welding, and the inherent strain shall be calculated separately. The line energy calculation of the flange and web can use the following equations [22]:

$$Q_f = \frac{2h_f}{2h_f + h_w} Q_{total} \tag{15}$$

$$Q_w = \frac{h_w}{2h_f + h_w} Q_{total} \tag{16}$$

where, Q_f and Q_w are the line energies obtained from the flange and web, respectively; h_f and h_w are the thicknesses of flange and web, respectively; Q_{total} is the total linear energy of the weld.

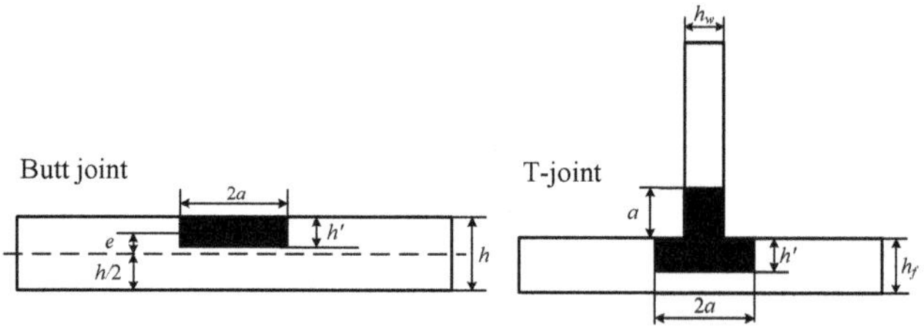

Fig. 2. Application methods of inherent strain for different joints.

Materials Research Forum LLC
https://doi.org/10.21741/9781644903254-27

Numerical simulation of welding deformation suppression

Double-layer shell structure and welded joints. The double-layer shell structure studied in this paper is the part of the carapace of the cyclotron, and its manufacturing accuracy will directly affect the subsequent multi-block assembly and welding. The structure is formed by welding, and the contour of the curved surface shape needs to be effectively guaranteed. The exploded view of the double-layer shell structure is shown in Fig. 3(a), which is composed of the outer shell, inner shells, toroidal ribs, and poloidal ribs, mainly including X-groove butt joint, double V-groove T-joints, and J-groove T-joints. To be more specific, the welding between inner shells is a X-groove butt joint. The welding between inner shells and ribs is double V-groove T-joints. The welding between ribs is double V-groove T-joints. The welding between the outer shell and ribs is J-groove T-joints, as shown in Fig. 3(b)-(d).

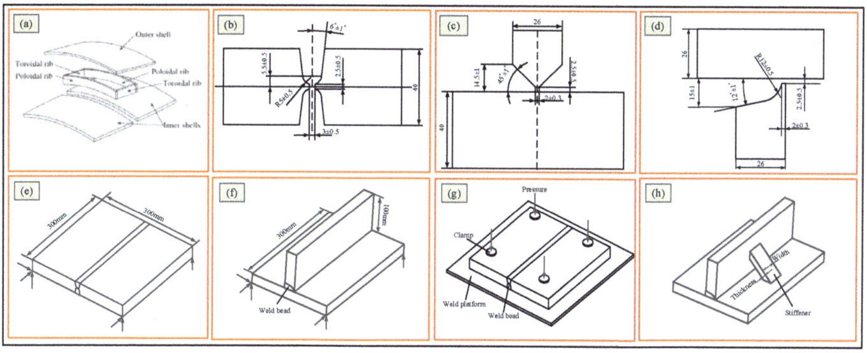

Fig. 3. Double shell structure and different welded joint: (a) Double shell structure, (b) X groove butt joint, (c) K groove T-joint, (d)J groove T-joint, (e) Unconstrained butt joint, (f) Unconstrained T-joint, (g) Restraint conditions of butt joint, (h) Restraint conditions of T-joint.

The typical joints of the shell structure are numerically simulated, and the influence of different constraints on the welding deformation is analyzed. The mitigation of deformation was the major focus of this work, and different constraint schemes were constructed utilizing the mechanical constraints. The first is the free state, which is bounded by three corners, as shown in Fig. 3(e)-(f). It is a necessary limitation to prevent inaccurate spatial motion trends. The butt joint is constrained by the clamping mechanism, and the T-joint is constrained by the reinforcing rib, as shown in Fig. 3(g)-(h). Considering the interaction between the fixture and the sample, the simulation model of the constraint fixture is established. To accurately simulate the thermodynamic behavior during the welding process, temperature-dependent material properties were used. The base metal is 304 stainless steel, and the welding wire is ER316. It is assumed that the material parameters of the welding material are the same as those of the base metal. The manual gas tungsten arc welding method was used to conduct experiments on welding joints to evaluate the reliability of the welding process. The welding process parameters are shown in Tab.1. In the multi-layer welding process, the interlayer temperature is controlled below 150°C.

Table 1. Welding parameters of the manual TIG welding.

Type	X-groove	K-groove	J-groove
Welding pass	20-25	12-15	12-15
Welding voltage (U/V)	8-12	9-12	8-12
Welding current (I/A)	125-220	160-293	150-274
Welding speed (v/mm·min^{-1})	80-120	80-120	80-360

Welding deformation suppression of butt joint. The out of plane deformation of butt joint after welding are shown in the Fig. 4. The results of welding deformation without constraints are shown in Fig. 4(a), from which the welding deformation in the vertical direction has a certain symmetry. It can be found that the transverse inherent strain has a greater effect on the welding deformation, resulting in an overall V-shaped deformation. The welding deformation constrained by the clamping mechanism is shown in Fig. 4(b). Due to the constraint of the welding platform, the deformation of the weld center is reduced, and the angular shrinking causes both ends being upward. In the free state, both sides of the edge by the initial restraint, the center region deformation to the downside, when set the restraint, the deformation to the bottom is significantly reduced, but the sides tend to upward deformation. Under the pressure constraint of the clamping mechanism, the out-of-plane deformation is obviously alleviated. Fig. 4(c) compares the distribution of welding deformation in the vertical direction, and with the application of constraints, the deformation of the weld joint decreases. The restraint scheme reduces out of plane deformation by 38.6%.

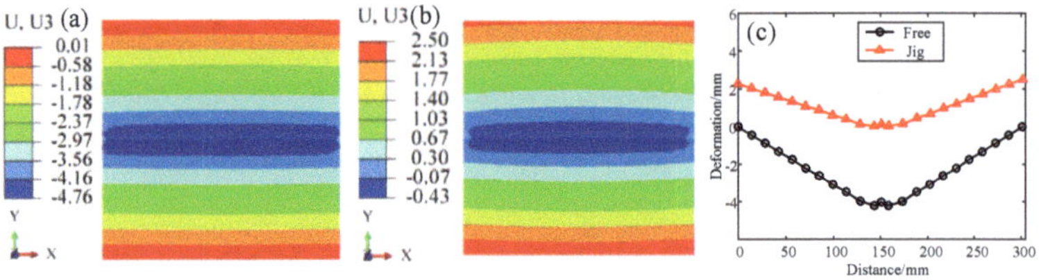

Fig. 4. Welding deformation simulation of butt joint. (a) Unconstrained state. (b) Restrained condition. (c) Comparison of angular deformation.

Welding deformation suppression of T-joint. The welding deformation of T-joint are shown in the Fig. 5. The result of welding deformation without constraint is shown in Fig. 5(a). Due to the boundary condition, edge region of the flange plate was fixed with almost none deflection, web plate bend downward with about 34.7 mm magnitude of welding angular distortion. The deformation result with stiffener is shown in Fig. 5(b). The stiffeners play a structural constraint role, which hinders the deformation trend and significantly reduces the welding deformation. The welding angle deformation of the web is reduced to 3.7 mm. This means that the structural constraints of the stiffeners reduce the welding deformation by 89.3 %. It can be determined that external mechanical constraints can significantly inhibit welding deformation.

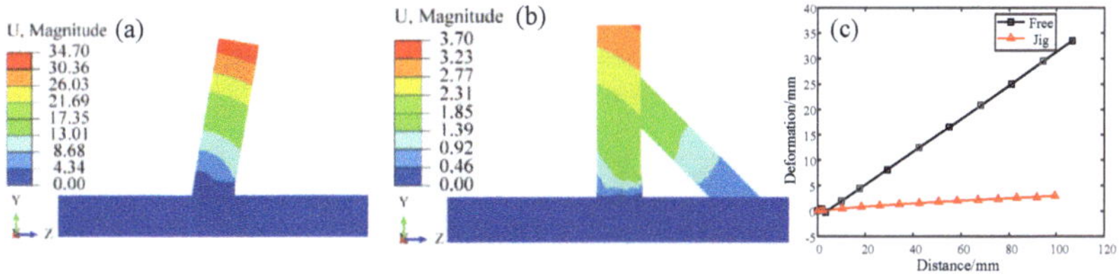

Fig. 5. Welding deformation simulation of T-joint: (a) Unconstrained state, (b) Restrained condition, (c) Comparison of angular deformation.

Welding deformation suppression of double-layer shell structure

Optimum welding sequence. The double-layer shell structure has multiple welded joints, and the welding sequence has a certain influence on the final welding deformation. The inner shell is the carrier of welding, so the first step is to weld the inner shell, then weld the four ribs on the inner shell, and finally weld the outer shell. According to the characteristics of the structure, we have developed the following possible welding sequences, as shown in Fig. 6. The welding of the outer shell has a variety of possibilities because it is a square weld, so it is necessary to analyze the influence of different welding sequences on the deformation of the outer shell. Four welding cases are designed and the optimum welding sequence should be analyzed.

Fig. 6. Design of different welding sequences.

In the numerical simulation analysis, in order to simulate the spot welding constraint in the welding experiment, four nodes are taken at both ends of the inner shell for constraint. Without other mechanical constraints, the welding deformation under unconstrained conditions is discussed. The parts are gradually added to the model according to the welding sequence, and the activation sequence is defined by the birth-death element. The welding sequence 1 and Case A simulation results in Fig. 6 are shown in Fig. 7, and the maximum deformation is 7.25 mm. When welding the inner shell, due to the shrinkage of the weld, the two sides are deformed upward. Then, four ribs are welded subsequently, which has little effect on the deformation of the inner shell. In addition, the maximum deformation of the longitudinal and transverse ribs is 4.2 mm and 3.9 mm. After welding, the outer shell deforms downward, with the maximum deformation of 6.15 mm.

The deformation in the middle area is the smallest, due to the mechanical interaction between the surrounding welds. From the deformation results, this is far beyond the tolerance requirements, so it is necessary to suppress the welding deformation.

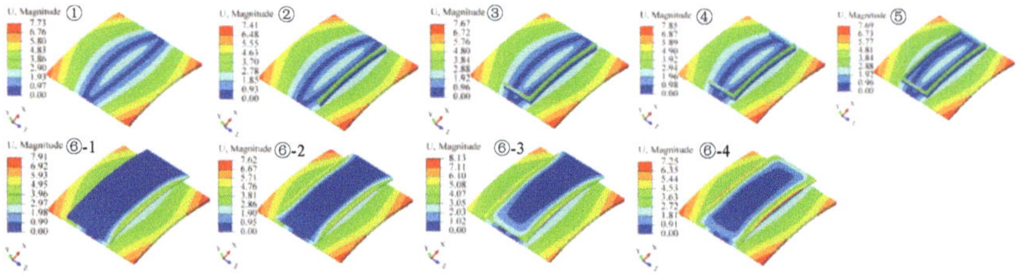

Fig. 7. Welding deformation results according to the welding sequence 1.

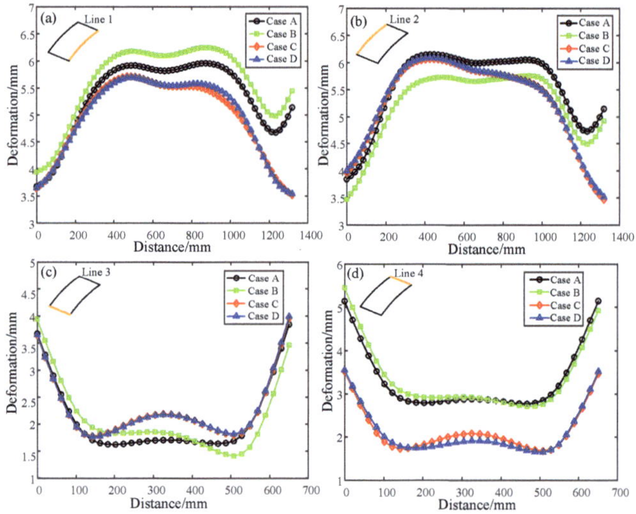

Fig. 8. Welding deformation of different lines of the outer shell: (a) Line 1, (b) Line 2, (c) Line 3, (d) Line 4.

As the weld bead is welded one by one, the stiffness of the weldment will change, and the shrinkage of the post weld bead will be constrained by the structure. The welding deformation of outer shell in Case A, Case B, Case C and Case D needs to be discussed in detail. Based on the welding sequence 1, the influence of different welding schemes on the deformation of the outer shell is analyzed by numerical simulation. Four lines were selected to compare the difference of welding deformation, as shown in Fig. 8. In summary, welding deformation of Case C and Case D is smaller than that of Case A and Case B. This is because Case C and Case D first weld short ribs to produce structural constraints, which reduces the deformation during subsequent welding of long ribs. Because of the stronger shrinkage tendency, the average deformation of the long ribs is obviously greater than that of the short ribs. In addition, the first welded rib always deforms significantly because it is more prone to deformation without the constraints of other ribs. After comparison, it can be found that Case C is the optimum sequence for welding the outer shell, that is, short ribs are first welded and than welding long ribs.

It has been determined that the welding scheme of Case C is adopted in the overall welding sequence. The numerical simulation process of welding deformation of different welding sequences in Fig. 6 is shown in Fig. 9. In the six welding sequences, the first step is to weld the inner shell, but the difference produced from the second step, which also leads to the difference of the final welding deformation. In summary, the maximum deformation is 7.40 mm and the minimum deformation is 7.27 mm in the six welding sequence. It can be concluded that the welding sequence has no obvious effect on the maximum deformation of the double-layer frame structure.

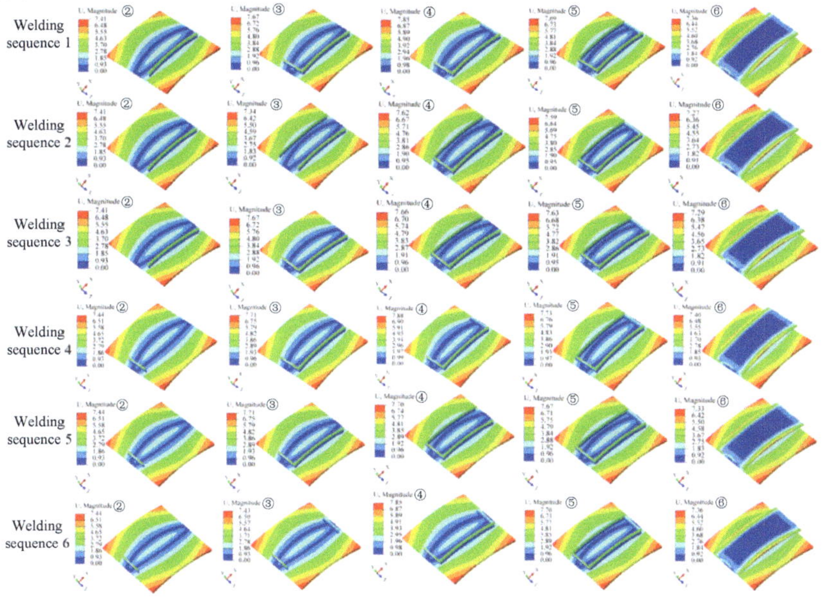

Fig. 9. Welding deformation under different welding sequences.

Welding deformation suppression. In the previous content, the method of suppressing welding deformation was studied. The external pressure is used to limit the deformation in the butt joint, and the stiffener is used to constrain the deformation in the T-joint. According to the mechanical constraints used in processing and converted into corresponding load in simulation. Specifically, when welding the inner shell, an external load is applied to suppress the deformation. When welding four ribs, the stiffeners are set to limit the deformation. When welding the outer shell, anti-deformation back ribs are applied to its back to resist the shrinkage. According to the unconstrained state, the deformation of the welding sequence 2 is the smallest, so the deformation is suppressed in accordance following the order. The welding deformation suppression results are shown in Fig. 10. Finally, The maximum deformation is is reduced to 2.89 mm.

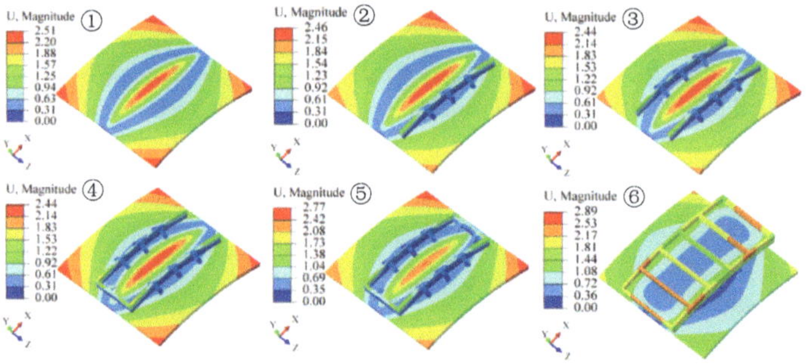

Fig. 10. Welding deformation suppression results following the welding sequence 2.

Welding experiment. The Jigs system in the numerical simulation process was manufactured for actual welding. Jigs were used to restrict welding distortion in all welding processes and add and remove the fixture in the appropriate process. Bolt clamping mechanisms were arranged around the inner shells, and stiffeners were used to fix ribs on the inner shells during welding. Finally, anti-deformation back ribs were used to reduce the deformation of the outer shell. The trend and magnitude of welding deformation were measured after welding. The double shell welded weldment is shown in Fig.11(a). The numerical calculation and experimental measurement results were compared, as shown in Fig. 11(b) and (c), which shows that the results of both are highly consistent. The deformation of numerical simulation is small because the jigs are fixed by spot welding for easy removal, so the degree of restraint is lower than that of numerical simulation.

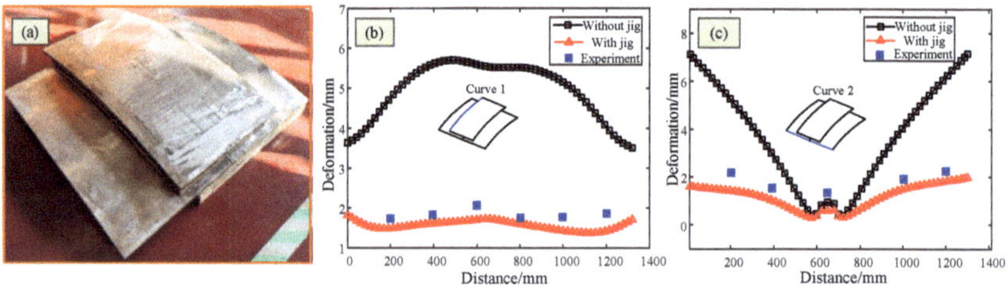

Fig. 11. Welding deformation suppression results: (a) Processed weldment, (b) Deformation of curve 1, (c) Deformation of curve 2.

Conclusions

In this paper, based on the inherent strain method, the welding deformation distribution of butt joints and T-joints is numerically simulated, and the influence of constraint schemes on the suppression of welding deformation is compared. Finally, the experimental study and numerical simulation of the welding deformation and its mitigation of the double-layer shell structure were conducted. The main conclusions are as follows:

(1) Numerical simulation models of welding deformation of butt joints and T-joints were established. Under the constraint scheme, the angular deformation of the butt joint is reduced by 38.6 %, and the angular deformation of the T-joint is reduced by 89.3 %, which verifies the effectiveness of the mechanical constraint to suppress the welding deformation.

(2) Six overall welding sequences of double-layer shell structure and four welding schemes of outer shell are designed. The distribution law of welding deformation is compared, and welding sequence 2 and Case C are determined as the optimum welding sequence.

(3) A jig system was designed to suppress the welding deformation of the double-layer shell structure, and the welding experiment was carried out. The influence of fixture on welding deformation distribution was analyzed. The maximum welding deformation was reduced from 7.27 mm down to 2.89 mm. The numerical simulation results are compared with the deformation experimental measurements, which verifies the effectiveness and provides guidance for engineering manufacturing.

Acknowledgements
The experimental part of this paper is supported by the Institute of plasma physics, Chinese academy of sciences and Hefei metal forming intelligent manufacturing Co., LTD. This work was supported by Anhui Province Key Project No. 202103a07020010, and No. 202304a05020074, flexible molding equipment for carbon fiber composite materials, JZ2023GQBK0070.

References

[1] Y.T. Kim, T.J. Kim, T.Y. Park, C.D. Jang, Welding distortion analysis of hull blocks using equivalent load method based on inherent strain, J. Ship Res. 56 (2012) 63-70. https://doi.org/10.5957/jsr.2012.56.2.63

[2] L. Xiu, J. Wu, Z. Liu, J. Ma, X. Fan, H. Ji, X. Xia, Y. Li, Weld distortion prediction of the CFETR vacuum vessel by inherent strain theory, Fusion Eng. Des. 121 (2017) 43-49. https://doi.org/10.1016/j.fusengdes.2017.03.175

[3] H. Ji, J. Wu, H. Wu, Z. Liu, J. Ma, X. Fan, Analysis of welding deformation on CFETR 1/32 vacuum vessel mockup, Fusion Eng. Des. 144 (2019) 160-163. https://doi.org/10.1016/j.fusengdes.2019.05.012

[4] Y. Kim, J. Kim, S. Kang, A study on welding deformation prediction for ship blocks using the equivalent strain method based on inherent strain, Appl. Sci. 9 (2019) 4906. https://doi.org/10.3390/app9224906

[5] J. Yi, J. Lin, Z. Chen, T. Chen, Prediction and controlling for welding deformation of propeller base structure, J. Ocean. Eng. Sci. 6 (2021) 410-416. https://doi.org/10.1016/j.joes.2021.06.002

[6] H. Murakawa, Y. Luo, Y. Ueda, Prediction of welding deformation and residual stress by elastic FEM based on inherent strain (first report) mechanism of inherent strain production, J. Soc. Nav. Archit. Jpn. 180 (1996), 739-751. https://doi.org/10.2534/JJASNAOE1968.1998.323

[7] H. Murakawa, D. Deng, N. Ma, Concept of inherent strain, inherent stress, inherent deformation and inherent force for prediction of welding distortion and residual stress, Trans. J.W.R.I. 39 (2010) 103-105. https://www.semanticscholar.org

[8] D. Deng, H. Murakawa, W. Liang, Numerical simulation of welding distortion in large structures. Comput. Methods Appl. Mech. Eng. 196 (2007) 4613-4627. https://doi.org/10.1016/j.cma.2007.05.023

[9] J. Wang, S. Rashed, H. Murakawa, Y. Luo, Numerical prediction and mitigation of out-of-plane welding distortion in ship panel structure by elastic FE analysis, Mar. Struct. 34 (2013) 135-155. https://doi.org/10.1016/j.marstruc.2013.09.003

[10] Y. Lu, C. Lu, D. Zhang, T. Chen, J. Zeng, P. Wu, Numerical computation methods of welding deformation and their application in bogie frame for high-speed trains, J. Manuf. Process. 38 (2019) 204-213. https://doi.org/10.1016/j.jmapro.2019.01.013

[11] Y. Zhong, Z. Liu, J. Wu, J. Ma, J. Tao, H. Ji, Optimization of welding distortion of vacuum vessel for nuclear fusion based on finite element analysis, Fusion Eng. Des.195 (2023) 113935. https://doi.org/10.1016/j.fusengdes.2023.113935

[12] N. Ma, H. Huang and H. Murakawa, Effect of jig constraint position and pitch on welding deformation, J. Mater. Process. Technol. 221 (2015) 154-162. https://doi.org/10.1016/j.jmatprotec.2015.02.022

[13] C. Zhang, S. Li, J. Sun, Y. Wang, D. Deng, Controlling angular distortion in high strength low alloy steel thick-plate T-joints, J. Mater. Process. Technol. 267 (2019) 257-267. https://doi.org/10.1016/j.jmatprotec.2018.12.023

[14] A.R. Gharib, F.R. Biglari, M. Shafaie, A.H. Kokabi, Experimental and numerical investigation of fixture time on distortion of welded part, Int. J. Adv. Manuf. Technol. 104 (2019) 1121-1131. https://doi.org/10.1007/s00170-019-03874-0

[15] X. Xia, H. Ji, J. Wu, H. Wu, Z. Liu, J. Ma, X. Fan, Research on welding deformation of CFETR 1/16 vacuum vessel mockup, Fusion Eng. Des. 151 (2020) 111411. https://doi.org/10.1016/j.fusengdes.2019.111411

[16] N. Ma, H. Huang, Efficient simulation of welding distortion in large structures and its reduction by jig constraints, J. Mater. Eng. Perform. 26 (2017) 5206-5216. https: //doi.org/10.1007/s11665-017-3000-4

[17] J. Goldak, A. Chakravarti, M. Bibby, A new finite element model for welding heat sources, Metal. Trans. B. 15 (1984) 299-305. https://link.springer.com/article/10.1007/BF02667333

[18] G. Fu, M.I. Lourenço, M. Duan, S.F. Estefen, Influence of the welding sequence on residual stress and distortion of fillet welded structures, Mar. Struct. 46 (2016) 30-55. https://doi.org/10.1016/j.marstruc.2015.12.001

[19] D. Deng, H. Murakawa, Prediction of welding distortion and residual stress in a thin plate butt-welded joint, Comput. Mater. Sci. 43 (2008) 353-365. https://doi.org/10.1016/j.commatsci.2007.12.006

[20] J. Xu, L. Chen, J. Wang, C. Ni, Prediction of welding distortion in multipass girth-butt welded pipes of different wall thickness, Int. J. Adv. Manuf. Technol. 35 (2008) 987-993. https://doi.org/10.1007/s00170-006-0782-1

[21] J. Wang, H. Lu, Prediction of welding deformations by FEM based on inherent strains, J Shanghai Jiaotong Univ. 2 (2000) 83-87. https://www.semanticscholar.org

[22] W. Liang, D. Deng, Influences of heat input, welding sequence and external restraint on twisting distortion in an asymmetrical curved stiffened panel, Adv. Eng. Softw. 115 (2018) 439-451. https://doi.org/10.1016/j.advengsoft.2017.11.002

Materials Research Proceedings 44 (2024) 254-260 https://doi.org/10.21741/9781644903254-28

Wrinkling behavior of the ultra-thin components with variable cross-sectional perimeter during bending process with local support

ZHUANG Yongpeng[1] and HAN Cong[1,2,a *]

[1]School of Materials Science and Engineering, Harbin Institute of Technology, Harbin 150001, China

[2]National Key Laboratory of Precision Hot Processing of Metals, Harbin Institute of Technology, Harbin 150001, China

[a]conghan@hit.edu.cn

Keywords: Bending, Ultra-Thin Component, Local Support, Wrinkling Behavior

Abstract: The wrinkling behavior of the ultra-thin components during the tube bending process was investigated by finite element analysis and confirmed by experiment and the causes of various typical defects were explored. The tube was bent through three local support methods: without support, hard support, soft support. The results indicate that severe wrinkling behavior of pipes occur during bending process. However, in the case of soft support, the wrinkling behavior of the pipe is suppressed and flattened during the subsequent bulging process.

Introduction

The metallic bent tubular parts play an extremely important role in the automotive and aerospace fields [1]. However, when forming bent pipes, wrinkles are easily formed on the inner side of the bend due to the shortening of the axis. Meanwhile, due to the axis not being in a straight line, it is prone to generate oblique wrinkles.

Conventional fabrication methods of tube bending mainly consist of roll bending [2], stretch bending, push bending [3], and rotary draw-bending [4]. The ratio of diameter to thickness (D/t) of the pipes formed by these methods are generally less than 100, and it is difficult to form pipes with relatively large ratio of diameter to thickness. M. Murata et al. prepare the bent tube with D/t of 21.1 and studied the effect of hardening exponent n on the springback, thickness strain distributions and flatness ratio [5]. Based on the knowledge of wall thickness changes and cross-sectional flattening, the bending limit of 14mm × 1.35mm TA18M titanium alloy pipes under aviation standards was determined, and bending experiments were conducted on the pipe specimens. Hong-wu Song et al. presents a new push bending method with granular media filler for thin-walled tubes with small relative bending radius [6]. The results showed that the granule media aided push bending process can be a feasible choice for thin-walled (D 70 × t 0.7, mm) tubes, which is difficult to achieve with other traditional bending processes for the tubes of small relative wall thickness (≤ 0.015). Saharnaz Montazeri et al. proposes a new compression bending method with internal pressure and prepared the tube with R/D ratio of 1 and D/t ratio of 20 [7].

In recent years, tube hydroforming has attracted widespread attention due to its weight reduction, good structural stiffness, good performance, and highly accurate dimensions, especially reducing the number of welds and subsequent processing machining passes [8]. Therefore, forming bent tubular parts through the tube hydroforming process has become a feasible method. Wang et al. investigated the effect of internal pressure on the axial stress and hoop stress [9]. The results indicate that Internal pressure decreased the axial compressive stress and wrinkling tendency was weakened. But the hoop tensile stress in the outer side of bent arc became larger with the increase of the internal pressure. Besides, the double-layered tube hydro bending method is proposed to solve the difficulties encountered in manufacturing ultrathin-walled elbow tubes by traditional

Materials Research Proceedings 44 (2024) 254-260 https://doi.org/10.21741/9781644903254-28

bending processes [10]. A bent tube with *D/t* ratio of 182 was prepared through this technology, and the wall thicknesses of the two layers decrease as the internal pressure increases.

However, wrinkle defect is one of the typical defects of the tube hydroforming process, which have been widely studied. Feng et al. studied the wrinkling behavior of overlapping tubular blanks during the hydroforming process [11]. Wrinkle defects can be effectively prevented by applying normal load, which is realized by covering a backing plate at the overlap of the outer layer. Liu et al. found that wrinkling is easy to occur in the middle of the side wall of the main tube within the thin-walled Tee-joint hydroforming process [12]. They also proposed that the critical wrinkling stress is proportional to the ratio of hoop stress to axial stress and material strength coefficients, but inversely proportional to work-hardening exponent and ratio of diameter to thickness when internal pressure is constant. T.F. Kong et al. minimize the wrinkling defects through the appropriate control of temperature distribution of the Mg alloy AZ31B tubular material in THF at evaluated temperatures [13]. The results showed that the most satisfactory component could be obtained when the average temperatures of axial-feeding and deformation regions were around 240 and 330°C, respectively.

Therefore, wrinkling behavior is the most important problem in the forming process of components, and researchers have done a lot of research to eliminate the occurrence of wrinkles. However, the wrinkling behavior of the ultra-thin components (D/t bigger than 500) during the tube bending process have not been deeply studied. In this paper, a novel method named Local Support Hydro-bending Process (LSHB) was proposed to restrain tube wrinkling during bending process. the wrinkling behavior of the LSHB was investigated by FE and confirmed by experiment and the causes of various typical defects were explored.

Materials and research methods

Material and Experiments. The GH3030 superalloy tube was used in the experiment. The tensile properties of GH3030 are shown in the Table 1. The working principle diagram of local support is shown in Fig. 1. During the bending process of pipes, the outer side of the bending is subjected to tensile stress, and the combined force points towards the inner side of the bending and acts on local supports, which gives the inner side of the bend an additional normal support force. The schematic diagram of bending process with soft support is shown in Fig. 2. It is mainly divided into four stages: placement stage, mold closing stage, liquid filling stage, and bulging stage. First, the soft support was put into the blank, and the position of the support was adjusted, and then put the blank into the mold, which is called the placement stage. When the upper mold goes down, the pipe contacts the mold and undergoes bending deformation due to the action of the mold. When the distance between the upper mold and lower mold is 0mm, the deformation completed, and it is called the mold closing stage. After the mold is closed, the support was taken out from the tube, the left and right punches were feed for sealing, and the tube was filled with liquid. This stage is the liquid filling stage. Finally, increase the pressure inside the pipe to reach p_2, so that the billet is formed by attaching to the mold, which is the bulging stage. For the tube bending process with hard support, the soft support should be replaced by hard support.

Table 1. Results of mechanical properties of the material.

Alloy	Yield strength (MPa)	Tensile strength (MPa)	Elongation (%)
GH3030	335	1035	30.1

 Materials Research Forum LLC
https://doi.org/10.21741/9781644903254-28

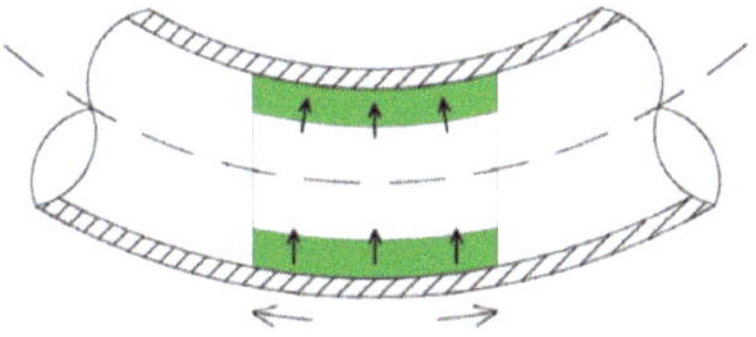

Fig. 1. Working principle diagram of local support.

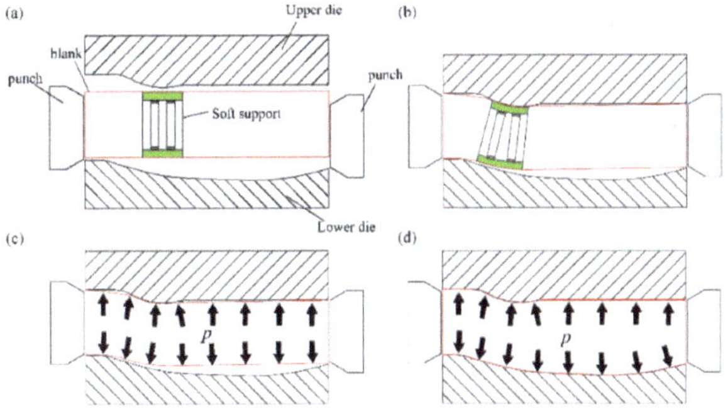

Fig. 2. The schematic diagram of bending process with soft support: (a) placement stage, (b) mold closing stage, (c) liquid filling stage, and (d) bulging stage.

Numerical modelling. In this study, Dynaform 5.9.4 was selected as the numerical simulation software to illustrate the wrinkling behavior of the material. The finite element models of three bending process were showed in Fig. 3. The dies are defined as rigid bodies, and the blank is defined as Belytschko-Tsay shell elements. The hard support is defined as 304 stainless steel, and the soft support is defined as polyurethane.

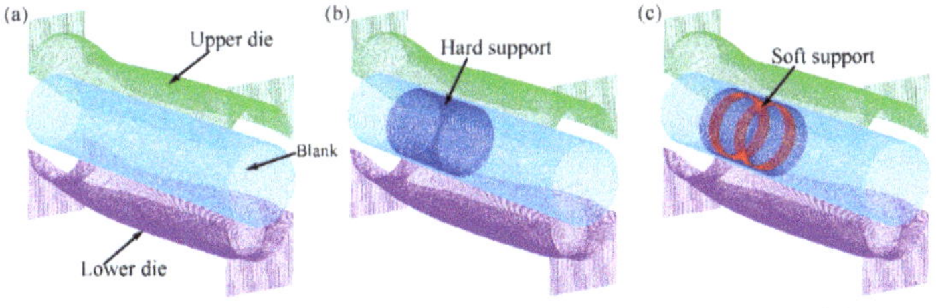

Fig. 3. Experimental setup of (a)the tube hydro-bending process, (b) Local Support Hydro-bending Process with hard support, (c) Local Support Hydro-bending Process with soft support.

Results and Discussion

Numerical simulation results. The numerical simulation results of three bending processes are analyzed, as shown in Fig. 4. It can be seen that when the tube bent without support, the wrinkles occur on the inner side and lateral side of the bending. This is mainly because the inner side is subjected to compressive stress due to the shortening of the axis, which easily generates compressive stress wrinkles. During the bending process, due to the force exerted by the upper and lower molds on the tube not being in a straight line, oblique wrinkles are formed at the lateral side of the tube. In addition, according to the cross-section at the bending of the tube, the length of cross-section at the upper half is 823.3 mm, and the length of cross-section at the lower half is

580.2 mm, which indicates that the length of the upper part is much greater than that of the lower part. Therefore, it is difficult to flatten the upper part during the bulging stage. This is another reason why the side edges of the tube wrinkle during the bending process. In order to suppress the appearance of oblique wrinkles at lateral side, local support is added to the bending process. As shown in Fig. 4, oblique wrinkles are significantly reduced due to the addition of local support. This is because due to the effect of the support, the cross-section distortion of the tube is less. The length of the upper and lower parts of the tube is similar, and there is no accumulation phenomenon in the upper part of the pipe, so it is easy to flatten during the bulging stage. However, for the hard support, the trend of transverse wrinkles on the inner side of the bend is increasing, as shown in Fig. 6(a). This is because during the bending process, the hard support cannot undergo bending deformation, resulting in stress concentration at the boundary attachment between the hard support and the tube.

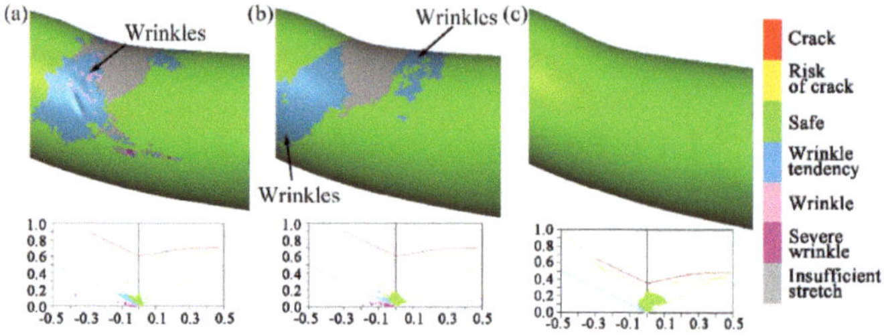

Fig. 4. Simulation results of (a)the tube hydro-bending process, (b) Local Support Hydro-bending Process with hard support, (c) Local Support Hydro-bending Process with soft support.

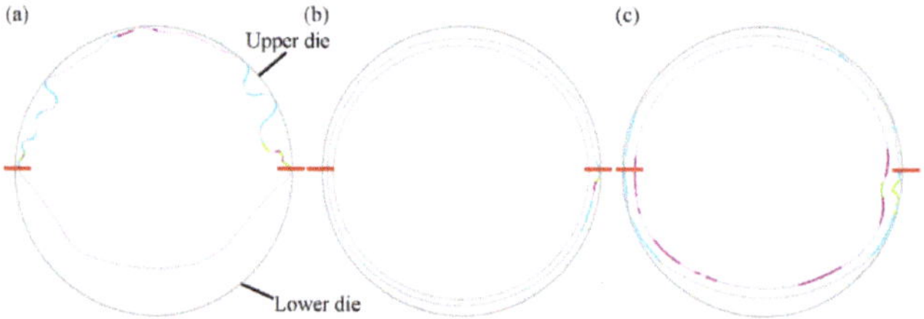

Fig. 5. The cross-section at the bending of the tube after mold close (a)the tube hydro-bending process, (b) Local Support Hydro-bending Process with hard support, (c) Local Support Hydro-bending Process with soft support.

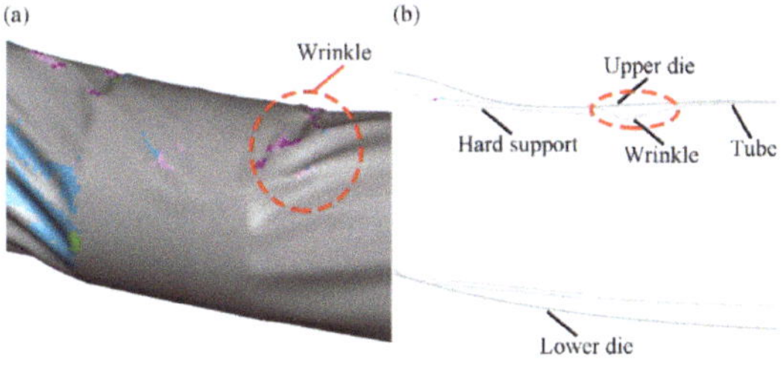

Fig. 6. (a) Simulation result of Local Support Hydro-bending Process with hard support after mold closing, (b) the axial section.

Experimental results. The experimental results are shown in Figs. 7-9. We can see that the experimental results are consistent with the simulation results. The oblique wrinkles are formed at the lateral side of the tube during the tube hydro-bending process. The transverse wrinkles generate on the inner side of the bend, as shown in figure 8. However, for the Local Support Hydro-bending Process with soft support, the workpiece with good surface quality are obtained. This indicates that soft support can effectively avoid oblique wrinkles on the side of the tube and transverse wrinkles on the inner side of the tube during the bending process.

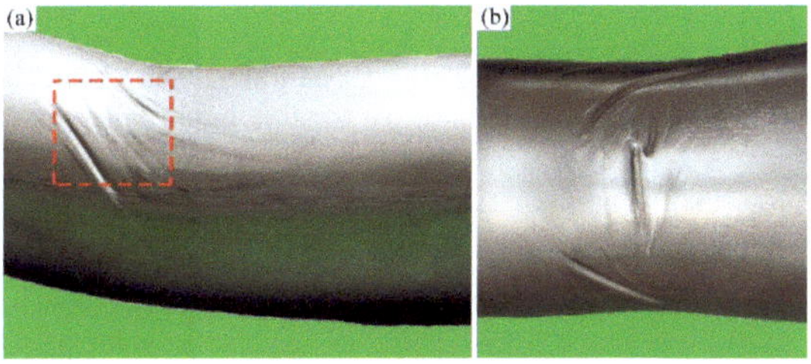

Fig. 7. Experimental piece of the tube hydro-bending process.

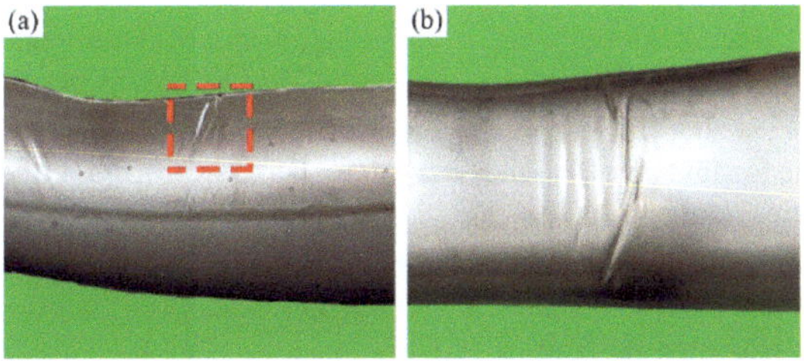

Fig. 8. Experimental piece of Local Support Hydro-bending Process with hard support.

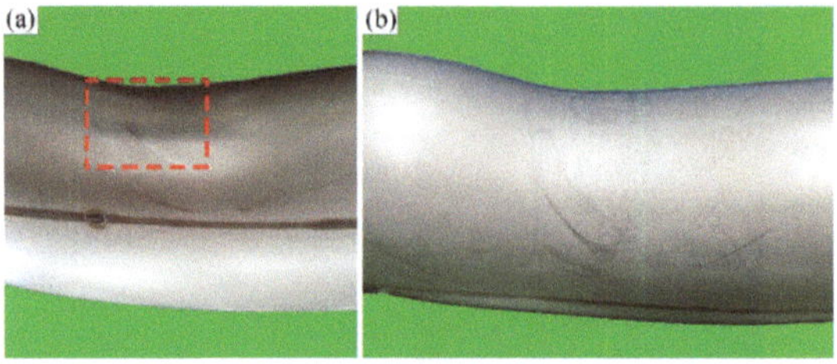

Fig. 9. Experimental piece of Local Support Hydro-bending Process with soft support.

Conclusions

This work investigated the wrinkling behavior of the ultra-thin components during the tube bending process and were confirmed by bending experiment. Besides, different support methods were added to bending process: hard support and soft support. The main conclusions could be summarized as follows:

(1) There are two typical wrinkle defects appearing on the part during the tube bending process: transverse wrinkles and oblique wrinkles. The transverse wrinkles are mainly caused by compressive stress. While, the oblique wrinkles are mainly due to shear stress.

(2) The soft support can effectively avoid oblique wrinkles on the side of the tube and transverse wrinkles on the inner side of the tube during the bending process.

Acknowledgements

This work was financially supported by the National Natural Science Foundation of China (project number: 51775136). The authors would like to take this opportunity to express their sincere appreciation.

Reference

[1] H. Yang, H. Li, Z. Zhang, M. Zhan, J. Liu, G. Li, Advances and Trends on Tube Bending Forming Technologies, Chin. J. Aeronaut. 25 (2012) 1–12. https://doi.org/10.1016/S1000-9361(11)60356-7

[2] M. Merklein, H. Hagenah, M. Cojutti, Investigations on Three-Roll Bending of Plain Tubular Components, Key Eng. Mater. 410–411 (2009) 325–334. http://dx.doi.org/10.4028/www.scientific.net/KEM.410-411.325

[3] C. Guo-qing, Y. Jian, L. Hong-xiang, F. Xue-song, T. Rui, Z. Wen-long, Push-bending Process and Forming Quality of Stainless-steel Tubes, MATEC Web Conf. 190 (2018) 04006.

[4] A. Mentella, M. Strano, R. Gemignani, A new method for feasibility study and determination of the loading curves in the rotary draw-bending process, Int. J. Mater. Form. 1 (2008) 165–168. https://doi.org/10.1007/s12289-008-0017-0

[5] M. Murata, T. Kuboki, K. Takahashi, M. Goodarzi, Y. Jin, Effect of hardening exponent on tube bending J. Mater. Process. Technol. 201 (2008) 189–192. https://doi.org/10.1016/j.jmatprotec.2007.11.286

[6] H. Song, W. Xie, S.-H. Zhang, W. Jiang, L. Lăzărescu, D. Banabic, Granular media filler assisted push bending method of thin-walled tubes with small bending radius, Int. J. Mech. Sci. 198 (2021) 106365. https://doi.org/10.1016/j.ijmecsci.2021.106365

[7] S. Montazeri, A. Gorji, M. Bakhshi, A new method for compression bending of thin-walled tubes in hydro-bending process, Int. J. Adv. Manuf. Technol. 85 (2016) 557–571. https://doi.org/10.1007/s00170-015-7910-8

[8] S. Yuan, Fundamentals and Processes of Fluid Pressure Forming Technology for Complex Thin-Walled, Components Eng. 7 (2021) 358–366. https://doi.org/10.1016/j.eng.2020.08.014

[9] X. Wang, P. Song, S. Yuan, Investigation on Corner Filling Process in Hydroforming of Thin-Walled Aluminum Alloy Tubular Part with Polygonal Sections, Mater. Trans. 53 (2012) 796–800. http://dx.doi.org/10.2320/matertrans.MF201114

[10] L. Hu, B. Teng, S. Yuan, Effect of internal pressure on hydro bending of double-layered tube, Proc. Inst. Mech. Eng. Part B J. Eng. Manuf. 226 (2012) 1717–1726. https://doi.org/10.1177/0954405412457517

[11] H. Feng, C. Han, Deformation behavior in the hydroforming of overlapping tubular blanks, Int. J. Mach. Tools Manuf. 158 (2020) 103624. http://dx.doi.org/10.1016/j.ijmachtools.2020.103624

[12] G. Liu, J. Peng, S. Yuan, B. Teng, K. Li, Analysis on critical conditions of sidewall wrinkling for hydroforming of thin-walled Tee-joint, Int. J. Mach. Tools Manuf. 97 (2015) 42–49. http://dx.doi.org/10.1016/j.ijmachtools.2015.06.004

[13] T.F. Kong, X.Z. Lu, L.C. Chan, Analysis and reduction of wrinkling defects for tube-hydroforming magnesium alloy components at elevated temperatures, Mater. Des. 173 (2019) 107761. https://doi.org/10.1016/j.matdes.2019.107761

Materials Research Proceedings 44 (2024) 261-270 https://doi.org/10.21741/9781644903254-29

Environmental impact assessment and comparative analysis of hot stamping and cold stamping processes: A cradle-to-gate lifecycle assessment study

BUCCONI Marco[1,a *], KARKASINAS Aineias[2,b], LIU Jun[1,c] and PARAMESWARAN Elangovan[1,d]

[1]Advanced Forming Research Centre AFRC, University of Strathclyde, 85 Inchinnan Dr, Inchinnan, Renfrew PA4 9LJ, United Kingdom

[2]National Manufacturing Institute Scotland (NMIS), University of Strathclyde, Netherton Sq, Paisley, Renfrew PA3 2EF, United Kingdom

[a]marco.bucconi@strath.ac.uk, [b]aineias.karkasinas@strath.ac.uk, [c]j.liu@strath.ac.uk, [d]elangovan.parameswaran@strath.ac.uk

Keywords: Lifecycle Assessment, Hot Stamping, Cold Stamping

Abstract. This manuscript presents the results of a cradle-to-gate lifecycle assessment (LCA) conducted on a component manufactured through two distinct process routes: Hot stamping of AA6082-T6 and cold stamping of AA5251-H22. The primary objective of this study is to provide a detailed understanding of the environmental impact associated with these processes and to conduct a comparative analysis of their environmental profiles. A comprehensive process map was developed for each manufacturing route, delineating all inputs and outputs at each step. Forming trials were executed during the LCA, capturing equipment energy consumption. When immediate data was unavailable from trials, the LCA model was supplemented with information from the Ecoinvent 3.6 database. The analysis demonstrates that the adoption of advanced near-net-shape manufacturing, specifically hot stamping, can significantly diminish the environmental impact compared to traditional cold stamping processes. Despite the additional energy requirements for heating in hot stamping, the overall environmental savings, supported by uncertainty analysis, are considerable. In the case of the examined demonstrator part, hot stamping showcased a noteworthy 35% reduction in CO2 equivalent emissions, equivalent to 6 kg CO2e per part. This reduction is primarily attributed to two key factors: the decreased material thickness achievable in hot stamping while preserving equivalent mechanical characteristics in the final part and the recycling of any material waste after forming. The results underscore the environmental advantages of embracing advanced manufacturing techniques, contributing valuable insights for environmentally conscious decision-making in the manufacturing industry.

Introduction

Cold stamping is a metalworking process where a sheet metal is precisely shaped into desired forms using rigid dies and mechanical force, all at ambient temperature. This process involves the use of pressure to deform the sheet metal, typically using a press. The metal is first placed into a die, and then pressure is applied through hydraulic cylinder by press to shape it into the desired component shape. Cold stamping can be used to create various products, from small parts such as fasteners and washers to larger components such as automotive panels and aerospace structural components [1]. This process offers numerous advantages over other metal shaping methods, including improved material properties, increased productivity at lower costs. However, this process may not be suitable for all materials or shapes. For instance, high strength materials may not be easily cold-formed, and highly complex shapes might require other manufacturing methods. Also, the springback of the material after forming can present a significant technical obstacle in certain cases. Furthermore, for some materials, an intermediate stress relieve heat treatment may also be required to reduce the residual stresses.

A novel hot stamping process, also known as Hot Form Quenching (HFQ) has been reported to manufacture high-strength aluminium alloys into complex and lightweight components [2]. It involves several key steps, including solution heat treatment, forming, in-die quenching and artificial ageing. The high temperature makes the sheet material more ductile and easier to deform, resulting in products with more precise and consistent dimensions. HFQ can create intricate part geometries that are difficult to achieve through traditional cold-forming processes. It is also beneficial in reducing part counts and assembly requirements by combining multiple components into one part to save significant cost and increase sustainability [3].

A Life Cycle Assessment (LCA) is a systematic and comprehensive methodology used to evaluate the environmental impact of a product, process or service throughout its entire life cycle. This paper aims to present a LCA on the environmental impacts of cold and hot stamping processes. For this analysis, an identical demonstrator part will be manufactured with both processes. The LCA of this study follow the ISO 14040 [4] and 14044 [5] standards, and it was carried out in the four following stages:

1. Goal and scope definition, where the objective, systems boundaries, functional unit, and assumptions are defined.
2. Life cycle inventory, which includes the identification of all the inputs and outputs associated with all process steps.
3. Life cycle impact assessment, where the inputs and outputs of the previous phase are used to evaluate the potential environmental impacts of the manufacturing steps.
4. Interpretation, in which the results of the LCA are discussed and compared.

Material and Methods

The study is designed as a ‘cradle-to-gate’ LCA; in other words, it includes the extraction and production of raw materials, converting processes, transports to the manufacturing site and the manufacturing processes. A schematic representation of this can be observed in Fig. 1.

In general, the study includes the following steps:

1. Production, transportation, and conversion of the raw materials
2. Transportation of the raw materials to the manufacturing site
3. Manufacturing of the demonstrator part using the selected process

Additionally, the study does not include the following steps:

1. Production and disposal of the infrastructure (machines, transport media, roads, etc.) and their maintenance as no significant impact is expected
2. Environmental effects from accidents
3. Transport of the demonstrator part from the manufacturing site to the customer
4. Use phase of the demonstrator part at the customer site
5. External manufacturing processes such as painting
6. Final disposal and recycling of the primary materials used and the demonstrator part

Below, Fig. 2 and Fig. 3 illustrate the simplified flowcharts and system boundaries for the cold-forming and hot-forming processes, respectively.

The function examined in this study is the manufacturing of a demonstrator part. Thus, the functional unit has been selected to be one demonstrator part. The reference flow is also one demonstrator part and has been used in the allocation of the tooling wear and production, the caption of the energy while the equipment is in standby and in some other auxiliary data.

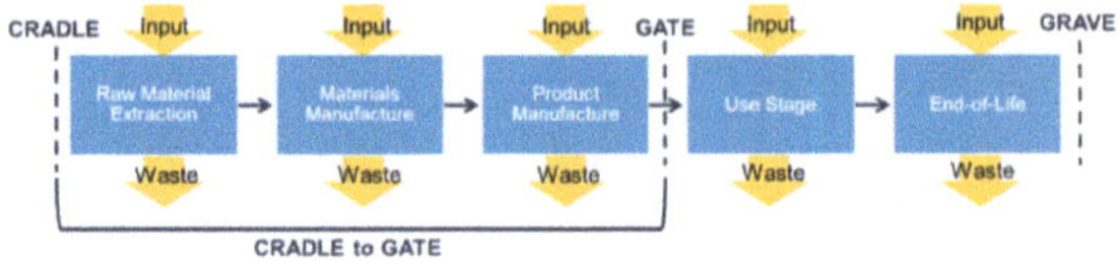

Fig. 1. LCA Stages.

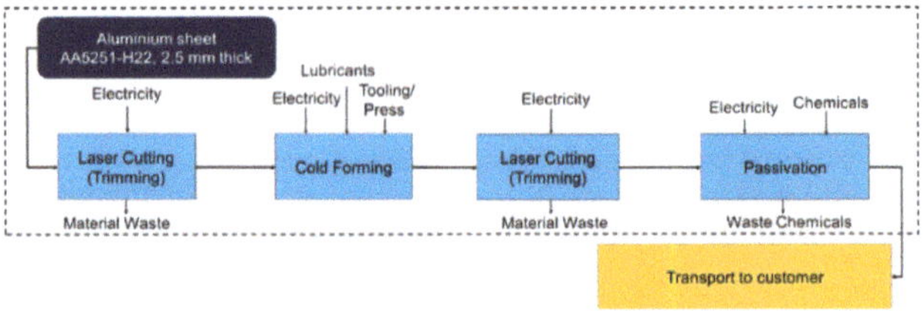

Fig. 2. Cold Stamping Process Map and System Boundaries.

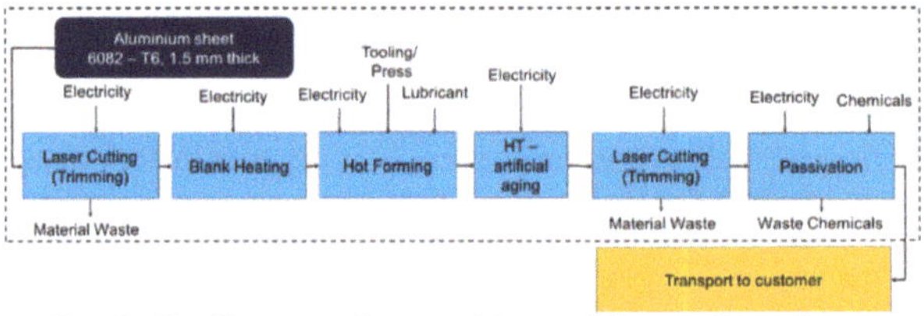

Fig. 3. Hot Stamping Process Map and System Boundaries.

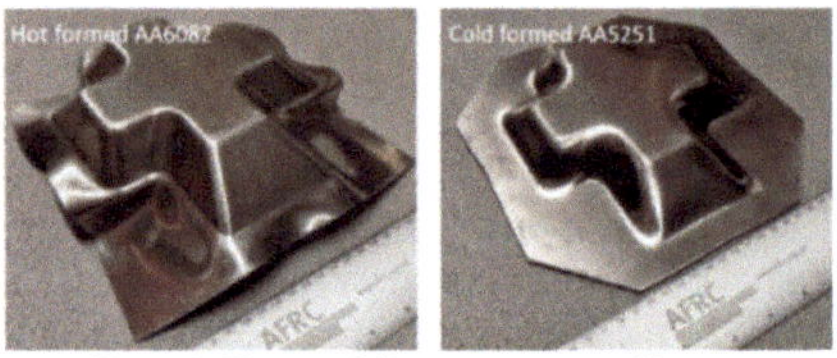

Fig. 4. Formed Demonstrator Parts.

To maintain the study within a feasible scope, it's necessary to limit the detail in system modelling using cut-off criteria. ISO standard [ISO 14044] states that cut-off criteria should consider mass, energy, or environmental significance. Prechains with an input material share less than 1% of the total mass input of a considered process are excluded from the study for mass-related cut-offs. However, the total cut-off cannot exceed 5% of input materials based on the functional unit. All energy inputs are considered, except for energy-related to material inputs from prechains that are cut off due to the mass-related rule.

In the next sections, more details are given on the assumptions and data used for modelling the environmental impact of the two manufacturing processes.

Data gathering and data quality

The data in this study have been meticulously chosen based on key criteria for collection. Specifically, data reliability is confirmed through validated sources and non-validated sources within selected databases. These data are deemed representative of all pertinent market sites over a sufficient timeframe to mitigate typical fluctuations, ensuring high completeness. Temporal correlation is robust, as primary data were acquired in the study, while secondary data originate from databases relevant to the study period. Geographically, the data exhibit a strong correlation, with industry experts consulted to guide regional data selection. Furthermore, primary data collected pertain closely to the study's objectives, encompassing equipment, technologies, and materials. Similarly, secondary data selection adheres to this principle. Overall, the data are of high quality. In Table 1 and Table 2, the primary and secondary data are illustrated.

Table 1. Cold-forming data.

	Resources	**Input**	**Output**
Freight, lorry, for raw materials in China	Other	secondary data	
Freight, sea, container ship, China to UK	Other	secondary data	
Initial Laser cutter	Resources- Material	primary data	primary data
	Resources- Other	secondary data	
	Waste		primary data
500T Press	Resources- Electricity	primary data	
	Resources- Other	secondary data	
	Resources- Other	secondary data	
	Waste		secondary data
Laser Cutting	Resources- Other	secondary data	
	Waste		primary data
Anodising, aluminium sheet	Resources- Other	secondary data	

Table 2. Hot-forming data.

	Resources	**Input**	**Output**
Freight, lorry, for raw materials in China	Other	secondary data	
Freight, sea, container ship, China to UK	Other	secondary data	
Initial Laser cutter	Resources- Material	primary data	primary data
	Resources- Other	secondary data	
	Waste		primary data
Furnace (blank preheating)	Resources- Material		
	Resources- Electricity	primary data	
500T Press	Resources- Electricity	primary data	
	Resources- Other	secondary data	
	Resources- Other	secondary data	
	Waste		secondary data
Furnace (Artificial ageing)	Resources- Electricity	primary data	
Laser Cutting	Resources- Other	secondary data	
	Waste		primary data
Anodising, aluminium sheet	Resources- Other	secondary data	

Starting Material. The demonstrator parts are fabricated using aluminium AA5251-H22 for cold stamping and AA6082-T6 for hot stamping. Due to the distinct mechanical properties of these alloys, with AA5251-H22 having a minimum tensile strength of 190 MPa and AA6082-T6 having a minimum of 310 MPa [6], a fair comparison between the two parts necessitates adjustments. It was assumed that for both parts to possess equivalent structural integrity, the AA5251-H22 part would need to be approximately 1.6 times thicker than the AA6082-T6 part. As such, for this study, the part thicknesses were standardised at 2.5 mm for AA5251-H22 and 1.5 mm for AA6082-T6. Also, based on data from the industry [6], the composition of the Alloys has been modelled using Global values, as seen in Table 4. The composition of the alloys is crucial for capturing the environmental impacts related to the extraction of the different raw materials, the refinement processes for the production of the elements, and any average travel in between stages.

Laser Trimming – electricity and waste material. To make the demonstrator part, two trimming operations are needed: one for cutting the blank to the required size and another one for trimming the formed component to the required shape.

Laser cutting is a widely used method across industries for precise material trimming. Utilising a high-powered laser beam, this technique offers accuracy in cutting various materials. Its precision enables the creation of intricate designs effortlessly, making it suitable for diverse applications. Laser cutting boasts advantages such as minimal waste, high accuracy, and reduced risk of material damage, making it a favoured choice for cutting materials. It is important to note that, the laser cutting eliminate the need of two additional trimming tools which makes the process cheaper, efficient, and sustainable.

For the manufacture of the demonstrator parts, a 4kW laser and a cycle time of 10 seconds were assumed. The literature [7] suggests that, on average, 35% of the sheet material is wasted during the initial and final laser cutting. In this work, it was assumed that 10% of the starting sheet thickness is trimmed off to make the starting blank, and 25% is cut after formation. The waste material, i.e. the off-cuts, is assumed to return to the production loop and be recycled. Table 3 summarises the weights of the starting material, blank, part, and off-cuts that can be recycled.

Forming – Electricity. To determine the quantity of electricity used by the two forming processes, manufacturing trials were run using a 500T hydraulic press. The electricity consumed from the equipment was collected for each process on four parts formed.

For the hot stamped parts, the press consumed, on average, 0.082kWh per part, including the forming operation and standby between parts.

For cold stamping this value increased to 0.085kWh per part.

Forming – Lubricant. The lubricant used in the manufacturing trials was the graphite lubricant Durcol W1010 – 01. The project team estimate an initial lubricant volume of 23.2 g/m^2 for best lubrication during the forming trials [8]. As a result, the weight of lubricant used per part has been calculated to be 2.5 g.

Forming – Tooling. For both the forming processes, the same forming tool was utilised. One of the model's assumptions for this LCA is that the tooling is considered consumable. This study was not aimed at conducting a detailed LCA on the tool but to capture the additional environmental impact from the normal wear of the tool used.

The tool used in the analysis is made of H13 steel, and its composition [9] is captured in Table 4. Vanadium (V) could not be modelled since data were not available and lithium was used as a replacement.

Table 3. Weights of input-output material.

	Hot stamping of AA6082-T6 1.5mm	Cold stamping of AA5754-H22 2.5 mm
Weight of initial sheet	400 g	670 g
Weight of blank	360 g	603 g
Weight of off-cuts after first trim	40 g	67 g
Weight of part	260 g	435 g
Weight of off-cuts after last trim	100 g	168 g
Total weight of off-cuts	**140 g**	**235 g**

Table 4. Aluminium alloy AA6082, AA5251 and H13 tool steel composition.

Aluminium alloys [6]			**Tool Steel**	
Element	Average in **Alloy 5251** [%]	Average in **Alloy 6082** [%]	Element	Average in **H13** [%] [9]
Silicon	0.20	1.00	Carbon	0.39
Magnesium-alloy	2.05	0.90	Chromium	5.13
Manganese	0.30	0.70	Molybdenum	1.43
Iron pellet	0.25	0.25	Vanadium	1.00
Chromium	0.075	0.125	Silicon	1.00
Zinc	0.075	0.10	Manganese	0.35
Titanium	0.075	0.075	Phosphorus	0.03
Copper	0.075	0.05	Sulphur	0.03
Aluminium	Balance	Balance	Steel	Balance

The exact manufacturing processes used for production of the tooling are not captured; instead, average global data for metal working have been selected from the database Ecoinvent 3.6. It was assumed that the tool could produce 150,000 parts and 100,000 parts in hot and cold stamping processes, respectively; the environmental impacts for production, wear, and waste management of the tool have been calculated and allocated per part.

Blank Heating – Electricity. The hot stamping method involves controlled heating of the blank before shaping. This heating occurs in a furnace, ensuring precise temperature and time control for uniformity. Energy usage in this step was evaluated through manufacturing trials. Parts were preheated to 535°C for 8 minutes, with energy consumption measured via an electricity meter connected to the furnace. Three key stages were identified during the preheating of the demonstrator part:

1. Direct energy. This is the energy used to heat blanks during an 8-minute heating phase. The study involved measuring energy usage for four parts and averaging the values.
2. Set up energy for furnace heat up. This is the energy required by the furnace to reach the necessary process temperature from room temperature, and it has to be split among all the parts that will be processed in the same batch before the furnace is cooled down again. For this study, it was assumed that a factory works on two 8-hour shifts a day Mon-Fri; a part is formed every 9 minutes; and the furnace is shut down after Friday's last shift and heated back up on Monday. Five hundred thirty parts will be heated up each week. In the trials, the energy consumption during the four hours required to reach a steady process temperature in the furnace was measured.
3. Stand by energy. This corresponds to the energy used by the furnace to maintain temperature during non-heating periods. The energy used during the 32 weekly standby

hours has to be divided by the 530 parts that will be preheated each week. Energy consumption was measured during one hour of steady-state furnace operation.

A model was built upon the assumptions above for each stage. In addition to these assumptions, primary data were gathered during manufacturing trials of the demonstration part. Table 5 presents these data, including energy consumption at key stages and total energy consumption per part.

Artificial Ageing – Electricity. In the hot stamping process, a post-forming heat treatment known as artificial ageing is employed for heat-treatable aluminium alloy components like AA6082. To quantify energy consumption in this stage, manufacturing experiments were conducted, with energy measured using an electricity meter connected to the furnace. Parts were subjected to heat treatment at 180°C for 3 hours, followed by air cooling. Three key stages in the furnace operation were identified during the heat treatment process:

1. Direct Energy: This is the energy utilised to heat parts during the 3-hour heat treatment. Given the production capacity of 530 parts per week and factory operation on two 8-hour shifts, Monday to Friday, each batch averages 21.2 parts. Therefore, energy allocation per part is calculated by dividing the total energy used for each heat treatment cycle by 21.2.
2. Set up energy for furnace heat up: This represents the energy needed by the furnace to reach process temperature from room temperature. This energy is distributed among all parts processed in the same batch before furnace cooldown, considering the production capacity of 530 parts.
3. Stand by energy. This corresponds to the energy used by the furnace to maintain temperature during non-heat treatment periods. The weekly standby energy, used in the 32 weekly standby hours, has to be divided by the 530 parts processed each week. Energy consumption was measured for one hour with the furnace in a steady state.

A dedicated model has been created to allocate emissions from each stage to the final products, utilising the described assumptions. In addition to these assumptions, primary data was gathered during the manufacturing trials of the demonstration part. Table 5 presents this data, including energy consumption for each key stage and the overall energy consumption for the part.

Anodising. This is a chemical process commonly used to protect aluminium surfaces from corrosion. It involves submerging the metal in an acid electrolyte bath and passing an electrical current through it. The process forms an aluminium oxide layer on the surface, known as the anodic layer, which provides protection. The quality of this layer depends on factors such as bath temperature, solution concentration, and current density. For both the hot- and cold-formed part, a current density of 8–12 mA/cm^2 is assumed, resulting in an energy consumption of 0.35–0.40 KWh/m^2 [10].

Table 5. Energy consumption blank preheating and part heat treatment.

Step	Phase	Average Energy measured (kWh)	Batch size	Energy used per part (kWh/part)
Blank Preheating	Direct energy	1	1	1
	Set up energy	42	530	0.08
	Stand by energy	160	530	0.3
	Total			**1.38**
Part Heat Treatment	Direct energy	7	21.2	0.33
	Set up energy	6	530	0.01
	Stand by energy	56.3	530	0.12
	Total			**0.46**

Materials Research Proceedings 44 (2024) 261-270 https://doi.org/10.21741/9781644903254-29

Results and Discussion

In this section the results for hot stamping and cold stamping are discussed.

Hot stamping. In Fig. 5, the network of hot stamping of AA6082-T6 for the demonstrator part is depicted, highlighting several key points. The main contributor to the environmental impact is identified as aluminium alloy AA6082, with each part having a total carbon footprint 11 kg CO2 equivalent. Laser-cutting processes are highlighted for their positive environmental impact despite generating waste, as they benefit from recycling and have relatively low energy consumption. The diagram selectively illustrates sources contributing more than 1.52% to the total impact, focusing on the most significant contributors to the manufacturing process's environmental footprint.

Cold stamping. The network diagram for cold stamping (Fig. 6) of the demonstrator part highlights key aspects of its environmental impact. Aluminium alloy AA5251 is identified as the primary contributor, resulting in a 17 kg CO2 equivalent carbon footprint per part. Similarly to hot stamping, laser cutting steps also contribute positively to the environment, thanks to recycling and low energy consumption. Notably, AA5251 stands out as the largest factor in the process's environmental impact, emphasizing its significant role in the manufacturing process's overall environmental footprint.

Comparison of hot and cold stamping processes. The following sub-sections will include a comparison between the two processes under the assumption that both will produce a part with the same (or very similar) properties.

Impact Assessment. The characterisation, as illustrated in Fig. 7, allows the reader to quickly identify how each process contributes to the total impact of each of the environmental categories. It becomes evident that hot stamping has a lower environmental impact in almost all environmental categories than cold stamping. The only exception is the ionising radiation category, where there is a small difference due to the extra carbon-14 emitted in the air due to the difference in composition of AA6082 and AA5251.

Uncertainty Analysis. In Figure 8, the uncertainty analysis is illustrated for this work. The results are based on 200 runs, with 58.7% of secondary data providing values that contain uncertainty-related data. For this data, lognormal distribution was used in 58.6% and undefined one for 41.3%. The primary data are considered of high quality with very limited uncertainty. Based on these assumptions, with a confidence interval of 95%, A- a part manufactured through hot forming, is always better environmentally than B- a part manufactured through cold forming.

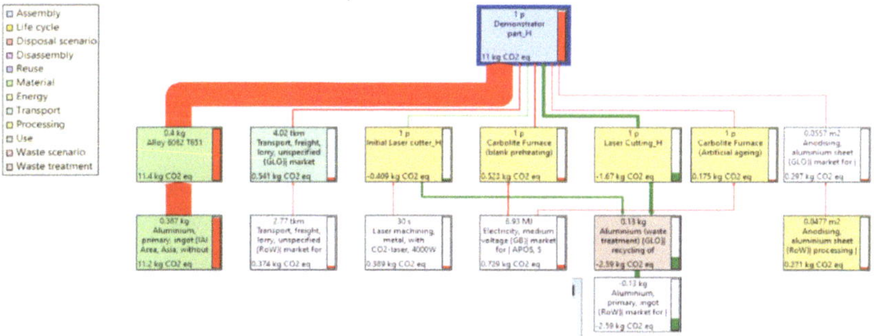

Fig. 5. NETWORK diagram for hot stamping of AA6082-T6.

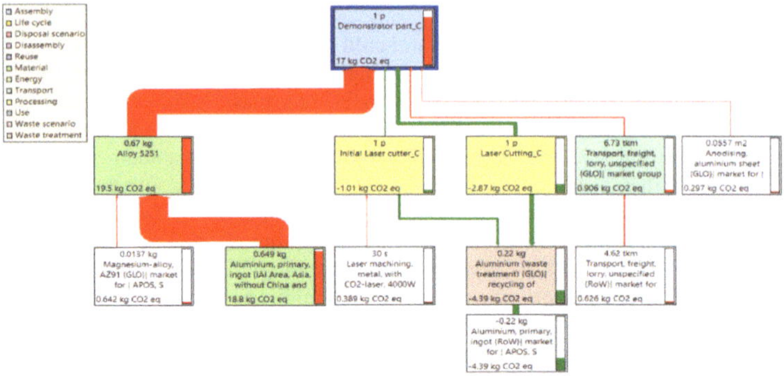

Fig. 6. NETWORK diagram for cold stamping of AA5251-H22.

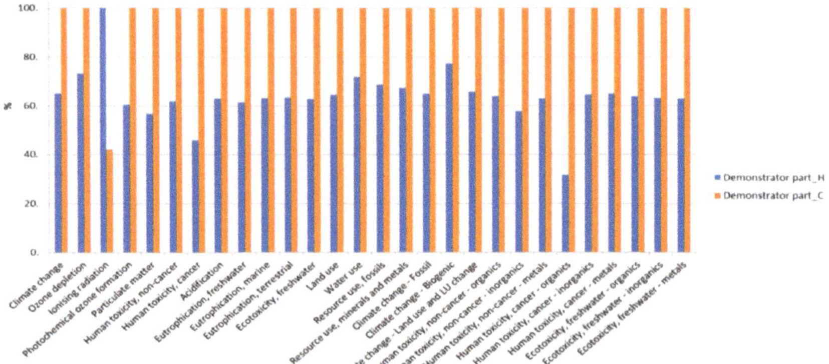

Fig. 7. Impact Assessment Characterisation Hot Stamping Vs Cold Stamping.

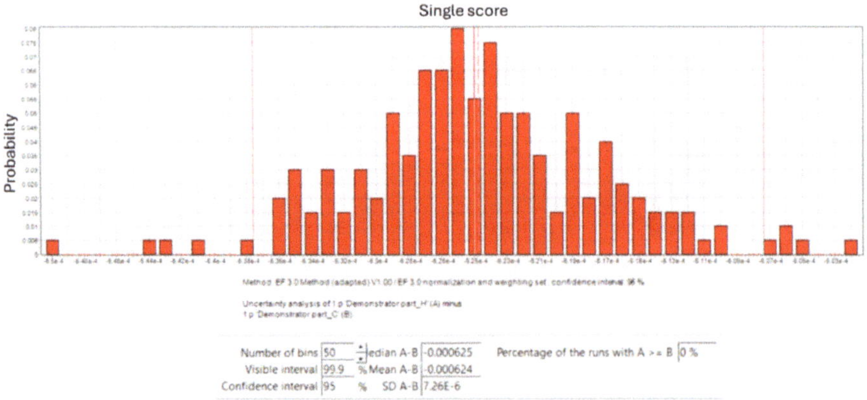

Fig. 8. Uncertainty Analysis.

Materials Research Proceedings 44 (2024) 261-270
https://doi.org/10.21741/9781644903254-29

Conclusions

The conclusions drawn from this 'cradle to gate' LCA study, designed to assess the environmental impacts of hot and cold stamping processes, have successfully met the goals. The study has generated updated insights into both forming processes' environmental strengths and weaknesses. Detailed process mapping and the development of initial LCA modelling has provided a clearer understanding of the environmental inputs and outputs associated with each process.

Moreover, the research has contributed significantly to the existing body of knowledge on the environmental performance of hot and cold forming processes. By conducting an in-depth LCA for each process and comparing the results, the study has highlighted the importance of specific data collection for accurately quantifying their environmental impacts.

Based on the 'cradle to gate' LCA, the total emissions for manufacturing the demonstrator part amount to 11 kg CO2 eq. for hot stamping, and 17 kg CO2 eq. for cold forming. It becomes evident that the difference between hot and cold stamping when a similar demonstration part is manufactured is equal to 6 kg CO2 eq. Hot stamping saves up to 35.3% for each manufactured part, compared with cold stamping. The observed difference is primarily attributed to material savings in the hot-formed parts. Specifically, hot forming requires 400g of AA6082-T6 per part, while cold forming necessitates 670g of AA5251-H22 for the same structure and strength, leading to a reduction of 40.3% in material usage for hot forming. The overall environmental savings are noteworthy despite the additional energy demand from furnaces in the hot stamping process. An uncertainty analysis was conducted to validate these results. The findings suggest that hot stamping is environmentally preferable, primarily due to material efficiency and, to a lesser extent, energy use, with the potential impact varying by the energy sources of the manufacturing country. This study underlines the importance of considering both material and energy efficiencies in reducing manufacturing processes' environmental footprint. The potential impact of the energy grid in the manufacturing country on this equilibrium is acknowledged.

Acknowledgement

Financial assistance from the High Value Manufacturing Catapult, project NMIS-CATX-06291, is acknowledged.

References

[1] T. Altan, A.E. Tekkaya, Sheet Metal Forming Process and applications, ASM International, 2012.

[2] J. Liu, L. Wang, J. Lin, Hot Stamping of Complex-Shaped High-Strength Aluminium Components, in Encyclopedia of Aluminum and Its Alloys, CRC Press, 2018, pp. 1244-1262.

[3] A. Wang, J. Liu, H. Gao, L. Wang, M. Masen, Hot stamping of AA6082 tailor welded blanks: Experiments and knowledge based cloud - finite element (KBC-FE) simulation, J. Mater. Process. Tech. 250 (2017) 228-238. http://dx.doi.org/10.1016/j.jmatprotec.2017.07.025

[4] Environmental management — Life cycle assessment — Principles and framework (ISO Standard No. 14040:2006), International Organization for Standardization, 2006.

[5] Environmental management — Life cycle assessment — Requirements and guidelines (ISO Standard No. 14044:2006), International Organization for Standardization, 2006.

[6] https://www.rightonblackburns.co.uk/products/metals/aluminium-alloys/commercial-aluminium-alloys.

[7] D. Cooper, K.E. Rossie, T.G. Gutowski, The energy requirements and environmental impacts of sheet metal forming: An analysis of five forming processes, J. Mater. Process. Tech. 244 (2017)116-135. http://dx.doi.org/10.1016/j.jmatprotec.2017.01.010

Segmented blank holder concept to influence the forming zone during incremental bending

PANICK Daniel[1,a], FROHN-SÖRENSEN Peter[1,b *] and ENGEL Bernd[1,c]

[1]University of Siegen, Chair of Forming Technology, Breite Straße 11, 57076 Siegen, Germany

[a]daniel.panick@uni-siegen.de, [b]peter.frohn@uni-siegen.de, [c]bernd.engel@uni-siegen.de

Keywords: Incremental Forming, Flexibility, Forming Zone, Adjustable Tools, Tool Segmentation, Profile Bending, Finite Element Method

Abstract. Decreasing batch sizes and a simultaneous increase in the number of variants are developments in production technology that can no longer be managed economically using conventional mould technology with rigid, product-specific functional surfaces for semi-finished products. Individual or even in-line reconfigurations of the tool's functional surfaces are being investigated to meet production requirements. Moreover, the goal is to control these surfaces during the process to compensate fluctuations, e.g. in semi-finished products. Incremental swivel bending (ISB) is a process for flexible bending forming in which the bending moment is transferred to the semi-finished product by friction by means of the clamping surfaces of the blank holder. In addition to the geometric and mechanical properties of the semi-finished product, the contact normal force and friction are responsible for the size and shape of the forming zone. In this paper, established analytics are utilized to present a tool concept that enables the clamping force distribution to be specifically adjusted by means of adjustable segmented hold-down surfaces with a vertical degree of freedom and, as a result, the size and shape of the forming zone to be influenced. The potential for extending the limits of forming technology is then discussed regarding the numerically determined local strain distribution through the dynamic loading and unloading of the clamping surface segments during incremental forming with the overall goal to achieve an in-process controllable material flow during sheet metal forming.

Introduction

Incremental forming processes offer a high degree of flexibility regarding the component geometry to be achieved, as the sequence of increments can be individually selected. At the same time, these processes fulfil current requirements, such as increasing variant diversity with simultaneously decreasing batch sizes, as one or more additional degrees of freedom allow components to be produced independently of the tool geometry [1-3].

Compared to continuous forming processes, incremental forming also has the advantage that higher degrees of deformation can be achieved due to the highly localised strain distribution [4-6]. Due to the freedom to sequence the forming increments, incremental processes offer an additional degree of flexibility compared to continuous processes. However, this requires a deep understanding of the forming zone and strategies to control the overlapping deformation zones [7].

For individualised and variant-intensive production, kinematic processes such as incremental swivel bending (ISB) have become established, which, for example, enables targeted adjustment of the strain curves by shifting the bending axis [8].

The sequential process repeats a sequence of production steps to line up individual bend increments to form a bend. Alternating bending directions or positioning the profile in opposite directions enables alternating bending and offers the potential to integrate geometric functions to produce side members for cars, for example (see Fig. 1.).

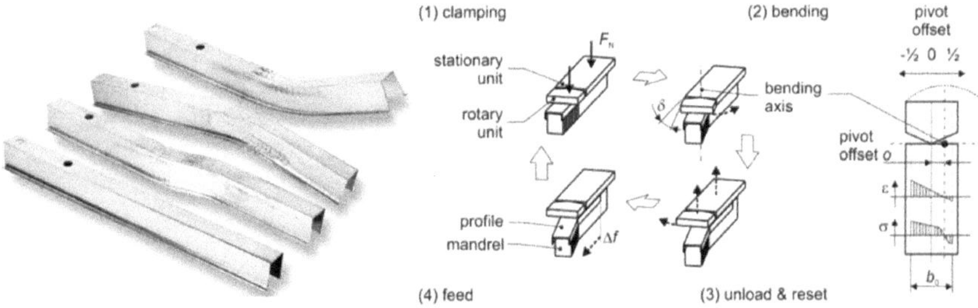

Fig. 1. left: hat profiles formed using the ISB process for the automotive industry (side members), right: process sequence of incremental swivel bending.

In the ISB process, the bending moment is transmitted by the clamping surfaces, which apply frictional tensile or compressive forces in the longitudinal direction of the semi-finished product. In addition to the geometric and mechanical properties of the semi-finished product, the contact normal force and the state of friction - which occurs between the clamping surface and the semi-finished product - are responsible for the size and shape of the forming zone [9].

The forming zone (see red shaded area Fig. 2a) develops after the yield point of the material is exceeded within the bending gap of the clamping tools rotating against each other and below the clamped contact surfaces. In the tensile stress range of the bending stress distribution, the material is plastically thinned within the forming zone angle γ due to the transverse strain (see Fig. 2b). The contact area A_c between the blank holder and the semi-finished product is reduced so that the contact normal stress increases at a constant press closing force and the frictional shear stress increases in direct proportion [10].

The prerequisite for a stable bending process is ensured by maintaining static friction, as slipping of the material under the clamping surfaces leads to an uncontrolled process sequence and - in the case of profile bending - to severe component twisting. Consequently, the extraction of the material must be ensured by maintaining the static friction in the form of a minimum requirement for the clamping force. In the further process sequence, the applied bending moment is only transferred to the remaining contact surface of the tool and solidifying materials also require a sharply increasing forming force. As a result, the minimum force requirement is expected to increase during the bending process [11].

There is therefore a requirement to design the functional surface for transmitting the clamping force and ensuring frictional locking in such a way that contact between the clamping surface and the semi-finished product is maintained and a contact pressure distribution that can be adjusted during the process can be realised.

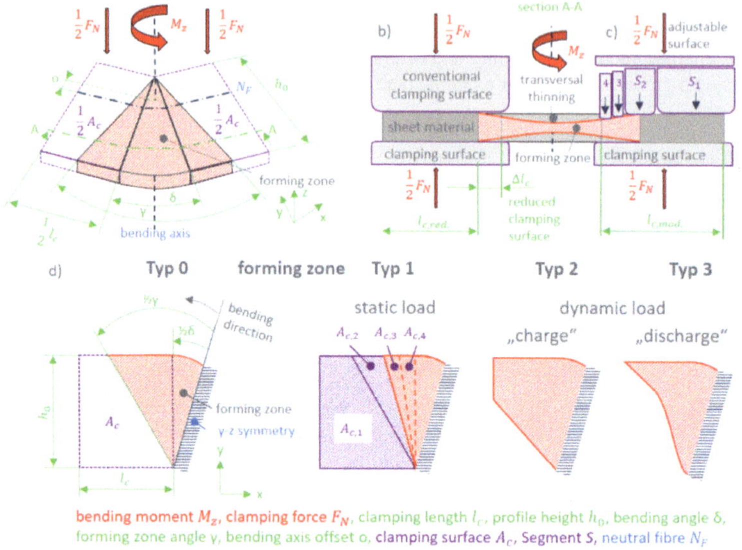

Fig. 2. a) Bending in the plane and formation of the forming zone, b) Sectional view with transverse thinning and reduced clamping surface, c) Adjustable surface follows the thinning, d) Predicted characteristics of different forming zones due to static or dynamic load.

Theoretical Background and Calculations

The ISB process has been demonstrated to work substantially under the principle of frictional engagement. In detail, the bending moment to form a single increment is transmitted by means of a firm fixation of the part in the heavily loaded clamping tools. Traction is a necessary premise for this forming principle since material must be prevented from slipping through the clamping tools to achieve a determined forming process. The central analytical description of the process relates the evolving forming zone to the applied process parameters as well as the mechanical material properties and geometrical relations. By the description of the forming zone, the corresponding strain gradient and the achieved curvature are determined for the bending operation to layout the process for a given application in production technology.

According to FROHN-SÖRENSEN [11], the forming zone angle γ is given by the relation

$$\gamma_{1/2} = -\frac{q_1}{2} \pm \sqrt{\left(\frac{q_1}{2}\right)^2 - q_2} \tag{1}$$

with the substitutions

$$q_1 = -\left(\frac{2}{3}\frac{E \cdot \delta}{R_e} + 2\delta + \pi\frac{l_{c,0}}{h_0} + \frac{\mu_0 \cdot F_N \cdot E \cdot \pi}{\sqrt{3} \cdot 2R_e \cdot m \cdot {h_0}^2}\right) \tag{2}$$

and

$$q_2 = \delta\left(\left(\frac{2}{3}\frac{E}{R_e} + 1\right) \cdot \left(\delta + \frac{\pi \cdot l_{c,0}}{h_0}\right) + \frac{\mu_0 \cdot F_N \cdot E \cdot \pi}{\sqrt{3} \cdot 2R_e \cdot m \cdot {h_0}^2}\right) \quad (3)$$

where E is YOUNGS's Modulus of Elasticity and R_e is the yield strength of the material. With respect to the process, F_N is the clamping force applied on both clamping tools and μ_0 is the static friction coefficient where tool surfaces and material are in contact. $l_{c,0}$ denotes the clamping length and h_0 the profile width. For the analytical process model, a bilinear hardening curve according to GERLACH [12] is applied for mathematical reasons where m is the inclination of the linear interpolation of hardening between yield strength and ultimate tensile strength R_m. Besides the linear expression in the elastic region according to HOOKE's law [13], the bilinear hardening curve is described in the plastic deformation region by

$$k_f = m \cdot \varepsilon + R_e \cdot \left(1 - \frac{m}{E}\right) \quad (4)$$

where m is expressed according to

$$m = \frac{R_m(1+A_G)-R_e}{A_G-\varepsilon_{el}}. \quad (5)$$

When comparing Eq. 1 with practical bending experiments and simulations, it has been observed that the forming zone angle strongly increases over the first few degrees of bending angle δ, which is imposed by the clamping tools. Given by the geometrical relation of the clamping length $l_{c,0}$ and the profile width h_0, γ converges until an ultimate forming zone angle γ_{ult} and would stay on this value with any further increase of bending angle. γ_{ult} is calculated according to

$$\gamma_{ult} = 2 * tan^{-1}\left(\frac{l_c}{h_0}\right). \quad (6)$$

Once the forming zone is determined, longitudinal strain ε_x in the centre of the forming zone is expressed by

$$\varepsilon_x = \frac{\Delta l}{l_0} = \frac{\delta}{\gamma-\delta}. \quad (7)$$

With the determination of strain, the unloaded bending angle δ_u is calculated from the bending angle given by the process δ minus the springback angle δ_{SB} which results from the elastic release of residual stresses. According to WITTEK *et al.* [14], the unloaded incremental bending angle δ_u is assumed as

$$\delta_u = \delta - \delta_{SB} \approx \delta\left(1 - 3 \cdot \frac{R_e}{E} \cdot \frac{1}{2\varepsilon_x}\right). \quad (8)$$

For incremental bending, multiple of these aforementioned forming zones γ are superposed. A number of n bending increments of angle δ are combined with incremental feed spacings with the distance Δf in an alternating process sequence. With the unloaded bending angles being calculated according to Eq. 8, a representative bending radius R can be assumed from the discontinuous manufacturing method ISB. By the relation of the curvature κ being the inverse expression of the radius, the arc manufactured by incremental bending processes is obtained by the following relation according to [15]

$$\kappa_{NF} = \frac{4}{\Delta f \cdot \left(\frac{1}{\sin\frac{\delta_u}{2}} + \frac{1}{\tan\frac{\delta_u}{2}}\right)}. \quad (9)$$

Materials Research Proceedings 44 (2024) 271-281 https://doi.org/10.21741/9781644903254-30

For the calculation of curvature at the strain-neutral fiber κ_{NF}, it is important mentioning that the unloaded bending angle must be justified by strain hardening in Eq. 8 since multiple forming zones overlap in the incremental sequence.

The number of superposed forming zones is denoted by the integer *SP*. With the successively increasing integer variable *i* for the incremental process sequence, strain obtained from incremental bending $\varphi_{x,s}$ is expressed by a sum formulation according to

$$\varphi_{x,s} = \sum_{i=1}^{SP} \ln\left(\frac{\delta \cdot \left(1 - 3 \cdot \frac{k_{f,i-1}(\varphi_{x,i-1})}{E} \cdot \frac{1}{2 \cdot \varepsilon_x}\right)}{\gamma - \delta \cdot \left(1 - 3 \cdot \frac{k_{f,i-1}(\varphi_{x,i-1})}{E} \cdot \frac{1}{2 \cdot \varepsilon_x}\right)} + 1\right) \tag{10}$$

where hardening $k_{f,i-1}$ is a function of strain $\varphi_{x,i-1}$. With the equations above, the forming potential of the herein presented novel tooling concept is predicted.

Tool Concept Development

A tool concept was developed with the overarching aim of influencing the forming zone in terms of its geometry and gradient. This considers the existing analysis to the effect that the maximum forming zone angle γ_ult converges asymptotically towards a limit value of 90° depending on the bending angle δ and assuming square hold-down surfaces. To ensure that the strain distribution is as localised as possible and thus higher degrees of deformation can be achieved [4,5], the approach of reducing the forming zone angle was pursued. The tool concept is based on the investigations from [16,17] and divides the blank holder surface in the upper part of the tool. This creates individual surface segments that allow the clamping force distribution and consequently the generated contact normal stress - which acts on the sheet metal strip - to be varied.

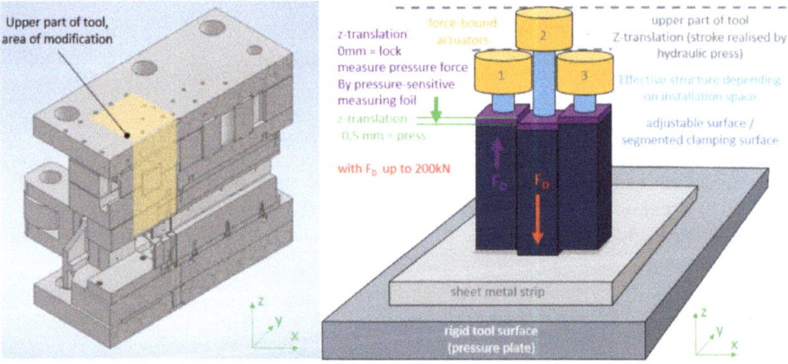

Fig. 3. left: ISB-experimental tool (CAD sectional view) with modification area in the upper part marked in orange, right: tool concept for static adjustment and dynamic control of the segmented blank holder (schematic).

The concept includes the following approaches (see Fig. 3):

- Geometry: The triangular geometry of the segments is based on the forming zone that forms in the event of bending under tensile stress. The geometries of the individual segments become finer the more angle pieces divide the area of the forming zone.
- Position: The symmetry plane (see Fig. 2. y-x symmetry) is referred to as the flow sheath. This is where the maximum strains are present in the current process design. The plane of symmetry and the centre of rotation of the bending axis lying on it are defined in the concept

(and for the case of bending under pure tensile stress) as the origin of the angle pieces. All segments meet here.

- Kinematics: The segments are supported by one vertical degree of freedom so that they can perform a translational movement orthogonal to the sheet plane. Due to the machine and tool compliance, the current design requires a translation of 0.5 mm to be able to follow the transverse thinning of the material in the thickness direction and thus maintain the condition of static friction between the clamping and surface of the semi-finished product.
- Measuring technology: To measure the process-related compressive forces, a surface pressure-sensitive film is mounted on the head side of each segment. To prevent the film from being destroyed by any shear stress, it is mounted between hardened plate inserts. Due to the fine subdivision of the hold-down surface in conjunction with the high pressures, this results in an installation space problem that severely limits the use of load cells.
- Actuators: Due to the different levels of complexity, two basic distinctions are made. On the one hand, the concept already enables static, manual adjustment of the compressive force without actuators, for example by using a thin sheet to individually support each segment, and on the other hand, dynamic control of the clamping force. The latter includes various methods for influencing the geometry and spread of the forming zone. This allows the hold-down segments to be loaded and unloaded over time and locally based on the subdivision. Due to the high energy density required, which results from the requirement for fine subdivision of the blank holder surface in combination with the clamping force of up to 200 kN to be realised, short-stroke hydraulic cylinders in the high-pressure range of 700 bar are used as actuators. The hydraulics enable force-based control and proportional measurement of the applied pressure force per segment via proportional valves. The pressure force is measured using pressure transmitters. The cylinders can be controlled individually and can move independently of each other with a stroke tolerance of 0.5 mm in a range of 0-5 mm. The cylinder force is then transferred to the segments.
- Active structure: Based on the preliminary work [18], a high-performance, i.e. the most compact, rigid, load bearing and load transferring active structure possible is developed, which realises the required kinematics with the desired accuracy. Furthermore, the active structure allows the actuators to be nested and thus enables the required compact design with sufficient accessibility.

Finite-Elemente-Simulation

A quasi-static, elasto-plastic deformation analysis of in-plane bending was implemented and calculated in the software ABAQUS CAE 2021. Based on the finite element method (FEM), the novel, segmented tool concept was implemented to study the effects of the process kinematics on the resulting forming zone. To this aim, a sufficient representation of the contact pressure distribution must be simulated since the process bases on the principle of frictional engagement. It was proven in preceding studies [11,16] and literature [19] that solid elements with elastic material properties are necessary to represent tool elasticity and thus obtain a valid contact pressure distribution and, in consequence, forming result.

Here, the sheet metal blank was discretized by a mesh of solid finite elements of 8-node linear brick type (C3D8) with a size of 1x1 mm in plane and 7 elements over thickness (see Fig. 4). A dual phase steel material with significant hardening behaviour – a HCT780X sheet metal steel of 1.5 mm thickness – was modelled in the simulation by means of a density of 7.85 kg/dm^3, a YOUNG´s modulus of 215 GPa and a Poisson Ratio of 0.3. Isotropic plastic deformation behaviour was assumed, and hardening was modelled by applying SWIFT/KRUPKOWSKY´s law. The segmented tool surfaces in contact are represented by the same type of hexagonal finite elements with a uniform mesh size of 2 mm and by using the same elastic parameters.

In the process model, the segments were loaded in a way proportionally to their contact surface so that the accumulated force from all segments would achieve the nominal press load which should be applied to the process for firm clamping conditions. In detail, a 45° and two 22.5° segments were represented in simulation where the first is loaded with half of the nominal clamping force and the latter segments are loaded with a quarter of the nominal load. Since a symmetric tool setup is suggested for the incremental bending process ISB, a central symmetry constraint is applied and only one of the bending tools is modelled. Under load, the clamping tool would then pivot around the bending axis thus applying the bending moment under the desired pressure application of the segments.

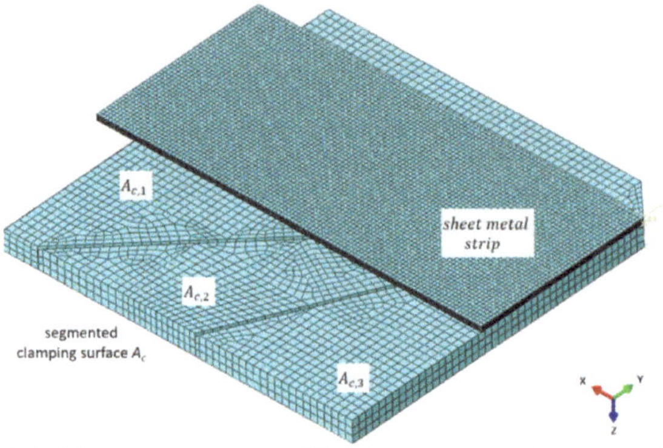

Fig. 4. The FE model of the incremental swivel bending process with segmented tooling concept, here three segments.

In this study, three different cases of contact pressure application were studied and evaluated with respect to the resulting forming zone. Firstly, a process operation with static application of load on each segment throughout the bending process is implemented in the FE simulation model. Next, dynamic operations of the segments were investigated. Either the inner segments are initially unloaded when the bending process initiates and loaded at a defined process point or, vice versa, they are loaded in the beginning and unloaded at a distinct instance of progress. It was shown in analytical considerations, that the forming zone of the in-plane bending process under frictional engagement expands in an angular way and reaches well within the clamping surfaces. Moreover, the forming zone angle γ would be significantly larger than the bending angle propagated by the clamping tools δ. In fact, γ would increase rapidly over the bending progression until reaching a certain threshold angle which is dictated by the tool geometry – here, $\gamma_{ult}/2$ is 45 ° when using quadratic tools. Therefore, the herein studied dynamic loading and unloading operations of the tool segments are operated in relation to the instance where γ reaches its threshold γ_{ult}.

Results

The herein suggested tooling concept would allow to restrict the forming zone to a prescribed angular value smaller than the ultimate forming zone angle γ_{ult} which is given by the tool geometry according to Eq. 6. Considering the unloaded bending angle δ_u according to Eq. 8 which depends on strain resulting from the bending angle δ and the forming zone γ as given in Eq. 7, the analytical consideration evidently shows the potential to limit elastic springback in incremental bending with the herein presented tooling concept, see Fig. 5 left.

According to Eq. 10, strain resulting from the incremental sequence is calculated. From the analytical equations, it is evident that larger strain results from a narrower forming zone angle. In turn, a lesser number of forming zones would overlap each other with such a tighter configuration (as given by superposition *SP*). Hence, in the range of forming fewer but more intense bending increments (i.e., lower values for *SP*), a potential to reduce overall strain in incremental bending is seen in the analytical consideration, cf. Fig. 5 right. Interestingly, a reduction of longitudinal strain in the incremental sequence is only seen for the first segmentation, i.e. from 45° to 22.5°. So far, these considerations only concern the static operation mode of the novel tooling concept, where the tool segments stay active or inactive throughout the bending process.

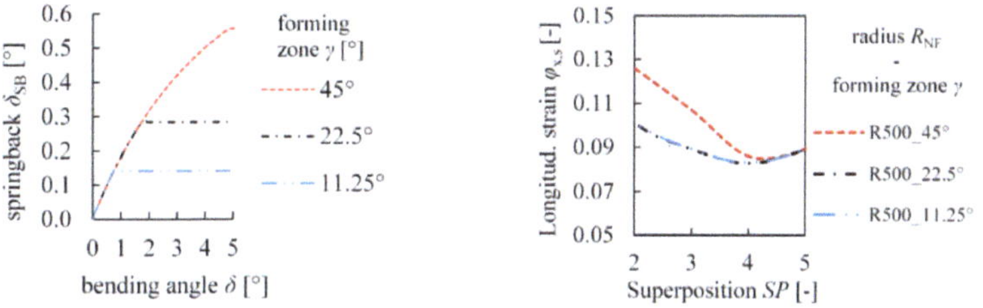

Fig. 5. Elastic springback (left) and strain resulting from bending a defined representative bending radius of 500 mm (right), both as a function of prescribed forming zone angles (45°, 22.5° and 11.25°). It is demonstrated based on analytical considerations, that the novel tooling concept potentially helps to limit elastic springback and strain in incremental bending.

With the help of the numerical process model, the dynamic operation of the segmented tool concept is simulated with the aim to evaluate the influence of an active loading or unloading of those tool segments which are in active contact with the forming zone. In Fig. 6 the numerical setup and the resulting strain maps are plotted for the three different process operations suggested in this paper for the segmented tool concept. In detail, a static operation of the tool concept is studied in simulation as well as a dynamic loading and unloading. The innermost segments are loaded/unloaded b) before, c) during or d) after the forming zone reaches its final angular extension γ_{ult}, respectively. All plots are extracted for the configuration at 0.2 seconds.

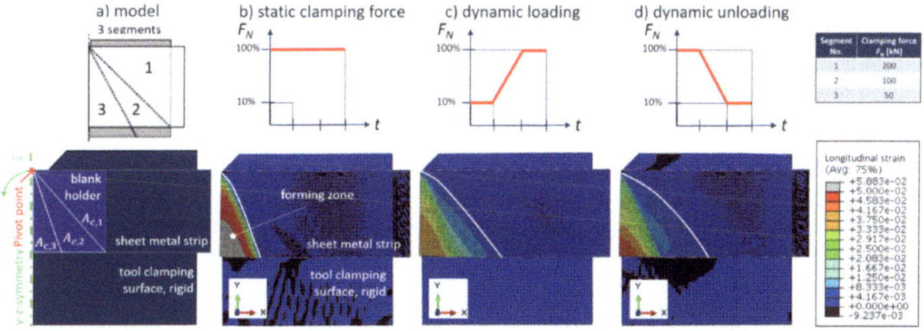

Fig. 6. Study on the evolution of forming zone by means of longitudinal strain with various static and dynamic concepts of tool operation. a) tool concept with three segments, b) static operation with the foremost segments under static load application, c) dynamic concept with the foremost segments being idle initially and loaded at δ = 2°, d) dynamic unloading of the foremost segments at δ = 2°. All plots contain a symmetry plane in the centre of the process.

From these simulation plots, evidently the static operation mode of segments is capable to keep the forming zone at a very narrow extension angle γ, which corresponds to the geometry of the foremost segment, i.e. 22.5° in this case. This concept has been presented in [16] however besides the option to restrict the forming zone, it delivers strong strain gradients and high maximum strain values (b). Evidently from the figure, the dynamic operation concepts deliver smoother strain transitions and softer strain maxima (c, d). By qualitative means, the dynamic loading concept initially lets the forming zone stretch out freely and then imposes a stronger straining in the central area after triggering the innermost segments (c). The dynamic unloading concept (d) delivers an intensely strained area in the central region since it operates during the first few degrees of bending like the static concept (b). However, after triggering the release of the innermost segments of concept (d), the forming zone is allowed to extend towards its final angle γ_{ult}, which results in an outer area of the forming zone that is much strained only weakly.

Conclusion

Advantages are predicted for the dynamic unloading concept, cf. Fig. 6 (d), since it initially plasticizes the sheet metal material sufficiently over the whole width h_0 but applies only limited strain values over the whole bending progression. In detail, the resulting forming zone is divided into two distinct areas, the inner area (the yield sheath) of higher strain application that develops during the loaded condition of the innermost segments, and the outer area that develops after unloading these segments. Two benefits might result in practical application: Firstly, the initial plasticizing helps to prevent uneven release of residual stress in springback, which results from a pronounced elastic deformation area in the central region of the forming operation. Undesired shape deviations are observed in practical experiments (twisting, buckling).

Moreover, this interpretation well agrees with the analytical calculations, see Fig. 5 left, which also predict reduced springback for narrower forming zone angles. Secondly, in an incremental process operation, the superposition of the outer area of the forming zone obtained from the dynamic unloading tool concept leads to fewer accumulation of strain from multiple forming zones. Failure due to cracking might be prevented from this concept. Again, this observation is confirmed by the analytical predictions, cf. Fig. 5 right, which suggest lower strain values for more narrow forming zones). However, besides the efforts in building such a tool in practice and the accompanying process control, such an operation might lead to pronounced coarseness of the bending geometry.

In practice, such a segmented tool concept would require a very high energy density to apply the desired contact pressure on relatively thin segments. Herein, we suggest hydraulics to actuate the individual tool segments like multi point forming [20]. Using the developed FE-model, the necessary segment forces might be studied in further approaches to define the minimum necessary load to achieve a process operation as suggested in the present work. As is, indeed considerable process forces are necessary on small tool segments which require high pressure hydraulic systems. Alternative tool concepts besides hydraulics are still under consideration by the authors.

References

[1] D.Y. Yang, M. Bambach, J. Cao, J.R. Duflou, P. Groche, T. Kuboki et al, Flexibility in metal forming, CIRP Annals 67 (2018) 743–765. https://doi.org/10.1016/j.cirp.2018.05.004

[2] R. Rauter, D. Globocnik, E. Perl-Vorbach, R.J. Baumgartner, Open innovation and its effects on economic and sustainability innovation performance, J. Innov. Knowledge 4 (2019) 226–233. https://doi.org/10.1016/j.jik.2018.03.004

[3] Y. Wang, H.-S. Ma, J.-H. Yang, K.-S. Wang, Industry 4.0: a way from mass customization to mass personalization production, Adv. Manuf. 17 (2017) 311–320. https://doi.org/10.1007/s40436-017-0204-7

[4] J.M. Allwood, D.R. Shouler, A.E. Tekkaya, The Increased Forming Limits of Incremental Sheet Forming Processes, KEM 344 (2007) 621–628. https://doi.org/10.4028/www.scientific.net/KEM.344.621

[5] L. Filice, L. Fratini, F. Micari, Analysis of Material Formability in Incremental Forming, CIRP Annals 51 (2002) 199–202. https://doi.org/10.1016/S0007-8506(07)61499-1

[6] K. Jackson, J. Allwood, The mechanics of incremental sheet forming, J. Mater. Process. Tech. 209 (2009) 1158–1174. https://doi.org/10.1016/j.jmatprotec.2008.03.025

[7] G. Hirt, J. Ames, M. Bambach, R. Kopp, Forming strategies and Process Modelling for CNC Incremental Sheet Forming, CIRP Annals 53 (2004) 203–206. https://doi.org/10.1016/S0007-8506(07)60679-9

[8] B. Engel, P. Frohn-Sörensen, M. Hillebrecht, A. Knappe, Incremental swivel bending for scalable lightweight structures, ATZ Worldw 119 (2017) 26–31. https://doi.org/10.1007/s38311-017-0023-2

[9] P. Frohn-Sörensen, L. Borchmann, B. Engel, Modelling the forming zone of force fitted bending processes, Procedia Manuf. 50 (2020) 411–417. https://doi.org/10.1016/j.promfg.2020.08.075

[10] P.J. Blau, Amontons' Laws of Friction. In: Wang QJ, Chung Y-W, editors, Encyclopedia of Tribology. Boston, MA: Springer US; 2013, p. 71.

[11] P. Frohn-Sörensen, Process Development of Incremental Swivel Bending – ISB [Dissertation], Siegen: Universität Siegen; 2023. http://dx.doi.org/10.25819/ubsi/10305

[12] C. Gerlach, Ein Beitrag zur Herstellung definierter Freiformbiegegeometrien bei Rohren und Profilen [Dissertation], Aachen: Shaker Verlag GmbH; 2010.

[13] R. Hooke, Lectures de Potentia Restitutiva, Or of Spring Explaining the Power of Springing Bodies, John Martyn; 1678.

[14] A.-M. Wittek, H.-C. Richter, B. Lazarz, Tubular stabilizer bars-calculations and construction, Transport Problems (2011) 17–25.

[15] P. Frohn-Sörensen, W. Hochstrate, D. Schneider, M. Schiller, B. Engel, Incremental bending of conic profiles on CNC hydraulic bending machines, Proceedings of the Institution of Mechanical Engineers, Part B: J. Eng. Manuf. 235 (2021) 1248–1268. https://doi.org/10.1177/0954405420982227

[16] P. Frohn-Sörensen, J. Reuter, B. Engel, Influencing the forming zone by altering the contact pressure in a bending process, ESAFORM 2021, 2021. https://doi.org/10.25518/esaform21.1870

[17] M. Gösling, S. Fruhstorfer, J. von Prondzinski, P. Frohn-Sörensen, B. Engel, Erweiterung der Prozessgrenzen für hochfestes mikrolegiertes Kaltband durch inkrementelles Profil-Schwenkbiegen, In: Transformation in der Biegetechnik. Universitätsbibliothek Siegen; 2023, p. 15–22. http://dx.doi.org/10.25819/ubsi/10390

[18] J. Reuter, P. Frohn-Sörensen, B. Engel, Segmentation Method for Bending Tools – Fundamental Investigation of Profile Forming by Segmented Tools, In: Mocellin K, Bouchard P-O, Bigot R, Balan T, editors, Proceedings of the 14th International Conference on the Technology of Plasticity - Current Trends in the Technology of Plasticity: ICTP 2023 - Volume 1, 1st ed. Cham: Springer Nature Switzerland; Imprint Springer; 2024, p. 547–559. https://doi.org/10.1007/978-3-031-41023-9_55

[19] M.C. Oliveira, D.M. Neto, J.L. Alves, L.F. Menezes, Considering the stiffness of the forming tools in the numerical analysis of the ironing process, Mater. Res. Forum LLC (2023). https://doi.org/10.21741/9781644902479-160

[20] K. Siegert, J. Hohnhaus, S. Wagner, Combination of Hydraulic Multipoint Cushion System and Segment-Elastic Blankholders, In: Williams D, Allen J, Hukkeri R, editors. Electronic control module network and data link development and validation using hardware in the loop systems, [Warrendale, PA]: SAE International; 2009.

Study influence of forming process on microstructure and properties of deep hole thick wall pipe blank

LEI Bingwang[1,a], LIU Haijiang[1,b*], TU Mingjin[1,c], WANG Jiaoqi[1,d], WANG Xing[1,e], QIN Ruiting[1,f], KOU Yanyan[1,g] and LIU Pengfei[1,h]

[1]Inner Mongolia North Heavy Industries Group Co., Ltd., Baotou, Inner Mongolia, P.R. China

[a]leibingwang@163.com, [b]13848522181@163.com, [c]tumingjin2004@163.com, [d]13847284477@163.com, [e]13940565250@163.com, [f]bzyunzhonglang@163.com, [g]15354903902@163.com, [h]liupengfei@163.com

Keywords: Forming Process, Heavy Duty Extrusion, Radial Forging, Super High Strength and Toughness, Large Diameter Thick Wall Pipe Blank

Abstract. According to the demand for high strength and high toughness of deep hole thick wall pipe blank, basing on 36,000 tons of vertical extrusion machine, 3,000 tons of free forging press and 1,600 tons of radial forging machine. Using electroslag remelting to smelt Cr-Ni-Mo-V steel ingot as raw material, three forming processes were studied, including free forging and radial forging, solid extrusion and radial forging, and hollow extrusion and radial forging. The physical quantity distributions of stress, strain and strain rate in the deformation zone of billet blank in different forming processes were analyzed via numerical simulation technology. Combined with the numerical simulation results, three forming processes were carried out in engineering trial production. The trial production results show that the structure of the pipe blank produced by the hollow extrusion + radial forging process is uniform, the mechanical properties of the pipe blank at both ends of the sampling $R_{P0.2}\geq1370$ MPa, -40°CKV≥28.0J, the mechanical properties of the full-length cross-section of the pipe blank $R_{P0.2}\geq1381$ MPa, -40°CKV≥28.2 J, the full-length yield strength is 28 MPa, and the comprehensive performance level is leading in the world.

Introduction

For a long time, people have actively explored new manufacturing methods (new processes, new equipment, etc.) of large forgings in order to economically obtain products with stable quality and excellent performance, and better serve aerospace, marine, national defense and other key areas of construction. As one of the most widely used weapons in various countries, barrel is the core component of artillery equipment and one of the important factors determining the performance and life of artillery [1-3]. The strength and toughness level of the barrel forgings have a decisive impact on the service life of the barrel, lightweight and the safety of the artillery [4-5]. With the rapid development of artillery to the path of large caliber, high bore pressure, high initial speed, long range and increasing muzzle energy, new requirements are put forward for the barrel. For the further improvement of the comprehensive performance of the barrel blank, such as strength and toughness, ablation resistance, integration of new generation of metallurgy, forming and heat treatment equipment, and further optimization of the manufacturing process to obtain ultra-high strength and toughness thick-walled billet, it is a major technical issue to support the development of future light large-caliber high bore pressure artillery and future high bore pressure main battle tank.

In order to improve the strength and toughness of thick-wall pipe blank for barrel, various countries have carried out in-depth research from the aspects of composition optimization [6-7], smelting process [8-9], forming process [8-9] and heat treatment process [10-11], and have achieved remarkable research results. Inner Mongolia North Heavy Industry Group Co., Ltd.

independently developed the world's largest ferrous metal vertical extrusion machine (Fig. 1) in July 2009, becoming the second enterprise in the world to master heavy extrusion technology. Continuously turning people's long-cherished wishes for heavy extrusion products into reality. Premium products such as extruding powder bar for large aircraft turbine disc, large caliber thick-wall seamless steel pipe of P91/P92 for thermal power, large caliber thick-wall seamless steel pipe of titanium alloy for marine construction, and TP316LN/TP316 series of large caliber pipes (rods) for fourth generation nuclear power have been mass produced, and the products performance are excellent. With our mature mastery of heavy extrusion technology, the integration of existing production equipment in our enterprise, developing "hollow extrusion + radial forging" composite forming process for medium alloy ultra-high strength and toughness barrel, and improving the ultra-high strength and toughness level of gun barrel, undoubtedly become a long-awaited technical research work in the field of artillery.

Based on 36,000 tons of heavy vertical extrusion machine, 3,000 tons of free forging and 1,600 tons of radial forging machine, this paper develops three kinds of composite forming process research on case1 "free forging + radial forging", case2 "solid extrusion + radial forging" and case3 "hollow extrusion + radial forging" for deep hole thick wall pipe billet, and makes beneficial exploration on the influence of forming process on the structure and performance of deep hole thick wall pipe billet.

Fig. 1. 36,000 tons heavy vertical extrusion production line.

Solid blanks and hollow blanks radial forging numerical simulation analysis

Radial forging machine with a unique high frequency pulse forging principle, realize short stroke, multi hammer, large tonnage, multi constraints of rapid forming, forging technology to bring a leap forward, but also for continuous research on metal forming process put forward new requirements. Using of deform11.0 simulation software (non MO module), for solid billet, hollow billet in the radial forging process of billet deformation area directly below the hammer strain, strain rate, stress field distribution and other physical quantities are analyzed by simulation technology. The rotary motion of manipulator and hammers movement synchronization, when hammers contact workpiece, the manipulator will stop rotating, feeding and other operations, four hammers began to hit the billet, when the hammer left the billet, the manipulator clamping billet rotating, feeding billet. The whole forging process of the manipulator, four hammers according to the above procedures cycle operation, until the end of the pass set task.

Numerical Simulation of solid blank radial forging process. Heads of the radial forging machine are distributed symmetrically along the circumference of 90°. The solid blank size of simulation in this section is Φ560×L. After the first radial forging to Φ490×L solid blank, the second radial forging to Φ410×L solid blank, the first and second forging using the same hammer head. The length of the hammer head is 850 along the Z-axis direction of radial forging, and the hammer head is placed symmetrically about the XOY plane. Fig. 2 shows the CAD model of radial forging simulation for solid blank. The numerical simulation model of solid blank radial forging process assumes the following:

(1) The elastic deformation of the material is ignored, and the material is regarded as rigid-plastic material;

(2) The friction model is shear friction;

(3) The hammer is regarded as a rigid body, and only heat exchange occurs between the hammer and the billet;

(4) The temperature inside and outside the billet is uniform;

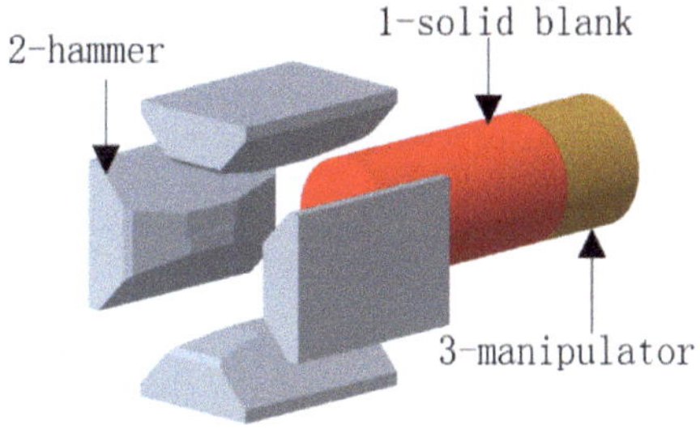

Fig. 2. Numerical simulation CAD model of four hammer head radial forging solid blank.

Initial temperature of solid blanks is 1230°C, initial temperature of hammer and other dies are 20°C, and the feeding speed of the billet is 30mm/s, and the frequency of the hammer is 3 times/s. Heat transfer coefficient between hammer and solid blanks is 1 N·(s·mm·°C^{-1}), the shear friction coefficient is 0.2, and rotation angle of solid blanks between the two hammering is 13.5°. Solid blanks is clamped by manipulator to rotate and feed. The CAE simulation model of the solid bar with four hammers is established in software of Deform11.0, as shown in Fig. 3. During the first four-hammer radial forging deformation, solid blanks equivalent deformation of the axial section was "X"-shaped distribution, as shown in Fig. 4.

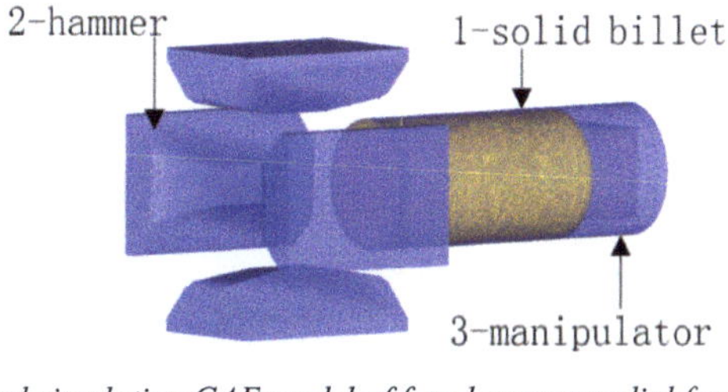

Fig. 3. Numerical simulation CAE model of four hammer radial forging solid blank.

Materials Research Proceedings 44 (2024) 282-296 https://doi.org/10.21741/9781644903254-31

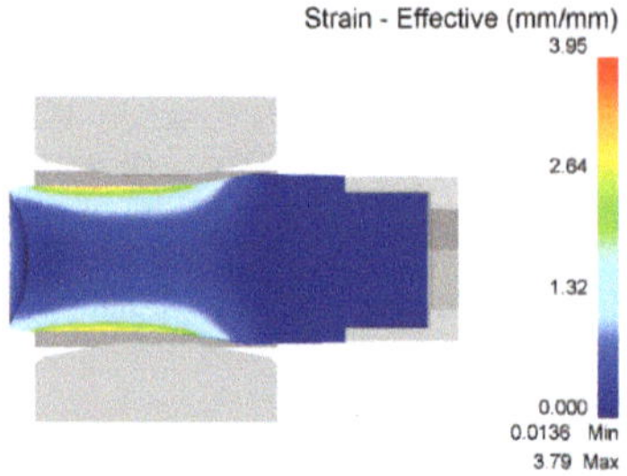

Fig. 4. Equivalent strain distribution of radial forging Φ560×L solid blanks in axial section.

On the basis of the first pass precision forging deformation, the second pass radial forging is entered. The outer diameter of blank is radial forged from 490 mm to 410 mm. In the process of second pass radial forging, the distribution of stress, strain and strain rate at the transverse interface of large deformation of blank during the second radial forging process are shown in Fig. 5 and Fig. 6.

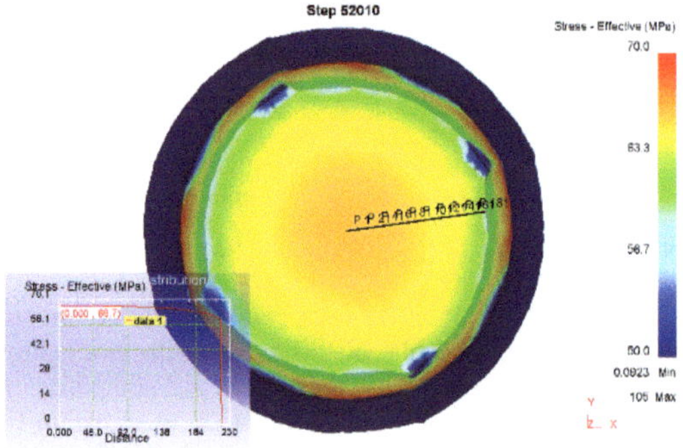

Fig. 5. Stress distribution of radial forging solid blank.

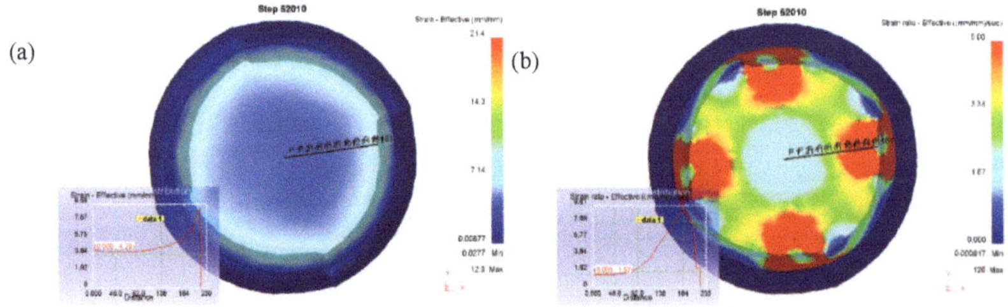

Fig..6 (a) Strain and strain rate (b) Distribution of radial forging solid blank.

Numerical simulation of hollow pipe blank radial forging process. For hollow radial forging, four special hammer heads are used. Four hammer heads are distributed at 90° along the circumference and are the same relative to the hammer heads. The same hammer heads are used in the first and second passes. The specification of hollow pipe blank is Φ560×Φ210×L in this section. The first pass is radial forged with Φ150mm mandrel, and the hollow pipe blank specification is radial forged to Φ490×Φ150×L; second pass is radial forged with Φ110 mandrel, and the hollow pipe blank specification is radial forged to Φ410×Φ110×L. Fig. 7 is the CAD model of the simulation of the precision forging of hollow pipe blank.

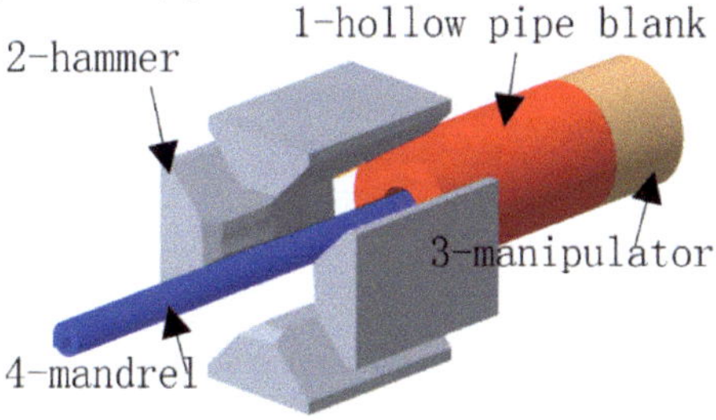

Fig. 7. Numerical simulation CAD model of four hammer head radial forging hollow pipe blank with mandrel.

The numerical simulation calculation model of hollow pipe blank radial forging process assumes:

(1) the elastic deformation of the material is ignored and the material is regarded as rigid-plastic material;

(2) the friction model is shear friction;

(3) the hammer and mandrel are regarded as rigid bodies, and only heat exchange occurs with the hollow pipe blank;

(4) the initial temperature hollow pipe blank t is uniform;

Initial temperature of hollow blank is 1230°C, initial temperature of hammer, mandrel and other molds are 20°C, manipulator holds hollow pipe blank to rotate and feed, rotation angle of hollow pipe blank between two hammers is 13.5°, the axial feeding speed is 30 mm/s, mandrel feeding speed is 0mm/s, and precision forging machine hits 3 times/s. Heat transfer coefficient between hammer and hollow pipe blank is 1 N·(s·mm·℃$^{-1}$), and heat transfer coefficient between extruded hollow pipe blank and mandrel is 0.4 N·(s·mm·℃$^{-1}$). The CAE simulation model of hollow pipe blank for radial forging with four hammers in the software of deform11.0 is shown in Fig. 8.

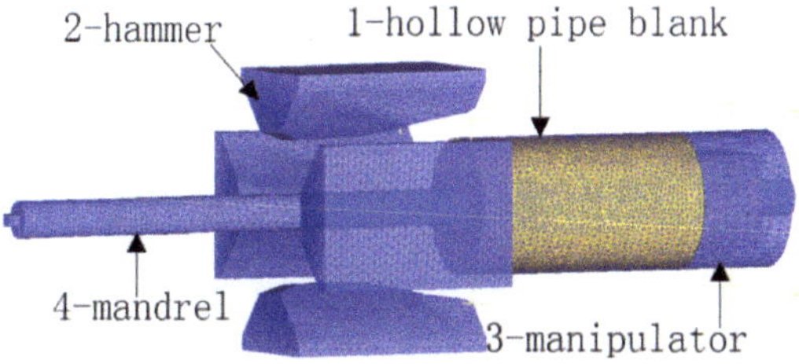

Fig. 8. Numerical simulation CAE model of four hammer head radial forging hollow pipe blank with mandrel.

In the radial forging process of Φ560×Φ210×L hollow pipe blank with Φ150 mm mandrel, the equivalent strain distribution of axial section of hollow pipe blank with mandrel in precision forging is shown in Fig. 9.

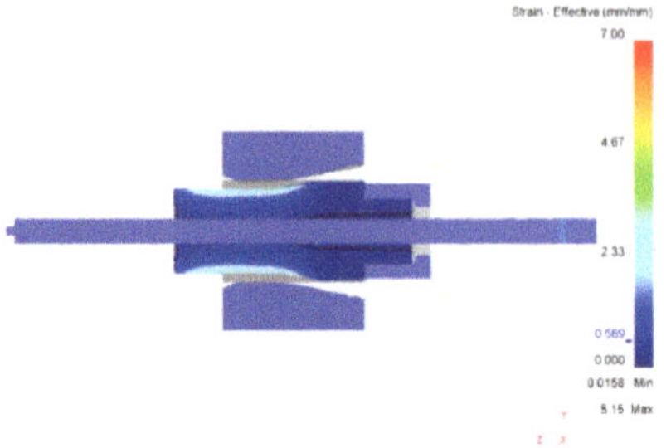

Fig. 9. Equivalent strain distribution of radial forging Φ560×Φ210×L hollow pipe blank with mandrel in axial section.

Fig. 10 and Fig. 11 show the stress, strain and strain rate distribution in the deformation zone of Φ490×Φ150×L hollow pipe blank radial forged to Φ410×Φ110×L using Φ110 mandrel in the second pass.

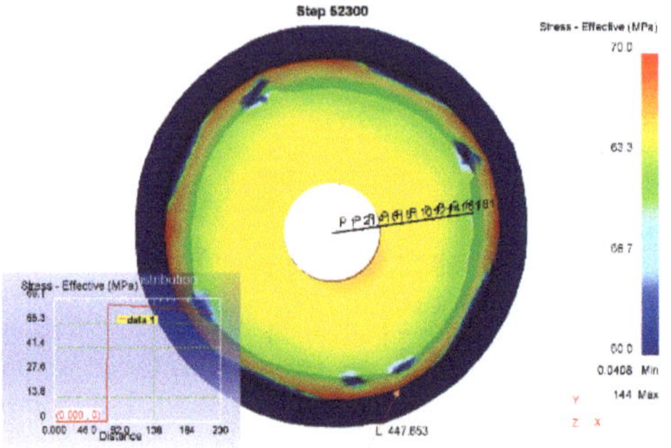

Fig. 10. Stress distribution of four hammer head radial forging hollow pipe blank with mandrel.

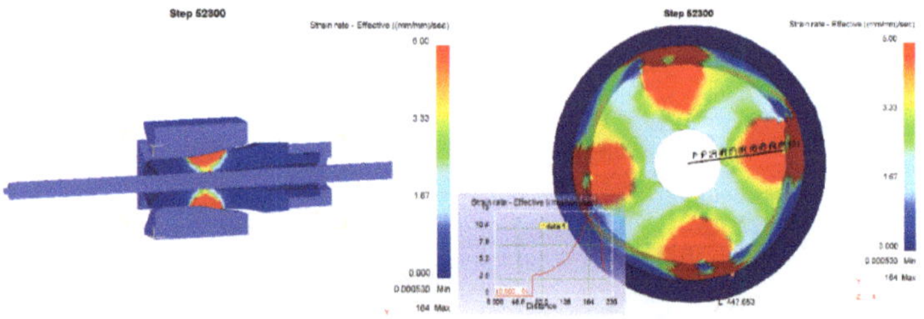

(a) (b)

Fig. 11. Strain rate distribution of radial forging hollow pipe blank: (a) Axial section (b) Transverse section.

Comparison of deformation parameters of solid radial forging and hollow radial forging. As can be seen from the above numerical simulation, the radial forging method with two pairs of hammers can deform the workpiece around the workpiece during forging, ensuring the precise control of the size of the forgings. Four hammers are divided into two groups in the same plane for radial forging.

The billet is pressed in two directions. The wide plastic deformation is limited by the tool, and the plastic deformation of the metal is mainly axial extension.

In the process of solid blank radial forging, solid blank is formed by the unformed billet, which rotates at a certain rotation angle and is sent directly below the hammer, and is then formed under the action of four hammers. The solid blank to be formed in the middle stage of mandrel precision forging is rotated at a certain rotation angle and is sent directly below the hammer, and is deformed under the action of the compound angle hammer surface. The hollow pipe blank directly below the hammer of hollow radial precision forging is usually divided into three states of the pipe blank: "sinking area", "precision forging area" and "plastic area", and the metal is formed in different areas at the same time. In the "sinking area", the inner wall of the tube blank decreases the distance from the axis of the billet, and the thickness of the pipe wall remains basically unchanged. As pipe blank rotates and is sent, the inner hole of the tube blank contacts the mandrel, and at this time the tube blank is in the main deformation area, namely the "precision forging area". Fig. 12 shows the equivalent stress curve from inside to outside and along the radial direction on the transverse section of large deformation of blank during the instantaneous precision forging forming.

Fig. 12 shows the equivalent stress curve of the transverse section of blank with large deformation from inside to outside and along the radial direction during the transient radial forging of blank. The equivalent stress of solid billet at the corresponding position on the cross section along the radial direction 0mm~193 mm is slightly greater than that of hollow billet, the equivalent stress of solid blank and hollow blank at 193 mm away from the axis is equal to 62.5 MPa, and the equivalent stress of solid blank at 193mm away from the axis to the outer surface of the pipe is slightly less than that of hollow pipe blank. The equivalent stress of solid blank along the radial direction from the axis is distributed between 59.49 MPa ~66.73 MPa, and the equivalent stress gradually decreases from 66.73 MPa to 62.5 MPa along the radial direction from the axis, and the equivalent stress value is 62.5 MPa at 193mm away from the axis, and the equivalent stress decreases from 62.5 MPa to 59.49 MPa from 193 m along the radial direction to the outer surface; the equivalent stress of hollow billet with mandrel precision forging is distributed between 60.13 MPa ~65.78 MPa, and the equivalent stress gradually decreases from 65.78 MPa to 62.5 MPa along the radial direction from the nearest point to the axis, and the equivalent stress decreases from 62.5 MPa to 60.13 MPa from 193 mm along the radial direction to the outer surface.

Many scholars have found that with the increase of detection strain rate, the strength and toughness of the material are improved accordingly, and the sensitivity of strength improvement to strain rate is better than that of elongation [12-15]. For example, for the ultra-high strength hot-formed 30MnB5 automotive steel, in the low strain rate range (10^{-3} s^{-1}-10^{-1} s^{-1}), the strain rate sensitivity of yield strength and tensile strength is low with the increase of strain rate; but in the high strain rate range (1 s^{-1}-103 s^{-1}), the yield strength increases from 1520 MPa to 1620 MPa, an increase of 6.58%. However, there are few reports on the sensitivity of strain rate to high strength steel in the process of hot forming. In this paper, the quantitative simulation analysis is carried out for the change of strain rate of material in the deformation area in the hot forming process of solid/hollow stepped blank, and then the influence of strain rate on the strength, impact toughness and other properties of the product in the process of hot forming is revealed.

Fig. 13 shows that at the moment of forming solid and hollow blanks, the strain rate on the transverse interface of the maximum deformation of the billet in the "precision forging zone" increases first and then decreases along the radial direction from the axis area to the outer near surface. The strain rate of solid precision forging blank along the radial direction is distributed between 1.56 s^{-1} and 9.1 s^{-1}, and the strain rate value increases continuously from 1.56s^{-1} at the axis to 9.1 s^{-1} at 180 mm away from the axis; then the strain rate value along the radial direction decreases continuously to 2.39 s^{-1}; the strain rate value of hollow radial forging blank with mandrel along the radial direction is distributed between 3.10 s^{-1} and 12.38 s^{-1}, and the strain rate value increases continuously from 3.10 s^{-1} near the mandrel surface to 12.38 s^{-1} at 193 mm away from the axis, and then decreases to 3.54 s^{-1} along the radial direction. It can be seen that in the process of four-hammer radial forging, the hollow radial forging blank with mandrel has a higher strain rate than the solid radial forging billet, which is more conducive to increasing the dislocation density of the metal in the thermal forming process, and provides a positive influence on improving the strength of the material.

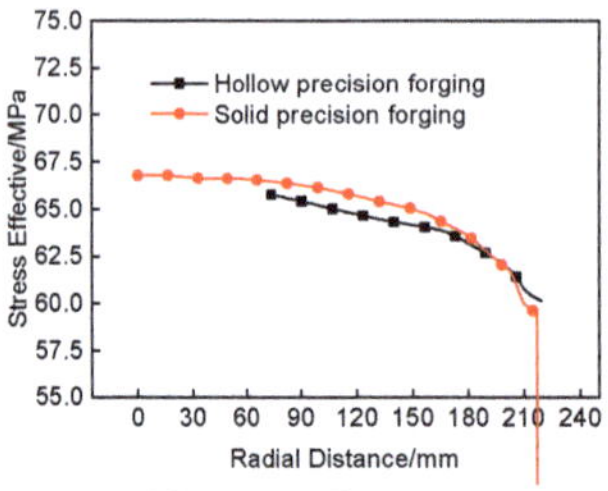

Fig. 12. Equivalent stress curve.

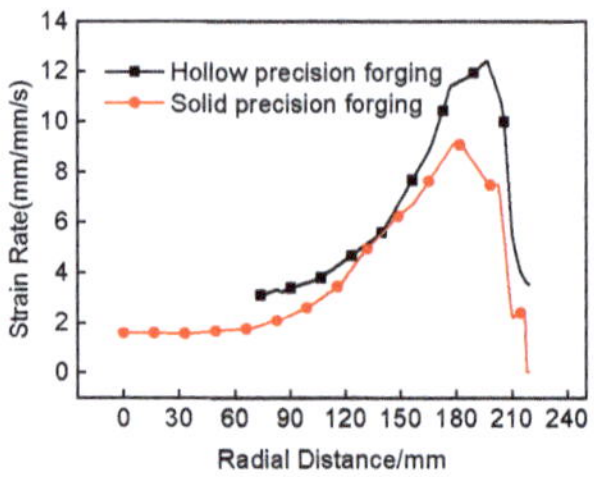

Fig. 13. Strain rate curve.

Numerical simulation analysis of moving force of mandrel in hollow radial forging. Considering the equipment capacity of 1600-ton four-hammer radial forging machine, the string force along the drawing direction is not more than 1×10^5N. Among the three combined forming schemes designed in this paper, there are problems in the radial forging process of four-hammer radial forging of hollow pipe blank with large aspect ratio and mandrel perforation. Therefore, four-hammer radial forging of hollow pipe blank with large aspect ratio is one of the key processes studied in this paper. In view of this problem, this paper carries out numerical simulation research. The CAD model established for numerical simulation analysis of four-hammer radial forging of hollow pipe blank with large aspect ratio is shown in Fig. 8. In Fig. 8, the specification of hollow radial forged pipe billet is Φ560×Φ210 × 1000, and four hammer radial forging with large deformation is adopted. The length of hammer along the pulling direction of extrusion blank is 850 mm, and the maximum diameter of mandrel is 150 mm (taper 0.0016°) and the length is 1900 mm. The simulated reduction amount is used to radial forge the extrusion blank with outer diameter of 560 mm to Φ390 mm, and the single pass deformation reaches the large deformation model of 170 mm. The temperature of the simulated blank is 1230°C, the initial temperature of the hammer, mandrel and other dies is 20°C, the feeding speed of mandrel is 0mm/s, the feeding speed of billet is 30 mm/s, and the frequency of radial forging machine is 3 times/s. The heat transfer coefficient between the extruded hollow pipe blank and the mandrel is 0.02, and the rotation angle is 13.5°. The friction coefficients between the inner surface of the extruded pipe blank and mandrel in the radial forging process were set as 0.3 and 0.05, and the force of mandrel when pulling and removing mandrel in the numerical simulation of the four hammer radial forging process is simulated.

Through simulation analysis, when the shear friction coefficient is 0.3, the mandrel needs 20 tons of force in the radial forging hollow pipe blank within 15 seconds after the start of four-hammer precision forging, which is much higher than the load of the radial forging machine. Moreover, with the increase of the feeding amount of the blank under the four-hammer, the maximum stringing force required by the mandrel in the blank reaches 8×10^5 N. When the friction coefficient is reduced to 0.05, the whole billet is fed directly under the hammer, and the maximum stringing force of the mandrel is 9.5×10^4N. Therefore, to realize the radial forging forming of the hollow pipe blank with large deformation and mandrel by four-hammer, the friction coefficient at high temperature should be less than 0.05. The force required for pulling and removing the mandrel with different friction coefficients between the mandrel and the hollow pipe blank is shown in Fig. 14.

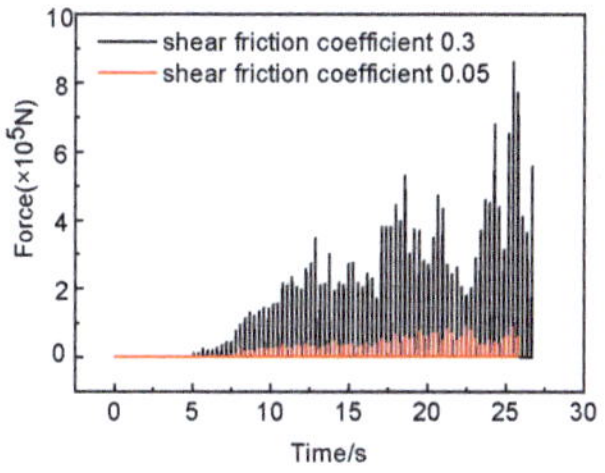

Fig. 14. Rod string pull-out load (friction coefficient 0.3, 0.05).

Trial-manufacture

Based on simulation results, three forming processes, including case1 "free forging + radial forging", case2 "solid extrusion + radial forging" and case3 "hollow extrusion + radial forging", were carried out. The electroslag remelted ingot of Cr-Ni-Mo-V medium and low carbon alloy was used in the trial-production.

Case 1"free forging + radial forging"

Technical process of case1: 3000 tons of free forging blanking - radial forging solid blank - annealing - rough machining - quenching and tempering heat treatment - performance detection.

(1) Free forging solid blank

Using 3000 tons free forging press for three upsetting and three drawing blank. For the first upsetting, the electroslag ingot is heated to $1210^{\pm10}$℃, and the electroslag ingot is upset to 700 mm by ratio of 2.15. Then the FM forging method is used to draw the long ingot. The sum of the upper and lower forging amount in each pass is distributed between 25% and 28%. After forming side A of the ingot, the ingot to be formed is rotated 90° along the axis and then adjacent side B is forged. After multiple draws, the size is $600^{\pm10}\times650^{\pm10}\times$L blank. For the second upsetting, the billet was returned to the furnace and heated to $1210^{\pm10}$℃, and the blank was upset to 700mm. After multiple upsetting, the specification was $600^{\pm10}\times650^{\pm10}\times$L. For the third upsetting, the billet was returned to the furnace and heated to $1210^{\pm10}$℃, and the blank was upset to 700mm. After multiple upsetting, the specification was $620^{\pm10}\times620^{\pm10}\times$L. Then the upper and lower flat anvil chamfering, and then using the upper flat anvil, the lower V-shaped anvil forging into the size of Φ620×L round rod.

(2) Radial forging solid stepped blank

The solid blank is freely forged, and then the blank is heated to $1150^{\pm10}$℃, and then radial forging is performed. The radial forging adopts R180 hammer head, blank feeding speed is 30mm/s, and radial forging frequency is 3 times/second. The solid billet of Φ315×L is forged by multi-pass pressing. During last pass, stepped blank is finished, forging ratio between 2.72 ~ 5.58.

Case 2"solid extrusion + radial forging"

Technical process of case 2: 15000t equipment upsetting blank - 36000t extrusion machine extruding solid blank - radial forging solid blank - annealing - rough machining - quenching heat treatment - performance detection.

(1) Extruding solid blank

The electroslag ingot is heated to $1260^{\pm10}$°C, the surface oxide skin is removed, and closed upsetting is completed on the 15000t equipment. The solid blank is heated to $1210^{\pm10}$°C, and 1100 series extrusion cylinder and Φ535 mm extrusion die are used on the 36000t extrusion machine to extrude Φ525×L solid blank, with an extrusion ratio of 4.62.The extruded blank is heated to $1150^{\pm10}$°C, and then radial forging is carried out. R180 hammer head is used for radial forging, with a billet feed rate of 30mm/s and a refined forging frequency of 3 times/second. The solid billet of Φ315×L is forged by multi-pass pressing. During last pass, stepped blank is finished, forging ratio between 2.72 ~ 5.58.

Case 3"hollow extrusion + radial forging"

Technical process of case3: upsetting and punching blank with 15000t equipment - 36000t extruding hollow pipe blank - radial forging hollow pipe blank - annealing - quenching heat treatment - performance detection.

(1) Extruding hollow pipe blank

The electroslag ingot is heated to $1260^{\pm10}$℃, the surface oxide skin is removed, and the closed upsetting and punching blank are completed on the 15000t equipment. The hollow pipe blank is heated to $1210^{\pm10}$℃, and 1100 series extrusion cylinder, Φ535 mm extrusion die, Φ155 mm extrusion mandrel are used on the 36000t extrusion equipment to extrude Φ525×Φ155×L hollow pipe blank, with an extrusion ratio of 4.75.

(2) Radial forging solid stepped blank

The radial forging process of extruded blank is completed by passes. The first pass is pre-forging: the extruded billet is heated to $1200^{\pm10}$°C, the inner hole is lubricated, and then the radial forging is carried out. R180 hammer head is used for radial forging, the billet feed rate is 30mm/s, and the precision forging frequency is 3 times/second. The extruded billet is radial forged into Φ380×Φ110, Φ420×Φ110×L hollow stepped billet. The second pass is final forging: the pre-forged billet is heated to $1200^{\pm10}$ ℃, and then the radial forging is carried out. R180 hammer head is used for radial forging, the blank feed rate is 30mm/s, and the radial forging frequency is 3 times/second. The pre-forged billet is radial forged into Φ255×Φ60, Φ295×Φ60, Φ315×Φ60 hollow stepped blank. The 2 solid stepped blanks trial-produced in case 1 and the 7 solid stepped blanks trial-produced in case 2 were machined into hollow stepped pipe blanks, and the same heat treatment process parameters were adopted to adjust the heat treatment of the 4 stepped hollow pipe blank trial-produced in case 3.

Performance testing

Test results of sampling at both ends of blank

A total of 13 stepped blanks were trial-produced in the three cases. Two stepped blanks were trial-produced in case1, serial numbered as 1～2., seven stepped blanks were trial-produced in case2, serial numbered as 3 ~ 9, and four pieces were trial-produced in case3, serial numbered as 10 ~ 13. Physical and chemical properties such as mechanical properties, microstructure and grain size were tested by sampling from both ends of each billet. Specific results are as follows:

(1) Mechanical property

The statistical results of mechanical properties of the 13 stepped blanks are shown in Table 1.

 Materials Research Forum LLC
https://doi.org/10.21741/9781644903254-31

Table 1. Statistical results of mechanical properties of the 13 stepped blanks.

case	$R_{P0.2}$/MPa		Z%		-40℃AKV/J	
	range	average	range	average	range	average
1	1381～1354	1370.7	55～51	52.8	30.1～19.0	25.5
2	1380～1354	1363.7	61～54	56.6	33.1～25.1	28.6
3	1386～1370	1378.2	61～53	55.0	33.3～28.0	30.1

(2) Microstructure
The microstructure and grain size of 13 stepped blanks are shown in Table 2.

Table 2. Microstructure and grain size of stepped blanks.

No.	microstructure		grain size		
	end1	end2	end1	end2	
1	T.P.	T.P.	9	9	case1
2	T.P.	T.P.	9	9	
3	T.P.	T.P.	8	8	case2
4	T.P.	T.P.	9	9	
5	T.P.	T.P.	7	8	
6	T.P.	T.P.	9	9	
7	T.P.	T.P.	7	7	
8	T.P.	T.P.	9	9	
9	T.P.	T.P.	7	7	
10	T.P.	T.P.	8	8	case3
11	T.P.	T.P.	8	8	
12	T.P.	T.P.	8	8	
13	T.P.	T.P.	7	7.5	

Note: 1. "end1" expresses large diameter end of stepped blanks; "end2" expresses small diameter end of stepped blanks.
2. T.P. expresses Tempered Martensite Pearlite.

Full-length cross-sectional test results of stepped pipe blank

The section test of No. 12 hollow stepped pipe blank was selected, and there were 1 test pieces in each area at 200mm from the end face of large-diameter of the pipe blank, 1/4, 1/2, 3/4 from the full length of the end face of large-diameter, and 200mm from the end face of small-diameter, that is, 5 test pieces in total. There are two test samples for tensile, impact, (-40℃) and grain size on the same test piece, and the test samples for each type are numbered 1 ~ 10 from large-diameter end to small-diameter end.

Fig. 15 shows sectional test results of tensile test at room temperature. $R_{P0.2}$: 1381 MPa ~ 1409 MPa, yield strength difference 28 MPa, indicating that the strength uniformity of stepped pipe blank reaches a relatively high level. Fig. 16 shows test results of low temperature impact toughness -40℃AKV: 28.2 J ~ 32.5 J, indicating that the total length of hollow stepped pipe blank meets requirement of -40℃AKV≥25 J.

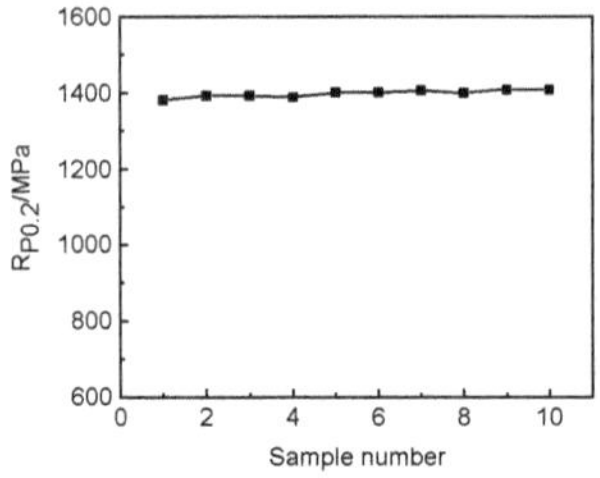

Fig. 15. Tensile strength and yield strength of cross sectional test.

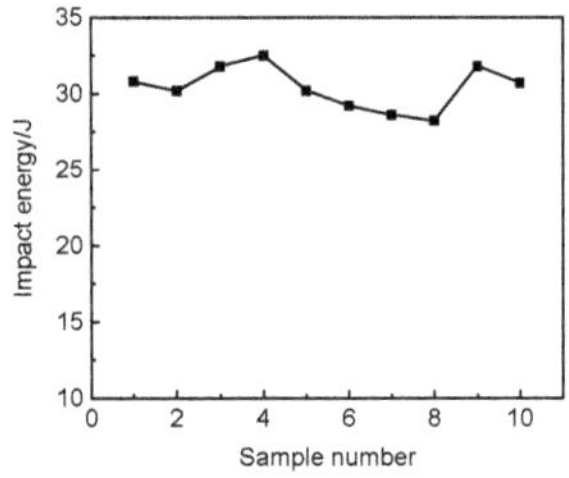

Fig. 16. Low temperature Impact toughness of cross sectional test.

Fig. 17 shows test results of cross-sectional grain size, and the grain size distribution ranges from 19.1um to 23.5um.

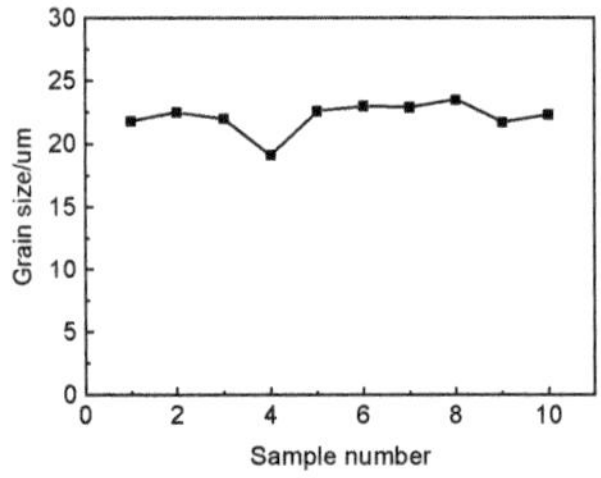

Fig. 17. Grain size.

Discussion

From the above physical and chemical test results, it can be seen that the mechanical properties of the stepped pipe blanks formed in case 2 "solid extrusion + radial forging" are significantly improved than those formed in case1 "free forging + radial forging". Using case1 "free forging + radial forging" process, the formed stepped blank $R_{P0.2}$:1354 MPa ~ 1381 MPa, -40℃AKV: 19 J ~ 30.1 J, impact fluctuation value is 11.1 J. Using Case1 "free forging + radial forging", due to the use of upper and lower hammer heads and 3000 tons of free forging press blanks, the deformation degree and deformation temperature of ingot in each region are very different, and there are risks such as large fluctuation value of mechanical properties difference and unstable performance. The lowest low temperature impact value of trial-produced blanks in this batch is 19 J and the fluctuation range is 11 J, that is, the fluctuation of -40℃AKV impact energy is 58.4%.

According to case 2 "solid extrusion blanking + four hammer head precision forging" process, the forming of 7 solid step billets is firstly closed upsetting and pressure maintaining in the 15,000 tons equipment of the 36,000 tons vertical extrusion production line. In the first stage of the deformation process of the ingot, three-dimensional compressive stress provides good conditions for the transformation of the organization of the ingot casting state to the organization of the plastic state, which is conducive to the uniformity of the subsequent products. In the subsequent extrusion forming process, the metal to be formed is extruded for the second time in the extrusion cylinder after three-dimensional compressive stress of 36,000 tons. Then solid blank is formed in the four-hammer radial forging under multi-directional constraints, which provides guarantee for the quality of blank. In this trial production of 7 to step billet $R_{P0.2}$:1354 MPa ~ 1380 MPa, -40℃ AKV: 25.1 J ~ 31.1 J, and the impact fluctuation range is 8 J, that is, the low temperature impact energy fluctuation of -40℃AKV is reduced from 58.4% in case1 to 31.8% in case 2 "solid extrusion + radial forging".

Case 3 "hollow extrusion + radial forging" process was used to form 4 stepped hollow pipe blanks. First, the closed upsetting and pressure holding were used in the 15,000 tons equipment of 36,000 tons vertical extrusion line. During the deformation of ingot in the first stage, the three-way compressive stress was applied to provide good conditions for the transformation of the cast structure to the plastic structure of ingot. Then the through hole billet is made by perforating needle. During the perforating process, the metal of the ingot axis is greatly plastic deformed to further improve the internal and external consistency of the through hole billet. In the subsequent through hole billet extrusion forming process, the metal to be formed is extruded in the extrusion cylinder for the second time under the action of 36,000 tons of three-way compressive stress. Due to the action of the mandrel on the hollow blank, the distribution of metal flow velocity at the extrusion die is more uniform and consistent than that in case2. Then the billet is formed under the multi-direction constraint condition of four-hammer radial forging, which provides guarantee for the quality of the extrusion billet. In this paper, the mechanical properties of 4 stepped hollow stepped pipe blanks formed by hollow extruding and radial forging with four hammer heads and mandrel. RP0.2:1370 MPa ~ 1386 MPa, -40℃AKV: From 28.0 J to 33.3 J, the impact fluctuation range is 5.3 J. On the basis of improving the yield strength, the impact fluctuation of -40°CAKV at low temperature decreased to 18.9%.

In addition, the sectional test of case 3 "hollow extrusion + radial forging" shows that the full-length performance of the stepped pipe blank meets the calculation requirements and the fluctuation range is small. Among them, $R_{P0.2}$:1381 MPa ~ 1409 MPa, difference 28 MPa, section shrinkage Z: 51% ~ 56%; -40℃AKV: 28.2 J ~ 32.5 J. In the tempering brittleness experiment, -40℃AKV: 28.3 J ~ 35.6 J.

Summary

Case 1 "free forging + radial forging", case2 "solid extrusion + radial forging" and case3 "hollow extrusion + radial forging" are three composite processes for manufacturing hollow pipe blank. The test results of yield strength are1354 MPa～1381 MPa, 1354 MPa～1380 MPa, 1370 MPa～1386 MPa, and the low temperature impact fluctuation value of -40°CAKV was 58.4%, 31.8%, 18.9%, respectively. The impact power of "extrusion + radial forging" is higher and more stable than that of "free forging + radial forging".

Hollow extrusion stepped hollow pipe blank is dissected, where $R_{P0.2}$:1381 MPa ~1409 MPa, section shrinkage Z: 51% ~ 56%; -40℃AKV: 28.2 J ~ 32.5 J, the tempering brittleness experiment

-40℃AKV: 28.3 J ~ 35.6 J, the results show that the full-length performance of hollow pipe blank meets technology requirements and the fluctuation range is small.

The solid blank Φ490×L was radial forged to Φ410×L with four hammer heads, and the hollow pipe blank Φ490×Φ150×L was radial forged to Φ410×Φ110×L with Φ110 mandrel. The strain rate on the transverse interface of blank in the two deformation processes was analyzed. The strain rates of solid radial forging blank ranged from 1.56 s^{-1}-9.1 s^{-1} along the radial direction, and those of hollow radial forging billets with mandrel ranged from 3.10 s^{-1} to 12.38 s^{-1} along the radial direction. Compared with solid radial forging blanks, hollow radial forging blanks with mandrel had higher strain rates and better comprehensive mechanical properties.

A four hammer radial forging CAE model with mandrel for pipe blank with outer diameter of 560 mm, inner hole of 210 mm, and length of 1000 mm was established in the simulation software of Deform-3D. For the calculation of 0.05 and 0.3 different shear friction coefficients between mandrel and hollow blank, the simulation results show that when the friction coefficient between mandrel surface and hollow blank deformation zone is controlled less than 0.05 in the radial forging process, the large-scale pipe blank with mandrel can be radial forged.

References

[1] Y.F. Xu, C.L. Shan, P.K. Liu, Review of the research on gun barrel life-span prolongation, Journal of gun launch & control (2019) 90-95. https://doi.org/10.19323/j.issn.1673-6524.2019.04.019

[2] P.K. Liu, D. Yang, J. Wang, Study on ablation and wear properties of barrel materials, Journal of gun launch & control (2021) 28-33. https://doi.org/10.19323/j.issn.1673-6524.2021.04.006

[3] M.L. Zhu, L.L. Liu, F.Z. Xuan, J. Effect of frequency on very high cycle fatigue behavior of a low strength Cr-Ni-Mo-V steel welded joint, Int. J. Fatigue (2015) 77-166. https://doi.org/10.3901//JME.2016.12.023

[4] Y.Q. Ou, C.G. Yu, Y.C. Zhang, Development of erosion and wear of gun barrel, Ordnance industry automation (2012) 44-46. https://doi.org/10.3969/j.issn.1006-1576.2012.06.014

[5]S.S. Zhang, Material process and service life of gun barrel, Journal of xi'an institute of technology (1989) 1-9. https://doi.org/10.16185/j.jxatu.edu.cn.1989.01.001

[6] Y. Lv, J. Hu, Z.N. Ren, Process in steel used for large-calibre thick-wall gun barrel, Ordnance material science and engineering (2013) 142-146. https://doi.org/10.14024/j.cnki.1004-244x.2013.02.011

[7]N. Zhang, C.R. Lv, L. Xu, Currents status and development trend of gun barrel steel, J. China metallurgy (2019) 6-9. https://doi.org/10.13228,j.boyuan.issn1006-9356.20190160

[8] X.D. Yuan, L.Q. Gao, H.F. Zhang, Development and prospect of gun barrel materials, Mater. Develop. Appl. (2017) 98-104. https://doi.org/10.19515/j.cnki.1003-1545.2017.05.019

[9] J.F. Huang, J. Zhang, J.Y. Chen, Failure mechanisms of gun barrels and the development of gun steel, Journal of gun launch & control (2023) 10-29. https://doi.org/10.19323/j.issn.1673-6524.2023.01.002

[10] D.B. Sun, Z.L. Jin, W. Li, Effect of tempering process on microstructure transformation of large-caliber 40CrNiMoV, Ordnance material science and engineering (2023) 72-76. https://doi.org/10.14024/j.cnki.1004-244x.20230516.004

Materials Research Forum LLC
https://doi.org/10.21741/9781644903254-31

[11] Y. Liu, M.Q. Wang, G.Q. Liu, Effects of tempering temperature on microstructure and mechanical properties of 40CrNiMoV steel, Journal of gun launch & control (2019) 90-95. https://doi.org/10.13251/j.issn.0254-6051.2014.06.012

[12] J.T. Liang, C.Z. Zhao, H.X. Yin, Strain rate sensitivity of ultra-high strength hot stamping steel, Chinese Journal of Engineering (2018) 1083-1090. https://doi.org/10.13374//j.issn2095-9389.2018.09.009

[13] H.J. Cai, J.F. Hu, R.B. Song, Plastic deformation microscopic mechanism of cold rolled dual phase steel DP1200 under high strain rate deformation, J. Mech. Eng. (2016) 23-29. https://doi.org/10.3901//JME.2016.12.023

[14] W. Zhang, Y. Pan, H.S. Liu, Effect of strain rate on properties of dual phase steel with high formability, Iron and steel (2022) 123-127. https://doi.org/10.13228/j.boyuan.issn0449-749x.20210619

[15] H.N. Liu, Y.D. Mei, L.B. Liu, Influence of strain rate on properties for low alloy high strength steel, Forg. Stamp. Tech. (2023) 253-257. https://doi.org/10. 13330/j.issn.1000-3940.2023.06. 034

Materials Research Proceedings 44 (2024) 297-304 https://doi.org/10.21741/9781644903254-32

Reduction of springback of Ti6Al4V alloy in ultrasonic vibration assisted deformation: Numerical simulation and experimental study

GAO Meng[1,a], WEN Dongxu[1,b,*], LI Jianjun[1,c] and HUANG Liang[1,d]

[1]State Key Laboratory of Materials Processing and Die & Mould Technology, School of Materials Science and Engineering, Huazhong University of Science and Technology, 1037 Luoyu Road, Wuhan, China

[a]gaomeng@hust.edu.cn, [b]dxwen@hust.edu.cn, [c]jianjun@hust.edu.cn, [d]huangliang@hust.edu.cn

Keywords: Ultrasonic Vibration, Titanium Alloys, Forming Process, Numerical Simulation

Abstract. The ultrasonic vibration assisted bending tests of Ti6Al4V are conducted under different forming temperatures and vibration amplitudes. The effects of forming temperature and vibration amplitude are investigated. Moreover, based on ABAQUS software and explicit solver, the ultrasonic vibration-assisted bending deformation process is simulated to explore the influence of temperature and amplitude on the springback of the plate. The results show that the springback of the plate decreases with the increase of vibration amplitude. Increasing the temperature can also reduce the springback of the plate. At 1000°C, increasing the amplitude shows the best effect on reducing the springback of the plate. The finite element simulation results show that increasing the amplitude can reduce the stress in the central deformation area and makes the stress distribution uniform. The increase of temperature reduces the overall deformation stress of the plate, thereby reducing the springback of the plate.

Introduction

Titanium alloys have been widely used in aerospace and aviation applications due to their attractive comprehensive properties such as high specific strength, excellent fracture toughness, and good heat resistance [1]. However, due to the poor formability of titanium alloys, it is difficult to form complex parts at room temperature. And the important reason for the poor formability of titanium alloy at room temperature is springback. Springback is one of the most serious defects in metal forming industry, because the shape of stamping parts after springback is related to the whole forming process [2]. Effective springback control is a key requirement for zero-defect metal forming and intelligent manufacturing.

Hot forming process is a traditional method to improve the formability of titanium alloy thin-walled plate [3]. However, this process has the disadvantages of surface oxidation, short mold life, complex microstructure evolution and high energy consumption [4]. As a new multi-energy field forming technology, ultrasonic vibration assisted forming has the advantages of green, energy saving and high efficiency [5]. During the deformation process, ultrasonic vibration assisted forming can effectively reduce the deformation resistance and improve the elongation and surface quality [6]. Therefore, it is expected to be an effective way to improve the forming ability of thin-walled plate of titanium alloy.

However, the mechanism of ultrasonic softening during plastic deformation is still unclear. The effect of ultrasonic vibration on the forming process is different for different materials such as copper, steel, aluminum alloys and titanium alloys. Liu et al. [7] investigated the effect of ultrasonic vibration compression on pure titanium, and found that ultrasonic vibration could reduce the yield and flow stresses in the plastic stage and the material showed residual softening after stopping the vibration. Kang et al. [8] studied the plastic deformation behavior of pure copper in ultrasonic-assisted micro-tensile experiments, and found that ultrasonic vibration promoted the reorientation of preferred grains and reduced the misorientation inside the grains. From the above studies, it can be concluded that the effect and inner mechanism of ultrasonic vibration on the forming process of alloys need be carefully studied.

Usually, the combination of finite element simulation and experiment is very important to study deformation. Sun et al. [9] combined simulation and experiment to study the ultrasonic vibration-assisted milling process of GH4169 superalloy, and the results show that increasing the ultrasonic amplitude can reduce the machining stress and improve the chipbreaking effect. Li et al. [10] investigated the principle of ultrasonic vibration assisted shear thickening polishing, and studied the influence of ultrasonic amplitude and ultrasonic frequency on polishing speed and polishing pressure based on simulation. However, the effect of ultrasonic vibration on the deformation process of Ti6Al4V alloy is rarely known.

In summary, the effect of ultrasonic vibration on springback during plastic deformation of titanium alloy is rarely known. Few researchers have studied the ultrasonic vibration assisted forming process by finite element simulation. Therefore, in this paper, the effect of ultrasonic vibration on the springback of plates of Ti6Al4V alloy after hot bending test is investigated. In addition, the whole process of ultrasonic vibration-assisted hot bending of Ti6Al4V alloy plates was constructed by finite element simulation model. The results illustrate the influence of forming temperature and vibration amplitude on the springback of Ti6Al4V alloy plate.

Material and experiments

The composition of Ti6Al4V alloy is shown in Table 1. The length of the plate is 110 mm, the width is 40 mm and the thickness is 1 mm. The principle diagram of ultrasonic vibration assisted bending is shown in Fig. 1. Firstly, the plate is heated by electricity, then the ultrasonic vibration equipment is turned on, and finally stamping is carried out. The stamping speed is set to 2mm / s. The holding time after stamping is 30 s. Then the plate is taken out and the springback of the plate is measured. Ultrasonic vibration equipment, mobile fixture and die are all equipped with insulation design. The time of energizing heating is set to about 8s, and the temperature of the plate can be adjusted by controlling the voltage. The forming temperature is set to 800°C, 900°C and 1000°C. The ultrasonic vibration frequency is set to 20khz, and the vibration amplitude (λ) is set to 0-16 μm.

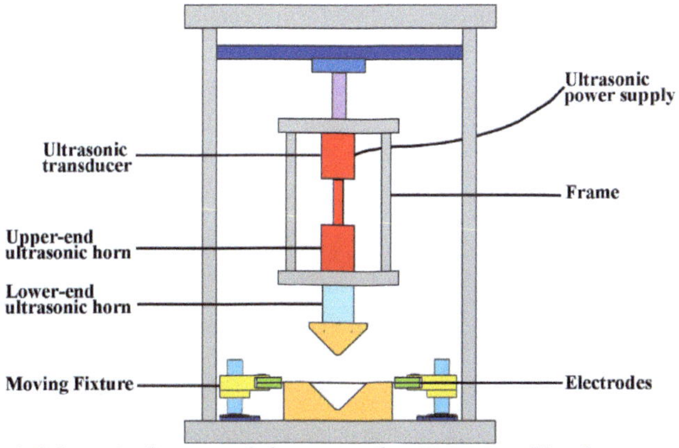

Fig. 1. Schematic diagram of ultrasonic vibration-assisted bending equipment.

Table 1. Chemical composition of Ti6Al4V alloy (wt.%).

Element	Al	V	Fe	Cu	Ni	Ti
wt.%	6.36	2.57	0.18	0.02	0.01	Bal.

The springback angle is shown in Fig. 2, and the change of the forming zone angle $\Delta\alpha$ is mainly studied. The angle to be measured after forming is δ, and α is the tool design angle. The following equation can be obtained:

$$\Delta\alpha = \frac{(\delta - \alpha)}{2} \tag{1}$$

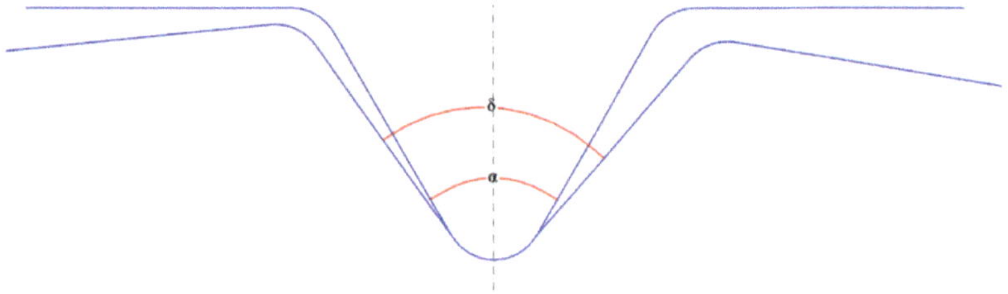

Fig. 2. Definition diagram of springback angle in the bending test.

Numerical modeling

In this study, ABAQUS software and explicit solver were used to simulate the ultrasonic vibration assisted bending process. The geometric model is shown in Fig. 3. Among them, the upper mold, the lower mold and the electrode are set as discrete rigid bodies, and the R3D4 element is used to divide the grid. The Ti6Al4V alloy plate is set as a deformable shell, and the S4RT element is used to divide the grid. The forming process is carried out by the display dynamics module, and the contact relationship of each component in the model is set by universal contact. The normal contact property is described by penalty function, and the friction coefficient is set to 0.15. The tangential contact property is defined as hard contact.

Fig. 3. Geometric model to simulate the ultrasonic vibration assisted bending test.

Results and discussion

Effect on springback angle

The relationship between springback angle, forming temperature and vibration amplitude is shown in Fig. 4. On the one hand, increasing the temperature leads to a significant reduction in the springback angle of the plate. When the temperature is 800°C, 900°C and 1000°C, the springback angles are 9.85°, 7.90° and 5.82° respectively without ultrasonic vibration. This is because the high temperature promotes the α-β phase transformation of Ti6Al4V alloy, and β phase with better plasticity improves the formability of Ti6Al4V. At the same time, the energy provided by the high temperature promotes the dislocation motion and promotes the plastic deformation of the plate.

On the other hand, at the same forming temperature, increasing the amplitude can effectively reduce the springback angle of plate. At 800°C, 900°C and 1000°C, the springback angle at the amplitude of 16 μm is respectively reduced by 0.80°, 1.04° and 1.15° compared with the amplitude of 0 μm. After the deformation, the ultrasonic vibration during the pressure holding process effectively eliminates the residual stress of the plate and reduces the springback of the plate.

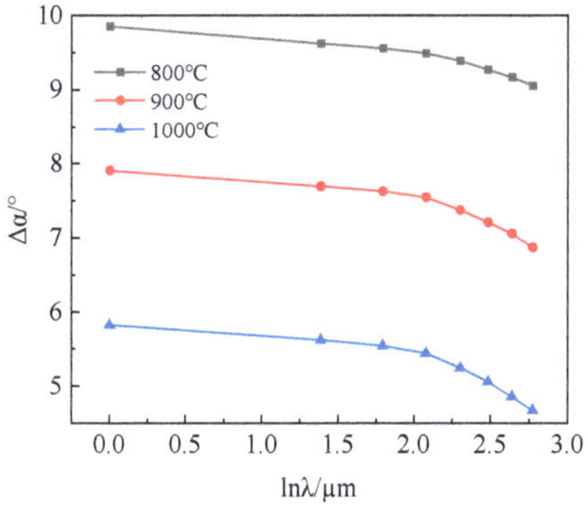

Fig. 4. Springback angle under different forming temperatures and vibration amplitudes.

Assessment of ultrasonic vibration assisted bending by FE simulation

Fig. 5 shows the stress distribution cloud diagram of Ti6Al4V alloy plate during bending under different forming temperatures. The central area of the plate and the outer area of the bending area are the areas with the largest deformation stress. Moreover, with the increase of temperature, the deformation stress decreases significantly, which leads to the reduction of springback of the plate.

Fig. 6 shows the stress cloud diagram of the plate under 800°C and different vibration amplitudes. It can be seen that as the amplitude increases, the stress in the deformation area gradually decreases and the stress distribution is more uniform. Therefore, the decrease of stress and the uniform distribution of stress caused by the increase of amplitude are the important reasons for the reduction of springback.

Materials Research Proceedings 44 (2024) 297-304 https://doi.org/10.21741/9781644903254-32

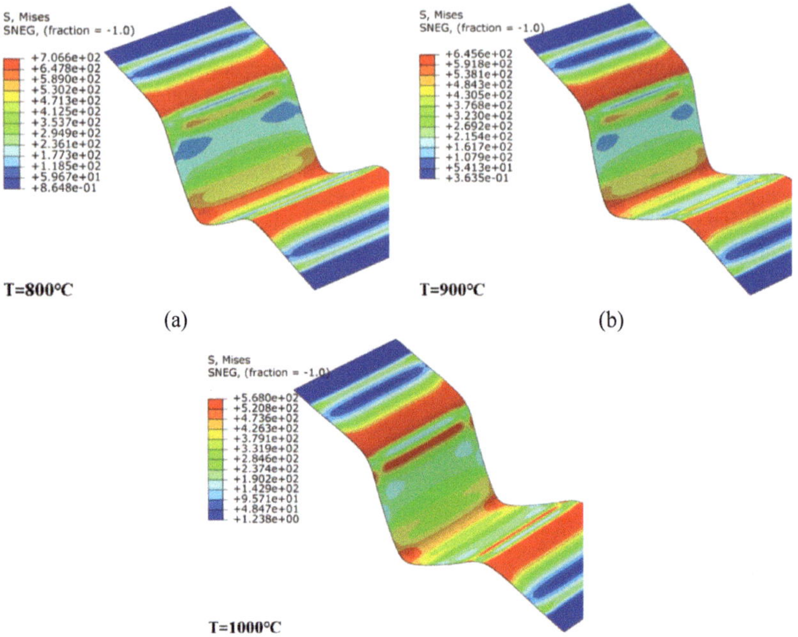

(c)

Fig. 5. Stress distribution cloud diagram under different forming temperatures: (a) 800℃, (b) 900℃, (c) 1000℃.

A path along the length direction is established at the midpoint of the plate to extract the stress distribution, and the equivalent stress distribution of the Ti6Al4V alloy plate along the length direction under different amplitudes is obtained. The results are shown in Fig. 7. The maximum stress in the middle of the plate reaches 635 MPa without ultrasonic vibration, and the maximum stress in the middle of the plate decreases to 554 MPa after ultrasonic vibration with amplitude of 10 μm is applied. When the applied ultrasonic vibration amplitude is increased to 16 μm, the maximum stress in the middle of the plate decreases to 488 MPa, and the stress distribution is more uniform.

Based on the deformed plate model, the springback angle of the plate after unloading can be derived. The results are shown in Table 2. The predicted springback angle is basically consistent with the experimental springback angle after comparison. The maximum relative error is 5.48%, indicating that the finite element simulation method is reasonable.

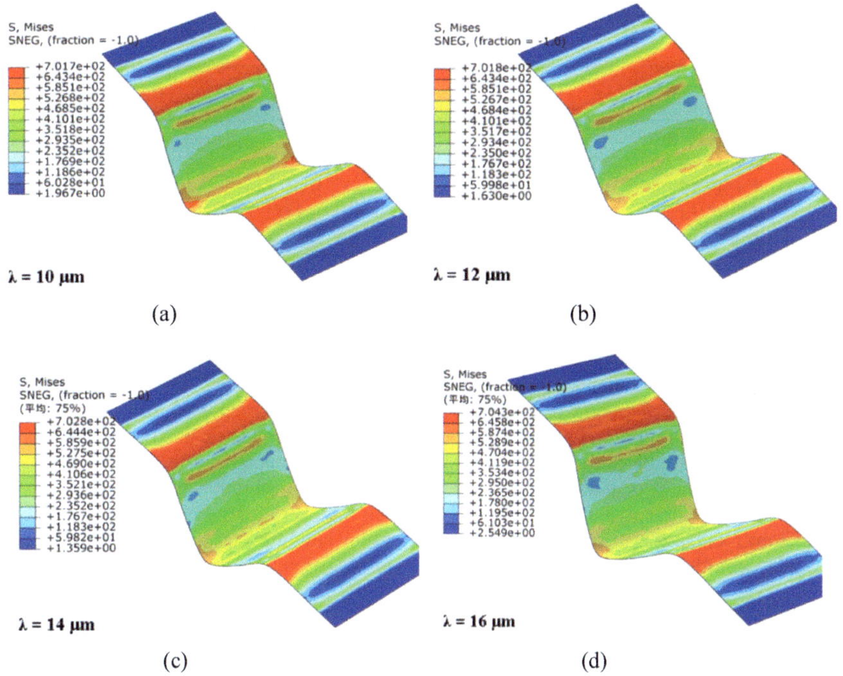

Fig. 6. Stress distribution cloud diagram of the plate under 800°C and different vibration amplitudes: (a) λ=10 μm, (b) λ=12 μm, (C) λ=14 μm, (d) λ=16 μm.

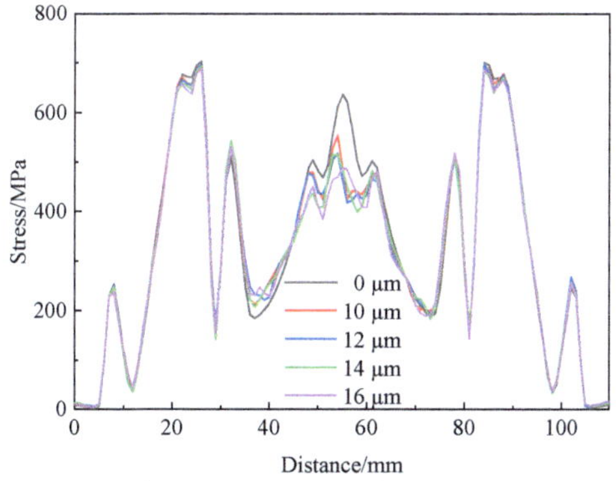

Fig. 7. Stress along the length direction under different vibration amplitudes at 800°C.

Table 2. Comparison of measured and predicted springback angle under different deformation parameters.

Deformation parameter	Measured springback angle	Predicted springback angle	Relative error
800°C/0 μm	9.85	10.39	5.48%
800°C/10 μm	9.39	9.80	4.32%
800°C/12 μm	9.27	9.01	-2.80%
800°C/14 μm	9.17	8.95	-2.37%
800°C/16 μm	9.08	8.86	-2.45%

Conclusion

In this study, the ultrasonic assisted bending test of Ti6Al4V alloy plate was carried out. The effects of temperature and ultrasonic amplitude on the springback of the plate after deformation were analyzed by finite element simulation. The conclusions are summarized as follows:

1. The springback of titanium alloy plate is significantly reduced due to the effect of ultrasonic vibration. As the amplitude increases, the springback angle further decreases.
2. Increasing the forming temperature can also reduce the springback. Increasing the amplitude can better reduce the springback of the plate at higher temperature.
3. The finite element simulation can well reflect the ultrasonic vibration assisted bending process. Increasing the temperature leads to a decrease in the overall deformation stress of the plate, thereby reducing springback. Increasing the vibration amplitude leads to a decrease in the stress in the central deformation area and a uniform stress distribution, thereby reducing springback.

References

[1] P.A. Grün, E.H. Uheida, L. Lachmann, D. Dimitrov, G.A. Oosthuizen, Formability of titanium alloy sheets by friction stir incremental forming, Int. J. Adv. Manuf. Technol. 99 (2018) 1993-2003. https://doi.org/10.1007/s00170-018-2541-5

[2] X.M. Yang, L.M. Dang, Y.Q. Wang, J. Zhou, B.Y. Wang, Springback prediction of TC4 titanium alloy V-bending under hot stamping condition, J. Cent. South Univ. 27 (2020) 2578-2591. https://doi.org/10.1007/s11771-020-4483-y

[3] G.D. Shao, H.W. Li, X. Zhang, M. Zhan, Z.Y. Xiang, Characteristics and mechanism in ultrasonic vibration-assisted deformation of Ni-based superalloy thin-walled sheet by quasi-in-situ EBSD, J. Alloys Compd. 908 (2022) 164591. https://doi.org/10.1016/j.jallcom.2022.164591

[4] S. Kumar, C.S. Wu, G.K. Padhy, W. Ding, Application of ultrasonic vibrations in welding and metal processing: a status review, J. Manuf. Process. 26 (2017) 295-322. https://doi.org/10.1016/j.jmapro.2017.02.027

[5] G. Shao, H. Li, M. Zhan, A review on ultrasonic-assisted forming: mechanism, model, and process, Chin. J. Mech. Eng. 34(1) (2021) 99. https://doi.org/10.1186/s10033-021-00612-0

[6] E. Malekipour, E. Sharifi, N.S. Majd, H. Heidary, Observation on the behavior of ultrasonic micro-hammer and its effects on the deep drawing process: numerical simulation and experimental study, Ultrasonics. 119 (2022) 106566. https://doi.org/10.1016/j.ultras.2021.106566

[7] T. Liu, J. Lin, Y. Guan, Z. Xie, L. Zhu, J. Zhai, Effects of ultrasonic vibration on the compression of pure titanium, Ultrasonics. 89 (2018) 26-33. https://doi.org/10.1016/j.ultras.2018.04.006

[8] J. Kang, X. Liu, M. Xu, Plastic deformation of pure copper in ultrasonic assisted micro-tensile test, Mater. Sci. Eng.: A. 785 (2020) 139364. https://doi.org/10.1016/j.msea.2020.139364

[9] J. Sun, P.Y. Li, S. Zhang, Y.S. Chen, H. Lu, G.Q. Chen, D. Shao, Simulation and experimental study of ultrasonic vibration-assisted milling of GH4169 high-temperature alloy, Alex. Eng. J. 73 (2023) 403-413. https://doi.org/10.1016/j.aej.2023.04.065

[10] X.H. Li, Z.H. Hu, Y. Tu, S.C. Wu, H. Zou, X.G. Chen, C.J. Qin, Y. Wang, Simulation and experimental study of ultrasonic-assisted shear thickening polishing of cemented carbide, J. Manuf. Process. 102 (2023) 106-118. https://doi.org/10.1016/j.jmapro.2023.07.042

Materials Research Proceedings 44 (2024) 305-319 https://doi.org/10.21741/9781644903254-33

The effect of web tension on calendering wrinkles and corrugations of coated foils

FU Zejun[1,a*], XU Zhutian[1,b] and PENG Linfa[1,c]

[1]Shanghai Jiao Tong University, No.800 Dongchuan Road, Minhang District, Shanghai, China

[a]fuzejun@sjtu.edu.cn, [b]zhutianxu@sjtu.edu.cn, [c]penglinfa@sjtu.edu.cn

Keywords: Lithium-Ion Batteries, Calendering Process, Wrinkles, Corrugations, Web Tension, Analytical Modelling

Abstract. Lithium-ion batteries have been widely used in energy storage for a range of applications from portable electronics, electric vehicles to power grids due to their high energy density, high power density and long cycle life. Calendering is a critical process in the manufacturing of lithium-ion batteries electrodes, which has a significant influence on volumetric energy density, long-term cycle stability, and electrochemical performance of lithium-ion batteries, etc. Wrinkles in the uncoated area and corrugations in the coated area of the electrode during the calendering process are serious problems in the industrial production of lithium-ion batteries, as shown in Fig. 1. Adjusting the web tensions has been a critical solution to mitigate these defects. However, large web tensions often lead to foil tearing, severe wrinkles, and high scrap rates, due to a lack of understanding of the wrinkling and corrugating behavior under different tension conditions. To cope with that, this stuedy aims to investigate the mechanism of wrinkling and the effect of web tension on the wrinkling of electrodes during calendering. Calendering experiments are carried out for lithium battery cathodes under different front and back tensions and the three-dimensional topographies of calendered electrodes are measured by a laser profiler. The corrugations of electrodes can be improved by the difference between the front and back tension, instead of the magnitude of the tension, and the tortuosity is reduced by 34.2% at a tension difference of 20 N, compared to that of 5 N. An increase of the tension difference leads to a linear increase of shear displacement δ, causing severe wrinkles in the uncoated area of electrodes. An analytical prediction model for the wrinkles during calendering is established based on shearing of the rectangular membrane and an electrode quality evaluation method for balancing wrinkles and corrugations is proposed. The determination of optimal web tension settings, with a tension difference of 17 N, is achieved to obtain the best quality calendered electrodes, which is in agreement with the experimental results. The method presented in this paper is helpful for the improvement of the production quality and efficiency of lithium battery electrodes.

Introduction

Lithium-ion batteries have been widely used in energy storage for a range of applications from portable electronics, electric vehicles to power grids due to their high energy density, high power density and long cycle life. The growing demand for high-performance and low-cost lithium-ion batteries results in challenges regarding refined understanding of engineering fundamentals and ingenuity in the electrode manufacturing process [1, 2]. The manufacturing process of lithium-ion batteries consists of slurry mixing, coating, drying, calendering, slitting, vacuum drying, and subsequent cell assembly and battery electrochemistry activation processes [3-5]. Calendering is a critical step in the fabrication of electrodes, which employs a high rolling load to compress the electrodes to a specified thickness. Calendering not only increases the volumetric energy density, but also enhances the electrical and thermal conductivity, long-term cycle stability, and electrochemical performance of the lithium-ion batteries [6-8].

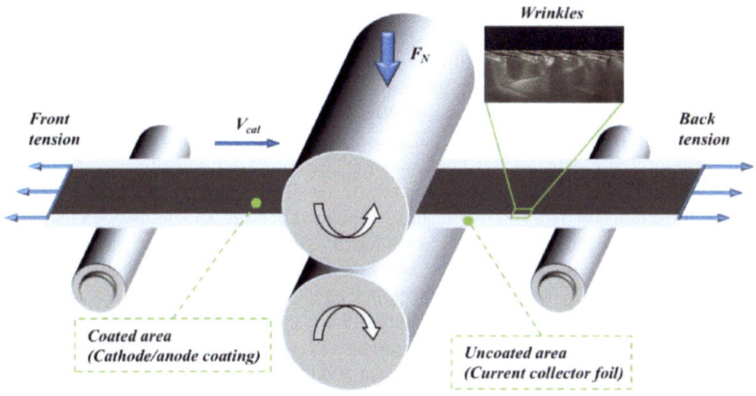

Fig. 1. Schematic of the calendering process including the appearance of electrode wrinkles.

The electrode is composed of a metal foil (aluminum for cathodes and copper for anodes) double-sided coated with a particulate composite (active ceramic powders for cathodes and graphite powders for anodes). Taking the cathode as an example, the thicknesses of the aluminum foil and single-side active coating are respectively 10-20 μm and 65-75 μm for commercial lithium-ion batteries, according to Zhu [9]. In the calendering process shown in Fig. 1, the electrodes are compacted to a required porosity by applying a line loads of more than 1000 N/mm between two rolls with a diameter of up to 1000 mm [10]. A high compaction pressure and thickness reduction rate are required to achieve a high volumetric energy density [11], which results in significant elongations of the electrodes. The elongation of the composite coating leads to evident corrugations in the coated area of the electrodes after calendaring. To eliminate those corrugations and maintain the flatness, web tension is applied to the front and back ends of the electrodes during the calendering process. In the existing processes, high web tension is used to reduce corrugations in the coated area, which at the same time leads to severe wrinkles in the uncoated area of the electrodes, as depicted in Figure 1. Those corrugations and wrinkles adversely affect the subsequent slitting, stacking, welding, electrolyte filling processes and cell performance [12, 13]. Besides, the increase of tension leads to tearing of ultra-thin foils more easily. Frequent failures thus occur on the production lines with the rolling speed up to 100 m/min, which significantly reduces productivity and increases costs.

A few studies have examined the wrinkles and corrugations of electrodes during calendering process. Guenther et al. divided calendering induced defects into three categories: geometrical, structural and mechanical ones, and evaluated the impact of them on the subsequent cell manufacturing processes. They found that corrugations in the coated area affect electrolyte filling, gripping and handling processes, while wrinkles in the uncoated area impair contacting and welding[12]. Mayer et al. investigated the corrugations and mechanical behavior of electrodes after calendering. The compaction rate, types of coating materials, and uncoated current collector foil were revealed to affect corrugations, while the web tension is identified to have little effect[13, 14]. This is attributed to the fact that, only cases with the same front and back tensions were involved in their study. Mayr et al. developed an in-line sensor-based process control method of the calendering process by applying in-line data acquisition for electrode thickness and surface topography, leading to a promising reduction of scrap and manufacturing costs [15]. Heating the rolls during calendering process has been investigated, the reduced line load and thus mechanical

 Materials Research Forum LLC
https://doi.org/10.21741/9781644903254-33

stresses applied throughout the process are thought to be able to reduce the wrinkles of electrodes, especially at high line loads and high compaction rates [16-19].

Analytical and numerical simulation models of the calendering process of the calendering process are used as additions to experimental analysis and provide insights into the actual deformation of electrodes between two rolls [20]. Zhang et al. conducted tensile and compression experiments to obtain the constitutive properties of porous electrodes and identified the mechanism of failure behavior of electrodes by implementing the proposed constitutive models into finite element analysis.[21, 22]. Gupta et al. characterized the stress-strain relationship for a dry cathode active layer in lithium-ion batteries by means of U-shaped bending of single-side coated aluminum foils. A finite element model, with an elastic-ideally plastic material for the aluminum and an elastic material for the cathode active layer, was established to validate the evaluation of properties of cathode active layer [23]. Meyer et al. developed an exponential model between the applied line load and the achieved coating density by introducing compaction resistance. The relationship between mass loading, active materials and compaction resistance was established [24, 25]. Zhu et al. obtained the mechanical properties of anode and cathode coatings by compression tests in a specially designed aluminum mold. The sequence of deformation and failure of various components of lithium-ion batteries under mechanical loadings were simulated by employing the Drucker-Prager/Cap plasticity model in finite element model [9, 26]. Sahraei et al. obtained the stress-volumetric strain curve from the uniaxial compression experiment of the cell and modeled the cell by finite element method using a crushable foam material and predicted the load-displacement response[27, 28]. Besides analytical and finite element modeling, the numerical simulation of the calendering process by means of the discrete element method (DEM) has given deeper insight into the compaction of electrodes[29-33]. However, the scale of simulation models for DEM is usually only a few hundred micrometers, whereas the battery modeling involves six orders of magnitude levels in length[27], which limits its applicability in the study of electrode manufacturing processes.

According to the above reviews, wrinkles and corrugations of the electrodes during calendering process are critical problems during the manufacturing of lithium-ion batteries, which attract both industrial and academic attentions. However, a detailed understanding of those defects and process design methods are still lacking. In practical production, the mitigation of wrinkles and corrugations relies on experiences, resulting in high labor intensity, long commissioning time, waste of electrode materials and high scrap rates. In this paper, Calendering experiments of electrodes for lithium-ion batteries under different tension settings are first carried out. Then, a laser sensor-based measurement and evaluation method of electrode wrinkles and corrugations is developed. The mechanism of electrode wrinkles and corrugations during calendering process and the effect of tension on them are explained. Finally, an analytical prediction model for the wrinkles and an evaluation method for the quality of the calendered electrodes is established. The selection of tension settings is accomplished using this model and an agreement with the experimental results is achieved. The method presented in this paper is helpful for the improvement of the production quality and efficiency of the calendering process for lithium-ion batteries.

Experiments

Materials. The uncalendered $LiNi_{0.5}Co_{0.2}Mn_{0.3}O_2$ (NCM523) cathode is manufactured by MTI KJ GROUP, containing 95.5% active material, 2.5% carbon black and 2% binder, with a total thickness of about 160 μm. The active material is coated on both sides of a 16 μm thick aluminum foil at a surface mass loading of 29.3 mg/cm^2. The width of the coated active material is 200 mm and the width of the uncoated current collector foil is 40 mm.

Experimental setup. In the calendering process, the electrodes are continuously compacted with a two-roller calender MSK-2300A-E as shown in Fig. 2(a). The diameter and the width of calender rolls are 200 mm and 300 mm respectively. The electrodes after coating and drying are calendered at ambient temperature and a rolling speed of 2 m/min. The applied line load of the calender is 1250 N/mm and the electrodes are compacted from an initial density of 2.03 g/cm^3 to a compaction one of 2.87 g/cm^3.

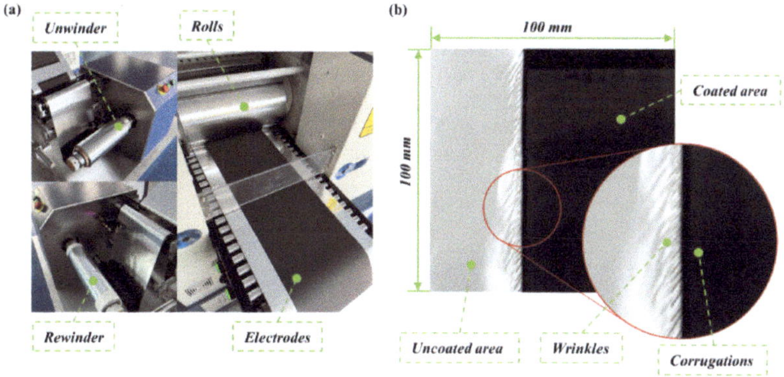

Fig. 2. Experimental procedures: (a) electrodes calendering systems; (b) images of electrodes after calendering.

As mentioned above, this study focuses on the effect of the front and back tensions on the electrode wrinkles and corrugations. An experiment with no tension applied is employed for comparison. After that, the complete experiments are designed to investigate the effect of front and back tensions. The front tension is designed from 5 to 20 N with an interval of 5 N, and the difference between the front and back tensions is designed from 0 to 20 N with an interval of 5 N. In addition, the uncoated current collector is thought to hinder electrode deformation according to Mayer [13], two different collector width, i.e. 0 and 40 mm, are discussed to investigate its effect on wrinkles and corrugations. The values of front and back tensions could be adjusted by the unwinder and rewinder presented in Fig. 2(a). Three electrodes are calendered for each experimental condition. As depicted in Fig. 2(b), all electrodes are cut into sheets with a length of 100 mm after calendering to enable the measurement of electrode wrinkles and corrugations.

Three-dimensional scanning of the calendered electrodes is performed using a Keyence LJ-X8020 laser profiler mounted on a three-axis motion stage. The laser sensor has a measuring range of 8 mm along the width and 4 mm along the height, and the accuracy of the sensor is 0.3 μm. The laser sensor acquired the height profiles of the electrodes along the x-direction (which is perpendicular to the running direction) at a frequency of 1000 Hz and the speed of the motion stage along the y-direction (which is in the running direction) is 10 mm/s. The three-dimensional topographies of the electrodes after calendering are obtained by stacking the acquired height profiles with the interval to 0.01 mm.

Results and discussion

Measurement of wrinkles and corrugations. Wrinkles in the uncoated area and corrugations in the coated area are geometric defects that coexist on calendered electrodes. The quantification of wrinkles and corrugations need to be conducted for the sake of evaluating the degree of geometrical defects of the calendered electrodes. As shown in Fig. 3, based on the three-dimensional scanning results of the calendered electrodes, two height profiles at a distance of ±1mm from the junction of the coated and uncoated areas are extracted along the rolling direction, as illustrated by the dash lines l_1 for the coated area and l_2 for the uncoated area, respectively.

The wavelength of corrugations in the coated areas is found in the centimeter range. The following data processing procedure is conducted to capture the major corrugation features after calendering: (1) smoothing: Savitzky-Golay filtering with a smoothing window of 300 data points is carried out to remove the minor fluctuations; (2) elimination of the overall electrode inclination after cutting: a straight line connecting the start and end points, which represents the overall inclination of the electrodes, is subtracted from the smoothed data. After those, the cross-section curve of corrugations in the coated area can be obtained as shown in Fig. 3(a). In order to avoid excessive random errors caused by the direct use of the maximum value of the height of corrugations, the degree of tortuosity ΔC, is proposed as an evaluation index of the corrugations in the coated area. ΔC is the difference between the total length C of the coated area in the calendered electrodes and the length C_0 scanned by the laser sensor along the rolling direction, and the value of C_0 is 100mm.

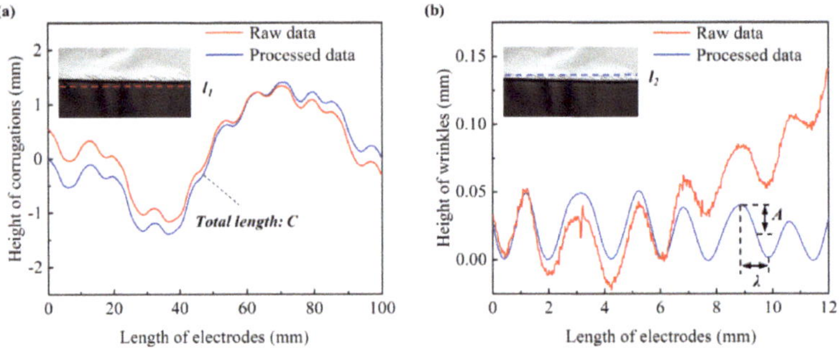

Fig. 3. Evaluation of calendered electrodes: (a) corrugations in the coated area; (b) wrinkles in the uncoated area of electrodes.

In the uncoated area, short and dense wrinkles in the collector foil are the main geometrical defects. As illustrated in Fig. 3(b), the wavelength and amplitude of wrinkles in this region are in the millimeter range. Both the overall inclination of the electrode and the corrugations in the coated area affect the measurement results. To capture the wrinkles data, the overall inclination of the electrode and the centimeter-range corrugations of the coated area need to be eliminated from the raw data. Therefore, the following processing procedure is conducted for the uncoated area: (1) smoothing: Savitzky-Golay filtering with a smoothing window of 100 data points is performed to smooth the data; (2) elimination of the overall inclination of the electrode and corrugations: a spline curve fitted by the valley points of each wrinkle, instead of a straight line, is subtracted from the smoothed data. After that, the cross-section curve of wrinkles can be obtained and the ratio R of amplitude A to half wavelength λ of the wrinkles is employed as an evaluation index aimed at evaluating the severity of wrinkles in the uncoated area.

Mechanisms of wrinkles and corrugations. To investigate the cause of wrinkles and corrugations in the calendering process of the electrodes, the results of experiments #1, #2, #8, and #11 are compared. The height profiles of the coated area and uncoated areas of the electrodes along the rolling direction under four different conditions, i.e. without uncoated current collector, no front and back tensions, equal front and back tensions, and greater back tension, are obtained.

As illustrated in Fig. 4(a), in the absence of the uncoated collector, the coated area of electrodes elongated after calendering and the coated area is not corrugated. In comparision with the case #2, where the uncoated collector is present, significant corrugations is observed in the coated area of electrodes. That is because the foil in the uncoated area is not in contact with the rolls and thus not elongated after calendering process. As a result, the foil hindered the elongation of the coated area, which results in out-of-plane bending and deformation of the coated area. Due to the elongation mismatch between the coated and uncoated areas, corrugations in the coated area occurred. However, if the uncoated area is removed, the elongation of the coated area is not restricted, leading to the elimination of the corrugations, as shown in Fig. 4(c).

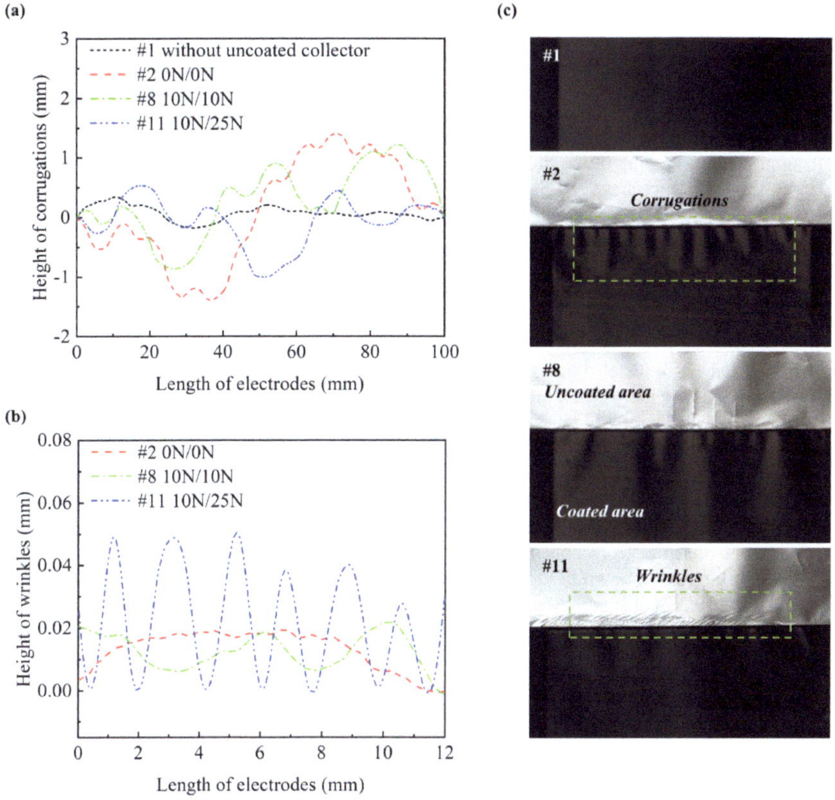

Fig. 4. Height profiles of electrodes under different conditions: (a) corrugations in coated area; (b) wrinkles in uncoated area; (c) the photographs of wrinkles and corugations after calendering.

In the presence of an uncoated collector, the front and back tensions have significant effect on the wrinkles and corrugaitons of the electrodes. As shown in Fig. 4(a), when no front and back tensions applied, evident out-of-plane corrugations can be observed in the coated area. Meanwhile, the millimeter-range wrinkles is not significant. By increasing both the front and the back tensions to 10 N, similar corrugations and little wrinkles is observed in the coated and uncoated areas, respectively. Those observations are consistent with the results reported in the literature[34].

On the other hand, when keeping the front tension at 10 N while increasing the back tension to 25 N, the geometrical defect patterns of the electrodes changed significantly. The tortuosity of the corrugations in the coated area, which can be evaluated by the value of ΔC as described above, is reduced from 0.36 to 0.27 mm, compared to the case that the front and back tensions are equal. Meanwhile, as shown in Fig. 4(b) and (c), evident wrinkles in the uncoated area also appeared. Short and dense wrinkles appeared in the junction area near the uncoated side at an inclination angle of approximately 30 degrees to the direction of rolling.

In summary, increasing the front and back tensions simultaneously has little improvement effect and only the difference between the front and back tensions is the main parameter affecting the corrugations in the coated area and the wrinkles in the uncoated area. That is because the coated area of the electrode is elongated in the rolling direction after passing through the roll gap, and the back tension acts on the uncoated area that is not elongated, resulting in the collector foil being stretched backward. Thus, the difference in displacement between the two areas decreased and the hindering effect of the uncoated area on the elongation of the coated area is alleviated, thereby the flatness of the coated area is improved. At the same time, the uncoated area is stretched backwards to form evident wrinkles at the coating edge. In the cases where no tension is applied, the front and back tensions are increased simultaneously or the front tension is greater than the back tension, there is no backward stretching effect on the uncoated collector, therefore there is almost no effect on the improvement of the geometric defects of the electrodes.

Effect of web tension. To elaborate the effect of tension on the wrinkles and corrugations of the electrodes, the evaluation index ΔC for the coated area and the ratio of amplitude to half wavelength R, are obtained for all experiments. The experimental results showed that the tension difference between the front and back tensions affected the geometrical defects, instead of the magnitude of them. As shown in Fig. 5(a), when the tension difference remained 0 N and the front tension is 5 N, 10 N, 15 N, and 20 N, respectively, the increase of the front tension had little effect on ΔC. On the other hand, by controling the front tension a constant of 5 N, the tortuosity of corrugations decreased with the increase of the tension difference. As depicted in Fig. 5(b), with the tension difference increased from 5 to 20 N, the value of ΔC decreased from 0.38 to 0.25 mm, for a reduction of 34.2%.

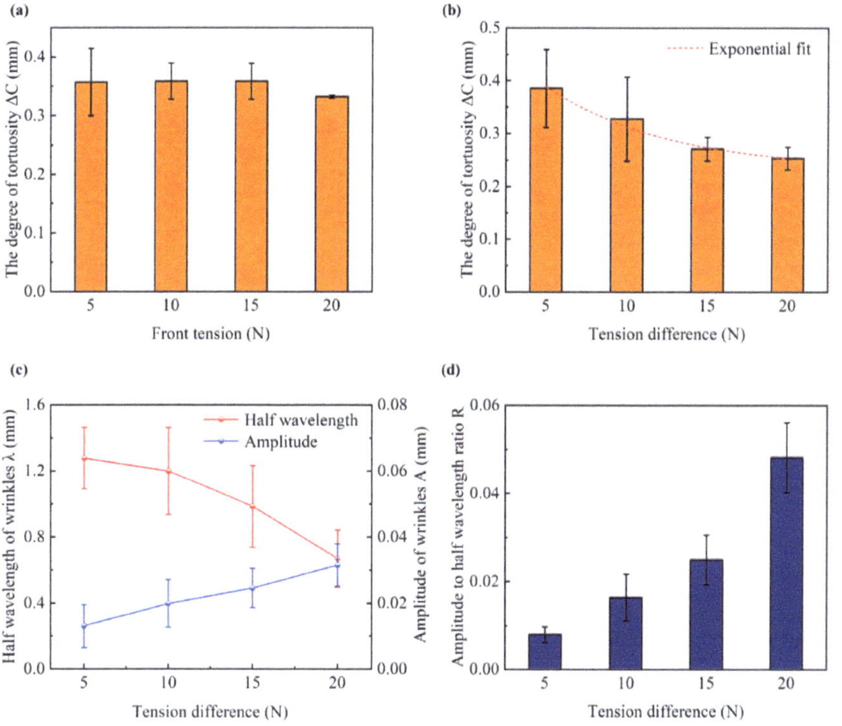

Fig. 5. Effect of the web tension, on corrugations: (a) front tension equals back tension; (b) front tension is 10 N, and the tension difference increased; on the half wavelength and amplitude (c), and the ratio of amplitude to half wavelength (d) of wrinkles at a front tension of 10 N.

As illustrated in Fig. 5(c), the height profiles of wrinkles in the uncoated area of the electrodes are obtained with different tension differences while making the front tension constant at 10 N. As the tension difference increased, the half wavelength of the wrinkles decreased while the amplitude increased. That leads to the difficulty of elastic recovery of the wrinkles and permanent deformation of the collector foil after the cutting process of the electrodes, which in turn affects subsequent processes of stacking, packaging and electrolyte filling in the production of lithium-ion batteries. As shown in Fig. 5(d), the effect of the tension difference on the amplitude-half wavelength ratio R of wrinkles is obtained. The increase in the difference between the front and back tensions led to a significant increase in the ratio R and more severe the wrinkles in the uncoated collector. That is due to an increased displacement difference between the coated and uncoated areas along the rolling direction induced by a larger tension difference.

As mentioned above, the corrugations and wrinkles of the electrodes after calendering are due to the mismatch of displacement between the coated and uncoated areas. To measure the difference in shear displacement, several grid lines are scribed on the surface of the uncalendered electrodes using a scribing needle and the 3D topographies of the electrodes are recorded before and after calenderging using a laser sensor described in Experimental setup. As shown in Fig. 6, the difference in shear displacement δ between the coated and uncoated areas is obtained by comparing

Materials Research Proceedings 44 (2024) 305-319 https://doi.org/10.21741/9781644903254-33

the deformation of the grid lines along the rolling direction before and after calendering. The shear displacement difference δ increased with the increasing tension difference T_d, which can be described by a linear relationship:

$$\delta = 0.01 + 0.00508T_d \quad (1)$$

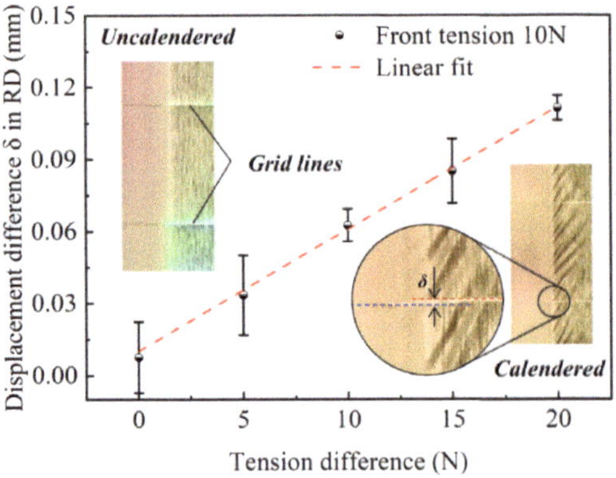

Fig. 6. Relationship between shear displacement and tension difference.

Modeling and optimization

Based on the results obtained above, the difference between the front and back tensions is the key parameter affecting the corrugations and wrinkles of the electrodes. The preferred tension settings need to be selected to balance the corrugations in the coated area and wrinkles in the uncoated area, to obtain a better quality of calendered electrodes. Thus, an analytical model for predicting wrinkles is developed to determine the optimal web tension parameters.

Materials Research Forum LLC
https://doi.org/10.21741/9781644903254-33

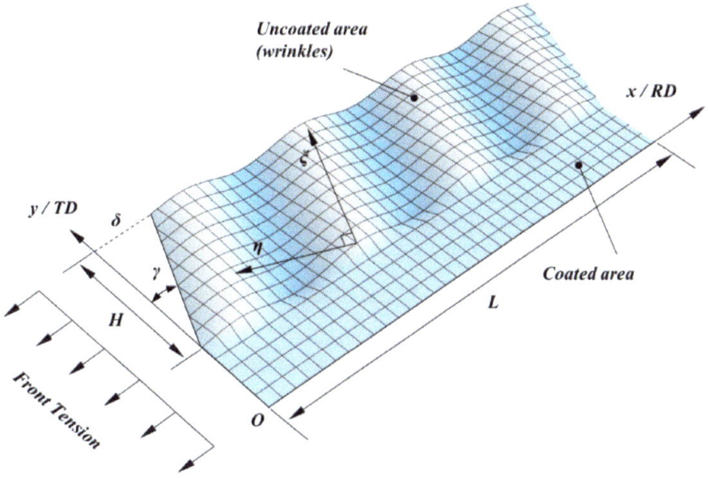

Fig. 7. Analytical model of shear wrinkling in the uncoated area of electrodes.

The following assumptions are made for the developed model: Firstly, linear elasticity, small deformations and homogeneity of the material are assumed. Secondly, the collector foil is assumed to be subjected to a uniform transverse in-plane displacement along the coating edge. Finally, the wrinkles are assumed to be parallel, evenly spaced, of uniform length, and inclined at 35° to the rolling direction, on the basis of experimental observations of wrinkles after calendering. The description of analytical model based on the bifurcation theory of thin-plate is shown in Fig. 7, the uncoated area of the electrode can be treated as a rectangular membrane with a width of H, the difference in displacement of coated and uncoated areas along the rolling direction is described by δ, and the shear angle γ, which can be calculated by $\gamma = \pi/180 \cdot arctan(\delta/H)$.

Referring to the analytical model for predicting wrinkles in rectangular membrane proposed by Wang [35], the initially flat collector foil deformed into a doubly curved shape, a model describing the wrinkled surface can be established using the coordinate systems ξ and η. ξ is parallel to the wrinkle direction and η is perpendicular to it.

The length of the model is kH, $k = 1/sin(35\pi/180)$ and the width is λ, the out-of-plane displacement function is configured as:

$$w = A \sin\frac{\pi\eta}{\lambda} \sin\frac{\pi\xi}{kH} \tag{2}$$

The uniaxial stress in the wrinkled region can be expressed as:

$$\sigma_\xi = E\gamma + \frac{\pi^2 E h^2}{12\lambda^2(1-v^2)} \tag{3}$$

where E is the Young's modulus with a value of 70 GPa, v is the Poisson's ratio with a value of 0.33 [28] and h refers to the thickness of the collector foil.

The critical wrinkling stress σ_η acting perpendicular to the wrinkles is assumed to equal the stress required to buckle a simply-supported, infinitely wide plate of length λ:

$$\sigma_\eta = -\frac{\pi^2 E h^2}{12\lambda^2(1-\nu^2)} \tag{4}$$

The force equilibrium relationship along the out-of-plane direction of the membrane can be expressed as:

$$\sigma_\xi k_\xi + \sigma_\eta k_\eta = 0 \tag{5}$$

where k_η and k_ξ are two curvatures of the surface and they are determined by Eq. (2). According to Eq. (2)-Eq. (5), the half wavelength λ of wrinkles of a rectangular collector foil subject to transverse in-plane shear displacement δ, as shown in Fig. 7, can be predicted by:

$$\lambda^2 = \frac{(kH\pi h)^2}{24\gamma(1-\nu^2)}[\sqrt{\frac{1}{(kH)^4} + \frac{48\gamma(1-\nu^2)}{(kH\pi h)^2}} - \frac{1}{(kH)^2}] \tag{6}$$

The total strain in the η direction is $-\gamma$, while the material strain in this direction is $-\nu\gamma$. The difference between the two is the "geometric strain":

$$\varepsilon_G = -\frac{\pi^2 A^2}{4\lambda^2} = -\gamma(1-\nu) \tag{7}$$

Thus, the amplitude of wrinkles can be solved from Eq. (6) and Eq. (7):

$$A = \frac{2\lambda\sqrt{\gamma(1-\nu)}}{\pi} \tag{8}$$

It should be noted that the accuracy of the amplitude prediction is not enough due to several simplified assumptions are used in the process of deriving the analytical model, such as linear elasticity, small deformations, etc. Based on the results of experiments, a modified value A_0 is subtracted, which has a value of 0.03, from the predicted value of amplitude to eliminate deviations resulting from simplified assumptions.

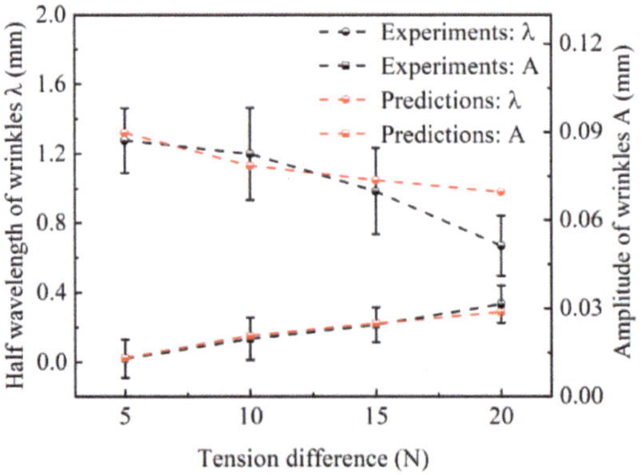

Fig. 8. Prediction of wrinkles: experimental and predicted results under different tension differences.

According to the analytical model for the prediction of wrinkles in the uncoated area of the electrodes, the half wavelength and amplitude of wrinkles are calculated using Eq. (6) and (8) at a front tension of 10 N and a tension difference of 5 N, 10 N, 15 N, and 20 N, as shown in Fig. 8. It should be noted that, the predicted results are in agreement with the experimental results, except at the point where T_d is 20 N. The uncoated area is no longer in the elastic range as assumed and thus the wavelength of wrinkles is overestimated when the great tension difference is too large. However, large tension differences should be avoided in production, which may lead to tearing of the electrodes.

As for corrugations in the coated area, as depicted in Fig. 5(b), the improvement effect of the tortuosity ΔC in the coated area gradually tended to be constant when the tension difference continued to increase, which can be described by an exponential curve:

$$\Delta C = 0.23 + 0.304\exp(-\frac{T_d}{7.62}) \tag{9}$$

where T_d is the difference between front and back tensions.

Both the corrugations in coated area and wrinkles in uncoated area have a detrimental effect on the quality of electrodes[12]. In order to find optimal settings of the web tension, an evaluation index I for the overall quality of calendered electrodes is established:

$$I = \alpha\Delta C + (1-\alpha)R \tag{10}$$

the relationship between ΔC and T_d can be obtained by Eq. (9). R is the ratio of amplitude to half wavelength, the relationship between R and T_d can be obtained by Eq. (1), Eq. (6) and Eq. (8). α is the weight, and its value is set to 0.2, based on expert consultations. The value of I is calculated for a range of tension differences T_d using Eq. (10). As illustrated in Fig. 10(a), the predicted minimum value of I is 0.0735, achieved at a T_d value of 17 N, at which point the best electrode quality is theoretically available.

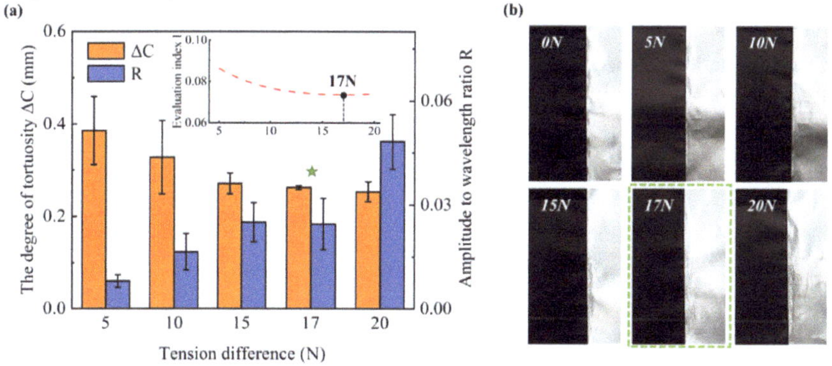

Fig. 9. Validation of the selected tension settings: (a) ΔC and amplitude to half wavelength ratio R of calendered electrodes; (b) photographs recording the appearance of calendered electrodes, under different tension differences.

To validate the previously chosen tension settings, experiments on electrode calendering at a tension difference T_d of 17 N are carried out additionally. As shown in Fig. 9(a), a comparison with the other four groups showed that, the corrugations in the coated area (described by the value of ΔC) are improved to a large extent and continued increase in the value of T_d did not improve it further, while the wrinkles in the uncoated area (described by the value of R) are limited to a relatively low level. From the appearance of the electrode after calendering, shown in Fig. 9(b), the flatness of the electrodes is optimized at T_d value of 17 N as compared to the other four groups, with both wrinkles in the uncoated area and corrugations in the coated area considered, which is consistent with previous predictions.

Conclusion

The effect of web tension is clarified via both experiments and analytical modeling to explore the wrinkles and corrugations of electrodes during calendering process. Calendering experiments of cathode electrodes of lithium-ion batteries under different tension settings are carried out. The measurement and data processing methods of corrugations in the coated area and wrinkles in the uncoated area are also developed. The optimal tension difference T_d is determined by employing the proposed evaluation index of geometrical defects of calendered electrodes. The following conclusive remarks can be drawn:

(1) The tension difference between the front and back tensions is the key parameter to determine the wrinkles and corrugations of calendered electrodes. Corrugations in the coated area can be mitigated while wrinkles in the uncoated area become more evident by increasing the tension difference.

(2) An analytical prediction model for the wrinkles after calendering is established based on shearing of the rectangular membrane and an evaluation method for the quality of the calendered electrode is developed to balancing the effects of wrinkles and corrugations.

(3) The determination of the web tension parameters to improve the quality of calendered electrodes is achived at T_d of 17 N and the predicted results are consistent with the experimental results.

Acknowledgements

This work was supported by National Key R&D Program of China (Grant No. 2022YFE0207500).

References

[1] W.B. Hawley, J. Li, Electrode manufacturing for lithium-ion batteries—Analysis of current and next generation processing, J. Energy Storage 25 (2019) 100862. https://doi.org/10.1016/j.est.2019.100862

[2] J. Li, J. Fleetwood, W.B. Hawley, W. Kays, From Materials to Cell: State-of-the-Art and Prospective Technologies for Lithium-Ion Battery Electrode Processing, Chem. Rev. 122 (2022) 903-56. https://doi.org/10.1021/acs.chemrev.1c00565

[3] J. Xiao, F. Shi, T. Glossmann, C. Burnett, Z. Liu, From laboratory innovations to materials manufacturing for lithium-based batteries, Nat. Energy 8 (2023) 329-339. https://doi.org/10.1038/s41560-023-01221-y

[4] J. Li, Z. Du, R.E. Ruther, S. Jin, L.A. David, K. Hays, M. Wood, N.D. Phillip, Y. Sheng, C. Mao, S. Kalnaus, C. Daniel, D.L. Wood III, Toward Low-Cost, High-Energy Density, and High-Power Density Lithium-Ion Batteries, JOM 69 (2017) 1484-1496. https://doi.org/10.1007/s11837-017-2404-9

[5] Y. Liu, R. Zhang, J. Wang, Y. Wang, Current and future lithium-ion battery manufacturing, iScience 14 (2021) 102332. https://doi.org/10.1016/j.isci.2021.102332

[6] X.K. Lu, S.R. Daemi, A. Bertei, M.D.R. Kok, K.B. O'Regan, L. Rasha, J. Park, G. Hinds, E. Kendrick, D.J.L. Brett, P.R. Shearing, Microstructural Evolution of Battery Electrodes During Calendering, Joule 4 (2020) 2746-2768. https://doi.org/10.1016/j.joule.2020.10.010

[7] H.X. Kang, C. Lim, T.Y. Li, Y. Fu, B. Yan, N. Houston, V. de Andrade, F. de Carlo, L. Zhu, Geometric and Electrochemical Characteristics of LiNi1/3Mn1/3Co1/3O2 Electrode with Different Calendering Conditions, Electrochim. Acta 232 (2017) 431-438. https://doi.org/10.1016/j.electacta.2017.02.151

[8] R. Sim, S. Lee, W.D. Li, A. Manthiram, Influence of Calendering on the Electrochemical Performance of LiNi0.9Mn0.05Al0.05O2 Cathodes in Lithium-Ion Cells, ACS Appl. Mater. Interfaces 13 (2021) 42898-42908. https://doi.org/10.1021/acsami.1c12543

[9] J. Zhu, W. Li, T. Wierzbicki, Y. Xia, J. Harding, Deformation and failure of lithium-ion batteries treated as a discrete layered structure, Int. J. Plast. 121 (2019) 293-311. https://doi.org/10.1016/j.ijplas.2019.06.011

[10] A. Kwade, W. Haselrieder, R. Leithoff, A. Modlinger, F. Dietrich, K. Droeder, Current status and challenges for automotive battery production technologies, Nat. Energy 3 (2018) 290-300. https://doi.org/10.1038/s41560-018-0130-3

[11] D.L. Wood, J. Li, C. Daniel, Prospects for reducing the processing cost of lithium ion batteries, J. Power Sources 275 (2015) 234-242. https://doi.org/10.1016/j.jpowsour.2014.11.019

[12] T. Guenther, D. Schreiner, A. Metkar, C. Meyer, Classification of Calendering-Induced Electrode Defects and Their Influence on Subsequent Processes of Lithium-Ion Battery Production, Energy Technol. 8 (2020). https://doi.org/10.1002/ente.201900026

[13] D. Mayer, A.-K. Wurba, B. Bold, J. Bernecker, A. Smith, J. Fleischer, Investigation of the Mechanical Behavior of Electrodes after Calendering and Its Influence on Singulation and Cell Performance, Processes 9 (2021). https://doi.org/10.3390/pr9112009

[14] D. Mayer, J. Fleischer, Concept for modelling the influence of electrode corrugation after calendering on stacking accuracy in battery cell production, Procedia CIRP 104 (2021) 744-749. https://doi.org/10.1016/j.procir.2021.11.125

[15] A. Mayr, D. Schreiner, B. Stumper, R. Daub, In-line Sensor-based Process Control of the Calendering Process for Lithium-Ion Batteries, Procedia CIRP 107 (2022) 295-301. https://doi.org/10.1016/j.procir.2022.04.048

[16] D. Schreiner, M. Oguntke, T. Günther, G. Reinhart, Modelling of the Calendering Process of NMC-622 Cathodes in Battery Production Analyzing Machine/Material–Process–Structure Correlations, Energy Technol. 7 (2019) 1900840. https://doi.org/10.1002/ente.201900840

[17] D. Schreiner, T. Zünd, F.J. Günter, L. Kraft, B. Stumper, F. Linsenmann, M. Schüßler, R. Wilhelm, A. Jossen, G. Reinhart, H.A. Gasteiger, Comparative Evaluation of LMR-NCM and NCA Cathode Active Materials in Multilayer Lithium-Ion Pouch Cells: Part I. Production, Electrode Characterization, and Formation, J. Electrochem. Soc. 168 (2021) 030507. https://doi.org/10.1149/1945-7111/abe50c

[18] E.N. Primo, M. Chouchane, M. Touzin, P. Vazquez, A.A. Franco, Understanding the calendering processability of Li(Ni0.33Mn0.33Co0.33)O2-based cathodes, J. Power Sources 488 (2021) 229361. https://doi.org/10.1016/j.jpowsour.2020.229361

[19] E.N. Primo, M. Touzin, A.A. Franco, Calendering of Li(Ni0.33Mn0.33Co0.33)O2-Based Cathodes: Analyzing the Link Between Process Parameters and Electrode Properties by Advanced Statistics, Batteries & Supercaps 4 (2021) 834-844. https://doi.org/10.1002/batt.202000324

[20] A. Diener, S. Ivanov, W. Haselrieder, A. Kwade, Evaluation of Deformation Behavior and Fast Elastic Recovery of Lithium-Ion Battery Cathodes via Direct Roll-Gap Detection During Calendering, Energy Technol. 10 (2022). https://doi.org/10.1002/ente.202101033

[21] C. Zhang, S. Santhanagopalan, M.A. Sprague, A.A. Pesaran, Coupled mechanical-electrical-thermal modeling for short-circuit prediction in a lithium-ion cell under mechanical abuse, J. Power Sources 290 (2015) 102-113. https://doi.org/10.1016/j.jpowsour.2015.04.162

[22] C. Zhang, J. Xu, L. Cao, Z. Wu, Constitutive behavior and progressive mechanical failure of

electrodes in lithium-ion batteries, J. Power Sources 357 (2017) 126-137. https://doi.org/10.1016/j.jpowsour.2017.04.103

[23] P. Gupta, İ.B. Üçel, P. Gudmundson, E. Olsson, Characterization of the Constitutive Behavior of a Cathode Active Layer in Lithium-Ion Batteries Using a Bending Test Method, Exp. Mech. 60 (2020) 847-860. https://doi.org/10.1007/s11340-020-00613-5

[24] C. Meyer, H. Bockholt, W. Haselrieder, A. Kwade, Characterization of the calendering process for compaction of electrodes for lithium-ion batteries, J. Mater. Process. Technol. 249 (2017) 172-178. https://doi.org/10.1016/j.jmatprotec.2017.05.031

[25] C. Meyer, M. Kosfeld, W. Haselrieder, A. Kwade, Process modeling of the electrode calendering of lithium-ion batteries regarding variation of cathode active materials and mass loadings, J. Energy Storage 18 (2018) 371-379. https://doi.org/10.1016/j.est.2018.05.018

[26] J. Zhu, W. Li, Y. Xia, E. Sahraei, Testing and Modeling the Mechanical Properties of the Granular Materials of Graphite Anode, J. Electrochem. Soc. 165 (2018) A1160-A1168. https://doi.org/10.1149/2.0141807jes

[27] E. Sahraei, R. Hill, T. Wierzbicki, Calibration and finite element simulation of pouch lithium-ion batteries for mechanical integrity, J. Power Sources 201 (2012) 307-321. https://doi.org/10.1016/j.jpowsour.2011.10.094

[28] E. Sahraei, E. Bosco, B. Dixon, B. Lai, Microscale failure mechanisms leading to internal short circuit in Li-ion batteries under complex loading scenarios, J. Power Sources 319 (2016) 56-65. https://doi.org/10.1016/j.jpowsour.2016.04.005

[29] D. Schreiner, A. Klinger, G. Reinhart, Modeling of the Calendering Process for Lithium-Ion Batteries with DEM Simulation, Procedia CIRP 93 (2020) 149-155. https://doi.org/10.1016/j.procir.2020.05.158

[30] R. Ge, A.M. Boyce, Y. Shui Zhang, P.R. Shearing, D.J. Cumming, R.M. Smith, Discrete element method and electrochemical modelling of lithium ion cathode structures characterised by X-ray computed tomography, Chemical Eng. J. 465 (2023) 142749. https://doi.org/10.1016/j.cej.2023.142749

[31] A. Lundkvist, P.-L. Larsson, E. Olsson, A discrete element analysis of the mechanical behaviour of a lithium-ion battery electrode active layer, Powder Technol. 425 (2023) 118574. https://doi.org/10.1016/j.powtec.2023.118574

[32] C. Sangrós Giménez, B. Finke, C. Nowak, et al. Structural and mechanical characterization of lithium-ion battery electrodes via DEM simulations, Adv. Powder Technol. 29 (2018) 2312-2321. https://doi.org/10.1016/j.apt.2018.05.014

[33] A.C. Ngandjong, T. Lombardo, E.N. Primo, M. Chouchane, A. Shodiev, O. Arcelus, A.A. Franco, Investigating electrode calendering and its impact on electrochemical performance by means of a new discrete element method model: Towards a digital twin of Li-Ion battery manufacturing, J. Power Sources 485 (2021) 229320. https://doi.org/10.1016/j.jpowsour.2020.229320

[34] D. Mayer, B. Schwab, J. Fleischer, Influence of Electrode Corrugation after Calendering on the Geometry of Single Electrode Sheets in Battery Cell Production, Energy Technol. 11 (2023) 2200870. https://doi.org/10.1002/ente.202200870

[35] C.G. Wang, H.F. Tan, X.W. Du, Z.M. Wan, Wrinkling prediction of rectangular shell-membrane under transverse in-plane displacement, Int. J. Solids Struct. 44 (2007) 6507-6516. https://doi.org/10.1016/j.ijsolstr.2007.02.036

Deformation laws and design method of near-net rolling for L-section rings with profile axial conical rolls

QIAN Dongsheng[1,2,3,a], GU Yixuan[1,2,b] , DENG Jiadong[2,3,c,*] and WANG Feng[1,2,3,d]

[1]School of Materials, Wuhan University of Technology, Luo Shi Road 122, 430070 Wuhan, China

[2]Hubei Key Laboratory of Advanced Technology for Automotive Components al. Luo Shi Road 122, 430070 Wuhan, China

[3]Hubei Engineering Research Center for Green Precision Material Forming al. Luo Shi Road 122, 430070 Wuhan, China

[a]qiands@whut.edu.cn, [b]gyx849757@163.com, [c]dengjd@whut.edu.cn, [d]wangfengwhut@163.com

Keywords: L-Section Rings, Near-Net Ring Rolling, Near-Net Rolling, FE Simulation, Ring Rolling Experiment

Abstract. L-section rings are widely used in key basic parts, such as flanges and rotary supports. When the radial length of the ring step is large while the axial depth is shallow, traditional ring rolling with profile mandrel may result in forming defects such as incomplete filling of the steps. This paper proposes a near-net rolling technology for L-section rings with profile axial conical rolls. Firstly, a finite element model for the rolling process of L-section ring was established. Then the revolution laws of the ring geometric dimension and the metal flowing behavior during the rolling process with profile axial conical rolls were studied. Further, the influence of key process parameters on the forming effects was revealed, including radial-to-axial deformation ratio, rolling ratio, and roller motion. Finally, the feasibility of the proposed process method was verified by FE simulation and rolling experiments. The research results provide a technical guidance for the near-net rolling forming of L-section ring, contributing to the high-performance, efficient, and cost-effective manufacturing of parts like L-section flanges and rotary supports.

Introduction

Ring rolling is an international mainstream technology for manufacturing high-performance large seamless rings, widely used in aviation, aerospace, automotive, engineering machinery, wind power, nuclear power and other fields, such as aviation casing rings, rocket storage tank rings, wind power bearing rings, etc. These ring components usually have complex structural shapes. However, in order to simplify the manufacturing process, rectangular or simple shaped section ring forgings are usually rolled, and then the target section shape is obtained through a large amount of cutting processing. This process not only has low material utilization and long processing cycles, but also weakens the mechanical properties of the product. Near-net rolling is an urgent need to achieve high-performance, high-efficiency, and low-cost manufacturing of such complex annular components, and it is also the forefront of research in the manufacturing technology of complex annular components for major equipment in aviation, aerospace, and wind power.

Compared to the rolling of rectangular section rings, the rolling process of irregular section rings is more complex, and there is relatively less research on it. D. S. Qian [1] designed four different ring blanks and determined the optimal blank design for the stepped section ring through theoretical analysis, finite element simulation, and experiments. Yu-Min Zhao [2] discussed the impact of rolling ratio on the forming outcomes of groove section rings and identified a suitable rolling ratio range.Lin Hua [3] first used constraining ring rolling technique to form a conical ring with thin walls and three high ribs, and studied the metal flow behavior during the constrained ring rolling process.Lei Lian [4] proposed and applied an advanced feeding strategy driven by staged growth velocity, which improved the rolling stability and forming quality of groove section ring rolling.

As a typical irregular section ring, the L-section ring has important applications in key components for wind turbine flanges and bearings. It consists of a large rectangular ring and a small rectangular ring, that combine to form an L-section ring[5]. When using shaped mandrel to directly roll and form L-shaped rings, as shown in Fig. 1, if the step length is large while the axial depth is shallow, forming defects such as incomplete filling and folding at the step are prone to occur [6].

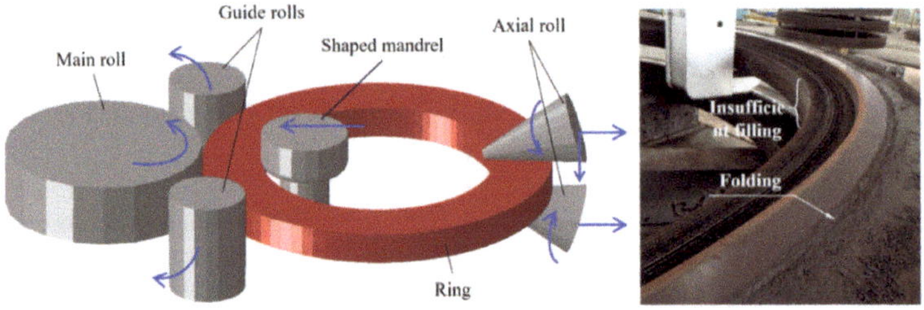

Fig. 1. L-section ring rolling with shaped mandrel.

For L-shaped section rings with large step lengths and shallow axial depths, profile axial conical roll can be used for near-net rolling, as shown in Fig. 2. This process utilizes an shaped axial roll with steps and utilizes the axial rolling pass to achieve direct rolling forming of L-rings [7]. During the rolling process, while the upper and lower cone rollers form the upper and lower surfaces of the ring axially, it is also necessary to ensure that the backward speed matches the growth speed of the outer diameter of the ring. By adding a pair of guide rolls on both sides of the axial rolls, the stability control of the ring rolling process is increased [8,9].

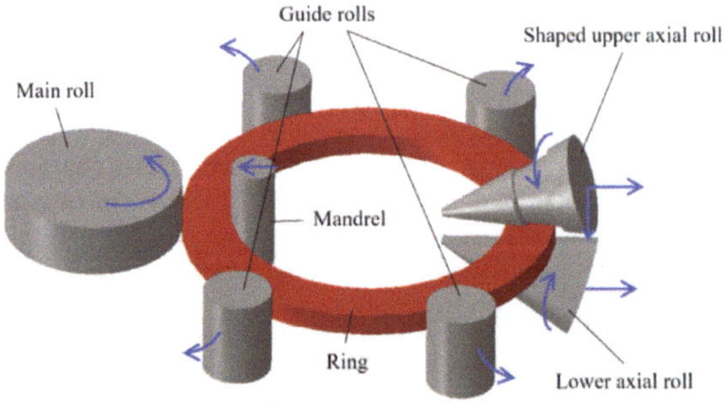

Fig. 2. Near-net shape rolling principle of L-section ring with profile axial conical rolls.

This article takes the L-section forging of a wind power yaw bearing as the research object. By establishing a finite element model for the rolling of L-section rings, the deformation law of the rings during the rolling process with profile axial conical rolls is analyzed, and the influence of key process parameters on the rolling forming is revealed, providing technical guidance for the near-net rolling of L-section rings.

Design of ring rolling process and FE modeling

Blank size design. The geometric relationship between the ring blanks and the forgings in the near-net rolling of L-section rings is shown in Fig. 3. Among them, H0 and B0 are the blank wall thickness and axial height of the ring, Bb and Bs are the axial height of the large hole section and the small hole section of the ring, Hb and Hs are the wall thickness of the large hole and the small hole of the ring, wherein the wall thickness and height of the large hole section of the ring are the key dimensions for the forming of the L-section ring. The shaped upper axial roll is pressed down to obtain the large hole part height Bb. And in order to ensure that the wall thickness of the large hole part of the L-section ring meets the design requirements, the shaped part of the upper axial roll should be kept away from the outer diameter of the ring Hb during the rolling process, so the backward speed of the cone roll VW and the growing speed of the outer diameter of the ring VD should meet the following definitions:

$$V_D = dD/dt. \tag{1}$$

$$V_W = V_D. \tag{2}$$

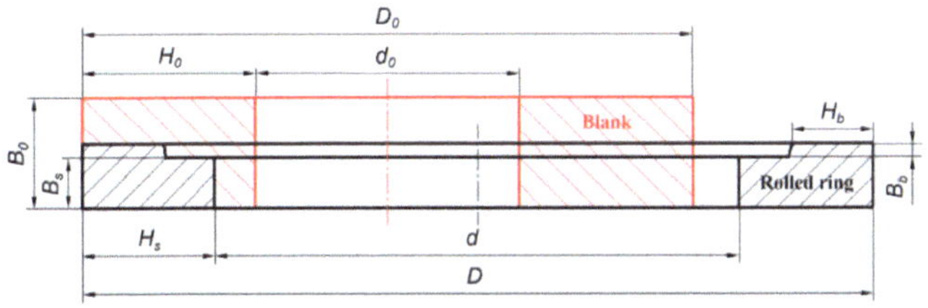

Fig. 3. Geometric relationship between ring blanks and forgings.

The specific parameters of L-section forging of wind power yaw bearing are shown in Table 1.

Table 1. Forgings dimension parameters (mm).

D_s	B_b	B_s	H_s	H_b
3711	30	114	306	186

In the rolling process of irregular section ring, the main factor affecting the filling of section profile is the shape and size of the blank. Reasonable design of the blank size is the key to ensure the metal flowing and filling of L-section forgings [10,11].

The rolling ratio is an important parameter in the blank size design, which determines the deformation degree of the ring. For non-rectangular section ring rolling, the selection of rolling ratio has a greater impact on the filling of the section profile, and the L-section ring rolling ratio λ is defined as follows:

$$\lambda = \frac{A_0}{A} = \frac{H_0 B_0}{H_b B_b + H_s B_s}. \tag{3}$$

In radial-axial ring rolling (RARR) of L-section ring, the reduction ratio of blank wall thickness and height of ring parts, that is, radial-to-axial deformation ratio η, has an important influence on the forming degree of step inside ring parts. η is determined by the formula:

Materials Research Forum LLC
https://doi.org/10.21741/9781644903254-34

$$\eta = \frac{\Delta h}{\Delta b} = \frac{H_0 - H_s}{B_0 - B_s} = \frac{B_b + B_s}{H_s}. \quad (4)$$

Where Δh=H0-Hs、Δb=B0-Bs are radial and axial feeds respectively.

In order to study the influence of ring blank size on the shaped axial roll near-net rolling, two groups of 9 ring blank were designed for finite element numerical simulation experiments. The size parameters of ring blank are shown in Table 2. Group 1: In order to study the effect of the ratio of radial and axial feeds on the near-net rolling with shaped axial roll, η values of ring blank 1, 2, 3, 4 and 5 were selected as 0.28, 0.47, 0.69, 0.96 and 1.31, respectively, and the outer diameters of ring parts were 2700mm, and the rolling ratio was basically unchanged. Group 2: In order to study the effect of rolling ratio on the near-net rolling with shaped axial roll, the radial feed Δh of ring blank 6, 7, 3, 8 and 9 were selected as 61 mm, 51 mm, 42 mm, 34 mm and 26 mm, the axial feed Δb were 61 mm, and the rolling ratio λ were 1.58, 1.54, 1.50, 1.47 and 1.43, respectively.

Table 2. Dimensions of ring blanks.

Blank number	D_0/mm	H_0/mm	B_0/mm	η	Mandrel feed speed Vc/mm s^{-1}	Upper axial roll feed speed Va/mms^{-1}	Rolling time /s	λ
1	2700	326	185	0.28	0.17	0.59	120	1.49
2	2700	337	180	0.47	0.26	0.55	120	1.50
3	2700	348	175	0.69	0.35	0.51	120	1.50
4	2700	360	170	0.96	0.45	0.47	120	1.51
5	2700	373	165	1.31	0.56	0.43	120	1.52
6	2600	367	175	1.00	0.51	0.51	120	1.58
7	2650	357	175	0.84	0.43	0.51	120	1.54
8	2750	340	175	0.56	0.28	0.51	120	1.47
9	2800	332	175	0.43	0.22	0.51	120	1.43

FE modeling. Based on the working principle of RARR, the 3D coupled thermo-mechanical FE model of L-section ring rolling with shaped axial roll was established by ABAQUS/Explicit program, as shown in Fig. 4. In the FE model, the ring part is set as a deformable body, the rolls are set as a rigid body, and the ring material is 42CrMo. The material constitutive equation and thermodynamic physical property parameters of the material are referred to paper[10]. The grid type is C3D8RT and dynamic explicit finite element algorithm is used to improve the computational efficiency. At the same time, ALE adaptive meshing is used to control mesh distortion in the deformation process. Other rolling parameters are shown in Table 3.

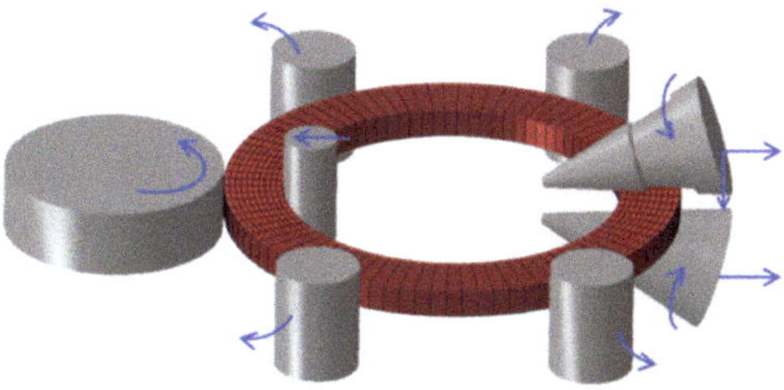

Fig. 4. FE model of L-section ring rolling with profile axial conical rolls.

Table 3. Simulated rolling parameters table.

Rolling parameter	value
Ring material	42CrMo
Driving roll working radius	650mm
Core roll working radius	160mm
Guide roll radius	250mm
Half cone Angle of cone roll	17.5°
Rolling time	120s
Initial ring temperature	1150°C
Tool temperature	200°C
Environment temperature	40°C
Main roll speed	19.10r/min
Friction coefficient	0.25

Result and discussion

Deformation law of the forging size. Four moments in the ring blank 3 forging process are depicted in Fig. 5: the beginning of rolling, the moment at which the upper axial roll is pressed down by 30 mm, the moment at which the large hole section of the ring comes into contact with the upper axial roll, and the end of rolling.

At the beginning of rolling, the upper surface of the blank only makes touch with the inner diameter side metal with the upper axial roll at the start of rolling, with a contact length of H0-Hb. In the early stage of rolling, the upper axial roll is pressed downwards, causing an axial height difference between the large and small hole parts of the ring, thereby obtaining the axial depth of the step. A tiny amount of deformation happens as a result of the axial deformation pulling on the metal at the large hole ring. As a result, the ring blank cross-section slopes when the upper axial roll is pressed down by 30 mm, and the step's axial depth is less than the upper axial roll's axial downward pressure. When rolling to D=3002mm, the axial depth of the step is equal to the downward pressure of the top cone roller shaft, and the height of the large hole part of the ring reaches Bb. In the latter stages of rolling, the upper axial roll and mandrel keep feeding, distributing the ring piece's volume in the direction of the large-hole ring, filling the axial roll cavity even more, and shortening the ring piece's upper surface slope. Rolling stops when the ring outer diameter D reaches 3711mm. Based on the simulation findings, it is evident that the key problem with the forgings' formation is insufficient metal filling at the step, which prevents the Hb value from reaching the design size. Fig. 6 displays the Hb values of two groups of ring blanks.

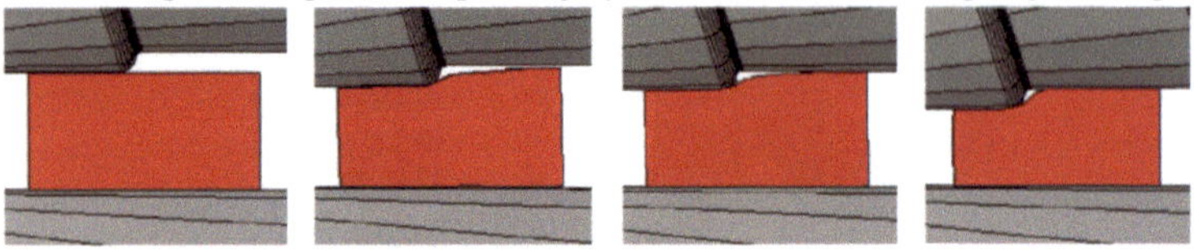

Fig. 5. The forming process of the forging cross-section.

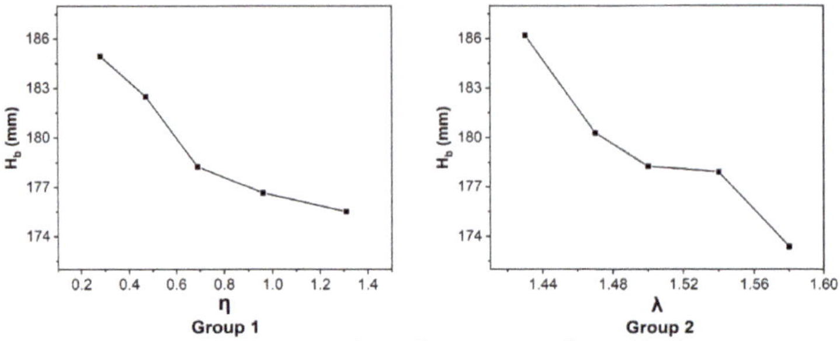

Fig. 6. The H_b values of two groups of ring blanks.

Factors affecting the forming size of the forging

Flow deformation of metals. It is known that when the height of the large hole part of the ring has just reached B_b, the ring needs to continue to grow and fill the step with metal. Fig. 7 shows the variation curve of the axial height of the forging step with the outer diameter of the ring during the rolling process of blank 3. According to the variation law of the axial height of the step, the curve can be divided into two parts: stage I and stage II.

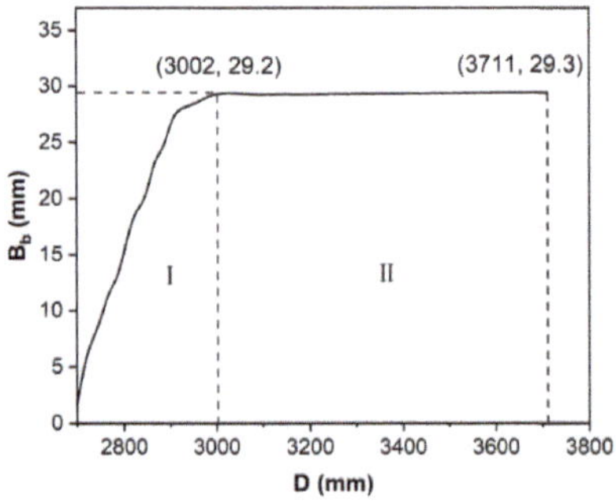

Fig. 7. Variation curve of step's axial height B_b.

As can be seen from the previous text, the proportion of stage II in the total curve has an impact on whether the step filling is sufficient. The proportion relationship is defined as follows:

$$\alpha = \frac{\Delta D_{II}}{\Delta D} = \frac{\Delta D_{II}}{D - D_0}. \tag{5}$$

Where ΔD_{II} is the change in the outer diameter stage II of the ring, and ΔD is the total change in the outer diameter of the ring. Fig. 8 shows the first group of ring blanks' α, It can be observed that the smaller α, the greater the difference between H_b and the design size of 186 mm. Fig. 9 analyzes the metal radial flow deformation behavior of forgings in stage II. It can be seen that the radial plastic strain direction of the forging in the cross-section is opposite, causing the inner and outer diameter metals to flow radially towards the core. At the same time, the direction of axial plastic deformation within the cross-section is consistent, but the amount of metal deformation on the outer diameter side is smaller than that on the inner diameter side, causing the forging metal to flow and fill towards the step. The α of the first group of ring blanks decreases with the increase of η, which is not conducive to the filling at the step, resulting in an increase in the difference between the final H_b and the design size.

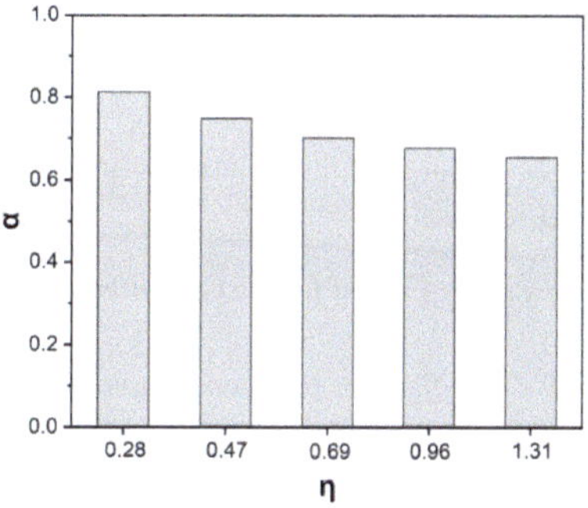

Fig. 8. The influence of radial-to-axial deformation ratio η on the value of α.

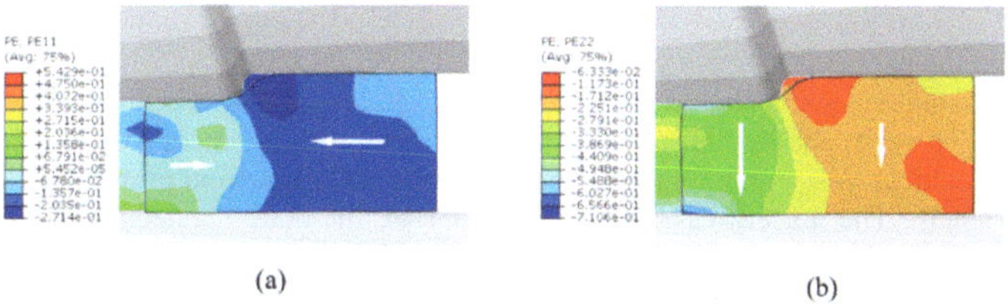

Fig. 9. Metal flow deformation in stage II: (a)radial plastic strain component; (b) axial plastic strain component.

Control of the roll motion. Compared with the first group of blanks, the second group has a larger variation in rolling ratio, with different ΔD, different deformation speeds, and a greater influence on the forming size by the motion of the rolls.

The motion of the rolls is the key to ensuring the forming size of the forging during rolling. Fig. 9 shows the V_D and V_W during the rolling process of blank 6 and 9. It was found that the matching between V_D and V_W was poor for blank 6 with larger λ. This is because in stage I, only the inner diameter side of blank is in contact with the upper axial roll, and the contact length changes with the wall thickness during the rolling process. In stage II, the axial roll can further adhere to the ring. This change in contact length leads to uneven force distribution on the forging, unstable rolling process, and susceptibility to eccentricity, resulting in a difference between the measured and the actual outer diameter of the ring, which affects the matching between V_D and V_W. Larger λ lead to larger deformation speeds as well, making the matching of V_D to V_W worse and leading to an increase in the difference between Hb and the design dimensions. This problem is also demonstrated by the difference between the setback distance of axial roll ΔL and ΔD for forgings of different λ in Table 5.

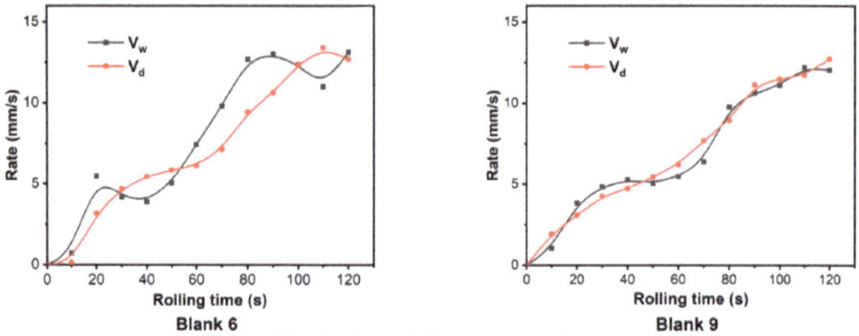

Fig. 9. V_D and V_W during rolling.

Table 5. The difference between ΔL and ΔD (mm).

λ	1.58	1.58	1.50	1.47	1.47
ΔD	1111	1061	1011	961	911
ΔL	1129	1066	1012	960	909
ΔD-ΔL	-19	-5	-1	1	2

Rolling experiment.
In summary, smaller λ and η should be selected for the rolling of L-section ring with profile axial conical rolls. To verify the feasibility of the near-net shape rolling process of L-section ring with profile axial conical rolls, rolling experiment was conducted on a wind power yaw bearing with larger step length. The η used was 0.22, and λ was 1.44, the outer diameter of the ring blank is 2860mm, the inner diameter is 2198mm, the height is 180mm, and the material is 42CrMo. The rolling experiment of the L-section forging is shown in Fig. 10. The rolling process is smooth, and the forming size is good, which verifies the feasibility of the rolling process with profile axial conical rolls.

Fig. 10. Rolling experiment with profile axial conical rolls.

Summary

This article takes an L-section wind power yaw bearing forging as the research object, and conducts research on its near-net rolling forming law and design method. The following conclusions are drawn:

(1) Two groups of blanks with differences η and λ was designed to study the influence of blank size on the near-net rolling forming size of L-section ring with profile axial conical rolls;

(2) When the η increases, the value of α decreases, which is not conducive to filling the step;

(3) The control of roll motion, especially the matching of V_D and V_W, has a significant impact on the forming size of L-section rings. Therefore, a smaller λ is beneficial for improving the matching between V_D and V_W.

Acknowledgment

This research work is supported by the National Key Research and Development Project (2018YFA0702900), Hubei Key Research and Development Project (2022EJD012), Jiangsu Transformation Projects of Scientific and Technological Achievements (BA2022106), China Postdoctoral Foundation Project (2023M742675), 111 Project (B17034) for the supports given to this research.

References

[1] D.S. Qian, L. Hua, Blank design optimization for stepped-section profile ring rolling, J. Sci. China-Technological Sciences 24 (2010) 1612-1619. https://doi.org/10.1007/s11431-010-3113-6

[2] Yu-Min Zhao, D.S. Qian, Effect of rolling ratio on groove-section profile ring rolling, J. Mech. Sci. Tech. 24 (2010) 1679-1687. https://doi.org/10.1007/s12206-010-0525-y

[3] L. Hua, D.Y. Tian, X.H. Han, W. Zhuang, Modelling and analysis of metal flowing behaviors in constraining ring rolling of tapered ring with thin wall and three high ribs, J. Chinese J. Aeronautics 36 (2023) 476-492. https://doi.org/10.1016/j.cja.2023.04.011

[4] L. Liang, L. Guo, Advanced feeding strategy driven by staged growth velocity for groove-section profiled ring rolling, J. Manuf. Process. 81 (2022) 907-921. https://doi.org/10.1016/j.jmapro.2022.07.050

[5] L. Hua, J.D. Deng, D.S. Qian, Recent development of ring rolling theory and technique, Int. J. Mater. Product Tech. 54 (2017) 65-87. https://doi.org/10.1504/IJMPT.2017.080566

[6] I.Y. Oh, T.W. Hwang, Y.Y. Woo, H.J. Yun, Y.H. Moon, Process-induced defects in an L-shape profile ring rolling process, Int. J. Mater. Forming 12 (2019) 727-740. https://doi.org/10.1007/s12289-018-1450-3

[7] P.Z. Zhou, L.W. Zhang, S.D. Gu, J. Ruan, L. Teng, Mathematic modeling and FE simulation of radial-axial ring rolling large L-section ring by shape axial roll, Int. J. Adv. Manuf. Tech. 72 (2014) 729-738. https://doi.org/10.1007/s00170-014-5705-y

[8] D.S. Qian, H Tian, J.D. Deng, Towards Extremely Large-scale Radial-axial Ring Rolling for Constant Ring Growth State with Accurate Closed-loop Control Method, J. Mech. Eng. 59 (2023) 85-95. https://doi.org/10.3901/JME.2023.10.085

[9] X. Li, L. Guo, F. Wang, On a plastic instability criterion for ultra-large radial-axial ring rolling process with four guide rolls, Chinese J. Aeronautics 35 (2022) 391-407. https://doi.org/10.1016/j.cja.2021.07.026

[10] D.S. Qian, B B Ma, J.D. Deng, Numerical simulation and experimental study on grain evolution in whole process of forging and rolling forming—waste heat quenching of 42CrMo ring, Chinese J. Aeronautics 29 (2022) 1-7. https://doi.org/10.3969/j.issn.1007-2012.2022.11.001

[11] L. Hua, X.G. Huang, C.D. Zhu, Theory and Technology of Ring Rolling, first ed., China Machine Press, 2001.

MATERIALS

Microstructure and texture evolution of Mg-Zn-Al-Ca (ZAX210) during groove rolling

STIRL Max[1,a *], ULLMANN Madlen[1,b] and PRAHL Ulrich[1,c]

[1]TU Bergakademie Freiberg, Bernhard-v.-Cotta Straße 4, 09599 Freiberg, Germany

[a]max.stirl@imf.tu-freiberg.de, [b]madlen.ullmann@imf.tu-freiberg.de, [c]ulrich.prahl@imf.tu-freiberg.de

Keywords: Microstructure, Magnesium, ZAX210, Groove Tolling

Abstract. This study introduces the ZAX210 magnesium alloy wire, produced using groove rolling. The starting material, initially extruded, was successfully rolled from 12.0 mm to 6.7 mm diameter in eight passes using square-oval grooves, finishing with an oval and round shape. High strain rates and an optimized rolling temperature of 265 °C were found beneficial, avoiding hot cracks and promoting recrystallization. Microstructural analysis using light and electron microscopy revealed a fully recrystallized, fine microstructure with uniform precipitate distribution. Texture development is being assessed with EBSD, where a strong dependence of the respective groove shape on the expression of the texture was observed. Mechanical properties of the final wire were determined, whereas $R_{p0.2}$ is 308 MPa, R_m is 337 MPa and A is 17.8 %.

Introduction

Magnesium alloys can play a key role as innovative lightweight materials for reducing CO_2 emissions. This is mainly due to its low density, high specific strength and high life cycle assessment potential. The limited formability of magnesium due to its hexagonal crystal structure is considered a challenge, which is why the forming processes are mainly carried out at elevated temperatures for a sufficient number of sliding systems. Furthermore, an understanding and specific manipulation of texture during the forming process is crucial to achieve ductile material behavior. Long products such as rod and wire material made of magnesium can be produced by casting, extrusion, or groove rolling (also known as caliber rolling), with the latter standing out in terms of economic efficiency due to high production throughput combined with promising mechanical properties [1]. Groove rolling is known from the processing of steel, where the material is formed into a defined groove shape by a large number of rolling passes. The precise design of the groove shape has an influence on the microstructure development due to the locally acting force and the material flow behavior. Groove rolling promises a weakening of the basal texture in contrast to, for example, extrusion [2]. Previous studies on the groove rolling of magnesium alloys have concentrated mainly on conventional alloy concepts like AZ31 [1–8]. In comparison to steel alloys, a stronger spreading behavior can be observed in magnesium alloys, which indicates other material flow mechanisms [8,9]. On the process side, other groove shapes are therefore necessary for optimum production. The improved mechanical properties of the groove-rolled samples are primarily attributed to the effects on the microstructure development and in detail the grain refining effect. Doiphode et al. observed a strength-increasing effect due to grain boundary consolidation, and the effect is also attributed to the texture and formation of twins [1]. Heat treatments to improve the mechanical properties only make sense in the low temperature range (373-473 K) for increased ductility [10]. To the authors best knowledge, there are only few investigations into the processing of this alloy system using groove rolling. In a Mg-Al-Ca-Mn-Sr alloy, which was processed using groove rolling, elongations of up to 37 % at a strength of 140 MPa were achieved in tensile tests [11].

Another promising alloying system is the use of calcium, which shows similar effects to the use of rare earths regarding the weakening of the basal texture and a grain-reducing effect through additional precipitate formation. This increases formability, especially at low temperatures [12]. The current state of research into the Mg-Zn-Al-Ca system relates primarily to the processing of flat products.

This paper is the first to investigate the behavior of a calcium-containing magnesium (ZAX210) alloy during groove rolling. After successfully determining the process parameters for processing the alloy, the focus is on the development of the microstructure. The development of the texture is of particular interest. Furthermore, the mechanical properties of the wire are determined.

Methodology

The starting material was an extruded ZAX210 wire with a diameter of 12 mm containing 2 % Zn, 1 % Al, and 0.3 % Ca. The samples are heated to forming temperature in a convection oven and subsequently formed on a three-high rolling mill at Institute of Metal Forming, TU Bergakademie, Freiberg, Germany. A total of 8 rolling steps are carried out, the corresponding groove pass design can be seen in Fig. 1. A square-oval groove pass is used, whereby 12 mm round bars will be fabricated to a final cross-section of 6.8 mm round. The degree of elongation reaches from 1.38 to 1.05, the overall area reduction is 68.8 %. Before each pass, the billet was rotated 90° clockwise along rolling direction. For the rolling process, no lubricant or reheating was used. Following preliminary tests, a rolling temperature of 265°C and a roll speed of 120 rpm (corresponds to a forming speed of approx. 3.5 m/s and 3.7 m/s for the present roll diameter of 297 mm and 278 mm) was choose. For the present alloy, a high forming speed shows positive effects as a result of the increasing recrystallized fraction. The same applies to an increased rolling temperature; because of the occurrence of hot cracks (possible melting of MgZn precipitates) from 350 °C and the heating due to the forming, a temperature window at 265°C was obtained.

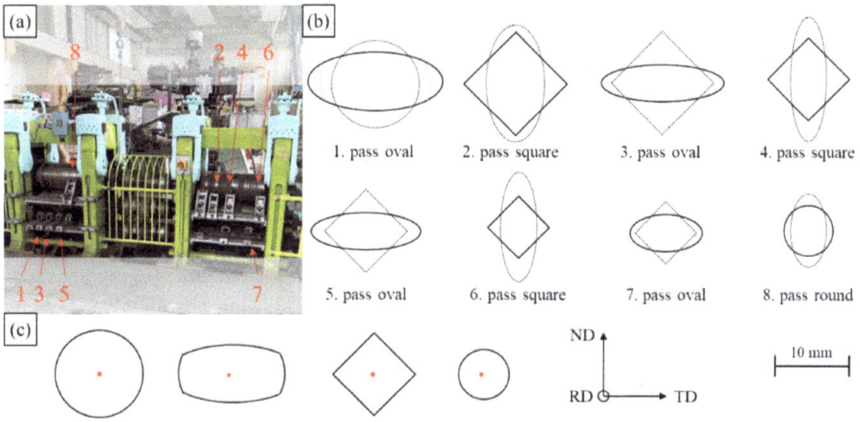

Fig. 1. Three high rolling mill (a) with the used square-oval groove pass design (b), where the dotted line is the input profile and the solid line groove pass. Blue Line in 1. pass represents an underfilling of the first groove pass. (c) Locations where microstructure is characterized (red dots).

Subsequently, microstructural images are taken from the initial extruded and rolled state using light and scanning electron microscopy to characterize the microstructure. For sample preparation, after cutting and grinding the samples with SiC-paper in five different steps, various polishing stages are carried out with MD-Chem OPS 300 magnesium polishing cloth and OP-chem, whereby attention is paid to regular rinsing with ethanol. The etching of the samples was conducted with a solution of picric acid (10 ml distilled water, 10 ml glacial acetic acid, 70 g ethanol and 4 g picric acid) for 15-30 seconds up to three times. The same samples were subsequently polished for scanning electron microscopy (SEM), whereby a ZEISS GeminiSEM 450 scanning electron microscope was used.

Texture analysis was performed using an electron backscatter diffraction detector (EBSD). The accelerating voltage was between 15 and 20 kV. A step size of 0.65 μm was selected. For the analysis of the EBSD data and the calculation of the pole figures, the free MTEX MATLAB toolbox [13] was used. A microstructural analysis was carried out on cross-sections in the center of the rod. Additionally, the samples were investigated by using energy dispersive X-ray spectroscopy (EDX).

Tensile tests are carried out at room temperature to determine the mechanical properties of the groove rolled samples. The specimen geometries correspond to DIN 50125 shape B with a specimen diameter of 4 mm and a test length of 20 mm.

Results and Discussion

Rolling process. The rolling could be carried out without cracking., it was found that the spreading behavior of the magnesium alloy was too strong for this schedule, resulting in overfilling. After adjusting the rolling gap, this was largely reduced, but overfilling still occurred in the final rolling pass. Furthermore, this was counteracted by deliberately underfilling the groove as the first rolling step. This can be seen in Fig. 1b with the blue line. The reason for the overfilling is the stronger spreading behavior in comparison to conventional steel alloys, for which the original groove pass design is intended. Compared to other magnesium alloys (see [9]), ZAX210 shows a similarly strong spreading behavior, which is, however, less dependent on temperature. Corresponding coefficients were included for the Freiberger model of spread [9] (Table 1) for the material coefficient $C_{W\beta}$ and the temperature coefficient, which is given by:

$$C_{T\beta} = a + b \cdot T \tag{1}$$

The Freiberg spreading model is used for the design of break down passes and includes all relevant influencing factors for groove rolling. The two coefficients for material and temperature behavior are decisive for transfer to other materials. Further explanations are given by [9].

Table 1. Coefficents of Freiberger model of spread [9] for different Mg wrought alloys.

Coefficient	$C_{W\beta}$ [ ]	a for $C_{T\beta}$ [ ]	b for $C_{T\beta}$ [ ]
ZAX210	1.4794	0.945894	1.38E-04
AZ31 (from [9])	1.0541	0.848512	5.13E-04
AZ81 (from [9])	1.0020	0.869500	4.28E-04
WE43 (from [9])	0.9119	0.886358	2.20E-04

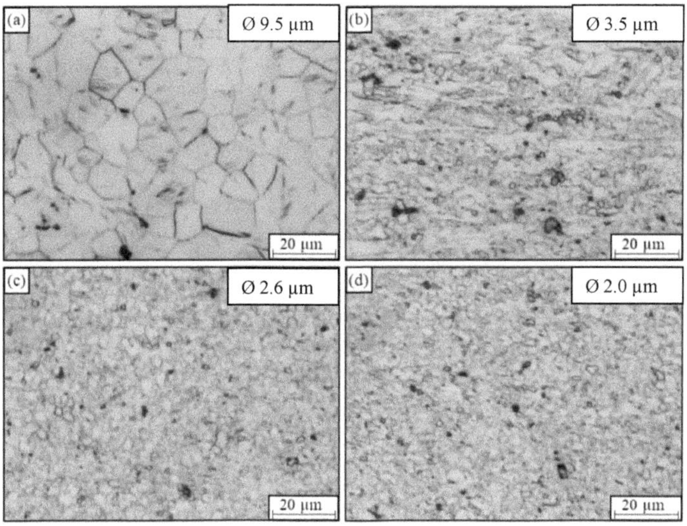

Fig. 2. Microstructure (center position) of the ZAX210 alloy in initial state (a), after 1. roll pass (b), 2. roll pass (c) and 8. roll pass (d). Average grain size via line intersection method on top right.

Microstructure analysis. The initial microstructure of the extruded state can be seen in Fig. 2a. The grain size distribution across the sample cross-section can be determined (Fig. 3a). There is also a partially recrystallized structure along the grain boundaries, which was produced by the forming process of extrusion. Fig. 2b - d show the microstructure after one, four and eight rolling passes respectively. A recrystallized microstructure can already be seen after the first rolling pass. This is also visible in the grain size distribution (Fig. 2b) with a decrease in grain sizes. The grain refining effect is mainly due to the dynamically recrystallized grains. The different force effect and thus degree of deformation distribution in the groove can be seen on the basis of light microscopic images of the first rolling pass at various points (Fig. 3). While a high proportion of recrystallization is already visible in the center of the sample, this cannot yet be observed in the edge areas. Initial recrystallization processes can be observed at the upper edge of the rolled sample, while in contrast there is hardly any change visible in the microstructure compared to the initial state at the edge without the effect of pressure. These differences can also be seen in the grain size distribution, whereby some larger grains in the area >10 µm exists. After the fourth or final eighth rolling pass, a completely recrystallized fine microstructure is present. Even after the fourth roll pass, no differences in the various position of the cross-section can be determined. Using EDX measurements, it was possible to determine that these are at least ternary intermetallic phases consisting of Al, Ca and Zn with different stoichiometric ratios (Fig. 4). The rolling processes have no influence on the number and distribution of the precipitates; similar precipitation tendencies could be observed in all samples at different points in the microstructure using EDX images. The exact determination of the precipitates present is difficult, as a large proportion of the precipitates are very fine and are therefore more difficult to detect using EDX.

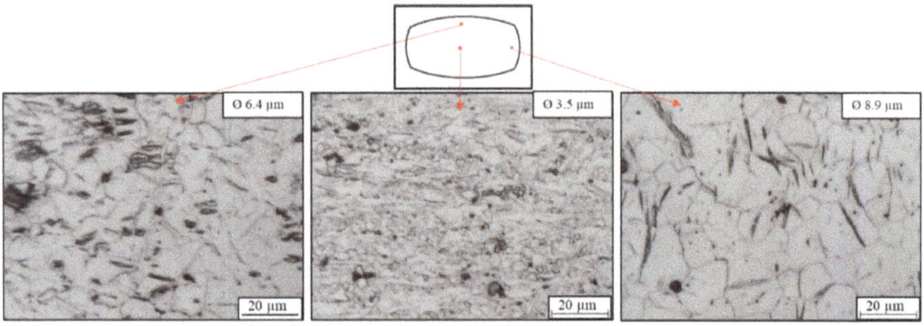

Fig. 3. Microstructure after the first roll pass at different positions. Average grain size via line intersection method on the top right.

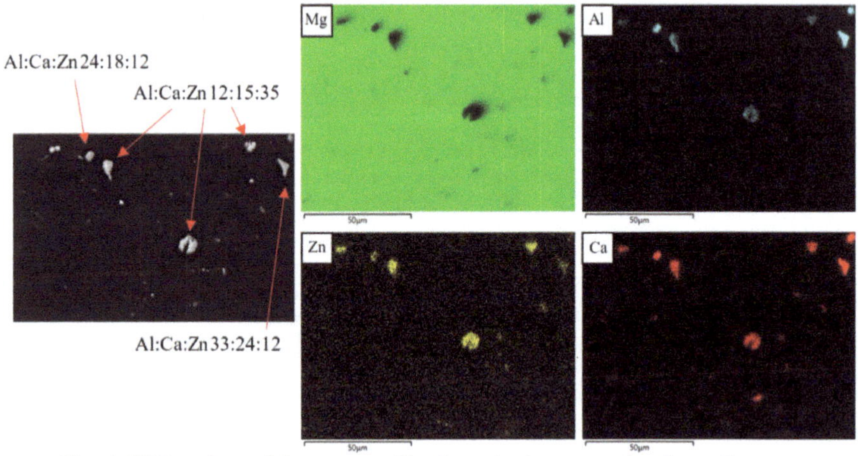

Fig. 4. EDX analysis of the intermetallic phases in the center after first roll pass.

Texture measurements. In the following, the EBSD pole figures calculated from the orientation distributions of grains in the rolled samples are discussed. An extruded texture is present prior to groove rolling, with a balanced orientation of the c-axes in the direction of the radial directions (here ND or TD) toward the outside. Furthermore, the peak of the misorientation angle at about 86° correspond to the $\{11\bar{2}0\}$ tensile twins. After the first rolling pass, a strong texture change is observed whereby the c-axis is oriented exclusively in the direction of the groove edges. These characteristics were confirmed by additional measuring points at the right edge of the rolled material and at the top contact point between the rolled material and the groove. A slight split form in the middle in the ND direction. A strong dependence of the texture on the corresponding groove shape is evident from the fourth pass. The sample was rolled from an oval to a square, with the first contact of the rolled material with the groove being off-center from the vertical line of symmetry. This can be seen in the texture as a slight shift of the maximum from the ND direction. The one-sided characteristic is technologically explained by a presumably slightly skewed entry of the rolled material, which shifted the maximum load to the right half of the groove. Similar observations can be made during the last rolling pass. Furthermore, the analysis of the twins shows that their importance in groove rolling appears to be secondary. Based on the misfit orientations, there is an increasing decrease in the twins, with slight peaks at 30° and 86°, which indicates $\{10\bar{1}1\}$-$\{10\bar{1}2\}$ double twinning with 38° <$11\bar{2}0$> rotation and $\{10\bar{1}2\}$ tensile twins with <$11\bar{2}0$> orientation. These texture investigations were made in the center part of the rolled samples. Further investigations should deal with the microstructural development in different areas of the rolled material cross-section.

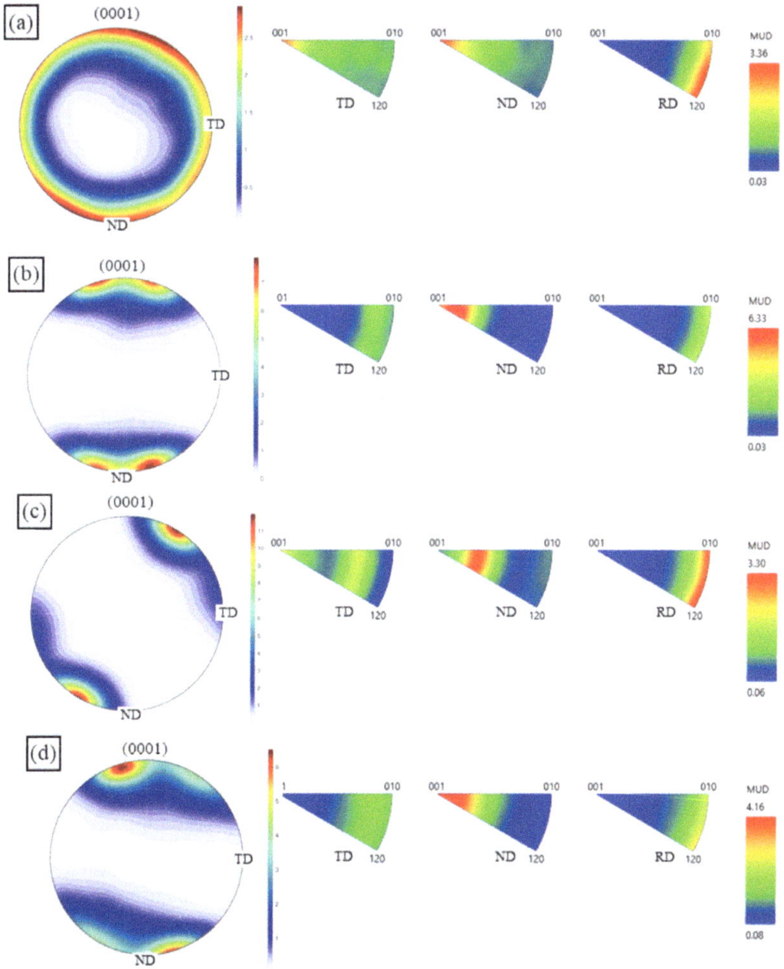

Fig. 5. {0001} and inverse pole figures of ZAX210 alloy in (a) initial extruded condition and after one (b), four (c) and eight passes (d) of groove rolling.

Materials Research Proceedings 44 (2024) 331-339 https://doi.org/10.21741/9781644903254-35

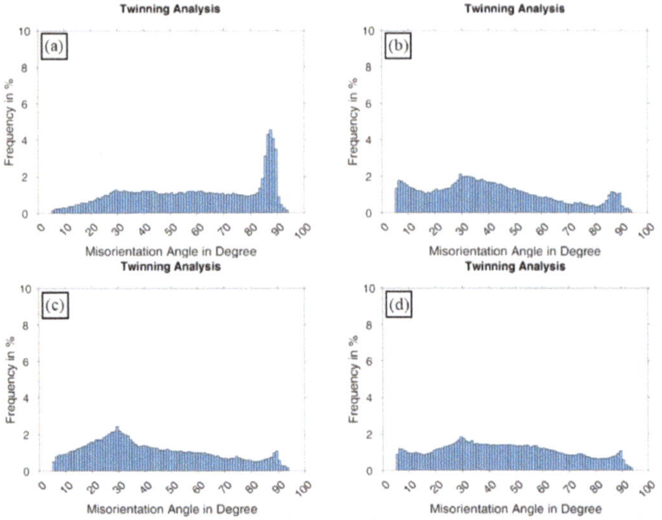

Fig. 6. Misorientation angle for (a) initial extruded condition and after one (b), four (c) and eight passes (d) of groove rolling.

Mechanical properties from tensile test. The results of the tensile tests are shown in table 2. The extruded material has a yield strength of 181 MPa, a tensile strength of 274 MPa and an ultimate strain of 17.2 %. After eight passes of groove rolling with a start rolling temperature of 265 °C, the material has a yield strength of 308 MPa, which is an improvement by 70.1 % compared to the extruded condition. For the tensile strength, an improvement of 23 % to 337 MPa can be determined. The elongation at break is almost identical to the extruded state at 17.8 %. The improvement of the tensile strength and especially the yield strength are due to the induced strain, which leads to dynamic recrystallization.

Table 2. Mechanical properties of the extruded and rolled alloy at room temperature.

Alloy	State	$R_{p0.2}$ [MPa]	R_m [MPa]	A [%]	Ref.
ZAX210	Extruded (initial)	181 ± 16.1	274 ± 6.4	17.2 ± 1.0	
ZAX210	Extruded and rolled (265 °C)	308 ± 13.0	337 ± 5.3	17.8 ± 1.3	
AZ31	Hot rolled (7 passes)	313	336	20.0	[10]
AZ80	Extruded and rolled (14 passes)	264 ± 3	365 ± 1	17 ± 1	[14]

Conclusion

In the present investigation, groove rolling of the calcium-containing magnesium alloy ZAX210 was carried out for the first time and microstructurally characterized and the mechanical properties determined. The following findings can be summarized:

a) The magnesium alloy ZAX210 containing calcium was groove rolled at 265°C without cracking, whereby a stronger spreading behavior was observed like other magnesium materials.

b) The finer microstructure improved the mechanical properties compared to the original material. After the 8th rolling pass, an average $R_{p0.2}$ of 308 MPa, R_m of 337 MPa and a total elongation of 17.8 % is achieved.

c) Groove rolling achieves significant grain refinement, with a completely recrystallized, fine-grained structure after the final rolling pass. Furthermore, a strong dependence of the respective groove shape on the expression of the texture was observed.

To gain a better understanding of the material during the groove rolling process, the microstructural development in different areas of the rolled material cross-section has to be investigated in the future. In particular, this will address the different stress and deformation states during groove rolling. Different texture development is also expected. This will be followed by further investigations into the material flow behavior in order to achieve the best possible groove filling and when designing the corresponding tools. In combination with an in-depth investigation of the process window, the effect and further improvement of the mechanical properties is also of interest.

Acknowledgement and Funding
The authors would like to thank the Federal Ministry for Economic Affairs and Climate Action for supporting this research work through the project "CLEAN-Mag: CO2-neutral production of lightweight magnesium components", project no. 03LB3080A, part of the Technologietransfer-Programm Leichtbau (TTP LB). Moreover, the authors would like to thank the colleagues from the IMF for operating the tensile tests (Mr. Winderlich) and the SEM/EBSD analysis (Ms. Korpala, Ms. Ueberschär).

References

[1] R.L. Doiphode, S.V.S. Narayana Murty, N. Prabhu, B.P. Kashyap, Effects of caliber rolling on microstructure and room temperature tensile properties of Mg–3Al–1Zn alloy, J. Magnes. Alloy. 1 (2013) 169–175. https://doi.org/10.1016/j.jma.2013.07.005

[2] J.H. Lee, B.J. Kwak, T. Kong, S.H. Park, T. Lee, Improved tensile properties of AZ31 Mg alloy subjected to various caliber-rolling strains, J. Magnes. Alloy. 7 (2019) 381–387. https://doi.org/10.1016/j.jma.2019.06.002

[3] A. Tripathi, S.V.S.N. Murty, P.R. Narayanan, Microstructure and texture evolution in AZ31 magnesium alloy during caliber rolling at different temperatures, J. Magnes. Alloy. 5 (2017) 340–347. https://doi.org/10.1016/j.jma.2017.07.001

[4] T. Mukai, H. Somekawa, T. Inoue, A. Singh, Strengthening Mg–Al–Zn alloy by repetitive oblique shear strain with caliber roll, Scripta Mater. 62 (2010) 113–116. https://doi.org/10.1016/j.scriptamat.2009.09.005

[5] T. Mukai, H. Somekawa, A. Singh, T. Inoue, Strengthening Mg-Al-Zn Alloy by Repetitive Oblique Shear Strain, in: W.H. Sillekens, S.R. Agnew, N.R. Neelameggham, S.N. Mathaudhu (Eds.), Magnesium Technology 2011, John Wiley & Sons, Inc., Hoboken, NJ, USA, 2011: pp. 211–214. https://doi.org/10.1002/9781118062029.ch40

[6] R.L. Doiphode, S.V.S.N. Murty, N. Prabhu, B.P. Kashyap, Grain growth in calibre rolled Mg–3Al–1Zn alloy and its effect on hardness, J. Magnes. Alloy. 3 (2015) 322–329. https://doi.org/10.1016/j.jma.2015.11.003

[7] H. Yu, D. Wang, Y. Liu, Y. Liu, L. Huang, B. Jiang, S. Park, W. Yu, F. Yin, Recrystallization mechanisms and texture evolution of AZ31 alloy by gradient caliber rolling, J. Mater. Res. Tech. 23 (2023) 611–626. https://doi.org/10.1016/j.jmrt.2023.01.044

[8] A. Stefanik, P. Szota, S. Mróz, T. Bajor, H. Dyja, Properties of the AZ31 Magnesium Alloy Round Bars Obtained in Different Rolling Processes / Własności Prętów Okrągłych Ze Stopu Magnezu AZ31 Otrzymanych W Różnych Procesach Walcowania, Arch. Metall. Mater. 60 (2015) 3001–3006. https://doi.org/10.1515/amm-2015-0479

[9] P. Adamyanets, M. Schmidtchen, R. Kawalla, Extension of the Freiberger Model of Spread for the Calculation of Material Flow during Rolling of Long Products to a New Material Group of Mg Alloys AZ31, AZ81, WE43, KEM 716 (2016) 677–684. https://doi.org/10.4028/www.scientific.net/KEM.716.677

[10] T. Kong, B.J. Kwak, J. Kim, J.H. Lee, S.H. Park, J.H. Kim, Y.H. Moon, H.S. Yoon, T. Lee, Tailoring strength-ductility balance of caliber-rolled AZ31 Mg alloy through subsequent annealing, J. Magnes. Alloy. 8 (2020) 163–171. https://doi.org/10.1016/j.jma.2019.11.005

[11] Y. Chino, Y. Nakaura, K. Ohori, A. Kamiya, M. Mabuchi, Mechanical properties at elevated temperature of a hot-deformed Mg–Al–Ca–Mn–Sr alloy, Mater. Sci. Eng. A 452–453 (2007) 31–36. https://doi.org/10.1016/j.msea.2006.12.118.

[12] M. Ullmann, K. Kittner, T. Henseler, A. Stöcker, U. Prahl, R. Kawalla, Development of new alloy systems and innovative processing technologies for the production of magnesium flat products with excellent property profile, Procedia Manuf. 27 (2019) 203–208. https://doi.org/10.1016/j.promfg.2018.12.065

[13] F. Bachmann, R. Hielscher, H. Schaeben, Texture Analysis with MTEX – Free and Open Source Software Toolbox, SSP 160 (2010) 63–68. https://doi.org/10.4028/www.scientific.net/SSP.160.63

[14] M. Moses, J. Luft, M. Ullmann, U. Prahl, R. Kawalla, Impact of Initial State during Calibre Rolling: Investigating Microstructure and Mechanical Properties of AZ80 Magnesium Alloy, MSF 941 (2018) 857–862. https://doi.org/10.4028/www.scientific.net/MSF.941.857

Forming of magnesium alloy cup using friction heated punch

HARADA Yasunori[1,a *] and TAKAHARA Taiki[2,b]

[1]Graduate School of Engineering, University of Hyogo, C229, 2167 Shosya, Himeji, Hyogo 6712280, Japan

[2]Graduate Student of Engineering, University of Hyogo, C230, 2167 Shosya, Himeji, Hyogo 6712280, Japan

[a]harada@eng.u-hyogo.ac.jp, [b]takaharak059@gmail.com

Keywords: Sheet Forming, Deep Drawing, Drawn Cup, Magnesium Alloy, Friction Heating, Thermal Conductivity, Dissimilar Materials

Abstract. In deep drawing of aluminum and magnesium alloys, it is known that many alloys are difficult to form at room temperature. In particular, many commonly used magnesium alloys are very difficult to form at room temperature. To improve formability, warm processing is used. However, the problem is that the mold requires heating and cooling equipment and its control equipment. In the present study, we attempted warm deep drawing using frictional heat generation. To heat the punch itself, a frictional heat generation structure was incorporated inside the punch. The materials of the friction heater jig were several kinds of stainless steel and alloy steel tools. The relationship between the heat generation temperature of the frictionally heated punch and the working time was investigated. In addition, the amount of wear due to friction of each jig was investigated. It was found that sufficient working temperatures were obtained in the friction-heated combination of austenitic stainless steel and alloy tool steel. It was also found that good formability of the deep-drawn cup was obtained in warm deep drawing using the prototype friction heated punch.

Introduction

In recent years, efforts have been made to improve the fuel efficiency of next-generation vehicles such as electric vehicles, hybrid vehicles, and fuel cell vehicles in order to reduce carbon dioxide emissions. In the case of automobiles, the weight of the vehicle body tends to increase due to the expansion of equipment such as electronic parts. In addition, rapid progress is being made in the combination of multiple materials with different properties, in other words, the use of multi-materials to achieve higher strength and lighter weight [1]. In the case of vehicle bodies for transportation machines such as automobiles and aircraft, the use of resins and light metals instead of steel materials is being promoted in order to reduce weight [2, 3]. Aluminum (Al) alloys and magnesium (Mg) alloys are attracting attention as representative light metals [4, 5]. Cast or extruded parts are used for many Al alloys and Mg alloys. Recently, from the viewpoint of mass production, material yield, and thinning of products, it has been desired to apply press processing to sheet materials [6]. However, it is known that some alloys are difficult to process at room temperature [7, 8]. For example, in the case of Mg alloys processed by rolling or extrusion, it is difficult to form them due to the characteristic texture formed during deformation [9, 10]. Currently, press processing is often used in the manufacture of automobiles using parts that make up automobiles such as bodies and doors [11]. Press forming has been used for a long time as a typical plastic working technique. This technique has relatively high productivity because it allows products to be produced in large quantities with fewer steps. Furthermore, by changing the mold, it can be processed into multiple shapes such as shearing and bending. One of these processing methods is deep drawing. This is a technology in which a blank metal sheet is placed in a metal mold called a die, and a punch tool is used to apply a predetermined load to form a seamless three-dimensional container. Although it is known as a severe processing method in press processing, it is relatively suitable for mass production. It is widely used in industrial products such as automotive parts such as sensor cases and battery cases, and in daily necessities such as beverage cans and cooking pots [12, 13]. As mentioned above, however, many Al alloys and Mg alloys have lower ductility and poor formability than steel materials. In particular, in the case of Mg alloys, many alloys are difficult to process at ambient temperature, with the exception of beta-type alloys [14] and microstructurally controlled alloys [15, 16]. For this reason, it is currently very difficult to form the material using cold deep drawing. One method to improve forming performance is warm deep drawing [17]. This process heats the flange before forming to reduce deformation resistance, thereby improving the formability of deep drawing. Formability is also improved by increasing deformation resistance and strength by cooling the part that contacts the punch shoulder. In the case of warm forming, however, the current problem is that a control device with a heating and cooling structure is required for the forming mold and the high cost associated with such equipment.

In the present study, we focused on the frictional heat generated during friction stir welding (FSW) and friction joining (FJ). In FSW and FJ technology, this is a technology in which materials softened by frictional heat are stirred and welded. Many studies have been reported on joining dissimilar materials [18, 19]. In the case of steel materials, it is known that the temperature generated by friction is over 900°C [20]. In order to heat the metal punch itself, a frictional heat generation structure was incorporated inside the punch. In other words, a friction heat generating jig made of different metals was built inside the punch. Warm deep drawing was attempted using the frictional heat generated by the rotation of the jig. The relationship between the frictionally heated punch temperature and working time was investigated. Warm deep drawing was also performed using a friction heated punch.

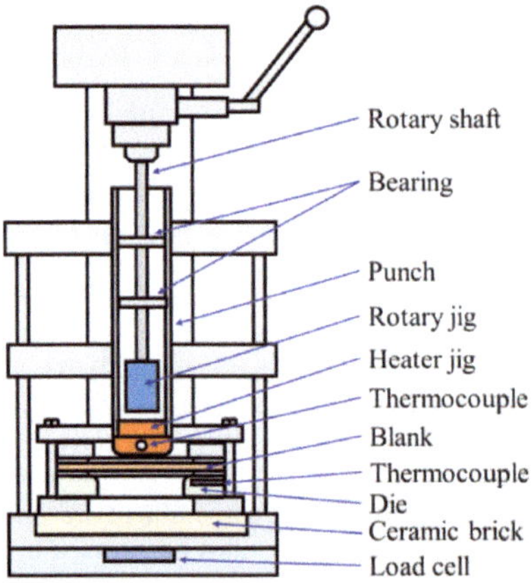

Fig. 1. Deep drawing method by using friction heating.

Experimental procedures

The prototype punch was heated by frictional heat. A deep drawing equipment consisting of punch and die was installed on a drilling machine. A schematic illustration of the warm deep drawing equipment using frictional heat is shown in Fig. 1. The punch is heated by the spindle of the drilling machine. At the tip of the punch, when a predetermined temperature is reached, the spindle is lowered by rotation of the drilling machine handle. The upper part of the rotary shaft connected to the rotary jig was attached to the gripping section of the drilling machine. The rotational speed ranged from 2400 to 2800 rpm. The rotating shaft was fixed by two bearings installed inside the punch. The temperature generated by frictional heating was measured by several thermocouples. The measurement locations were near the heater jig and the die shoulder radius.

The material of the rotary jig was tool steel or stainless steel. The heater jig was the same alloy tool steel used for the die. These jigs used are material grades and designation defined in JIS (Japanese Industrial Standards) standards. In the deep drawing process, the blanks used were Al alloy A5052 and Mg alloy AZ31. The blanks were 35 mm in diameter and 0.5 mm thick. The punch was a commercially available stainless steel SUS304 pipe. SUS304 is chromium-nickel austenite stainless steels. This steel is a material grade and designation defined in JIS G4303 standard. It was 160 mm long, with outer and inner diameters of 20 mm and 16 mm, respectively. The punch had a diameter of 20 mm and a shoulder radius of 3 mm. The die had a hole diameter of 21 mm and a shoulder radius of 4 mm. The lubricant was a heat-resistant molybdenum disulfide-based work oil. The clearance between the punch and the die was 0.5 mm. Table 1 shows working conditions in friction heating.

Table 1. Working conditions in friction heating.

Rotary jig	Alloy steel	SUP9 (spring steel, Cr-Mg steel), 299 HV
		SUJ2 (bearing steel, high-carbon Cr steel), 318 HV
		SCM440 (Cr-Mo steel), 326 HV
		SKS3 (Mn-Cr-W steel), 210 HV
	Alloy tool steel	SKD11 (cold work tool steel), 232 HV
		SKD61 (hot work tool steel), 220 HV
		SKH51 (high-speed tool steel), 231 HV
	Stainless steel	SUS303 (austenite stainless steel), 300 HV
		SUS304 (austenite stainless steel), 188 HV
		SUS310S (austenite stainless steel), 237 HV
		SUS316L (austenite stainless steel), 249 HV
		SUS403 (martensitic stainless steel), 206 HV
		SUS430 (ferritic stainless steel), 188 HV
		SUS327L1 (duplex phase stainless steel), 273 HV
		SUS630 (precipitation hardening steel), 415 HV
Heater jig	Alloy tool steel	SKD11 (cold work tool steel), 232 HV
		SKD61 (hot work tool steel), 220 HV
		SKH51 (high-speed tool steel), 231 HV
Rotating velocity		2400 - 2800 rpm
Surface roughness of worked surface (by polishing paper, particle size: d)		Sa*: 0.33 mm, Sz**: 5.97 mm (abrasive paper count: #240, d: 0.08 mm)
Load force of rotary jig		200 - 400 N

* Sa (ISO 25178) is a parameter that extends an arithmetic mean height Ra to a surface.
** Sz (ISO 25178) is the extension of a maximum height Rz to a surface.

After friction processing, the surface state of the rotary jig and heater jig that generated heat was observed with a digital microscope. The wear depth was also measured for the rotary jig and the heater jig after processing. The wear depth was defined as the dimensional change in the depth direction from the surface before and after processing. The maximum length before and after friction heating was measured with a digital micrometer in the rotary and the heater jigs.

Results and Discussions

The relationship between the punch temperature heated by friction and the processing time of friction was investigated. The temperature of the heater jig was measured when the rotary jig was in contact with the heater jig while rotating. The relationship between the temperature and working time of heater jig SKD11 is shown in Fig. 2. The rotary jig was alloy steel type (a) and stainless steel type (b). The broken line in the figure indicates the temperature at which the rotary jig stopped rotating, which is 350°C. In the warm forming of light metals, it is generally known that the forming limit is greatly improved when the metal is heated to approximately 300°C [21, 22]. In the preliminary test, the heater jig reached a temperature of more than 500°C after a few minutes due to friction with the rotary jig. Since a heating temperature of approximately 300°C was sufficient for forming light metals, the rotation of the rotary jig was stopped when the heating temperature reached 350°C. The temperature of heat generated by friction was 350°C or higher in all rotary jigs. After the rotary jig stopped rotating, the heat temperature increased. When the rotary jig was made of stainless steel, the maximum temperature of the heater jig was approximately 400°C . Moreover, the time during which the temperature was maintained at 350°C or higher was long, i.e., 100 s or longer. This is because the thermal conductivity of stainless steel is lower than that of other steel materials [23].

The effect of the material of the heater jig on heat generation was investigated. When the material of the heater jig was changed, there was little effect on the relationship between the frictional heat-generating temperature and the processing time. The maximum temperature of the stainless steel type rotary jig was found to be higher, and the retention time above 350°C was also longer. The relationship between heat generation temperature and rotation time is shown in Figs. 3 and 4, respectively.

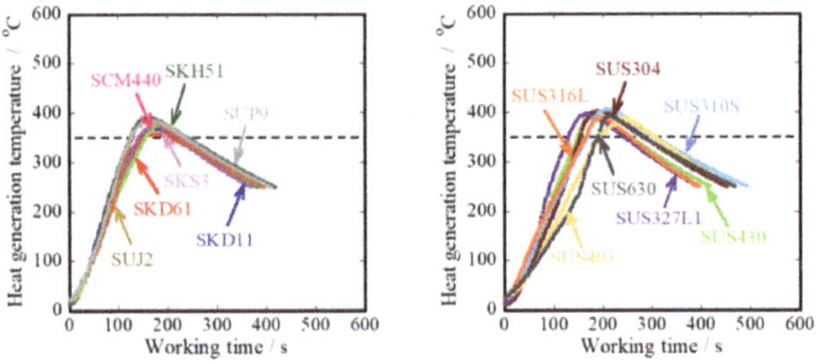

(a) Rotary jig: alloy steel type (b) Rotary jig: stainless steel type

Fig. 2. Variations of heat generation temperature with working time for heater jig SKD11.

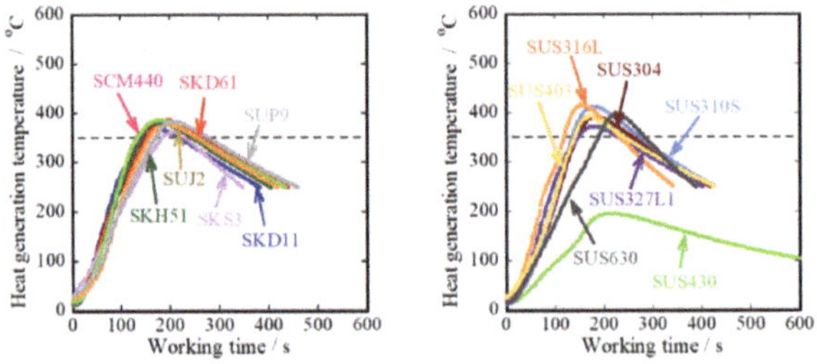

(a) Rotary jig: alloy steel type (b) Rotary jig: stainless steel type

Fig. 3. Variations of heat generation temperature with working time for heater jig SKD61.

Materials Research Proceedings 44 (2024) 340-349 https://doi.org/10.21741/9781644903254-36

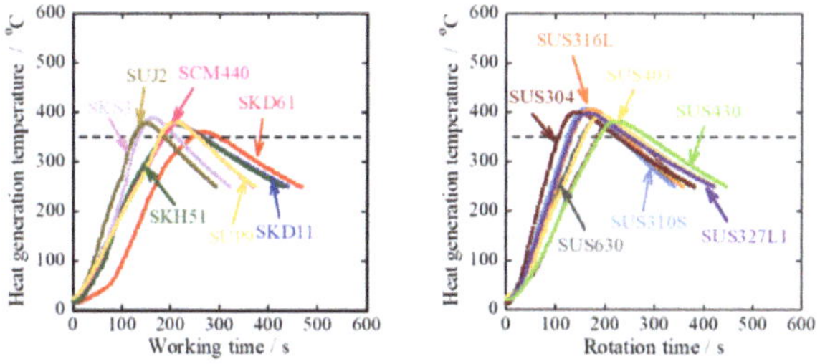

(a) Rotary jig: alloy steel type (b) Rotary jig: stainless steel type

Fig. 4. Variations of heat generation temperature with working time for heater jig SKH51.

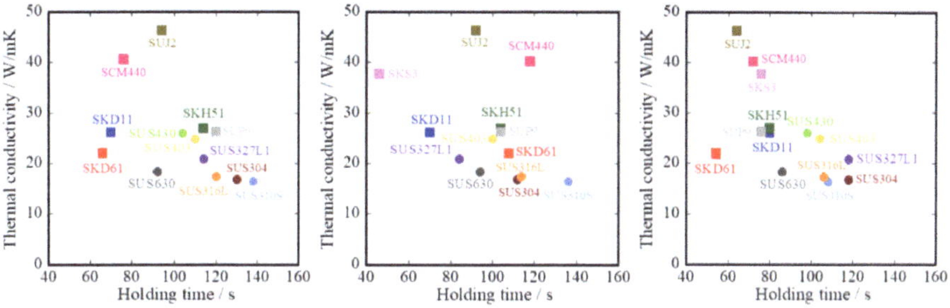

(a) Heter jig: SKD11 (b) Heater jig: SKD61 (c) Heater jig: SKH51

Fig. 5. Relationship between thermal conductivity and holding time of over 350°C.

Differences in time were observed between different types of rotary jigs in terms of holding time at heating temperatures above 350°C. In the rotary jigs made of stainless steel, the time to hold the heating temperature above 350°C was long. This is because the thermal conductivity of stainless steel is smaller than that of other steel materials. Therefore, the relationship between the time for which the temperature is maintained at 350°C and the thermal conductivity was investigated. The relationship between holding time and thermal conductivity in various rotary jigs is shown in Fig. 5. The heater jigs are SKD11 (a) , SKD61 (b) and SKH51 (c). Especially, the holding time of SUS304 and SUS310S was 120 s or longer. In other words, the thermal conductivity of these materials is low. The thermal conductivity of tool steel is approximately 45.1 W/(m-K), while that of stainless steel SUS304 is approximately 16.0 W/(m-K) [24]. Therefore, the use of stainless steel as the rotary jig was found to be effective for heating the punch.

After frictional heating, the surface condition of the rotary and heater jig was observed. The surfaces of the rotary jig SUS304 and heater jig SKH51 after friction processing are shown in Fig. 6. Although traces of friction were observed, no seizure was observed. Similarly, almost no seizure was observed in other combinations of rotary jigs and heater jigs.

The amount of wear on the rotary jig and heater jig after frictional heating was investigated. The amount of wear was defined as the wear depth obtained by measuring the dimensional change of the jig. The wear depth after friction tests on alloy tool steel type (a) and stainless steel type rotary jigs is shown in Fig. 7. The heater jig was SKD11. In the case of rotary jigs made of alloy tool steel, the wear depth for all rotary jigs was less than 0.02 mm. In the case of rotary jigs made of stainless steel, the wear depth was more than 0.02 mm. Due to the low thermal conductivity of stainless steel, heat tends to accumulate in the surface layer. The softening of the surface layer is thought to have increased the amount of wear. In stainless steel rotary jigs, the wear depth of SUS310S, SUS403, and SUS630 was 0.04 mm or more. The exact reason why the wear amount of the three types of steel was higher is currently unknown.

Similarly, when the heater jig was changed to SKD61 or SKH51, the amount of wear of the stainless steel type jig increased compared to that of the alloy tool steel jig. The wear depth after friction tests on alloy tool steel type (a) and stainless steel type rotary jigs is shown in Fig. 8. The heater jig was SKD61.

(a) Rotary jig: SUS304 (b) Heater jig: SKH51

Fig. 6. Surfaces of rotary jig and heater jig after friction processing.

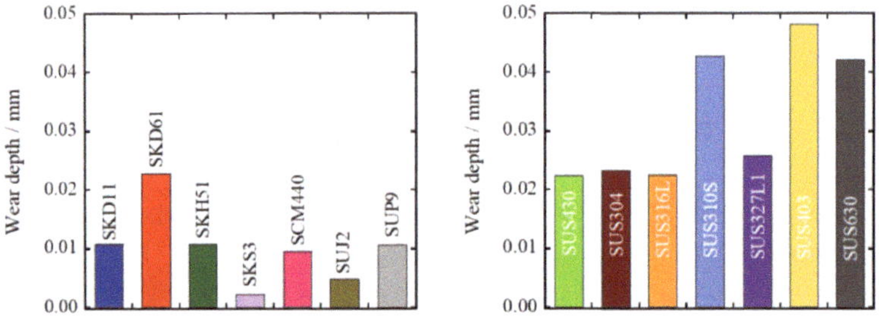

(a) Rotary jig: alloy tool steel type (b) Rotary jig: stainless steel type

Fig. 7. Wear depth of rotary jig relative to heater jig SKD11.

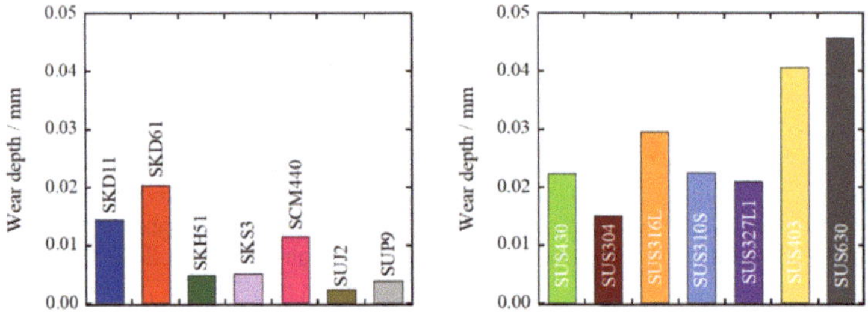

(a) Rotary jig: alloy tool steel type (b) Rotary jig: stainless steel type

Fig. 8. Wear depth of rotary jig relative to heater jig SKD61.

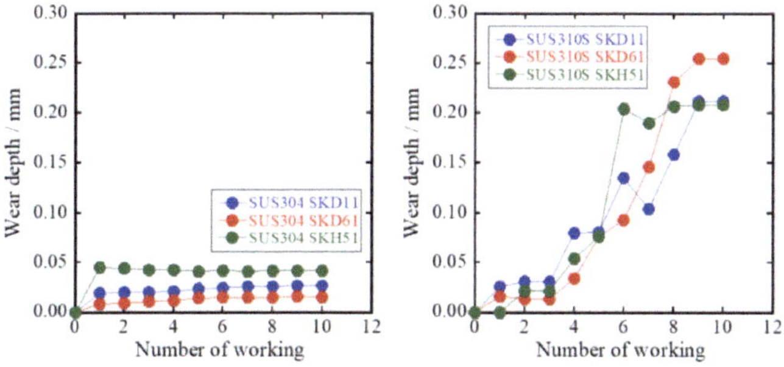

(a) Rotary jig: SUS304 (b) Rotary jig: SUS310S

Fig. 9. Variations of wear depth of rotary jig with number of working.

The effect of the number of friction tests on the wear depth was investigated. The friction test was performed multiple times in succession to measure the amount of wear. The wear depth after multiple friction tests on the stainless steel type rotary jig is shown in Fig. 9. The number of steps in the wear test was 10 times. In the second and subsequent wear depth measurements, the length of the jig before the test was used as the standard. The rotary jig was made of stainless steel SUS304 (a) and stainless steel SUS310S (b). The heating jigs were SKD11, SKD61, and SKH51. In both rotary jigs, the rotary jig was worn in the first test. In the second and subsequent tests, the depth of wear for SUS304 did not change much from the depth of wear in the first test. In the tenth test, the jig wear depth was less than 0.05 mm for the three types of heater jigs. On the other hand, when the rotary jig was made of SUS310S, the wear depth increased significantly after the second test; at the 10th test, the wear depth ranged from 0.20 mm to 0.26 mm. In austenitic stainless steels, the addition of Cr and Ni in SUS310S is higher. The Cr and Ni content in SUS304 is 18 wt.% and 8 wt.%, and that in SUS310S is 20 wt. % and 25 wt. %. When the Ni content is high, the work hardening rate decreases; the work hardening coefficients for SUS304 and SUS310S are approximately 0.44 and approximately 0.34. The reason why the wear of SUS310S jigs is greater than that of SUS304 jigs is because the hardness of the surface layer during friction is lower.

The effect of the material type of the heater jig on the wear depth after multiple friction tests was investigated. There was almost no effect of material on the wear depth of SUS304 and SUS310S rotary jigs.

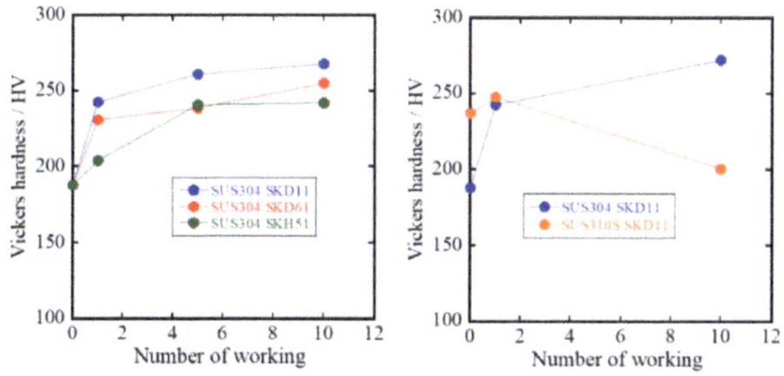

(a) Rotary jig: SUS304 (b) Rotary jig: SUS304, SUS310S

Fig. 10. Variations of Vickers hardness with number of working.

R.T. (4.8 kN) 133 °C (4.2 kN) R.T. (fractured) 210 °C (2.9 kN)
(a) Aluminum alloy A5052 (b) Magnesium alloy AZ31

Fig. 11. Appearances of drawn cups by using friction heated punch.

Differences in the depth of wear after multiple friction tests were observed for the SUS304 and SUS310S rotary jigs. The hardness of the surface layer of both jigs after the multiple friction test was examined. The relationship between the hardness of the surface layer of the rotary jig and the number of friction tests is shown in Fig. 10. Each figure shows the change in hardness of the rotary jig SUS304 with respect to three different heater jigs (a) and the change in hardness of two different rotary jigs with respect to the heater jig SKD11 (b). In the rotary jig SUS304 (a), as the number of tests increased, the hardness for the three types of heater jigs also increased. When the number of tests was 10 times, the hardness of the rotary jig was approximately 250 HV. The hardness of the rotary jig was approximately 1.3 times higher than the pre-test hardness. Microstructural observation of the cross section showed that fine microstructures were observed in the surface layer. After the rotation of the rotary jig stopped, recrystallization is thought to have occurred in the worked layer. As the number of tests increased, the area of fine structure expanded. Regarding the rotary jig SUS304 and SUS310S for the heater jig SKD11 (b), the hardness of the rotary jig SUS310S decreased as the number of tests increased. In the surface layer of the rotary jig SUS310S, few areas of fine structure were observed.

Warm deep drawing was performed using the punch heated by frictional heating. The appearances of the drawn cup A5052 (a) and drawn cup AZ31 (b) are shown in Fig. 11. The temperature of the process was measured inside near the shoulder of the die. In the A5052 blank (a), formability was good. The cup could be formed at room temperature, and the warm forming load was reduced. The maximum loads were 4.8 kN and 4.2 kN at room temperature and warm forming. In the AZ31 blank (b), fracture occurred at room temperature due to poor ductility. At warm temperatures, however, the formability was improved. In warm forming, the maximum load was 2.9 kN. The thickness distribution of the warm drawn cups was measured; the maximum reduction in thickness for the A5052 and AZ31 cups was approximately 5 % for both cups.

Summary

Friction heating was performed on several combinations of steel materials using the punch with a built-in friction heater jig. The following results were obtained: (1) When the rotary jig was made of alloy tool steel or stainless steel, a heat generation temperature of 370 to 400°C was obtained at a rotation time of 180 s; (2) When the rotary jig was made of austenite stainless steel, a holding time of more than 350°C was obtained for more than 100 s due to low thermal conductivity; (3) No trace of burning was observed on the friction surfaces where the rotary jig and heater jig contact each other; (4) When the rotary jig was made of stainless steel, the depth of wear increased compared to other grades of rotary jigs, but the maximum value was less than 0.1 mm; (5) The proposed method was capable of heating to the punch. Therefore, it was found that there was a possibility to improve the workability of Mg alloys by warm deep drawing.

Acknowledgment

The authors would like to thank Mr. H. Togawa, Material and Manufacturing Center, for advice on experimental equipment. This research was supported in part by a grant from AMADA foundation under grant No. AF-2020016-B3.

References

[1] D. Li, X. Fang, Lightweight design approach of an LFT-metal multi-material vehicle door concept, Automot. Engine Tech. 7 (2022) 385-407. https://doi.org/10.1007/s41104-022-00121-9

[2] I. Turkmen, N.S. Koksal, Investigation of mechanical properties and impact strength depending on the number of fiber layers in glass fiber-reinforced polyester matrix composite materials, Mater. Testing 56 (2014) 472-478.

[3] H. Li, X. Sun, J. Wang, L. Wang, Study of aging stability and precipitation kinetic of the Al-Zn-Mg alloy applied for auto body structures, J. Mater. Res. Tech. 9 (2020) 15575-15584. https://doi.org/10.1016/j.jmrt.2020.11.025

[4] P. Shurkin, G. Scamans, N. Barekar, L. Hou, T. Subroto, C. Barbatti, Phase composition and microstructure of high strength AA6xxx aluminium alloys with nickel additions, Materials Trans. 64 (2023) 398-405. https://doi.org/10.2320/matertrans.MT-LA2022047

[5] A. Kielbus, R. Jarosz, Gating system optimization for EV31A magnesium alloy engine body sand casting, Materials 15 (2022) 4620. http://dx.doi.org/10.3390/ma15134620

[6] K.S. Bhandari, S. Aziz, W.N. Chen, S.J. Li, D.W. Jung, Deformation evaluation of A5052 sheet metal in SPIF process, Mater. Sci. Forum 1084 (2023) 91-95. http://dx.doi.org/10.4028/p-e6768o

[7] R. Comanei, L. Zaharia, R. Chelariu, Damaging prediction of difficult-to-work aluminum alloys during equal channel angular pressing, J. Mater. Eng. Perform. 213 (2012) 287-297. https://ui.adsabs.harvard.edu/link_gateway/2012JMEP...21..287C/doi:10.1007/s11665-011-9904-5

[8] S. Fritsch, S. Hunger, M. Scholze, M. Hockauf, M.F.-X. Wagner, Optimisation of thermo mechanical treatments using cryogenic rolling and aging of the high strength aluminium alloy AlZn5.5MgCu (AA7075), 42 (2011) 573-579. http://dx.doi.org/10.1002/mawe.201100835

[9] E. Tolouie, R. Jamaati, Effect of rolling reduction on the microstructure, texture, and mechanical behavior of AZ91 alloy, J. Mater. Res. Tech. 26 (2023) 7947-7957. https://doi.org/10.1016/j.jmrt.2023.09.119

[10] S.M. Fatemi, H. Paul, A. Hemmati-Hosuri, Microstructure, texture, and instability during superplastic deformation of ECAP-treated AZ31 Mg alloy, Mater. Sci. Tech. 39 (2023) 1661-1669. http://dx.doi.org/10.1080/02670836.2023.2180594

[11] P. Kaczynski, M. Skwarski, K. Jaskiewicz, Development of the technology for press-forming of energy-absorbing elements made of 7075 aluminum alloy, J. Manuf. Process. 50 (2020) 676-683. https://doi.org/10.1016/j.jmapro.2020.01.023

[12] F. Han, R. Radonjic, New approach for wrinkle prediction in deep drawing process, Key Eng. Mater. 639 (2015) 459-466. https://doi.org/10.4028/www.scientific.net/KEM.639.459

[13] C. Lee, G. Park, C. Kim, Design of Type 3 High-Pressure Vessel Liner (Al 6061) for Hydrogen Vehicles, J. Pressure Vessel Technology 144 (2022). https://doi.org/10.1115/1.4054366

[14] F.R. Cao, H. Ding, Y. Li, G. Zhou, I.Z. Cui, Superplasticity, dynamic grain growth and deformation mechanism in ultra-light two-phase magnesium-lithium alloys, Mater. Sci. Eng. A 527 (2010) 2335-2341. https://doi.org/10.1016/j.msea.2009.12.029

[15] S.-J. Kim, C. Lee, J. Koo, J. Lee, Y. Lee, D. Kim, Improving the room-temperature formability of a magnesium alloy sheet by texture control, Mater. Sci. Eng. A 724 (2018) 156-163. http://dx.doi.org/10.1016/j.msea.2018.03.084

[16] K. Suzuki, Y. Chino, X. Huanng, M. Yuasa, M. Mabuchi, Enhancement of Room Temperature Stretch Formability of Mg-1.5 mass%Mn Alloy by Texture Control, Mater. Trans. 54 (2013) 392-398. https://doi.org/10.2320/matertrans.M2012312

[17] Y. Nakayama, T. Naka, T. Uemori, I. Shimizu, Temperature and Processability of Magnesium Alloy AZ31 on Rectangular Cup Deep Drawing, Key Eng. Mater. 535/536 (2013) 326-32. https://doi.org/10.4028/www.scientific.net/KEM.535-536.326

[18] K. Bhardwaj, P. Sandeep, T. Saurabh, B. Santosh, Corrosion and Mechanical Aspects of Friction Stir Welded AA6061 Joints: Effects of Different Backing Plates, J. Mater. Eng. Perform. 32 (2023) 10817-10833.
https://ui.adsabs.harvard.edu/link_gateway/2023JMEP...3210817K/doi:10.1007/s11665-023-07900-x

[19] B. Wang, P. Zhu, Y. Cao, L. Zhou, P. Xue, L. Wu, Effects of different friction stir welding processes on residual stress and deformation of Ti62A alloy joints, J. Mater. Res. Tech. 26 (2023) 6096-6107. https://doi.org/10.1016/j.jmrt.2023.08.308

[20] T. Li, X. Xie, J. Xu, R. Li, K. Qi, X. Zhang, H. Yue, Y. Zhao, L. Lei, Research on AZ31 Mg alloy/22MnB5 steel pinless friction stir spot welding process and interfacial temperature field simulation, J. Mater. Res. Tech. 26 (2023) 3710-3725. https://doi.org/10.1016/j.jmrt.2023.08.169

[21]Y. Lee, S. Kim, S.-Y. Park, J. Yoo, Y. Moon, Friction effect of surface treated tools used for warm forming of Mg alloy sheets, Int. J. Precision Eng. Manuf. 15 (2014) 2631-2637. https://doi.org/10.1007/s12541-014-0637-x

[22] F. Tamashiro, H. Hamasaki, F. Yoshida, Equi-Plastic Work Locus of 5000 Series Aluminum Alloy Sheet at Warm Temperature, Key Eng. Mater. 725 (2017) 695-699. http://dx.doi.org/10.4028/www.scientific.net/KEM.725.695

[23] P. Muthuswamy, Investigation on sustainable machining characteristics of tools with serrated cutting edges in face milling of AISI 304 Stainless Steel, Procedia CIRP 105 (2022) 865-871. https://doi.org/10.1016/j.procir.2022.02.143

[24] Information on https://www.jssa.gr.jp/english/

[25] M. Toit, H. G. Steyn, Comparing the Formability of AISI 304 and AISI 202 Stainless Steels, J. Mater. Eng. Perform. 21 (2012) 1491-1495. http://dx.doi.org/10.1007/s11665-011-0044-8

Effect of surface characteristics on strain distribution in air- bending

KAIJALAINEN Antti[1,a *], POKKA Aki-Petteri[1,b], JASKARI Matias[2,c], HUUKI Juha[3,d], HINTSALA Tommi[1,e] and KÖMI Jukka[1,f]

[1]Materials and Mechanical Engineering, Centre for Advanced Steels Research, University of Oulu, Pentti Kaiteran katu 1, 90570 Oulu, Finland

[2]Kerttu Saalasti Institute, University of Oulu, Pajatie 5, 85500 Nivala, Finland

[3]Department of Mechanical Engineering, Aalto University, Puumiehenkuja 3, 02150 Espoo, Finland

[a]antti.kaijalainen@oulu.fi, [b]aki-petteri.pokka@oulu.fi, [c]matias.jaskari@oulu.fi, [d]juha.huuki@aalto.fi, [e]tommi.hintsala@oulu.fi, [f]jukka.komi@oulu.fi

Keywords: Roughness, Hardness, Bendability, DIC, Ultra-High Strength Steel

Abstract. This study is a continuation of previous studies; therefore, the idea is to determine the influence of surface roughness and hardness on the bendability of a UHSS grade, and it is investigated with 3-point bending tests, utilizing Digital Image Correlation (DIC) for measuring the strain distributions on the outer curvature. Investigated bending samples of a 4 mm thick commercial bainitic-martensitic steel sheet were tested in different surface conditions: (1) as-rolled, (2) dry electropolished, (3) ultrasonic burnishing and (4) a combination of dry electropolishing and ultrasonic burnishing. Ultrasonic burnishing increased the surface hardness and decreased the surface roughness. Bending results showed that smoother surface did not lead to improved bending angles and higher hardness in the subsurface decreased the bending capacity, i.e., reduced the maximum bending angle and critical strain.

Introduction

For ultra-high strength strip steels (UHSS), cold bending is the most important method of forming for applications such as containers and crane booms. Bendability is commonly determined as the minimum ratio of the punch radius to the strip thickness that the steel can tolerate without the appearance of surface defects during bending to an angle of 90° [1]. However, in this study, the bendability of a UHSS grade is investigated with 3-point bending tests, utilizing Digital Image Correlation (DIC) for measuring the strain distributions on the outer curvature [2, 3, 4, 5] and not into angle of 90°. Earlier bending studies [6, 7] have shown that strain localization is the precursor to failure in the air bending of ultrahigh-strength strip steels and shear bands have been observed to initiate at depths in the range of 1 – 6 % of the total strip thickness from the surface at angles of ~45° to the surface, i.e. in the maximum shear stress directions. Thus, it is reasonable to suppose that the near-surface properties of the strip govern the bendability.

Therefore, the ultrasonic burnishing is selected to for the surface treatment, as it strongly affects the surface quality [8], by increasing the surface hardness and decreasing the surface roughness. During the burnishing the surface layer of a workpiece plastically deforms under the combined effect of the static force created by pressing the tool onto the workpiece and the dynamic force created by the ultrasonic oscillation system [9]. Ultrasonic transducers convert high frequency electrical power into ultrasonic mechanical vibration. The equipment forges the surface of the workpiece with a hard metal point more than 20000 impacts per second. There is no removal of the material surface, unlike in grinding or polishing.

As has been shown in earlier studies [3, 10], the shot blasting affects the surface quality and decreases bendability, as well as increases the hardness at the subsurface layer. This study is a continuation of previous studies; therefore, the main aim of the present paper is to establish an understanding of how bendability can be influenced by modifying the surface characteristics using the polishing and ultrasonic burnishing to achieve the smooth and harder surface for bending trials. The differences in the extent of multi-breakage and the bend shapes are also studied, and these observations are correlated with the findings from the bending force, strain measurements and surface roughness and hardness on the bendability of a UHSS grade.

Experimental

Investigated bending samples of 4 mm thick commercial hot-rolled bainitic-martensitic steel (0.08C-0.2Si-1.0Mn, in wt.%) were tested in different surface conditions: (1) as-rolled, (2) dry electropolished (P), (3) ultrasonic burnishing (U) and (4) a combination of dry electropolishing and ultrasonic burnishing (PU), which are seen in Fig. 1a.

Surface Treatments. Dry electropolishing was utilized using a commercial electropolishing equipment (DLyte 100 HF+). The machine is equipped with a 16-litre container which can be used to polish several samples simultaneously. Instead of liquid electrolyte typical for conventional electropolishing, polymer media containing mild acids (labelled Steel 04) was used in this study, supplied by the manufacturer. Used alternating voltage was 24 V, and maximum of 2 hours polishing time was used, until roughly 20 µm was removed from the sample surface (example of as-rolled and polished surfaces are shown in Fig. 1a).

The burnishing was performed together with the Mazak FH 480 milling center and HIQUSA ultrasonic burnishing equipment (Fig. 1b), with a wolfram-carbide ball 6 mm diameter. The used milling center table feed was 2000 mm/min and the side shift 0.025 mm, while the ultrasonic burnishing forging was performed at a frequency of 19 kHz. The spring preload was 1.0 mm for the workpiece. A cutting fluid (filtered 5 % oil-water emulsion) was used to cool the tool in the process.

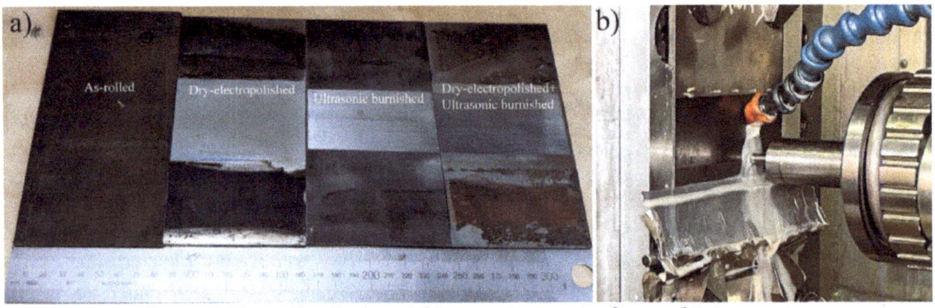

Figure 1. *Setup for the burnishing equipment in the machining center.*

Surface Roughness and Hardness Measurements. Surface roughness was measured using laser scanning confocal microscope (LSCM) according to ISO 4287:1997. Subsurface hardness profiles were measured using a micro-hardness tester (CSM, Switzerland) under a 0.2 (HV0.02) and 0.5 N (HV0.05) load.

Bending Trials. The investigated materials are tested in the transverse orientation, referring to the bend axis being perpendicular to the rolling direction, and they are bent until fracture. In the bending samples, the naming "b" means repeat of the test. Rectangular specimens with width (bendlength) of 80 mm and thickness of 4 mm were bent in room temperature using a Zwick 100 kN universal tensile test machine with purpose-build three-point bending tools (Fig. 2a) with DIC camera setup. The setup allows measuring of bending force, punch stroke and most importantly, continuous DIC measurement throughout the test through the openings in the lower tool. The punch is also stationary in this setup to ensure that the measurement area stays in focus throughout the tests. As the punch is stationary, the necessary punch stroke is produced with the lower tool moving upwards at a speed of 1 mm/s. For this study, punch radius of 4 mm and a die shoulder radius of 6 mm were used, with a 70 mm gap (die width) between the centers of the shoulders. Die shoulders are freely rotating and lubricated with PTFE lubricant. More information of the test setup can be found in Refs. [2, 3].

The specimen width was chosen small enough to fit the available tools but large enough to ensure plane strain condition at the center of the bend. The bending force is expressed as force per width (N/mm) to compensate for the small variation in the specimen widths. The punch stroke data was used to calculate the bending angles throughout the test according to ISO 7438:2016. The punch stroke data was adjusted for the vertical elasticity of the setup (combined rigidity of the tensile test machine and the bending tools), measured at 51.9 kN/mm.

The DIC-system used for measuring deformation on the outer surface of the bend was Strainmaster by Lavision, equipped with two monochrome CCD cameras with a resolution of 2456 × 2058 pixels. The specimens were painted with a black-and-white speckle pattern to enable accurate tracking of points on the measured surface by the DIC-software (DaVis 8.4). The DIC recording and processing parameters are presented in Table 1. Fig. 2b provides an example of a strain map of the outer surface of a bent specimen, obtained through DIC measurement. From the obtained strain maps, the principal strain values were extracted from three lines (sections A – C) positioned at the centre of the bend at 10 mm intervals as can be seen in Fig. 2b. The peak strain ε_{b_max} at each point in time was calculated as the average of the peak values of the sections A – C.

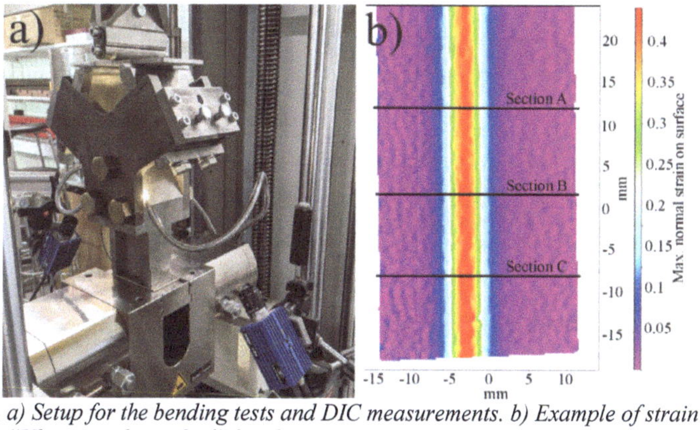

Figure 2. *a) Setup for the bending tests and DIC measurements. b) Example of strain map of "Ultrasonic burnished" bend specimen at 60° angle and sections A – C.*

Materials Research Proceedings 44 (2024) 350-357 https://doi.org/10.21741/9781644903254-37

Table 1. *Recording and processing parameters for the DIC system.*

DIC system and software	LaVision Strainmaster (Stereo DIC), DaVis 8.4
Sensor, digitization, imaging rate	2456 × 2058, 12-bit, 2 Hz
Lens, imaging distance, image scale	35 mm C-mount, 0.43 m, 25.7 pixels/mm
Subset size, step size, VSG	15 × 15 pixels, 5 pixels, 35 pixels
Strain window, smoothing method	5 × 5 data points, 2nd order polynomial fit
Standard deviation of principle strain	380 – 460 microstrain
Interpolation, shape function, algorithms LSM	6th order spline interpolation; affine shape function; (iterative least squares matching) algorithm based on optical flow estimation

Results and Discussion

Surface roughness and hardness. Surface roughness measurements are presented in Table 2 and roughness profiles of as-rolled, dry electropolished, ultrasonic burnishing and a combination of dry electropolishing and ultrasonic burnishing are shown in Fig. 3. R_a means the arithmetical mean deviation of the assessed profile, R_z is the maximum peak to valley height of the profile, and R_v is the max. valley depth below the mean surface. Dry electropolishing reduced the surface roughness as the R_a value decreased from 1.62 μm (as-rolled) to 0.88 μm (dry electropolished). The ultrasonic burnishing smoothed the surface even more to 0.56 μm. As previously discussed in Ref. [3, 10], R_a and R_z values might not be the most suitable values as the crack initiation during bending starts from the sharp pits or scratches, so R_v value is better to use when the correlation between the bendability and roughness is evaluated. The as-rolled material has the deepest valleys on the surface and an R_v value of 19.90 μm. While the rest of the surfaces (P, U and PU) have similar level R_v values of 3.6 – 4.6 μm.

Table 2. *Investigated surface roughness values.*

Material	R_a (μm)	R_z (μm)	R_v (μm)
As-rolled	1.62	30.30	19.90
Dry electropolished (P)	0.88	9.81	4.58
Ultrasonic burnished (U)	0.56	8.16	3.63
Polished + burnished (PU)	0.64	8.43	3.75

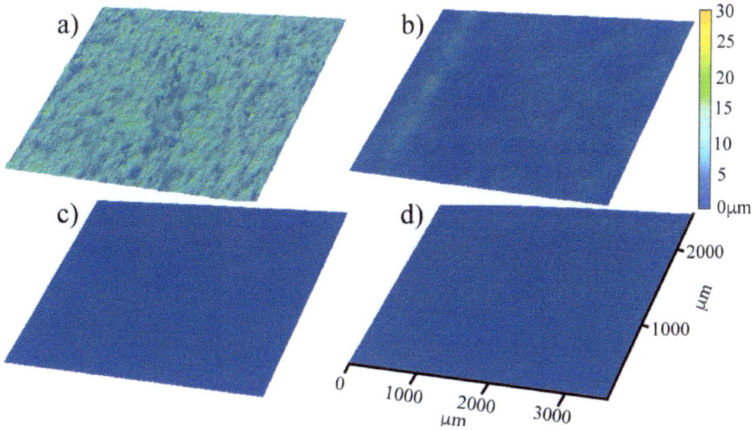

Figure 3. Topographical 3D surface profiles of a) as-rolled (after HCl pickling), b) dry electropolished, c) ultrasonic burnishing and d) a combination of dry electropolishing and ultrasonic burnishing.

Subsurface microhardness profiles are presented in Fig. 4. The hardnesses measured using a 200 mN load (at 25 μm below the surface) show larger scatter and higher hardness values compared to the rest of the measurements made with a 500 mN load. This means that Vickers hardness values measured using different loads are not comparable. All the samples show that the hardness generally increases throughout the thickness from the surface to 100 μm and after 100 μm the hardness is stabilized. As shown in previous studies [3, 10], the surface treatments affect the hardness values near the surface. Dry electropolishing does not seem to affect the hardness values as those are similar to the as-rolled surface hardness. However, the hardness at 25 μm below surface is increased from ~355 HV (as-rolled) to 380 HV with ultrasonic burnishing (U) or even to 405 HV with the combination of polishing and burnishing (PU). Same behavior of increased hardness is observed at 50 μm below surface. Microstructural investigation showed that the microstructure was homogenous in the measurement area.

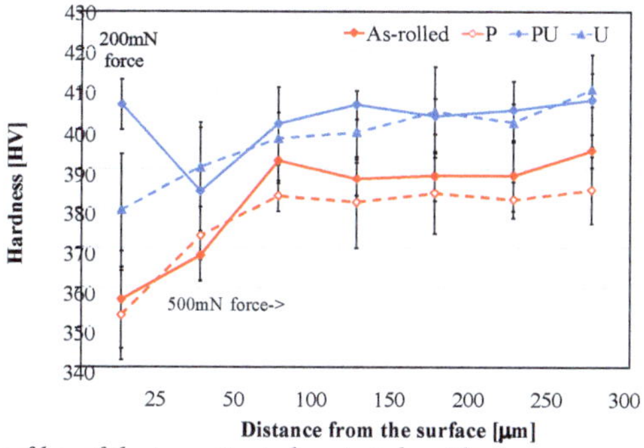

Figure 4. Hardness profiles of the investigated materials with six measurements of each. Notice: 25 μm is measured with a 200 mN load and 50 – 300 μm is measured with a 500 mN load. Therefore, the results are not comparable.

Bending. Examples of bend samples are shown in Fig. 5. The bending tests with DIC-measurement showed that the 60° bend angle is the last point where all specimens were still intact. Therefore, the strain distributions at 60° bend angle are presented in Fig. 6. The distributions were extracted from section B, illustrated in Fig. 2b. The distributions of all the specimens are almost identical at this point, and the small differences are a combination of random noise in the DIC measurement and actual variance in the material properties as seen between the as-rolled S and Sb samples (Fig. 6a) and ultrasonic burnished U and Ub samples (Fig. 6b).

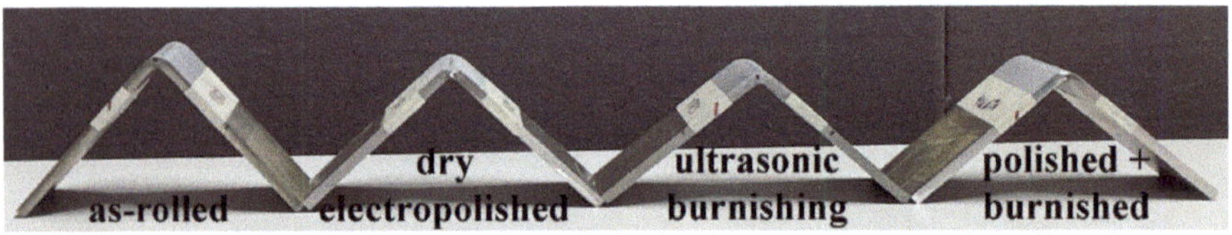

Figure 5. *Examples of the bent samples with the bend axis transverse to the rolling direction.*

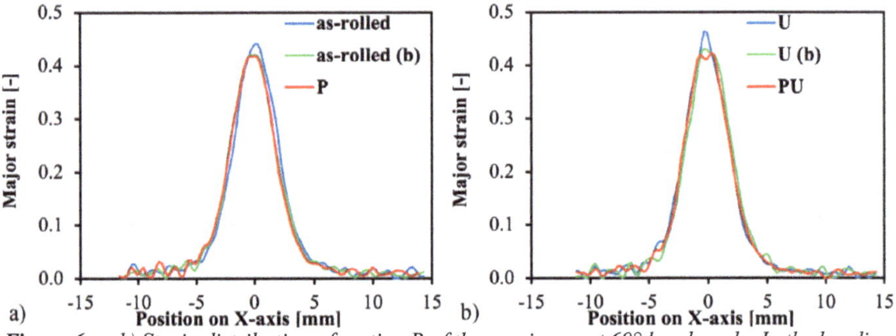

Figure 6. *a, b) Strain distribution of section B of the specimens at 60° bend angle. In the bending samples, the naming "b" means repeat of the test.*

The measured bending force and the average peak strain of sections A – C are plotted against the bend angle in Fig. 7. Since there were some small variations between the repeat tests of the as-rolled and ultrasonic burnished specimens, both repeats for these specimen types are included in Fig. 7. The development of both the force and the strains are near identical between all the specimens until 60° bend angle, when the first specimens start to fracture. The critical angle and strain, at which damage on the specimen surface starts to appear, can be determined from the sharp increase in the peak strain. The determined critical bending angles and peak strains of DIC-bending trials are listed in Table 3.

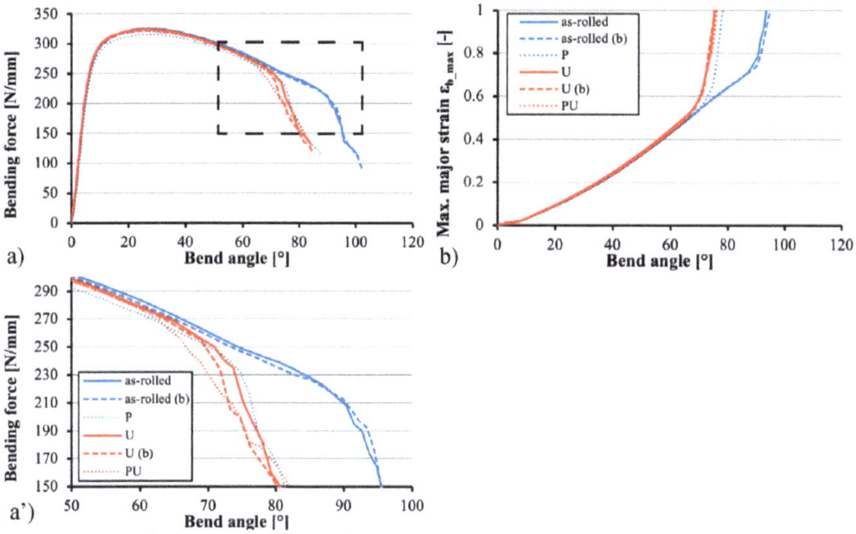

Figure 7. *Development of a) the bending force and b) the peak strain on the outside surface. a') Magnification of bending force from bend angle 50° to 100°.*

Table 3. *Critical bending angles ($\alpha_{crit.}$) and peak strains ($\varepsilon_{crit.}$) of the investigated specimens.*

Results of the repeat tests in brackets.

Material	$\alpha_{crit.}$ (°)	$\varepsilon_{crit.}$ (-)
As-rolled	87 (90)	0.71 (0.74)
Dry electropolished (P)	69 (70)	0.53 (0.54)
Ultrasonic burnished (U)	65 (63)	0.50 (0.47)
Polished + burnished (PU)	66 (66)	0.50 (0.51)

Bending results showed only minor variation between repeat tests, thus it can be stated that the DIC bending test is a solid method for studying the differences between the surfaces of materials. The as-rolled tests have clearly the highest critical angle (87 – 90°) and strain (0.71 – 0.74). The polished and burnished specimens started to fracture first at approx. 65° angle or 0.50 strain. Interestingly, no major difference can be seen between the dry polished (P) and ultrasonic burnished (U or PU) samples even though the surface hardnesses were higher on the U and PU samples compared to the dry polished sample (P). Therefore, it can be stated that in this study, smoothness did not improve the bending properties and that the dry electropolishing and ultrasonic burnishing decreased the bending capacity. These treatments affect the surface, not only by increased hardness, but possibly also by increasing the dislocation densities and residual stresses. However, these factors are not determined in this study, but could be investigated in the future.

Summary

Dry electropolishing and ultrasonic burnishing were found to decrease the surface roughness. Ultrasonic burnishing increased the hardness of the surface. Both surface treatments affected the critical angle and strain significantly but had no effect on the strain distribution. Based on this, it could be concluded that the strain distribution in bending is determined by the properties of the whole cross-section, while the critical strain is more affected by the properties near the surface. Dry electropolishing seemed to have no effect on the surface hardness, however surprisingly it still decreased the critical bending angle. The fractures may well be initiated from below the surface, in which case the surface roughness would be irrelevant. The main effect of the surface treatment (i.e., polishing or burnishing) might come from the work-hardening of the material near the surface, where the hardness is increased. Therefore, more detailed surface characterization is needed, e.g., residual stress and dislocation density investigations.

Acknowledgements
Financial assistance of the Business Finland, project FOSSA– Fossil-Free Steel Applications, are acknowledged.

References

[1] SFS-EN ISO 7438, Metallic materials – Bend test, (2016).

[2] A.-P. Pokka, A.-M. Arola, A. Kaijalainen, V. Kesti, and J. Larkiola, Strain distribution during air bending of ultra-high strength steels, ESAFORM 2021, (2021). https://doi.org/10.25518/esaform21.2509

[3] A. Kaijalainen, A.-P. Pokka, M. Jaskari, and J. Kömi, Effect of surface roughness on strain distribution during bending, IOP Conf. Ser. Mater. Sci. Eng., 1270 (2022) 012080. https://doi.org/10.1088/1757-899x/1270/1/012080

[4] M. Kaupper and M. Merklein, Bendability of advanced high strength steels—A new evaluation procedure, CIRP Ann. Manuf. Technol., 62 (2013) 247–250. https://doi.org/10.1016/j.cirp.2013.03.049

[5] K. Cheong, C. Butcher, and J. Dykeman, The influence of the through-thickness strain gradients on the fracture characterization of advanced high-strength steels, SAE Int. J. Mater. Manuf., 11 (2018) 2018-01–0627. https://doi.org/10.4271/2018-01-0627

[6] A.-M. Arola, A. Kaijalainen, and V. Kesti, The effect of surface layer properties on bendability of ultra-high strength steel, AIP Conf. Proc., 1769 (2016) 200024. https://doi.org/10.1063/1.4963642

[7] A. J. Kaijalainen, P. P. Suikkanen, L. P. Karjalainen, and D. A. Porter, Influence of subsurface microstructure on the bendability of ultrahigh-strength strip steel, Mater. Sci. Eng. A, 654 (2016) 151–160. https://doi.org/10.1016/j.msea.2015.12.030

[8] J. Huuki, R. Ullah R, and E. Fangnon, Effects of ultrasonic burnishing on the surface quality of corrosion-resistant tool steel using a hard-carbon-coated burnishing tool, Mater. Res. Proc., 28 (2023) 1775–1780. https://doi.org/10.21741/9781644902479-192

[9] J. Huuki and S. V. Laakso, Integrity of surfaces finished with ultrasonic burnishing, Proc. Inst. Mech. Eng. B: J. Eng. Manuf., 227 (2013) 45–53. https://doi.org/10.1177/0954405412462805

[10] A. J. Kaijalainen, V. Kesti, J. Heikkala, R. Ruoppa, D. A. Porter, and J. I. Kömi, Bendability of shot blasted ultrahigh-strength steels, Mater. Sci. Forum, 941 (2018) 510–515. https://doi.org/10.4028/www.scientific.net/MSF.941.510

Hole expansion performance of a medium manganese advanced high-strength steel after hot rolling and intercritical annealing

PLOSILA Pekka[1,a*], KANTANEN Pekka[1,b], HANNULA Jaakko[1,c], JAVAHERI Vahid[1,d], KÖMI Jukka[1,e] and KAIJALAINEN Antti[1,f]

[1]Centre for Advanced Steels Research, Materials and Mechanical Engineering, University of Oulu, Pentti Kaiteran katu 1, 90570 Oulu, Finland

[a]pekka.plosila@oulu.fi, [b]pekka.kantanen@oulu.fi, [c]jaakko.hannula@oulu.fi, [d]vahid.javaheri@oulu.fi, [e]jukka.komi@oulu.fi, [f]antti.kaijalainen@oulu.fi

Keywords: Medium Manganese Steel, Annealing, Retained Austenite, Formability, Stretch-Flangeability

Abstract. This paper investigates the hole expansion behavior of a medium manganese advanced high-strength steel. The study began with vacuum-cast 0.3C–1Si–6Mn–2Al steel, which was subsequently laboratory hot-rolled to a thickness of 4 mm. Intercritical annealing treatments (IAT) were then carried out at temperatures of 650 °C and 700 °C to produce varying fractions of tempered martensite/ferrite and retained austenite (RA). Tensile tests were performed in the longitudinal direction, while hole expansion tests following ISO 16630 were conducted to assess the stretch-flangeability of the IAT materials. Test holes were prepared using punching and wire electrical discharge machining (W-EDM) to examine the effects of different edge conditions. Tensile strength of 1027 MPa and 911 MPa corresponding to a microstructure consisting of RA fraction of 7.5% and 43.3% were obtained at IA temperatures of 650 °C and 700 °C, respectively. Mean hole expansion ratios (HER) with punched holes were relatively low for the given strength levels (3% and 17% for IAT 650 °C and IAT 700 °C, respectively). W-EDM edge condition considerably improved HER but not as much as expected, potentially due to the high inhomogeneity of the microstructure. Further research on optimization of processing and microstructure is required to enhance the poor hole expansion performance of intercritically annealed medium manganese steels.

Introduction

In the automotive sector, utilizing advanced high-strength steels, which offer an enhanced combination of strength and ductility, can result in improvements in cost-effectiveness, weight reduction, and fuel efficiency. This is particularly evident in chassis components. Production of sheet metal components often involves blanking followed by different cold-forming processes. During these forming operations, the blank edges are subjected to various stress and strain conditions. Consequently, there is a potential risk of failure at the cut edge during subsequent stages of production due to the damage and strain hardening present in the sheared edge [1]. The formability of sheared edge becomes an increasingly critical concern as material tensile strength rises.

The ISO 16630 hole expansion test [2] is the most commonly used method for assessing the suitability of sheet metals for forming processes involving stretching of cut edges. In this test, a conical expansion tool is forced into a punched hole until a through-thickness crack occurs. Diameter of the expanded hole after failure is then compared to the initial hole diameter. The test results typically exhibit significant variability and can be greatly influenced by testing conditions, such as cutting clearance, tool shape and wear [3]. Alternative hole preparation methods, such as wire electrical discharge machining (W-EDM) [4,5], demonstrate significant improvement in the hole expansion performance of high-strength steels. This enhancement is attributed to the prevention of the formation of a highly deformed zone near the edge of the material.

Advanced high-strength steels (AHSS) consist of diverse microstructures, including different fractions of martensite, bainite, ferrite, and retained austenite (RA), which result in varying local ductility. In AHSSs with complex microstructures containing considerable amount of retained austenite, increase in the fraction of RA [6] and fresh martensite [7] have been reported to decrease hole expansion performance.

Medium (4–10%) manganese steel is one of the possible candidates to achieve the requirements of third generation AHSS grades [8]. Beyond desired properties of AHSSs, considerations of industrial manufacturability are crucial in developing new steel grades. In the case of medium manganese steels, the IAT process leverages a substantial fraction of metastable austenite to enhance ductility. The amount of RA is primarily controlled by the intercritical annealing time and temperature at which austenite and ferrite coexist allowing for alloying element partitioning to occur. During IAT carbon (C) and substitutional elements like manganese (Mn) partition from martensite/ferrite to austenite. Both C and Mn are strong austenite stabilizers and an increase in their concentration can result in increasing austenite stability at room temperature [9,10].

Hence, this study aims to evaluate hole expansion performance of a medium manganese advanced high-strength hot-rolled steel after intercritical annealing treatments producing different fractions of tempered martensite/ferrite and particularly different amount of retained austenite. Additionally, effects of shear cutting on the stretch-flangeability of the IAT materials are investigated by comparing punched and wire electrical discharge machined hole edges.

Experimental

Research Materials and Processing. Experimental medium manganese steel containing 0.3C–1Si–6Mn–2Al was designed and vacuum cast into a 65 kg ingot. After the vacuum casting, pieces of around 40 × 80 × 90 mm were cut from the ingot and soaked in a furnace at 1200°C for 2 hours. Subsequently, the steel pieces were laboratory hot-rolled to a thickness of 4 mm with a final rolling temperature (FRT) of 920°C followed by air cooling to room temperature (RT).

After the hot rolling, the steels were annealed in a furnace at intercritical region with temperatures of 650°C and 700°C to produce two test materials with different microstructures; especially with low and high fraction of retained austenite. Holding times at IA temperatures were 10 min, and after IAT the materials were air-cooled to RT.

Base Material Microstructural Investigation. Microstructures of the materials were investigated utilizing a Zeiss Sigma field emission scanning electron microscope (FESEM). The samples were prepared from the rolling direction – normal direction (RD-ND) cross-sections and polished with the standard metallographic sample preparation. Samples were etched with 2% Nital before investigation.

Detailed microstructural features were characterized by employing electron backscatter scanning diffraction (EBSD) detector. EBSD investigations were made with JEOL JSM-7900F FESEM equipped with an Oxford EBSD detector. All samples for EBSD and x-ray diffraction (XRD) were polished to mirror finish using a silica suspension (0.04 μm) in the final step. EBSD samples (RD-ND cross-section) were investigated at quarter thickness of the materials with acceleration voltage of 20 kV and a step size of 0.02 μm. The captured data was post-processed using Oxford AZtecCrystal software.

Retained austenite (RA) fractions were quantified utilizing XRD analysis using a 9 kW Rigaku SmartLab XRD unit operated on a Co-Kα source. XRD analysis was facilitated by PDXL2 analysis software. A de-texturing method was adopted in the XRD measurements, and a Rietveld refinement technique was utilized in the determination of the phase fractions. Volume fractions of RA (XRD) were measured at the material quarter thickness on rolling direction – transverse direction (RD-TD) cross-sections.

Material bulk Vickers hardness was also measured from the FESEM samples using a Zwick ZHU 2.5 universal hardness testing machine. The testing was conducted at material quarter thickness with five measurement points and 98.07 N testing force.

Tensile Testing. Uniaxial tensile tests were performed at room temperature with a Zwick Z100 materials testing machine in accordance with EN ISO 6892-1 [11] utilizing sheet thickness flat specimens with gauge length of 50 mm, parallel length of 75 mm, and width of 12.5 mm. Tensile testing was conducted in the longitudinal direction with 3 repeat specimens. Strain hardening exponents (n_{4-6}) were determined from the tensile test data in a plastic strain range between 4% and 6%.

To evaluate local ductility of the investigated materials, thickness strains at fracture of the tensile specimens were investigated. Measurements were conducted on the fracture surfaces of the tensile specimens with a digital microscope (Dino-Lite Edge). The microscope was focused on the thinnest point of the fracture surface. The true plastic thickness strain ($\varepsilon_{p}\_a_{min}$) was calculated with Eq. 1 [12]

$$\varepsilon_{p}\_a_{min} = \ln(a_0/a_{min}), \qquad (1)$$

where a_0 is the initial specimen thickness and a_{min} is the specimen thickness at minimum thickness point of the fracture surface.

Hole Expansion Testing. To evaluate stretch-flangeability of the materials, hole expansion tests according to ISO 16630 [2] were performed. 10 mm diameter test holes for the hole expansion testing were prepared using two different methods: punching according to the standard and wire electrical discharge machining (W-EDM) to obtain edges with shear-affected and essentially pre-damage free conditions, respectively. Prior to hole preparation, rolling surfaces of all hole expansion test pieces were sandblasted to remove excess furnace and rolling scale. Punching was conducted according to the ISO 16630 with a 12.5% cutting clearance. Four punched and two W-EDM holes were produced for each material. One unexpanded hole for both punched and W-EDM conditions was reserved for hole edge quality investigations.

Hole expansion tests were conducted at Lapland University of Applied Science in Kemi with an Erichsen universal sheet metal testing machine. For each material, three punched and one W-EDM hole was tested. A conical expansion tool with a 60° tip angle was forced into the hole of the specimen at rate of 15 mm/min. The burr side of the specimen was set facing away of the expansion tool and no lubrication was used. The movement of the expansion tool was stopped as a through-thickness crack was observed in the hole edge by the operator. Hole diameters were determined after the tests with caliper measurements and the limiting hole expansion ratio (HER) was calculated with Eq. 2 [2]

$$HER = \frac{D_h - D_0}{D_0} \times 100, \qquad (2)$$

where D_h is the hole diameter after the failure and D_0 is the original hole diameter. The diameters were measured in two perpendicular directions. Expanded specimen edges were also investigated with a digital microscope.

Unexpanded Hole Edge Investigation. To investigate differences between the initial hole edge conditions, visual inspection was carried out on all hole edge quality specimens. Hole surfaces of the materials were imaged with laser scanning confocal microscopy. Cross-sectional samples (RD-ND) of material near the hole edge were cut, polished, and etched with 2% Nital. Vickers microhardness profiles inwards from the hole edge of the cross-sectional samples were measured with 3 measurements at a given distance from the hole surface. The measurement location in all samples was at quarter thickness from the top surface, and the utilized measurement force was 0.981N.

Results and discussion

Base Material Microstructural Investigation. Determined fractions of RA and material bulk hardness measurement results are presented in Table 1. Significantly higher RA fraction was achieved with the higher IA temperature. Detected local RA fractions were considerably lower in the EBSD measurements compared to the XRD measurements. IAT 650°C material showed higher bulk material hardness accordingly to the lower RA fraction.

Table 1. Fractions of retained austenite and material bulk hardness measurement results. Standard deviations of hardness measurements are given.

Material	Retained austenite content [vol.%]		Hardness [HV10]
	XRD	EBSD	
IAT 650°C	7.5	2.1	331 ± 2
IAT 700°C	43.3	32.4	305 ± 1

FESEM images of material microstructures at quarter thickness are presented in Fig. 1a–b. IAT 650°C material consisted mostly of fully tempered martensite/ferrite with RA and cementite between the laths in addition to carbide precipitation. In IAT 700°C material, the microstructure was a mixture of fully tempered martensite/ferrite and RA with some cementite precipitation.

EBSD phase maps of material microstructures at quarter thickness are presented in Fig. 1c–d. In IAT 650°C material, most of the map consist of martensite/ferrite with some visible small pools of RA. In IAT 700°C material, the fraction of RA in the map is high. Some unindexed spots were left in the maps, which most likely correspond to carbon-enriched constituents such as cementite.

In IAT 650 C material, with high probability, most of the RA is finely divided as interlath austenite between tempered martensite laths, whereas small fraction of RA existed as micro-size pools as shown in Fig. 1c. From Table 1, it can clearly be perceived that the RA fractions differ between the XRD and EBSD methods. These differences in RA fractions are partly due to the resolution limit of EBSD measurements. Finely divided interlath films are undetectable in EBSD and are detectable in XRD. Measurement location can also affect the results of the local determinations, as there is a significant difference between the RA fractions of IAT 700°C material.

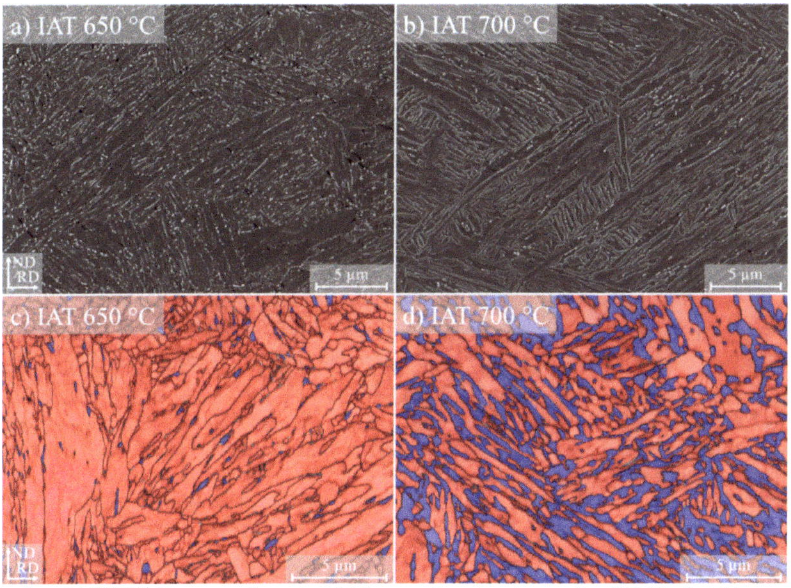

Fig. 1. a–b) FESEM images of quarter-thickness microstructures after 2% Nital etching. c–d) EBSD band contrast, phase, and grain and phase boundary maps of quarter thickness microstructures. Red color corresponds to martensite/ferrite, blue to austenite, black to ≥15° grain and phase boundaries, and grey to unindexed areas.

Tensile Testing. Mean results of tensile testing are presented in Table 2. Lower IA temperature resulted in higher yield and tensile strengths, and simultaneously in lower uniform and total elongations. In IAT 650°C material $n_{4\text{-}6}$ values were also higher indicating greater strain hardening capability, however, yield-to-tensile strength ratios were similar between the materials.

Table 2. Mean tensile and true thickness strain measurement results tested in the longitudinal direction. Standard deviations of the measurements are given.

Material	$R_{p0.2}$ [MPa]	R_m [MPa]	$R_{p0.2}/R_m$ [-]	$n_{4\text{-}6}$ [-]	A_g [%]	A_{50} [%]	$R_m \times A_{50}$ [GPa%]	$\varepsilon_{p}\_a_{min}$ [-]
IAT 650°C	841±8	1027±13	0.82±0.00	0.18±0.00	10.4±0.6	15.8±2.2	16.3±2.4	0.40±0.04
IAT 700°C	774±17	911±10	0.85±0.01	0.11±0.00	16.7±0.4	27.1±0.8	24.7±0.9	0.68±0.03

Thickness strain measurement results are presented in Table 2. IAT 700°C material had higher $\varepsilon_{p}\_a_{min}$ values implying higher tolerance to local deformation, which is also reflected in the higher post-uniform elongation. The fracture surfaces of IAT 700°C specimens were visibly smoother and had more pronounced localization of thinning at the midpoint of the fracture surface compared to IAT 650°C material. Slight local delaminations were observable in the fracture surfaces of IAT 650°C specimens.

Engineering stress-strain curves of the IAT materials tested in the longitudinal direction are presented in Fig. 2a. A yield point plateau is observed in IAT 650 °C material prior to increased strain hardening. In IAT 700°C material, yielding behavior is more constant.

 Materials Research Forum LLC
https://doi.org/10.21741/9781644903254-38

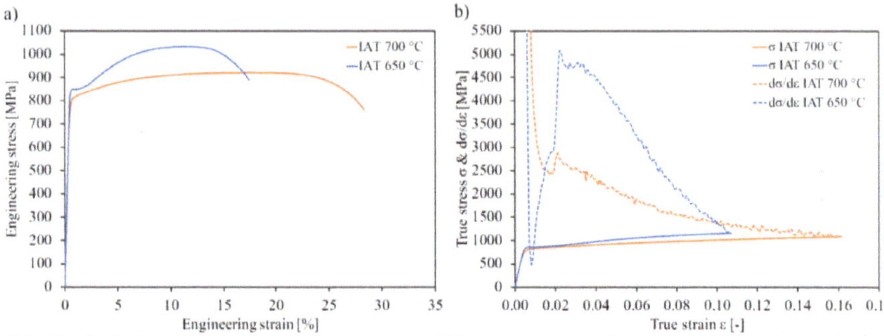

Fig. 2. a) Engineering stress-strain curves and b) true stress-strain curves and strain hardening rates of the IAT materials tested in the longitudinal direction.

True stress-strain curves and strain hardening rates of the IAT materials are presented in Fig. 2b. IAT 650°C material had a significant drop in the strain hardening rate prior to sharp increase to higher values, displaying a pronounced TRIP-effect (transformation-induced plasticity) after the yielding indicating to less stable retained austenite owe to lower IA temperature. In IAT 700°C material, the decrease in the strain hardening rate was steadier. The differences indicate that the stability of the RA is higher in IAT 700 °C material and strain-induced martensite transformation did not occur during the straining the sample. Deformation of ductile austenite enables the high uniform and total elongations in the IAT 700 °C material.

Hole Expansion Testing. Results of hole expansion testing are summarized in Table 3. With the punched hole edge condition, IAT 650°C material had extremely low HER values. In IAT 700°C, the mean HER was slightly higher. With the present testing, it is difficult to decisively state whether this difference is due to increased tensile formability introduced by the higher RA fraction or by the decreased tensile strength, as higher tensile strengths tend to lead to lower mean HER values.

W-EDM edge condition improved HER of both materials, especially IAT 700°C. However, the increase in HER was not as substantial as expected. In hot-rolled complex-phase steels with over 1000 MPa tensile strength, W-EDM HER values over 100% are possible to obtain [5]. High inhomogeneity of the microstructure could have limited the W-EDM HER even when the effects of shear-affected zone are removed.

Table 3. Mean hole expansion test results. Standard deviations of tests with punched hole edge condition are given.

Material	Punched HER [%]	W-EDM HER [%]
IAT 650 °C	4±3	18*
IAT 700 °C	17±4	62

*Failure occurred in an irregular zone in the edge.

In IAT 650°C W-EDM specimen, the failure occurred from a point of unevenness in the edge resulting from the starting/ending point of the hole circumference cutting. Without it, the W-EDM HER value could have been higher, and this should be considered when comparing the HER results. In IAT 700°C specimen, a smaller propagating crack was observed in the similar location, however, the through-thickness crack was located near the material rolling direction. This could indicate that IAT 700°C material is less sensitive to edge defects. As the number of tested specimens was low due to limited amount of laboratory-rolled material, confirmation would require additional investigation.

In Fig. 3a–d, top-down images of the expanded hole expansion specimens are presented. In the punched specimens, especially for IAT 650°C material, the through-thickness failures occurred prior to proper flange formation. In the IAT 700°C W-EDM specimen with higher HER, flanging was more pronounced. The poor surface quality of the laboratory hot-rolled material is also observable in Fig. 3a–d, with visible spots of furnace scale rolled into the material as scale breaking was not possible between annealing and hot rolling. The surface quality could have had an influence on the hole expansion performance, as the defects can function as strain localizers near the hole edge.

Fig. 3. a–d) Top-down images of hole expansion specimens after the testing. e–h) Close-up images of the through-thickness failures in the hole edges.

In Fig. 3e–f, close-up images of the through-thickness failure locations in the hole edges are presented. Comparing the hole surfaces of the punched specimens, IAT 650°C specimens had smaller number of propagating cracks from the top surface of the test compared to IAT 700°C specimens. In Fig. 3g, small cracks can also be seen propagating from the expanding tool side of the hole edge of the IAT 700°C specimen, and delamination of a surface layer in the IAT 650°C specimen is visible in Fig. 3e. The hole edges of W-EDM specimens were smoother with only a few propagating cracks. Fig. 3f shows the through-thickness crack originating from the irregular zone in the edge of the IAT 650°C specimen. In Fig. 3h, some amount of necking is observable prior to the through-thickness failure of the IAT 700°C specimen. In all specimens of Fig. 3e–f, based on the angle of the through-thickness crack, the final failure is occurring through shearing after the crack initiation.

Unexpanded Hole Edge Investigation. Fig. 4 shows the differences between the hole edge conditions with the different preparation methods prior to the hole expansion testing. Laser microscope surface scans of punched edges are presented in Fig. 4a–b. Burnish-to-fracture zone ratios were low in both materials, 15% for IAT 650°C material and 11% for IAT 700°C material. Fig. 4a–b also shows the fracture surface of IAT 700°C material being comparably rougher. Essentially no burring was observed in the specimens. Slight oxidation was visible in the surfaces of the W-EDM specimens.

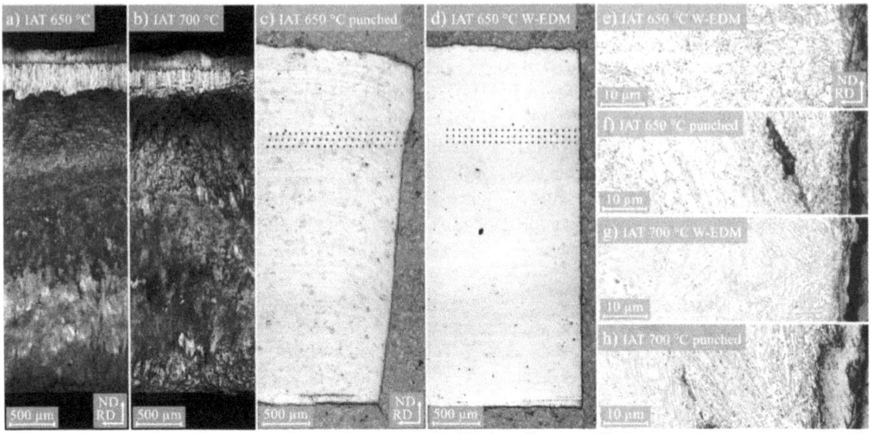

Fig. 4. Images from investigations of unexpanded hole edges: a–b) Laser microscope scans of punched hole surfaces. c–d) Cross-sections of material close to the hole edges. e–f) Laser microscope images of microstructures close to hole edges, 2% Nital etching.

The microhardness measurement lines are visible in the example hole edge cross-section presented in Fig. 4c–d. Nital etching in Fig. 4c also shows deformation of the material towards the sheared edge. Fig. 4d shows the W-EDM edge being considerably smoother compared to rolling surfaces. In Fig. 4c–d, initial delamination of a bottom surface layer is observable. Fig. 4e and g show essentially undeformed microstructure close to the W-EDM edges. Fig. 4f and h show intensive deformation of the microstructure close to the punched edges. Additionally, void formation along the material flow lines can be observed.

Results of microhardness testing of unexpanded hole edges are presented in Fig. 5. No change in the material hardness was observed in the W-EDM edge for both investigated materials. In the punched hole edges, there was a significant difference in the strain hardening of the material near the hole edges between the materials. In IAT 700°C material, the maximum hardness near the edge was higher, and depth of the shear-affected zone was deeper compared to IAT 650°C material. IAT 700°C base material had a major fraction of retained austenite (43.3%) compared to IAT 650°C base material (7.5%). Strain-induced martensite transformation because of high strain rate of the punching process could be a contributing factor to the major increase in the hardness of the punched IAT 700°C sample. The stability of retained austenite is known to be influenced by strain rate dynamics. Therefore, while RA remains stable under tensile loading for the IAT 700°C material, it exhibits instability under applied punching condition. Consequently, while TRIP was not observed during tensile testing, a significant increase in hardness is noted in the punched sample.

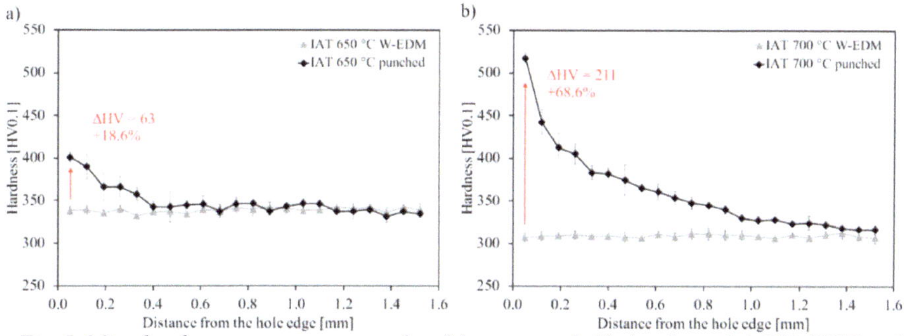

Fig. 5. Microhardness measurement results of the unexpanded hole edges of a) IAT 650°C and b) IAT 700°C materials.

The closest microhardness measurement points were located 50 μm from the hole surfaces. The deformation is more severe closer to the edge as can be seen from the Fig. 4f and h microstructures, thus the actual hardness maximums near the edges can be even higher. For future investigation, it could be beneficial to further examine the fracture surfaces in detail to confirm whether the failure in punching occurs in a more brittle manner in the IAT 650°C material. Additionally, comparisons to edge hardness after the hole expansion testing could provide more insight on the failure behavior of the IAT medium Mn steels with significant RA fraction.

Summary

In this paper hole expansion of a 0.3C–1Si–6Mn–2Al medium manganese advanced high-strength steel was investigated. The steel was laboratory hot-rolled and intercritically annealed (IA) with two IA temperatures to produce different fractions of tempered martensite/ferrite and retained austenite (RA). Tensile tests were performed, and hole expansion tests according to ISO 16630 were conducted to obtain hole expansion ratios (HER) with punched and wire electrical discharge machined (W-EDM) edge conditions.

The results led to following conclusions:

- With the higher IA temperature, material with increased RA fraction and improved tensile strength – total elongation balance was achieved.
- High RA fraction did not lead to drastic improvement of HER in punched materials.
- Punched HER values were poor due to inhomogeneity of microstructure and high amount of strain hardening in the edge.
- W-EDM edge condition improved HER due to removal of shear-affected zone for both materials, the effect was more significant in the material with the higher RA fraction.
- Investigated materials are sensitive to strain rate, which is high in the hole punching process. Stability of RA in shearing should be investigated in further detail.
- Limitations of the laboratory hot rolling compared to industrial production, such as poor surface quality, scale, and temperature variations, could have reduced the hole expansion performance.
- Further optimization of processing and microstructures are required to improve the hole expansion performance of IA treated medium Mn steels.

Acknowledgements

The authors are grateful for financial assistance of the Business Finland research project FOSSA – Fossil-Free Steel Applications. The authors would also like to thank Jane and Aatos Erkon säätiö

(JAES) and Tiina ja Antti Herlinin säätiö (TAHS) for their financial support on Advanced Steels for Green Planet Project.

References

[1] B.S. Levy, C.J. Van Tyne, Review of the shearing process for sheet steels and its effect on sheared-edge stretching, J. Mater. Eng. Perform. 21 (2012) 1205–1213. https://doi.org/10.1007/s11665-011-9997-x

[2] International Organization for Standardization, ISO 16630:2017 Metallic materials – Sheet and strip – Hole expanding test, (2017)

[3] P. Larour, J. Hinterdorfer, L. Wagner, J. Freudenthaler, A. Grünsteidl, M. Kerschbaum, Stretch flangeability of AHSS automotive grades versus cutting tool clearance, wear, angle and radial strain gradients, IOP Conf. Ser. Mater. Sci. Eng. 1238 (2022) 012041. https://doi.org/10.1088/1757-899x/1238/1/012041

[4] A. Karelova, C. Krempaszky, E. Werner, P. Tsipouridis, T. Hebesberger, A. Pichler, Hole expansion of dual-phase and complex-phase AHS steels - Effect of edge conditions, Steel Res. Int. 80 (2009) 71–77. https://doi.org/10.2374/SRI08SP110

[5] W. Cho, B.S. Jeong, K. Jeong, S.H. Lee, H. Kim, J. Lee, S. Kim, H.N. Han, New approach to hole-expansion ratio in complex phase and martensitic steels: Understanding the role of punching damage, J. Mat. Res. Tech. 26 (2023) 837–849. https://doi.org/10.1016/j.jmrt.2023.07.253

[6] J.H. Kim, E.J. Seo, M.-H. Kwon, S. Kang, B.C. De Cooman, Effect of quenching temperature on stretch flangeability of a medium Mn steel processed by quenching and partitioning, Mat. Sci. Eng. A 729 (2018) 276-284. https://doi.org/10.1016/j.msea.2018.05.083

[7] Y.-R. Im, E.-Y. Kim, T. Song, J.S. Lee, D.-W. Suh, Tensile Properties and Stretch-Flangeability of TRIP Steels Produced by Quenching and Partitioning (Q&P) Process with Different Fractions of Constituent Phases, ISIJ Int. 61 (2021) 572-581. https://doi.org/10.2355/isijinternational.ISIJINT-2020-388

[8] N. Fonstein, Advanced High Strength Sheet Steels, first ed., Springer International Publishing, Cham, 2015, XXII 396. https://doi.org/10.1007/978-3-319-19165-2

[9] S. Lee, S.-J. Lee, B.C. De Cooman, Austenite stability of ultrafine-grained transformation induced plasticity steel with Mn partitioning, Scr. Mater. 65 (2011) 225-228. https://doi.org/10.1016/j.scriptamat.2011.04.010

[10] P. Kantanen, S. Anttila, P. Karjalainen, R. Latypova, M. Somani, A. Kaijalainen, J. Kömi, Microstructures and mechanical properties of three medium-Mn steels processed via quenching and partitioning as well as austenite reversion heat treatments, Mater. Sci. Eng. A 847 (2022) 143341. https://doi.org/10.1016/j.msea.2022.143341

[11] International Organization for Standardization, EN ISO 6892-1:2019 Metallic materials. Tensile testing. Part 1: Method of test at room temperature, (2019)

[12] VDA 238-110. Verband der Automobilindustrie e. V., VDA 238-110 Ermittlung der Lokalen Duktilität aus der Bruchfläche von Zugproben / Determination of Local Ductility by Fracture Surface Analyses of Tensile Test Specimens 08/2023, (2023)

Improvement of formability of silicon-containing recycled wrought aluminum alloy by hot stamping after rapid heating

KAWANA Ryohei[1,a *], HIROSAWA Shoichi[1,b *], OOTAKI Mitsuhiro[1,c], MAENO Tomoyoshi[1,d], SUZUKI Yasushi[2,e] and YABUKI Yuuji[2,f]

[1]Graduate School of Engineering Science, Yokohama National University, 79-5 Tokiwadai, Hodogaya-ku, Yokohama, Kanagawa 240-8501, Japan

[2]R&D Division, G-TEKT Corporation, 4-7-28 Shinmeidai, Hamura, Tokyo 205-0023, Japan

[a]kawana-ryohei-fv@ynu.jp, [b]hirosawa@ynu.ac.jp, [c]ootaki-mitsuhiro-kb@ynu.ac.jp, [d]maeno-tomoyoshi-yf@ynu.ac.jp, [e]yasushi_suzuki.nv@g-tekt.jp, [f]yuuji_yabuki@g-tekt.jp

Keywords: Al-Si Alloy, Recycled Aluminum, Rolling, Hot Stamping, Formability, V-Bending

Abstract. To clarify the detrimental effect of Si content on the fabrication of wrought products with recycled aluminum ingots, the formability of cold-rolled Al-Si alloy sheets with different Si contents was evaluated and improved by a V-bending test after rapid heating. It was shown that rapid heating to a certain temperature can improve the formability of Al-Si alloy sheets, and the higher heating temperatures, the more severe forming can be accomplished even for difficult-to-process Al-12%Si alloy sheet. High temperature tensile tests after rapid heating and observation of fracture surfaces indicated that rapid heating improves the ductility of aluminum matrix, and relieves stress accumulation around Si particles, resulting in the suppressed initiation of cracks.

Introduction

To improve fuel efficiency of automobiles, light-weight Al-Si alloy (die)castings are utilized for engine parts, while 5000 series (Al-Mg) and 6000 series (Al-Mg-Si) alloy sheets have been adapted for panel components. However, when the vehicles are disposed, the (die)castings and alloy sheets are mechanically crushed and melted together without sorting, resulting in the production of secondary ingots with various alloy compositions of Si, Mg, Cu, Fe, Mn and so on [1, 2]. In the present recycling system, such recycled ingots are exclusively utilized for fabricating low-purity casting materials (i.e. cascade recycling), but electrification of automobiles forecasts the decreased demand of engine parts. Therefore, application of secondary ingots to wrought products (i.e. upgrade recycle) is highly anticipated by improving their formability.

Recently, the authors successfully improved the formability of difficult-to-process A7075-T6 alloy sheets by hot stamping after rapid heating to medium temperatures [3]. The proposed forming method appears to be applicable to recycled wrought aluminum because short heating time of 10-20 s decreases the deformation resistance but increases the elongation to fracture at the temperatures. In this study, rapid heating to 200, 250 or 300°C was applied to cold-rolled Al-Si alloy sheets with different Si contents, and then their formability was evaluated by 90° V-bending test with punch radii $R = 2$ and 9 mm. The purpose of this study is to optimize heating temperatures for applying recycled aluminum sheets with various Si contents (3, 7 and 12%Si) to wrought materials.

Experiment

Al-3, 7 and 12%Si binary alloys prepared from 99.7%-purity aluminum and Al-19.5%Si mother ingot were melted in a graphite molten crucible by a resistance furnace below 720°C, degasified by Ar gas flow, and then casted into an iron mold preheated to 150°C. The compositions of the fabricated alloys were evaluated by a handheld XRF analyzer (HITACHI X-MET range) and listed in Table 1. The ingots with a dimension of 200×80×20 mm^3 were subjected to surface milling, homogenization treatment at 540°C for 4 h, hot rolling to 10 mm thickness and cold rolling to 2 mm thickness (i.e. reduction ratio of cold rolling is 80%).

Table 1. Compositions of the fabricated Al-Si binary alloys [%].

	Si	Fe	Al
Al-3%Si	3.43	0.22	Bal.
Al-7%Si	6.78	0.22	Bal.
Al-12%Si	12.73	0.28	Bal.

Mechanical properties of the as-cast, as-homogenized and cold-rolled samples were evaluated by room-temperature tensile tests. Tensile test specimens with a dimension shown in Fig. 1 were fabricated by punching with a die. Microstructural observation using optical microscopy (OM) and scanning electron microscopy (SEM) was conducted to clarify the relationship between microstructures and mechanical properties. Specimens for the microscopy observation were mechanically polished by waterproof papers, and then etched with a Keller solution. Size distribution and area fraction of Si particles were quantitively evaluated by analyzing OM images using ImageJ software.

To evaluate the formability of the fabricated Al-Si cold-rolled sheets, 90° V-bending test with punch radii $R = 2$ and 9 mm was conducted at room temperature or, if failed, after rapid heating to 200, 250 or 300°C as schematically illustrated in Fig. 2. A square blank with a dimension of 150×20×2 mm^3 was sandwiched between hot plates at 300°C maximum, and then heated to the target temperature within 10-20 s.

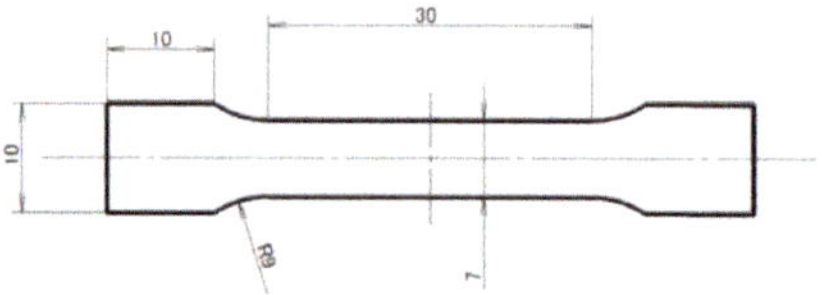

Fig. 1. The dimension of tensile test specimens.

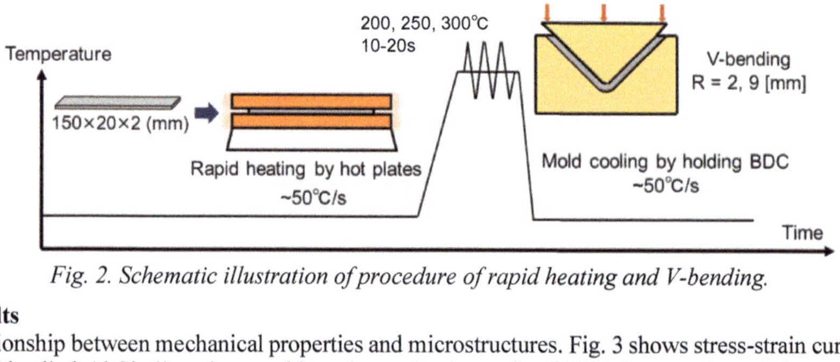

Fig. 2. Schematic illustration of procedure of rapid heating and V-bending.

Results

Relationship between mechanical properties and microstructures. Fig. 3 shows stress-strain curves of cold-rolled Al-Si alloy sheets subjected to reduction ratio of 80%. Irrespective of Si content, ultimate tensile stress (UTS) of the three alloys reached 200-210 MPa although the elongation to fracture decreased from 8% to 4% with increasing Si content. The corresponding OM and SEM microstructures are shown in Fig. 4. It was found from SEM images that Si particles are fractured through rolling process, and sharp edges cause stress concentration during tensile deformation, resulting in lower ductility. The decrease in ductility is greater in the highest Si-containing Al-12%Si alloy because Si particles are coarser and more angular as illustrated by the size distribution in Fig. 5.

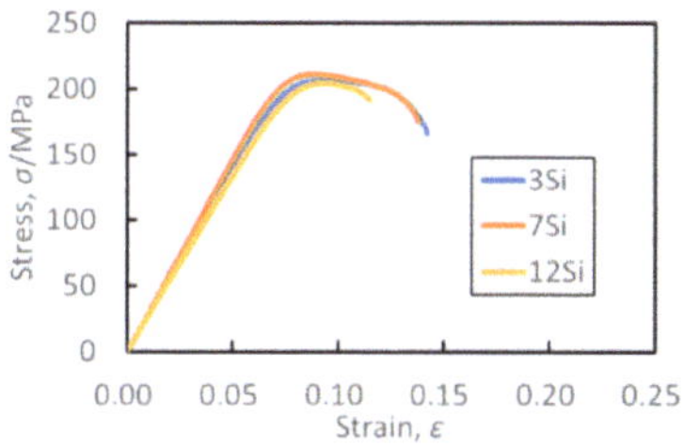

Fig. 3. Stress-strain curves of cold-rolled Al-Si alloy sheets.

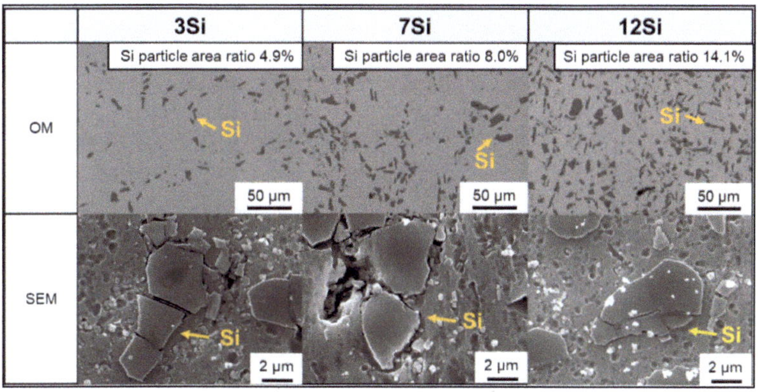

Fig. 4. OM and SEM microstructures of cold-rolled Al-Si alloy sheets.

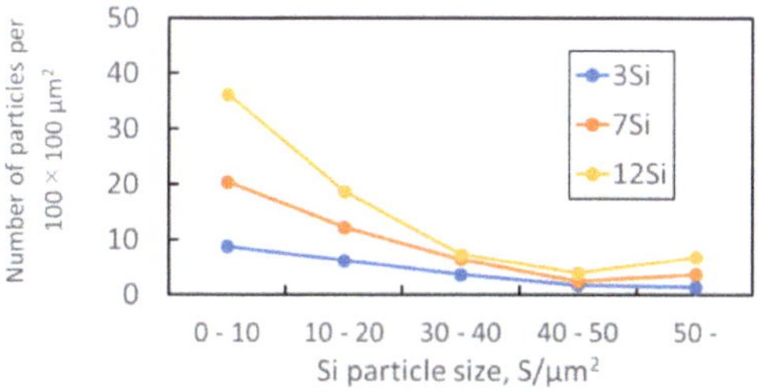

Fig. 5. Size distribution of Si particles in cold-rolled Al-Si alloy sheets.

V-bending tests. Results of 90° V-bending tests with punch radii of $R = 9$ mm and 2 mm are shown in Fig. 6. At $R = 9$ mm, V-bending of Al-3%Si and Al-7%Si alloy sheets was successful even without rapid heating, indicating that the amount of Si particles has less influence on the degradation of their formability (Fig. 6(a)). In V-bending of more Si containing Al-12%Si alloy, on the other hand, V-bending could not be accomplished at room temperature, and therefore rapid heating to 250°C became quite effective for improving the formability of Al-12%Si alloy.

The effectiveness of such a hot stamping method was clarified even in the case of $R = 2$ mm (Fig. 6(b)); Al-3%Si alloy could be V-bent at room temperature, whereas rapid heating to 250 and 300 °C was required for Al-7%Si and Al-12%Si alloys, respectively. Therefore, it was revealed that higher Si content degrades the formability of the Al-Si alloy sheets, but rapid heating to a certain temperature can improve the formability of recycled wrought aluminum.

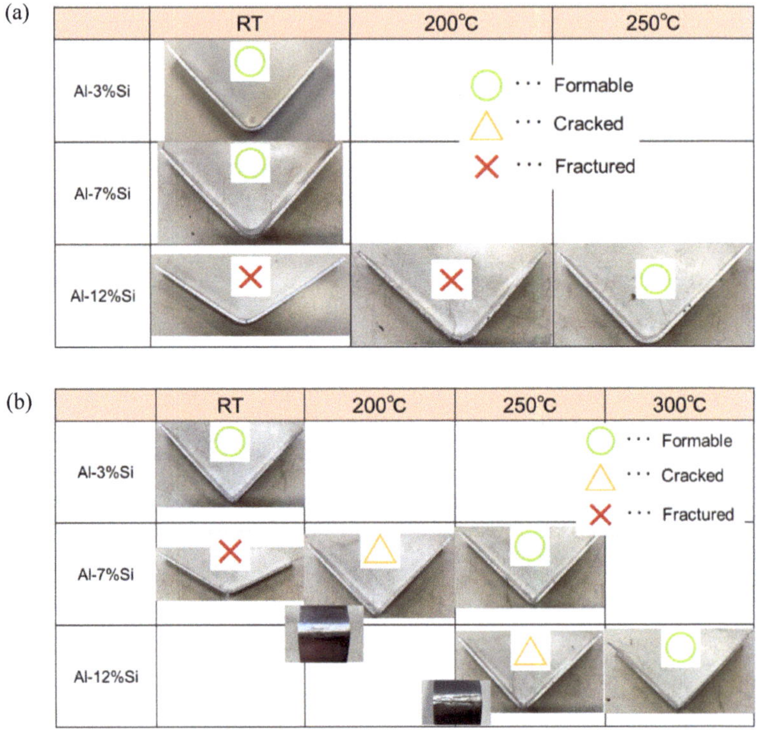

Fig. 6. Results of 90° V-bending tests of cold-rolled Al-Si alloy sheets with punch radii of (a) 9 mm and (b) 2 mm.

Discussion

To discuss the reason why rapid heating before forming improves the formability of cold-rolled Al-Si alloy sheets, the influence of rapid heating on mechanical properties of Al-12%Si alloy was evaluated by high-temperature tensile tests after rapid heating (Fig. 7(a)). The obtained stress-strain curves confirmed that heating to 300°C decreases deformation resistance to ~50 MPa, but improves the elongation to fracture above 20% (Fig. 7(b)).

The corresponding tensile fracture surfaces of cold-rolled Al-12%Si alloy sheets are shown in Fig. 8. It was revealed from the SEM images that coarse Si particles are exposed from the matrix with fine dimples at room temperature, but larger dimples at higher temperatures. This suggests that cracks were propagated along Si particles, but the fracture was retarded by the deformation of the surrounding matrix with increasing heating temperatures. Therefore, rapid heating before forming can improve the formability of even difficult-to-process Al-12%Si alloy because relieved stress concentration between Si particles and Al matrix delays the propagation of cracks.

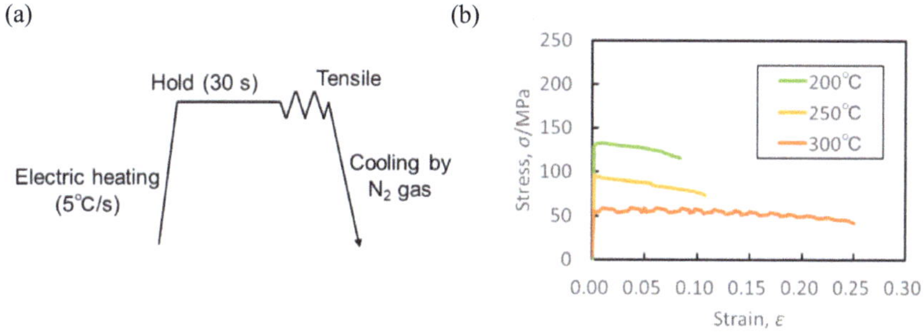

Fig. 7. Schematic illustration of procedure of high-temperature tensile test after rapid heating (a). The obtained stress-strain curves at 200, 250 or 300°C was also shown in (b) for cold-rolled Al-12%Si alloy.

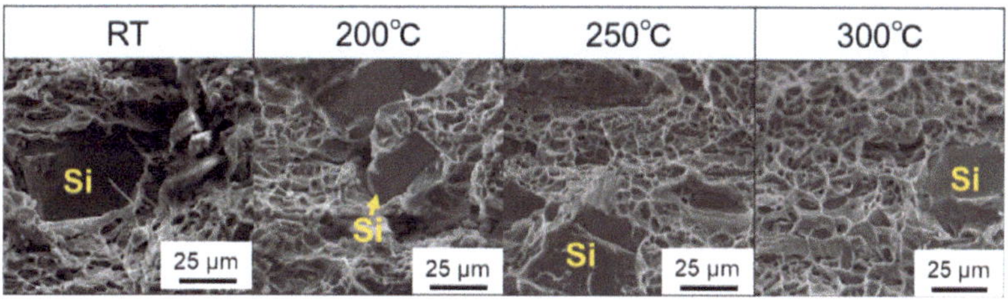

Fig. 8. Tensile fracture surfaces of cold-rolled Al-12%Si alloy at different temperatures.

Summary

Al-3, 7 and Al-12%Si binary alloys were subjected to cold rolling by 80%, and then V-bent at room temperature or, if failed, after rapid heating to 200, 250 or 300°C for evaluating their formability. The obtained results are summarized below.

1. Irrespective of Si content, higher ultimate tensile stress (UTS) of 200-210 MPa were obtained, but the elongation to fracture decreased from 8% to 4% with increasing Si content, resulting from stress concentration caused by sharp edges of fractured Si particles.
2. From the results of 90° V-bending tests, it was found that the formability of cold-rolled Al-Si alloys is degraded with increasing Si content, but improved by rapid heating to a certain temperature. In the case of punch radius $R = 9$ mm, for example, Al-3%Si and Al-7%Si alloy sheets were successfully V-bent at room temperature, but V-bending of more Si containing Al-12%Si alloy could not be accomplished at room temperature. Therefore, rapid heating to 250 °C become quite effective for improving the formability of Al-12%Si alloy.
3. The effectiveness of rapid heating was clarified even in more severe case of $R = 2$ mm. Al-3%Si alloy could be V-bent at room temperature, but rapid heating to 250 and 300 °C was required for Al-7%Si and Al-12%Si alloys, respectively. Therefore, rapid heating before

forming can improve the formability of even difficult-to-process Al-12%Si alloy because relieved stress concentration between Si particles and Al matrix delays the propagation of cracks.

References

[1] J. Cui, H.J. Roven, Recycling of automotive aluminum, Trans. Nonferrous Met. Soc. China, 20 (2010) 2057–2063. https://doi.org/10.1016/S1003-6326(09)60417-9

[2] S. Kumai, Role and potential of aluminium and its alloys for a zero-carbon society, Mater. Trans. 64 (2023) 319-333. https://doi.org/10.2320/matertrans.MT-LA2022009

[3] S. Hirosawa, T. Maeno, J. Kim, K. Nariya, Y. Suzuki, International patent, WO2023/053845 A1.

Development of X40CrSiMo10-2 steel series for heavy-duty in hot environment of marine diesel engines

EL-BITAR Taher[1,a], EL-MELIGY Maha[1,b *] and GAMIL Mohammed[2,c]

[1]Plastic Deformation Department, Central Metallurgical R&D Institute (CMRDI), Cairo, 11422, Egypt

[2]Department of Mechanical Engineering, Faculty of Engineering at Shoubra, Benha University, Cairo, 11629, Egypt

[a]elbitart@yahoo.com, [b]dr.mahaelmeligy@gmail.com, [c]mohammed.gamil@feng.bu.edu.eg

Keywords: Exhaust Valve, Marine Diesel Engine, Heavy-Duty, Hot Temperature Environment, Induction Melting, Sand Casting, 5-Successive Forging Steps, Scattered Cr-Mo Carbides

Abstract. Different X40CrSiMo10-2 Steel series are developed to serve as a heavy-duty exhaust valve for marine diesel engines. The alloys contain 0.35-0.50 % Carbon combined with 8.5-10.0 % Chrome, 2.0-2.5 % Silicon and 1.0-1.3 % Molybdenum. The steel series were processed in an induction melting furnace and cast as round cylinders with 105 mm mean diameter, and 360 mm height. Melt steel was bottom poured with 27 mm average diameter runner. Each Cast ingot was weighting 26.5 Kg. The thermodynamic calculations ensured formation of strong-stable Cr-Mo carbides containing $Cr_{23}C_6$ and Mo_2C at the specified working temperatures. Cr enhances oxidation resistance of the alloy, while Si besides Cr are working to secure a ferrite phase matrix at room temperature and up to Ac1(855 °C). On the other hand, Mo beside the strengthening effect, it is fighting against coagulation of carbides within the matrix during working in the hot environment. A microstructure of the as – cast condition exhibits coarse ferrite grains due to slow rate of cooling during solidification in the sand mold. Cr and Mo- carbides are precipitated as thin layers enveloping the solidified ferrite grains, while the majority of carbides are precipitated as clusters scattered over the grains. On heating the steel to a temperature higher than Ac1, considerable portion of the ferrite matrix is transformed to austenite phase and the scattered carbides are then dissoluted. Multi-step hot forging is a complementary process, where the matrix would be finer with homogeneously distributed embedded carbides. Forging would be carried out by 5- successive controlled steps to obtain 50 mm diameter round bars.

Introduction

Carbon-steels containing Chromium -Molybdenum (Cr-Mo) are considered as the most suitable candidates for heavy duty service in hot-temperature environment. The alloy should possess high adhesive wear resistance and hot corrosion. The working environment of exhaust valves of marine diesel engines are typical to hot temperature (300 °C) and adhesive wear action. Furthermore, the combustion gases contain considerable amount of Sulphur, as the diesel fuel composed of numerous Sulphur compounds. Consequently, the exhaust valve should posse good hot oxidation and corrosion resistance to withstand against the sulfuric acid (H2SO4) formed during combustion stroke.

The behavior of valve alloy at high temperature, and oxidation characteristics should be carefully designed [1]. Due to non-uniform cross-section of the valve, considerable temperature variations and thermal loads are maintained causing excessive stresses, which are considered as one of the reasons for premature valve damage [2, 3].

Considerable microstructure changes and alloy hot oxidation are maintained due to high temperature environment. Sulfidation reaction accelerates the oxidation phenomenon especially in marine internal combustion engines operating with high-sulfur diesel. Hot corrosion attack is characterized by a loosely adhering oxide scale separated from the alloy-affected region. Sodium chloride that forming sodium sulfate at marine environment with the high sulfur contain diesel. is the most aggressive of alkali

Hot corrosion is obviously met with a diesel engine exhaust environment, where an increasing demand for extensive engine output has raised the valve temperature. Diesel fuels contain varying amounts of sulfur so the combustion product can promote sulfidation. Sodium chloride have been recognized as the most potent of alkali salts to accelerate the hot corrosion, where a marine environment is playing very active roles [4].

On crude oil refining, two main desulfurization processes are recently implemented for sulfur compounds removal from the crude oil. Hydrodesulfurization (HDS) is one of the catalytic processes, which aims at turning organic sulfur compounds into H2S using H2 as the reactant in the presence of metal catalysts operating at high temperature and pressure. Furthermore, a recent process is bio desulfurization (BDS), which is a new bio-catalytic desulfurization method, where sulfides in crude oil can be turned into elemental sulfur that can be removed by 3 steps. S-content after the first removal step would be 0.58%, then S is lowered to 0.11% after the second step, and lastly S reaches to 0.09% at the end of the process [5].

Investigating deformation behavior of alloy X45CrSi93, which is not containing Mo, at temperatures between 900 and 1000°C. The microstructural analysis showed a pronounced phase transformation from α-ferrite dissolution to austenite and carbide [6].

In the absence of alloying with Mo at high working temperature, the austenitic matrix is decomposed to form the lamellar structure consisting of $\gamma + Cr_{23}C_6$, where carbides were existed at the grain boundary and within the γ grains. The lamellar structure nucleates at the boundaries and grows at the expense of the parent matrix, and known as cellular decomposition. The separated lamellar structure from the matrix decreases strength, toughness, loading ability of the matrix are leading to early fatigue failure and hot corrosion [7]. The amount of the lamellar structure ($Cr_{23}C_6 + \gamma$) is recommended to be less than 15% [7].

Molybdenum additions improve resistance to pitting and crevice corrosion in chloride containing environments and sulfuric, phosphoric, and hydrochloric acids. Mo Improves toughness and wear resistance of steel, where Mo has a solid solution strengthening effect on Ferrite, in addition to its role of forming stable Mo-carbides (Mo_2). Mo effectively inhibits the accumulation of cementite at 450~600°C and promotes the precipitation of Cr and Mo carbides, and thus becoming the most effective alloy element to improve the thermal strength of steel. The development of high-performance materials for ultra-high temperature applications represents one of the biggest challenges of the last decades. molybdenum-based alloys have been considered as their high temperature strength, and creep resistance [8, 9]. Mo containing steel can be fair forged and rolled. Mo can be dissolved in ferrite, austenite and forming carbide, and is an element to reduce Austenite phase zone.

Recently, a work was dealing with simulation of different forms of carbide precipitation and mechanisms of crack propagation using advanced Finite Element approaches [10, 11]. The work proposed that carbides could be precipitated in the form of any of net, band, cluster, and random structures. Each of these structures could be either coarse or fine. Fig. 1 presents the proposed forms of carbide precipitation [10, 11].

Materials Research Proceedings 44 (2024) 374-380 https://doi.org/10.21741/9781644903254-40

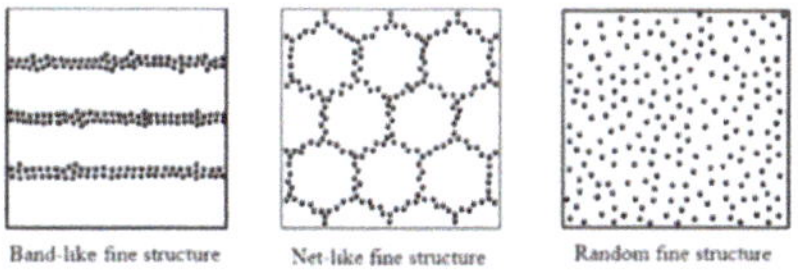

Fig. 1. Forms of carbide precipitation.

Crack propagates in the net-like carbide following the carbide network. However, if the matrix grains are too big, the crack goes through. On the other hand, in band-like carbides, cracks jump from one band to another. However, crack jumps through the random carbide structure from one carbide to another, leading to intensive crack branching [10, 11].

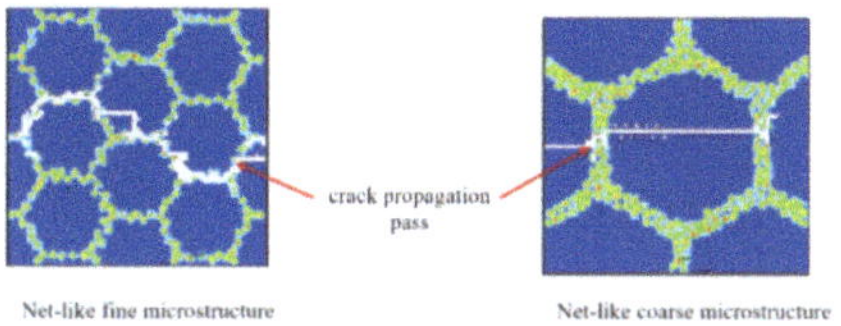

Fig. 2. Crack propagation modes at net like carbide structures [10, 11].

Fig. 3 presents values of calculated energy needed to fracture each of the different carbide precipitation mode.

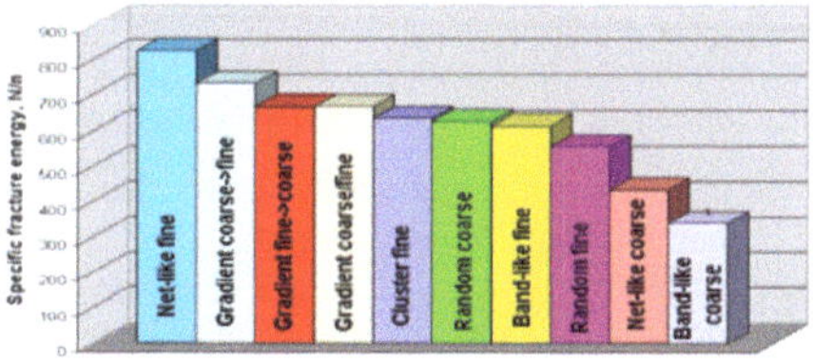

Fig. 3. Calculated energy needed to fracture different carbide structures [10, 11].

The steel alloys under investigation, undergoes microstructure phase transformations by heating. At room temperature, the steel composes of ferrite matrix scattered with different size carbides (mainly Cr and Mo carbides). On heating to higher than Ac_1, considerable portion of the ferrite matrix is transformed to austenite phase and dissolution dimensioning the volume fraction of the scattered carbides. The fraction of transformed austenite increases with the increase of heating temperature. Continuous increase of transformed austenite with few scattered carbides prevail on temperature higher than Ac_3 [6]. However, the flow stress (σf) suffers and decreases appreciably on heating from the room temperature to the Ac_1, after which σf recovers and increase as a consequence austenite phase creation and carbides dissolution.

Experimental Procedure

A steel series was cast containing 0.35-0.50 % carbon in combination with 8.5-10 % Chrome, 2-2.5% Silicon 0.99-1.28 % Molybdenum. Melting was carried out in an induction furnace. Melt steel was bottom poured and cast as round cylinder with a 105 mm mean diameter, and 360 mm height. The castings were weighting 26.5 Kg. Detailed chemical composition of processed steel is presented in Table 1.

Materials Research Proceedings 44 (2024) 374-380 https://doi.org/10.21741/9781644903254-40

Table 1. Chemical composition of procced steel alloys for Exhaust Valve.

Element, wt%	C	Si	Mn	P	S	Cr	Mo	Ni
EXH 2	0.51	2.57	0.441	0.0283	0.0115	8.45	1.28	0.226
EXH 3	0.36	2.44	0.341	0.0222	0.0104	9.15	0.986	0.202

Fig. 4 presents a schematic dimensional drawing and photo of the cast ingot in combination with the liquid steel feeding system (runner). The feeding system and raiser blocks are trimmed for further processing.

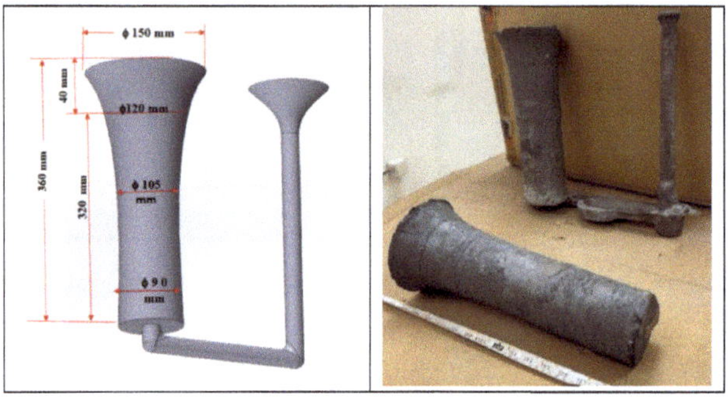

Fig. 4. presents schematic dimensional drawing and photo of the cast ingots.

Representative samples were taken and suitable prepared for microstructure investigations of the as cast structure and Brinell hardness measurements.

The cast ingots are then forged by 5-successive steps open die-controlled forging for grain size and carbide refining and homogenizing. During the 1st forging step, 105 mm diameter round cross section ingots are hot deformed to a rectangle with a cross section 118 mm x 59 mm. The 2nd step forging results in 73 mm edge square, while the 3rd step is forging to Octagon with 29 mm side length. The 4th and 5th forging steps are round bars with ϕ 61 mm and ϕ 51 mm respectively. Fig. 5 presents the mean deformation temperatures which are ranging between 1100°C and 810°C. Fig. 5 additionally shows the cross-sectional area reduction per each forging step.

Materials Research Proceedings 44 (2024) 374-380 https://doi.org/10.21741/9781644903254-40

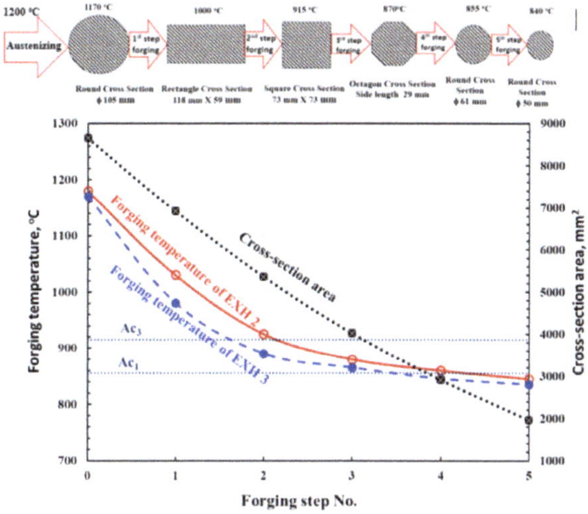

Fig. 5. Cross section reduction and forging temperature during successive 5 – steps hot deformation.

Forging stock schedule are schematically presented for successive 5-steps forging whole necessary processing data are clearly stated within the Fig. 6.

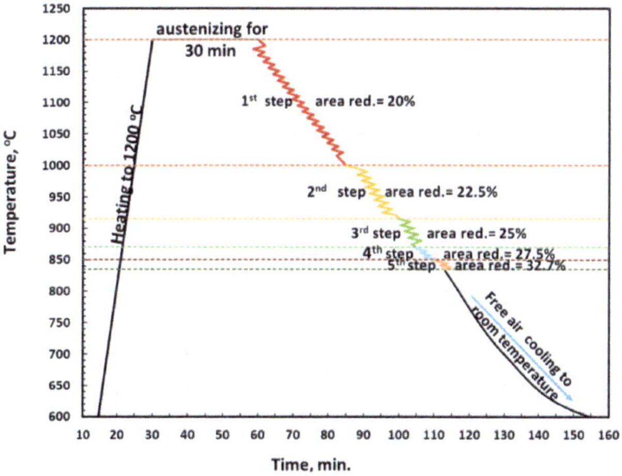

Fig. 6. Forging stock Schedule for successive 5 – steps hot deformation.

Calculated and measured results

Thermodynamic calculations of coexisting triple system Chromium (Cr) and Molybdenum (Mo) with Carbon (C) at different working temperatures ensure formation of strong (stable) carbides at temperatures up to 1100°C [12]. Fig. 10 presents calculated Gibbs free energy (ΔGo) for Cr and

Mo carbides formations at different environment temperatures. Both $Cr_{23}C_6$ and Mo_2C are stable at the specified working temperature range. However, $Cr_{23}C_6$ possesses the most -ve ΔGo, indicating highest stability. However, importance of carbides depends mainly (in high temperature environment) on fighting against coagulations of carbides by time [6]. The phenomena are usually ruled by the Mo content in the alloy, where Mo delays and suppresses coagulations.

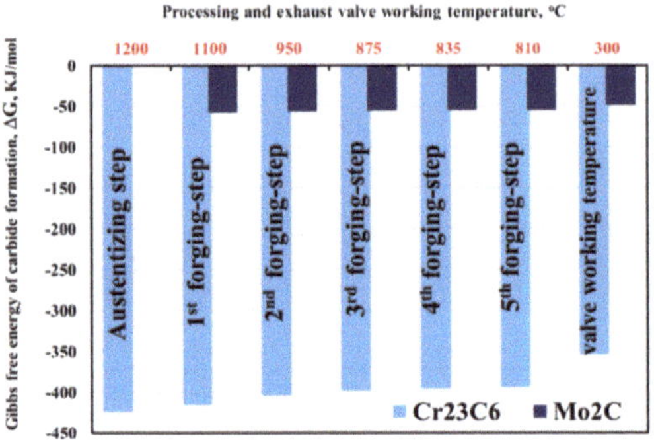

Fig. 7. Gibbs free energy of formation of the carbides $Cr_{23}C_6$, Mo_2C.

Fig. 8 contains representative microstructure of alloys EXH 2 and EXH 3 at magnifications 50X and 100X respectively. Both microstructures exhibit coarse ferrite grains due to slow rate of cooling during solidification in the sand mold. Furthermore, Chromium and Molybdenum carbides are precipitated as thin layers enveloping the solidified ferrite grains, while the majority of carbides are precipitated as clusters scattered over the grains, which are the worst case for crack initiation and propagations as presented previously in Fig. 2 [10, 11]. Further processing by multi step forging are essentially needed as previously described in Fig. 5.

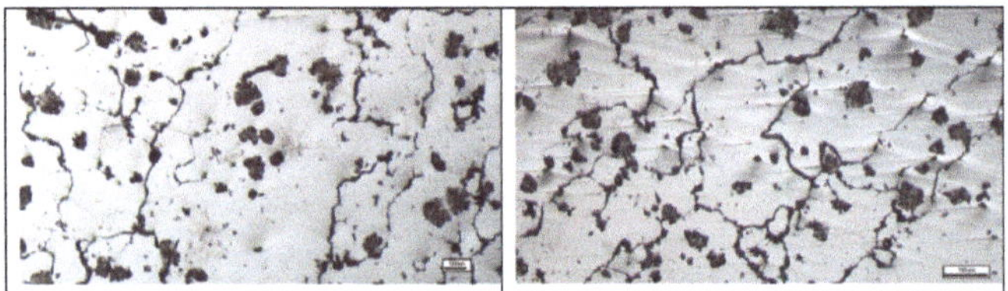

EXH 2, 50 X	EXH 3, 100X

Fig. 8. Representative microstructure of alloys EXH 2 and EXH 3.

Summary

1. Cr- carbides ($Cr_{23}C_6$) possess the most -ve ΔGo, indicating highest stability
2. Mo is ruled on delays and suppresses coagulations of carbides during heating.
3. The as cast microstructures exhibit coarse ferrite grains due to slow rate of cooling during solidification in the sand mold.
4. Cr and Mo- carbides are precipitated as thin layers enveloping the solidified ferrite grains, while the majority of carbides are precipitated as clusters scattered over the grains.
5. On heating the steel to a temperature higher than Ac1, considerable portion of the ferrite matrix is transformed to austenite phase and the scattered carbides are dissolute.

Announcement
Casting and forging processes in the current work were executed at the pilot plant of CMRDI-Egypt in the fram work of finiancial support provided by the Egyptian goverment-Ministry of Higher Education and Scientific Research.

References

[1] R. Prasad, N.K. Samria, Transient heat transfer studies on a diesel engine, Int. J. Mech. Sci. 33 (1991) 179-195. https://doi.org/10.1016/0020-7403(91)90045-5

[2] S. Seralathan, T. Naga Raju, I. Guru Venkat, V. Hariram, S. Dinesh, Thermal analysis on different exhaust valve materials of compression ignition engine, Mater. Today Proc. 33 (2020) 4105-4111. https://doi.org/10.1016/j.matpr.2020.06.550

[3] Y.S. Wang, S. Narasimhan, J.M. Larson, J.E. Larson, G.C. Barber, The effect of operating conditions on heavy duty engine valve seat wear, Wear 201 (1996) 15-25. https://doi.org/10.1016/S0043-1648(96)06945-1

[4] A. Chaudhuri, Hot Corrosion of Diesel Engine Exhaust Valves, SAE Trans. 82(1973) 2478-2486. www.jstor.org/stable/4472130

[5] L. Lin, L. Hong, Q. Jianhua, X. Jinjuan, Progress in the Technology for Desulfurization of Crude Oil, China Petroleum Processing and Petrochemical Technology 12 (2010) 1-6.

[6] Y. Wu, M. Zhang, X. Xu, Investigations on hot deformation behaviors and abnormal variation mechanisms of flow stress at elevated temperature for X45CrSi93 valve steel, J. Mater. Res. 30 (2015) 1715–1726. https://doi.org/10.1557/jmr.2015.98

[7] Z.W. Yu, X.L. Xu, Failure analysis and metallurgical investigation of diesel engine exhaust valves, Eng. Fail. Anal. 13 (2006) 673-682. https://doi.org/10.1016/j.engfailanal.2004.10.018

[8] V. Petrusha, G. Hasemann, R.S. Touzani, V. Bolbut, I. Bogomol, M. Krüger, Microstructure Formation of Cast and Directionally Solidified Mo-Ti-B Alloys, Metals 12 (2022) 916. https://doi.org/10.3390/met1206091

[9] R. Wan, F. Sun, L. Zhang, A. Shan, Effects of Mo on high-temperature strength of fire-resistant steel, Mater. Des. 35 (2012) 335-341. https://doi.org/10.1016/j.matdes.2011.09.009

[10] L.M. Jr., N. Lippmann, S. Schmauder, Computational Design Of Multiphase Materials At The Mesolevel, IMECE International Mechanical Engineering Congress and Exposition, 11-16 Nov., New York, USA, 2001.

[11] T. El-Bitar, A. Ismail, A. El-Morsy, A. Amer, Deformation of Special Steels, Steel Grips 2 (2004) 364-371.

[12] T. El-Bitar, M. El-Melig, M. Khedr, Investigation of exhaust valve failure in a marine diesel engine, Eng. Fail. Anal. 114 (2020) 1-6. https://doi.org/10.1016/j.engfailanal.2020.104574

Influence of particle characteristics on mechanical properties of particle reinforced tungsten alloys

XIAO Fangnao[1,a], CHENG Gang[2,b] *and BARRIERE Thierry[3,c]

[1]New Engineering Industry College, Putian University, 1133, Middle Xueyuan Street, Chengxiang District, 351100 Putian, China

[2]INSA Centre Val de Loire, LaMé, 3, Rue de la Chocolaterie, BP 3410, 41034 Blois, France

[3]Université de Franche-Comté, CNRS, Institut FEMTO-ST, 26, Rue de l'Épitaphe, 25000 Besançon, France

[a]xfnaolw@163.com, [b]gang.cheng@insa-cvl.fr, [c]thierry.barriere@univ-fcomte.fr

Keywords: Tungsten Alloy, Oxide Particle Strengthening, Microscale Simulation

Abstract. The manufacturing processing and the mechanical properties of particle reinforced metal matrix composites are strongly dependent on their microstructural characteristics. In this research, 3D models in microscale of tungsten alloys reinforced by $Zr(Y)O_2$ particles ($W-Zr(Y)O_2$) were established to investigate their uniaxial compression deformation behaviours. The effects of particles contents, size and their distribution on the compressive properties of $W-Zr(Y)O_2$ alloy were discussed. The mechanical behaviours of the reinforced W alloys were improved by increasing the content of $Zr(Y)O_2$ particles. With the same particles content, the strength of the reinforced alloys increased with smaller size particles. With the same particle size and content, the stress concentration was reduced with more homogeneous distribution of the reinforced particles. The predicted strengths with 3D models are compared with the experiment data, which exhibits high prediction accuracy.

Introduction

Particle reinforced metal matrix composites (PRMMCs) show distinct merits over unreinforced metals and alloys, such as high stiffness and strength, outstanding wear resistance, attractive fatigue properties, and superior thermal and electrical characteristics [1, 2]. Designing new PRMMCs and predicting their physical and mechanical performance are challenging. The advanced and reliable numerical models of PRMMCs are essential to optimise and predict their mechanical properties [2, 3].

With the continuous development of numerical methods, the prediction of mechanical properties of materials with simulation has become an important scientific issue [4]. Compared with the traditional experimental methods, the numerical simulation exhibits the following advantages: 1). It is difficult to determine the deformation processing parameters for the materials, such as tungsten, due to their brittleness during the recrystallization. The numerical simulation can predict the mechanical properties over a large temperature range, in which the experimental tests are difficult to carry out; 2). The parameters in numerical simulation could be adjusted easily without the wasting of time and cost; 3). The test conditions, including temperature, duration time and pressure, could be set up smoothly in the numerical simulation. In the past decades, more and more researchers developed the numerical simulation tools to investigate the mechanical properties in room and high temperatures, microstructure evaluation and fracture law of W-based materials.

The mechanical properties of particle reinforced metal matrix composites (PRMMCs) depend on their microstructure characteristics, such as size, fraction, distribution and morphology [3]. The experimental, analytical and numerical simulation methodologies have been employed to investigate the microstructure and physical propriety of PRMMCs. The macroscopic simulation of the composite materials is difficult to reveal the interface evolution, strengthening mechanism and damage. The effective properties and damage mechanism of PRMMCs are strongly dependent on the properties of matrix, particles, interface, and the microstructure of the composites. Various representative finite element models in microscale have been developed to estimate the effects of microstructural features of PRMMCs to optimize their mechanical properties.

The effect of mesoscopic characteristics on mechanical performance of PRMMCs are investigated based on the 2D plane model, which exhibits some disadvantages as follows: 1) the simplified 2D plane model cannot truly characterize actual damage until the whole process of the fracture damage of concrete material and evolutionary regularity of crack extension; 2) it cannot accurately reflect the macroscopic mechanical properties of PRMMCs. In the actual production and life, the mechanical problems and the structure of the composite materials relate to dimensional entities. It is necessary to establish a 3D numerical model to study the mechanical properties and failure mechanism of PRMMCs more truly and intuitively. 3D model has been proposed to analyze the influence of particle size [5], volume fraction [6], morphology [7], distribution [8], interface damage [9], crack and void evolution [10] on the mechanical performance of composite [8, 9, 11]. Compared with 2D model, 3D model is more realistic in the characterization of the microstructure.

Zhang et al. [7] investigated the effects of particles' aspect ratio on the deformation of Al-SiC composite by using microscale models. The representative volume element-based microscale model of the composites with different particle aspect ratio was employed. An elastoplastic model was employed to describe the relation between the stress and the plastic strain of the Al matrix [12]. The simulation results show that in the longitudinal direction, the stress in the SiC particles increases with their aspect ratio and the stress in the metal decreases. The effects of the particles' shape and size on the mechanical properties of the reinforced composites are essential to achieve the numerical simulation with W alloys. Williams et al. [11] studied the effects of particle and matrix interface debonding in Al-SiC composites. Different microstructures containing angular and idealized ellipsoidal SiC particles were analyzed. During plastic deformation, an isotropic matrix hardening model was introduced to describe their mechanical behaviour. The results indicated that the angular particles exhibit higher degree of load transfer and are more sensitive to interfacial debonding during tensile deformation. The geometrical shape of the reinforced particles is important to investigate the improvement of the mechanical property of the reinforced alloys.

In this study, the microscale model of particles reinforced tungsten alloy is established. The complete stress-strain field in microscale of PRMMCs is obtained to reflect their mechanical characteristics. The influence of material parameters such as $Zr(Y)O_2$ particle contents, size and distribution on the mechanical properties of W alloy is investigated, which may be necessary for the mechanical design of structural materials in computational materials science.

This paper is structured as follows: In Section 2, manufacturing processes are presented for the elaboration of $W\text{-}Zr(Y)O_2$ alloy. In section 3, 3D simulation model is introduced with the associated boundary conditions. Section 4 consists the numerical simulation results. A comparison with experimental investigations is conducted in Section 5. The main conclusions were summarized in Section 6.

Presentation of $W\text{-}Zr(Y)O_2$ alloys

Four kinds of doped tungsten powders, 0 wt% $Zr(Y)O_2$, 0.25 wt% $Zr(Y)O_2$, 0.5 wt% $Zr(Y)O_2$, and 0.75 wt% $Zr(Y)O_2$, have been used in this study. The manufacturing processes concerning the fabrication of $W\text{-}Zr(Y)O_2$ alloys was described as follows [13]: Firstly, $ZrOCl_2{\cdot}8H_2O$, $Y(NO_3)_3{\cdot}6H_2O$ and $(NH_4)_6H_2W_{12}O_{40}{\cdot}5H_2O$ were dissolved in distilled water. A certain amount of $ZrOCl_2{\cdot}8H_2O$ and $Y(NO_3)_3{\cdot}6H_2O$ were homogeneously mixed. The mixture was added slowly to the $(NH_4)_6H_2W_{12}O_{40}{\cdot}5H_2O$ solution with stirring. HNO_3 was adopt to decrease the solution's pH value to lower than 1.0. The mixed solution was put into a stainless-steel autoclave with a

Materials Research Forum LLC
https://doi.org/10.21741/9781644903254-41

polytetrafluoroethylene lining. The autoclave was sealed and then heated at 276.5 K min^{-1} until the reaction temperature was reached to 120°C. The holding time of the hydrothermal reaction was 17 h. The product was stirred for 120 min using an electric mixer with some distilled water. Through drying at 90°C, the precursor powder was obtained. The obtained powder was calcined at 550°C for 4 h in air and reduced through the hydrogen reduction processes 550°C (2 h) + 900°C (2 h). The milled powders were pressed by the cold isostatic pressing process at a pressure of 250 MPa. The green compacts were sintered at 2400°C for 4 h in a hydrogen atmosphere. Then the sintering samples were hot isostatic pressed at 1400°C and 200 MPa for 1h. Over 90% of particles range from 100 nm to 650 nm in diameter without the coarse particle of more than 1 μm. The average particle size is about 410 nm.

Numerical simulation in 3D microscale of W-Zr(Y)O2 alloy

3D model is representative of the experimental compression specimen, with the size of 5×8 μm. In the numerical simulation, it is assumed that all $Zr(Y)O_2$ particles in W matrix are homogenous and considered as spherical with a constant diameter, which is calculated based on their volume fraction. The coordinates of the particles were generated by using the random distribution function. They should satisfy two conditions: firstly, the particles should not exceed the boundary of the model, it means that the radius of particles should be smaller than the distance between the center of particle and the boundary; Secondly, the particles cannot overlap with each other.

The matrix and particles are merged together in the numerical model. The contact conditions between the matrix and circles are not considering in the simulation. The lower and upper mold tools are created as rigid body in the numerical model: one is placed on the top surface of the W alloy and the other one is placed on its bottom surface. The lower tool is fixed, while an imposed displacement of 1.3 μm is applied on the upper tool. The imposed displacement is calculated according to the sample height and strain in the experimental tests. The friction between the tools and W alloy is not considered in the simulation. The top and bottom surfaces of the W alloy are free to move in radial direction. The 3D cylindrical geometry model is 5 μm in diameter and 8 μm in height. The tetrahedral element C3D10 is used in the modelling.

Simulation results of compression tests with W-Zr(Y)O2 alloy

Validation of mesh size. In order to improve the calculation accuracy of the numerical simulation, different mesh conditions are applied on the 3D model. The mesh size, including 0.3 μm, 0.4 μm, 0.6 μm and 0.8 μm, corresponding to element number of 777915, 138712, 93730 and 79978 are employed in the simulation, as shown in Fig. 1.

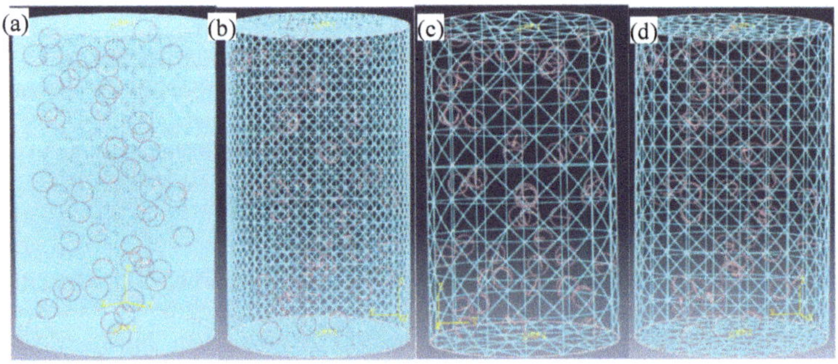

Fig.1. 3D models with different mesh size: (a) 0.3 μm; (b) 0.4 μm; (c) 0.6 μm; (d) 0.8 μm.

The 3D models with different mesh size are created in the simulation to study the convergence of mesh. The evolution of stress in function of mesh size is illustrated in Fig. 2. The stress decreased with the mesh size and becomes constant until mesh size 0.3 μm. The mesh size 0.3 μm is selected to achieve a high calculate accuracy with reasonable computational time.

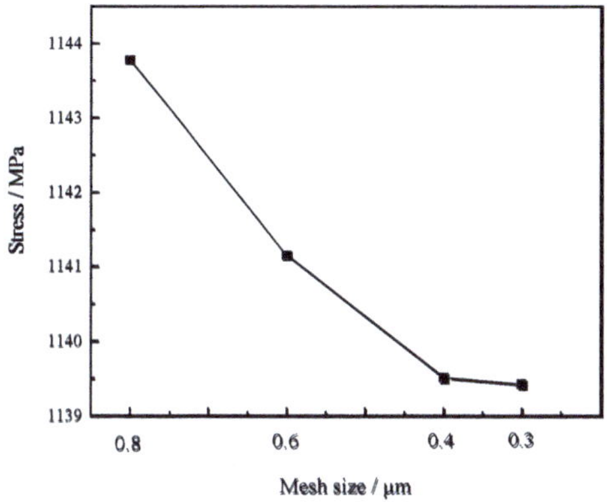

Fig. 2. Evolution of the von Mises stress in function of mesh size.

$Zr(Y)O_2$ mass fraction. The mass fraction of $Zr(Y)O_2$ has a significant effect on the mechanical properties of reinforced W alloy. The effects of particle mass fraction on mechanical behaviors, including yield and compressive strength, are calculated by using the 3D model. 3D model with different particle mass fraction are created as shown in Fig. 3. The particle number in each model is 50.

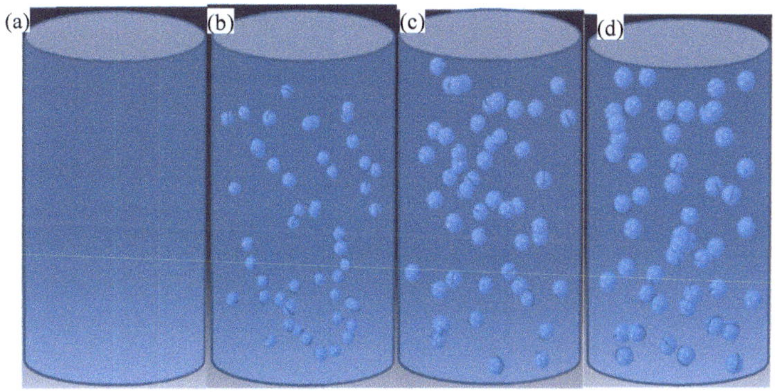

Fig. 3. 3D models embedded with 50 particles with different mass fraction of $Zr(Y)O_2$: (a) 0; (b) 0.25 wt. %; (c) 0.50 wt. %; (d) 0.75 wt. %.

Based on Fig. 4a, the ultimate compressive stresses obtained by 3D models increase slightly with the particle mass fraction. Fig. 4b shows that, when the particle mass fraction increases from 0 to 0.75 %, the ultimate compressive stress of the material increases from 1117.7 MPa to 1140.1 MPa, which is 2.77 % higher than that of the undoped alloy. When the particle mass fraction increases from 0 to 0.75 %, the yield strength increases 3.91 %. Fig. 5 presents the stress

Materials Research Proceedings 44 (2024) 381-387 https://doi.org/10.21741/9781644903254-41

distribution of axial section in loading direction of 3D models. As the particles randomly distributed in the whole 3D model, a small number of visible particles are in the whole 3D model compared to corresponding 2D model. As the mass fraction of particle increases, the maximum stress in loading direction observed does not vary linearly. This is attributed to the various particle number existed in or near axial section in different models. This comparison shows the stress in the matrix increases with particle content. It indicates the reinforcement effect of particle on the matrix.

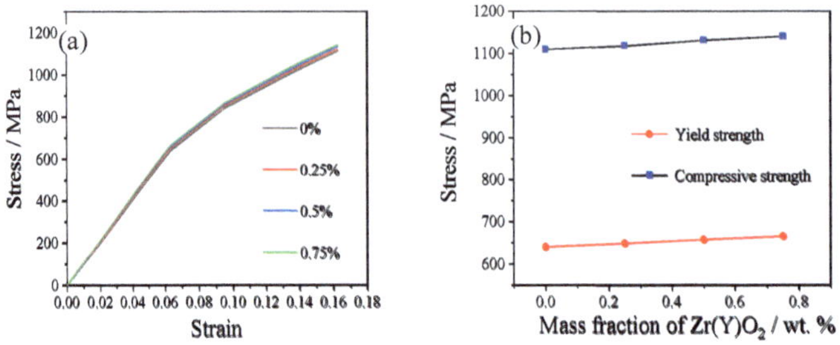

Fig. 4. Mechanical properties obtained by 3D models with different particle mass fraction: (a) Stress-strain curves; (b) Yield and compressive strength.

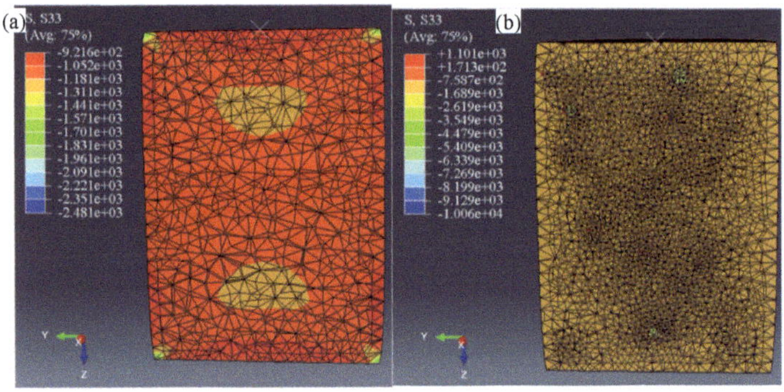

Materials Research Proceedings 44 (2024) 381-387 https://doi.org/10.21741/9781644903254-41

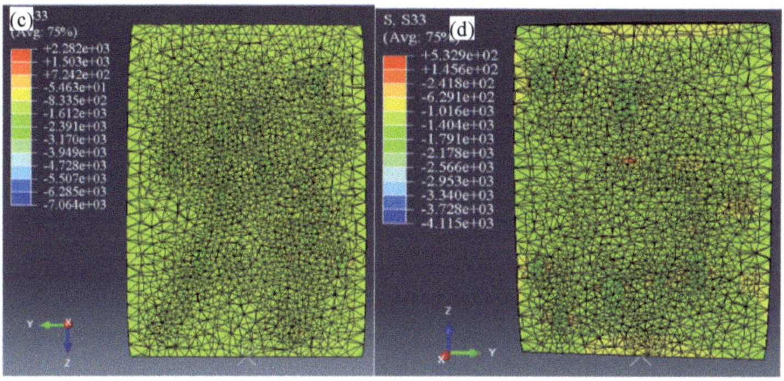

Fig. 5. Stress distribution in loading direction of 3D models embedded with 50 particles with different mass fraction of Zr(Y)O₂: (a) 0; (b) 0.25 wt. %; (c) 0.50 wt. %; (d) 0.75 wt.

Comparison of simulation results with experimental investigations

The mass fraction of the particles in the two models affects the yield strength. The simulated values are compared with the experimental values of the room-temperature compressive properties of W-$Zr(Y)O_2$ alloy by HIP. The yield strength of the reinforced alloys increases with the mass fraction of the particles, both in the numerical simulation and experimental investigation, as shown in Fig. 6. The numerical results in the 3D simulation are more important than these in the 2D simulation and closer to the experimental data. The difference between the numerical and experimental results is mainly due to no consideration of the interaction conditions between particles and matrix, and the grain refinement in the alloys.

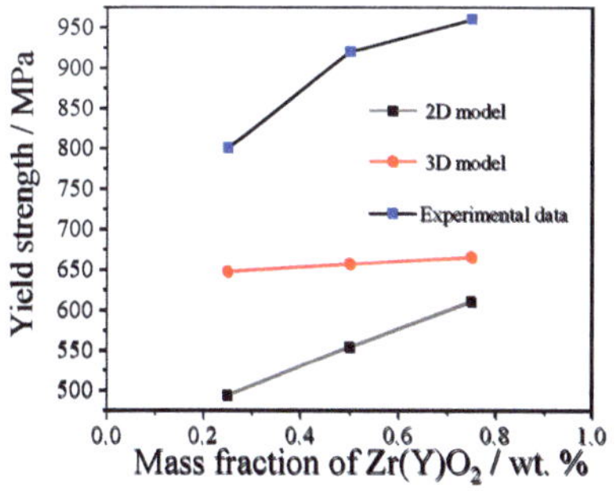

Fig. 6. Effect of $Zr(Y)O_2$ mass fraction on yield strength of reinforced alloy in numerical models and experimental investigations.

Conclusions

1. The random distribution of particles in W matrix is generated and implemented with finite element software Abaqus.
2. 3D models in microscale are created by considering of the material properties of W-$Zr(Y)O_2$. The influences of the particle mass fraction on their behaviours are investigated. The yield increases by 3.91 % when the particle mass fraction increases from 0 to 0.75 %. The predicted strengths from 3D simulation are higher than those from experiment data.

3. The proposed numerical approach will be employed to develop the innovative particle reinforced materials with high performance at various temperatures. The simulation models enlarge its applications for other solicitations, such as tensile, shear and torsion.

Funding
This work is supported by National Natural Science Foundation of China, China [No. U2341258], Fujian Province Young and Middle-aged Teacher Education Research Project, China [JZ230040] and EIPHI Graduate School (contract ANR-17-EURE-0002), Université Bourgogne Franche-Comté, Besancon.

References

[1] S. Ma, X. Zhuang, X. Wang, Particle distribution-dependent micromechanical simulation on mechanical properties and damage behaviors of particle reinforced metal matrix composites, J. Mater. Sci. 56 (2021) 6780-6798. https://doi.org/10.1007/s10853-020-05684-2

[2] M. Saravana Kumar, C.I. Pruncu, P. Harikrishnan, S.R. Begum, M. Vasumathi, Experimental investigation of in-homogeneity in particle distribution during the processing of metal matrix composites, Silicon 14 (2022) 629-641. https://doi.org/10.1007/s12633-020-00886-4

[3] J. Geng, Y. Li, F. Wang, C. Zhang, D. Chen, M. Wang, H. Wang, Revealing the complex effects of particle bands on fatigue crack growth in an extruded aluminium matrix composite, Int. J. Fatigue 157 (2022) 106720. https://doi.org/10.1016/j.ijfatigue.2022.106720

[4] R. Kocich, L. Kunčická, D. Dohnalík, A. Macháčková, M. Šofer, Cold rotary swaging of a tungsten heavy alloy: Numerical and experimental investigations, Int. J. Refract. Metal. Hard Mater. 61 (2016) 264-272. https://doi.org/10.1016/j.ijrmhm.2016.10.005

[5] F. Zhang, Y. Huang, K.C. Hwang, S. Qu, C. Liu, A three-dimensional strain gradient plasticity analysis of particle size effect in composite materials, Mater. Manuf. Process. 22 (2007) 140-148. https://doi.org/10.1080/10426910601062032

[6] J. Zahr Viñuela, J.L. Pérez Castellanos, Elastic constants and isotropy considerations for particulate metal-matrix composites. A multi-particle, cell-based approach, Compos. Part A: Applied Sci. Manuf. 42 (2011) 521-533. https://doi.org/10.1016/j.compositesa.2011.01.011

[7] J.F. Zhang, X.X. Zhang, Q.Z. Wang, B.L. Xiao, Z.Y. Ma, Simulation of anisotropic load transfer and stress distribution in sicp/Al composites subjected to tensile loading, Mech. Mater. 122 (2018) 96-103. https://doi.org/10.1016/j.mechmat.2018.04.011

[8] A. Abedini, C. Butcher, Z.T. Chen, Numerical simulation of the influence of particle clustering on tensile behavior of particle-reinforced composites, Computat. Mater. Sci. 73 (2013) 15-23. https://doi.org/10.1016/j.commatsci.2013.02.021

[9] Y. Su, Q. Ouyang, W. Zhang, Z. Li, Q. Guo, G. Fan, D. Zhang, Composite structure modeling and mechanical behavior of particle reinforced metal matrix composites, Mater. Sci. Eng. A 597 (2014) 359-369. https://doi.org/10.1016/j.msea.2014.01.024

[10] Z. Yang, W. Ren, R. Sharma, S. McDonald, M. Mostafavi, Y. Vertyagina, T.J. Marrow, In-situ X-ray computed tomography characterisation of 3D fracture evolution and image-based numerical homogenisation of concrete, Cement and Concrete Composites 75 (2017) 74-83. https://doi.org/10.1016/j.cemconcomp.2016.10.001

[11] J.J. Williams, J. Segurado, J. LLorca, N. Chawla, Three dimensional (3D) microstructure-based modeling of interfacial decohesion in particle reinforced metal matrix composites, Mater. Sci. Eng. A 557 (2012) 113-118. https://doi.org/10.1016/j.msea.2012.05.108

[12] C.-W. Nan, D.R. Clarke, The influence of particle size and particle fracture on the elastic/plastic deformation of metal matrix composites, Acta Mater. 44 (1996) 3801-3811. https://doi.org/10.1016/1359-6454(96)00008-0

[13] F. Xiao, Q. Miao, S. Wei, T. Barriere, G. Cheng, S. Zuo, L. Xu, Uniform nanosized oxide particles dispersion strengthened tungsten alloy fabricated involving hydrothermal method and hot isostatic pressing, J. Alloy. Compd. 824 (2020) 153894. https://doi.org/10.1016/j.jallcom.2020.153894

Materials Research Proceedings 44 (2024) 388-392 https://doi.org/10.21741/9781644903254-42

Research of MF series glass protective lubricant for titanium alloy precision forging process

FENG Chi[1,a *] and DUAN Su-jie[1,b]

[1]Beijing TLC Glass Tech Co., Ltd

Block C, Bldg.21, No.2 Huankezhong Road, Tongzhou Dist, Beijing, China

[a]fengchi@tlcglass.com.cn, [b]duansujie@tlcglass.com.cn

Keywords: Glass Protective Lubricant, Titanium Alloy, Precision Forging

Abstract. Titanium alloy has high specific strength, high temperature resistance, corrosion resistance, excellent comprehensive performance and wide application, and most titanium alloy parts need to be forged, among which titanium alloy parts with complex structure need to be formed by precision forging. However, titanium alloy is prone to oxidation and gas permeation at high temperature, with large deformation resistance and narrow forging temperature range, so it is very important to solve the problems of protection, lubrication and heat preservation and heat insulation in titanium alloy precision forging. This paper mainly studies the working principle and preparation method of MF series glass protective lubricant for titanium alloy precision forging process, and its protective, lubricating and heat insulation effects on titanium alloy during heating and forging. It is found that the use of MF series glass protective lubricant for titanium alloy precision forging process can significantly improve the quality of titanium alloy precision forgings, showing good protective, lubricating and heat insulation effects.

Introduction

Precision forgings such as titanium alloy aircraft engine blades and artificial joints have complex shapes, and the machining allowance after forging is small. Even some parts need to be precision forged without allowance, and strict three-coordinate testing and mechanical properties testing are required after forging, so the quality of forgings is very high. It is found that the quality of titanium alloy precision forgings can be significantly improved by using glass protective lubricant during the forging process. Specifically, the glass protective lubricant can produce good protective, lubricating and isolating effects in the whole process of forging heating and forging, which can prevent the billet from oxidation and hydrogen absorption at high temperature, delay the temperature drop of forging and reduce the deformation resistance during forging, and isolate it from the die, thus prolonging the service life of the die[1-8].

Formulation design of MF series glass protective lubricant for titanium alloy precision forging process

The glass protective lubricant used in titanium alloy precision forging process is a suspension prepared by ball milling with a variety of amorphous composite glasses as the main solid base and a certain amount of crystalline fillers, adhesives, additives and water. Before the titanium alloy blank is heated in the furnace, the coating is evenly stirred and sprayed on its surface, and the coating is formed after drying. When designing the coating formula, we should consider the specific thermal system of titanium alloy, that is, heating temperature, holding time and furnace atmosphere, and design a coating with good compatibility and appropriate high-temperature viscosity and expansion coefficient, so as to achieve a perfect match with the corresponding thermal system.

(1) Amorphous composite glass powder

As the most important substance in coatings, glass powders with different softening temperatures, high-temperature viscosities and different particle sizes are used in coatings to form amorphous composite glass powders. Among them, the glass powder with low softening temperature will quickly melt and form a film when the alloy is heated in the furnace, so as to isolate the oxidation of the alloy matrix by harmful gases such as O_2. With the increase of alloy temperature, the glass powder with higher softening temperature gradually melts to form a film, so that the coating can remain dense, uniform and stable at high temperature and isolate harmful gases. During forging, the glass powder that has melted into a dense and closed film on the surface of the blank can play a good role in lubrication, reduce the deformation resistance of titanium alloy, facilitate the forming of forgings, isolate it from the die and prolong the service life of the die.

At present, the realization of glass components with different softening temperatures and high-temperature viscosities is mainly achieved by adjusting the chemical composition of glass. For example, increasing SiO_2, Al_2O_3, MgO and other compounds in glass components can increase the high-temperature viscosity of glass, while increasing TiO_2, Na_2O, K_2O, B_2O_3 and other compounds in glass components can reduce the high-temperature viscosity of glass. The viscosity changes of glass powder at different temperatures are detected by a high-temperature viscometer to determine the optimal viscosity range.

(2) Crystalline filler

Specifically, crystalline fillers mainly include SiO_2, Al_2O_3, ZrO_2, kaolin and other substances with high temperature resistance, stable chemical properties and no high-temperature thermal corrosion to titanium Al_2O_3, so as to enhance the high-temperature stability of the coating.

(3) Binder

The binder in the coating directly affects the adhesion of the coating on the blank after spraying or brushing, and good adhesion can make the coating not easy to fall off after drying. At the same time, a good binder should also ensure that the coating is easy to wash after drying, so that the coating can be easily removed when there is an error in coating and the blank will not be scrapped. The selected binder should also be non-toxic and non-irritating to ensure the environment of the spraying site. According to the above requirements, this coating adopts environmentally-friendly waterborne acrylic emulsion binder, which can ensure the adhesion of the coating on the blank, ensure that the coating is easy to wash, and help to enhance the suspension performance of the coating, and avoid the problems of precipitation and delamination of the coating after long-term storage.

(4) Additives

This coating draws lessons from the design concept of architectural coating formula, and adds multifunctional additives such as wetting agent, defoamer, film-forming additive, thickener and dispersant, which adjusts the viscosity of the coating to a more suitable range and improves the comprehensive performance of the coating.

Adhesion of MF series glass protective lubricant for titanium alloy precision forging process

The MF-T21 glass protective lubricant for titanium alloy precision forging process was prepared through the above ideas. MF-T21 glass protective lubricant with different thickness was sprayed on TC4 titanium alloy with the same surface state, and its adhesion was tested by PosiTest AT-M drawing coating adhesion tester in the United States. According to the test experience, the coating adhesion can meet the requirements when it reaches more than 0.6MPa, that is, it can ensure that the blank will not fall off during the transportation after spraying and drying. The test results are shown in Fig. 1.

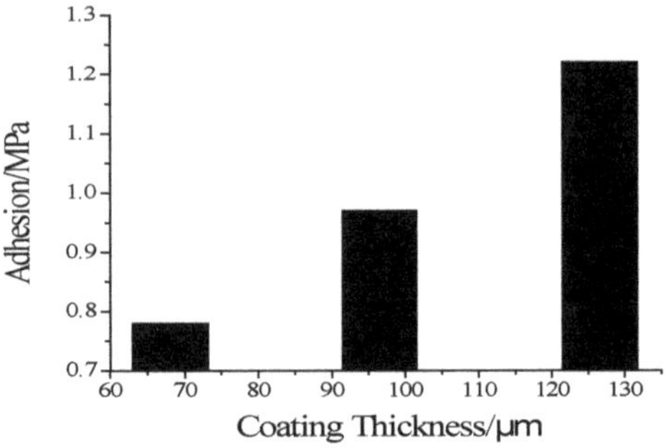

Fig. 1. Adhesion of glass protective lubricant for MF-T21 titanium alloy precision forging process.

As can be seen from Fig.1, the adhesion of the binder added in MF-T21 glass protective lubricant increases with the increase of coating thickness, and both are greater than 0.6MPa. It can be seen that MF-T21 glass protective lubricant can be well adapted to spraying with different thicknesses in the production site, and can be firmly attached to the surface of titanium alloy after spraying and drying.

Protective performance of MF series glass protective lubricant for titanium alloy precision forging process

In order to verify the high-temperature protection effect of glass protective lubricant for MF-T22 titanium alloy precision forging process on titanium alloy, TC4 titanium alloy was selected for the high-temperature protection effect experiment. Fig. 2 shows the micro-morphology of TC4 titanium alloy bare material and TC4 titanium alloy sample coated with MF-T22 glass protective lubricant for titanium alloy precision forging process after 950°C for 90min.

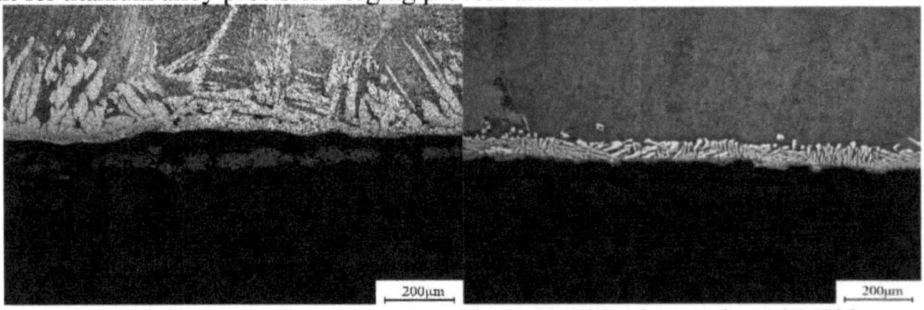

(a). 950°C/90min, uncoating; (b) 950°C/90 min, coating MF-T22

Fig. 2. Microstructure of oxide layer of TC4 titanium alloy.

As can be seen from Fig. 2, the thickness of α oxide layer on TC4 titanium alloy sample coated with MF-T22 glass protective lubricant is obviously reduced compared with that of bare material, and the anti-oxidation effect is obvious. At the same time, the coating can still firmly adhere to the surface of titanium alloy after the sample is cooled, which shows that the expansion coefficient of the coating is well matched with that of TC4 titanium alloy.

Lubrication performance of MF series glass protective lubricant for titanium alloy precision forging process

In order to determine the lubrication effect of MF-T22 glass protective lubricant on titanium alloy at high temperature, the friction coefficient (μ) difference between TC4 titanium alloy surface coated with MF-T22 glass protective lubricant for precision forging process at different temperatures was measured by ring upsetting experiment on a 630-ton hydraulic press.

The specific experimental conditions are as follows: the upper and lower backing plates with roughness of 0.6μm are installed on a 630-ton hydraulic press, and the backing plates are heated to 400±20°C before forging, and the dimensions of titanium alloy rings are-20±0.2 mm in outer diameter, 10±0.2 mm in inner diameter and 7±0.2mm in height. The TC4 titanium alloy sample is heated by a box-type electric furnace, and the pressing speed of the press is 3mm/s during forging. The coated and uncoated titanium alloy samples were forged at three temperature sections of 930°C, 950°C and 970°C respectively, in which five parallel samples were set at each temperature section. The experimental results are shown in Table 1 and Table 2. According to the experimental data, the average compression ratio and inner diameter reduction ratio of TC4 titanium alloy samples can be calculated, and then the corresponding friction coefficient μ can be found by the theoretical correction curve of measuring friction coefficient by ring upsetting method.

Table 1. Experimental results of ring upsetting of TC4 titanium alloy samples without MF-T22 glass protective lubricant coating.

Heating temperature(°C)	930	950	970
Deformation (%)	31	28	32
Height after compression (mm)	5.4	5.3	5.2
Compression ratio (%)	23	24	26
Internal diameter after compression (mm)	8.2	8.3	8.4
Inner diameter reduction rate (%)	18	17	16
Friction coefficient(μ)	0.57	0.53	0.48

Table 2. Experimental results of ring upsetting of TC4 titanium alloy samples coated with MF-T22 glass protective lubricant.

Heating temperature(°C)	930	950	970
Deformation (%)	34	32	31
Height after compression (mm)	4.8	4.7	4.6
Compression ratio (%)	31	33	34
Internal diameter after compression (mm)	8.5	8,6	8.7
Inner diameter reduction rate (%)	15	14	13
Friction coefficient(μ)	0.26	0.19	0.20

After calculation and analysis, the friction coefficient of the sample without glass lubricant is 0.48 ~ 0.57 when forging at high temperature. However, the friction coefficient decreases to 0.19~0.26 with glass protective lubricant. This is mainly because the glass protective lubricant coating forms a uniform, dense and closed molten glass film at high temperature, which can separate the forging from the die, thus changing the solid-solid contact into the solid-liquid contact, that is, changing the external friction between the forging and the die into the internal friction between the molten glass, and playing a good lubrication effect. Furthermore, the influence of lubrication on forging production is comprehensive, not limited to reducing the friction between workpiece and die contact surface, reducing deformation resistance and improving die life. Good lubrication can also save equipment energy and forging materials, improve equipment life and forging performance.

Conclusion

To sum up, MF-T22 series glass protective lubricants for titanium alloy precision forging process can form a uniform, dense and closed glass film during the heating process of titanium alloy blank, which can protect titanium alloy at high temperature and play a good lubrication role in the forging process of titanium alloy blank. At the same time, due to the low thermal conductivity of glass, it can play a good role in heat preservation of the blank, ensuring that the titanium alloy blank maintains a low deformation resistance for a long time during forging, which is a powerful guarantee for obtaining excellent quality titanium alloy precision forgings.

References

[1] C.C. Cojihiieb, A.T. Tymahob, Protection Coating for Metals Heating (Beijing: China Machine Press), 1979.

[2] C. Ding-chun, C. Jiang, Effect of Glass-protection Lubricants on Oxidation Resistance Behavior of TC6 Titanium Alloy, Hot Working Technol. 4 (2016) 121-123.

[3] W. Lian-jun, M. Chang-gong, The Study of Protection Coatings for Titanium Alloys at High Temperature, Rare Metal. Mater. Eng. 32 (2003) 151-153.

[4] C. Ming-hui, S. Ming-li, Effect of Sand Blasting and Glass Matrix Composite Coating on Oxidation Resistance of a Nickel-based Superalloy at 1000°C, Corrosion Sci. 73 (2013) 331-341. https://doi.org/10.1016/j.corsci.2013.04.022

[5] C. Ming-hui, L. Wen-bo, Glass Coatings on Stainless Steels for High-temperature Oxidation Protection: Mechanisms, Corrosion Sci. 82 (2014) 316-327. https://doi.org/10.1016/j.corsci.2014.01.033

[6] W. Le-an, Manufacturing Engineering for Hard Wrought Alloy Forgings (Beijing: Nation Defense Industry Press), 2005.

[7] W. Shu-yun, D. Su-jie, Study of Glass-protection Lubricants for Stainless Steel Forging, Forging & Stamping Technol. 34 (2009) 8-10.

[8] D. Su-jie, L. Xi-chun, T281 Environment-friendly and Air Resistant Glass Lubricants for Precision Forging of Titanium Alloys, Forging & Stamping Technol. 35 (2010) 114-116.

The role of thermomechanical processing in controlling the microstructure inhomogeneity as a way to create special properties of $(FeMnNiCo)_{1-x}Mo_x$ high entropy alloy - a practical approach

CICHOCKI Kamil[1,a *], BAŁA Piotr[1,b], HENSCHEL Sebastian[2,c], KRÜGER Lutz[2,d] and MUSZKA Krzysztof[1,f]

[1]AGH University of Krakow, Department of Metal Forming, and Metallurgical Engineering, al. A. Mickiewicza 30, 30-059 Kraków, Poland

[2]TU Bergakademie Freiberg, Institute of Materials Engineering, Gustav-Zeuner-Str. 5, 09599 Freiberg, Germany

[a]cichocki@agh.edu.pl, [b]pbala@agh.edu.pl, [c]sebastian.henschel@iwt.tu-freiberg.de, [d]krueger@ww.tu-freiberg.de, [e]muszka@agh.edu.pl

Keywords: High Entropy Alloy, Strain Rate Effect, Temperature Effect, Compression Test, Strain Hardening, Microstructure, Twinning

Abstract. From modern alloys, increasingly demanding combinations of properties are expected, enabling their use as structural materials in new areas of application such as the energy industry (Liquid Natural Gas and hydrogen tanks), military, or space. One of the fundamental expectations for constructional materials used in these sectors is high energy absorption capacity while maintaining high strength (often at low temperatures where e.g. conventional steels exhibit brittleness). Simultaneously, they require appropriate plastic deformation susceptibility (formability) and other additional parameters (e.g., corrosion resistance, weldability, fatigue resistance). Designing manufacturing technologies for the metal forming of structural elements from modern engineering materials for the aforementioned applications requires a holistic approach and understanding of the relationships between process parameters (time, temperature, deformation, strain rate), microstructural phenomena (recovery/strengthening mechanisms, phase transformations, precipitation processes), and resulting properties (strength, ductility, toughness). In this work I will provide a practical example of designing specific properties of $(CoNiFeMn)_{1-x}Mo_x$ high entropy alloys by controlling microstructure development at each stage of the metal forming process. The role of recrystallisation, grain growth, and the precipitation of the phase in controlling grain size - a key parameter defining the susceptibility to twinning/nanotwinning in this alloy at cryogenic temperatures - will be discussed.

Introduction

High-entropy alloys are a family of alloys that has been intensively studied over the last few years. The first alloy was the so-called Cantor alloy with an equilibrium chemical composition[1]. These alloys are characterised by four effects: slow diffusion, cocktail effect, high entropy and lattice distortion[2]. Further research has shown that high-entropy alloys have a number of unique properties such as, low stacking fault energy, very good mechanical properties at cryogenic conditions.

The susceptibility to hot deformation in high-entropy alloys has been studied over recent years. These studies mainly concern alloys in the FCC lattice. Studies for Cantor alloys have shown that high-entropy alloys are characterised by a relatively high activation energy of dynamic recrystallisation [3]. Studies have shown that the addition of molybdenum increases the activation energy of recrystallisation in the $CrFeCoNiMo_{0.3}$ alloy [4]. Investigating the possibility of deforming these alloys by twinning is also a popular research direction. Kaushik [5] showed that Cantor's alloy tends to deform by twinning during cold rolling.

This paper presents a scheme for the preparation, in terms of microstructure, of high-entropy alloys of composition $(CoNiFeMn)_{1-x}Mo_x$ for mechanical testing under cryogenic conditions at high strain rates. The desired initial microstructure was achieved by hot deformation analysis, cold rolling and appropriate heat treatment.

Materials and Methods

Ingots were prepared in a VIM LAB 20-50 type induction vacuum furnace and cast in a controlled atmosphere. For the study, ingots of 1 kg were prepared for three alloys with the following chemical compositions CoNiFeMn, $(CoNiFeMn)_{95}Mo_5$ and $(CoNiFeMn)_{90}Mo_{10}$.

The ingots were then hot forged into flat bars at 1150° C using a 500T ZAMET servo-hydraulic press. The initial microstructure for the compression tests is shown in Figure 1. Microstructure analysis was performed after heating to 1150°C and holding for 300 s.

CoNiFeMn (CoNiFeMn)₉₅Mo₅ (CoNiFeMn)₉₀Mo₁₀

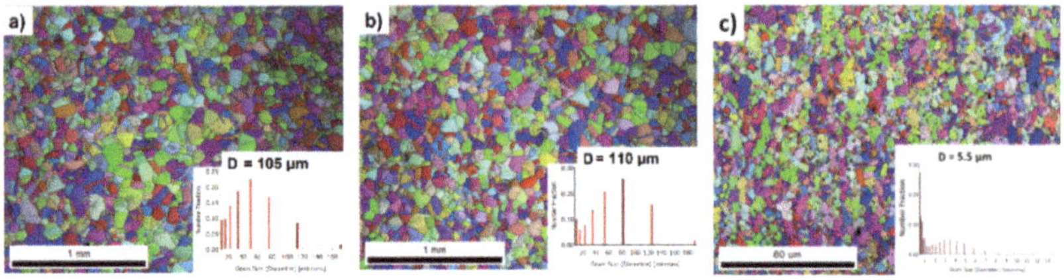

Fig. 1. Initial microstructure for compression tests of the tested alloys.

Grain size measurements were made after the annealing twins were removed. The materials are characterised by a stable solid solution in the FCC lattice. The significantly smaller grain size for the alloy $(CoNiFeMn)_{90}Mo_{10}$ is due to the presence of the intermetallic μ phase, which limits grain growth [6].

Methods. Compression tests were performed on an ASP servo-hydraulic thermomechanical simulator equipped with an induction furnace. Cylindrical specimens measuring 8x8 mm were prepared for compression tests. The specimens were cut from pre-plasticised material. The specimens were heated at a rate of 5°C/s to a temperature of 1150°C and held for 300 s. The choice of temperature was dictated by the fact that materials oxidise rapidly at higher temperatures. Then cooled at a cooling rate of 4°C/s to the deformation temperature. Deformation was carried out at temperatures of 900° C, 1000° C, 1100°C with strain rates of 0.1 s^{-1} , 1 s^{-1} , 10 s^{-1} . The final strain was set at 0.7 of the true strain. A more accurate analysis of the hot deformation behaviour of the materials can be based on the determination of the deformational consolidation:

$$\theta = \frac{d\sigma}{d\varepsilon} \tag{1}$$

The method of Najafizadeh and Jonas was used to determine the values of stress σ_C and strain ε_C critical [7]. The critical stress is determined from a third-degree polynomial function in θ - σ, determined from zero to the peak stress value according to equation:

$$\theta = A\sigma^3 + B\sigma^2 + C\sigma + D \tag{2}$$

where: A, B, C, D - are constants. The second derivative of Eq. 2 as a function of σ can be written as:

$$\frac{d^2\theta}{d\sigma^2} = 6A\sigma + 2B \tag{3}$$

The critical stress can be determined at point $\frac{d^2\theta}{d\sigma^2} = 0$. Then:

$$6A\sigma + 2B = 0 \rightarrow \sigma_C = \frac{-B}{3A} \tag{4}$$

Cold rolling was carried out on a laboratory 4-high mill with a working roll diameter of 100 mm.

SEM/EBSD analysis was performed on a FEI Nova NanoSEM 450 scanning microscope (SEM) (STEM) using an EDAX EBSD Velocity detector. The specimens were prepared by grinding and mechanical polishing. EBSD analysis was performed in the centre of the compressed sample, with an accelerating voltage of 20 kV, a spot size of 6 and a step of 0.5 μm. The processing of the obtained results was performed in OIM-TSL 8 software.

Results and Discussion

The resulting flow curves for the tested high-entropy tested, obtained at different temperatures and strain rates, are shown in Fig. 2. During hot deformation of the material, the characteristic plateau is the result of the phenomenon of recovery (DRV) whereas the flow softening results from occurrence of the dynamic recrystallisation (DRX).

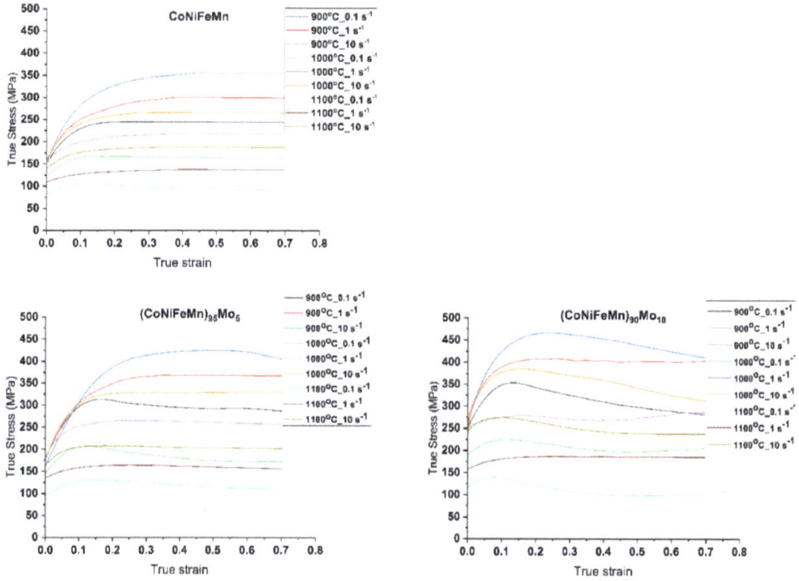

Fig. 2. Flow curves for the high-entropy alloys tested.

Materials Research Proceedings 44 (2024) 393-401 https://doi.org/10.21741/9781644903254-43

From the obtained flow curves it can be seen that, the addition of molybdenum causes an increase in flow stress. Comparing CoNiFeMn and $(CoNiFeMn)_{95}Mo_5$ alloys, this increase is mainly due to solution strengthening, due to the similar grain size of the materials studied. In the case of the $(CoNiFeMn)_{90}Mo_{10}$ alloy, the increase in stress is additionally due to the presence of the μ-phase precipitates, which cause the effect of precipitation strengthening. It should also be noted that with the addition of molybdenum the peak deformation as well as the subsequent softening is much more pronounced.

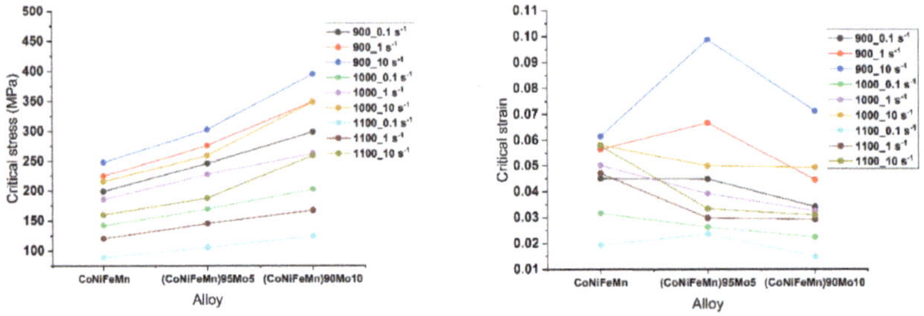

Fig. 3. a) critical stress, b) critical strain.

Analysing the results obtained for critical and peak stress, the stress values increase with the addition of molybdenum. In the case of critical stress values, the strongest, inter-alloy relationship is observed at 900°C. At 1100°C the relationship is linear. In the case of critical strain values at 900° C in the alloy $(CoNiFeMn)_{95}Mo_5$ there is an increase in critical strain values, followed by a decrease in $(CoNiFeMn)_{90}Mo_{10}$. This phenomenon may be due to a strain-induced precipitation process where, as will be discussed in the following sections, in the case of the deformed material, μ-phase precipitates appear at the temperatures up to 1000°C [8]. The observation of a decrease in critical strain confirms that the addition of molybdenum promotes the occurrence of dynamic recrystallisation, occurring earlier. Comparing CoNiFeMn and $(CoNiFeMn)_{95}Mo_5$ alloys this is also due to the fact that molybdenum reduces the stacking fault energy which favours the occurrence of DRX. In the $(CoNiFeMn)_{90}Mo_{10}$ alloy there are additionally μ-phase precipitation in the temperature range of hot forming. The larger decrease in strain values is due to the higher stored energy associated with the additional separation strengthening. Another aspect is the grain size, which is much smaller compared to other alloys. A smaller grain size, increases the volume of grain boundaries, which are privileged nucleation sites for new grains which means that dynamic recrystallisation can start at lower strains. In addition, the smaller grain size increases the proportion of strengthening from the grain boundaries, which also translates into a shift in the onset of recrystallisation towards smaller strains.

The activation energies of dynamic recrystallisation were determined from the slope factor of the linear function at ln(sinh(ασp)) - 1000/T. The values are included in the graphs in Fig. 4.

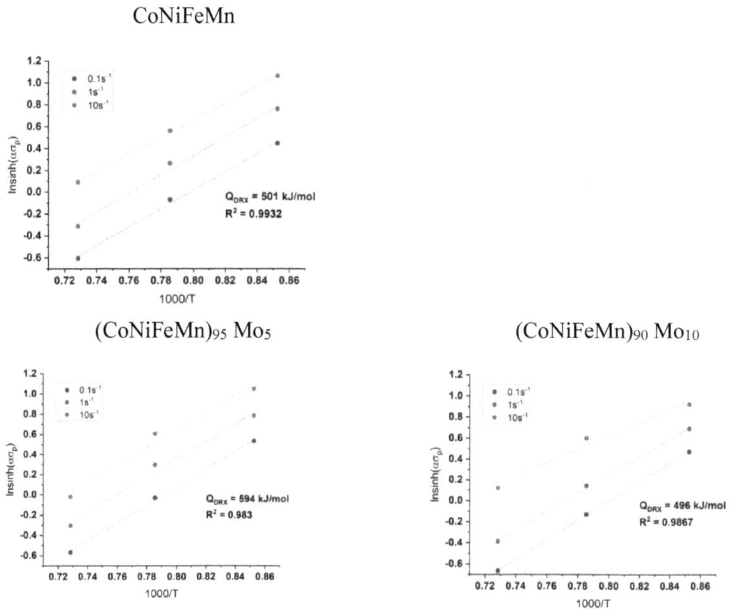

Fig. 4. Activation energy of dynamic recrystallisation for the alloys studied.

The above analysis shows that the addition of molybdenum increases the activation energy of dynamic recrystallisation. For alloys containing pure FCC solid solution, the addition of 5% atomic molybdenum results in an increase in Q_{DRX} of 93 kJ/mol. The decrease in Q_{DRX} values is dictated by the previously described effects from the existing precipitates as well as the smaller grain size. The values obtained are comparable to those obtained for other high-entropy alloys[3][9]. High values of activation energies, close to or slightly higher than the diffusion activation energies of the individual elements, indicate the effect of sluggish diffusion and hinder the occurrence of diffusion processes such as dynamic recrystallisation [10][11].

Investigations into the behaviour of materials at elevated temperatures made it possible to design the hot-rolling process and prepare the material accordingly for subsequent cold rolling.

Cold rolling, microstructure evolution. The hot-rolling process parameters were developed based on the results obtained on the ASP thermomechanical simulator. The materials were plastically processed according to the scheme shown in Fig. 5. The total given strain was 0.74. The material was then relaxed and cooled in air. The relaxation allowed the completion of the meta-dynamic and static recrystallisation processes. The selection of appropriate process parameters allowed the material to be fully recrystallised, with equiaxial grains, and with the characteristic annealing twins present (Fig. 5).

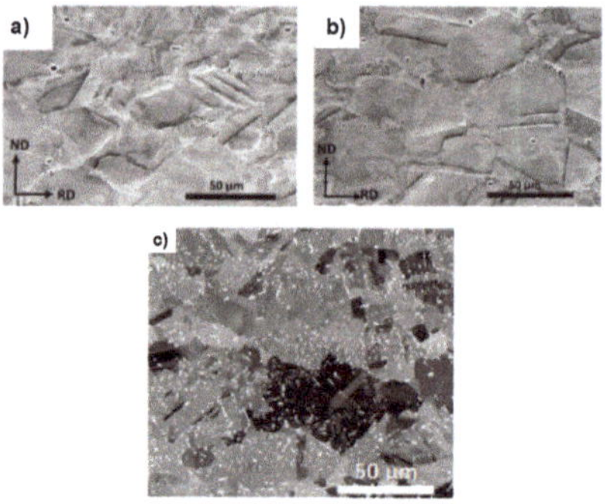

Fig. 5. SEM-BSE microstructure images after hot rolling of (a) CoNiFeMn, (b) (CoNiFeMn)95Mo5, (c) (CoNiFeMn)90Mo10 alloys.

By analysing the microstructures, it can be seen that the (CoNiFeMn)90Mo10 alloy exhibits μ-phase precipitates with varying morphologies. In addition, there are small areas free of precipitates in the microstructure (Fig. 5c). This may be due to the fact of uneven deformation during hot rolling, as well as the occurrence of a heat dissipation gradient during cooling. The microstructure photographs show the initial microstructure for cold rolling. The final actual deformation was taken as $\varepsilon = 0.7$. Fig. 6 shows the IPF after cold rolling after various deformations for the tested materials.

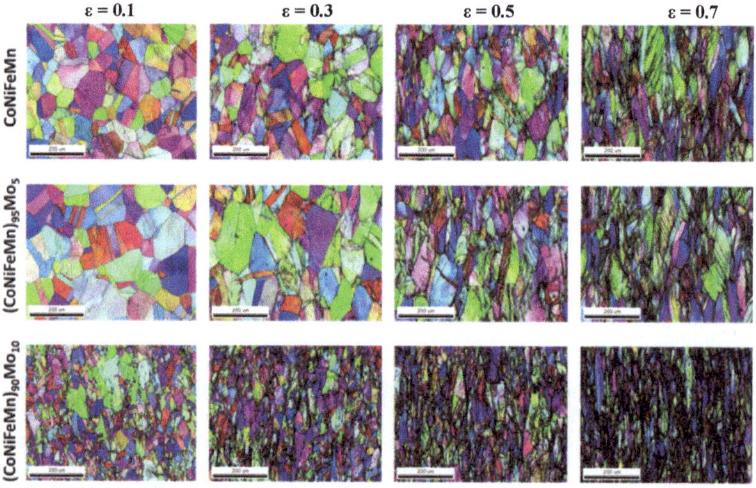

Fig. 6. IPF maps obtained by EBSD analysis after cold rolling with different degrees of deformation.

In the microstructure after deformation $\varepsilon = 0.1$, it is still possible to observe equiaxial grains with visible annealing twins. In this case, the difference in grain size between the materials can be seen. In the alloys CoNiFeMn and $(CoNiFeMn)_{95}Mo_5$, the grain size is similar. In the case of the $(CoNiFeMn)_{90}Mo_{10}$ alloy, the grain size is smaller as a result of the blocking of grain growth after

Materials Research Proceedings 44 (2024) 393-401 https://doi.org/10.21741/9781644903254-43

hot forming by μ-phase precipitation. The beginning of substructure formation can be observed, suggesting the appearance of low angle grain boundaries (LAB). After deformation $\varepsilon = 0.3$, the grains begin to elongate in the rolling direction. Increasing the degree of deformation also increases the number of low angle grain boundaries. The microstructure after a strain of 0.5 is already characterised by elongated grains in the rolling direction. The microstructure of the alloy $(CoNiFeMn)_{90}Mo_{10}$ already reveals thick bands extending along the rolling direction. The formation of a band structure begins. For the other materials, these microstructural phenomena occur at strain $\varepsilon = 0.7$. In addition to the high- and low-angle grain boundaries, the appearance of strain twins was observed in the microstructure (Fig. 7). At strain $\varepsilon = 0.1$, annealing twins are still visible. Deformation twins appear in the CoNiFeMn alloy at a strain of $\varepsilon = 0.7$. In the case of the alloy $(CoNiFeMn)_{95}$ Mo_5 twins can already be observed at lower strains.

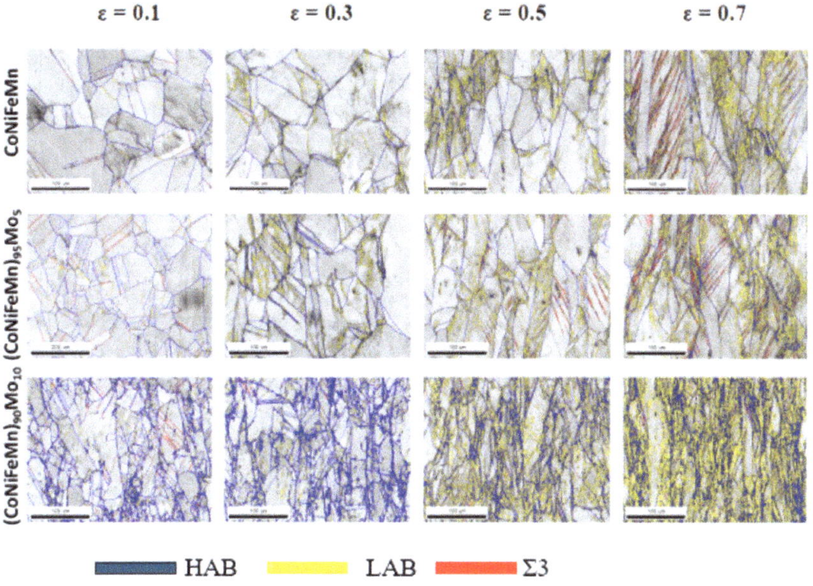

Fig. 7. High-angle, low-angle and twin boundaries evolution during deformation progresses.

The presence of twins is also confirmed by the linear disorientation measurement presented in Fig. 8. The linear disorientation measurement was performed inside the grain in the <110> direction. From the analysis presented in the graph, increase in disorientation up to 60° can be observed, which confirms the rotation of the crystal during deformation and thus indicates the appearance of deformation twins.

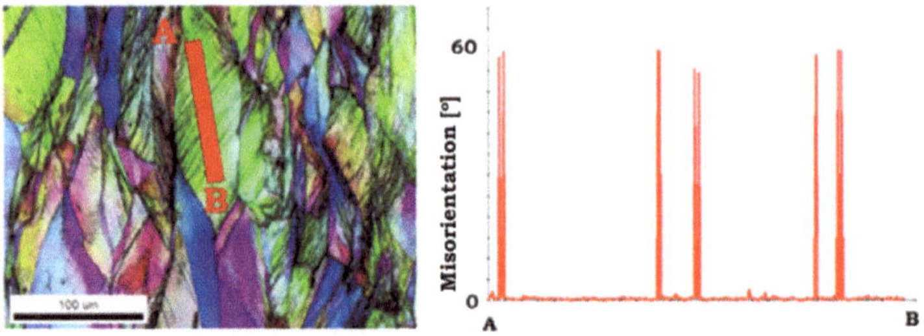

Fig. 8. Linear grain disorientation in the <110> direction.

Static recrystallisation and grain growth after cold rolling is therefore a very important factor in controlling the ability to absorb the impact energy through twinning mechanism. A detailed analysis of the static recrystallisation process and grain growth has been discussed in detail in our previous work [8]. Fig. 9 shows the grain size as a function of temperature after cold rolling.

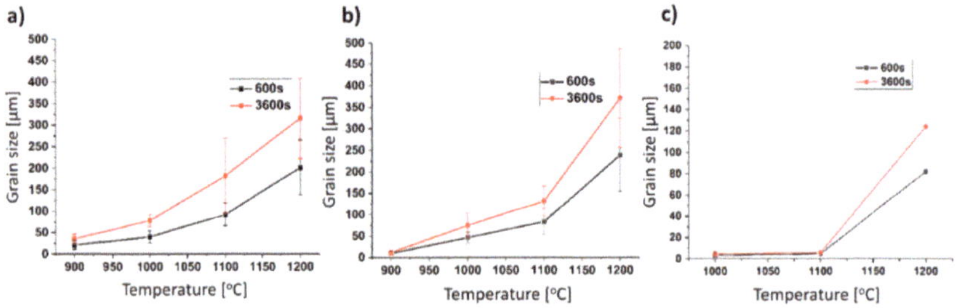

Fig. 9. a) CoNiFeMn [8], (b) $(CoNiFeMn)_{95}Mo_5$ [8], (c) $(CoNiFeMn)_{90}Mo_{10}$.

Based on the presented research it can be concluded that by knowing the mechanisms controlling the grain size both during hot and cold deformation, makes it possible to prepare a material with a suitable microstructure for possible applications under both quasi-static or dynamic load–bearing applications – especially at low temperature where nano-twinning and twinning is expected as a main mechanism that increasing the toughness of these new group of materials.

Summary

In this study, the hot deformation behaviour of the proposed alloys was studied, as well as the microstructure evolution during cold rolling and heat treatment. Based upon the presented research, the following conclusion can be drawn:

- The addition of 5% molybdenum to the CoNiFeMn alloy increases the activation energy of dynamic recrystallisation. In the alloy with 10% Mo, the DRX activation energy is lower due to the increased energy accumulated by the smaller grain size and the presence of µ-phase precipitates.
- The evolution of the microstructure after rolling shows that strain twins can be observed in the alloy with 5% Mo at lower strains than in the alloy without Mo. This demonstrates the reduction of alignment error energies by the addition of Mo to the alloy.

Acknowledgement

Research project supported by the program "Excellence initiative—research university" from the AGH University of Krakow is acknowledged.

References

[1] B. Cantor, I.T.H. Chang, P. Knight, A.J.B. Vincent, Microstructural development in equiatomic multicomponent alloys, Mater. Sci. Eng. A 375-377 (2004) 213-218. https://doi.org/10.1016/j.msea.2003.10.257

[2] D. B. Miracle, High-Entropy Alloys: A Current Evaluation of Founding Ideas and Core Effects and Exploring 'Nonlinear Alloys, JOM 69 (2017) 2130-2136. https://doi.org/10.1007/s11837-017-2527-z

[3] J.M. Park, J. Moon, J.W. Bae, M.J. Jang, J. Park, S. Lee, H.S. Kim, Strain rate effects of dynamic compressive deformation on mechanical properties and microstructure of CoCrFeMnNi high-entropy alloy, Mater. Sci. Eng. A 719 (2018) 155-163. https://doi.org/10.1016/j.msea.2018.02.031

[4] J. Wang, Y. Liu, B. Liu, Y. Wang, Y. Cao, T. Li, R. Zhou, Flow behaviour and microstructures of powder metallurgical CrFeCoNiMo0.2 high entropy alloy during high temperature deformation, Mater. Sci. Eng. A 689 (2017) 233-242. https://doi.org/10.1016/j.msea.2017.02.064

[5] L. Kaushik, M.-S. Kim, J. Singh, J.-H. Kang, Y.-U. Heo, J.-Y. Suh, S.-H. Choi, Deformation mechanisms and texture evolution in high entropy alloy during cold rolling, Int. J. Plast. 141 (2021) 102989. https://doi.org/10.1016/j.ijplas.2021.102989

[6] K. Cichocki, P. Bała, T. Kozieł, G. Cios, N. Schell, K. Muszka, Effect of Mo on Phase Stability and Properties in FeMnNiCo High-Entropy Alloys, Metall. Mater. Trans. A Phys. Metall. Mater. Sci. 53 (2022) 1749-1760. https://doi.org/10.1007/s11661-022-06629-x

[7] A. Najafizadeh, J.J. Jonas, Predicting the critical stress for initiation of dynamic recrystallization, ISIJ Int. 46 (2006) 1679-1684. https://doi.org/10.2355/isijinternational.46.1679

[8] K. Cichocki, P. Bała, M. Kwiecień, M. Szymula, K. Chrzan, C. Hamilton, The influence of Mo addition on static recrystallization and grain growth behaviour in CoNiFeMn system subjected to prior deformation, Arch. Civ. Mech. Eng. 24 (2024). https://doi.org/10.1007/s43452-024-00888-8

[9] S.A. Sajadi, M.R. Toroghinejad, A. Rezaeian, G.R. Ebrahimi, Dynamic recrystallization behavior of the equiatomic FeCoCrNi high-entropy alloy during high temperature deformation, J. Mater. Res. Technol. 20 (2022) 1093-1109. https://doi.org/10.1016/j.jmrt.2022.07.055

[10] W. Kucza, J. Dąbrowa, G. Cieślak, K. Berent, T. Kulik, M. Danielewski, Studies of 'sluggish diffusion' effect in Co-Cr-Fe-Mn-Ni, Co-Cr-Fe-Ni and Co-Fe-Mn-Ni high entropy alloys; determination of tracer diffusivities by combinatorial approach, J. Alloy. Compd. 731 (2018) 920-928. https://doi.org/10.1016/j.jallcom.2017.10.108

[11] K.Y. Tsai, M.H. Tsai, J.W. Yeh, Sluggish diffusion in Co-Cr-Fe-Mn-Ni high-entropy alloys, Acta Mater. 61 (2013) 4887-4897. https://doi.org/10.1016/j.actamat.2013.04.058

Microstructure, strain rate, and temperature effects on the compressive loading behavior of $(FeMnNiCo)_{1-x}Mo_x$ high entropy alloy

KRÜGER Lutz[1,a *], CICHOCKI Kamil[2,b], HENSCHEL Sebastian[1,c], CHULIST Robert[3,d], BAŁA Piotr[2,e] and MUSZKA Krzysztof[2,f]

[1]TU Bergakademie Freiberg, Institute of Materials Engineering, Gustav-Zeuner-Str. 5, 09599 Freiberg, Germany

[2]AGH University of Science and Technology, Department of Metal Forming and Metallurgical Engineering, al. A. Mickiewicza 30, 30-059 Kraków, Poland

[3]Institute of Metallurgy and Materials Science, Polish Academy of Sciences, ul. W. Reymonta 30, 30-059 Kraków, Poland

[a]krueger@ww.tu-freiberg.de, [b]cichocki@agh.edu.pl, [c]sebastian.henschel@iwt.tu-freiberg.de, [d]r.chulist@imim.pl, [e]pbala@agh.edu.pl [f]muszka@agh.edu.pl

Keywords: High Entropy Alloy, Strain Rate Effect, Temperature Effect, Compression Test, Strain Hardening, Microstructure, Twinning

Abstract. High entropy alloys based on FeMnNiCoMo offer a potential for high energy absorption and high strength, even at low temperatures. In this work, the effects of chemical composition and grain size on the mechanical properties were investigated. Compression tests at a wide range of strain rates and temperatures have been performed. For a strain rate of about 10^2 s^{-1} an instrumented drop weight tower was applied. Furthermore, cryogenic temperature tests at −196 °C were carried out. The microstructure was analyzed by means of scanning electron microscopy and transmission Kikuchi diffraction. Hence, the role of grain size, precipitations, and twinning-induced plasticity (TWIP) on the yield strength and the further strain hardening were investigated.

Introduction

High entropy alloys (HEAs) are under focus since the investigations of Cantor et al. [1] and Yeh et al. [2] in 2004. Many alloy systems have been proposed. For some systems, the high ductility and fracture toughness especially at cryogenic temperatures are notable features. A reason for this behavior is the high strain hardening due to solid solution strengthening and deformation twinning [3]. Li et al. [4] suggested dual-phase HEAs which use additional hardening by martensitic phase transformation. A smaller grain size will not only result in a higher strength due to Hall-Petch effect. The stress to induce twinning will also increase [5].

Basu et al. [6] observed a low strain rate sensitivity for the dual-phase $Fe_{50}Mn_{30}Co_{10}Cr_{10}$ alloy at strain rates between 10^{-4} to 5×10^{-3} s^{-1}. Huang et al. [7] reviewed the results of several HEAs at a wide range of strain rates, i.e. 10^{-4} to 10^4 s^{-1}. For the CoCrFeMnNi alloy, an increase in strain rate (10^{-4} to 5×10^3 s^{-1}) leads to a strong increase in strength and work hardening rate. They also investigated the adiabatic heating and determined a rather low Taylor-Quinney coefficient of about 0.5 to 0.6 for strain rates of 10^0 to 3×10^3 s^{-1}. This finding is in contrast to observations of Meyer et al. [8] at a 1045 steel. According to Li et al. [9] the deformation of fcc HEAs at higher strain rates and lower temperatures is mainly based on twinning. This behavior was confirmed in CoCrFeMnNi by Park et al. [10]. However, they did not study the material behavior at intermediate strain rates of about 10^2 s^{-1}. Due to the high work hardening rate, the resistance to adiabatic shear band localization is high [9].

In this work, the mechanical behavior of the system $(FeMnNiCo)_{1-x}Mo_x$ was studied under compressive loading. The effects of Mo content, strain rate and temperature were investigated. In addition to quasi-static and high strain rate tests the mechanical behavior at intermediate strain rates was also measured.

Materials and Methods

Materials. The chemical compositions of the investigated high entropy alloys are given in Table 1. Ingots were produced in a Bühler Arc Melter AM. Further details can be found in a recent study on the phase stability of these alloys [11].

Table 1. Chemical composition of investigated alloys, in atom-%.

Alloy	Fe	Mn	Ni	Co	Mo
FeMnNiCo	24.7	24.2	25.5	25.6	0
$(FeMnNiCo)_{90}Mo_{10}$	22.2	21.2	23.5	23.8	9.3

The ingots were hot rolled at 1100°C with a thickness reduction of 40 %. After air cooling, cold rolling with 6 passes and a total thickness reduction of 10 % was performed. In order to statically recrystallize the microstructure, annealing was done at different temperatures and times. Details are given by Cichocki et al. in this journal issue. The initial microstructure is shown in Fig. 1.

0 % Mo

10 % Mo

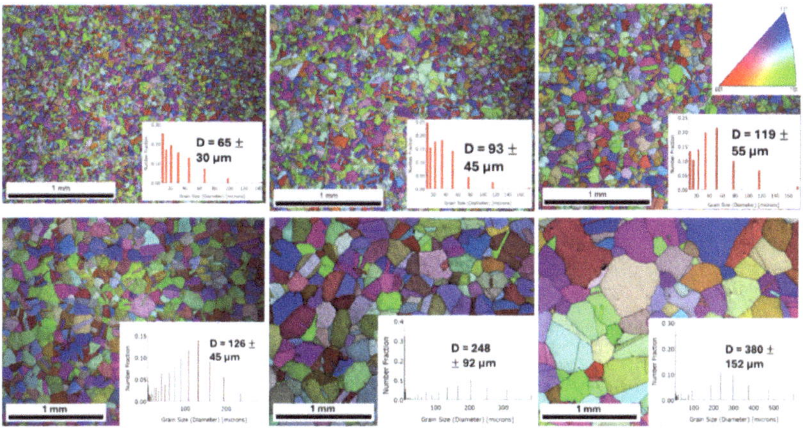

Fig. 1.: Initial microstructure of materials with 0 % Mo (first row) and 10 % Mo (second row).

The different heat treatment led to different average grain sizes and grain size distributions. In all cases, a polygonal grain shape was achieved. Only some annealing twins are visible. The material basically consists of a single phase with fcc lattice. However, it should be noted, that the hard µ phase is present in small amounts in the materials with 10 % Mo [11]. The average grain sizes of the materials are given in Table 2.

Table 2. Average grain sizes and standard deviations of different materials. Grain sizes are referred to by their class name.

Material / Grain size in µm	class A	class B	class C	class D	class E
FeMnNiCo	65 ± 30	93 ± 45	119 ± 55		
$(FeMnNiCo)_{90}Mo_{10}$			126 ± 45	248 ± 92	360 ± 152

To compare the materials, classes of grain sizes were defined. In class C, there are two materials, i.e. 0 % Mo and 10 % Mo. The other classes of grain sizes contain either the 0 % Mo material or the material with 10 % Mo.

Methods. Compression tests were performed at 4 different strain rates: 10^{-1}, 10^{0}, 10^{2} and 10^{3} s^{-1}. A bath with liquid nitrogen was applied for the low temperature tests. Tests at strain rates of 10^{0} and lower were performed in a universal testing machine. An instrumented drop tower was applied for tests at approximately 10^{2} s^{-1}. The drop tower consists of a 190 kg drop weight with a punch that is instrumented with calibrated strain gauges and a 4,000 kg anvil. Strain rates of about 10^{3} s^{-1} were achieved in a split Hopkinson pressure bar. The setup consists of incident and transmitted bars, each 2 m long and 20 mm in diameter and made of high-strength maraging steel, which are instrumented with strain gauges. The striker bar had a length of 0.5 m and was made of the same bar material as the other bars. Pulse shaping was used to obtain early force equilibrium and relatively constant strain rates. More details on pulse shaping can be found elsewhere [12]. Interrupted tests at strain rates > 10^{-1} s^{-1} were achieved by stop rings made of high-speed steel with a defined thickness. The samples can only be deformed as long as there is a gap between punch and the stop ring.

In addition to true compressive strain ε and true compressive stress σ, strain hardening coefficient Θ was calculated:

$$\Theta = \frac{d\sigma}{d\varepsilon}. \quad (1)$$

SEM/EBSD analysis of the initial microstructure was performed on a FEI Nova NanoSEM 450 scanning microscope (SEM) using the Velocity EBSD detector from EDAX. The samples were prepared by grinding and mechanical polishing. The EBSD analysis was performed in the center of the compressed sample, with an accelerating voltage of 20 kV and a spot size of 6 with and step size 0.2 µm. HR EBSD was performed in selected samples with a step size of 0.05 µm with an accelerating voltage of 20 kV. Specimens from interrupted compressive tests were analyzed in a Tescan Mira 3 XMU SEM at 15 kV with a Digiview EBSD detector from EDAX and a step size of 0.25 µm. The obtained results were processed with OIM-TSL 8.

Results and Discussion

Effect of Mo content and grain size. The dependence of true stress and strain hardening behavior on the Mo content and the grain sizes is shown in Fig. 2 for room temperature tests at a strain rate of 10^{-1} s^{-1}. It was observed that an increase in Mo results in a higher yield strength, cf. Fig. 2a. This behavior is explained by the solid solution hardening effect. Furthermore, the HEAs with 0 % Mo exhibited a continuously decreasing strain hardening coefficient. The materials with 10 % Mo exhibited a local maximum in strain hardening at a true strain of approximately 0.2.

Within the analyzed range of grain sizes, the materials show an increase in yield strength with decreasing grain size. This is explained by the Hall-Petch effect. However, there is no difference in further strain hardening behavior for the materials with 0 % Mo, cf. Fig. 2b. The material with

10 % Mo and the smallest grain size (class C) exhibited the highest local maximum of the strain hardening coefficient Θ, cf. Fig. 2c.

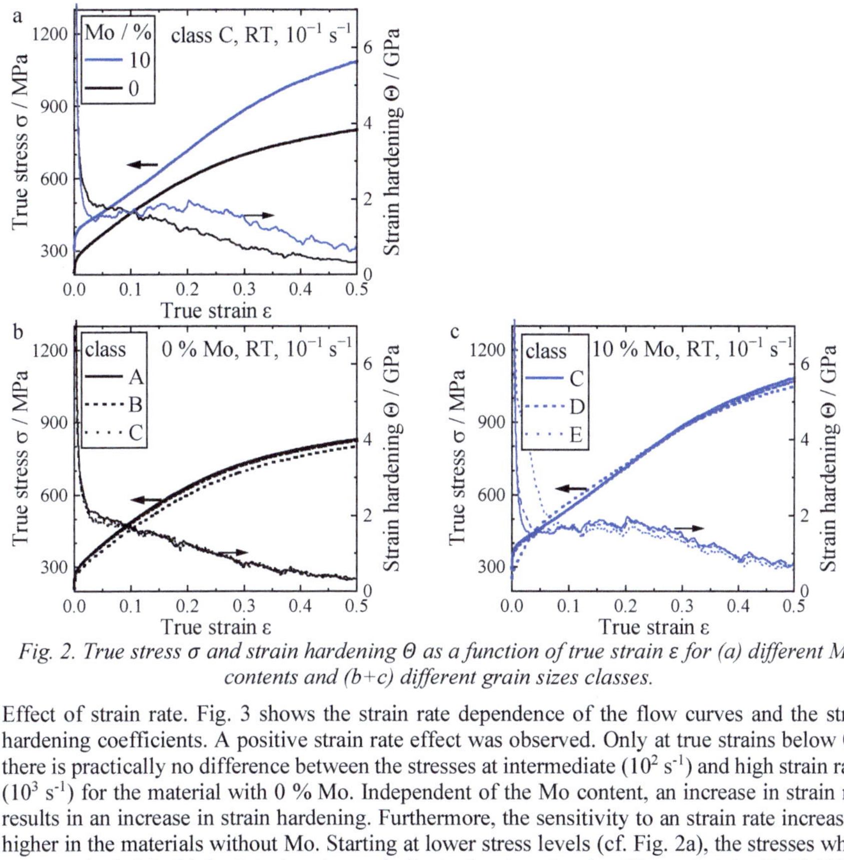

Fig. 2. True stress σ and strain hardening Θ as a function of true strain ε for (a) different Mo contents and (b+c) different grain sizes classes.

Effect of strain rate. Fig. 3 shows the strain rate dependence of the flow curves and the strain hardening coefficients. A positive strain rate effect was observed. Only at true strains below 0.1, there is practically no difference between the stresses at intermediate (10^2 s^{-1}) and high strain rates (10^3 s^{-1}) for the material with 0 % Mo. Independent of the Mo content, an increase in strain rate results in an increase in strain hardening. Furthermore, the sensitivity to an strain rate increase is higher in the materials without Mo. Starting at lower stress levels (cf. Fig. 2a), the stresses which were reached at the highest strain rate are similar to the stress levels of the material with 10 % Mo.

Even at the highest strain rate of 10^3 s^{-1}, the material with 0 % Mo shows no local maximum of the strain hardening coefficient, c.f. Fig. 3a. This local maximum was intensified for the material with 10 % Mo at higher strain rates. Furthermore, the maximum of the strain hardening coefficient was shifted to lower strains, c.f. Fig. 3b.

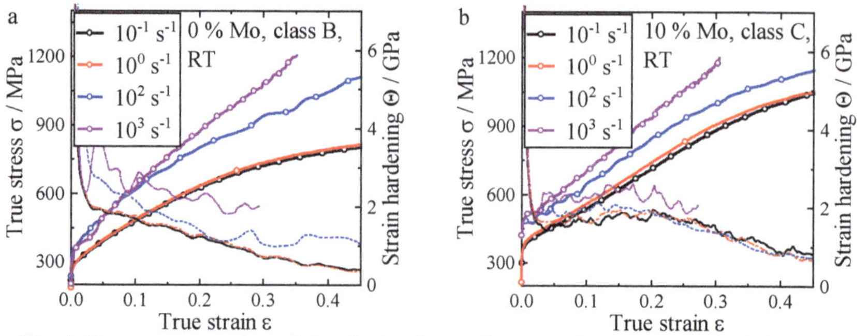

Fig. 3. True stress σ and strain hardening Θ as a function of true strain ε for different strain rates (a) with 0 % Mo and (b) 10 % Mo.

Effect of temperature. At a cryogenic temperature of −196°C the yield strength and strain hardening are higher than at ambient temperature, cf. Fig. 4a. The temperature effect on the flow stress at an intermediate strain rate of about 10^2 s^{-1} is more pronounced for the material with 10 % Mo addition, cf. Fig. 4b. The addition of 10 % Mo has a similar effect on the strain hardening as the decrease of temperature to −196°C. Like the tests at ambient temperature, the strain hardening behavior of the material with 10 % Mo is characterized by a local maximum, with highest strain hardening coefficient at −196°C.

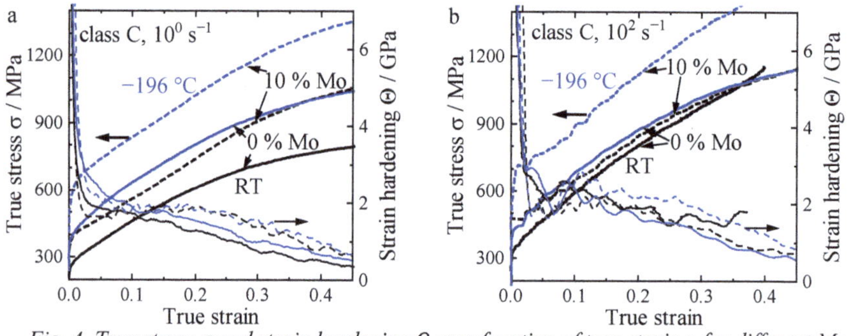

Fig. 4. True stress σ and strain hardening Θ as a function of true strain ε for different Mo contents and temperatures at strain rates (a) 10^{-1} s^{-1} and (b) 10^{2} s^{-1}.

Microstructural development. To explain and discuss the mechanical behavior, the microstructure of selected tests was analyzed at different degrees of deformation. Firstly, interrupted compressive tests are discussed. Secondly, the microstructure at a true compressive strain of about 0.4–0.5 is evaluated.

From the EBSD measurements in Fig. 5 it was derived that the formation of twins is affected by temperature and strain rate. The basic relationships from other studies like [9] are also found here. At the highest strain rates in the present study, multiple twinning was observed even at relatively low strains of about 0.02. This phenomenon was not observed at tests below strain rates of 10^2 s^{-1}, even at cryogenic temperatures.

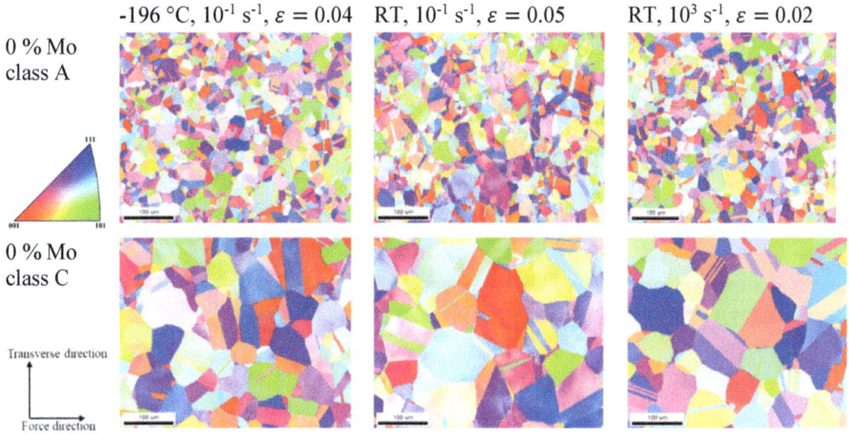

Fig. 5. Inverse pole figure visualizations of the microstructure after small deformation in materials with 0 % Mo: First row: smallest grain size, Second row: largest grain size. Effect of temperature from first to second column. Effect of strain rate from second to third column.

The EBSD results at large degrees of deformation are shown in Fig. 6. Due to the large deformation ($\varepsilon \approx 0.5$), the elongation of the grains is clearly visible. Moreover, there was further twinning within favorably oriented grains. A quantitative analysis of the $\Sigma3$ grain boundaries, i.e. the twin planes, reveals a significantly higher amount of twinning after testing at cryogenic temperature. Furthermore, with the addition of 10 % Mo the effect of deformation twinning was already observable largely at room temperature and at relatively small strain rates.

The microstructural observations in Fig. 6 are well related to the mechanical behavior in Fig. 4. The material with 10 % Mo shows a relatively large strain hardening even at room temperature. To achieve a similar strain hardening for the materials without Mo, a temperature of -196 °C is needed. The pronounced effect of twinning formation leads to small free paths for dislocation motion and, hence, for high strength and high strain hardening.

The twins are already formed after a short amount of deformation, e.g. after reaching a threshold stress. Hence, the material with 10 % Mo exhibiting higher stress levels due to solid solution hardening shows a higher amount of strain hardening due to easier formation of twins. Similarly, lower temperatures and higher strain rates will increase the strain hardening in fcc materials, cf. Fig. 4. Hence, higher stress levels are reached after smaller deformation.

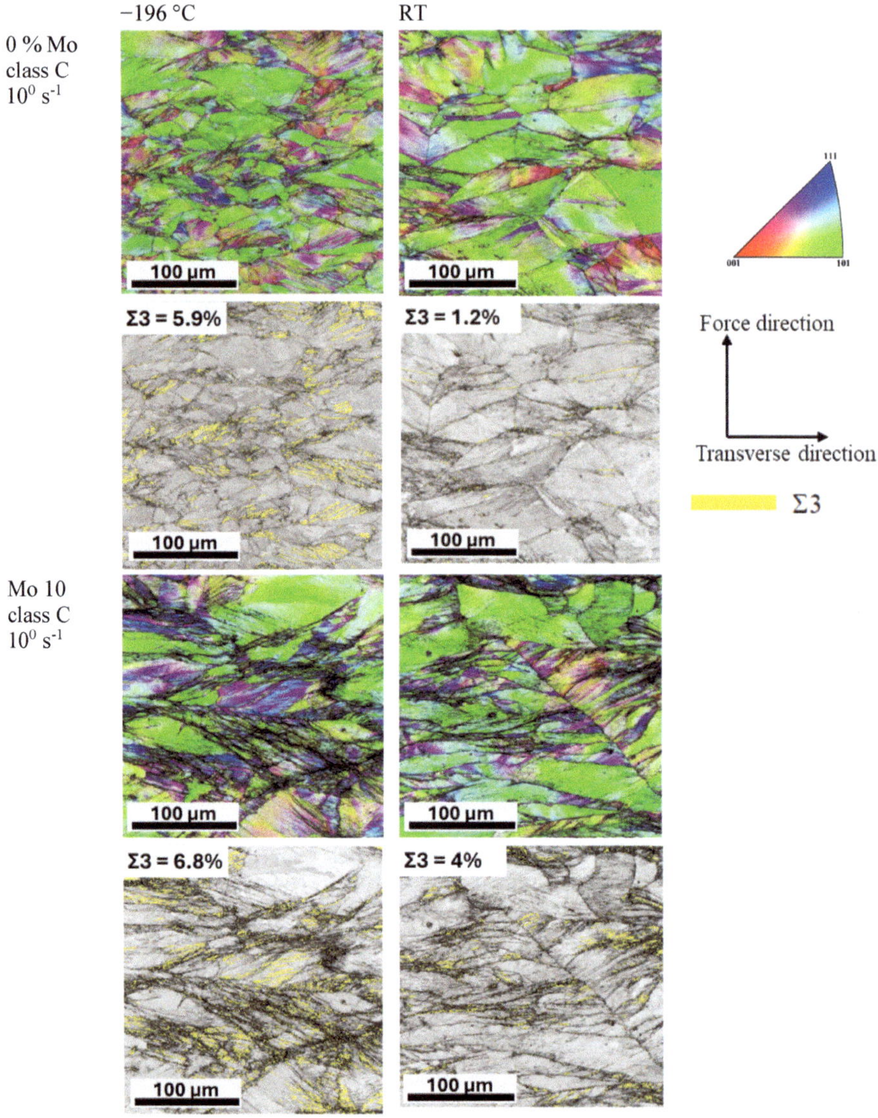

Fig. 6. Inverse pole figure visualization and image quality map with Σ3 grain boundary of the microstructure after deformation: Temperature effect from left to right column; Effect of Mo addition from upper to lower part of the figure.

Summary and Conclusions

The compressive loading behavior of high entropy alloys based on FeMnNiCo with and without Mo addition was investigated regarding grain size, strain rate, and temperature effects. To reveal the interdependencies between microstructure and mechanical properties, microstructural analysis of different deformation degrees was performed by means of SEM/EBSD. The following conclusions can be drawn:

- The addition of 10 % Mo significantly increases the susceptibility to form twins during deformation. This observation leads to higher strain hardening coefficients. Furthermore, the strain hardening shows a local maximum which is affected by temperature and strain rate.
- Higher strain rates and a cryogenic temperature result in higher stress levels and higher strain hardening. These observations are explained both by typical behavior of fcc metals and by pronounced twinning at lower temperatures and higher strain rates.
- Within the analyzed range, the mechanical behavior is relatively weakly affected by the grain size. Nevertheless, an decreased grain size will increase the flow stress.
- In all material conditions a damage tolerant compressibility without cracks on the surface of the samples was observed.

Acknowledgement

The support of Astrid Leuteritz, Diane Hübgen, Katrin Becker and Robert Dehne during specimen preparation is gratefully appreciated.

References

[1] B. Cantor, I. Chang, P. Knight, A. Vincent, Microstructural development in equiatomic multicomponent alloys, Mat. Sci. Eng. A 375-377 (2004) 213–218. https://doi.org/10.1016/j.msea.2003.10.257

[2] J.-W. Yeh, S.-K. Chen, S.-J. Lin, J.-Y. Gan, T.-S. Chin, T.-T. Shun, C.-H. Tsau, S.-Y. Chang, Nanostructured High-Entropy Alloys with Multiple Principal Elements: Novel Alloy Design Concepts and Outcomes, Adv. Eng. Mater. 6 (2004) 299–303. https://doi.org/10.1002/adem.200300567

[3] E.P. George, W.A. Curtin, C.C. Tasan, High entropy alloys: A focused review of mechanical properties and deformation mechanisms, Acta Mater. 188 (2020) 435–474. https://doi.org/10.1016/j.actamat.2019.12.015

[4] Z. Li, K.G. Pradeep, Y. Deng, D. Raabe, C.C. Tasan, Metastable high-entropy dual-phase alloys overcome the strength-ductility trade-off, Nature 534 (2016) 227–230. https://doi.org/10.1038/nature17981

[5] M.A. Meyers, O. Vöhringer, V.A. Lubarda, The onset of twinning in metals: a constitutive description, Acta Mater. 49 (2001) 4025–4039. https://doi.org/10.1016/S1359-6454(01)00300-7

[6] S. Basu, Z. Li, K. G. Pradeep, D. Raabe, Strain Rate Sensitivity of a TRIP-Assisted Dual-Phase High-Entropy Alloy, Front. Mater. 5 (2018). https://doi.org/10.3389/fmats.2018.00030

[7] A. Huang, S.J. Fensin, M.A. Meyers, Strain-rate effects and dynamic behavior of high entropy alloys, J. Mater. Res. Technol. 22 (2023) 307–347. https://doi.org/10.1016/j.jmrt.2022.11.057

[8] L.W. Meyer, N. Herzig, T. Halle, F. Hahn, L. Krüger, K.P. Staudhammer, A basic approach for strain rate dependent energy conversion including heat transfer effects: An experimental and numerical study, J. Mater. Process. Tech. 182 (2007) 319–326. https://doi.org/10.1016/j.jmatprotec.2006.07.040

[9] Z. Li, S. Zhao, R.O. Ritchie, M.A. Meyers, Mechanical properties of high-entropy alloys with emphasis on face-centered cubic alloys, Prog. Mater Sci. 102 (2019) 296–345. https://doi.org/10.1016/j.pmatsci.2018.12.003

[10] J.M. Park, J. Moon, J.W. Bae, M.J. Jang, J. Park, S. Lee, H.S. Kim, Strain rate effects of dynamic compressive deformation on mechanical properties and microstructure of CoCrFeMnNi high-entropy alloy, Mat. Sci. Eng. A 719 (2018) 155–163. https://doi.org/10.1016/j.msea.2018.02.031

[11] K. Cichocki, P. Bała, T. Kozieł, G. Cios, N. Schell, K. Muszka, Effect of Mo on Phase Stability and Properties in FeMnNiCo High-Entropy Alloys, Metall. Mater. Trans. A 53 (2022) 1749–1760. https://doi.org/10.1007/s11661-022-06629-x

[12] S. Henschel, L. Krüger, Dynamic crack initiation measurements in a four-point split Hopkinson bending device, Eng. Fract. Mech. 133 (2015) 62–75. https://doi.org/10.1016/j.engfracmech.2015.05.020

Investigation of mesoscale deformation and damage behaviour in high-carbon bearing steel via multiscale simulations and experiments

WANG Huiling[1,a], QIAN Dongsheng[1,b*], WANG Feng[1,c], and HUA Lin[1,d]

[1]West Courtyard, Mafanshan Campus, No. 122, Luoshi Road, Hongshan District, Wuhan City, Hubei Province, China

[a]wanghlwhut@163.com, [b]qiands@whut.edu.cn, [c]wangfengwhut@163.com, [d]hualin@whut.edu.cn.com

Keywords: High-Carbon Steel, In Situ SEM, Digital Image Correlation, Multiscale Simulations, Damage and Fracture, Representative Volume Element, Molecular Dynamics

Abstract. High-carbon steel is susceptible to micro-defects owing to the heterogeneous microstructures during forming, which seriously deteriorates the fatigue life. However, the investigation of the damage behavior during cold deformation remains largely undisclosed in these steels, which hinders further control of the microstructure and forming process. The current work investigates the deformation and damage characteristics of high-carbon steel composed of soft ferrite and hard cementite particles. A methodology coupled with multiscale simulations and experiments is applied to analyze large plastic deformation and damage characteristics. The simulation utilizes nanoscale molecular dynamics simulation to obtain matrix-particle interface strength properties. The Rice-Tracey fracture model and Weibull distribution capture experimental fracture characteristics of matrix and particles. The above failure criteria are incorporated within a real microstructure-based representative volume element (RVE) to conduct the mesoscale deformation and damage process in the spheroidized ferrite-cementite steel. Additionally, in-situ SEM uniaxial tensile tests are carried out to assess the damage mechanism and validity of the mesoscale simulation. The numerical simulation exhibits a well coincidence with the experimental trends in damage evolution of the individual particles, matrix, and matrix/particle interfaces. It is also observed that damage is a function of inherent particle properties, particle morphological features, and matrix strain localization characteristics. Larger-size and long-striped particle undergoes fracture at an earlier stage. Consequently, the incompatibility and stress concentration between matrix and particles affect the strain localization characteristics. As a result, higher stresses inside strain localization bands results in the increase of void damage initiation and growth along the interfaces. Overall, the way of matrix/particle decohesion should be marginally higher compared to particle fracture, which primarily dominates the final fracture of the high-carbon bearing steel.

Introduction

High-carbon bearing steels have been widely used in the aerospace, precision machine tool, and automotive industries because of its the relatively high machinability, good ductility, and excellent surface finish [1-3]. However, as a dual-phase polycrystalline alloy, micro-defects are highly susceptible to be produced in these steels owing to the mismatched deformation between the ferrite and cementite during cold-forming [4, 5]. These defects act as crack sources and reduce the impact toughness [6] and fatigue life [7]. This phenomenon is mainly influenced by the interleaving of length-scales caused by the deformation geometry, loading conditions and/or heterogeneous microstructure of the material. From the early studies [8-12], clearly the microstructural factors such as grain size, particle size, distribution, mechanical properties of the constituent phases, should be considered in analysis of deformation and fracture behavior. This complex

Materials Research Forum LLC
https://doi.org/10.21741/9781644903254-45

microstructure with different phases produces a highly non-uniform stress and strain distribution in the localized microstructure. However, the complexity of the microstructure makes it difficult to capture this strain distribution and damage response. There have been some studies on the damage mechanism of dual-phase steels numerically and experimentally, but less research has been done on the damage and strain distribution phenomena of this microstructure of cementite steel. Therefore, it is necessary to study the local deformation and damage mechanism of these steels in detail, as well as the correlation between them and the microstructure parameters.

In recent years, numerical simulation has been the most popular method to investigate the ductile damage and fracture behavior of metals, such as macroscale finite element (FE) simulation [13-15], nanoscale molecular dynamics (MD) [16, 17] simulation, and mesoscale representative volume element (RVE) methods [18-22]. From a macroscale level, the information about the failure properties of the components can be predicted through macroscale FE simulations, such as the stress and strain evolution of the components during the complex molding and manufacturing process. From the crystal and phase level, the correlation between the mesoscopic mechanical properties in the material with the microstructural characteristics can be accessed from RVE simulations. From the atomic level, the intrinsic properties can be described by MD simulation. Based on these numerical calculations, the damage and fracture mechanisms can be detected that have not yet been revealed only through experiments. In this way, a variety of damage models can be developed that are suitable for application in different materials and forming processes. Thus, the combination of micro-, macro-, and nanoscale simulation greatly enriches the understanding of damage and fracture processes in metals and other engineering materials.

In addition, the latest advancement of in situ experiment coupled with micro Digital Image Correlation (μDIC) has been also an effective method for observing the damage mechanism and non-uniform strain distribution in multiphase microstructure. For instance, [23] studied the neighborhood effects on micromechanical behaviors of multi-phase TRIP steel by employing in situ SEM/EBSD and in situ synchrotron X-ray diffraction tensile tests. [24] analyzed the evolution of the whole damage process of SiCp/Al composites through in situ tensile testing. [25] provided a unique set of data including twin/detwin characteristics and twin area fraction of Mg alloys by the in situ SEM-DIC method. [26] investigated the strain localization in commercially pure titanium under uniaxial tension by means of in situ digital image correlation (DIC) at different length scales. [19] observed the dominant damage nucleation mode changed from ferrite grain boundary (F/F) decohesion to ferrite/martensite interface (F/M) decohesion and finally to martensite cracking by in-situ tensile testing in a SEM. The microstructure of high-carbon bearing steels is a mixture of a mechanically soft phase (ferrite matrix) and hard phases (cementite particles), and as a result of their distinct deformation behaviors, the amount of plastic strain partitioned into these phases varies significantly [4]. On top of these points about strain partitioning, there is also microstructural strain path development that needs to be taken into account.

In this work, the multiscale simulations and the in-situ SEM-based μDIC were combined to investigate the damage evolution and failure behavior of high-carbon bearing steel under the uniaxial tensile loading. First, the microstructure characteristics and damage mechanism of the as-received material was characterized by the in-situ SEM-based μDIC. Second, macroscale tensile model was conducted based on the experimental sample, and the GTN plasticity model was used in the model. Third, a microstructure-based RVE from the in-situ SEM image was reconstructed to analyze the micromechanical behavior, and estimate the failure model parameters of two phases. In particular, the zero-thickness cohesive elements were inserted at ferrite/carbide interfaces. Thereafter, the isotropic elastoplastic model, damage criterion, and CZM were assigned to the ferrite, cementite, and interface elements, whereas the elastic property was considered for the carbides. Fourth, the nanoscale MD model under different angle loading was performed by using the LAMMPS software. For the purpose of identifying the model parameters, the RSM method was applied to evaluate the GTN parameters, and then MD simulations were used to obtain the CZM parameters. The μDIC analysis from in-situ results was used to estimate the damage model parameters of two phases. Finally, the validity of the model was evaluated by comparing with the in-situ tensile results. Specifically, the ductile fracture, micromechanical behavior, damage evolution of the high-carbon bearing steel were explored.

Methodology

Experiments. The as-received material used in this study was the annealed 1.0C-1.5Cr steel bar with a size of Ø50 × 110 mm. The chemical composition is listed in Table 1. The initial microstructure was characterized by a Quanta FEG 450 Scanning Electron Microscope (SEM) and a JEOL-2100F high resolution transmission electron microscopy (HRTEM). The specimen for SEM observation was mechanically polished and then etched in an alcohol solution containing 4% nitric acid (volume fraction) for 30 s. The specimen used for HRTEM experiment was prepared by mechanical polishing into the disk with the thickness of 50 μm, and then thinned by twin-jet electrolytic polishing apparatus. To determine the OR and coherency of ferrite/cementite interface, electron diffraction pattern and HRTEM images of the selected region were evaluated by means of Digital Micrograph software tool and fast Fourier transformation (FFT), respectively.

Table 1. Chemical composition of the as-received material.

Element	C	Cr	Si	Mn	Mo	Ni	P	S	Cu	Fe
Wt [%]	0.967	1.38	0.30	0.23	0.02	0.07	0.02	<0.023	0.05	Bal.

In-situ SEM tensile tests were conducted to observe the damage evolution and strain field at room temperature using a Gemini 300 SEM equipment. The dimensions and geometry of the tested specimens are shown in Fig. 1(a). Before the tensile tests, the specimens were mechanically polished, followed by ultrasonic cleaning with acetone, and electropolishing in an alcohol solution containing 4% nitric acid. During the tensile test, the specimen was deformed at with a velocity of 0.033 mm/min, and microstructures were captured by the repeated interruptions at the local strains of 2%, 5%, 10%, 15% and 20%. Fig. 1(b) shows the force-displacement curve for in situ tensile test. In addition, evolution of the equivalent plastic strain field (PEEQ) from undeformed to deformed in-situ SEM images was captured based on μDIC technology.

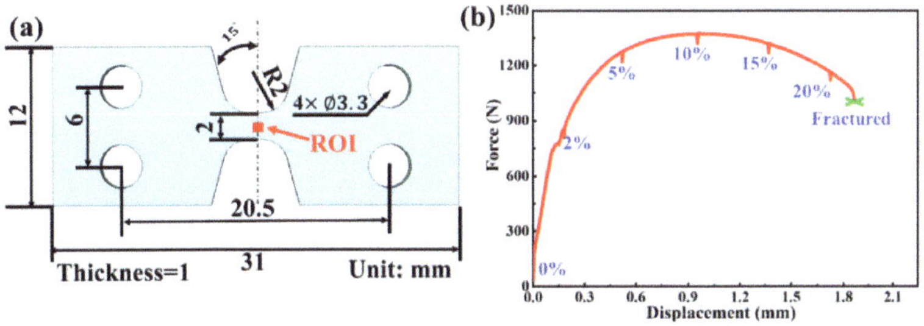

Fig. 1. (a) Dimensions of the specimen for in situ tensile tests. (b) Force-displacement curve.

The uniaxial tensile tests were conducted at room temperature on an AG-IC universal material testing machine. Standard tensile specimens were machined from adjacent locations on the as-

Materials Research Proceedings 44 (2024) 411-424 https://doi.org/10.21741/9781644903254-45

received bar to ensure similarly of the microstructure. Fig. 2 shows the shape and dimension of the tensile specimen with a gauge distance of 15 mm. The tests were repeated three times, and the strain rate was 10-3 s^{-1}. The macro digital image correlation (DIC) tests were performed to capture the local strain distribution during the tension. The specimen surfaces were sprayed with black and white scatter dots, and the speckle images of the specimen were collected at each deformation stage using high precision camera. Subsequently, the speckle images were processed by commercial software XTDIC to obtain the local strains at the different macroscopic strains.

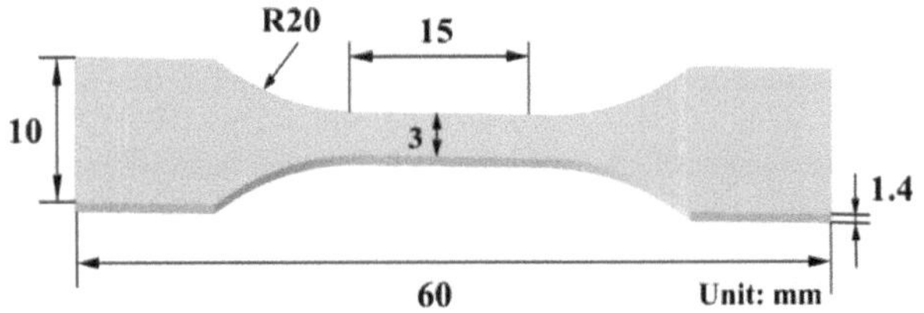

Fig. 2. Geometrical size of the uniaxial tensile specimen.

Multiscale modelling - Macroscale FE modelling. A uniaxial tensile model of the as-received steel was conducted as a macroscale finite element (FE) analysis, as showed in Fig. 3. The displacements of the plane z=0 and y=-30 mm have been constrained in the z-direction and y-direction, while the planes y=30 mm has been imposed uniform time-dependent displacement amounting to 5 mm, respectively. All other boundary surfaces of the domain have been considered traction-free. To make the simulation more accurate, the mesh size of the tensile model is divided into three sections, among which the middle segment is 0.2 mm, the circular segment is 0.5 mm, and the holding segment at both end is 1 mm. The mesh type was the eight-node linear hexahedron (C3D8R) with linear interpolation and selective reduced integration 8670 hexahedra elements, which was built via the ABAQUS/Explicit software. In addition, the Ludwik law and GTN porous plasticity model [27-29] were used as the elastic-plastic behavior and damage behavior of as-received steel, respectively. The laws are shown in following:

$$\sigma = \sigma_0 + K(\varepsilon_p)^n \tag{1}$$

$$\Phi = (\frac{\sigma_{eq}}{\sigma})^2 + 2f^* q_1 \cosh(\frac{3q_2}{2}\frac{\sigma_m}{\sigma}) - (1 + q_3 f^{*2}) = 0 \tag{2}$$

$$f^* = \begin{cases} f, & f > f_c \\ f_c + \frac{f_u - f_c}{f_f - f_c}(f - f_c), & f_c \leq f \leq f_f \\ f_u, & f \geq f_f \end{cases} \tag{3}$$

$$\dot{f} = \frac{f_N}{s_N\sqrt{2\pi}} exp\left[-\frac{1}{2}\left(\frac{\bar{\varepsilon}_{eq}^p - \varepsilon_N}{S_N}\right)^2\right]\dot{\bar{\varepsilon}}_{eq}^p + (1 - f)\dot{\varepsilon}_{eq}^p \tag{4}$$

where σ is the flow stress of undamaged material. σ_0 is the yield stress. K and n are the hardening modulus and exponent, respectively. σ_{eq} is the von Mises equivalent stress. σ_m is the hydrostatic pressure, which can be calculated as $\sigma_m = \frac{1}{3}\sigma_{ij}\delta_{ij}$. q_1, q_2, q_3 are the Tvergaard parameters, which are equal to 1, 1.5, and 2.25. $f_N, f_c, f_f, \varepsilon_N$ and S_N are GTN model parameters, which can be obtained by inverse extrapolation method in comparison with experimental data.

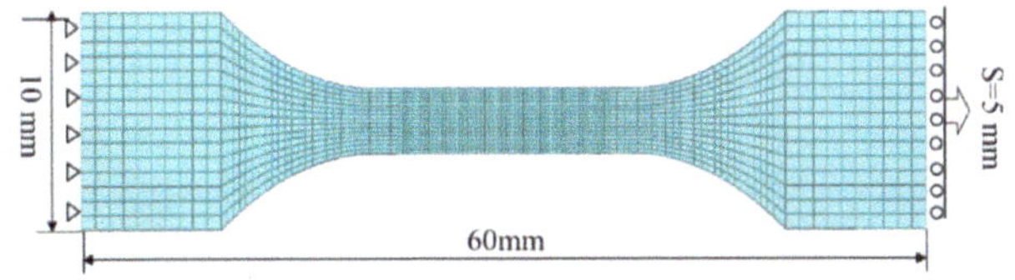

Fig. 3. The schematic of macroscale FE model: (a) tensile model; (b) zoom over the middle segment of tensile model.

Multiscale modelling - Mesoscale RVE modelling. In this section, mesoscale RVE is reconstructed based on the undeformed in-situ SEM micrographs by the image processing method [30, 31]. The generated RVE model, consisting of 456 particles, is presented in Fig. 4. The volume fraction of spheroidized cementite in the RVE was 20.12%. Subsequently, the generated RVEs were meshed in the commercial software ABAQUS (2020) with the 3-node linear plane stress triangle (CPS3) elements. The total number of elements and nodes are 151040 and 148007, respectively. To simulate the tensile deformation, symmetric boundary condition (SBC) was used by constraining all nodes of each element along the left edge in the X direction and move freely in the Y direction, and all nodes on the bottom edge to move in the Y direction but were free to move in the X direction. In addition, the cementite particles were assumed to follow linear elasticity with a Young's modulus of 230 GPa. Linear elasticity with a Young's modulus of 205 GPa and Ludwik law were used to describe the elastic-plastic behavior of ferrite matrix. According to the inverse analysis method based on genetic algorithm (GA), the model parameters for ferrite phase was obtained.

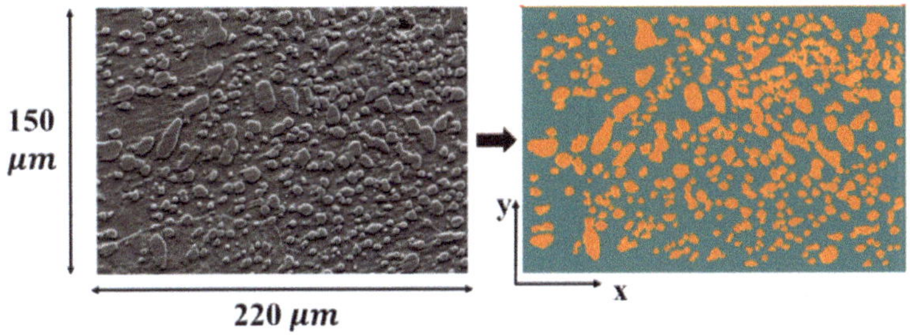

Fig. 4. A composite view of the reconfigured RVE model from the undeformed in-situ SEM image.

Multiscale modelling – Nanoscale MD modelling. To analyze the interfacial debonding between ferrite and cementite under the tension, MD simulations were conducted by using the atomic/molecular Massively Parallel Simulator (LAMMPS) package. Due to irregular morphology of cementite particles, the interface between ferrite matrix and cementite particles are subjected to mixed-mode loading [32]. The size of box is 300 Å in the x direction, 200 Å in the y direction, and 150 Å in the z direction, for a total of 848249 atoms. The orientation relationship between the ferrite and cementite was based on the transmission electron microscopy (TEM) test. Therefore, a ferrite-cementite interface model was generated using the Atomsk software, which is shown in Fig. 5.

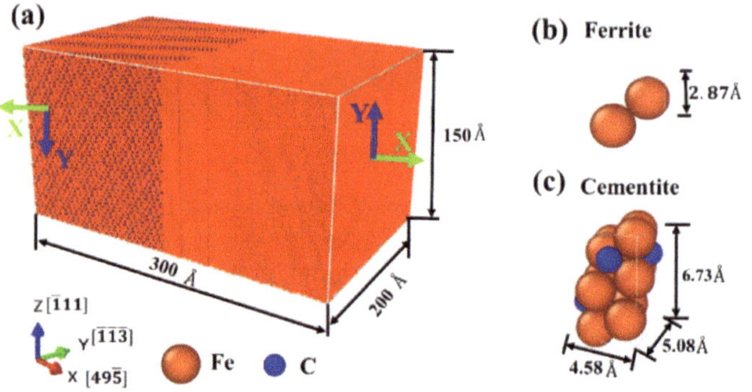

Fig. 5. A schematic of MD simulations to obtain the traction-separation (T-S) behavior at the matrix/particle interface.

Additionally, an embedded atom method (EAM) potential for Fe-C system was utilized to describe the atomic interaction, which has been verified to be reliable [33]. Before loading, the model was relaxed for 100 ps under the isothermal-isobaric (NPT) conditions to allow the system to reach most stable state. Afterwards, the relaxed model was carried out at a strain rate 10^9 s^{-1} along x and y directions (corresponded to the normal and shear directions) until fracture, respectively. The mixed mode loading conditions are determined by the shear and normal strain rate, respectively. The two strain rates are proportional to the rates of boundary displacements in the x and y directions, $\triangle \dot{X}$ and $\triangle \dot{Y}$. The relative normal and shear loading condition is expressed by a loading angle, $\theta = ArcTan(\frac{\triangle \dot{X}}{\triangle \dot{Y}})$. In order to obtain sufficient data for traction-separation relationships, MD simulations were carried out at five selected loading angles ($\theta = 0, 30°, 45°, 60°, 90°$). Finally, the stress-strain curves were obtained to represent the traction(τ)-separation (δ) behaviors.

Results and discussion

Microstructural characterization. The typical microstructure and statistical result are presented in Fig. 6. As presented in Fig. 6(a)~(b), the microstructure of as-received steel was consisted of ferrite matrix (dark zone, α) and randomly distributed cementite particles (bright zones, θ), in which the morphologies of particles were spherical. The particle volume fraction is about 20%, and the particle diameters range from 0.1~1 μm. As illustrated in Fig. 6(c)~(f), the ferrite/cementite interface is semi-coherent with a lattice misfit of 7.6%. The zone axes of ferrite and cementite were not exactly parallel, and the orientation relationship (OR) of the ferrite/cementite interface was $[\bar{1}0\bar{1}]_\alpha \parallel [\bar{2}11]_\theta$.

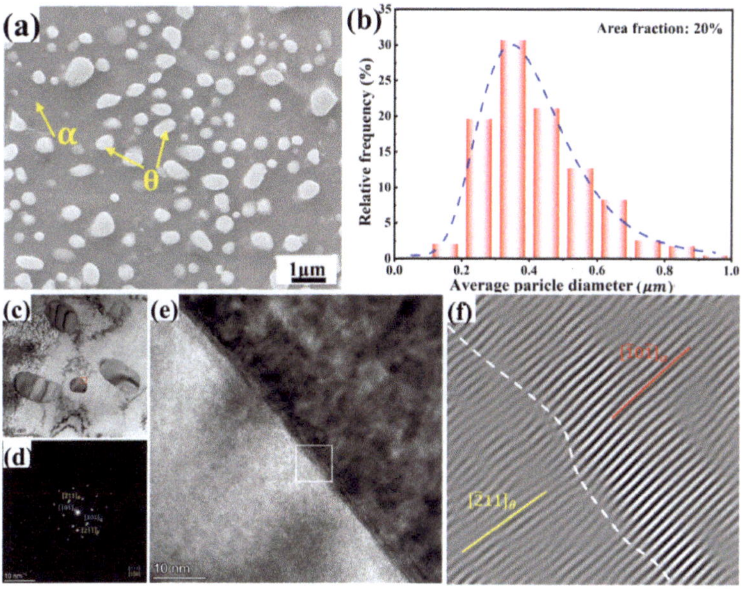

Fig. 6. Initial microstructural observations from SEM and TEM tests: (a) SEM image, (b) size distribution histogram of cementite particles. (c) Bright-field TEM image of ferrite and cementite. (d) Corresponding selected area diffraction pattern of the interface along $[001]_\alpha$ *zone axis. (e) HRTEM image of the interface observed from the red-boxed area in (c). (f) FFT image of the region labeled by the white box in (e).*

In situ SEM images, corresponding to the different global strains marked in Fig. 1(b), are shown in Fig. 7. As shown in Fig. 7, the as-received steel was consisted of the ferrite matrix and cementite particles. In the strain of 2%, no obvious variation occurred in the microstructure. When the strain was increased to 5%, several plastic slip bands appeared in the ferrite phase (marked by yellow dotted lines), and a bar-shaped cementite particle was broken (marked by red dotted box). At a strain of 10%, numerous plastic slip bands were observed in the ferrite phase, which showed a 45° angle to the direction of loading. As the strain reached to 15%, significant plastic deformation was observed in the ferrite phase. A large number of micro-voids (marked by green dotted circles) were detected at the interface between ferrite and cementite particles. Once the strain further increased to 20%, the volume fraction of voids between ferrite and cementite increased. In addition, voids in the ferrite phase were also observed (marked by pink triangle). From the above damage process, three types of damage mechanisms can be summarized: (1) Interfacial debonding; (2) cementite cracking; (3) ferrite cracking. Among them, the final fracture mainly occurred from the interface-debonding voids. In addition, the particles with irregular shapes were more likely to undergo fracture, rather than eventual failure.

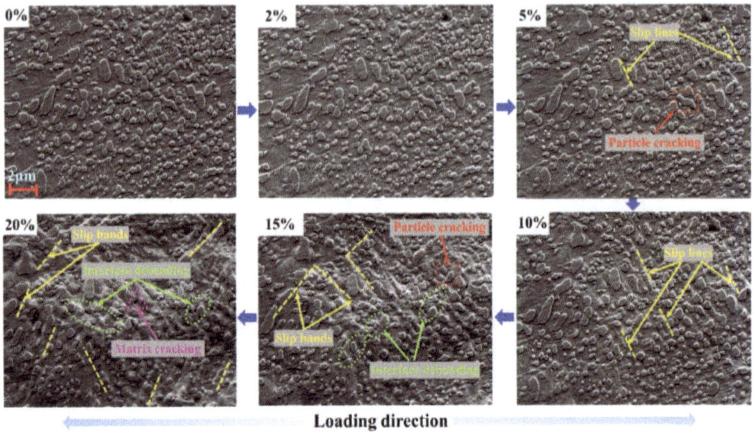

Fig. 7. SEM images at magnification of 5000×, corresponding to the strains of 0%, 2%, 5%, 10%, 15% and 20% for in situ tensile test.

Macroscale mechanical properties. Fig. 8 shows the experimental stress-strain curves and strain distributions by macro DIC analysis. As shown in Fig. 8(a), the yield stress and elongation are 407.26 MPa and 24.47%, respectively. As presented in Fig. 8(b), the strain distribution of as-received steel is extremely heterogeneous. When the engineering strain increases to the 0.15 (marked by the point A), the local strain ($\boldsymbol{\varepsilon_{xx}}$) in the central region of specimen is 0.205. As the engineering strain reaches the necking stage of the specimen (marked by the point B), the local strain ($\boldsymbol{\varepsilon_{xx}}$) in the central region increases to 0.508.

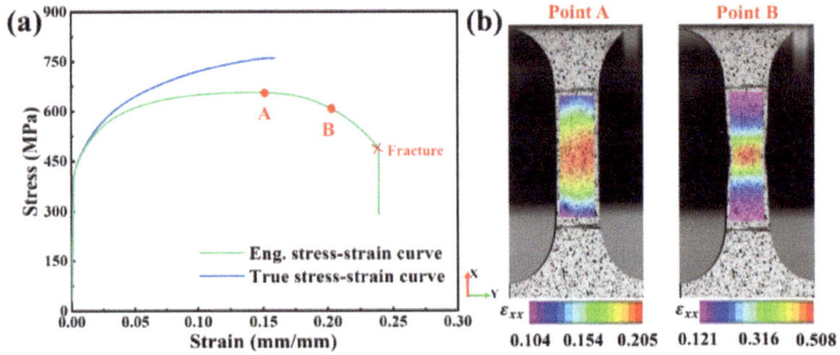

Fig. 8. (a) Stress-strain curves of uniaxial tensile experiment. (b) Strain distribution captured by macro DIC analysis.

To identify the model parameters in the macroscale FE simulation, the inverse extrapolation method was used to identify the unknown parameters in combination with experimental and simulated results under uniaxial tensile loading. First, the Young's modulus (E) was obtained by fitting the experimentally elastic stage of the true stress-strain curve. Second, the Ludwik model parameters were determined from fitting Eq. (1) to the true stress versus true plastic strains curves of the uniaxial tensile test. Third, the unknown GTN model parameters (ε_N, f_c, f_N, f_f) are calibrated by response surface method (RSM) [34]. Fig. 9 shows the simulated results in comparison with experimental result. As shown in Fig. 9(a), the simulated force-displacement curve under the optimized damage parameters were also well fitted to the experimental results. As shown in Fig. 9(b), the strain distribution at the necking stage was well match to the experimental result (Fig. 8(b)). In presented in Fig. 9(c), the void volume fractions of central point at the necking stage are 0.02. The optimal model parameters for macroscale simulation are listed in Table 2.

Table 2. Identified material parameters for the macroscale tensile simulation.

Parameters	$E(GPa)$	σ_0	K	n	ε_N	f_N	f_c	f_f
Value	205	262	830	0.25	0.2	0.01	0.05	0.15

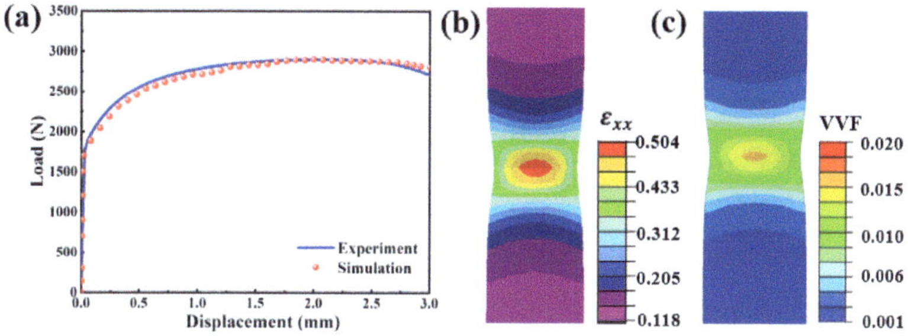

Fig. 9. (a) Stress-strain curves of uniaxial tensile test. (b) Strain distribution captured by DIC test during the tensile loading. (c)Void volume fractions of central point at the necking stage.

Mesoscale mechanical properties. Fig. 10 shows strain maps by using μDIC analysis during in situ tension. It is worth noting that strain maps could not be approached due to the excessive plastic deformation after the strain of 10%. Therefore, only the strain fields at the global strains of 2% and 5% were analyzed. In these two stages, the material damage was driven by the particle cracking. It shows that the distribution of local strains was more homogenously distributed in the ferrite matrix, while the larger strain gradients occurred close to the cementite particles. The highest localized strains were both distributed in the ferrite phase close to the irregular elongated particles. It is assumed that the particle strain in μDIC map (ε_θ^{DIC}) was equal to an average value of PEEQ of matrix surrounding to the fractured particle that was observed in in-situ SEM images. Similarly, the matrix strain (ε_α^{DIC}) was calculated by averaging the PEEQ at four surrounding points in the matrix from μ-DIC strain maps. As showed in Fig. 10(a), the local strains of ROI ranged from 0 to 0.095 at a global strain of 0.02. The averaged strain of ferrite matrix is ε_α^{DIC}= (0.048+0.035+0.008+0.064)/4=0.039 at the stage before particle fracture. In Fig. 10(b), the local strains of ROI ranged from 0.039 to 0.132 at a global strain of 0.05. The matrix strain is ε_α^{DIC}= (0.109+0.118+0.128+0.101)/4=0.114 at the stage after particle fracture.

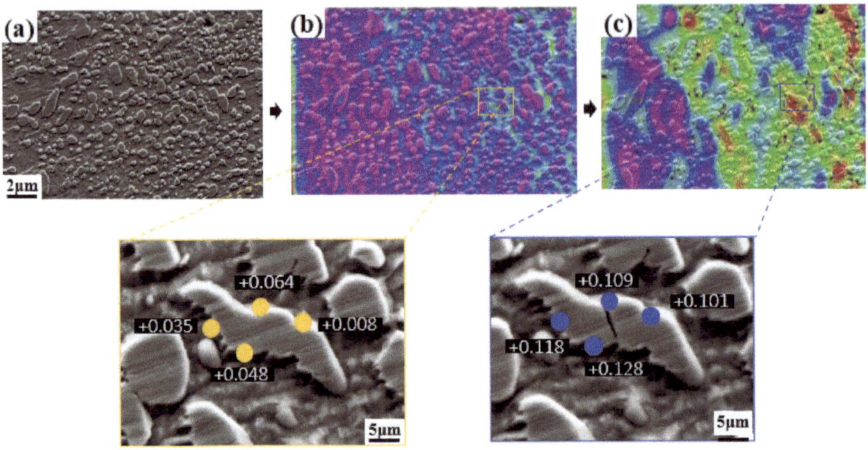

Fig. 10. Strain maps and matrix strain points by using μ-DIC technology from the in-situ tensile test with the global strain of (a) 0%, (b) 2%, (c) 5%.

Fig. 11 shows the micromechanical behaviors of as-received steel were analyzed using microstructure-based RVE simulation. It can be seen from Fig. 11(a) that the stress distribution in the ferrite phase was heterogeneous, and increased preferentially along the ferrite phase around the large particle. The stress localization became more pronounced with increasing strain, localizing along a direction 45° to the loading direction. Similarly, the stress distribution in the cementite phase exhibited the same inhomogeneity, as shown in Fig. 11(b). In addition, the stresses along the tips of the particles or in the depressed regions with a small radius of curvature were significantly higher than in other locations with higher roundness. In Fig. 11(c), the strain distribution in the ferrite and cementite phases is also inhomogeneous, and strain localization gradually develops with increasing strain. These results indicate that the mismatched properties between the cementite and ferrite led to an inhomogeneous distribution of the mesoscale stresses and strains between the two phases. The internal stresses and strains of the ferrite and cementite phases also showed inhomogeneous distributions. The inhomogeneous profiles were closely related to the shape, size, and distribution of the particles, and to the level of plastic slip of the ferrite.

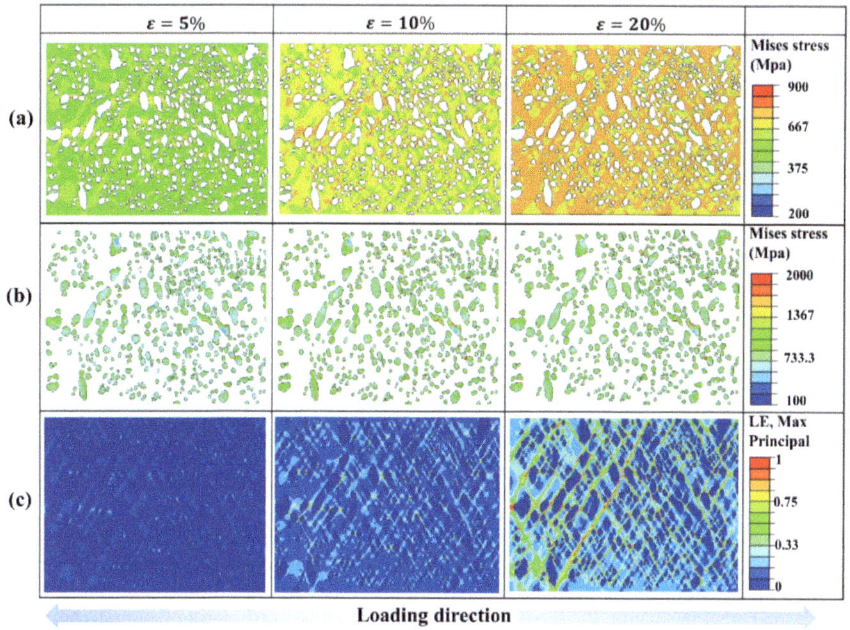

Fig. 11. The strain and stress distributions of the local microstructure under the in-situ tensile loading: (a) mises stress in ferrite phase, (b) mises stress in cementite phase, and (c) maximum principal strain.

Nanoscale MD results. Fig. 12 shows the interfacial decohesion of ferrite-cementite interface by MD simulation under the normal loading. As shown in Fig. 12(a), decohesion preferentially appears at the interface of the two phases , as well as fracture with increasing strain. In presented in Fig. 12(b), a large plastic slip occurred in the ferrite, while the cementite phase remaind essentially elastically deformed. This also indicated the interfacial debonding caused by the large strain gradient difference between the ferrite and cementite phases. As illustrated in Fig. 12(c), the maximum force and total energy to separation for interfacial debonding under different angles of loading are varied. The maximum traction of interface debonding was 3.26 GPa under pure shear loading (marked with purple line), and the corresponding failure displacement to separation was 80Å. Under uniaxial tension, the maximum stress and displacement were 7.41 GPa and 45 Å, respectively. Interfacial debonding occurred preferentially under shear loading compared to tensile loading. Conversely, complete fracture was delayed much more under shear loading than under tensile loading. Comparing the displacements for the onset of interface debonding and the displacements for the final failure under different loading, there is a gradual transfer from shear to tensile with a gradual delay.

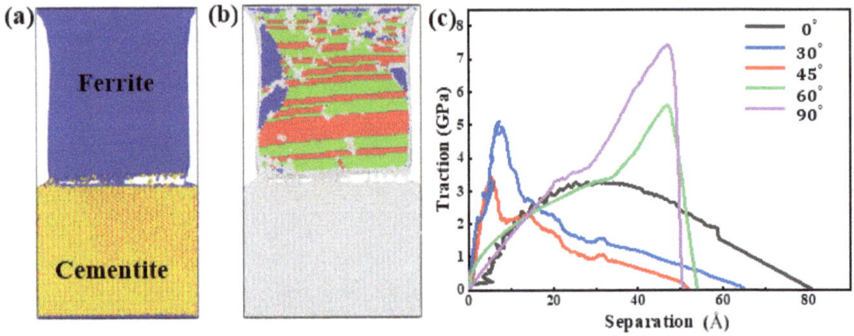

Fig. 12. Traction-separation beahvior of ferrite-cementite interface by MD simulation: (a) interfacial debonding; (b) Comoin neighbor analysis (CNA); (c) traction-separation curves under different angle loading. Noting that the CAN can be used to reflect the capacity for plastic deformation.

Summary

The damage and fracture behavior of high-carbon steel containing ferrite and cementite particles are systematically investigated through multiscale finite element simulation and in situ experiment. The experimental results demonstrated that the damage mechanism of high-carbon steel under tensile loading primarily involved strip-shaped particle cracking, interfacial debonding at the two phases, and cracking within the ferrite matrix. Moreover, it is worth noting that material fracture ultimately occurs due to interfacial debonding. The multiscale simulations indicated a large plastic deformation of the ferrite under large deformations, while the particles maintained the elastic deformation. This large strain localization led to the ultimate failure of the material, and the morphology, size and distribution of the particles strongly influence the void initiation. The driving forces for the three types of damage mechanism were equivalent stresses, maximum principal stresses, and equivalent plastic strains, respectively. In addition, interfacial debonding occurred preferentially under shear loading and is most difficult to occur under tensile loading. This work provides significant guidance for the subsequent modulation of the microstructure and forming parameters of this steel.

References

[1] T.A. Zubkova, I.L. Yakovleva, L.E. Kar Kina, I.A. Veretennikova, Study of the hardness and elastic modulus of cementite in the structure of granular pearlite by the nano-indentation method, Met. Sci. Heat Treat. 56 (2014) 330-335.https://doi.org/10.1007/s11041-014-9756-3

[2] H.J.P.i.m.S. Bhadeshia, Steels for bearings, 57 (2012) 268-435.

[3] C. Prasad, P. Bhuyan, C. Kaithwas, R. Saha, S. Mandal, Microstructure engineering by dispersing nano-spheroid cementite in ultrafine-grained ferrite and its implications on strength-ductility relationship in high carbon steel, Mater. Des. 139 (2018) 324-335. https://doi.org/10.1016/j.matdes.2017.11.019

[4] H.N. Wu, D.S. Xu, H. Wang, R. Yang, Institute, Of, Metal, Research, Chinese, Academy, Molecular Dynamics Simulation of Tensile Deformation and Fracture of γ-TiAl with and without Surface Defects, J. Mater. Res. Technol. 10 (2016) 49-58. https://doi.org/10.1016/j.jmst.2015.12.001

[5] W. Li, T. Sakai, Q. Li, L.T. Lu, P. Wang, Reliability evaluation on very high cycle fatigue property of GCr15 bearing steel, Int. J. Fatigue 32 (2010) 1096-1107. https://doi.org/10.1016/j.ijfatigue.2009.12.008

[6] M. Yang, L. Zhou, C. Wang, P. Jiang, F. Yuan, E. Ma, X. Wu, High impact toughness of CrCoNi medium-entropy alloy at liquid-helium temperature, Scr. Mater. 172 (2019) 66-71. https://doi.org/10.1016/j.scriptamat.2019.07.010

[7] Q. Deng, N.A. Bhatti, X. Yin, M. Abdel Wahab, The effect of a critical micro-void defect on fretting fatigue crack initiation in heterogeneous material using a multiscale approach, Tribol. Int. 141 (2020) 105909. https://doi.org/10.1016/j.triboint.2019.105909

[8] A. Inoue, T. Ogura, T.J.T.o.t.J.I.o.M. Masumoto, Transmission electron microscope study on deformation and fracture of cementite in cold-rolled steels, 17 (1976) 149-157.

[9] S.P. Rawal, J. Gurland, Observations on the effect of cementite particles on the fracture toughness of spheroidized carbon steels, Metall. Trans. A 8 (1977) 691-698. https://doi.org/10.1007/BF02664778

[10] K. Shibanuma, S. Aihara, K.J.E.F.M. Suzuki, Prediction model on cleavage fracture initiation in steels having ferrite–cementite microstructures–Part I: Model presentation, 151 (2016) 161-180.

[11] A. Nagao, K. Hayashi, K. Oi, S. Mitao, Effect of uniform distribution of fine cementite on hydrogen embrittlement of low carbon martensitic steel plates, ISIJ Int. 52 (2012) 213-221. https://doi.org/10.2355/isijinternational.52.213

[12] K. Shibanuma, S. Aihara, K.J.E.F.M. Suzuki, Prediction model on cleavage fracture initiation in steels having ferrite–cementite microstructures–Part II: Model validation and discussions, Eng. Fract. Mech. 151 (2016) 181-202. https://doi.org/10.1016/j.engfracmech.2015.03.049

[13] F.-F. Li, G. Fang, Stress-state dependency of ductile fracture in an extruded magnesium alloy and its underlying mechanisms, Int. J. Plast. 152 (2022) 103258. https://doi.org/10.1016/j.ijplas.2022.103258

[14] M. Brünig, S. Gerke, M. Schmidt, Damage and failure at negative stress triaxialities: Experiments, modeling and numerical simulations, Int. J. Plast. 102 (2018) 70-82. https://doi.org/10.1016/j.ijplas.2017.12.003

[15] H. Wu, X. Zhuang, Z. Zhao, Extended GTN model for predicting ductile fracture under a broad range of stress states, Int. J. Solids Struct. 239-240 (2022) 111452. https://doi.org/10.1016/j.ijsolstr.2022.111452

[16] Q. Deng, L. Xiong, Y. Chen, Coarse-graining atomistic dynamics of brittle fracture by finite element method, Int. J. Plast. 26 (2010) 1402-1414. https://doi.org/10.1016/j.ijplas.2010.04.007

[17] V.S. Krasnikov, A.E. Mayer, Plasticity driven growth of nanovoids and strength of aluminum at high rate tension: Molecular dynamics simulations and continuum modeling, Int. J. Plast. 74 (2015) 75-91. https://doi.org/10.1016/j.ijplas.2015.06.007

[18] K. Paveebunvipak, V. Uthaisangsuk, Microstructure based modeling of deformation and failure of spot-welded advanced high strength steels sheets, Mater. Des. 160(DEC.) (2018) 731-751. https://doi.org/10.1016/j.matdes.2018.09.052

[19] A. Tang, H. Liu, R. Chen, G. Liu, Q. Lai, Y. Zhong, L. Wang, J. Wang, Q. Lu, Y. Shen, Mesoscopic origin of damage nucleation in dual-phase steels, Int. J. Plast. 137 (2021) 102920. https://doi.org/10.1016/j.ijplas.2020.102920

[20] K. Ismail, A. Perlade, P.J. Jacques, T. Pardoen, L. Brassart, Impact of second phase morphology and orientation on the plastic behavior of dual-phase steels, Int. J. Plast. 118 (2019) 130-146. https://doi.org/10.1016/j.ijplas.2019.02.005

[21] U.B. Asim, M.A. Siddiq, M.E. Kartal, A CPFEM based study to understand the void growth in high strength dual-phase titanium alloy (Ti-10V-2Fe-3Al), Int. J. Plast. 122 (2019) 188-211. https://doi.org/10.1016/j.ijplas.2019.07.002

[22] H.-J. Guo, C. Ling, E.P. Busso, Z. Zhong, D.-F. Li, Crystal plasticity based investigation of micro-void evolution under multi-axial loading conditions, Int. J. Plast. 129 (2020) 102673. https://doi.org/10.1016/j.ijplas.2020.102673

[23] J. Kang, N.S. Pottore, H. Zhu, C.C. Tasan, An in situ investigation of neighborhood effects in a ferrite-containing quenching and partitioning steel: Mechanical stability, strain partitioning, and damage, Acta Mater. 254 (2023). https://doi.org/10.1016/j.actamat.2023.118985

[24] Z. Yang, J. Fan, Y. Liu, Z. Yang, Y. Kang, J. Nie, Effect of combination variation of particle and matrix on the damage evolution and mechanical properties of particle reinforced metal matrix composites, Mater. Sci. Eng. A 806 (2021). https://doi.org/10.1016/j.msea.2021.140804

[25] M. Yaghoobi, Z. Chen, A.D. Murphy-Leonard, V. Sundararaghavan, S. Daly, J.E. Allison, Deformation twinning and detwinning in extruded Mg-4Al: In-situ experiment and crystal plasticity simulation, Int. J. Plast. 155 (2022). https://doi.org/10.1016/j.ijplas.2022.103345

[26] X.H. Gong, Z.D. Feng, D. Fan, L. Lu, S.N. Luo, Strain localization in titanium investigated via in situ digital image correlation with multiscale speckles, Mater. Charact. 189 (2022). https://doi.org/10.1016/j.matchar.2022.111940

[27] Gurson, L. A, Continuum Theory of Ductile Rupture by Void Nucleation and Growth: Part I—Yield Criteria and Flow Rules for Porous Ductile Media, J. Eng. Mater. Technol. 99 (1977) 297-300. https://doi.org/10.1115/1.3443401

[28] A. Needleman, V. Tvergaard, An analysis of ductile rupture in notched bars, J. Mech. Phys. Solids 32 (1984) 461-490. https://doi.org/10.1016/0022-5096(84)90031-0

[29] V. Tvergaard, Influence of voids on shear band instabilities under plane strain conditions, Int. J. Fract. 17 (1981) 389-407. https://doi.org/10.1007/BF00036191

[30] C. Shen, C. Wang, M. Huang, N. Xu, S. van der Zwaag, W. Xu, A generic high-throughput microstructure classification and quantification method for regular SEM images of complex steel microstructures combining EBSD labeling and deep learning, J. Mater. Res. Technol. 93 (2021) 191-204. https://doi.org/10.1016/j.jmst.2021.04.009

[31] H. Wang, F. Wang, D. Qian, F. Chen, Z. Dong, L. Hua, Investigation of damage mechanisms related to microstructural features of ferrite-cementite steels via experiments and multiscale simulations, Int. J. Plast. 170 (2023). https://doi.org/10.1016/j.ijplas.2023.103745

[32] X.W. Zhou, J.A. Zimmerman, E.D. Reedy, N.R. Moody, Molecular dynamics simulation based cohesive surface representation of mixed mode fracture, Mech. Mater. 40 (2008) 832-845. https://doi.org/10.1016/j.mechmat.2008.05.001

[33] M. Guziewski, S.P. Coleman, C.R. Weinberger, Atomistic investigation into the atomic structure and energetics of the ferrite-cementite interface: The Bagaryatskii orientation, Acta Mater. 119 (2016) 184-192. https://doi.org/10.1016/j.actamat.2016.08.017

[34] X. Li, M. Xu, Z. Zhang, Hot damage evolution in a high strength aluminum alloy during hot forming: a study using the Gurson–Tvergaard–Needleman model, J. Mater. Res. Technol. 14 (2021) 1366-1376. https://doi.org/10.1016/j.jmrt.2021.07.066

Effects of network-distributed TiB on dynamic recrystallization of TiB/(TA15-Si) composites during the deformation in β phase region

LI Liting[1,3,a], WANG Kehuan[1,2,3,b *], WANG Dongjun[1,2,3,c] and LIU Gang[1,2,3,d]

[1]National Key Laboratory for Precision Hot Processing of Metals, Harbin Institute of Technology, Harbin, 150001, China

[2]Institute of High Pressure Fluid Forming, Harbin Institute of Technology, Harbin, 150001, China

[3]School of Materials Science and Engineering, Harbin Institute of Technology, Harbin, 150001, China

[a]liliting1022@126.com, [b]wangkehuan@hit.edu.cn, [c]dongjunwang@hit.edu.cn, [d]gliu@hit.edu.cn

Keyword: Network-Structured Composite, Hot Deformation, Microstructure Evolution, Dynamic Recrystallization

Abstract. To study the effects of network distributed TiB on dynamic recrystallization of high temperature β phase of near α titanium matrix composites, the TiB/(TA15-Si) composites were compressed under 1000°C with strain rates ranging from 1 s^{-1}~0.01 s^{-1}. Results show that the flow stress was sensitive to strain rates. During hot compression, the microstructure evolution included single TiB rotation, network structure changing from circular to oval, and dynamic recrystallization (DRX) of high temperature β phase. The mechanism of DRX was continuous and discontinuous DRX mixed mechanism. Attributed to the accelerative effects of TiB by providing nucleation sites and improving storage energy, DRX preferentially occurred around TiB. After compression, equiaxed grains of high temperature β phase were formed in TiB rich region, while elongated grains were formed in TiB lean region. The dislocations density in elongated grains at higher strain rate was higher than that at lower strain rate. Under high strain rate, high temperature β phase was dominated by DRX in TiB rich region and dynamic recovery in TiB lean region. Under lower strain rates, high temperature β phase was dominated by the growth of DRX grains.

Introduction

With the rapid development of aero engines, it is urgent to develop lightweight and heat-resistant structural materials with service temperatures above 650°C. Due to the excellent mechanical performance and low manufacturing cost, titanium composites with network structure have received extensive concern [1,2]. Compared with matrix alloy, the service temperature of network structured composites could be improved by 100°C~200°C [3]. Therefore, network structured TiB/(TA15-Si) composites are expected to replace nickel-based alloys to be serviced at 650°C.

Generally, metal materials undergo hot deformation to optimize the microstructure and performance [4,5]. For titanium matrix composites with network structure, the deformation in β phase region has more research value and practical significance. Firstly, the deformation resistance of titanium alloy in α+β phase region is greater than that in β phase region, and the addition of reinforcements could increase the deformation resistance, which makes the deformation of the composites in α+β phase region require higher her tonnage of the press machine. Thus, the composites have more anvantages as deformed in β phase region. Secondly, for near α titanium alloys, it is usually prohibited to forge in β phase region as finished forging of products, which is because fast growth of β phase results to β brittleness. For network structures composites, TiB restricts the grains size within the size of a titanium alloy powder [6]. As well, during the following cooling process, the continuous α phase at the grain boundary can also be avoided to be formed because of TiB, thus it is more likely to prevent the β brittleness. Thirdly, in practice, to avoid β brittleness of near α titanium alloys, the hot deformation is required to be completed in a short time as deformed in β phase region. The total deformation needs to be completed through small deformation in pass with multi-pass, which reduces the production efficiency and wastes energy. For network distributed composites, it has obvious advantages by TiB limiting the growth of high temperature β phase, decreasing the requirements for deformation time, increasing the pass deformation, and then reducing the number of passes. Above all, the deformation of network structured TiB/(TA15-Si) composites in β phase region is necessary to be studied and some investigation has been done [7,8,9]. The results show that the performance was enhanced after deformation in β phase region, and the strength increased with deformation passes. However, those researchers paid close attention to the strength improvement. The deformation behavior and microstructure evolution of the composites in β phase region needs to be further studied.

In this paper, the hot deformation behavior of the composites was investigated through hot compression at 1000°C with different strain rates. The microstructure evolution after compression was studied by scanning electron microscope (SEM) and electron backscattering diffraction (EBSD) characterization, and the effects of TiB on microstructure were analyzed.

Material and experimental procedures

Initial material. TiB/(TA15-Si) composites with network structure were utilized in the present study. The as-received composites were sintered billet with 3.5% volume fraction of TiB and 0.5% weight fraction of Si. The microstructure of the as-sintered sample is shown in Fig. 1(a), which consisted of lamellar α phase, β phase and network distributed TiB. In order to study the deformation behavior of the composite in single β phase region, the β transformation temperature was determined to be 995°C by metallographic method. As the sample was heated to 1000°C and held for 20 min, the α phase disappeared and transformed to β phase completely, as shown in Fig. 1 (b). The high temperature β phase of the as-sintered sample was reconstructed, as shown in Fig. 1 (c). The grain boundary of the reconstructed β phase coincides with the position of network-structured TiB, and the size of the prior β grain is about 120 μm.

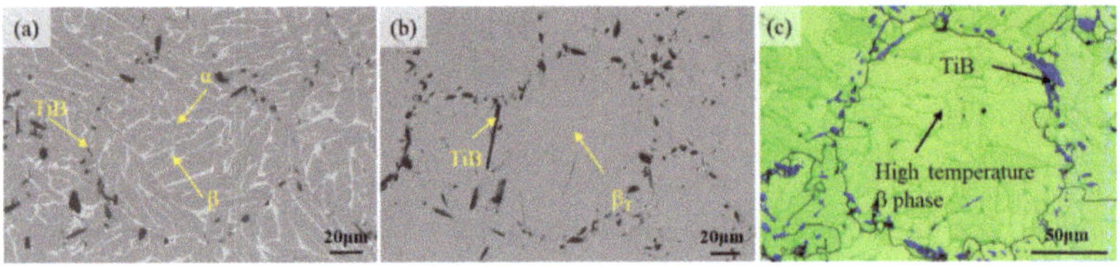

Fig. 1. SEM of the sample with different conditions: (a) as-sintered; (b) after heat treatment at 1000°C for 20 min; (c) phase maps of reconstructed β phase of as-sintered.

Experimental procedures. The cylindrical samples with ϕ8 × 12 mm were compressed on a universal material testing machine (Japan Shimadzu AG-X Plus 250 kN/50 Kn). To reduce friction, the samples' surface was uniformly coated with the antioxidant of glass protective lubricant, and a graphite sheet was placed between the samples and indenters. The samples were held at 1000°C for 20 min before deformation and then were compressed with a reduction of 60% at strain rates of 1 s^{-1} and 0.01 s^{-1} respectively. After the compression, the samples were water quenched immediately.

To investigate the microstructure evolution during the deformation, the microstructure was characterized by SEM (Hitachi Scanning Electron Microscope SU 5000) and EBSD. The deformed samples were ground mechanically then electrolytic polished for 90 s at a voltage of 30V and current of 0.68A ~ 0.72A. The electrolytic solution was composed of 60% methanol, 34% butyl alcohol and 6% perchloric acid and kept at -30°C. The polished samples were etched in Kroll solution for 5 seconds for TiB observation. AZtecCrystal software was used to analyze the EBSD data. The reconstruction of high temperature β phase was accomplished using the function of Parent Grain Analysis in AZtecCrystal software. The main parameter setting were as follow: the parent phase and child phase was selected as β phase and α phase of titanium based on the acquired data information, the orientation relation was the Burgers relationship including $\{110\}_{\beta}//\{0001\}_{\alpha}$ and $<\text{-}11\text{-}1>_{\beta}//<2\text{-}1\text{-}10>_{\alpha}$, the refinement range was 10°, and the window size, step width, voting thresh, and parent thresh were 20 pix, 25%, 5° and 10°, respectively.

Results and Discussion

Hot deformation behaviors. Fig. 2 shows the true stress-strain curves of the samples compressed at 1000°C with different strain rates. It is found the flow stress increases rapidly at the initial stage of the deformation due to work hardening effect, then the increase rate of stress decreases, and the flow stress gradually increases until the peak stress is reached. After that, softening of flow stress decreases gradually. The softening behavior may be related to the DRX of matrix [10] and evolution of the reinforcements [11]. It is noted that when the strain rate is $0.01 s^{-1}$, the flow stress shows a slow increase trend in the late stage of the deformation. By comparing the curves under different strain rates, the flow stress is positively correlated with the strain rate when the composite is formed in β phase region.

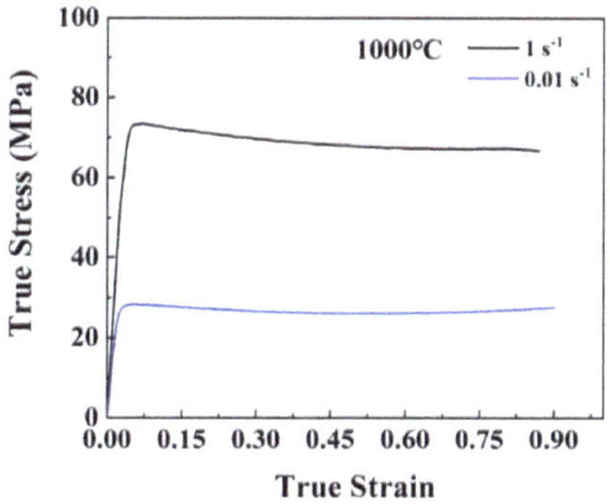

Fig. 2. Stress-strain curves of the samples compressed at 1000°C with different strain rates.

Microstructure evolution. Fig. 3 shows the SEM microstructure of the deformed samples in the center position under different strain rates. It is found that the TiB is still distributed in a circular network structure in the center of the cross section, while in a oval network structure of the longitudinal section. Moreover, the direction of a single TiB is perpendicular to the compression direction. The reason is the lower resistance of high temperature β phase to deformation, resulting in the rotation of TiB with the flow of matrix. In addition, fine β grains are found, as marked in yellow circle in Fig. 5(c) and (f), which indicates the occurrence of dynamic recrystallization of high temperature β phase during the compression.

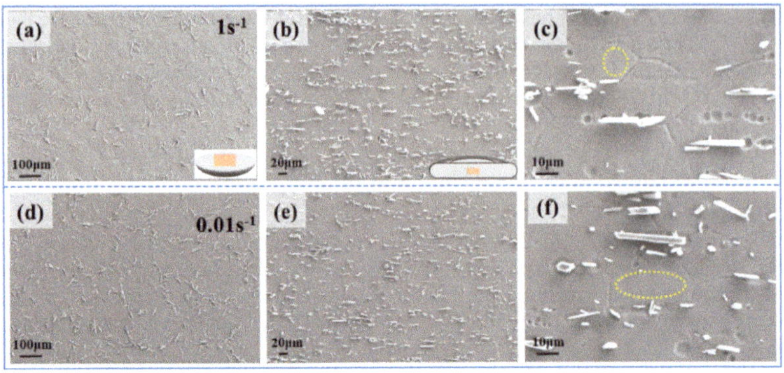

Fig. 3. SEM microstructure of the deformed samples in the center position under different strain rates: (a, b and c) 1 s^{-1}; (d, e and f) 0.01 s^{-1}; (a, d) in the cross section; (b, e) in the longitudinal section; (c, f) high magnification of figure (b) and (e).

Fig. 4 shows the EBSD results of the reconstructed β phase of the deformed samples under different strain rates. As the strain rate is $1s^{-1}$, fine β grains can be seen around TiB. As marked in yellow arrows in Fig. 4(a), the fine grains exhibit a distinct misorientation with the near and elongated grains, which is a typical characteristic of discontinuous dynamic recrystallization (DDRX). Additionally, sub-grains with low-angle grain boundaries are found within the deformed grains, as shown in the white arrows in Fig. 4(a). The misorientation of the sub-grains is close to that of the surrounding grain, which is a characteristic of continuous dynamic recrystallization (CDRX). For near α titanium alloy, the microstructure evolution during the deformation in β region is dominated by the recovery (DRV) of the β phase. However, TiB can promote the DRX of β phase by supplying the nucleation sites [12]. In this paper, the DRX of β phase occurs in TiB rich region. In the TiB lean region, it can be found that a slight difference in color within the grains, as shown in the yellow frame in Fig. 4(a), which suggests there are certain dislocations inside the grain. Thus, the elongated grain is dominated by DRV. Thus, at the strain rate of $1s^{-1}$, DRX of high temperature β phase occurs in TiB rich region, while DRV of high temperature β phase in TiB lean region.

When the strain rate is $0.01s^{-1}$, the size of the DRX grains around TiB increases as shown in yellow arrows in Fig. 4(b). Moreover, the color within the elongated β grains in TiB lean region is consistent, as shown in the yellow frames in in Fig. 4(b). It means a lower density of dislocations within the elongated β grains, which suggests the elongated β grains at the strain rate of $0.01s^{-1}$ are DRX grains. This is because the migration of grain boundaries reduces dislocation density. Thus, at the strain rate of $0.01s^{-1}$, the DRX grains around TiB grow to a certain extent along the direction to TiB rich region due to a limited effect of TiB, but the DRX grains grow rapidly along the direction to TiB lean region fastly due to the absence of the restriction of TiB.

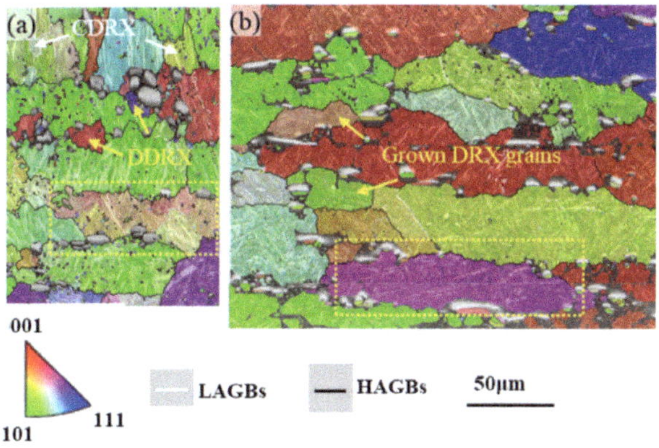

Fig. 4. EBSD results of the reconstructed β phase of the deformed samples under different strain rates: (a) 1 s^{-1}; (b) 0.01 s^{-1}.

The microstructure evolution of the composites during the deformation in β phase region with different strain rates is summarized, as shown in Fig. 5. The network distribution of TiB changes from circular to oval, accompanied by rotation of the single TiB, whose direction is gradually perpendicular to the compression direction. At the initial stage of deformation, a large number of dislocations are generated in the high temperature β phase. Due to TiB providing sufficient nucleation sites and higher storage energy, the nucleus of crystallization is preferentially formed around TiB. At a high strain rate (1s^{-1}), the nucleus of crystallization develops into refined DRX grains in TiB rich region. In TiB lean region, since the nucleus of crystallization has no sufficient time to grow to TiB lean region, the high temperature β phase is mainly dominated by the DRV process, and a certain density of dislocations are still distributed. At a low strain rate (0.01s^{-1}), the nucleus of crystallization has sufficient time to grow. In TiB rich region, it can only grow to a certain extent due to a restriction of TiB and elongate along the direction perpendicular to the compression direction, but it can grow fastly to TiB lean region due to the absence of the restriction of TiB. Thus, the dislocation density in TiB lean region is lower, and the corresponding high temperature β phase mainly occurs DRX process.

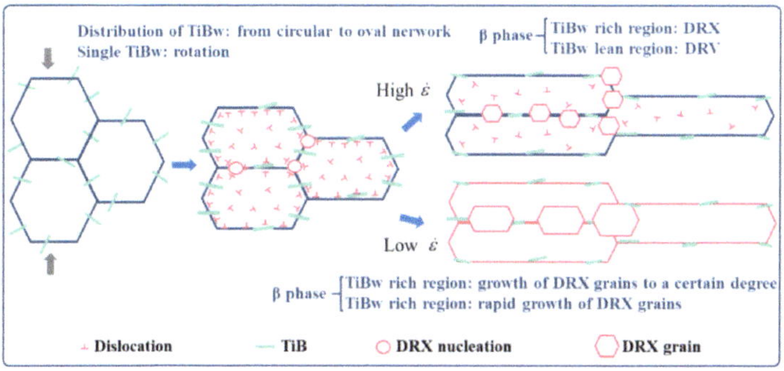

Fig. 5. Schematic diagram of the microstructure evolution during the deformation in β phase region.

Conclusions

In summary, the as-sintered TiB/(TA15-Si) composites with network structure have been compressed with different strain rates in β phase region. The effects of network distributed TiB on dynamic recrystallization of high temperature β phase was investigated. The main conclusions were as follows:

1. After compression, the single TiB rotated, network structure evolved from circular to oval on longitudinal section, and high temperature β phase around TiB occurred dynamic recrystallization preferentially due to TiB supplying nucleation sites and higher storage energy. The mechanism of dynamic recrystallization included continuous and discontinuous dynamic recrystallization.
2. After compression, equiaxed grains of high temperature β phase were formed in TiB rich region due to dynamic recrystallization, while elongated grains were formed in TiB lean region. The dislocations density in elongated grains at strain rate of 1 s^{-1} was higher than that at strain rate of 0.01 s^{-1} due to their different envolution mechanism.
3. At the strain rate of 1 s^{-1}, high temperature β phase was dominated by dynamic recrystallization in TiB rich region and dynamic recovery in TiB lean region. At the strain rate of 0.01 s^{-1}, it was dominated by the growth of dynamic recrystallization grains with a certain extent in TiB rich region and rapid growth in TiB lean region.

Acknowledgments

The authors acknowledge Prof. L. J. Huang from Harbin Institute of Technology for providing the network-structured titanium matrix composites. This project is supported by the National Natural Science Foundation of China (Grant No. 52375324), and the Hei Long Jiang Postdoctoral Foundation (LBH-TZ2107).

Declaration of interests

The authors declare that they have no known competing financial interests or personal relationships that could have appeared to influence the work reported in this paper.

References

[1] L. Huang, Q. An, L. Geng, S. Wang, S. Jiang, X. Cui, R. Zhang, F. Sun, Y. Jiao, X. Chen, C. Wang, Multiscale Architecture and superior high-temperature performance of discontinuously reinforced titanium matrix composites, Adv. Mater. 33 (2021) 2000688. https://doi.org/10.1002/adma.202000688

[2] L.J. Huang, L. Geng, H.X. Peng, Microstructurally inhomogeneous composites: is a homogeneous reinforcement distribution optimal?, Progress Mater. Sci. 71 (2015) 93-168. https://doi.org/10.1016/j.pmatsci.2015.01.002

[3] Y. Jiao, L.J. Huang, Q. An, et al, Effects of Ti5Si3 characteristics adjustment on microstructure and tensile properties of in-situ (Ti5Si3+ TiBw)/Ti6Al4V composites with two-scale network architecture, Mater. Sci. Eng. A 673 (2016) 595-605.

[4] Q. Peikun, Y. Han, G. Huang, L. Jianwen, L. Lei, L. Xiao, W. Lu, Texture evolution and dynamic recrystallization behavior of hybrid-reinforced titanium matrix composites: enhanced strength and ductility, Metall. Mater. Trans. A 51 (2020) 2276-2290. https://doi.org/10.1007/s11661-020-05677-5

[5] L. Jianwen, Y. Han, J. Xiang, L. Lei, G. Huang, W. Lu, Understanding the Role of Multi-scale Reinforcements on Severe Plastic Deformation of Titanium Matrix Composites, Metall. Mater. Trans. A 51 (2020) 1732-1743. https://doi.org/10.1007/s11661-020-05630-6

[6] J. Wei, F. Yang, M. Qi, C. Zhang, C. Chen, Z. Guo, The “pinning” effect of in-situ TiB whiskers on the hot deformation behavior and dynamic recrystallization of (10 vol%) TiB/Ti–

6Al–4V composites, J. Mater. Res. Tech. 22 (2023) 1695-1707. https://doi.org/10.1016/j.jmrt.2022.12.048

[7] S. Wang, Q. An, R. Zhang, S. Jiang, L.J. Huang, R. Chen, B. Liu, Y. Jiao, L. Geng, Microstructure characteristics and enhanced properties of network-structured TiB/(TA15-Si) composites via rolling deformation at different temperatures, Mater. Sci. Engi. A 829 (2022) 142176. https://doi.org/10.1016/j.msea.2021.142176

[8] Z. Li, Q. Kang, G. Wang, X. Sui, Y. Liu, S. Luo, Microstructure evolution during hot-packed rolling and mechanical properties anisotropy of as-rolled network-structured TiBw/TA15 composites, Mater. Sci. Eng. A 849 (2022) 143518. https://doi.org/10.1016/j.msea.2022.143518

[9] R. Zhang, L. Huang, Q. An, L. Geng, B. Wang, Y. Jiao, The hyperbolic constitutive equations and modified dynamic material model of TiBw/Ti-6.5 Al-2.5 Zr-1Mo-1V-0.5 Si composites, Mater. Sci. Eng. A 766 (2019) 138329. https://doi.org/10.1016/j.msea.2019.138329

[10] Z.B. Yang, C.J. Zhang, X. Ji, S. Zhang, H. Feng, J. Han, P. Peng, T. Wang, Microstructural evolution and silicide precipitation behavior of TiCp/near-α titanium matrix composite during hot compression, Mater. Charact. 189 (2022) 111933. https://doi.org/10.1016/j.matchar.2022.111933

[11] R. Zhang, D.J. Wang, L.J. Huang, S.J. Juan, L. Geng, Deformation behaviors and microstructure evolution of TiBw/TA15 composite with novel network architecture, J. Alloy. Compd 722 (2017) 970-980. https://doi.org/10.1016/j.jallcom.2017.06.197

[12] H.B. Zhang, B. Wang, Y.T. Zhang, Y. Li, J.L. He, Y.F. Zhang, Microstructure evolution and deformation mechanisms of β phase in TiBw/Ti60 composites studied by reconstructing grain during the hot deformation, J. Alloys Compd. 825 (2020) 153872. https://doi.org/10.1016/j.jallcom.2020.153872

MODELLING

Materials Research Proceedings 44 (2024) 433-443 https://doi.org/10.21741/9781644903254-47

Semisolid deposition of metallic material by extrusion-base analytical and simulative methodologies

BRUNI Carlo

Università Politecnica delle Marche -DIISM, Via Brecce Bianche 3, 60100 Ancona, Italia

c.bruni@univpm.it

Keywords: FDM, Analytical Modeling, Finite Element Simulation, Magnesium Alloy

Abstract. The objective of the paper is the analytical and finite element simulation study of the material behaviour at temperatures of 400 - 450°C of AZ31 and ZM21 magnesium alloys tested for bulk metal working. In the extrusion, representing the main phase of the fused deposition modeling, the metal work material considered was a bulk magnesium alloy billet. The stress and deformation fields are reported in detail using analytical modeling to describe the material in a very small volume of a extruder. It is observed that the increase in the extrusion temperature also determined by the deformation and friction heat produces an increase in the thermal load particularly visible in the high deformed zone. This allows the obtaining of the semisolid condition for the material described by the temperature levels reported by the numerical simulations. Such technology results very useful for ribs deposition over flat surfaces.

Introduction

Today working of metal or polymeric materials requires the understanding of the material plastic flow in order to avoid all of the possible defects related with uncontrolled conditions getting voids and something similar in the bulk of the workpiece [1]. The material processes already known as additive manufacturing represent up today the most rapid way to realize physical models in different materials. When approaching metals the temperature managing represents the most important factor. That is why high temperature means shrinkage, warpage and oxide phenomena requiring subsequent treatments [2].

However the direct realization of a near net shape object under the additive manufacturing philosophy represents a useful task in order to realize fully metal parts with an effective layer by layer approach instead of the binder [3] or composite based ones. The latter requiring complex methodologies when separated materials are used.

But, the net shape realization of metal workpieces represents the real objective in such field. Different are the methods used that are SLS, SLM, jetting powder, fused filament and fused metal deposition to produce the desired metal object [4-8]. Even if the extrusion based one is the more simple to realize a semisolid filament to generate ribs and something similar [7-9]. Some studies appear also on microextrusion of material in which the tool is made by micro- EDM [10].

Of course the realization of extrusion trials at dimensions comparable with the millimiter and sub-millimiter order of magnitude could be sensitive to the friction conditions and subsequent heat generated. In general, the surface and bulk characteristics of the extruded object are function of the tool material contact interface [11-13]. Such contact in general increases the temperature of the extruded material by the nominal. The modeling of such behaviour can be found in literature and able to represent the micro-mesoscale. In particular, the thermal conditions for magnesium alloys as well as the modeling approaches whose variables, included the thermal-physical properties with temperature, is described by some researchers in [14-18].

Up today the FDM of metal material can be feasible not only for reduced complexity of tools and models typical of other technologies [19-20] in realizing components for high precision assemblies, but also for a traditional approach in working metals consisting of in temperature deformation and semisolid manufacturing in order to obtain the desired geometry of the filament. The pressure to material to feed can be imposed by a screw or a drum or a feed forward system in which a solid filament is heated by induction coil and deposited under semisolid condition.

Of course the process conditions need to be strictly monitored in order to avoid lack of respect of the filament requirements. Among the different variables such as deformation, strain rate, geometry of the container and of the tool, the most important is the temperature and temperature variation during working. The possibility to realize metal ribs by a semisolid filament could be a valid and sustainable solution in all of the situations in which metal powder is difficult to treat directly.

To do that the modeling of the extruding apparatus represents useful for understanding and controlling the manufacturing tools.

In literature different scientific works related to material behaviour [8-9,12] under extruding conditions and material flow description using analytical and non analytical methods such as numerical simulations in different conditions in terms of strain, strain rate, temperature and friction can be found. Some of them are concentrated on uniform and non uniform deformation, friction and subsliding effects in order to obtain the effective stress – strain conditions and the effective temperature values for load evaluation purposes.

The extrusion of lightweight alloys was investigated in the past by the author and other researches and the studied completed with the material behaviour knowledge [21, 22]. But the extrusion at millimeter and sub-millimiter order of magnitude, that is the filament metal deposition, at high temperatures is still under investigation being a promising technique to realize additive manufactured metal components. Under such conditions the understanding of the effect of the deformation and friction on the temperature represents still today an objective of the current research above all when low amount of material needs to be treated for filament deposition in particular for magnesium alloys.

The objective of the paper is the study through the analytical and finite element simulation of two magnesium alloys that are AZ31 and the ZM21 under extrusion conditions at elevated temperatures taking into consideration the effect of deformation and friction heating at 400-450°C. The stress and load obtained by an analytical modeling on both the materials and the temperature increase during forming is described. The thermal effect on the working tools studied by means of finite element simulation. It results that such methodology is very useful for determining the semisolid status for rib applications under the additive manufacturing philosophy.

Analytical and simulative procedures

The modeling of FDM for metal deposition of ribs is investigated by the realization of a proper procedural tool. Considering that the critical component of the system is represented by the extruder in which the material flow needs to be modeled two types of approaches are followed in such investigation. Fig. 1 shows the one based on the analytical modeling and the other one on the finite element simulation in order to describe both the tensional and the thermal behaviours inside the extruder. The billet of the initial radius of 3.5 mm was extruded to 2.5 mm over a length of about 14 mm included the cylindrical part of 10 mm.

By the analytical modeling the material flow during extrusion is described with the typical equations given by the slab method:

$$\sigma = \bar{\sigma} \cdot \frac{1+\mu cotg(\alpha)}{\mu cotg(\alpha)}\left(1 - \left(\frac{R_0}{R_f}\right)^{\mu \cot(\alpha)}\right). \quad (1)$$

$$\sigma_L = \sigma \cdot exp\left(\frac{2\mu}{R_0}\Delta l\right) \quad (2)$$

where:

σ: extrusion stress/tension

$\overline{\sigma}$: material flow

α : half cone angle

μ: friction coefficient

R_0: initial billet radius

R_f : final billet radius

σ_L : tension at the stem of the container

Δl: travel covered by the stem

The material behaviour is described by two models related to two different magnesium alloys that are the AZ31 magnesium and the ZM21 magnesium alloys. They are described by the two following equations proposed in [21, 22]:

$$\bar{\sigma} = K\varepsilon^n\dot{\varepsilon}^m \quad (3)$$

$$\bar{\sigma} = \beta_0 + \beta_1\varepsilon + \beta_2 \ln(\dot{\varepsilon}) + \beta_3 T^2 + \beta_4\left[exp\left(\frac{\varepsilon}{T}\right)\right]^2 + \beta_5\left[exp\left(\frac{\dot{\varepsilon}}{T}\right)\right]^2 + \beta_6\varepsilon^2 + \beta_7\dot{\varepsilon}^4 \quad (4)$$

with:

K: Material coefficient depending on Temperature

n: material coefficient depending on temperature

m: strain rate sensitivity depending on temperature

ε: deformation

$\dot{\varepsilon}$: strain rate

β_1, β_n : constant coefficients of equation.

By the previous equations the extrusion load can be calculated by using a 1.5/2 coefficient to consider the distortion effect with angles approaching 45°. The extrusion load calculated by the following:

$$L = \sigma_L \cdot \pi \cdot R_0^2 \quad (5)$$

The temperature increase due to deformation is obtained by balancing heat exchange as follows:

$$\Delta T_{def} = \frac{\beta\bar{\sigma}\Delta\bar{\varepsilon}}{\rho c} \quad (6)$$

Where $\beta=\beta_{hg}\beta_{hr}$ with β_{hg} is the percent of heat generated by deformation and β_{hr} is the percent of heat due remaining in the workpiece. $\Delta\varepsilon$ is deformation increment in the conical zone.

Materials Research Proceedings 44 (2024) 433-443 https://doi.org/10.21741/9781644903254-47

The friction heating is calculated by the following equation:

$$\Delta T_{attr} = \frac{\tau 2\pi R_0 \Delta l \cdot v \cdot t}{\rho V c} + \frac{\tau\pi(R_0^2 - R_f^2)\sqrt{\frac{tg\alpha^2+1}{tg\alpha^2}} \cdot \left(\frac{R_0 - R_f}{sin\alpha}\right)}{\rho V c} \tag{7}$$

while the temperature of the billet during cooling by tool contact calculated as:

$$T = T_{sup} + \left(T_0 - T_{sup}\right) exp\left(\frac{-\alpha A(t-t_0)}{\rho c V}\right). \tag{8}$$

Finally the total temperature of the billet due to different contributions given by:

$$T_f = \Delta T_{def} + \Delta T_{attr} + T \tag{9}$$

where:

t_0: initial time

v: velocity of the stem

t: time of displacement

ρ: density of material

c: specific heat

V: volume of the material

ΔT: temperature increment

T_{sup}: surface temperature of container and stem

T_0: initial temperature of the billet

ΔT_{def}: actual temperature increment of the billet due to deformation

ΔT_{attr}:actual temperature increment of the billet for the friction

T: actual temperature due to tool contact

T_f : total actual temperature of workpiece.

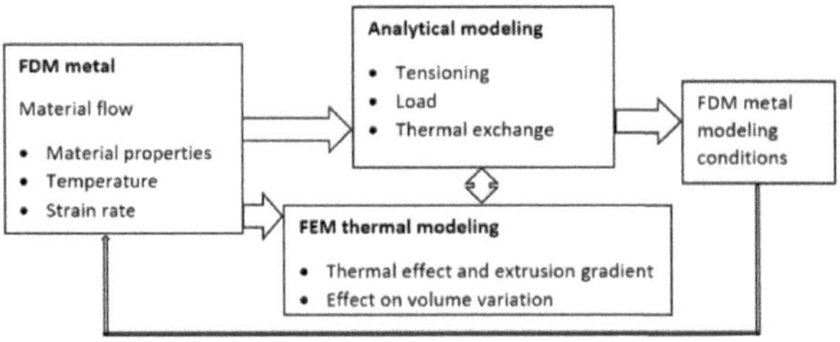

Figure 1 Procedure.

The previous equations by (1) to (9) are used to quantify the stress of the material inside the container and under the action of the stem. By those the temperature increase, under two nominal temperatures investigated that are 400° and 450°C, due to the deformation and friction heating was calculated and computed in a finite element simulation model.

The finite element analysis of the extruder behaviour was performed mainly in terms of the thermal load obtained by the imposed and during working generated temperature. The built model is characterized by an adaptive re-meshing. The heat flow and the thermal dilatation of the material and of the container in order to take chance to quantify elastic deformation and its effect on the extrusion are obtained. The material used for the stem and for the container was conventional tool steel. In Fig.2 a scheme of the extruding model is given.

Figure 2 Extruding system model.

Results

For the temperatures considered in the study the equation (1) and (2) using the material models (3) and (4) proposed in previous works produce the results in terms of stress reported in Figs. 3 and 4. At 400° and 450°C for AZ31 magnesium alloy a slight decrease and then an increase in the stress vs. displacement curve is given in particular in the conical zone of the extruder in which the main deformation appears. In detail Fig. 3 reports the values at different levels of the displacement, while the Fig. 4 shows the same values increased to consider at the beginning the final stress values at the end of the extrusion phase. This provides useful information for project design evaluations at the extruder level.

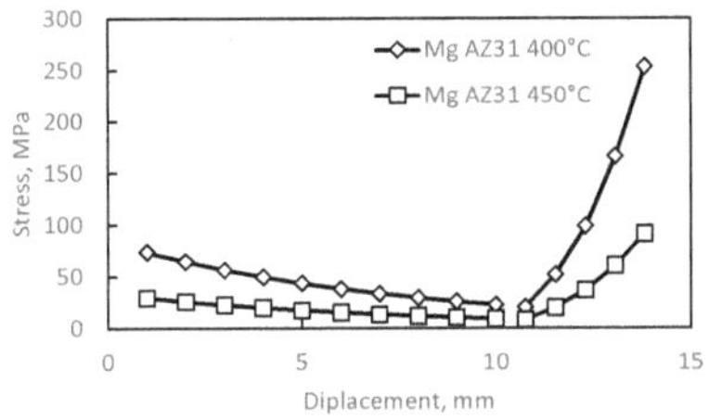

Figure 3 Stress behaviour for AZ31 Mg alloy during extrusion at 400° and 450°C.

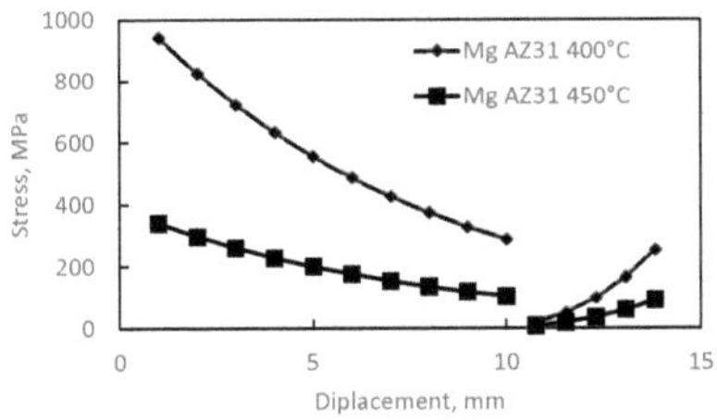

Figure 4 Stress behaviour for AZ31 Mg alloy during extrusion at 400° and 450°C increased to consider at the stem level the stress after the entire deformation.

Under the two temperature conditions also for the ZM21 magnesium alloy a decrease in tension until the beginning of the conical zone and then an increase can be observed as reported in Figs. 5 and 6.

The tensioning at 400°C is higher than the results obtained by the same equations (1) and (2) at 450°C. It means that under the conditions of the present investigation the high values of tension in the cylindrical part of the extruder depend on the friction while those obtained in the conical one are related mainly to the deformation.

The eq. (5) allows the calculation of the highest load that could be of about some tens of kN.

By those values of stress the temperature increase can be analysed in detail. In particular, by equation (6) and (7) the temperature contribution due to deformation and mainly that due to friction can be calculated, respectively. In figs. 7 and 8 the temperature increase in the cylindrical and conical zone for both magnesium alloys at 400° and 450°C is shown previously (condition 1) and after deformation (condition 2).

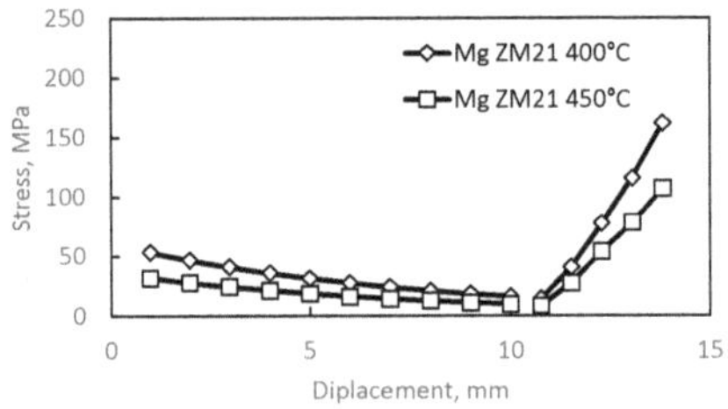

Figure 5 Stress behaviour for ZM21 Mg alloy during extrusion at 400° and 450°C.

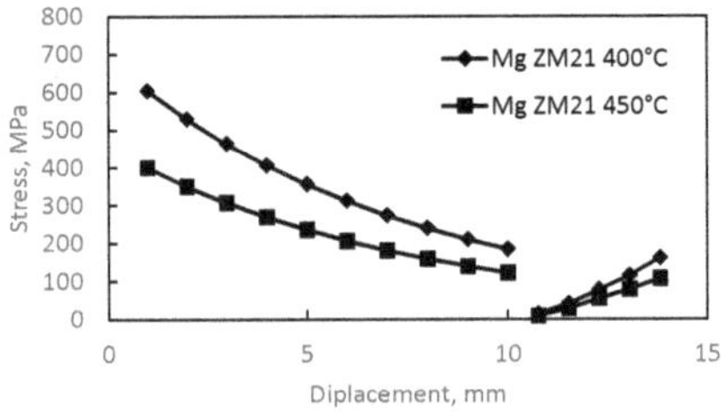

Figure 6 Stress behaviour for ZM21 Mg alloy during extrusion at 400° and 450°C increased to consider at the stem level the stress after the entire deformation.

Instantaneusly the local temperature can rise up to about 550°C for both AZ31 and ZM21 magnesium alloys due to friction and deformation at the conical zone and up to a temperature lower than the former in the cylindrical. If the cooling effect is not considered the temperature rise is able to produce the beginning of semisolid behaviour of the material under investigation notwithstanding the nominal temperature imposed. The material behaviour under similar conditions in terms of temperature is also reported in [8,9].

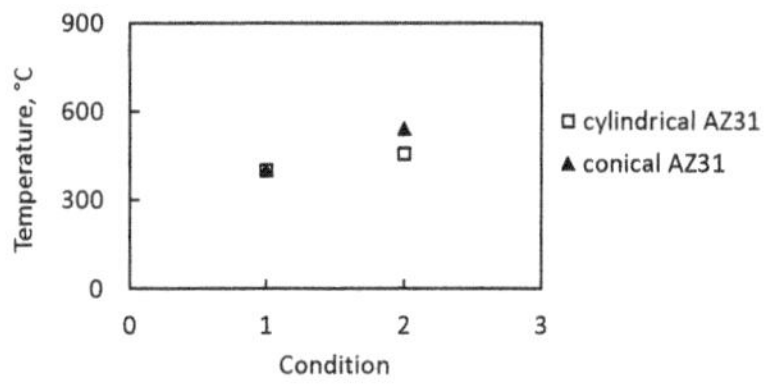

a)

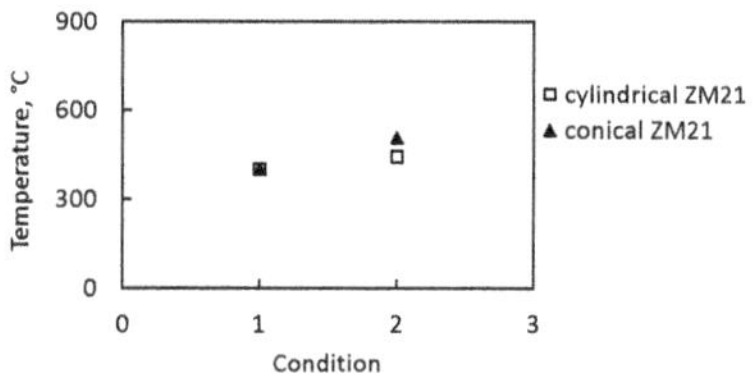

b)

Figure 7 Temperature increase for AZ31 Mg (a) and ZM21 Mg (b) alloy during extrusion 400°C increased for deformation and friction heating.

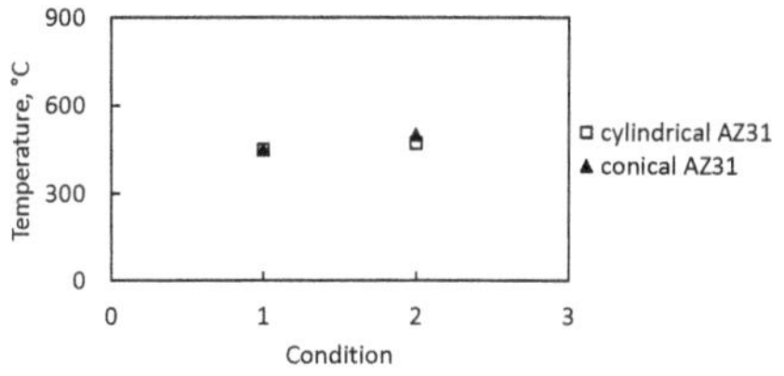

a)

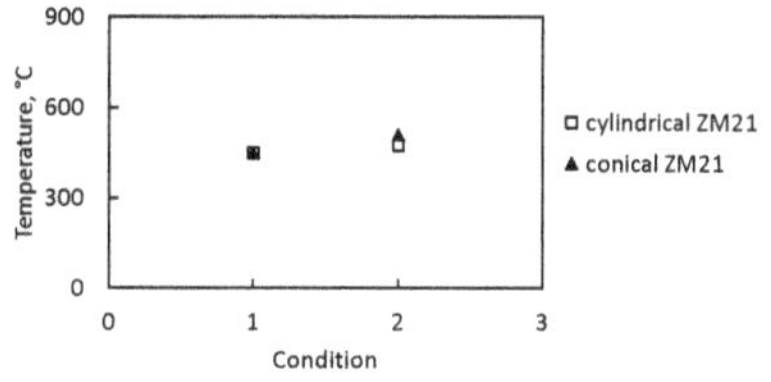

b)

Figure 8 Temperature increase for AZ31 Mg (a) and ZM21 Mg (b) alloy during extrusion 450°C increased for deformation and friction heating.

The temperature of AZ31 and of ZM21 mg alloys obtained by equation (8) related to the nominal at 450°C is reported in Fig. 9a) for the entire volume and the cylindrical part. The vacuum cooling condition that is 0 m/s air velocity is considered. The decrease is due mainly in the extruding system at the tool-material contact interface.

Due to the effect of the very low deforming volume considered that is of about some hundreds of cubic millimetres the real temperature rising is affected by a factor of 10^3 respect to conventional extruding problems. The cooling effect could reduce the temperature increase to maintain the above values. Such results are similar to those found for aluminium alloys under similar conditions [16, 23-25]. Fig. 9b) reports the temperature behaviour given by the equation (9) representing the whole temperature of the alloy considering all of the effects represented by the deformation and friction heating for the nominal temperature of 400°C, included the effect of cooling at the tool-material contact interface at the same temperature without such effects.

The temperatures determine an increase in the volume that produces deformations, displacements and tensions in the container with a given thermal flow at the specimen tool contact interface as can be observed in Figs. 10, 11 and 12. They report the real temperatures in the whole volume and in the conical zone approximated to 600°C in order to overestimate the effect of pressure on the tool, the elastic deformation amount at the extruder and on the material with related displacements.

The Fig. 13 reports the extruder behaviour under thermal load when the volume of material inside the container is less than that effectively available. At the deformation zone the pressure is equal to 50-65 MPa with a load not higher than 3.5-4.5 kN. Under such condition the analytical modeling is largerly able to describe the behaviour of the magnesium alloys under investigation.

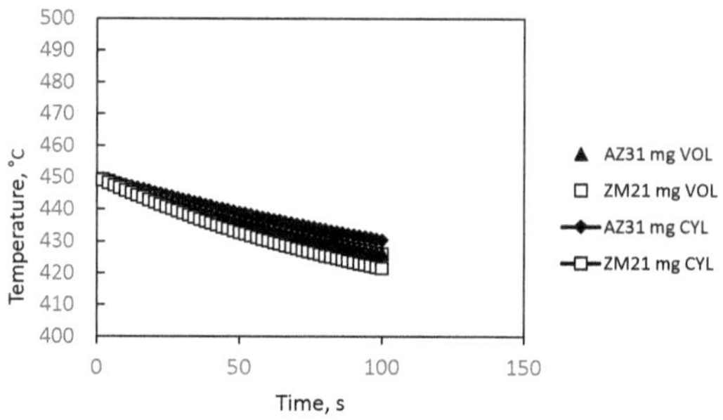

a)

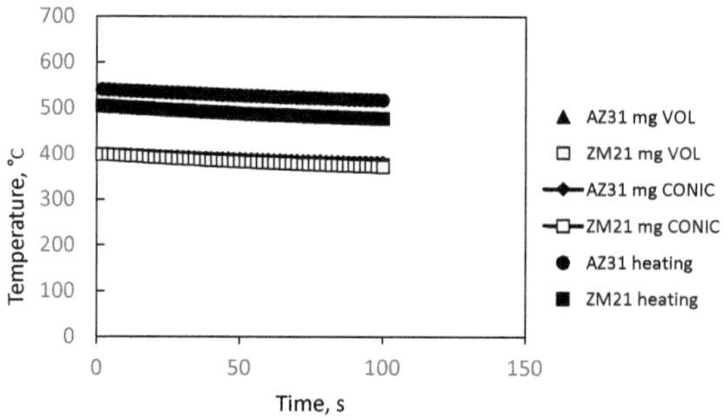

b)

Figure 9 Whole temperature behaviour for AZ31 Mg and ZM21 Mg alloys during extrusion at 450°C in the cylindrical and conical region: a) nominal temperature and b) temperature after deformation and friction heating.

a) b)

Figure 10 Extruding system (a) and temperature of magnesium alloy during extrusion at 450°C with the contribution of deformation and friction heating (b).

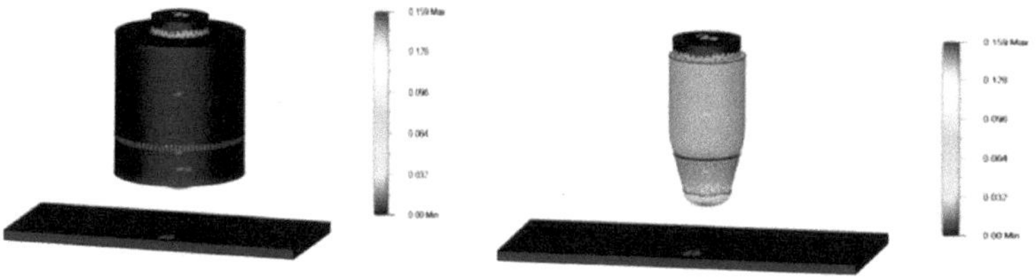

a) b)

Figure 11 Deformation of the extruder (a) and of the material magnesium alloy (b) during extrusion at 450°C with the contribution of deformation and friction heating.

Figure 12 Displacements in mm of the extruder wall at 450°C plus the contribution of deformation and friction heating.

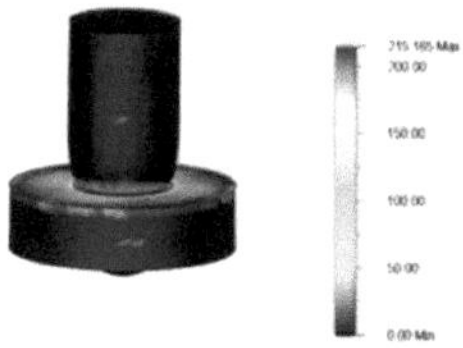

Figure 13 Pressure contact at the conical zone for temperature increase.

Summary

The study of the extrusion of magnesium AZ31 and ZM21 alloys in order to obtain semisolid condition useful for ribs deposition is performed.

The analytical modeling allows to determine the stress values of the materials under investigation knowing the behaviour for given conditions of temperature and strain rate. In detail, the flow stress distribution along the container represented by the cylindrical and the conical zone in which effectively takes place the deformation. Two different conditions were considered in terms of initial nominal working temperature that are 400° and 450°C. For both the conditions the effect of deformation and friction heating produces an increase in temperature up to a value not higher than 600°C. Such temperature is the most useful for determining the semisolid status for rib applications under the additive manufacturing philosophy. In particular when higher conicity angles than 15° do not allow to neglect distortion effects in the conical zone.

The increase in temperature determines a thermal load on the extruder evaluated by means of the numerical simulation getting an acceptable elastic deformation on the tools with a reacting force at least one order of magnitude less than the extruding force.

References

[1] S. Singh, G. Singh, C. Prakash and S. Ramakrishna, Current status and future directions of fused filament fabrication, Journal of Manufact. Proc. 55, 2020, 288. https://doi.org/10.1016/j.jmapro.2020.04.049

[2] M. Mousapour, M. Salmi, L. Klemettinen and J. Partanen, Feasibility study of producing multi-metal parts by Fused Filament Fabrication (FFF) technique, Journal of Manufact. Proc. 67 (2021) 438. https://doi.org/10.1016/j.jmapro.2021.05.021

[3] N. Kumar, V. Gupta, P. Saxena, A. Bajpai, C. Lahoda, J. Polte, Filament fabrication and subsequent additive manufacturing, debinding, and sintering for extrusion-based metal additive manufacturing and their applications: A review, Composites Part B 264 (2023) 110915. https://doi.org/10.1016/j.compositesb.2023.110915

[4] M. B. Jabłońska, Effect of the conversion of the plastic deformation work to heat on the behaviour of TWIP steels: a review, Archives of Civil and Mechanical Engineering 23 (2023) 136 https://doi.org/10.1007/s43452-023-00656-0

[5] S. Roshchupkin, A. Kolesov, A. Tarakhovskiy, I. Tishchenko, A brief review of main ideas of metal fused filament fabrication, Materials Today: Proceedings 38 (2021) 2063-2067. https://doi.org/10.1016/j.matpr.2020.10.142

[6] F. W. C. Farias, V. R. Duarte, I. Oliveira, J. da Cruz P. Filho, N. Schell, E. Maawad, J.A. Avila, J.Y. Li , Y. Zhang, G. Santos, J.P. Oliveira, In situ interlayer hot forging arc based directed energy deposition of Inconel® 625: process development and microstructure effects, Additive Manufacturing 66 (2023) 103476. https://doi.org/10.1016/j.matpr.2020.10.142

[7] A. I. Nurhudan, S. Supriadi, Y. Whulanza A. S. Saragih, Additive manufacturing of metallic based on extrusion process: A review. Journal of Manufacturing Processes 66 (2021) 228–237. https://doi.org/10.1016/j.jmapro.2021.04.018

[8] Y. Qi Li, F. Li, F. Wei Kang, H. Q. Du, Z. Y. Chen, Recent research and advances in extrusion forming of magnesium alloys: A review, Journal of Alloys and Compounds, 953 (2023) 170080. https://doi.org/10.1016/j.jallcom.2023.170080

[9] J. Feng, D. Zhang, H. Hu, Y. Zhao, X. Chen, B. Jiang, F. Pan, Improved microstructures of AZ31 magnesium alloy by semi-solid extrusion, Materials Science & Engineering A 800 (2021) 140204. https://doi.org/10.1016/j.msea.2020.14020

[10] S. A. Parasız , B. L. Kinsey, N. Mahayatsanun, J. Cao, Effect of specimen size and grain size on deformation in microextrusion, Journal of Manufacturing Processes 13 (2011) 153–159. https://doi.org/10.1016/j.jmapro.2011.05.002

[11] B. Meng, M. Wan, R. Zhao, Z. Zou, H. LIU, Micromanufacturing technologies of compact heat exchangers for hypersonic precooled airbreathing propulsion: A review. Chinese Journal of Aeronautics (2021) 34 79-103. https://doi.org/10.1016/j.cja.2020.03.028

[12] N. Krishnan, J. Cao, K. Doh, Study of the Size Effect on Friction Conditions in Microextrusion—Part I: Microextrusion Experiments and Analysis, J. Manuf. Sci. Eng. 2007, 129(4): 669-676. https://doi.org/10.1115/1.2386207

[13] Yu Y. , Hisanao K., Tsuyoshi, F, Effect of forming conditions on microstructure and room-temperature mechanical characterization of Zn–22Al superplastic microtubes fabricated by direct extrusion, Materials Science & Engineering A 844 (2022) 143160. https://doi.org/10.1016/j.msea.2022.143160

[14] M. Schewe, H. Wilbuer, A. Menze, Simulation of wear and effective friction properties of microstructured surfaces, Wear 464–465 (2021) 203-491. https://doi.org/10.1016/j.wear.2020.203491

[15] S. Li, X. Yang, J. Hou, W. Du, A review on thermal conductivity of magnesium and its alloys,

Journal of Magnesium and Alloys, 8 (2020) 78–90. https://doi.org/10.1016/j.jma.2019.08.002

[16] B. Liu, Y. Wang , Z. Lin, T. Zhang, Creating metal parts by Fused Deposition Modeling and Sintering, Materials letter 263 (2020) 127252. https://doi.org/10.1016/j.matlet.2019.127252

[17] N. Bontcheva , G. Petzov , L. Parashkevova, Thermomechanical modelling of hot extrusion of Al-alloys,followed by cooling on the press, Computational Materials Science 38 (2006) 83–89. https://doi.org/10.1016/j.commatsci.2006.01.009

[18] M. S. Kesler, M. L. Neveau, W. G. Carter, H. B. Henderson, Z. C. Sims, D. Weiss, T. T. Li, K. McCall, M. Glicksman, O. Rios, Liquid direct reactive interface printing of structural aluminum alloys, Applied Materials Today 13 (2018) 339–343. https://doi.org/10.1016/j.apmt.2018.10.005

[19] Huang, W., Chen, S., Xiao, J., Jiang, X., Jia, Y. Laser wire-feed metal additive manufacturing of the Al alloy. Opt. Laser Technol. (2021) 134, 10662. https://doi.org/10.1016/j.optlastec.2020.106627

[20] L. E. S. Paes, H. Santos Ferreira, M. Pereira, F. A. Xavier, W. L. Weingaertner, L.O. Vilarinho, Modeling layer geometry in directed energy deposition with laser for additive manufacturing, Surface & Coatings Technology 409 (2021) 126-897. https://doi.org/10.1016/j.surfcoat.2021.126897

[21] C. Bruni, M. El Mehtedi, and F. Gabrielli, Flow curve modelling of a ZM21 magnesium alloy and finite element simulation in hot deformation, Key Engineering Materials Vols. 622-623 (2014) 588-595. https://doi:10.4028/www.scientific.net/KEM.622-623.588

[22] C. Bruni et al., Constitutive models for AZ31 Magnesium alloys, Key Engineering Materials, 367 (2008) 87-94. https://doi.org/10.4028/www.scientific.net/KEM.367.87

[23] R. R. Yadav, Y. Dewang, J. Raghuwanshi,V. Sharma, Finite element analysis of extrusion process using aluminum alloy, Materials Today: Proceedings 24 (2020) 500–509. https://doi.org/10.1016/j.matpr.2020.04.302

[24] N. Fietier, Y. Krahenbuhl, M. Vialard, New methods for the fast simulations of the extrusion

process of hot metals, Journal of Mat. Proc. Technology 209 (2009) 2244–2259. https://doi.org/10.1016/j.jmatprotec.2008.05.022

[25] A. García-Domínguez, J. Claver, A.M. Camacho, M. Sebastián. Comparative Analysis of Extrusion Processes by Finite Element Analysis, Procedia Engineering 100 (2015) 74. https://doi.org/10.1016/j.proeng.2015.01.344

Metal Forming 2024 Materials Research Forum LLC
Materials Research Proceedings 44 (2024) 444-455 https://doi.org/10.21741/9781644903254-48

Roll pass design for round and square sections using an informed artificial neural network

OVERHAGEN Christian[1,a *] and MARTIN Robert[2,b]

[1]University of Duisburg-Essen, Faculty of Engineering, Institute for Technologies of Metals, Chair of Metallurgy and Metal Forming, Forsthausweg 2, 47057 Duisburg, Germany

[2]University of Duisburg-Essen, Faculty of Engineering, Institute for Technologies of Metals, Forsthausweg 2, 47057 Duisburg, Germany

[a]christian.overhagen@uni-due.de, [b]robert.martin@uni-due.de

Keywords: Rod Rolling, Bar Rolling, Neural Networks, Roll Pass Design, Machine Learning, Custom Loss Function

Abstract. Nowadays, the roll pass design process for hot rolling mills of full sections like round and square sections is carried out by FEM simulations or analytical calculations. Both techniques require a highly iterative method with numerous iterations required to determine a satisfying pass design solution. In the present work, a neural network is trained for the process of roll pass design for round and square sections. The steps of data generation and training are combined by the formulation of a custom loss function for the underlying physical problem. In this way, the neural network is informed about the metal forming problem, enabling the generation of the necessary training data to be carried out implicitly during training. This technique increases the flexibility and extensibility of the current approach, eliminating the necessity of external training data generation. The metal forming equations for the pass design problem are presented for the pass sequences round-oval-round, as well as square-diamond-square, which are included in custom loss functions for the neural networks. The prediction results are compared to analytical iterative calculations, indicating very good agreements. Therefore, multiple iterations and computation time can be saved when the new approach of pass design is applied.

Introduction

Roll Pass Design of full sections like round and square sections is a task which is carried out to a big extent within the mill building and rolling mill industries. Therefore, we seek for methodologies which allow this process to be completed fast and accurately. Classical approaches require numerous iterations, starting from simple initial conditions for the groove design, which are refined step by step to finally yield a satisfactory solution. Balan et al. presented surrogate models for the flat rolling process based on neural networks [1] to increase the efficiency of numerical simulations. A new surrogate method for roll pass design was presented in [2], where an artificial neural network was trained with classically precalculated roll pass designs. There, the task was split into two parts, the data generation, and the training of a suitable neural network. The present approach follows a different route. Instead of utilizing a classical neural network with a default loss function provided by the machine learning framework (here: PyTorch [3]), we leverage the metal forming equations governing the pass design procedure as a custom loss function during training of the neural network. Therefore, the tedious and computationally expensive procedure of an iterative pre-calculation of the final pass design is no longer necessary. Instead, the synthetic data is generated on-the-fly during the training procedure. In the beginning, we only need to define the data range in which the neural network should be trained. Everything else will be handled by the loss function.

A roll pass design of square or round sections consists of a series of two-pass sequences, each consisting of an oval or diamond groove in the first pass (the so-called minor groove), and a round or square groove in the second pass (the so-called major groove). Usually, the sequence of major grooves in a pass design is predefined, and the task of the pass designer is to find the necessary geometries of the minor grooves. This is a highly iterative technique, requiring multiple evaluations of the same equations. In the present approach, we introduce a method to replace this procedure by direct prediction of the minor groove geometries by an artificial neural network.

The analytical formulation of the pass design problem

Since optimization techniques used for training of neural networks are generally based on gradient descent methods, the loss function of the neural network must be and differentiable. Therefore, we must find analytical expressions for the lateral spreading problem of section passes.

To calculate the groove filling by lateral spread of given groove geometries, a suitable spreading model for flat passes must be applied in conjunction with an equivalent pass method as to enable the spread calculation of section passes. For the spread calculation, we will use Roux's spreading model as given in [5, 6]. For the transformation between the section passes and equivalent flat passes, we use an extension of the equivalent pass method by A.E. Lendl [7], which will be shown in the following.

Lendl's method defines the mean entry and exit heights of the equivalent flat pass according to the relations

$$h_{0L} = \frac{A_{0L}}{w_i} \quad \text{and} \quad h_{1L} = \frac{A_{1L}}{w_i} \,. \tag{1}$$

The areas A_{0L} and A_{1L} are the parts of the entry and exit cross sections A_0 and A_1 which are subjected to the direct pressure exerted by the top and bottom roll.

A_{0L} and A_{1L} can be constructed graphically or calculated from geometrical considerations of the pass geometries. Additionally, we must define a working roll diameter, which is given by

$$d_{work} = d_{nom} - h_{1L} + s \tag{2}$$

In Eq. 2, d_{nom} is the nominal barrel diameter of the rolls and s is the roll gap of the groove.

The pass sequence round-oval-round. Fig. 1 shows the pass round-to-oval with the geometrical parameters d_0 (entry diameter), h_{ov} (height of the oval and, r_{ov} (radius of the oval).

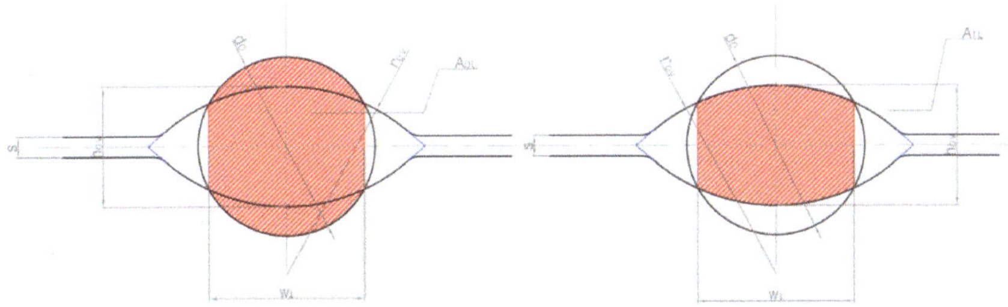

Fig. 1. The construction of the equivalence areas A_{0L} (left) and A_{1L} (right) for the pass round-to-oval.

 Materials Research Forum LLC
https://doi.org/10.21741/9781644903254-48

In the present approach, we restrict our considerations to the single-radius oval, noting that the approach can be extended to multi-radius ovals as long as their geometry can be described analytically. The intersection width w_i is the distance between the intersection points of the entry section and roll contours (see Fig. 1). From geometrical considerations, it follows that

$$w_i = 2\sqrt{r_{ov}^2 - \left(\frac{\frac{d_0^2}{4} - 2r_{ov}^2 + h_{ov}r_{ov} - \frac{h_{ov}^2}{4}}{h - 2r_{ov}} \right)^2} \,. \tag{3}$$

The areas A_{0L} and A_{1L} can then be calculated via

$$A_{0L} = w_i d_0 - 2w_i \frac{d_0}{2} + w_i \sqrt{\frac{d_0^2}{4} - \frac{w_i^2}{4}} + \frac{d_0^2}{2} \arcsin\left(\frac{w_i}{d_0}\right) \tag{4}$$

and

$$A_{1L} = w_i h_{ov} - 2w_i r_{ov} + w_i \sqrt{r_{ov}^2 - \frac{w_i^2}{4}} + 2r_{ov}^2 \arcsin\left(\frac{w_i}{2r_{ov}}\right) \,. \tag{5}$$

The mean heights of the equivalent flat pass are then calculated from Eq. 1. From the spread model, the exit width of the oval section $w_1 = w_{os}$ is then calculated, from which the output oval contour is constructed.

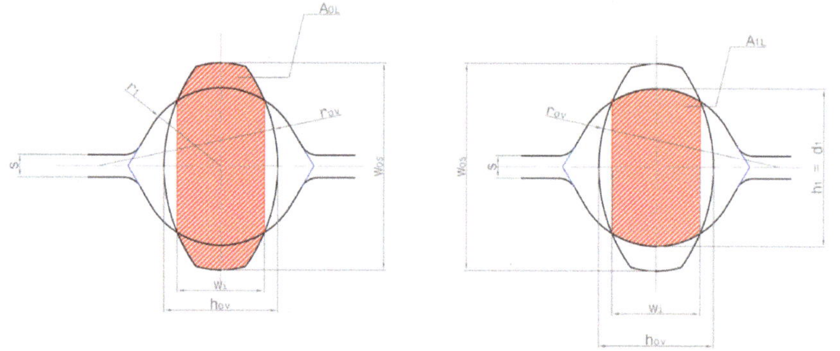

Fig. 2. Construction of the equivalence areas A_{0L} (left) and A_{1L} (right) for the pass oval-to-round.

For the second pass oval-to-round, the oval section will be rotated by 90° to an upright position and fed into a round groove. It should be noted that due to the upright rotation, the former width of the oval w_{os} now becomes the initial height, whereas the former height of the oval h_{ov} becomes the initial width. For clarity, we retain the definitions shown in Figs. 1 and 2.

In the round pass, the height of the section is reduced from w_{os} to $h_1 = d_1$ (see Fig. 2). The final radius of the round section is r_1. We shall note that the round groove is tangentially opened to prevent rolling faults due to spreading variations. The intersection width w_i of this pass is now given as

 Materials Research Forum LLC
https://doi.org/10.21741/9781644903254-48

$$w_i = \frac{h_{ov}}{2} - r_{ov} + \sqrt{r_{ov}^2 - y_i^2}, \tag{6}$$

where

$$y_i = \sqrt{r_{ov}^2 - \left(\frac{\frac{d_1^2}{4} - 2r_{ov}^2 + h_{ov}r_{ov} - \frac{h_{ov}^2}{4}}{h_{ov} - 2r_{ov}}\right)^2} \tag{7}$$

is the y-coordinate of the intersection point. Therefore, the area A_{0L} follows to (cf. Fig. 2):

$$\begin{aligned} A_{0L} &= w_{os} - h_{ov} - 2w_{os}r_{ov} + w_{os}\sqrt{r_{ov}^2 - \frac{w_{os}^2}{4}} + 2r_{ov}^2 \arcsin\left(\frac{w_{os}}{2r_{ov}}\right) \\ &- \left[2y_i\left(h_{ov} - w_i\right) - 4r_{ov}y_i + 2y_i\sqrt{r_{ov}^2 - y_i^2} + 2r_{ov}^2 \arcsin\left(\frac{y_i}{r_{ov}}\right)\right] \end{aligned} \tag{8}$$

and the area A_{1L} is:

$$A_{1L} = w_i\sqrt{\frac{d_1^2}{4} - \frac{w_i^2}{4}} + 2\left(\frac{d_1^2}{4}\right)\arcsin\left(\frac{w_i}{d_1}\right). \tag{9}$$

The equivalent heights h_{0L} and h_{1L} are again calculated by Eq. 1. Now, with the working roll diameter calculated according to Eq. 2, we can carry out the spread calculation for the second pass oval-to-round using Roux's spreading model, yielding the final width w_2 of the round section.

As a result of the spread calculation, the final width of the round section is found. To assess the suitability of the present oval geometry, we express the filling condition of both grooves (oval and round) in terms of a filling ratio given as f_{fill} = *actual width / nominal width*. The nominal width is defined differently depending on the groove type.

For round grooves, we usually take the diameter of the intended round section as the nominal width. For minor grooves, the width on face of the minor groove is used as the nominal width. The width on face *wof* of a single-radius oval groove is calculated according to

$$wof = 2\sqrt{r_{ov}^2 - \left(r_{ov} - \frac{h_{ov} - s}{2}\right)^2} . \tag{10}$$

For non-final round sections, target filling ratios between *0.95* and *0.98* are used, while for single-radius ovals, the target filling ratios area usually in the range between *0.8* and *0.9*.

The pass sequence square-diamond-square. To work out the geometrical relations leading to the definition of the equivalence areas A_{0L} and A_{1L} on the general pass diamond-to-diamond, we shall first treat the definition of the section area on a rolled diamond section, as given in Fig. 3.

 Materials Research Forum LLC
https://doi.org/10.21741/9781644903254-48

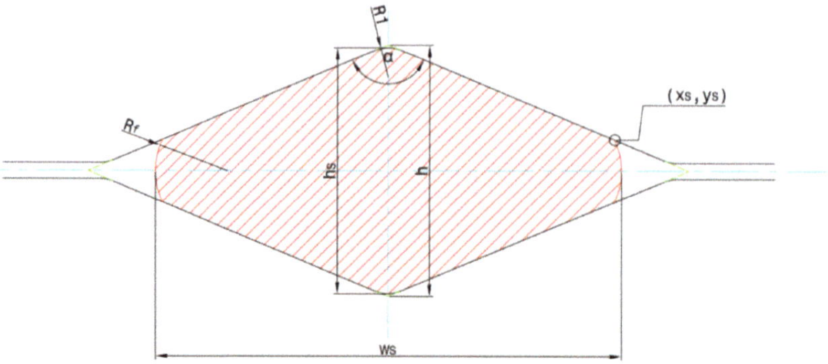

Fig. 3. Typical cross section rolled in a diamond groove.

The diamond has the theoretical height h. Practically, the groove tips are rounded with a main radius R_1, leading to a smaller height of the exit section h_S. Under lateral spread, a final width w_S of the section is generated with the sides of the rolled section being curved at a filling radius R_f. The total area of this cross section A_S can now be calculated as follows:

$$A_S = w_S \cdot \left(h - \frac{1}{2} w_S \tan\gamma\right) - 2\left[R_1^2 \sin^2\gamma \tan\gamma - \frac{R_1^2}{2}(2\gamma - \sin 2\gamma)\right]$$
$$-\left[\left(2y_S + (h - w_S \tan\gamma)\right)\left(\frac{w_S}{2} - x_s\right) - R_f^2 (2\varphi - \sin 2\varphi)\right] \tag{11}$$

Here, x_S and y_S are the coordinates of the intersection point between the arc of the filling radius and the inclined groove contour in the first quadrant (see Fig. 3). The angle φ corresponds to the chord of the filling radius and is given by $\tan\varphi = y_S / (x_S - x_M)$, where $x_M = w_S/2 - R_f$ is the x-coordinate of the center of the radius R_f. The angle γ is always $\gamma = 90° - \alpha/2$. Note that for this construction, R_f cannot be less than

$$R_{f,\min} = \frac{\frac{h}{2} - \frac{w_s}{2}\tan\gamma}{\cos\gamma - \tan\gamma(1 - \sin\gamma)} \tag{12}$$

Similarly, we can determine the equivalence areas A_{0L} and A_{1L} as shown in Fig. 4.

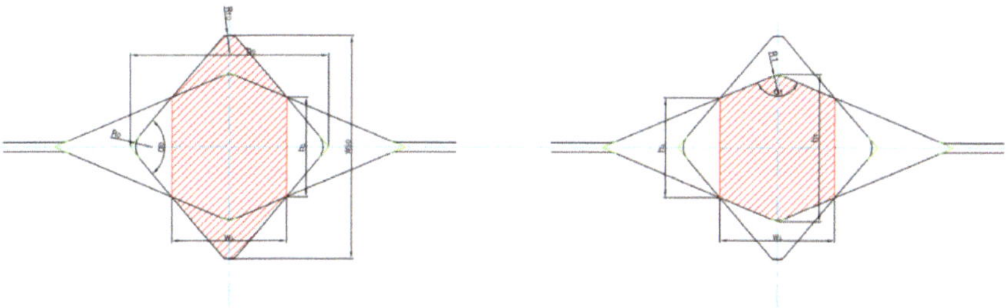

Fig. 4. The construction of the areas A_{0L} (left) and A_{1L} (right) at the generalized pass diamond-to-diamond.

In Fig. 4, the general pass diamond-to-diamond is considered, which also applies to the passes square-to-diamond and diamond-to-square as special cases. Note that the groove opening angles α_0 on the entry section, as well as α_1 of the exit section are not equal to $90°$ in Fig. 4 for generality of the considerations.

By comparing Fig. 4 (left) to Fig. 3 and subtracting the triangular regions left and right to the intersection distance, we can calculate the shaded area A_{0L} in Fig. 4 (left) via

$$A_{0L} = A_{0S} - \frac{h_i \cdot (h_0 - w_i)}{2} + 2R_0^2 \left[\frac{\cos^3 \frac{\alpha_0}{2}}{\sin \frac{\alpha_0}{2}} - \frac{1}{2}\left((\pi - \bar{\alpha}_0) - \sin \alpha_0\right) \right], \tag{13}$$

where R_0 and α_0 are the original tip radius and opening angle of the entry diamond, respectively. The area A_{1L} in Fig. 4 (right) is given by

$$A_{1L} = \frac{w_i \cdot (h_1 - h_i)}{2} - 2R_1^2 \left[\frac{\cos^3 \frac{\alpha_1}{2}}{\sin \frac{\alpha_1}{2}} - \frac{1}{2}\left((\pi - \bar{\alpha}_1) - \sin \alpha_1\right) \right] + w_i h_i \tag{14}$$

The height and width values of the region bounded by the intersection points (see Fig. 4) are

$$w_i = \frac{h_0 \tan \frac{\alpha_0}{2} - h_1}{\tan \gamma_0 - \tan \gamma_1}, \tag{15}$$

$$h_i = h_1 - w_i \tan \gamma_1 \tag{16}$$

The passes square-to-diamond and diamond-to-square are special cases of the general case treated here, with $\alpha_0 = 90°$ for square-to-diamond and $\alpha_1 = 90°$ for diamond-to-square.

For the target filling criterion, we again define the width on face *wof* of the diamond and square grooves at the roll gap *s* and the groove height *h* by

$$wof = \tan \frac{\alpha}{2}\left(h - \frac{s}{2}\right). \tag{17}$$

In this pass sequence, we usually require a well-filled diamond groove with filling ratios of 0.95 to 0.98.

Structure of the neural network

As the roll pass design problem can be split into several units, each of which comprises a minor groove and a major groove, we design the neural network to predict the minor groove dimensions which are optimal for the transition between a given entry section (round or square) into the first pass and a given exit section out of the second pass (round or square).

We use three input features, which comprise the entry size s_0 or d_0, the exit size s_1 or d_1 and the nominal roll diameter d_{nom}. The output features comprise the height h and width w of the minor groove to be used for the intended deformation.

It was found that a deep neural network with 3 hidden layers provided the best prediction capabilities for the current task. The first hidden layer uses 90, the second one 40 and the third hidden layer 20 neurons.

In all hidden layers, the Rectified Linear Unit (ReLU) is used as the activation function. In the output layer, we use the linear activation function as not to restrict the output data range.

Training procedure

Prior to the training, we define ranges of the input values which should be covered. Table 1 gives an overview of the training data ranges which were used for the pass design of round sections.

Table 1. Overview of the training data for the round-oval-round pass design.

	Minimum value	Maximum value	Number of values
Initial diameter d_0	15 [mm]	30 [mm]	10
Mean reduction per pass	18 [%]	25 [%]	10
Roll diameter d_{nom}	210 [mm]	500 [mm]	10

Therefore, a total of 1000 combinations of rolling parameters are considered during the training procedure. The target filling ratios are always set to 0.85 for the ovals and 0.95 for the round grooves.

The input data is normalized to a range of [0,1] for each input feature. Out of the 1000 samples, 150 are selected for the calculation of the validation loss.

A training batch size of 16 samples is used. For each batch, the neural network will propose 16 corresponding minor grooves for the intended deformation task. Based on the equations presented above, the loss function calculates the groove fillings of both passes (minor and major) and compares these values to the target filling ratios. Then, the loss function is calculated as

$$L = \left[w_1(h,w) - w_1^* \right]^2 + \left[w_2(h,w) - w_2^* \right]^2, \tag{18}$$

where w_1 and w_2 are the section widths of both passes calculated for the predicted (guessed) minor groove characterized by its height h and width w, while w_1* and w_2* are the target widths of the two passes.

In the beginning of the training process, the neural network will present random guesses of groove geometries to the loss function, which might lead to a non-defined loss value. Therefore, we guide the neural network towards a technically feasible range by a normalization of the output data to a range of [0,1]. However, the outputs are not constrained to this range since we use linear activation in the output layer.

The training is carried out similarly for the square-diamond-square pass design. Table 2 provides an overview of the data ranges and the sample sizes used in the training.

Table 2. Overview of the training data for the square-diamond-square pass design.

	Minimum value	Maximum value	Number of values
Initial sidelength s_0	30 [mm]	[60] mm	10
Mean reduction per pass	20 [%]	25 [%]	10
Roll diameter d_{nom}	350 [mm]	650 [mm]	10

As a result, two trained neural networks were developed, one for the pass sequence round-oval-round and one for the pass sequence square-diamond-square. For each network training, the Adam optimization algorithm [4] is applied for 500 epochs at a learning rate of 0.0001.

Results of the neural networks and comparison to analytical calculations

The round-oval-round pass design. The neural network, trained as described above, can now be used to predict the necessary width and height values of a single-radius oval in a two-pass sequence round-oval-round.

To examine the precision of the prediction and to prove to usability of the present approach, testing data was generated using the classical iterative method on the data range given in Table 3.

Table 3. Testing data for the round-oval-round pass design neural network.

	Minimum value	Maximum value	Number of values
Initial diameter d_0	16 [mm]	29 [mm]	16
Mean reduction per pass	18 %	23 %	16
Roll diameter d_{nom}	250 [mm]	420 [mm]	16

For each of these combinations (16*16*16 = 4096 samples in total), the necessary oval groove was calculated classically using the equations (1) to (9). The calculation was repeated iteratively, until a filling error of less than 0.05 mm occurred in both passes. The neural network (trained previously using the data ranges given in Table 1) was then applied to the input data given in Table 3. The results of oval heights and widths are shown in comparison in Fig. 5.

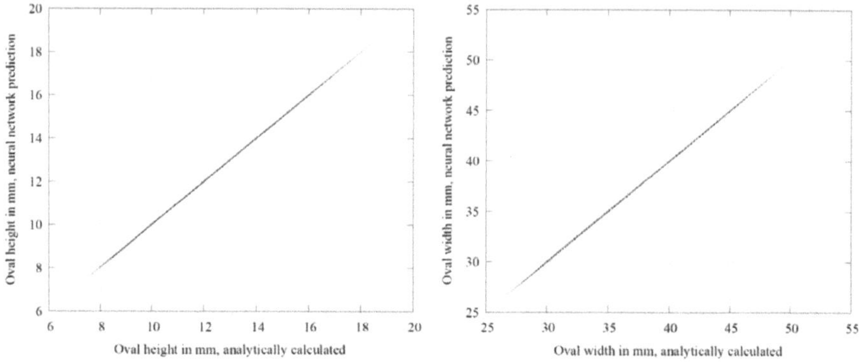

Fig. 5. Comparison of analytically calculated test data to the corresponding neural network predictions for the oval width and height.

The left-hand part of Fig. 5 shows the comparison between predicted and analytically calculated oval widths, whereas the right part of the figure shows the corresponding results for the oval heights. As can be seen, the predicted oval geometries are in very good agreement to the direct iterative calculation.

The square-diamond-square pass design. The similar procedure was carried out for the pass sequence square-diamond-square on the testing data range given in Table 5.

Table 4. Testing data for the square-diamond-square pass design neural network.

	Minimum value	Maximum value	Number of values
Initial sidelength s_0	30 [mm]	60 [mm]	16
Mean reduction per pass	20 [%]	25 [%]	16
Roll diameter d_{nom}	350 [mm]	650 [mm]	16

In Fig. 6, the prediction results of the neural network are shown in comparison to the direct analytical calculations of the pass design. Similarly as for the round-oval-round pass design, the prediction results are very close to the ones calculated analytically.

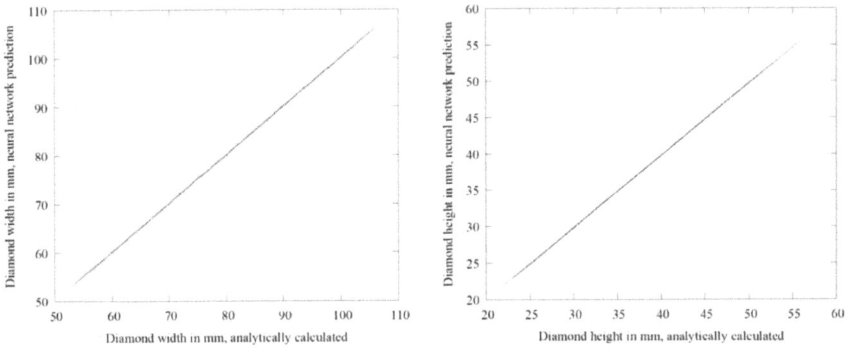

Fig. 6. Comparison of analytically calculated test data to the corresponding neural network predictions for the diamond width and height.

Checking the predictions using a numerical pass-design software. A more practical evaluation of the prediction results can be carried out when the predicted grooves are checked with a simulation and optimization software for roll pass designs. Such software solution called *MPC – Mill Process Calculations* is maintained at the Chair of Metallurgy and Metal Forming [8]. The algorithm used by the software to calculate the groove fillings is based on the same principles as the one employed as a loss function in the current approach, but not in its analytical form. Instead, a numerical treatment of the section and groove contour geometries is applied which provides a higher flexibility at the geometrical section formation.

As a study, we want to consider a pass sequence round-oval-round for an initial section with a diameter of 27 mm, to be rolled down to a final diameter of 16 mm in 4 passes at a nominal roll diameter of d_{nom} = 300 mm. As we see from Table 1, this data is well inside the training range of the neural network. The intermediate round diameter needed in the second pass was d_2 = 20.66 mm for a homogeneous deformation distribution. The two oval grooves in passes 1 and 3 were predicted by the neural network, and the groove fillings were computed using the numerical model. The cross-sectional results for the first two passes are shown in Fig. 7.

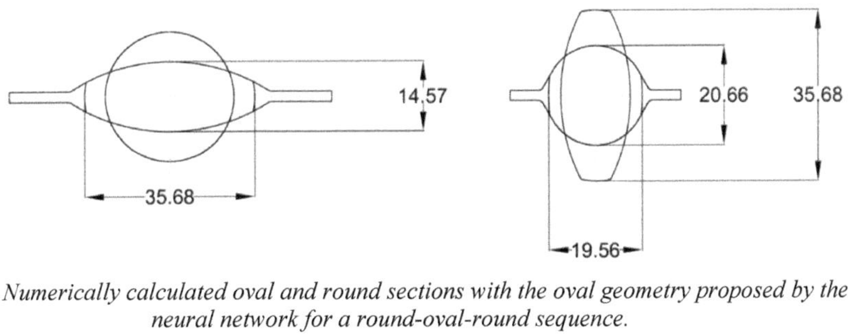

Fig. 7. Numerically calculated oval and round sections with the oval geometry proposed by the neural network for a round-oval-round sequence.

Table 5 gives an overview of the calculated results for all four passes. For the oval grooves, we define the filling ratio as the width of the section of the width on face of the groove, where the neural network was trained to design the grooves for a value of 0.85. For the round grooves, the filling ratio is defined as the section width divided by the section height. Here we trained the neural network to reach a value of 0.95. This slight underfilling of the round grooves is intended for a good contact condition in the following oval groove within a pass sequence.

Table 5. Results of the numerical validation calculations for round-oval-round with the oval geometries proposed by the neural network.

Pass	Section shape	Height of section [mm]	Width of section [mm]	Width on face of the oval grooves [mm]	Actual filling ratio
1	Oval	14.57	35.68	41.96	0.850
2	Round	20.66	19.56		0.947
3	Oval	10.89	27.90	33.24	0.839
4	Round	16.00	15.12		0.945

Summary

In the present work, it is shown how a neural network can be informed about the metal forming equations governing the pass design process of round and square sections. The analytical model is used as a custom loss function to train a neural network for the task of roll pass design of two-pass sequences. Therefore, no pre-generated target data is used for the training, but the targets are generated by the neural network itself during the training process, based on a prescribed input data range. The data range of the network can easily be extended without the necessity of external data generation. Once trained, the neural network can predict the necessary groove geometries without the necessity of multiple iterations in close agreement to the iterative analytical calculations.

In the current usage example, we showed that the data-driven model predicts the groove geometries close to the results of analytical and numerical models. The neural network has the advantage of its easy portability compared to analytical and numerical models. Therefore, this technique is suitable to create a general method of rapid roll pass design.

Generally, artificial neural networks with custom loss functions are an interesting technique to solve systems of nonlinear equations, which can otherwise only be solved iteratively. The present roll pass design problem is only one example. However, one limitation of the approach is that the underlying physical model must be given in its analytical form, as to ensure differentiability with respect to the input parameters.

References

[1] K. Slimani, M. Zaaf, T. Balan, Accurate surrogate models for the flat rolling process, Int. J. Mater. Form. 16 (2023) 23. https://doi.org/10.1007/s12289-023-01744-5

[2] C. Overhagen, K. Fu, Data-Driven Roll Pass Design of Wire Rod Mills, Key Eng. Mater. 926 (2022) 589-601. http://dx.doi.org/10.4028/p-dwm928

[3] A. Paszke, S. Gross, F. Massa, A. Lerer, J. Bradbury, G. Chanan, T. Killeen, Z. Lin, N. Gimelshein, L. Antiga, A. Desmaison, A. Köpf, E. Yang, Z. DeVito, M. Raison, A. Tejani, S. Chilamkurthy, B. Steiner, L. Fang, J. Bai, S. Chintala, PyTorch: An Imperative Style, High-Performance Deep Learning Library, arXiv: 1912.01703v1. https://doi.org/10.48550/arXiv.1912.01703

[4] D.P. Kingma, J. Ba, Adam: A Method for Stochastic Optimization, arXiv: 1412.6980v9. https://doi.org/10.48550/arXiv.1412.6980

[5] M.J. Roux: Etude sur le phenomene de l'elargissement dans les laminoirs, Revue de metallurgie 36 (1939).

[6] P.J. Mauk, R. Kopp, Spread under Hot Rolling, The roll pass designer 37 (39) 3-68.

[7] A.E. Lendl, Rolled bars – part I, Iron and Steel 21 (14) 397-402.

[8] C. Overhagen, Numerical assessment of interstand tensions in continuous hot rolling processes of flat and long products, Int. J. Adv. Manuf. Tech. 127 (2023) 4677-4696. https://doi.org/10.1007/s00170-023-11584-x

Materials Research Proceedings 44 (2024) 456-464 https://doi.org/10.21741/9781644903254-49

Effect of ductile cast iron tool with grooved polishing and die coatings on friction behaviour in flat strip drawing of A7075-T6 alloy strip with quick heating

HARADA Sohdai[1,a], MAENO Tomoyoshi[2,b] *, NOMURA Atsuya[1,c], HIROSAWA Shoichi[2,d], YABUKI Yuji[3,e] and SUZUKI Yasushi[3,f]

[1]Department of Mechanical Engineering, Materials Science, and Ocean Engineering, Graduate School of Engineering Science, Yokohama National University, Yokohama, 240-8501, Kanagawa, Japan

[2]Division of Systems Research, Faculty of Engineering, Yokohama National University, Yokohama, 240-8501, Kanagawa, Japan

[3]R&D Division, G-TEKT Corporation, 4-7-28 Sinmeidai, Hamura, Tokyo 205-0023, Japan

[a]harada-sodai-dg@ynu.jp, [b]maeno-tomoyoshi-yf@ynu.ac.jp, [c]nomura-atsuya-gc@ynu.jp, [d]hirosawa@ynu.ac.jp, [e]yuuji_yabuki.iq@g-tekt.jp, [f]yasushi_suzuki.nv@g-tekt.jp

Keywords: Aluminium, Hot Stamping, Warm Forming, Friction, Die Material, Surface Texture, Seizure, Lubricant

Abstract. A7075 aluminium alloy, a heat-treatable aluminium alloy, is expected to be applied to automobile frame parts from the viewpoint of high strength and light weight. To form this aluminium alloy, hot stamping using quick contact heating with a hot plate has been developed. To prevent seizure caused by hot stamping, the frictional properties of different tool coatings under warm and dry conditions were investigated by drawing tests of aluminium alloy strips using quick contact heating. Diamond like carbon (DLC) showed a low coefficient of friction and prevented seizure. The warm forming process allows the use of liquid lubricants. The frictional properties of cast iron and a simple surface texturing technique, grooved polishing, were investigated under warm and liquid lubrication conditions. The cast iron with grooved polishing showed the reduction in friction equivalent to DLC and prevented seizure as well.

Introduction

The weight of vehicles has been increasing due to improved collision safety and environmental performance. Therefore, it is important to improve fuel efficiency. Reducing the weight of the automobile body is effective for low fuel efficiency. Conventionally, increasing the strength of steel strips and reducing their thickness have been considered. Among them, high-strength steel strips are widely used. However, reducing the strip thickness reduces rigidity and reduces collision safety and driving performance. Therefore, it is important to replace the material with one that has a lower specific gravity without reducing the strip thickness. Especially, A7075 aluminium alloy, a heat-treatable alloy, has particularly high strength and is expected to be used for automobile frame parts.

Forming of heat-treated aluminium alloys under heated conditions has been investigated, and hot stamping (HFQ® (Hot Form Quench)) with die quenching has been developed [1]. For HFQ, the blank is heated to above the solution treatment temperature in the furnace and is stamped and simultaneously held at the bottom dead centre for quenching of solution heat treatment. Heating not only improves the formability but also eliminates the solution heat treatment process after forming. However, artificial ageing after stamping is required for precipitation hardening and is reduces a productivity.

To omit not only solution heat treatment but also artificial ageing after forming, warm hot stamping using quick heating was developed [2]. Although high temperature or long heating reduces the strength of age-hardened aluminium alloy by over ageing and re-solid solution, the precipitation hardening is maintained by quick heating such as contact or Joule heating.

Since even cold stamping of aluminium alloys easily causes adhesion, friction is more essential in hot stamping of aluminium alloy. For high-temperature hot stamping of aluminium alloys such as HFQ, the properties of solid lubricants and tool coatings have been investigated [3]. For warm hot-stamping, liquid lubricants can be used because the heating temperature is around 200°C. In addition, a combination of surface textures of tools and lubricants can also be considered. Spherical graphite cast iron has been confirmed to exhibit friction-reducing effects using liquid lubricants [4]. In addition, surface dimples have been confirmed to function as micro reservoirs for lubricant retention using liquid lubricants [5].

In this study, we investigated the frictional properties of A7075-T6 aluminium alloy in a drawing test using quick contact heating at warm temperatures. Under warm (200 to 300°C) dry conditions, the effects of heating temperature and heating time on friction properties were researched. Under the same conditions, the friction properties of tool surface coatings were also researched. 200°C liquid lubricant was used to investigate the friction properties of spheroidal graphite cast iron and a grooved polishing proposed as a simple surface texturing method.

Warm and hot flat strip drawing test using contact heating and tool polishing

Fig. 1 shows the sequence of the aluminium alloy flat strip drawing test with quick contact heating. The system is composed of hot plates, a fixed tool, and a movable tool by a pneumatic cylinder and is installed in a universal testing machine. The shoes imitate a stamping tool. The A7075-T6 aluminium alloy strips having 2 mm in thickness are cut into 20×300 mm and clamped with the chuck. The aluminium strip is sandwiched with the hot plates and is heated to target temperature for almost 30 s. The strip is ascended by immediately moving the crosshead when reaching target temperature. When the heated area of the strip reaches the shoe position, the strip is caught between shoe tools by a pneumatic cylinder and is drawn.

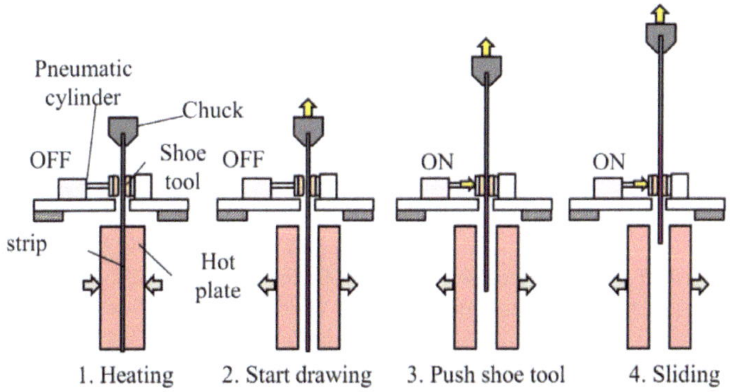

Fig. 1. Sequence of warm and hot flat strip drawing test using quick contact heating.

The drawn force is measured as frictional force. The coefficient of friction μ was determined by Eq. 1.

$$\mu = F/2N \quad (1)$$

F is a drawing load of a strip and *N* is a vertical load of the shoe by the pneumatic cylinder. The vertical load of the shoe was measured in advance by a load cell, and the universal testing machine measured the drawing load. Since the frictional force is generated on both sides of the strip, the drawing force of strip F is double the frictional force on each side.

The detailed layout of aluminium alloy strip in the warm and hot flat strip drawing testing machine is shown in Fig. 2(a). The temperature of the hot plates were set at 50 °C higher than the target temperature. The temperature of the strip was measured by a K-type thermocouple inserted into a hole on the thickness side of the strip. Since the temperature within the centre 110 mm of the hot plates is uniform, the bottom 110mm of the strip is uniformly heated to the target temperature. This part is caught between shoes, and the friction is evaluated. The shoe dimensions is illustrated in Fig. 2(b). The flat area slidden with the strip is 36×20 mm and the length in drawn direction is 20 mm. The appearance of shoe is shown in Fig. 2(c).

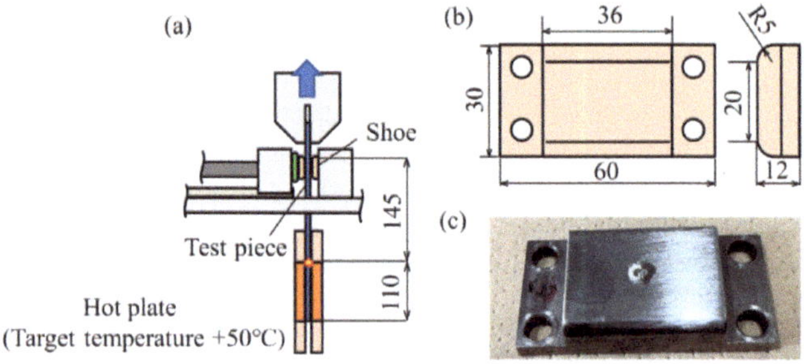

Fig. 2. (a) Detailed layout of aluminium alloy strip in warm and hot flat strip drawing testing machine, (b) shoe dimensions and (c) used shoe.

Table 1 shows the test conditions for the hot flat strip drawing test. The shoe materials used were SKD61 (hot work tool steel in JIS) and FCD600 (spheroidal graphite cast iron in JIS). For SKD61, tests were conducted with and without surface coatings. The coatings were AlSiCrN, vanadium carbide (VC), and DLC. The AlSiCrN and DLC were formed by the PVD process, VC was formed by thermo-reactive deposition and diffusion process. The coatings were processed after miller polishing and heat treatment for SKD material shoe.

The "conventional polishing" is composed of rough polishing (#600), intermediate polishing (#1000), and finish polishing (#3000). In contrast, the intermediate polishing is omitted for "grooved polishing". To keep the lubricant, grooves are formed with rough polishing and remained by omission of intermediate polishing. Finish polishing smooths the sharp surface edge. The grooved polishing was performed only for FCD600 shoe. Shoe was polished 90° angle to drawing direction. Experiments were investigated mainly at heating temperatures of 200°C. To compare the heating effects of furnace heating and quick heating, 30 s and 600 s were used. The average contact pressure between the shoe and the strip is 0.75 MPa. The liquid lubricant was an insoluble oil with a kinematic viscosity of 40 mm^2/s and was applied with a micro-pipette, the film thickness to be 1.5 μm. The drawing distance after the catching with the shoe tools was 80 to 96 mm, and drawing speed is 8.3 mm/s.

Table 1. Experimental conditions for hot flat strip drawing test.

Parameter	Value
Die Materials	SKD61(tool steel) FCD600(ductile cast iron)
Die Coatings (for SKD61)	None, AlSiCrN, VC, DLC
Polishing method	•conventional polishing • grooved polishing
Polishing direction [°] (0° is parallel to the drawing direction)	90 (0, 45)
Heating temprature T [°C]	200, 300
Heating time [s]	20 - 30, 600
Average contact surface pressure [Mpa]	0.75
Drawing distance s[mm]	88±8
Lubricant film thickness [μm]	1.5
Drawing speed [mm/s]	8.33

Fig. 3(a) and (b) show the tool surface profiles of only the rough polished and the grooved polished ones, respectively. In the case of grooved polishing, the plateau area is flatter, and the corners of that are rounded compared to that of only rough polishing. This surface shape is less scratching against to aluminium alloy sheet during friction.

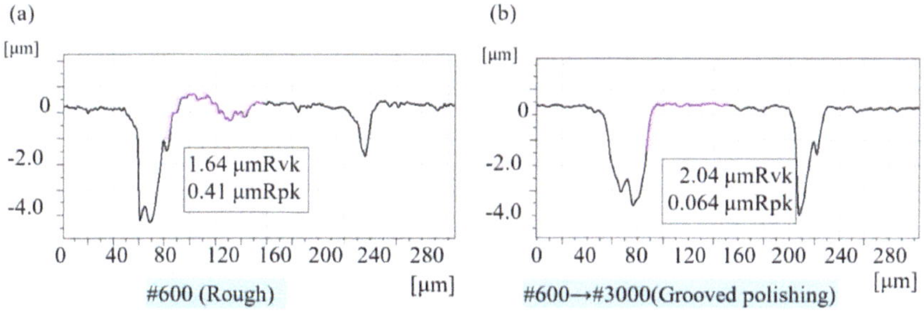

Fig. 3. Surface profiles of (a) only rough polished and (b) grooved polished tools.

Influence of heating temperature and time on frictional behaviour

Aluminium alloy flat strip drawing tests were conducted using quick contact heating at heating temperatures of 200°C, and 300°C and a heating time of about 20-30 s. For comparison, contact heating at 200°C for 600 s is also performed, which simulates long-time heating in a furnace. Fig. 4(a) shows the relationship between an average coefficient of friction and an average contact pressure of shoe at $T = 200$°C and 300°C for quick heating. The friction at 200°C is approximately half that of 300°C, and a difference of only 100°C significantly reduces friction. Fig. 4(b) shows the friction coefficient-stroke curves for different heating time for flat strip drawing test of contact heated aluminium alloy strip for dry condition at 200°C. The coefficient of friction increased with heating time, indicating that quick heating is better for friction. This is thought that during quick heating, the slight amount of pollution such as oil remaining on the surface of the strip acted as a lubricant.

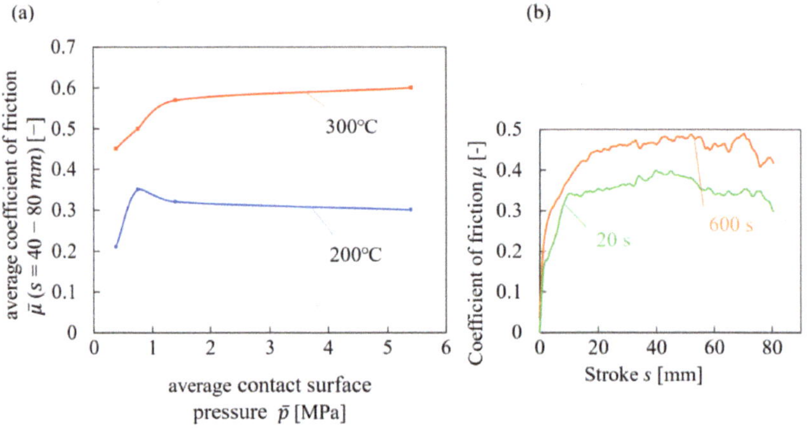

Fig. 4. (a) Relationship between average coefficient of friction and average contact pressure of shoe in flat strip drawing test of aluminium alloy strip for dry condition, (b) Friction coefficient-stroke curve for different heating time for flat strip drawing test of contact heated aluminium alloy strip of $T = 200$ °C for dry condition.

Influence of tool surface coating on friction behaviour

Fig. 5 shows the friction coefficient-stroke curves obtained by the flat strip drawing test using coated shoe for dry condition. Each coating exhibited lower friction in the initial stage of drawing than the without coating. However, the friction coefficients of AlSiCrN and VC increased with an increase in stroke, and AlSiCrN reached to that of no coating at $s = 80$ mm. This is thought to be because the Al contained in AlSiCrN easily reacts with the strip specimen, causing adhesion to occur. DLC exhibited stable low friction despite warm sliding.

Fig. 6 shows the seizure on the surface of each shoe. The degree of seizure is severe for the case with no coating. For AlSiCrN and VC, seizures are also observed, although more moderate than without coating. There is no seizure for DLC-coated shoe.

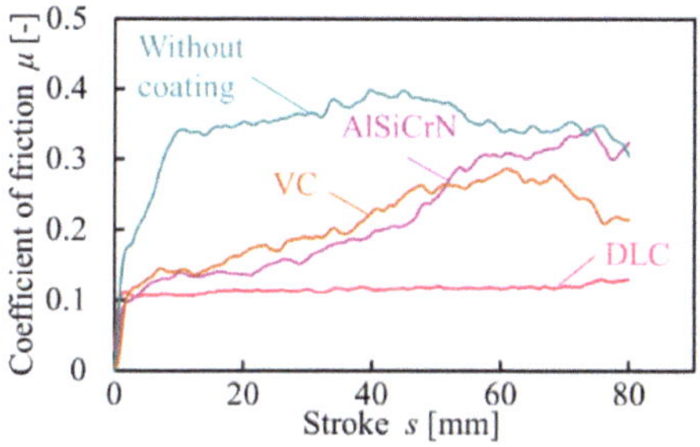

Fig. 5. Friction coefficient-stroke curve for different tool surface coatings for flat strip drawing test of contact heated aluminium alloy strip of T = 200°C for dry condition.

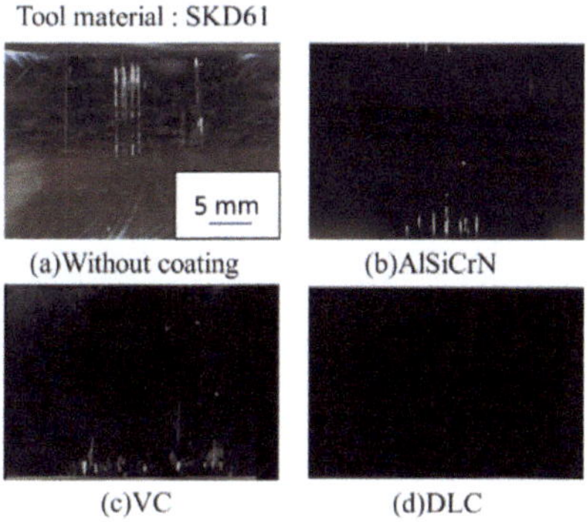

Fig. 6. Surface appearances of tools after flat strip drawing test of contact heated aluminium alloy strip of T = 200 °C for dry condition.

Influence of tool material and grooved polishing of shoe surface on friction behaviour

Although the DLC coating has low friction and prevents seizures, the coating process costs. Applying spheroidal graphite cast iron to warm hot-stamping of aluminium alloy is attractive as an inexpensive tool. A reduction of friction by the spheroidal graphite as a lubricant and holding of liquid lubricant at the graphite escaped pit are expected. Fig. 7 shows the friction coefficient-stroke curves of an aluminium alloy strip drawing test using quick contact heating with a conventional polished SKD61, a conventional polished FCD600 and a grooved polished FCD600 shoe using a liquid lubricant. Both SKD61 and FCD600 showed low friction in the initial stage of drawing. However, the friction coefficient for conventionally polished SKD61 and FCD600 increased at $s = 20$ mm and $s = 40$ mm, respectively. When conventional polished, FCD600 showed lower friction for a longer distance than SKD61 when liquid lubricant was used, but both were finally seized.

The friction coefficient for groove polishing of FCD600 is constant, equivalent to that of DLC for dry. Fig. 8 shows that no seizure was observed for the groove-polished FCD600, whereas seizure occurred when the FCD600 was conventionally polished.

Materials Research Proceedings 44 (2024) 456-464 https://doi.org/10.21741/9781644903254-49

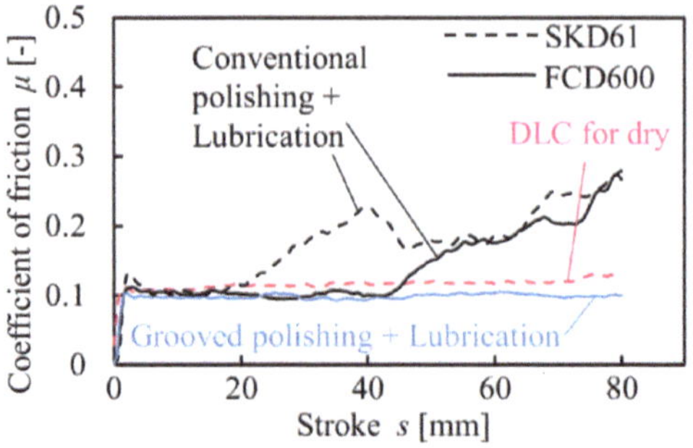

Fig. 7. Friction coefficient-stroke curves for conventionally polished SKD61 and FCD600 and grooved polished FCD600 in flat strip drawing test at T = 200°C with liquid lubricant.

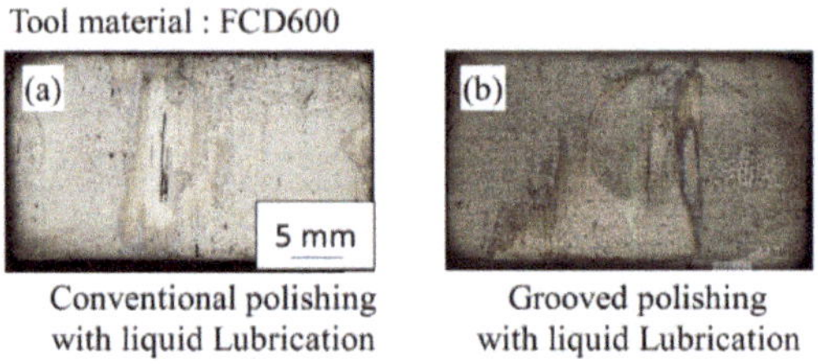

Fig. 8. Surface appearances of the FCD600 shoes of conventionally polished and grooved polished, after flat strip drawing test at T = 200°C with liquid lubricant.

Rvk of the surface roughness parameter was used to represent the value of the valley's depth on the shoe surface. The relationship between an area of adhesion on the strip and Rvk of the shoe is shown in Fig. 9. The area of adhesion decreased with increasing Rvk, indicating that seizure scratches did not occur when Rvk ≈ 2 μm. This suggests that for a given drawing distance, Rvk represents a threshold value for the occurrence of a seizure.

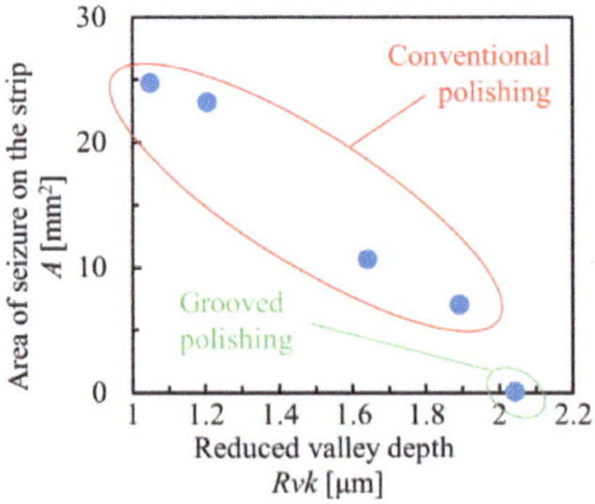

Fig. 9. Area of adhesion on aluminium strip for different Rvk with liquid lubricant obtained by flat strip drawing test using FCD600 shoe for T = 200°C.

Influence of groove polishing direction on friction behaviour

It has been shown that the combination of groove-polished FCD600 of spheroidal graphite cast iron and liquid lubricant keeps low friction and prevents the occurrence of seizures. Next, the optimal polishing direction was investigated. Fig. 10 shows the coefficient of friction-stroke curves for a spheroidal graphite cast iron tool surface polished in three different directions. The difference in the polishing direction on the friction coefficient is slight. Whereas the effect on the occurrence of seizures is observed. Although no seizure was observed at 90°, as the polishing direction was closer to the direction parallel to the drawing direction, the stroke without seizure became shorter.

We expect the phenomenon shown in Fig. 11 to occur. The left figure shows the case where the polishing direction of the tool surface is orthogonal to the drawing direction, and the right figure shows the case where it is parallel to the drawing direction. In the case of drawing direction and groove's direction are orthogonal, the liquid lubricant accumulated in the groove leaches into the boundary lubrication area as the strips draw, preventing contact between the material and the tool. On the other hand, if the grooves are parallel, the liquid lubricant does not leach into the boundary lubrication area, and tool and material contact at the boundary lubrication area, resulting in a seizure.

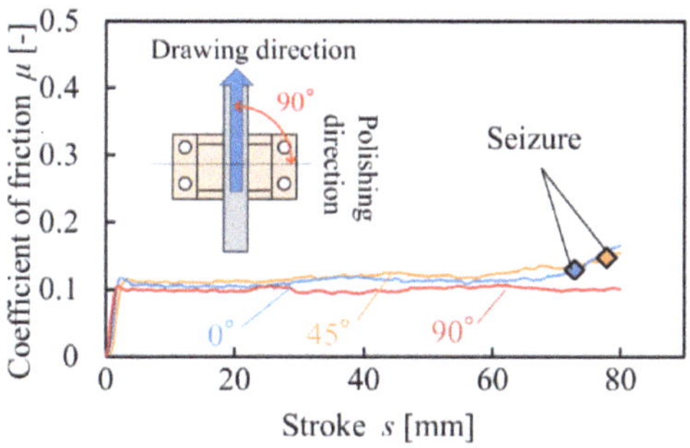

Fig. 10. Friction coefficient-stroke curves for different Grooved polishing directions of ductile cast iron tool obtained by flat strip drawing test.

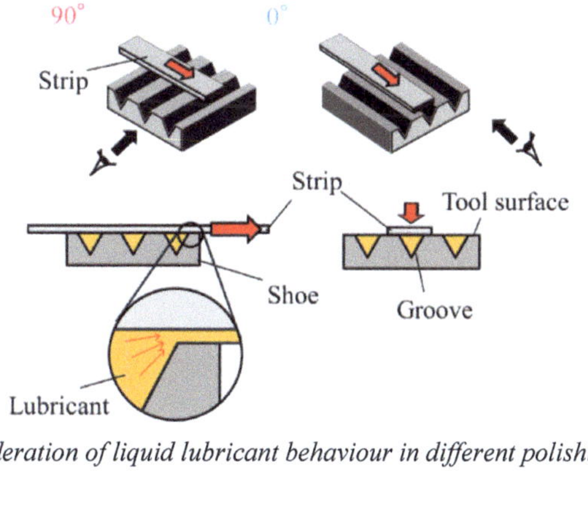

Fig. 11. Consideration of liquid lubricant behaviour in different polishing directions.

Summary
A strip drawing test was conducted under warm conditions to investigate the frictional properties of an age-hardened A7075 aluminium alloy strip during warm heating using quick contact heating. The following results were obtained.

Friction for quick heating at 200°C becomes mild compared with the high-temperature and long heating time. Therefore, warm stamping using quick heating has the advantage of friction against to hot stamping including solution heat treatment and warm stamping with a furnace.

A DLC coating showed very stable and small friction without a seizure, even in dry conditions. The DLC coating is very suitable for hot and warm stamping of aluminium alloy although the price is high cost.

Conventional-polished ductile cast iron showed superior low friction compared to tool steel under liquid lubrication. However, it was not enough against the seizure. By combining the ductile cast iron and grooved polishing, lubricant is kept and seizure was prevented. The friction was small and stable like that of a DLC coating. A combination of ductile cast iron tool and grooved polishing is ideal for the warm forming of aluminium alloy, where liquid lubricants can be used. In addition, Rvk was also shown to be a useful parameter for expressing the frictional properties of the surface using grooved polishing.

References

[1] R.P. Garret, J. Lin, T.A. Dean, Solution heat treatment and cold die quenching in forming AA 6xxx sheet components: Feasibility study, Adv. Mater. Res. 6-8 (2005) 673-680. https://doi.org/10.4028/www.scientific.net/AMR.6-8.673

[2] T. Maeno and T. Taniguchi, Japan Patent, JP 2019-188453. (2019).

[3] Y. Dong, K. Zheng, J. Fernandez, X. Li, H. Dong, and J. Lin, Experimental investigations on hot forming of AA6082 using advanced plasma nitrocarburised and CAPVD WC: C coated tools: J. Mater. Process. Technol. 240 (2017) 190-199. https://doi.org/10.1016/j.jmatprotec.2016.09.023

[4] D. Gowda, Tribological Characterization of Centrifugally Cast Graphite Cast Iron under Dry and Wet conditions, Materials Today: Proceedings 5 (2018) 145–151. https://doi.org/10.1016/j.matpr.2017.11.065

[5] W. Tang, Y. Zhou, H. Zhu, H. Yang, The effect of surface texturing on reducing the friction and wear of steel under lubricated sliding contact, Appl. Surf. Sci. 273 (2013) 199-204. https://doi.org/10.1016/j.apsusc.2013.02.013

Materials Research Proceedings 44 (2024) 465-474 https://doi.org/10.21741/9781644903254-50

Lightweight battery enclosure design

GANGULI Tanmay[1,a*] and DATE Prashant P.[1,b]

[1]Dept. of Mech. Eng. Indian Institute of Technology, Bombay Powai, Mumbai, India

[a]tanmayganguli09@gmail.com, [b]ppdate@iitb.ac.in

Keywords: Light Weight Battery Enclosure, Taguchi Orthogonal Arrays, Design of Experiments, Optimization

Abstract: The battery box is the heaviest part of an electric vehicle and may account for 20-30 per cent of its weight. Therefore, the design of a lightweight battery box is an important step towards the lightweight design of the entire vehicle. In this paper, a systematic procedure has been followed to reduce the weight of a given battery enclosure structure, while ensuring that the structure can withstand the stresses during acceleration, braking and turning of the vehicle, and the compressive stresses on the cells inside the enclosure do not exceed the maximum limit. The important design parameters were identified, and an optimal parameter set with respect to lightweight design and structural stability was determined. The Taguchi method of orthogonal arrays was used for the analysis to optimize the calculation efforts. The stress distribution over the body of the enclosure was analyzed and material was removed from the regions having low stress concentration. Subsequently, the structure was modified and forced air cooling was introduced to ensure that the temperature rise during operation was within acceptable limits. The stress, thermal and modal analysis of the resultant structure was then carried out to demonstrate its feasibility.

Introduction

The use of batteries for various applications has become extremely important to replace conventional fossil fuels. This is specifically pertinent in the case of vehicles, where the design of batteries and the battery pack enclosure (BPE), is of great importance not just for the safety of operation of the batteries but also to reduce the weight and consequently the efficiency of the complete system. It is estimated that ~ 20-30% of the total weight of a battery electric vehicle is contributed by the BPE structure. Thus, novel designs and their implementation to reduce the weight of BPEs has been an important area of research by numerous groups in the recent past. [1,2,3]. Another approach is to combine the design novelties for weight reduction with the use of lightweight and high strength materials for the construction of the BPE structures. Qiu-Sheng Chen et al. have used Aluminum alloy in place of mild steel to design the BPEs [4]. Chen et. al have reported the use of carbon fiber reinforced plastics as battery case and Aluminum alloy as frame structure [5]. Battery operation also generates a lot of heat and hence, thermal management is also an important issue for the safe operation of batteries. Passive management techniques have been reported by Parsons et al. [6] and active cooling by the flow of forced air has also been reported [7]. Thus, in this work, an approach to reduce the weight of a given battery pack enclosure is presented, with a resulting optimized structure design that is structurally robust (shown by stress and modal analysis) that also allows for efficient heat transfer out of the BPE.

Procedure overview

We have used two main approaches towards optimizing a given structure with regards to weight reduction: topology optimization and parametric optimization. Topology optimization aims at removing material from the given body from low stress areas, while ensuring its mechanical and structural robustness to applied forces and stresses, etc. Parametric optimization on the other hand identifies certain parameters in the body design (e.g. thickness of a battery box) and aims to tune those parameters to obtain the desired objective, which in this case is weight reduction. First, we identified relevant parameters and optimized them. However, the resulting battery box structure was found to be thermally infeasible as the temperature of the cells was found to reach very high values in a steady state thermal analysis. Thus, the original structure had to be modified to account for a thermal management system. The battery box structure was suitably modified to allow for air cooling and parametric optimization was repeated. The feasibility of the new battery box structure both mechanically and thermally was checked through simulations. Once the structure was finalized, low stress regions in the enclosure were identified by inspection of the stress distribution and material was removed from those regions (topology optimization). Structural simulations were repeated to ensure mechanical robustness. Further practical feasibility was ensured through modal analysis of the BPE. In all the simulations, only the cells and enclosure were considered. All other components of the battery pack like battery management system, thermal management apparatus, etc. were not considered. All simulations were carried out using ANSYS 2024 (student) and 2016 versions. The CAD models were designed in SOLIDWORKS 2019.

Parametric Optimization of Given Structure

Initial Structure. The initial battery box structure is given in Fig. 1. The properties of the battery box and pouch cells used for the simulations are listed in Table 1. The bulk properties of pouch cells have been reported in previous works [8]. The boundary conditions applied on the battery box are summarized in Table 2. The box was subjected to an acceleration of g (gravity) in -Y (vertical) and 3g in either X or Z axis (signifies acceleration, braking, rapid turning), where the axes are as shown in Fig. 1. The bottom edges were assumed to be attached to the surface on which the box was kept, by a fixed joint. Similar boundary conditions are considered by other authors [4]. Five parameters of the battery box design were identified: enclosure wall thickness, cell gap, cell-separator wall height, middle partition thickness and base plate thickness, which were optimized in this work.

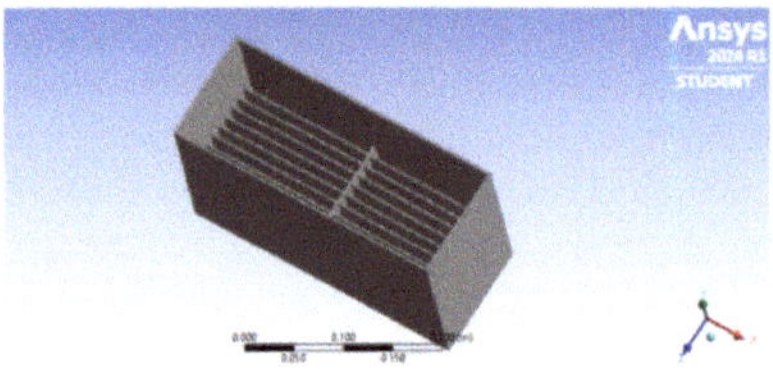

Fig. 1. Initial battery box structure.

Table 1. Properties of the initial battery box and Properties of each pouch cell.

Mass	7.67 Kg
Material	Structural steel
Size	20 x 2 cells
Enclosure thickness	2 mm

Size	200 x150 x10 mm^3
Mass	500 g
Eff. Young's modulus	4.2 GPa
Poisson's ratio	0

Table 2. Boundary conditions.

Accelerations considered in the calculations	a) 30 m/s^2 along X and -10 m/s^2 along -Y b) 30 m/s^2 along Z and -10 m/s^2 along -Y
Support	Bottom 4 edges fixed
Displacement	No Y displacement

Taguchi method of Design of Experiments

Techniques like multiobjective genetic algorithm (MOGA), particle swarm optimization (PSA) etc. have been used for parametric optimization [9]. A simpler approach is to consider various ‘levels’ of each parameter and try out different combinations to reach an optimal parameter set. A full factorial design method where the complete parameter space is explored is extremely time consuming. A faster way to design experiments is by using orthogonal arrays [10]. For our case, three levels for the five parameters were identified (Table 3). Thus, an L-18 Taguchi orthogonal array can be constructed, reducing the number of experiments needed from 3^5 to just 18 [11]. For each case, the mass of the box, maximum and average stress, and maximum and average deformation for both acceleration directions were tabulated (Table 4).

Table 3. Optimization parameters and their levels.

Enclosure thickness (a)	2 mm (1)	1 mm (2)	0.5 mm (3)
Base thickness (b)	2 mm (1)	1 mm (2)	0.5 mm (3)
Cell gap (c)	2 mm (1)	1 mm (2)	0.5 mm (3)
Middle partition thickness (d)	2 mm (1)	1 mm (2)	0.5 mm (3)
Cell separator height (e)	100 mm (1)	80 mm (2)	50 mm (3)

Results

We find that the maximum stress and deformation values for all the simulations are within acceptable bounds, much lower than the yield strength of structural steel, which is around 240 MPa. Experiment 16 gives the lightest battery box weight (1.7 kg). The results for all the combinations are summarized in Table 4. The stress distribution in the structure for 4 cases, namely: a) Stress distribution in the original battery box for acceleration in +Z direction, b) Stress distribution for acceleration in +X direction, c) Stress distribution for Experiment 16, acceleration in + X direction, and d) Stress distribution for Experiment 16, acceleration in +Z direction; are shown in Figs. 2a, 2b, 2c and 2d respectively. Thus, we can get the optimal lightweight structure from the combination of these parameters. This weight is around 1.7 kg, thereby reducing the weight by over 80% from the initial assumed structure.

Table 4. The summary of the results of max stress, max deformation and mass for each factor-level combination of the L-18 orthogonal array.

Expt. No	t_ side(a)	t_ bottom(b)	t_ cell(c)	t_ cent(d)	h_ part(e)	max stress (MPa) (x accl, z accl)	max deform (mm) (x , z)	Mass (kg)
1	2	2	2	2	100	3.27,5.52	0.017,0.043	7.667
2	2	1	1	1	80	4.68,18.93	0.02,0.085	4.63
3	2	0.5	0.5	0.5	50	9.18,37.6	0.057,0.18	3.245
4	1	2	2	1	80	4.81,13.42	0.023,0.08	5.3
5	1	1	1	0.5	50	12.98,43	0.06,0.183	2.64
6	1	0.5	0.5	2	100	5.87,7.75	0.014,0.06	2.63
7	0.5	2	1	2	50	8,9.43	0.055,0.18	2.31
8	0.5	1	0.5	1	100	6.5,10.76	0.017,0.05	2.03
9	0.5	0.5	2	0.5	50	7.1,27.64	0.056,0.17	2.94
10	2	2	0.5	0.5	80	8.44,26.12	0.23,0.084	3.97
11	2	1	2	2	50	5.46,9.54	0.056,0.16	5.17
12	2	0.5	1	1	100	3.15,13.76	0.016,0.046	4.94
13	1	2	1	0.5	100	3.7,21.32	0.017,0.048	3.98
14	1	1	0.5	2	80	9.2,13,82	0.022,0.085	2.51
15	1	0.5	2	1	50	8,20.14	0.057,0.165	3.63
16	0.5	2	0.5	1	50	12.16,25.46	0.058,0.18	1.7
17	0.5	1	2	0.5	100	3.8,15.47	0.018,0.047	5.23
18	0.5	0.5	1	0.5	80	5.23,20.2	0.023,0.087	2.5

Thermal simulations

Apart from the structural simulations, the thermal feasibility of this structure also must be checked. It is unlikely that conduction along with natural convection in the surrounding air will be sufficient to keep the cells within the desired temperature limit. The desired temperature limit of a pouch cell is 40-50°C [12]. Pouch cell thermal properties such as specific heat, thermal conductivity and heat generation rate have been reported [13]. The parameters used in our simulations are summarized in Table 5. Steady state thermal analysis was performed, with only natural convection in the ambient air and conduction into the battery enclosure as the modes of heat dissipation. It was observed that temperatures of the cells became too high for the original design (Fig. 1), as shown in Fig. 3, and the simulation failed to converge for many of the cases with thinner cell separation. Obviously, the structure by itself cannot maintain the cell temperatures in the desired range. Thus, thermal management is required, along with weight reduction.

Fig. 2a. Stress distribution of original battery box for acceleration in +Z direction.

Fig. 2b. Stress distribution for acceleration in +X direction.

Materials Research Proceedings 44 (2024) 465-474 https://doi.org/10.21741/9781644903254-50

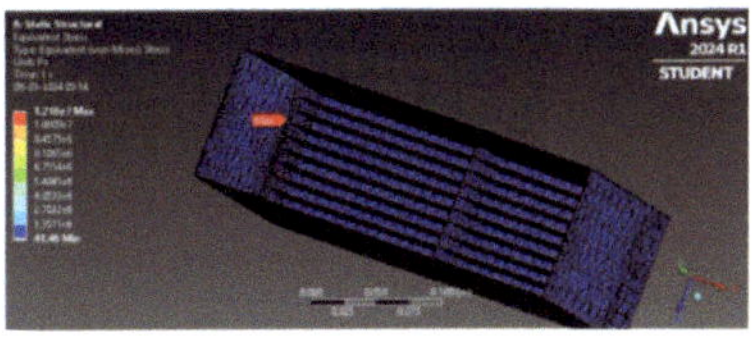

Fig. 2c. Stress distribution for Expt. 16, acceleration in +X direction.

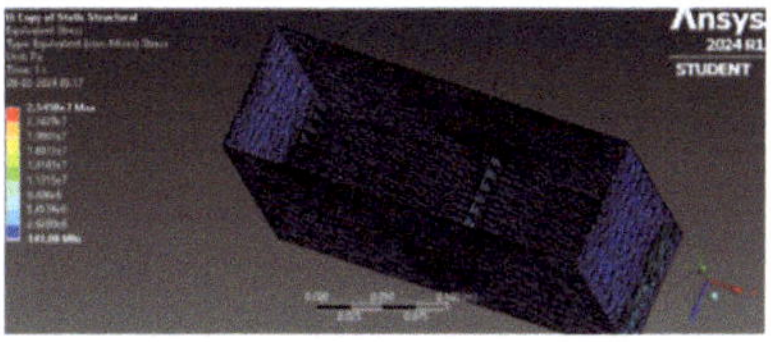

Fig. 2d. Stress distribution for Expt.16, acceleration in +Z direction.

Table 5. Properties of pouch cells used in thermal simulations.

Specific heat	550 J/kg-K
Thermal conductivity	X, Z : 25 W/m-K; Y : 0.5 W/m-K
Heat generation rate	5000 W/m^3

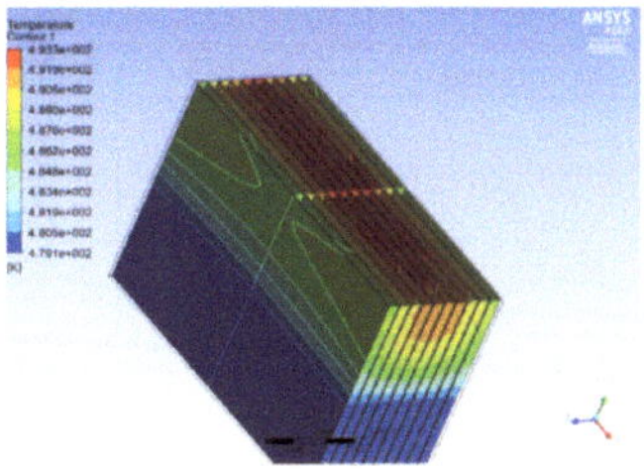

Fig. 3. Temperature distribution for the original battery enclosure without thermal management.

Thermally feasible design

New design. One possible change in design considered is to have a gap between two cells and the system being cooled by blowing air through the gap. Thus, the design of the box shown in Fig. 1 was modified as shown in Fig. 4.

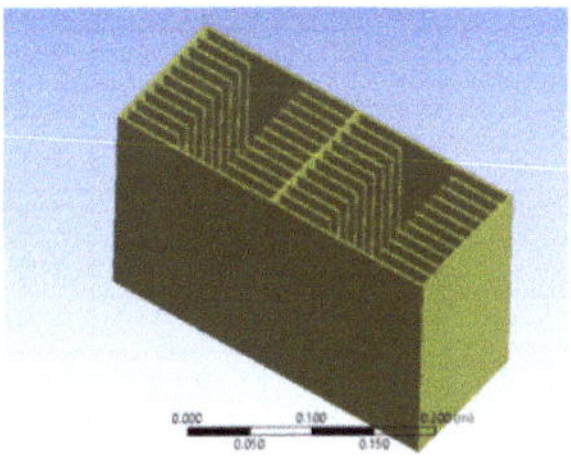

Fig. 4a. Modified battery box design.

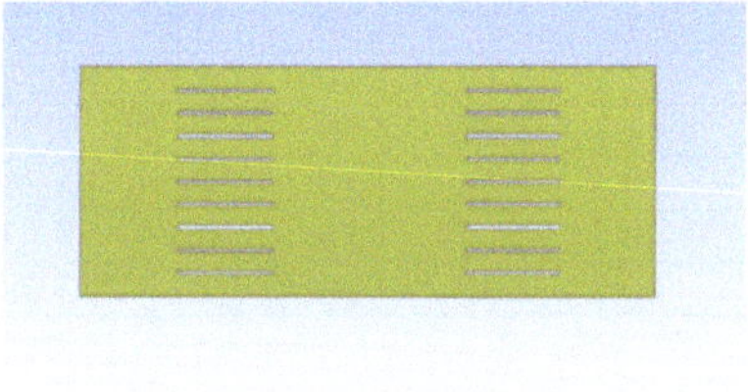

Fig 4b. Gaps for air flow.

Steady state thermal analysis of the structure in Fig. 4 (performed on a small group of 4 cells due to computation limitations) shows a maximum temperature of about 41°C at an inlet air velocity of 1 m/s (Fig. 5). It is observed that the steady state temperature predicted by the simulation is within acceptable limits, and the air inlet speed required is also not so high as to demand huge amount of power from the battery pack.

Materials Research Proceedings 44 (2024) 465-474 https://doi.org/10.21741/9781644903254-50

Parametric Optimization. Based on the above, the parameters and their levels for the new design are summarized in Table 6. Considering four factors with three levels each, we have an L-9 orthogonal array. The previous analysis for the two acceleration conditions was repeated and Table 7 is populated accordingly.

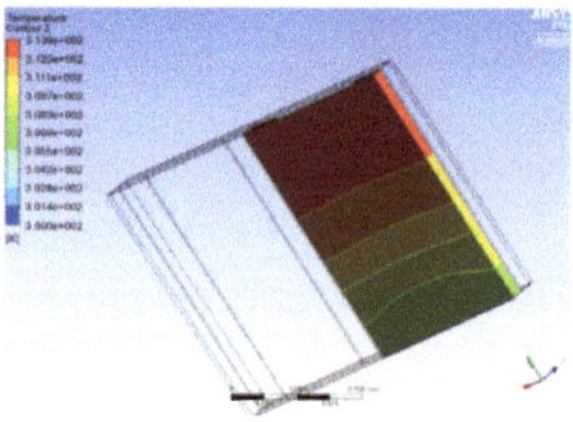

Fig. 5. Temperature profile of the cell.

Table 6. Parameters and their levels for the second optimization.

Enclosure thickness (mm)(a)	2 (1)	1 (2)	0.5 (3)
Cell gap(mm) (b)	2 (1)	1 (2)	0.5 (3)
Air flow space width(mm) (c)	50 (1)	30 (2)	10 (3)
Middle partition thickness (mm) (d)	2 (1)	1 (2)	0.5 (3)

Table 7. Mass, stress and deformation values.

Exp no.	a (mm)	b (mm)	c (mm)	d (mm)	max stress (MPa) (Z, X-accn)	max def. (mm) (Z, X-accn)	Mass (kg)
1	2	2	50	2	2.22, 2.17	0.0044, 0.0055	8.96
2	2	1	30	1	2.88, 5.1	0.0064, 0.0128	4.72
3	2	0.5	10	0.5	5.94, 10.042	0.0125, 0.032	3.183
4	1	2	30	0.5	2.58, 9.167	0.0087, 0.0349	2.45
5	1	1	10	2	2.73, 9.326	0.0087, 0.0346	2.45
6	1	0.5	50	1	2.97, 5.47	0.0092, 0.0095	3.1
7	0.5	2	10	1	3.36, 6.34	0.0078, 0.0287	2.22
8	0.5	1	50	0.5	3.63, 6.04	0.0076, 0.0078	3.8
9	0.5	0.5	30	2	3.08, 6.34	0.0078, 0.015	2.06

We find that maximum stress and deformation values are again within acceptable bounds, in fact resulting in a better performance than the original design. This shows that the new design (Fig. 4) is also structurally more robust.

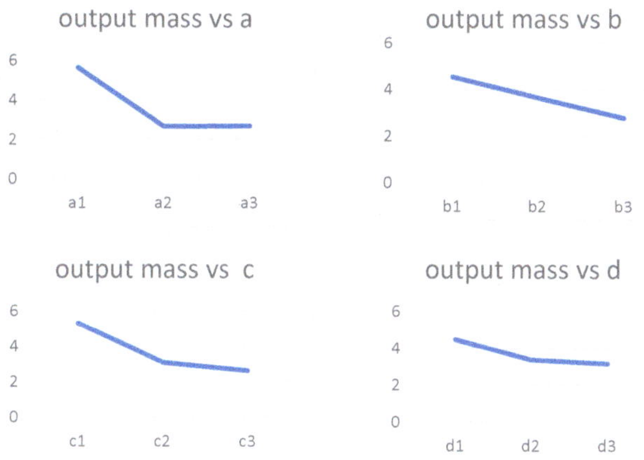

Fig. 6. Variation of mass with each parameter separately. The parameter details are indicated in Table 6.

Thermal analysis. Using the graphs (Fig. 6), we can find the optimal parameter set and hence the structurally optimal battery box design, with the least weight. Steady state thermal analysis of this design was performed with the same cell parameters as in Table 4. Again only 4 adjacent cells were simulated to reduce computational cost. This approach is expected to be valid as the significant heat flow is expected to be by the flow of air which is perpendicular to the axis of the stacked cells. It is seen that with an air inlet velocity of 5 m/s, the maximum cell temperature is around 47°C (Fig. 7). Thus, the temperature of the entire 20 X 2 pack can also be managed, with a similar air inlet between each cell. This proves that this structure is indeed feasible both with respect to mechanical integrity and thermal management.

Topology Optimization. Once the parametric optimization is complete, topology optimization can give a further weight reduction. It can be observed that the regions near the top of the enclosure wall experience little to no stresses and are hence suitable candidates for mass removal. Further removal of material gives the structure in Fig. 8, with a final weight of 1.48 kg. Since we are just removing some material from the enclosure, we do not compromise on thermal management, instead the cells near the edge will be cooler than before as they are also exposed to natural convection from the sides where material has been removed.

Materials Research Proceedings 44 (2024) 465-474 https://doi.org/10.21741/9781644903254-50

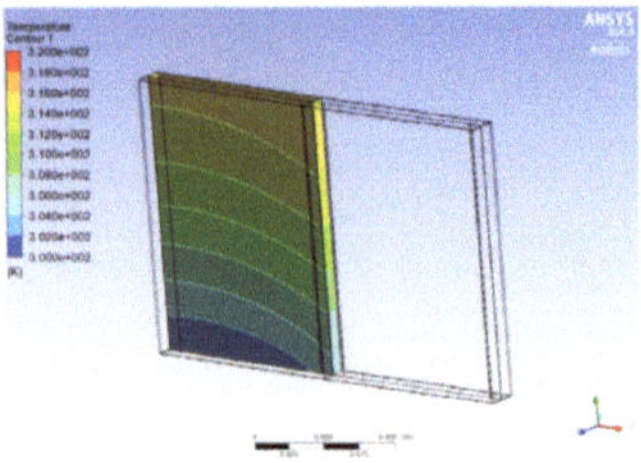

Fig. 7: Temperature profile of the cell in the final optimized structure

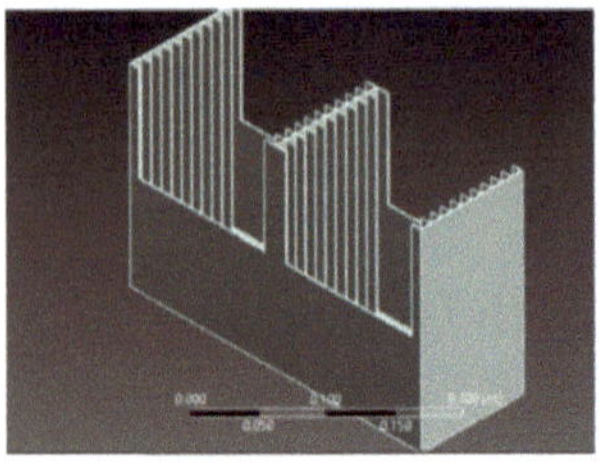

Fig. 8: Structure after topology and parametric optimization

Stress Analysis. Structural simulations were performed on this structure and maximum stresses and deformations on the battery enclosure are found to be within acceptable bounds, as shown in Fig. 9. The stresses on the individual cells are also found to be negligible, as shown in Fig. 10. Cells and enclosure are part of the same simulation, with the cells hidden in Fig. 9 and the enclosure and other cells hidden in Fig. 10.

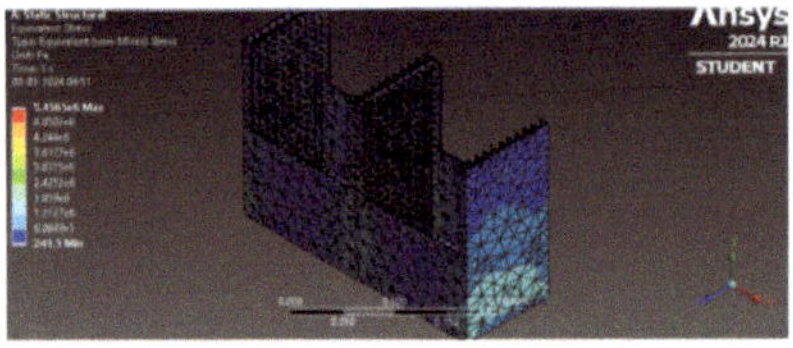

Fig. 9a. Stress dist. in final struct. for Z accel.

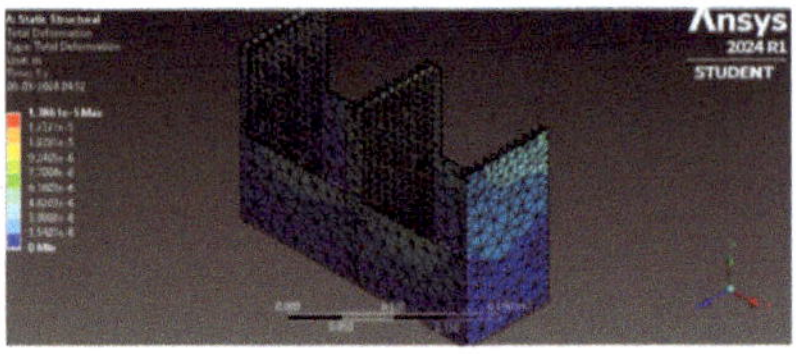

Fig. 9b. Deformation distribution for Z accel.

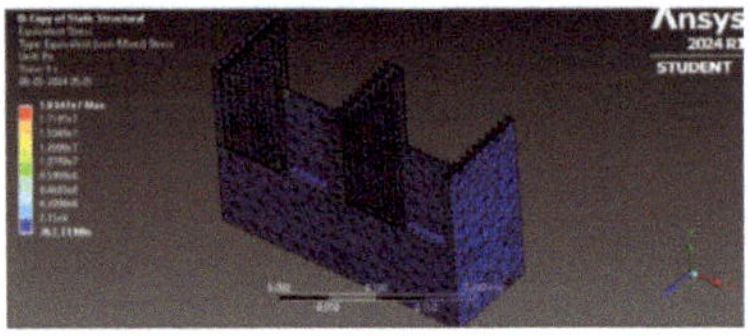

Fig. 9c. Stress distribution for X accel.

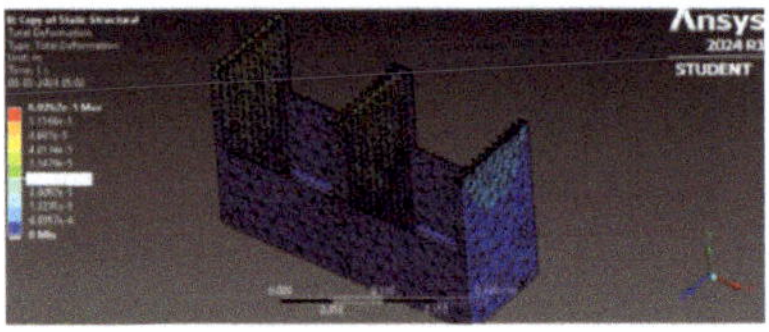

Fig. 9d. Deformation distribution for Z accel.

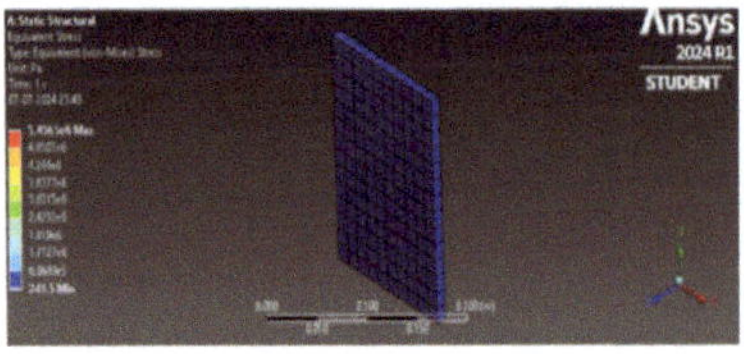

Fig. 10a. Stress Dist. for a cell, Z accel.

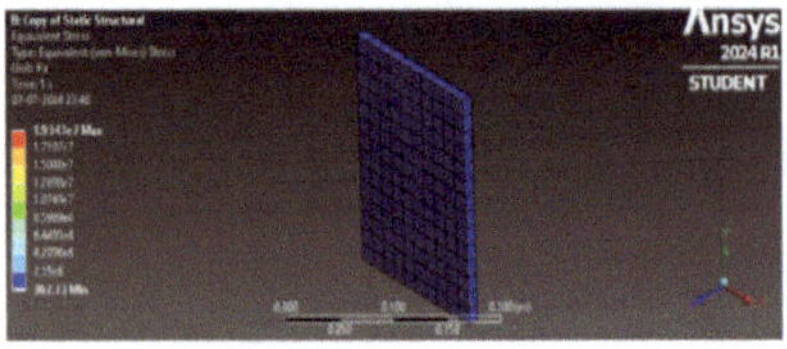

Fig. 10b. Stress Dist. for a cell, X accel.

Mode	Frequency (Hz)
1	359.94
2	374.4
3	391.94
4	614.67
5	691.11

Table 8. Frequency modes of battery box.

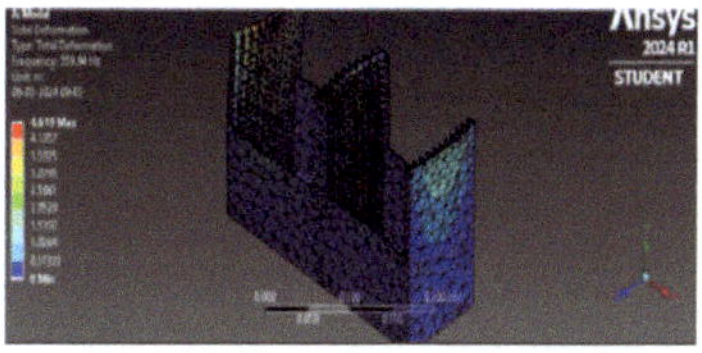

Fig. 11. First frequency mode

Modal Analysis. Another important aspect of structural design is modal analysis. The battery box and cells will be exposed to vibrations as the vehicle moves on the ground. The frequency modes of the battery box should be higher than the normal frequencies of vibration during the vehicle's motion to avoid resonance. Based on previous reports, we want the frequency modes to be higher than ~ 80 Hz [14]. The first 5 frequency modes for the battery box are tabulated in Table 8, and we find that the modes are much higher than the lower frequencies encountered during driving.

Conclusions and Future work

The given battery enclosure structure was optimized, to make it lightweight and at the same time able to resist forces during braking, acceleration and turning. The structure was also shown to be thermally feasible and temperature of the cells during steady state was found to be within their functional range. In all the simulations, the stresses on the cells were negligible, ensuring that the cells were not damaged. In the process, weight reduction from 7.67 kg to 1.48 kg was achieved, a reduction by almost 80%. Thus, significant reduction in weight could be achieved by first identifying and optimizing relevant parameters of the structure and later by topology optimization. The model above can be made more realistic by considering the exact properties of different parts of the pouch cells (electrodes, electrolyte, etc.), instead of volume averaged properties, as used here. The cells are assumed to be nearly rigid, thus can be constrained by two short separators (Fig. 8). A more realistic model will encompass the non-rigidity of the cell. The boundary conditions of the battery enclosure can also be modified to nuts/bolts at specific locations, instead of fixed joints along the entire edge. The methodology presented in this paper can be used as per the system under consideration to make the system lightweight and reliable. Further lightweighting of battery packs is proposed to be attempted in the future by material substitution for some or all portions.

References

[1] Y. Chen, G Liu, Z.Y. Zhang, S.J. Hou, Integrated design technique for materials and structures of vehicle body under crash safety considerations, Struct. Multidiscip. O. 56 (2017) 455. https://doi.org/10.1007/s00158-017-1674-8

[2] L. Chen, X.Y. Zhao, Proc. 2019 Int. Conf. Robotics, Intelligent Control and Artificial intelligence, Association for Computing Machinery, Shanghai, China.

[3] Y. Pan, Y. Xiong, L. Wu, K. Diao, W. Guo, Lightweight Design of an Automotive Battery-Pack Enclosure via Advanced High-Strength Steels and Size Optimization, Int. J. Automot. Tech. 22 (2021) 1279. https://doi.org/10.1007/s12239-021-0112-5

[4] Q.-S. Chen, H. Zhao, L.-X. Kong, K.-W. Chen, Research on Battery Box Lightweight Based on Material Replacement, Adv. Eng. Res. 141 (2017) 346. Proc. of 5th Int. Conf. on Mechatronics, Materials, Chemistry and Computer Engg (ICMMCCE 2017). https://doi.org/10.2991/icmmcce-17.2017.72

[5] X. Chen, M. Li, S. Li, J. Jin, C. Zhang, Machine and Deep Learning for Digital Twin Networks: A Survey, Society of Automotive Engineers (SAE)-China Cong., Singapore. https://doi.org/10.1109/JIOT.2024.3416733

[6] K.K. Parsons, T.J. Mackin, Design and Simulation of Passive Thermal Management System for Lithium-Ion Battery Packs on an Unmanned Ground Vehicle, J. Therm. Sci. Eng. Appl. 9 (2017) 011012. https://doi.org/10.1115/1.4034904

[7] H. Sun, R. Dixon, Development of cooling strategy for an air cooled lithium-ion battery pack, J. Power Source. 272 (2014) 404-414. https://doi.org/10.1016/j.jpowsour.2014.08.107

[8] J. Lian, M. Koch, W. Li, T. Wierzbicki, J. Zhu, Mechanical Deformation of Lithium-Ion Pouch Cells under in-plane Loads—Part II: Computational Modeling, J. Electrochem. Soc. 167 (2020) 090556. https://doi.org/10.1149/1945-7111/ab9eee

[9] L. Cheng, A. Garg, A.K. Jishnu, L. Gao, Surrogate based multi-objective design optimization of lithium-ion battery air-cooled system in electric vehicles, J. Energy Stor. 31 (2020) 101645. https://doi.org/10.1016/j.est.2020.101645

[10] Ranjit Roy, A Primer on the Taguchi Method, Society of Manufacturing Engineers, ISBN 13: 978-0-87263-864-8

[11] https://www.me.psu.edu/cimbala/me345/Lectures/Taguchi_orthogonal_arrays.pdf

[12] D. Agwu Daberechi, F.K. Opara, N. Chukwuchekwa, D.O. Dike, L. Uzoechi, 2017 IEEE 3rd Int. Conf. on Electro-Tech. for Nat. Dev. (NIGERCON)

[13] G. Vertiz, M. Oyarbide, H. Macicior, O. Miguel, I. Cantero, P. Fernandez de Arroiabe, I. Ulacia, Thermal characterization of large size lithium-ion pouch cell based on 1d electro-thermal model, J. Power Source. 272 (2014) 476. https://doi.org/10.1016/j.jpowsour.2014.08.092

[14] M. Hartmann, M. Roschitz, Z. Khalil, Enhanced Battery Pack for Electric Vehicle: Noise Reduction and Increased Stiffness, Mater. Sci. Forum 765 (2013) 818. https://doi.org/10.4028/www.scientific.net/MSF.765.818

Materials Research Proceedings 44 (2024) 475-481 https://doi.org/10.21741/9781644903254-51

Residual stress and part distortion prediction in L-PBF of Ti-6Al-4V using layer-by-layer FEM simulation

POLLARA Gaetano[1,a*], PALMERI Dina[1,b], BUFFA Gianluca[1,c] and FRATINI Livan[1,d]

[1]Dipartimento di Ingegneria, Università Degli Studi di Palermo, Viale delle Scienze, Palermo, 90128, Italy

[a]gaetano.pollara@unipa.it, [b]dina.palmeri@unipa.it, [c]gianluca.buffa@unipa.it, [d]livan.fratini@unipa.it

Keywords: Numerical Simulation, Laser Powder Bed Fusion, Ti-6Al-4V

Abstract. Due to its ability to accommodate customer demands and produce objects with complex shapes, Laser Powder Bed Fusion (LPBF) has been widely adopted in numerous industry areas, including biomedical, automotive, and aerospace. Even with all the benefits that LPBF has to offer, its use may be limited by the development of residual stress according to the strong thermal gradients produced throughout the process. Residual stresses within the samples can result in part distortion after the removal from the built platform or even in part failure during the process if the residual stresses are excessive. In order to save time and costs, numerical simulation can be an effective tool to predict residual stress and part distortion in opposition to the trial-and-error approach which involves an expansive and time-consuming experimental campaign. To this aim a finite element method (FEM) together with a layer-by-layer approach was used in this study. Numerical simulations were performed on the commercial software DEFORM-3D™ with which different values of laser power were investigated. Moreover, the influence of the voxel mesh on the FEM model accuracy was also investigated.

Introduction

Laser power bed fusion (L-PBF) thanks to the possibility to produce complex shapes and to obtain highly customized products, is widely adopted in the aerospace, automotive, and biomedical industries [1]. L-PBF is mostly adopted in the case of metallic components, especially for the production of Ti-6Al-4V parts that can be difficult to work with traditional manufacturing technology. Moreover, Ti-6Al-4V is widely used in aerospace and biomedical applications thanks to its higher strength, low density, and high corrosion resistance [2]. Due to the high thermal gradients and cooling rates developed during the process, residual stresses are usually presented in the produced part at the end of the process [3]. Depending on the amount of residual stresses generated in the L-PBF process, distortion of the part or job failure can occur due to the collision of the warped part with the recoater [4,5]. The trial-and-error approach is not suitable for this technology due to the elevated production time and materials costs [6]. For this reason, numerical simulation can help predict the outcome of the printing process in terms of residual stress and part distortion in order to save time and money [7]. There are many numerical simulation approaches depending on the scale of the problem being issued. The microscale approach can be used when the study is focused on process parameters optimization and the scope is to analyze the behavior of the melt pool. In this case, a Computational Fluid Dynamics (CFD) numerical method is preferred [8]. In the mesoscale approach, one or more hatches are deposited simultaneously. In this way, it is possible to study the effect of the scan strategy on the residual stress distribution within the layer [9]. When the aim is to analyze the macroscopic effect of the process parameters, a part-scale approach is usually adopted in which one or more layers are heated up together. This approach allows for shortening the computational cost and so the simulation time [10].

In this paper, a Finite Element Method (FEM) was used for predicting residual stress and part distortion of Ti-6Al-4V parts during the L-PBF process. In detail, the effect of laser power on residual stress and part distortion was investigated. Moreover, the influence of the voxel mesh and the number of computational layers on the model accuracy was also analyzed. The results of this study can help engineers in the design phase for the L-PBF process of Ti-6Al-4V alloy.

FEM model and experimental details

Numerical model. In this study, a Finite Element Method (FEM) was used for predicting residual stress and part distortion of Ti-6Al-4V parts during the L-PBF process. To this aim, FEM analyses were carried out through the commercial software DEFORM-3D™ v12.0 (V12.0, SFC, Columbus, OH, USA) in which a part-scale approach was adopted. In the part-scale approach, the power input is applied to the whole layer simultaneously and can be expressed as:

$$\dot{q} = \frac{\eta P}{V_{pool}} \tag{1}$$

Where η is the laser efficiency, P is the laser power [W], and V_{pool} is the heat-affected volume. In detail, V_{pool} represents the elements being activated by the heat source in the considered time step [11]. The elements are activated according to the born-dead-elements technique in which the elements affected by the heat source are identified through a search algorithm depending on the used voxel mesh. The reader can refer to [12] for further details regarding the voxel mesh. The dog bone geometry of external dimensions equal to 46 x 8 x 3 mm^3 and was designed in Autodesk Fusion 360 and the STL file was imported in DEFORM-3D™ v12.0 (Fig. 1).

Fig. 1. Sample geometry used in this study to evaluate part distortion in L-PBF.

The specimen was placed on a build plate of 80 x 80 x 25 mm^3 with which sticking conditions were applied in order to avoid detachment during the simulation of the printing process. The build chamber and the build plate temperatures were set at 40°C and 200°C, respectively. Only the heat exchange with the argon was considered in the numerical model (h_{conv} = 1 W/m2 °C) to reduce the calculation time. The thermomechanical problem was solved with a 12th Gen Intel (R) Core (TM) i9-12900 2.40 GHz processor by using a Newton-Rapson iteration coupled with a MUMPS (MUltifrontal Massively Parallel Sparse) solver. In this study, three different values of laser power were investigated by using three voxel mesh dimensions to investigate their effect on the model accuracy. Particularly, the voxel mesh dimension on the XY plane was kept constant and equal to 0.4 x 0.4 mm^2, while the dimension along the thickness of the sample was varied by using 5, 10, 15 and 20 elements for the voxel meshes v_z=5, v_z=10, v_z=15 and v_z=20 respectively. In this way, the voxel mesh along the Z-direction will result in 0.6 mm, 0.3 mm, and 0.15 mm for v_z=5, v_z=10, v_z=15 and v_z=20 respectively. As a consequence, by considering a layer thickness of 60 μm, the number of real layers in the computational one will be equal to 10 for v_z=5, 5 for v_z=10, 3.3 for v_z=15 and 2.5 for v_z=20.

Experimental Details. Numerical simulations were validated through an experimental campaign where samples with three different values of laser power (300W, 350W, and 400W) were fabricated with an SLM280HL 3D printing machine using Ti-6Al-4V spherical powder with a Gaussian size distribution of 20–63 μm and mass density of 4.43 g/cm3. During the L-PBF process of the dog bone specimens the build orientation, hatch distance (h), scan strategy (s), and layer thickness (l_t) were kept constant and equal to 0°, 100 μm, 0°, and 60 μm, respectively. The oxygen level during the printing process was maintained lower than 1% by filling the build chamber with argon. Moreover, in order to minimize the build-up of residual stress during the printing process, the platform was preheated up to 200 °C.

To validate the result from the numerical simulation, the CAD model of the dog bone specimen geometry was compared to the acquired printed geometry after the removal from the build platform. In detail, the printed geometry was acquired according to the fringe projection approach through a 3D COMET V system (Steinbichler, Neubeuern, Germany). In this way, it was possible to measure the difference between the CAD model and the acquired printed geometry in terms of the Z-direction displacement of the top right corner (Fig. 2).

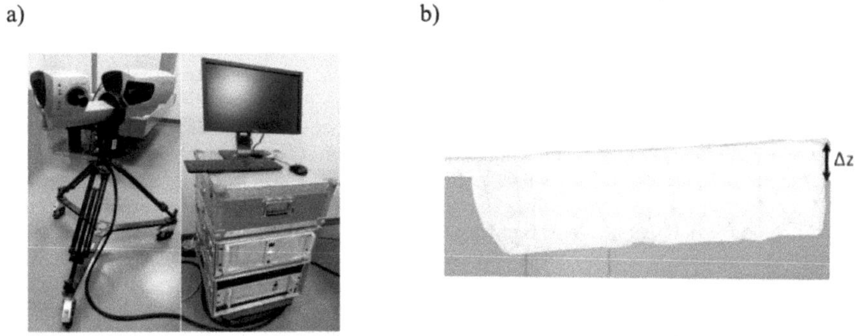

Fig. 2. Distortion measurement a) 3D acquisition system and b) comparison between the CAD model and the acquired geometry in terms of Z-displacement (Δz).

This value was compared with the Z displacement of the top right corner obtained with the numerical simulations.

Results and discussion

The accuracy of the numerical model, by varying the voxel mesh simulation, was investigated. To this aim, real distortions in terms of Δz displacement of the top right corner of the sample were measured as previously explained. The results from the real distortion measurements are shown in Fig. 3 together with the simulated values obtained with different voxel mesh elements and laser power values.

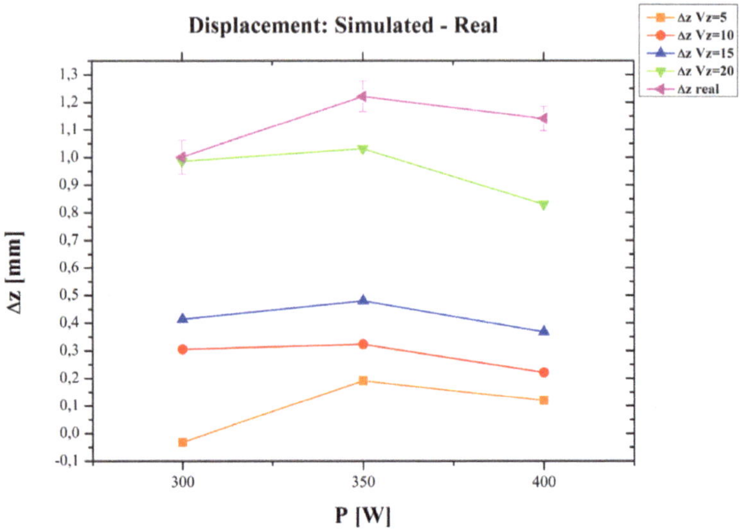

Fig. 3. Distortion measurements from numerical simulation with different voxel meshes (v_z=5, v_z=10, v_z=15 and v_z=20) compared to experimental ones (Δz real).

It can be observed, for the real distortion measurements, how they first increase by increasing the laser power and then decrease. In detail, for laser power below P = 350 W, there is an increase in the thermal gradients, along the build direction (z), that directly affects the residual stress profiles and distortion. For laser power higher than 350 W it can be noted how the gradient of residual stress, along the build direction (z), is lower than that one for the P = 350 W. This is because of the greater remelting area interested by the higher value of laser power as shown in the temperature profiles obtained with the numerical simulations (Fig. 4). Moreover, this laser power value can result in gas porosity acting as a stress relief factor.

By considering the distortion values shown in Fig. 3, for different voxel mesh elements along the z-direction (v_z), it can be stated that for $v_z = 20$ it is possible to have the best accuracy of the numerical simulation, especially for P = 300 W. Overall, by increasing v_z from 5 to 20 the accuracy of the FEM model increases. This is because increasing the number of voxel mesh elements along the z-direction means reducing the number of real layers embodied in the computational one. Using a few elements, as in the case of $v_z = 5$, is not enough to well approximate the L-PBF process when a sample of the geometry used in this study is considered. In this case, where a layer thickness of 60 μm is adopted, using a $v_z = 5$ means including 10 real layers in the computational ones.

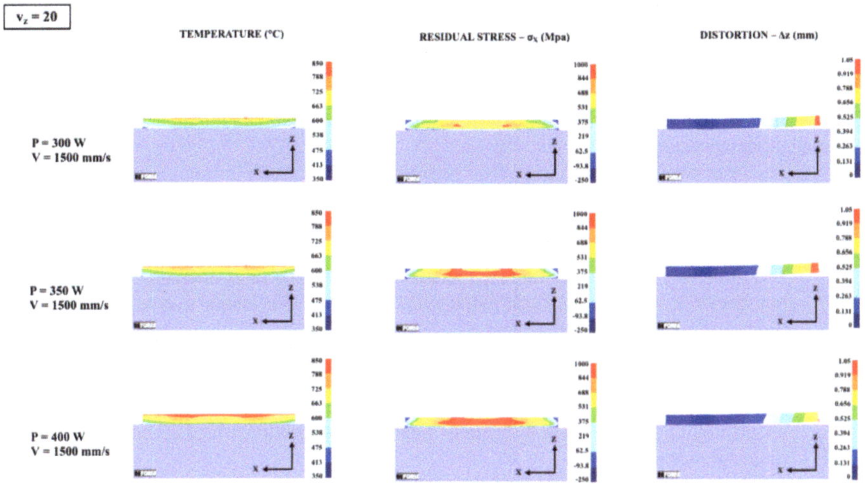

Fig. 4. Temperature profiles, residual stress profile, and distortion obtained from the numerical simulation with voxel mesh $v_z = 20$ and different laser power (300 W, 350 W, and 400 W).

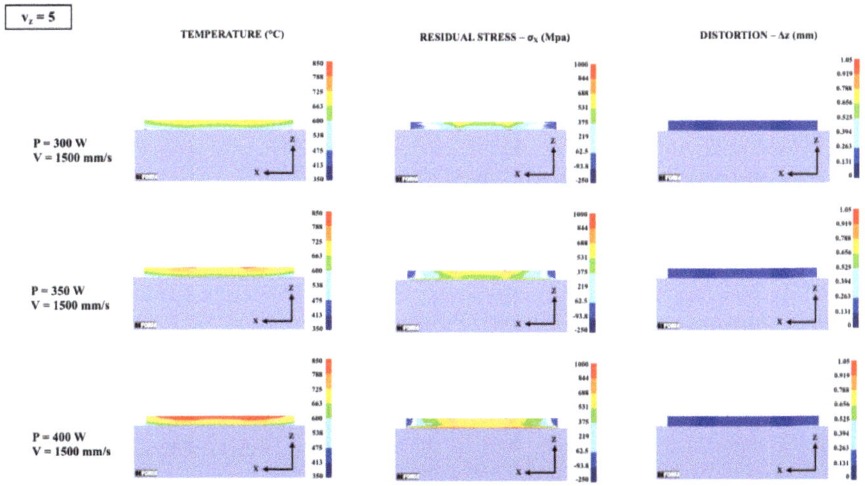

Fig. 5. Temperature profiles, residual stress profile, and distortion obtained from the numerical simulation with voxel mesh $v_z = 5$ and different laser power (300 W, 350 W, and 400 W).

In this way, the numerical model cannot reproduce with a good approximation the thermal exchange condition that occurs in the L-PBF process.

Fig. 5 shows how using $v_z = 5$ does not allow the correct development of residual stresses because there is no time for their development. The low residual stress values in fact result in almost no distortions.

Conclusions

In this paper, the commercial FEM software DEFORM 3D™ whit a layer-by-layer approach was used to perform numerical simulations of the Ti-6Al-4V L-PBF process. The effect of laser power on residual stresses and part distortion was studied. Moreover, the model accuracy by varying the voxel mesh elements along the build direction was investigated. The main findings of this study can be summarized as follows:

- The laser power has a big impact on part distortion due to its influence on thermal gradients and residual stress profiles.
- The accuracy of the numerical model is strictly correlated to the number of voxel mesh elements used along the build direction.
- The number of real layers included in the computational ones decreases by increasing the number of voxel mesh elements in the build direction. This results in a better approximation of the L-PBF process.
- Using a few numbers of voxel mesh elements ($v_z = 5$) leads to the activation of several elements simultaneously without giving enough time for the residual stresses to develop.
- For the analyzed geometry, a good accuracy of the numerical model can be obtained only by employing 20 voxel mesh elements along the z-direction which results in a voxel mesh dimension of 0.15 mm.
- The best results in terms of model accuracy were obtained in the case of laser power P = 300 W when a $v_z = 20$ is considered.

Acknowledgments

This research was funded by Regione Sicilia under the scheme "Azione 1.2.1_03 del PO FESR SICILIA 2014-2020" – Progetto PON03 PE_00206_1 AMELIE "Advanced framework for Manufacturing Engineering and Product Lifecycle Enhancement." CUP G76I20000060007.

References

[1] T. DebRoy, H.L. Wei, J.S. Zuback, T. Mukherjee, J.W. Elmer, J.O. Milewski, A.M. Beese, A. Wilson-Heid, A. De, W. Zhang, Additive manufacturing of metallic components – Process, structure and properties, Prog. Mater. Sci. 92 (2018) 112–224. https://doi.org/10.1016/j.pmatsci.2017.10.001

[2] G. Buffa, D. Palmeri, G. Pollara, F. Di Franco, M. Santamaria, L. Fratini, Process parameters and surface treatment effects on the mechanical and corrosion resistance properties of Ti6Al4V components produced by laser powder bed fusion, Progress in Additive Manufacturing (2023). https://doi.org/10.1007/s40964-023-00440-9

[3] S. Liu, Y.C. Shin, Additive manufacturing of Ti6Al4V alloy: A review, Mater. Des. 164 (2019) 8–12. https://doi.org/10.1016/j.matdes.2018.107552

[4] X. Lu, M. Chiumenti, M. Cervera, M. Slimani, I. Gonzalez, Recoater-Induced Distortions and Build Failures in Selective Laser Melting of Thin-Walled Ti6Al4V Parts, J. Manuf. Mater. Process. 7 (2023). https://doi.org/10.3390/jmmp7020064

[5] G. Buffa, A. Costa, D. Palmeri, G. Pollara, A. Barcellona, L. Fratini, A new control parameter to predict micro-warping-induced job failure in LPBF of TI6AL4V titanium alloy, Int. J. Adv. Manuf. Tech. 126 (2023). https://doi.org/10.1007/s00170-023-11179-6

[6] A. Razavykia, E. Brusa, C. Delprete, R. Yavari, An overview of additive manufacturing technologies-A review to technical synthesis in numerical study of selective laser melting, Materials 13 (2020) 1–22. https://doi.org/10.3390/ma13173895

[7] L. Bertini, F. Bucchi, F. Frendo, M. Moda, B.D. Monelli, Residual stress prediction in selective laser melting: A critical review of simulation strategies, Int. J. Adv. Manuf. Tech. 105 (2019) 609–636. https://doi.org/10.1007/s00170-019-04091-5

[8] E.L. Li, L. Wang, A.B. Yu, Z.Y. Zhou, A three-phase model for simulation of heat transfer and melt pool behaviour in laser powder bed fusion process, Powder Technol. 381 (2021) 298–312. https://doi.org/10.1016/j.powtec.2020.11.061

[9] M. Masoomi, S.M. Thompson, N. Shamsaei, Laser powder bed fusion of Ti-6Al-4V parts: Thermal modeling and mechanical implications, Int. J. Mach. Tool. Manuf. 118–119 (2017) 73–90. https://doi.org/10.1016/j.ijmachtools.2017.04.007

[10] N. Bastola, M.P. Jahan, N. Rangasamy, C.S. Rakurty, A Review of the Residual Stress Generation in Metal Additive Manufacturing: Analysis of Cause, Measurement, Effects, and Prevention, Micromachines (Basel) 14 (2023). https://doi.org/10.3390/mi14071480

[11] C. Chen, Z. Xiao, H. Zhu, X. Zeng, Deformation and control method of thin-walled part during laser powder bed fusion of Ti–6Al–4V alloy, Int. J. Adv. Manuf. Tech. 110 (2020) 3467–3478. https://doi.org/10.1007/s00170-020-06104-0

[12] D. Palmeri, G. Pollara, R. Licari, F. Micari, Finite Element Method in L-PBF of Ti-6Al-4V: Influence of Laser Power and Scan Speed on Residual Stress and Part Distortion, Metals (Basel) 13 (2023). https://doi.org/10.3390/met13111907

Modelling of dynamic scale layer growth considering temperature, time and alloying elements

BERGELT Tim[1,a]*, GRAF Marcel[1,b], HUNZE-TRETOW Jan[2,c], BEHRENS Bernd-Arno[2,d] and LAMPKE Thomas[1,e]

[1]Materials and Surface Engineering Group, Institute of Materials Science and Engineering, Chemnitz University of Technology, 09107 Chemnitz, Germany

[2]Institute for Forming Technology and Forming Machines, Leibniz University Hannover, 30823 Garbsen, Germany

[a]tim.bergelt@mb.tu-chemnitz.de; [b]marcel.graf@mb.tu-chemnitz.de, [c]hunze@ifum.uni-hannover.de, [d]behrens@ifum.uni-hannover.de, [e]thomas.lampke@mb.tu-chemnitz.de

Keywords: Scale Layer, Microstructure, Thermophysical Properties, Modelling

Abstract: During hot forming of steel oxide scaling occurs at higher temperatures caused by reactions with oxide containing atmospheres. Three characteristic iron oxides exist for steel at temperatures above 570°C: Wustite (FeO), magnetite (Fe_3O_4) and hematite (Fe_2O_3). Scale layer formation is influenced by various process parameters, such as temperature, process time and furnace atmosphere. Additionally, the base material with different alloying elements (e.g. C, Cr, Si and Ni) also affects the scale layer formation. Therefor oxide scales are very difficult to handle in the entire manufacturing process. The aim of this work is to examine and evaluate the influence of temperature, time and the alloying elements C and Cr in association with the layer growth, layer composition and thermophysical properties for scale layers. Based on the achieved correlations, a model is developed, which is able to predict the scale formation and scale properties, depending on temperature, time and alloying elements. With rising temperature and time increasing layer thicknesses were observed. Further, the additional Cr ensured lower layer thicknesses compared to the unalloyed steels. The iron oxide distribution changed with rising temperature to higher oxide containing phases like magnetite and hematite. The mathematical model, developed based on this results, is able to calculate the resulting layer structure, thickness and thermophysical properties depending on temperature, time and chemical composition of the material.

Introduction

During the hot forming of metals, such as forging or rolling, the material surface reacts with oxide-containing atmospheres at temperatures above 570°C. Consequently, a new solid phase is formed at the material's surface: the oxide scale. These scale layers usually consist of three iron oxides: Wustite FeO, magnetite Fe_3O_4 and hematite Fe_2O_3. The scale layer formed during hot forming processes causes material loss and damages the forming tools and workpieces. Hence, a better understanding of scale layer growth and scale layer properties could reduce material losses and damage, enabling more cost and resource efficient production processes.

It is well known from the literature that scale layer formation depends on different process parameters, such as temperature, time and alloying composition of the base material [1,2]. *Graf et al.* [3] determined the influence of time and temperature on the scale development. The scale layer thickness increases with higher temperatures and longer process times due to the enhanced diffusion processes with rising temperatures. Furthermore, an increase in carbon content leads to a decrease in scale layer thickness. Besides carbon, other alloying elements, notably chromium, influence the scale layer formation. In case of Cr steels, the microstructure comprises additional chromium precipitates (chromium spinels and chromium(III) oxide), which inhibit the surface oxidation processes, thereby leading to a decreased scale layer formation [4,5].

Therefore, detailed knowledge of the thermophysical properties of the scale layer is required as they differ from the base material to optimize the hot forming processes. The scale layer may act as a barrier for the workpiece and influence the thermal behaviour of the entire system due to the different thermophysical properties [2]. To model a thermomechanical process, the material's mechanical properties and stress collective, as well as the thermal properties of the modelling materials, need to be known. For thermo-mechanical process modelling, the important properties are thermal conductivity λ, thermal expansion coefficient α and specific heat capacity c_p. Since the fractions of iron oxides within the scale layer can vary based on the process conditions, it is necessary to determine the thermophysical properties of each iron oxide to predict the thermophysical properties of the complete scale layer. *Dubey et al.* [6] developed a numerical heat conduction model to predict the transient three-dimensional temperature field in billets, considering the scale layer growth on the billet surfaces during the reheating simulations. However, the scale layer was only described using the iron oxide wustite without considering the other two oxide types. In addition, the layer growth was only modelled using the parabolic law, as the study focused on longer application times of the temperature. The resulting models can predict process influences on the temperature field and the scale growth of the billets. Especially for scale growth, different growth rates can be determined depending on the temperature of the furnace gas. The maximum scale layer thicknesses were identified at the locations with the highest temperature on the billet. The hot rolling model developed by *Krzyzanowski et al.* [7] describes the influences of the scale layer composition, crack formation and propagation during the rolling process and the material flow of the formed material. To simplify the model, *Krzyzanowski et al.* chose a 2D model. The scale layer consists of individual oxide fragments connected to each other. The size of the individual fragments is based on the experimentally determined crack distances, and it is less than the smallest detected crack distance, allowing a prediction of the crack distances and predictions about the crack distribution over the entire scale layer. However, this model does not simulate the scale layer growth during the heating process; instead, it employs a predefined scale layer as an initial step. As a result, the model cannot simulate various process conditions during the heating.

Materials and experimental

Material. In this work, the influence of carbon and chromium on the scale formation is investigated by a separate and comparative analysis of the element's influences. Therefore, seven different steels were used. The chemical composition of the steels, determined by melt analysis, is listed in Table 1.

Table 1. Chemical composition of the seven analysed steels.

Steel	C	Si	Mn	P	S	Cr	Mo	Ni	Cu	Al
C15	0.16	0.21	0.42	<0.001	<0.001	0.1	0.045	0.087	0.108	0.017
15Cr3	0.16	0.21	0.42	<0.001	<0.001	0.6	0.045	0.087	0.108	0.017
C45	0.49	0.2	0.42	<0.004	<0.001	0.1	0.034	0.081	0.096	0.021
45Cr3	0.49	0.2	0.42	<0.004	<0.001	0.6	0.034	0.081	0.096	0.021
45Cr4	0.49	0.2	0.42	<0.004	<0.001	1.15	0.035	0.076	0.1	0.017
C60	0.63	0.2	0.42	<0.001	<0.001	0.1	0.035	0.076	0.1	0.017
60Cr3	0.63	0.2	0.42	<0.001	<0.001	0.6	0.035	0.076	0.1	0.017

Three different C steels (C15, C45 and C60) were chosen to analyse the influence of carbon. Further, four different Cr steels (15Cr3, 45Cr3, 45Cr4 and 60Cr3) were selected to determine the influence of chromium. Using C45, 45Cr3 and 45Cr4 three different chromium contents are available for the same carbon concentration of 0.49 wt. %. The influence of the chromium content will be focussed predominantly on this steel series.

Scale formation and analysis. To determine the influence of temperature and time on the scale layer growth, each steel was subjected to scaling at temperatures of 900, 1050 and 1200°C for 20, 40 and 120 s durations, respectively. These scaling tests were performed on a quenching dilatometer (DIL805 A/D, Bähr) with cylindrical samples of 10 mm in height and 10 mm in diameter. The samples were inductively heated, and the temperature was monitored and controlled by using thermocouples. The closed test chamber of the dilatometer allowed a defined atmosphere containing argon during heating and cooling to prevent pre- and post-scaling. During the scaling process, the test chamber was flooded with atmosphere. The heating and cooling rates, and the gas flow, were set identically over the entire series of tests. After scaling, the samples were embedded in epoxy resin to protect the scale layer from further damage. Besides, cross sections were metallographically prepared for all scaled samples to identify the scale layer thickness and the iron oxide distribution depending on the temperature and time regime. The cross-sections were then analysed using light microscopy (OLYMPUS GX51 inverted optical microscope).

Dilatometer, DSC and LFA analysis. In addition to the scale layer growth and scale layer structure, the thermophysical properties of thermal expansion, specific heat capacity and thermal conductivity were determined for a complete scale layer. To account for the influence of the base materials and the layer growth, the base material was scaled to generate the pure scale samples. This contrasts with most other studies, in which classic iron oxide powder represents the scale layer. To produce pure scale samples, the scale layer requires a thickness of 3–4 mm. Therefore, cylindrical samples were scaled in the furnace at 900, 1050 and 1200°C up to 96 hours. For 900°C a thickness of 2 mm could not be achieved and for 1200 °C the scale layer showed intense cracking and high porosity compared to the scale layers produced at 1050°C. For further analyses, the material was scaled at 1050°C for 96 h. The 3–4 mm thick scale layer was separated from the substrate using a cutting machine and reduced to the required dimensions by manual grinding. The investigation of the thermophysical properties was focused on C45, 45Cr3 and 45Cr4 to reduce the number of necessary tests.

Further, thermal expansion was determined by using a dilatometer (*DIL402E, NETZSCH GmbH & Co. Holding KG*). The pure scale sample was measured from RT (room temperature) up to 1200 °C at a heating rate of 5 K/min and adjusted with the reference measurement. The specific heat capacity was determined using Differential scanning calorimetry (DSC) on *STA449J* (*NETZSCH GmbH & Co. Holding KG*). For the experiments, pure scale samples were pulverised and tested between RT and 1200 °C with a heating and cooling rate of 5 K/min. Based on preliminary investigations, it was impossible to determine the thermal conductivity of the pure scale samples using laser flash analysis. Due to the porous and heterogeneous layer structure of the entire scale layer, no comparative results could be obtained. Therefore, the modelling steps will use comparative literature data [8–13].

Modelling

To utilise the obtained data for a prediction model, all results and their dependencies need to be described by mathematical equations. To reduce the complexity of all experimental data, the layer growth, layer composition and thermophysical properties are considered separately and combined using a *Fortran-77* script. By choosing *Fortran-77* as the programming language, the integration to the Solver: *MSC.Marc* within *Simufact Forming* can be used in future works to implement the scale layer model in the simulations of hot forming processes. To describe the experimental data obtained in Scale formation and analysis and Dilatometer, DSC and LFA analysis for scale layer growth, scale layer structure and thermophysical properties as a function of temperature and time it is necessary to describe and fit the data via mathematical equations. For the iron oxide distribution of the scale layer, a classic polynomial fit is used, allowing a sufficiently high coefficient of determination for the experimental data. Due to the characteristics of the experimental data, a separate fit is necessary for temperatures up to 1050°C and above 1050°C, as the curve characteristics change significantly at this point and a fit over the entire temperature range provides poor results. To describe the iron oxide evolution up to 1050°C for wustite and magnetite as well as for hematite, a polynomial of 7th degree (Eq. 1) is used. For temperatures above 1050°C, a polynomial of 3rd degree (Eq. 2) is accurate enough. In the polynomial equations, B0–B7 are coefficients and y is the volume fraction of the respective iron oxide from 100 % volume fraction.

$$y = B0 + B1 \cdot x^1 + B2 \cdot x^2 + B3 \cdot x^3 + B4 \cdot x^4 + B5 \cdot x^5 + B6 \cdot x^6 + B7 \cdot x^7 \quad (1)$$

$$y = B0 + B1 \cdot x^1 + B2 \cdot x^2 + B3 \cdot x^3 \quad (2)$$

For scale layer growth, the ARRHENIUS approach (Eq. 3) is used as a combined linear and parabolic layer growth model to describe the resulting layer thickness as a function of temperature and time for all steels.

$$d_{oxid} = -\frac{k_{p0} * \exp\left(\frac{-Q_p}{R*T}\right)}{2*k_{l0} * \exp\left(\frac{-Q_l}{R*T}\right)} + \sqrt{\left(\frac{k_{p0} * \exp\left(\frac{-Q_p}{R*T}\right)}{2*k_{l0} * \exp\left(\frac{-Q_l}{R*T}\right)}\right)^2 + k_{p0} * \exp\left(\frac{-Q_p}{R*T}\right) * t_{oxidation}} \quad (3)$$

Where d_{oxid} is the scale layer thickness, R is the universal gas constant and T is the scaling temperature. The model coefficients are k_{p0}, k_{l0}, Q_p and Q_l. The coefficients were determined using the least squares method for all process parameter variations to achieve a high coefficient of determination. For the thermophysical properties (thermal expansion, specific heat capacity and thermal conductivity), it is only possible to measure experimentally fixed scale layer compositions (see Thermophysical properties). Further, to obtain a dynamic model, it is necessary to know the thermophysical properties of each iron oxide as a function of temperature. Only then it is possible to predict the thermophysical properties for each scale layer composition by using a weighting function combined with the iron oxide distribution model. The required data for the individual iron oxides is taken from the literature [8–13] and validated experimentally using the pure scale samples.

Results and Discussion

Iron oxide distribution. To characterise the scale layers, light microscopy images were taken of all scaled samples. The microscopic scale layers at 1200°C for 120 s for C15 and 15Cr3 are shown in Fig. 1. In both images, the three-layer system is noticeable, which allows the identification of the different iron oxide types through various grey colours [14]. The dark grey area on the bottom of the scale layer is wustite (FeO), the slightly brighter layer on top is magnetite (Fe_3O_4) and the brightest layer near to the surface is hematite (Fe_2O_3).

The resulting iron oxide distribution was determined based on the light microscopy images of all parameter constellations for the seven steels. The measured distribution is described as a proportion of 100 % with the entire layer. For C and Cr steels, similar tendencies could be achieved for 900, 1050 and 1200°C. At scaling temperatures at 900 and 1050°C, wustite dominates with over 80 % volume fraction. Magnetite occurs at less than 20 % and hematite is formed at a maximum of up to 1 % (see Fig. 2: B and C). For 1200 °C, the proportions of wustite and magnetite vary between 40–50 % and hematite exhibits a constant volume fraction of 10 % (see Fig. 2: D). Generally, an increase in temperature leads to a change in the iron oxide distribution inside the scale layer. With higher temperatures, more hematite forms at the expense of wustite and magnetite. Also, with temperatures over 1050°C, wustite and magnetite get more equally balanced up to 1200°C. However, it can be assumed that with a further temperature increase, wustite will

Materials Research Forum LLC
https://doi.org/10.21741/9781644903254-52

continue to decrease in favour of magnetite and hematite. The volume fraction of wustite, magnetite and hematite show no significant change with rising time. This distribution of the oxides is also described in literature [15].

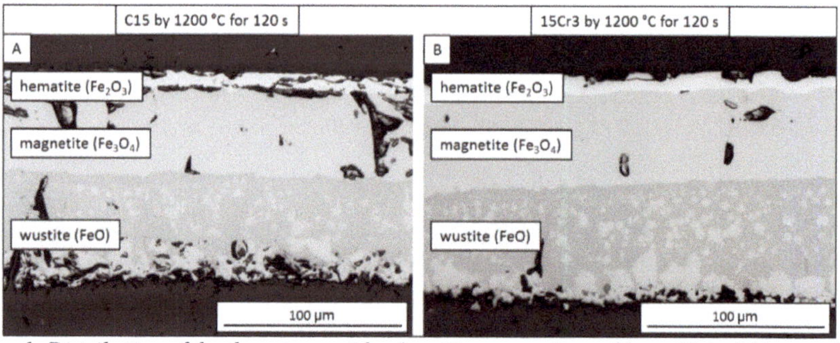

Fig. 1. Distribution of the three iron oxides (wustite, magnetite and hematite) within a scale layer produced at 1200 °C for 120 s for the material C15 (A) and 15Cr3 (B).

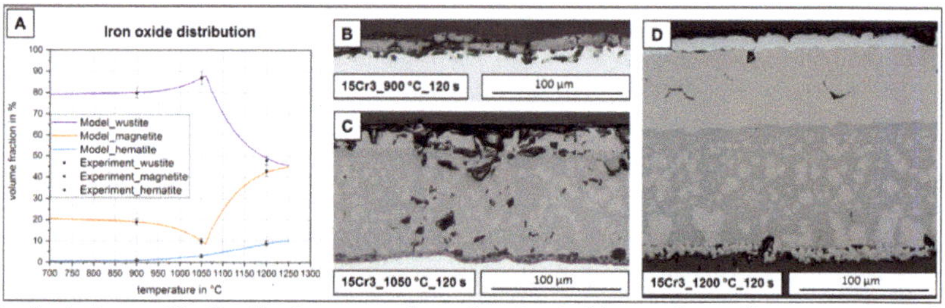

Fig. 2. A: Iron oxide distribution as a volume fraction in % over the scaling temperature for each iron oxide (wustite, magnetite and hematite). The colored curves display a polynomial fit based on the experimental data (black points). B: scale layer for 15Cr3 after 120 s at 900°C. C: scale layer for 15Cr3 after 120 s at 1050°C. D: scale layer for 15Cr3 after 120 s at 1200°C.

To describe the layer composition via mathematical equations, only the influence of various temperatures is considered. The time variation is too small and will be cut out for simplification. Furthermore, no differences were found in the iron oxide distribution in the resulting scale layer between the different steel alloys, so that one model can be used for all steel compositions. The mathematical approach, described in chapter Modelling, using a polynomial fit is shown in Fig. 2: A. The determined coefficients B0–B7 for the polynomial fits are listed in Table 2.

Materials Research Proceedings 44 (2024) 482-492 https://doi.org/10.21741/9781644903254-52

Table 2. Coefficients B0–B7 for each iron oxide for temperatures up to 1050°C and from 1050°C.

Phase		B0	B1	B2	B3	B4	B5	B6	B7
FeO	up to 1050 °C	73.992	0.0413	-3.32e-4	1.52e-6	-3,79e-9	5.26e-12	-3.8e-15	1.11e-18
	from 1050 °C	6030	-13.913	1.08e-2	-2.8e-6	-	-	-	-
Fe_3O_4	up to 1050 °C	26.094	-0.0437	3.58e-4	-1.65e-6	4.1e-9	-5.8e-13	4.22e-15	-1.2e-18
	from 1050 °C	-5135	12.019	-0.0093	2.4e-6	-	-	-	-
Fe_2O_3	from 900 °C	1.7651	-0.0388	3.14e-4	-1.24e-6	2.64e-9	-3.1e-12	1.81e-15	-4.2e-19

The proportion of wustite (purple curve) rises to a maximum of 1050°C and then decreases to approximately 45 % at 1200°C. The proportion of magnetite initially decreases until 1050°C and begins to increase from 1050°C. The scale layer contains almost no hematite at temperatures around 900°C. However, with temperatures above, the amount of hematite increases continuously with increasing temperature. The black points represent the experimental data, and the model curves are in very good agreement with this data. However, the model can only be considered as a first approximation, since there are not enough reference points. Therefore, only a coarse extrapolation between the reference points can be applied. In future works, the model shown in Fig. 2: A must be supported with further data and need to be updated.

Influence of temperature, time and material. Based on the light microscopy images shown in chapter Iron oxide distribution, the resulting layer thicknesses were evaluated as a function of the temperature and time. The below graphs (Fig. 3) illustrates the variation in layer thickness over time for 900, 1050 and 1200°C. The selected curves compare various C content (Fig. 3: A) and Cr content (B). The layer growth of each steel corresponds to the established √t–law for layer growth kinetics.

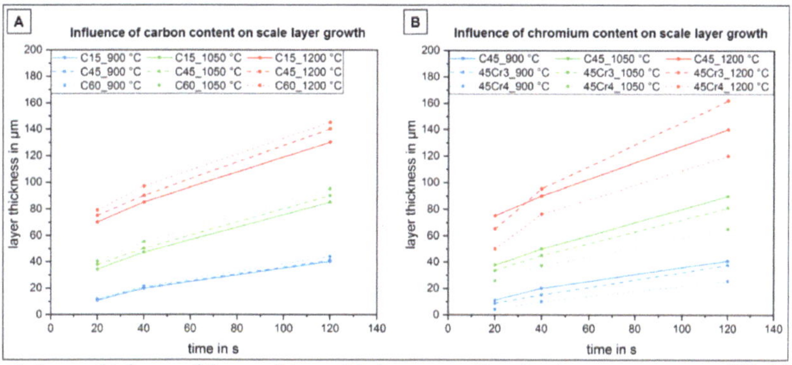

Fig. 3. Layer thickness of the resulting oxide layer over time for three temperatures 900, 1050 and 1200°C. A: comparison between C15, C45 and C60. B: comparison of two selected Cr-steels 45Cr3 and 45Cr4 and the chromium-free C45.

The oxide scale growth will slow down with increasing time, but never stop completely. Nevertheless, the interesting area is within the first few minutes, due to the point of highest oxide scale growth and can be extrapolated for longer periods according to the parabolic law. In addition to time, an increasing scaling temperature ensures thicker layers. As the temperature increases the activation energy increases, which is directly associated with enhanced diffusivity of the reactants. The layer growth is completely diffusion-controlled, so an increase in temperature results automatically in higher layer thicknesses. It is observed that as carbon content rises, layer thickness increases (Fig. 3: A). This tendency corresponds to the development of the austenitic grain size over the carbon content. It was determined that the average grain size increases with higher carbon content. For the scale layer, it can be concluded that a higher number of grain boundaries in the microstructure lead to a thinner layer thickness. As an explanation, the grain boundaries act as slight diffusion barriers and seem to affect the layer growth. This effect can be observed for all three temperatures.

Further, the influence of the different Cr content on the layer thickness of C45, 45Cr3 and 45Cr4 (Fig. 3: B) shows an increase in layer thickness with rising temperature, comparable to the C steels. Moreover, it could be determined that the layer thickness decreases with increasing Cr content. Between 45Cr3 and 45Cr4, a significantly higher influence of the chromium content on the layer thickness is observed compared to C45 and 45Cr3. The chromium-free C45 shows the highest layer thicknesses, but only slightly above that for the 45Cr3. This effect is most likely the result of chromium precipitations, which act as slight diffusion barriers during layer formation and hinder the layer growth. With rising Cr content, this effect seems to increase. However, for 45Cr3 at 1200 °C, a pronounced increase in layer thickness was observed, which cannot be explained by literature currently.

To describe the correlations between temperature and time on the layer growth in a single equation, the ARRHENIUS-approach, described in Modelling, was used. The four unknown coefficients k_{p0}, k_{l0}, Q_p and Q_l were determined using a least square fit (see Table 3) and allow a description of the scale layer thickness based on the temperature and time as inputs. In the current state it was impossible to establish factorial relationships between the coefficients to describe the influence of the chemical composition in the base material. Therefore, the coefficients for each material must be defined separately in the scale layer model and the layer thickness can only be predicted accurately for the analysed steels.

Table 3. Coefficients k_{p0}, k_{l0}, Q_p and Q_l from the ARRHENIUS-approach for each steel to describe the scale layer growth dependent on temperature and time.

Steel	k_{p0} in µm²/s	Q_p in J/mol	k_{l0} in µm/s	Q_l in J/mol
C15	1.76e+06	1.14e+05	9.89e+06	1.47e+05
15Cr3	1.37e+07	1.35e+05	2.11e+07	1.69e+05
C45	2.15e+06	1.14e+05	2.11e+07	1.57e+05
45Cr3	2.14e+07	1.40e+05	2.10e+07	1.63e+05
45Cr4	4.03e+06	1.26e+05	2.12e+07	1.72e+05
C60	1.85e+06	1.11e+05	2.58e+07	1.59e+05
60Cr3	9.97e+06	1.29e+05	2.10e+07	1.65e+05

Fig. 4 shows the fits for C45 (A), 45Cr3 (B) and 45Cr4 (C) as examples for all seven steels. Identical to Fig. 3, the scale layer thickness is plotted over time. The coloured points indicate the experimentally measured layer thickness at the respective temperature and time. The solid lines represent the fit using the ARRHENIUS-approach for temperatures at 900, 1050 and 1200°C. For most of the curves, a coefficient of determination (R^2) with a value over 0.9 was achieved, with only a few curves falling between 0.8 and 0.9. Based on the obtained R^2 between the fit and the experiments, the coefficients (k_{p0}, k_{l0}, Q_p and Q_l) for each tested steel are validated.

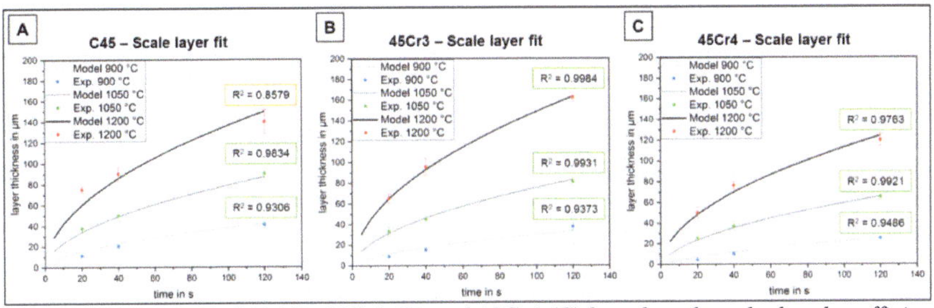

Fig. 4. Scale growth for C45 (A), 45Cr3 (B) and 45Cr4 (C) based on the calculated coefficients of the ARRHENIUS-approach (solid lines) for 900, 1050 and 1200 °C. Compared with the experimental data (colored points), a sufficiently high coefficient of determination was achieved for each curve.

Thermophysical properties. As stated in chapter Iron oxide distribution, the distribution of the iron oxides inside the scale layer changes as a function of temperature. Due to the distinct thermophysical properties of every iron oxide, the properties of the resulting scale layers also vary at different temperatures. It was impossible to produce samples consisting of only one iron oxide (wustite, magnetite or haematite), therefore pure scale samples of a complete layer system were used. These scale samples were produced at 1050°C in a furnace, resulting in an iron oxide distribution of 80 % wustite, 19 % magnetite and 1 % hematite. Nevertheless, in all three experiments to determine the thermophysical properties, an oxidation reaction was detected in the first heating cycle up to 1200 °C. In all further test cycles from RT up to 1200°C, no further oxidation peaks were observed. The oxidation in the first heating cycle caused a change in the iron oxide distribution to 0 % wustite, 30 % magnetite and 70 % hematite. The lack of bonding between the scales and the base material results in the consumption of the wustite phase and the high temperature phase haematite with remaining magnetite is formed. The assumption is that the 0-30-70 distribution is always achieved with pure scale samples, regardless of the scale layer structure before the first heating cycle. Based on this result, the determined experimental data will be validated using a model based on literature data for exactly this 0-30-70 distribution.

For the thermal expansion α, data was taken from the publication by *Beygelzimer* [16] for all three iron oxides and the α for the 0-30-70 distribution was calculated using a weighting function. The experimentally measured thermal expansion coefficients for the pure scale samples based on the steels C45 (green), 45Cr3 (yellow) and 45Cr4 (orange) is illustrated in Fig. 5: A. In addition, the black curve shows the expansion model of the 0-30-70 distribution based on the literature data. The experimental curves of the three scale samples are slightly below the model curve and show a peak at approximately 630 °C where the model detects it around 590°C. These slight differences can be explained by the additional alloying elements such as chromium and silicon, which are also present in the pure scale samples based on the base material. However, this assumption cannot be confirmed currently by experimental data. The slightly higher curve of 45Cr4 compared to C45 and 45Cr3 is also attributed to the influence of the additional chromium in the scale samples. Nevertheless, it can be concluded that the thermal expansion model can represent real conditions with sufficient accuracy, so it can be used further in the scale layer model.

For the specific heat capacity c_p, the same procedure was chosen as for the thermal expansion. The experimental data were determined using pure scale samples for DSC testing and the heat capacity model was developed, based on the literature data of the individual iron oxides according to *Beygelzimer* [17].The experimental data of the pure scale samples based on C45 (green), 45Cr3 (yellow) and 45Cr4 (orange) with the model curve (black) for the 0-30-70 distribution is shown in Fig. 5: B. For the specific heat capacity, the model curve shows a very good correlation with the experimental data for the 45Cr3 and 45Cr4. The C45 appears to be clearly higher, compared to the model curve. These differences can be explained by the additional alloying elements such as chromium and silicon, which are also present in the pure scale samples based on the base material. However, currently it is unexplainable why the samples with higher chromium closer to the model curve. Therefore, further investigations are needed. Both peaks at 575°C and 690°C matches exactly for each material. Only at temperatures above 1000°C, the experimental curve increases significantly more compared to the literature data. In the temperature range from RT to 1000°C, the heat capacity model is validated based on the experimental data. In accordance with [18,19], the literature data will be used from a temperature of 1000°C and above.

To determine the thermal conductivity λ, pure scale samples were analysed using LFA (Laser Flash Analysis). Due to the partially high porosity and strong heterogeneity of the scale layer, it was impractical to generate reproducible and valid results. Therefore, the thermal conductivity will be determined solely based on the literature data. The thermal conductivity model is developed according to the data from *Beygelzimer* [13] for each iron oxide and was validated using further literature [8–12]. The thermal conductivity for three exemplary layer compositions are displayed in Fig. 5: C. The 70-30-0 distribution represents a scale layer at 900°C and the 45-45-10 distribution represents a scale layer at 1200°C. The 0-30-70 distribution represents the known distribution after further oxidation without bonding to the substrate material. The three model curves demonstrate clearly that scale formed at different heating temperatures has different thermophysical properties and this fact has to be taken into account in the modelling of scale layers.

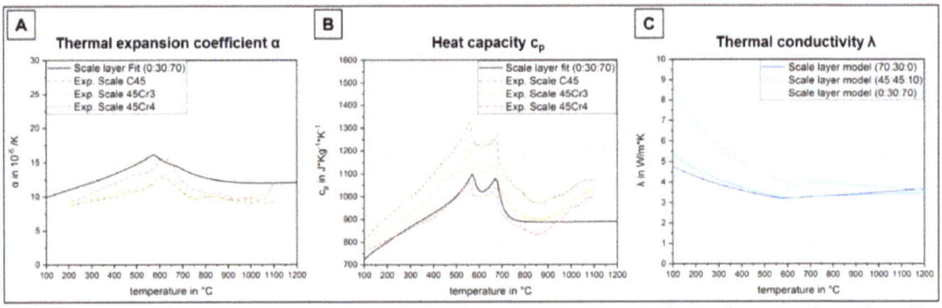

Fig. 5: A: Calculated thermal expansion coefficient using the weighting function based on the literature data of Beygelzimer [16] for a 0-30-70 distribution. Compared with the experimental data (colored lines), the model shows good agreement. B: Calculated heat capacity using the weighting function based on the literature data of Beygelzimer [17] for a 0-30-70 distribution. Up to a temperature of 850°C, model and experimental curves show very good agreement. After that, the experimental curves differ greatly from the literature. C: Calculated thermal conductivity using the weighting function based on the literature data of Beygelzimer [13] for a 70-30-0, 45-45-10 and 0-30-70 distribution.

All the equations and coefficients determined for the iron oxide distribution, the scale layer growth and the thermophysical properties were finally implemented in a *Fortran-77* script to predict the expected scale layer thickness, layer composition and the resulting properties by entering the temperature and time. As all sub-models (iron oxide distribution, scale layer growth and thermophysical properties) were considered to be validated, this also applies for the complete scale layer model. In future work, this model will be used in hot forming simulations with the FE program *Simufact Forming*. This enables the detection of the influence of different scale layers on heating and forming, as well as optimisation to reduce or systematically adjust scale layers.

Conclusions

The aim of this work was to create a scale layer model that can predict the resulting scale layer thickness and structure with associated thermophysical properties as a function of temperature, time and chemical composition. To accomplish this goal, seven different steels with varying carbon and chromium contents were scaled.

- Higher carbon content increases layer thickness, while additional chromium content leads to smaller layer thicknesses.
- High agreement between ARRHENIUS-fit and experimental data, allowed the use of the layer growth model outside the reference points.
- Carbon and chromium have a relatively small influence on the iron oxide composition of the scale layer. With higher temperatures significantly more hematite is formed and wustite and magnetite are evenly distributed.
- The thermophysical properties of the scale layer focussed on thermal expansion α, and specific heat capacity c_p could be determined using pure scale samples
- The models for thermal expansion, heat capacity and thermal conductivity were created using data from the literature and experimentally validated.
- In *Fortran-77* a Scale-Model was developed that can predict the scale layer thickness, iron oxide distribution and thermophysical properties for any temperature and time.

In future works, the scale layer model will be used in hot forming simulations with the FE programme *Simufact Forming* to consider scale formation during heating and forming processes. Another field of application could be the optimisation of furnace strategies with the aim of reducing scale formation or for the systematic adjustment of required scale layer properties.

Funding

The authors would like to thank the German Research Foundation (DFG) for funding the research project "Generally modelling of material behavior and surface modifications for FEM analysis of die forging of carbon steels" (grant number 316273316).

Author Credits

The author would like to thank Stephan Daniel Schwöbel from TU Chemnitz/Germany for the support in developing the fits, Max Müller from TU Chemnitz/Germany for the work at the dilatometer and metallographic preparation.

References

[1] R.M. Pineda Huitron, P.E. Ramírez López, E. Vuorinen, P. Nazen Jalali, L. Pelcastre, M. Kärkkäinen, Scale Formation on HSLA Steel during Continuous Casting Part I: The Effect of Temperature–Time on Oxidation Kinetics, Metals 10 (2020) 1243. https://doi.org/10.3390/met10091243

[2] E. Wielgosz, T. Kargul, J. Falkus, Comparison of experimental and numerically calculated thermal properties of steels, METAL 2014 - 23rd International Conference on Metallurgy and Materials, Conference Proceedings (2014).

[3] M. Graf, M. Ullmann, G. Korpala, H. Wester, B. Awiszus, R. Kawalla, B.-A. Behrens, Forming and Oxidation Behavior During Forging with Consideration of Carbon Content of Steel, Metals 8 (2018) 996. https://doi.org/10.3390/met8120996

[4] D.J. Young, J. Zurek, L. Singheiser, W.J. Quadakkers, Temperature dependence of oxide scale formation on high-Cr ferritic steels in Ar–H2–H2O, Corrosion Science 53 (2011) 2131–2141. https://doi.org/10.1016/j.corsci.2011.02.031

[5] F.H. Stott, G.C. Wood, J. Stringer, The influence of alloying elements on the development and maintenance of protective scales, Oxid Met 44 (1995) 113–145. https://doi.org/10.1007/BF01046725

[6] S.K. Dubey, P. Srinivasan, Development of three dimensional transient numerical heat conduction model with growth of oxide scale for steel billet reheat simulation, International J. Therm. Sci. 84 (2014) 214–227. https://doi.org/10.1016/j.ijthermalsci.2014.05.022

[7] M. Krzyzanowski, J.H. Beynon, Modelling the Behaviour of Oxide Scale in Hot Rolling, ISIJ Int. 46 (2006) 1533–1547. https://doi.org/10.2355/isijinternational.46.1533

[8] M. Takeda, T. Onishi, S. Nakakubo, S. Fujimoto, Physical Properties of Iron-Oxide Scales on Si-Containing Steels at High Temperature, Mater. Trans. 50 (2009) 2242–2246. https://doi.org/10.2320/matertrans.M2009097

[9] T. Akiyama, H. Ohta, R. Takahashi, Y. Waseda, J. Yagi, Measurement and Modeling of Thermal Conductivity for Dense Iron Oxide and Porous Iron Ore Aggliomerates in Stepwise Reduction, ISIJ Int. 32 (1992) 829–837. https://doi.org/10.2355/ISIJINTERNATIONAL.32.829

[10] J. Slowik, G. Borchardt, C. Köhler, R. Jeschar, R. Scholz, Influence of oxide scales on heat transfer in secondary cooling zones in the continuous casting process, part 2: determination of material properties of oxide scales on steel under spray-water cooling conditions, Steel Res. 61 (1990) 302–311. https://doi.org/10.1002/srin.199000353

[11] J. Mo/lgaard, W.W. Smeltzer, Thermal Conductivity of Magnetite and Hematite, J. Appl. Phys. 42 (1971) 3644–3647. https://doi.org/10.1063/1.1660785

[12] Y. Grosu, A. Faik, I. Ortega-Fernández, B. D'Aguanno, Natural Magnetite for thermal energy storage: Excellent thermophysical properties, reversible latent heat transition and controlled thermal conductivity, Solar Energy Materials and Solar Cells 161 (2017) 170–176. https://doi.org/10.1016/j.solmat.2016.12.006

[13] E. Beygelzimer, Y. Beygelzimer, Thermal conductivity of oxide scale and its components in the range from 0 C to 1300 C: Generalized estimates with account for movability of phase transitions, 2021.

[14] R.Y. Chen, W.Y.D. Yuen, A Study of the Scale Structure of Hot-Rolled Steel Strip by Simulated Coiling and Cooling, Oxid. Met. 53 (2000) 539–560. https://doi.org/10.1023/A:1004637127231

[15] G. Korpała, M. Ullmann, M. Graf, H. Wester, A. Bouguecha, B. Awiszus, B.-A. Behrens, R. Kawalla, Modelling the influence of carbon content on material behavior during forging, Penang, Malaysia, Author(s), 2017, p. 190013

[16] E. Beygelzimer, Y. Beygelzimer, Generalized estimates for thermal expansion of oxide scale in the range from 0C to 1300C with account for movability of phase transitions in its components, 2021.

[17] E. Beygelzimer, Y. Beygelzimer, Heat Capacity of oxide scale in the range from 0 C to 1300 C: Generalized estimates with account for movability of phase transitions, 2021.

[18] F. Grønvold, A. Sveen, Heat capacity and thermodynamic properties of synthetic magnetite (Fe3O4) from 300 to 1050 K. Ferrimagnetic transition and zero-point entropy, J. Chemic. Thermodynamic. 6 (1974) 859–872. https://doi.org/10.1016/0021-9614(74)90230-4

[19] J.P. Coughlin, E.G. King, K.R. Bonnickson, High-temperature Heat Contents of Ferrous Oxide, Magnetite and Ferric Oxide 1, J. Am. Chem. Soc. 73 (1951) 3891–3893. https://doi.org/10.1021/ja01152a098

Neural surrogate-driven modelling, optimisation, and generation of engineering designs: A concise review

CHEN Siyi[1,a *], DING Jiangfeng[1,b], SHAO Zhutao[1,c], SHI Zhusheng[1,d] and LIN Jianguo[1,e]

[1]Imperial College London, Department of Mechanical Engineering, London SW7 2AZ, UK

[a]siyi.chen18@imperial.ac.uk, [b]j.ding22@imperial.ac.uk, [c]z.shao12@imperial.ac.uk, [d]zhusheng.shi@imperial.ac.uk, [e]jianguo.lin@imperial.ac.uk

Keywords: AI-Driven Design, Feature Engineering, Neural Surrogate Modelling, Surrogate-Driven Design Optimisation, Generative Design, Text-to-CAD, Digital Twin

Abstract. Synergies between neural networks and traditional surrogate modelling techniques have emerged as the forefront of data-driven engineering. Neural network-based surrogate models, trained on carefully selected experimental data or high-fidelity simulations, can predict behaviours of complex systems with remarkable speed and accuracy. This review examines the current state and recent developments in neural surrogate technologies, highlighting their expanding roles in engineering design optimisation and generation. It also covers various feature engineering methods for representing 3D geometries, the principles of neural surrogate modelling, and the potential of emerging AI-driven design tools. While feature engineering remains a challenge, especially in parameterising complex designs for machine learning, recent advancements in code/language-based representations offer promising solutions for digitalising various design scenarios. Moreover, the emergence of AI-driven design tools, including text-to-CAD models powered by large language models, enables engineers to rapidly generate and evaluate innovative design concepts. Neural surrogate modelling has the potential to transform engineering workflows. Continued research into geometric feature engineering, along with the integration of AI-driven design tools, will speed up the use of neural surrogate models in engineering designs.

Introduction

Engineering design is the systematic process of conceiving, developing, and optimising products or services to fulfil specific needs and requirements [1]. Traditionally, this process relies on physical prototyping, repetitive simulations and human intuition, often leading to a time-consuming and potentially suboptimal outcomes. Advancements in machine learning (ML) technologies, particularly neural networks, are addressing longstanding challenges in this field by automating repetitive tasks [2], optimising designs for specific goals [3,4], and generating entirely new design concepts [5].

ML refers to the automatic mapping of underlying function(s) between input features and the desired output by optimising model parameters using observed data. ML models leverages probability theory to identify (learn) and approximate (fit) data distributions [6], enabling them to make predictions without explicit programming [7]. Feature engineering is a critical process in ML where raw data is transformed into meaningful features (inputs) to improve the performance and accuracy of the model [8]. It encompasses feature abstraction, which extracts meaningful patterns from complex raw data, and parameterisation, the process of defining and adjusting parameters of a model for optimal performance. In engineering contexts, feature engineering focuses on the abstraction and parametrisation of geometric information and performance-related attributes from relevant systems, processes or machines.

Digital twins are virtual replicas of the real world, serving as a key driver of the Fourth Industrial Revolution. They use real-time data to simulate physical systems, identifying issues and enhancing real-world performance [9]. For instance, Google Maps serves as the digital twin of the planet's surface, enabling navigation and virtual exploration [10]. Engineering digital twins typically leverage existing computer-aided design (CAD), computer-aided engineering (CAE) and computer-aided manufacturing (CAM) tools, which are tailored to various stages of the product lifecycle. Surrogate modelling, which creates simplified mathematical approximations of complex systems, has emerged as a vital component of digital twins [11]. The superior speed and data efficiency of neural networks make neural surrogate modelling ideal for industrial design applications [12,13]. Its computational efficiency enables rapid exploration of high-dimensional design spaces, a significant advantage over traditional simulation-heavy design approaches [14].

Emerging AI-driven design tools are further transforming the engineering design cycle [2]. Cloud-based ML platforms are streamlining the development of neural surrogates for engineers [15,16]. Generative design software autonomously creates design options, guided by performance requirements and manufacturing constraints [17]. Additionally, text-to-CAD (TTC) models can convert natural language descriptions into CAD-ready geometries [18,19]. The synergy of these technologies enables rapid generation, evaluation and optimisation of engineering designs.

This concise review aims to elucidate the frontier of neural surrogate technologies and their expanding applications in engineering design optimisation and generation. It explores different geometric feature engineering techniques, advancements in surrogate modelling for industrial design tasks, their integration with optimisation algorithms, and finally, capabilities of emerging AI-driven design tools. By examining the latest research developments in neural surrogate-driven engineering design, this paper offers insights into the future of this exciting field.

Geometric Feature Engineering

Geometric feature engineering involves the extraction, manipulation, and encoding of geometric properties into formats compatible with ML algorithms [8]. However, the lack of standardised, broadly applicable geometric feature engineering techniques has hindered the widespread adoption of ML in engineering.

Both numerical and categorical scalar data can be directly fed into scalar-based machine learning methods like support vector machines without extensive feature engineering [7]. In contrast, 2D information, typically in the form of pixel images, is typically processed through machine vision techniques such as convolutional neural networks (CNNs) [7]. The digital representation of 3D data is more complex, with diverse methods employed without a standard approach. A common technique involves projecting 3D objects into 2D profiles, enabling the application of image-based machine learning techniques [3,14]. While 3D projections of CAD models were utilised in several deep learning studies [20,21], this image-based representation of 3D shapes is often restricted to simple objects without internal features.

Other computational techniques frequently used for representing 3D geometry include voxel-, point cloud-, mesh-based, and implicit representations. Fig. 1 illustrates these four 3D representation techniques' depictions of the Stanford Bunny. Point clouds (Fig. 1(a)) are collections of points in 3D space, defined by their spatial positions (x, y, z) and sometimes other attributes (like colour) [16]. Voxel grids (Fig. 1(b)), on the other hand, resemble 3D versions of pixels and can be considered as quantised and fixed-sized point clouds [22]. As shown in Fig. 1(c), polygon meshes connect vertices with polygonal faces to form surfaces, often with additional 2D information mapped onto the triangular surfaces of the polygons [23,24]. In contrast to the explicit representations that capture geometric information in discretised formats, implicit representations use continuous, infinite-resolution functions (e.g., signed distance functions) to define a volume's occupancy field [3,5], as illustrated in Fig. 1(d). The recent use of neural networks to learn compact latent representations of implicit functions has made implicit neural representations more memory-efficient and better suited for numerical optimisation [25]. Despite their prevalent use, those position-based representations are incompatible with modern CAD software that models engineering components using sequential and parametric commands [5,26].

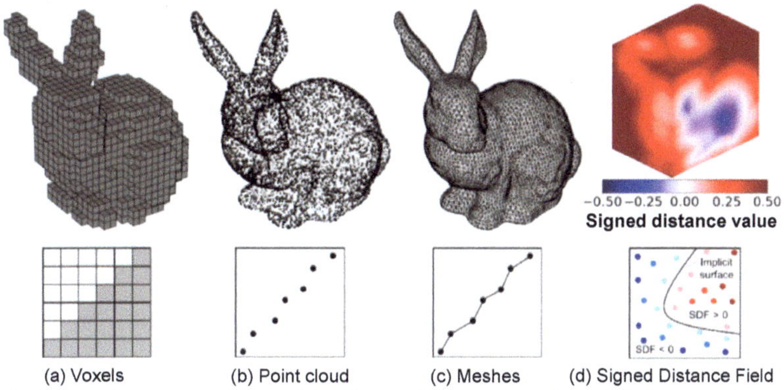

Fig. 1. Geometric representations of the Stanford bunny through explicit forms such as voxels (a), point cloud (b) and meshes (c), and implicit representations like signed distance fields (d) [25].

Given the rising prominence of large language models (LLMs) like ChatGPT, recent research in geometric feature engineering is pivoting towards code/language-based representations of 3D geometries [5]. These methods mainly encompass the boundary representation (BREP) techniques and CAD sequence-based representations, as shown in Fig. 2(a) and 2(b), respectively. The BREP format defines the boundary between the interior and exterior of a solid's volume using a collection of connected surface segments (Fig. 2(a)), complemented by topological information that explains their adjacency relationships [27]. In contrast, most CAD programmes use command (construction) sequences to create models, as shown in Fig. 2(b). This parametric representation allows one to model complex geometries using several parameters, without manually building it up from scratch [26]. The BREP serves as an abstraction of the CAD command sequences [26]. While viewable in most CAD software, BREP files lack the parametric history needed for direct modification using those tools. Additionally, the vocabulary formed by CAD command sequences is analogous to natural language. This similarity has encouraged the exploration of Autoencoder (AE)-based Transformer networks to produce command sequences of novel 3D CAD designs, which can be directly edited by users [26]. Furthermore, fine-tuned LLMs can generate construction sequences of CAD models based on user prompt inputs. These sequences can then be imported into existing CAD software to construct the final 3D model [19].

AEs are neural-network architectures designed for unsupervised learning tasks, particularly focusing on dimensionality reduction and feature extraction [7]. During the encoding process, AEs automatically identify a user-specified number of features from the input data, transforming them into latent encodings. These encodings are then used to reconstruct the output information in the decoder stage [28]. In engineering design, AEs can function as geometric feature detectors, providing more insightful or nuanced ways to describe geometries than human-designed methods [7,29]. For example, a latent parameter extracted from an input 3D surface mesh might represent a combination of lean and sweep of a compressor blade in turbomachinery [28].

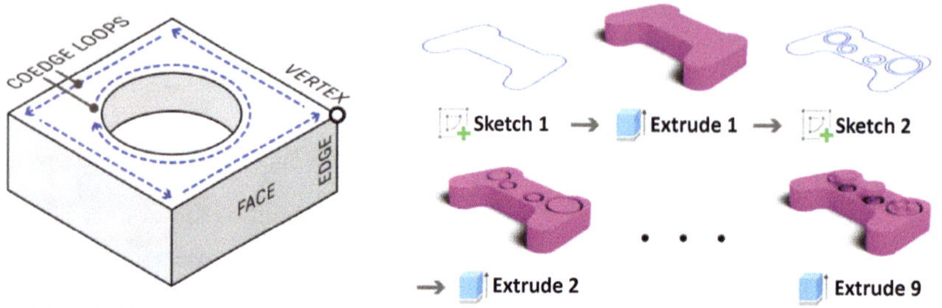

(a) Modified from Lambourne et al. [27] (b) Adapted from Willis et al. [30]

Fig. 2. Code/text-based representation of 3D shapes – Boundary representation with its face-adjacency graph (a); And CAD command (construction) sequence (b).

Surrogate Modelling

Surrogate modelling is a computational strategy that utilises intelligently sampled data to approximate complex mathematical models in unobserved regions of the design space [11]. Unlike purely data-driven methods, surrogate-based design optimisation can leverage data generated from simulations, thus overcoming the data scarcity issue common in engineering design. These models solely rely on the input-output behaviour of the true function, without consideration for its internal workings [1]. Modern design of experiment (DoE) methods, such as the Latin hypercube sampling, were utilised to ensure efficient data collection and maximised sampling uniformity across the design space [11,31].

Classic surrogate modelling techniques such as polynomial response surface model, Kriging, and radial basis functions incorporate statistical learning principles, but they do not engage in "learning" as comprehensively as modern ML algorithms [11]. Conversely, ML-based surrogates, particularly neural networks, are less confined by predefined mathematical structures and excel at dimensionality reductions for very high-dimensional and non-linear problems [12]. Multi-layer perceptron (MLP), a class of feed-forward neural networks, is especially suitable for surrogate modelling of complex non-linear problems in engineering design [12]. These networks excel at predicting and optimising both scalar performance metrics and 2D/3D physical fields [12,14]. Owing to their rapid evaluation speed, neural surrogates can significantly accelerate certain engineering design processes by eliminating the need for the repetitive evaluation of time-consuming CFD or FEA codes [3,13].

Neural surrogate models have become instrumental in predicting the performance of parameterised engineering designs [13]. Parametric design variables, such as those defined for the aeroengine nacelle of a civilian airplane (Fig. 3), were commonly used as inputs for the neural networks that predict performance metrics [32,33]. Tejero et al. modelled the nacelle drag characteristics across the design space, and under different operating conditions, using MLP networks [32]. Similar MLP networks were also employed to estimate the coefficients of lift and drag, pitching moment, and the lift-to-drag ratio based on the angle of attack and flap configurations for experimental aeroplane designs [33].

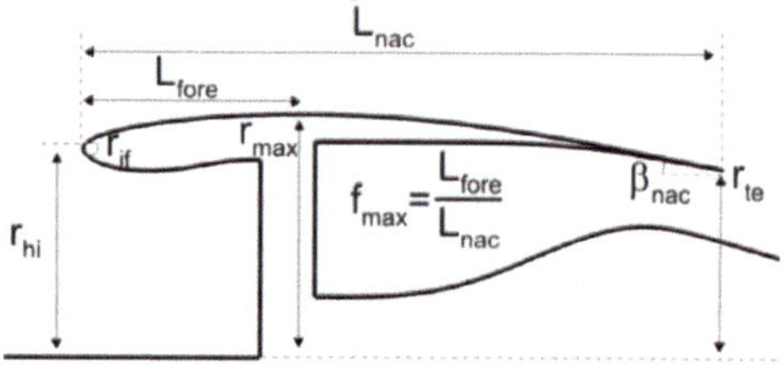

Fig. 3. Design variables used in the parametrisation of an aeroengine nacelle [32].

Beyond parametric design inputs and outputs, surrogate models are increasingly embracing image-based representations [14,34]. Fig. 4 depicts an image-to-image surrogate model employing AEs [3], which maps input images to corresponding output images [28]. Attar et al. [14] and Zhou et al. [34] used Res-SE-U-Net based image-to-image models to predict blank's thinning fields (as 2D image) from die geometry and blank shape images. Comparative analyses indicate that image-based neural surrogate models outperform their scalar-based counterparts in prediction accuracy, data efficiency, informativeness and generalisability [4,34,35].

In addition to images, neural surrogates can also process 3D input and output data [28,36]. For example, Pongetti et al. deployed a multi-level neural surrogate model to predict various quantities of interest, such as the static pressure field on turbomachinery compressor blades based on their 3D surface mesh input [28]. Moreover, Petrik et al. developed neural surrogate models for optimising open die forging processes [36]. These models accept the input of a 3D voxel mesh representing the pre-stroke geometry of the workpiece, along with a forging vector that defines the forging path. They output predictions of the recrystallisation and shape deformation of the corresponding workpiece in 3D space, both represented as voxel meshes. Results suggest that this system can accurately mimic simulations throughout the entire interpolation space [36].

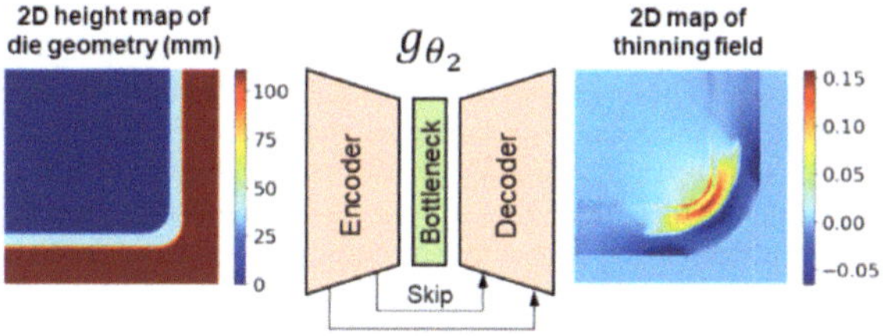

Fig. 4. Image-to-image surrogate modelling driven by an autoencoder [3].

Neural Surrogate-Driven Design Optimisation

Industrial and scientific endeavours are fraught with design problems that require the development of novel systems or processes for improved performance and efficiency [37]. Traditional design optimisation methods adopt a "top-down" approach where designers iteratively adjust design variables to achieve pre-defined performance metrics [13], often relying on trial-and-error or empirical strategies [4]. Given that modern engineering is primarily driven by expensive computer simulations [38], this process typically involves a non-exhaustive exploration of the design space, frequently resulting in suboptimal design outcomes [13]. In comparison, neural surrogate-assisted design optimisation incurs significantly lower computational costs, allowing near-instant evaluations of designs [4,38]. The fast surrogate modelling enables the optimisation algorithm to iteratively search for the optimal point(s) within the design space starting from an initial configuration [1]. As the outcomes of these optimisation processes are unforeseen, this constitutes a "bottom-up" approach [13]. Moreover, for optimisation problems, neural surrogate modelling helps imposing various boundary conditions, such as manufacturing and geometric constraints, by directly modelling them within the surrogate [3,4]. Ultimately, highly accurate and generalisable neural surrogates pave the way to be integrated within full-function digital twins, enabling real-time, multi-objective design optimisation [4].

Surrogate-driven design optimisation has been prevalent in aerodynamic design, encompassing aeroplane components [32,33], turbomachinery [28], flight envelopes [39], etc. Norgaard et al. determined the optimal flap settings and flap schedule for a research airplane by employing neural networks to predict performance metrics [33]. Leveraging predictions of quantities of interest from 3D-to-3D neural surrogate models, Pongetti et al. iteratively refined the optimal geometry of turbine compressor blades with an evolutionary algorithm to solve the targeted optimisation problem [28]. Additionally, the optimal landing trajectory for a spacecraft was determined using gradient descent optimisation based on aerodynamic predictions from neural surrogates [39].

Neural surrogate modelling also finds extensive application in the metal forming sector, particularly for optimising tooling design and process parameters vital to the cost and quality. Due to the intricate nature of these manufacturing processes, slight adjustments can result in significant changes in manufacturability metrics, rendering the traditional design process slow and error-prone [4]. Sheet metal forming processes rely heavily on optimal blank shapes, but scalar-based surrogate models, limited by accuracy, robustness, generalisability and the data inefficiency, struggled in the modelling and optimising blank shapes [4,34]. Attar et al. leveraged non-parametric neural surrogates capable of handling intricate morphing geometries, and gradient-based algorithms, to optimise geometries of hot stamping dies and blank shapes for corners and bulkheads. These geometric adjustments driven by manufacturability predictions resulted in notable enhancements in component quality [3]. Furthermore, CrystalMind's surrogate models for optimising open-die forging processes are coupled with a dual annealing-based optimisation algorithm, which iteratively adjusts the forging vector to achieve deformation and recrystallisation patterns that closely match target outcomes. The rapid feedback of these neural surrogates enables the optimisation process to finish much faster than with FEA simulations [36].

Emerging AI-driven Design Tools

Digital transformation in engineering industries has primarily focused on surrogate modelling of engineering products and processes, frequently leveraging raw CAD and CAE data [16]. This approach has enabled faster design iterations and higher product quality [16], effectively addressing the slowness of traditional CAD and CAE tools [15]. Web-based platforms such as NeuralConcept and Monolith AI offer streamlined software solutions to engineers that facilitate surrogate-driven design evaluation and optimisation [15,16]. These enterprise AI infrastructures integrate the cutting-edge deep learning algorithms with a user-friendly interface into a modular platform, enabling users to rapidly create and scale up various types of surrogate models using cloud computing resources [15]. By minimising coding requirements and providing comprehensive guidance on ML, these end-to-end platforms make powerful machine learning tools accessible to mainstream engineers [16]. Industrial applications of these AI platforms demonstrate significant saving in engineering time. For instance, a car manufacturer slashed the development period of a new vehicle from five years to just one after adopting this solution [15].

Additionally, tools for generative topology optimisation, also known as generative design, have been introduced in leading CAD programmes [17]. This technology automates the creation and iterative modification of a component's topology, based on user-defined performance requirements and manufacturing constraints [17]. Topology optimisation tools work by iteratively refining the geometry of a design until it converges to an optimal solution that satisfies all objectives and constraints. In comparison, generative topology optimisation produces multiple design choices that satisfy specified design requirements. More advanced versions even rank generated design options to facilitate user evaluation [17]. These tools are often integrated with CAE software for design evaluation, a process that neural surrogate modelling can greatly expedite. Some of them also incorporate CAM functionalities to impose design constraints [17].

Text-guided generative AI models that can produce 2D images, 3D geometries and sequences of programming languages, are quickly emerging in both academic and industrial sectors [18,40,41]. Text-to-image generators like the DALL-E series, Imagen and Midjourney, have been utilised to produce high-quality images of design concepts, especially in architecture [42]. Similarly, text-to-3D models generate various parameterised 3D geometries (point cloud [43], textured mesh [43], implicit neural radiance fields [44], etc.) from text input. These models are widely applicable in 3D content creation for gaming, virtual reality and industrial design [43]. TTC models, on the other hand, produce parametric command (construction) sequences that adhere to existing CAD file formats, making them directly importable into most CAD software [18,19,41]. As shown in Fig. 5(a), one TTC app facilitates parametric editing of CAD objects based on user's command prompts and drawing inputs [40], while another supports the generation of complex, fully-mated assemblies (Fig. 5(b)) [41]. Moreover, some TTC models offer "copilot" features like the autocompletion of CAD files [18]. These functionalities are derived from the models' deep understanding of both CAD file syntax and its correlation to 3D geometry. By training on open-source or proprietary datasets of engineering designs, these models learn to effectively leverage historical design information to synthesise new designs [19]. Certain platform even allows for fine-tuning of the pre-trained TTC models to users' specific database [19], further bolstering the benefit of this AI infrastructure. These text-to-design tools enable novice users to exploit insights from past design data and assist experienced designers to quickly visualise and explore innovative design concepts. In near future, the capabilities of TTC models are expected match the LLMs that power them. This will unlock extraordinary geometric inference capabilities, transforming the way designers interact with complex shapes. With the incorporation of neural surrogate modelling, these generative AI models can instantly assess created designs, thereby intelligently guiding the optimisation of complex design cases.

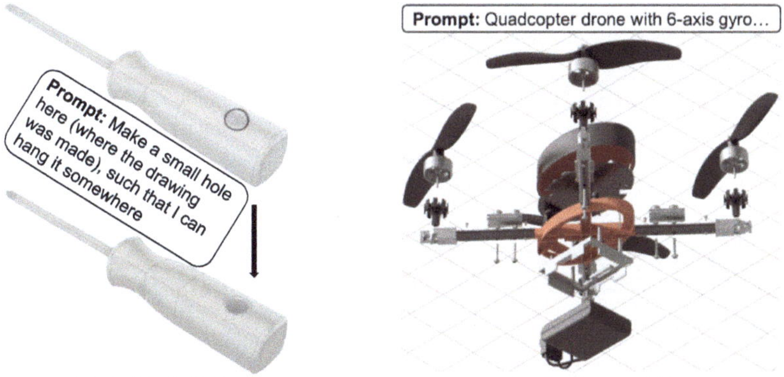

(a) CadifyAI [40] *(b) GetLeo.ai* [41]

Fig. 5. AI-powered generation and modification of CAD parts and assemblies: Text prompts and drawing inputs allow for real-time model modification (a); Complex, fully-mated assemblies of engineering designs are generated from scratch using simple text instructions (b).

As AI-powered design tools become more accessible, they are poised to transform numerous engineering practices. These tools are devised to augment the capabilities and efficiency of human designers, who remain the principal decision maker, instead of replacing them [45]. Looking ahead, future developments incline towards merging these technologies to create full-function digital twins that will intelligentise the design and manufacturing of engineering products.

Summary

This review represents a pioneering survey of how neural surrogate technologies can revolutionise reduced-order modelling, optimisation and generation of engineering designs. Despite challenges in geometric feature engineering for complex hardware designs, neural surrogate modelling stands out as a substantial advancement over traditional CAE methods. Neural surrogates empower the broader use of optimisation algorithms and enable the exploration of extensive design variations, potentially shortening design cycles and improving product performance. Furthermore, emerging AI-driven design tools and infrastructures are reshaping engineering practices, supporting engineers throughout the design process. Integration of neural surrogate modelling with these AI-driven tools brings the realisation of comprehensive digital twins, a cornerstone of Industry 4.0

initiatives, closer to reality. The creation of comprehensive and accurate digital twins necessitates powerful modelling methods, thus underscoring the critical need to advance neural surrogate technologies. Future research into ingenious geometric feature engineering techniques, and the strategic integration of novel AI-driven design tools, is imperative for facilitating the widespread adoption of neural surrogate modelling across various engineering domains.

Acknowledgement
We thank Minbo Ma for his constructive suggestions on the clarity and organisation of our review.

References
[1] A. Keane, A. Forrester, A. Sobester, Engineering design via surrogate modelling: A practical guide, John Wiley & Sons, 2008. https://doi.org/10.2514/4.479557

[2] L. Regenwetter, A. Nobari, F. Ahmed, Deep generative models in engineering design: a review, J. Mech. Des. 144 (2022) 071704. https://doi.org/10.1115/1.4053859

[3] H. Attar, A. Foster, N. Li, Development of a deep learning platform for sheet stamping geometry optimisation under manufacturing constraints, Eng. Appl. Artif. Intell. 123 (2023) 106295. https://doi.org/10.1016/j.engappai.2023.106295

[4] H. Zhou, N. Li, Image-based Artificial Intelligence empowered surrogate model and shape morpher for real-time blank shape optimisation in the hot stamping process. https://doi.org/10.48550/arXiv.2212.05885

[5] Information on https://shorturl.at/efjSV

[6] S. Liu, Development of an Intelligent Flexible Sheet Metal Forming Technique, PhD Thesis, Imperial College London, 2022

[7] A. Géron, Hands-on machine learning with Scikit-Learn, Keras and TensorFlow: concepts, tools, and techniques to build intelligent systems, Oreilly, Newton, 2019. https://www.oreilly.com/library/view/hands-on-machine-learning/9781098125967/

[8] Information on https://domino.ai/data-science-dictionary/feature-engineering

[9] V. Jovanovic, M. Kuzlu, U. Cali, D.H. Utku, F.O. Catak, S. Sarp, N. Zohrabi, Digital Twin in Industry 4.0 and Beyond Applications, in: Digital Twin Driven Intelligent Systems and Emerging Metaverse, Springer, Singapore, 2023, pp. 155–174. https://doi.org/10.1007/978-981-99-0252-1_7

[10] Information on https://shorturl.at/bfF15

[11] Z. Han, K. Zhang, Surrogate-Based Optimization, in: Engineering Design Optimization, InTech, 2012, pp. 343–362. https://doi.org/10.5772/36125

[12] F. Viana, C. Gogu, T. Goel, Surrogate modeling: tricks that endured the test of time and some recent developments, Struct. Multidiscipl. Optim. 64 (2021) 2881–2908. https://doi.org/10.1007/S00158-021-03001-2

[13] G. Sun, S. Wang, A review of the artificial neural network surrogate modeling in aerodynamic design, Proc. Inst. Mech. Eng. G. J. Aerosp. Eng. 233 (2019) 5863–5872. https://doi.org/10.1177/0954410019864485

[14] H. Attar, H. Zhou, A. Foster, N. Li, Rapid feasibility assessment of components to be formed through hot stamping: A deep learning approach, J. Manuf. Process. 68 (2021) 1650–1671. https://doi.org/10.1016/j.jmapro.2021.06.011

[15] Information on https://www.neuralconcept.com/

[16] Information on https://www.monolithai.com/products

[17] Information on https://shorturl.at/twGPU

[18] Information on https://www.visendi.ai/

[19] Information on https://zoo.dev/text-to-cad

[20] C. Zhang, R. Pinquié, A. Polette, G. Carasi, H. De Charnace, J.P. Pernot, Automatic 3D CAD models reconstruction from 2D orthographic drawings, Comput. Graph. 114 (2023) 179–189. https://doi.org/10.1016/J.CAG.2023.05.021

[21] J. Yet, C. Lau, K. Thang, 2D Orthographic Drawings to 3D Conversion using Machine Learning, ICOSEC (2022) 1308–1312. https://doi.org/10.1109/ICOSEC54921.2022.9952009

[22] R. John, S. Peter, J. Alfred, R.J. Spick, S. Demediuk, J. Alfred Walker, Naive mesh-to-mesh coloured model generation using 3D GANs, Proc. ACSW (2020) 1–6. https://doi.org/10.1145/3373017.3373067

[23] K. Murdock, Advanced Materials, in: 3ds Max 2011 Bible, Wiley Publishing, Indianapolis, 2010: pp. 759–815. https://www.wiley.com/en-gb/3ds+Max+2011+Bible-p-9780470908372

[24] T. Mullen, Working with Textures and Materials, in: Mastering Blender, Second Edition, John Wiley & Sons, Inc., Indianapolis, 2013: pp. 33–84. https://www.wiley.com/en-cn/Mastering+Blender,+2nd+Edition-p-9781118275405

[25] H. Attar, A. Foster, N. Li, Implicit Neural Representations of Sheet Stamping Geometries with Small-Scale Features, Eng. Appl. Artif. Intell. 123 (2023) 106482. https://doi.org/10.1016/j.engappai.2023.106482

[26] R. Wu, C. Xiao, C. Zheng, DeepCAD: A deep generative network for computer-aided design models, Proc. IEEE Int. Conf. Comput. Vis. (2021) 6752-6762. https://doi.org/10.1109/ICCV48922.2021.00670

[27] J. Lambourne, K. Willis, P. Jayaraman, A. Sanghi, P. Meltzer, H. Shayani, BRepNet: A topological message passing system for solid models, IEEE/CVF CVPR (2021) 12768–12777. https://doi.org/10.1109/CVPR46437.2021.01258

[28] J. Pongetti, T. Kipouros, M. Emmanuelli, R. Ahlfeld, S. Shahpar, Using autoencoders and output consolidation to improve machine learning models for turbomachinery applications, Proc. ASME Turbo Expo. (2021) V02DT36A018. https://doi.org/10.1115/GT2021-60158

[29] T. Rios, B. Van Stein, T. Back, B. Sendhoff, S. Menzel, Multitask shape optimization using a 3-D point cloud autoencoder as unified representation, IEEE TEVC 26 (2022) 206–217. https://doi.org/10.1109/TEVC.2021.3086308

[30] K. Willis, Y. Pu, J. Luo, H. Chu, T. Du, J. Lambourne, A. Solar-Lezama, W. Matusik, Fusion 360 Gallery: A dataset and environment for programmatic CAD construction from human design sequences, ACM Trans. Graph. 40 (2021) 1–24. https://doi.org/10.1145/3450626.3459818

[31] S. Kucherenko, D. Albrecht, A. Saltelli, Exploring multi-dimensional spaces: a comparison of Latin hypercube and quasi monte carlo sampling techniques, (2015). https://doi.org/10.48550/arXiv.1505.02350

[32] F. Tejero, D. MacManus, F. Sanchez-Moreno, C. Sheaf, Neural network-based multi-point, multi-objective optimisation for transonic applications, Aerosp. Sci. Technol. 136 (2023) 108208. https://doi.org/10.1016/J.AST.2023.108208

[33] M. Norgaard, J. Ross, M. Ncrgaard, C. Jorgensen, Neural network prediction of new aircraft design coefficients, (1997). https://ntrs.nasa.gov/citations/19970023478

[34] H. Zhou, Q. Xu, N. Li, A study on using image based machine learning methods to develop the surrogate models of stamp forming simulations, J. Manuf. Sci. Eng. 144 (2022) 021012. https://doi.org/10.1115/1.4051604

[35] H. Zhou, H. Attar, Y. Pan, X. Li, P. Childs, N. Li, On the feasibility of small-data learning in simulation-driven engineering tasks with known mechanisms and effective data representations, in: NeurIPS-AI4Science Poster, 2021. https://openreview.net/forum?id=qvxJBCp7aji

[36] J. Petrik, S. Ali, M. Feistle, M. Bambach, CrystalMind: A surrogate model for predicting 3D models with recrystallization in open-die hot forging including an optimization framework, Mech. Mater. 189 (2024) 104875. https://doi.org/10.1016/J.MECHMAT.2023.104875

[37] B. Shahriari, K. Swersky, Z. Wang, R.P. Adams, N. De Freitas, Taking the human out of the loop: A review of Bayesian optimization, Proc. IEEE 104 (2016) 148–175. https://doi.org/10.1109/JPROC.2015.2494218

[38] N. Loka, I. Couckuyt, F. Garbuglia, D. Spina, I. Van Nieuwenhuyse, T. Dhaene, Bi-objective Bayesian optimization of engineering problems with cheap and expensive cost functions, Eng. Comput. 39 (2023) 1923–1933. https://doi.org/10.1007/s00366-021-01573-7

[39] E. Ruff, R. Russell, M. Stoeckle, P. Miotto, J.P. How, Surrogate neural networks for efficient simulation-based trajectory planning optimization (2023). https://doi.org/10.48550/arXiv.2303.17468

[40] Information on https://www.cadify.ai/

[41] Information on https://www.getleo.ai/cad

[42] F. Bie, Y. Yang, Z. Zhou, A. Ghanem, M. Zhang, Z. Yao, X. Wu, C. Holmes, P. Golnari, D.A. Clifton, Y. He, D. Tao, S.L. Song, RenAIssance: A survey into AI text-to-image generation in the era of large model, (2023). https://doi.org/10.48550/arXiv.2309.00810

[43] A. Nichol, H. Jun, P. Dhariwal, P. Mishkin, M. Chen, Point-E: A system for generating 3D point clouds from complex prompts, (2022). https://doi.org/10.48550/arXiv.2212.08751

[44] C. Lin, J. Gao, L. Tang, T. Takikawa, X. Zeng, X. Huang, K. Kreis, S. Fidler, M. Liu, T. Lin, Magic3D: High-resolution text-to-3D content creation, Proc. IEEE CVPR 2023-June (2023) 300–309. https://doi.org/10.1109/CVPR52729.2023.00037

[45] Information on https://www.solidworks.com/media/wild-imagination-artificial-intelligence

Electroshock plastic constitutive modelling of high strength aluminum alloy based on GA-BP neural network

SONG Yanli[1,a *], XU Hainan[1,b], CHEN Long[1,c], LU Jue[1,d] and HUA Lin[1,e *]

[1]Hubei Key Laboratory of Advanced Technology for Automotive Components, Wuhan University of Technology, Wuhan 430070, China

[a]ylsong@whut.edu.cn, [b]xhn0801@whut.edu.cn, [c]335194@whut.edu.cn, [d]lujue@whut.edu.cn, [e]hualin@whut.edu.cn

Keywords: 7075 Aluminum Alloy, Genetic Algorithm, BP Neural Network, Electroshock Plastic Constitutive Model, Electrically-Assisted Forming

Abstract. Electrically-assisted forming is an effective way to improve the difficult-to-form materials. On basis of electro- and magneto- plastic effects, our team put forward a novel electroshock treatment (EST) process, which can not only improve the formability of the material but also improve the service performance of the formed parts. However, electrically-assisted forming involves nonlinear deformation with multi-physics coupling, the microstructural evolution and plastic mechanical response of materials under the coupling of electric current field and stress field are highly intricate, posing significant challenges in predicting macroscopic deformation behavior. The study of the characterization and precise control of plastic deformation's constitutive relation is therefore highly significant. In this work, the stress-strain curves of AA7075 under, different current densities, periods and power duration are obtained by means of the pulse electric current assisted tension test, and the intelligent prediction of the electroshock plastic constitutive relationship of AA7075 is realized based on the Genetic Algorithm (GA) optimized Back Propagation (BP) (GA-BP) neural network. The results show that the proposed algorithm can reduce the input parameters of the constitutive model to current, temperature and strain, and improve the computational efficiency. GA is used to optimize the initial weight and threshold of BP neural network, and the best hidden node number was selected as 12 by error verification to train the network. Compared with the traditional BP neural network, the neural network optimized by GA has higher accuracy, and in the plastic deformation stage, the coefficient of determination R^2 between the predicted results and the experimental results is basically 99%, which is about 20% higher than that of the BP neural network. At the same time, it solves the data fluctuation in the long-term prediction due to the standard of satisfying the error in the prediction.

Introduction

In recent years, the application of electrically-assisted forming in improving the formability of metal materials has received widespread attention, since it can improve the plasticity and forming performance of the material and reduce flow stress and springback [1]. During the electrically-assisted forming process, the material is subject to the coupling effect of multiple physical fields, and its macroscopic mechanical response and microstructure evolution are very complex. Hao et al. [2] studied the effect of extremely low frequency electric pulse on the mechanical properties of AA7075-T6, and found that the tension strength of AA7075-T6 was reduced by the pulse electric current assisted tension, the elongation of the specimen after fracture can be increased by 32.3%, and a linear relationship can be established between the current density, duration and the stress drop. Conrad et al. [3] explored the influence of an electron wind on each of the parameters in the equation for the thermally-activated motion of dislocations, and found that the pre-exponential impact was the greatest and the derived electron wind push coefficient was one or more orders of magnitude larger than the value normally accepted for the electron drag coefficient. Ross et al. [4] studied the material property response of Ti-6Al-4V alloy under the action of electric current and the electromigration theory of metals. They applied the EAF process to the forming process and found that the electric current can be used at lower temperatures. Reduce the deformation resistance of the material, improve the formability of the metal during deformation and reduce springback after deformation.

Traditionally, material constitutive equations are established by multiple linear regression analysis of experimental results. However, many factors affecting flow stress are highly nonlinear, which makes the establishment of theoretical models difficult and easy to deviate from the actual situation. There are three main methods for establishing constitutive models: phenomenological models fitted by empirical formulas, models based on physical internal variables and statistical models based on test data. The test data statistical model has fast operation speed and high accuracy. Therefore, the neural network was used to establish the constitutive relationship of the material, and has been widely used. Commonly used intelligent algorithms are usually predicted by neural networks such as Back Propagation, Artificial, and Elman. Wan et al. [5] used Back Propagation (BP) neural network to establish a Zr-4 alloy constitutive model. The correlation coefficient is 0.9981, the average absolute relative error is 4.497, and 93.75% of the data points deviate from the predicted value within 10%. Yu et al. [6] established an Artificial neural network constitutive model of zr based the bulk metallic glass composites based on Genetic Algorithm (GA) and particle swarm optimization (PSO). The study found that compared with the traditional constitutive model, the constitutive model established through neural network is more accurate and more suitable for (Zr55Cu30Al10Ni5)94Ta6. Neural network is a complex nonlinear system with high adaptability, self-learning and fault tolerance capabilities. When it is applied to complex nonlinear prediction, its effect is better. The GA has the ability to quickly seek optimization.

A novel electroshock plasticizing forming process has been proposed by our team, which can not only improve the formability of the material but also improve the service performance of the formed parts [2]. However, there is still a lack of constitutive equations to quantitatively describe its periodic zigzag stress-strain behavior. This paper uses GA to optimize the weights and thresholds of BP neural network, and establishes a constitutive relationship prediction model of AA7075-T6 under electroshock based on GA-BP neural network. This model can be used to predict AA7075-T6 rheological behavior of alloys under electroshock treatment (EST).

Experimental process and mechanical behavior analysis

Experimental Materials. This paper uses AA7075-T6 as the test material, and the specimen thickness is 1.6mm. Its nominal chemical composition is shown in Table 1. The tension specimen required for the pulse electric current assisted tension test was prepared by laser cutting along the rolling direction of the sheet, and then the flash on the edge of the specimen was polished with 800-grit and 1500-grit sandpaper. The specific dimensional drawings and physical pictures are as follows as shown in Fig. 1. In order to facilitate the power supply, both ends of the specimen were additionally processed into terminals. The specimen and cable are reliably connected using bolts and nuts.

Table 1. Nominal chemical compositions of AA7075-T6 sheets (wt. %).

Si	Fe	Cu	Mn	Mg	Cr	Zn	Ti	Al
0. 40	0. 35	1. 52	0. 092	2. 66	0. 22	5. 28	0. 029	Bal.

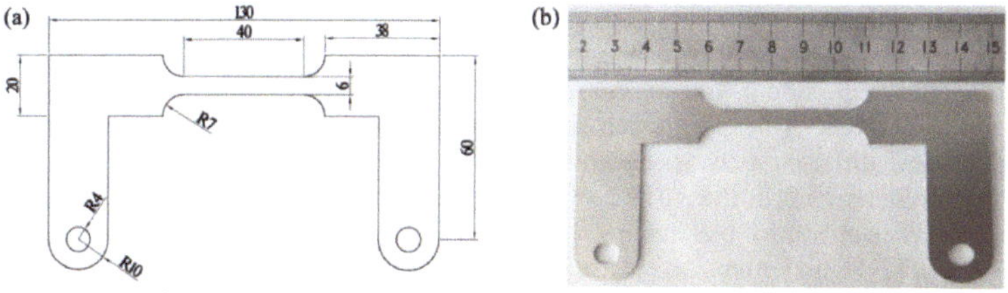

Fig. 1. The pulse electric current assisted quasi-static unidirectional tension test specimens: (a) dimensional drawing; (b) photograph (Unit of length: mm).

Experimental method. The physical diagram and schematic diagram of the pulse electric current assisted tension equipment are shown in Fig. 2. It consists of a DC power supply, the electronic universal tension testing machine and a Fotric 226 infrared thermometer. The electronic universal tension testing machine is a CMT of the SANS MTS brand. Model -5205, the maximum test force is 200kN, the load accuracy is ±0.5%, and the data sampling interval is 0.01s. During the pulse electric current assisted tension test, electric pulses and quasi-static tension loads are applied to the specimen simultaneously. An insulation conversion head is installed between the chuck and chuck base of the testing machine to ensure sufficient insulation between the universal testing machine and the tension specimen. At the same time, the temperature of the specimen surface is measured with an infrared thermometer, and a layer of black paint is sprayed on the specimen surface to increase the energy emissivity, thereby improving measurement accuracy.

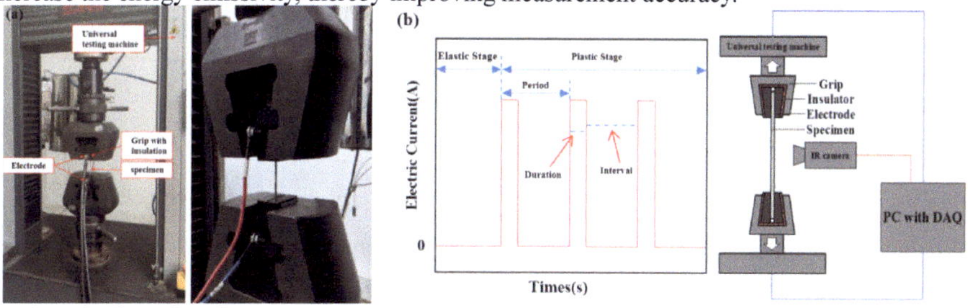

Fig. 2. The pulse electric current assisted quasi-static unidirectional tension platform: (a) schematic diagram; (b) photograph.

The pulse electric current assisted quasi-static unidirectional tension test of AA7075-T6 was carried out with different current density (J), pulse duration (td) and pulse period (ω) as process parameters, and the pulse electric current waveform is a unidirectional wave with constant current density and pulse period. Each set of tests starts to energize after the material enters plastic deformation.

The current density was selected as 50 A/mm^2 and 60 A/mm^2, the pulse time is 1s, 3s, and the pulse period is 30 s, 40 s. Among them, the current density J is the ratio of the current value output by the pulse power supply to the cross-section of the specimen gauge section, and the unit is A/mm^2. In the pulse electric current assisted quasi-static unidirectional tension test, the medium speed was maintained at 2.4 mm/min. The relationship between the specimen strain rate and the displacement speed of the tension testing machine beam is:

$$\dot{\varepsilon}=\frac{V}{60L_c}. \quad (1)$$

where $\dot{\varepsilon}$ is the strain rate of the specimen (s^{-1}); V is the beam displacement speed of the tension testing machine mm/min; L_C is the gauge length.

During the experiment, the microcomputer processing system automatically collects, calculates and corrects the experimental data, and outputs the flow stress data under different current densities, different pulse periods and different pulse durations. At the same time, the infrared thermometer records the surface stress of the specimen temperature.

Analysis of electroshock plasticizing mechanical behavior. Fig. 3. shows the true stress-strain curves of AA7075-T6 under tension effects of 50A/mm^2-3s-40s, 60A/mm^2-1s-30s, and 60A/mm^2-3s-40s. During the plastic deformation stage, after the current flows during the pulse duration, the stress value of the aluminum alloy material will suddenly drop. When the current flows and enters the pulse interval, the stress value rises in a slope, then enters the strengthening stage, completing a local stress-strain curve, and then continues this cycle after the arrival of the second electroshock until the specimen breaks. It can be seen from the blue curve and black curve in Fig. 3 that under the same pulse period and different current densities, the stress drop value after each electrical pulse shock increases with the increase of current density. While for the blue curve and the red curve, it can be seen that under the same current density and different pulse periods, the longer the current duration, the more obvious the stress drop, and the overall trend of stress is downward in a longer duration.

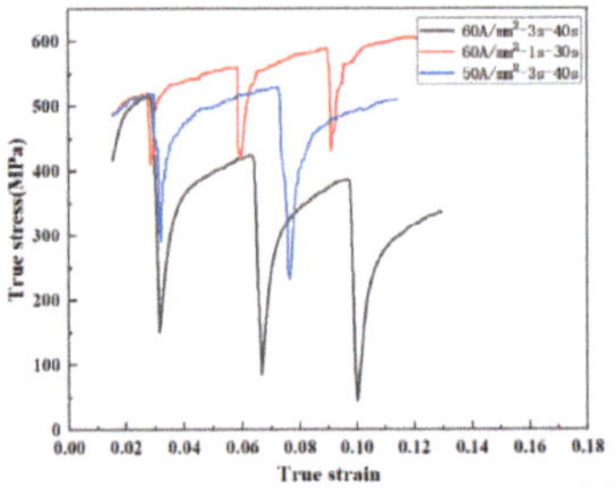

Fig. 3. The true stress-strain curves of AA7075-T6 under three different states of the pulse electric current assisted quasi-static unidirectional tension.

Neural network constitutive model based on neural network

BP neural network theory. The nonlinear mapping relationship of the BP neural network model is superior to the linear relationship of the multiple linear regression model in processing variables, and is widely used in various complex system analysis, especially in aspects such as simulation, fitting, and prediction that are difficult to quantitatively analyze [7]. Its structure is shown in Fig. 4.

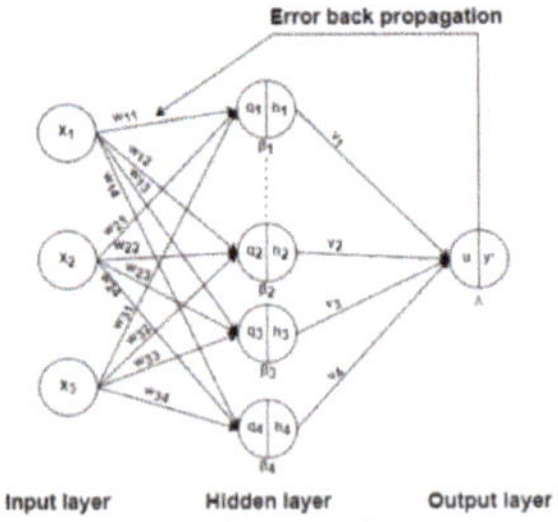

Fig. 4. BP neural network structure.

In this work, the strain, temperature and current obtained during the pulse electric current assisted quasi-static unidirectional tension test of AA7075-T6 are used as input, and the stress is used as output. Therefore, the number of neurons in the input layer is 3, and the number of neurons in the output layer 1. First, normalize the data [8]:

$$x' = \frac{x - x_{min}}{x_{max} - x_{min}}. \tag{2}$$

where x' is the normalized data; x is the original input data, x_{max} and x_{min} respectively correspond to the maximum and minimum values of the original input data.

Genetic algorithm. GA is a stochastic global search optimization method that simulates phenomena such as replication, crossover, and mutation that occur in natural selection and inheritance, allowing the population to evolve to better and better areas in the search space [9], thereby obtaining high-quality solutions to the problem. The steps of the GA are shown in Table 2:

Table 2. GA steps.

Number	Name	Definition	Methods
1	Encoding	GA mapping in space and mapping traits to genes.	Encoding method: number encoding.
2	Population	Randomly generate n initial individuals to form a population.	/
3	Fitness	The higher the fitness, the better the individual is.	/
4	Selection	Select individuals with higher fitness to enter the next generation.	Selection method: roulette.
5	Crossover	Select individuals from two parents and exchange part of the genes of the parents to produce new offspring.	Crossover method: the two-point crossover, the probability is 0.8.
6	Mutation	One or several genes on a chromosome mutate, forming new chromosomes.	The probability is 0.01.

According to the empirical formula, the larger the population size, the better is not generally selected. Although the search ability is stronger if the size is too large, the algorithm time will be too long, and the search ability of smaller scales is limited. Considering the complexity of the problem, through experience and experiments, the earliest Determine and select the initial population size as 50, and set the parameters of the GA as follows:

Table 3. The setting of experimental parameters.

Parameter settings	Value
Population size	50
Crossover probability	0. 8
Mutation probability	0. 01
Number of iterations	50

Analysis of prediction results of constitutive model

The prediction effect of the constitutive model of the plastic deformation stage of the AA7075-T6 obtained by the BP neural network is shown in Fig. 5, where the coefficient of determination R^2 in the Fig. (a), (b), (c) is 0.75, 0.85 and 0.84. From a mathematical point of view, the BP algorithm is a local search method. When solving the global extreme value of some complex nonlinear functions, the algorithm may fall into a local minimum, causing training to fail. And the training of neural network is the process of solving the optimal weights and biases. Different initial weights and biases have a great impact on the training of the network. It is necessary to try different initial values many times to obtain the optimal network, so the task is heavy and causes greater randomness to the results. And because the BP neural network prediction stops network training based on whether the target error meets the conditions, this also leads to the fact that although the BP neural network prediction result is within the error range, it may fluctuate up and down, leading to prediction of the actual result. Trend generates errors [10], Fig. 5(d) shows a partial enlargement of Fig. 5(c).

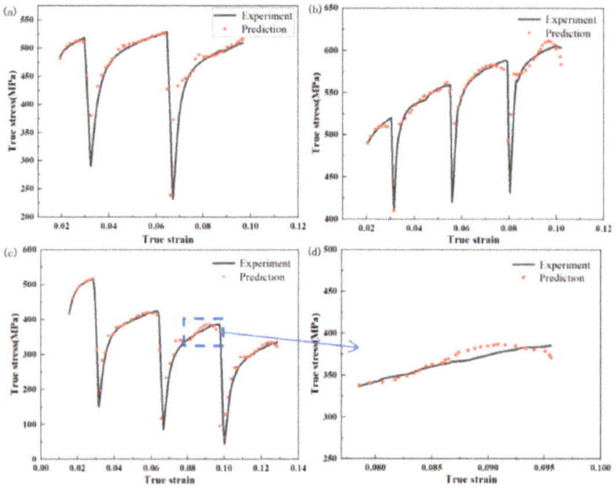

Fig. 5. The true stress-strain of AA7075-T6 based on BP model under various conditions: (a) 50A/mm²-3s-40s, (b) 60A/mm²-1s-30s, (c) 60A/mm²-3s-40s, (d) partial enlargement of c.

After optimizing the BP neural network using GA, the prediction effect of the plastic deformation stage of the AA7075-T6 is obtained, as shown in Fig. 6, and the prediction accuracy of the optimized neural network reaches 99%, and the fluctuations of the predicted values are also in line with the actual trend.

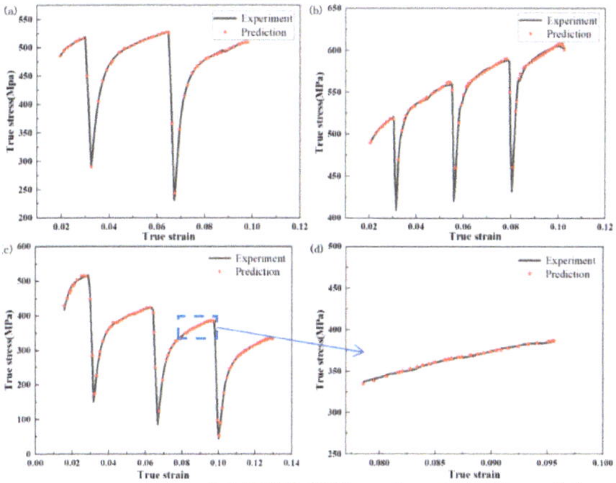

Fig. 6. The true stress-true strain of AA7075-T6 based on GA-BP model under various conditions: (a) 50A/mm²-3s-40s, (b) 60A/mm²-1s-30s, (c) 60A/mm²-3s-40s, (d) partial enlargement of c.

The Fig. 7 is the error diagram between the predicted results and the actual results. It can be seen from the figure that the error fluctuations are small, and the maximum error does not exceed 25 MPa. The larger error fluctuations exist in the turning point of stress decline and rise. Because the number of specimens here is small, and the neural network is more sensitive to small changes in specimen data, the prediction results will have slightly larger fluctuations. The fluctuations at the end and front end in Fig. 7(b) and c are due to sampling, due to sudden small changes in the data detected by the microcomputer. Besides, the setting neural network threshold is too large, resulting in large fluctuations. Subsequent improvement work needs to use a greater frequency in the turning point of stress decline and rise during data sampling to obtain more data, and at the same time reduce the threshold of the neural network, thereby reducing the error.

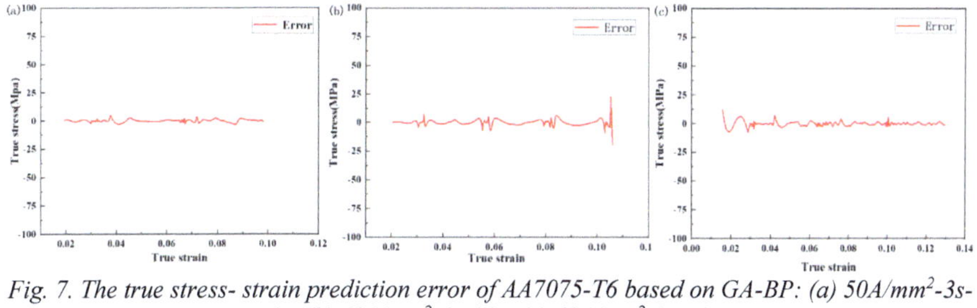

Fig. 7. The true stress- strain prediction error of AA7075-T6 based on GA-BP: (a) 50A/mm²-3s-40s, (b) 60A/mm²-1s-30s, (c) 60A/mm²-3s-40s.

Conclusions

This paper uses a GA to optimize the BP neural network to predict the constitutive relationship of AA7075-T6 during the pulse electric current assisted quasi-static unidirectional tension:

(1) Build a BP neural network model to predict constitutive relationships. Experimental data show that the fitting accuracy of BP neural network does not meet the requirements, and the predicted data trend is inconsistent with the actual situation. Therefore, in view of the above situation, the GA is used to optimize the initial weight threshold of the BP neural network to obtain

the best prediction results.

(2) Using the global search capability of the GA, the optimal initial weight is found. The optimized neural network has greatly improved the prediction accuracy. The determination coefficient R^2 of its prediction results has reached 99%. The error ranges of the three prediction results are all within 25MPa, which meets the prediction requirements. Compared with the BP neural network The prediction accuracy is increased by about 20%, and the inaccuracy of the BP neural network in predicting trends is improved.

(3) It is found that the neural network prediction has large error fluctuations, which are more obvious when the stress decreases and rises suddenly. The reason is that there is less sampling data at the fluctuation point and the threshold setting is too large. The improvement measure is to increase the fluctuation. Larger sampling frequency and set smaller threshold.

Acknowledgements
This work was financially supported by the National Key Research and Development Project of China (Grand No. 2020YFA0714900), and the National Natural Science Foundation of China (51975440).

References

[1] S. Wang, A. Xiao, Y. Lin, Effect of induced pulse current on mechanical properties and microstructure of rolled 5052 aluminum alloy, J. Materials Characterization. 185 (2022) 111757. https://doi.org/10.1016/j.matchar.2022.111757

[2] Y. Song, C. Hao, Effect of Extremely Low Frequency Pulse Current Treatment on the Mechanical Properties of Al-Zn-Mg-Cu Aluminum Alloy, J. Mech. Eng. 58 (2022) 68-77. https://doi.org/10.3901/JME.2022.10.068

[3] H. Conrad, Thermally activated plastic flow of metals and ceramics with an electric field or current, Mater. Sci. Eng. A 322 (2002) 100-107. https://doi.org/10.1016/S0921-5093(01)01122-4

[4] C.D. Ross, T.J. Kronenberger, J.T. Roth, Effect of DC on the formability of Ti-6Al-4V, J. Eng. Mater. Tech.-Transactions of the ASME 131 (2009). https://doi.org/10.1115/1.3078307

[5] P. Wan, H. Zou, K. Wang, Research on hot deformation behavior of Zr-4 alloy based on PSO-BP artificial neural network, J. Alloy. Compd. 826 (2020) 154047. https://doi.org/10.1016/j.jallcom.2020.154047

[6] G. Yu, X. Bao, X. Xu, Constitutive modeling of Ta-rich particle reinforced Zr-based bulk metallic composites in the supercooled liquid region by using evolutionary artificial neural network, J. Alloy. Compd. 938 (2023) 168488. https://doi.org/10.1016/j.jallcom.2022.168488

[7] R. Chen, J. Song, M. Xu, Prediction of the corrosion depth of oil wel cement corrdded by carbon dioxide using GA-BP neural network, Constr. Build. Mater. 394 (2023) 132127. https://doi.org/10.1016/j.conbuildmat.2023.132127

[8] J. Shi, J. Wang, D.D. Macdonald, Prediction of primary water stress corrosion crack growth rates in Alloy 600 using artificial neural networks, Corros. Sci. 92 (2015) 217-227. https://doi.org/10.1016/j.corsci.2014.12.007

[9] K.G. Srinivasa, K.R. Venugopal, L.M. Patnaik, A self-adaptive migration model genetic algorithm for data mining applications, Inform. Sci. 177 (2007) 4295-4313. https://doi.org/10.1016/j.ins.2007.05.008

[10] Y. Zhou, S. Li, BP neural network modeling with sensitivity analysis on monotonicity based Spearman coefficient, Chemometr. Intell. Lab. 200 (2020) 103977. https://doi.org/10.1016/j.chemolab.2020.103977

Materials Research Proceedings 44 (2024) 511-518 https://doi.org/10.21741/9781644903254-55

A review of deep neuron network applications in extrusion die design

DING Jiangfeng[1,a*], CHEN Siyi[1,b], SHAO Zhutao[1,c], SHI Zhusheng[1,d] and LIN Jianguo[1,e]

[1]Imperial College London, Department of Mechanical Engineering, South Kensington, London SW7 2AZ, UK

[a]j.ding22@imperial.ac.uk, [b]siyi.chen18@imperial.ac.uk, [c]z.shao12@imperial.ac.uk, [d]zhusheng.shi@imperial.ac.uk, [e]jianguo.lin@imperial.ac.uk

Keywords: Metal Forming, Extrusion Die Design, Machine Learning, Neuron Network

Abstract. Ensuring the quality of extrusion product necessitates meticulous die design, typically achieved through simulation iterations and/or experimental trials. However, this process is not only time-consuming but also costly. Despite substantial research utilizing historical data and finite element analysis (FEA) to elucidate design guidelines and principles, and the existence of numerous empirical equations guiding die design, it remains more of an art reliant on the designer's experience. In contrast, Deep Neural Networks (DNNs) have the capability to capture design experience with appropriately defined inputs and outputs, transforming it into abstract features for further application. With the advancement of DNNs, the automatic generation of precise die designs has become achievable. Several research studies have been undertaken to enhance die design through the application of DNNs, particularly Convolutional Neural Networks (CNNs). CNNs, a machine learning method commonly applied to extract information from images, have been utilized due to the intricate nature of die design. Given the inherent characteristics of DNNs, a significant challenge in incorporating DNNs into die design lies in devising a scheme to abstract 3D die designs for defining inputs without loss of information. Various methods exist for handling 3D objects, such as point clouds or projecting 3D objects into 2D depth graphs. Nonetheless, most of these methods prove challenging to implement effectively in the realm of die design. Another challenge stems from the overall complexity of the extrusion die. While most research has focused on automatically designing specific features of the die, such as the location or shape of portholes, there have also been data-driven studies attempting to generate entire die designs using historical data. This paper aims to review the status of the application of DNNs in hot extrusion die design and explore the further potential in this field.

Introduction

Extrusion stands out as one of the most iconic metal forming techniques in modern industries, finding widespread applications in architecture, construction, display equipment, electrical systems, and various sectors of industrial and transportation fields [1]. This metal-forming process involves pushing a long cylindrical billet within a closed cavity through a die with the desired cross-section. The precision of the die is a critical factor influencing the quality of the final product. While other essential parameters like pressure and temperature play a role, die design emerges as the foremost consideration for ensuring product quality.

Currently, finite element analysis (FEA) remains the mainstream approach for die design. Once a die design is successfully simulated, experiments can be conducted, and adjustments to the die are made based on the results. However, this process involves several iterations, proving to be both time-consuming and costly. Despite extensive research leveraging historical data and FEA to establish design guidelines and principles, die design still relies heavily on the experience of the designer, making it more of an art than an analytical science.

In contrast, DNNs have shown promise in capturing design experience by translating it into abstract features for practical applications [2]. With advancements in DNNs, the automatic generation of precise die designs has become achievable. Numerous research studies have explored enhancing die design through the application of DNNs, with a particular focus on CNNs, a machine learning method commonly used for extracting information from images [3]. CNNs prove beneficial due to the intricate nature of die design.

Despite the potential advantages, incorporating DNNs into die design presents significant challenges. One major hurdle involves devising a scheme to abstract 3D die designs for defining inputs without losing information. While various methods exist for handling 3D objects, such as point clouds or projecting 3D objects into 2D depth graphs, effectively implementing them in the realm of die design has proven challenging. Additionally, the overall complexity of the extrusion die poses another challenge. Most research has concentrated on automatically designing specific features of the die, such as the location or shape of portholes [4]. Still, there have been data-driven studies attempting to generate entire die designs using historical data [5].

This article aims to review the current status of the application of DNNs in hot extrusion die design and explore their further potential in this field.

Die Design Challenges

The crux of a successful extrusion process lies in the intricacies of die design. While variables such as temperature and pressure also wield significant influence, these factors are often predetermined in practice. Achieving a straight final product or ensuring a uniform metal flow at a steady state necessitates the precision of die design.

Presently, the prevailing method for die design unfolds as follows: Designers begin by tailoring an initial design to the product's profile using empirical equations and expertise, sometimes adjusting from existing designs if applicable. They then employ FEA to test structural integrity, iterating if necessary. Validation through real-world experimentation follows, with any disparities prompting further refinement. Once the flow is achieved uniform at steady state, the design process concludes, readying the die for implementation.

The described method is notably laborious and time intensive, often stretching over months, particularly for inexperienced designers. Moreover, the execution of multiple sets of experiments adds to the overall costliness of this approach. The intricacies of die design give rise to a variety of guidelines and principles tailored to specific scenarios. However, many of these principles may conflict with each other, underscoring the experiential aspect of die design. Consequently, the quality of the final product is heavily reliant on the designer's experience and expertise.

The knowledge possessed by die designers is often subjective and lacks a systematic structure, leaning more towards intuition than analytical knowledge. For instance, study suggests that die designers commonly reduce the bearing length at the tip of a leg to between 50% and 75% of the bearing in the rest of the leg to account for the greater relative perimeter [6]. However, the exact percentage reduction is subjective and varies among designers. Designers often intuitively determine an optimal percentage based on their experiences with similar dies. Therefore, finding effective way to leverage this experiential knowledge to guide future designs would be invaluable.

Deep Neural Networks in Die Design

The continuous advancement of Deep Neural Networks (DNN) is shaping a significant shift in uncovering latent knowledge, while remaining grounded in academic principles. Essentially, DNN can be understood as a detailed approach using stochastic gradient descent (SGD) to minimize errors [7]. This method has earned DNN a reputation for its ability to uncover hidden features and simplify complex knowledge, leading to their widespread adoption across various industries. Particularly, it finds notable applications in areas like strain analysis and crack prediction [8].

The burgeoning potential of DNN is especially conspicuous in the domain of metal forming. Pioneering contributions by Zhou et al. [9] in successfully developing a surrogate model for the hot stamping process and by Liu et al. [10] in effectively harnessing DNN in sheet metal forming, underscore the expanding utility and applicability of DNN methodologies across various facets of industrial practice.

Considering the intrinsic nature of extrusion dies, where the product profile is represented as a 2D graph, the utilization of neural networks becomes imperative for handling graph-based data. Consequently, the CNN emerges as a pivotal player in contemporary automatic die design systems. Tailored for image processing, CNN employs filters and convolutional layers to autonomously learn hierarchical features, showcasing exceptional efficacy in tasks such as image recognition.

However, the integration of DNNs into the realm of extrusion die design introduces unique challenges. Unlike dies used for stamping or sheet metal forming, extrusion dies involve multiple parts, complicating efforts to simplify their intricate 3D properties into 2D or 1D representations. This intricacy necessitates further exploration and refinement in the seamless integration of DNN methodologies to address the distinct challenges posed by extrusion die design, a pivotal avenue for future research and advancement in the field.

Challenges in Incorporating DNNs into Die Design

Although DNNs hold significant potential in the field of extrusion die design, they face an inevitable challenge: digitization. To effectively utilize DNNs, all aspects of a die design must be captured and represented in a suitable format. Currently, there are two primary methods for extrusion die design: one method involves representing all information in a Multiview graph, while the other employs computer-aided design (CAD) techniques. Fig. 1a and 1b provide examples for both cases, respectively.

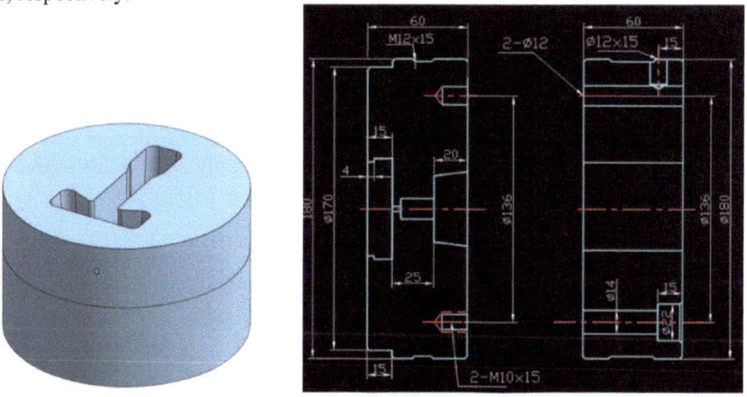

a) 3D representation of Die design b) Multiview representation of Die design

Fig. 1. Two different representations of die design.

There are two potential strategies for digitization 3D design emerge:

1. Finding an abstract representation of 3D design.
2. Finding a 3D representation that can capture most attributes.

While these strategies have been extensively studied, primarily not for the analysis of extrusion die design, they offer valuable insights into the digitization of CAD models.

Various methods exist to represent a 3D object as abstract vectors. For instance, Wu et al. [11] successfully translated CAD models into semi-natural language sequences (CAD sequences). They broke down the process of drawing a CAD model into step-by-step sequences, as illustrated in Fig. 2. This approach, based on the logic of the CAD platform Onshape, was first introduced by Willis et al. [12] in Fusion 360 Gallery. A CAD model is formed step by step through "sketch" followed by "extrude". By generating such sequences for a dataset, a transformer-based autoencoder is trained to capture the embedded abstract features of CAD designs, translating them into a 256*1 latent vector. Inspired by this work, Jobczyk and Homann [13] successfully created CAD models from multi-view images. The process involves generating CAD sequences, producing latent vectors using DeepCAD as ground truth, generating Multiview graphs from CAD models, and training a CNN-based DNN to return a 256*1 vector as output. This method has proven effective in generating CAD sequences (representing CAD models) from Multiview graphs.

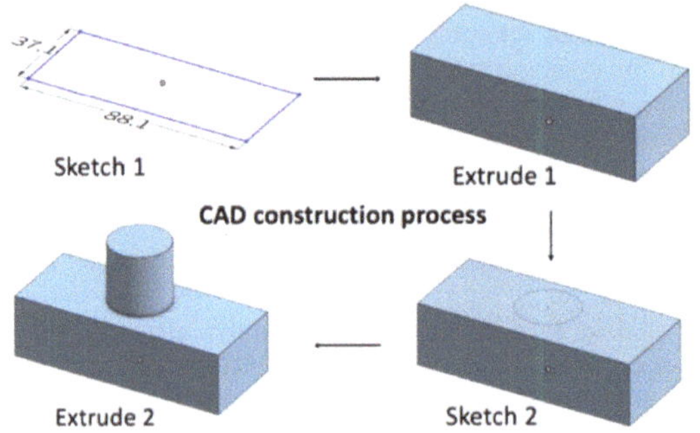

Fig. 2. The CAD construction process used in DeepCAD (unit: mm).

Medial axis transformation (MAT) is another technique employed in extrusion die design. The medial axis of an object refers to a set of points within the object, where each point serving as the center of a circle tangent to the object's boundary at two nonadjacent points [14]. Prior to the emergence of DNN, MAT was applied in automatic extrusion die design, two representative product's profiles are illustrated in Fig. 3. Lin [15] successfully utilized this method to analyze geometric aspects using MAT.

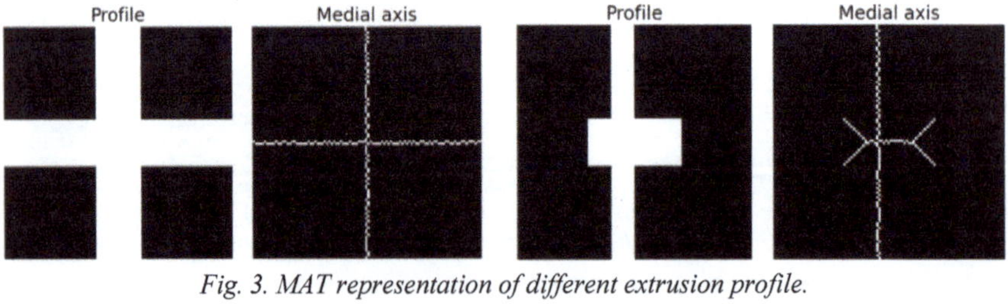

Fig. 3. MAT representation of different extrusion profile.

While MAT provides a valuable means of transforming the profile into a lower-dimensional representation, it does not offer a comprehensive solution for extrusion die design due to its limitation in addressing the entire 3D object. Despite this drawback, MAT has historically contributed to understanding and reasoning about certain geometric aspects relevant to extrusion die design.

An alternative widely employed approach involves sampling points across the entire 3D object, known as a point cloud. This method utilizes a collection of data points in a 3D coordinate system, where each point within the cloud signifies a specific spatial position, typically defined by its x, y, and z coordinates. In contrast to triangle meshes, point clouds do not necessitate the storage or maintenance of polygonal-mesh connectivity or topological consistency [16].

This method has demonstrated effectiveness in various studies; for instance, Lin et al. [17] utilized point clouds to represent 3D objects and train a surrogate model, a type of DNN designed to replace FEA in certain applications. Lin's approach proved capable of reconstructing 3D objects, however, achieving a more accurate representation often requires a substantial number of points, typically ranging from 2000 to 3000 for a solid representation. This quantity of points can be substantial as depicted in Fig. 4, leading to prolonged training times.

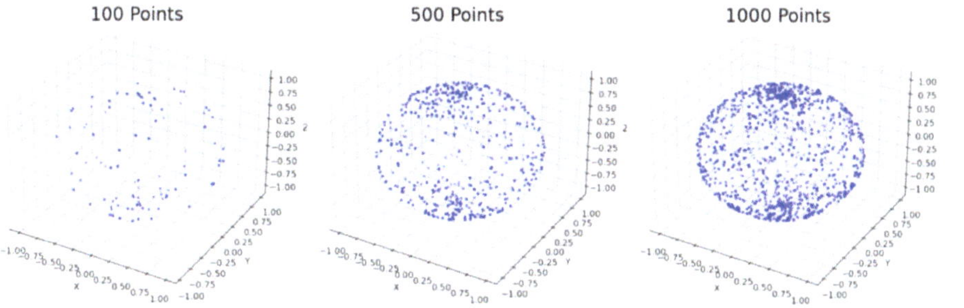

Fig. 4. Different number of points represent a same ball.

An alternative approach involves the use of 3D CNN to directly process 3D CAD data. Lee et al. [18] have employed this method to recognize features of 3D CAD models. While the results are promising in terms of feature recognition, it is noteworthy that existing literature predominantly focuses on classification and recognition tasks rather than generation using this method.

Status of DNNs in Hot Extrusion Die Design

Despite these aforementioned challenges, DNNs are beginning to make an impact in extrusion die design. However, owing to the regression-like attributes of DNNs, their current capability is primarily limited to providing predictions based on available data. Given the precision requirements of die design, existing literature offers guidance for specific parts of the die or provides a preliminary design for further refinement.

Llorca-Schenk et al. (2023) successfully analyzed the geometry of ports, including parameters such as port area, port perimeter, and die center to port center distance, utilizing various machine learning methods. Some variables they used to decompose a port is shown in Fig. 5. However, this analysis relied on a substantial amount of data from successfully designed porthole dies, specifically those made of H13 hot work steel with aluminum alloy 6063 billet material, which can be hard to collect.

Zangara et al. [19] decomposed critical attributes of extrusion die design into 12 parameters for digitization, e.g., Pocket width is called Rpc, Bridge length is called Hbr. The whole die is then parametrized with such decomposition and then can be used for training. Authors analyzed the correlations among these parameters using a support vector machine (SVM).

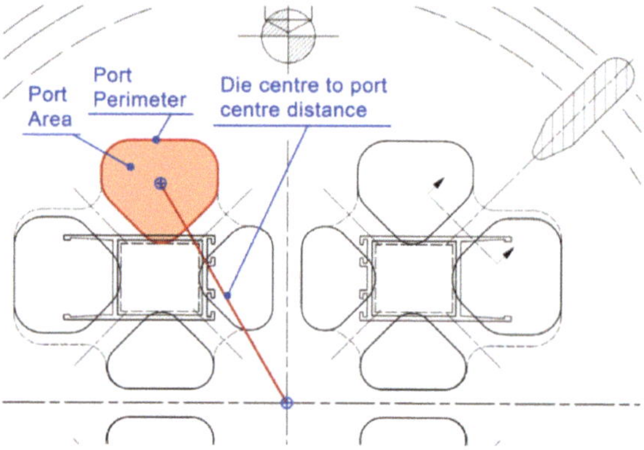

Fig. 5. Variables that been used to describe a porthole die [4].

In contrast to previous studies that focused predominantly on specific parts (ports) or aimed to assist in die design, Yu et al. (2023) successfully employed DNNs to design entire extrusion dies. By analyzing a large number of successful die designs and incorporating prior analytical studies on porthole die design, they identified porthole geometries and bearing length as the most crucial parameters. Porthole geometries were categorized based on shape and the number of portholes. For bearing length, an empirical equation with several constants was employed, with the values of these constants varying with changes in the profile's geometries. Different scenarios were categorized, and CNN was used to detect and locate them, reporting a set of constants. Bearing length was then calculated using these constants. Although numerous parameters require consideration, their results provide a robust outcome at steady state, demonstrating the effectiveness of their DNN-based approach in designing entire extrusion dies.

Potential of DNNs in Hot Extrusion Die Design

Incorporating DNNs into die design poses significant challenges, since emulating the expertise of experienced designers demands a substantial dataset of successful die designs for supervised learning. Yet, the intricacies of die design, considered a company's proprietary knowledge, hinder the collection of such data. Compounded by the messy nature of industrial data, primarily tailored for in-house understanding, the lack of a standardized approach across the industry further complicates the process. Consequently, the current study is constrained to specific die features or a narrow profile range due to these limitations.

The prospect of constructing a large-scale neural network akin to ChatGPT for die design becomes feasible if a standardization effort succeeds. However, establishing such a standard proves challenging. An alternative approach involves exploring semi-unsupervised learning, akin to control theory or reinforcement learning with feedback. While this method requires FEA to provide feedback, potentially time-consuming in practice, the application of surrogate models presents a viable solution to meet time constraints. Attar et al. [20] has successfully developed an automated platform for stamping die design. This approach acknowledges the intricate nature of die design while leveraging DNNs for improved automation and innovation.

Summary

Until now, extrusion die design has retained its status as more of an art form, relying heavily on designers' experience to create initial editions. Subsequent iterations involve the use of FEA and experiments to test design solidity, a process notorious for its time and cost intensiveness. Despite meeting analytical standards, designs frequently fall short of delivering robust results.

In efforts to mitigate design costs, DNNs are increasingly adapted in the realm of die design. Presently, studies predominantly concentrate on optimizing specific die components, such as porthole positioning or geometry, and bearing length design. Some endeavors even attempt to holistically design the entire die by segmenting it into models, selected based on product profiles.

However, integrating DNNs into extrusion die design encounters numerous challenges. The intricate nature of extrusion dies, comprising multiple complex parts, complicates the acquisition of concise parameterizations encompassing all geometric information for intricate 3D objects. Therefore, devising methods to digitize die designs for training purposes becomes imperative. Additionally, training DNNs necessitates vast quantities of consistent data, a hurdle exacerbated by the customized nature of die designs, often tailored for in-house comprehension. The establishment of a standard for extrusion die design could facilitate the creation of a comprehensive database for training purposes.

References

[1] Metal Extrusion - Engineering Ideas Clinic - Confluence. (n.d.). Uwaterloo.atlassian.net. Retrieved December 3, 2023, from https://uwaterloo.atlassian.net/wiki/spaces/EIC/pages/31727390596/Metal+Extrusion#handytip-7517022545358001

[2] J. Heaton, I. Goodfellow, Y. Bengio, A. Courville, Deep learning, Genet Program Evolvable Mach. 19 (2018) 305-307. https://doi.org/10.1007/s10710-017-9314-z

[3] Z. Li, F. Liu, W. Yang, S. Peng, J. Zhou, A Survey of Convolutional Neural Networks: Analysis, Applications, and Prospects, IEEE T. Neur. Net. Lear. 33 (2022) 6999-7019. https://doi.org/10.1109/TNNLS.2021.3084827

[4] J. Llorca-Schenk, J.R. Rico-Juan, M. Sanchez-Lozano, Designing porthole aluminium extrusion dies on the basis of eXplainable Artificial Intelligence, Expert Syst. Appl. 222 (2023) 119808. https://doi.org/10.1016/j.eswa.2023.119808

[5] Y.B. Yu, Y.R. Lai, Q.C. Hsu, T.T. Truong, Deep Convolutional Neural Network to Assist Die Design for Flow Balance of Aluminum Hollow Extrusion, in: K. Mocellin, P.O. Bouchard, R. Bigot, T. Balan (Eds.), Proceedings of the 14th International Conference on the Technology of Plasticity - Current Trends in the Technology of Plasticity, ICTP 2023, Lecture Notes in Mechanical Engineering (2024). https://doi.org/10.1007/978-3-031-42093-1_7

[6] N. Miles, G. Evans, A. Middleditch, Bearing lengths for extrusion dies: rationale, current practice and requirements for automation, J. Mater. Process. Tech. 72 (1997) 162-176. https://doi.org/10.1016/S0924-0136(97)00150-7

[7] Y. LeCun, Y. Bengio, G. Hinton, Deep learning, Nature 521 (2015) 436-444. https://doi.org/10.1038/nature14539

[8] N.S. Gulgec, M. Takáč, S.N. Pakzad, Structural sensing with deep learning: Strain estimation from acceleration data for fatigue assessment, Comput.-Aided Civ. Infrastruct. Eng. 35 (2020) 1349-1364. https://doi.org/10.1111/mice.12565

[9] H. Zhou, Q. Xu, Z. Nie, N. Li, A study on using image-based machine learning methods to develop surrogate models of stamp forming simulations, J. Manuf. Sci. Eng. 144 (2022) 021012. https://doi.org/10.1115/1.4051604

[10] S. Liu, Z. Shi, J. Lin, Z. Li, Reinforcement learning in free-form stamping of sheet-metals, Procedia Manuf. 50 (2020) 444-449. https://doi.org/10.1016/j.promfg.2020.08.081

[11] R. Wu, C. Xiao, C. Zheng, DeepCAD: A deep generative network for computer-aided design models, in: Proceedings of the IEEE/CVF International Conference on Computer Vision (2021) 6772-6782. https://doi.org/10.48550/arXiv.2105.09492

[12] K.D.D. Willis, Y. Pu, J. Luo, H. Chu, T. Du, J.G. Lambourne, A. Solar-Lezama, W. Matusik, Fusion 360 gallery: A dataset and environment for programmatic CAD construction from human design sequences, ACM Transactions on Graphics 40 (2021) 1-24. https://doi.org/10.48550/arXiv.2010.02392

[13] H. Jobczyk, H. Homann, Automatic Reverse Engineering: Creating Computer-Aided Design (CAD) Models from Multi-view Images, in: U. Köthe, U., Rother, C. (eds.) Pattern Recognition, DAGM GCPR 2023, Lect. Notes Comput. Sc. 14264 (2024). https://doi.org/10.1007/978-3-031-54605-1_16

[14] H.I. Choi, S.W. Choi, H.P. Moon, Mathematical theory of medial axis transform, Pacific J. Math. 181 (1997) 57-88. https://msp.org/pjm/1997/181-1/p02.xhtml

[15] C. Lin, Optimal design of aluminium extrusion dies using a novel geometry based approach, Swansea University (2005). https://cronfa.swan.ac.uk/Record/cronfa42734

[16] X.F. Han, J.S. Jin, M.J. Wang, W. Jiang, L. Gao, L. Xiao, A review of algorithms for filtering the 3D point cloud, Signal Process. Image Commun. 57 (2017) 103-112. https://doi.org/10.1016/j.image.2017.05.009

[17] C.H. Lin, C. Kong, S. Lucey, Learning efficient point cloud generation for dense 3D object reconstruction, in: Proceedings of the AAAI Conference on Artificial Intelligence 32 (2018). https://doi.org/10.48550/arXiv.1706.07036

[18] J. Lee, H. Lee, D. Mun, 3D convolutional neural network for machining feature recognition with gradient-based visual explanations from 3D CAD models, Sci. Rep. 12 (2022) 14864. https://doi.org/10.1038/s41598-022-19212-6

[19] G. Zangara, F. Gagliardi, L. Filice, G. Ambrogio, A Coupled Approach Based on Statistical Methods and Machine Learning Techniques to Improve Porthole Die Design, in: K. Mocellin, P.O. Bouchard, R. Bigot, T. Balan (Eds.), Proceedings of the 14th International Conference on the Technology of Plasticity - Current Trends in the Technology of Plasticity, ICTP 2023, Lecture Notes in Mechanical Engineering (2024). https://doi.org/10.1007/978-3-031-42093-1_6

[20] H.R. Attar, H. Zhou, A. Foster, N. Li, Rapid feasibility assessment of components to be formed through hot stamping: A deep learning approach, J. Manuf. Process. 68 (2021) 1650-1671. https://doi.org/10.48550/arXiv.2104.13199

The influence of process parameters on the axial force of the continuous rolling section in the synchronous forming process of shape and inner hole for hollow axles

YE Caoqi[1,a], SHU Xuedao[1,b*], WANG Jitai[1,c], XIA Yingxiang[1,d], LI Zixuan[1,e] and XU Haijie[1,f]

[1]Ningbo University No. 818 Fenghua Road, Jiangbei District, Ningbo, Zhejiang Province, China

[a]yecaoqi925@gmail.com, [b]shuxuedao@nbu.edu.cn, [c]wangjitai0708@gmail.com, [d]xiayingxiangnbu@163.com, [e]lizixuan@nbu.edu.cn, [f]xuhaijie@nbu.edu.cn

Keywords: Process Parameters, FEM, Synchronous Forming, Axial Stress, Hollow Axle

Abstract. To verify the feasibility of the synchronized forming of shape and inner hole for hollow axles and the influence of process parameters on the axial force in the continuous rolling section, a finite element model of the synchronized forming of shape and inner hole for hollow axles was established using finite element analysis software, and the forming process of hollow axle shafts was simulated. The results indicate that synchronized forming with shaped holes for hollow axle shafts is feasible. The axial stress in the continuous rolling section increases with the feed angle of the piercing roll. With the increase in the speed of the piercing roll, the axial stress in the continuous rolling section shows a trend of rapid increase followed by a slow increase. Moreover, as the conical angle of the disc-shaped rolls increases, the axial stress in the continuous rolling section generally increases.

Introduction

The hollow axle is a core component of high-speed trains. Compared with traditional solid axles, it has advantages such as reducing the mass below the spring, reducing the force between the wheel and the rail, and facilitating ultrasonic flaw detection. However, its higher production cost and technical barriers have led to a single manufacturing technology for hollow axles, cumbersome production processes, and key technologies being monopolized by a few companies. In recent years, many scholars have conducted extensive research on new hollow axle production technologies. Zheng Shuhua et al. analyzed the stable rolling conditions of the cross wedge rolling of hollow axles and obtained criteria for determining the instability of cross wedge rolling of hollow axles. They analyzed the effects of spread angle and forming angle on the thickness uniformity and obtained the regularity of the influence of process parameters on the thickness of hollow axles rolled by cross wedge rolling [1]. Shuexuedao et al. studied the influence of die process parameters on the energy parameters of cross wedge rolling of hollow axles [2]. Xu Chang et al. simulated the forming process of hollow axles by three-roll skew rolling using finite element software. They analyzed the metal flow law, strain field, temperature field, and variation law of rolling force in this process, elucidating the forming mechanism of hollow axles by three-roll skew rolling and the influence of process parameters on mechanical parameters [3,4]. Pater et al. successfully rolled reduced-scale solid axles on a CNC three-roll skew rolling machine, confirming that axle shafts of rail vehicles can be formed by the three-roll skew rolling process. They also explored the influence of process parameters on the occurrence of spiral thread defects in the transition section of hollow axles and discussed the impact of spiral threads on the quality of axle cores, finding that this defect does not significantly affect the core quality [5,6]. Romanenko V.P. et al. proposed a manufacturing technology for hollow axles using rotary tube piercing to form inner holes and radial forging technology to form outer shapes. They tested the mechanical properties of hollow axles manufactured using this process [7-8].The introduction of these new hollow axle forming technologies has enriched the manufacturing technology of hollow axles and provided new ideas for their production.

The above-mentioned new technologies cover plastic forming methods such as cross wedge rolling, skew rolling and forging, exploring exploratory forming of hollow axles by utilizing the strengths of various processes. Inspired by previous research, this study innovatively proposes the hollow axle synchronous forming process, as shown in Fig. 1. The rolling process of this technology is as follows: firstly, the billet reaches the deformation zone of the perforation section under the pushing of the plunger and is bitten by the piercing rolls. After passing through the perforation deformation zone, the billet becomes a tube, which then enters the rolling deformation zone and is bitten by the disc rolls. The tube is rolled into a stepped shaft with certain dimensions by the radial movement of the disc rolls. After the entire billet passes through the piercing zone, the clamp behind the disc rolls bite the rolled piece and moves axially at a certain speed coordinated with the radial speed of the disc roll until the rolling is completed.

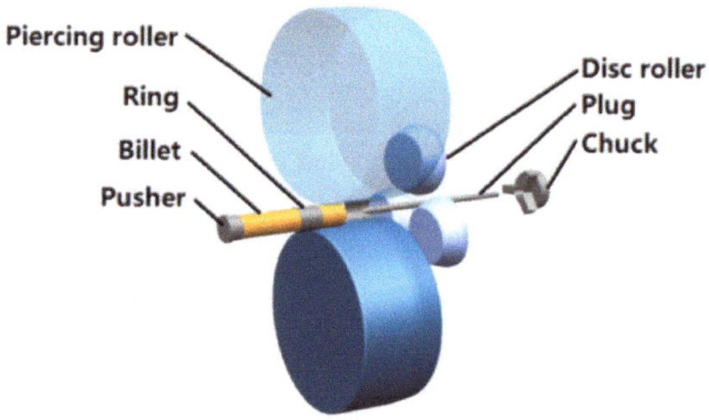

Fig. 1. Principle of synchronous forming of hollow axle.

Compared with traditional technologies for producing hollow axles, this technology improves the forming efficiency and material utilization of hollow axles, and reduces the footprint of equipment, as both external and internal shaping can be performed on a single mill. As this process shares characteristics with continuous rolling processes, analyzing the stress state of the hollow axle's rolling segment and the influence of process parameters on its stress state is a key issue in this technology. Therefore, this study conducts finite element simulations of the process and investigates the effects of piercing roll speed, feed angle $\alpha 1$ of piercing roll , and cone angle β of disc roll (as shown in Fig. 2) on the axial force of the rolling segment of hollow axles.

Materials Research Forum LLC
https://doi.org/10.21741/9781644903254-56

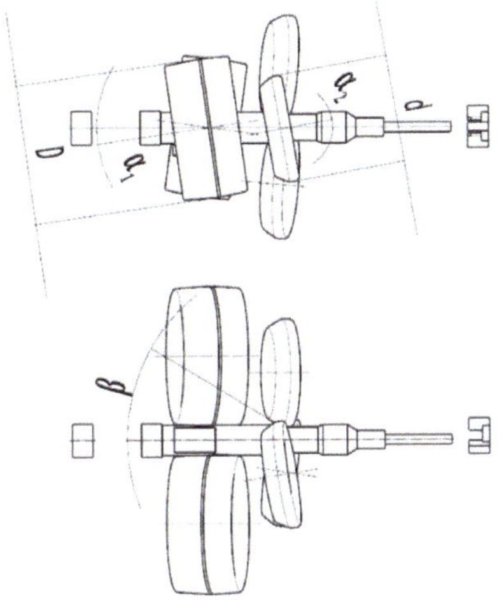

Fig. 2. Projection diagram of this process.

Establishment of Finite Element Model and Analysis of the Process

Hot Compression Behavior of 30NiCrMoV12

Hot Compression Test. To investigate the high-temperature flow behavior of 30NiCrMoV12 during thermal plastic deformation and to construct a more reliable numerical simulation model, a hot compression test plan for 30NiCrMoV12 was developed in conjunction with the process characteristics of skew rolling forming. High-temperature flow stress curves under different conditions were obtained, and the influence of deformation parameters on the high-temperature deformation behavior of 30NiCrMoV12 was analyzed. The high-temperature uniaxial compression tests were conducted on a Gleeble-3500 thermal simulation machine, using cylindrical specimens with dimensions of φ8 mm × 12 mm. Before deformation, tantalum chips were attached to both ends of the specimen to reduce friction between the specimen and the compression head. After securing the specimen, it was heated to the preset temperature at a rate of 5 °C/s and held for 3 minutes to reach thermal equilibrium. Subsequently, the specimen was compressed to a deformation of 50% at the preset strain rate, followed by immediate water quenching upon completion of compression deformation. The process of high-temperature uniaxial compression testing is illustrated in Fig. 3.

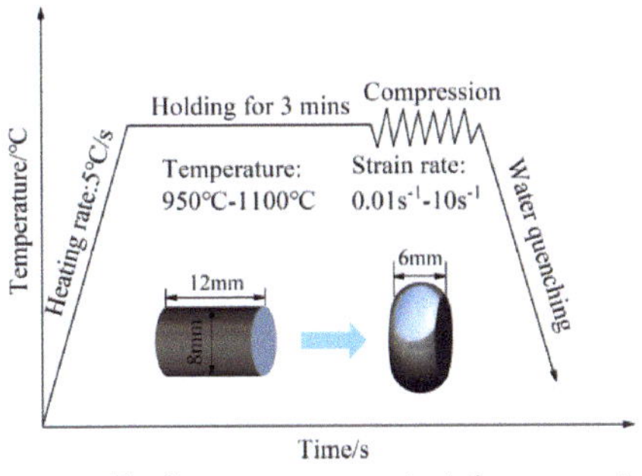

Fig. 3. Process of high-temperature uniaxial compression testing.

Hot Compression Test Results. Fig. 4 shows the stress-strain curves of 30NiCrMoV12 under different high-temperature deformation conditions. It can be observed from the graph that the stress under different deformation conditions increases sharply and reaches a peak within a very small strain range. This is attributed to the generation, intersection, and entanglement of dislocations causing work hardening, leading to a rapid increase in flow stress. As deformation continues to increase, dynamic recovery rearranges, slips, and annihilates accumulated dislocations, resulting in a decreasing trend or reduction in the increase of flow stress [9]. Moreover, when microstructural changes occur within the material, such as recrystallization and dynamic recrystallization, the flow softening rate caused can further counteract the hardening rate, leading to a decrease in flow stress or reaching a steady state [10]. Therefore, plastic deformation of the material is a process where hardening and softening mechanisms coexist and compete, and this process exhibits strong temperature and strain rate sensitivity.

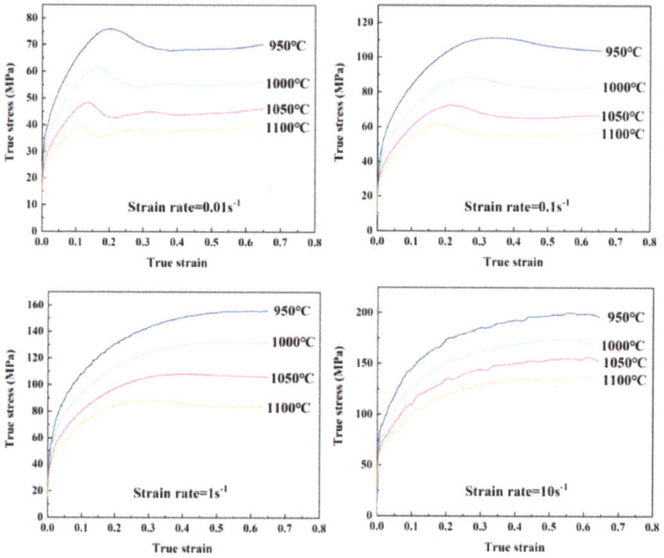

Fig. 4. Stress-strain curves of 30NiCrMoV12 under different high-temperature deformation conditions.

Establishment of the Constitutive Model for 30NiCrMoV12. Based on the Arrhenius constitutive model, the measured flow stress data from the hot compression experiments of 30NiCrMoV12 are subjected to multivariate regression analysis, resulting in the following constitutive equation:

$$\dot{\varepsilon} = 7.05\times10^{11}[\sinh(0.009832\sigma_{\mathrm{p}})]^{4.66}\exp\left(-\frac{314505}{RT}\right) \tag{1}$$

Where: $\dot{\varepsilon}$ is the strain rate, σ_p is the yield stress, R is the gas constant, and T is the temperature.

Finally, the obtained constitutive equation is imported into the material library of finite element software using secondary development software to provide a more accurate material model for finite element analysis.

Materials Research Proceedings 44 (2024) 519-527
https://doi.org/10.21741/9781644903254-56

Establishment of Finite Element Model. Fig. 5 illustrates the finite element model of synchronous forming of 30NiCrMoV12 axle-shaped holes established in this study using the FORGE NxT 1.1 software. The parameter settings and boundary conditions of this model are as follows: During the rolling process, the workpiece undergoes severe plastic deformation, and its elastic deformation can be neglected. The rolling rolls and guides undergo minimal deformation, thus the workpiece is set as rigid-plastic, and the rolling rolls and guides are set as rigid bodies. The rolling rolls rotate about the inner diameter center axis, driven by a motor, and the rotational speed is set according to the rolling conditions. A shear friction model commonly used to describe volume forming friction conditions is selected to describe the friction between the workpiece and the rolling rolls during contact. The friction factor between the workpiece and the rolling rolls is set to 0.8.The contact thermal conductivity coefficient between the workpiece and the rolling rolls/guides is set to 25,000 W/(m·K), and the convective heat transfer coefficient between the workpiece and air is set to 20 W/(m·K).

Fig. 5. Finite element model of the process.

The finite element simulation process is illustrated in Fig. 6. From the diagram, it can be observed that the billet is first pierced to form a hollow tube in the piercing section. Subsequently, the hollow tube is rolled into a stepped shaft with certain dimensions in the continuous rolling section. This process eliminates the need for reheating the hollow tube, saving energy and time, which is highly beneficial for metals with a narrow rolling temperature range. From the final formed shape, it can be seen that apart from the material head at the tail end of the axle and the spiral threads on the surface of the long axle section, which need to be removed by machining, the dimensions of the remaining parts are very close to those of standard axles. Therefore, forming a hollow axle using this process is feasible.

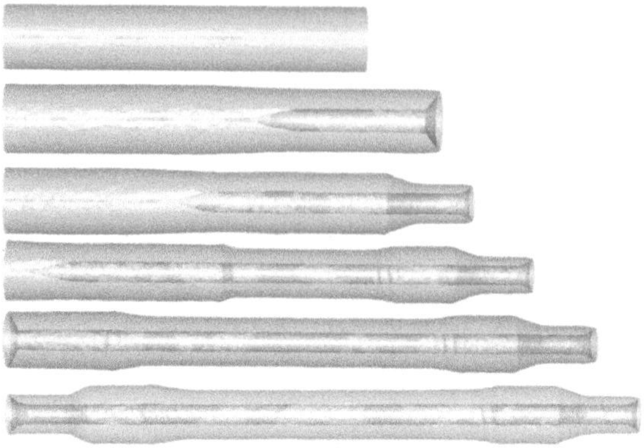

Fig. 6. Finite element simulation process.

The Influence of Process Parameters on Axial Force in the Continuous rolling section

The synchronized forming of shape and inner hole for hollow axles process can shape both the inner hole and the outer shape of the axle on a single rolling mill. During this process, two sets of rolls simultaneously act on the billet. Therefore, the working conditions of these two sets of rolls need to satisfy certain continuous rolling relationships to ensure the smooth progress of the rolling process. The stress state of the billet between the two sets of rolls during the rolling process is the key issue we need to study. The billet in the continuous rolling section is subjected to circumferential forces from the piercing roll and the disc roll, which cause it to rotate, as well as axial forces that induce axial movement. The magnitude of these two forces determines whether the rolling process can proceed smoothly. Therefore, in the following discussion, the axial force is selected as the research object, and the influence of the piercing roll feed angle, piercing roll speed, and disc roll cone angle on the axial force is investigated. The selection of process parameters is shown in Table 1.

Table 1. Selection of process parameters.

Condition	piercing roll feed angle[°]	piercing roll speed[rpm]	disc roll cone angle[°]
1	4/5/6/7	20	20
2	5	17/20/23/26	20
3	5	20	15/20/25/30

In order to obtain the magnitude of the axial stress in the continuous rolling section during the rolling process, three tracking points are selected at the exit of the piercing section. The method for selecting the tracking points is shown Fig. 7. The axial stress of these three points from the exit of the piercing section to the entrance of the continuous rolling section is recorded. Then, the average axial stress of the three points is taken as the axial stress value of the continuous rolling section under this process parameter.

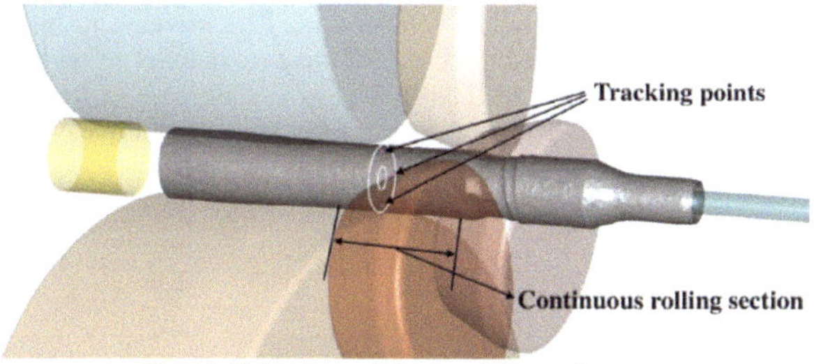

Fig. 7. Selecting of the tracking points.

The influence of piercing roll feed angle. The variation in axial stress magnitude in the continuous rolling section under different piercing roll feed angles is shown in Fig. 8. From the graph, it can be observed that the axial stress experienced by the continuous rolling section is compressive stress. This indicates that the axial velocity imparted to the billet by the piercing roll is greater than the axial velocity imparted by the disc roll during the rolling process. This is because, with the same circumferential speed of the rolls, the axial component of the rolling force exerted by the piercing roll on the billet is greater than the axial component of the rolling force exerted by the disc roll. Furthermore, the graph shows that the axial stress in the continuous rolling section increases with an increase in the piercing roll feed angle. This is attributed to the increase in the axial velocity component of the rolls with the increase in the piercing roll feed angle. Consequently, the axial velocity imparted to the billet by the piercing roll increases due to frictional forces. However, under this circumstance, the axial velocity imparted to the billet by the disc roll does not increase, directly leading to an increase in the axial stress in the continuous rolling section.

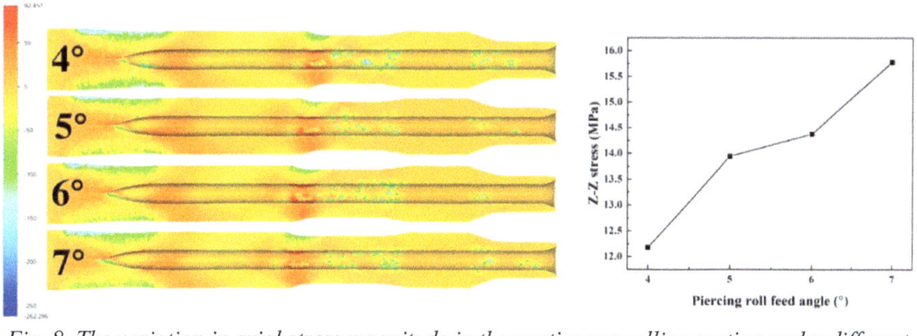

Fig. 8. The variation in axial stress magnitude in the continuous rolling section under different piercing roll feed angles.

The influence of piercing roll speed. Fig. 9 depicts the variation in axial stress in the continuous rolling section under different piercing roll speeds. From the graph, it can be observed that the axial stress experienced by the continuous rolling section during the rolling process remains compressive. Moreover, with an increase in the piercing roll speed, the axial stress in the continuous rolling section exhibits a trend of rapid increase followed by a gradual increase. The primary reason for the increase in axial stress is the increasing difference in axial velocities between the billet in the piercing section and the continuous rolling section. However, after the piercing roll speed reaches 20 rpm, the rate of increase in axial stress slows down. This phenomenon can be attributed to the following reasons: With the increase in roll speed, the material's strain rate increases, and the friction between the rolls and the billet becomes more intense. This leads to a rise in the billet's temperature. The increase in billet temperature has a softening effect on the material, causing the rate of increase in axial stress to decrease. Therefore, while the difference in axial velocities contributes to the increase in axial stress, the softening effect induced by the rise in billet temperature mitigates the rate of increase in axial stress beyond a certain roll speed.

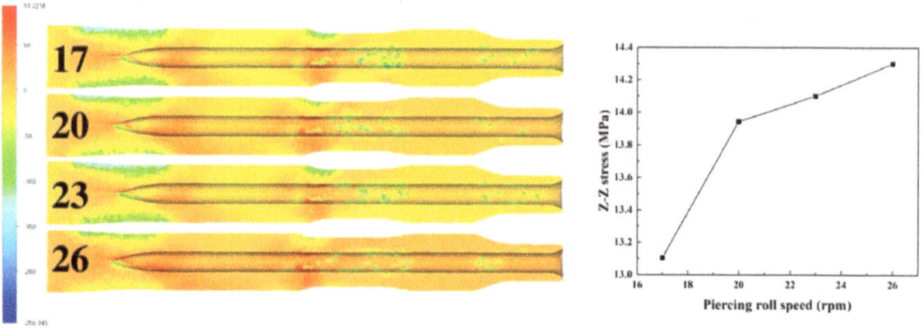

Fig. 9. The variation in axial stress magnitude in the continuous rolling section under different piercing roll speed.

The influence of disc roll cone angle. Fig. 10 illustrates the influence of different cone angles of the disc rolls on the axial stress in the continuous rolling section. From the graph, it can be observed that as the cone angle of the disc rolls increases, the overall trend of the axial stress in the continuous rolling section is an increase. This is because, with the increase in cone angle, the contact area between the disc rolls and the billet decreases. Consequently, this directly leads to a reduction in the rolling force exerted on the continuous rolling section. As a result, the axial velocity of the continuous rolling section decreases. This, in turn, increases the difference in axial velocities between the billet in the piercing section and the continuous rolling section, thereby causing an increase in axial stress in the continuous rolling section.

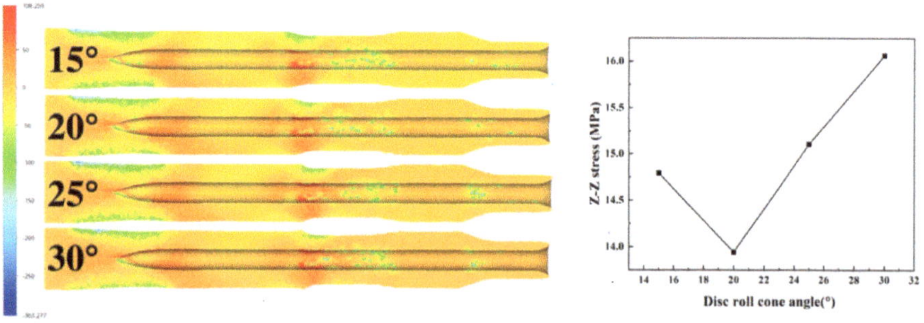

Fig. 10. The variation in axial stress magnitude in the continuous rolling section under different disc roll cone angles.

Conclusion

1. The synchronized forming of shape and inner hole for hollow axles process is feasible. It can effectively increase material utilization and rolling efficiency.
2. The axial stress in the continuous rolling section increases with an increase in the piercing roll feed angle. Additionally, with an increase in piercing roll speed, the axial stress in the continuous

rolling section exhibits a trend of rapid increase followed by a gradual increase. Moreover, as the cone angle of the disc rolls increases, the axial stress in the continuous rolling section shows an overall increasing trend.

Funding
This study was funded by the Natural Science Foundation of Zhejiang, China (Grant Number: LZ22E050002), the National Natural Science Foundation of China (Grant Number: 51975301), and the Major Project of Science and Technology Innovation 2025 in Ningbo City, China (Grant Number: 2022Z064, 2022Z009, 2022Z015).

References

[1] S.H. Zheng, X.D. Shu, B.S. Sun, W.F. Peng, Wall thickness uniformity of railway hollow shafts by cross-wedge rolling. Chin. J. Eng, 37 (2015) 648-654. https://doi.org/10.13374/j.issn2095-9389.2015.05.017

[2] X.D. Shu, C. Liu, B.S. Sun, W.F. Peng, P.H. Yu, Influence of mold technological parameters on the forming force parameters in multi-wedge rolling of the railway hollow shafts. J. Plast. Eng. 23 (2016) 23-28. https://doi.org/10.3969/j.issn.1007-2012.2016.03.005

[3] C. Xu, X.D. Shu, Y. Zhu, Simulation and analysis of three-roll skew rolling forming of the railway hollow shaft. J. Ningbo Univ. NSEE 31 (2018) 16-19.

[4] C. Xu, X.D. Shu, Influence of process parameters on the forming mechanics parameters of the three-roll skew rolling forming of the railway hollow shaft with 1 : 5, Metalurgija 57(2018) 153 -156.

[5] Z. Pater, J. Tomczak, K, Lis, T. Bulzak, X.D. Shu, Forming of rail car axles in a CNC skew rolling mill, Arch. Civ. Mech. Eng. 20 (2020) 1-13. https://doi.org/10.1007/s43452-020-00075-5

[6] Z. Pater, J. Tomczak, T. Bulzak, Problems of forming stepped axles and shafts in a 3-roller skew rolling mill, J. Mater. Res. Technol. 9 (2020) 10434-10446. https://doi.org/10.1016/j.jmrt.2020.07.062

[7] V.P. Romanenko, B.A. Romantsev, G.P. Illarionov, Billet Preparation Method for Railcar Hollow Axle Production, Metallurgist 58 (2014) 684-688. https://doi.org/10.1007/s11015-014-9977-0

[8] V.P. Romanenko, P.P. Stepanov, S.M. Kriskovich, Production of Hollow Railroad Axles by Screw Piercing and Radial Forging, Metallurgist 61 (2018) 873-877. https://doi.org/10.1007/s11015-018-0579-0

[9]Z.L. Liu, S. Ren, W. Liu, Hot deformation behavior and microstructure evolution of GH141 nickel-based superalloy, The Chinese J. Nonferrous Met. 33 (2023) 2577-2592. https://doi.org/10.11817/j.ysxb.1004.0609.2022-43577

[10]W.W. Zhang, X.G. Liu, H.Z. Li, Study on microstructure evolution of GH4742 alloy underthermo-mechanical coupling based on LAM technology, J. Mech. Eng. 58 (2022) 110-120. https://doi.org/10.3901/JME.2022.16.110

Optimizing deep drawing parameters for power battery shells through the integration of feature-weighted SVM and genetic algorithm

WANG Ruoda[1,a], SUN Yu[1,b*], WU Kai[1,c] and WANG Yu[1,d]

[1]Nanjing University of Science and Technology, Xiaolingwei No. 200 Street, 210094 Nanjing, China

[a]ruoda_wang@njust.edu.cn, [b]sunyu@njust.edu.cn, [c]wukai@njust.edu.cn, [d]wy@njust.edu.cn

Keywords: Deep Drawing, Finite Element Analysis, Feature-Weighted Support Vector Machine, Optimization, Power Battery Shells

Abstract. Deep drawing is one of the main forming processes for battery shells, and the rational setting of its process parameters directly affects the quality of the formation. Selecting the optimal deep drawing process parameters requires repeated trials and simulations, which increases the cost, reduces the efficiency, and poses a significant challenge to enterprises. To address this challenge, we focus on the first deep drawing process of the battery shell, proposing a parameter optimization method for battery shell deep drawing based on a Feature-Weighted Support Vector Machine (FWSVM) and Genetic Algorithms (GA). Our aim is twofold: on the one hand, to establish an agent model for finite element analysis of the deep drawing process using the FWSVM technique to enhance prediction accuracy and save simulation time; on the other hand, to derive optimal deep drawing process parameters using GA. The experimental results indicate that the FWSVM can accurately establish the relationship between process input/output parameters, and the optimized process parameters achieved through this method can realize the minimum thinning rate and convex mold contact force.

Introduction

The forming process of the rectangular power battery shell requires strict control over the process parameters, as the quality of forming directly affects the safety and stability of the battery [1]. Due to its unique geometric structure, the molding process for battery shells involves multiple deep drawings, stamping, and other complex processes [2]. Precise control of these processes is crucial for maintaining product consistency and quality. In the molding process, the battery shell's rounded corners undergo mainly tensile deformation, while its straight edges primarily experience bending deformation, due to varying forces [3]. The uneven forces and deformations complicate material flow, increasing the likelihood of defects like wrinkles and cracks. The first deep drawing stage is particularly prone to defects due to the substantial material deformation. Therefore, precise control of the process parameters at first deep drawing step becomes crucial to ensure the quality of the molding [4].

Modern numerical modeling and simulation techniques offer an efficient alternative to the costly and inefficient traditional trial-and-error methods. Simulating the molding process makes it possible to predict material behavior under specific parameters, enabling optimization of process parameters and mold design prior to production. This approach reduces trial-and-error iterations, cuts production costs, and accelerates product development [5].

Over the past few decades, researchers have focused on the optimization of deep drawing process. Park et al. [6] conducted finite element analysis on a multistage deep drawn punch featuring a high aspect ratio rectangle to focus on process design. Numerical methods were employed for analysis, which was then experimentally validated. To minimize material waste in the trimming process, simulations were performed in LS-DYNA3D using brick elements and explicit time integrals to optimize the billet shape. Yaghoubi et al. [7] investigated how die geometry parameters affect the hydro-mechanical deep drawing process of 2024 aluminum alloy at high temperatures. Initially, they trained a neural network with the grouped data processing method (GMDH) to analyze process behavior, then optimized process variables using the Honeybee algorithm. The simulation results, validated against experimental data, demonstrated a strong correlation. Özek et al. [8] investigated the effects of die/blank holder angles, blank holder force, and die/punch diameters on the drawing limits, punch force, and wall thickness of DIN EN 10130–1999 steel sheets in angular squared deep drawing. Data was collected using profiles machined at angles ranging from 0° to 15° and 8, 12, and 16 mm diameters through a complete factorial design. Analysis techniques such as regression, signal-to-noise ratio, and ANOVA showed that increased angles and diameters extend drawing limits while reduced forces enhance surface uniformity. Padmanabhan et al. [9] explored optimizing die radius, blank holder force, and friction coefficient in deep drawing stainless steel cups via the finite element method and Taguchi technique. Findings indicate the die radius as the most critical parameter, followed by blank holder force and friction coefficient in terms of impact. Furthermore, adjusting the blank holder force and applying local lubrication significantly enhance the quality of formed parts, underscoring these parameters' key role in reducing manufacturing costs.

Despite extensive research on the deep drawing process by many scholars, there remains room for improvement. Firstly, in the design of the experiments method for agent model training, data is often assigned equal weights, neglecting the impact of feature weights on the model's performance. Secondly, relying on statistical analysis for optimization restricts process variable settings to a predetermined range of values. To address these issues, this study introduces an optimization method based on Feature-Weighted Support Vector Machine and Genetic Algorithms (FWSVM-GA) for the deep drawing process, specifically focusing on optimizing parameters in the battery shell's initial deep drawing process.

Methodology

The optimization framework of battery shell deep drawing process parameters by integrating FWSVM and GA:

The framework of FWSVM-GA method is shown in Fig. 1. Firstly, we designate the blank holder force (*BHF*), the punch fillet radius (R_p), the die fillet radius (R_d), the die gap (*DG*), and the friction conditions (*FC*) as input parameters. The thinning rate (*TR*) is chosen as an evaluation indicator and process output parameter. Subsequently, we use these input parameters as experimental factors, select four levels for each factor, and generate 111 experiments using the D-optimal experimental design method. Considering the different weights of input parameters, we determine the weight of each parameter through the entropy method. Then, these weights are applied to the classical support vector machine kernel function, resulting in the development of a FWSVM model. This model aims to establish a correlation between process parameters and output characteristics. Following this, we train and test the FWSVM model using input/output data from finite element analysis. We employ the FWSVM as a fitness function in a GA for optimizing process parameters. Lastly, the effectiveness of the finite element and agent models is verified through experiments.

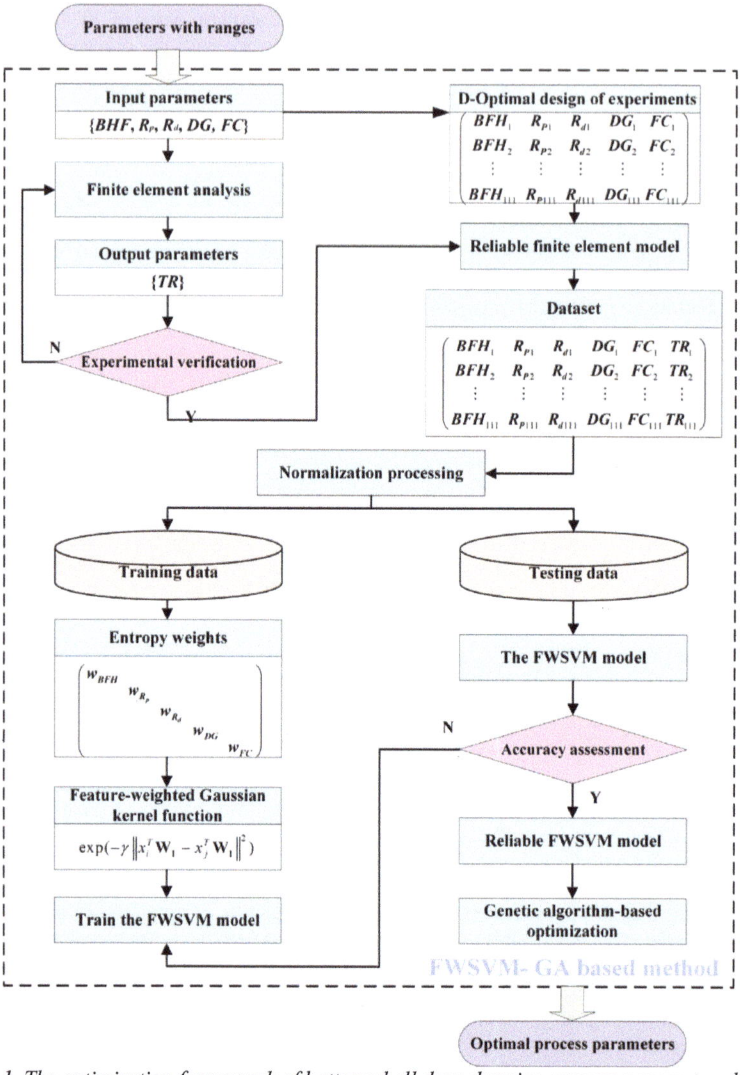

Fig. 1. The optimization framework of battery shell deep drawing process parameters by integrating FWSVM and GA.

Finite-element modeling. The power battery shell, characterized by its tall rectangular shape, requires multiple deep drawing processes for formation. This study primarily focuses on the initial deep drawing process of the battery shell. Fig. 2 shows the 3-D model after molding. The blank is made from a 1.5 mm thick rolled aluminum alloy sheet (AL3003H14), featuring a density of 2.7×103 kg/m^3, an elasticity modulus of 69 GPa, a yield strength of 125 MPa, and a tensile strength of 150 MPa.

Based on experience, the die's cross-section is elliptical, with a long semi-axis of 120.85 mm and a short semi-axis of 83.84 mm. The punch, die, part, and blank holder's 3-D models were initially crafted in SolidWorks and then converted to IGS format. For simulation, the part is initially imported into DYNAFORM software. Using the one-shot expansion method generates the blank shape, followed by the sequential import of the remaining models. In the meshing process, the punch, die, and blank holder are considered rigid bodies, with a cell size of 3mm and a thickness of 5 layers. The finite element model is depicted in Fig. 3. The initial deep drawing height is 80 mm, with a virtual drawing speed of 2 m/s. Parameters are configured and submitted to LS-DYNA for analysis, with post-processing conducted using eta/POST.

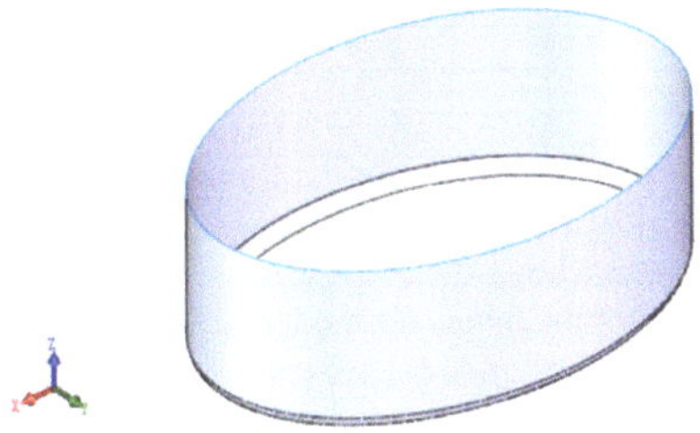

Fig. 2. 3-D model after drawing.

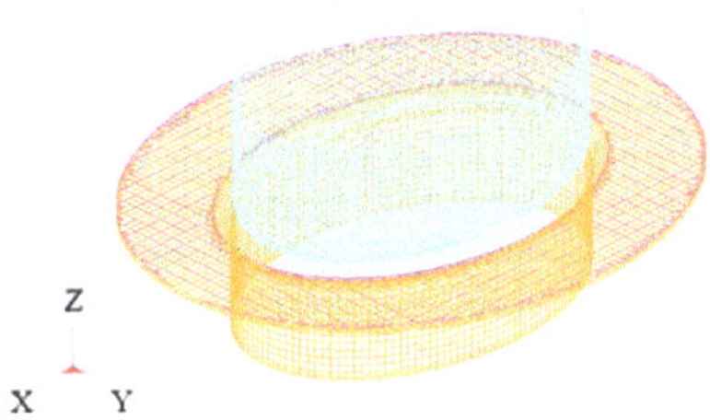

Fig. 3. Finite element model.

Orthogonal experimental design. One of the main objectives of this study is to predict the product quality under different process parameters. Finite element modeling, while useful for this purpose, is time-intensive due to the lengthy duration of each FE model run. Consequently, FWSVM is utilized as a more efficient technique to estimate process responses for specific sets of process parameters. Required data was collected through a series of simulation runs using the design of experiments (DOE) methodology.

Several process parameters, such as blank holder force, punch fillet radius, die fillet radius, die clearance, friction conditions, and blank size, jointly influence the quality of the battery shell's first deep drawing. Blank size has a minimal impact on forming. Using a larger blank optimizes material use, while a smaller blank can result in insufficient trimming allowance, impacting accuracy. Blank dimensions are derived from the part shape in DYNAFORM in this study. the blank holder force (*BHF*), the punch fillet radius (R_p), the die fillet radius (R_d), the die gap (*DG*), and the friction conditions (*FC*) as inputs, with each factor having four levels, as detailed in Table 1. For a complete factorial design, the number of possible designs is $N=L^m$, where L is the number of levels for each factor and m is the number of factors. Consequently, the full factorial design in this study encompasses 576 experiments. To further reduce simulation runs, a D-optimal experiment design was implemented. As a result, only 111 runs are necessary to gather the data needed for the FWSVM. The dataset was generated by obtaining the maximum thinning rate (*TR*) from all 111 experiments via FEA. In engineering applications, the thinning rate should not exceed 25%; Table 2 presents some of the datasets.

Table 1. Variable parameters with their levels.

Levels	*BHF* (mm)	R_p (mm)	R_d (mm)	*DG* (mm)	*FC*
1	30	10	11	1.65	0.13
2	40	11	12	1.8	0.15
3	50	12	13	1.95	0.17
4	60	13	14	2.1	0.19

Table 2. Dataset.

Levels	*BHF* (mm)	R_p (mm)	R_d (mm)	*DG* (mm)	*FC*	*TR*
1	50	10	11	2.1	0.13	19.268
2	40	11	13	2.1	0.13	12.242
3	60	12	12	1.8	0.19	14.798
⋮	⋮	⋮	⋮	⋮	⋮	⋮
110	60	11	11	1.95	0.17	17.941
111	30	11	14	1.8	0.13	14.574

The proposed FWSVM. Since different input parameters have different effects on the results during drawing, we propose to use weighted support vector machines for modeling. Let $D_{train} = (x_i, y_i)$ be the training dataset, where $x_i \in R^n$ is the i-th input feature, $y_i \in \{-1, +1\}$, $i = 1, 2, \ldots, l$, is the class label. Suppose the hyperplane correctly divides the D_{train} into two classes, in that case, the optimal hyperplane should maximize the sum of the minimum distances from the two classes of samples to the hyperplane.

The optimal hyperplane is obtained by solving the following optimization problem:

$$\min_{\omega,b,\xi_i} : \frac{1}{2}\|\omega\| + C\sum_{i=1}^{n}\xi_i$$
$$\text{suject to:}\begin{cases} y_i[\omega\cdot\phi(x\mathrm{W})+b]\ge 1-\xi,\ i=1,2,\cdots,n \\ \xi_i \ge 0,\ i=1,2,\cdots,n \end{cases} \tag{1}$$

where C is a positive constant parameter used to control the tradeoff between the margin and the classification error. ξ_i denotes slack variables, ϕ is the feature mapping function, which maps the input space to a usually high dimensional feature space where the data points become linearly separable, $\mathrm{W}_{n\times n}$ represents a feature weighted matrix.

Since Eq. 1 is a convex quadratic programming problem, using the Lagrange multiplier method, its Lagrange function is shown as follows,

$$L(\omega,b,\alpha,\xi,\mu) = \frac{1}{2}\|\omega\| + C\sum_{i=1}^{n}\xi_i + \sum_{i=1}^{n}\alpha_i[1-\xi_i - y_i(\omega\cdot\phi(x_i\mathrm{W})+b)] - \sum_{i=1}^{n}\mu_i\xi_i \tag{2}$$

where α_i and μ_i are Lagrange multipliers. Letting the partial derivatives of $L(\omega, b, \alpha, \xi, \mu)$ with respect to ω, b, ξ_i be 0 yields:

$$\omega = \sum_{i=1}^{n}\alpha_i y_i x_i \tag{3}$$

$$0 = \sum_{i=1}^{n}\alpha_i y_i \tag{4}$$

$$C = \alpha_i + \mu_i \tag{5}$$

Bringing Eq. 3-5 into Eq. 2 to get the dual problem of Eq. 1,

$$\begin{aligned}&\max_{\alpha} : \sum_{i=1}^{n}\alpha_i - \frac{1}{2}\sum_{i=1}^{n}\sum_{j=1}^{n}\alpha_i\alpha_j y_i y_j \phi(x_i\mathrm{W})\phi(x_j\mathrm{W})\\ &\text{subject to: } \sum_{i=1}^{n}\alpha_i y_i = 0\\ &\qquad\qquad 0 \le \alpha_i \le C\end{aligned} \tag{6}$$

For the solution of Eq. 6, the Karush–Kuhn–Tucker (KKT) conditions [10] require:

$$\begin{cases}\alpha_i \ge 0,\ \mu_i \ge 0,\\ y_i(\omega\cdot\phi(x_i\mathrm{W})+b)-1+\xi_i \ge 0,\\ \alpha_i[y_i(\omega\cdot\phi(x_i\mathrm{W})+b)-1+\xi_i]=0,\\ \xi_i \ge 0,\ \mu_i\xi_i=0.\end{cases} \tag{7}$$

Then b can be calculated as:

$$b = y_j - \omega\cdot\phi(x_j\mathrm{W}) \tag{8}$$

Since the value of b is not unique, its average value is taken,

$$\bar{b} = \frac{1}{N}\sum_{0\le\alpha_j\le C}[y_j - \sum_{i=1}^{n}\alpha_i y_i(\phi(x_i\mathrm{W})\phi(x_j\mathrm{W}))] \tag{9}$$

where N is the number of support vectors. Then the corresponding decision function is denoted as,

$$f(x) = sign(\omega\cdot\phi(x\mathrm{W})+\bar{b}) \tag{10}$$

In order to express FWSVM, The feature weighted kernel function $K_f(x_i, x_j)$ is defined as follows:

$$K_f(x_i,x_j) = \phi(x_i\mathrm{W})\cdot\phi(x_j\mathrm{W}) = K_p(x_i\mathrm{W},x_j\mathrm{W}) \tag{11}$$

Only the individual features were weighted in this study, so W is a diagonal matrix of order n. At this point $(\mathrm{W})_{ii} = e_i$ ($1 \leqslant i \leqslant n$) represents the weight of the i-th feature, and in general e_i is not all equal.

$$\mathrm{W} = \begin{pmatrix} e_1 & & & \\ & e_2 & & \\ & & \ddots & \\ & & & e_n \end{pmatrix} \tag{12}$$

Therefore, the non-linear classification decision function is,

$$f(x) = sign(\sum_{i=1}^{n}\alpha_i y K_f(x_i,x_j) + \frac{1}{N}\sum_{0\le\alpha_j\le C}[y_j - \sum_{i=1}^{n}\alpha_i y_i K_f(x_i,x_j)]) \tag{13}$$

the Feature-weighted Gaussian radial basis function is adopted as kernel function,

$$K_f(x_i,x_j)=[\gamma(x_i\mathrm{W})\cdot(x_j\mathrm{W})+r]^d=(\gamma x_i\mathrm{W}\mathrm{W}^T x_j^T+r)^d,\ \gamma>0 \tag{14}$$

This study uses the entropy weight method to determine the weight of each input parameter, which is an objective assignment method commonly used in multi-indicator comprehensive evaluation problems. It reflects the dispersion of the indicators by calculating the entropy value of each indicator, and then determines the weight of each indicator. The smaller the entropy value, the greater the variability of the indicator, the more information it carries, so it should be given a larger weight.

Firstly, we construct the input parameters into a matrix V,

$$\mathrm{V}=\begin{pmatrix} p_{11} & p_{11} & \cdots & p_{1m} \\ p_{21} & p_{22} & \cdots & p_{2m} \\ \vdots & \vdots & \ddots & \vdots \\ p_{k1} & p_{k2} & \cdots & p_{km} \end{pmatrix} \tag{15}$$

where p_{ij} represents the value of the j-th index in the i-th case (i = 1, 2, 3, …, k, j = 1, 2, 3, …, m).

Using Eq. 16 to normalize the elements in matrix V to obtain the matrix Q = $(q_{ij})_{k\times m}$.

$$q_{ij}=\frac{p_{ij}}{\sum_{i=1}^{m} p_{ij}} \tag{16}$$

The entropy weight is obtained via:

$$e_{oj}=-\frac{1}{\ln m}\sum_{j=1}^{m} q_{ij}\ln q_{ij}$$
$$g_j=1-e_{oj}$$
$$w_{oj}=\frac{g_j}{\sum_{j=1}^{m} g_j} \tag{17}$$

where e_{oj} represents the entropy value of index j, g_j is the variation coefficient, the smaller the value of e_{oj}, the larger the value of g_j, and the greater the contribution of the attribute, w_{oj} denotes the weight. The objective weight vector W can be written as,

$$\mathrm{W}=\left(w_{o1},\ w_{o2},\ \ldots,\ w_{oj},\ \ldots,\ w_{om}\right) \tag{18}$$

To validate the superiority of FWSVM, we compare it with other commonly used agent models (SVM, BPNN). We use leave-one-out cross-validation (LOO-CV) to estimate the performance of comparative methods and employ the mean absolute percentage error (MAPE) to measure the discrepancy between computed and FEM values. The accuracy for each LOO-CV iteration can be calculated as follows:

$$\mathrm{acc}=1-MAPE=1-\frac{1}{n}\sum_{i=1}^{N}\left|\frac{TR_i-y_i}{TR_i}\right| \tag{19}$$

where n is the number of samples, TRi is the finite element simulation value and yi is the predicted value.

Optimization of deep drawing process parameters. The objective of this section is to identify optimal process parameters through Genetic Algorithms (GA). Drawing on Darwin's principles of natural selection and genetics, GA is a search algorithm that emulates the natural evolutionary process to tackle optimization and search challenges.GA is capable of conducting a global search across the search space, addressing complex issues with numerous objectives and constraints to identify the global optimal solution or one nearing the global optimum.

In this study, FWSVM serves as the GA's fitness function, aiming to identify optimal deep drawing parameter values. During the optimization search, outputs below 25% receive a fitness value that increases inversely with the output value. Concurrently, a substantial negative penalty is applied for unsatisfactory solutions to steer the algorithm away from these options. For predicted outputs under 25%, a fitness value is assigned using the formula - (25 - output). This approach aims to increase fitness (in absolute terms) as output diminishes, maintaining a negative fitness to identify parameter combinations yielding the smallest possible FWSVM model outputs under 25%. The input parameters were constrained to BHF (kN): [30, 60], Rp (mm): [10, 13], Rd (mm): [11, 14], DG (mm): [1.65, 2.1], FC: [0.13, 0.19]. After 50 iterations, the algorithm concludes, with optimal parameters for each iteration displayed in Fig. 4. Final optimal parameter combinations is detailed in Table 3.

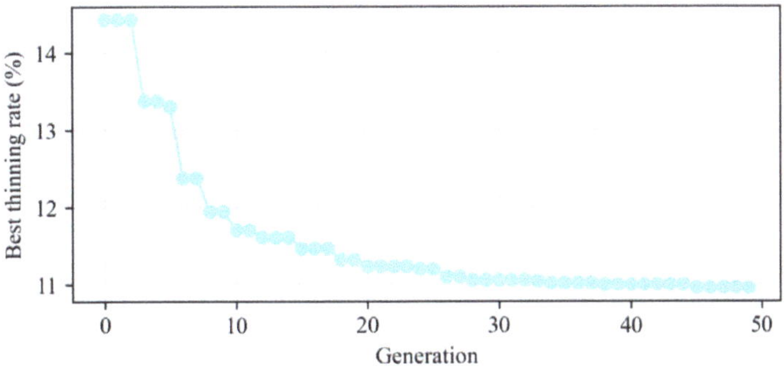

Fig. 4. Iteration process.

Table 3. Optimal parameter combination.

BHF (mm)	R_p (mm)	R_d (mm)	*DG* (mm)	*FC*	*TR (%)*
52.153	13.012	13.995	2.037	0.171	10.945

Result and discussion

As shown in Fig. 5, the weights of the five input parameters were calculated using Eq. 15-18. These weights were used to construct the FWSVM. Using Eq. 19, we computed the accuracy of the FWSVM with respect to the SVM, and the BPNN (default values were used for the parameters of the SVM and the BPNN), and the results are shown in Fig. 6.

Materials Research Proceedings 44 (2024) 528-537 https://doi.org/10.21741/9781644903254-57

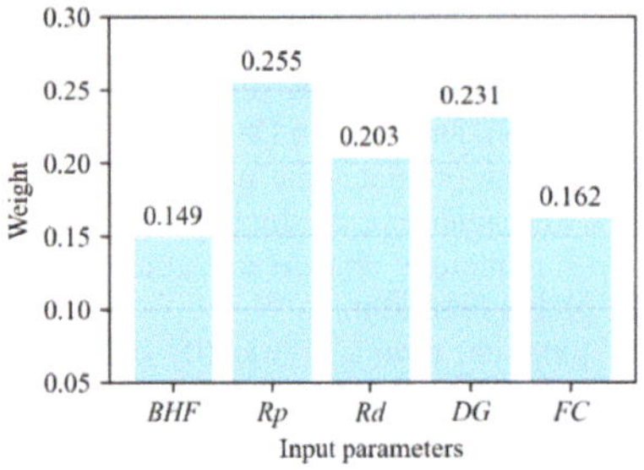

Fig. 5. Weighting distribution.

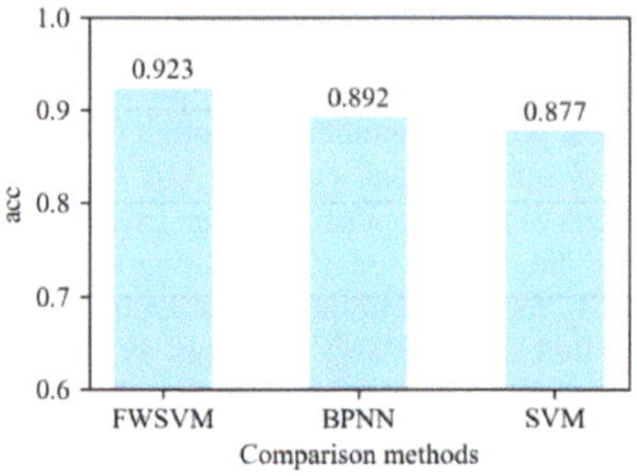

Fig. 6. Accuracy comparison.

According to Fig. 5, in constructing the FWSVM for enhanced prediction accuracy, R_p holds the highest weight, while *BHF* has the lowest, with the overall weight ranking as $R_p > DG > R_d > FC > BHF$. However, this ranking is based on the dataset's dispersion and does not reflect the parameters' significance for deep drawing molding quality. From Fig. 6, it can be seen that all three methods achieve good prediction results, Specifically, FWSVM leads in accuracy at 0.923, followed by BPNN at 0.892, with SVM trailing at 0.877. These results demonstrate FWSVM's superiority.

We also performed finite element analysis and experimental verification on the optimal process parameter combinations in Table 3. The results of the finite element analysis are shown in Fig. 7, and the experimental parts are shown in Fig. 8.

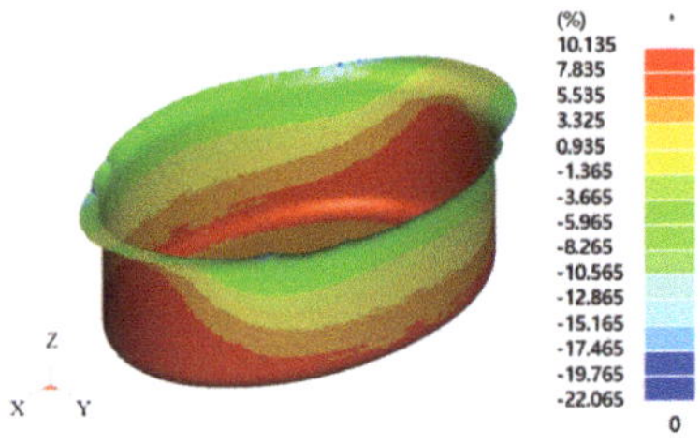

Fig. 7. The finite element analysis results.

Fig. 8. The experimental result.

From Fig. 7, it can be seen that the maximum thinning rate of the part under the optimal parameter combination is 10.135%, which is much less than the limiting value of 25%. The difference between the finite element simulation results and the predicted results (10.945%) is only 0.81%, which further illustrates the accuracy of FWSVM. The experimental results Fig. 8 show that the parts are free from defects such as rupture and wrinkle and possess excellent quality.

Summary

This study employs the FWSVM technique to develop an agent model for finite element analysis of the deep drawing process, enhancing prediction accuracy and reducing simulation time. Subsequently, GA is applied to identify the deep drawing process's optimal parameters, thereby elevating the molding quality of the battery shell's initial deep drawing. As input parameters, we selected blank holder force (*BHF*), punch fillet radius (R_p), die fillet radius (R_d), die gap (*DG*), and friction conditions (*FC*). Thinning rate (*TR*) is chosen as the output parameter. The FWSVM dataset is generated through the design of experiments method and finite element analysis. The trained FWSVM serves as the fitness function for GA, leading to the identification of the optimal process parameter combination. Experimental evidence shows that battery shells' initial deep drawn parts exhibit outstanding quality under this set of process parameters.

References

[1] T.D. Nguyen, J. Deng, B. Robert, W. Chen, T. Siegmund, Deformation Behavior of Single Prismatic Battery Cell Cases and Cell Assemblies Loaded by Internal Pressure, J. Electrochem. Energy Conversion Storage 18 (2021). https://doi.org/10.1115/1.4050101

[2] Q. Yu, Y. Nie, S. Peng, Y. Miao, C. Zhai, R. Zhang, J. Han, S. Zhao, M. Pecht, Evaluation of the safety standards system of power batteries for electric vehicles in China, Appl. Energy 349 (2023) 121674. https://doi.org/10.1016/j.apenergy.2023.121674

[3] S. Yaghoubi, F. Fereshteh-Saniee, An investigation on the effects of the process parameters of hydro-mechanical deep drawing on manufacturing high-quality bimetallic spherical-conical cups, Int. J. Adv. Manuf. Technol. 110 (2020) 1805–1818. https://doi.org/10.1007/s00170-020-05985-5

[4] R. Dwivedi, G. Agnihotri, Study of Deep Drawing Process Parameters, Materials Today: Proceedings 4 (2017) 820–826. https://doi.org/10.1016/j.matpr.2017.01.091

[5] M. Manoochehri, F. Kolahan, Integration of artificial neural network and simulated annealing algorithm to optimize deep drawing process, Int. J. Adv. Manuf. Technol. 73 (2014) 241–249. https://doi.org/10.1007/s00170-014-5788-5

[6] C.S. Park, T.W. Ku, B.S. Kang, S.M. Hwang, Process design and blank modification in the multistage rectangular deep drawing of an extreme aspect ratio, J. Mater. Process. Technol. 153–154 (2004) 778–784. https://doi.org/10.1016/j.jmatprotec.2004.04.306

[7] S. Yaghoubi, F. Fereshteh-Saniee, Optimization of the geometrical parameters for elevated temperature hydro-mechanical deep drawing process of 2024 aluminum alloy, Proceedings of the Institution of Mechanical Engineers, Part E: J. Process. Mech. Eng. 235 (2021) 151–161. https://doi.org/10.1177/0954408920949364

[8] C. Özek, E. Ünal, Optimization and Modeling of Angular Deep Drawing Process for Square Cups, Mater. Manuf. Process. 26 (2011) 1117–1125. https://doi.org/10.1080/10426914.2010.532526

[9] R. Padmanabhan, M.C. Oliveira, J.L. Alves, L.F. Menezes, Influence of process parameters on the deep drawing of stainless steel, Finite Elements in Analysis and Design 43 (2007) 1062–1067. https://doi.org/10.1016/j.finel.2007.06.011

[10] S.P. Boyd, L. Vandenberghe, Convex optimization, Cambridge University Press, Cambridge, UK ; New York, 2004

Materials Research Proceedings 44 (2024) 538-547 https://doi.org/10.21741/9781644903254-58

Knowledge-guided particle swarm optimization of multi-link systems for cold and warm forging presses

ZHENG Han[1,a], SUN Yu[1,b *], NI Jun[1,c] and DING WuXue[1,d]

[1]Nanjing University of Science and Technology No. 200 Xiaolingwei Street, 210094 Nanjing, China

[a]zhenghan@njust.edu.cn, [b]sunyu@njust.edu.cn, [c]junnee@njust.edu.cn, [d]wuxuexie@njust.edu.cn

Keywords: Forging Press, Elbow-Bar Mechanism, Knowledge-Guided, Size Optimization

Abstract. The performance of the multi-link system for mechanical press directly affects the performance of the press and the quality of the forgings. In order to improve the quality of forgings and enhance the design efficiency, a knowledge-guided particle swarm optimization design framework for multi-link systems is proposed, taking the cold and warm forging press as the research object. A knowledge database consisting of historical cases, rule knowledge and other knowledge was created. A reasoning machine consisting of retrieval and adoption was then built to determine the configuration of the multi-link system through case similarity calculations. A multi-link system optimization model was established for the selected configuration. To solve the optimization model, a knowledge-guided improved particle swarm optimization algorithm is developed, from which we obtain the optimized dimensions of the multi-link system. Comparing the results before and after optimization, we found that all the performance indexes have been improved to a certain extent, which proves the effectiveness of the knowledge-guided particle swarm optimization design framework for multi-link systems. Moreover, the framework facilitates the storage and reuse of design knowledge, which improves design efficiency and promotes design automation.

Introduction

Mechanical presses play a crucial role in forging and stamping processes. With the development of forging technology, in order to obtain near-net shape forgings, the development and use of cold forging technology has become the consensus of the manufacturing powerhouse. In the cold forging conditions, the fluidity of the metal is much lower than it in the hot forging conditions, so the cold forging forming is more sensitive to the forming speed, which requires “forming in accordance with the trend”, that is, the forming speed should be as low as possible, and the forming force should be much higher. Traditional crank-slide mechanical presses are difficult to meet the needs of current industrial production due to problems such as high speeds and difficulty in adjusting them [1]. Multi-link mechanical presses have received widespread attention and application in recent years due to their ability to approach the material being processed at slower, more uniform speeds and their quick-return characteristics. Their performance is mainly determined by the structural and dimensional parameters of the multi-link system. However, the multi-link systems for presses are complex and various. The conventional design methods depend on the experience of designers and a trial-and-error process, leading to low precision, extended design cycle and other problems.

With the advancement of computer technology, data-based optimization design methods [2,3] and knowledge-based design methods [4,5] are gradually maturing. In recent years, there have been many researchers using data-based optimization design methods combined with computer simulation for the optimal design of multi-link systems. Dong [6] proposed an improved optimization algorithm based on virtual prototype technology by considering the kinematics and dynamics performance indexes of the elbow-bar drive mechanism, and determined the optimal dimensions of the elbow-bar drive mechanism. Kütük [7] established an optimization model of a two-degree-of-freedom seven-link mechanism using a genetic algorithm to determine the optimal link lengths and applied kinetostatic approach to analyse crank torques and all the forces on the links. However, data-based optimization design methods require a large amount of data to support them. With the growing complexity of products, conventional intelligent optimization algorithms encounter challenges such as low efficiency and local optimization. Moreover, most of the studies have neglected the selection of the transmission configuration of the multi-link system, which is a difficult step to establish an optimization model. Selecting the most suitable transmission configuration requires more experience and theoretical knowledge of the designers.

Knowledge-based design methods are good at aggregating and reusing design knowledge (e.g., expert experience, design manuals, industry specifications, etc.) for reasoning, judging, and decision-making to achieve automation, and intelligence in product design. However, they are seldom applied to the design of multi-link system of mechanical presses, and knowledge-based design methods are difficult to apply in practice due to incomplete knowledge, difficulties in knowledge acquisition and characterization, and limited adaptability. Therefore, it is imperative to integrate and improve these two methods for better application in practice [8,9].

Aiming at the above problems, this paper builds a design knowledge base and a reasoning machine to realize the rapid selection of transmission configuration for cold and warm forging presses through the case similarity calculation, and establishes an optimization model based on the elbow-bar transmission configuration determined by the above. A knowledge-guided improved particle swarm optimization algorithm is proposed to solve the optimization model. Then the optimized dimensional parameters of the multi-link system are obtained. The purpose of improving design efficiency, press performance and product quality is achieved.

Rapid Selection of Transmission Configuration

The selection of the transmission configuration is not only the first step of press design, but also the most critical step. The types of configurations are complex and varied. In order to select the most suitable transmission configuration, the theoretical knowledge and design experience of the designers are required to a higher level. In this paper, knowledge engineering technology is used to create a knowledge base and a reasoning machine to store and reuse historical cases and design experience to realize the rapid selection of transmission configurations.

Knowledge base is to store the knowledge in some form of representation in the computer for subsequent recall. Multi-link systems design knowledge is varied and complex in form, which can be divided into case knowledge, rule knowledge and related information knowledge. The above knowledge is stored in an appropriate way to build a knowledge base for multi-link system design. The structure of knowledge base is shown in Fig. 1.

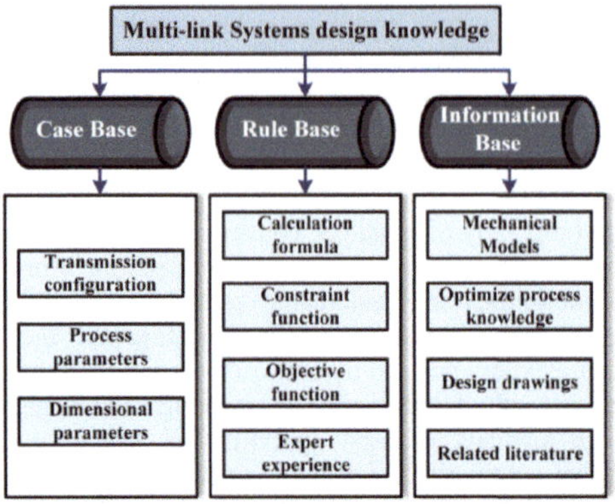

Fig. 1. Structure of multi-link systems design knowledge base.

The reasoning machine is a bridge between the knowledge base and the design requirements. In this paper, firstly, input the design demand parameters, the reasoning machine retrieves whether there are similar cases. If there is, then adopt the case with the highest similarity, follow its transmission configuration, and improve the design of partial size parameters according to the design demand to form a new case. If an identical case exists, the case is used directly. The reasoning process for the transmission configuration of the multi-link systems is shown in Fig. 2.

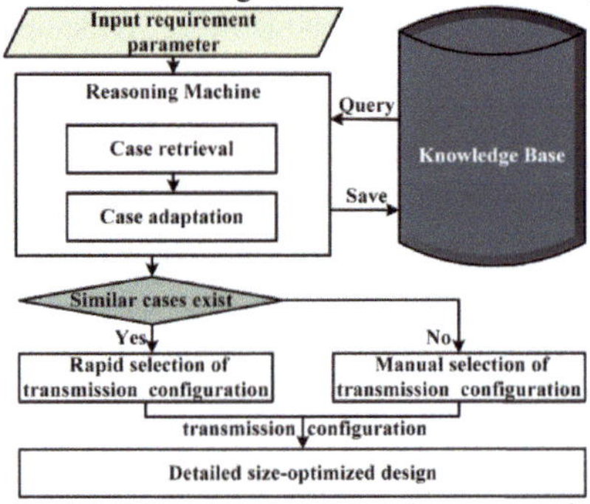

Fig. 2. Reasoning process for the transmission configuration of the multi-link systems.

The main technical parameters to be entered for case retrieval include: machine type, nominal pressure, slide stroke, nominal pressure stroke and process parameters. The case similarity formula based on the above parameters is as follows

$$S(A,B)=\sum_{i=1}^{n}\frac{\mu_i}{1+|a_i-b_i|}. \tag{1}$$

Where S represents the similarity of cases, A, B represents different cases, the set of attributes of these two cases is (a_1, a_2, ..., a_n), (b_1, b_2, ..., b_n), n represents the similarity of cases, and μ_i represents the weights, the effect of each attribute on the multi-link systems.

Considering the richness of the cases in the case base, we set the similarity threshold to 0.6. If there are cases with similarity more than 0.6, they are considered similar. If the similarity of all cases is lower than 0.6, it is considered that there is no similar case. So, the transmission configuration needs to be selected manually. With the continuous enrichment of the knowledge base, the accuracy and automation of the reasoning machine will be continuously improved.

Take the main technical parameters of a cold forging press for near net shape as an example: Nominal pressure, 12500 kN; Nominal pressure stroke, 10mm; slide stroke, 400 mm; Maximum stroke rate of slide is 25SPM. After the case retrieval and adoption, according to the highest similarity case, the reasoning obtained the transmission configuration should be the elbow-bar mechanism of bias driving, which is the same as the original transmission configuration of this press, proving the validity of this reasoning mechanism. The transmission sketch of the elbow-bar driving mechanism is shown in Fig. 3.

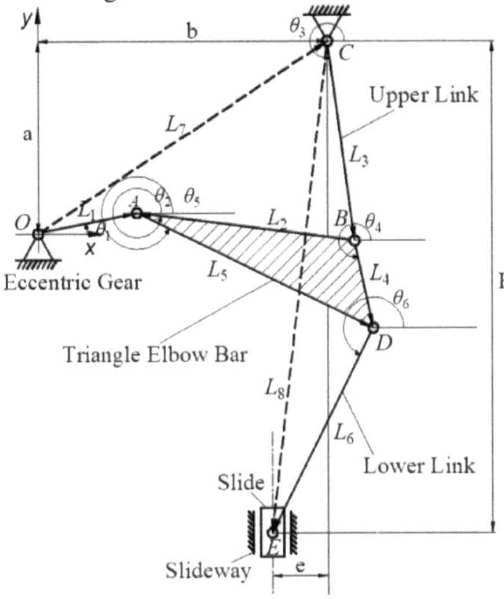

Fig. 3. Transmission sketch of the elbow-bar driving mechanism.

Optimization Model of Elbow-Bar Driving Mechanism

The kinematic model of the elbow-bar driving mechanism needs to be established firstly. Establish the coordinate system shown in Fig. 3. L_1~L_6 is the length of each link, and θ_1~θ_6 is the turning angle of each link. According to the closed vector method, three closed loops are selected as OABCO, ABDA and CBDEC, which can be obtained as:

$$\begin{cases} L_1\cos\theta_1 + L_2\cos\theta_2 = a + L_3\cos\theta_3 \\ L_1\sin\theta_1 + L_2\sin\theta_2 = b + L_3\sin\theta_3 \\ L_2\cos\theta_2 + L_4\cos\theta_4 = L_5\cos\theta_5 \\ L_2\sin\theta_2 + L_4\sin\theta_4 = L_5\sin\theta_5 \\ L_3\cos\theta_3 + L_4\cos\theta_4 + L_6\cos\theta_6 = e \\ L_3\sin\theta_3 + L_4\sin\theta_4 + L_6\sin\theta_6 = H \end{cases} . \tag{2}$$

From Eq. 2, Under the condition that the size of links is fixed, there exist θ_1, θ_2, θ_3, θ_4, θ_5, θ_6, H a total of 7 variables, of which θ_1 is the independent variable. In the limit of 6 equations, cancelling θ_2, θ_3, θ_4, θ_5, θ_6, we can get the equation of H about θ_1, denoted as $H(\theta_1)$.

Then we can get the kinematic equations of the elbow-bar driving mechanism are

$$\begin{cases} s = H_{\max} - H \\ v = \dfrac{\mathrm{d}s}{\mathrm{d}t} \\ a = \dfrac{\mathrm{d}^2 s}{\mathrm{d}t^2} \end{cases} . \tag{3}$$

where s, v, a are the displacement, velocity and acceleration of the slider, respectively.

Once the kinematic model of the mechanism is complete, the corresponding optimization model can be established. The optimization vector is as follows:

$$X = [x_1, x_2, x_3, x_4, x_5, x_6, x_7, x_8, x_9]^{\mathrm{T}} = [L_1, L_2, L_3, L_4, L_5, L_6, a, b, e]^{\mathrm{T}} . \tag{4}$$

In order to establish the constraint equations and objective functions, it is necessary to take into account the structure of the multi-link system, the transmission performance and the characteristics of the cold and warm forging process.

As can be seen from Fig. 3, there is a crank-rocker mechanism in the driving mechanism. The constraint functions for establishing the crank rock mechanism are

$$\begin{cases} g_1(X) = x_1 - x_2 \le 0 \\ g_2(X) = x_1 - x_3 \le 0 \\ g_3(X) = x_1 - x_7 \le 0 \\ g_4(X) = x_1 + x_7 - x_2 - x_3 \le 0 \\ g_5(X) = x_1 + x_2 - x_7 - x_3 \le 0 \\ g_6(X) = x_1 + x_3 - x_2 - x_7 \le 0 \end{cases} . \tag{5}$$

In order to ensure good transmission performance of the mechanism, it is necessary to limit the range of transmission angle between two limit positions. Establish the constraint functions $g_7(X)$, $g_8(X)$ for the pole drive angle.

$$\begin{cases} g_7(X) = x_2^2 + x_3^2 - (\sqrt{x_7^2 + x_8^2} - x_1)^2 - 2x_2 x_3 \cos\gamma_{\min} \le 0 \\ g_8(X) = -x_2^2 - x_3^2 + (\sqrt{x_7^2 + x_8^2} + x_1)^2 + 2x_2 x_3 \cos\gamma_{\max} \le 0 \end{cases} . \tag{6}$$

Where $\gamma_{\min}$, $\gamma_{\max}$ is the allowable minimum and maximum transmission angle between two limit positions, generally take $\gamma_{\min} = 45°$ and $\gamma_{\max} = 135°$.

Slider stroke is one of the main technical parameters of the press, which should be controlled within a certain error range at the design stage. Therefore, in this paper, the slider stroke error is constrained to be within ±1%, and the constraint functions $g_9(X)$, $g_{10}(X)$ are established.

$$\begin{cases} g_9(X) = H_{\max} - H_{\min} - 1.01s \le 0 \\ g_{10}(X) = 0.99s - H_{\max} + H_{\min} \le 0 \end{cases} . \tag{7}$$

The establishment of the optimization objective function needs to take into account the motion curve of the slider and the maximum input torque of the eccentric gear.

The low speed and smooth motion of the slide can make the material uniformly stressed during the forming process and improve the quality of forgings. In addition, lower slider speeds mean less input torque for the eccentric gear. Therefore, the minimization of the maximum speed and maximum acceleration of the slider within the nominal pressure stroke is taken as optimization objectives. The objective functions are

$$f_1(X) = |v|_{s=\mathrm{h}} \tag{8}$$

$$f_2(X) = \max|a|_{s\in[0,\mathrm{h}]}. \tag{9}$$

Where h is the displacement of the slide when it first contacts the workpiece, that is, the nominal pressure stroke.

At the same time, in order to increase the work efficiency, the return time should be reduced, that is, the quick-return characteristic should be strengthened. The objective function is

$$f_3(X) = t_{\mathrm{up}}. \tag{10}$$

Where t_{up} is the return time of slider.

Combining all objective functions and taking the constraint function as penalty terms, the total objective function is established as:

$$\min f(X) = \min\{\sum_{i=1}^{3} \mathrm{w}_i \frac{f_i(X)}{f_{i\min}} + \sum_{j=1}^{10} \gamma_j g_j(X)\}. \tag{11}$$

Where w_i are the weights of each optimization objective, and $\mathrm{w}_1+\mathrm{w}_2+\mathrm{w}_3=1$; $f_{i\min}$ are the estimated optimal values for each optimization objective, so that the values of the three optimization objectives are of the same order of magnitude; γ_j is the penalty factor

Any kinematic and optimization models of the transmission configuration will be stored as knowledge in the knowledge base. This avoids repeated modeling and improves the design efficiency in the subsequent similar case designs.

Knowledge-Guided Particle Swarm Optimization

In this paper, the particle swarm optimization algorithm is used to solve the optimization model, in order to improve the optimization efficiency, this paper proposes a knowledge-guided improved particle swarm optimization algorithm. And the following optimization strategies are proposed.

Firstly, this paper adopts a hierarchical case knowledge-guided particle population initialization strategy by making full use of historical similar cases. The aim is to reduce the number of iterations and to improve the local optimization accuracy while making the algorithm capable of global search.

For the characteristics of the transmission configuration combined with the actual production situation, set the upper and lower limits of each size parameter $x_{i\max}$, $x_{i\min}$ (i=1,2,3...9), to determine the global optimization space for the

$$D = [x_{1\min}, x_{1\max}; x_{2\min}, x_{2\max}; x_{3\min}, x_{3\max}; x_{4\min}, x_{4\max}; x_{5\min}, x_{5\max}; \\ x_{6\min}, x_{6\max}; x_{7\min}, x_{7\max}; x_{8\min}, x_{8\max}; x_{9\min}, x_{9\max};]^{\mathrm{T}}. \tag{12}$$

According to the similar historical cases (similarity greater than 0.6), obtained in the transmission configuration reasoning stage, the maximum and minimum values of each design variable in the similar historical cases with the same transmission configuration are taken as $sx_{i\max}$, $sx_{i\min}$ (i=1,2,3...9), and thus the local optimization space is determined as

$$D_1 = [sx_{1\min}, sx_{1\max}; sx_{2\min}, sx_{2\max}; sx_{3\min}, sx_{3\max}; sx_{4\min}, sx_{4\max}; sx_{5\min}, sx_{5\max}; sx_{6\min}, sx_{6\max}; sx_{7\min}, sx_{7\max}; sx_{8\min}, sx_{8\max}; sx_{9\min}, sx_{9\max};]^{T} \quad (13)$$

If the number of similar historical cases is m and the particle swarm population size is n, the above m historical cases are used as m population initial solutions, and 0.6n-m population initial solutions are taken to be randomly generated in the local optimization space D_1, and the remaining 0.4n population initial solutions are randomly generated in D-D_1. So that the algorithm has global search capability while improving the local optimization accuracy.

If there are few similar cases in the case base to generate local optimization intervals or lack of practical reference value, it is necessary to judge whether to adjust the global optimization space by oneself, and carry out random initialization of all particle `populations in this global optimization space.

Then, the particle iteration process is guided by the sensitivity knowledge of each optimization variable. In the particle swarm optimization algorithm, the particles find the global optimal solution by continuously updating their position and velocity, and its particle iteration formula is

$$\begin{cases} v_i(t+1) = \lambda v_i(t) + c_1 r_1 (pb_i - x_i(t)) + c_2 r_2 (gb_i - x_i(t)) \\ x_i(t+1) = x_i(t) + v_i(t+1) \end{cases} \quad (14)$$

Where v, x are the velocity and position of the particle, respectively; t is the number of iterations; λ is the inertia weight; c_1 is the self-learning factor; c_2 is the social learning factor; r_1, r_2 are the random numbers between (0,1); pb_i and gb_i are the positions of the individual and global guides for particle i.

According to the objective function and global optimization space of elbow-bar driving mechanism, the sobol method is applied to perform parameter sensitivity analysis. The parameter sensitivity vector is obtained as

$$V_{sen} = [S_{x1}, S_{x2}, S_{x3}, S_{x4}, S_{x5}, S_{x6}, S_{x7}, S_{x8}, S_{x9}]. \quad (15)$$

where S_{xi} is the sensitivity of the objective function to each optimization variable x_i (i=1,2,3...9).

Based on the above knowledge of parameter sensitivity, the particle iteration formula is dynamically updated to improve the search direction and step size. Firstly, the particles are guided to iterate in the direction of high sensitivity variables by updating the important parameters λ, c_1, c_2 as

$$\begin{cases} \lambda = \lambda_{\max} - \dfrac{V_{sen} \cdot v_i(t)}{\|V_{sen}\| \times \|v_i(t)\|} (\lambda_{\max} - \lambda_{\min}) \\ c_1 = c_{1\min} + \dfrac{V_{sen} \cdot (pb_1 - x_i(t))}{\|V_{sen}\| \times \|pb_1 - x_i(t)\|} (c_{1\max} - c_{1\min}) \\ c_2 = c_{2\min} + \dfrac{V_{sen} \cdot (gb_1 - x_i(t))}{\|V_{sen}\| \times \|gb_1 - x_i(t)\|} (c_{2\max} - c_{2\min}) \end{cases} \quad (16)$$

Where λ_{max}, λ_{min} are the maximum and minimum values of the inertia weights; c_{1max}, c_{1min}, c_{2max}, c_{2min} are the maximum and minimum values of the corresponding learning factors, respectively.

Meanwhile, the particles should be made to search quickly in the direction of low sensitivity variables with a larger step size while reducing the step size in the direction of high sensitivity variables in order to improve the algorithm's optimization ability during iteration. the iteration method is as follows

Materials Research Forum LLC
https://doi.org/10.21741/9781644903254-58

$$x_i(t+1) = x_i(t) + (1.5 - \frac{V_{sen} \cdot v_i(t+1)}{\|V_{sen}\| \times \|v_i(t+1)\|}) v_i(t+1) \cdot \tag{17}$$

Fig. 4 shows the knowledge-guided particle swarm optimization process. The optimization process is as follows.

Step 1: Determine the global and local optimization intervals, and initialize the particle population hierarchically in different intervals, using the historical similar cases as a guide;

Step 2: Iterate according to the particle iteration strategy guided by the sensitivity knowledge to update the position and velocity of each particle. Then, calculate the fitness of each particle based on the knowledge of constraints and objective functions;

Step 3: update the historical optimal positions of particles themselves and population, as well as the main parameters of the algorithm, such as the inertia weight λ;

Step 4: Stop iteration when the end condition is met, the optimization results are processed and outputted. The design results are stored in the example knowledge base to further enrich and improve the knowledge base.

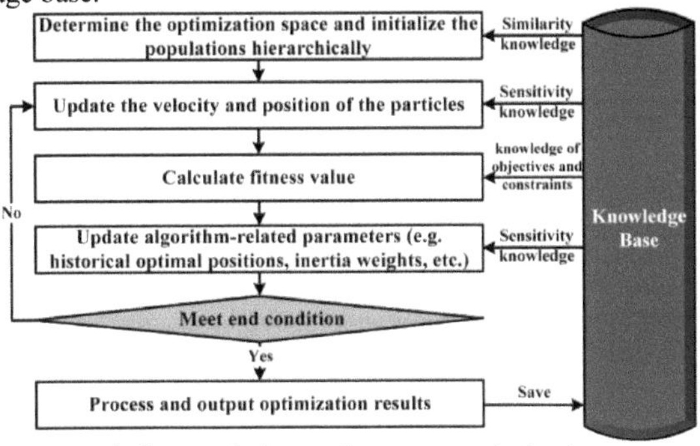

Fig. 4. Knowledge-guided particle swarm optimization process.

Optimization Result

Based on the above optimization model and the knowledge-guided particle swarm optimization algorithm, an optimization program is written in MATLAB to perform the optimization calculations. In order to facilitate the manufacturing and accuracy testing of the link system in engineering practice, the optimization results are rounded off and the optimized links system parameters are compared with the original links system parameters, as shown in Table 1 below.

Table 1. Optimization results of the elbow-bar driving mechanism [mm].

	L_1	L_2	L_3	L_4	L_5	L_6	a	b	c
Initial	205	1720	610	720	1930	845	570	1500	-15
Optimal	260	1725	705	545	1900	920	760	1550	-28

Through kinematic analysis, we can get different curves of the slider. Comparing the slider curves before and after optimization, the optimized results are better than the original parameters as shown in Fig. 5.

Materials Research Proceedings 44 (2024) 538-547 https://doi.org/10.21741/9781644903254-58

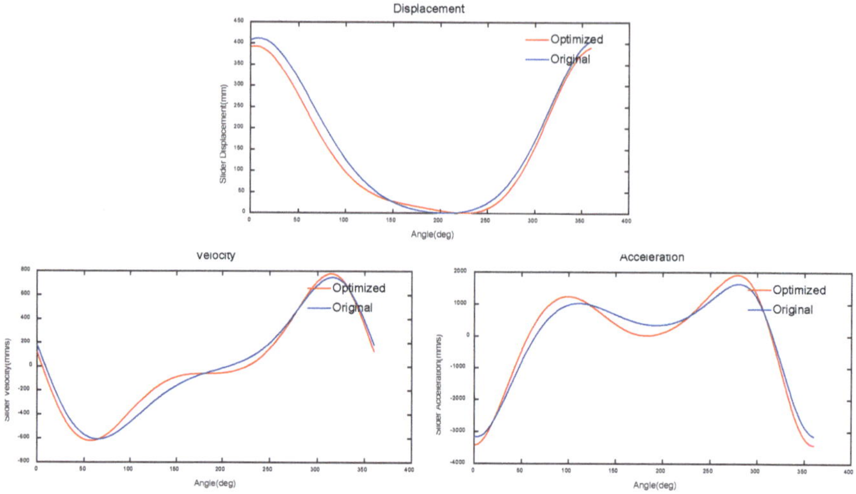

Fig. 5. Kinematics curve of slider.

The position of the lower dead center is shifted to the right, the motion curve is smoother in the forming stage. It meets the process requirements of low and smooth forming speed and long holding time for cold and warm forging, which is conducive to improving the quality of forgings.

The main technical indexes of the elbow-bar driving mechanism of the cold and warm forging press before and after optimization are shown in Table 2.

Table 2. Main technical indexes of the elbow-bar driving mechanism .

Main technical indexes	Initial	Optimal	optimzation rate
Stroke [mm]	411.11	392.12	29.07%
Total height [mm]	2212	2164	2.17%
Velocity on the height of 25 mm [mm/s]	152.58	79.06	47.83%
Maximum acceleration during the nominal stroke [mm/s^2]	645.40	576.93	10.61%
Eccentric gear input torque [N·m]	728584	377515	48.19%
Slider return time [s]	0.71	0.62	12.68%

As can be seen from Table 2, after the optimization of the knowledge-guided particle swarm algorithm, not only the motion curve of the mechanism is improved, but also its total height, the input torque of the eccentric gears is reduced, and the quick-return characteristics are also improved. The above analysis and data show that the knowledge-guided particle swarm algorithm can effectively optimize the size of the multi-link system to achieve the purpose of improving the performance of the press and the quality of forgings.

Summary

Based on the knowledge engineering technology, the knowledge base of multi-link systems design and the related reasoning machine are built, and the rapid selection of the transmission configuration for the designed press is realized. According to the elbow-bar driving mechanism obtained by reasoning, its kinematic model and optimization model are established, and the knowledge-guided improved particle swarm optimization algorithm is proposed to solve the optimization model. After optimization, all the kinematic performance indexes of the slider are optimized to different degrees and the maximum input torque is reduced. Meanwhile, the method presented in this paper facilitates the accumulation and transfer of design knowledge. The models and equations established for each design will be classified and stored in the knowledge base so that they can be used for subsequent design of similar cases to avoid repetitive work. With the gradual enrichment of the knowledge base, the accuracy and efficiency of the optimization design method proposed in this paper will be gradually improved.

References

[1] K. Osakada, K. Mori, T. Altan, Mechanical servo press technology for metal forming, CIRP annals 60 (2011) 651-672. https://doi.org/ 10.1016/j.cirp.2011.05.007

[2] D. Kang, Z. Chen, Y.H. Fan, C. Li, C. Mi, Optimization on kinematic characteristics and lightweight of a camellia fruit picking machine based on the Kriging surrogate model, Mech. Ind. 22 (2021) 16. https://doi.org/10.1051/meca/2021017

[3] C. Balasubramanyam, AB. Shetty, KR. Spandana, Analysis and optimization of an 8 bar mechanism, Int. J. Mach. Learn. 6 (2015) 655-666. https://doi.org/10.1007/s13042-015-0368-z

[4] F. Dworschak, P. Kügler, B. Schleich, S. Wartzack, Model and knowledge representation for the reuse of design process knowledge supporting design automation in mass customization, Appl. Sci. 11 (2021) 9825. https://doi.org/10.3390/app11219825

[5] X. Long, H. Li, Y Du, E Mao, J Tai, A knowledge-based automated design system for mechanical products based on a general knowledge framework, Expert Syst. Appl. 178 (2021) 114960. https://doi.org/10.1016/j.eswa.2021.114960

[6] ME. Kütük, M. Artan, Hybrid seven-bar press mechanism: link optimization and kinetostatic analysis, Tehnički glasnik 12 (2018) 181-187. https://doi.org/10.31803/tg-20180203202102

[7] X. Dong, Y. Sun, Multi-objective Sizing Optimization of Elbow-Bar Driving Mechanism of Cold Forging Press, Forming the Future: Proceedings of the 13th International Conference on the Technology of Plasticity (2021) 2899-2907. https://doi.org/10.1007/978-3-030-75381-8_240

[8] D. Wu, G.G. Wang, Knowledge-assisted optimization for large-scale design problems: A review and proposition, J. Mech. Design 142 (2020) 010801. https://doi.org/10.1115/1.4044525

[9] C. Li, D. Wang, Integrated knowledge-based system for containership lashing bridge optimization design, 13th international Marine design conference, (2018) 429-438.

Development of unified constitutive model for hot deformation behavior of TC4 with and without diffusion bonding

CAI Zhongman[1,a], CHEN Haosheng[1,b], LI Can[1,c], CHEN Jiarui[1,d], LI Xiaoxing[1,e] and LI Yong[1,f*]

[1]Beihang University XueYuan Road No.37, Haidian District, Beijing, China

[a]czmncst@163.com, [b]chenhs@buaa.edu.cn, [c]lican9504@163.com, [d]jiaruic@buaa.edu.cn, [e]li.xiaoxing@buaa.edu.cn, [f]liyong19@buaa.edu.cn

Keywords: TC4, Hot Deformation, Modeling, Diffusion Bonding

Abstract. The hot deformation behavior of TC4 with and without diffusion bonding (DB) is studied. Firstly, the hot tensile testes of TC4 specimens with and without DB were carried out at temperature of 750°C and strain rate of 0.1 ~ 0.0001 s^{-1}. It was found that the peak stress and fracture strain of DB specimens are lower than those of base metal (BM) specimens. The quasi-cleavage fracture appeared in the DB specimens at 750°C, and the mechanism could be explained as straight grain boundary formed by incomplete recrystallization and insufficient grain boundary migration energy and the joints weak left over from the DB process causing rapid voids nucleation at the DB interface. The shorter distance of voids coalescence within the interface leads to quasi-cleavage fracture. In addition, a unified constitutive model based on internal variables was developed by introducing the weld-dependent fracture coefficients and damage tolerance coefficient. The model can accurately predict the hot deformation behavior of TC4 with and without diffusion bonding, describing the influence behavior of DB area on hot deformation, as well as the evolutions of internal variables, including dislocation density, reserve fraction, dynamic recrystallization (DRX) fraction and damage. It provides a theoretical basis for further accurate simulation of the hot forming of titanium alloy with diffusion bonding.

Introduction

TC4 titanium alloy has been widely used in aerospace engineering, vehicle engineering and medical fields [1, 2]. TC4 hollow parts have low material density and higher hollowness, which highlights its lightweight characteristics. The main manufacturing processes of hollow parts are integral casting, additive manufacturing, and diffusion bonding (DB). Compared with integral casting and additive manufacturing, DB has the advantages of easy control of process parameters, small deformation, higher dimensional accuracy and mechanical properties of welded joints that are equivalent to those of base metal (BM). At present, superplastic forming and diffusion bonding (SPF/DB) or hot forming and DB hybrid processes are often used for hollow blade manufacturing [3]. On the basis of DB, subsequent forming is carried out to obtain the final part. In the process of hot forming, the deformation is large and the relationship of microstructure evolution is more complicated. The alloy is very sensitive to the initial state [4]. Different initial states also lead to different deformation behaviors in subsequence processing. Therefore, it is important and necessary to characterize and realize the hot deformation characteristics of materials after DB.

Guo and Peng [5] studied the welding and microstructure evolution during TC4 diffusion bonding. Two states will be formed at the interface: one is that the recrystallized grains pass through the original interface and replace, and the other is that the recrystallized grains form a new straight grain boundary at the interface. Li and Imran [5] conducted tensile tests on TC4 with DB specimens and found that the elongation of TC4 with DB was lower than BM at elevated temperatures. Li [3] studied the microstructure evolution of tc4 at hot temperatures and believed that the temperature required for DRX to occur at the DB interface was higher. The characterization of this difference in hot deformation characteristics is of great significance for process optimization and precision manufacturing.

E. Alabort [8] established a unified viscoplastic constitutive model to simulate the superplastic diffusion bonding process of Ti-6Al-4V. It is proved that deformation behavior prediction with processing history information and internal microstructure state is more accurate. Lin[18] applied the internal variable constitutive model of dislocations, grain size and recrystallization variables to TC4 alloy, and achieved good prediction results. On this basis, Bai [19] established a unified constitutive of dislocation and spheroidization behavior and applied it to TC4 alloy. However, the above studies are all based on homogeneous TC4, which does not have the ability to describe the changes of hot deformation characteristics caused by specific structures at the same time.

Therefore, in this work, TC4 in DB and BM states are selected as the original state to study. Firstly, the hot tensile test is carried out at 750°C and 0.0001 ~ 0.1 s^{-1}, and the hot deformation behavior is analyzed. On this basis, a set of internal variable constitutive model is developed to describe the hot deformation characteristics of both states at the same time. The parameters of the equation will be calibrated by the Particle Swarm Optimization (PSO) algorithm. Finally, the prediction effect is evaluated by statistics.

Material and experimental

The as-received material is prepared by diffusion welding of two TC4 alloy plates, as shown in Fig. 1(a). On the welding material, DB and BM round bar tensile specimens with a diameter of 3 mm and a gauge distance of 15mm are made along the thickness direction and length direction, respectively, as shown in Fig. 1(b). The temperature is selected as 750°C, TC4 conventional thermoforming processing temperature. The strain rates are 0.1,0.01,0.001, 0.0001s^{-1}.

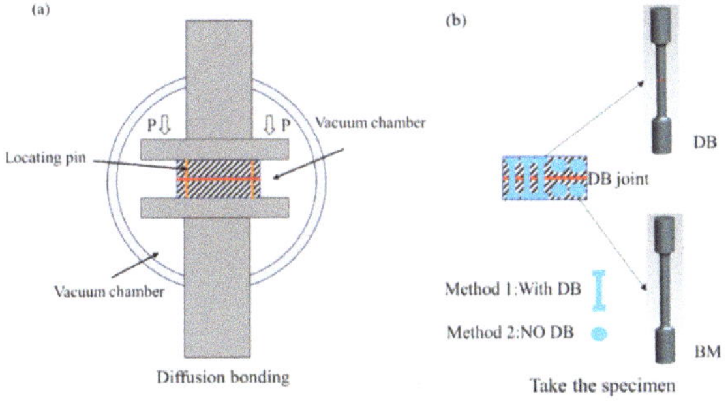

Fig. 1. (a) Billet diffusion welding diagram and (b) specimen preparation diagram.

Experimental results and discussions

The flow stress-strain curves comparison of TC4 between BM and DB states is shown in Fig. 2 (a). Comparing the two can be found that the stress-strain curves of BM have a higher peak stress at the beginning of deformation, as shown in the pink dotted circle. The peak stresses are plotted as shown in Fig. 2(b) It can be found that with the increase of strain rate, the difference of peak stress between BM and DB becomes smaller. The DB process of TC4 is studied in reference [5]. The results show that the DRX will occur at the interface during DB, which will lead to a decrease of dislocation density. The dislocation density of the BM specimen is higher, which makes the dislocation accumulation and entanglement more intense. Work hardening is strengthened, making the stress required for deformation greater. So the peak stress of BM is greater. With the increase of strain rate, the deformation time is shorter and the dislocation increment rate is larger. Therefore, the effect of the original dislocation density is weakened. This results in the phenomenon that the peak stress difference decreases with the increase of strain rate.

As the deformation continues, the stress of BM softens faster, and the stress gradually approaches. The higher dislocation density allows the DRX behavior to occur earlier and more fully resulting in the softening effect stronger. When the stress is close, such as the position indicated by the red arrow, the DB specimen breaks and the stress is 0. However, the BM specimen continued to deform and gradually decreased to 0. Fig. 2(c) shows the comparison of fracture strain between DB and BM specimens. Due to the rapid fracture of the DB specimen, the fracture strain is lower than that of the BM specimen. The results show that the DB specimen has worse formability than the BM specimen, and it is necessary to pay more attention to its damage behavior.

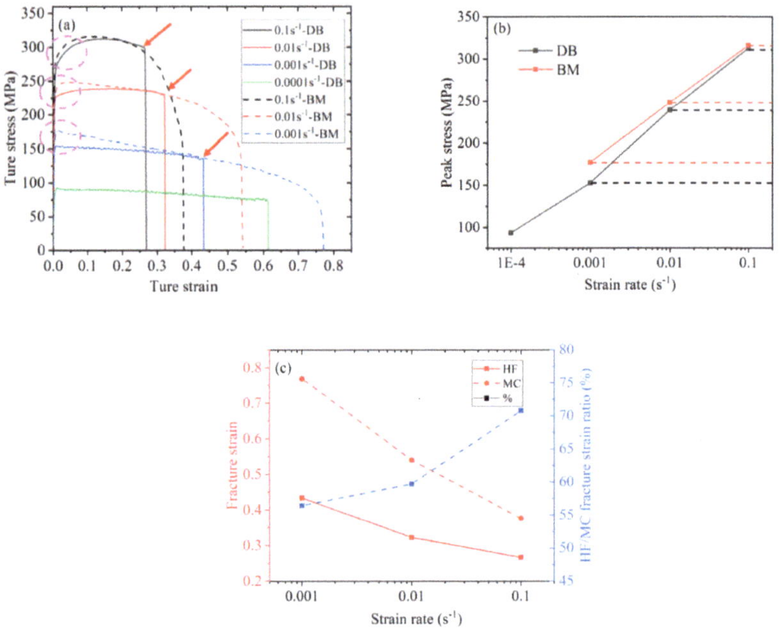

Fig. 2. (a) The flow stress-strain curves of TC4 comparison between BM and DB; (b) peak stress and (c) comparison of fracture strain between DB and BM.

Fig. 3(a) shows the fracture morphology of BM at a strain rate of $0.1s^{-1}$. We can clearly see a necking phenomenon at the fracture. By observing the microstructure of the fracture, a large number of dimples were found. It shows that the fracture of BM is a typical ductile fracture. Fig. 3(b) and (c) show the fractures morphology of DB with strain rates of $0.1s^{-1}$ and $0.0001s^{-1}$, respectively. The whole fracture is straight, and there is no obvious necking phenomenon. By observing the microstructure of the fracture, it is found that dimple and transgranular fracture are the typical characteristics of quasi-cleavage fracture. This indicates that at 750°C, 0.1 ~ 0.0001 s^{-1}, the DB specimen will have quasi-cleavage fracture during hot deformation. Fig. 3 (d) shows the forming mechanism of the fracture at the DB interface on a microscopic scale. Some scholars [5] have shown that during the DB process, the inadequate deformation resulted in incomplete recrystallization and grain boundary migration, the storage energy difference on both sides of the interface was not enough to drive the grain boundary migration process. Therefore, the straight grain boundary formed. The voids preferentially nucleate at the grain boundary [9], and damage is

Materials Research Forum LLC
https://doi.org/10.21741/9781644903254-59

more likely to occur at the interface. In the process of deformation, the voids are preferently nucleated at the joints weak, impurities, and grain boundaries at the DB interface Resulting in the rapid formation of a large number of voids in the DB interface. Moreover, the voids coalescence distance at the interface is shorter, and it is easier to coalesce with continuous deformation. This results in a quasi-cleavage fracture. This also indicates that the damage of DB specimens is more inclined to coalescence at the interface of DB, resulting in a decrease in fracture strain. In order to simulate this kind of specimen, the damage tolerance is proposed. The hot deformation and fracture behavior of DB specimens is simulated by controlling damage tolerance.

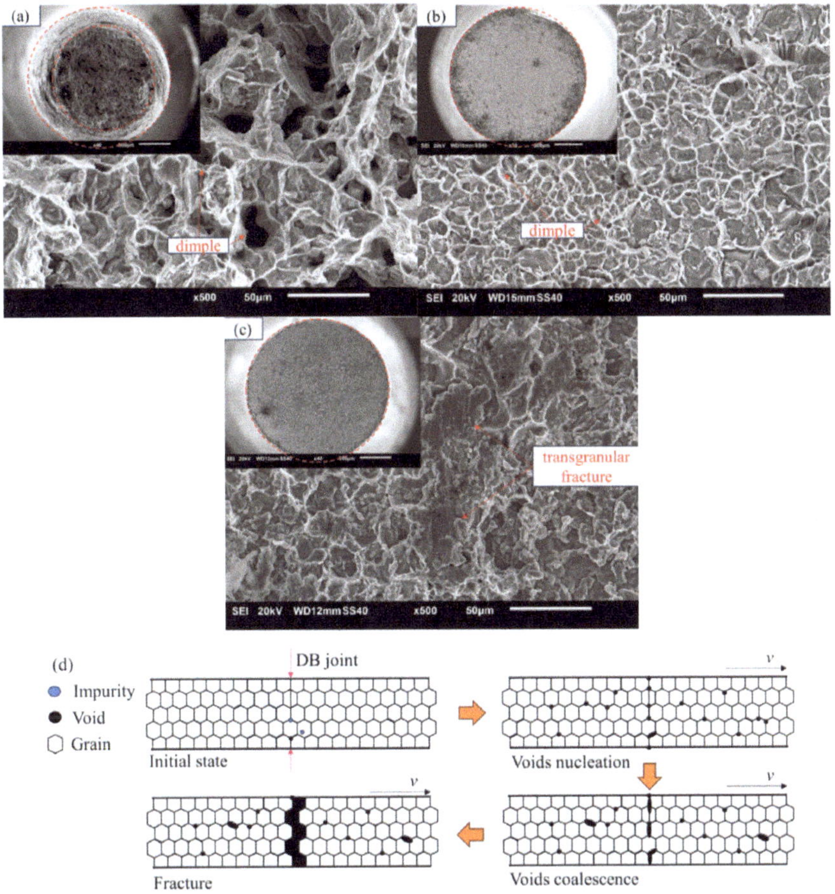

Fig. 3. Fracture (a) BM specimen 0.1s^{-1}; (b) DB specimen 0.1s^{-1}; (c) DB specimen 0.001s^{-1} and (d) schematic diagram of forming mechanism.

Development of viscoplastic constitutive equations

Viscoplastic flow rule. Lin [10] proposed that the flow stress of a material is mainly composed of three parts: initial yield stress, viscoplastic stress, and hardening stress. The material undergoes deformation behavior under the synergistic action of the two phases. Therefore, the total plastic strain rate can be expressed as [11]:

$$\begin{cases} \dot{\varepsilon}_{\mathrm{p},\alpha} = \left(\dfrac{\sigma - k_\alpha - H}{K_\alpha}\right)^n \\ \dot{\varepsilon}_{\mathrm{p},\beta} = \left(\dfrac{\sigma - k_\beta - H}{K_\beta}\right)^n \end{cases} . \tag{1}$$

k_α and k_β are the initial yield stress of the α and β phases, respectively. K_α and K_β are the material parameters of the α and β phases, respectively. Based on BAI et al. 's study [11], the relationship between α and β phases can be defined as $k_\alpha = 1.25\,k_\beta$, $K_\alpha = 1.11\,K_\beta$.

The total strain rate is expressed as:

$$\dot{\varepsilon}_p = (1 - f_\beta)\dot{\varepsilon}_{p,\alpha} + f_\beta \dot{\varepsilon}_{p,\beta} . \tag{2}$$

Hardening mechanism. H is the hardening stress due to the evolution of dislocation density during plastic deformation. After extensive experimental verification, its value is known to satisfy the Taylor model. The normalized dislocation density has a value between 0 and 1. Eq. (3) can be obtained by simplifying:

$$H = B\overline{\rho}^{0.5} . \tag{3}$$

B is the isotropic hardening constant, and $\overline{\rho}$ is the relative dislocation density.

Dislocation evolution and softening mechanisms. The dislocation density is elevated, which in turn drives the onset of DRV and DRX behavior, and the reciprocal cancellation of heterogeneous dislocations leads to a decrease in dislocation density. Therefore, the dislocation density evolution equation is expressed as [12]:

$$\frac{d\rho}{d\varepsilon_p} = k\sqrt{\rho} - k_1 \rho . \tag{4}$$

where ε_p is the plastic strain, and k and k_1 are material constants. The first term is related to the dislocation multiplication as a result of plastic straining. The second term models the reduction of dislocation density due to DRV. Since the effects of static recovery and recrystallization on the dislocation density during deformation should not be neglected. The dislocation accumulation and ablation are different for different strain rates at hot temperatures [13]. The strain rate influence term is introduced and expressed in simplified form:

$$\dot{\overline{\rho}} = A\,\dot{\varepsilon}_p^{\delta_1}\left|\dot{\varepsilon}_p\right| - m_1\overline{\rho}\dot{\varepsilon}_p^{\delta_2}\left|\dot{\varepsilon}_p\right| - m_2\overline{\rho}^{\delta_3} - \left[m_3\overline{\rho}/(1-S)\right]\dot{S} . \tag{5}$$

The results of Alabort et al [8] define the critical value of dislocation density for DRX behavior occurs. In addition, Lin [14] suggested that there is a reserve period for DRX. There is a relationship between the reserve period and the dislocation density. In summary, the evolution of the DRX fraction is described as:

Materials Research Forum LLC
https://doi.org/10.21741/9781644903254-59

$$\begin{cases} \dot{S} = q_1[x\bar{\rho} - \bar{\rho}_c(1-S)](1-S)^{q_2}. \\ \dot{x} = q_3(1-x)\bar{\rho}. \\ \bar{\rho}_c = q_4\dot{\varepsilon}_p^{q_5}. \end{cases} \tag{6}$$

where q_1, q_2, q_3, q_4 and q_5 are the material parameter, x is the DRX reserve fraction, and $\bar{\rho}_c$ is the critical dislocation density.

The DRX softening effect be introduced based on Zhao's findings [15], modifying Eq. (2) to read:

$$\dot{\varepsilon}_p = ((1-f_\beta)\dot{\varepsilon}_{p,\alpha} + f_\beta\dot{\varepsilon}_{p,\beta})*1/(1-S). \tag{7}$$

Damage evolution. Some scholars believe that the plastic damage mechanism of α, β duplex titanium alloy is the duplex deformation incoordination [16]. The damage evolution equation is expressed as

$$\dot{D} = d_1(1-D)|\dot{\varepsilon}_{p,\beta} - \dot{\varepsilon}_{p,\alpha}|^{d_2} + \frac{d_6\cosh(d_3\varepsilon_p)}{(1-D)^{d_4}}\dot{\varepsilon}_p^{d_5}. \tag{8}$$

where d_1, d_2, d_3, d_4, d_5 and d_6 are material constant.

The damage is introduced into Eq. (1) and corrected to:

$$\begin{cases} \dot{\varepsilon}_{p,\alpha} = \left(\dfrac{\sigma/(1-D) - k_\alpha - H}{K_\alpha}\right)^n. \\ \dot{\varepsilon}_{p,\beta} = \left(\dfrac{\sigma/(1-D) - k_\beta - H}{K_\beta}\right)^n. \end{cases} \tag{9}$$

Based on the concept of damage tolerance mentioned above, a function is set. When the damage reaches the tolerance level, the material undergoes quasi-cleavage fracture. And set the weld-dependent fracture coefficients to control the fracture behavior. Ultimately, the equation is expressed as:

$$F = \mathrm{sign}(f + \mathrm{abs}(\ln(\mathrm{fix}(1000*D)/\mathrm{fix}(\omega)))). \tag{10}$$

where F is the weld-dependent fracture coefficient and ω is the damage tolerance coefficient. f is the material state coefficient: when f is 1, the material state is determined to be BM; when f is 0, the material state is determined to be DB.

The coupled the weld-dependent fracture coefficients, the stresses can be expressed as:

$$\sigma = (1-D)E(\varepsilon_T - \varepsilon_p)*F \tag{11}$$

Determination of model parameters. There are many parameters, which are difficult to realize by conventional solution methods. In this paper, the PSO algorithm is used to optimize the parameters. According to the research results of LIN [17] is used as the optimization objective. First, assuming an initial value of dislocation density is 0.1, The experimental data of BM are fitted to the determining parameter. The initial dislocation density is then set to 0 and the damage tolerance coefficient is controlled to obtain accurate DB fracture behavior. The damage tolerance coefficients at different strain rates are plotted as shown in Fig. 4. A linear relationship is met. The

Materials Research Forum LLC
https://doi.org/10.21741/9781644903254-59

coefficient is obtained by linear fitting. The final expression for the damage tolerance coefficient is obtained as Eq.12. Finally, the parameters are shown in Table 1.

$$\omega = w1 + w2 * \dot{\varepsilon}_T \quad (12)$$

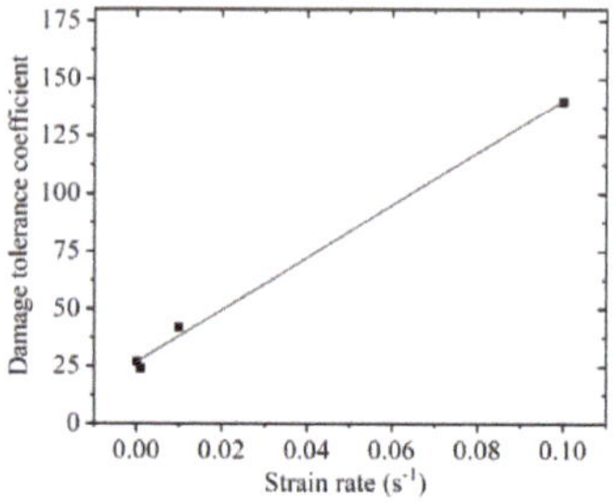

Fig. 4. Damage tolerance coefficient ω value fitting.

Table 1. Optimized values of equation parameters.

E	6.5671e+4	m_2	7.0570	d_4	1.0000	δ_1	0.2495
k_α	1.3736e-7	m_3	1.7013	d_5	1.7322	δ_2	8.1352
K_α	397. 0950	d_1	0.0024	d_6	1.0000	δ_3	6.3147
n	6.9572	d_2	5.5277	A	2.2699	f_β	0.1214
m_1	0.0521	d_3	10.0000	B	93.4351	q_1	0.0363
q_2	1.2703	q_3	6.1014	q_4	4.0477	q_5	3.0832
w_1	26.6744	w_2	1136.8338				

Model application and discussions

Fig. 5(a) shows the comparison between the BM prediction curve (line) and the experimental value (symbol) of BM. The predicted stress values are in good agreement with the experimental values. To quantitatively describe the predictive effect of the model, it was evaluated by statistical methods. The correlation coefficient (R), the average absolute relative error (AARE), and the root mean square error (RMSE) are 0.9933, 3.4%, and 8.1 MPa, respectively. Fig. 5(c) shows the comparison of fracture strain experiment and prediction, and the maximum error is -5.23%. It shows that the model has high accuracy in predicting fracture behavior under MC parameters.

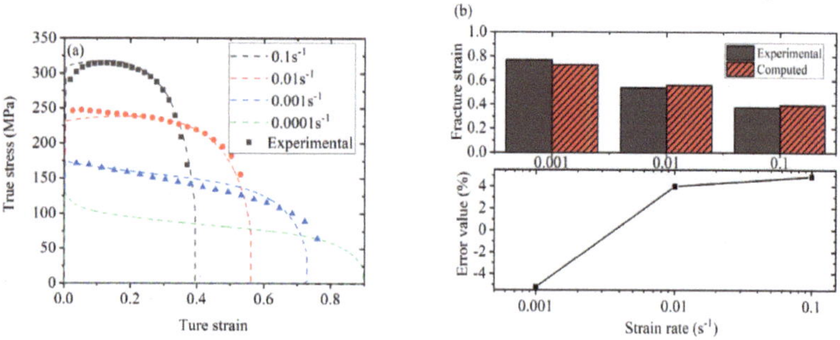

Fig. 5. (a) Stress prediction of BM specimen and (b) fracture strain prediction.

Fig. 6(a) shows the comparison between the BM prediction curve (line) and the experimental value (symbol) of DB. R, AARE, and RMSE are 0.9951,38.1%, and 12.3MPa, respectively. It is shown that the model is also highly accurate in predicting the stresses for the DB state. Fig. 6(c) shows the comparison of fracture strain experiment and prediction, and the maximum error is -3.34%.

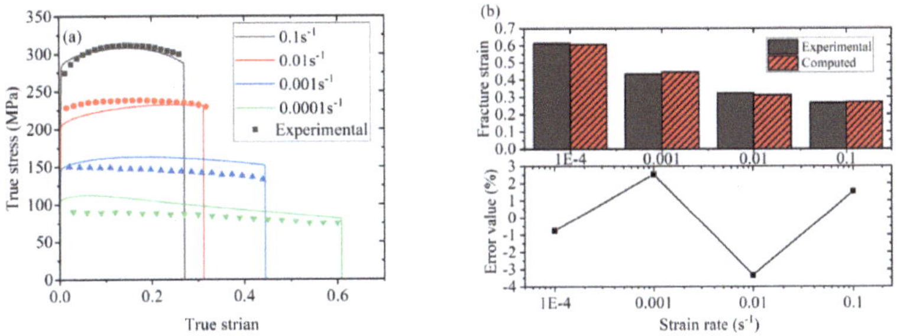

Fig. 6. (a) Stress prediction of DB specimen and (b) fracture strain prediction.

Fig. 7 shows the evolution of the internal variables for BM and DB at 0.001s^{-1}. It is worth noting that differences in the initial dislocations make a difference in the evolution of the dislocation. It is shown that the higher initial dislocation density of the BM leads to earlier completion of the reserve fraction. This also contributes to earlier and fuller DRX. BM has a higher DRX fraction relative to DB, which is also consistent with the results of the previous stress-strain curves analysis.

At the beginning of deformation, the damage increases relatively slowly and voids nucleate. When the deformation reaches a certain level, a large number of voids begin to coalesce. This causes the damage to grow dramatically until the material breaks. For DB material, the voids are more inclined to the nucleate and coalescence at the DB interface. Most of the damage is concentrated at the DB interface, leading to fracture behavior at small damage values. This is consistent with the experimental performance obtained. According to the above analysis, the prediction of internal variable evolution in the model is also effective.

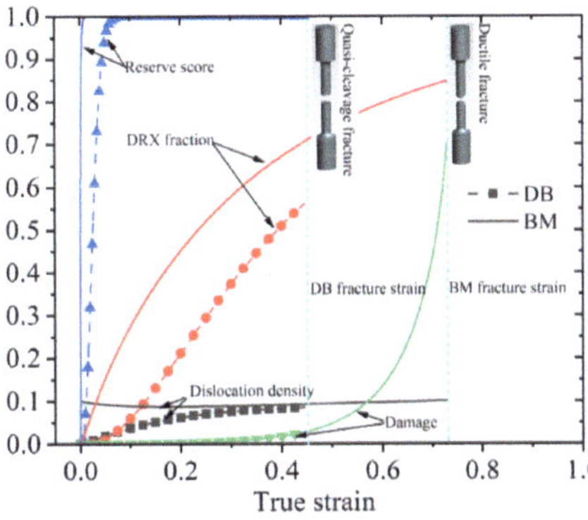

Fig. 7. Evolution of internal variables at 0.001s^{-1}.

Summary

The thermal deformation behavior of TC4 with and without DB is studied by tensile experiments. The influence of DB structure on thermal deformation behavior is analyzed. A set of internal variable constitutive model considering DB structure is developed and its prediction accuracy is verified. The main conclusions are:

At 750°C, 0.1 ~ 0.0001 s^{-1}, TC4 with DB shows smaller peak stresses during tensile tests than the one without DB, while the fracture strain is smaller, which would greatly affect the forming properties of the DB material.

At 750°C, 0.1-0.0001s^{-1}, the fracture type of TC4 without DB is a ductile fracture and the fracture type of TC4 with DB is a quasi-cleavage fracture. Only a dimple was found in TC4 without DB fracture, while TC4 with DB fracture consists of two structures: transgranular fracture and dimple. The formation mechanism of TC4 quasi-cleavage fracture was expounded as a straight grain boundary formed by incomplete recrystallization and insufficient grain boundary migration energy and the joints weak left over from the DB process caused rapid voids nucleation at the DB interface. The shorter distance of voids coalescence within the interface leads to quasi-cleavage fracture.

By introducing the weld-dependent fracture coefficients and damage tolerance coefficient setting, a unified constitutive model was constructed to describe the hot deformation characteristics of both TC4 with and without DB. The model is capable of describing the evolution of dislocation density, reserve fraction, DRX fractions and damage variables. It can accurately predict TC4 with and without DB stress-strain curves. Within the predicted fracture strain, the prediction accuracy of TC4 without DB statistics: R is 0.9933. The prediction accuracy of TC4 with DB statistics: R is 0.9955. The fracture strain accuracy of TC4 with and without DB is within 3.5,5.5% respectively. It shows that the constitutive model can accurately predict the hot deformation characteristics of TC4 with and without DB at the same time.

References

[1] K. Zheng, D. Li, H. Chen, S. Qu, Z. Zhao, Y. Zhang, Y. Li, Effect of cooling rate on the phase transformation and post strength of Ti-6Al-4V under hot forming conditions: Experiments and modelling, J. Alloy. Compd. 972 (2024) 172868. https://doi.org/10.1016/j.jallcom.2023.172868

[2] Y. Li, H. Chen, L. Du, F. Yang, Y. Zhang, D. Li, Characterization and unified modelling of creep and viscoplasticity deformation of titanium alloy at elevated temperature, Int. J. Plast. (2024) 103892. https://doi.org/10.1016/j.ijplas.2024.103892

[3] C. Li, Y. Li, D. Zhang, X. Li, W. Zhao, Experimental investigation of the mechanical properties and microstructures of Ti–6Al–4V solid-state diffusion bonding joint under high temperature conditions, J. Mater. Res. Technol. 26 (2023) 4042-4058. https://doi.org/10.1016/j.jmrt.2023.08.172

[4] S.V.S. Narayana Murty, B. Nageswara Rao, B.P. Kashyap, Processing maps for hot deformation of α 2 aluminide alloy Ti-24Al-11Nb, J. Mater. Sci. 37 (2002) 1197-1201. https://doi.org/10.1023/A:1014323723598

[5] Y. Peng, J. Li, Z. Li, S. Li, W. Guo, X. Gao, J. Xiong, Interfacial voids and microstructure evolution, bonding behavior and deformation mechanism of TC4 diffusion bonded joints, J. Manuf. Process. 81 (2022) 837-851.

[6] S.M. Imran, C. Li, L. Lang, Y. Guo, H.A. Mirza, F. Haq, S. Alexandrova, J. Jiang, H. Han, An investigation into Arrhenius type constitutive models to predict complex hot deformation behavior of TC4 alloy having bimodal microstructure, Materials Today Commun. 31 (2022)

103622. https://doi.org/10.1016/j.mtcomm.2022.103622

[7] C. Li, I. Sardar Muhammad, L. Lang, Y. Guo, X. Li, S. Alexandrova, D. Zhang, Hot deformation behavior and strain compensation constitutive model of equiaxed fine grain diffusion-welded micro-duplex TC4 titanium alloy, Chinese J. Aeronautics 36 (2023) 510-522. https://doi.org/10.1016/j.cja.2022.07.025

[8] E. Alabort, D. Putman, R.C. Reed, Superplasticity in Ti–6Al–4V: Characterisation, modelling and applications, Acta Mater. 95 (2015) 428-442. https://doi.org/10.1016/j.actamat.2015.04.056

[9] D. Wu, K. Wang, B. Liu, H. Qu, X. Wang, G. Liu, Damage evolution of titanium alloy laser-welded joint during hot deformation: Experiment and modeling, J. Mater. Res. Technol. 26 (2023) 6375-6388. https://doi.org/10.1016/j.jmrt.2023.09.023

[10] B.H. Cheong, J. Lin, A.A. Ball, Modelling of hardening due to grain growth for a superplastic alloy, J. Mater. Process. Technol. 119 (2001) 361-365. https://doi.org/10.1016/S0924-0136(01)00929-3

[11] Q. Bai, J. Lin, T.A. Dean, D.S. Balint, T. Gao, Z. Zhang, Modelling of dominant softening mechanisms for Ti-6Al-4V in steady state hot forming conditions, Mater. Sci. Eng. A 559 (2013) 352-358. https://doi.org/10.1016/j.msea.2012.08.110

[12] Y. Estrin, Dislocation theory based constitutive modelling: foundations and applications, J. Mater. Process. Technol. 80-81 (1998) 33-39. https://doi.org/10.1016/S0924-0136(98)00208-8

[13] Y. Wu, D. Wang, Z. Liu, G. Liu, A unified internal state variable material model for Ti2AlNb-alloy and its applications in hot gas forming, Int. J. Mech. Sci. 164 (2019) 105126. https://doi.org/10.1016/j.ijmecsci.2019.105126

[14] Y. Huo, J. Lin, Q. Bai, B. Wang, X. Tang, H. Ji, Prediction of microstructure and ductile damage of a high-speed railway axle steel during cross wedge rolling, J. Mater. Process. Technol. 239 (2017) 359-369. https://doi.org/10.1016/j.jmatprotec.2016.09.001

[15] L. Zhao, T. Yasmeen, P. Gao, S. Wei, Z. Bai, J. Jiang, J. Lin, Mechanism-based constitutive equations for superplastic forming of TA15 with equiaxed fine grain structure, Procedia Eng. 207 (2017) 1874-1879. https://doi.org/10.1016/j.proeng.2017.10.954

[16] M.A. Greenfield, H. Margolin, The mechanism of void formation, void growth, and tensile fracture in an alloy consisting of two ductile phases, Metall. Mater. Trans. B 3 (1972) 2649-2659. https://doi.org/10.1007/BF02644241

[17] J. Cao, J. Lin, A study on formulation of objective functions for determining material models, Int. J. Mech. Sci. 50 (2008) 193-204. https://doi.org/10.1016/j.ijmecsci.2007.07.003

A customized design tool incorporating web and client functions for new mechanical presses

NI Jun[1,a *], SUN Yu[1,b], WANG Yu[1,c] and WU Kai[1,d]

[1]Nanjing university of science and technology, Xiaolingwei Street No. 200, Xuanwu District, Nanjing 210094, P.R. China

[a]junnee@njust.edu.cn, [b]sunyu@njust.edu.cn, [c]wy@njust.edu.cn, [d]wukai@njust.edu.cn

Keywords: Mechanical Press, CAD, Knowledge-Guided, Multi-Link System, Structural Consistency

Abstract. To improve the design efficiency and consistency of many new mechanical presses developed by a few engineers, an application with two tabs was proposed to incorporate a web browser and a client window, respectively, for the press modification and the new press design. The press modification is based on the standardized press data in server while the new press design adopts the functions of Computer-Aided Design (CAD) and Word processing tools in local computers. For the press modification: (1) a case retrieval function was developed to find similar features of standardized press information that was then adapted to meet the new custom order; (2) some structural parameters and component selections were adjusted and checked by the calculation algorithms and material properties; (3) within the browser tab, design document was eventually reported. For the new press design: (1) besides the BOM templates, two dimensional (2D) sketch templates were summarized for non-standard parts or assemblies including the machine frame, crank, links, slider, work-table, etc.; (2) the slider movement design and dimensional allocation of press components were performed according to a series of interactive dialogs, answers and views with regard to the custom order; (3) the dot net framework was adopted to integrate Solidworks, Auto CAD and Word functions to generate parts, assemblies, drawings and the Bill Of Materials (BOM) from the 2D sketch and BOM templates using the dimensional allocations and the design results. Examples validated the feasibility of the proposed tool.

Introduction

Mechanical presses play a crucial role in manufacturing the parts made by steel or aluminum alloys that are widely used in automotive, rail transit, household appliance, aircraft and aero-engine industries, accounting for 76% the global market amount of metal forming machines [1]. In response to various specific working scenarios, the customized designs of new presses are not uncommon. For a large number of design tasks, many problems restrict the rapid and accurate response of metal forming machine manufacturer, such as the growing human cost, high-strength workload, and the inevitable diversity of design results [2]. Especially, every design result needs to be confirmed to meet the custom order and structural safety. To deal with this challenge, it requires to normalize the custom design that uses the knowledge of standardized presses, and performs interactive designs and analyses, from which the parts, assemblies, drawings, and Bill of Materials (BOM) will be yielded.

Over decades, web applications are usually applied to database managements while desktop applications are able to perform the Computer-Aided Design / Engineering / Manufacturing (CAD / CAE / CAM). For example, Siemens team center product lifecycle management is good choice for one source of standardized mechanical press (parts, assemblies, drawings, and bill of materials) shared by all the designers within a company [3]. These standardized mechanical presses are cases that would be used for new customer orders. In many theoretical researches, the case-based reasoning methods, which separately adopt the case retrieval [4], case adaptation [5], or expert experience [6], are usually applied to find an appropriate case or adjust a case to meet the new plan requirement. For the customized design of a new mechanical press, some components of different standardized mechanical presses are selected, adjusted and assembled together; this is a complex case retrieval and adaptation using a series of standardized mechanical presses with many components because of the complexities of various hierarchical component features within a product; it is seldom solved either by the case retrieval or by the case adaptation. To combine both the case retrieval and adaptation of standardized mechanical presses with their components, the new hybrid feature weights with the weight mean of design parameters are proposed, which make true the press modification for the customized design.

Recently about 10 years after the concept of Industry 4.0 was introduced [7], and despite that the cloud-based finite element modeling was developed and open [8], this still is a time-consuming and error-prone procedure when the manufacturers use the remote database or a solver in a server far away to perform the design generating two more solutions for each customer. Hence the desktop application is effective for the design of a brand-new new press. For the design of a brand-new new press, the kinematics and dynamics of the crank, link and slider system are analyzed firstly, the dimensional allocation are then performed which support the component design and check using CAD and CAE tools, and finally the CAD tool [9] outputs the parts, assemblies, drawings and the Bill Of Materials (BOM) for the further manufacture and test. A rapid product design system based on 3D CAD that uses the part templates and BOM templates is proposed with the application to a vehicle steering gear [10]; this system is based on the secondary development technology, XML data interface technology, and parameterized component design technology, but the hierarchical component features are not addressed. A rule-based and feature-based approach is employed in 3D CAD modeling to create a parametric design platform with integrated evaluation features, which is able to automatically determine the dimensions of spherical joints, effectively boosting their design efficiency and safety [11]; so it would be more efficient if the hierarchical component features were handled in assembly templates. Recently, an explicit template is referred to internal structure of functional geometries, subtractions, combination operations, replication features, and the localized modifications would significantly improve the robustness and reusability of feature-based CAD models, particularly in response to changes [12]. In light of the above advances, the templates of the 2D part sketches and the axis and point assembly templates with explicit functions of subtractions, combination operations, replication features, and the localized modifications are proposed for new press design.

Methodology

A New Architecture Incorporating Web and Client Functions. As shown in Fig. 1, it is comprised of five lines. The top line is the functions of the press modification and new press design. The second line is the workflow having two operational directions. The first operational direction is to modify a press, which starts from case retrieval and adaptation, and then involves the modification, calculation, save and report. And the other one is to perform dimensional allocation, 3D modeling, save and report. The third line is the user interface made up of web view, window files, dialogs and views, which accommodates the workflow. The fourth line is the subroutine functions written in Vue, Oracle, Java Script, C++, C#, and dot Net Framework. The subroutine functions written in Java Script provide services for the case retrieval, case adaptation, press data revision, design calculations for press forces and the dimensions of press components, the selection of motors, clutch and brakes, and the report. The subroutine functions written in C++ provide the calculation algorithms in DLL files and the subroutine functions written in C# and dot Net Framework provide the service of dimensional allocation and 3D modeling. The fifth line is for the recourses embedded in some of the upper tool line or stored in Oracle database. The recourses embedded in tool line are the work flow of the design, the rules of case retrieval and adaptation, the calculation algorithms, the templates of BOM and sketches, and related API of Solidworks, auto CAD and

Word. The recourses stored in Oracle database comprise the standardized presses and components, material properties, and the user information.

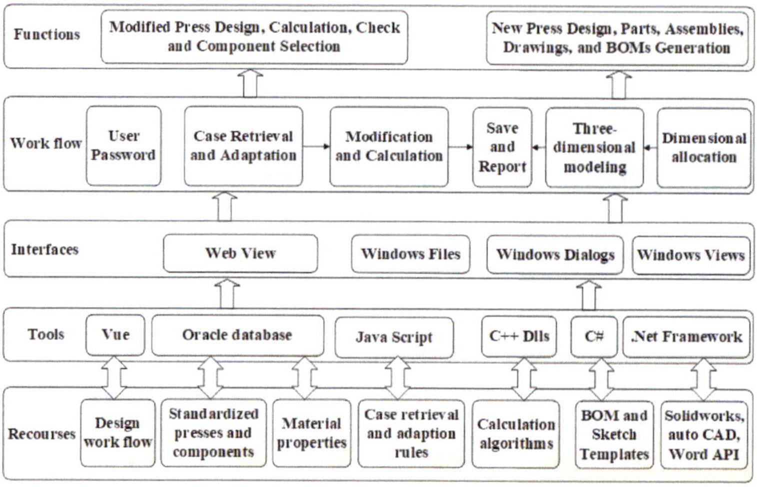

Fig. 1. The architecture with web and client functions.

For the top two functions, the press modification is based on the Oracle database in server while the new press design adopts the functions of Computer-Aided Design (CAD) and Word processing tools and 2D sketch templates in local computers. The novelty of this new architecture lies in the case retrieval and adaptation as well as the parametric assembly design of mechanical presses from the templates. This is due to the factors that the calculation algorithms and BOM templates can be found in any textbook or obtained from the experience summarized by any press manufacturer.

Case Retrieval and Adaptation of Standardized Mechanical Presses. Each case is represented as a vector containing m question and n solution attributes, thus the cases can be represented as Eq. 1.

$$\mathbf{C}_k = \{p_{k1}, p_{k2}, \ldots, p_{ki}, \ldots, p_{km}, s_{k1}, s_{k2}, \ldots, s_{kj} \ldots, s_{kn}\} \quad 1 \le i \le m;\ 1 \le j \le n \quad . \tag{1}$$

where $\mathbf{C}_k$ is the k-th case, p_{ki} is the i-th problem attribute, s_{kj} is the j-th solution.

Let $\mathbf{C}_0$ be the vector of a new case, thus its vector of question attributes is $\{p_{01}, p_{02}, \ldots, p_{0i}, \ldots, p_{0m}\}$. According to Eq.1, a series of similarities can be calculated by the comparison between the new case question attributes $\{p_{01}, p_{02}, \ldots, p_{0i}, \ldots, p_{0m}\}$ and one of $\mathbf{C}_k$, i.e., Eq. 2.

$$sim_i = \sum_{j=1}^{m}\left[1-\left|q_{ij} - p_{0i}\right|\right] \times w_{hj} \Big/ \sum_{j=1}^{n} w_{hj} \quad , \tag{2}$$

where sim_i is the similarity that take a higher value to demonstrate a higher similarity, p_{0j} is the new problem attributes, q_{ij} that satisfies Eq. 3 is the element of the normalized problem matrix $(q_{ij})_{k\times m}$, and w_{hj} of is the element of the proposed hybrid weight matrix $(w_{hi})_{k\times m}$.

$$q_{ij} = p_{ij} \Big/ \sum_{i=1}^{m} p_{ij} \quad , \tag{3}$$

where p_{ij} is the problem attributes in Eq. 1.

Element w_{hj} combines both the subjective and objective weights, i.e., w_{sj} and w_{oj}, is obtained by the optimization in form of Eq. 4.

$$\min \sum_{i=1}^{m}\sum_{j=1}^{n}\left[\left(w_{hj}-w_{sj}\right)p_{ij}\right]^2+\sum_{i=1}^{m}\sum_{j=1}^{n}\left[\left(w_{hj}-w_{oj}\right)p_{ij}\right]^2$$
$$\text{s.t. } \sum_{j=1}^{n} w_{hj}=1,\ w_{hj}\geq 0. \tag{4}$$

As shown in Eq. 5, element w_{sj} in $(w_{si})_{k\times m}$ is generated from the judgment matrix $(\alpha_{ij})_{k\times m}$ defined by experts according to the analytic hierarchy process (AHP) [13].

$$w_{sj} = \overline{w_{sj}} \Big/ \sum_{j=1}^{n} \sqrt[n]{\prod_{j=1}^{n} \alpha_{ij}} \quad , \tag{5}$$

where n is the number of features, w_{sj} is the j-th weight of features, α_{ij} represents the importance of criterion i compared to criterion j, $\alpha_{ij} > 0$, $\alpha_{ii} = 1$, and $\alpha_{ij} = 1 / \alpha_{ji}$.

As shown in Eq. 6, element w_{oj} in $(w_{oi})_{k\times m}$ is yielded from the entropy method [14] and the normalized problem matrix $(q_{ij})_{k\times m}$.

$$w_{oj} = \left(1+\frac{1}{\ln m}\sum_{j=1}^{m} q_{ij}\ln q_{ij}\right) \Bigg/ \sum_{j=1}^{m}\left(1+\frac{1}{\ln m}\sum_{j=1}^{m} q_{ij}\ln q_{ij}\right). \tag{6}$$

According to Eq. 2, a number of top-ranked cases would be selected for the case adaptation. Without lose generality, k top-ranked cases are selected and denoted by the superscript in form of Eq. 7.

$$\mathrm{C}_s = \left\{p_{s1}^{(1)},\ p_{s2}^{(1)},\ \ldots,\ p_{si}^{(1)},\ \ldots, p_{sm}^{(1)},\ s_{s1}^{(1)},\ s_{s2}^{(1)},\ \ldots, s_{sj}^{(1)}\ \ldots,\ s_{sn}^{(1)}\right\}\ 1\leq i\leq m;\ 1\leq j\leq n\ . \tag{7}$$

As shown in Eq. 8, a hybrid weighted mean w_{kj} is proposed to covers similar case problems p_i, the utility of each candidate case u_i, the gray relationship between each candidate case problem and solution r, and the feature adaptability of each case solution $a(C_{k,j}^{sol})$.

$$w_{kj} = a_{kj}\times\sum_{i=1}^{m} p_{ki}\times\left(\sum_{s=1}^{s}\frac{u_i}{\sum_{i=1}^{k}u_i}r\left(\frac{p_{si}^{(1)}}{p_{1i}^{(1)}},\frac{s_{sj}^{(1)}}{s_{1j}^{(1)}}\right)\right). \tag{8}$$

Specifically, r is the coefficient of the gray relationship between similar problem $p_{si}^{(1)}/p_{1i}^{(1)}$ and solution $s_{sj}^{(1)}/s_{1j}^{(1)}$,

$$u_i = \sum_{j=1}^{m} w_{hi} p_{si}^{(1)}, \tag{9}$$

$$a(C_{k,j}^{sol}) = w^S sim(case^P, case^C) + w^A a(f_p), \tag{10}$$

where w^S and w^A represent the weights that are both set to 0.5.

The solution to the new case $\mathbf{C}_0$ is adapted by the hybrid weighted mean and the k similar cases as shown in Eq. 11, which would be efficient and effective for the customized press design with a little modification compared to standardized presses.

$$s_{0j} = \sum_{j=1}^{n} w_{kj} s_{kj}^{(1)}. \tag{11}$$

Once the similar press is found and modified, the calculation algorithms will check whether the press modification is acceptable or not. If it is acceptable, it will be able to be saved and written in a report. Otherwise, calculation algorithms will provide some recommendation of improvement. In brief, Fig. 2 is the flowchart for the above procedure.

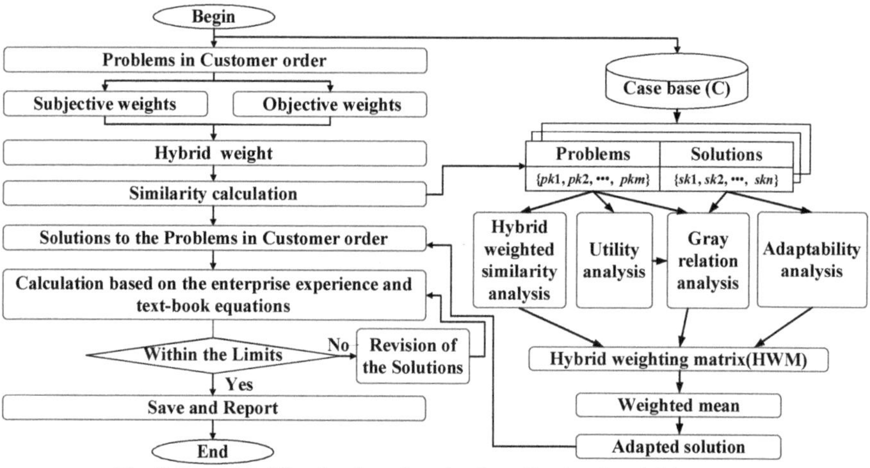

Fig. 2. Press Modification based on the Case Retrieval and Adaptation.

Parametric Assembly Design of Mechanical Presses from the Templates of Modeling. A set of parameters in 2D sketch templates are revised according to the dimensional allocation yielded from the kinematic analysis of multi-links that represent the dimension of crank, link, slider and wheel. Some points and axis in these 2D sketch templates that are never changed are defined especially for the location of parts in the assembly. On one hand, subroutine functions written in C# and dot Net Framework copy the 2D sketch templates in software template directory to a new part in working directory, and use the parameters in the result of dimensional allocation to generate all the features of a part in form of the shaft, link, slider, wheel, work-table, etc. On the other hand, the unchanged axis and points in the 2D sketches of the generated shaft, link, slider, wheel or work-table are helpful to locate the related parts with respect to the assembly platform. By combining the two folds, the press assembly would be able to obtained and save to a series of files as well as the 2D drawing and BOMs. In a nutshell, the proposed parametric assembly design methodology of mechanical presses from the templates is summarized in Fig. 3.

In addition, Fig. 3 shows that the assembly would be able to be saved as STEP files that could be meshed and applied to related stress analysis, supporting a comprehensive structural validation before the production using the 2D drawings and BOMs.

Examples

The Customized Design Tool of New Mechanical Presses. According to the methodology, the customized design tool is developed as shown in Fig. 4. This design tool is an application with two tabs that incorporate a web browser (Fig. 4a) and a client window (Fig. 4b), respectively, for the press modification and the new press design.

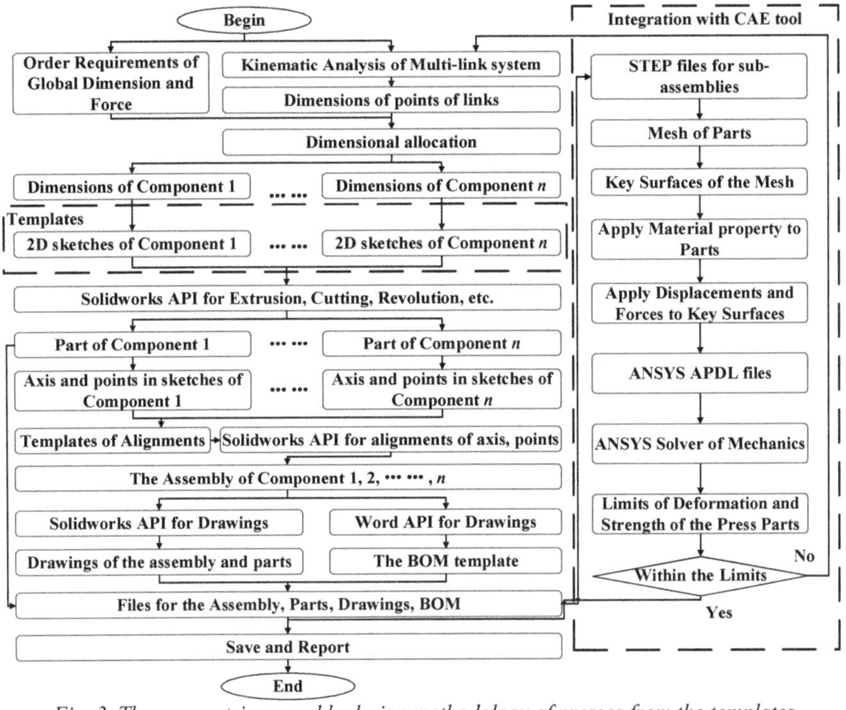

Fig. 3. The parametric assembly design methodology of presses from the templates.

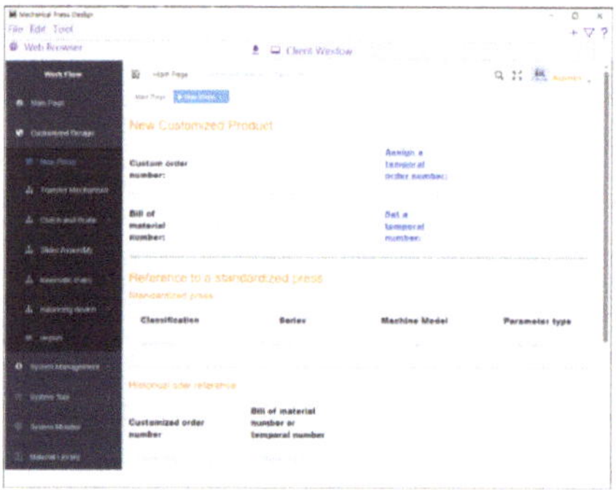

(a) the web browser tab

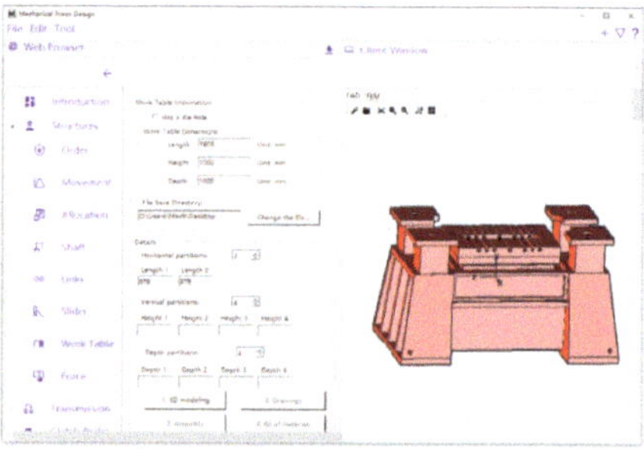

(b) the client window tab

Fig. 4. The web browser tab and client window tab of the proposed application.

The web browser is for the press modification. Especially the client window is not visible until the authorized user enters the correct password. In the web browser, the user could perform the case retrieval and adaptation for all the components in the standardized mechanical presses stored in the server database. The left side of the web browser comprises the modification for components in transfer mechanism, clutch and brake, slider assembly as well as the kinematic pairs. The left side also provides the report function if the modification is validated by the calculations performed in the web browser and server. The report function outputs a PDF file of all the design information of the new press. The administrator could see the system management, tool and monitor in the left side, and some managers are able to add, delete and search data in the material library and standardized mechanical press library.

The client window is for the new press design. It comprises three views, i.e., the left tree view, the middle form view and the right graphic view. For each node in the left tree view, there is a form view in the middle view. In the middle form view, the user could input related parameters, and click confirm and start the functions in Fig. 2. If the functions in Fig. 2 generate the assembly, part or drawings, the model would be displayed in the right graphic view.

The following sections is two specific examples of press modification and design using the customized design tool.

The Knowledge-guided Press Modification. As shown in Fig. 5, the knowledge-guided press modification is illustrated by the modification of a crank shaft.

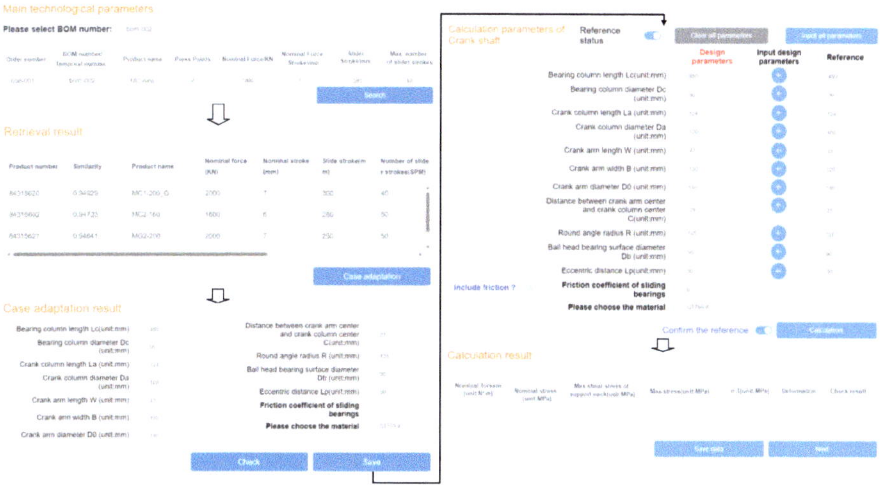

Fig. 5. Case retrieval, adaptation, calculation for a crank shaft of a customized press.

The top-left view in Fig. 5 is main technological parameters of a press required by a customer order, including press points, the nominal force, the stroke of nominal force, maximum number of slider strokes per minutes. When we click the button Search, the middle-left view is shown, and a set of similar cases are retrieved from the database of knowledge that contains many parameters of standardized presses. Then, if the button Case adaptation is clicked, the case adaptation result is shown as the bottom-left view. Then, as shown in top-right figure, if the button Input all parameters is clicked, all the adaptation result will be put into the textboxes of design parameters of the crank shaft. If we confirm the reference parameters and click the button Calculation, some equations of mechanics that are obtained from textbooks and experience of engineering practice will check the nominal torsion, stress and deformation, the result of which are listed in the bottom-right view of Fig. 5. Especially, the last column of Check result in the bottom-right view of Fig. 5 will show whether the new parameters of the crank shaft are acceptable or not.

The Template-driven New Multi-Link System Design. In this client window, a multi-link mechanism of six links is shown as example. In Fig. 6, a variation of eccentric wheel-link mechanism is calculated for the input coordinates of six points. After the length of crank shaft is allocated according to the length and width of the work-table required by the customer, the feature-based 3D modeling is performed for the crank shaft.

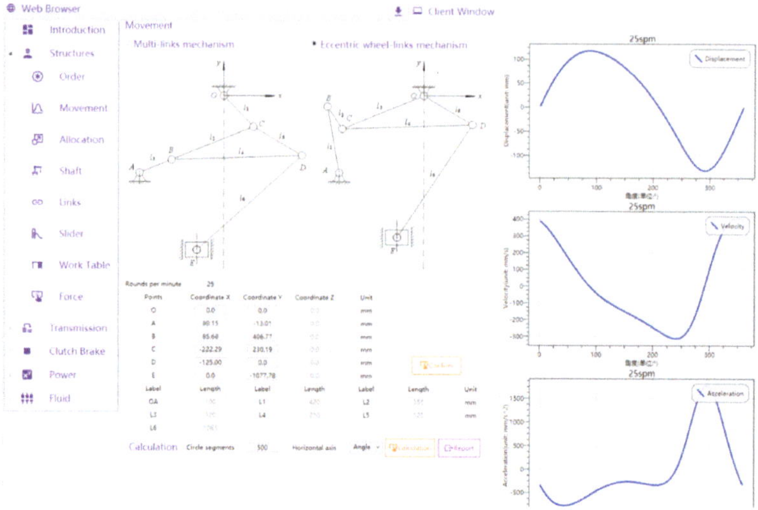

Fig. 6. Stroke curves of multi-link mechanism.

Fig. 7 shows the 3D modeling of the crank shaft starting from a series of 2D sketch templates by the modeling, assembly and drawing functions.

Fig. 7. Sketch templates and the parametric design view of crank shaft.

The left-top side figure in Fig. 7 is the designed sketch templates, the left bottom side figure in Fig. 7 lists modeling, assembly and drawing functions that take use of the solidworks API, and the right view shows the user interface made of a form view and a graphic view.

The modeling is achieved by the following process. The form view has many textboxes for each column of the shaft. Some of the textboxes are column and arm parameters especially for the position, length and diameter of all the columns as well as the position, length, width and diameter of the arms. Specifically, the bearing column will contact with the sliding bearings that support the shaft. The number of the bearing columns can be changed by the textbox with the up and down triangles according to the length of the press worktable, the simplest case of which is 2 bearing columns. The least crank column number is one, which will contact with the link. One crank column has two crank arms, the number of which can be automatically set when that of crank column is set. The textboxes of column and arm parameters increase as the number of columns and arms increase, i.e., the value of textboxes with up and down triangles. When all the column and arm parameters as well as the round angle and end taper angle are all confirmed, the shaft parametric design can be executed. By clicking the button 3D Modeling, we will obtain the part of the shaft as shown in the right graphic view, and the related part is saved in working directory.

Similar to the above process, the parts of all links and wheel components can be generated from the parameters set by the user. By automatically making some of the axis and points of different parts coincide with each other, the assembly of wheel, shaft and links will be completed as shown in Fig. 8. Similar to the assembly procedure of wheel link mechanism, the whole mechanical press will be assembled.

Fig. 8. The assembly of eccentric wheel-link mechanism.

Summary

This paper proposes a new architecture incorporating web and client functions for the modification and design of mechanical presses. The press modification is done in the web browser using the case retrieval, adaptation and calculating validation. The new press design is finished in the client window using modeling, drawing and assembling functions that integrate the Solidworks API. The cases used by the web browser is the knowledge of standardized presses and related BOMs and drawings stored in the server. Many templates including BOM templates and 2D sketches with mating feature (points, axis and faces) are defined for the report, modeling, drawing and assembling. By the proposed architecture, a customized design tool of mechanical presses is developed, the feasibility of which is validated by the modification and design of a multi-link system.

Acknowledgements

This work was financially supported by Jiangsu Provincial Department of Science and Technology, project BA2021067.

References

[1] Q.Y. Research Centre, Press machine report 2024, global revenue, key companies market share & rank, Abstract, 2023, Information on https://www.qyresearch.com.cn/reports/press-machine-p3072233.html

[2] W.P. Neumann, S. Winkelhaus, E.H. Grosse, C.H. Glock, Industry 4.0 and the human factor – A systems framework and analysis methodology for successful development, Int. J. Prod. Econ. 233 (2021) 107992. https://doi.org/10.1016/j.ijpe.2020.107992

[3] S. Herbst, A. Hoffmann, Product Lifecycle Management (PLM) mit Siemens Teamcenter, Carl Hanser Verlag GmbH & Co. KG Press, Göttingen, 2018. https://doi.org/10.3139/9783446446496.fm

[4] R. Lopez De Mantaras, D. Mcsherry, D. Bridge, D. Leake, B. Smyth, S. Craw, B. Faltings, M.L. Maher, M.T. Cox, K. Forbus, M. Keane, A. Aamodt, I. Watson, Retrieval, reuse, revision and retention in case-based reasoning, Knowl. Eng. Rev. 20 (2005) 215–240. https://doi.org/10.1017/S0269888906000646

[5] H. Munoz-Avila, M.T. Cox, Case-Based Plan Adaptation: An Analysis and Review, IEEE Intell. Syst. 23 (2008) 75–81. https://doi.org/10.1109/MIS.2008.59

[6] Y. Huo, J. Liu, J. Xiong, W. Xiao, J. Zhao, Machine learning and CBR integrated mechanical product design approach, Adv. Eng. Inform. 52 (2022) 101611. https://doi.org/10.1016/j.aei. 2022.101611

[7] H. Liu, S. Dhawan, M. Shen, M. Shen, K. Chen, V. Wu, L. Wang, Industry 4.0 in metal forming industry towards automotive applications: a review, Int. J. Automot. Mech. Eng. 1 (2022) 2-12. https://doi.org/10.53941/ijamm0101002

[8] V.A. Ciubotariu, C. Grigoras, C.M. Radu, C.N. Tampu, V. Zichil, The opportunity of using cloud-based computing in numerical simulations on structural analysis - case study, J. Eng. Stud. Res. 28 (2022) 38-42. https://doi.org/10.29081/jesr.v28i1.004

[9] M. Shubhit, K. Vineet, G. Pradeep, An analogy between conventional and unorthodox cad methods, Mater. Today: Proc. 43 (2021) 2098-2104. https://doi.org/10.1016/j.matpr.2020.11.889

[10] F. Liu, Fast industrial product design method and its application based on 3D CAD system, Comput. Aided Des. Appl. 18 (2021) 118-128. https://doi.org/10.14733/cadaps.2021. S3.118-128

[11] N. Geren, O. Oktay Akçalı, E. Unver, J. Allport, Automated sizing of automotive steering ball joints in parametric CAD environment using expert knowledge and feature-based computer-assisted 3D modelling, Adv. Eng. Inform. 52 (2022) 101630. https://doi.org/10.1016/j.aei.2022.101630

[12] Aranburu, J.D. Camba, D. Justel, M. Contero, An improved explicit reference modeling methodology for parametric design. Comput. Aided Des. 161 (2023) 103541. https://doi.org/10.1016/ j.cad.2023.103541

[13] T.L. Saaty, J.S. Shang, An innovative orders-of-magnitude approach to AHP-based mutli-criteria decision making: prioritizing divergent intangible humane acts, Eur. J. Oper. Res. 214 (2011) 703-715. https://doi.org/10.1016/j.ejor.2011.05.019

[14] J. Zhao, J. Tian, F. Meng, M. Zhang, Q. Wu, Safety assessment method for storage tank farm based on the combination of structure entropy weight method and cloud model, J. Loss Prev. Process Ind. 75 (2022) 104709. https://doi.org/10.1016/j.jlp.2021.104709

Slider movement planning based on improved S-curve acceleration algorithms for servo presses

WANG Jun[1,a], WU Kai[1,b,*] and DING Wuxue[1,c]

[1]Nanjing University of Science and Technology, No. 200 Xiaolingwei Street, 210094 Nanjing, China

[a]wang-jun@njust.edu.cn, [b]wukai@njust.edu.cn, [c]wuxuexie@njust.edu.cn

Keywords: Slider Movement Planning, Servo Press, S-Curve, Deep Drawing Process

Abstract. Servo press is a forming equipment that uses a servo motor to drive a crank slider for stamping and deep drawing process. The servo drive can adjust the motor speed in real time and dynamically, which provides an important support for the dynamic adjustment of the crank slider movement. To realize the smooth reciprocating movement of the large-mass and high-inertia slider, it is necessary to develop a motor speed adjustment algorithm to achieve a good match between the movement load of the slider and the torque that the motor can bear, to ensure the operation safety of the whole machine. S-curve algorithms are often used for the planning of slider movement curves, which is easy to produce flexible impact on the motor. Therefore, to improve this situation, an improved S-curve algorithm is proposed based on the above common methods, which realizes the motion control method of smooth speed change. In this paper, a movement planning algorithm for servo press slider with improved S-curve is proposed based on the 7-phase S-curve acceleration algorithm. Firstly, the kinematic calculation of the crank slider mechanism is carried out. Secondly, 4 control points are set up to divide the slider movement curve of the typical drawing process, including the idle feed stage, deep drawing stage, idle return stage, and loading/unloading stage. Furthermore, the uniform acceleration stage and the uniform deceleration stage are removed, and the acceleration cusp was smoothed, and the improved S-curve algorithm was obtained. Finally, the large-scale crank slider servo press and its deep drawing parameters in a literature are taken as an example to solve the slider movement curves by using the improved S-curve. The results show that the improved S-curve algorithm can be used for servo press slide motion profile planning, which can realize efficient deep drawing process mode.

Introduction

With the rapid development of the manufacturing industry, the traditional mechanical press working mode is fixed, the motion characteristics cannot be adjusted, poor process adaptability and other defects, cannot meet the forming requirements of different materials and different stamping processes; on the other hand, with the development of servo control technology, servo motors are widely used. Compared with the traditional mechanical press, servo press has the advantages of flexibility, intelligence, high precision, low noise and high productivity [1-3]. By controlling the rotational speed of the servo motor, the motion characteristics of the slider can be effectively controlled, so that it can meet the requirements of different materials and different forming processes. The focus of the trajectory planning of the main drive mechanism of the servo press is the acceleration control of the main drive servo motor.

The current acceleration control algorithms mainly include S-curve acceleration [4-6], polynomial acceleration [7], trigonometric function acceleration [8], NURBS curve [9], Bezier curve [10,11], B-spline curve [12,13]. Among them, S-curve acceleration has abrupt changes in acceleration, and trigonometric and polynomial acceleration satisfy the requirements of acceleration and acceleration continuity, avoiding the flexural impact of machining [14]. Q. Hao [4] used S-curve to optimize the maximum acceleration of the trajectory. The maximum driving force and the requirements of the motor were reduced. Z. Liu [5] used a 5-phase S-curve to optimize the amplitude of acceleration to suppress the torsional vibration of a synchronous belt-driven printing press system. X. Zhang [6] designed the traditional S-curve acceleration control model of the acceleration law change rule as a triangular shape, to solve the algorithm itself exists in the acceleration of the step of the problem. Q. Song [15] used the superposition of trigonometric functions of 2 different periods for acceleration model construction, thus reducing the requirement of motor power, but the algorithm fixed the time share of the variable speed interval.

In this paper starts from the common control algorithm of slider motion curve planning and proposes a servo press slider motion planning algorithm with improved S-curve. (1) Firstly, the kinematics calculation of crank slider mechanism is carried out to get the displacement, speed, acceleration over time curves and their relation equations; (2) Then, 4 control points are set to divide the typical deep drawing process curve into the idle feed stage, deep drawing stage, idle return stage, and loading/unloading stage, and the trend is illustrated; (3) Uniform acceleration stage and uniform deceleration stage in the S-curve algorithm are removed, and the acceleration cusps are smoothed, and an improved S-curve algorithm is obtained further to get the acceleration. The acceleration cusp is smoothed to obtain the improved S-curve algorithm, and the expressions of acceleration, speed and displacement over time are further obtained; (4) the parameters of crank-slider servo press in the reference paper are taken as an example, and the improved S-curve algorithm is used to plan and solve the slider motion curve.

Analysis of typical servo drawing process

Kinematic analysis of crank slider servo press. Taking the crank-slider press transmission system as an example, the kinematic analysis of the slider, its transmission mechanism and sketch are shown in Fig. 1(a) and 1(b), respectively, in which the eccentric crankshaft is the active part, and the connecting rod and the slider are the follower parts.

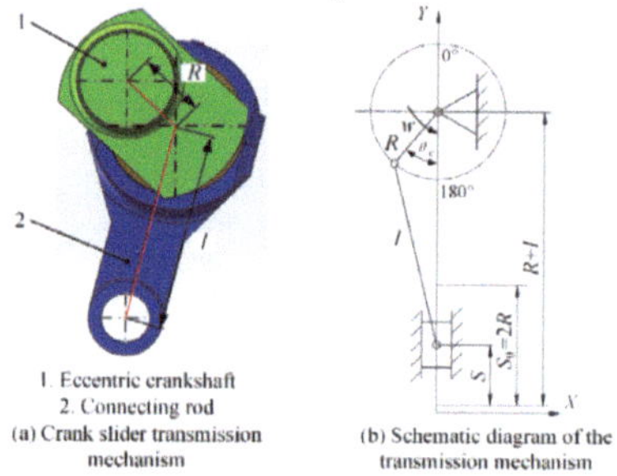

Fig. 1. Crank press drive system.

From Fig. 1(b), the relationship between the slider position, speed and acceleration and the crank angle can be established as follows:

$$S = R\left[(1-\cos\theta_c)+\frac{R}{4l}(1-\cos 2\theta_c)\right] \tag{1}$$

$$v = \frac{\mathrm{d}S}{\mathrm{d}t} = \omega R(\sin\theta_c + \frac{R}{2l}\sin 2\theta_c) \tag{2}$$

$$a = \frac{\mathrm{d}v}{\mathrm{d}t} = \frac{\mathrm{d}^2 S}{\mathrm{d}t^2} = -\omega^2 R(\cos\theta_c + \frac{R}{l}\cos 2\theta_c) \tag{3}$$

Where, S is the stroke of the slider, mm; θ_c is the angle of the crank, °; R is the radius of the crank, mm; l is the length of the connecting rod, mm; v is the speed of the slider, mm/s; ω is the angular speed of the crank, rad/s; a is the acceleration of the slider, mm/s^2. When θ_c=0° (top dead centre, TDC)and θ_c=180° (bottom dead centre, BDC), there is a maximum acceleration $a_{max}=-w^2R(1+R/l)$. Further kinematic analysis using ADAMS yields the relationship between crank angle and slider displacement, speed and acceleration versus crank angle[16-17], as shown in Fig. 2.

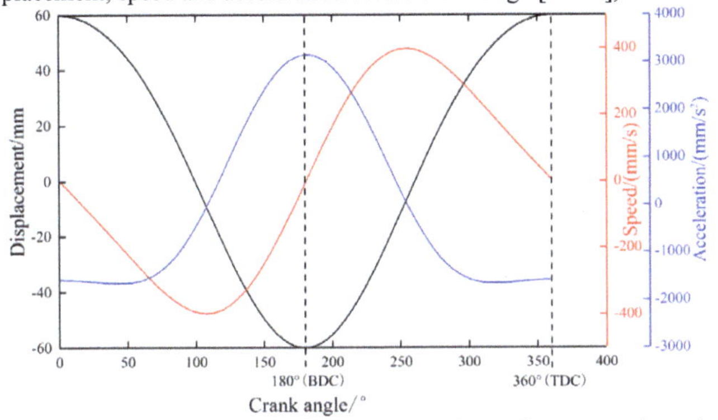

Fig. 2. Relation between slider displacement, speed, acceleration and crank angle.

Crank slider press, due to its structural parameters to determine, when the crank input constant speed, the slider movement parameters are also determined, when the slider is at the dead center, the acceleration will be a sudden change, easy to impact on the motor, affecting the stability of its normal function. Servo press can realize the dynamic adjustment of the output speed, to a certain extent, can avoid the slider in the dead position at the sudden change, so the crank slider servo press servo motor acceleration control, to realize the smooth transition of the slider speed, acceleration.

Analysis of servo drawing process curve. Taking a servo drawing process curve as an example, 4 control points are set up to divide the slider movement curve of the typical drawing process, including the idle feed stage I, deep drawing stage II, idle return stage III, and loading/unloading stage IV, as shown in Fig. 3. Where the slider runs to the idle feed stage I, the slider first accelerated and then decelerated, there is a maximum speed; when the slider runs to the deep drawing stage II, the slider is a uniform speed; when the slider runs to the idle return stage III, the slider has a sharp return characteristics, the speed is gradually reduced; when the slider runs to the loading/unloading stage IV, the slider runs to the top dead centre near the lower speed, for the loading and unloading to set aside a longer period of time for the manipulator to carry out loading and unloading operation.

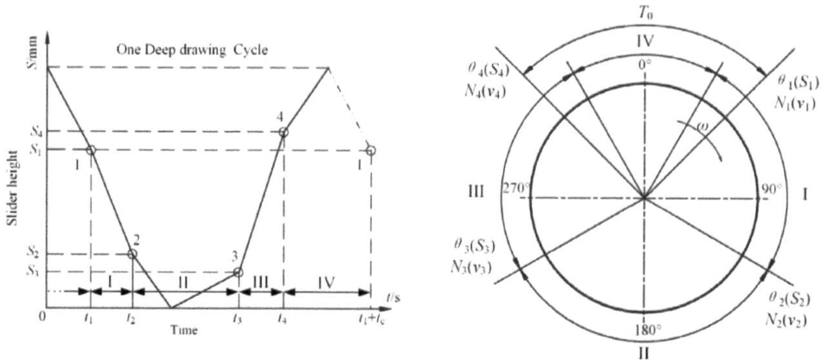

Fig.. 3. Typical deep drawing process for servo presses.

Key point 1: Slider closing angle θ_1, according to the transmission system can determine the slider position S_1, θ_1 is the corresponding crank angle when the robot transfers the blank to the forming and then safely leaves the press; S_1 is the corresponding slider safety height; N_1 is the corresponding crank instantaneous speed; v_1 is the corresponding slider instantaneous speed. Key point 2: the stamping start angle θ_2 and stamping start rotational speed N_2, according to the transmission system, the height of the slide S_2 and speed v_2 can be calculated, where θ_2 is the crank angle when the press starts stamping; N_2 is the corresponding crank rotational speed; S_2 is the initial height of the part when the press starts stamping; v_2 is the speed of the slide when the part starts

stamping. Key point 3: the stamping end angle θ_3 and stamping end rotational speed N_3, according to the transmission system can be calculated slider height S_3 and speed v_3, where θ_3 is the crank angle at the end of the die pulling; N_3 is the corresponding crank rotational speed at the end of the die pulling; S_3 is the height of the die pulling/slider contacting the height of the part; v_3 is the speed of slider at the end of the die pulling. Where N_2=N_3. Key point 4: the slider opening angle θ_4, according to the transmission system can be calculated slider height S_4 and speed v_4, where θ_4 is the corresponding crank angle when the manipulator leaves the press after picking up the workpiece; S_4 is the height of the corresponding slider; N_4 is the instantaneous rotational speed of the corresponding crank; and v_4 is the instantaneous speed of the corresponding slider.

The T_0 is the safety time required for the press manipulator to remove the workpiece and place the blank, the time required for the crank to run from θ_4 to θ_1, during which the manipulator is required to complete the removal of the workpiece, the transfer of the blank, and leave the press safely.

Principle of improved S-curve acceleration algorithm

Principle of 7- phase S-curve acceleration algorithm. S-curve acceleration algorithm is a commonly used algorithm for controlling the object under test to realize smooth acceleration, uniform speed and deceleration process, to ensure the smoothness and stability of the mechanical system when the motor starts, stops, and moves at a constant speed. S-curve acceleration consists of 7 segments, acceleration, uniform acceleration, deceleration, uniform speed, acceleration, uniform deceleration, and deceleration segments, shown in Fig. 4.

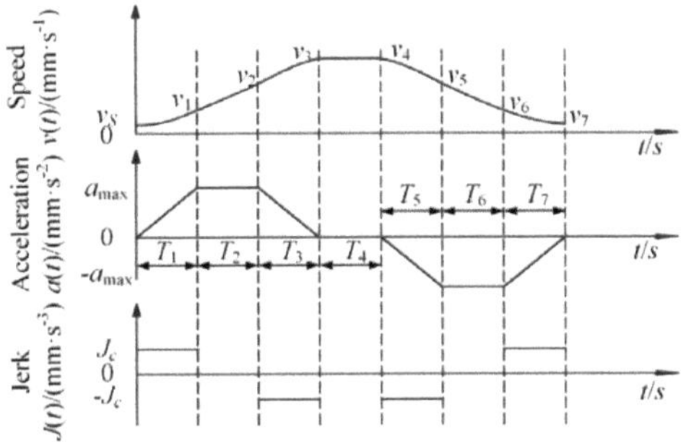

Fig. 4. S-curve acceleration principle.

Define the following variables:

$$T_k = t_k - t_{k-1}\,(k = 1,2,3,4,5,6) \tag{4}$$

$$\tau_k = t - t_{k-1}\,(k = 1,2,3,4,5,6) \tag{5}$$

$$T_1 = T_3 = T_5 = T_7; T_2 = T_4 = T_6 \tag{6}$$

$$JT_1 = a_{max} \tag{7}$$

Combining Fig. 4 and Eq. 4 to Eq. 7, the equations for acceleration, speed and time in the S-curve acceleration algorithm can be clarified as Eq. 8 to Eq. 9.

$$a(t) = \begin{cases} J_C\tau_1, & 0 < t \le t_1 \\ J_C T_1, & t_1 < t \le t_2 \\ J_C T_1 - J_C\tau_3, & t_2 < t \le t_3 \\ 0, & t_3 < t \le t_4 \\ -J_C\tau_5, & t_4 < t \le t_5 \\ -J_C T_1, & t_5 < t \le t_6 \\ -J_C T_1 + J_C\tau_7, & t_6 < t \le t_7 \end{cases} \tag{8}$$

$$v(t) = \begin{cases} v_S + \frac{1}{2}J_C\tau_1^2, & 0 < t \le t_1,\; t = t_1,\; v_1 = v_S + \frac{1}{2}J_C T_1^2 \\ v_S + \frac{1}{2}J_C T_1^2 + J_C T_1\tau_2, & t_1 < t \le t_2,\; t = t_2,\; v_2 = v_S + \frac{1}{2}J_C T_1^2 + J_C T_1 T_2 \\ v_S + \frac{1}{2}J_C T_1^2 + J_C T_1 T_2 + J_C T_1\tau_3 - \frac{1}{2}J_C\tau_3^2, & t_2 < t \le t_3,\; t = t_3,\; v_3 = v_S + J_C T_1 T_2 + J_C T_1^2 \\ v_S + J_C T_1^2 + J_C T_1 T_2 & t_3 < t \le t_4,\; t = t_4,\; v_4 = v_S + J_C T_1 T_2 + J_C T_1^2 \\ v_S + J_C T_1^2 + J_C T_1 T_2 - \frac{1}{2}J_C\tau_5^2, & t_4 < t \le t_5,\; t = t_5,\; v_5 = v_S + \frac{1}{2}J_C T_1^2 + J_C T_1 T_2 \\ v_S + \frac{1}{2}J_C T_1^2 + J_C T_1 T_2 - J_C T_1\tau_6, & t_5 < t \le t_6,\; t = t_6,\; v_6 = v_S + \frac{1}{2}J_C T_1^2 \\ v_S + \frac{1}{2}J_C T_1^2 - J_C T_1\tau_7 + \frac{1}{2}J_C\tau_7^2, & t_6 < t \le t_7,\; t = t_7,\; v_7 = v_S \end{cases} \tag{9}$$

By setting the time of acceleration, deceleration, acceleration and deceleration segments in the 7-phase S-curve acceleration algorithm equal to the time of acceleration, deceleration, acceleration and deceleration segments, and the time of uniform acceleration, uniform speed and uniform

deceleration segments, the solution of speed can be simplified, and it can also ensure that the initial speed v_0 is equal to the final speed v_7 after the acceleration algorithm is run as shown in Eq. 9.
Principle of improved S-curve acceleration algorithm. Due to the specific machining process, the slider by the acceleration algorithm run by the beginning and end of the speed is not equal, and it may introduce many different time divisions of the S-curve acceleration algorithm, which will have many parameters. Therefore, the 7-phase S-curve acceleration algorithm is improved, the uniform acceleration section and acceleration/deceleration section are removed, while the uniform speed section in the middle section is moved to the two ends of the acceleration section, deceleration section next to the curve, and at the same time the curve obtained is smoothed by using a quadratic parabola at the acceleration mutation cusp, and the schematic diagram is shown in Fig. 5.

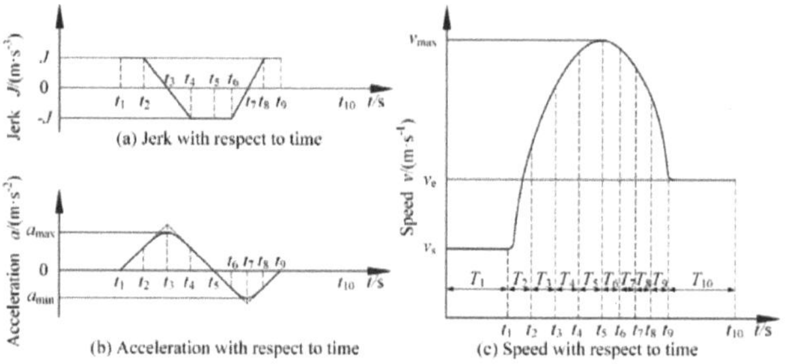

Fig. 5. Improved S-curve acceleration algorithm principle schematic diagram.

To simplify the solution, the following conditions are assumed here: $T_2=T_3=T_4=T_5$, $T_6=T_7=T_8=T_9$, then $v_S \neq v_e$; If $T_2=T_3=T_4=T_5=T_6=T_7=T_8=T_9$, then $v_S=v_e$. Therefore, taking $v_S \neq v_e$ as an example, the relations between acceleration, acceleration, speed and time of the improved S-curve acceleration algorithm are given, as shown in Eq.10 to Eq.12.

$$J(t)=\begin{cases} 0, & 0<t\le t_1 \\ J, & t_1<t\le t_2 \\ J-\dfrac{J}{T_3}\tau_3, & t_2<t\le t_3 \\ -\dfrac{J}{T_4}\tau_4, & t_3<t\le t_4 \\ -J, & t_4<t\le t_5 \\ -J, & t_5<t\le t_6 \\ \dfrac{J}{T_7}\tau_7-J, & t_6<t\le t_7 \\ \dfrac{J}{T_8}\tau_8, & t_7<t\le t_8 \\ J, & t_8<t\le t_9 \\ 0, & t_9<t\le t_{10} \end{cases} \tag{10}$$

Further integration yields an expression for the acceleration -time relationship as follows.

$$a(t)=\begin{cases}0, & 0<t\le t_1\\ J\tau_2, & t_1<t\le t_2, t=t_2, a_2=JT_2\\ J(T_2+\tau_3-\frac{\tau_3^2}{2T_3}), & t_2<t\le t_3, t=t_3, a_3=\frac{3}{2}JT_2\\ J(\frac{3}{2}T_2-\frac{\tau_4^2}{2T_2}), & t_3<t\le t_4, t=t_4, a_4=JT_2\\ J(T_2-\tau_5), & t_4<t\le t_5, t=t_5, a_5=0\\ -J\tau_6, & t_5<t\le t_6, t=t_6, a_6=-JT_6\\ J(-T_6+\frac{\tau_7^2}{2T_6}-\tau_7), & t_6<t\le t_7, t=t_7, a_7=-\frac{3}{2}JT_6\\ J(-\frac{3}{2}T_6+\frac{\tau_8^2}{2T_6}), & t_7<t\le t_8, t=t_8, a_8=-JT_6\\ J(-T_6+\tau_9), & t_8<t\le t_9, t=t_9, a_9=0\\ 0, & t_9<t\le t_{10}, t=t_{10}, a_{10}=0\end{cases} \tag{11}$$

Further integration yields an expression for the speed-time relationship as follows.

$$v(t)=\begin{cases}v_s, & 0<t\le t_1, t=t_1, v_1=v_s\\ v_s+\frac{1}{2}J\tau_2^2, & t_1<t\le t_2, t=t_2, v_2=v_s+\frac{1}{2}JT_2^2\\ v_s+\frac{1}{2}JT_2^2+J(T_2\tau_3+\frac{1}{2}\tau_3^2-\frac{\tau_3^3}{6T_3}), & t_2<t\le t_3, t=t_3, v_3=v_s+\frac{11}{6}JT_2^2\\ v_s+\frac{11}{6}JT_2^2+J(\frac{3}{2}T_2\tau_4-\frac{\tau_4^3}{6T_2}), & t_3<t\le t_4, t=t_4, v_4=v_s+\frac{19}{6}JT_2^2\\ v_s+\frac{19}{6}JT_2^2+J(T_2\tau_5-\frac{1}{2}\tau_5^2), & t_4<t\le t_5, t=t_5, v_5=v_s+\frac{11}{3}JT_2^2\\ v_s+\frac{11}{3}JT_2^2-\frac{1}{2}J\tau_6^2, & t_5<t\le t_6, t=t_6, v_6=v_s+\frac{11}{3}JT_2^2-\frac{1}{2}JT_6^2\\ v_s+\frac{11}{3}JT_2^2-\frac{1}{2}JT_6^2-J(T_6\tau_7-\frac{\tau_7^3}{6T_6}+\frac{1}{2}\tau_7^2), & t_6<t\le t_7, t=t_7, v_7=v_s+\frac{11}{3}JT_2^2-\frac{11}{6}JT_6^2\\ v_s+\frac{11}{3}JT_2^2-\frac{11}{6}JT_6^2-J(\frac{3}{2}T_6\tau_8-\frac{\tau_8^3}{6T_6}), & t_7<t\le t_8, t=t_8, v_8=v_s+\frac{11}{3}JT_2^2-\frac{19}{6}JT_6^2\\ v_s+\frac{11}{3}JT_2^2-\frac{19}{6}JT_6^2-J(T_6\tau_9-\frac{1}{2}\tau_9^2), & t_8<t\le t_9, t=t_9, v_9=v_s+\frac{11}{3}JT_2^2-\frac{11}{3}JT_6^2\\ v_s+\frac{11}{3}JT_2^2-\frac{11}{3}JT_6^2, & t_9<t\le t_{10}, t=t_{10}, v_e=v_s+\frac{11}{3}JT_2^2-\frac{11}{3}JT_6^2\end{cases} \tag{12}$$

The above parameters $a_{\max}$、$a_{\min}$、T_f、$v_{\max}$、v_s、v_e are the key parameters in the process curve, which can be used to express the control parameters in the process of S-curve acceleration algorithm, such as: time interval T_i, the jerk J and other parameters, which can simplify the acceleration algorithm in the solution, according to the boundary conditions can be obtained in part of the parameters of the relationship between the Eq.13 to Eq. 21.

$$v_{\max}=v_5=v_s+\frac{11}{3}JT_2^2 \tag{13}$$

$$J=\frac{2a_{\max}}{3T_2}=-\frac{2a_{\min}}{3T_6}=\frac{44a_{\max}^2}{27(v_{\max}-v_s)}=-\frac{44a_{\min}^2}{27(v_{\max}-v_e)} \tag{14}$$

$$T_2=T_3=T_4=T_5=\frac{9(v_{\max}-v_s)}{22a_{\max}} \tag{15}$$

$$T_6 = T_7 = T_8 = T_9 = -\frac{9(v_{max} - v_e)}{22a_{min}} \tag{16}$$

$$a_{min} = a_7 = -\frac{3}{2}JT_6 = -\frac{T_6}{T_2}a_{max} \tag{17}$$

$$T_1 = T_{10} = \frac{T_f}{2} - \frac{2(a_{max} - a_{min})T_2}{a_{max}} = \frac{T_f}{2} - \frac{9a_{min}(v_{max} - v_s) - 9a_{max}(v_{max} - v_e)}{11a_{max}a_{min}} \tag{18}$$

$$v_e = v_s + \frac{11}{3}JT_2^2 - \frac{11}{3}JT_6^2 \tag{19}$$

$$T_f = \sum_{i=1}^{10} T_i, \quad i = 1,2,3,\ldots,10 \tag{20}$$

$$a_{max} = a_3 = \frac{3}{2}JT_2 \tag{21}$$

Trajectory planning of servo motor press

Trajectory planning model establishment. According to the previous analysis of the servo press drive system and the deep-drawing process curve, the angular speed, angular acceleration and angular jerk curve of the servo motor in the complete cycle of the deep-drawing process are given, as shown in Fig. 6. The relevant variables in the improved S-curve acceleration control algorithm can be determined according to the process parameters in the actual case. The selected range of angular speed extreme angles, x_1, x_2, x_3, and x_4, is determined based on θ_1, θ_2, θ_3, and θ_4. Where θ_1, θ_2, θ_3, and θ_4, and w_1, w_2, and w_3 are known variables, angular speed extremes angles x_1, x_2, x_3, and x_4, angular speed extremes w_{m1}, and w_{m2}, angular acceleration extremes angles y_1, y_2, y_3, and y_4 and known angles θ_1, θ_2, θ_3, and θ_4 in the interval time/angle T_0, T_1, T_2, T_3, T_4, T_5, T_6, T_7, T_8, T_9, T_{10}, T_{11}, T_{12}, T_{13}, T_{14}, T_{15}, T_{16}, and T_{17} are unknown parameters and need to be solved. The following relationship exists for the interval time/angle: $T_1=T_2=T_3=T_4$; $T_5=T_6=T_7=T_8$; $T_{10}=T_{11}=T_{12}=T_{13}$; $T_{14}=T_{15}=T_{16}=T_{17}$. There is also $w_2=w_3$.

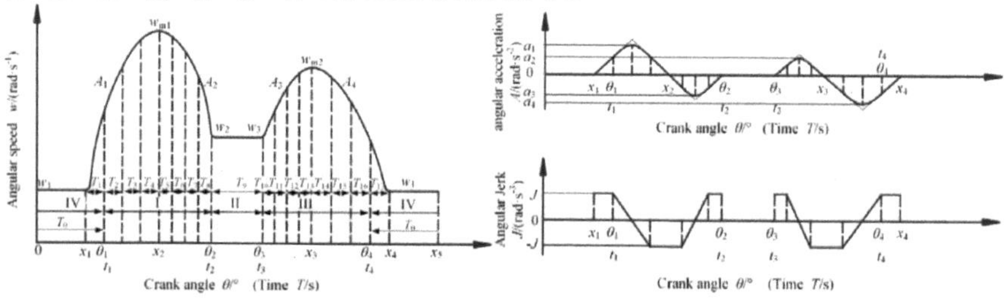

Fig. 6. Angular speed curve and key points of servo motor throughout the whole week.

From the Fig. 6, in the idle feed stage, the crank is first accelerated by acceleration A_1 from w_1 to the angular speed extremes w_{m1}, and then decelerated by acceleration A_2 from w_{m1} to the starting speed of the stamping w_2; into the deep drawing stage, the crank is deep-drawn at a constant speed of w_2 until the end of the θ_3; and then into the idle return stage, the crank is first accelerated by acceleration A_3 from w_3 to the angular speed extremes w_{m2}, and then decelerated by acceleration

A_4 from w_{m2} to w_1 in order to achieve the upper dead center of the low-speed operation, and so on to realize the continuous stamping operation.

Slider motion trajectory planning solution. Taking the relevant parameters of a certain type of servo press in the references as an example, solving the speed trajectory planning and compare the results with those in the references [18]. The relevant parameters of the servo press are as follows: crank radius R=450 mm, connecting rod length L=1200 mm, and slider stroke S=900 mm, and the process control parameters are set up as an example for the deep drawing forming of a certain model's side panel, in which the forming height of the part is 200 mm, and the speed of the start of the deep drawing process is required to be less than 300 mm/s, and the height of the safety space required by the manipulator is not less than 700 mm, and the transfer time is not less than 2 s, the speed of die pulling is less than 80 mm/s when the height of die pulling is 10 mm, and the maximum acceleration of slider is less than 3500 mm/s^2. The key point control parameters are set as follows: θ_1=66°, θ_4=294°, S_1=S_4=705.7 mm, T_0=2s; θ_2=131.5°, S_2=200.1 mm, w_2=w_3=6.7 rpm, v_2=297.7 mm/s, θ_3=190°, S_3=9.4 mm, v_3=75.1 mm/s.

As shown in Fig. 6, the following relationships exist for some parameters: $x_1+T_1=t_1/\theta_1$; $t_1/\theta_1+T_1=y_1$; $x_1+4T_1=x_2$; $x_2+2T_5=y_2$; $x_2+4T_5=t_2/\theta_2$; $t_2/\theta_2+T_9=t_3$; $t_3/\theta_3+2T_{10}=y_3$; $t_3/\theta_3+4T_{10}=x_3$; $x_3+2T_{14}=y_4$; $x_3+3T_{14}=t_4/\theta_4$; $t_4/\theta_4+T_{14}=x_4$. Where the solution parameters include X=[x_1, x_2, x_3, x_4, x_5, w_{m1}, w_{m2}, T_1, T_5, T_9, T_{10}, T_{14}, J], and hence a set of solutions X_0=[0.9, 1.3, 3.24, 4.14, 5.14, 22.5, 24.5, 0.1, 0.11, 1.4, 0.125, 0.1, 313.63], and the corresponding crank and slider kinematic parameter plots can be obtained, as shown in Figs. 7, 8, and 9.

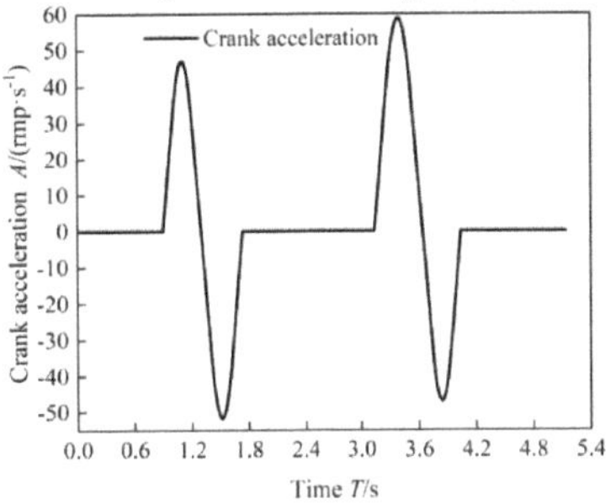

Fig. 7. Slider height and crank speed curve.

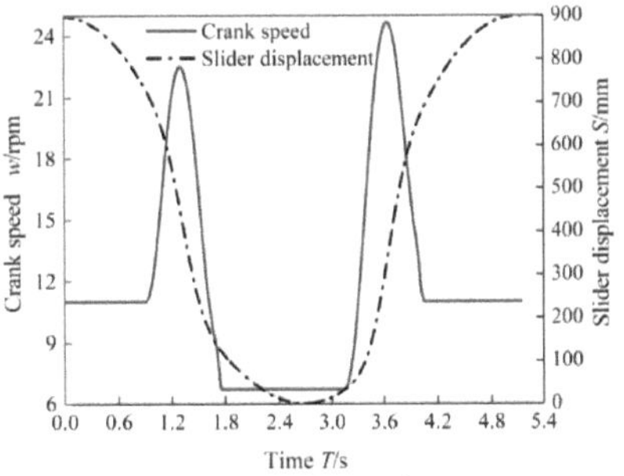

Fig. 8. Crank acceleration curve.

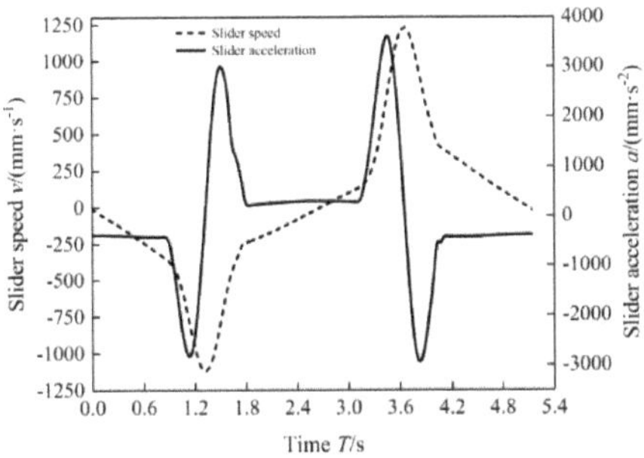

Fig. 9. Slider speed and acceleration curve.

Fig. 7, 8, and 9 illustrate that while satisfying the spatial and temporal constraints of the transfer manipulator and the maximum acceleration of the press slider, the maximum production rate can reach 11.7 times/min. This significantly enhances production efficiency. The continuous variation of crank acceleration depicted in Fig. 8 and 9 ensures reliable equipment operation with heightened flexibility. These findings underscore the efficacy of the improved S-curve acceleration algorithm for planning the motion curve of the servo press slider.

Summarize

An efficient deep drawing process mode of servo press is established, and the servo press machining process trajectory planning can be realized by setting 4 key point control parameters, which is simple and convenient to set up and easy to operate on site. The servo motor flexible acceleration and deceleration control algorithm with improved S- curve is proposed, and the crank speed, acceleration and acceleration changes are continuous, which avoids the impact of acceleration and deceleration process on the equipment, and it can be used for servo press deep drawing trajectory planning.

References

[1] F. Peng, C. Wang, H. Chai, Z. Shao, S. Wang, B. Wang, Lightweight slider design for a servo press based on its layered structure, J. Tsinghua Univ. (Sci. Technol.) 12 (2020) 1016-1022. https://doi.org/10.16511/j.cnki.qhdxxb.2020.22.016

[2] F. Peng, P. Zhang, L. Wang, Z. Shao, D. Yang, K. Yang, Energy management of servo press lines based on flywheel speed adaptive planning, J. Tsinghua Univ. (Sci. Technol.) 11 (2020) 927-933. https://doi.org/10.16511/j.cnki.qhdxxb.2020.22.014

[3] L. Yang, S. Xu, W. Shi, H. GE, Research on S Type Acceleration and Deceleration Time Planning Algorithm with Beginning and End Speed Non-zero, J. Mech. Eng. 23 (2016) 199-206. https://doi.org/10.3901/JME.2016.23.199

[4] Q. Hao, L. Guan, L. Wang, Intelligent acceleration/deceleration control algorithm for drive force for a heavy duty hybrid machine tool, J. Tsinghua Univ. (Sci. Technol.). 11 (2020) 1770-1773, 1778. https://doi.org/10.3321/j.issn:1000-0054.2009.11.009

[5] Z. Liu, S. Zhang, L. Cai, J. Yang, B. Xu, P. Xu, Five-phase S-curve Control Method Based on Particle Swarm Optimization, J. Beijing Univ Technol. 5 (2015) 641-648. https://doi.org/10.11936/bjutxb2014060067

[6] X. Zhang, D. Zhu, F. Pan, Research on S-curve flexible Acc/Dec control algorithm in NC machining, Manuf. Autom. 12 (2020) 68-73. https://doi.org/10.3969/j.issn.1009-0134.2020.12.014

[7] J. Chen, S.S. Yeh, J. Sun, An S-curve Acceleration/Deceleration Design for CNC Machine Tools Using Quintic Feedrate Function, Comput - Aided Des & Appl. 4 (2011) 583-892. https://doi.org/10.3722/cadaps.2011.583-592

[8] X. Guo, C. Li, A New Flexible Acceleration and Deceleration Algorithm, Journal of Shanghai Jiaotong University. 2 (2003) 205-207. https://doi.org/10.3321/j.issn:1006-2467.2003.02.016

[9] Y. Chen, W. Guo, F. Gao, NURBS-Based Feature Modeling and Trajectory Planning for Ram Motion of Servo Mechanical Presses, J. Shanghai Jiaotong Univ. 1 (2009) 138-142. https://doi.org/10.3321/j.issn:1006-2467.2009.01.031

[10] X. Qu, Design and simulations of non-uniform speed motion curves for a slider-crank servo press, Assem. Autom. 3 (2018) 336-346. https://doi.org/10.1108/AA-10-2016-131

[11] W. Shang, S. Zhao, Servo Press Processing Bezier-Model with Optimum, J. Xi'an Jiaotong Univ. 3 (2012) 31-35.

[12] J. Gu, T. Wu, C. Li, B. Zhang. Y. Zhang, Trajectory Fairing Algorithm for Shrub Pruning Based on Improved Cubic B-spline, Trans. Chin. Soc. Agric. Mach. S1 (2021) 89-97. https://doi.org/10.6041/j.issn.1000-1298.2021.S0.012.

[13] P. Cheng, P. Chen, Y. Lin, Small mechanical press with double-axis servo system for forming of small metal products, Int. J. Adv. Manuf. Technol. 68 (2013) 2371-2381. https://doi.org/10.1007/s00170-013-4851-y

[14] H. Luo, The Research on Interpolation Algorithm of NURBS Curve Based on Milne-Hamming Method and Acceleration/Deceleration Control Algorithm, Nanjing: Nanjing University of Aeronautics and Astronautics, 2020.

[15] Q. Song, B. Guo, J. Li, W. Yin, Flexible Acceleration and Deceleration Control Algorithm for Servo Press, Trans. Chin. Soc. Agric. Mach. 6 (2013) 269-273. https://doi.org/10.6041/j.issn.1000-1298.2013.06.047

[16] H Lin, D Ye, X Wang, Y Liu, Motion planning of main drive system of high speed fine blanking press by double servo drive, J. Huazhong Univ. Sci. Technol. 4 (2018) 6-11. https://doi.org/10.13245/j.hust.180402

[17] Y. Sun, J. Hu. Y. Cheng, W. Ruan, Characteristics Analysis of Working Mechanisms for Servo Mechanical Presses, Appl. Mech. Mater. 220-223 (2012) 762-767. https://doi.org/10.4028/www.scientific.net/AMM.220-223.762

[18] W. Yin, J. Li, Q. Song, Planed and Optimized Drawing Trajectory for Heavy Servo Presses, CFHI Technol. 1 (2015) 1-5. https://doi.org/10.3969/j.issn.1673-3355.2015.01.001

Influence of hardening model on draw-bending springback prediction of DP980 dual-phase steel

ZHENG Xuebin[1,2,a], HAN Longshuai[2,b], Hongwei E[2,c], LIU Yu[2,d], LI Xuetao[2,e], WU Xiangdong[1,f] and WAN Min[1,g]*

[1]School of Mechanical Engineering & Automation, Beihang University, Xueyuan Road, Haidian District, Beijing, China

[2]Research Institute of Technology, Shougang Group Co., Ltd., Yangzhuang Street, Shijingshan District, Beijing, China

[a]zhengxb@shougang.com.cn, [b]hanlongshuai@shougang.com.cn, [c]ehw0635@shougang.com.cn, [d]liuy0911@shougang.com.cn, [e]lxt@shougang.com.cn, [f]xdwu@buaa.edu.cn, [g]mwan@buaa.edu.cn

Keywords: DP980 Dual-Phase Steel, Y-U Model, Draw-Bending, Elastic Modulus, Springback Prediction

Abstract. Springback is a key factor affecting the dimensional and shape accuracy of cold-formed high-strength steel parts, and the accuracy of high-strength steel springback prediction depends on the accurate material constitutive model. Therefore, it is important to study the influence of the constitutive model on springback prediction accuracy. In this paper, the parameters of the Yoshida-Uemori (Y-U) dynamic hardening model of DP980 dual-phase steel were determined through tension-compression tests with two different strain levels and multicycle tension-compression tests. Based on the draw-bending test platform, the influence of roll radius and normalized back force on the springback of DP980 was investigated. Additionally, the effect of different hardening models (Swift model, Y-U model) on the prediction of draw-bending springback was studied by using Abaqus finite element analysis software. The results show that increasing the bending radius and normalized back force can reduce the springback angle and side wall crimp. The tension-compression testing strategy shows little effect on the calibrated Y-U model parameters of DP980 dual-phase steel. The Y-U model can achieve better prediction accuracy for draw-bending springback angle and curvature compared to the Swift isotropic hardening model.

Introduction

Advanced high strength steels (AHSS) are highly promising materials for lightweight vehicle structures. In recent years, advanced high strength steels (AHSS) are being used in the automotive industry in an effort to reduce vehicle weight and increase fuel efficiency without sacrificing passenger safety [1]. DP (Dual Phase) steels exhibit continuous yielding behavior, a low yield to tensile strength ratio, a high work-hardening rate, relatively high elongation, bake hardenability and cost efficiency, the use of DP steels has increased [2-3]. However, the practical application of these materials has been hampered by the technical challenges of understanding and accounting for springback behaviour after forming [4]. AHSS components are often modified via trial-and - error, which is not only costly but also inefficient. The application of finite element (FE) methods to predict springback of sheet products and compensate for the design of stamping dies can effectively reduce the expense of die tryout and improve the dimensional accuracy of parts.

There are many examples of predicting springback after simple forming operations such as cylindrical tool bending, V-die bending, U-channel forming and stretch bending[5]. These test methods can characterise test conditions such as pure bending or bending with a small tensile force.

It was concluded that springback increased with the increase of material yield strength and the tool radius, but decreased with increasing elastic modulus for an elastic perfectly plastic material after pure bending. However, among all the process variables that affect springback, sheet tension has the most prominent effect in reducing springback [6]. While stretch bending test is unable to represent the complex deformations that occur in the practical forming processes in which a sheet metal slides over a rigid tool surface as it is drawn into a die cavity. However, the draw bending test overcame this shortcoming and provided a better way of studying springback behavior [7-9].

The accuracy of high-strength steel springback prediction largely depends on the accurate material constitutive model [10]. In recent studies, attention has been drawn to the importance of material constitutive models for FE formability and springback analysis. Yoshida and Uemori [11-12] implemented a back stress idea to develop a theory (the Y-U model) for cyclic plasticity that describes the Bauschinger effect as well as an effect of hardening stagnation that has been observed in some materials upon reverse loading. Lin et al. [13] investigated the effect of the Y-U model on the springback prediction accuracy of MP980 advanced high strength steel and AA6022-T4 aluminium alloy in U-bending. For MP980 advanced high strength steel exhibiting a clear Bauschinger effect but insignificant texture anisotropy, the springback prediction accuracy can be significantly improved by using the Y-U model in combination with the Hill48 yield criterion, and for the AA6022-T4 aluminium alloy with significant texture anisotropy, it is more important to improve the springback prediction accuracy by using the YLD2000-2D yield criterion. Min et al. [14] investigated the effect of tension-compression testing strategy on calibration of Y-U model parameters and on the springback prediction accuracy of advanced high-strength steels. For QP980 advanced high-strength steel, which has a phase transition induced plasticity (TRIP) effect, the tensile-compression strategy has a significant effect on the parameters of the Y-U model and on the springback simulation results. The HAH model developed by Barlat et al [3] was able to describe the complex material behavior such as Bauschinger effect, transient hardening and permanent softening and was in good agreement with experimental data for both cases with and without pre-strain. Barlat et al. [15] developed the HAH model was able to describe the complex material behavior such as Bauschinger effect, transient hardening and permanent softening and was in good agreement with experimental data for both cases with and without pre-strain. Lee et al. [16] applied the HAH model for springback predictions in U draw of as-received and pre-strained DP780 steel sheets. Chaboche [17] developed the Chaboche model and proposed the decomposition of the nonlinear kinematic hardening into several kinematic rules used for the description of special ranges of strain.

In this work, the parameters of the Yoshida-Uemori (Y-U) dynamic hardening model of DP980 dual-phase steel were determined by tension-compression tests. The influence of roll radius and normalized back force on the springback of DP980 was investigated using a draw-bending test platform. The influence of different hardening models on drawing-bending springback prediction was studied by using Abaqus finite element analysis software.

Experimental details

Materials. The material used in this study was hot-dip galvanized DP980 dual-phase steel sheets with a thickness of 1.2 mm. The typical microstructure of dual-phase steel sheets is ferrite and martensite islands. The strength of DP steels is achieved by the presence of martensite, and the ductility is due to the presence of ferrite. The chemical composition of the DP980 dual-phase steel used in this study is given in Table 1.

Table 1. Chemical composition of DP980 steel.

Steel	C	Si	Mn	S	Alt	Cr	Mo	Ti
DP980	0.095	0.4	2.4	0.003	0.035	0.5	0.2	0.025

Uniaxial tensile tests. The mechanical properties of the materials were characterized by uniaxial tensile tests. The tests were carried out on standard GB/T228.1-2010 with a strain rate of 0.01/s. The specimens were tested along the rolling direction (RD), diagonal direction (DD) and transverse direction (TD) with a gauge length of 80 mm. The directional r-values were measured by two extensometers. The true stress-true strain curves are plotted in Fig. 1 and the corresponding mechanical properties of the DP980 dual phase steel are summarized in Table 2.

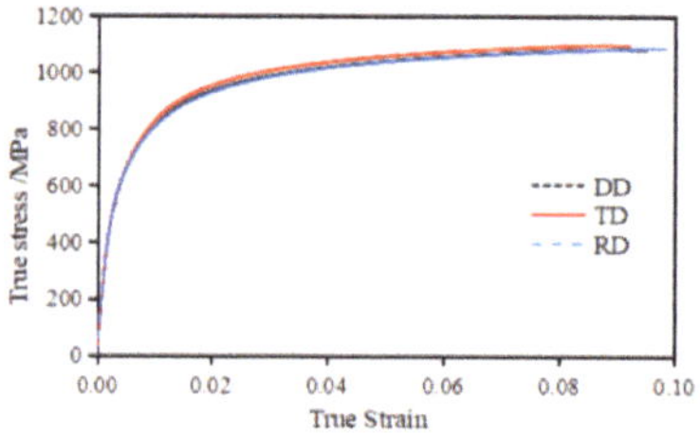

Fig. 1. True stress-true strain curves in RD, TD and DD directions for DP980 steel.

Table 2. The mechanical properties of DP980 steel.

Direction	Yield stress (MPa)	Tensile strength (MPa)	Elongation (%)	n value	r value
RD	638	1008	10.5	0.094	0.96
TD	638	1008	10.5	0.094	0.96
DD	640	993	11	0.096	0.94

Tension-compression tests. It has been shown that the Bauschinger effect and cyclic hardening behavior are significant factors in the prediction of springback for automotive sheet materials. To determine the parameters of the Y-U model, the experimental method in reference [13] was used for the tension compression test of DP980 dual phase steel. Strain was measured using the digital image correlation (DIC) method at room temperature. The crosshead speed was set at 2 mm/min for both tension and compression tests. A lateral force fixture with a nominal force of approximately 2 kN was designed to prevent specimen buckling. Polytetrafluoroethylene (Teflon) sheets 0.8 mm thickness were adhered to both surfaces of each specimen with lubricating grease to minimize friction between the specimen and the fixture. The dimension of the specimen for the tension compression test was shown in Fig. 2.

Materials Research Proceedings 44 (2024) 580-588 https://doi.org/10.21741/9781644903254-62

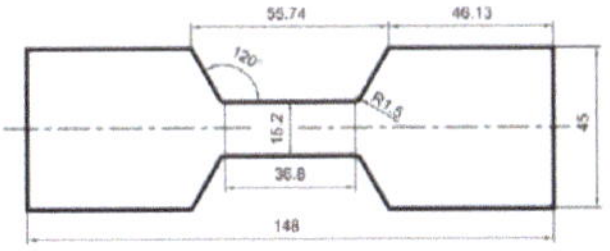

Fig. 2. Dimensions of specimen for tension-compression test (mm).

Draw-bending tests. The draw-bending test and schematics of the test procedure are shown in Fig. 2. During the test, the strip was drawn over a fixed or rotating cylindrical roller at a speed of 40 mm/s for a total travel of 127 mm, while the back force was maintained constant. A typical unloaded specimen is shown schematically in Fig. 3(b), from which four distinct regions of deformation were identified. The material in regions 1 and 4 only experienced pure stretching and these two regions were not considered in this experiment. Regions 2 was in contact with the forming tool before unloading, and its radius of curvature changes from R to R' after springback. The material in region 3 has undergone sequential bending and unbending with superposed stretching and has a radius of curvature r' after unloading. The parameter $\Delta\theta$ has been used to characterize the amount of springback for draw bending tested specimens.

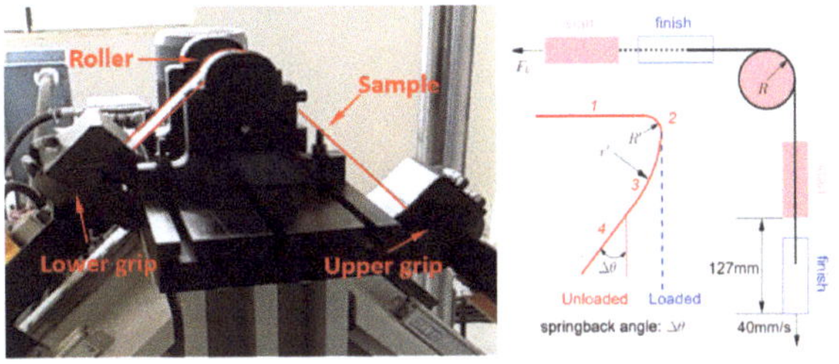

(a) (b)

Fig. 3. Draw bending experiment: (a) equipment, (b) schematics of test procedure.

FE modeling of springback

Y-U model parameter fitting. The Y-U model uses two surfaces to describe the hardening rule, wherein the yield surface moves within the bounding surface [11]. In this work, a single-element model was built to optimize the Y-U model parameters [14] . An sequence optimization response surface method (SRSM) was used to minimize the difference between experimental and calculated stresses with a mean square function as follows,

$$\delta = \frac{1}{P}\Sigma_{p=1}^{P} \mathrm{W_p}\left[\frac{f_p(x) - G_p}{S_p}\right]^2 = \frac{1}{P}\Sigma_{p=1}^{P} \mathrm{W_p}\left[\frac{\delta_p(x)}{S_p}\right]^2 \tag{1}$$

where G_p, $p = 1 \cdots P$ is the number of the experimental curve, G, fp(x) is the corresponding component value of the curve f calculated by the response surface, x is the vector of design variables and W_p is the weight coefficient. In addition, an identification process based on the feedforward neural network method was carried out to investigate the influence of the optimization method on the parameter identification.

Springback prediction with Abaqus. The finite element analyses for the tension-bending tests were performed in ABAQUS. The draw-bending and springback simulations were conducted using ABAQUS/Explicit and ABAQUS/implicit, respectively. The Y-U model was implemented into ABAQUS through the user material subroutines UMAT and VUMAT. The Coulomb friction law with a friction coefficient of 0.1 was used for the contact between the tools and the blank. For the simulation, the forming tools were defined as rigid analytical surfaces and four-node shell elements (S4R) with 8 integration points through the thickness were used for the blank. The Hill48 yield criterion was used for the simulated material, and the Swift and Y-U models were compared to determine their effect on the accuracy of springback prediction.

Fig. 4. The 3D-FE model of Draw bending test and mashing.

Results and Discussion

Tension-Compression test results. To investigate the influence of the tension-compression test strategy on the Y-U model parameters, the tension-compression-tension (TCT) tests under two pre-strains of 0.03 and 0.05, and the multi-cycle tension-compression-tension (MTCT) tests under three pre-strains of 0.02, 0.035, and 0.05 were conducted on the DP980 dual phase steel. Experimental true stress vs. true strain curves of DP980 are shown in Fig. 5. It can be illustrated that both methods can well describe the deformation behaviour of DP980. The Y-U model can describe the transient Bauchinger effect and permanent softening phenomenon of the material during cyclic loading by introducing the parameters C and h. Therefore, the Y-U model has a high fitting accuracy compared with the test curve. Table 3 shows the parameters of the Y-U model obtained by fitting single-cycle tensile-compression test curves and multi-cycle tensile-compression test curves, respectively, with fitting errors of 0.2%.

Materials Research Proceedings 44 (2024) 580-588 https://doi.org/10.21741/9781644903254-62

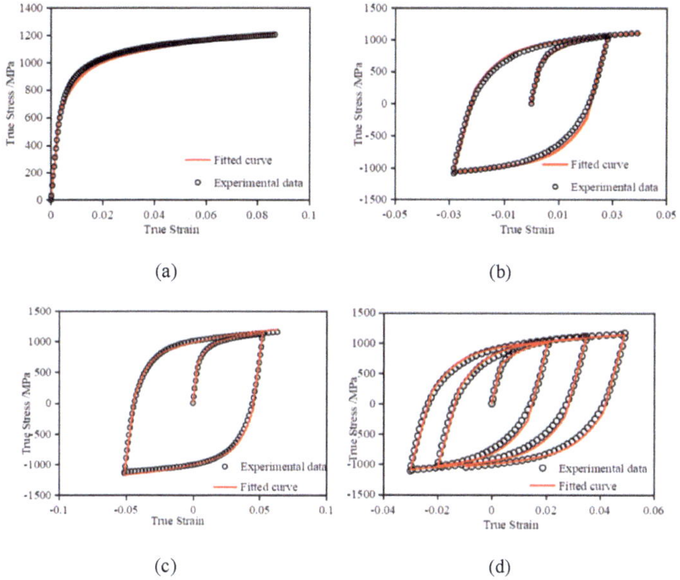

Fig. 5. Experimental and curve fit data of DP980: (a) tensile test, (b) 0.03 pre-strain, (c) 0.05 pre-strain, (d) multi-cycle pre-strain of 0.02, 0.035, and 0.05.

Table 3. Fitted parameters of the Y-U model for DP980.

Strategy	Y (MPa)	B (MPa)	Rsat (MPa)	m	h	b (MPa)	C_1	C_2
TCT	583.7	862.4	604.5	31.1	179.1	378.0	0.87	0.0233
MTCT	637.9	941.6	810.2	15.8	142.0	376.3	0.69	0.0002

Draw-bending test results. Since the sheet tension played a dominant role in reducing springback, multiple tests were conducted with the normalized back forces ranging from 0.2to 1.2, at an increment of 0.05-0.1 to improve resolution.The normalized back force is calculated using equation as follows,

$$\overline{F}_b = \frac{F_b}{\sigma_s \times a_0 \times b_0} \tag{2}$$

Where σ_s is the yield strength of the material, a_0 is the thickness of the specimen, b_0 is the width of the material.

Fig. 6 shows the results of the tests with tool radius of 6.2 and 12.7 mm, respectively, from which it can be seen that the stretching force plays an significant role in reducing the springback. For the same bending radius, the springback angle $\Delta\theta$ decreases with the increase of Normalized back force. Therefore, in the actual stamping process, the method usually used to reduce springback is to increase the binder force. For the same stretching force, the total springback angle decreases as the bending radius increases. By increasing the bending radius and stretching force, both side wall curling and springback can be reduced. Accordingly, R' has the same tendency as the springback angle, while r' has the opposite tendency. In addition to this commonly observed trend, a dramatic drop in the springback angle occurs near F_b = 1.1- 1.2, which is associated with a sudden increase in the anticlastic curvature. It is noted that the critical back force, at which the fast decline of springback angle occurs, weakly depends on tool radius.

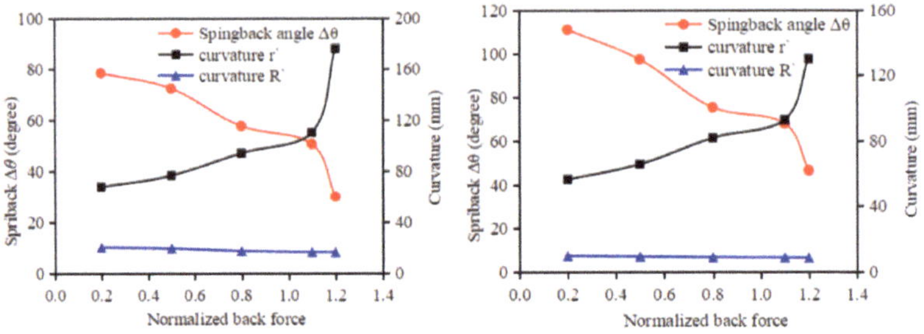

Fig. 6. Effect of normalized back force on the springback angle and the curvature: (a) R6.2 mm, (b) R12.7 mm.

Springback prediction. The effect of the Swift isotropic hardening model and the Y-U kinematic hardening model on the springback is investigated by using the bending radius of 6.2 mm in the draw-bending test, and the comparison of the predicted results with the test results is shown in Fig. 7. It can be seen that a better prediction accuracy of springback angle Δθ and radius of curvature r' can be obtained using the Y-U model, with the corresponding prediction errors of 5.3% and 4.6% respectively. The prediction errors of the Swift hardening model for springback angle Δθ and radius of curvature r' are 28.7% and 22.8% respectively. This suggests that the prediction of springback in draw-bending can be significantly improved by using the Y-U model. The elastic modulus decreases with the increasing of cyclic load strain, and the Y-U model uses an exponential function to describe the relationship between the elastic modulus and the equivalent plastic strain, which improves the accuracy of the springback prediction.

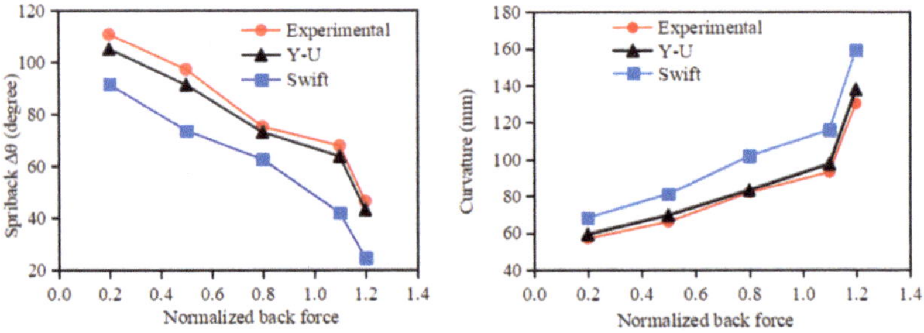

Fig. 7. Comparison of the simulation results with the experimental data: (a) the springback angle, (b) the unloaded anticlastic curvature

Conclusions

(1) The springback angle and side wall crimp decrease with increasing normalized back force, and there is a dramatic decrease in the springback angle at normalized back force close to 1.1- 1.2, which is associated with a sudden increase in the anticlastic curvature.
(2) The tension-compression testing strategy shows little effect on the calibrated Y-U model parameters of DP980 dual-phase steel.
(3) The Y-U model can achieve better prediction accuracy for springback angle and curvature compared to the Swift isotropic hardening model.

References

[1] S. Li, Y. Zhang, L. Qi, et al., Effect of single tensile overload on fatigue crack growth behavior in DP780 dual phase steel, Int. J. Fatigue 106 (2018) 49-55. https://doi.org/10.1016/j.ijfatigue.2017.09.018

[2] D.-K. Kim, E.-Y. Kim, J. Han, et al., Effect of microstructural factors on void formation by ferrite/martensite interface decohesion in DP980 steel under uniaxial tension, Int. J. Plast. 94 (2017) 3-23. https://doi.org/10.1016/j.ijplas.2017.04.019

[3] A.S. Khan, M. Baig, S.-H. Choi, et al., Quasi-static and dynamic responses of advanced high strength steels: Experiments and modeling, Int. J. Plast. 30-31 (2012) 1-17. https://doi.org/10.1016/j.ijplas.2011.08.004

[4] J.-Y. Lee, F. Barlat, M.-G. Lee, Constitutive and friction modeling for accurate springback analysis of advanced high strength steel sheets, Int. J. Plast. 71 (2015) 113-135. https://doi.org/10.1016/j.ijplas.2015.04.005

[5] Wang Jianfeng, Principles of the draw-bend springback test, The Ohio State University, Columbus, 2004.

[6] X. Li, H. Dong, C. Yu, H. Wang, Z. Zang, B. Song, Y. Wang, D. Li, Influence of yield criteria and hardening model on draw-bending springback prediction of DP780, J. Mech. Eng. 45 (2020) 110-115. https://doi:10.3901/JME.2020.12.042

[7] Carden W D, Geng L M, Matlock D K, et al., Measurement of Springback, International Journal of Mechanical Sciences, 44 (2002) 79-101. https://doi.org/10.1016/S0020-7403 (01)00082-0

[8] Wang Jianfeng, Wagoner R H, Matlock D K., Anticlastic curvature in draw bend springback, International Journal of Solids and Structures, 42 (2005) 1287-1307. https://doi.org/10.1016/ j.ijsolstr.2004.08.017

[9] Hocine Chalal, Sever-Gabriel Racz, Tudor Balan., Springback of thick sheet AHSS subject to bending under tension, International Journal of Mechanical Sciences, 59 (2012) 104-114. https://doi.org/10.1016/j.ijmecsci.2012.03.011

[10] Song Bingyi, Meng Bao, Wan Min. Research progress of cyclic plastic hardening model and experiment for metal Sheetss, Journal of netshape forming engineering, 11 (2019) 28-41. https://doi.org/10.3969/j.issn.1674-6457.2019.03.003.

[11] Yoshida F, Uemori T., A model of large-strain cyclic plasticity describing the Bauschinger effect and workhardening stagnation, International Journal of Plasticity, 18 (2002) 661-686. https://doi.org/10.1016/S0749-6419(01)00050-X

[12] Yoshida F, Uemori T., A model of large-strain cyclic plasticity and its application to springback simulation, International Journal of Mechanical Sciences, 45 (2003) 1687-1702. https://doi.org/10.1016/j.ijmecsci.2003.10.013

[13] Lin Jianping, Hou Yong, Min Junying, et al., Effect of constitutive model on springback prediction of MP980 and AA6022-T4, International Journal of Material Forming, 13 (2020) 1-13. https://doi.org/10.1007/s12289-018-01468-x.

[14] Min Junying, Guo Nan, Hou Yong, et al., Effect of tension-compression testing strategy on kinematic model calibration and springback simulation of advanced high strength steels, International Journal of Material Forming, 14 (2021) 435-448. https://doi.org/10.1007/ s12289-020-01583-8

[15] Barlat F, Gracio J, Lee M G, et al., An alternative to kinematic hardening in classical plasticity, International Journal of Plasticity, 27 (2011) 1309-1327. https://doi.org/10.1016/ j.ijplas.2011.03.003

[16] Lee J Y, Lee J W, Lee M G, et al., An application of homogeneous anisotropic hardening to springback prediction in pre-strained U-draw/bending, International Journal of Solids and Structures, 49 (2012) 3562-3572. https://doi.org/10.1016/j.ijsolstr.2012.03.042

[17] Chaboche J L., Constitutive equations for cyclic palsticity and cyclic viscoplasticity, International Journal of Plasticity. 5 (1989) 247-302. https://doi.org/10.1016/0749-6419 (89)90015-6

Eccentricity-resistant process design and finite element analysis of deep hole cylindrical parts

LI Shuo[1,a], ZHAI Yuewen[1,b*], HE Xiaomao[1,c], QIN Sixiao[1,d] and JIN Hong[1,e]

[1]China Academy of Machinery Beijing Research Institute of Mechanical & Electrical Technology Co., Ltd., No.18 Xueqing Road, Haidian District, Beijing, P.R. China

[a]957729591@qq.com, [b]zhaiyuewen@163.com, [c]hexiaomaos@163.com, [d]sxgcxb_qsx@126.com, [e]jinhong2020@hotmail.com

Keywords: Deep Hole Cylindrical Parts, Double-Process Hot Extrusion, Extrusion Off-Center

Abstract: In this paper, finite element analysis software was used to simulate and compare the deep-hole cylindrical parts formed by single process hot extrusion and double-process hot extrusion under the condition of uneven initial circumferential temperature of billet. The results show that the deformation resistance of the material is affected by the uneven temperature, and the strength of the single process hot extrusion parts is weakened because of the long punch length, which is easy to cause the eccentricity problem of the deep-hole cylindrical parts, and the wall thickness difference of the deep-hole cylindrical parts reaches 1.23 mm. The double-process hot extrusion can greatly reduce the influence of eccentricity on the deep hole cylindrical parts, make the wall thickness distribution more uniform, and the wall thickness difference is 0.38 mm. At the same time, the double-process hot extrusion reduces the forming force by 900 T and improves the feasibility of the process.

Introduction

Deep hole cylindrical components are widely used in fields such as automobile manufacturing, electronic communication, etc. Previously, the manufacturing of deep hole cylindrical components used forging processes, combined with cutting treatment, which could meet the needs of early small-scale experiments. However, this method produces large amounts of waste material and affects the mechanical properties of the product, making it unsuitable for mass production [1,2]. Although the use of casting processes reduces cutting and improves material utilization, the structure of the deep hole cylindrical components produced by casting is loose, and the products cannot meet design requirements [3,4].

Compared to forging and casting, the backward extrusion process can improve the organizational properties of materials during the forming process, refine grains, and enhance their strength and plasticity; it achieves high processing precision and improves material utilization [5,6]; it reduces production costs and is suitable for batch production [7]. However, when producing deep hole cylindrical components by backward extrusion, eccentricity problems can easily occur. Especially when the cylindrical component is longer, the more serious the eccentricity problem becomes, leading to reduced material utilization; cutting of machining metal flow lines; and the long-term biased load on the mold affecting its service life [8], thus deep hole cylindrical components require multi-process backward extrusion forming. Factors causing forging eccentricity during backward extrusion forming include uneven blank temperature, mold position eccentricity, and inaccurate blank positioning, etc. The mold position and blank positioning can be adjusted by improving machining and mold assembly precision, so most eccentricity problems with deep hole cylindrical components in actual production are due to uneven temperature. During the forming process, the blank does not receive continuous heating, and there is a temperature difference between the punch and the die, with the mold also working on the blank. There is heat transfer during contact with the blank, leading to uneven reduction and distribution of its own temperature, especially in the circumferential direction of the blank, where there is a significant temperature distribution difference. This results in different metal deformation resistance at different parts, with the metal on the higher temperature side having better fluidity and lower deformation resistance [9]. Furthermore, due to the long design of the punch, it is prone to deformation during the forming process due to uneven force, leaning towards the higher temperature side of the blank, thus causing eccentricity issues. This article proposes a dual-process thermal extrusion forming process for typical products, which can effectively suppress the generation of eccentricity problems during the manufacturing process of deep hole cylindrical components.

With the help of finite element analysis software, the effects of different parameters can be efficiently analyzed, reducing trial-and-error costs, optimizing process paths, and minimizing the impact of eccentricity problems of deep hole cylindrical components [10].

Establishment of finite element model

Introduction of typical deep hole cylindrical parts product. The typical product studied in this article, as shown in Fig.1, is an irregular-shaped deep cylindrical component with a bottom.

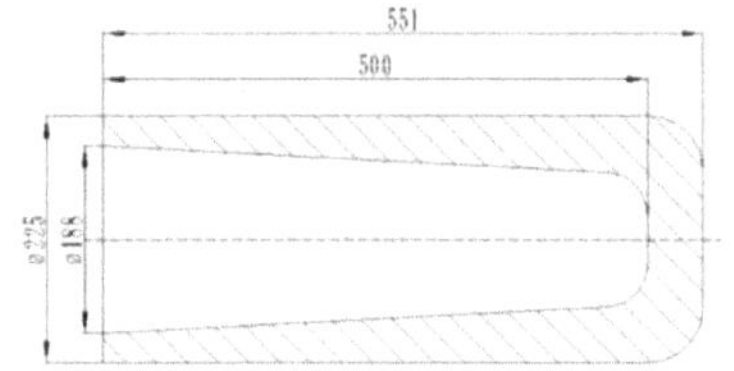

Fig.1. Drawing of forgings with bottom barrel.

Fig. 1 depicts a forging of a cylindrical component with a base. Its material is AISI-1045, with a maximum outer diameter of φ225 mm, a maximum inner diameter of φ188 mm, and a minimum wall thickness of 18.5 mm. The inner and outer shapes are similar, with the bottom being slightly thicker. It features an internal blind hole with a depth of 500 mm and a total length of the product is 551 mm.

Establishment of the geometric model. In the 3D software, the blank and the mold are modeled to generate 3D object format files. Entering the pre-processing module of the finite element analysis software, import the mold models for each process, and the blank, punch, and die form a complete model. The blank is set as a plastic body. Since the uneven temperature distribution of the blank will cause bending elastic deformation of the punch, the punch is set as an elastic body. The die is surrounded by a pre-tightening sleeve on the outside, with very little deformation, hence it is set as a rigid body.

Setting of simulation parameters. Select AISI-1045 for the blank material, and AISI-H13 for the punch and die materials. During finite element analysis, a non-uniform temperature field can be set to cause uneven deformation resistance in the blank during deformation. Different parts of the blank are set to different temperatures: the left side of the blank is 1000℃, the middle part is 1015 ℃, and the right side is 1030℃, as shown in Fig. 2. The temperature of the punch and die is set to 300℃, and the external environment is set to room temperature of 20℃.

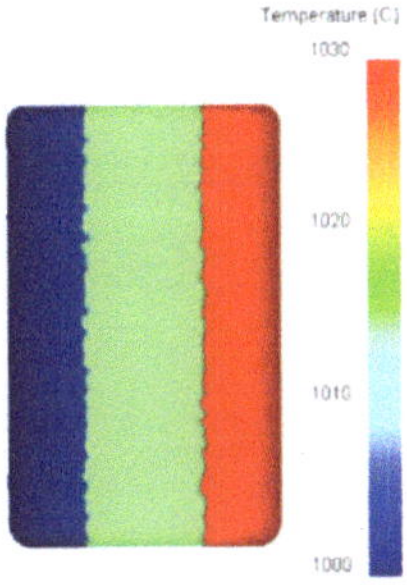

Fig. 2. Temperature distribution of different parts of the blank.

The change in the temperature of the blank during the backward extrusion process has a significant impact on the quality of the deep hole cylindrical components. It is necessary to define the heat exchange surface between the blank and the punch and die, and set an appropriate heat exchange coefficient. In actual production, there is also heat exchange during the transfer of the blank, so at the beginning of the simulation, the blank is not in contact with the mold for a period of time. The blank, being the main object of study, is divided into a dense grid. When dividing the grid for the punch and die, the grid needs to be refined at local details to better fit reality. When setting boundary conditions, the degrees of freedom in the actual motion direction are released, the movement speed of the blank is set to 0mm·s^-1, and the punch is set to a downward speed of 50 mm·s^-1. After setting up and checking for errors, the analysis is carried out using finite element software.

Simulation results analysis

Analysis of single-process thermal extrusion process. The overall forming effect of the forging is good, with the lowest temperature at the end of forging being 702℃ as shown in Fig. 3.

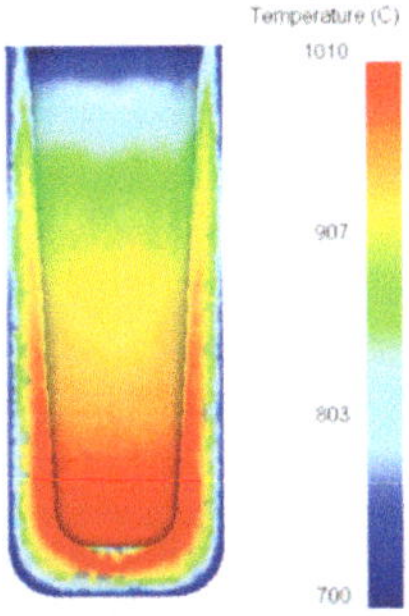

Fig. 3. Final forging temperature distribution of single process hot extrusion.

Using this scheme, the forging is at a reasonable extrusion temperature. The forming force, as shown in Fig. 4, peaks at 2570T. The forging forming force increases sharply after the end of the compound extrusion in the initial stage. The reason is that after the end of the compound extrusion, the lower end of the forging has been completely filled, the extrusion method changes from compound extrusion to backward extrusion, and the change in stress state leads to an increase in forming force. Moreover, the internal hole of the forging is a blind hole with a slope. As the extrusion continues, the projection area of the upper mold contact increases, with the projection area at the topmost end being the largest, thus requiring the greatest forming force. Therefore, it is necessary to avoid using backward extrusion to form the upper end of the forging.

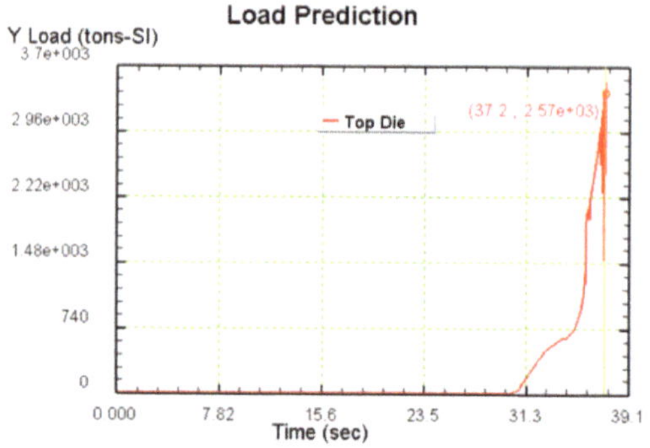

Fig. 4. Single-process thermal extrusion forming force curve.

The simulation results of the single-process thermal extrusion, as shown in Fig. 5, indicate that when there is a temperature difference of 30℃ on both sides of the blank, the wall thickness difference at 250 mm above the bottom surface is 1.23 mm.

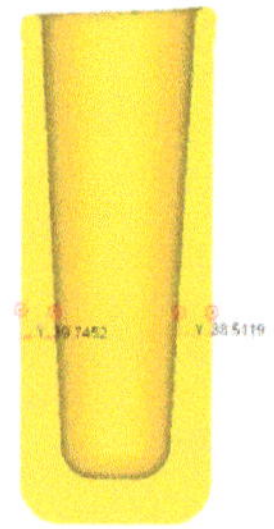

Fig. 5. Wall thickness at the bottom surface starting height 250 mm of the single process hot extrusion forging.

Analysis of double-process thermal extrusion process. In the single-process thermal extrusion, the forming force of the forging increases sharply after the initial compound extrusion ends, resulting in a high final forming force. This is because after the compound extrusion ends, the lower end of the forging is completely filled, and the extrusion method changes from compound extrusion to backward extrusion, causing the stress state to change and increase the forming force. In addition, the internal hole of the forging is a blind hole with a slope. As the extrusion continues, the projection area of contact of the upper mold becomes larger, with the largest projection area at the uppermost end, which requires the most significant forming force. Therefore, it is necessary to avoid using backward extrusion to form the upper end of the forging.

The double-process thermal extrusion scheme uses a first process compound extrusion to form the upper end of the forging, as shown in Fig. 6. By forming the upper end of the forging through a first process compound extrusion (the red box indicates the same part), the required forming force can be effectively reduced. Moreover, the punch needed for forming the upper end is thicker and shorter, which enhances rigidity. It can withstand the maximum forming force and also resist the biased load caused by temperature differences, lubrication, etc., effectively reducing the eccentricity of the forging. The second process uses a thinner punch to form a smaller inner hole.

The upper part extruded in the first process can guide the second process punch, ensuring the alignment of the punch and lateral support during the extrusion process.

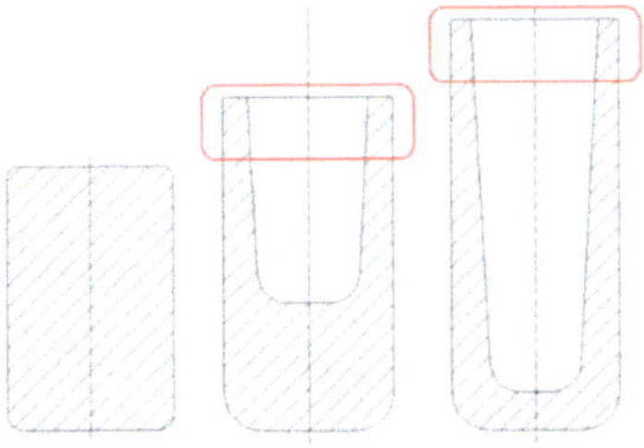

Fig. 6. Double-process thermal extrusion process diagram Blank diagram (b) First process forging diagram (c) Second process forging diagram.

The forging is well-formed, with no tendency for defects during the process, and the final forging temperature distribution is shown in Fig. 7.

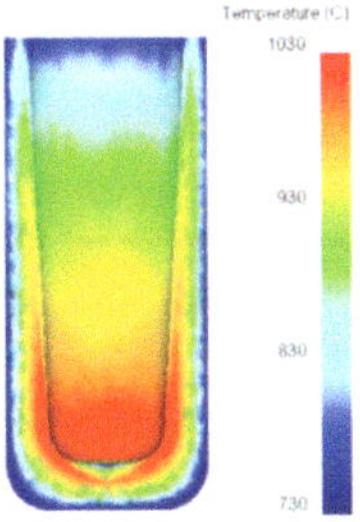

Fig. 7. Double-process extrusion final forging temperature distribution map.

The first 30 seconds of the simulation are for transfer time, with no extrusion force generated. The maximum forming force for the first extrusion is 1670T, and for the second extrusion, it is 1260T, with the maximum forming force for the double-process extrusion being 1670T. The double-process thermal extrusion forming force curve is shown in Fig. 8.

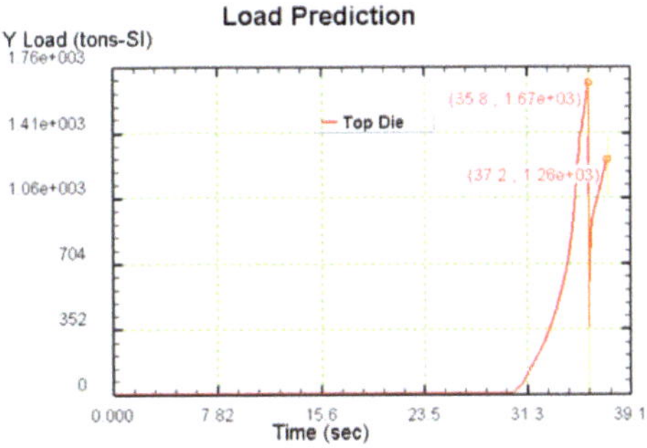

Fig. 8. Double-process extrusion forming force curve.

The overall forming effect of the forging is good, with reasonable metal flow and no tendency for defects. The lowest final forging temperature is 738℃, and the forging is at a reasonable extrusion temperature. The double-process scheme effectively reduces the extrusion forming force.

The double-process thermal extrusion scheme uses a first process compound extrusion to form the upper end of the forging, with a thicker and shorter punch needed for forming the upper end, increasing rigidity. It can withstand the maximum forming force while also resisting the biased load caused by temperature differences, lubrication, etc., effectively reducing the eccentricity of the forging. The second process uses a thinner punch to form a smaller inner hole. The upper part extruded in the first process can guide the second process punch, ensuring the punch's alignment and lateral support during the extrusion process.

The simulation results of the double-process thermal extrusion, as shown in Fig. 9, indicate that when there is a temperature difference of 30℃ on both sides of the blank, the wall thickness difference at 250 mm above the bottom surface is 0.38 mm.

Fig .9. Double-process thermal extrusion forging bottom surface raised 250mm wall thickness map.

The double-process thermal extrusion process scheme can significantly reduce the eccentricity problem caused by the temperature difference in the blank compared to the single-process thermal extrusion scheme. Without considering other factors, when the temperature difference in the blank reaches 30℃, the double-process thermal extrusion can control the forging wall thickness difference within a very small range.

The temperature distribution of the forging in the process of extrusion is shown in Fig. 10.

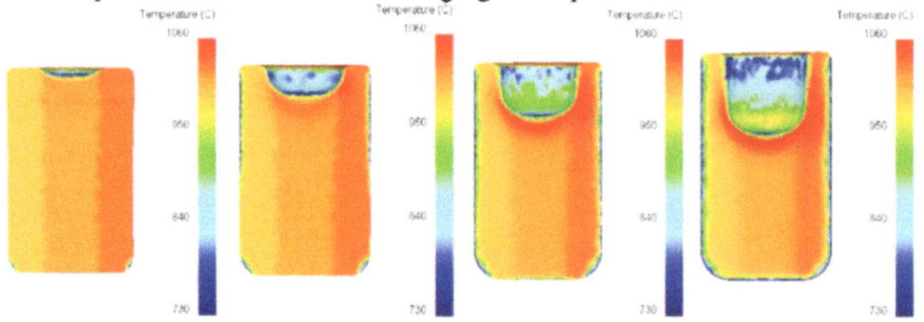

Fig. 10. Temperature distribution of forgings in the middle step of the first process of double process hot extrusion.

The temperature distribution of the forging in the second process of extrusion is shown in Fig. 11.

Materials Research Proceedings 44 (2024) 589-596 https://doi.org/10.21741/9781644903254-63

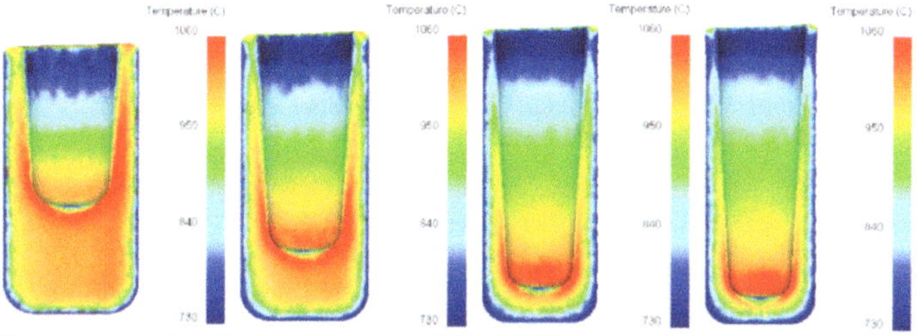

Fig. 11. Temperature distribution diagram of the middle step of the second process of double process hot extrusion.

During the first process of extrusion, there is a clear problem with the temperature distribution inside the blank, with the left side being cooler and the right side being hotter. During the second process of extrusion, due to the heat generated by the extrusion and the long period of heat exchange, the circumferential temperature distribution inside the forging is relatively uniform. During the first process, the side with the lower temperature has greater deformation resistance, and the punch at this time will be subjected to a force towards the right side, as shown in Fig. 12.

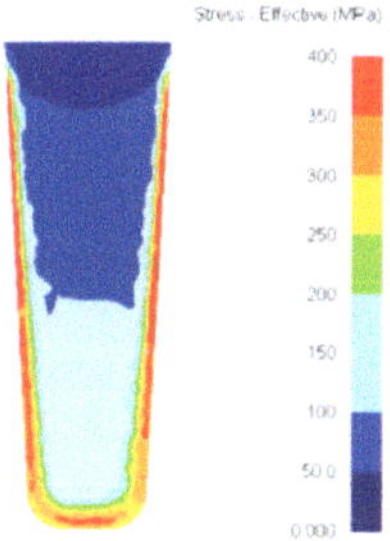

Fig. 12. Double-process thermal extrusion first process punch equivalent stress map.

In the stress field of the punch, it can be seen that the punch has a tendency to bend towards the right side due to the lateral force received, causing greater stress on the right side above. However, at this time the punch is shorter and has a larger diameter, which means its overall rigidity is strong, effectively preventing significant bending. In the second process of extrusion, since the temperature of the forging is more uniform, the punch receives less lateral force. Even if the punch is long and slender, it will not produce significant bending.

Conclusion

1) The uneven circumferential temperature change of the blank during the extrusion forming process is the main cause of eccentricity issues, affecting the wall thickness distribution of the formed parts. The double-process approach resolves the alignment problem with two differently designed punches and also improves the wall thickness distribution of the formed parts, increasing production efficiency.

2) The single-process thermal extrusion does not allocate the deformation volume reasonably, resulting in a large forming force. During the forming process, the uneven cooling of the blank's temperature and the long punch length, but insufficient strength, can easily cause eccentricity issues in the formed parts, affecting quality. The double-process extrusion scheme shows good forming effects with a smaller forming force, and the double-process setup can also reduce the eccentricity of the forging inner hole, improving forming quality.

Materials Research Proceedings 44 (2024) 589-596 https://doi.org/10.21741/9781644903254-63

References

[1] C. Zhang, H.X Sun, J.X Zhang, Research on the molding process of deep hole cylindrical parts with bottom flanges, Hot Work. Tech. 51 (2022) 82-86. https://dx.doi.org/10.14158/j.cnki.1001-3814.20220080

[2] J.B Zhang, Y. Hao, K.W. Li, Influence of hot reverse extrusion punch structure on extrusion forming force of nitrogen spring cylinder, Forging Tech. 46 (2022) 109-115. https://dx.doi.org/10.13330/j.issn.1000-3940.2021.05.017

[3] X.H. Ye, Numerical simulation and experimental research on hot precision forging of press cylinder block, Anhui University of Technology, Maanshan, 2017. https://dx.doi.org/cdmd/j.cnki.2.1017.246907

[4] Y.H. Qiao, C. Song, Z.S. Wang, Casting process analysis of engine cylinder block, Hot Work. Tech. 43 (2014) 65-67. https://dx.doi.org/10.14158/j.cnki.1001-3814.2014.07.019

[5] L.L. Jia, S.Z. Zhang, Extrusion Process and Die, China Machine Press, Beijing, 2004.

[6] H.J. Choi, J.H. Choi, B.B. Hwang, The forming characteristics of radial–backward extrusion, J. Mater. Process. Tech. 113 (2001) 141-147. https://doi.org/10.1016/S0924-0136(01)00703-8

[7] S.J. Zhou, Study on precision isothermal forming process of typical structure of 7A04 aluminum alloy, Aerospace Manuf. Tech. 3 (2010) 34-37. https://dx.doi.org/sun htgy/j.cnki.0.2010-03-012

[8] F. Wang, C.W. Cui, F. Fang, Research on shrink forming process of 7A04 alloy cylinder, Hot Work. Tech. 40 (2011) 98-100. https://dx.doi.org/10.14158/j.cnki.1001-3814.2011.11.036

[9] C.C. Zhao, H.B. Hao, L.M. Hu, Influence of uneven billet temperature on quality of high-pressure gas cylinder back-extruding parts, Forg. Stamp. Tech. 44 (2019) 88-97. https://dx.doi.org/10.13330/j.issn.1000-3940.2019.06.013

[10] R.S. Sharma, V. Upadhyay, K.H. Raj, Neuro-fuzzy modeling of hot extrusion process, Indian J. Eng. Mater. Sci. 16 (2009) 86-92.

Metal Forming 2024 | Materials Research Forum LLC
Materials Research Proceedings 44 (2024) 597-605 | https://doi.org/10.21741/9781644903254-64

Integrating Numerical and experimental analyses for the realization of pre-detection of delayed fractures on punched surfaces of ultra-high-strength steel sheets

UEDA Yuya[1,a], EGUCHI Jin[1,b], and MATSUNO Takashi[1,c *]

[1]Faculty of Engineering, Tottori University 4-101 Koyama-cho-minami, Tottori-city, Tottori 680-8552, Japan

[a]yuya.flower0307@gmail.com, [b]vintagejkj@gmail.com, [c]matsu@tottori-u.ac.jp

Keywords: Punching, Finite Element Simulation, Residual Stress

Abstract. A key challenge encountered in finite element (FE) punching simulations is accurately capturing the arrest or delay in crack propagation during punch penetration. Conventional methods relying on element deletion demonstrate inadequacies in representing crack arrest, particularly complicating the simulation of the punched surface states in scenarios with minute clearances. This study introduces a novel approach for fracture representation in FE punching simulations, wherein cracks were integrated as perimeter lines during the remeshing process. Utilizing observation-driven crack propagation paths and arrest behaviors, we analyzed the residual stress on punched surfaces under minimal clearance conditions. Results indicate that for 22MnB5 hot-stamped steel, the residual stresses within the interior side of the burnished region can exceed 1 GPa in cases with a 10-μm clearance.

Introduction

Punched surfaces of ultra-high-strength steel sheets (UHSSs) exhibit significant residual tensile stresses [1, 2] and reduced ductility compared to their as-received conditions [3]. Consequently, in automotive applications, concerns arise as the tensile strength of UHSSs exceeds 1.5 GPa, leading to increased apprehensions regarding delayed fractures induced by hydrogen embrittlement on punched surfaces [1, 2, 4]. When conventional punch-die clearances surpass 10%t, where %t represents the ratio of the clearance relative to the sheet thickness, these delayed fractures can be attributed to pronounced tensile residual stresses on the punched surfaces [1, 2, 4]. The process of actual press forming emphasizes monitoring the potential for delayed fractures and comprehending the associated residual stress on the punched edges.

Matsuno et al. [5] revealed that delayed fractures are absent in die-quenched steel sheets with a tensile strength of 1.5 GPa on punched surfaces with minute clearances of less than 10 μm. The prevailing theory posits that this absence results from the compressive residual stress applied to the burnishing process within these minute clearances. However, another report have suggested that UHSSs may exhibit tensile residual stresses within the burnished surfaces [6], introducing ambiguity about the definitive influence of the compressive residual stress at such minute clearances.

Finite element (FE) simulations are crucial for assessing residual stress on punched surfaces, yet they encounter obstacles in accurately capturing the fracture process in cases involving minute clearances and repeated crack progression and arrest. Currently, a ductile fracture model applicable to such conditions of compression has not been sufficiently formulated. Our group endeavored to inversely derive ductile fracture loci from FE punching simulations [7], revealing that the shapes of the identified ductile fracture loci (design curves) significantly deviate from those traditionally recognized. Moreover, in several commercially available FE codes that employ the element deletion technique, fractures invariably propagate instantly through the entire sheet thickness without pausing, due to the inability of these codes to replicate crack arresting behavior that leads to the formation of secondary burnished zones [8].

In this study, a FE punching simulation employing a crack line update method [9] is introduced. This method, distinctively devoid of ductile fracture loci, heuristically updates crack lines to closely mimic the actual observed profile of the punched surface. Simulations performed on quenched 22MnB5 steel sheets with a minute clearance of 10 μm revealed an internal residual stress distribution exceeding 1 GPa. Such elevated stress levels are posited as a potential cause for delayed fractures on punched surfaces. Intriguingly, punched holes in this study exhibited delayed fractures under bending conditions, even though JSC1180Y steel, boasting a tensile strength of 1180 MPa, previously showed no signs of delayed fractures on surfaces punched with similar minute clearance [5].

Material

The test material was a quenched 22MnB5 steel sheet employed in the automobile hot-stamping process, whose mechanical properties are detailed in Table 1. The tensile strength of this material surpasses 1.5 GPa.

Table 1. Mechanical properties of 22MnB5.

Yield stress [MPa]	Tensile strength [MPa]	Uniform elongation [%]	Total elongation [%]
951.4	1600.9	4.78	7.19

For the shearing simulation, the flow stress curve within the large deformation region must be established. Consequently, this study employed a real-time diameter measurement type tensile test combined with FE simulation [10-12] to acquire the flow stress curve extending to fracture. The stress curve was characterized by the modified Swift law [10-12] as follows:

$$Y(\bar{\varepsilon}_p) = (1-a)Y_s(\bar{\varepsilon}_p) + a\{Y_l(\bar{\varepsilon}_p) - 7.134\}, \tag{1}$$

where

$$Y_s(\bar{\varepsilon}_p) = 1672(\bar{\varepsilon}_p + 1.460 \times 10^{-3})^{0.06}, \tag{2}$$

$$Y_l(\bar{\varepsilon}_p) = 1574(\bar{\varepsilon}_p + 2.503 \times 10^{-4})^{0.04}, \tag{3}$$

$$a = 0.5\left\{\tanh\left(10(\bar{\varepsilon}_p - 0.06)\right) + 1\right\}. \tag{4}$$

In Eq. 1, $Y_s(\bar{\varepsilon}_p)$ represents the Swift work-hardening law for the lower strain region, while $Y_l(\bar{\varepsilon}_p)$ represents the Swift work-hardening law for the larger strain region. For the yield stress $Y(\bar{\varepsilon}_p)$ in the modified Swift law, $Y_s(\bar{\varepsilon}_p)$ and $Y_l(\bar{\varepsilon}_p)$ are smoothly connected at an equivalent plastic strain $\bar{\varepsilon}_p$ of 0.06 using a switching function a.

Shearing simulation with crack-line update method

This analysis was performed using the commercial Abaqus/Standard FE software. Fig. 1 illustrates the FE model configured for the shearing simulation. The specimen was modeled using axisymmetric two-dimensional (2D) quadrilateral elements. In regions undergoing substantial deformation, a dense mesh with an element side length of 0.0025 mm was applied, while other areas were meshed with the larger size of 0.2 mm. The punch and die edges were modeled with 0.02 mm radii. The clearance between the punch and die was 0.8%t.

A remeshing sequence was deemed essential for the accuracy of the shearing simulation owing to the significantly large expected strain, which could lead to element distortion. Accordingly, remeshing was performed at each increment of 0.01 mm of punch penetration to maintain the integrity of the mesh throughout the simulation process.

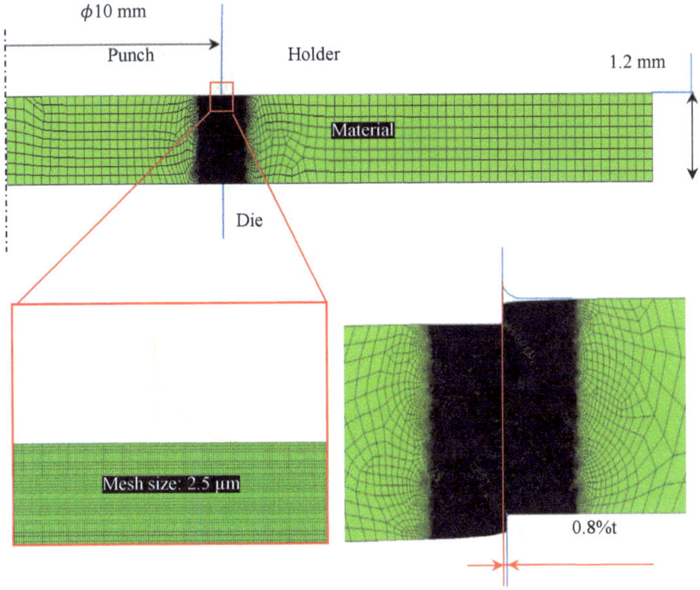

Fig. 1. Boundary conditions and initial meshes for punching simulation.

A notable aspect of this study is the strategic addition of crack lines originating close to the edges of the punch and die. These crack lines were systematically updated along the periphery of the workpiece with each remeshing sequence, as illustrated in Fig. 2. This technique, designated as the "crack-line update method," [9] is a key feature of our simulation approach.

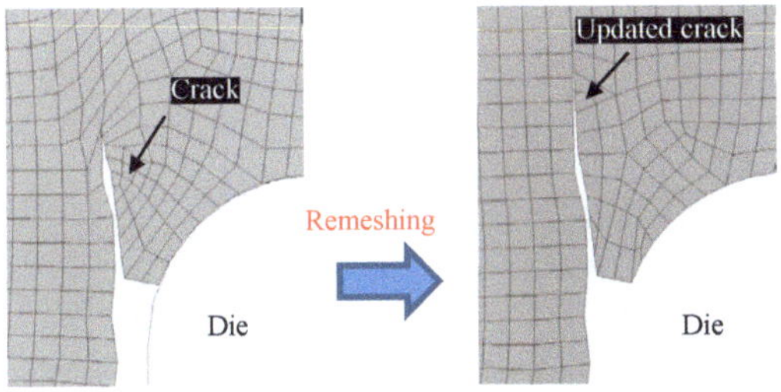

Fig. 2. Fracture expression in FE simulation with crack-line update method.

Materials Research Proceedings 44 (2024) 597-605
https://doi.org/10.21741/9781644903254-64

Assimilation approach for crack-line update method

We devised an assimilation technique precisely capturing the actual profile of the fracture zone observed on the cross-section of punched specimens in our residual stress analysis, separate from prediction methodologies. Fig. 3 illustrates the observation of the actual punched surface for this analysis, with the reference profile derived from the cross-section (Fig. 3(a)). This technique mandated periodic updates to the crack lines to accurately mirror the progression of observed fractures, emulating the fracture zone profile.

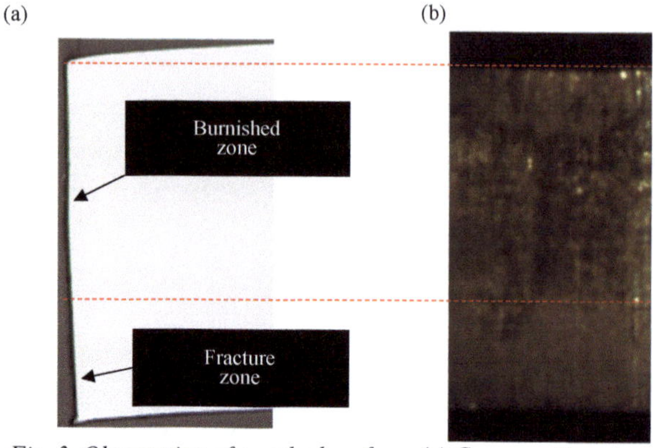

Fig. 3. Observation of punched surface. (a) Cross-section view, (b) Front view.

The progression of these crack lines was determined using empirical data from half-punched specimens, as depicted in Fig. 4. Notably, no cracks are observed around the punch edges. Consequently, only the cracks initiated around the die edge were included in the FE simulation. Internal cracks, such as those in Fig. 4, were omitted from evaluating crack length, considering their origin from propagation in the circumferential direction around the hole, distinct from the assessed primary crack growth.

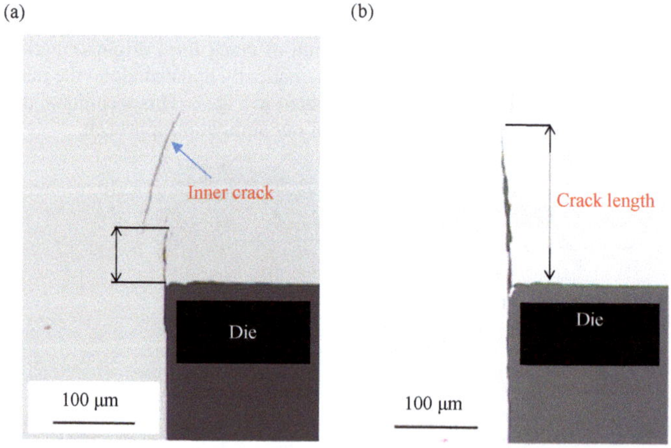

Fig. 4. Cracks observed on the cross-section of a half-punched specimen: punch penetration of (a) 0.24 mm and (b) 0.36 mm.

Materials Research Forum LLC
https://doi.org/10.21741/9781644903254-64

Fig. 5 demonstrates a linear correlation between crack length and punch penetration up to a 0.6 mm threshold. Beyond this penetration depth, the crack growth pattern became erratic, with non-uniform expansion around the circumference of the hole. Importantly, some specimens, subjected to punch penetrations beyond 0.6 mm, fractured during preparation, indicating complete crack penetration. These half-punched specimens contributed only to the incremental extension of crack length, not determining the directionality of crack-line updates.

Observational crack-propagation uncertainty implies variability in crack length due to crack arrest phenomena, particularly prevalent under conditions of minute clearance. Consequently, certain approximations were necessary when defining the additional crack length in each update sequence to integrate the observed crack progression into our simulation accurately.

This study performed crack-length determination through two different approaches. The first approach envisioned cracks growing continuously and uninterrupted, providing a clear view of progression without the complexities of arrest. This method used the correlation between punch penetration and crack length (Fig. 5) to determine crack lengths through linear interpolation. The second approach adopted a more nuanced view by incorporating the concept of crack arrest, which was achieved by introducing stopping points in crack growth rather than a straightforward linear interpolation.

In the second approach, the update sequence for the crack lines paused when the propagation trajectory approached the boundary defined by the punch-die clearance space. In the first approach, crack lines tended to follow the material surfaces as they rotated because of punch penetration, often causing them to extend beyond the designated clearance area. Hence, a crack-arrest sequence was employed in the second approach for its proactive prevention. Following the arrest phase, the crack lines were updated to align with the punch penetration and crack length relationship, as depicted in Fig. 5. This process resulted in the arrest of crack-line propagation within the steps corresponding to punch penetration increasing from 0.37 mm to 0.76 mm, an effective increment of 0.31 mm. When the punch penetration reached 0.76 mm, the material completely separated.

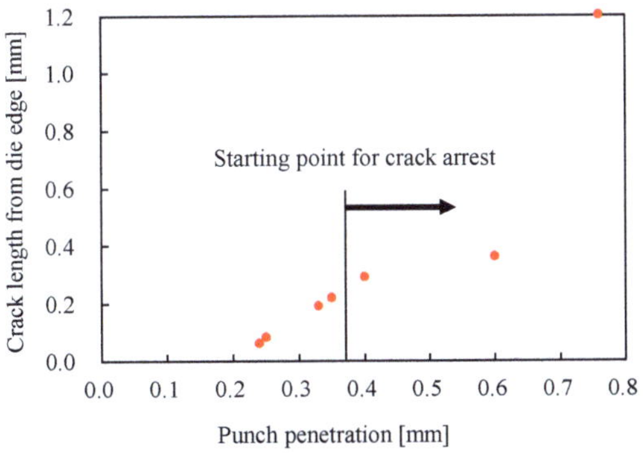

Fig. 5. Relationship between crack length and punch penetration for half-punched specimens.

Materials Research Forum LLC
https://doi.org/10.21741/9781644903254-64

Results and discussion

Fig. 6 depicts the FE simulations (circumferential stress map) with the actual cross-sectional image. Fig. 6(a) represents the case without crack arrest, while Fig. 6(b) represents the case with crack arrest.

A significant deviation from the actual cross-sectional profile is observed in the simulation case without crack arrest (Fig. 6(a)). A large and irregular protrusion is formed due to the orientation of crack lines during punch penetration. Without crack arrest, the cracks excessively expand, deviating from the clearance region and extending into the interior of the punch side. The deviated crack is subsequently pressed by the punch and rotated along the direction of punch penetration, resulting in an abnormal protrusion. This protrusion significantly influences the residual stress distribution, inducing a compressive stress state around it. It leads to high tensile residual stress in the burnished region as a response to the localized compressive stress concentration.

Conversely, when crack arrest is incorporated into the simulation (Fig. 6(b)), protrusion formation is notably absent, resulting in a closer approximation of the actual punched surface profile. However, the simulation did not completely replicate the effects of an insufficient blank holder force, resulting in a slight discrepancy between the simulated and actual raised punched surface profiles. Therefore, the simulation provided a significant first step in understanding cases with minute clearances. Particularly notable was the simulation revelation of tensile stresses exceeding 1 GPa in the interior of the burnished zone, contrasting sharply with the low or compressive stresses on the surface, providing crucial insights for delayed fracture analysis. This outcome emphasizes the need for further investigations into the mechanisms governing such stress distributions, which are crucial for predicting the service life of punched components.

To validate the circumferential residual stress as numerically simulated in the FE analyses, an X-ray diffraction (XRD) measurement was conducted at the center of the punched surface. This measurement was performed using a μ360s instrument (Pulstec Industry Co., Ltd.) equipped with a 0.8-mm beam spot and 0.5-mm collimator size; it yielded a circumferential residual stress of -149 MPa. The circumferential stress simulated in the case with crack arrest was approximately 10 MPa, averaged over the corresponding surface area in alignment with the measurement.

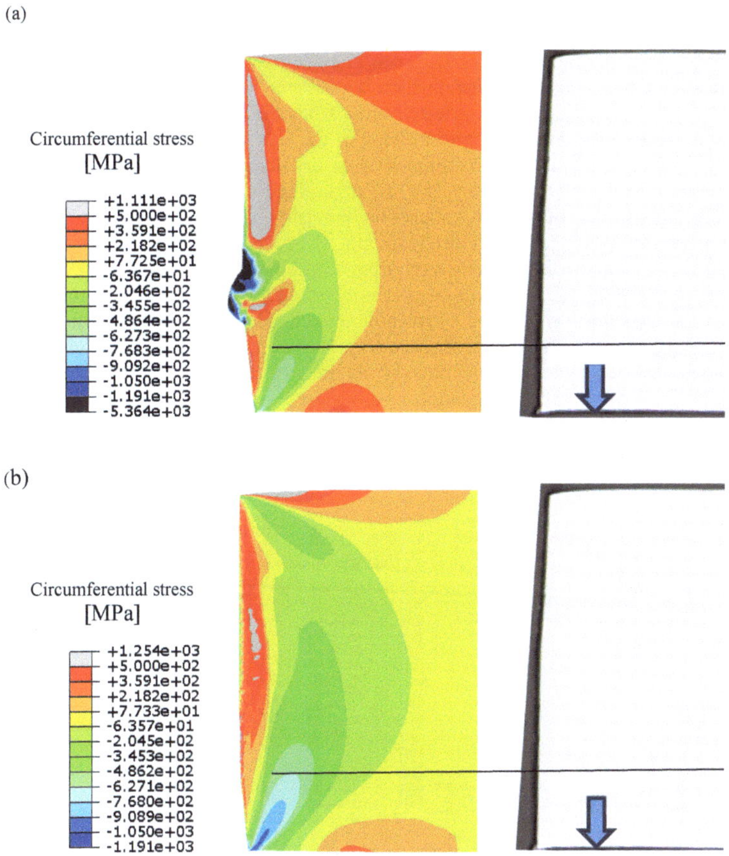

Fig. 6. Circumferential residual stress FE-simulated with crack-line update method. The actual punched surface shown in Fig. 3 is also shown for reference. (a) Case without crack arrest, (b) Case with crack arrest.

The simulation indicated higher residual stress levels than the actual measurements. This discrepancy can be largely attributed to the flattening process designed to eliminate concavities during the creation of the burnished zone [8]. The simulated residual stresses were considered valid despite this error source. This is particularly true because the high tensile residual stresses, critical for actual delayed fracture, were broadly distributed and more pronounced. Essentially, the localized residual stress near the surface was found to have a limited impact. As our methodology advanced, the precision of replicating the fracture surface improved significantly, enhancing the accuracy of the residual stress distribution analysis.

The validation of the FE-simulated stresses in the thickness direction remained underexplored in our study, primarily due to the significant scattering observed in XRD measurements of residual stress. This variability constrained our ability to validate the analytical results related to this aspect robustly. Consequently, we present our stress distribution analysis outcomes in Fig. 7 as illustrative data rather than definitive conclusions.

A notable concentration of compressive stress is evident around the protrusion in the absence of crack arrest (see Fig. 7(a)). This phenomenon is attributed to the residual stress induced by the punch penetration, persisting in the affected area. Conversely, such pronounced compressive stresses are not observed with crack arrest owing to the absence of protrusions (Fig. 7(b)). The punched surface generally exhibits compressive or low-tensile residual stress, excluding the protrusion region. Notably, residual tensile stresses ranging from 200 to 300 MPa are detected in certain areas within the burnished zone. The precision of this analysis warrants further investigation, presenting an area for future research.

Simulating crack progression exclusively from the die side and subsequent residual stress analysis was challenging due to the absence of established ductile fracture criteria under low or negative stress triaxiality conditions. Typically, punch-side cracking was inevitable in FE simulations using conventional ductile fracture loci. However, the proposed assimilation method facilitates such analyses without relying on a ductile fracture model. This significant advancement is the primary focus of our rapid publication.

In the future, we aim to further refine our assimilation technique by incorporating actual punched surface irregularities, such as surface tilting and extensive burr formation, to improve the fidelity of the simulation to real-world conditions.

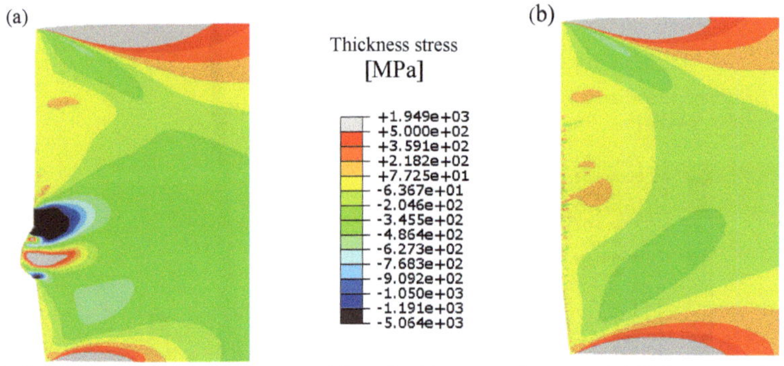

Fig. 7. Comparison of thickness stress with and without crack arrest. (a) Without crack arrest, (b) With crack arrest.

Conclusion

In this study, we introduced a FE simulation technique capable of accurately replicating the punched surface profile of hot-stamped 22MnB5 UHSS without utilizing a damage model. This approach allowed successful visualization of residual stresses in cases of minute clearance.

During the FE simulations, straightforward linear interpolation of crack propagation behavior led to the development of a protrusion on the punched surface. Consequently, the residual stress analyzed in this case differed from the actual residual stress observed on the punched surface. Nevertheless, by implementing a crack arrest strategy, we were able to prevent the formation of this protrusion. The residual stresses determined were in close agreement with empirical measurements. Future efforts should focus on further refining the replication of crack halting behavior, as well as integrating considerations of ductile fracture models.

The insights gained from this study contribute to the broader discussion on the delayed fracture stress threshold of punched surfaces in UHSSs, augmenting separate investigations into delayed fractures on punched surfaces carried out by the authors. To gain a more comprehensive understanding of the findings of this paper, readers are encouraged to consult our previous publications.

References

[1] K.-i. Mori, Y. Abe, K. Sedoguchi, Delayed fracture in cold blanking of ultra-high strength steel sheets, CIRP Annals, 68 (2019) 297-300. https://doi.org/10.1016/j.cirp.2019.04.111

[2] T. Matsuno, Y. Sekito, E. Sakurada, T. Suzuki, K. Kawasaki, M. Suehiro, Resistance of hydrogen embrittlement on hot-sheared surface during die-quench process, ISIJ Int. 54 (2014) 2369-2374. https://doi.org/10.2355/isijinternational.54.2369

[3] T. Matsuno, N. Kinoshita, T. Matsuda, Y. Honda, T. Yasutomi, Microvoid formation of ferrite-martensite dual-phase steel via tensile deformation after severe plastic shear-deformation, ISIJ Int. 63 (2023) 941-949. https://doi.org/10.2355/isijinternational.ISIJINT-2023-012

[4] M. Yoshino, Y. Toji, S. Takagi, K. Hasegawa, Influence of sheared edge on hydrogen embrittlement resistance in an ultra-high strength steel sheet, ISIJ Int. 54 (2014) 1416-1425. https://doi.org/10.2355/isijinternational.54.1416

[5] T. Matsuno, Y. Ueda, T. Takahashi, T. Hama, T. Hojo, Y. Shibayama, M. Ridha, Y. Okitsu, M. Takamura, Delayed fracture stress thresholds degraded by shear punching process in ultra-high-strength steel sheets: Analysis using an in-plane bending test with numerical assimilation, J. Manuf. Process. 119 (2024) 1005-1021. https://doi.org/10.1016/j.jmapro.2024.04.013

[6] T. Matsuno, Y. Ochiai, Y. Okitsu, M. Iga, A. Kohri, T. Mikami, Surface and interior residual stress analysis of a deep-drawn 1180-MPa class ultra-high strength steel sheet with scratch marks, Int. J. Adv. Manuf. Tech. 116 (2021) 2873-2884. https://doi.org/10.1007/s00170-021-07675-2

[7] T. Matsuno, K. Nakagiri, T. Matsuda, T. Tanaka, T. Yasutomi, H. Shoji, M. Ohata, Identification of ductile fracture design curve for hardened quasi-brittle AISI-D2 tool steel to predict shearing tool failure, J. Mater. Process. Tech. 307 (2022). https://doi.org/10.1016/j.jmatprotec.2022.117680

[8] T. Matsuno, A. Seto, M. Suehiro, Y. Kuriyama, H. Murakami, FEM analysis of high strength steel sheets up to fracture and its application to piercing process, Proceedings of IDDRG2012, (2012) 134-140.

[9] T. Matsuno, M. Takamura, S. Mihara, T. Koda, K. Tsujioka, Development of the crack-line-update method for two-dimensional piercing simulations, Journal of Physics: Conference Series, 1063 (2018) 012157. https://doi.org/10.1088/1742-6596/1063/1/012157

[10] T. Matsuno, T. Fujita, T. Matsuda, Y. Shibayama, T. Hojo, I. Watanabe, Unstable stress-triaxiality development and contrasting weakening in two types of high-strength transformation-induced plasticity(TRIP) steels: Insights from a new compact tensile testing method, J. Mater. Process. Tech. 322 (2023) 118174. https://doi.org/10.1016/j.jmatprotec.2023.118174

[11] T. Matsuno, D. Kondo, T. Hama, T. Naito, Y. Okitsu, S. Hayashi, K. Takada, Flow stress curves for 980MPa- and 1.5GPa-class ultra-high-strength steel sheets weakened under high-stress triaxiality, Int. J. Mech. Sci. 261 (2024) 108671. https://doi.org/10.1016/j.ijmecsci.2023.108671

[12] T. Matsuno, T. Hojo, I. Watanabe, A. Shiro, T. Shobu, K. Kajiwara, Tensile deformation behavior of TRIP-aided bainitic ferrite steel in the post-necking strain region, Sci. Tech. Adv. Mater.: Methods 1 (2021) 56-74. https://doi.org/10.1080/27660400.2021.1922207

Multi-scale numerical modeling of the multi-pass rolling texture evolution by using crystal plasticity and finite element method

BYRSKA-WÓJCIK Dorota[1,a *], WIERZBANOWSKI Krzysztof[1,b], WRÓBEL Mirosław[1,c] and BŁONIARZ Remigiusz[1,d]

[1]AGH University of Krakow, al. Mickiewicza 30, 30-059 Krakow, Poland

[a]byrska@agh.edu.pl, [b]Krzysztof.Wierzbanowski@fis.agh.edu.pl, [c]mwrobel@agh.edu.pl, [d]bloniarz@agh.edu.pl

Keywords: Multiscale Model, Multi-Pass, Finite Element Method, Crystal Plasticity, Asymmetric Rolling, Aluminum, Texture, Product Homogeneity

Abstract. The finite element method (FEM) combined with the crystal plasticity (CP) model was used to predict crystallographic texture evolution during multi-pass asymmetric rolling of the polycrystalline aluminum with tilted entry into the rolling gap. The FEM macro-scale model of the rolling process with various asymmetry was performed using Abaqus commercial software. The asymmetry was introduced by difference in the diameters of cooperating rolls and various tilt angles of the rolled bar entry into the rolling gap. Predicted stress distribution over sample volume was used as the boundary conditions in the crystal plasticity meso-scale model of the plastic deformation. The CP meso-scale model developed by Leffers and Wierzbanowski [1], [2] was used as a post-processing procedure. The predicted evolution of the rolling texture distribution on the rolled bar thickness turned out to be in good agreement with that measured for specimens rolled with the same parameters as the ones used for modeling. Impact of the rolling parameters on the texture evolution was found.

Introduction

Rolling is one of the most used processes in metal forming due to its simplicity and economic aspects. Traditionally, symmetric rolling has been used. However, it has been shown that the use of appropriate and controlled asymmetry can bring tangible benefits. Thus, in this way it is possible to significantly homogenize the microstructure and crystallographic texture (shortly texture) distribution throughout the product thickness. Therefore, asymmetric rolling of various metals has been recently studied, and a vast literature on this subject is available (e.g., [3]–[5]). According to the Web of Science, approximately 160 publications on this topic were published annually in 2022 and 2023, and during this period the average annual number of citations was over 5,000. The texture effect has been depicted in numerous papers (e.g. [2], [4], [6]–[9]) and improvement in texture gradient by tilt rolling has been patented [10]. The influence of rolling asymmetry on mechanical properties and microstructure has also been frequently characterized (e.g. [3], [6], [7], [11], [12]). However, most of these papers focused on experimental studies, and numerical analysis is usually limited to single-pass rolling processes and often involve only the horizontal introduction of the rolled strip into the rolling gap. The present paper is devoted to development of a multi-scale numerical model of multi-pass asymmetric rolling for which the asymmetry is caused by both a difference in the diameters of cooperating rolls and various tilt angles of the rolled bar entry into rolling gap.

Experiment and material

The 6-pass rolling process for pure aluminum was done in laboratory conditions using a 4-high rolling mill. The asymmetry was introduced by a difference in rolls diameters ($\boldsymbol{d_1}$ and $\boldsymbol{d_2}$ in Fig. 1) and by tilting the rolled bar into rolling gap (the tilt angle $\boldsymbol{\alpha}$ in Fig. 1). The diameter of the upper roll was 100 mm while the diameter of the lower one was equal to 100 or 120 mm. Three values of the tilt angle were used, i.e., $\alpha = 0°$, $30°$ or $-30°$ where the positive sign refers to the tilt towards the smaller, upper roll and negative one towards the bigger, lower one (see Fig. 1). The initial sample thickness was 5 mm. The roll distance set for each pass (see table 1) did not change significantly during rolling.

Table 1. Distance between rolls in each pass of rolling.

Pass number	1	2	3	4	5	6
Distance between rolls [mm]	4.2	3.51	2.8	2.09	1.43	0.99

Multi-scale numerical model

The developed model combines a macro-scale finite element (FE) description with a meso-scale crystal plasticity (CP) approach. Thus, the stress/strains fields predicted by the macro-scale simulations of multi-stage asymmetric rolling with the parameters used in the experiment were applied to determine the boundary conditions for the meso-scale CP numerical model of plastic deformation. This way it was possible to calculate the rolling texture evolution. The scheme of the multi-scale approach can be seen in Fig. 1.

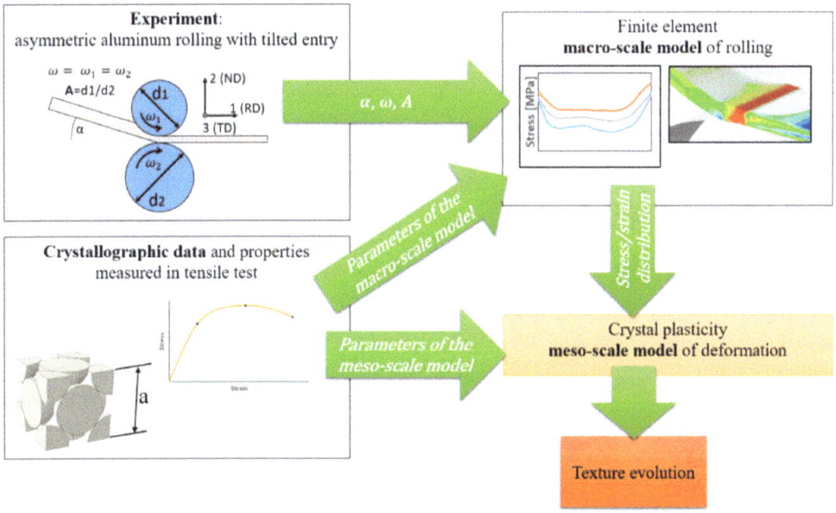

Fig. 1. The idea of the multi-scale approach used.

Macro-scale model

The three-dimensional macro-scale model of rolling was prepared by using commercial Abaqus software. The reference sets, rolling geometry and finite element mesh are presented in Fig. 2. The mesh consisted of 59 004 elements of type C3D8R (linear brick element, with integration reduced to one point).

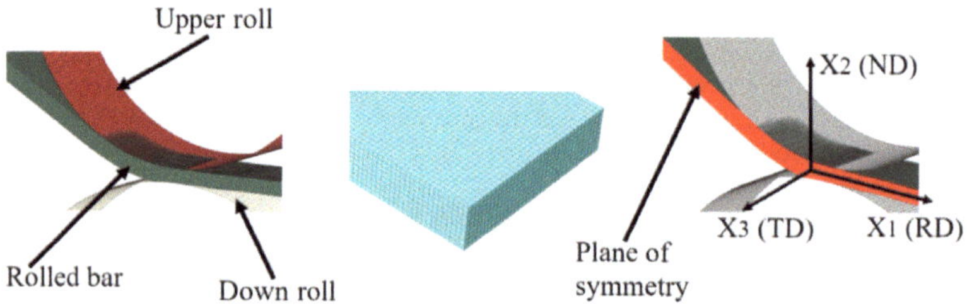

Fig. 2. Geometry of the Abaqus model (on the left), FE mesh (in the middle) and plane of symmetry (on the right).

Modelling of the multi-stage process (6 passes) requires transferring the deformed mesh and resulting stresses and strains fields after each pass to the initial conditions of the next one. Rolls are simulated as rigid body with constant angular velocity. No lubrication during rolling experiment was applied, thus the friction coefficient between the rolls and rolled bar was taken as equal to 0.3. The mirror symmetry at the center of the rolled bar along the rolling direction was assumed (symmetry plane perpendicular to X_3 = TD, see Fig. 2), thus only half of bar width was taken into account. All variants' simulations were performed in ACK Cyfronet using supercomputer Ares, but calculations can be also done on standard PC computer. The estimated time of calculations for one variant of a 6-pass rolling should not be longer than few days, thus this model could be recommended for usage in industry as tool for rolling asymmetry optimization.

Meso-scale model

The utilized meso-scale model was developed by Leffers and Wierzbanowski [1], [2], [13]. It is based on crystal plasticity theory. The relationship between the stress and strain in the sample and in the grain is the following:

$$\dot{\sigma}_{ij} = \dot{\Sigma}_{ij} + \alpha G\left(\dot{E}^p_{ij} - \dot{\epsilon}^p_{ij}\right), \tag{1}$$

where σ_{ij} and ϵ^p_{ij} are the local stress and plastic strain of a grain and Σ_{ij} and $\mathrm{E}^{\mathrm{p}}_{\mathrm{ij}}$ are their global equivalents. G is a shear modulus of the material, and α is the elasto-plastic accommodation parameter. The Einstein summation notation is used in this and following formulas. In a face-centered-cubic (fcc) metal with a high stacking fault energy (like aluminum) deformed at room temperature, crystallographic slip of the perfect dislocation with Burgers vector of (*a*/2) <110> on the {111} planes is dominant (*a* – is the lattice parameter). According to Schmid's law [14], slip occurs when the resolved shear stress τ reaches the critical value τ_{cr}:

$$\tau = m_i n_j \sigma_{ij} = \tau_{cr}, \tag{2}$$

where m is the slip direction and n is the slip plane normal vector (in our case <110> and <111> vectors, respectively). The value of critical resolved shear stress τ_{cr} increases with deformation due to the material hardening which is anisotropic. Thus, hardening of the slip system i depends on shear glides on other slip systems j ($d\gamma^j$):

$$d\tau^i_{cr} = H^{ij} d\gamma^j, \tag{3}$$

where H^{ij} is the strain hardening matrix.

The total plastic strain of a grain I is the sum of glides on the active slip systems (s) combination:

$$\Delta\epsilon_{ij}^{p(I)} = \Sigma_s \left[\frac{1}{2}\left(m_i^s n_j^s + m_j^s n_i^s\right)\delta\gamma^s\right]. \tag{4}$$

Thus, the plastic strain increment of the polycrystalline sample can be expressed as the average of its grains' strains:

$$\Delta E_{ij}^{p} = \frac{1}{V_0}\Delta\epsilon_{ij}^{p(I)} V^I, \tag{5}$$

where V_0 is the sample volume and V^I is the volume of the grain I.

Rotation of the grain lattice can be calculated by:

$$\delta\omega_{ij} = -\frac{1}{2}\left(m_i n_j - m_j n_i\right)\delta\gamma. \tag{6}$$

This rotation results in an increase in the share of preferential crystallographic orientations (i.e., in development of the crystallographic texture). During simulation the texture developed from 1000 initially randomly oriented grains. It was presented as the Orientation Distribution Function (ODF) in the Euler orientation space. Bunge convention of the Euler angles rotations was used [15]. Description of the mathematical details of the model used can be found elsewhere [1], [13].

Values of the stress tensor components evolution obtained from the macro-scale FEM model for movement of the selected integration point through the deformation zone in 6 passes were introduced in the equation (1). The first rolling pass evolution of the stress components can be seen in Tables 3 and 4, as an example. The final textures both models predicted and measured are shown in Tables 3 and 4 as the $\Phi_2 = 0°$ ODF sections. The m3m (O_h) cubic symmetry of the crystal and no sample symmetry were assumed during ODF calculations.

Results

The von Mises stress field and selected stress field components for the first pass of symmetric and asymmetric rolling can be seen in Table 2. Changes in selected stress components on the rolled bar lower surface in the rolling gap of the first pass can be also seen in Tables 3 and 4. They clearly show a trend typical for all rolling passes. The rolling asymmetry effect on the stress field isclearly visible.

Differences in the evolution of the stress and strain fields caused by rolling asymmetry result in differences in the development of the crystallographic texture, as expected. One can see this in the ODFs cross sections from Tables 3 and 4 as a value of the maxima shift along Φ. This effect is visible for both simulated and measured textures._Although quantitative differences between these results are visible, the qualitative nature is the same. Considering the simplicity of the model used to simulate the texture development, the simulation results can be considered satisfactory.

Table 2. Stress distribution for the first pass of the rolling.

Rolling variant	Mises stress [MPa] 0 11 22 32 43 54 65 76 87 98 108 119 130	σ_{12} [MPa] -71 -59 -47 -36 -24 -12 0 12 24 35 47 59 71	σ_{22} [MPa] -285 -255 -224 -194 -163 -133 -102 -72 -42 -11 19 50 80
$A = 1.0$ $\alpha = 0°$	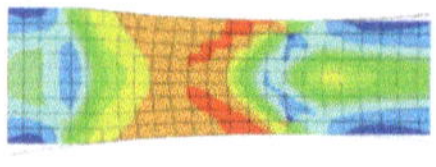		
$A = 1.0$ $\alpha = 30°$	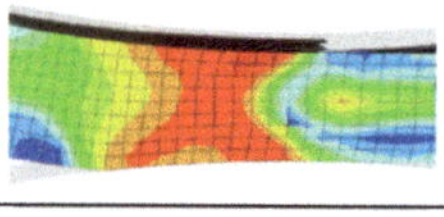	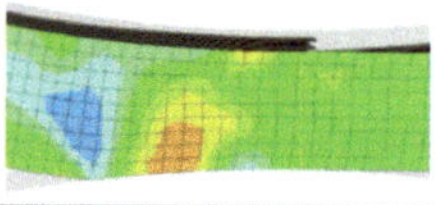	
$A = 1.2$ $\alpha = 0°$	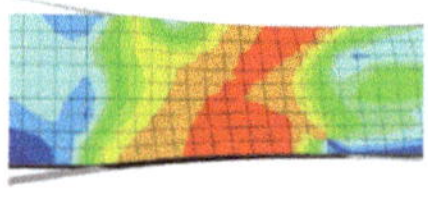	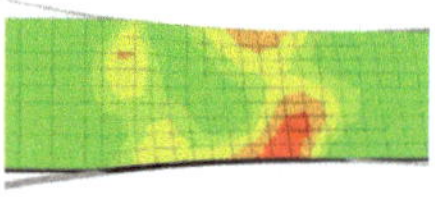	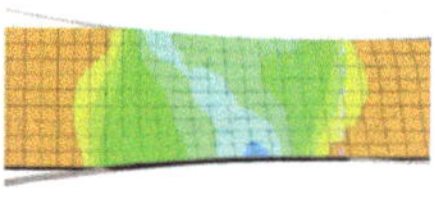
$A = 1.2$ $\alpha = 30°$	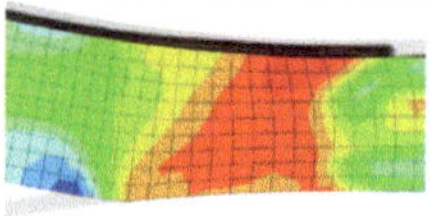	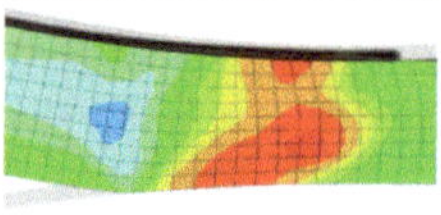	
$A = 1.2$ $\alpha = -30°$	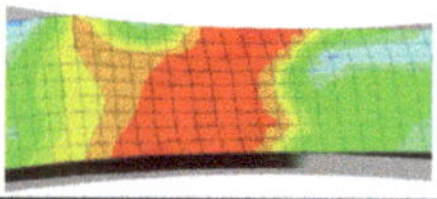	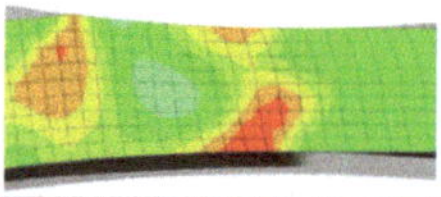	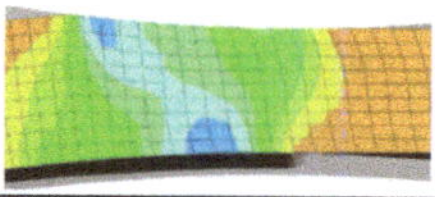

Significant changes in the rolled material flow due to rolling asymmetry can also be clearly discerned in the FE mesh deformation. Selected examples for the third pass of symmetric ($A = 1.0$), and asymmetric ($A = 1.2$) rolling with different tilt angle values can be seen in Fig. 3. For symmetric rolling (Fig. 3a), the two-fold symmetry axis parallel to RD and lying in the thickness center of the rolled strip is visible. The two-fold symmetry axis parallel to ND attached to the half thickness plane describe symmetry of the upper and lower part of the bar rolled with $A = 1.0$ and $\alpha = 30°$ (Fig. 3b). The lack of this type of symmetry is visible for the remaining cases. However, differences in the way the material flows on the rolling bar thickness are minimized for cases shown in Figs. 3c and 3e.

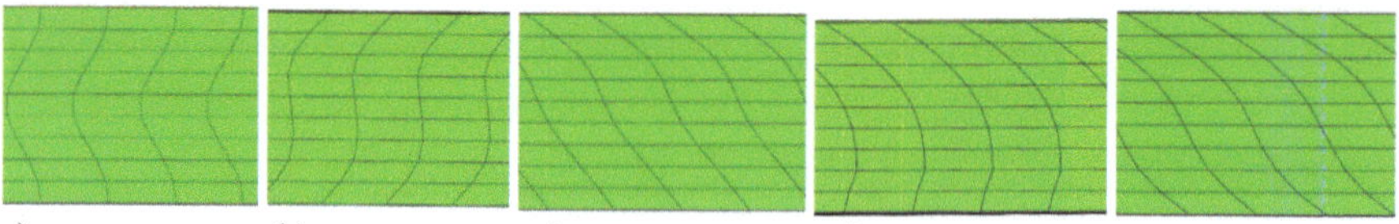

a) b) c) d) e)

Fig. 3. The deformed mesh after third pass of rolling in different variants: a) $A = 1.0$, $\alpha = 0°$, *b)* $A = 1.0$, $\alpha = 30°$, *c)* $A = 1.2$, $\alpha = 0°$, *d)* $A = 1.2$, $\alpha = 30°$, *e)* $A = 1.2$, $\alpha = -30°$.

https://doi.org/10.21741/9781644903254-65

Table 3. Stress distribution during the first rolling pass and experimental and simulated texture after sixth pass at down surface of the sample for symmetric flat and tilt rolling. The ODF cross-sections at $\varphi_2 = 0°$ are shown.

	$A = 1.0$ $\alpha = 0°$	$A = 1.0$ $\alpha = 30°$
Normal stress components	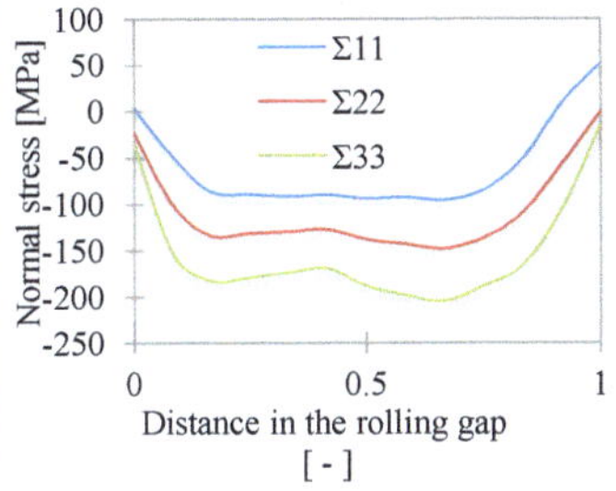	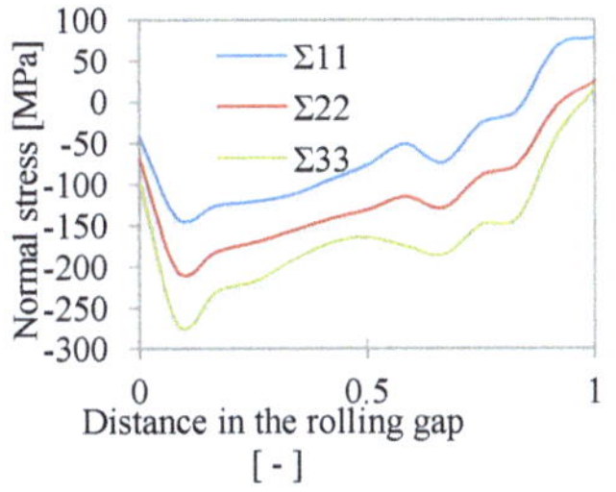
Shear stress components	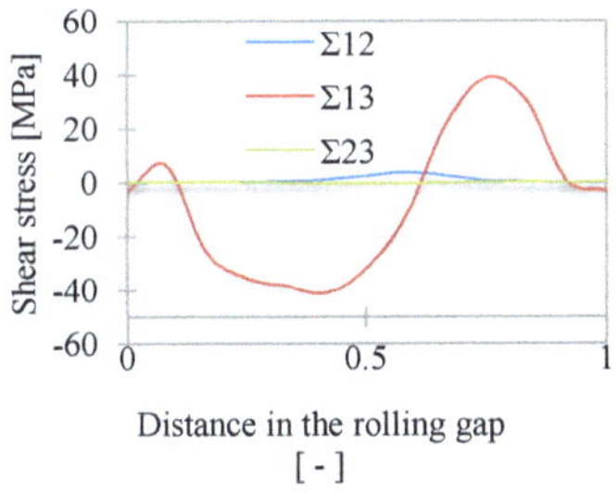	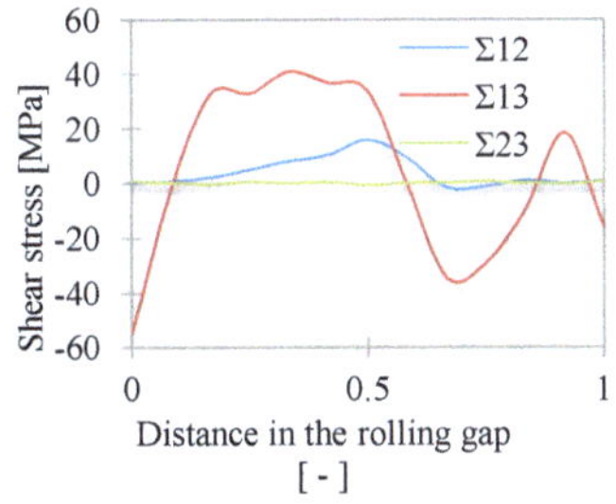
Experimental texture	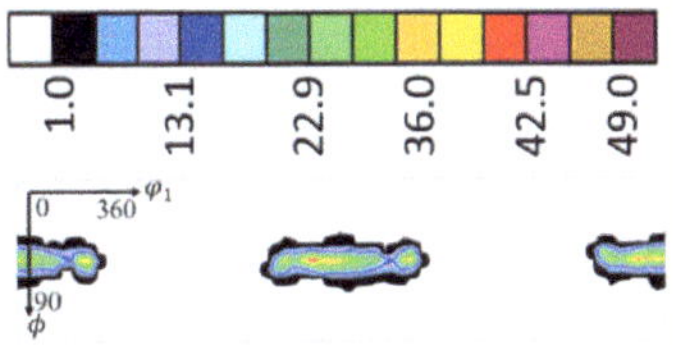	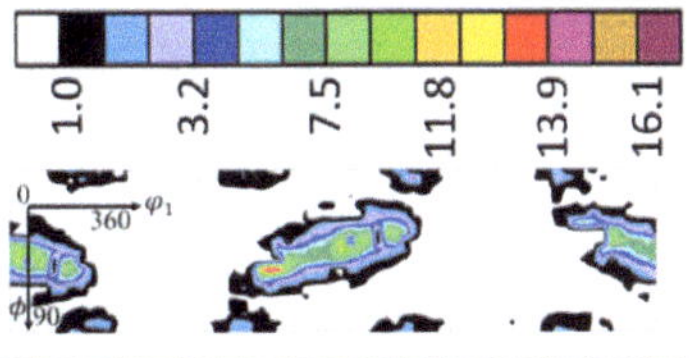
Simulated texture	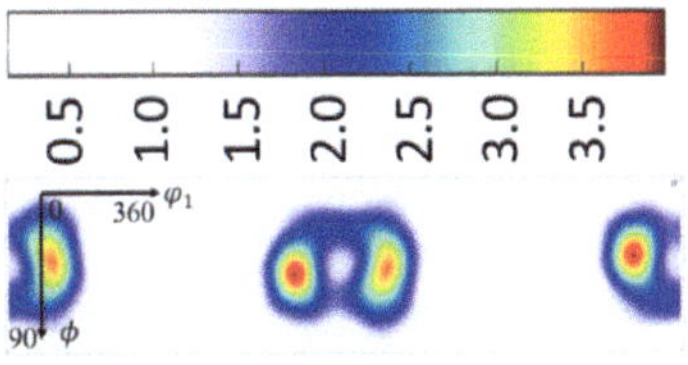	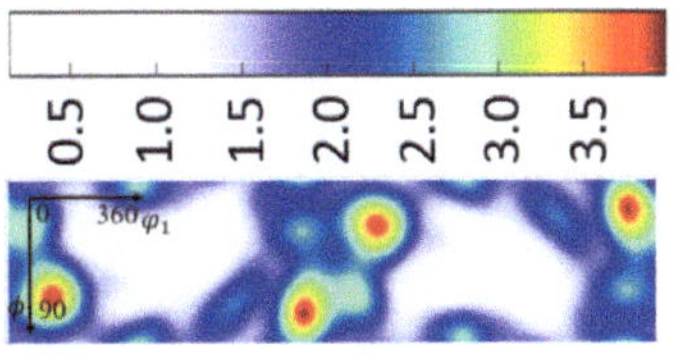

Table 4. Stress distribution after the first pass and experimental and simulated texture after sixth pass at down surface of the sample for symmetric and asymmetric 6-pass flat rolling. The ODF cross-sections at $\varphi_2 = 0°$ are shown.

	$A = 1.0$ $\alpha = 0°$	$A = 1.2$ $\alpha = 0°$
Normal stress components	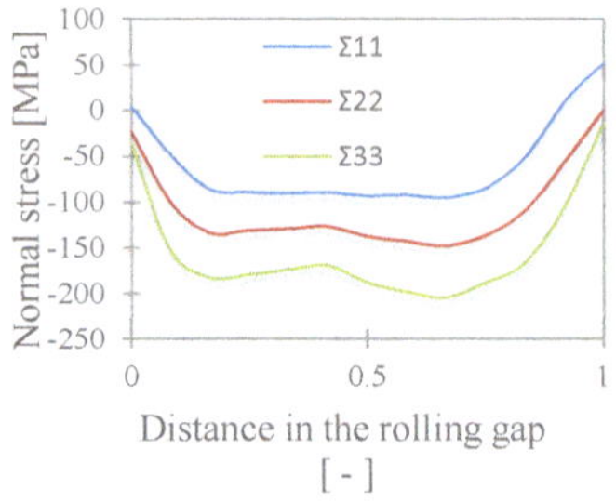	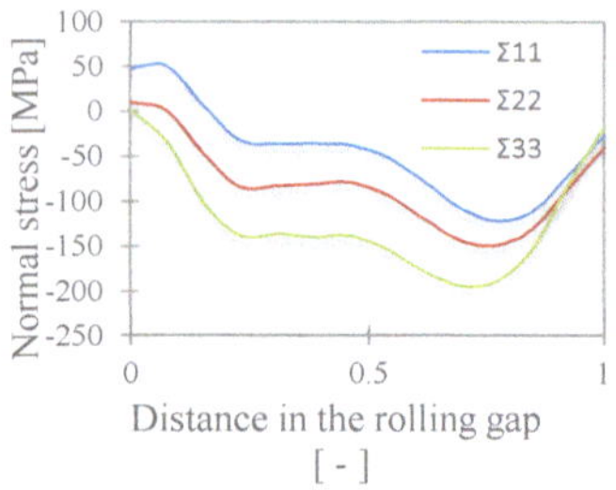
Shear stress components	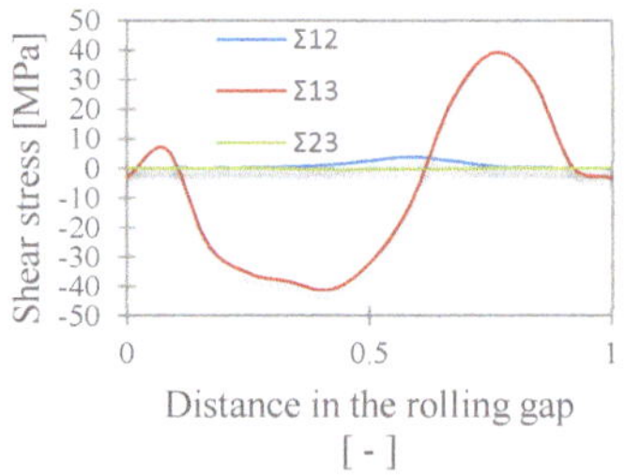	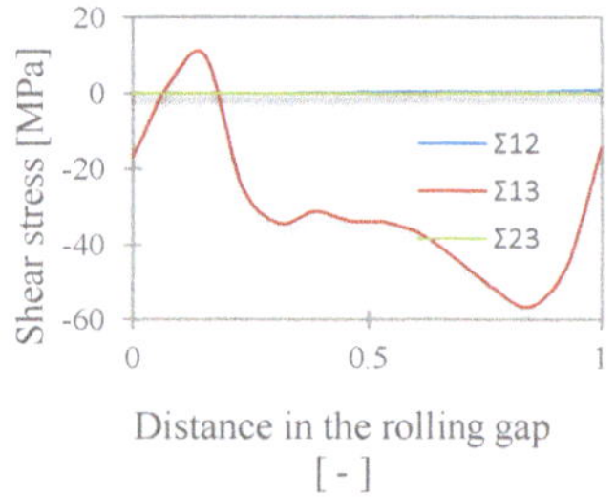
Experimental texture	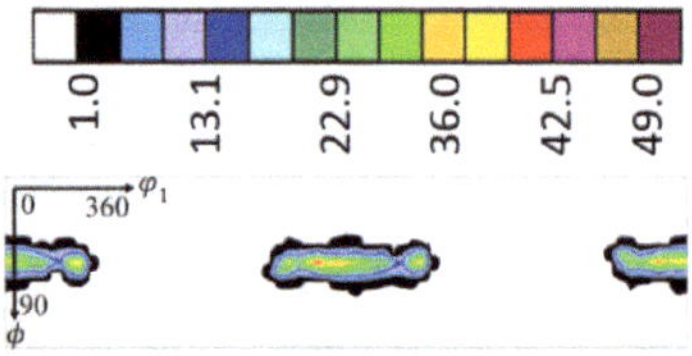	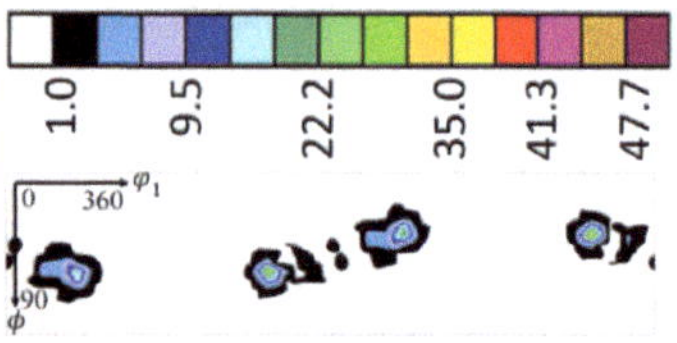
Simulated texture	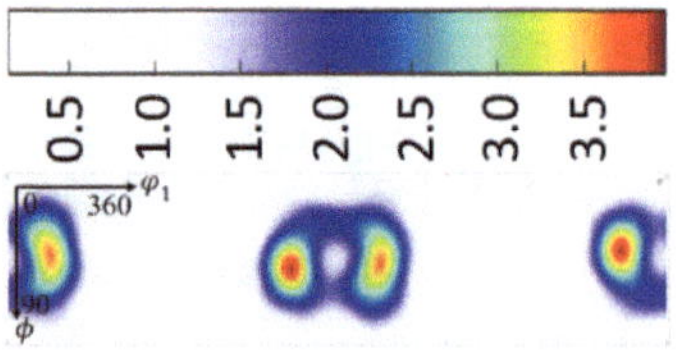	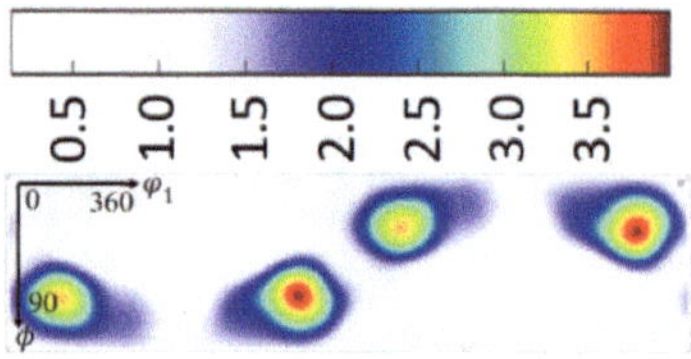

It is evident that the appropriate selection of the rolling asymmetry (i.e., A and α values) can significantly homogenize the distribution of strains in the volume of the rolled flat bar.

Conclusions
The developed multi-scale model describes the evolution in the deformation field and in the crystallographic texture development of aluminum flat products. This model can be used in industry to select asymmetry conditions that enable to produce material with controlled properties through the rolled strip thickness.

Acknowledgments
We gratefully acknowledge Polish high-performance computing infrastructure PLGrid (HPC Center: ACK Cyfronet AGH) for providing computer facilities and support within computational grant no. PLG/2023/016602.

References
[1] K. Wierzbanowski, M. Wroński, T. Leffers, FCC rolling textures reviewed in the light of quantitative comparisons between simulated and experimental textures, Crit. Rev. Solid State Mater. Sci. 39 (2014) 391–422. https://doi.org/10.1080/10408436.2014.899485

[2] S. Wronski, M. Wrobel, A. Baczmanski, K. Wierzbanowski, Effects of cross-rolling on residual stress, texture and plastic anisotropy in f.c.c. and b.c.c. metals,Mater. Charact. 77 (2013) 116–126. https://doi.org/10.1016/j.matchar.2013.01.005

[3] A. Uniwersał, M. Wróbel, K. Wierzbanowski, S. Wroński, A. Baczmański, Mechanical and microstructural characteristics of polycrystalline copper rolled asymmetrically to a high deformation level, Mater. Charact. 148 (2019) 214–223. https://doi.org/10.1016/j.matchar.2018.12.017

[4] A. Uniwersał, M. Wroński, M. Wróbel, K. Wierzbanowski, A. Baczmański, Texture effects due to asymmetric rolling of polycrystalline copper, Acta Mater. 139 (2017) 30–38. https://doi.org/10.1016/j.actamat.2017.07.062

[5] M.Y. Amegadzie, D.P. Bishop, Effect of asymmetric rolling on the microstructure and mechanical properties of wrought 6061 aluminum, Mater. Today Commun. 25 (2020) 101283. https://doi.org/10.1016/j.mtcomm.2020.101283

[6] A. Uniwersał, M. Wróbel, K. Wierzbanowski, S. Wroński, M. Wroński, I. Kalemba-Rec, T. Sak, B. Bacroix, Microstructure, texture and mechanical characteristics of asymmetrically rolled polycrystalline copper, Mater. Charact. 118 (2016) 575–583. https://doi.org/10.1016/j.matchar.2016.07.004

[7] S. Wroński, K. Wierzbanowski, B. Bacroix, M. Wróbel, M. Wroński, Crystallographic textures variation in asymmetrically rolled steel, Mater. Sci. Forum 638–642 (2010) 2811–2816. https://doi.org/10.4028/www.scientific.net/MSF.638-642.2811

[8] S. Wroński, K. Wierzbanowski, B. Bacroix, T. Chauveau, M. Wróbel, Effect of asymmetric rolling on mechanical characteristics, texture and misorientations in ferritic steel, J. Cent. South Univ. 20 (2013) 1443–1455. https://doi.org/10.1007/s11771-013-1633-5

[9] H. Wang, Y. Li, G. Xu, H. Tang, Y. Wang, J. Li, B. Li, T. Zhang, W. Yu, Y. Wang, Y. Liang, The effect of asymmetric rolling on the microstructure and properties of Al–Cu–Li–TiC/TiB2 alloys, Mater. Sci. Eng. A 897 (2024) 146333. https://doi.org/10.1016/j.msea.2024.146333

[10] "United States Patent US008250895B2," Aug. 2008.

[11] A. Uniwersał, M. Wróbel, K. Wierzbanowski, S. Wroński, A. Baczmański, Rolling asymmetry effects on recrystallization process and on properties and microstructure of annealed copper, Mater. Charact. 153 (2019) 136–147. https://doi.org/10.1016/j.matchar.2019.05.001

[12] M. Wroński, K. Wierzbanowski, M. Wróbel, S. Wroński, B. Bacroix, Effect of rolling asymmetry on selected properties of grade 2 titanium sheet, Met. Mater. Int. 21 (2015) 805–814. https://doi.org/10.1007/s12540-015-5094-2

[13] K. Wierzbanowski, M. Wroński, A. Baczmański, B. Bacroix, P. Lipinski, A. Lodini, Problem of lattice rotation due to plastic deformation. Example of rolling of F.C.C materials, Arch. Metall. Mater. 56 (2011) 575–584. https://doi.org/10.2478/v10172-011-0061-5

[14] E. Schmid, W. Boas, Kristallplastizität. Springer Berlin Heidelberg, 1935.

[15] G. Wassermann, H.-J. Bunge. Mathematische Methoden der Texturanalyse Akademie-Verlag Berlin 1969, 330 Seiten Geb. M 68.–, Krist. und Tech. 5 (1970) K23–K23. https://doi.org/10.1002/crat.19700050319

PROCESSES

Effect of counter punch on formability in two-step compression forming for producing extremely deep cup

KAJIKAWA Shohei[1,a *], UMEHARA Yutaro[1,b], KUBOKI Takashi[1,c] and IIZUKA Takashi[2,c]

[1]Department of Mechanical and Intelligent Systems Engineering, The University of Electro-Communications, 1-5-1 Chofu Gaoka, Chofu-shi, Tokyo, 182-8585, Japan

[2]Faculty of Mechanical Engineering, Kyoto Institute of Technology, Matsugasaki Goshokaido-cho, Kyoto-shi, Kyoto, 606-8585, Japan

[a]s.kajikawa@uec.ac.jp, [b]umehara@mt.mce.uec.ac.jp, [c]kuboki@mce.uec.ac.jp, [d]tiizuka@kit.ac.jp

Keywords: Sheet Metal, Compression, Deep Cup, Sheet Forging, Aluminum, FEM

Abstract. This paper presents two-step compression forming with a counter punch for producing an extremely deep cup with good shape accuracy. In the proposed method, the bottom shape is formed in the first step of the boss forming, and then the side wall is formed in the second step of the flange compression. The influence of the counter punch position in the boss forming was investigated to improve the shape accuracy of the bottom portion in this study. Appropriate counter punch position was predicted in finite element method (FEM) analysis. When the counter punch position was not appropriate, the forming defects, which were the local thinning and the backflow of the material, occurred. Based on the FEM result, the experiments were carried out. The experimental result agreed with the analysis result, and a deep cup with good shape accuracy was successfully formed under the optimum condition.

Introduction

Deep cup-shaped parts made from sheet metal are generally used for various industrial fields such as automobiles, household electronics, precision equipment, and so on. For example, battery cases for electric vehicles are conventionally produced by multi-step deep drawing from sheet metal [1]. This is because the productivity of deep drawing is high compared to the other methods such as spinning [2]. The deep case is formed gradually by multi-step deep drawing because the material could fracture at the punch shoulder portion due to tensile force if the deep cup is attempted to be formed by single-step deep drawing.

To further reduce the processing time, a reduction in the number of steps in the deep-drawing process has been tackled by various methods. It is possible to reduce the drawing stage by optimizing the punch, die, and blank shape [1]. The use of the blank with localized hard portions that contact the punch shoulder is effective in preventing crack occurrence [3,4], but the blank with localized hard portions needs to be formed as a pre-process. Incremental forming in the deep drawing was proposed to reduce the frictional force between the die and the blank holder [5,6]. However, these methods need the press machine with motion controllable, or specialized equipment.

Authors propose a flange compression forming method to form an extremely deep cup by only one process with simple tools and a conventional press machine [7-9]. In the proposed method, the sheet does not fracture with the effect of hydrostatic pressure because the sheet is stretched by compression in the thickness direction. However, there remains the problem of low shape accuracy at the cup bottom. In this study, we propose two-step compression forming with a counter punch to produce an extremely deep cup with good shape accuracy. Appropriate forming conditions were investigated, and the effectiveness of the proposed method was verified by a series of finite element method (FEM) analyses and experiments.

Two-step Compression Forming

Fig. 1 shows the schematic illustration of two-step compression forming for producing the extremely deep cup. A punch with a pin, a die, and a counter punch are used. In the first step of boss forming, the bottom portion of the cup is formed. The center portion of the blank is compressed between the pin and the counter punch until the clearance between the punch and the die at the side is equal to the initial blank thickness as shown in Fig. 1 (b). In the second step of flange compression, the side wall of the cup is formed. In this step, the counter punch is removed, and then the material is extruded from the clearance c_{pd} between the pin and the die by compressing the flange portion between the punch and the die as shown in Fig. 1 (b).

Pin length l_p is geometrically determined based on the previous study [8]. Fig. 2 shows a schematic illustration of the boss forming at the beginning and end. l_p should be larger than the total value of the initial blank thickness t_0, the die corner radius r_{cd}, and the pin corner radius r_{cp} for forming the side wall with uniform thickness. This is because the clearance c_{pd} between the pin and the die changes at the initial stage of the flange compression if l_p is smaller than the total value of t_0, r_{cd}, and r_{cp}, and then the side wall thickness becomes thick near the bottom portion. However, the local thinning appeared at the portion that contacts the pin shoulder during the boss forming when l_p is too long. Therefore, l_p was determined by the following equation in this study.

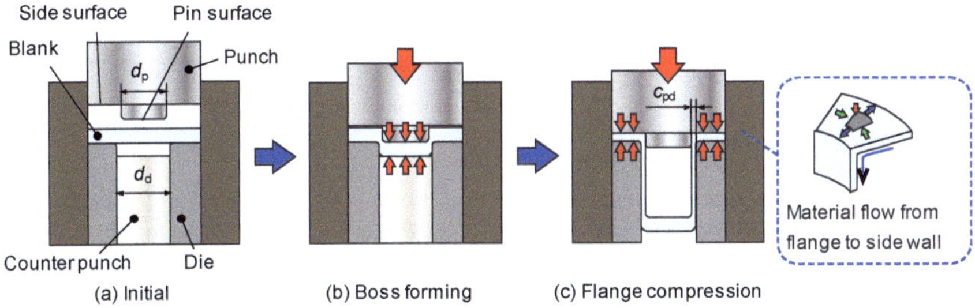

Fig. 1. Schematic illustration of two-step compression forming for producing extremely deep cup.

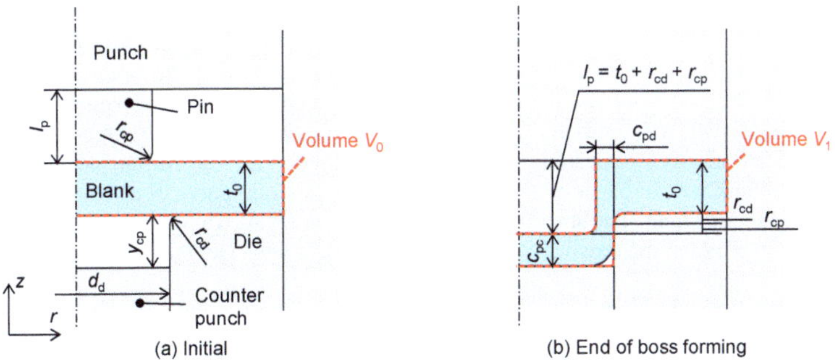

Fig. 2. Schematic illustration of the boss forming at the beginning and end.

$$l_\mathrm{p} = t_0 + r_\mathrm{d} + r_\mathrm{p} \tag{1}$$

The clearance c_pc between the pin and the counter punch at the end of the boss forming, which becomes the target bottom thickness t_bt of the formed cup, could be important. This is because the volume V_1 of the space enclosed by the punch, die, and counterpunch at the end of the boss forming changes with c_pc as shown in Fig. 2 (b). V_1 must be larger than the initial volume V_0 of the blank. If V_1 is smaller than V_0 due to small c_pc, excessive load could be applied to the device because the material cannot escape from the space in the device. Therefore, c_pc, when V_1 is equal to V_0, is considered to be the limit clearance c_pcl. When V_1 is equal to V_0, the following equation is obtained.

$$\pi \left(\frac{d_\mathrm{d}}{2}\right)^2 c_\mathrm{pcl} + V_\mathrm{a} + V_\mathrm{b} = \pi \left(\frac{d_\mathrm{p}}{2}\right)^2 t_0 \tag{2}$$

where t_0 is the initial blank thickness, and d_d and d_p are the diameters of the die and the punch pin. V_a and V_b were calculated by the following equations.

$$V_\mathrm{a} = \pi \left(\frac{d_\mathrm{d}}{2}\right)^2 r_\mathrm{cp} - \int_0^{r_\mathrm{cp}} \pi \left(\sqrt{r_\mathrm{cp}^2 - \left(z - r_\mathrm{cp}^2\right)^2} + \frac{d_\mathrm{p}}{2} - r_\mathrm{cp}\right)^2 dz \tag{3}$$

$$V_\mathrm{b} = -\pi \left(\frac{d_\mathrm{p}}{2}\right)^2 r_\mathrm{cd} + \int_0^{r_\mathrm{cp}} \pi \left(-\sqrt{r_\mathrm{cd}^2 - z^2} + \frac{d_\mathrm{d}}{2} + r_\mathrm{cd}\right)^2 dz \tag{4}$$

Therefore, c_pcl was obtained from Eq. 2, 3, and 4 as follows.

$$c_\mathrm{pcl} = \left(\frac{d_\mathrm{p}}{d_\mathrm{d}}\right)^2 t_0 - \frac{4(V_\mathrm{a}+V_\mathrm{b})}{\pi d_\mathrm{d}^2} \tag{5}$$

On the other hand, local thinning could appear if c_pc is too large. Therefore, an optimum value of c_pc was investigated for each pin and the die shoulder corner radius r_cp and r_cd to form a deep cup with good shape accuracy in this study. The optimum value of c_pc was predicted by FEM analysis for the boss forming. The optimum c_pc, which was predicted by FEM analysis, was verified by the experiment.

Material and Methods

An elastic-plastic analysis was carried out by using commercial code "Elfen" produced by Rockfield Software Ltd., Swansea, UK. Table 1 shows the analysis conditions. The analysis was conducted for the first step of the boss forming in order to predict the optimum value of the clearance c_pc between the pin and the counter punch at the end of the boss forming. The model is two-dimensional with axisymmetry. An explicit scheme was adopted. Three-node triangular elements and an adaptive remeshing scheme were applied. The von Mises yield criterion was selected. The blank was an elasto-plastic body, and a stress-strain curve of aluminum alloy 1100-O was applied to the blank. The stress-strain curve for the analysis was obtained by approximating the stress-strain curve in the tensile test according to Voce's law. The punch, the die, the counter

Table 1. FEM analysis conditions.

Blank	Material	Aluminum alloy 1100-O σ=90 - (90 - 35)e$^{-50\varepsilon}$ according to Voce's law
	Diameter d_0 [mm]	24
	Thickness t_0 [mm]	3
Punch	Material	Rigid body
	Pin diameter d_p [mm]	9.0
	Pin shoulder corner radius r_{cp} [mm]	0.2 ~ 1.0
	Pin length l_p [mm]	$t_0 + r_{cp} + r_{cd}$
Die	Material	Rigid body
	Diameter d_d [mm]	11.0
	Shoulder Corner radius r_{cd} [mm]	0.2 ~ 1.0
Counter punch	Material	Rigid body
	Diameter d_c [mm]	11
Clearance between pin and die c_{pd} [mm]		1.0
Clearance between pin and counter punch at end of boss forming c_{pc} [mm]		1.5 ~ 3.0
Friction coefficient μ		0.1

Table 2. Experimental conditions.

Punch pin shoulder corner radius r_{cp} [mm]	0.6
Die shoulder Corner radius r_{cd} [mm]	0.2
Clearance between pin and counter punch at end of boss forming c_{pc} [mm]	1.9, 2.2, 2.5
Lubricant	White petrolatum (Petrolatum HG, Taiyo Pharmaceutical Co. Ltd., Tokyo, Japan) on both sides of blank

punch, and the container were rigid bodies. The clearance c_{pc} between the pin and the counter punch at the end of the boss forming, the pin, and the die corner radius r_{cp} and r_{cd} were changed as the forming parameters. The local thinning ratio γ was calculated by the following equation to evaluate the local thinning of the formed cup.

$$\gamma = \frac{t_{st} - t_{min}}{t_{st}} \tag{6}$$

t_{st} is the target side wall thickness, which is equal to the clearance c_{pd} between the pin and the die, and t_{min} is the minimum wall thickness.

An experiment was carried out by using the device shown in Fig. 1. A universal testing machine (Autograph AG-300kNXplus, Shimadzu Corporation, Kyoto, Japan) was used for pressing the punch. Table 2 shows the experimental conditions. The other conditions for the experiment were the same as that for the analysis shown in Table 1. In the experiment, the deep cups were formed by conducting the first and the second steps, which are the boss forming and the flange compression. The pin and the die shoulder corner radius r_{cp} and r_{cd} were set to 0.6 and 0.2 mm, and the influence of the clearance c_{pc} between the pin and the counter punch at the end of the boss forming was investigated. In the second step, the flange was compressed until the punch load achieved 200 kN. The distribution of the cup diameter D from the bottom in the height direction was measured to evaluate the shape accuracy of the cup. The radius R of the cup was obtained by dividing D by 2.

Influence of Clearance c_{pc} between Pin and Counter Punch on Formability in FEM Analysis
Fig. 3 shows the influence of the clearance c_{pc} between the pin and the counter punch at the end of the boss forming when the pin and the die corner radius r_{cp} and r_{cd} were 0.6 and 0.2 mm. In this condition, the limit clearance c_{pcl} is 1.7 mm according to Eq. 5. The boss without any defects was formed under the condition of c_{pc} = 1.9 mm, while any defects, which were the backflow and the local thinning, occurred when c_{pc} was too small or large.

When the clearance c_{pc} was 1.6 mm, which was smaller than the limit clearance c_{pcl}, the backflow of the material occurred as shown in Fig. 3 (a). This is because the space between the pin and the counter punch is small compared to the initial volume of the blank under the pin. Fig. 4 shows the relationship between punch stroke S and punch load P. Under conditions where c_{pc}

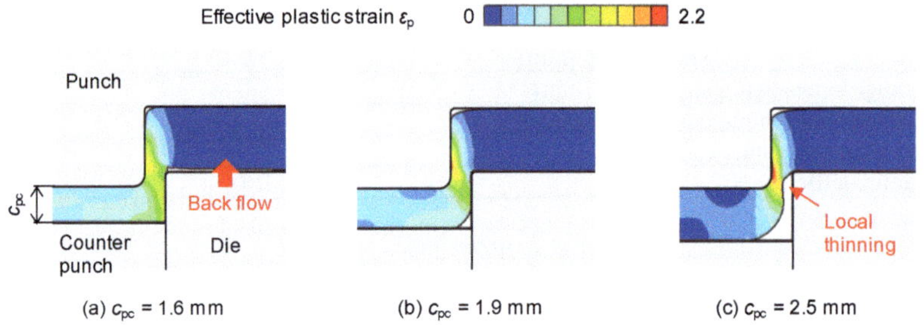

Fig. 3. Influence of clearance c_{pc} between pin and counter punch at the end of boss forming (Pin shoulder corner radius r_{cp} = 0.6 mm, die shoulder corner radius r_{cd} = 0.2 mm).

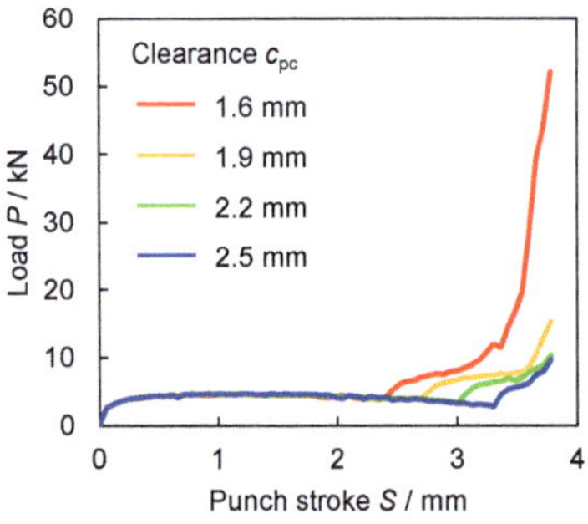

Fig. 4. Relationship between punch stroke S and punch load P (Pin shoulder corner radius r_{cp} = 0.6 mm, die shoulder corner radius r_{cd} = 0.2 mm).

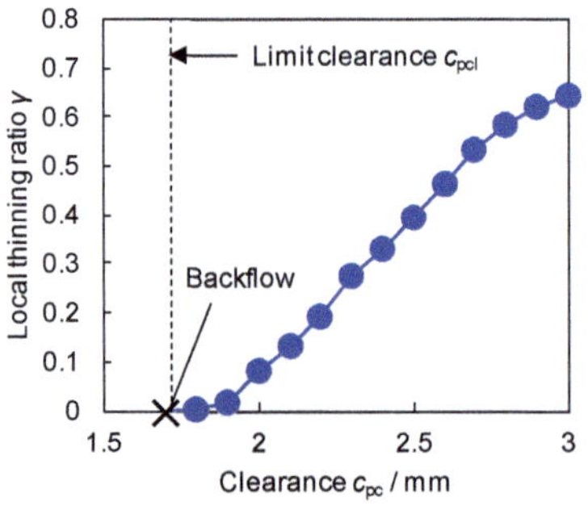

Fig. 5. Influence of clearance c_{pc} between pin and counter punch on local thinning ratio γ (Pin shoulder corner radius r_{cp} = 0.6 mm, die shoulder corner radius r_{cd} = 0.2 mm).

was 1.6 mm with the backflow occurring, P increased rapidly at the end of boss forming. This is because the space surrounded by the punch, die, and counterpunch was filled with material, but

Materials Research Forum LLC
https://doi.org/10.21741/9781644903254-66

the material could not escape from the die. c_{pc}, where the backflow occurred, should not be selected because the large load may cause tool rupture.

When the clearance c_{pc} was too large such as c_{pc} = 2.5 mm, the local thinning appeared at the portion near the pin shoulder as shown in Fig. 3 (c). The local thinning should be suppressed because the local thinning causes the material's fracture. Fig. 5 shows the influence of c_{pc} on the local thinning ratio γ. γ was 0 under the condition of c_{pc} = 1.8 mm, and γ increased with the increase in c_{pc}. When the material did not contact the counter punch under the condition of c_{pc} = 3 mm, γ was 0.64. Therefore, the local thinning was prevented by compressive stress from the counter punch.

Influence of Shoulder Corner Radius r_{cd} and r_{cp} of Die and Punch on Appropriate Clearance c_{pc} between Pin and Counter Punch in FEM Analysis

The appropriate value of the clearance c_{pc} between the pin and the counter punch could change with the die and the pin shoulder corner radius r_{cd} and r_{cp} because the pin length l_p changes with r_{cd} and r_{cp} according to Eq. 1. Therefore, the influence of c_{pc} on the local thinning ratio γ was investigated for each r_{cd} and r_{cp} as shown in Fig. 6. The limit clearance c_{pcl}, which was calculated by Eq. 5, was also shown in Fig. 6. c_{pcl} decreases with the increase in r_{cd} or r_{cp} because the volume V_1 of the space enclosed by the punch, die and counterpunch at the end of boss forming increases with the increase in r_{cd} or r_{cp}.

Fig. 6 (a) shows the influence of the clearance c_{pc} and the die shoulder corner radius r_{cd} on the local thinning ratio γ when r_{cp} was 0.6 mm. γ decreased with the decrease in c_{pc} and the backflow occurred when c_{pc} was near the limit clearance c_{pcl} in the case that r_{cd} was 0.2 mm. However, γ increased with the increase in r_{cd}, and there was no c_{pc} for preventing the local thinning when r_{cd} was too large. This is considered to be because the pin length l_p is needed to increase with r_{cd}. When l_p is large, the pin is pressed deeper into the blank, and then the local thinning easily occurs in the boss forming. Under the condition that r_{cd} was 0.8 and 1.0 mm, the backflow occurred with the local thinning when c_{pc} was small. For this reason, c_{pc} of which the backflow occurred did not

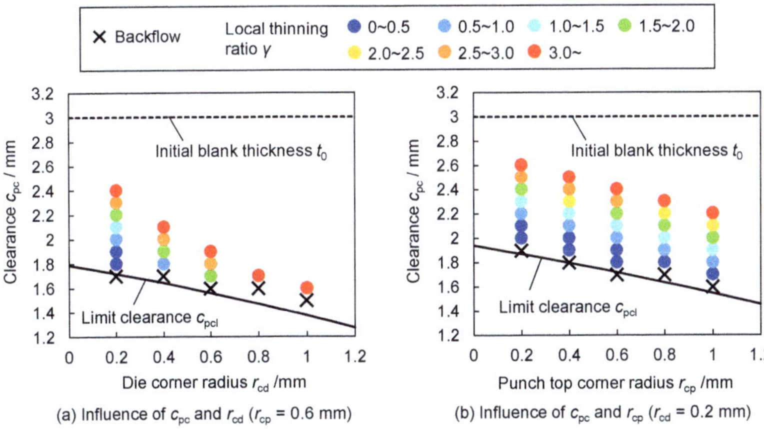

(a) Influence of c_{pc} and r_{cd} (r_{cp} = 0.6 mm)   (b) Influence of c_{pc} and r_{cp} (r_{cd} = 0.2 mm)

Fig. 6. Influence of clearance c_{pc} between pin and counter punch, die shoulder corner radius r_{cd}, and pin shoulder corner radius r_{cp} on the local thinning ratio γ.

agree with c_{pcl} under the condition that r_{cd} was 0.8 and 1.0 mm. From the above result, r_{cd} should be small to prevent local thinning.

Fig. 6 (b) shows the influence of the clearance c_{pc} and the pin shoulder corner radius r_{cp} on the local thinning ratio γ when r_{cd} was 0.2 mm. γ decreased with the decrease in c_{pc} for each r_{cp}, and the backflow occurred when c_{pc} was near the limit clearance c_{pcl}. γ increased with the increase in r_{cp} under the same conditions of c_{pc}, but there was c_{pc} for reducing the local thinning for each r_{cp}. Large r_{cp} could suppress the local thinning because the stress concentration is relieved at the pin shoulder. Therefore, the pin could be pushed in deeper without the local thinning when the pin length l_p was large due to large r_{cp}. The above result indicated that the influence of r_{cp} on the local thinning is small, and the bottom thickness of the formed cup could be controlled by adjusting r_{cp}.

Influence of Clearance between c_{pc} Pin and Counter Punch on Formability in Experiment

The influence of the clearance c_{pc} on the formability was investigated in the experiment in order to verify the FEM result. c_{pc} was selected to 1.9, 2.2, and 2.5 mm based on the FEM results. c_{pc} = 1.9 mm was the optimum condition for reducing the local thinning in the FEM analysis. The condition of c_{pc} = 2.2 and 2.5 mm could cause the local thinning or the crack because the γ for c_{pc} = 2.2 and 2.5 mm were 0.19 and 0.40 in the FEM analysis.

Fig. 7 shows the cross-sectional view of the formed cup after the second step. When the clearance c_{pc} was 1.9 mm, the cup was successfully formed without any defects. However, a crack appeared near the bottom corner when c_{pc} was 2.2 and 2.5 mm. Fig. 8 shows a magnified view of the bottom corner. In the case of c_{pc} = 2.2 mm, there was a small dent on the outer side, and a crack appeared on the inner side of the cup. In the case of c_{pc} = 2.5 mm, there was a large dent on the outer side, and a crack propagated from the inner side. Fig. 9 shows the distribution of the cup radius R in the height direction for the experiment and the FEM analysis. The experimental results qualitatively agreed with the FEM results. This result indicates that the optimum conditions could be predicted by the FEM analysis.

(a) c_{pc} = 1.9 mm (b) c_{pc} = 2.2 mm (c) c_{pc} = 2.5 mm

Fig. 7. Cross-sectional view of formed cup for each clearance c_{pc} after second step.

(a) c_{pc} = 1.9 mm (b) c_{pc} = 2.2 mm (c) c_{pc} = 2.5 mm

Fig. 8. Magnified view of bottom corner of formed cup for each clearance c_{pc} after second step.

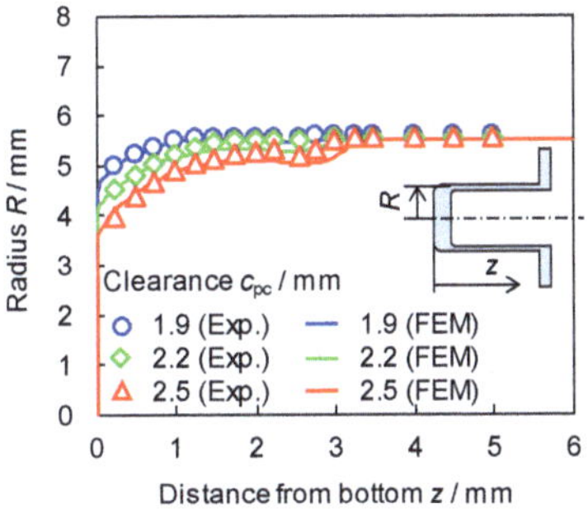

Fig. 9. Distribution of the cup radius R in height direction.

Conclusions

This paper proposed two-step compression forming, which is composed of the boss forming and the flange compression, for producing an extremely deep with high shape accuracy. In the first step of the boss forming, the bottom shape is formed by compressing the sheet between the pin and the counter punch. In the second step of the flange compression, the side wall is formed by compressing the flange portion between the punch and the die. The influences of the tool position and shape, which is the clearance c_{pc} between the pin and the counter punch, and the shoulder corner radius r_{cp} and r_{cd} of the pin and the die, were investigated for forming the bottom shape without any defects by FEM analysis. The bottom shape of the cup was successfully formed by setting c_{pc}, r_{cp} and r_{cd} appropriately. When c_{pc} was too small, the backflow of the material occurred, and then the forming load became too large. When c_{pc} was too large, the local thinning appeared due to the stress concentration at the pin shoulder. A small value of r_{cd} was appropriate because the range of c_{pc} for reducing the local thinning was wider with the decrease in r_{cd}, and there was no appropriate value of c_{pc} in the case that r_{cd} was too large. On the other hand, there were appropriate values of c_{pc} for each r_{cp}, while the appropriate value of c_{pc} decreased with the increase in r_{cp}. To verify the analysis results, the experiment was carried out. The deep cup with good shape accuracy was successfully formed under the appropriate conditions, and the experimental results agreed with the analysis results.

Acknowledgement

This work was supported by Sango Monozukuri Foundation, grant number 2022-RD-04.

References

[1] S. Kajikawa, T. Takasan, T. Kuboki, K. Tamaoki, K. Onishi, A. Gunji, A. Yamauchi, Finite element analysis of multi-stage deep drawing for forming deep rectangular case with high aspect ratio, Proc. COMPLAS XVI (2021), Barcelona, Spain, Paper No. 556.

[2] O. Music, J.M. Allwood, K. Kawai, A review of the mechanics of metal spinning, J. Mater. Process. Tech. 210 (2010) 3-23. https://doi.org/10.1016/j.jmatprotec.2009.08.021

[3] K. Mori, Y. Abe, K. Osakada, Plate forging of tailored blanks having local thickening for deep drawing of square cups, J. Mater. Process. Tech. 211 (2011) 1569-1574. https://doi.org/10.1016/j.jmatprotec.2011.04.010

[4] C.S. Namoco Jr., T. Iizuka, K. Narita, N. Takakura, K. Yamaguchi, Effects of embossing and restration process on the deep drawability of aluminum alloy sheets, J. Mater. Process. Tech. 187-188 (2007) 202-206. https://doi.org/10.1016/j.jmatprotec.2006.11.182

[5] M.A. Hassan, R. Suenaga, N. Takakura, K. Yamaguchi, A novel process on friction aided deep drawing using tapered blank holder divided into four segments, J. Mater. Process. Tech. 159 (2005) 418-425. https://doi.org/10.1016/j.jmatprotec.2004.06.006

[6] Y. Okude, T. Iwaoka, I. Nakamura, T. Muraoka, T. Katagiri, Development of warm and cold forming of Ti-6Al-4V alloy by press forming, Proc. 14th ICTP (2023), French Riviera, France, 213-222.

[7] T. Iizuka, N. Hatanak, N. Takakura, Deep drawing of high cup of magnesium alloy AZ31 by compressive force, Steel Res. Int. 81 (2010) 1247-1250.

[8] S. Kajikawa, T. Kuboki, T. Iizuka, Effect of tool design on formability in deep drawing by applying compressive force on flange, Proc. COMPLAS XIV (2017), Barcelona, Spain, 54-62.

[9] S. Kajikawa, T. Kuboki, T. Iizuka, Flange compression using stepped punch for forming extremely deep cup with flange from aluminum alloy sheet, J. Mater. Process. Tech. 288 (2021) 116835. http://dx.doi.org/10.1016/j.jmatprotec.2020.116835

Development of a Mg-6.8Y-2.5Zn-0.4Zr alloy (WZ73) under varying twin roll casting conditions

UEBERSCHÄR Franziska[1,a *], ULLMANN Madlen[1,b] and PRAHL Ulrich[1,c]

[1]Insititute of Metal Forming, Technische Universität Bergakademie Freiberg, Bernhard-von-Cotta-Str. 4, 09599 Freiberg, Germany

[a]franziska.ueberschaer@imf.tu-freiberg.de, [b]madlen.ullmann@imf.tu-freiberg.de, [c]ulrich.prahl@imf.tu-freiberg.de

Keywords: WZ73, Twin Roll Casting, Microstructure, Texture, LPSO

Abstract. Twin roll casting (TRC) has a big potential to produce thin strip material from magnesium alloys and enables the production in an economic manner. However, the final properties of TRC strips, such as microstructure and texture, are influenced by the twin roll casting conditions. In this work the development of a Mg-6.8Y-2.5Zn-0.4Zr alloy (WZ73) during different twin roll casting conditions, varying the twin roll casting speed, were studied. As a reference the strain and strain rate were determined. After twin roll casting the microstructure is inhomogeneous over the strip thickness and consists of a network-like structure of the LPSO phases and the α-Mg matrix. The α-Mg matrix is made up of dobulites (flake-like structures), which is already known for this alloy. [1,2] Typical defects of the twin roll cast strips were observed as well. It was also revealed that the twin roll casting conditions have a big influence on the precipitation, morphology, and phase fraction of the LPSO phases. For example, the phase fraction increases with the strain decreasing whilst the thickness of the precipitated phases increases with an increased strain. In all samples kink bands and yttrium enriched precipitations within the network like structures were detected. No dynamic recrystallization nor grain boundaries were detected. The resulting textures revealed the activation of basal slip and non-basal slip, but the intensities are small, regardless of the twin roll casting conditions.

Introduction

In recent decades, twin roll casting has become an essential method in the sheet production of light metals. This technology has proven to be especially beneficial for the fabrication of magnesium alloys. Conventional systems for forming thin strips of magnesium alloys through sheet metal involve a series of steps like continuous casting, a surface pretreatment, homogenization, pre-rolling in multiple passes, hot rolling, and multiple heat treatments. In contrast, the twin roll casting (TRC) method allows for the direct production of thin strips with thicknesses ranging from 3 to 7 mm from the molten state. This approach combines casting and rolling into a single process, making it possible to create near-net-shape semi-finished products. By leveraging the combination of TRC and strip rolling, the production of magnesium strips can be carried out in an economically and energy-efficient manner. Currently, TRC is primarily used for producing low alloy magnesium alloys like AZ31 or ZAX210. [3–6]

Over the last years magnesium alloys containing of rare earth elements became of much interest due to their good mechanical properties. [7,8] Within certain alloying systems, such as Mg-Y-Zn, the presence of Long Period Stacking Ordered (LPSO) structures significantly contributes to the enhancement of both strength and ductility. These LPSO structures exhibit a stacking sequence along the c-axis that manifests in a 10-, 14-, 18-, or 24-fold increase. [9,10] The formation of LPSO structures has been observed under various conditions, particularly in conventionally cast, heat treated, and/or hot deformed samples. [10–13]

Available literature mainly focuses on extrusion of Mg-Y-Zn alloys, different compositions or on the further processes after the initial production of sheets like the heat treatment or hot rolling. [2,14,15] Therefor in this work the focus is for the first time on the investigation of the different twin roll casting parameters and how they affect the microstructure and texture of the magnesium alloy WZ73. The investigated twin roll casting process is unique to the Institute of Metal Forming at the TU Bergakademie Freiberg.

Material and Methods

In the current study, the pilot-plant scale implementation of the horizontal TRC process was carried out using a TRC plant at an industrial scale at the Institute of Metal Forming (IMF, TU Bergakademie Freiberg). [2] The rolls had a diameter of 840 mm and were water cooled. A protective gas atmosphere was maintained while melting of the Mg-6.8Y-2.5Zn-0.4Zr (wt.%) (WZ73) magnesium alloy ingots. The chemical composition of the WZ73 alloy is provided in Table 1. The melting of WZ73 alloy ingots occurred within a steel crucible in a melting furnace, also under a protective gas atmosphere (SF_6), set at 730°C. The melting furnace was connected to a preheated launder and the adjacent nozzle through a riser. A pump facilitated the continuous supply of molten magnesium, which was then fed into the roll gap (4.5 mm) via the casting nozzle. The contact between the melt and the top and bottom rolls enabled heat transfer at the contact surfaces. Consequently, a meniscus-shaped solidification zone formed, leading to delayed solidification in the mid-section of the strip. By employing this method, it was feasible to produce a magnesium strip with a width of 710 mm and an average thickness between 5.3 mm and 5.5 mm. The TRC speed was adjusted between 1.5 and 2.1 m/min.

Table 1. Chemical composition of Mg-6.8Y-3.0Zn-0.4Zr [wt.%] alloy.

Mg [wt.%]	Y [wt.%]	Zn [wt.%]	Zr [wt.%]	Others [wt.%]
Balance	6.8	3.0	0.4	0.01

To conduct the applied equivalent strain during the twin roll casting the solidification model by Weiner et al. [16] was used. This model is based on the Freiberg Layer Model and is able to simulate the strain φ and compressed length l_d. These were then used to calculate the equivalent strain φ_v and equivalent strain rate $\dot{\varphi}_v$ via the Eq. 1 and 2 using the twin roll casting speed v_{TRC}.

$$\varphi_v = \frac{2}{\sqrt{3}} \cdot \varphi \tag{1}$$

$$\dot{\varphi}_v = \frac{v_{TRC}}{l_d} \cdot \varphi_v \tag{2}$$

Samples for microstructure analysis were prepared using traditional grinding and polishing methods. Two different acids were used for etching. In a first step a mixture of distilled water, glacial acetic acid, ethanol, and picric acid was used and for the second step nitric acid. The microstructure analysis was conducted using optical microscopy with the Keyence VHX 6000 at the IMF in Freiberg, Germany. The specimens were measured in the longitudinal direction. Additionally, a scanning electron microscope evaluation was carried out using a ZEISS GeminiSEM 450 at the same location. X-ray diffraction analysis was performed on a Seifert-FPM RD 7 instrument at the Institute of Materials Science at the TU Bergakademie Freiberg, utilizing CuKα radiation (λ = 1.540598 Å) to identify the phases. The diffraction patterns were measured within a 2θ-range of 20° to 150°, with a step size of 0.02° and a step time of 25 s. Furthermore, texture analysis was conducted using electron backscattering diffraction (EBSD) analysis on a ZEISS GeminiSEM 450 instrument at the IMF. EBSD was performed with a voltage

Materials Research Proceedings 44 (2024) 625-634 https://doi.org/10.21741/9781644903254-67

of 15 kV and a step size of 2.5 μm. The EBSD data was analyzed and the pole figures were calculated using the MTex MATLAB Toolbox (version 5.9.0, MTex, Ralf Hielscher, TU Bergakademie Freiberg, Germany). [17]

Results and Discussion

Calculations. For a better comparison of the different microstructures the Freiberg layer model developed by Weiner et al. [16] was used to estimate the equivalent strain φ_v and equivalent strain rate $\dot{\varphi}_v$. The results are presented in Table 2. The TRC conditions influence the strain, because due to the higher TRC speed, the material solidifies later. The reason for this is the shorter contact with the water-cooled rolls, which leads to a lower heat flow and a later material solidification. In addition, the temperature of the strip being produced is higher due to the shorter contact time with the rolls. The higher temperature delays solidification as well and provides more time for segregation processes. [18,19] A comparison of the points of solidification shows that the material solidifies approx. 10 mm closer to the roll gap outlet at the twin roll casting speed of 2.1 m/min. The subsequent solidification in return leads to a shorter compressed length and consequently a lower strain. The dependence of the strain on the TRC speed leads to the change of the equivalent strain, but the equivalent strain rate is nearly the same. Therefor for further investigation only the dependence of the equivalent strain will be considered.

Table 2. Results of the calculations based on the Freiberg layer model developed by Weiner et al. [16] regarding twin roll casting of the WZ73 alloy.

TRC speed v_{TRC} [m/min]	compressed length l_d [mm]	strain φ [-]	equivalent strain φ_v [-]	equivalent strain rate $\dot{\varphi}_v$ [s^{-1}]
1.5	46.0	0.6	0.7	0.4
1.6	44.2	0.5	0.6	0.4
2.1	38.3	0.4	0.5	0.5

Microstructure. Fig. 2 shows an overview of the microstructure over the whole strip thickness. Typical twin roll casting defects were observed and are shown in Fig. 1. For example, surface bleeds, centerline segregations, an inhomogeneous microstructure and some cracks could be detected. These are defects which have been reported before for twin roll cast WZ73 alloy and others. [19–22] The centerline segregations are the most dominant defect. It can be seen that with a decreased equivalent strain the thickness of the centerline segregation increases. This is the result of the higher deformation, the higher TRC speed and the higher temperatures of the strip as explained before. The higher speed and related phenomena lead to a bigger volume of liquid being squished into the middle, because the melt does not solidify as quickly and the contact with the rolls is shorter resulting in a lower heat rate. [18,19] The segregations observed in Fig. 2 can be described as deformation segregations. They appear due to rapid deformation process according to Lockyer et al. [23] The melt and the solid material are deformed together and small liquid regions are formed between the solid grains. [23,24] Centerline segregations in general occur due to the rapid cooling of the material which is in contact with the water-cooled rolls. The solidification front migrates towards the center of the strip which results in an alloying element enriched residual melt solidifying in the center. Furthermore, the microstructure does not consist of large dendrites which grow from the surface into the middle of the strip. These are characteristics of the microstructure observed for AZ31 or ZAX210 after twin roll casting. [4,20]

 Materials Research Forum LLC
https://doi.org/10.21741/9781644903254-67

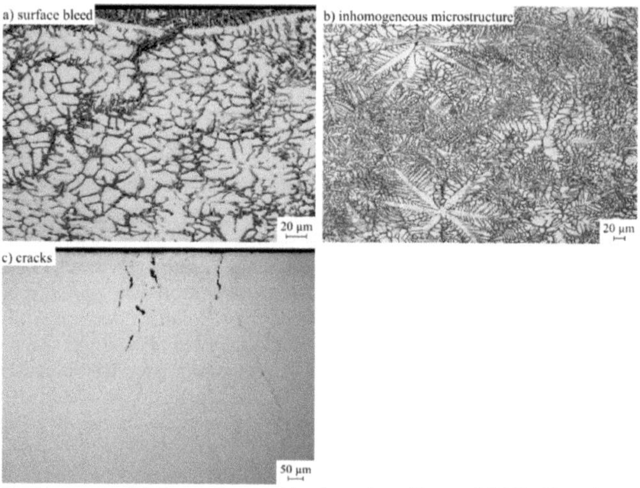

Fig. 1. Optical images of defects observed in the twin roll cast WZ73 alloy a) surface bleed; b) inhomogeneous microstructure; c) cracks.

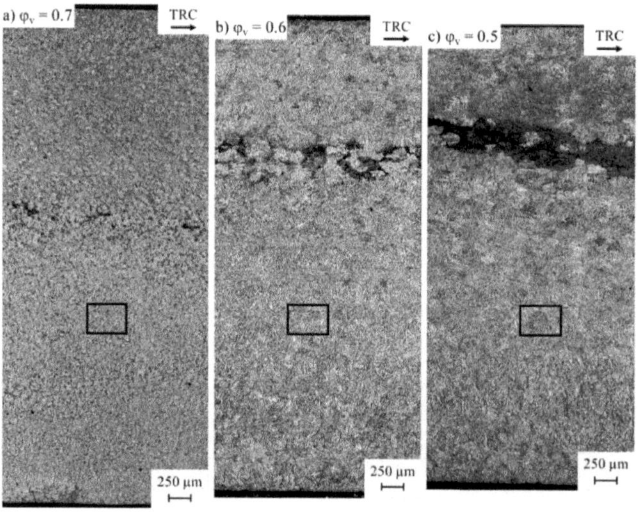

Fig. 2. Optical micrographs of the twin roll cast WZ73 alloy showing the entire strip thickness a) $\varphi_v = 0.7$; b) $\varphi_v = 0.6$ and c) $\varphi_v = 0.5$ including marked areas for Fig. 3.

Regarding the more detailed microstructure images in Fig. 3 it is visible that the microstructure consists of two parts: the α-magnesium matrix and an LPSO phase. The matrix (light parts in the optical images) is made up of dobulites (flake-like structures) whereas the LPSO phase has a network-like structure precipitated at the grain boundaries (dark parts in the optical images), which is similar to previous reported microstructure of a twin roll cast WZ73 alloy. [2,25] The phase fraction of the LPSO phase increases slightly when the equivalent strain decreases as can be seen in Table 3. In parallel the thickness of the participated LPSO phase decreases with the equivalent strain decreasing. The increasing phase fraction is consistent with other observation in AZ31 alloys which show more precipitates when the solidification rate rises. [26] Furthermore no fine grains were detected which would indicate a dynamic recrystallisation due to the deformation process. Ullmann et al. [14] discovered that the recrystallisation behavior of the WZ73 alloy during hot rolling depends more on the equivalent strain rate than the equivalent strain. [14] The during twin roll casting applied equivalent strain rate is rather low which can be a reason for the missing sites of dynamic recrystallisation. In all samples kink bands could be identified.

Table 3. Phase fraction and thickness of the LPSO phase depending on the equivalent strain φ_v.

equivalent strain φ_v [-]	phase fraction [%]	thickness [µm]
0.7	33 ± 1.5	1.7 ± 0.7
0.6	36 ± 1.8	1.5 ± 0.7
0.5	41 ± 2.5	1.1 ± 0.4

Fig. 3b,d,f show SEM images of the twin roll cast WZ73 alloy including points of EDS measurements. In addition to the dark matrix and the LPSO phase (light grey) some white points are visible. Those were identified as yttrium rich precipitates which are too small for further identification. Other measuring points have been indicated in Fig. 3 with numbers 1 to 4. The points 1 show the α-magnesium matrix. Other yttrium and zinc rich phases have been indicated with numbers 2 and 4. The composition measured via EDS can be compared to the $Mg_{12}YZn$ phase. In previous papers this phase was noted as an LPSO phase which can be either 18R or 14H like ordered. But the precipitation along the grain boundaries indicates more likely the presence of the 18R phase. [27–29] Points 3 show zirconium rich phases. The precise composition could not be identified, because of the small size. But it can be noted, that zirconium was only found in precipitates which favorably where in the middle of dobulite structures as seen for B3 and C3. Neither in the matrix nor the in the network like structures zirconium was detected.

Table 4. Chemical composition of structural constituents according to marking in SEM images Figure 3b,d,f determined via EDS analysis (energy dispersive X-ray spectroscopy) of the twin roll cast WZ73 alloy.

		Mg [wt.%]	Y [wt.%]	Zn [wt.%]	Zr [wt.%]
φ_v = 0.7	A1	97.7	1.5	0.8	-
	A2	77.4	13.2	9.4	-
	A3	56.5	24.4	3.8	15.3
	A4	73.9	15.9	10.2	-
φ_v = 0.6	B1	98.2	1.3	0.5	-
	B2	58.7	23.3	18.0	-
	B3	61.4	29.6	1.5	7.5
	B4	80.9	12.3	6.8	-
φ_v = 0.5	C1	96.8	2.3	0.9	-
	C2	58.5	23.0	18.5	-
	C3	48.7	40.5	1.4	9.4
	C4	74.3	15.7	10.0	-

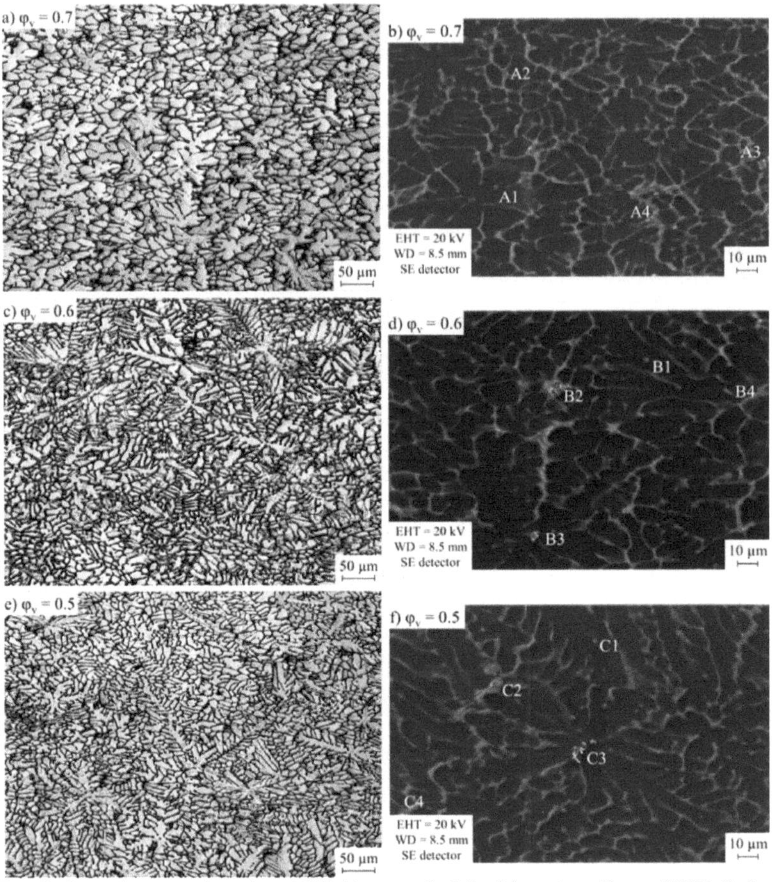

Fig. 3. Optical images (a,c,e) and SEM images (b,d,f) of the twin roll cast WZ73 (below mid-thickness as marked in Fig. 2) a,b) $\varphi_v = 0.7$; c,d) $\varphi_v = 0.6$; e,f) $\varphi_v = 0.5$ with measuring points of the EDS analysis (point 1: magnesium matrix, points 2 to 4: precipitates).

The results of the X-ray diffraction (XRD) analysis in Figure 4 show the presence of the following phases: α-Mg, $Mg_{12}YZn$, $Mg_3Y_2Zn_3$ and LPSO phases. This confirms some of the via EDS identified phases. But some of the peaks of the XRD are assigned to the $Mg_3Y_2Zn_3$ phase which usually is described as the W-phase in WZ73 alloys and has been detected in twin roll cast Mg-Y-Zn alloys before. [13,29] The little peaks at 31° and 33° are assigned to LPSO phases according to literature. [2,21,30] But a clear identification of either the 14H or 18R phase is not possible.

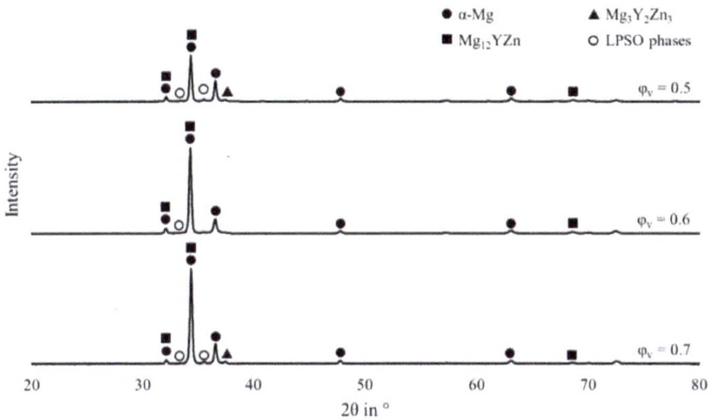

Fig. 4. Results of the XRD analysis of the twin roll cast WZ73 alloy along the cross section.

Texture. Fig. 5 shows the (0001) and $(10\bar{1}0)$ pole figures of the twin roll cast WZ73 alloys (below mid thickness) depending on the different equivalent strain rates. Regardless of the equivalent strain, the pole figures show a central intensity maximum, which is slightly stretched in the transverse direction and rather low intensities. In general, however, the texture components are pronounced in the TRC direction. Only the pole figure of the samples with $\varphi_v = 0.5$ shows a second intensity maximum, which is pronounced in the twin roll cast direction. The $(10\bar{1}0)$ pole figures show a so-called sixfold symmetry regardless of the equivalent strain. This symmetry and the broadened maxima of the (0001) pole figure indicate the activity of non-basal slip systems like pyramidal slip systems. This sixfold symmetry was already observed by Ullmann et al. [14] during hot rolling of WZ73 alloys. [1,14] Comparable textures were also described for twin roll cast ZAX210 or AZ31 alloys and hot rolled WE43 alloys. [4,6,31]

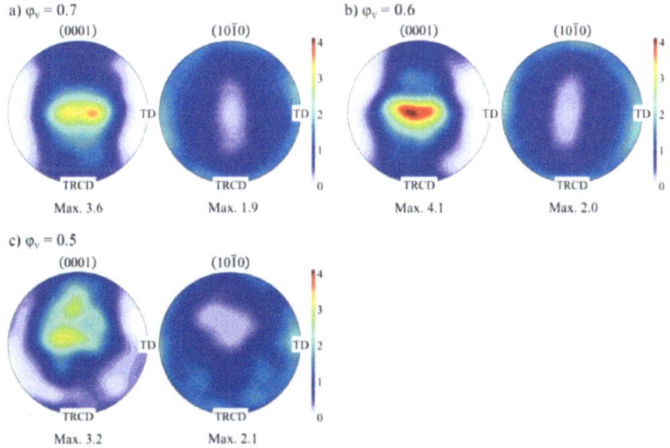

Fig. 5. (0001) and $(10\bar{1}0)$ pole figure of the twin roll cast WZ73 alloy below mid-thickness a) φ_v = 0.7; b) φ_v = 0.6; c) φ_v = 0.5.

Conclusions

The present work shows the influence of twin roll casting parameters on the microstructure and texture evolution based on the equivalent strain and equivalent strain rate. The results can be summarized as follows:

1. Different TRC speeds varying from 1.5 m/min up to 2.1 m/min were tested. The solidification model was used to calculate the strain and compressed length. For all the investigated TRC speeds it was found that the equivalent strain rate is similar whilst the equivalent strain decreases with an increasing TRC speed.
2. As typical twin roll casting defects centerline segregations, an inhomogeneous microstructure, surface bleeds and cracks could be detected.
3. The microstructure consists of the α-magnesium matrix and the network-like shaped LPSO phases which participated at the grain boundaries. The phase fraction of the LPSO phases increases with the equivalent strain decreasing. The $Mg_{12}YZn$ phase could be assigned to the LPSO phase via XRD measurements. The specific ordering could not be identified clearly. But the precipitation along the grain boundaries indicate the 18R LPSO phase. Some yttrium and zirconium rich precipitates were also observed but are too small in size to clearly identify a chemical composition.
4. The texture is characterized by basal pole which is elongated in transverse direction and a low intensity. Together with the sixfold symmetry observed in the $(10\bar{1}0)$ pole figure an activation of pyramidal slip systems is indicated. Those findings are the same for all investigated equivalent strains.

Funding

This research was funded by the Deutsche Forschungsgemeinschaft (DFG, German Research Foundation) – Project number 497456843.

References

[1] M. Ullmann, K. Kittner, T. Henseler, C. Krbetschek, D. Rafaja, R. Kawalla, U. Prahl, Dynamic recrystallization and texture evolution of Mg-6.8Y-2.5Zn-0.3Zr alloy during hot rolling, Procedia Manuf. 50 (2020) 809–816. https://doi.org/10.1016/j.promfg.2020.08.146

[2] K. Kittner, M. Ullmann, F. Arndt, R. Kawalla, U. Prahl, Microstructure and Texture Evolution during Twin-Roll Casting and Annealing of a Mg–6.8Y2.5Zn–0.4Zr Alloy (WZ73), Crystals 10 (2020) 513. https://doi.org/10.3390/cryst10060513

[3] K. Neh, M. Ullmann, R. Kawalla, Twin-Roll-Casting and Hot Rolling of Magnesium Alloy WE43, Procedia Eng. (2014) 1553–1558. https://doi.org/10.1016/j.proeng.2014.10.189

[4] K. Kittner, M. Ullmann, T. Henseler, R. Kawalla, U. Prahl, Microstructure and Hot Deformation Behavior of Twin Roll Cast Mg-2Zn-1Al-0.3Ca Alloy, Materials (Basel, Switzerland) 12 (2019). https://doi.org/10.3390/ma12071020

[5] R. Kawalla, M. Ullmann, C. Schmidt, J. Dembińska, H.P. Vogt, Properties of Magnesium Strips Produced by Twin-Roll-Casting and Hot Rolling, Mater. Sci. Forum 690 (2011) 21–24. https://doi.org/10.4028/www.scientific.net/MSF.690.21

[6] G.T. Bae, J.H. Bae, D.H. Kang, H. Lee, N.J. Kim, Effect of Ca addition on microstructure of twin-roll cast AZ31 Mg alloy, Met. Mater. Int. 15 (2009) 1–5. https://doi.org/10.1007/s12540-009-0001-3

[7] Y.C. Wan, S.N. Jiang, C.M. Liu, B.Z. Wang, Z.Y. Chen, Effect of Nd and Dy on the microstructure and mechanical property of the as extruded Mg–1Zn–0.6Zr alloy, Mater. Sci. Eng. A (2015) 158–163. https://doi.org/10.1016/j.msea.2014.12.003

[8] Y.C. Wan, C. Liu, H. Xiao, Y. Gao, S. Jiang, Z.Y. Chen, Improving the Ductility of Mg-Gd-Y-Zr Alloy through Extrusion and a Following Rolling, Adv. Eng. Mater. 20 (2018) 1701041. https://doi.org/10.1002/adem.201701041

[9] A. Ono, E. Abe, T. Itoi, M. Hirohashi, M. Yamasaki, Y. Kawamura, Microstructure Evolutions of Rapidly-Solidified and Conventionally-Cast Mg97Zn1Y2 Alloys, Mater. Trans. 49 (2008) 990–994. https://doi.org/10.2320/matertrans.MC200763

[10] K. Hagihara, A. Kinoshita, Y. Sugino, M. Yamasaki, Y. Kawamura, H.Y. Yasuda, Y. Umakoshi, Effect of long-period stacking ordered phase on mechanical properties of Mg97Zn1Y2 extruded alloy, Acta Mater. 58 (2010) 6282–6293. https://doi.org/10.1016/j.actamat.2010.07.050

[11] K. Hagihara, Z. Li, M. Yamasaki, Y. Kawamura, T. Nakano, Strain-rate dependence of deformation behavior of LPSO-phases, Mater. Letters 214 (2018) 119–122. https://doi.org/10.1016/j.matlet.2017.11.117

[12] J.-B. Liu, K. Zhang, J.-T. Han, X.-G. Li, Y.-J. Li, M.-L. Ma, J.-W. Yuan, G.-L. Shi, Microstructure and texture evolution of Mg–7Y–1Nd–0.5Zr alloy sheets with different rolling temperatures, Rare Metals 39 (2020) 1273–1278. https://doi.org/10.1007/s12598-016-0740-5

[13] K.-H. Kim, J.G. Lee, G.T. Bae, J.H. Bae, N.J. Kim, Mechanical Properties and Microstructure of Twin-Roll Cast Mg-Zn-Y Alloy, Mater. Trans. 49 (2008) 980–985. https://doi.org/10.2320/matertrans.MC200752

[14] M. Ullmann, K. Kittner, U. Prahl, Hot Rolling of the Twin-Roll Cast and Homogenized Mg-6.8Y-2.5Zn (WZ73) Magnesium Alloy Containing LPSO Structures, Metals 11 (2021) 1771. https://doi.org/10.3390/met11111771

[15] K. Suzawa, S. Inoue, S. Nishimoto, S. Fuchigami, M. Yamasaki, Y. Kawamura, K. Yoshida, N. Kawabe, High-strain-rate superplasticity and tensile behavior of fine-grained Mg97Zn1Y2 alloys fabricated by chip/ribbon-consolidation, Mater. Sci. Eng. A 764 (2019) 138179. https://doi.org/10.1016/j.msea.2019.138179

[16] M. Weiner, M. Schmidtchen, U. Prahl, Extension of the Freiberg Layer Model by Means of Solidification for Roll Casting, Adv. Eng. Mater. 24 (2022). https://doi.org/10.1002/adem.202101546

[17] F. Bachmann, R. Hielscher, H. Schaeben, Texture Analysis with MTEX – Free and Open Source Software Toolbox, Semi-Solid Processing of Alloys and Composites X 160 (2010) 63–68. https://doi.org/10.4028/www.scientific.net/SSP.160.63

[18] M.A. Wells, A. Hadadzadeh, Twin Roll Casting (TRC) of Magnesium Alloys – Opportunities and Challenges, MSF 783-786 (2014) 527–533. https://doi.org/10.4028/www.scientific.net/MSF.783-786.527

[19] C. Gras, M. Meredith, J.D. Hunt, Microdefects formation during the twin-roll casting of Al–Mg–Mn aluminium alloys, J. Mater. Process. Technol. 167 (2005) 62–72. https://doi.org/10.1016/j.jmatprotec.2004.09.084

[20] A. Hadadzadeh, M.A. Wells, Inverse and centreline segregation formation in twin roll cast AZ31 magnesium alloy, Mater. Sci. Technol. 31 (2015) 1715–1726. https://doi.org/10.1179/1743284714Y.0000000750

[21] K. Kittner, M. Ullmann, F. Arndt, S. Berndorf, T. Henseler, U. Prahl, Analysis of defects in a twin roll cast Mg-Y-Zn magnesium alloy, Engineering Reports 4 (2022). https://doi.org/10.1002/eng2.12394

[22] Y. Li, C. He, J. Li, Z. Wang, Di Wu, G. Xu, A Novel Approach to Improve the Microstructure and Mechanical Properties of Al-Mg-Si Aluminum Alloys during Twin-Roll Casting, Materials (Basel, Switzerland) 13 (2020). https://doi.org/10.3390/ma13071713

[23] S.A. Lockyer, M. Yun, J.D. Hunt, D.V. Edmonds, Micro- and macrodefects in thin sheet twin-roll cast aluminum alloys, Materials Characterization 37 (1996) 301–310. https://doi.org/10.1016/S1044-5803(97)80019-8

[24] M. Yun, S. Lokyer, J.D. Hunt, Twin roll casting of aluminium alloys, Mater. Sci. Eng. A 280 (2000) 116–123. https://doi.org/10.1016/S0921-5093(99)00676-0

[25] G. Kurz, J. Wendt, J. Bohlen, D. Letzig, Microstructure Evolution of Different Magnesium Alloys During Twin Roll Casting 465–470. https://doi.org/10.1002/9781119093428.ch86

[26] A. Javaid, J. Hanke, C.H. Simha, M.S. Kozdras, Twin Roll Casting of Magnesium Strip at Canmet Materials — Modeling and Experiments, in: M.V. Manuel, A. Singh, M. Alderman, N.R. Neelameggham (Eds.), Springer eBook Collection Chemistry and Materials Science, Magnesium Technology 2015, Springer, Cham, 2016, pp. 461–464.

[27] J.F. Nie, Y.M. Zhu, A.J. Morton, On the Structure, Transformation and Deformation of Long-Period Stacking Ordered Phases in Mg-Y-Zn Alloys, Metall and Mat Trans A 45 (2014) 3338–3348. https://doi.org/10.1007/s11661-014-2301-6

[28] C. Krbetschek, R. Trân, H. Wemme, M. Ullmann, U. Prahl, D. Rafaja, Hot crack susceptibility of cast Mg 97 Y 2 Zn 1, Engineering Reports 4 (2022). https://doi.org/10.1002/eng2.12380

[29] J.Y. Yang, W.J. Kim, Effect of I(Mg3YZn6)-, W(Mg3Y2Zn3)- and LPSO(Mg12ZnY)-phases on tensile work-hardening and fracture behaviors of rolled Mg–Y–Zn alloys, J. Mater. Res. Technol. 8 (2019) 2316–2325. https://doi.org/10.1016/j.jmrt.2019.04.016

[30] M. Saadati, R.A. Khosroshahi, G. Ebrahimi, M. Jahazi, Formation of precipitates in parallel arrays on LPSO structures during hot deformation of GZ41K magnesium alloy, Mater. Charact. 131 (2017) 234–243. https://doi.org/10.1016/j.matchar.2017.07.007

[31] H. Azzeddine, D. Bradai, Texture and Microstructure of WE54 Alloy after Hot Rolling and Annealing, MSF 702-703 (2011) 453–456. https://doi.org/10.4028/www.scientific.net/MSF.702-703.453

Suppression of non-uniform deformation using die shape in cylindrical deep drawing

MURAOKA Tsuyoshi[1,a,*], OKUDE Yusuke[1], NAKAMURA Isao[1] and KATAGIRI Takashi[1]

[1]Tokyo Metropolitan Industrial Technology Research Institute, 2-4-10 Aomi, Koto-ku, Tokyo, Japan

[a]muraoka.tsuyoshi@iri-tokyo.jp

Keywords: Deep Drawing, Anisotropy, Die Shape, Ear Height, Stainless Steel

Abstract. This study presents a cylindrical deep drawing method that suppresses non-uniform deformation by die shape. In the cylindrical deep drawing, the material inflow in the circumferential direction varies due to the material anisotropy, resulting in unevenness at the edges of the cup, called the ear, after forming. Since the ears are usually trimmed after forming, there are concerns that the ear reduces yield rate and productivity due to the increased number of processes. The main approach study to the suppression of non-uniform deformation due to anisotropy has been the development of materials with suppressed material anisotropy. However, material development has been difficult to introduce because of the increased costs involved, and there has been a lack of development from the press process side. Therefore, this study clarified the effect of die shoulder radius on forming shapes in "Multiple R die" for cylindrical deep drawing by experiments and finite element analysis. Ear height can be suppressed if forming is performed in such a way that the effects of anisotropy and die shape cancel out each other by experimental and analysis results.

Introduction

Deep drawing, shown in Fig. 1, is used for automotive parts and beverage cans as a basic sheet press forming method. In recent years, electric vehicles and other transportation machinery have become increasingly electrified, requiring higher output and lighter-weight motors and batteries [1]). Therefore, further expansion of demand for high-quality and highly functional housings and cases is expected. In press forming, studies have been reported on the application of high-performance materials and improved formability by utilizing forming temperatures and press motions [2, 3]. On the other hand, these studies often require special press machines, which are often restrictive from an equipment standpoint.

From the viewpoint of formability, one of the issues is that ears occur in the formed cups due to non-uniformity of deformation caused by material anisotropy in deep drawing. Thus, there are concerns about yield reduction due to the occurrence of ear and difficulty in forming as designed. The main approach study to the suppression of non-uniform deformation due to anisotropy has been the development of materials with suppressed material anisotropy [4]. Several studies on press forming analysis considering anisotropy have also been reported and its practicality has been examined [5, 6]. However, material development has been difficult to introduce because of the increased costs involved, and there has been a lack of development from the press process side. For suppression of non-uniform deformation, "Multiple R die" in which the die shoulder R shape is varied in the circumferential direction has also been proposed in contrast to the conventional "Single R die" in which the die shoulder R dimensions are uniform [7]. The effect of "Multiple R die" on forming behavior is shown in Fig. 2. In "Multiple R die", the blank folder force can be relatively changed in the circumferential direction by changing the die shoulder R in the circumferential direction, and material flowing into the die can be expected to be controlled.

However, previous study has been limited to "Multiple R die" shapes and have not adequately examined the forming shape relative to the die shape.

In this study, cylindrical deep drawing experiments using a "Multiple R die" were conducted on stainless steel, and the ear height and wall thickness distribution were measured to clarify the basic machining characteristics in the "Multiple R die". Finite element analysis was also conducted to clarify the relationship between material anisotropy and "Multiple R die".

Deep Drawing by "Multiple R die"

Die shape. The determination of the method in the die shoulder shape of "Multiple R die" is shown in Fig. 3. The die R shape of "Multiple R die" can have various possible shapes. In this study, the die R shape is defined as a die shoulder R dimension function and determined by the following procedure.

(1) Set R_{Dmin} and R_{Dmax} to be the minimum and maximum die shoulder R dimensions.

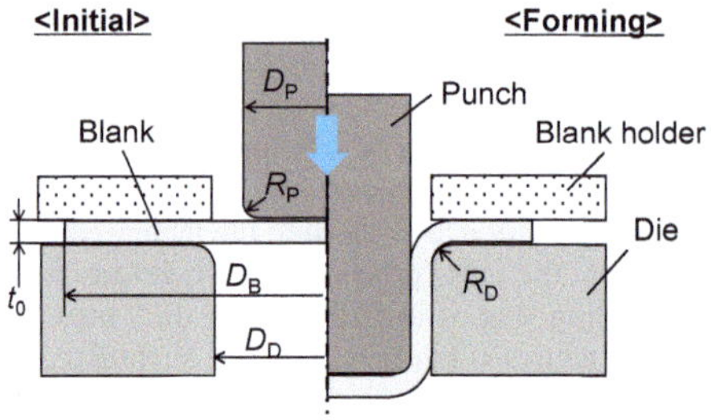

Fig. 1. Cylindrical deep drawing.

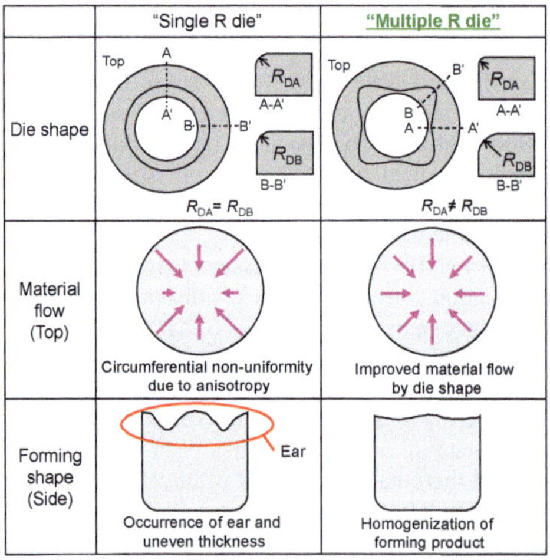

Fig. 2. Effect of "Multiple R die" forming.

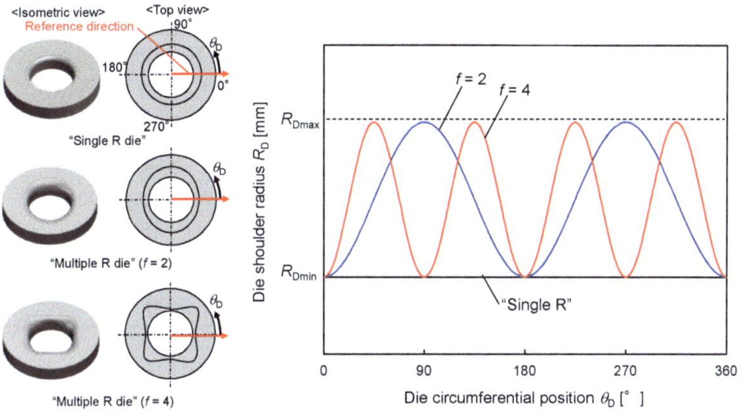

Fig. 3. The determination of the method in die shoulder shape of "Multiple R die".

(2) Set a shape factor f.

(3) Eq. 1 determines the die shoulder R dimension R_D at the circumferential die angle position θ_D.

$$R_D(\theta_D) = \frac{R_{Dmax}+R_{Dmin}}{2} - \frac{R_{Dmax}-R_{Dmin}}{2}\cos(f \cdot \theta_D) \tag{1}$$

In this study, the dies were formed using $f = 2$ and 4 dies shown in Fig. 3. The direction in which $\theta_D = 0$ ° will be referred to as the die reference direction.

Material anisotropy. To investigate the relationship between material anisotropy and die shape, material anisotropy was measured beforehand. The anisotropy of the material was obtained by tensile test to obtain Lankford values. The tensile test conditions and specimen shape are shown in Table 1 and Fig. 4. In this study, SUS304 (Japan Industrial Standard, JIS), an austenitic stainless steel, was used as the work material. JIS 13B specimens with rolling directions $\theta_B = 0$, 45, and 90 ° were cut from the same lot of sheet as the work material. The Lankford value (r-value), an anisotropy parameter, was calculated from the change in the width direction strain and the longitudinal strain of the parallel section of the specimen [7]. The results of the r-value measurement are shown in Table 2. The highest r-value was obtained at $\theta_B = 45$ °.

Forming experiment

Experimental condition. The experimental conditions for the cylindrical deep drawing test are shown in Table 3. Each dimension of the die is shown in Fig. 1. "SERVOPULSER universal testing machine (SHIMADZU CORPORATION)" was used for deep drawing. In this study, the forming results depend on the material anisotropy and the die position. Hence, the angle between the rolling direction and the die reference direction was defined as the material installation angle α, as shown in Fig. 5. The forming experiment was performed with $\alpha = 0$, 45, and 90 ° for $f = 2$ and $\alpha = 0$ and 45 ° for $f = 4$.

Table 1. Tensile test condition.

Material	SUS304 (JIS)
Specimen shape	JIS13B test specimen
Thickness t_0 [mm]	0.5
Tensile direction (Rolling direction) θ_B [°]	0, 45, 90
Test temperature	Room temperature
Test velocity V_t [mm/min]	20
Strain measurement equipment	Extensometer
Pre-strain ε_{pre} [%]	15

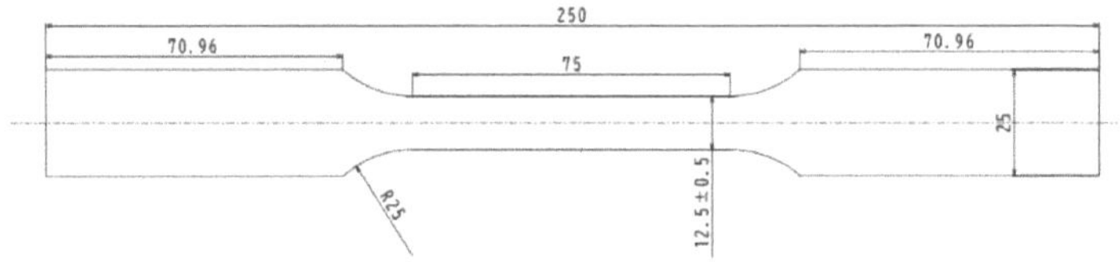

Fig. 4. Specimen shape.

Table 2. Tensile test result.

Rolling direction θ_B [°]	Lankford value $r_{\theta B}$
0	0.852
45	1.388
90	0.781

Formed products were measured for cup height $H_C(\theta_B)$ and external and internal surface shape. A CMM (Coordinate Measuring Machine) was used for the measurements. Ear height ΔH_E was obtained from the difference between the maximum and minimum cup heights $H_C(\theta_B)$. The external and internal surface shapes were measured circumferentially at two heights h = 7, 19 mm from the cup bottom, compared to the approximate height of the formed product of 30 mm. The circumferential wall thickness distribution $t(\theta_B)$ was obtained from the coordinate positions of the outer and inner surfaces. The difference between the maximum and minimum values of the wall thickness distribution was used to determine thickness deviation Δt.

Experimental result. The appearance of the formed product is shown in Fig. 6. Compared to "Single R die", "Multiple R die" changes the ear height under all conditions. Therefore, material flow can be changed by changing the die shoulder R shape.

The cup height distribution is shown in Fig. 7. The cup rim shapes are changed by using "Multiple R die" and the change in material flow can be confirmed. For f = 2, cup height was generally lower than "Single R die", but the cup height distribution itself did not change. For f = 4, the cup height distribution changes abruptly at α = 0 °, and the distribution flattens out at α = 45 °. Therefore, the material inflow can be well controlled by setting f = 4 for "Multiple R die". A comparison of ear heights calculated from cup heights is shown in Fig. 8. For f = 2 of "Multiple R die", the ears were larger than "Single R die" under all conditions. On the other hand, for f = 4 in

Table 3. Experimental condition.

Blank	Material	SUS304 (JIS)
	Outer diameter D_B [mm]	60
	Thickness t_0 [mm]	0.5
Punch	Material	SKD11 (JIS)
	Outer diameter D_P [mm]	28.4
	Shoulder radius R_P [mm]	3
Die (Common)	Material	SKD61 (JIS)
	Inner diameter D_D [mm]	30
"Single R die"	Shoulder radius R_D [mm]	3
"Multiple R die"	Shape factor f	2, 4
	Minimum shoulder radius R_{Dmin} [mm]	3
	Maximum shoulder radius R_{Dmax} [mm]	5, 7 (only f = 4)
Punch speed V_P [mm/s]		5
Blank holder pressure P_B [MPa]		2.94
Lubricant		G-3060M (NIHON KOHSAKUYU CO., LTD.)

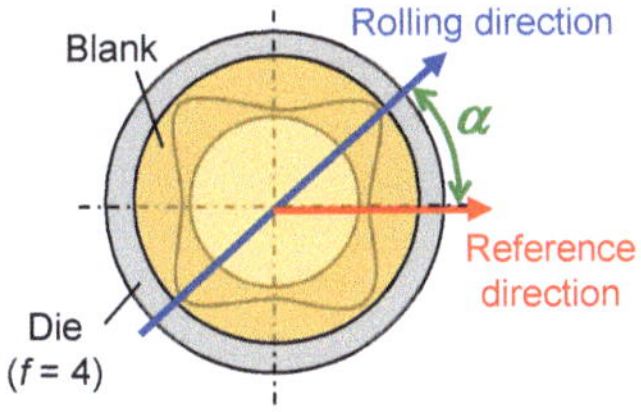

Fig. 5. Material anisotropy and die positioning.

"Multiple R die", the ear heights tended to be smaller at $\alpha = 45$ °.

The circumferential wall thickness distribution of the formed product is shown in Fig. 9. First, in "Single R die" at $h = 7$ mm, near the cup bottom, there was a tendency for the wall thickness to decrease at $\theta_B = 0$ ° and increase at $\theta_B = 45$ °, with a reversed relationship at $h = 19$ mm. Fig. 9(a) and Fig. 9(c) show the forming results at $f = 2$. At both position angles α, "Multiple R die" produced more wall thinning and thickening than the "Single R die". Fig. 9(b) and Fig. 9(d) are the results at $f = 4$. At $\alpha = 0$ °, there is almost no thickness change at $h = 7$ mm, but at $h = 19$ mm, the wall thickness increases at $\theta_B = 0$ ° and decreases at $\theta_B = 45$ °.On the other hand, at $\alpha = 45$ °, the wall thickness distribution was similar to "Single R die" but there was no significant increase or decrease in wall thickness compared to $\alpha = 0$ °. The thickness deviation of the circumferential direction is shown Fig. 10. At $f = 2$, the thickness deviation increased more than that of the "Single R die" under all forming conditions. For $f = 4$, the thickness deviation could be suppressed at

Materials Research Forum LLC
https://doi.org/10.21741/9781644903254-68

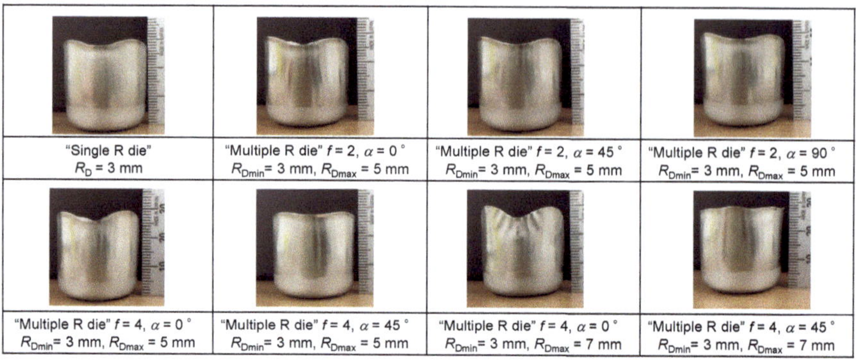

Fig. 6. Formed product.

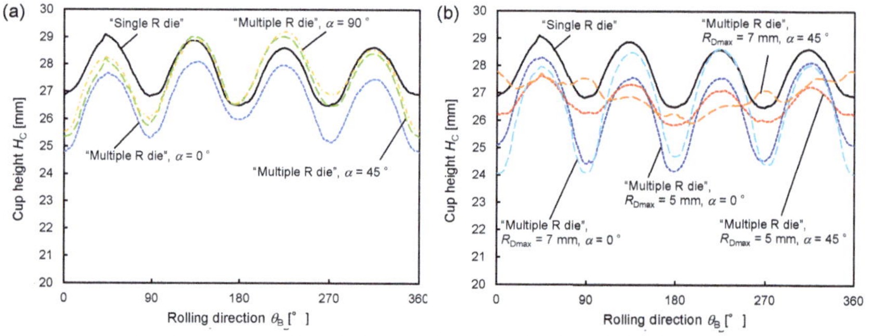

Fig. 7. Effect of die shape on cup height shape (a) "Multiple R die", f = 2 (b) "Multiple R die", f = 4.

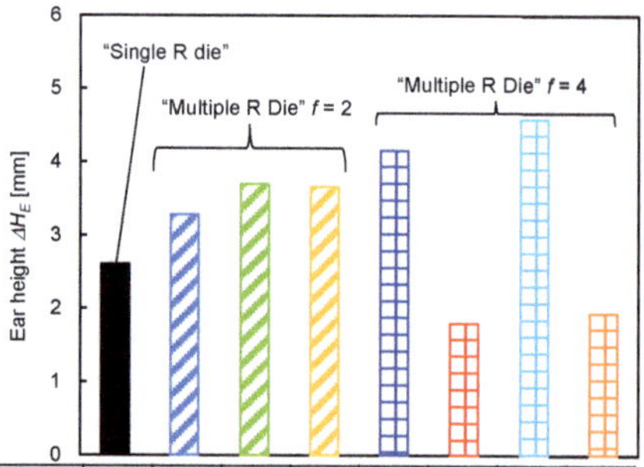

R_{Dmin} [mm]	3	3	3	3	3	3	3	3
R_{Dmax} [mm]	3	5	5	5	5	5	7	7
α [°]	-	0	45	90	0	45	0	45

Fig. 8. Effect of die shape on ear height.

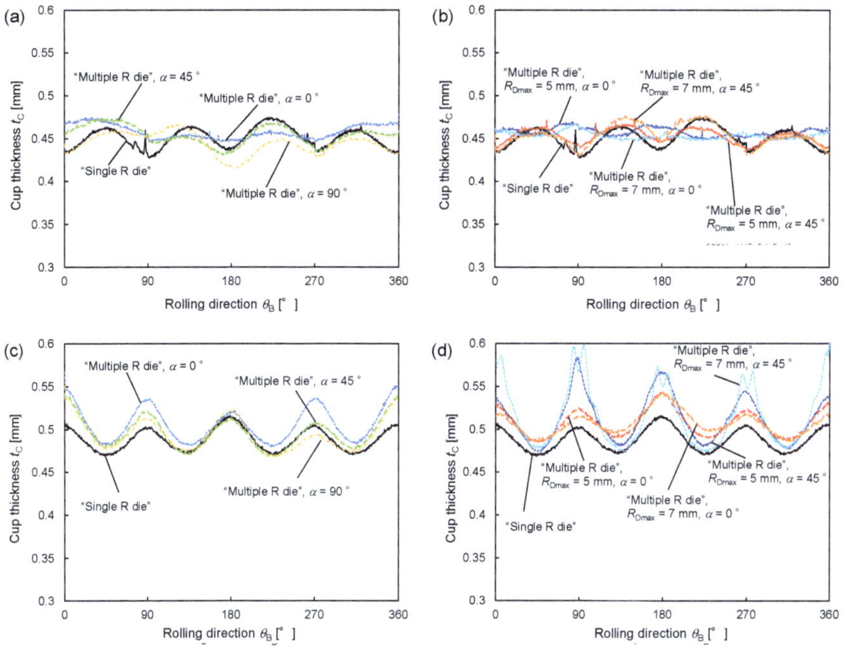

Fig. 9. Effect of die shape on thickness distribution (a) h = 7 mm, "Multiple R die", f = 2 (b) h = 7 mm, "Multiple R die", f = 4 (c) h = 19 mm, "Multiple R die", f = 2 (d) h = 19 mm, "Multiple R die", f = 4.

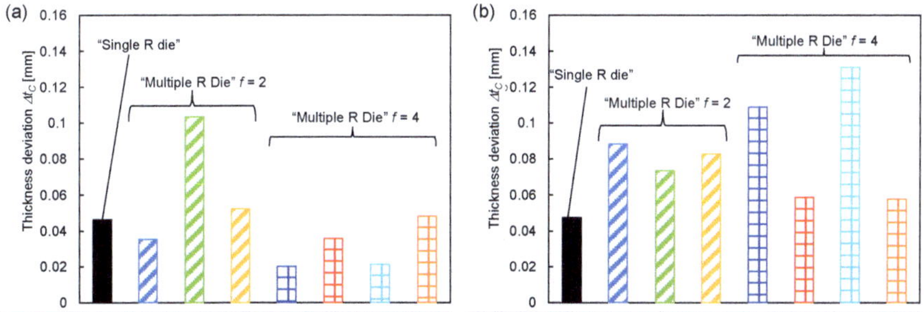

(a)

R_{Dmin} [mm]	3	3	3	3	3	3	3	3
R_{Dmax} [mm]	3	5	5	5	5	5	7	7
α [°]	-	0	45	90	0	45	0	45

(b)

R_{Dmin} [mm]	3	3	3	3	3	3	3	3
R_{Dmax} [mm]	3	5	5	5	5	5	7	7
α [°]	-	0	45	90	0	45	0	45

Fig. 10. Effect of die shape on thickness deviation (a) h = 7 mm (b) h = 19 mm.

h= 7 mm and α=0°, but the thickness deviation increased markedly at h = 19 mm. At α = 45 °, the trend of the thickness deviation was slightly larger than that of "Single R die". Therefore, if "Multiple R die" is formed with f = 4 and α = 45 °, it is possible to form less ear height and the same thickness deviation as "Single R die".

Forming analysis

Analysis condition. Numerical analysis using the finite element method was conducted to discuss the forming experiment results. The analysis condition and analysis model are shown in Table 4 and Fig. 11. In this study, the analysis was carried out using the dynamic explicit method of the general-purpose finite element analysis code "LS-Dyna". The work-hardening characteristics of the work material were approximated by Swift's equation by obtaining the stress-strain relationship from tensile tests. The quadratic Hill yield criterion [9] was used for the anisotropic yield function, and Lankford values from Table 2 were used for the anisotropic hardening parameters. In the forming of "Multiple R die", dies at $f = 4$ which had the greatest effect on the ear height suppression in the experiment were used. In this study, isotropic materials were also analyzed to investigate the effects of pure die shapes only.

Table 4. Analysis condition.

Blank	Element type	Solid
	Outer diameter D_B [mm]	60
	Thickness t_0 [mm]	0.5
	Young's modulus E [GPa]	210
	Material hardening (σ_T: True stress, ε_p: Plastic strain)	$\sigma_T = 1535(\varepsilon_p+0.029)^{0.41}$
Punch	Element type	Shell
	Outer diameter D_P [mm]	28.4
	Shoulder radius R_P [mm]	3
Die (Common)	Element type	Shell
	Inner diameter D_D [mm]	30
"Single R die"	Shoulder radius R_D [mm]	3
"Multiple R die"	Shape factor f	4
	Minimum shoulder radius R_{Dmin} [mm]	3
	Maximum shoulder radius R_{Dmax} [mm]	5, 7
Blank holder pressure P_B [MPa]		2.94
Coefficient of friction μ		0.05

Analysis result. The appearance of typical analysis results is shown in Fig. 12. It is similar to the experimental results shown in Fig. 6, confirming the validity of the analysis results.

The analysis results of the cup rim shape at a punch stroke S of about 15 mm are shown in Fig. 13. The vertical axis is the distance r from the cup center to the cup rim, and the horizontal axis is the angle θ_B from the rolling direction. The trend of the cup rim shape during the forming process is very similar to the results for the height of the formed cup in Fig. 7. Increasing R_{Dmax} also results in a significant change in cup rim shape. When the analysis is performed on isotropic material, the cup rim distribution is qualitatively consistent with the cup rim shape in the "Single R die". Thus, it can be seen that the effect of anisotropy in "Single R die" and the effect of die shape in "Multiple R die" $f = 4$ show similar trends. Therefore, if the directionality of the effects of material anisotropy and die shape is matched ($\alpha = 0$ °), the ears will be larger, and if they are placed so that material and die cancel out each other (α = 45 °), the effect of ear shape will be smaller. In this study, stainless steel was used as the material, and the highest r-value was obtained at θ_B = 45 °. The higher the r-value, the greater the circumferential shrinkage deformation of the flange section and the more the sidewall section becomes a mountain shape [10]. The same effect can be obtained by "Multiple R die" if the materials show the same tendency of anisotropy. On the other hand, a slight increase in thickness deviation is observed when "Multiple R die" is used, as shown in Fig. 9 and Fig. 10.

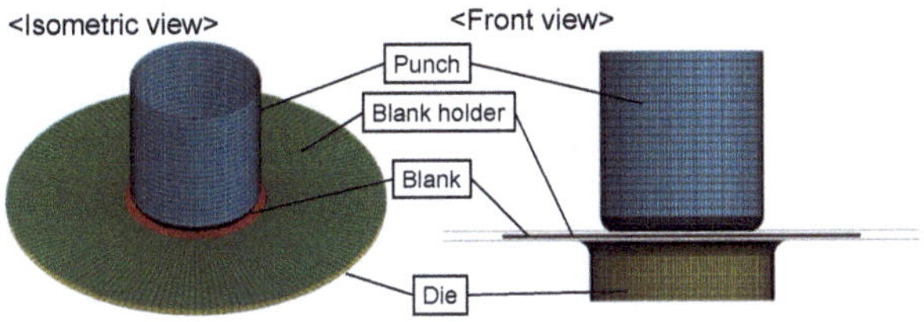

Fig. 11. Analysis model ("Single R die").

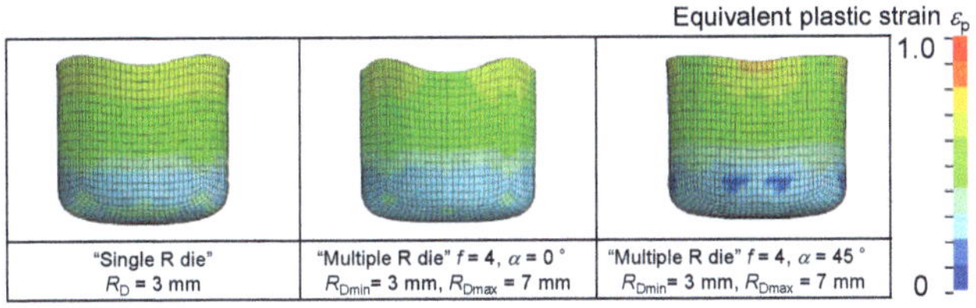

Fig. 12. Analysis result of appearance.

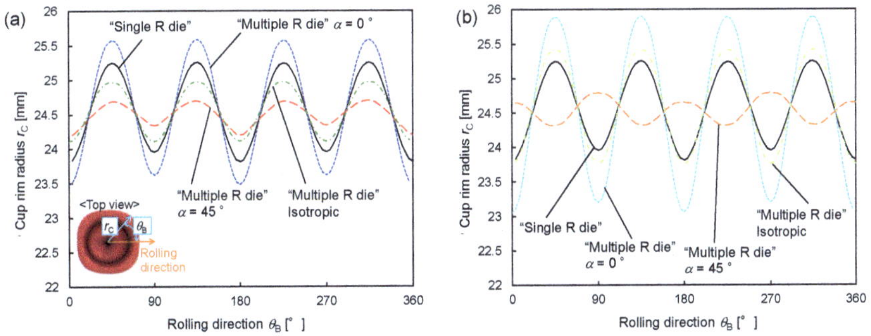

Fig. 13. Analysis results of cup rim shape during forming (Stroke S = approx. 15 mm) (a) f = 4, R_{Dmax} = 5 mm (b) f = 4, R_{Dmax} = 7 mm.

Conclusion

In this study, "Multiple R die" in which the die shoulder R shape was changed in the circumferential direction was proposed for cylindrical deep drawing. Forming experiments and finite element analysis were conducted to investigate the suppression effect of die shape on non-uniform deformation caused by anisotropy. The results are shown below.

- By using a "Multiple R die" the amount of material circumferential flow can be controlled.
- When stainless steel SUS304 (JIS), which has the highest Lankford value in the rolling direction θ_B = 45 °, is formed, the ear height can be suppressed when the shape factor f = 4 and the material installation angle α = 45 °. On the other hand, the thickness deviation in the side wall of the cup tends to be larger than that of "Single R die".
- The results of the finite element analysis showed qualitative agreement between the deformation behavior due to the anisotropic material in the "Single R die" and the deformation behavior due to the isotropic material in the "Multiple R die" at f = 4. Therefore, it was found that ear height can be suppressed if the material and die positions are installed so that the effects of non-uniform deformation due to material anisotropy and the die shape due to "Multiple R die" cancel out each other.

References

[1] T. Ohba, Present Status and Future Prospects of Automotive Weight Reduction -Expectations to Future Plastics and Adhesive Technology, J. Adhesion Soc. Jpn. 49-7 (2013) 260-266. https://doi.org/10.11618/adhesion.49.260

[2] N. Kotkunde, H. N. Krishnamurthy, S. K. Singh and G. Jella, Experimental and Numerical Investigations on Hot Deformation Behavior and Processing Maps for ASS 304 and ASS 316, High Temp. Mater. Proc. 37 (2018) 873-888. http://dx.doi.org/10.1515/htmp-2017-0047

[3] Y. Okude, T. Iwaoka, I. Nakamura, T. Katagiri, T. Muraoka, Development of Method of Cold Incremental Drawing and Ironing Ti-6Al-4V Alloy Sheets, JSTP 62-729 (2021) 133-138. https://doi.org/10.9773/sosei.62.133

[4] Y. Inoue, Masahiro Yamaguchi, Effect of Si content on earing behavior in deep drawing of 3104 aluminum alloy sheets, JILM 73-1 (2023) 17-23. http://dx.doi.org/10.2464/jilm.73.17

[5] H. Hamasaki, Anisotropic yield function of aluminum alloy sheet for accurate press forming simulation, JILM 65-11 (2015) 536-541. https://doi.org/10.2464/jilm.65.536

[6] T. Kuwabara, K. Hashimoto, E. Iizuka, J. W. Yoon, Effects of anisotropic yield functions on the accuracy of forming simulations of bore expansion, JSTP 50-585 (2009) 925-930. https://doi.org/10.9773/sosei.50.925

[7] W. Phanitwong, S. Thipprakmas, Multi draw radius die design for increases in limiting drawing ratio, Metals 10-870 (2020) 1-17. https://doi.org/10.3390/met10070870

[8] W. T. Lankford, S. C. Snyder, J. A. Bausher, New criteria for predicting the press performance of deep drawing sheets, Trans. ASM 42 (1950) 1197.

[9] R. Hill, A theory of the yielding and plastic flow of anisotropic metals, Proc. Roy. Soc. A193 (1948) 281. https://doi.org/10.1098/rspa.1948.0045

[10] T. Ohwue, K. Sato, Y. Kobayashi, Analysis of Earring in Circular-Shell Deep-Drawing Test, Trans. Jpn. Soc. Mech. Eng. A 79-801 (2013), 595-608. https://doi.org/10.1299/kikaia.79.595

Twin roll casting of ZAX210 magnesium wire: Processing, microstructure, texture and mechanical properties

KITTNER Kristina[1,a], ULLMANN Madlen[1,b *] and PRAHL Ulrich[1,c]

[1]Technische Universität Bergakademie Freiberg, Institute of Metal Forming, Bernhard-von-Cotta-Straße 4, 09599 Freiberg, Germany

[a]kristina.kittner@imf.tu-freiberg.de, [b]madlen.ullmann@imf.tu-freiberg.de, [c]ulrich.prahl@imf.tu-freiberg.de

Keywords: Twin Roll Casting, Wire, Texture, Twinning, Dynamic Recrystallization

Abstract. Wires of magnesium alloys are appropriate candidates for applications as filler materials, as biomaterials due to their good biocompatibility and excellent mechanical properties but also as materials for joining applications, such as welding wires or screws. Twin roll casting (TRC) provides an innovative and efficient technology for the production of semi-finished wires with a good property profile. TRC of wire is a comparatively new technology that has so far only been performed at the Institute of Metal Forming at Technische Universität Bergakademie Freiberg, Germany. In recent studies, the process was combined with the promising calcium-containing Mg alloy ZAX210 (Mg-2Zn-1Al-0.3Ca). The wire was produced with an oval cross section under variation of twin roll casting speed in order to investigate the influence of twin roll casting speed (2.5 m/min to 4.5 m/min) on microstructure and texture evolution. Microstructure after TRC consists of fine dendrites and locally deformed microstructure. Twinning and small globular recrystallized grains can be detected as well. Furthermore, it can be observed that higher twin roll casting speeds (4.5 m/min) result in more pronounced formation of intermetallic phases and segregations but also in a higher amount of recrystallized grains, which arise due to dynamic recrystallization during twin roll casting. Texture analysis reveals the development of strong basal textures at lower (2.5 m/min) twin roll casting speed and weakened textures with prismatic character at increased twin roll casting speed of 4.5 m/min.

Introduction

The twin-roll casting (TRC) processes, that combines both solicitation and rolling in a single step, is known to allow the production of near net shaped products. In the last decades TRC for magnesium alloys was widely investigated in laboratory as well as in industrial scale. Compared to conventional processing routes for magnesium semi-finished products, twin-roll casting can save energy and capital costs and reduce material losses [1–3]. The research activities are mainly focussed on the production of sheets and strips. However, in 2017, a pilot plant for twin-roll casing of magnesium wire was installed at the Institute of Metal Forming (IMF) at Technische Universität Bergakademie Freiberg. The TRC wire pilot plant enables the production of wires with an oval cross sections 20 mm in width and 9.2 mm in height. First investigations were done with the magnesium alloy AZ31 [4].

Conventional production processes for magnesium wire consist of continuous casting combined by several steps of extrusion and heat treatment. It is well known, that extrusion can result in fine-grained microstructures that lead to high strength values of the rods or wires. Tong et al. (2015) [5] present a Mg-5.3Zn-0.6Ca alloy with an ultimate tensile strength (UTS) of 279 MPa after extrusion at 300 °C. Further results are shown for example by Bazhenov et al. (2021) [6] with 352 MPa ultimate tensile strength of a Mg-4Zn-1Mn-0.7Ca alloy or Du et al. (2013) [7] with 378 MPa UTS of a Mg-4.5Zn-1.13Ca alloy. However, deformation processes like extrusion contribute to the development of strong basal textures having a negative impact on ductility and formability of magnesium alloys [8–10]. Strong basal textures result in poor ductility at room temperature due to insufficient number of available slip systems.

In order to overcome those drawbacks, research is focussed on addition of alloying elements to magnesium, which lead to weaker basal textures, and suitable processing routes, that also contribute to texture weakening by dynamic recrystallization (DRX). Tong et al. (2015) [5] showed the deviation and weakening of conventional basal textures after extrusion of a Mg-5.3Zn-0.6Ca magnesium alloy. DRXed grains exhibited inclined c-axis about 30° to 40° to extrusion direction. Zhang et al. (2012) [11] also showed texture weakening by Ca addition to a Mg-1Zn magnesium alloy during extrusion. A for Ca addition characteristic basal pole split developed, which leads to enhanced ductility and formability. Wang et al. (2017) [12] reported a split texture with double peaks of misorientation angle distribution by 30° to 60° in extrusion direction. The authors suggest that Ca addition reduced the generalized stacking fault energies of basal slip and prismatic slip and therefore promoted the activity of prismatic slip systems. Former investigations of Ullmann et al. (2019) [13] of twin roll cast and hot rolled sheet of the ZAX210 alloy also reveals that a texture with basal pole split and lower intensity develops. The texture weakening was attributed to the occurring recrystallization mechanisms, which were mainly identified as twin-induced dynamic recrystallization (TDRX).

In this study, twin roll casting of wire was performed on the unique pilot plant at the IMF. The Ca-containing magnesium alloy ZAX210 (Mg-2Zn-1Al-0.3Ca) was used in order to investigate the influence of twin roll casting speed on microstructure and texture development. The responsible mechanisms, which may arise during TRC, were discussed.

Experimental Procedure

In this study, ingots of the magnesium alloy ZAX210 (Mg-2Zn-1Al-0.3Ca) with a chemical composition according to Table 1 was used. Twin roll casting was conducted at a pilot plant for TRC of magnesium wire at the IMF (Fig. 1a). The ingots were melted in an electrical furnace at temperatures from 690°C to 715°C under protective gas atmosphere. The melt was transferred into a headbox using a dosing tube and finally, the melt was forwarded into a casting nozzle, which feds the melt into the water-cooled closed groove. Further information can be found in [14,15,4]. The twin roll casting speed was varied between 2.5 m/min to 4.5 m/min in 0.5 increments. The twin roll cast wire exhibits an oval cross section with 20 mm width and 10.5 mm to 11.7 mm height (Fig. 1b). The edges of the TRC wire exhibit burrs, which occur, when the melt flows into the narrow gap between the upper and the lower roll. If further processing is intended, burrs must be removed from the surface of the wire.

Table 1. Chemical composition of the ZAX210 magnesium alloy in mass percent measured via optical emission spectrometry.

Zn	Al	Ca	Si	Cu	Fe	Ni	Mg
1.89	0.78	0.28	0.011	<0.001	0.006	0.003	Bal.

The twin roll casting process was accompanied by the development of finite element-based model using Abaqus. This enables information to be obtained on the temperature profile within the rolling gap and the mechanisms during solidification. The following assumptions were made prior to the simulation: Melt is a Newtonian fluid, melt domain (nozzle and the region between rollers) tundish is a closed channel and the boundary condition at the walls is non-slip/shear. For density 1,750 kg/m³ was used. The viscosity varies linearly with the solid fraction. Velocities for inlet and outlet were set to 2.5 m/min and 4.5 m/min with regard to the experimental procedure. The heat flux to the rollers, the air and the nozzle were estimated to be 2,000,000 W/m², 50 W/m² and 0 W/m^2. Melt domain and boundary conditions as well as the used meshing systems are shown in Fig. 2.

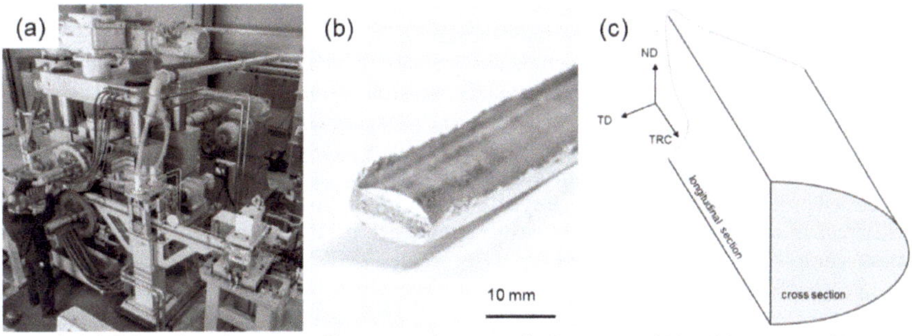

Fig. 1. (a) Pilot plant for twin roll casting of wire at the Institute of Metal Forming, (b) twin roll cast wire of the ZAX210 magnesium alloy and (c) schematic view of the wire for microstructure analysis.

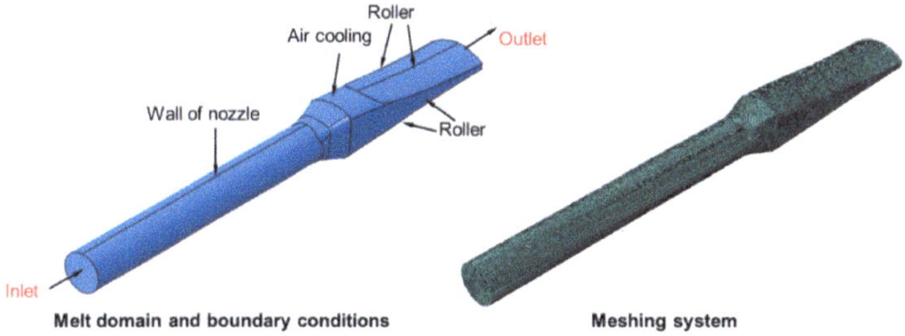

Fig. 2. Melt domain and boundary conditions and meshing systems of the applied twin roll casting system for the production of wire at the Institute of Metal Forming.

After twin roll casting samples for microstructural analysis were cut from the wire according to Fig. 1c with regard to the cross and the longitudinal section. Samples were metallographically prepared by several steps of grinding and polishing with oxide polishing suspension. After polishing and cleaning in an ultrasonic bath, the samples were etched with a picric acid solution for microstructural characterization. Optical and scanning electron microscopy (SEM) was conducted with the help of Zeiss GeminiSEM 450 in order to analyse microstructural characteristics with regard to the twin roll casting speed. Energy dispersive X-ray spectroscopy (EDX) was used to determine the chemical composition of individual microstructural constituents. Acceleration voltage of 20 kV and a working distance of 8.5 mm were applied. To investigate orientation relationships of the grains, electron backscatter diffraction (EBSD) analysis was used. The EBSD operating conditions were 15 nA probe current with 20 kV accelerating voltage and a step size of 0.5 µm to 2 µm. For analysis of the EBSD data and the calculation of pole figures the free MTEX MATLAB toolbox [16] was used.

Tensile tests at room temperature were performed to determine the mechanical properties in TRC direction. Sample shape and dimension were selected according to DIN 50125-A5x25. Traverse speed was 0.625 mm/min.

 Materials Research Forum LLC
https://doi.org/10.21741/9781644903254-69

Results and Discussion

Fig. 3 shows the microstructure of the twin roll cast wire in the polished and the etched state, exemplarily for the wire twin roll cast with a twin roll casting speed of 4.5 m/min. The polished state of the longitudinal and the cross section reveals, that fine and coarse intermetallic phases are evenly distributed, with fewer phases only occurring in the peripheral area. After etching a homogenous fine dendritic microstructure can be observed in the macroscopic micrographs. Strong centre line segregation, as it is reported for twin roll cast magnesium sheet [17,18], cannot be found in the twin roll cast wire of the ZAX210 magnesium alloy. Furthermore, the three characteristic areas, that can be found in TRC sheet (edge zone, dendritic columnar zone and central equiaxed zone) [19], are not obvious in the longitudinal section of the TRC wire. Because of the closed groove, cooling takes place evenly from all sides, so that the solidification conditions are different from those for the twin roll casting of sheet metal and consequently, the resulting microstructures differ from each other [19].

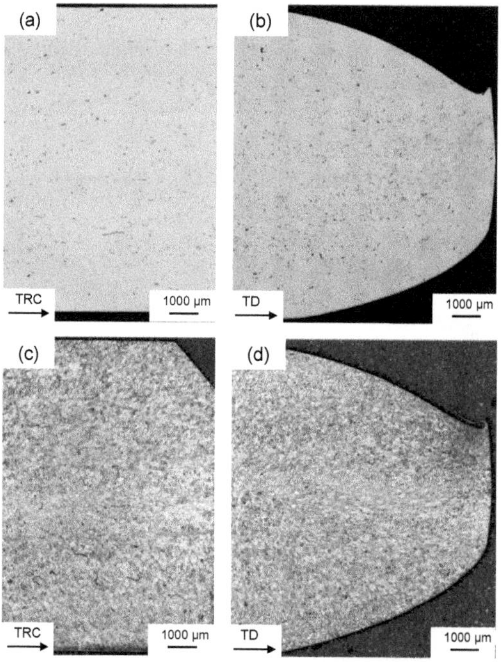

Fig. 3. Optical micrographs of the twin roll cast wire, exemplarily shown for a twin roll casting speed of 4.5 m/min in the polished state (a) longitudinal and (b) cross section and in the etched state (c) longitudinal and (d) cross section.

Detailed views of the microstructure are shown in Figure 4 for twin roll casting speeds 2.5 m/min, 3.5 m/min and 4.5 m/min, respectively. The micrographs refer to the middle and the edge of the cross section of the wire. Especially in the middle section, high amounts of twins can be observed in all samples. However, lower twin roll casting speeds seem to lead to a more pronounced formation of twins. Beetles et al. (2012) [20] reported that twins can be easily activated in Ca-containing magnesium alloys and may be responsible for an accelerated dynamic recrystallization (DRX) due to twin-induced dynamic recrystallization (TDRX). Basis for this mechanism is the formation of double twins. Double twinning is defined as the simultaneous or partly nucleation of secondary twinning at parent twin-twin interfaces. Most commonly reported double twin structures are parental {1011} or {1013} primary twins and secondary {1012} twins [21,22]. For the magnesium alloy ZAX210, the formation of double twins after hot deformation was presented by Kittner et al. (2022) [23].

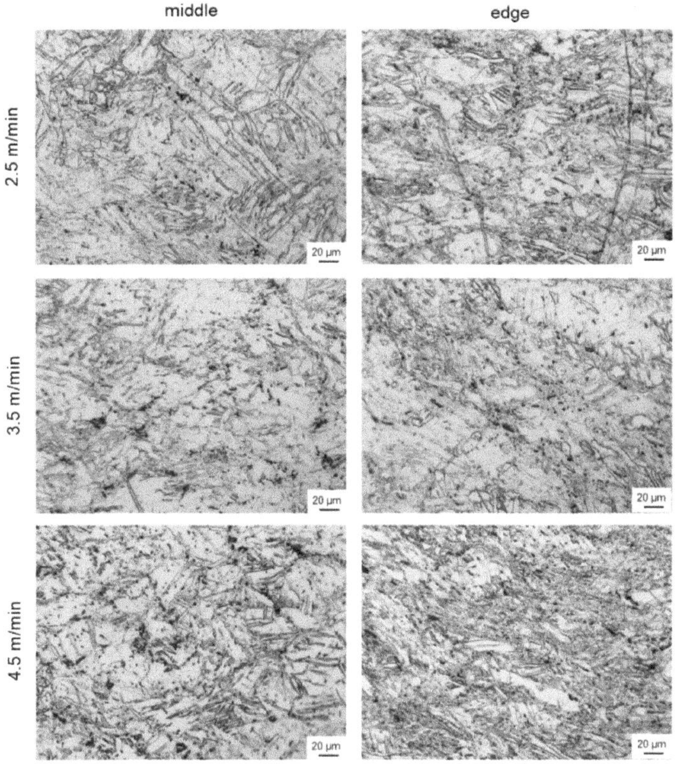

Fig. 4. Optical micrographs in detailed view of the middle and the edge of the cross section of the twin roll cast wire at twin roll casting speed 2.5 m/min, 3.5 m/min and 4.5 m/min.

Twinning analysis in Fig. 5 shows the frequency of misorientation angles in a range of 5° to 95°. There were basically three recognizable peaks. These can be assigned to double twinning (30°), twinning with twin plane {-2111} (75°) and extension twinning with twin plane {10-12} (86°). Commonly reported double twin types are primary twins of the {10-11} or {10-13} planes with {10-12} twinning within the original twin interior [24]. The mentioned twin types have already been identified in magnesium alloys by other research groups [25]. The occurrence of double twinning is known to have a favourable effect on DRX of Ca-containing magnesium alloys [23]. Higher number of twins at 2.5 m/min twin roll casting speed can be assigned to the existing solidification conditions.

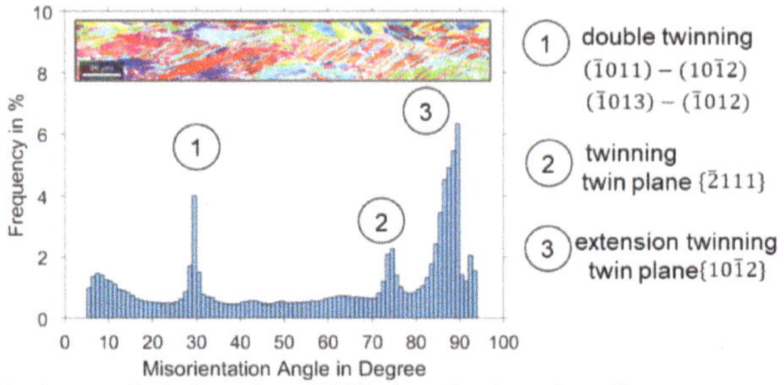

Fig. 5. Twinning analysis based on EBSD data for the twin roll cast magnesium wires.

The temperature distribution was simulated at a low (2.5 m/min) and a higher (4.5 m/min) twin roll casting speed (Fig. 6). At both speeds, it is clear that the edge area cools rapidly as the melt contacts the rolls. The temperature profile shows a more uniform cooling of the melt at the slower twin roll casting speed, which in turn leads to a more uniform solidification. Increasing twin roll casting speeds result in a delayed cooling of the melt at the centre of the wire and therefore to a change in the solidification and deformation conditions. From the FE-based model, it can be suggested, that at lower twin roll casting speeds (2.5 m/min), the maximum temperature of the wire when leaving the roll gap is nearly 200°C in the centre and 100°C at the edges. Consequently, an almost uniform temperature distribution over the cross-section can be achieved. Twinning can easily be activated at lower temperatures owning to the low critical resolved shear stress (CRSS) [26,27].

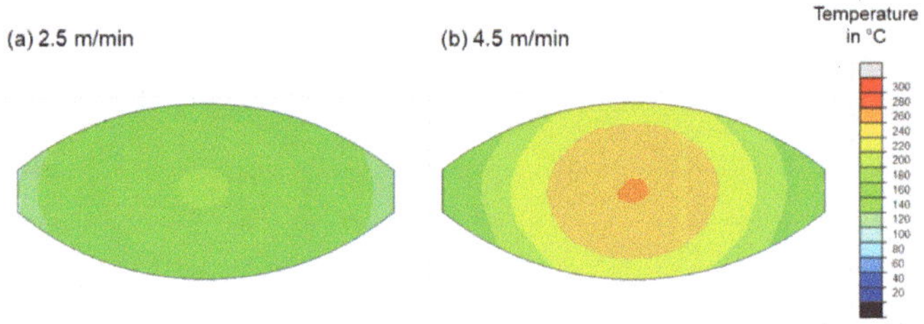

Fig. 6. Temperature distribution of the twin roll cast wires at 2.5 m/min and 4.5 min at the outlet of the roll gap.

With increasing twin roll casting speed, the temperature difference form edge to centre also increases and the maximum centre temperatures rise to values of 260°C to 280°C. A temperature of 220°C extends to the very edge of the wire. Those temperatures during the deformation, when the wire passes the closed groove, contribute to the initiation of dynamic recrystallization processes. For a twin roll casting speed of 4.5 m/min the formation of fine grains in the centre as well as at the edges can be observed. The amount of fine grains at the edges was determined to be 18 %, 29 % and 43 % for twin roll casting speeds of 2.5 m/min, 3.5 m/min and 4.5 m/min, respectively and is therefore higher compared to the core of the wire (< 10 %). This can be attributed to the higher strain that arises during deformation at the top and the bottom of wire, where the largest reduction takes place. The ZAX210 alloy exhibits a promoted recrystallization, especially at high strains [23,19]. Dynamic recrystallized grains are mainly located along the grain boundaries of the original microstructure or at twin boundaries. Therefore, continuous dynamic

Materials Research Forum LLC
https://doi.org/10.21741/9781644903254-69

recrystallization (CDRX) and TDRX are assumed to be the main softening mechanisms during twin roll casting of ZAX210 wire at higher twin roll casting speeds.

Scanning electron microscopy (SEM) and energy dispersive X-ray spectroscopy (EDX) analysis provide information about the composition of the microstructure. Fig. 7 presents SEM images of the twin roll cast wire at 2.5 m/min and 4.5 m/min twin roll casting speed. At lower twin roll casting speeds, the intermetallic phases (light grey) are finely distributed in the interdendritic areas. Because of their small size (< 2 µm) determination of the chemical composition via EDX is difficult and not reliable. X-ray diffraction (XRD) analysis of twin roll cast ZAX210 sheets conducted by Kittner et al. (2019) [19] indicates the occurrence of Mg_2Ca, MgZn-phases and $Ca_2Mg_6Zn_3$. Predominantly, the α-magnesium matrix was detected. Due to the more homogenous solidification, the intermetallic phases are finely distributed especially compared to the wire cast at higher twin roll casting speeds. Time for precipitation and diffusion processes is restricted. Locally, enrichment of residual melt with alloying elements can occur and small segregation areas developed. Overall, the intermetallic phases are unevenly distributed. In the segregation areas, the size of the intermetallic phases is larger than 2 µm and the EDX analysis provides a chemical composition of 11 % Ca, 35 % Zn and 54 % Mg, which corresponds to the phase composition of $Ca_2Mg_6Zn_3$ [28]. The chemical composition of the α-magnesium matrix exhibits no variations as a function of twin roll casting speed. Alloying elements contain a mass fraction of 1.4 % to 1.65 % Zn, 0.6 % Al and <0.1 % Ca.

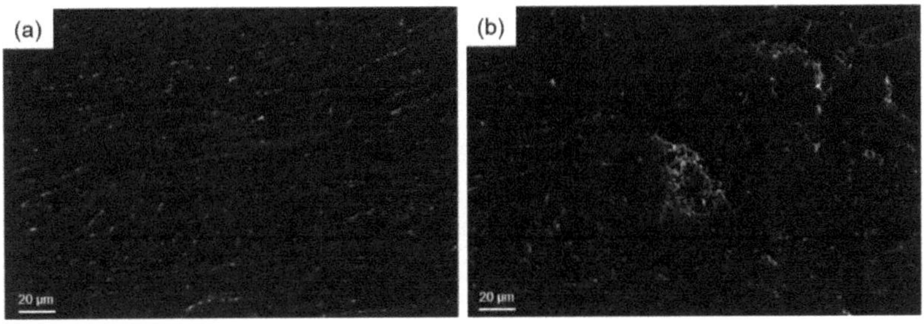

Chemical composition of the magnesium matrix in wt.%

Mg	Zn	Al	Ca	Mg	Zn	Al	Ca
97.92	1.41	0.59	0.08	97.70	1.64	0.60	0.06

Fig. 7. SEM figures and chemical composition of the magnesium matrix, determined via EDX, of the twin roll cast wires at (a) 2.5 m/min and (b) 4.5 m/min.

EBSD analysis was performed to determine orientation relationships of the grains and to calculate pole figures. Fig. 8 presents the microstructure of the twin roll cast wires in inverse pole figure notation together with the calculated (0001) and (10-10) pole figures. It can be observed, that increasing twin roll casting speeds leads to the formation of small recrystallized grains, which exhibit a non-basal orientation (Fig. 8c). At low twin roll casting speed (2.5 m/min) no characteristics of DRX can be found. Resulting texture is dominated by a basal character with a maximum core intensity of 7.5 mrd (multiples of a random distribution) (Fig. 8a). Those textures also develop during deformation processes like rolling or extrusion. Especially deformed microstructures without dynamic recrystallization exhibit strong basal textures [9]. With increasing twin roll casting speed, the maximum core intensity is split in multiple maxima, which are tilted away from ND to TRC direction and texture with prismatic character is developed. The maximum intensity decreased to 4.5 mrd at 3.5 m/min and 3.6 mrd at 4.5 m/min twin roll casting speed. At 4.5 m/min the prismatic character of the texture is more pronounced. The texture weakening by increasing twin roll casting speed of the Ca-containing ZAX210 magnesium alloy is attributed to the dynamic recrystallization. Comparable results were reported for magnesium alloys with Ca addition [11,12].

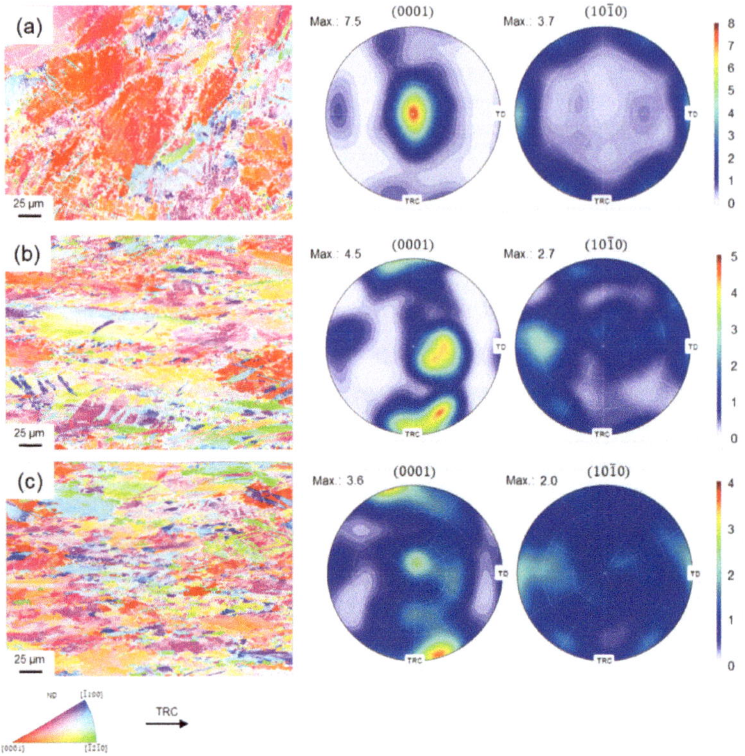

Fig. 8. Inverse pole figures and calculated (0001) and (10-10) pole figures of the twin roll cast ZAX210 wires produced with twin roll casting speed of (a) 2.5 m/min, (b) 3.5 m/min and (c) 4.5 m/min.

In order to determine the mechanical properties of the twin roll cast wires of the ZAX210 magnesium alloy, tensile testing was performed at room temperature. The results of proof stress, ultimate tensile strength (UTS) and elongation at fracture are listed in Table 2 It seems that twin roll casting speed has significant impact on the strength values of the twin roll cast wires. Proof stress and UTS range from 130 MPa to 135 MPa and 245 MPa to 257 MPa respectively. Strong basal textures may lead to limited plasticity because of operative basal slip and extension twinning [27]. However, at higher twin roll casting speed the UTS is lower. This may be due to the softening processes that occur during twin roll casting. The drop of the UTS seems to be less pronounced, as the small new recrystallized grains contribute to an increase in strength according to the Hall-Patch-relationship [29]. However, a significant increase in elongation at fracture can be observed with increasing the content of DRXed grains. Grain refinement can also effectively improve the plasticity of the material. Texture weakening promotes the activity of basal < a > slip systems as well as non-basal slip systems, which can enhance the plasticity significantly [27].

Table 2. Mechanical properties of the twin roll cast ZAX210 wires determined via tensile testing at room temperature.

Twin roll casting speed in m/min	Proof stress in MPa	Ultimate tensile strength in MPa	Elongation at fracture in %
2.5	135	257	9.2
3.5	133	250	13.1
4.5	130	245	21.6

Conclusions

In this study, twin roll casting of the Ca-containing magnesium alloy ZAX210 was conducted to produce wire under variation of the twin roll casting speed from 2.5 m/min to 4.5 m/min. Microstructural and texture development as well as mechanical properties were analysed with regard to the twin roll casting speed. The main results can be summarized as follows:

(1) Microstructures of twin roll cast wires exhibit several types of twinning: double twinning (30°), twinning with twin plane {-2111} (75°) and extension twinning with twin plane {10-12} (86 °). Twinning formation is less pronounced when twin roll casting speed rises.
(2) Increasing twin roll casting speed leads to initiation of dynamic recrystallization. Because of the occurring solidification kinetics, higher temperatures remain in the solidified wire before passing the closed groove. Deformation during TRC results in DRX via CDRX and TDRX.
(3) Because of the occurring DRX processes, texture weakening can be observed at high twin roll casting speed. The texture changes from a strong basal texture with high core intensity (2.5 m/min) to a texture with a prismatic character and basal pole split with lower intensities (4.5 m/min).
(4) The twin roll cast wires offer good mechanical properties with 135 MPa proof stress and 257 MPa UTS as maximum strength values at 2.5 m/min. Increasing twin roll casting speed results in a slight decrease of the UTS to 245 MPa. However, due to the recrystallization processes, which lead to grain refinement and texture weakening, plasticity can be improved. Finally, the elongation at fracture enhances from 9.2 % to 21.6 % when twin roll casting speed was increased from 2.5 m/min to 4.5 m/min.

Funding

The authors would like to thank the Federal Ministry for Economic Affairs and Climate Action for supporting this research work through the project "CLEAN-Mag: CO2-neutral production of lightweight magnesium components", project no. 03LB3080A, part of the Technologietransfer-Programm Leichtbau (TTP LB).

References

[1] R. Kawalla, M. Ullmann, T. Henseler, U. Prahl, Magnesium Twin-Roll Casting Technology for Flat and Long Products - State of the Art and Future, Mater. Sci. Forum 941 (2018) 1431–1436. https://doi.org/10.4028/www.scientific.net/MSF.941.1431

[2] D. Liang, C.B. Cowley, The twin-roll strip casting of magnesium, JOM 56 (2004) 26–28. https://doi.org/10.1007/s11837-004-0122-6

[3] L. Löchte, H. Westengen, J. Rødseth (Eds.), An efficient route to magnesium alloy sheet: Twin roll casting and hot rolling, 2005.

[4] M. Moses, C. Kawalla, R. Kawalla, M. Höck, Development of an Innovative and Quality-Focused Production Technology for Magnesium Wire, MSF 918 (2018) 34–39. https://doi.org/10.4028/www.scientific.net/MSF.918.34

[5] L.B. Tong, M.Y. Zheng, L.R. Cheng, D.P. Zhang, S. Kamado, J. Meng, H.J. Zhang, Influence of deformation rate on microstructure, texture and mechanical properties of indirect-extruded Mg–Zn–Ca alloy, 8th STERMAT on Stereology and Image Analysis in Materials Science 104 (2015) 66–72. https://doi.org/10.1016/j.matchar.2014.09.020

[6] V.E. Bazhenov, A.V. Li, A.A. Komissarov, A.V. Koltygin, S.A. Tavolzhanskii, V.A. Bautin, O.O. Voropaeva, A.M. Mukhametshina, A.A. Tokar, Microstructure and mechanical and corrosion properties of hot-extruded Mg–Zn–Ca–(Mn) biodegradable alloys, J. Magnes. Alloy. 9 (2021) 1428–1442. https://doi.org/10.1016/j.jma.2020.11.008

[7] Y.Z. Du, M.Y. Zheng, X.G. Qiao, K. Wu, X.D. Liu, G.J. Wang, X.Y. Lv, M.J. Li, X.L. Liu, Z.J. Wang, Y.T. Liu, The effect of double extrusion on the microstructure and mechanical properties of Mg–Zn–Ca alloy, 14th International Conference on the Strength of Materials 583 (2013) 69–77. https://doi.org/10.1016/j.msea.2013.06.054

[8] M.G. Jiang, C. Xu, T. Nakata, H. Yan, R.S. Chen, S. Kamado, High-speed extrusion of dilute Mg-Zn-Ca-Mn alloys and its effect on microstructure, texture and mechanical properties, 14th International Conference on the Strength of Materials 678 (2016) 329–338. https://doi.org/10.1016/j.msea.2016.10.007

[9] T. Al-Samman, G. Gottstein, Deformation Conditions and Stability of the Basal Texture in Magnesium, Mater. Sci. Forum 539-543 (2007) 3401–3406. https://doi.org/10.4028/www.scientific.net/MSF.539-543.3401

[10] C. Schmidt, R. Kawalla, Decomposing the basal texture in rolled az31 magnesium sheets, Mg2012 (2012).

[11] B. Zhang, Y. Wang, L. Geng, C. Lu, Effects of calcium on texture and mechanical properties of hot-extruded Mg–Zn–Ca alloys, 14th International Conference on the Strength of Materials 539 (2012) 56–60. https://doi.org/10.1016/j.msea.2012.01.030

[12] G. Wang, G. Huang, X. Chen, Q. Deng, A. Tang, B. Jiang, F. Pan, Effects of Zn addition on the mechanical properties and texture of extruded Mg-Zn-Ca-Ce magnesium alloy sheets, 14th International Conference on the Strength of Materials 705 (2017) 46–54. https://doi.org/10.1016/j.msea.2017.08.036

[13] M. Ullmann, K. Kittner, T. Henseler, A. Stöcker, U. Prahl, R. Kawalla, Development of new alloy systems and innovative processing technologies for the production of magnesium flat products with excellent property profile, Procedia Manuf. 27 (2019) 203–208. https://doi.org/10.1016/j.promfg.2018.12.065

[14] F. Arndt, S. Berndorf, M. Moses, M. Ullmann, U. Prahl, Microstructure and Hot Deformation Behaviour of Twin-Roll Cast AZ31 Magnesium Wire, Crystals 12 (2022) 173. https://doi.org/10.3390/cryst12020173

[15] M. Moses, M. Ullmann, R. Kawalla, U. Prahl, Improving Mechanical Properties of Twin-Roll Cast AZ31 by Wire Rolling, Mater. Sci. Forum 1016 (2021) 957–963. https://doi.org/10.4028/www.scientific.net/MSF.1016.957

[16] F. Bachmann, R. Hielscher, H. Schaeben, Texture Analysis with MTEX – Free and Open Source Software Toolbox, SSP 160 (2010) 63–68. https://doi.org/10.4028/www.scientific.net/SSP.160.63

[17] M. Zimina, M. Šlapáková, J. Bohlen, D. Letzig, G. Kurz, S. Zaunschirm, J. Kastner, M. Cieslar, Center Line Segregation in Twin-Roll Cast AZ31 Magnesium Alloy, Acta Phys. Pol. A 134 (2018) 774–778. https://doi.org/10.12693/APhysPolA.134.774

[18] K. Kittner, M. Ullmann, F. Arndt, S. Berndorf, T. Henseler, U. Prahl, Analysis of defects in a twin roll cast Mg-Y-Zn magnesium alloy, Eng. Reports 17 (2021) 177. https://doi.org/10.1002/eng2.12394

[19] K. Kittner, M. Ullmann, T. Henseler, R. Kawalla, U. Prahl, Microstructure and Hot Deformation Behavior of Twin Roll Cast Mg-2Zn-1Al-0.3Ca Alloy, Materials 12 (2019) 1020.

[20] C. Beetles, M.R. Barnett (Eds.), Advances in Wrought Magnesium Alloys: Fundamentals of Processing, Properties and Applications, 2012.

[21] F. Mokdad, D.L. Chen, D.Y. Li, Single and double twin nucleation, growth, and interaction in an extruded magnesium alloy, Mater. Des. 119 (2017) 376–396. https://doi.org/10.1016/j.matdes.2017.01.072

[22] I.J. Beyerlein, L. Capolungo, P.E. Marshall, R.J. McCabe, C.N. Tome, Statistical analyses of deformation twinning in magnesium, Philos. Mag. 90 (2010) 2161–2190. https://doi.org/10.1080/14786431003630835

[23] K. Kittner, M. Ullmann, U. Prahl, Microstructural and Textural Investigation of an Mg-Zn-Al-Ca Alloy after Hot Plane Strain Compression, Materials (Basel, Switzerland) 15 (2022). https://doi.org/10.3390/ma15217499

[24] M.R. Barnett, Twinning and the ductility of magnesium alloys Part II. “Contraction” twins, Mater. Sci. Eng. A 464 (2007) 8–16. https://doi.org/10.1016/j.msea.2007.02.109

[25] J. Wang, I.J. Beyerlein, Atomic Structures of $$ [0\bar{1}10] $$ Symmetric Tilt Grain Boundaries in Hexagonal Close-Packed (hcp) Crystals, Metall. Mater. Trans. A 43 (2012) 3556–3569. https://doi.org/10.1007/s11661-012-1177-6

[26] C. Xie, Q.H. Fang, X. Liu, P.C. Guo, J.K. Chen, M.H. Zhang, Y.W. Liu, B. Rolfe, L.X. Li, Theoretical study on the {1¯012} deformation twinning and cracking in coarse-grained magnesium alloys, Int. J. Plast. 82 (2016) 44–61. https://doi.org/10.1016/j.ijplas.2016.02.001

[27] L. Wang, Y. Li, H. Zhang, Z. Zhang, Q. Yang, Q. Zhang, H. Wang, W. Cheng, K.S. Shin, M. Vedani, Review: Achieving enhanced plasticity of magnesium alloys below recrystallization temperature through various texture control methods, J. Mater. Res. Tech. 9 (2020) 12604–12625. https://doi.org/10.1016/j.jmrt.2020.09.002

[28] S. Wasiur-Rahman, M. Medraj, Critical assessment and thermodynamic modeling of the binary Mg–Zn, Ca–Zn and ternary Mg–Ca–Zn systems, Fourth International Conference on Bulk Metallic Glasses 17 (2009) 847–864. https://doi.org/10.1016/j.intermet.2009.03.014

[29] H. Somekawa, T. Mukai, Hall–Petch relation for deformation twinning in solid solution magnesium alloys, 14th International Conference on the Strength of Materials 561 (2013) 378–385. https://doi.org/10.1016/j.msea.2012.10.040.

Effect of feed pitch on formability in tube expansion using ball spin forming

OGURI Soma[1,a]*, KAJIKAWA Shohei[1,b] and KUBOKI Takashi[1,c]

[1]Department of Mechanical and Intelligent Systems Engineering, The University of Electro-Communications, 1-5-1 Chofu Gaoka, Chofu-shi, Tokyo, 182-8585, Japan

[a]s.oguri@mt.mce.uec.ac.jp, [b]s.kajikawa@uec.ac.jp, [c]kuboki@mce.uec.ac.jp

Keywords: Tube Forming, Expansion, Ball Spin Forming, Spinning, Aluminum

Abstract. This study proposes a simple and localized forming method using a train of four balls in a tube-hoop direction that expands tube diameter at arbitrary axial positions. The balls travel in the tube direction while rotating around the tube axis. Compared with conventional methods, the expansion mechanism is compact as only one input-rotational drive forms taper and parallel sections. This paper particularly focuses on the effect of feed pitch for the parallel section forming and the ball gap to the cylinder on dimensional precision. An outer cylinder was used for securing the outer diameter as the inner surface of the cylinder can suppress the overshoot of the tube diameter. The experimental results verified the effect of the cylinder. When the ball gap to the cylinder was small enough to push the tube wall to the cylinder-inner surface, the tube diameter was stable at the target value regardless of the feed pitch for the parallel section forming. The tube thickness t decreased to 0.5 mm with increase of the clearance between the balls and the cylinder-inner surface just as designed. However, the thickness could not be less than 0.5 mm.

Introduction

Tubes are used in various fields, such as automotive parts, aircraft parts, heat exchangers, and construction materials, and there are various forming methods to manufacture according to their purpose[1]. In forming a tube, tube end forming is simple. The die can be pressed against the tube end to form the tube end[2]. In the case of tube nosing, a method has been proposed to increase productivity and forming limits by pressing a die with a rotating relief against the tube end[3]. In the case of tube expansion, a method of forming by inserting and moving the plug into the inside of the tube has been proposed[4]. Forming the center of the tube tends to complicate the structure of the machine due to the radial movement of the tool. In the case of diameter reduction forming, spinning has been used as an effective means [5]. Tube expanding in the centers of the tube is more difficult and the structure of the forming machine is more complex. Wines et al. proposed a forming method where the radial displacement of the ball train is realized by a tapered plug inserted inside the ball train [6]. According to this method, the machine is complex because the rotation of the tube, the movement of the taper plug, and the movement of the ball train are controlled independently of each other. This tube expansion processes require large forming equipment and cannot easily expand tubes at workplaces for plant engineering and others.

This study proposes a simple and localized forming method using a train of four balls in a tube-hoop direction that expands tube diameter as shown in Fig. 1. The balls travel in the tube direction while rotating around the tube axis. Compared with conventional methods, the expansion mechanism is compact as only one input-rotational drive forms taper and parallel sections. In a previous study, a prototype machine based on this proposal was built and experiments were conducted for investigating the effect of the feed pitch for the parallel section on the outer diameter and wall thickness. The previous study did not apply a cylinder or other tools as a diameter constraint, and then the outer diameter and thickness were unstable and changed in the tube-axial direction. The results showed that overshoot occurred at all feed pitches and that the amount of overshoot increased as the feed pitch increased. It was also found that the feed pitch did not affect the thickness.

This paper investigates the effect of feed pitch for the parallel section P_2 forming and the ball gap to the cylinder γ_α on the tube-outer diameter and thickness. An outer cylinder was used for suppressing the overshoot of the outer diameter and improving the diameter precision.

Fig. 1. Expanded tube.

Machine configuration and forming process

Fig. 2(a) shows a longitudinal section of the expansion machine and Fig. 2(b) shows a 3D model of the yellow part in Fig. 2(a). Note that parts [A] and [B] are screws. Fig. 3 shows a schematic diagrams of the forming process. The forming process can be divided into the first process for forming tapered section and the second process for forming expanded parallel section. A rotary drive is applied to Bolt 1 in the first and second steps. In the first step of taper forming, Bolt 1 moves forward while rotating along the sliding part [A] of the screw. The ball-pushing jig gives the balls a circumferential orbital motion so that the balls move forward along the plug slope with an increasing orbital radius. This results in taper forming (Fig. 3 (a)). The second step, forming the expanded parallel section, starts when Bolt 1 encounters Bolt 2, and the sliding part transitions from the screw [A] to [B]. The orbital radius of the balls is constant, so parallel sections are formed (Fig. 3 (b)). In this way, the tapered and parallel sections are formed simply by applying a rotary drive to Bolt 1. The amount of expansion γ is expressed by the following Eq. 1.

$$\gamma = 2L_{\mathrm{m}}\tan\alpha - 2G \tag{1}$$

Where L_{m}: clearance between Bolt 1 and Bolt 2, α: the plug half angle, and G: clearance between the tube and the balls at the initial position. The feed pitches P_1 for 1st step and P_2 for 2nd step of ball per revolution are determined by the screw pitches of Bolt 1 and Bolt 2.

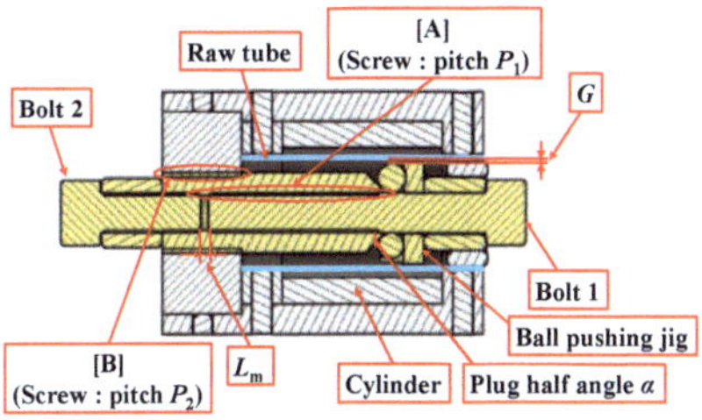

(a)Longitudinal section of expansion machine

(b) 3D model

Fig. 2. Schematic diagrams of the expansion machine.

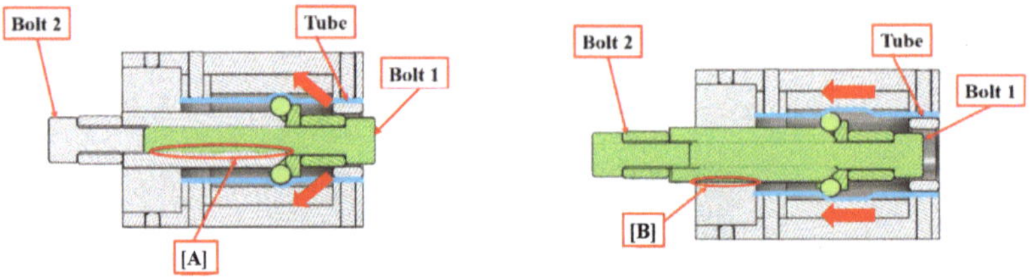

(a) 1st step (b) 2nd step

Fig. 3. Schematic diagrams of forming (Green part: moving tools).

Experimental condition

Table 1 shows the conditions of this experiment. The experiment investigated the effect of feed pitch for the parallel section P_2 on dimensional precision. The effect of the ball gap to the cylinder γ_α on the outer diameter D and thickness t was also studied. The ball gap to cylinder γ_α is defined as shown in Fig. 4. If the gap γ_α is smaller than thickness t, the tube wall is ironed by the balls.

Table 1. The condition of this experiment.

Plug half angle α [°]	45
Feed pitch P_1 [mm·rev^{-1}]	1.5
Feed pitch P_2 [mm·rev^{-1}]	1.25, 1.0, 0.75, 0.5
Ball gap to cylinder γ_α [mm]	1.0, 0.75, 0.5, 0.25
Ball diameter [mm]	7.5
Cylinder inner diameter [mm]	32.03
Material of raw tube	A1070-H14
Length of raw tube l_0 [mm]	30
Outer diameter of raw tube D_0 [mm]	28
Thickness of raw tube t_0 [mm]	1.0

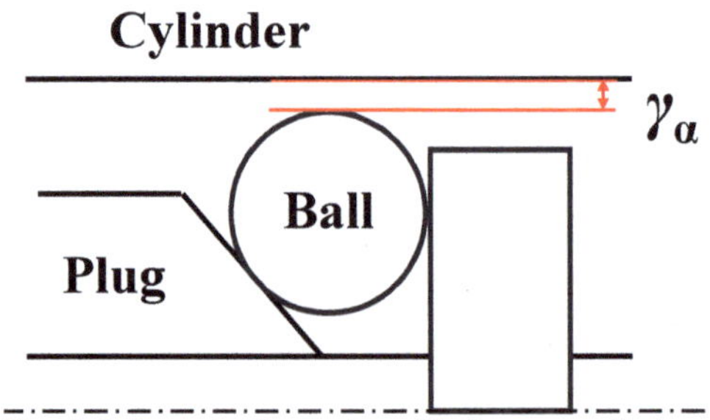

Fig. 4. Schematic illustration of ball gap to cylinder γ_α.

Evaluation of dimensional precision of expanded tube

The axial distribution of the outer diameter D and thickness t of the expanded tube was measured. As shown in Fig. 5, eight lines were scribed at 45° pitch in the circumferential direction, and four lines were scribed at 2 mm pitch in the longitudinal direction from the starting point of the parallel section of the expanded tube, and the outer diameter D was measured using a digital caliper and the wall thickness t using a micrometer at the intersection points. The data obtained were averaged in the circumferential direction to obtain the axial distribution of outer diameter D and wall thickness t.

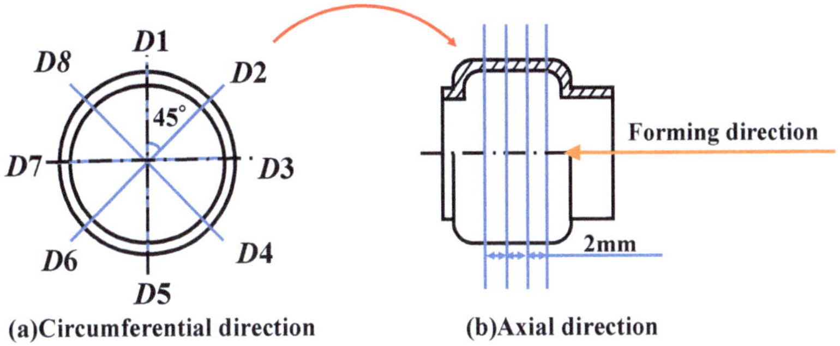

Fig. 5. Position for evaluating thickness t and outer diameter D.

Results and discussion

(1) Outer diameter. Fig. 6(a) shows the effect of feed pitch for the parallel section forming P_2 on outer diameter D when the ball gap to the cylinder γ_α = 1 mm. If overshoot does not occur, the tube-outer surface does not contact the cylinder-inner surface as the gap γ_α is equal to the initial tube thickness t_0 and the thickness after expansion t must be smaller than t_0. When P_2 = 1.0 - 0.5 mm·rev^{-1}, D and increased as P_2 increased. This would be attributed to the overshoot which must increase with increase of P_2. It is noteworthy that the diameter D is equal to the cylinder-inner diameter when P_2 is 1.25 mm·rev^{-1}, and this would be attributed to a large overshoot.

Figs. 6(b), (c), and (d) show the effect of P_2 on D for γ_α = 0.75 mm, 0.5 mm, and 0.25 mm. Under all three conditions, the error of D to the cylinder-inner diameter was within 0.1 mm. This is because the ball gap to the cylinder γ_α was smaller than the naturally-expanded thickness t_n which was an assumed thickness after expansion without the constraints by the cylinder. Therefore, the tube wall must have been ironed between the ball and the cylinder-inner surface. It should be noted that the diameter error to the cylinder diameter increased with decrease of the gap γ_α. This would be due to excessive ironing which might cause unstable deformation.

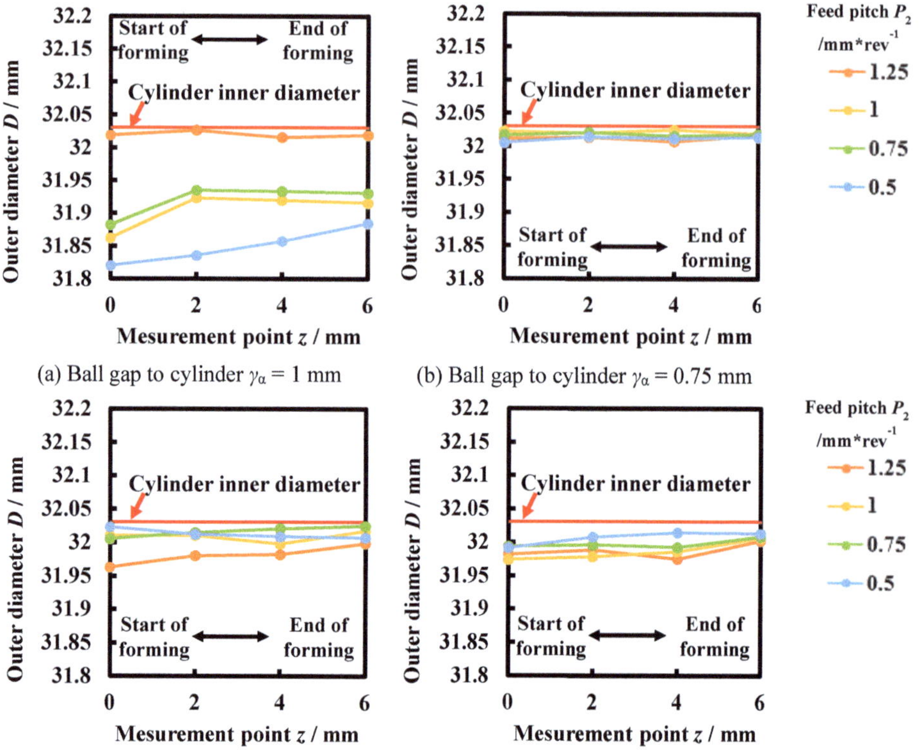

(a) Ball gap to cylinder $\gamma_\alpha = 1$ mm (b) Ball gap to cylinder $\gamma_\alpha = 0.75$ mm

(c) Ball gap to cylinder $\gamma_\alpha = 0.5$ mm (d) Ball gap to cylinder $\gamma_\alpha = 0.25$ mm

Fig. 6. Effect of feed pitch P_2 on outer diameter D.

(2) Thickness. Fig. 7(a) shows the effect of the feed pitch P_2 on the wall thickness t at the ball gap to the cylinder $\gamma_\alpha = 1$ mm. The thickness t was around 0.8 mm regardless of P_2. This is because the gap γ_α was more than the naturally-formed thickness t_n without the cylinder and the thickness was not ironed by the ball.

Figs. 7(b), (c), and (d) show the effect of P_2 on t for $\gamma_\alpha = 0.75$ mm, 0.5 mm, and 0.25 mm, respectively. The target thickness was the ball gap to the cylinder γ_α as the gap γ_α was small enough to iron the tube wall with the balls. The thickness t was close to the target thickness when $\gamma_\alpha = 0.75$ mm, while t was larger than the target thickness when $\gamma_\alpha = 0.5$ and 0.25 mm. This is thought to be due to unstable deformation of the tube wall and elastic deformation of the supporting tools as the ironing deformation was very severe. No clear effect of P_2 on t was observed.

Materials Research Forum LLC
https://doi.org/10.21741/9781644903254-70

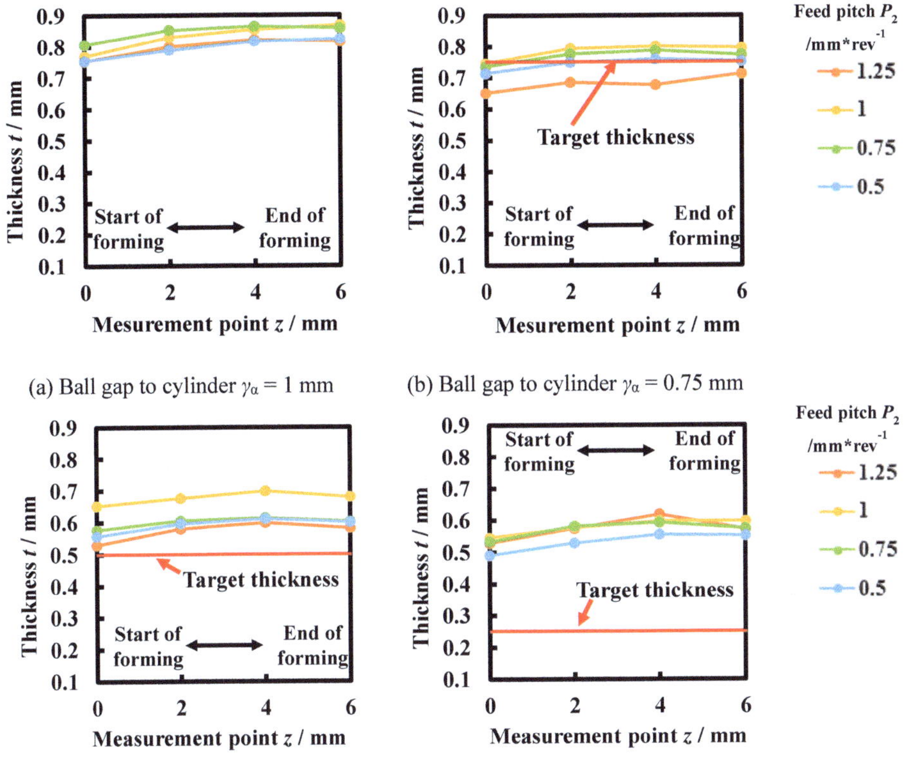

(a) Ball gap to cylinder γ_α = 1 mm (b) Ball gap to cylinder γ_α = 0.75 mm

(c) Ball gap to cylinder γ_α = 0.5 mm (d) Ball gap to cylinder γ_α = 0.25 mm

Fig. 7. Effect of feed pitch P_2 on thickness t.

Conclusion

This paper proposed a simple and localized forming method using a train of four balls in a tube-hoop direction that expands tube diameter at arbitrary axial positions. A prototype tool-set was built up based on the proposed method and experiments were conducted to verify the effectiveness of the proposed method and check the dimensional precision of the formed parts by changing the feed pitch P_2 and ball gap to cylinder γ_α for forming of the expanded parallel section with a cylinder placed on the outer surface of the tube. The following conclusions were obtained.

1) The proposed method successfully expanded the tube diameter in the middle of the longitudinal direction with a concise mechanism using only one input drive. The expansion process is composed of step 1 of taper forming and step 2 of parallel forming.
2) Setting an outer cylinder was effective for securing the outer diameter of the parallel section.
3) The cylinder-inner diameter should be the target diameter, while the ball gap to the cylinder should be the target thickness.
4) The dimensional precision changes depending on the ball gap to cylinder γ_α, and the most appropriate value was clarified for the gap γ_α.
5) When the gap γ_α is equal to the initial tube thickness t_0, the feed pitch P_2 affects the expanded diameter D. When the feed pitch P_2 was large enough, the overshoot was large enough,

expanding the tube diameter to the cylinder-inner diameter. However, the tube thickness t is smaller than the gap γ_α.

6) When the gap γ_α was 0.75 mm for the tube thickness t_0 = 1.0 mm, the tube must have been ironed between the balls and the cylinder-inner surface appropriately, and the dimensional precision was highest, i.e. diameter D was almost the same as the cylinder-inner diameter and thickness t was almost the same as the gap γ_α.
7) When the gap γ_α was smaller than 0.75 mm, the tube must have been ironed severely, resulting in a smaller diameter than the cylinder-inner diameter and a thicker wall than the gap.

References

[1] S. Izumi, K. Kaita, H. Takeda, K. Nisawa, Effect of manufacturing condition of hot rolled steel sheet of formability of Steel Tube for Automobile, Iron and Steel 14 (1974) 2173-2184.

[2] M. Alves, B. Almeida, P. Rosa, P. Martins, End forming of thin-walled tubes, J. Mater. Process. Technol. 177 (2006), 183-187. http://dx.doi.org/10.1016/j.jmatprotec.2006.04.040

[3] T. Kuboki, M. Abe, Y. Yamada, M. Murata, Flexible rotary reduction of tube tips by dies with relief surfaces for attaining high forming limit and productivity, CIRP Annals – Manuf. Tech. 64 (2015) 269-272. http://dx.doi.org/10.1016/j.cirp.2015.04.088

[4] S. Zhang, T. Kuboki, M. Akiyama, S. Kajikawa, Effect of material on strain direction in tube expansion drawing process, Proc. 14th Int. Conf. Techn. Plast. (ICTP2023), (2023), French Riviera, France, 594-602.

[5] M. Murata, T. Kuboki, T. Murai, Compression spinning of circular magnesium tube using heated roller tool, J. Mater. Process. Tech. 162-163 (2005) 540-545. https://doi.org/10.1016/j.jmatprotec.2005.02.199

[6] W. Homberg, T. Rostek, E. Wiens, Internal Flow-Turning - An Innovative Technology for the Manufacture of Tailored Tubes, Key Eng. Mater. 639 (2015) 65-70. http://dx.doi.org/10.4028/www.scientific.net/KEM.639.65

Metal Forming 2024 | Materials Research Forum LLC
Materials Research Proceedings 44 (2024) 663-673 | https://doi.org/10.21741/9781644903254-71

Evaluation of the friction and wear behavior of a-C:H coatings for lubricant-reduced sheet metal forming

REBLITZ Jonas[1,a*], ZETTL Bastian[1,b], SEYNSTAHL Armin[2,c], ORGELDINGER Christian[2,d], TREMMEL Stephan[2,e] and MERKLEIN Marion[1,f]

[1]Institute of Manufacturing Technology, Friedrich-Alexander-Universität Erlangen-Nürnberg (FAU), Egerlandstraße 13, 91058 Erlangen, Germany

[2]Engineering Design and CAD, Universität Bayreuth, Universitätsstraße 30, 95447 Bayreuth, Germany

[a]jonas.reblitz@fau.de, [b]bastian.zettl@fau.de, [c]armin.seynstahl@uni-bayreuth.de, [d]christian.orgeldinger@uni-bayreuth.de, [e]stephan.tremmel@uni-bayreuth.de, [f]marion.merklein@fau.de

Keywords: Physical Vapor Deposition (PVD), DLC Coatings, Sheet Metal, Tribology

Abstract. Industrial companies are confronted with increasing demands for sustainability, challenges in the availability of raw materials as well as rising energy and labor costs. Thus, technological measures for improving the ecologic and economic potential of production are required. An important production stage for the manufacturing of metal-based products is sheet metal forming which consists of several production steps. Initially, the semi-finished sheets are lubricated with oil to reduce friction in the actual forming stage and afterwards they have to be cleaned. One approach for the decrease of production costs and increase of sustainability is the avoidance or reduction of oil-based lubricants and hence subsequent cleaning of the sheet metal parts. However, reducing the amount of lubrication leads to higher friction and wear. Consequently, the part quality as well as tool lifetime are significantly decreased. To meet these challenges, surface modifications for the reduction of wear can be applied. In this study, diamond like carbon (DLC) tool coatings were combined with minimum quantities of a solid lubricant on semi-finished parts to analyze the potential of the coatings for beneficial friction conditions. In this context, suitable a-C:H coating systems were needed to be identified. The associated friction and wear behavior was evaluated as a function of the lubricant amount for different variants of semi-finished parts. Initially, several a-C:H coatings were generated with varying deposition parameters. Promising variants regarding layer adhesion were analyzed for sheet metal forming in terms of tribological behavior. The resulting coefficients of friction (CoF) and wear resistance were investigated with strip drawing tests. For the comparison of the results over different industrially relevant sheet metals, two aluminum alloys as well as one steel material with different zinc coatings were evaluated. For all materials, the amount of solid lubricant was varied in a range below 0.8 g/m^2 up to a completely dry condition without lubrication.

Introduction

Increasing demands for sustainability and the reduction of dependence on fossil energy are current challenges for the industry. Thus, imports of gas are diversified, and renewable energy is focused for power generation in the European Union [1]. Consequently, energy costs in the industry are increasing. In 2022, the price of electricity in Germany rose by around 50% [2] and the price of gas by 150% [3] compared to 2021. Furthermore, the labor costs are 44% higher than the EU average [4]. To remain competitive as a business location ecologically and economically efficient manufacturing processes are essential. One approach in forming technology is the decrease of lubricant amount to enable a reduction of process steps and resource consumption. Conventional steps for processing sheet metal in the automotive sector are lubrication of the sheets, forming, cleaning, joining and coating [5]. By reducing the lubricant amount, the lubrication and cleaning steps become more efficient or can even be omitted. The reduction of lubricant amount can be realized by minimum quantity lubrication or local oiling. However, removing or reducing the lubricant increases friction and wear, which leads to lower component quality and tool lifetime. To meet these challenges, tool modifications can be suitable. These include the application of oxide layers [6], macroscopic [7] and microscopic [8] surface textures as well as DLC coatings [9]. In this study, the tribological behavior of DLC coated tools combined with minimum quantities of solid lubricants on the sheet material are investigated. This should enable the processing of industrial quantities in future applications. For this purpose, a-C:H coatings with and without SiO are deposited with varying coating parameters. Suitable coatings regarding adhesive strength are tested via scratch tests and Rockwell-C indentation tests. Promising coating systems for metal forming are tribologically analyzed with the aid of strip drawing tests. To ensure industry-related conditions three in the automotive sector common materials are used. These include the aluminum alloys AA5182 and AA6014 with the dry lubricant Drylube E1 in quantities of 0.2 to 0.8 g/m^2. Galvanized DC04 (GI) with and without Prime Lubrication Treatment (PLT) is analyzed as well. This enables the derivation of the required lubricant amount as a function of sheet material and tool coating. The tribological target values are the CoF as well as the surface topography and roughness of the friction jaws and sheets.

Material and methods

Coating deposition. To characterize the DLC coatings, they were deposited to the same substrate material as the friction jaws, tool steel 1.2379, hardened to 60 ± 2 HRC appropriate to the following coating process. The specimens were flat round discs with a diameter of 30 mm and a height of 5 mm. Prior to deposition, the surfaces were fine polished to $R_a \approx 0.013$ µm with a 1 µm diamond suspension (Saphir 550.3, ATM Qness GmbH) and then cleaned in an ultrasonic bath in acetone and isopropanol for 10 min each. An industrial-scale PVD coating unit (TT 300 K4, H-O-T Härte- und Oberflächentechnik GmbH & Co. KG) was used for coating production. For this study, two coating systems with hydrogen-containing amorphous carbon (a-C:H) and alternating SiO-doped (a-C:H:SiO) functional layers were deposited under 2-fold substrate rotation. Doping of the SiO content was realized by evaporation of the liquid precursor hexamethyldisiloxane (HMDSO). Before the actual coating process, the coating chamber was evacuated to high vacuum conditions with an initial pressure below 5.0×10^{-4} Pa and subsequently heated to 250 °C for 40 min. Furthermore, the surface of the specimens was plasma etched and cleaned by argon (Ar+)-ion plasma etching for 40 min working with an argon (Ar purity 99.999 %) gas flow of 500 sccm and a bipolar pulsed bias voltage of −500 V (pulse frequency 40 kHz, reverse recovery time 5 µs). Subsequently, a sputtering process was carried out for 3 min with closed shutters to remove impurities from the chromium (Cr) and tungsten carbide (WC) targets (both a purity of 99.9 %) with dimension of 267.5 × 170 mm. The basic coating architecture was characterized by a thin Cr adhesion layer, a graded CrWC intermediate layer and a WC support layer, which were deposited by PVD using unbalanced magnetron (UBM) sputtering in an argon atmosphere. Afterwards, the different functional layers (a-C:H and a-C:H:SiO) were applied in a Plasma Enhanced Chemical Vapor Deposition (PECVD) process using a mixed Acetylene-Ar (C_2H_2-Ar) plasma. The relevant deposition parameters are listed in Table 1. To determine favorable coating parameters, the a-C:H functional layer was deposited on a Box-Behnken DoE, varying the bias voltage (−450 V; −525 V; −600 V), C_2H_2/Ar-ratio (156/104 sccm; 202/58 sccm; 220/40 sccm) and substrate rotation speed (1 rpm; 3 rpm; 6 rpm). Based on the Box-Behnken DoE, the best coating with respect to adhesion behavior was identified with the parameters (bias voltage −450 V, C_2H_2/Ar-ratio 220/40 sccm, substrate rotation speed 3 rpm). A second full-factorial DoE was defined for doping with SiO. The a-C:H and a-C:H:SiO layers were deposited alternately and both the HMDSO flow (2 sccm;

5 sccm; 8 sccm) and the layer changes (2, 25, 50) were varied. In all cases, the functional layer was deposited with a coating time of 8580 s and a chamber temperature of 100°C, whereby the time of the alternating layers was adjusted accordingly. For the coating of the friction jaws, one coating from each of the two coating systems was selected based on the adhesion tests.

Table 1. Relevant deposition parameters for the DLC coatings considering the basic coating architecture without the functional layer.

Layer	Power	Pulse frequency	Reverse recovery time	Duration	Chamber temperature	Bias voltage	Ar flow
Cr	5.0 kW	70 kHz	4 µs	240 s	140°C	−100 V	180 sccm
CrWC	5.0 | 0.3 ↗ 1.2 kW	40 kHz	5 µs	1260 s	140°C	−100 V	180 sccm
WC	1.2 kW	40 kHz	5 µs	1080 s	120°C	−100 V	195 sccm

Characterization of the coatings. The indentation hardness of the coatings was measured by means of microhardness measurement according Oliver and Pharr [10] on a nanoindenter (Picodentor HM500, Helmut Fischer GmbH). A test force of 18 mN was used and 15 indentations were evaluated for both coating systems.
Rockwell indentations (HRC method) were generated (DuraJet 10G5, EMCO-TEST Prüfmaschinen) to characterize the coating adhesion. The adhesive strength was then evaluated under a light microscope (DM4000 M, Leica Microsystems) in accordance with DIN 4856. Five Rockwell indentations were analyzed for each coating variant. In addition, the adhesive strength was evaluated using a scratch test in accordance with DIN EN ISO 20502 with a scratch tester (RTG-2, KTmfk) with a normal force increasing linearly to 100 N over a scratch path of 10 mm. The scratch was optically analyzed with the same light microscope in accordance with the standard and the critical load values *Lc1*, *Lc2* and *Lc3* were determined. Five scratch tests were carried out for each layer variant and the resulting load values were averaged. The statistical evaluation between a-C:H and a-C:H:SiO coatings was performed using the Welch Two Sample t-test in R.

Sheet materials and lubrication. To enable close to industry conditions three common sheet materials and associated solid lubricants were chosen. These include the two aluminum alloys AA5182 and AA6014 with mill finish texture. For both variants, the lubricant Drylube E1 (Zeller + Gmelin GmbH & Co. KG) was used in varying amounts. Tests without tool coating and with completely decreased semi-finished parts were performed as a reference. Additionally, galvanized DC04 with EDT texture was investigated. In this context, variants with and without Prime Lubrication Treatment (PLT) were analyzed. An overview of the investigated sheet materials, lubrication and tool coatings is given in Table 2.

Table 2. Variation of sheet material, lubrication and tool coating.

Sheet material	AA5182	AA6014	DC04
Surface texture	Mill Finish		EDT
Lubricant	Drylube E1		Prime Lubrication Treatment (PLT)
Lubricant amount in g/m^2	0.0, 0.2, 0.4, 0.8		With/without PLT
Tool coating	Uncoated, a-C:H, a-C:H:SiO		

Tribological characterization of DLC coatings. Tribological investigations for different combinations of tool coatings, sheet material and lubricant amount were performed using strip drawing tests, as shown in Fig. 1. The material of the friction jaws is 1.2379, which is hardened to 60 HRC. They have an ideal contact surface of 100 x 55 mm^2 with a bevel.

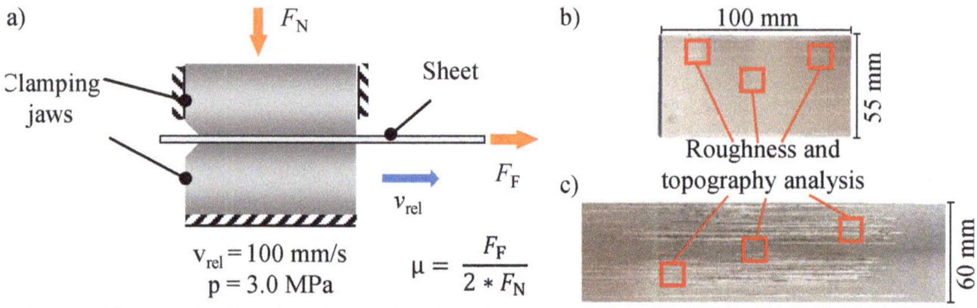

Fig. 1. a) Setup of strip drawing test and surface analysis of b) friction jaws and c) sheets.

The friction jaws were polished to reach a surface roughness of R_a = 0.003 µm before they were tested uncoated or with a-C:H coating and a-C:H:SiO coatings. By coating the jaws, the average surface roughness increases to R_a = 0.012 µm. To avoid a further expensive process step in industrial application, post-processing of the surfaces was not performed. Sheet strips with dimensions 60 x 500 mm^2 were then clamped with a contact pressure of 3 MPa between two friction jaws and drawn at a constant speed of 100 mm/s, as shown in Fig. 1 a). The tensile force was measured using a load cell so that the resulting CoF can be determined between a drawing length of 100 mm and 170 mm. For the evaluation of the surface, a laser-scanning microscope (VK-X200, Keyence) was used. The topography and roughness measurements were analyzed transverse to the drawing direction. Five strip drawing tests were performed with one pair of friction jaws and evaluated in terms of CoF. Three surface measurements per friction jaw were carried out for each parameter set. The measurements represent one area with smooth and two areas with rough surfaces. Regarding the sheet surface, the fourth and fifth test was analyzed.

Results and discussion

Mechanical properties of the a-C:H and a-C:H:SiO coatings. The determined indentation hardness for the a-C:H coatings averaged with H_{IT} = 25.8 GPa and indentation modulus E_{IT} = 200 GPa which is similar or even higher compared to literature [11,12]. For the SiO-doped variants H_{IT} = 21.5 GPa and E_{IT} = 187 GPa are measured which is also higher compared to [13,14]. The SiO-doped coatings therefore achieved slightly lower mechanical properties.

Adhesion strength of the a-C:H and a-C:H:SiO coatings. The adhesive strength of all the coating variants examined using Rockwell indentation could be classified between HF2 and HF4. No systematic differences were identified regarding the used coating parameters. There were also no indications that the SiO-doped coatings exhibited better or worse overall coating adhesion than the pure a-C:H coatings. However, some of the a-C:H coatings exhibited larger contiguous flaking in some variants, presumably due to the slightly higher indentation hardness and thus a brittle fracture of coating flakes. Fig. 2 a) and b) show examples of the best coatings for a-C:H (bias voltage −450 V, C_2H_2/Ar-ratio 220/40 sccm, substrate rotation speed 3 rpm) and a-C:H:SiO (HMDSO flow 2 sccm, 50 layer changes) the corresponding Rockwell indentations. These coating systems achieved an adhesive strength class of HF2, which is comparable to literature [15,16]. The results of the scratch tests are shown in Fig. 2 c). Similar to the Rockwell adhesion strength, there were no differences that can be explained by the coating parameters used for the a-C:H and a-C:H:SiO coatings. On the other hand, the SiO-doped coatings differed significantly from the pure a-C:H coatings regarding the critical loads Lc1 (p = 5.7e-12) and Lc3 (p = 0.0006). Averaged over all coating systems, the value for the critical load Lc1 was 15.0 N for the a-C:H coatings and 10.9 N for the SiO-doped coatings. At 43.8 N, Lc3 for the SiO-doped coatings was also higher than the value of 37.4 N for the pure a-C:H coatings. In contrast, no difference was determined for the critical load Lc2 ($p = 0.95$); the critical load was 25.6 N in both cases. The increased resistance in the scratch test is assumed to be due to the higher indentation hardness of the a-C:H coatings. In comparison to literature values, it must be considered that these can vary considerably depending on the exact coating system and can depend on the device used and the person evaluating it. However, the values calculated are quite common for a-C:H coating systems [16].

The coatings tested showed a coating adhesion typical for a-C:H coatings in this application [15] with the best variants of HF2 in the Rockwell test. As the coating adhesion seemed to be most important for the chosen application, the best a-C:H and a-C:H:SiO coatings in terms of adhesion were selected for further deposition on the friction jaws.

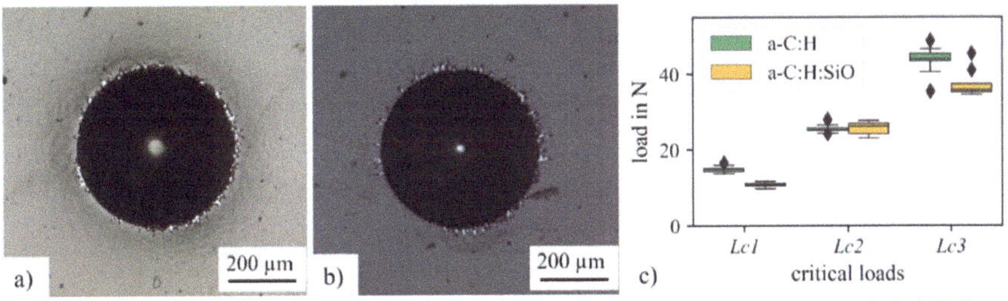

Fig. 2. Exemplary Rockwell indentations of the best a) a-C:H (HF2) and b) a-C:H:SiO (HF2) coatings and c) results of the scratch tests with n = 13 for a-C:H and n = 9 for a-C:H:SiO.

Characterization of friction. As a reference, tests without tool coating and lubrication were performed for both aluminum alloys. As expected, the CoF showed high values of $\mu = 0.41$ for AA5182 and $\mu = 0.55$ for AA6014. Due to high abrasion together with adhesion and thus plastic deformations of the strips, only one valid test could be performed for AA6014. Analogous to the studies of Krachenfels et al. [17], AA6014 seemed to exhibit a higher adhesion tendency under dry conditions. According to Fig. 3 a significant decrease of the CoF could already be reached by the smallest lubricant amount of 0.2 g/m^2 Drylube E1. The CoF was then reduced to $\mu = 0.11$ for AA5182 and $\mu = 0.12$ for AA6014. With increasing lubrication, the friction reduced continuously down to $\mu = 0.03$ for AA5182. Maximum values for CoF were detected for a-C:H:SiO coatings and AA6014 sheets, which might be explained by the higher surface roughness of the semi-finished parts, as shown in Fig. 6. Consequently, the higher roughness of the tools and the sheets led to an enhanced interlocking of the surfaces and thus higher CoF. For galvanized DC04 the friction was significantly lower under dry conditions than for both aluminum alloys. In the case of uncoated friction jaws, the CoF was $\mu = 0.12$. This might be attributed to the zinc coating that was already observed by Krachenfels et al. [17]. They assumed that the hexagonal densely packed lattice structure of zinc exhibited a lower adhesion tendency than the cubic face-centered lattice structure of aluminum. Friction jaws with the a-C:H coating led to a higher CoF of $\mu = 0.16$, which might also be caused by the higher surface roughness. SiO-doped coatings showed a similar friction level as uncoated tools with a CoF of $\mu = 0.12$. Unlike the aluminum alloys, the SiO-doped coatings seemed to reduce friction for DC04. Semi-finished parts with PLT exhibited a slightly lower CoF in general. The lowest value of $\mu = 0.10$ was reached for a-C:H:SiO coatings.

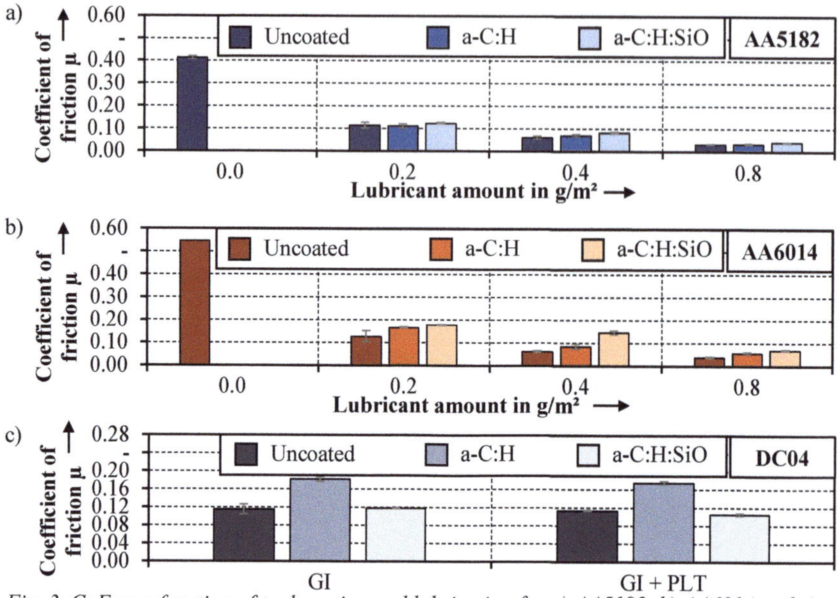

Fig. 3. CoF as a function of tool coating and lubrication for a) AA5182, b) AA6014 and c) DC04.

Characterization of the friction jaw surfaces. Initially, the characterization of the topography of the friction jaws was performed. These are shown in Fig. 4 as a function of tool coating and the lowest and highest lubricant amount. a-C:H coated friction jaws had surface defects. Strong adhesion was observed for AA5182 with 0.2 g/m^2 Drylube E1 and uncoated tools. It seemed that there was still an insufficient surface separation. Otherwise, there was no visible wear for AA5182.

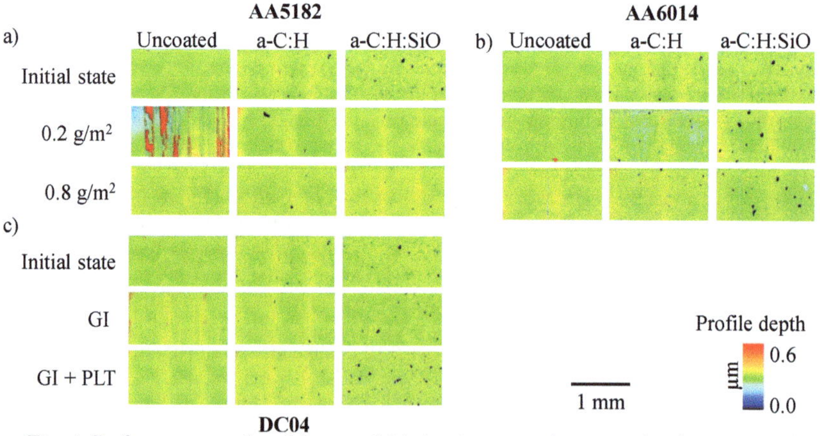

Fig. 4. Surface topography of the tested friction jaws as a function of tool coating and lubrication for a) AA5182, b) AA6014 and c) DC04.

AA6014 exhibited a similar effect, whereas there was only a quiet small area with adhesion at the lowest lubricant amount. Tests with DC04 GI and uncoated friction jaws indicated a slightly increased surface roughness. This might be caused by the direct metallic contact without intermediate medium. Since there were almost no visible signs of roughening, the surface roughness is also evaluated, as shown in Fig. 5.

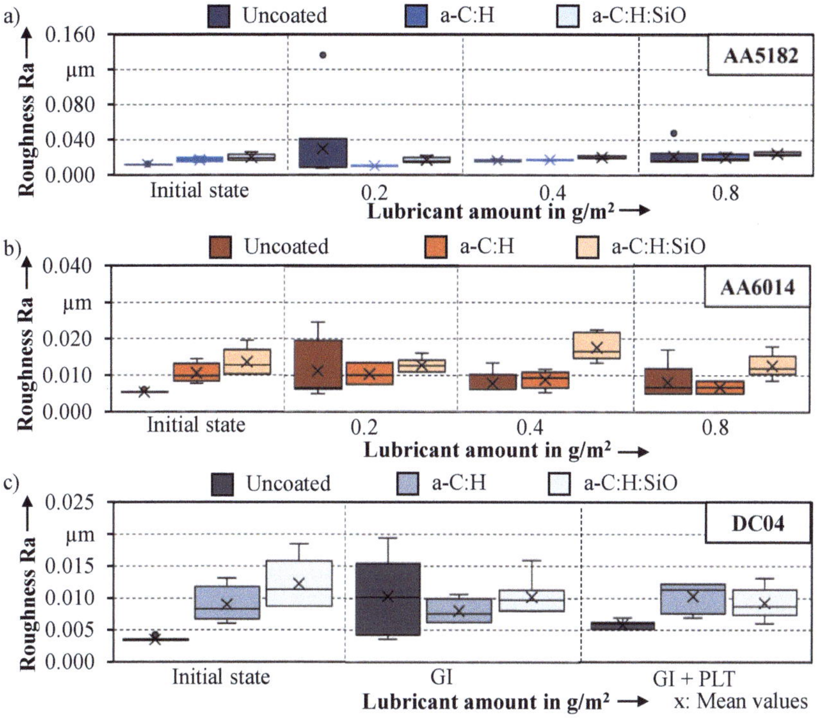

Fig. 5. Surface roughness of the tested friction jaws as a function of tool coating and lubrication for a) AA5182, b) AA6014 and c) DC04.

To derive correlations between the testing conditions and the tool surface a comparison of the roughness with the initial state was necessary. Uncoated friction jaws had an average roughness of $R_a = 0.003$ µm. a-C:H-coated tools showed a roughness of $R_a = 0.009$ µm and $R_a = 0.012$ µm with SiO. For uncoated friction jaws and without lubrication the tests with aluminum led to a high surface roughness of $R_a = 0.127$ µm for AA5182 and $R_a = 0.474$ µm for AA6014 due to strong adhesion. The higher adhesion tendency of AA6014 under dry conditions was clearly visible. By the application of lubricants, a significant decrease of adhesion was achieved. Analogous to Fig. 4a) high surface roughness of $R_a = 0.028$ µm was measured for AA5182 with 0.2 g/m^2 lubrication. When increasing the lubricant amount to 0.4 g/m^2 the surface roughness was on a similar level as in the initial state. A sufficient separation of the surfaces seemed to be reached. Using the a-C:H-coated friction jaws, the surface roughness decreased from $R_a = 0.009$ µm to $R_a = 0.005$ µm and showed no significant change with increasing lubricant quantity. Possibly a smoothing of the friction jaws occurred, which was independent of the lubricant amount. The surface roughness of the SiO-doped a-C:H coatings was similar to the initial state. Thus, the lowest lubricant amount led to acceptable surface protection in context of the strip drawing test. A slight increase in surface roughness was observed for AA6014 with uncoated friction jaws. In contrast to AA5182, almost no adhesion was visible at the smallest amount of lubricant, as shown in Fig. 4. According to Zhao et al. [18] AA6014 had a low tendency to build loose wear particles in contrast to AA5182. This might be one reason why the adhesion tendency was lower for AA6014 under lubricated condition. Only slight differences could be found for a-C:H-coated friction jaws compared to the initial state. However, the surface roughness tended to be lower after the strip drawing tests. Due to the high hardness of the coatings only marginal flattening occurred. For a-C:H:SiO coatings no trend was visible for different lubricant amounts. The surface roughness was at the level of the initial state except for 0.4 g/m^2. Hence there was no significant influence for AA5182 as well, SiO-doped a-C:H layers seemed to be a suitable option to reduce surface roughening for both aluminum alloys. Tests with DC04 without tool coating and lubrication led to an increase of surface roughness up to $R_a = 0.011$ µm. This effect could be decreased by the application of PLT on the semi-finished parts, which resulted in values of $R_a = 0.006$ µm. Compared to the roughness of the friction jaws for both aluminum alloys under dry conditions DC04 caused a low surface roughening. As already discussed for the CoF the zinc coating might lead to low friction and wear. Variants with coated friction jaws exhibited no significant change in surface roughness. In combination with the low CoF for a-C:H:SiO coatings, provided suitable surface protection of the friction jaws for galvanized DC04.

Characterization of the sheet surfaces. The sheets were evaluated in terms of topography and surface roughness, according to Fig. 6. In initial state the surface roughness of the sheet material was between $R_a = 0.168$ µm and 0.309 µm. Because of adhesion, for uncoated friction jaws and the lowest lubricant amount the roughness of the AA5182 sheets increased up to 0.715 µm. By increasing the lubricant amount, the sheet roughness decreased. For 0.8 g/m^2 Drylube E1 AA5182 showed only weak adhesion in the topography images and a roughness of $R_a = 0.240$ µm. Tests with a-C:H coatings tended to higher values than in initial state, but the difference was not significant. However, roughness peaks seemed to be slightly removed. This was also observed for SiO-doped coatings but with a higher surface roughness of $R_a = 0223$ µm than in initial state. The higher roughness of the friction jaws might be the reason for this. For AA6014 a significantly higher surface roughness was measured in the initial state with $R_a = 0.309$ µm. No significant change in roughness was observed for uncoated friction jaws with a value of $R_a = 0.335$ µm. The surface roughness of coated friction jaws was below the initial value, which meant that the higher hardness of the a-C:H coatings results in smoothing due to abrasion or plastic deformation. This was also indicated by the lack of smoothing on the friction jaws in contrast to AA5182. The analyzed AA6014 sheets with SiO-doped coatings on the tools showed small grooves and adhesion as well as an increased roughness. A higher lubricant amount tended to a lower surface roughness on the sheets. In case of DC04 GI, uncoated friction jaws caused a significant increase in surface roughness from $R_a = 0.198$ µm in initial state to $R_a = 0.331$ µm. This could be also seen in the topography images and might be explained by adhesion. By the application of PLT on the semi-finished parts, this effect decreased and resulted in smaller values for the surface roughness. a-C:H coated friction jaws might induce strong abrasion and adhesion. It seemed as if the roughness peaks were removed and then accumulated in the roughness valleys and welded together. It is likely that this was due to the zinc layer, which was removed by the harder a-C:H coatings. This effect decreased by the application of PLT on the sheets, which might be explained by the better separation of the surfaces. Furthermore, PLT could impeded the adhesion of already abraded particles. Using a-C:H:SiO coatings small grooves appeared on the surface of DC04 sheets without PLT. Additionally, zinc abrasion and adhesion were significantly reduced compared to the coatings without SiO. When PLT was applied to the sheets, the formation of grooves was inhibited, and the topography was similar to the tests without tool coating. Thus, slight adhesion occurred on the surface. The elimination of excessive abrasion and adhesion could cause the significant lower CoF compared to a-C:H:SiO coatings. In general, the coating of the sheet material with PLT led to less abrasion and adhesion. Therefore, it was well suited as surface protection. Regarding the sheet surfaces, SiO-doped a-C:H coatings were most suitable for galvanized DC04 sheets.

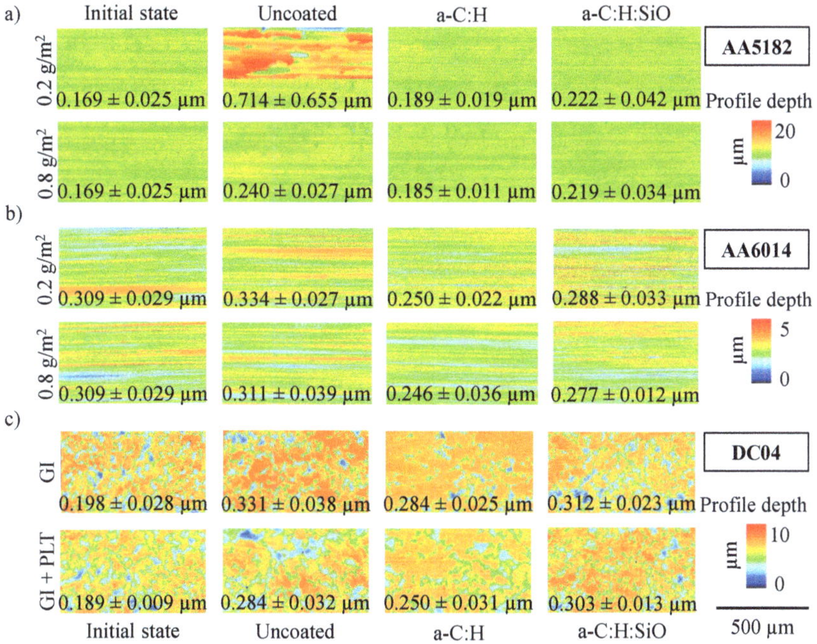

Fig. 6. Surface topography and surface roughness R_a of the tested sheets as a function of tool coating and lubrication for a) AA5182, b) AA6014 and c) DC04.

Summary

In general, the reduction of lubricant amount leads to an increase of friction and wear in metal forming. To decrease this influence, a-C:H coatings with and without SiO doping were applied to tool surfaces. Using Rockwell C indentation, the adhesive strength class HF2 was achieved for both coatings systems. In scratch tests, the selected a-C:H coating showed higher critical loads *Lc1* and *Lc3* compared to a-C:H:SiO. In addition to the tool coatings, dry lubricants were applied to the semi-finished products in small quantities to enable tool lifetimes suitable for large-scale production and thus industrial applications. The selected coatings and lubricant variants were then tested regarding their tribological behavior for forming technology in a strip drawing test. Topography and roughness measurements were used to analyze the surface of friction jaws and semi-finished products. As a reference for the effectiveness of the coatings, all tests were also carried out with uncoated friction jaws. It was found that the application of lubricant resulted in a strong reduction of the CoF for both aluminum alloys tested. The CoF decreased further as the amount of lubricant increased. The use of coated friction jaws led to an increase in friction, which could be due to the higher surface roughness. With DC04, the lubricant only caused a slight reduction in the CoF. However, the values were significantly lower than for aluminum, possibly due to the zinc coating. The a-C:H coating increased friction; with SiO doping, the CoF was like that of an uncoated friction jaw. Adhesion on uncoated friction jaws was observed for all materials. a-C:H-coated tools tended to be slightly smoothed, while no significant change in roughness was measured for SiO-doped friction jaws. Similar tendencies were found for the sheet surfaces. In general, adhesion and thus the amount of lubricant could be reduced by coating the friction jaws. When doped with SiO, abrasion was also reduced. Consequently, this variant appeared to be the most suitable for all tested materials.

References

[1] Information on https://www.consilium.europa.eu/en/policies/energy-prices-and-security-of-supply/

[2] Information on https://www.statista.com/statistics/595803/electricity-non-household-price-germany/

[3] Information on https://ec.europa.eu/eurostat/statistics-explained/index.php?title=Natural_gas_price_statistics#Natural_gas_prices_for_non-household_consumers

[4] Information on https://www.destatis.de/EN/Themes/Labour/Labour-Costs-Non-Wage-Costs/_node.html

[5] F. Vollertsen, F. Schmidt, Dry Metal Forming: Definition, Chances and Challenges, Int. J. Precis. Eng. Manuf.-Green Tech. 1 (2014) 59-62. https://doi.org/10.1007/s40684-014-0009-0

[6] S. Schöler, F. Özkaya, C. Kock, H. J. Maier, B.-A. Behrens, Selective thermally oxidised tool surfaces for dry deep drawing, Dry Met. Forming OAJ FPR 6 (2020) 001–029.

[7] T. Jähnig, A. Mousavi, T. Roch, E. Beyer, A.F. Lasagni, A. Brosius, Macro and microstructuring of deep drawing tools for dry forming processes, Dry Met. Forming OAJ FPR 6 (2020) 030–068.

[8] T. Jähnig, A.F. Lasagni, Laser interference patterned ta-C-coated dry forming tools, Industr. Lubric. Tribol. 72/8 (2020) 1001–1005. http://dx.doi.org/10.1108/ILT-08-2019-0314

[9] J. Henneberg, B. Rothammer, R. Zhao, M. Vorndran, J. Tenner, K. Krachenfels, T. Häfner, S. Tremmel, M. Schmidt, M. Merklein, Analysis of tribological behavior of surface modifications for a dry deep drawing process, Dry Met. Forming OAJ FMT 5 (2019) 013–024.

[10] W.C. Oliver, G.M. Pharr, Measurement of hardness and elastic modulus by instrumented indentation: Advances in understanding and refinements to methodology, J. Mater. Res. 19 (2004) 3–20. https://doi.org/10.1557/jmr.2004.19.1.3

[11] C. Chouquet, C. Ducros, S. Barrat, A. Billard, F. Sanchette, Mechanical properties of a-C:H/Si-containing a-C:H multilayered coatings grown by LF-PECVD, Surf. Coat. Tech. 203 (2008) 745 – 749. https://doi.org/10.1016/j.surfcoat.2008.08.008

[12] E. Barba, A. Claver, F. Montalà, J.F. Palacio, C.J. Luis-Pérez, N. Sala, C. Colominas, J.A. Garcia, Study of the Industrial Application of Diamond-Like Carbon Coatings Deposited on Advanced Tool Steels, Coatings 14 (2024) 159. https://doi.org/10.3390/coatings14020159

[13] A.S. Grenadyorov, K.V. Oskomov, N.F. Kovsharov, A.A. Solovyev, Effect of precursor flow rate on physical and mechanical properties of a-C:H:SiOx films deposited by PACVD method, J. Phys. 1115 (2018) 042046. https://doi.org/10.1088/1742-6596/1115/4/042046

[14] A.S. Grenadyorov, A.A. Solovyev, K.V. Oskomov, M.O. Zhulkov, Dependence of Mechanical and Tribological Properties of a-C:H:SiOx Films on the Bias Voltage Amplitude of

the Substrate, Technic. Phys. 66 (2021) 1111 – 1117. http://dx.doi.org/10.1134/S1063784221080089

[15] J. Henneberg, B. Rothammer, R. Zhao, M. Vorndran, J. Tenner, K. Krachenfels, T. Häfner, S. Tremmel, M. Schmidt, M. Merklein, Lubricant free forming with tailored tribological conditions, Dry Met. Forming OAJ FPR 6 (2020) 228–261.

[16] Z. Huang, Z. Chen, W. Lang, X. Wang, Adhesion Studies of CrC/a-C:H Coatings Deposited with Anode Assisted Reactive Magnetron Sputtering Combined with DC-Pulsed Plasma Enhanced Chemical Vapor Deposition, Materials 14 (2021) 2954. https://doi.org/10.3390/ma14112954

[17] K. Krachenfels, B. Rothammer, S. Tremmel, M. Merklein, Experimental investigation of tool-sided surface modifications for dry deep drawing processes at the tool radii area, Procedia Manuf. 29 (2019) 201-208. https://doi.org/10.1016/j.promfg.2019.02.127

[18] R. Zhao, J. Steiner, K. Andreas, M. Merklein, S. Tremmel, Investigation of tribological behaviour of a-C:H coatings for dry deep drawing of aluminuium alloys, Tribol. Int. 118 (2018) 484-490. http://dx.doi.org/10.1016/j.triboint.2017.05.031

Research on material deformation in double-sided shearing with preventing enlargement of sheet material on the die using digital image correlation

UEDA Kouki[1,a,*], SASADA Masahiro[1,b] and TANAKA Tatsuya[1,c]

[1]1-3 Miyakodani, Tatara, Kyotanabe-city, Kyoto 610-0394, Japan

[a]ctwj0589@mail4.doshisha.ac.jp, [b]msasada@mail.doshisha.ac.jp, [c]tatanaka@mail.doshisha.ac.jp

Keywords: Shearing, Digital Image Correlation, Preventing Enlargement, Crack, Strain

Abstract. In shearing, the effect of preventing the enlargement of sheet material on increasing burnished surface was reported. In this study, double-sided shearing with a suppression plate for preventing the enlargement of sheet material was carried out to investigate the effect of preventing the enlargement of sheet material on the material deformation and crack initiation. During shearing, the material deformation was observed using high speed camera and the distribution of displacement and strain of material were determined from moving images using digital image correlation. The following results were obtained from the observation of the cut surface obtained by shearing and the material deformation during shearing. When the suppression plate prevented the enlargement of the sheet material, the occurrence of cracks in the sheet material around the punch and die cutting edge was delayed. The length of burnished surface increased. For the same reason, the length of burnished surface increased as the trimming allowance of the sheet material was decreased when the enlargement of sheet material was prevented. It was confirmed that preventing the enlargement of sheet material affects the material flow. It is clarified that the equivalent strain around the punch and die cutting edge until crack initiation was increased when the enlargement of sheet material was prevented.

Introduction

Shearing is the cutting of a sheet material with a punch and a die. The cut surface obtained by shearing consisted of a rollover, burnished surface, fractured surface and burr. The relationship between the crack initiation and hydrostatic pressure was discussed [1]. Influence of fine blanking conditions on the quality of sheared surface have been experimentally investigated [2]. The finite element analysis was performed to investigate the effect of the blank holder and counter punch forces on the material flow features and the cut surfaces in conventional blanking and fine blanking [3]. The FEM analysis was performed to investigate the mechanism of the V-ring indenter. The material flow and stress distribution calculated using FEM were reported [4]. As described above, there has been a useful report on the cut surface by fine blanking. The effect of preventing the enlargement of sheet material on increasing burnished surface was reported using experiment and finite element methods. It has been reported that the preventing the enlargement of sheet material on the die during shearing has an effect on the cut surface [5]. Based on the results of FEM and observation of the product obtained by shearing, the discussion on the formation of the cut surface was reported. In order to further investigate, it is desirable to clarify the material flow by taking moving images during shearing, since cracks occur in the shearing process.

On the other hand, in recent years, the displacement and strain have been determined from digital images of sheet material deformation during experiment [6]. The relationship between the material flow and the formation of rollover was confirmed experimentally using image processing [7]. Measurement of strain, strain rate and crack evolution in shear zone based on the recorded

 Materials Research Forum LLC
https://doi.org/10.21741/9781644903254-72

high speed images was reported [8]. It is expected to clarify the effect of preventing the enlargement of sheet material on the displacement using digital image correlation. The purpose of this study is clarifying that the effect of preventing enlargement of sheet material on the material deformation and crack initiation based on the moving images about material deformation during shearing process. In this report, the results of observation of material deformation during experiments show that the effect of preventing enlargement on the punch stroke in which cracks occurred. In addition, the results of clarifying material flow and strain using digital image correlation method are reported.

Experiment

Double-sided shearing with counter punch was carried out using the experimental equipment shown in Fig.1. The horizontal enlargement of the sheet material due to shearing is constrained by the suppression plate. Shearing without the suppression plate also was carried out. The width of the punch was 14.8 mm and the clearance was 5% of the thickness of the sheet material. The radius of punch cutting edge was 0.01mm and the radius of the die cutting edge was 0.2 mm. The sheet material was aluminum (JIS A5052P-O) with length of 19 mm or 23 mm, a width of 20 mm and a thickness of 2 mm. When the specimen with length of 19 mm was used, the width of trimming allowance width (L_t) is 2 mm, which is 100% of thickness of sheet material. For sheet material with length of 23 mm, the trimming allowance width (L_t) is 4 mm, which is 200% of thickness of the sheet material. The reinforced glass had a length of 40mm, a width of 63 mm and a thickness of 10.2 mm. The counter punch force of 1380 N was applied to a sheet material with a counter punch. Blank holding force was also applied to the sheet material with a blank holder to prevent warpage of the sheet material.

During shearing, the deformation of sheet material was observed through reinforced glass using a high speed camera. The resolution of the high speed camera was 1280×1024 pixels. The frame rate was 50fps. The shearing speed was set to 1mm/min. In this experiment, shearing was carried out without lubrication. The observation surface of sheet material was polished with #800 sandpaper to apply a pattern to the observation surface. The punch stroke (P_s) in shearing was measured from moving images. Images at P_s 0.01mm intervals were created from moving images, and the distribution of displacement and strain were determined using the digital image correlation software MatchID 2023.2.

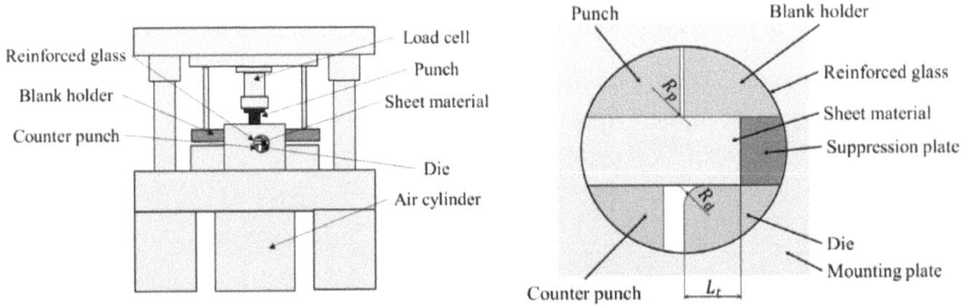

(a) Entire experimental apparatus

(b) Enlarged view of area around the cutting edge

Fig. 1. Experimental apparatus with the suppression plate.

Materials Research Forum LLC
https://doi.org/10.21741/9781644903254-72

Result and discussion

Cut surface. The cut surface of sheet material under the punch obtained by shearing for L_t=4 mm are shown in Fig. 2. The fractured surface of sheet material with the suppression plate is smaller than that without the suppression plate. The cut surface of sheet material under the punch obtained by shearing for L_t=2 mm are shown in Fig. 3. Similarly, the fractured surface of sheet material with the suppression plate is smaller than that without the suppression plate. As the trimming allowance decreases, the fractured surface of the sheet material with the suppression plate becomes smaller.

The cut surface of sheet material on the die obtained by shearing for L_t=4 mm are shown in Fig. 4. The fractured surface of sheet material with the suppression plate is smaller than that without the suppression plate. The cut surface of sheet material on the die obtained by shearing for L_t=2 mm are shown in Fig. 5. Similarly, the fractured surface of sheet material with the suppression plate is smaller than that without the suppression plate. As the trimming allowance decreases, the fractured surface of the sheet material with the suppression plate becomes smaller.

The composition of the cut surface near the surface in contact with the reinforced glass is shown in Fig. 6. When the horizontal enlargement of sheet material was prevented, the length of the fractured surface decreased, and the length of the burnished surface increased both on the product and scrap side. It was also observed that the length of the burnished surface increased as the trimming allowance decreased when the horizontal enlargement of sheet material was prevented.

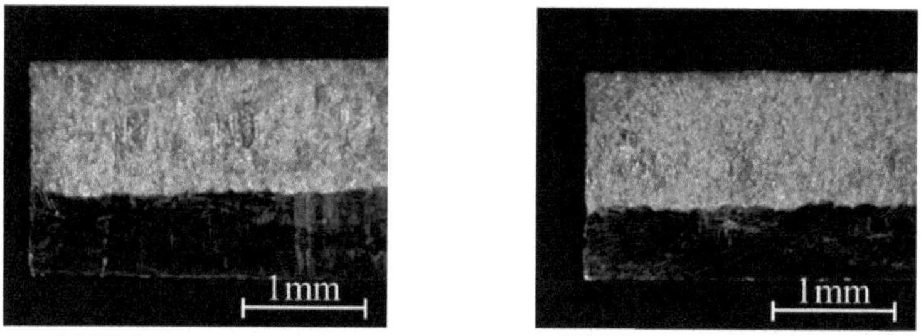

(a) With suppresion plate (b) Without suppresion plate

Fig. 2. Cut surface of the sheet material under the punch obtained by shearing for L_t=4 mm.

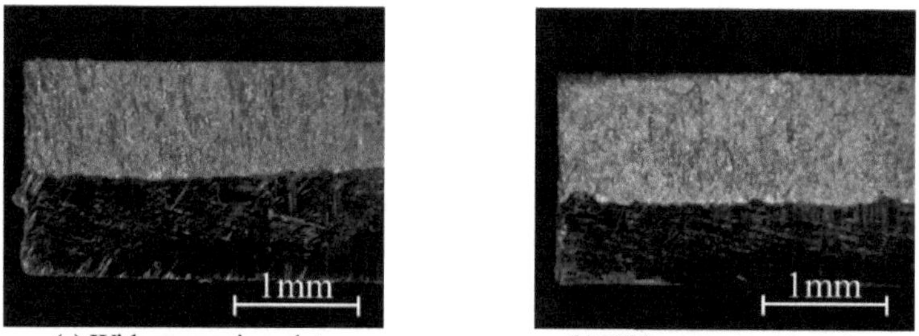

(a) With suppresion plate (b) Without suppresion plate

Fig. 3. Cut surface of the sheet material under the punch obtained by shearing for L_t=2 mm.

Materials Research Forum LLC
https://doi.org/10.21741/9781644903254-72

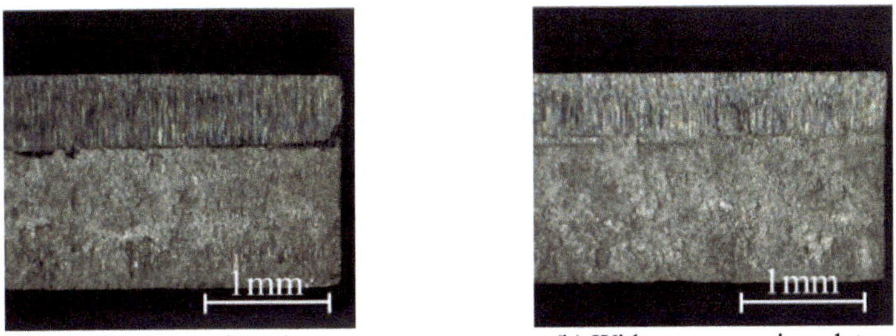

(a) With suppresion plate (b) Without suppresion plate

Fig. 4. Cut surface of the sheet material on the die obtained by shearing for L_t=4.mm.

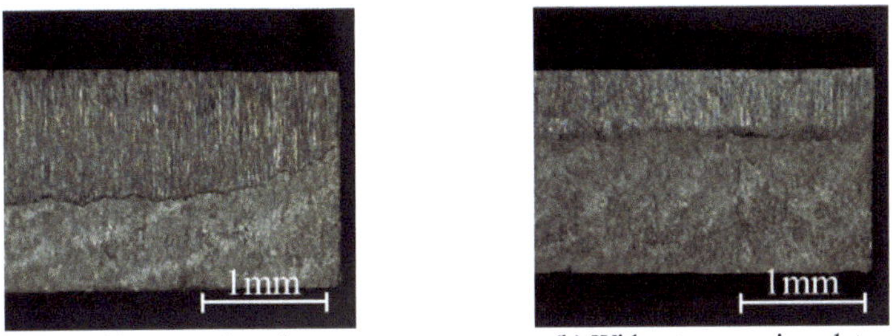

(a) With suppresion plate (b) Without suppresion plate

Fig. 5. Cut surface of the sheet material on the die obtained by shearing for L_t=2 mm.

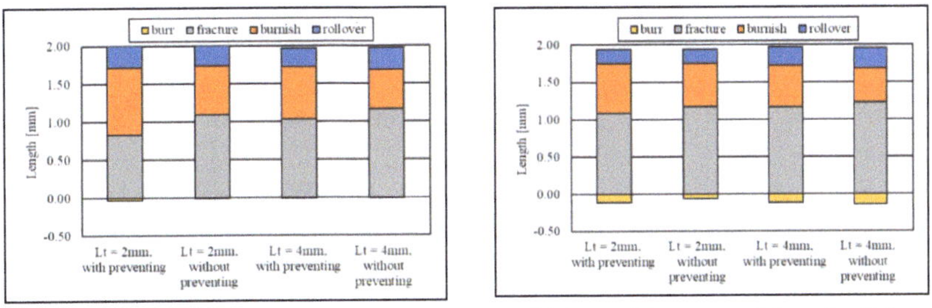

(a) Product (sheet material under the punch) (b) Scrap (sheet material on the die)

Fig. 6. Measurement of cut surface (a) product, (b) scrap.

Crack initiation. In order to investigate the effect of preventing enlargement of sheet material, the punch stroke at crack initiation was confirmed. Two different methods were used to confirm the punch stroke in which the crack occurred.

First, the punch stroke at crack initiation around tool cutting edge was calculated from composition of the cut surface. Punch stroke at crack initiation around punch cutting edge is the sum of the length of the rollover of scrap, the burnished surface of scrap and the burr of product. Similarly, punch stroke at crack initiation around die cutting edge is the sum of the length of the rollover of product, the burnished surface of product and the burr of scrap. From the above, punch stroke at crack initiation around punch and die cutting edge was calculated from the length of cut surface of Fig. 6. The punch stroke at crack initiation by composition of the cut surface is shown https://doi.org/10.21741/9781644903254-72

in Fig. 7. It is shown in the blue bar. Another method was to find the punch stroke at crack initiation from the moving images by visual observation. The punch stroke at crack initiation around punch and die cutting edge was confirmed visually from moving images taken by the high speed camera. The punch stroke at crack initiation by moving images is also shown in Fig. 7. It is shown in orange bar. There is no significant difference between the results measured by two methods. It is clarified that crack in the sheet material around the punch cutting edge occurs faster than cracks in the sheet material around die cutting edge. Punch stroke at crack initiation increased when the enlargement of sheet material was prevented. From the above, it is considered that the reduction of fractured surface by preventing the enlargement of sheet material is an effect of the delay in the occurrence of crack near the die cutting edge.

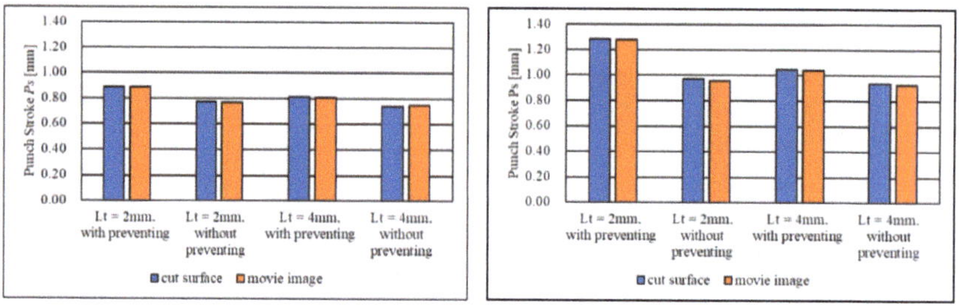

(a) Around the punch cutting edge (b) Around the die cutting edge

Fig. 7. Punch stroke at crack initiation (a) punch, (b) die.

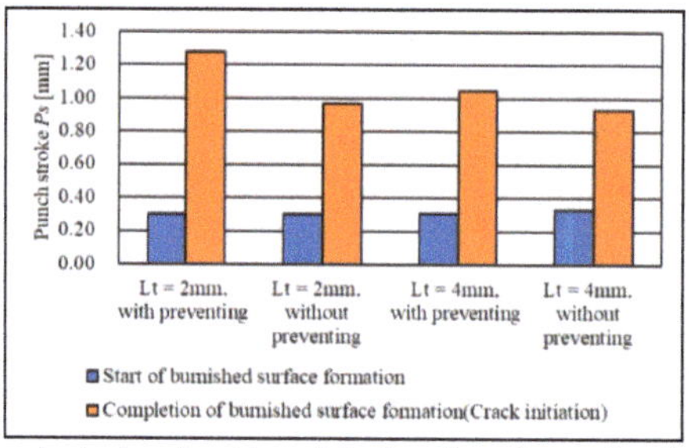

Fig. 8. Punch stroke about burnished surface formation for product.

Burnished surface formation. The duration of the formation of the burnished surface was confirmed from the moving images taken by the high speed camera. The punch strokes in which the formation of the burnished surface was started on the product and the punch strokes in which the formation of the burnished surface was completed were visually confirmed, and the results are shown in Fig. 8. It should be noted that the punch stroke at which the formation of the burnished surface begins is the punch stroke at which the sheet material begins to contact with the side of die. The punch stroke at which the formation of the burnished surface is completed is the punch stroke when a crack occurs near the die cutting edge. Fig. 8 shows that the punch stroke at which the formation of the burnished surface started did not vary much according to the shearing conditions. On the other hand, the punch stroke at which the formation of the burnished surface was completed is different according to shearing conditions. The increase in the length of the burnished surface due to preventing horizontal enlargement of sheet material may be attributed to the suppression of crack initiation around die cutting edge.

Distribution of displacement of material. From moving images taken by the high speed camera, distribution of displacement of sheet material was determined using digital image correlation. Fig. 9 shows distribution of horizontal displacement and displacement vector for L_t=2 mm. The material flow in the horizontal direction away from the die cutting edge on the die was also reduced when the enlargement of sheet material was prevented. In the case of the L_t=2 mm without the suppression plate, the lower part of the sheet material on the die was displaced significantly away from the die cutting edge. This is due to the warping of the material on the die.

Fig. 10 shows distribution of displacement after crack initiation around punch cutting edge for L_t=2 mm. It should be noted that the region of interest for digital image correlation was set only around the die cutting edge, where no cracks had occurred in order to prevent crack from affecting the determination of displacement. The material flow in the horizontal direction away from the die cutting edge was reduced on the die even after crack initiation around punch cutting edge when the enlargement of sheet material was prevented. Therefore, in shearing with the suppression plate, the material flow on the die may be drawn into the die hole.

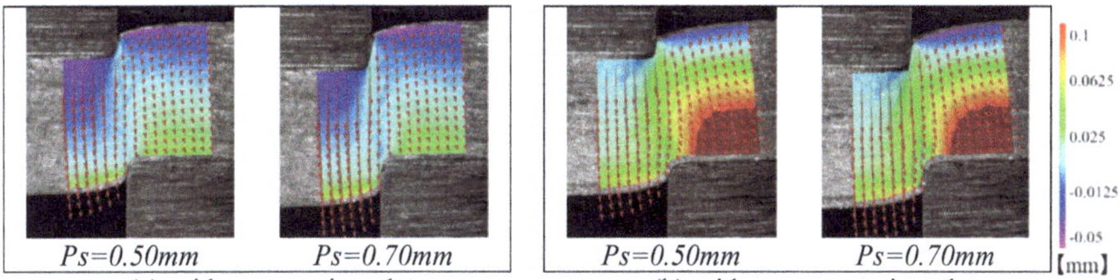

(a) with suppression plate (b) without suppression plate

Fig. 9. Variation of distribution of horizontal displacement and material flow for L_t=2 mm.

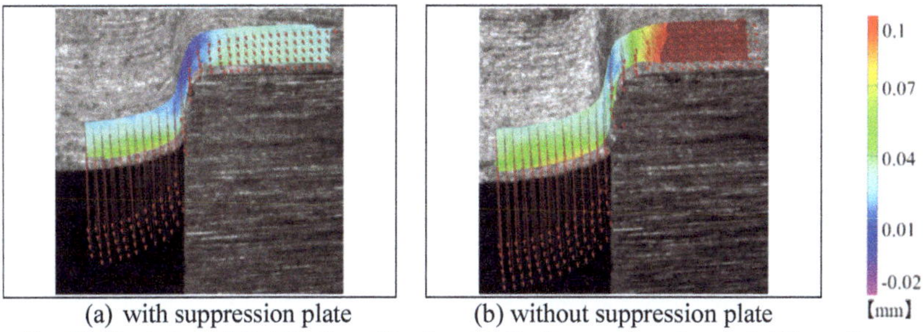

(a) with suppression plate (b) without suppression plate

Fig. 10. Variation of distribution of horizontal displacement and material flow after crack initiation around punch cutting edge for L_t=2 mm (P_s =0.90 mm).

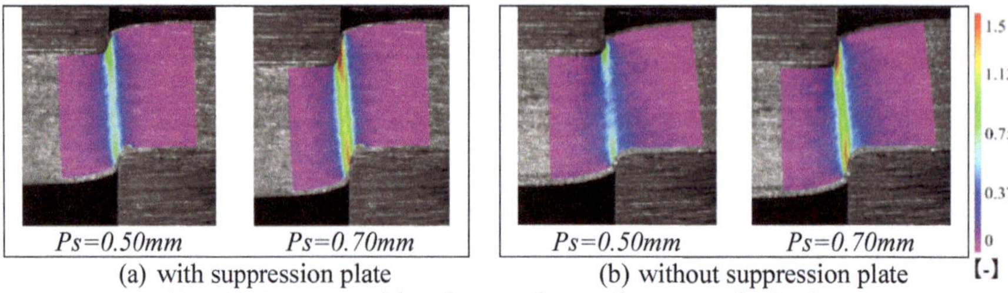

Fig. 11. Variation of distribution of equivalent strain for L_t=2 mm.

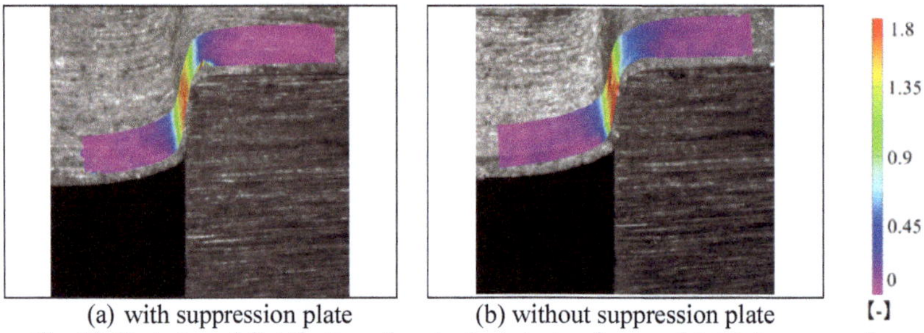

Fig. 12. Variation of distribution of equivalent strain after crack initiation around punch cutting edge for L_t=2 mm (P_s =0.90 mm).

Distribution of equivalent strain. From moving images taken by the high speed camera, distribution of strain of material was determined using digital image correlation. Fig. 11 shows distribution of equivalent strain for L_t=2 mm. At the same punch stroke, there was no significant difference in equivalent strain regardless of preventing enlargement of sheet material.

Fig. 12 shows distribution of equivalent strain after crack initiation around punch cutting edge for L_t=2mm. There was no significant difference in equivalent strain regardless of preventing enlargement of sheet material even after crack initiation around punch cutting edge.

Relationships between equivalent strain and punch stroke. The relationships between equivalent strain and punch stroke for the location where the equivalent strain was maximum around tool cutting edge in the punch stroke at crack initiation shown in Fig. 6 were determined using digital image correlation. The relationships between equivalent strain and punch stroke are shown in Fig. 13. It should be noted that the results were determined until the punch stroke at crack initiation. The equivalent strain at crack initiation increased when material enlargement was prevented. The equivalent strain in the punch stroke at crack initiation increased as the trimming allowance was decreased when enlargement of sheet material was prevented. Similarly, Fig. 13(b) shows that the equivalent strain at crack initiation around die cutting edge increased when material enlargement was prevented. Hence, it is clarified that preventing enlargement of sheet material had little effect on the trend of increasing equivalent strain and increased the equivalent strain until crack initiation.

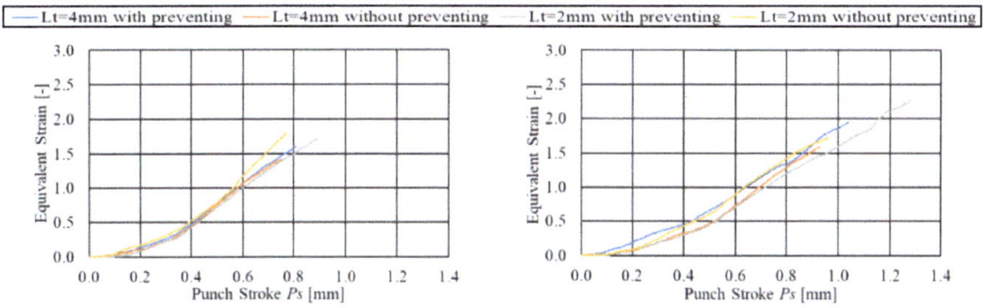

(a) Punch (b) Die

Fig. 13. History of equivalent strain on crack initiation point (a) punch, (b) die.

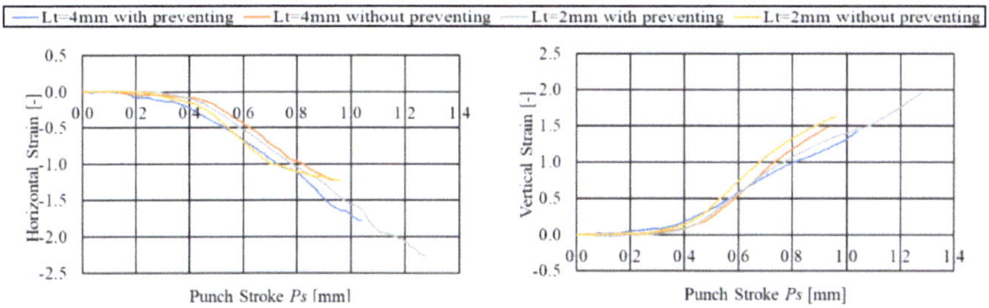

(a)Horizontal strain (b)Vertical strain

Fig. 14. History of normal strain on crack initiation point around die cutting edge.

Normal strain along the horizontal and vertical direction. The relationships between normal strain along the horizontal or vertical direction determined using digital image correlation and punch stroke, at the point of Fig. 13(b) are shown in Fig. 14. Fig. 14(a) shows that compressive strain along the horizontal direction of the material around the die cutting edge increased when material enlargement was prevented for L_t=4 mm. For L_t=2 mm, the compressive strain along the horizontal direction of the sheet material around the die cutting edge was larger when material enlargement was not prevented. Further investigation is needed about this, but the sheet material on the die warped up for L_t=2 mm without the suppression plate according to section Distribution of displacement of material, and the material deformation was different from other conditions. Further investigation of the strain for L_t=2 mm is considered to be necessary in the future. Fig. 14(b) shows that tensile strain along the vertical direction of the material around the die cutting edge decreased when material enlargement was prevented for both L_t=2mm and L_t=4mm.

Conclusion

In order to clarify that the effect of preventing the horizontal enlargement of sheet material on the die on the material deformation in double-sided shearing, deformation of the material was observed in shearing. The distribution of the displacement and strain were also determined from moving images taken by high speed camera using digital image correlation. The following conclusion were drawn.

1) When the enlargement of sheet material on the die was prevented in the shearing, the length of the burnished surface increased. It was also observed that the length of the burnished surface increased as the trimming allowance was decreased when the enlargement of sheet material was prevented.

Materials Research Forum LLC
https://doi.org/10.21741/9781644903254-72

2) The increase in the length of the burnished surface due to preventing horizontal enlargement of sheet material was attributed to the delay of crack initiation around the die cutting edge.
3) At the same punch stroke, there was no significant difference in equivalent strain regardless of preventing enlargement of sheet material. It is clarified that the equivalent strain around the punch and die cutting edge until crack initiation was increased when the enlargement of sheet material was prevented.
4) From the distribution of strain determined using digital image correlation, it was confirmed that tensile strain along the vertical direction of sheet material around the side of the die decreased when the enlargement of sheet material on the die was prevented.

References

[1] K. Kondo, Mechanism of the Shearing Processes of Ductile Sheet Metals (1st Report) -Effect of the Mechanical Properties of Material and Friction of the Tool Surfaces-, J. Japan Soc. Precision Eng. 31 (1965) 476-483.

[2] T. Maeda, T. Nakagawa, Experimental Investigations on Fine Blanking -1st Report-, J. Japan Soc. Tech. Plast. 9 (1968) 618-626.

[3] S. Thippeakmas, M. Jin, M. Murakawa, An investigation of material flow analysis in fine blanking process, J. Mater. Process. Tech. 192 (2007) 237-242. http://dx.doi.org/10.1016/j.jmatprotec.2007.04.065

[4] S. Thipprakmas, Finite element analysis of V-ring indenter mechanism in fine blanking process, Mater. Des. 30 (2009) 526-531. https://doi.org/10.1016/j.matdes.2008.05.072

[5] H. Hoshi, M. Sasada, I. Aoki, Influence of constraint of material on shearing characteristics, Trans. Japan Soc. Mech. Eng. Series C 74 (2008) 206-211. http://dx.doi.org/10.1299/kikaic.74.206

[6] I. Aoki, T. Takahashi, Material flow analysis on shearing process by applying Fourier phase correlation method—analysis of piercing and fine-blanking, J. Mater. Process. Tech. 134 (2003) 45-52.

[7] M. Sasada, T. Togashi, Measurement of rollover in double-sided shearing using image processing and influence of clearance, Procedia Eng. 81 (2014) 1139-1144. https://doi.org/10.1016/j.proeng.2014.10.248

[8] C. Hartmann, H.A. Weiss, P. Lechner, W. Volk, S. Neumayer, J.H. Fitschen, G. Steidl, Measurement of strain, strain rate and crack evolution in shear cutting, J. Mater. Process. Tech. 288 (2021) 116872. http://dx.doi.org/10.1016/j.jmatprotec.2020.116872

Materials Research Proceedings 44 (2024) 683-690
https://doi.org/10.21741/9781644903254-73

Example of the methodology for producing heterostructural systems using plastic deformation

PABICH Bartłomiej[1,a *], MAJTA Janusz[1,b], KWIECIEŃ Marcin[1,c], CICHOCKI Kamil[1,d] and BŁONIARZ Remigiusz[1,e]

[1]AGH University of Krakow al. Mickiewicza 30, 30-059 Krakow, Poland

[a]bpabich@agh.edu.pl, [b]majta@agh.edu.pl, [c]mkwiecie@agh.edu.pl, [d]cichocki@agh.edu.pl, [e]bloniarz@agh.edu.pl

Keywords: Heterostructured Materials, Multilayer Composite, Multi-Stage Wire Drawing

Abstract. The study specifically examines the changes in the properties of the investigated systems as metal-to-metal composites and their constituent components. These heterostructural systems were crafted from microalloyed steel, reinforced with rheological non-uniformity due to the presence of Ti or Mg hcp phases. The influence of total deformation on the continuity of hcp layers and the degree of refinement of the matrix microstructure, i.e., microalloyed steel obtained through continuous recrystallization, was examined. The analysis delves into various strengthening mechanisms within the studied heterostructural systems, with particular emphasis on the strengthening effect resulting from the presence of low-plasticity hcp phases and their interaction with the entire system. The results of rheological studies conducted using channel compression tests and deep wire drawing (DWD) are presented. Microstructural studies have shown not only changes in the microstructure of the constituent materials but also the mutual interaction of these changes, directly affect the mechanical properties of the investigated heterogeneous materials.

Introduction

The demand for new structural materials that would meet the expectations of designers of new structural components in devices that have to operate under increasingly demanding operating conditions is a continuous challenge facing materials engineering. One of the commonly used solutions to optimize mechanical and operational properties is the use of metal-to-metal type composites [1,2]. This work proposes an innovative methodology for the fabrication of heterostructural materials characterized by an appealing combination of mechanical properties resulting from the interaction of multiple strengthening mechanisms and their synergistic effect. Heterostructured materials are defined as materials that contain heterogeneous zones that have dramatically different constitutive properties in the case of structural metallic materials or alternatively, very different physical properties in the case of functional materials [3]. According to the above definition, the materials with the following structures can be classified as typical heterostructured materials [1]: heterogeneous lamella structures, gradient structures, laminate structures, dual/multi-phase structures, harmonic (core shell) structures, multimodal structures, etc. The purpose of the present study is to outline the new strategies for structural materials development offered by new degrees of freedom and by their combination: not only playing with the microstructure or with the macroscopic shape, but allowing a new scale for materials organization, the "architecture", and controlling a new degree of freedom, the "spatial heterogeneity,, [4]. The presented research discusses the rheological properties and the results of microstructural analysis of multilayer systems produced through the channel test and deep wire drawing (DWD). The analysis of results obtained in channel die compression tests allowed for the identification of conditions for conducting deep drawing processes of wires, to take into account

Materials Research Proceedings 44 (2024) 683-690 https://doi.org/10.21741/9781644903254-73

all interactions arising from both the inhomogeneity of strengthening processes and microstructural changes.

Experimental

The presented studies aimed to produce mechanically and structurally inhomogeneous materials and investigate the microstructural changes and mechanical response of such multilayered systems. Two deformation processes were employed for the multilayer systems composed of microalloyed steel and titanium (St/Ti) or magnesium (St/Mg): channel test and deep wire drawing (DWD). Both applied deformation methods allowed the accumulation of significant plastic deformation energy in the deformed materials, resulting in non-uniform strengthening through an increased density of geometrically necessary dislocations (GND) revealed using electron backscatter diffraction *(EBSD)* based on kernel average misorientation *(KAM)*. Experimental research included tests of packet compression in the channel test and processes of wire drawing with very large total deformations. In Fig. 1, the experimental setup for channel tests is shown, with samples constructed with a packet arrangement of layers and an illustrative diagram of the drawing process of inhomogeneous wires with the concept of creating the structure of the matrix material (microalloyed steel, bcc) reinforced with macro and meso particles of a less ductile phase (Ti, Mg, hcp). Published earlier studies [5, 6] have shown that the microalloyed steel with the addition of Nb, used in the presented research, is a suitable material for very large deformations and for the generation of Severe Plastic Deformation (SPD) effects in the examined heterostructural wires. In the case of channel compression tests, packages consisting of two layers of microalloyed steel were compressed, with a layer of Ti or Mg between them (Fig. 1b). All samples had a length of 15mm and a width of 10mm. Two thicknesses of the steel layer were used, namely 4.0 mm and 2.0 mm, along with spacers of Ti or Mg with a thickness of 1 mm. These prepared packages underwent the compression process with two total strain magnitudes of 0.25 and 0.5. For the DWD processes, the procedure involved wire drawing a system with a circular cross-section, constructed by placing a rod of Ti or Mg with a diameter of 3 mm in a hollowed microalloyed steel cylinder with a diameter of 6.5 mm (Fig. 1c, d). The multi-stage wire drawing process was carried out with a unit reduction of cross-sectional area equal to 15%, until the maximum possible reduction in the diameter of the investigated multilayer system was achieved. Additionally, wire drawing processes were carried out in the same multi-pass scheme using a "homogeneous" microalloyed steel rod with an initial diameter of 6.5mm. In this case, the wire drawing process was executed until achieving a diameter reduction to 2.5mm. Some wires were drawn to a diameter of 0.75mm. Intermediate annealing was not applied between drawing operations. The resulting research materials underwent microstructural analysis and mechanical property testing i.e. tensile tests and hardness measurements.

The research materials used had the following characteristics:

- *Microalloyed steel*: bcc, Shear Modulus: G = 78 GPa, Chemical composition, (wt%): 0.07C/0.29Si/1.36Mn/0.067Nb/0.03Ti/0.16Cu/0.009N//Fe – bal. Initial grain size: 22 μm.
- *Ti, grade 2*: hcp, Shear Modulus: G = 43 GPa, Chemical composition, (wt%): 0.01Fe/0.03H/ 0.04O/0.01N/Ti-bal. Initial grain size: 28 μm.
- *Mg, AZ31*: hcp, Shear Modulus: G = 17 GPa, Chemical composition, (wt%): 3.27Al/0.98Zn/0.28Mn/0.0027Fe/0.00078Ni/0.11Si/Mg- bal. Initial grain size: 20-25 μm.

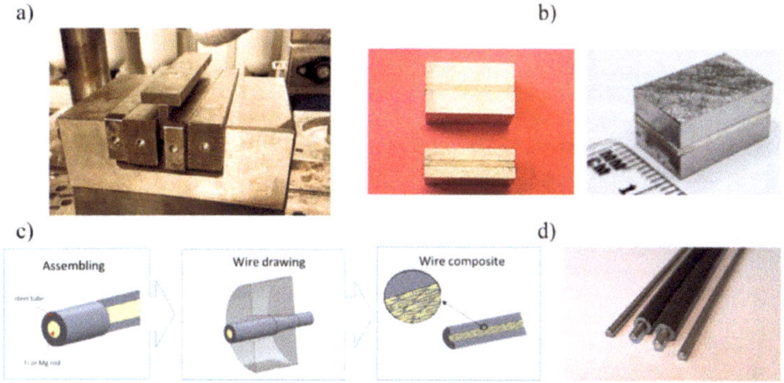

Fig. 1. Research methods employed in the production of the studied multilayer and heterostructural systems. Test stand for channel testing (a); constructed multilayer samples for channel compression (b). Diagram of heterostructural material production using the deep wire drawing (DWD) process (c). Inhomogeneous wire samples for DWD tests (d).

The impact of the scale of microstructure elements participating in the strengthening of the examined systems was considered, taking into account the following factors: i) the degree of reinforcement achieved by developing the matrix microstructure (grain refinement and perlite colonies) and the second phase (particles of fragmented Ti or Mg layers produced by wire drawing and deep wire drawing, ii) the introduction of new deformation mechanisms due to the presence of incoherent second phase particles on the micro-meso scale, iii) changes in competitive processes, such as the loss of coherence and healing processes, cracking events, or recovery processes, accompanying the very strong deformation of the studied heterostructural multilayer systems.

Results and discussion

The first stage of the research involved testing by the channel compression tests of multilayer systems. The aim of these tests was to obtain information necessary for the proper design of the second stage of experimental research, namely the deep drawing processes of heterostructured wires. Example results of computer simulations and microstructure analysis, using EBSD technique, of the multilayered sample (steel/Ti/steel) are presented in Fig. 2. Suitable rheological and microstructural models have been proposed, the application of which can rationalize the behavior of the studied materials by considering different dimensional scales and allowing for interactions between incoherent phases of materials used in the studied metal-to-metal composites. As seen in Fig. 2a, the deformation of the compressed sample and the non-uniform distribution of effective strain translate into the degree of grain deformation in the layers of steel, i.e., stronger effects are visible in the upper layer, which was influenced by the moving hammer. This observation is significant from the planning-optimization perspective of wire drawing processes for heterostructured wire systems, where the forces in the deformation zone originate from the cone of the drawing die, causing typical non-uniformity in plastic deformation zone with all the consequences in microstructural changes. In Fig. 2b, the intensity of strain energy accumulation is shown, represented by the degree of grain deformation and distributions of geometrically necessary dislocation densities as an image of KAM (Kernel Average Misorientation). For the applied thickness of the steel layers (4.0 mm) and the deformation magnitude of the studied system (0.25), a noticeably stronger work hardening is evident in the Ti layer. It can be presumed that the direct cause of this is the lower plasticity of Ti, due to its hcp structure.

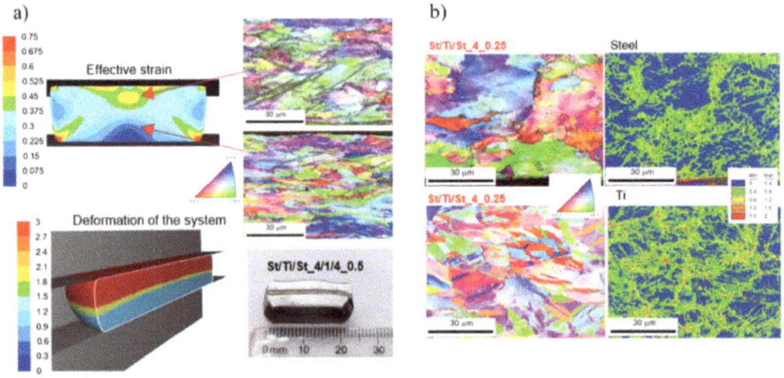

Fig. 2. Example results of computer simulations and microstructure analysis of the multilayered sample (steel/Ti/steel), using EBSD technique (a). Representation of strain hardening in adjacent layers of steel and Ti, showing, for a given package (steel layer thickness 4.0 mm) and strain magnitude (0.25), a faster increase in the density of geometrically necessary dislocations, represented by KAM, in the titanium layer (b).

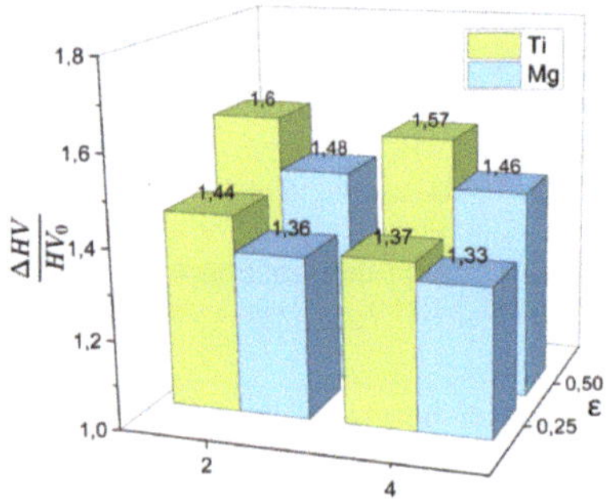

Fig. 3. Relative hardness increases of Ti and Mg as a function of the thickness of the steel layers and the total deformation of the packages produced in the channel compression tests.

Understanding the characteristics of the strengthening processes in Ti or Mg layers is crucial for designing the heterostructural systems targeted in the current research. Increased work hardening (relative hardness increase) was significantly observed in the Ti layers compared to Mg, especially when higher deformation was applied (Fig. 3). In-depth studies in this area have also been presented in previous publications by the authors [7,8]. From these observations, a direct conclusion can be drawn that Ti, due to more effective accommodation of deformation effects, may be a more effective layer in creating dispersed particles resulting from the loss of coherence in a highly deformed heterostructural system (see Fig. 1c).

In the second stage of the research, heterostructural materials produced through deep multistage wire drawing were investigated. In this process, deformation effects corresponding to reductions in diameters to 4.0 mm and 2.5 mm were accumulated. Wires with structurally non-uniform characteristics were examined, consisting of a layer of steel and a titanium core (see Fig. 1c,d). Mechanical property studies and microstructural analysis were conducted, allowing for the identification of the fundamental causes behind the observed mechanical responses of the investigated materials. In Table 1, the mechanical properties of heterostructural wire systems with diameters of 2.5 mm, featuring cores made of Ti and Mg, are presented. For comparison, the results of tensile tests on a homogeneous wire made of microalloyed steel with the same deformation history are also included. Significant differences in the strength and plastic properties of the investigated wires can be observed. The ability to formulate more general conclusions from the research on the mechanical properties of produced heterostructural wires requires further investigations with an increased degree of deformation. This is necessary to achieve the expected microstructural effects in the form of Ti core fragmentation, which is the overarching goal of the presented studies.

Table 1. Mechanical properties of the investigated "homogeneous" steel wires and heterostructural wire systems of Steel/Ti and Steel/Mg.

Type of system, ϕ 2.5 mm	YS [MPa]	TS [MPa]	EL [%]	RA [%]
Steel	1153	1236	1.90	37.00
St/Ti	1007	1227	5.51	21.56
St/Mg	800	818	3.85	5.66

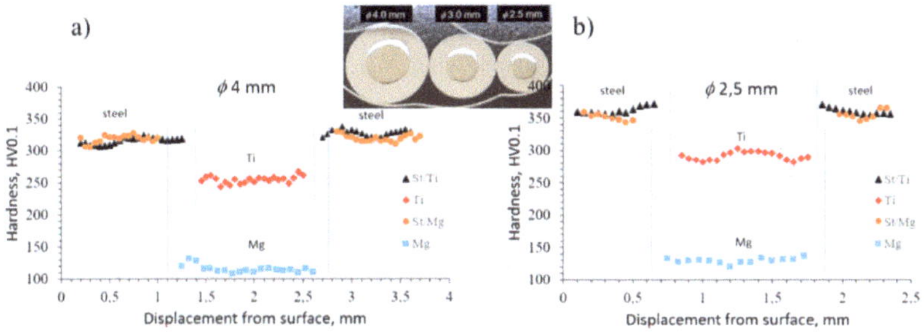

Fig. 4. Hardness distributions in the layers of steel and Ti or Mg on the cross-section of heterostructural wires after reducing their diameter to 4.0 mm (a) and 2.5 mm (b).

Undoubtedly, a crucial aspect of the discussion regarding the results obtained at this stage of the research is the assessment of the non-uniformity of strain strengthening in the examined heterostructural systems. The direct representation of the strain strengthening characteristic is hardness distributions on cross-sections of produced wires. In Fig. 4, the hardness measurement results for steel/Ti systems deformed by reducing diameters to 4.0 and 2.5 mm are presented. The presented hardness distributions in the layers of steel, Mg, and Ti confirm earlier observations from the first stage of the study, specifically the channel die compression tests, where an increase in hardness was noted in the steel layers near the contact with the Ti layer, especially for higher effective strains. No such effect was observed in the case of the Mg system.

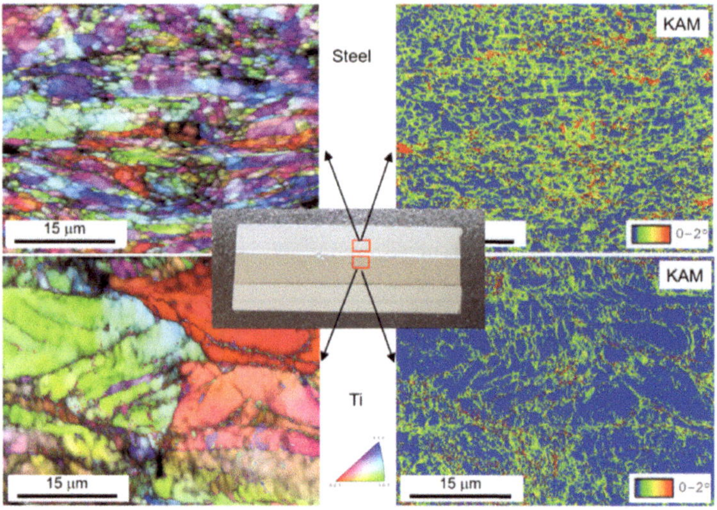

Fig. 5. Results of microstructural investigations using EBSD and KAM analysis in the contact zone of the steel and Ti layers on the longitudinal cross-section of the heterostructural wire after reducing the diameter to 4.0 mm.

The results of microstructural analysis presented in Fig. 5 show that even after reducing the diameter to 4.0 mm, a distinct separation of work hardening is observed between adjacent materials, characterized by different ductility. The ability to accommodate deformation is represented in this case by KAM maps showing the geometrically necessary dislocation densities. In the case of the steel layer, the effects of initiated continuous recrystallization can be observed, resulting in clear colonies of a fine-grained structure. It is expected that this *in situ* recrystallization process, as a result of strong accumulation of plastic deformation energy, will progress with increasing levels of deformation in subsequent passes. In the presented microstructures, it is evident that in adjacent regions of the steel and Ti system, the former exhibits a higher degree of work hardening. This means that in the case of wire drawing heterogeneous material systems, such as the investigated steel/Ti samples, after reaching a certain degree of hardening, the Ti core begins to function as a specific tool for the more ductile steel layer. Confirmation of this is also found in observations of the relative change in the thickness of the steel layer to the diameter of the Ti core. It was noticed that the reduction in the diameter of the entire heterogeneous steel/Ti wire, with the increase in total reduction, occurred to an increasing extent at the expense of reducing the thickness of the steel layer. From this, the conclusion can be drawn that, in order to ensure the possibility of further deformation through the drawing process of the wire until achieving a core thinning of Ti to the point where its fragmentation begins, leading to the formation of fine particles, it will be necessary to introduce interoperative annealing.

In Fig. 6, microstructure images of etched longitudinal sections of steel/Ti wires deformed by reducing the diameter to 2.5 mm are presented. Clear deformed grains of both steel and Ti can be observed. From these observations, it can be inferred that the steel/Ti wires constitute a suitable research material for further studies on intense deformation of heterostructural systems, until the expected strengthening effect by Ti particles formed from its fragmentation due to the loss of coherence after utilizing the entire plasticity is achieved. The subject of further research will also be the condition of the connection between the layers of steel and Ti at various stages of total deformation in consecutive sequences, down to reducing the diameter below 0.75 mm. Previous studies [6] have shown that it is possible to produce in homogeneous wires from the investigated microalloyed steel, nanostructures with an average grain size of about 250 nm as a result of the in-situ recrystallization process during Severe Plastic Deformation (SPD).

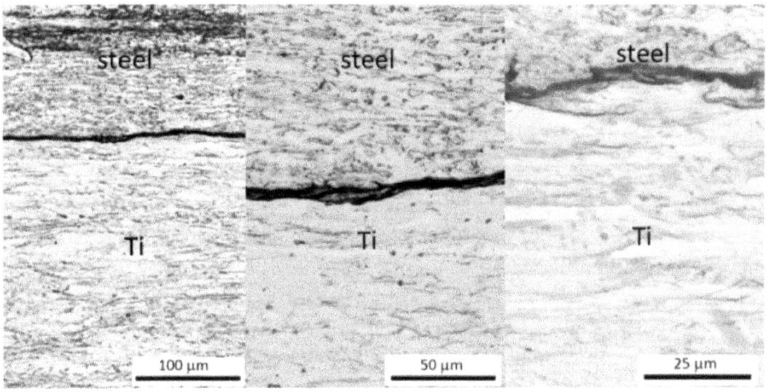

Fig. 6. Microstructures at the interface of the steel layer and Ti of the heterostructural wire after reducing the diameter to 2.5 mm, at various magnification scales. Cross-section.

Summary

The conducted plastometric and microstructural studies have shown that there is a significant variation in work hardening effects not only in adjacent layers of microalloyed steel and Ti or Mg, but also as an effect of the magnitude of the total applied strain and the positioning of these layers in a multilayer stack. The obtained results were used to plan and carry out Deep Wire Drawing (DWD) processes for the production of heterostructural systems. The experimental studies were supported by computer simulations, allowing for the assessment of mechanical states accompanying the investigated deformation processes, such as channel die compression tests and wire drawing processes. The effects of the mutual interaction of the constituent layers of the studied heterogeneous systems have been presented.

The main cognitive effects of this work were achieved through exploring the possibilities of new processes for manufacturing multilayer systems based on microalloyed steels, which can ultimately be reinforced with particles of hcp phases with low ductility and varying degrees of dispersion. The research results will be utilized to design subsequent stages of the study, aiming to produce new multilayer drawn products with attractive combinations of strength and ductility.

Acknowledgment

Financial assistance of the National Science Center, Poland, project 2022/45/B/ST8/01383, is acknowledged.

References

[1] X.L. Wu, Y.T. Zhu, Heterogeneous materials: a new class of materials with unprecedented mechanical properties, Mater. Res. Lett. 5 (2017) 527–532. https://doi.org/10.1080/21663831.2017.1343208

[2] Y. Kim, G.H. Gu, R. E. Kim, M. H. Seo, H. S. Kim. Deformation behavior of lightweight clad sheet: Experiment and modeling, Mater. Sci Eng A852 (2022) 1-9. https://doi.org/10.1016/j.msea.2022.143666

[3] Y. Zhu , K. Ameyama , P.M. Anderson , I.J. Beyerlein , H. Gao , H.S. Kim , E. Lavernia, S. Mathaudhu , H. Mughrabi , R.O. Ritchie , N. Tsuji, X. Zhang, X. Wu, Heterostructured materials: superior properties from hetero-zone interaction, Mater. Res. Lett. 9 (2021) 1-31. https://doi.org/10.1080/21663831.2020.1796836

[4] O. Bouaziz, Y. Bréchet, J.D. Embury, Heterogeneous and architectured materials: a possible strategy for design of structural materials, Adv. Eng. Mater. 10 (2008) 24-36. https://doi.org/10.1002/adem.200700289

[5] L.Y. Sun, X. Liu, F. Xu, S.W. Lei, H. Li, Q.J. Zhai, Review on niobium application in microalloyed steel, J. Iron Steel Res. Int. 29 (2022) 1513-1525. https://doi.org/10.1007/s42243-022-00789-1

[6] J. Majta, M. Kwiecien, P. Lisiecka-Graca, R. Bloniarz, E. Kopytek, K. Muszka, Precipitation -based Strengthening Mechanisms in Ultrafine-grained and Nano-Structures in Microalloyed Steel Wires, 11th International Conference on Advanced Computational Engineering and Experimenting ACE-X, Vienna, 2017, 3-6 July.

[7] J. Majta, B. Pabich, Effects of work hardening gradients in incoherent metal-to-metal composites, Materials Info 2023: 6th World Congress on Materials Science and Engineering: theme “Materials innovation and impact”, Rome, 2023 18-19 September.

[8] B. Pabich, J, Majta, M Kwiecień, K. Cichocki, Ł. Lisiecki, Heterostructures produced through severe plastic deformation of multilayered systems: steel-Ti and steel-Mg, Materials Research Proceedings, 27th International ESAFORM Conference on Material Forming, Toulouse, 2024, 24-26 April.

Development of joining technology between FeCrAl alloys and fusion reactor piping materials

KANAI Ryui[1,a *], NAOKO Oono[1,b,] INAO Daisuke[2,c], NAGASAKA Takuya[3,d] and HOKAMOTO Kazuyuki[2,e]

[1]Yokohama National University, 79-5 Tokiwadai, Hodogaya-ku, Yokohama 240-8501, Japan

[2]Kumamoto University, 2-39-1 Kurokami, Chuo-ku, Kumamoto-shi 860-8555, Japan

[3]National Institute for Fusion Science,322-6 Oroshicho, Toki 509-5202, Japan

[a]kanai-ryui-nr@ynu.ac.jp, [b]oono-naoko-yh@ynu.ac.jp, [c]inao@tech.kumamoto-u.ac.jp, [d]nagasaka.takuya@nifs.ac.jp, [e]hokamoto@mech.kumamoto-u.ac.jp

Keywords: Clad Materials, Intermediate Products, Bonding Distance

Abstract. The use of FeCrAl alloys as an inner lining for piping materials for fusion reactors is being considered. The FeCrAl alloy/SUS316L joints were explosion welded to avoid the influence of an oxide film at the joint interface. The joint length could be increased by increasing the charge ratio or reducing the material thickness. Together with the microstructure observation, hardness test, 4-point bending test and heat treatment test results, it is clear that explosion welding is suitable for joining FeCrAl-ODS and SUS316L.

Introduction

FeCrAl alloys have excellent corrosion resistance due to the formation of a protective alumina film by pre-oxidation treatment and are therefore suitable for the liners used in highly corrosive environments, such as liquid metal breeder blanket piping with operating temperatures above 500°C [1]. However, since fusion reactor designs require low activation elements, the addition of non-low activation Al is problematic [2]. In addition, the high temperature strength of the tubes cannot be maintained with ferritic FeCrAl alloys alone [3], and powder metallurgy strengthened versions of FeCrAl alloys are expensive to produce. We are considering clad materials of FeCrAl alloys with current structural material candidates to provide answers to the above issues. Welding is undesirable because it alters the microstructure of the thin pipe material (a few mm to 10 mm is assumed in some designs [2,4]) in fusion reactors under thermal effects. Solid phase diffusion bonding is difficult to apply to FeCrAl alloys due to the formation of a protective alumina film at high temperatures. Explosion welding is a joining method that produces very few intermediate products and can join a wide range of dissimilar metals without losing the characteristics of each metal, because a gaseous metal splash (metal jet) produced in front of the point of impact blows away oxides films and allows contact between the newly formed metal surfaces [5]. In this study, FeCrAl alloys are bonded to SUS316L fusion reactor pipe material by explosion welding and their bondability is tested.

Experimental Methods

Two types of FeCrAl alloys were selected: Kanthal® APMT, a commercial FeCrAl alloy, and FeCrAl based oxide dispersion strengthened alloy (FeCrAl-ODS). Table 1 shows the composition of these alloys and SUS316L. For the FeCrAl alloys, test specimens for explosion welding were cut out from the ingots. For SUS316L, pre-cut pieces were purchased. The dimensions of APMT and FeCrAl-ODS were 40 mm long x 11 mm wide x 3 mm thick and 40 mm long x 14.9 mm wide x 3 mm thick respectively; the dimensions of SUS316L were 180 mm long x 100 mm wide with varying thicknesses of 2 mm and 3 mm. The specimen surfaces were polished to a roughness of less than 1 μ m using an automatic polishing machine (Buehler EcoMet250) prior to the explosion welding.

Table 1. Composition of APMT, SUS316L and FeCrAl-ODS [wt.%].

Elements	Fe	Cr	Al	Ni	Mo	Ti	Zr	Y_2O_3
APMT[11]	balance	21	5	—	3.0	—	—	—
FeCrAl-ODS	balance	15	7	—	—	0.5	0.4	0.5
SUS316L	balance	18	—	12	2.5	—	—	—

Explosion welding was performed using facilities at the Institute of Industrial Nanomaterials, Kumamoto University. The test layout is shown in Fig. 1. A SUS316L flyer plate was installed on the upper side, and FeCrAl-ODS or APMT as a cover plate on the lower side. The explosives used was ammonium nitrate-based explosive (PAVEX manufactured by Kayaku Japan Co., Ltd., whose explosion velocity, and density are approximately 2.4 km/s and 530 kg/m3, respectively), and acrylic resin (13 mm) spacers were inserted at the four corners to ensure impact velocity. The details of the test conditions are given in Table 3. The charge ratio is expressed as the weight of the explosive per the weight of flyer plate. For example, in the case of a 3 mm SUS316L plate, the charge ratio of 0.81 corresponds to an explosive weight of 350 g.

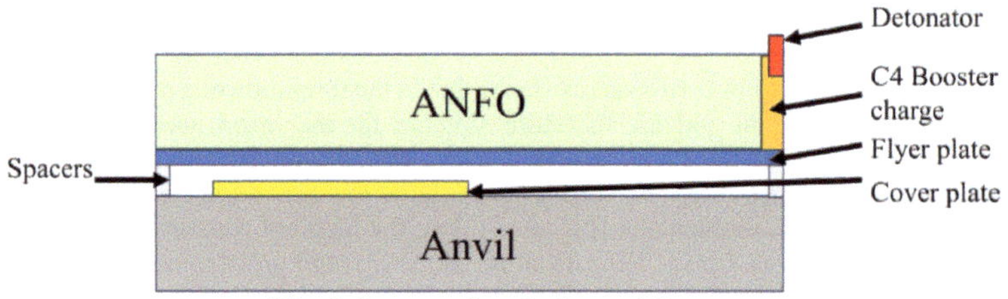

Fig. 1. Layout of explosion welding.

Table 2. Charge ratio and SUS316L thickness.

		Thickness of SUS316L [mm]	Charge ratio	Number of sheets	Number of tests
(a)	APMT	3	0.81	1	2
(b)	APMT	3	0.95	1	4
(c)	FeCrAl-ODS	2	1.43	1	2
(d)	FeCrAl-ODS	2	1.63	1	2
(e)	FeCrAl-ODS	3	0.95	1	1
(f)	FeCrAl-ODS	3	1.09	1	2

After explosion welding, the specimens were evaluated by cross-sectional microstructure observation, hardness testing and four-point bend testing. For the cross-sectional observation, the specimen cross-sections were polished to a roughness of less than 1μm as described above and observed using an optical microscope to measure the length of connection sections (hereafter referred to as "bonding length"). Scanning electron microscope (SEM) observation was also used to determine the morphology and bonding state of the interface. A micro-Vickers hardness test was carried out using the Akashi MVK-E to investigate the hardness distribution near the interface with a test load of 9.8 N and a holding time of 15 sec. The distance from the interface and the diagonal length of the imprints were measured using a polarizing telescope (Nicon ECLIPSE ME600). The four-point bending test was performed using equipment at the National Institute for Fusion Science (NIFS). Three bending specimens with the dimensions shown in Fig. 2 were obtained from one APMT explosion welding specimen. The specimens were tested at a displacement rate of 0.20 mm/min in the bending direction at room temperature.

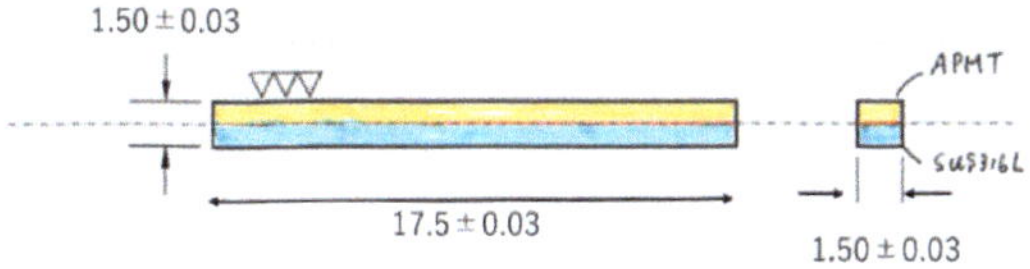

Fig. 2. 4-point bending specimen dimensions.

Results and Discussion

Fig. 3 shows the relationship between the bonding length and the number of explosives (listed as charge ratio). The bonding length of 0.0 refers to the one that was peeled off after the test.

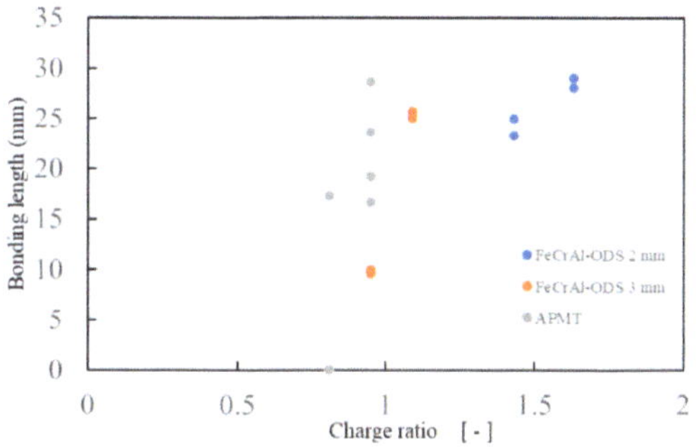

Fig. 3. Relation between charge ratio and bonding length.

As shown in Fig. 3, the bonding length increases with increasing the charge ratio and decreasing the thickness of the flyer plate (SUS316L). It can be interpreted that the charge ratio and plate thickness contribute to an increase in the velocity of the impact. For the same charge ratio, bonding length becomes shorter in the APMT/SUS316L joints than in the FeCrAl-ODS/SUS316L joints. The Vickers hardness of SUS316L is 200 HV [6], that of APMT is 250 HV [7], and that of FeCrAl-ODS is 326 HV [8]. During explosion welding, the metal jets produced cause an enormous deformation at very high strain rates on the surfaces being joined, and this is the origin of the bonding. The stress required for this deformation depends on the yield stress and work hardening coefficient of the materials being welded [9]. The stress is provided by the pressure of the jet generated between the joined materials; in other words, stiffer (harder to deform) materials require higher impact velocities. The exact work hardening coefficients of the FeCrAl alloys used in this work are not available, but the 0.2% proof stress of APMT and FeCrAl-ODS at room temperature are 510 to 600 MPa [7] and 887MPa [10] at room temperature, respectively. It can therefore be said that FeCrAl-ODS requires an even higher charge ratio or a thinner cover plate than APMT.

Cross-sectional views of the bonded interfaces by optical microscope are shown in Fig. 4. In all images, the upper side of the figure shows APMT or FeCrAl-ODS, and the lower side shows

SUS316L. Comparisons between (a) and (b), (c) and (d), (e) and (f), the shape of the wave tip (surrounded by the red lines) grows into a hook shape for larger charging ratios. The tips of the waves appear to have areas of different color from both FeCrAl and SUS316L. From the SEM energy-dispersive spectroscopy (EDS), there were no oxide films at the interfaces (map not shown).

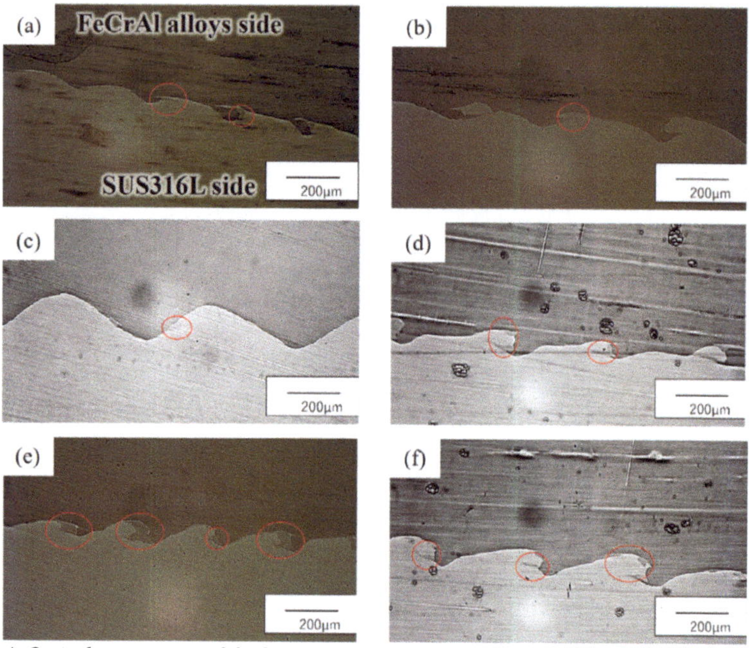

Fig. 4. Optical microscope of the bonding interfaces: (a) APMT charge ratio 0.81 (b) APMT charge ratio 0.95 (c) FeCrAl-ODS SUS316L 2mm charge ratio 1.43 (d) FeCrAl-ODS SUS316L 2mm charge ratio 1.63 (e) FeCrAl-ODS SUS316L 3mm charge ratio 0.95 (f) FeCrAl-ODS SUS316L 3mm charge ratio 1.09.

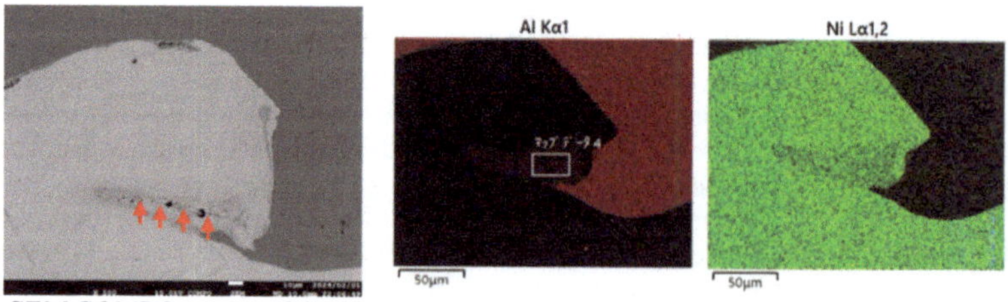

Fig. 5. SEM COMPO image of the bonding interface under the condition (f) and EDS elemental maps of Al and Ni at the same position.

Fig. 5 shows the detailed SEM COMPO image at the wave tip of the joint under the condition (f) in Fig. 4 and its Al and Ni elemental maps obtained by EDS. Al or Ni diffusion near the interface was negligible. From the EDS maps, Al and Ni coexist in intermediate products in the area of different color in the optical microscope (corresponding to the area surrounded by a rectangle in the Al map), indicating the molten portions of the FeCrAl-ODS and SUS316L. In addition, cracks were observed in this area (indicated by arrows in the SEM COMPO image). The same phenomenon occurred under conditions (c) and (e). In the molten area, Al, a ferrite-stabilizing element, and Ni, an austenite-stabilizing element, coexist, and the crystal structure changes from the austenite phase at high temperature immediately after solidification to the ferrite or martensite phase as the temperature decreases, which may cause local cracking. On the other hand, in condition (d), the presence of a molten zone was also confirmed but no cracks were observed (image not shown).

Fig. 6 shows the Vickers microhardness distribution across the bonding interface under condition (b) and (f). As a comparison, the original hardness of SUS316L, APMT, and FeCrAl-ODS are drawn as lines on the graph. In both conditions, the hardness increase nearby the bonding interface and become even higher than the original materials. This trend was observed to varying degrees in all conditions. The main possible cause for this is severe deformation at the interface during explosion welding: as can be seen in Fig. 4, the joint interface has a severe wave shape, clearly indicating that significant plastic deformation has occurred during joining. Martensitic transformation in the molten area, as described above, can also affect the hardness of the interface. Fig.7 shows the layout of the indentation.

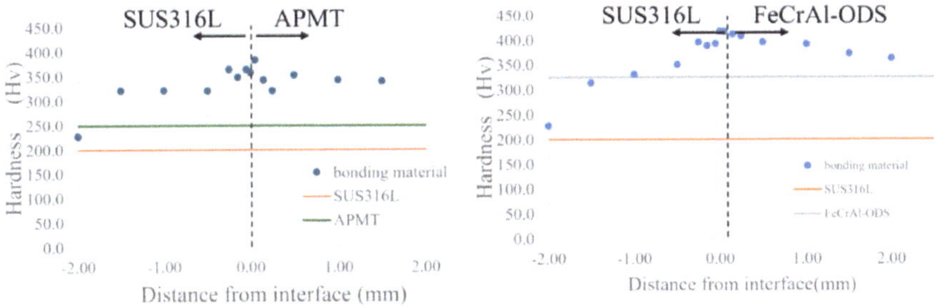

Fig. 6. Hardness changes in the vicinity of the bonding interface under condition (b) (left) and (f) (right) in Table 2.

Fig. 7. The layout of the indentation.

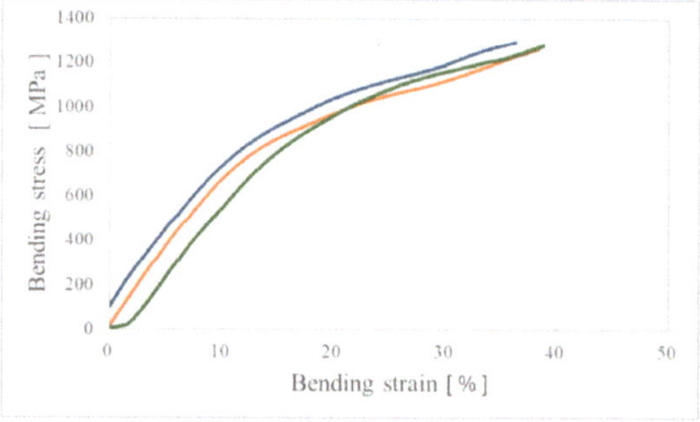

Fig. 8. Stress-strain diagram of bending test.

Fig. 8 shows stress-strain diagrams from the 4-point bending test. Stresses were calculated from the equation in the literature [11]. No interface failure was observed after the test. From Fig. 7, the first yield was observed at a bending stress of approximately 800 MPa. The bending strength of SUS316L is reported to be 896 MPa [12], which is similar to the results of this experiment. As the bending strength of APMT is not known, a comparison is made using 0.2% proof stress. The proof stress of SUS316L is 175 MPa [8] at room temperature. Since SUS316L is weaker than APMT (the proof stress of APMT was described above), the first yielding should be caused by SUS316L. It can be said that the joint strength is sufficient for explosion welding, although the possibility that the interface strength up to specimen fracture needs to be investigated cannot be ruled out. As measurement is impossible due to the design of the 4-point bending test, this will be a future issue, including the development of a suitable test.

Conclusions

Explosion welding was performed between two types of FeCrAl alloys (APMT and FeCrAl-ODS) and SUS316L. The following conclusions were drawn from the analysis.

(1) The joint length of the explosion weld joint depends on the amount of explosive and the thickness of the flyer plate. In the case of FeCrAl-ODS, a good joint length and a crack-free joint interface were obtained as the thickness of SUS316L decreased and the charge ratio increased.

(2) Cross-sectional observation at the bonding interface showed intermediate products at the interface; SEM-EDS analysis confirmed that these products were molten portions of the base metal and mating material. Cracks were observed in the molten portions in some of the joint specimens.

(3) Hardening was observed in the vicinity of the interface, but this did not affect the fracture origin during 4-point bending test. No fracture was observed at the interface, and the yield was observed at about 800 MPa. Therefore, it can be said that a good joint material was obtained in this test range.

Acknowledgements

Explosion welding was funded and supported by the Joint Usage/Research, the Institute of Industrial Nanomaterials, Kumamoto (K2023-19). The SEM microanalysis was supported by the Instrumental Analysis Center at Yokohama National University. 4-point bending test was funded and supported by the NIFS Collaboration Research program (NIFS23KIEF047) and NINS program of Promoting Research by Networking among Institutions (Grant Number 01412302).

References

[1] M. Kondo, S. Hatakeyama, N. Oono, T. Nozawa, Corrosion-resistant materials for liquid LiPb fusion blanket in elevated temperature operation, *Corros. Sci.* **197** (2022) 110070. https://doi.org/10.1016/j.corsci.2021.110070

[2] T. Tanaka, T. Chikada, T. Hinoki, T. Muroga, Electrical insulation performances of ceramic materials developed for advanced blanket systems under intense radiations, J. Nucl. Mater. 569 (2022) 153917. https://doi.org/10.1016/j.jnucmat.2022.153917

[3] M.S. Burton, Metallurgical Principles of Metal Bonding, Welding-J 11 (1954) 1051.

[4] R. Nishio, T. Tanaka, N. Oono, M. Kondo, Reduction of MHD pressure drop by electrical insulating oxide layers in liquid breeder blanket of fusion reactors, Nucl. Mater. Energy 34 (2023) 101382. https://doi.org/10.1016/j.nme.2023.101382

[5] T. Fujio, Explosion Welding, Japan Weld. Soc. 41 (1972) 28-33.

[6] https://www.toyo-success.co.jp/product/characteristic_s.html

[7] https://www.kanthal.com/en/products/material-datasheets/tube/kanthal-apmt/

[8] S. Ukai, K. Sakamoto, S. Ohtsuka, S. Yamashita, A. Kimura, Alloy design and characterization of a recrystallized FeCrAl-ODS cladding for accident-tolerant BWR fuels: An overview of research activity in Japan, J. Nuclear Mater. 583 (2023) 154508. https://doi.org/10.1016/j.jnucmat.2023.154508

[9] Y. Ma, T. Wang, G. Wang, X. Fang, C. Chu, Numerical and experimental studies of the interface characteristics and wave formation mechanism of Hastelloy/stainless steel explosive welding composite plate, *Mater. Today Commun.* **36** (2023) 106880. https://doi.org/10.1016/j.mtcomm.2023.106880

[10] P. Dou, Z.-X. Xin, W. Sang, A. Kimura, Age-hardening mechanisms of 15Cr ODS ferritic steels with 5, 7 and 9 wt.% Al at 475°C for 9000 h, J. Nucl. Mater. 540 (2020) 152368. https://doi.org/10.1016/j.jnucmat.2020.152368

[11] H. Fu, T. Nagasaka, T. Muroga, W. Guan, S. Nogami, H. Serizawa, S. Geng, K. Yabuuchi, A. Kimura, Plastic deformation behavior and bonding strength of an EBW joint between 9Cr-ODS and JLF-1 estimated by symmetric four-point bend tests combined with FEM analysis, Fusion Eng. Des. 102 (2016) 88-93. https://doi.org/10.1016/j.fusengdes.2015.11.037

[12] https://www.silicolloy.co.jp/characteristic/mechanical2/

Hemming for improvement of joint strength in aluminium alloy and carbon fibre-reinforced thermoplastic sheets

ABE Yohei[1,a]* and KIMURA Kensuke[1,b]

[1]Toyohashi University of Technology 1-1 Hibarigaoka, Tempaku, Toyohashi, 441-8580, Japan

[a]abe@me.tut.ac.jp, [b]k-kimura@plast.me.tut.ac.jp

Keywords: Metal/Polymer, Joining, Hemming, Carbon Fibre-Reinforced Thermoplastic, Hot Pressing, Aluminium Alloy Sheet

Abstract. To increase the joint strength in hemmed aluminium alloy and carbon fibre-reinforced thermoplastic (CFRTP) sheets, two kinds of hemming processes were proposed. In a hemming process, hemming was combined with protrusions with structures on the surface on the aluminium alloy sheet. The protrusions with structure were firmly joined to the surface on the sheet by laser cladding with fine ceramic powders, and then the heated and softened CFRTP sheet material was compressed and infiltrating into the protrusions in hot pressing. In the other hemming process, hemming was combined with the aluminium alloy sheet with drilled holes. The heated and softened CFRTP was flowed into the holes in hot pressing. Each hot press condition was investigated. And then, the strength in the pull-out direction and the strength in the peel direction were measured and compared. The joint strengths by the proposed two hemming processes were larger than that by hemming only.

Introduction

To reduce carbon dioxide emissions, the reduction weight in automobile body parts is a method. To reduce the weight, using high specific materials such as ultra-high strength steel, aluminium alloy and carbon fibre-reinforced plastic (CFRP) is effective. It is not easy for automobile bodys to use single material because of the different requirements of each body part. Therefore, many kinds of joining processes are used for joining these materials in assembling automobile body. For joining overlapped two steel sheets, resistance spot welding [1] is generally used. For joining overlapped two aluminium alloy sheets, self-piercing riveting [2] tends to be used. For joining overlapped steel and aluminium alloy sheets, self-piercing riveting and a screw turning tend to be used [3]. The joining process is selected depending on not only sheet combination but also the requirements of joints.

For joining CFRTPs and metals, adhesive bonding [4] and mechanical fastening are conventionally used. Although adhesive bonding is effective for joining CFRTPs and metals, not only adhesive material, but also an adhesive application process and a curing process including temporarily fixing are required. In mechanical fastening, self-piercing riveting without pre-hole is possible to join CFRP and aluminium alloy sheets [5]. A clinching process, which is a method for joining aluminium alloy and CFRP as mechanical fastening without rivets, is placed two sheets in high temperature are pressed by a punch, and the sheets are joined with interlocking [6]. A hole-clinching process, which is using CFRP having a hole, is possible to join in room temperature [7]. On the other hand, CFRTPs and metals are possible to be joined by heating CFRTPs. Friction spot joining [8] and friction stir spot welding [9] are a joining method using the thermal energy generated by the friction between the rotating tool and the metal. Heat is transferred from the metal heated by friction to the CFRTP, and the plastic near the joint interface is heated and joined.

Hemming joints are one of the joints that using plastic deformation [10]. Hemming joint is a joining method in which the edges of the sheet metal that will become the outer sheet are folded back and the sheet that will be the middle sheet is sandwiched between sheets and it is used with bonding in joining automobile parts [11]. Hemming joints can be made by using metal sheet on the bending side and CFRTP on the sandwiching side. However, simply hemmed joint has the problem that the joining force is low because the only the frictional force between the sheets occurs.

In this study, two kinds of hemming processes for joining the aluminium alloy and CFRTP sheets were proposed to increase the joint strength. In these processes, hemming and hot press were used. The heating conditions in hot pressing was investigated to join without defects. And then, the strength in the pull-out direction and strength in the peel direction were measured and compared.

Hemming for improvement of joint strength

The hemming processes for improvement of the joint strength in the aluminium alloy and CFRTP sheets are shown in Fig. 1. The strength of the hemmed joint in the tension shearing and the peel directions are measured by a pull-out testing and a peel testing, respectively. The load of the hemmed joint is only generated by the frictional force between the aluminium alloy and CFRTP sheets. Therefore, two kinds of methods were proposed to improve joint strength using the anchor effect. The first hemming method was combined with protrusions with structures on the surface on the aluminium alloy sheet. The anchoring effect of the protrusions and carbon fibres improves the joining load [12, 13]. In this process, the protrusions with structure were firmly joined to the surface on the sheet by laser cladding with fine ceramic powders, and then the sheet was hemmed, finally the heated and softened CFRTP material was compressed and infiltrating into the protrusions in hot pressing. The second hemming method was combined with drilled holes in the aluminium alloy sheet. The anchoring effect of the flowed CFRTP material into the holes improves the load. In this process, the holes in the aluminium alloy sheet were obtained by drilling, and then the sheet was hemmed, finally the heated and softened CFRTP material was flowed into the hole cavities in hot pressing.

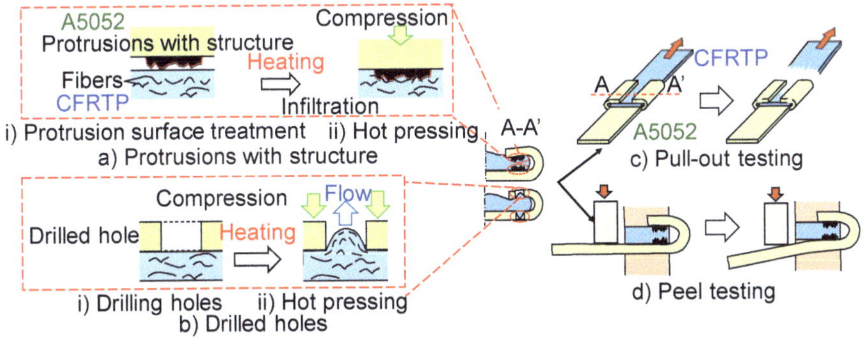

Fig. 1. Hemming processes for improvement of joint strength in aluminium alloy and CFRTP sheets.

Conditions for pull-out and peel tests

The used sheet materials were a 2 mm in thickness of aluminium alloy A5052-H sheet and a 3 mm in thickness of carbon fibre reinforced thermoplastic (CFRTP) sheet. The base material in CFRTP sheet was polyamide MXD6 with short fibres woven in one direction. The material properties are shown in Table 1.

In a pull-put testing, the tensile shear stress is measured by pull-out test specimen in Fig. 2(a). The CFRTP sheet were joined in both edges by hemming, and then the upper sheet was pulledout. In the peel testing, the peel stress was measured by pushing the hemmed specimen with a punch in Fig. 2(b). The CFRTP sheet were joined in an edge by hemming. In the both specimens, the sheet widths and the hemmed length were 20 mm and 6 mm, respectively.

Table 1. Material properties of sheets.

Sheet	Material	Thickness [mm]	Tensile strength [MPa]	Melting point [℃]	Fibre length [mm]
Outer	A5052-H	2.0	242	-	-
Inner	CFRTP (PAMXD6)	3.0	151 (Injection direction)	243	< 0.5

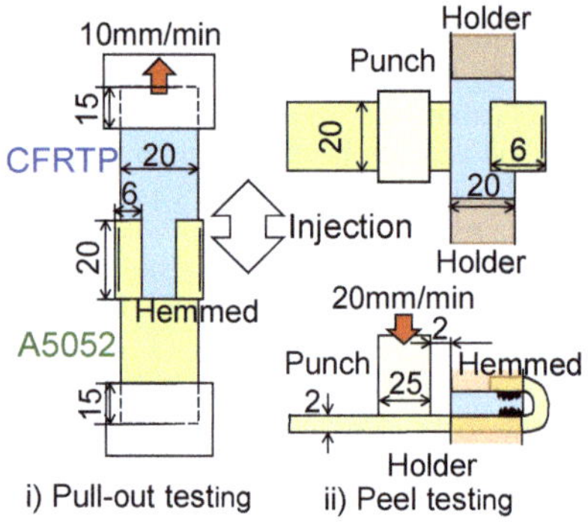

Fig. 2. Conditions for pull-out testing and peel testing.

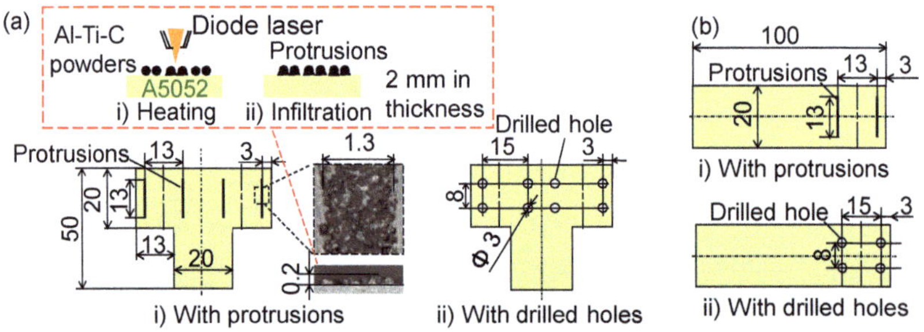

Fig. 3. Specimens for pull-out testing and peel testing.

The specimens for the pull-out testing and peel testing are shown in Fig. 3. The specimens by hemming with the protrusions and by hemming with the drilled holes were prepared for each testing. For hemming specimen with the protrusions, the protrusions, which was made of Al-Ti-C powders, were infiltrated four portions on the aluminium alloy surface. The length and width were 13 mm and 1.3 mm, respectively. For hemming specimen with the drilled holes, total eight holes with 3 mm in diameter were drilled in aluminium alloy sheet. The hemming specimens for peel testing in Fig. 3(b) were used with two portions of the protrusions and four holes, respectively.

Materials Research Forum LLC
https://doi.org/10.21741/9781644903254-75

Conditions for pull-out and peel tests

The joining process for improvement of the joint strength is shown in Fig. 4. The joining process consists of four-stage bending and hot press. In three-stage bending, aluminium alloy sheet was bent to 90°, 110°, 135° and 180°. Although A5052 was enough ductility to bend to 180°, a punch having sidewall to contact with the outer corner of the bending sheet was used[14]. In 180°-bending, the CFRTP sheet was inserted in bent aluminium alloy sheet. In hot pressing, the sheets were heated by the heated blocks on the aluminium alloy top and bottom surfaces. The heating temperature T_h in the heated blocks and holding force F_s were investigated to join the sheets. The holding time was 10 minutes, and heating temperature was reduced under -2C°/min to 180°C after holding time. After 180°C, the specimens were cool to room temperature under holding force applied.

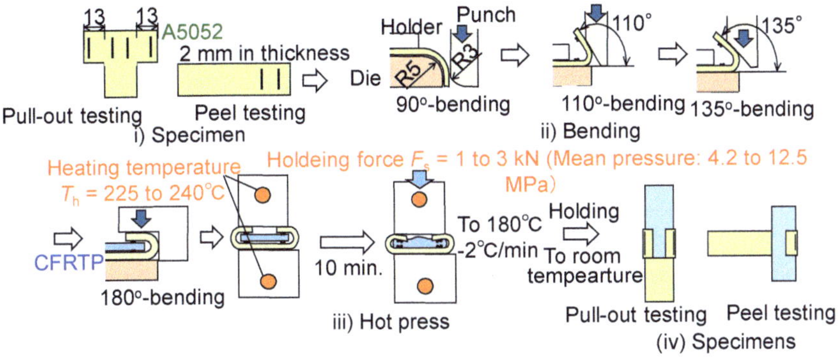

Fig. 4. Joining process for improvement of joint strength.

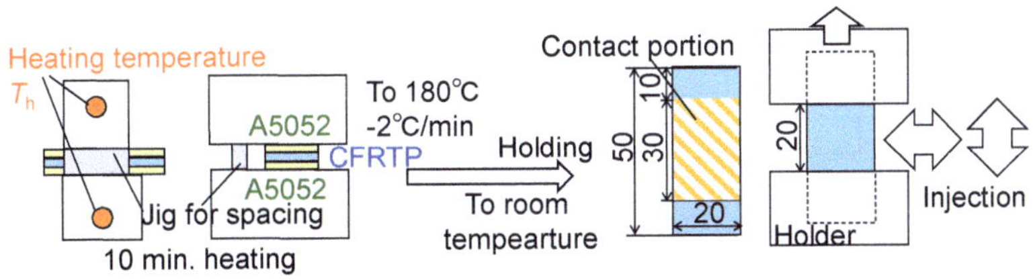

Fig. 5. Tension test conditions for CFRTP.

To investigate the material properties of CFRTP sheet after hot pressing, tension testing in room temperature of once heated CFRTP sheet was performed. The conditions of tension testing for CFRTP sheet are shown in Fig. 5. The CFRTP sheet was 50 mm in length and 20 mm in width, and then the sheet was sandwiched between two aluminium alloy sheets. The sheets were heated by the heated steel blocks on the aluminium alloy top and bottom surfaces as same as in hot pressing. After cooling to room temperature, tensile testing was carried out. The effects of the heating temperature T_h and injection direction were investigated.

The tensile force and stress of CFRTP sheet at room temperature without heating are shown in Fig. 6. The tensile force was increased with increasing of stroke. When the force in both the directions reached at a peak, the force suddenly dropped with fracture. The force in the injection parallel direction was higher than that of the injection vertical direction. Therefore, the injection parallel direction in the specimens was used in this study.

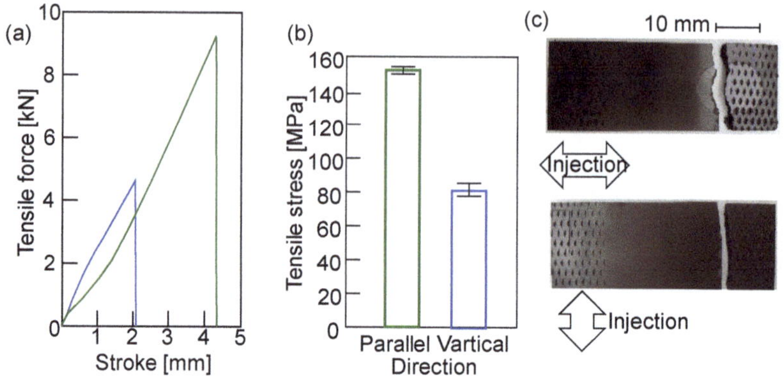

Fig. 6. Tensile force and stress of CFRTP at room temperature.

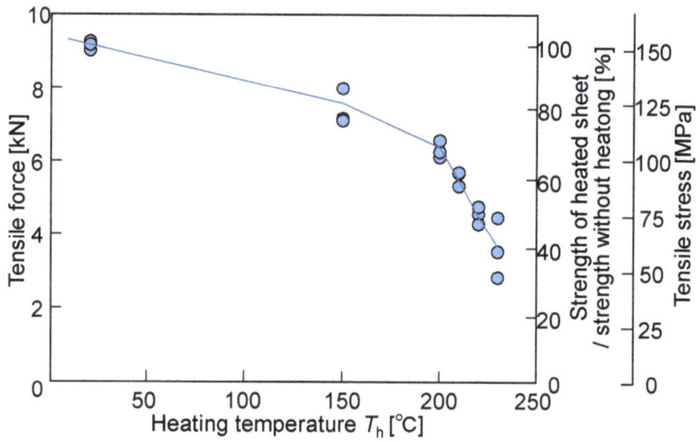

Fig. 7. Effect of heating temperature on tensile force and stress of CFRTP.

The effect of heating temperature on the tensile force and nominal stress of the CFRTP sheet is shown in Fig. 7. The tensile force and nominal stress of the CFRTP sheet were reduced by heating. At 230°C in hot pressing, the strength of the heated sheet was reduced to about 40% of the strength without heating.

Results in hot pressing

The aluminium alloy sheet was successfully bend in four-stage bending without defects. The effect of the heating temperature on the joinability in hot pressing to flow the CFRTP into protrusions with structures is shown in Fig. 8. In high holding force, the cracks were observed on the surface of CFRTP sheet because of large material flow in the longitudinal direction. In insufficient heating temperature, CFRTP material was unfilled into the protrusions. The CFRTP material was infiltrating into the protrusions in appropriate holding force and appropriate heating temperature.

Materials Research Forum LLC
https://doi.org/10.21741/9781644903254-75

The effect of the heating temperature on the joinability in hot pressing to flow the CFRTP into the holes is shown in Fig. 9. In high holding force, cracks were observed on the surface of CFRTP sheet. Although the voids in all CFRTP cross-section were observed, we classified cross-section having CFRTP flow into the holes without cracks as joined. The joining range including the voids in CFRTP sheet was higher temperature and was wider than that by hot pressing to flow CFRTP into protrusions with structures.

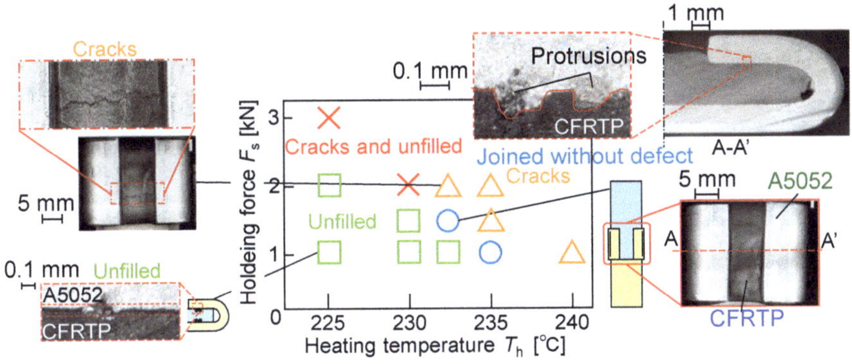

Fig. 8. Effect of heating temperature on joining ability in hot pressing to flow CFRTP into protrusions with structures.

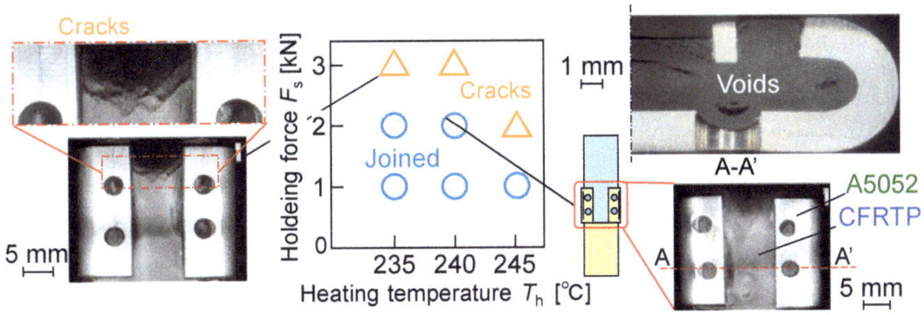

Fig. 9. Effect of heating temperature on joining ability in hot pressing to flow CFRTP into holes.

Joint strength

The pull-out load-stroke curves are shown in Fig. 10. All pull-out loads increased sharply, and then, suddenly dropped. In the load by hemming with the drilled holes, the CFRTP was fractured at the peak load. In the other loads after the peak loads, the loads reduced gradually with long stroke by pull-out in the hemmed portions.

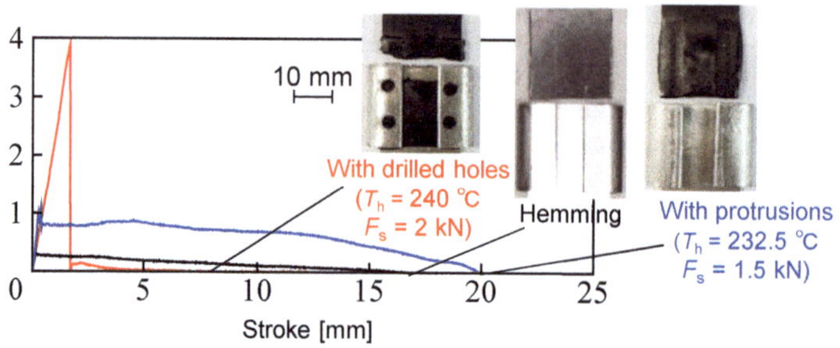

Fig. 10. Pull-out load-stroke curves.

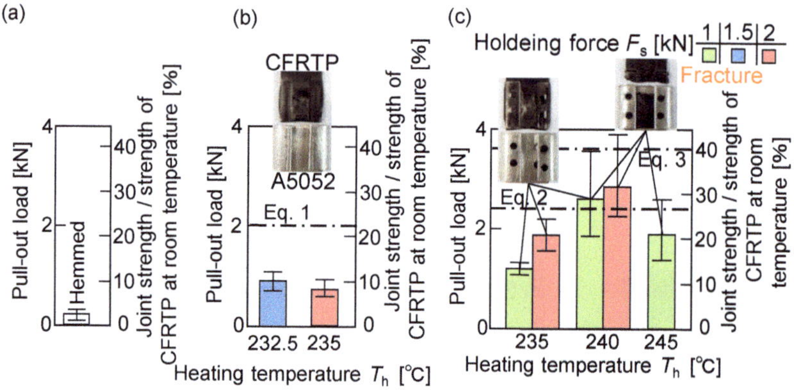

Fig. 11. Pull-out loads for (a) hemmed, (b) hemmed with protrusions and (c) hemmed with drilled holes.

The pull-out loads for joints by hemming, hemming with protrusions and hemming with the holes were summarized in Fig. 11. The estimated pull-out load for joints by hemming with protrusions is

$$P_p = \sigma A_p, \tag{1}$$

where, σ and A_p are joint strength in the protrusions and area of the protrusions, respectively. σ = 30 MPa was used from reference [13]. The estimated pull–out load for joints by hemming with holes is

$$P_h = C\,\sigma_b\,A_h, \tag{2}$$

where, C, σ_b and A_h are constant, tensile strength of the CFRTP and the area of holes, respectively. C = 0.8 and σ_b at 230°C were used. The estimated tensile load of CFRTP is

$$P_c = \sigma_b\,A_c, \tag{3}$$

where, A_c is the cross-sectional area of the CFRTP sheet. The load by the hemmed joint was very small. The loads by hemming with protrusions in T_h= 232.5°C and 235°C were similar. However, the both loads do not reach to the load from Eq. 1. It seems that the material flowing into the structure may not be enough. Although the loads by hemming with holes in high temperature tended to show fracture in the CFRTP sheet, the CFRTP material in holes was fractured in the load in lower temperature. The loads at 240°C were almost in the estimated loads between Eq. 2 and Eq. 3.

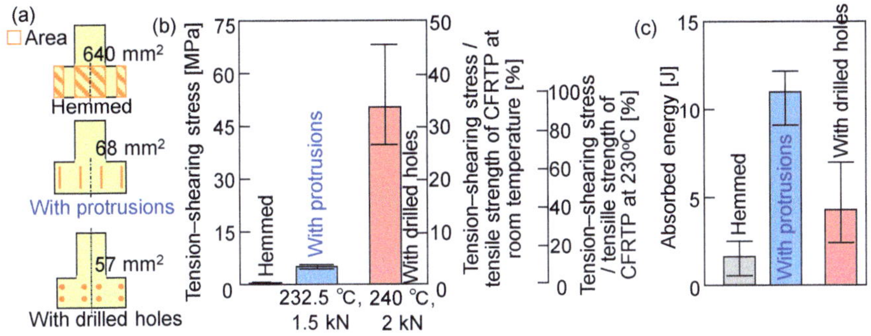

Fig. 12: (a) Joint area, (b) tension–shearing stress and (c) absorbed energy of joints.

The tension–shearing stress and absorbed energy of joints are shown in Fig. 12. The load was divided by the area to show as tension–shearing stress. The area of hemmed sheets was defined the flange portions. The area of hemming with protrusions was defined the area of the all protrusions. The area of hemming with holes was defined the area of all holes. The stress of joint by hemming with holes was the highest. Although the ratio of the tension–shearing stress and the strength at room temperature was about 34%, the ratio of the stress and the strength of CFRTP at 230°C was about 92%. The absorbed energy in hemming with protrusions was the highest because of medium pull-out load and long stroke as shown in Fig. 10.

The peeling load-stroke curves are shown in Fig. 13. The peeling load of the hemmed sheets was not measured because of the unfixed sheets. The loads of the joint by hemming with protrusions and hemming with the holes increased with increasing the stroke. In the both joints, the loads dropped rapidly by occurrence of fracture on the lower surface. Finally, the fracture occurs on the upper surface, and then the sheets were separated.

The peeling stress and absorbed energy of joints are shown in Fig. 14. The load was divided by the area to show as peeling stress. The peeling stress and tension–shearing stress of joints by hemming with protrusions were same in Fig. 12, although the peeling stress of joints by hemming with holes was about 10% of tension–shearing stress. It seems that the protrusions with structures were effective for loading in the both pull-out and peel directions.

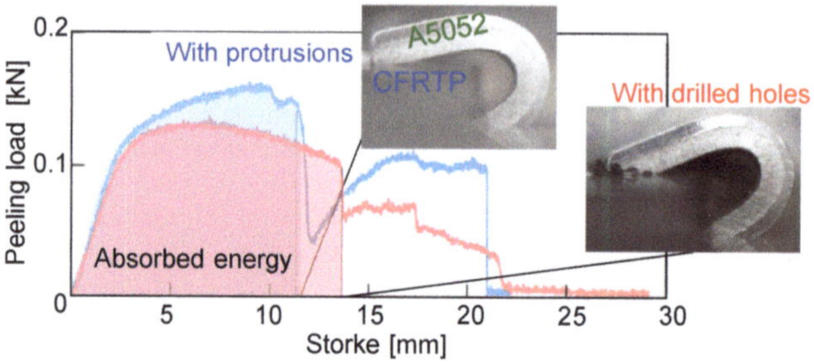

Fig. 13. Peeling load-stroke curves.

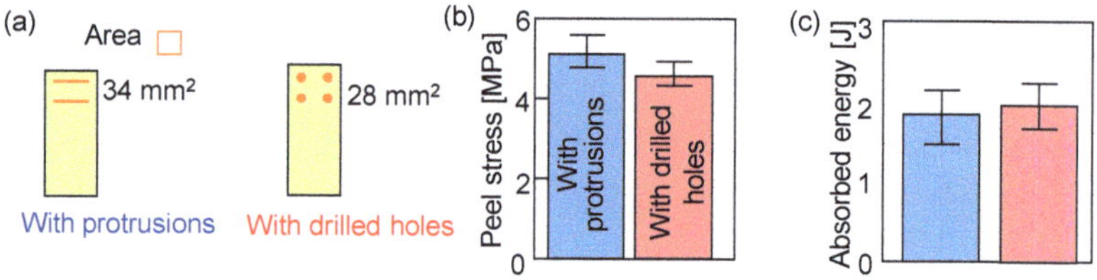

Fig. 14. (a) Joint area, (b) peeling stress and (c) absorbed energy of joints.

Conclusion

Two hemming processes for joining the aluminium alloy and CFRTP sheets were proposed to increase the joint strength. The main conclusions of this study are as follows:

1) In the process of hemming with the protrusions with structure, the CFRTP material was infiltrating into the protrusions in appropriate holding force and appropriate heating temperature.
2) In the process of hemming with the holes, the joining range was wider, although CFRTP sheet included the voids.
3) The tension–shearing stress of joint by hemming with holes was the highest. The ratio of the stress and the strength of CFRTP at 230°C was about 92%.
4) The peeling stress and tension–shearing stress of joints by hemming with protrusions were same. It seems that the protrusions with structures were effective for loading in the both pull-out and peel directions although the stresses were about 5 MPa.

This work was partly supported by "Knowledge Hub Aichi" which is a Priority Research Project of the Aichi Prefectural Government and by Foundation of Public Interest of Tatematsu. We would like to thank Tomohiro Maeda (Kisoh Co., Ltd) for his support and comments.

References

[1] M. Pouranvari, S.P.H. Marashi, Critical review of automotive steels spot welding: Process, structure and properties, Sci. Tech. Weld. Join. 18-5 (2013) 361 – 403. https://doi.org/10.1179/1362171813Y.0000000120

[2] D. Li, A. Chrysanthou, I. Patel, G. Williams, Self-piercing riveting-a review, Int. J. Adv. Manuf. Tech. 92-5-8 (2017) 1777 – 1824. https://doi.org/10.1007/s00170-017-0156-x

[3] T. Sakiyama, Y. Naito, Y. Miyazaki, T. Nose, G. Murayama, K. Saita, H. Oikawa, Dissimilar metal joining technologies for steel sheet and aluminum alloy sheet in auto body,

Nippon Steel Technical Report 103 (2013) 91 – 98.
https://www.scopus.com/inward/record.uri?eid=2-s2.0-84879387161&partnerID=40&md5=82c16e8027334a80781738cc86d97690

[4] J.M. Arenas, C. Alía, J.J. Narbón, R. Ocaña, C. González, Considerations for the industrial application of structural adhesive joints in the aluminium–composite material bonding, Compos. Part B: Engineering 44 (2013) 417–423. https://doi.org/10.1016/j.compositesb.2012.04.026

[5] G. Di Franco, L. Fratini, A. Pasta, V. Ruisi, On the self-piercing riveting of Al blanks and carbon fiber composite panels, Int. J. Mater. Forming 31 (2010) 1035–1038. https://doi.org/10.1007/s12289-011-1067-2

[6] F. Lambiase, D.C. Ko, Feasibility of mechanical clinching for joining aluminum AA6082-T6 and carbon fiber reinforced polymer sheets, Mater. Design 107 (2016) 341–352. https://doi.org/10.1016/j.matdes.2016.06.061

[7] C.J. Lee, J.M. Lee, H.Y. Ryu, K.H. Lee, B.M. Kim, D.C. Ko, Design of hole-clinching process for joining of dissimilar materials – Al6061-T4 alloy with DP780 steel, hot-pressed 22MnB5 steel, and carbon fiber reinforced plastic, J. Mater. Process. Tech. 214–10 (2014) 2169–2178. https://doi.org/10.1016/j.jmatprotec.2014.03.032

[8] K. Nagatsuka, S. Yoshida, A. Tsuchiya, K. Nakata, Direct joining of carbon-fiber–reinforced plastic to an aluminum alloy using friction lap joining, Compos. Part B: Engineering 73 (2015) 82–88. https://doi.org/10.1016/j.compositesb.2014.12.029

[9] S.M. Goushegir, J.F. dos Santos, S.T. Amancio-Filho, Friction Spot Joining of aluminum AA2024/carbon-fiber reinforced poly (phenylene sulfide) composite single lap joints: Microstructure and mechanical performance, Mater. Design 54 (2014) 196–206. https://doi.org/10.1016/j.matdes.2013.08.034

[10] S. Kalpakjian, S. Schmid, Manufacturing Engineering & Technology, 7th ed.; Pearson Education South Asia Pte Ltd.: Singapore (2013) 410.

[11] W. Woyke, F. Jesche, S. Menzel, D. Landgrebe, Designing a flawless hem flange bonding process. adhesion ADHESIVES + SEALANTS, 14 (2017) 26–29. https://doi.org/10.1007/s35784-017-0020-2

[12] T. Maeda, Direct joining technology for metal and plastics by positive anchor effort, Journal of the Japan Society for Precision Engineering 84–5 (2018) 419–422 (in Japanese). https://doi.org/10.2493/jjspe.84.419

[13] K. Iwata, A. Suzuki, S.G. Kim, N. Takata, M. Kobashi, Enhancing the solid-state joinability of A5052 and CFRTP via an additively manufactured micro-structure, J. Mater. Process. Tech. 306 (2022) 117629. https://doi.org/10.1016/j.jmatprotec.2022.117629

[14] Z. Hamedon, K. Mori, Y. Abe, Hemming for joining high strength steel sheets, Procedia Eng. 81 (2014) 2074–2079. https://doi.org/10.1016/j.proeng.2014.10.288

α'-martensite grading techniques in reverse flow forming of AISI 304L

ARIAN Bahman[1,a *], HOMBERG Werner[1], KERSTING Lukas[2,b], TRÄCHTLER Ansgar[2], ROZO VASQUEZ Julian[3,c] and WALTHER Frank[3]

[1]Paderborn University, Forming and Machining Technology (LUF), Germany

[2]Fraunhofer Institute for Mechatronic Systems Design (IEM), Germany

[3]TU Dortmund University, Chair of Materials Test Engineering (WPT), Germany

[a]ba@luf.upb.de, [b]lukas.kersting@iem.fraunhofer.de, [c]julian.rozo@tu-dortmund.de

Keywords: Reverse Flow Forming, α'-Martensite, Grading Strategies, Property-Control

Abstract. Manufacturing processes benefit from property-control enabling reproducibility, application-oriented outcomes, and efficient part production. In reverse flow forming, state-of-the-art practices focus primarily on geometry-control, neglecting property-control. Given the intricacies of the process involving the interaction of tool and machine behavior, process parameters, properties of semi-finished products and temperatures, incorporating process-control becomes an imperative for producing components with predefined properties. The property controlled within this reverse flow forming process is the local α'-martensite content. Therefore, process strategies to actively influence the α'-martensite content must be implemented. In this study seamless AISI 304L steel tubes are used, where α'-martensite formation is strain- and/or temperature-induced through phase transformation within the process. This paper presents innovative process strategies, methods, and specially developed mechanical and thermal actuator systems to locally increase or suppress the α'-martensite content. The use and implementation of these approaches and tools allows the creation of unique optically invisible microstructure profiles containing 3D-gradings, implying a radial grading of α'-martensite. The locally implemented α'-martensite, forming these 3D-gradings, offers potential applications for functional or sensory purposes. This paper extends beyond theoretical concepts, providing tangible component outcomes.

Introduction

Flow formed tubes, exemplifying outstanding surface quality and precise dimensional accuracy, find widespread application in the automotive and aerospace industries [1]. Besides the visible effects on components, metal forming processes have a profound effect on the microstructure of formed parts [2]. The resulting properties of components in forming processes are an outcome of the collective influence of input parameters, such as the selected process parameters of the forming machine or the choice of the semi-finished material, along with the concurrent process variables that prevail during the deformation, including factors like the temperature within the forming zone and the applied forming forces. Thus, significant property-control within flow forming, which is not yet state of the art, requires the ability to control both, process parameters like the feed rate or the infeed as well as the forming temperature. This is particularly crucial when using TRIP steels such as AISI 304L, where phase transformation from austenite to α'-martensite is strain- and/or temperature-induced [3]. Although achieving geometrical reproducibility in flow formed AISI 304L parts is feasible, each produced component exhibits an arbitrary three-dimensional microstructure profile, indicating a corresponding arbitrary distribution of α'-martensite. A promising approach for the production of flow formed parts with defined geometry and property is closed-loop property-control. Numerous experiments were carried out as part of a priority program funded by the DFG, aiming for the development of innovative process strategies to

Materials Research Proceedings 44 (2024) 708-717 https://doi.org/10.21741/9781644903254-76

successfully produce flow formed parts with locally enhanced or reduced α'-martensite contents. These grading techniques can then be used to create invisible microstructure profiles, in the sense of an invisible "barcode", within the forming process of the tubes. This proposes a pioneering method for achieving tamper-proof part production in reverse flow forming, requiring inventive measurement concepts, as the fundamental aspect of property-control involves observing the pertinent properties or their associated system states [4]. Hence, a softsensor has been devised and implemented within the flow forming process to detect local variations in magnetic component properties and to correlate these with the estimated α'-martensite content [5]. The measurement concept also includes a laser system that ensures the component geometry by monitoring plastic deformation and provides real-time information to the control system. Building upon the accomplished local grading and control of α'-martensite in axial (1D) and axial-angular (2D) directions through solely mechanical or thermomechanical process strategies, this paper advances further by presenting innovative and intricate approaches for achieving α'-martensite grading in radial direction (3D). The aim is to experimentally validate whether the indention depth of the α'-martensite can be influenced by thermal actuators and/or mechanical forming strategies, for example by generating radial temperature gradients or varying the number of flow forming passes. In Fig. 1, various grading types are showcased through a color-coded tube. The red color highlights the 3D-grading, the primary focus of this study, with the radial direction representing the thickness of the tube.

Fig. 1. Differentiation of grading types.

Reverse Flow Forming

The incremental process of flow forming offers in particular significant advantages in flexibility and efficiency [6]. A key feature of flow forming is the intentional reduction in wall thickness Δw of semi-finished tubes. This reduction results in a component elongation Δl caused by volume constancy, accompanied by excellent shape, dimensional accuracy and superior surface qualities [7]. Flow formed tubes adhere to stringent standards, particularly those established by the aerospace industry, and find widespread application in high-performance components such as drive shafts for jet engines and helicopters [8]. This paper focuses on reverse flow forming, as illustrated in Fig. 2, where the material flow occurs counter to the axial motion of the roller tools. These versatile roller tools, capable of both axial and radial movement, rotate around their own axis due to contact friction with the rotating tube.

Materials Research Forum LLC
https://doi.org/10.21741/9781644903254-76

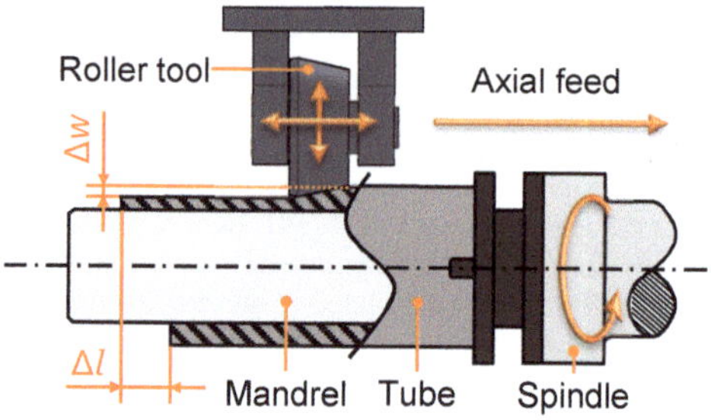

Fig. 2. Reverse flow forming process principle.

Experimental Setup

All the experiments for the study were carried out on a PLB 400 spinning machine from Leifeld Metal Spinning GmbH (Ahlen, Germany) with a drive power reaching up to 11 kW and a maximum spindle speed of 950 rpm. The forming machine is equipped with a hydraulically driven cross support, featuring two machining axes custom-developed at Forming and Machining Technology (LUF) in Paderborn University. Each axis (X/Y-axis) can generate a maximum force of 35 kN. The experimental setup, depicted in Fig. 3, incorporates a single-roller tool configuration, crucial to create sufficient space to accommodate thermal actuator systems and sensors dedicated to process control.

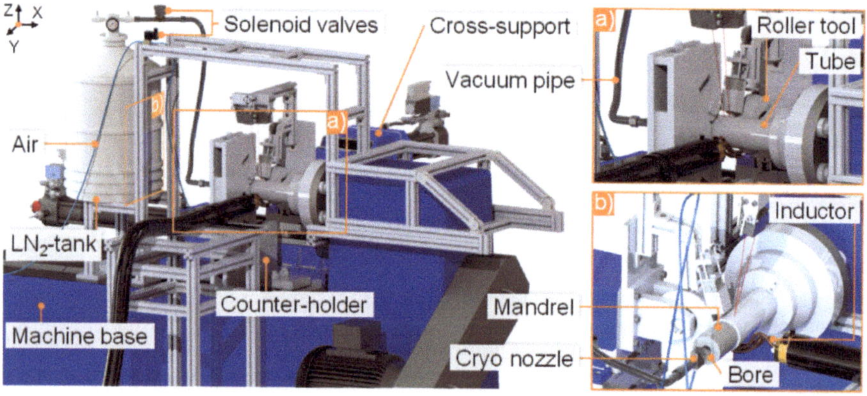

Fig. 3. Machine and actuator configuration with a) and b) showing different viewing angles.

Initially introduced in [9] and later optimized in [10], the sensor concept developed at Forming and Machining Technology (LUF) will not be the primary focus of this work. Moreover, the setup includes a robust counter-holder specifically designed to support the mandrel, along with an absolute value encoder enabling angle-dependent operations within the process. The thermal heating actuator comprises a custom-developed inductor that conforms to the shape of the semi-finished tube, allowing for localized heating in a specific angular area. In addition to the thermal heating actuator, the setup also includes its counterpart, a thermal cooling actuator. The custom-developed automated cryogenic system (Fig. 4) can be used for global or local cooling.

Materials Research Forum LLC
https://doi.org/10.21741/9781644903254-76

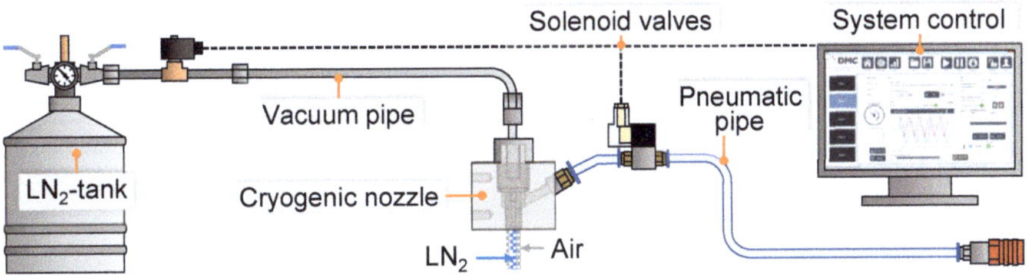

Fig. 4. Automated cryogenic cooling system.

Previously detailed in [11], it includes a liquid nitrogen tank, a flexible stainless steel vacuum pipe, pneumatic pipes, a set of solenoid valves and a two-component cryogenic nozzle bundling liquid nitrogen surrounded by compressed air. The use of solenoid valves, coupled with an absolute value encoder, enables automated and reproducible cooling through computer assistance. To achieve efficient cooling, the cryogenic nozzle focuses a stream of liquid nitrogen at a temperature of -196°C, which creates significant temperature gradients in a short period of time. Utilizing the nozzle allows for the atomization of liquid nitrogen, preventing the Leidenfrost phenomenon [12] that typically occurs when pouring liquid nitrogen in its liquid form. During the conducted experiments, a specially designed hollow mandrel with a diameter of 72 mm and a centric bore of 25 mm was employed. Despite its bore, the mandrel exhibits adequate stability and facilitates indirect cooling of the tube by cooling through the bore. Temperature measurements during the experiments were carried out using a set of tactile thermocouples, chosen for their superior reliability in comparison to optical methods, such as thermographic measurements. A spring-loaded thermocouple, positioned above the tube, measured its surface temperature in proximity to the forming zone (the contact area between the roller tool and the rotating tube). The second thermocouple was positioned inside the tube, between the mandrel, to measure the temperature of the tube's inner surface. Due to the rotation of both the tube and the mandrel, the second thermocouple could not be connected to a standard measuring device via cable. Consequently, a Bluetooth Mini Datalogger was integrated into the rotating spindle, enabling cable-free transmission of temperature values. The semi-finished material used for the experiments was AISI 304L (X2CrNi18-9, 1.4307), a TRIP steel with strain- and/or temperature-induced α'-martensite formation [13]. The detection method for α'-martensite included a precision measurement setup, featuring a Feritscope for reproducible results.

3D-Grading Strategies

The novel introduced grading strategies aim for 3D-graded structures with locally adjusted α'-martensite content useful for functional or sensory purposes in high-performance components.

Temperature gradient strategy. Controlling α'-martensite formation in the radial (thickness) direction of tubes encompasses the management of various factors, including temperature. To establish the desired temperature gradient, cryogenic cooling and induction heating with thermal actuators are employed. One thermal strategy involves combining tube induction heating with indirect mandrel cooling of its inner surface. Copper foil, acting as magnetic induction shielding, can be placed between the tube and mandrel to block heat transfer to the tube's interior and enhance the temperature gradient across the wall thickness (Fig. 5).

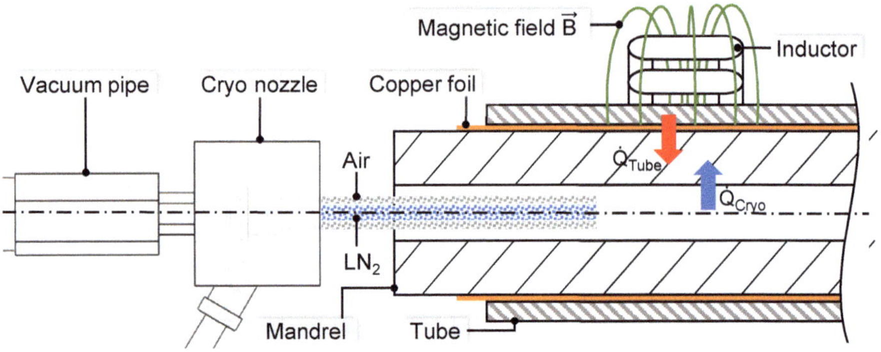

Fig. 5. Temperature gradient strategy: Combining induction heating and indirect cooling.

As previously described, the tube's inner surface cools indirectly through contact with the cooled mandrel. In isothermal flow forming, α'-martensite content peaks on the outer surface due to roller tool contact, gradually diminishing towards the inner surface. To counter this, cooling the inner surface and heating the outer surface during flow forming could normalize or even reverse α'-martensite formation in the radial direction.

Knead strategy. This isothermal knead strategy leverages the phenomenon of increased α'-martensite formation with successive passes for the same amount of wall thickness reduction. Whether inducing a 2 mm reduction in wall thickness with a single pass or with multiple passes in increments of e.g. 0.2 mm, yields significantly different α'-martensite content on the outer surface of the tube, as demonstrated in [14]. Further investigation is required to ascertain if this discrepancy extends to influence α'-martensite formation in the radial direction, potentially making it a viable strategy for influencing such formation.

Results and Discussion

For comparison purposes, a flow formed reference sample was produced isothermally, utilizing a 2 mm infeed (*r*), an axial feed rate of 0.1 mm/s (*f*), and a rotational speed of 30 rpm (*n*). The sample served to investigate the impact of thermal and kneading strategies on radial α'-martensite formation.

Temperature gradient strategy. Fig. 6, captured with the Smartzoom 5 automated digital microscope from ZEISS (Oberkochen, Germany), compares cross-sections of the reference sample in a) with samples produced via b) induction heating (tube surface temperature of 100°C) and c) cryogenic cooling (tube surface temperature of -40°C) using identical process parameters and wall thickness reductions. The microstructure comparison reveals that lower temperatures result in a higher concentration of α'-martensite needles (indicated by black areas), making the microstructure in c) appear darkest. Additionally, the concentration is observed to be highest at the surface, diminishing radially.

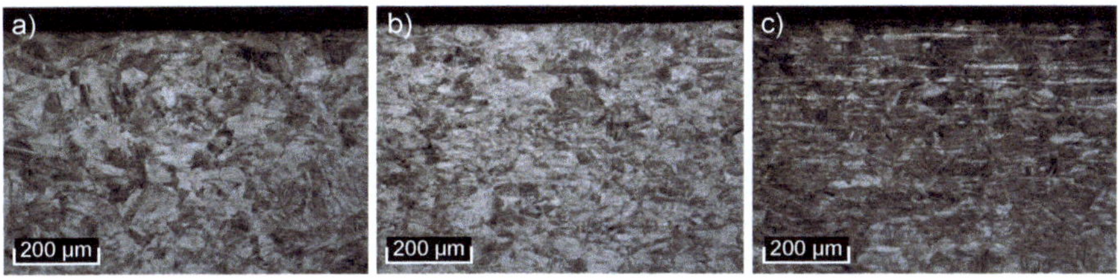

Fig. 6. Microscopic cross-sections: a) ref. sample, b) induction heating, c) cryogenic cooling.

As this type of evaluation is limited to a very localized area and only allows qualitative statements, the entire cross-section (Fig. 7 d) was measured precisely and reproducibly using a specially developed measurement setup. For this, a Feritscope was used measuring the α'-martensite content along the cross-section in the form of a grid with intervals of 0.375 mm/0.25 mm in the Y-(radial) direction and 3 mm in the X-direction (Fig. 7).

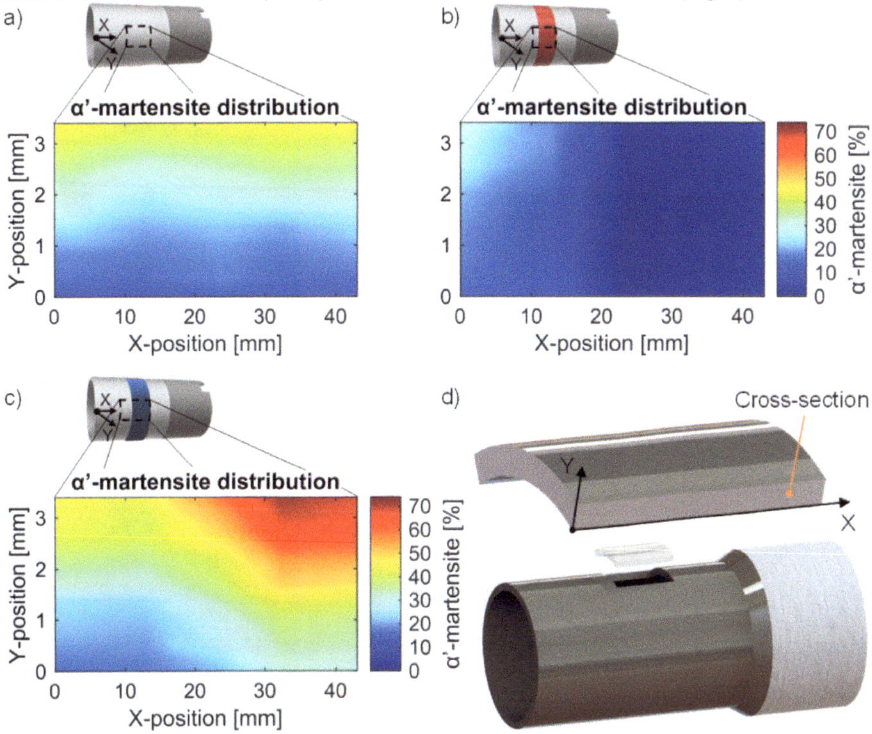

Fig. 7. Cross-section measurements: a) ref. sample, b) induction heating, c) cryogenic cooling, d) highlights the focused cross-section.

The measurements confirm the qualitative evaluation using microscopy. Additionally, cooling in Fig. 7 c) results in deeper penetration of α'-martensite formation compared to the reference sample in Fig. 7 a). The comparison of the grading resolution/sharpness between tempered and non-tempered areas (transition area) shows significant differences between induction heating (Fig. 7 b) and cryogenic surface cooling (Fig. 7 c). Induction heating creates a vertical grading line in radial direction, indicating an abrupt temperature change. Cryogenic surface cooling results in a diagonal grading line and a transition length of approx. 20 mm, implying a gradual temperature change due to radial heat conduction dependency. The temperature gradient for cryogenic surface cooling amounted approx. 20 K ($T_{surface} \approx$ -40°C, $T_{inner\ surface} \approx$ -20°C) while for induction heating no temperature gradient was observed ($T_{surface} \approx$ 100°C, $T_{inner\ surface} \approx$ 100°C). Further experiments compared radial α'-martensite formation between direct surface cooling at -40°C (Fig. 8 b) and indirect cooling (Fig. 8 c) via the mandrel (with both tube surface and inner surface at -40°C), as illustrated in Fig. 8. Indirect cooling from the tube's interior results in greater depth and α'-martensite content compared to surface cooling. The deformation process during surface cooling heats deeper material layers reducing α'-martensite formation.

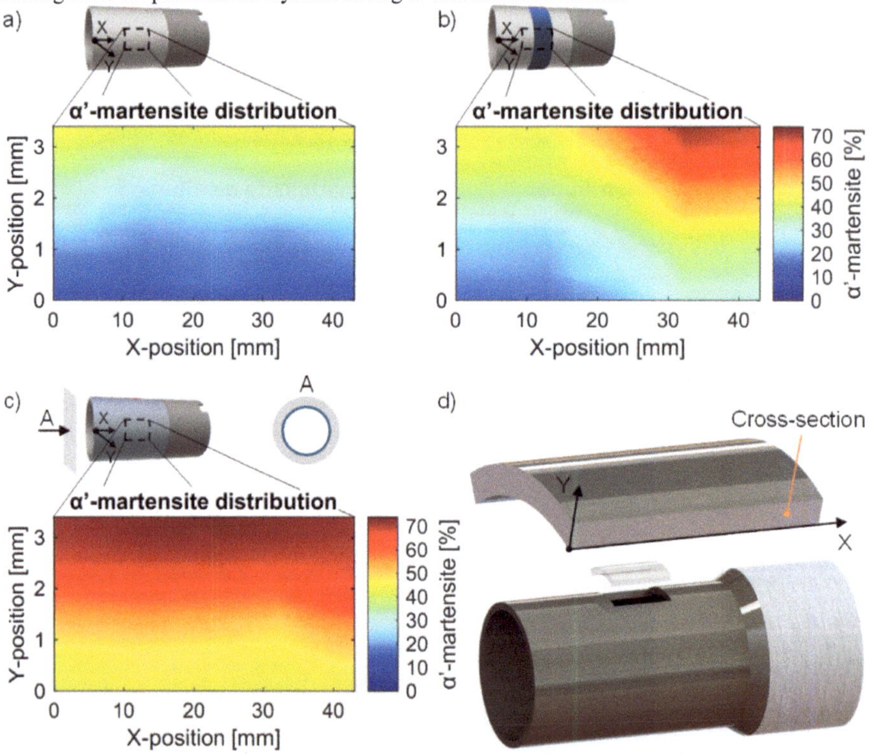

Fig. 8. Cross-section measurements: a) ref. sample, b) direct surface cooling, c) indirect cooling, d) highlights the focused cross-section.

The next strategy involves combining tube induction heating with indirect mandrel cooling of its inner surface, which was described in Fig. 5. Fig. 9 compares radial α'-martensite formation among the reference sample (Fig. 9 a), the combined heating and cooling strategy without (Fig. 9 b), and with copper foil (Fig. 9 c). Without copper foil, α'-martensite formation is completely suppressed, while its use effectively shields the magnetic field, keeping the mandrel cool and preventing heating of the tube's interior. This leads to a homogenization of the α'-martensite content throughout the tube's wall thickness. The use of copper foil enables investigations of the grading resolution/sharpness of the transition area between indirectly cooled and cooled as well as heated areas. The result is a diagonal grading line and a transition length of approx. 13 mm, implying a gradual temperature change. The max. temperature gradient for the combined indirect cryogenic cooling and induction heating amounted approx. 50 K ($T_{surface} \approx 25$°C, $T_{inner\ surface} \approx -25$°C).

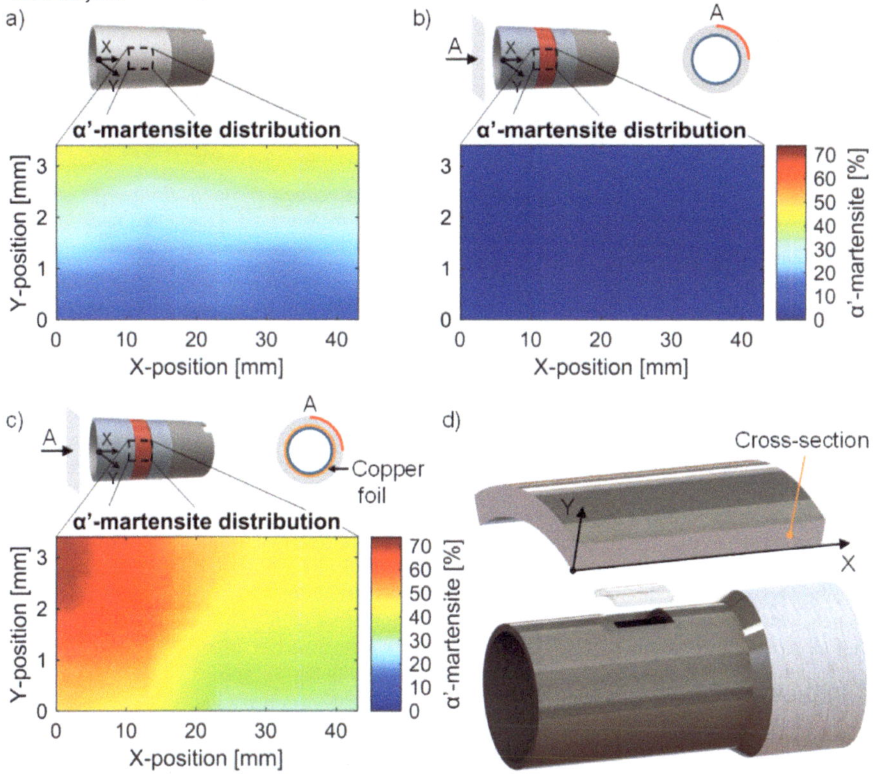

Fig. 9. Cross-section measurements: a) ref. sample, combination of indirect cooling and induction heating in b) without and c) with copper foil, d) highlights the focused cross-section.

Knead strategy. This isothermal strategy uses the number of passes instead of temperature to affect the α'-martensite content. Fig. 10 compares the cross-section of two samples with a wall thickness reduction of 2 mm each. Fig. 10 a) depicts radial α'-martensite distribution for five passes using a 2 mm infeed (*r*). In contrast, Fig. 10 b) shows the distribution of α'-martensite resulting from five passes using a 1 mm infeed followed by four passes with 2 mm. The comparison demonstrates that the radial distribution of α'-martensite is influenced by the number of passes. More passes yield increased depth and higher, more homogenized concentration of α'-martensite for the same 2 mm wall thickness reduction, making the Knead strategy suitable for 3D-grading.

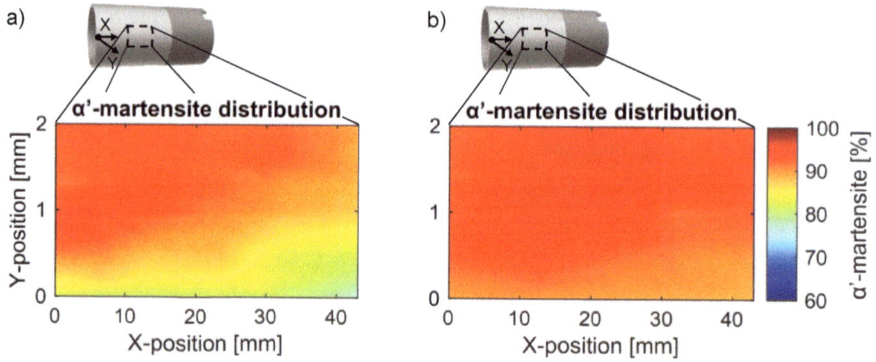

Fig. 10. Cross-section measurements: 2 mm thickness reduction with a) five passes and r = 2 mm, b) five passes and r = 1 mm followed by four passes and r = 2 mm.

Conclusions and Outlook

In this study, innovative strategies for 3D-grading in reverse flow forming have been demonstrated, notably the "Temperature gradient strategy" and the isothermal "Knead strategy". These strategies have demonstrated remarkable effectiveness in influencing the α'-martensite formation and distribution in radial direction over the thickness of tubes, while maintaining constant wall thickness reduction. The "Temperature gradient strategy", leveraging specially developed cryogenic and induction heating systems, proved instrumental in creating controlled temperature gradients during the reverse flow forming process. Of particular significance is the integration of induction heating alongside indirect cooling, augmented by the strategic application of copper foil to neutralize undesired magnetic field induction. This refined method not only facilitates temperature management but also empowers manipulation of α'-martensite formation and distribution in radial direction. Furthermore, it promotes deeper penetration of α'-martensite formation and addresses inherent inhomogeneities typically observed in conventional flow forming processes by homogenizing α'-martensite distribution. Furthermore, the isothermal "Knead strategy" showcased the potential for manipulating α'-martensite distribution through variations in flow forming passes. These innovative approaches offer precise control over microstructural features, promising advancements in material processing and component manufacturing, with potential applications in automotive and aerospace industries.

Acknowledgements

The authors would like to thank the German Research Foundation (Deutsche Forschungsgemeinschaft, DFG) for their support of the depicted research within the priority program SPP 2183 "Property controlled deformation processes", through project no. 424335026 "Property control during spinning of metastable austenites".

References

[1] M. Haridas, G. Gopal, A. Ramesh, R.K. Katta, Modelling and simulation of single and multi-pass flow forming to investigate the influence of process parameters on part accuracy, IJMR 11 (2016) 79473. https://doi.org/10.1504/IJMR.2016.079473

[2] D.A. Hughes, N. Hansen, The microstructural origin of work hardening stages, Acta Mater. 148 (2018) 374–383. https://doi.org/10.1016/j.actamat.2018.02.002

[3] A. Weidner, Deformation Processes in TRIP/TWIP Steels (295). Cham: Springer International Publishing, 2020.

[4] M. Bambach, T. Meurer, W. Homberg, S. Duncan, Editorial to special issue "Property-controlled forming processes", 2022. https://doi.org/10.3929/ethz-b-000530958

[5] W. Homberg, B. Arian, V. Arne, T. Borgert, A. Brosius, P. Groche, C. Hartmann, L. Kersting, R. Laue, J. Martschin, T. Meurer, D. Spies, A.E. Tekkaya, A. Trächtler, W. Volk, F. Wendler, M. Wrobel, Softsensors: key component of property control in forming technology, Prod. Eng. Res. Devel. 18 (2023) 603-614. https://doi.org/10.1007/s11740-023-01227-1

[6] M. Runge, Spinning and Flow forming: spinning and flow forming technology, product design, equipment, control systems. Landsberg/Lech: Moderne Industrie, 1994.

[7] M. Sivanandini, S.S. Dhami, B.S. Pabla, Flow forming of tubes-A review, Int. J. Sci. Eng. Res. 3 (2012) Art. no. 5. [Online]. Available: https://www.ijser.org/researchpaper/Flow-Forming-Of-Tubes-A-Review.pdf

[8] J. Savoie, M. Bissinger, Case studies and applications of flow forming to aircraft engine component manufacturing, KEM 344 (2007) 443–450. https://doi.org/10.4028/www.scientific.net/KEM.344.443

[9] M. Riepold, B. Arian, J. Rozo Vasquez, W. Homberg, F. Walther, A. Trächtler, Model approaches for closed-loop property control for flow forming, Adv. Industr. Manuf. Eng. 3 (2021) 100057. https://doi.org/10.1016/j.aime.2021.100057

[10] B. Arian, W. Homberg, L. Kersting, A. Trächtler, J. Rozo Vasquez, F. Walther, Produktkennzeichnung durch lokal definierte Einstellung von ferromagnetischen Eigenschaften beim Drückwalzen von metastabilen Stahlwerkstoffen: Ideen Form geben: 36. Aachener Stahlkolloquium Umformtechnik: 26.-27. Oktober 2022, Eurogress Aachen: Tagungsband, 1st ed. Aachen: Verlagshaus Mainz GmbH, 2022.

[11] B. Arian, W. Homberg, J. Rozo Vasquez, F. Walther, L. Kersting, A. Trächtler, Cryogenic reverse flow forming of AISI 304L, Mater. Res. Forum LLC (2023). https://doi.org/10.21741/9781644902479-219

[12] Y. Guo, X. Liu, J. Ji, Z. Wang, X. Hu, Y. Zhu, T. Zhang, T. Tao, K. Liu, Y. Jiao, Delayed leidenfrost effect of a cutting droplet on a microgrooved tool surface, Langmuir: the ACS journal of surfaces and colloids, early access. https://doi.org/10.1021/acs.langmuir.3c00592

[13] M. Jambor, T. Vojtek, P. Pokorný, M. Šmíd, Effect of solution annealing on fatigue crack propagation in the AISI 304L TRIP steel, Materials, early access. https://doi.org/10.3390/ma14061331

[14] B. Arian, W. Homberg, J. Rozo Vasquez, F. Walther, M. Riepold, A. Trächtler, Forming of metastable austenitic stainless steel tubes with axially graded martensite content by flow-forming, ESAFORM 2021, 2021. https://doi.org/10.25518/esaform21.2759

Microstructure evolution of semi-solid billet fabricated by semi-solid isothermal heat treatment of wrought AlSi7Mg aluminum alloy

JIANG Jufu[1,a *], SONG Tao[2], WANG Ying[2,b *], ZHANG Ying[1], ZHU Liang[1] and DONG Jian[1]

[1]School of Materials Science and Engineering, Harbin Institute of Technology, Harbin, 150001, P.R. China

[2]School of Mechatronics Engineering, Harbin Institute of Technology, Harbin, 150001, P.R. China

[a]jiangjufu@hit.edu.cn, [b]wangying1002@hit.edu.cn

Keywords: Semi-Solid, Isothermal Heat Treatment, Microstructure, AlSi7Mg Aluminum Alloy

Abstract: In this paper, the as-cast AlSi7Mg aluminum alloy billet was upsetted above recrystallization temperature to obtain wrought aluminum alloy with 50% deformation degree, and semi-solid isothermal treatment method was employed to achieve semi-solid billet of the wrought AlSi7Mg aluminum alloy. The microstructure observation and evolution law were investigated via optical microscope and scanning electron microscope. The results show that the primary α-Al phase can be changed from dendrite to spherical structure by semi-solid isothermal treatment of the wrought alloy (SSITWA). The complete transformation of the primary α-Al phase from dendrite to spherical structure can be achieved by increasing the isothermal temperature and prolongation of holding time. However, when the holding time is too long, the spheroidized grains will grow abnormally and eventually become irregularly shaped grains. High isothermal temperature can increase the primary α-Al size and reduce the time of dendrite to spheroidal grains. The average grain size of AlSi7Mg aluminum alloy semi-solid billet fabricated by SSITWA varies in a range of from 15 μm to 65 μm when the temperature and holding time change. The content of Al and Si elements in primary α-Al phase is obviously higher than that in liquid eutectic phase consisting of Al and Si elements, and the distribution of Mg element is uniform in the microstructure of semi-solid billet fabricated by SSITWA.

Introduction

AlSi7Mg aluminum alloy is a hypoeutectic alloy with silicon and magnesium as the main alloying elements. Because of the large amount of Al-Si eutectic, large temperature interval of solidus and liquidus and good fluidity, it is a desirable alloy for semi-solid metal forming [1]. It is attractive in the automotive and aerospace industries for its light weight and good corrosion resistance [2-5]. AlSi7Mg aluminum alloy also has good mechanical strength, ductility, fatigue properties, and machinability [6-8].

Semi-solid processing technology is a kind of advanced forming technology combining liquid metal casting and solid metal plastic processing [9-12]. It is widely used in aluminum alloy processing [13, 14]. The primary process of semi-solid processing is to prepare semi-solid slurry (billet) with spherical grains and thixotropic property [15]. At present, there are many preparation methods of slurry, including mechanical stirring (MS) method, electromagnetic stirring (ES) method, ultrasonic treatment (UT) method, strain induced melt activation method (SIMA) and recrystallization and partial remelting method (RAP) [16-18]. Many scholars have carried out relevant research based on fabrication of semi-solid slurry of aluminum alloys. Jiang et al.[19] proposed a novel method for fabricating semi-solid billet of aluminum alloy, named semi-solid isothermal treatment of wrought alloy (SSITWA). It involves a partial melting of industrial wrought aluminum alloy (hot-rolled plate or hot-extrusion bar). Four kinds of deformed aluminum alloys were prepared and then the thixoforming experiments were carried out on four typical parts. The results further confirmed the feasibility of thixoforming route of short process. Guan et al.[20] studied the microstructure and mechanical properties of rheological castings of 7075 alloy treated by annular electromagnetic stirring melt. The results showed that the degree of refinement and uniformity of solidification structure could be significantly improved under such conditions. The slurry was used for subsequent rheological casting, and the hot crack defects in the casting process were significantly alleviated. Liu et al. [21] prepared a high strength Al-6Si-6Cu semi-solid billet by liquid phase reaction sintering. The results showed that Cu nanoparticles precipitated in Al grains of the Al-6Si-6Cu (wt%) semi-solid billet and showed strong pinning effect. Under this condition, the Al-6Si-6Cu semi-solid billet yield strength (YS) and ultimate strength (US) increased by 120% and 67.4%. Zhang et al. [22] studied the microstructure evolution of an Al-Si alloy during shear-vibration coupling sub-rapid solidification and the effects of three key parameters, i.e., pouring temperature, vibration frequency, and tilting angle of the slope plate on the microstructure. The results showed that compared with the traditional solidification on the cooling slope plate, more dissociated grains can be produced rapidly by the debonding with the plate surface. The additional vibration would make the dissociated grains more susceptible to subsequent diffusion and would also generate smaller, more uniform primary grains. Yu et al. [23] prepared rare-earth praseodymium cerium (Pr/Ce) reinforced semi-solid ADC12 slurry by high-energy ultrasonic method. The effects of solid fraction, ultrasonic power and Pr/Ce addition on the steady-state apparent viscosity of slurry at a certain shear rate were investigated.. The results showed that the apparent viscosity and solid phase fraction of semi-solid ADC12 slurry increased with the addition of Pr/Ce, but the ultrasonic power was opposite. At the same time, high energy ultrasound also played a role in grain refinement. Semi-solid isothermal heat treatment of wrought aluminum (SSITWA) is that the alloy billet is first subjected to a certain thermal deformation process, and then the semi-solid isothermal treatment is carried out to transform the dendrite structure into spheroidized structure. Then the semi-solid forming process is carried out. The process of non-dendrite ingot billet preparation is omitted as compared to MS, ES and UT methods, and it reduces the cold working process at room temperature as compared to SIMA method. It also reduces the resistance to formation of material as compared to RAP. Therefore, this method has the advantages of short process flow, simple equipment and high operability, and it provides a new idea for the preparation of aluminum alloy semi-solid billet.

In this study, AlSi7Mg semi-solid aluminum alloy billets were prepared by semi-solid isothermal heat treatment of wrought alloy (SSITWA) in order to shorten the working hours of semi-solid billets. The wrought AlSi7Mg alloy was prepared by upsetting with a severe plastic deformation above recrystallization temperature, and isothermal treatment experiments were carried out at different isothermal temperatures and holding time. By microstructure observation and energy spectrum analysis of the semi-solid billet, the influence of process parameters on microstructure of the semi-solid billet was quantitatively analyzed with the average grain size and average shape factor. The microstructure evolution of AlSi7Mg aluminum alloy under the semi-solid temperature range isothermal treatment was determined.

Experimental procedure

In this Study, AlSi7Mg aluminum alloy semi-solid billets were prepared by SSITWA. First, the hot upsetting experiment was carried out on AlSi7Mg alloy, then the intermediate billets were prepared under the selected parameters, and the isothermal treatment experiments were carried out at different isothermal temperatures and holding time.

Materials Research Forum LLC
https://doi.org/10.21741/9781644903254-77

Experimental material. The commercial AlSi7Mg aluminum alloy bar was used as the feedstock, with a chemical composition shown in Table 1. The solidus and liquidus temperatures were determined by STA449F3 differential scanning calorimeter (DSC). 3 mm diameter and 1.5 mm height disc sample was heated in an argon atmosphere at a rate of 10°C min^{-1} from room temperature to 700°C. According to the DSC result [24], the solidus temperature was 574.5°C, and liquidus temperature was 636.2°C, indicating a wide range of semi-solid state. The solid fractions at semi-solid isothermal temperatures 590°C, 595°C, 600°C, 605°C and 610°C were 0.64, 0.55, 0.491, 0.433 and 0.367, respectively [24].

Table 1. Chemical composition of AlSi7Mg alloy (wt.%).

Si	Mg	Ti	Fe	Mn	Cu	Zn	Al
6.930	0.490	0.054	0.020	0.033	0.011	0.043	Bal.

Semi-solid isothermal treatment of wrought AlSi7Mg alloy. The equipment used in the experiment was 50000 kN hydraulic press and box-type resistance heating furnace. 120 mm×160 mm original ingot was heated to 150°C in a heating furnace and sprayed with a lubricant of water-based graphite. After the billet was heated to 350°C, it was upsetted on a press at a deformation rate of about 80 mm/min, and the upsetting process was stopped until the deformation reached 50%. Finally, the drum-shaped billet was obtained. The cuboid specimens with 10*10*12 mm^3 for semi-solid isothermal treatment experiment were cut from the upsetted billet. The semi-solid isothermal temperatures were 590°C, 595°C, 600°C, 605°C and 610°C, and the holding time was 1 min, 5 min, 10 min, 15 min and 20 min, respectively. 0.5% HF solution was used for corrosion treatment, and the corrosion time was 5~10 s. There were 25 groups of experimental parameters. The samples were isothermal treated with a box-type resistance heating furnace (temperature accuracy ± 1 °C). The samples were heated to a specified temperature and held for a set time. Then the samples were taken out from the box-type resistance furnace and quenched in water to retain the microstructure under this parameter. The specific parameters of semi-solid isothermal treatment are shown in Table 2. The experimental flow chart is shown in Fig. 1.

Table 2. Experimental parameters of AlSi7Mg alloy isothermal treatment.

NO.	temperature / ℃	holding time /min	NO.	temperature / ℃	holding time /min
1	590	1	16	605	1
2	590	5	17	605	5
3	590	10	18	605	10
4	590	15	19	605	15
5	590	20	20	605	20
6	595	1	21	610	1
7	595	5	22	610	5
8	595	10	23	610	10
9	595	15	24	610	15
10	595	20	25	610	20
11	600	1			
12	600	5			
13	600	10			
14	600	15			
15	600	20			

 Materials Research Forum LLC
https://doi.org/10.21741/9781644903254-77

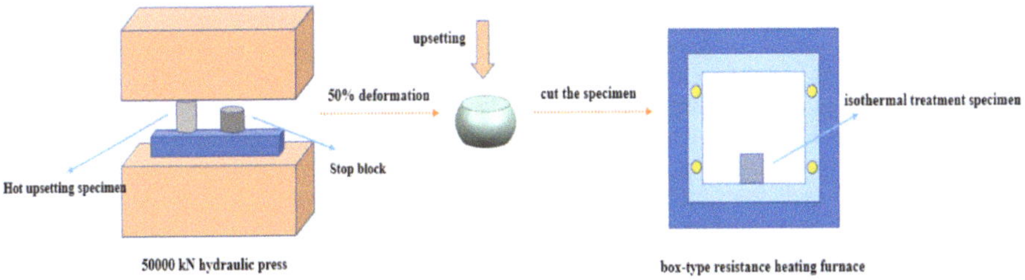

Fig. 1. Experimental flow chart.

After the isothermal treatment, the samples were ground with 200, 400, 800, 1200, 1500, and 2000 grit SiC sandpapers, and they were polished and corroded by polishing machine and 0.5% HF etchant. The metallographic images were taken by an optical microscopy (Olympus GX71). Image Pro Plus software was used to analyze and process the metallographic images (no less than 300 grains were measured in order to reduce statistical error). The area and perimeter of the grains were counted, and the average grain size was calculated. The average shape factor was used to describe the roundness of the grains. The calculation method was shown in formulas (1) and (2), respectively [25]:

$$D=\frac{1}{N}\sum_{i=1}^{N}\sqrt{\frac{4A_i}{\pi}} \quad (1)$$

$$f=\frac{1}{N}\sum_{i=1}^{N}\frac{4\pi A_i}{P_i^2} \quad (2)$$

Where, D——Average grain size/μm;
f——Average shape factor;
A_i——The area of the ith grain or grain measured;
P_i——The circumference of the ith grain or grain measured;
N——The total number of grains or grains measured.

Results and Discussion

Microstructure evolution of AlSi7Mg Al alloy during semi-solid isothermal treatment. Fig. 2 to Fig. 6 show the metallographic images of AlSi7Mg alloy in different temperature groups with an increase of soaking time. The grey colour part represents primary α-Al grain and the black colour part represents liquid phase, i.e. low-melting-point eutectic structure. Fig. 2 shows the microstructure of AlSi7Mg alloy held at 590°C for 1 min, 5 min, 10 min, 15 min and 20 min. Fig. 2(a) shows the microstructures of AlSi7Mg alloy held at 590°C for 1 min. Due to the short holding time, the dendrites in the original as-cast structure still existed. After holding at 590°C for 1 min, the dendrite grains were relatively coarse and the intergranular liquid phases were less. Only a small number of eutectic phases melted, and no liquid phases appeared in the grains. The morphology of eutectic phases did not change much. This is because that solid-liquid transition depends on solute diffusion and the solute diffusion requires enough time at a constant temperature. In the early stage of semi-solid isothermal treatment, the soaking time is too short to ensure enough time of solute diffusion. As a result, the microstructure still exhibits a dendrite morphology. Fig. 2(b) shows the microstructures with a holding time of 5 min. With the increase of holding time, the grain size became uniform, and the particles became fine and spheroidization occurred. A small number of liquid phases began to appear in the grain, most of the eutectic phases

 Materials Research Forum LLC
https://doi.org/10.21741/9781644903254-77

on the α phase grain boundary had melted into liquid phases. But it is found that a small amount of granulated eutectic phases had not melted. The α-Al phase was basically spheroidized, and the adjacent grains merged and grew up. The increase of holding time was beneficial to the formation of spherical grains and the melting of eutectic phases.

Fig. 2(c) shows the microstructures of AlSi7Mg alloy after holding for 10 min. It could be seen that the grains had obvious signs of growth. With the increase of holding time, the grains became larger and larger after merging, and the liquid phases in the grains also became more and more, which tended to melt from the inside. Grains began to break down at grain boundaries to form small grains, which became new grains and grew up. Under the joint action of interface curvature and interface energy, small grains were gradually engulfed by large grains and the large grains kept growing. Fig. 2(d) shows the microstructures of AlSi7Mg alloy with a holding time of 15 min. The ratio of liquid phases was relatively high, the distance between adjacent large grains was far. A large number of fine grains generated by the grain breakage in the previous stage were distributed between the liquid phases. The rose-shape appeared between grains in the region with less liquid phases due to their close grain orientation and merger growth. Because of the liquid phase wrapping, the large grains were concave inward to form rosette and had the tendency of breaking into new grains. Fig. 2(e) shows the microstructures of AlSi7Mg alloy after holding for 20 min. Most of the rose-like dendrites were melted. Traces of original rose-like grains could be seen from the microstructure. The burrs previously formed by grain and grain boundary fragmentation had grown and spheroidized and new burrs were also being formed. Grown burr grains had smooth rounded edges.

As can be seen from the changes of the microstructures at different temperatures with changing time, when the isothermal temperature was 590°C and the holding time was less than 5 min, the microstructures first grew from dendrites and gradually broke into clumps or particles at the root with the increase of the holding time. With the further increase of holding time, Al grains grew from granular or massive, and part of them became rose-shaped and then fragmented into spherical and massive again. In this process, the distribution of eutectic structure became more uniform. It can be concluded that the microstructures evolution of semi-solid isothermal treatment process was as follows: (1) dendrite growth, (2) massive or granular, (3) rosette, (4) fragmentary massive or granular, (5) spheroidization, (6) grain growth.

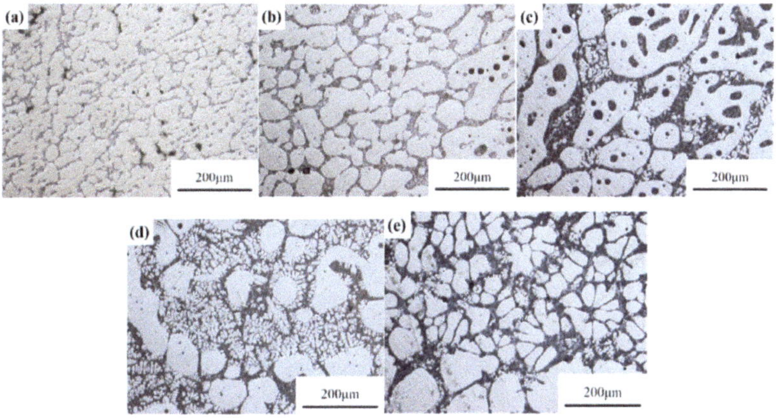

Fig. 2. Microstructure images of AlSi7Mg alloy held at 590°C for different time: (a) 1 min; (b) 5 min; (c) 10 min; (d) 15 min; (e) 20 min.

Fig. 3 shows the microstructures of AlSi7Mg alloy after holding at 595°C for different time. Fig. 3(a) shows the microstructures of AlSi7Mg alloy after holding at 595°C for 1 min. The grains were coarse, with more coarse dendrites and less intergranular liquid phases, and there was bond between grains. Fig. 3(b) shows the microstructures of AlSi7Mg alloy after holding at 595°C for 5 min. Combined growth of adjacent grains appeared, and some of the dendrite melted into an independent dendrite arm due to their proximity to each other. Some of the dendrites were broken to form new grains with small size and good roundness. Fig. 3(c) shows the microstructures of AlSi7Mg alloy after holding at 595°C for 10 min. Intergranular liquid phases increased, and ingrain liquid phases also began to appear. The new grains generated by dendrite fragmentation in the previous stage gradually grew into massive or spherical grains, and the grain boundaries of grains began to break and form a large number of burr grains. Fig. 3(d) shows the microstructures of AlSi7Mg alloy after holding at 595°C for 15 min. The liquid phases ratio tended to be stable and mainly distributed in the intergranular area. The internal melted grains became small massive or spherical grains and grew continuously. The process of burr formation and burr growth was carried out dynamically at grain boundaries. Two or more grain boundaries of adjacent grains were connected, and there was a tendency to form rose grains. Fig. 3(e) shows the microstructures of AlSi7Mg alloy after holding at 595°C for 20 min. At this time, the number of large grains had been greatly reduced and separated from each other, and the liquid phases were filled with small size burr grains and newly grew burr grains. The edges of large grains were rounded and smooth, and the two adjacent grains grew in "8" shape. It could be found that when the holding time was short, the microstructures and grain size were not uniform, while when the holding time was enough, the distribution of grains and eutectic phases became uniform, but the grains were coarse and round. With the increase of time, a large number of recrystallized grains (burrs) appeared at the grain edge and were in the process of dynamic generation and gradual growth. The microstructures evolution was as follows: (1) dendrite growth, (2) massive or granular, (3) rosette, (4) fragmentary massive or granular, (5) spheroidization, (6) grain growth. This was the same as the semi-solid microstructure evolution at 590 °C for different holding time.

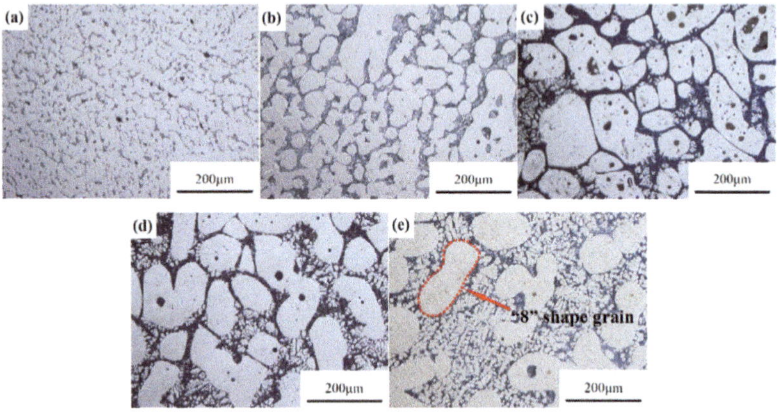

Fig. 3. Microstructure images of AlSi7Mg alloy held at 595℃ for different time: (a) 1 min; (b) 5 min; (c) 10 min; (d) 15 min; (e) 20 min.

Fig. 4 shows the microstructures of AlSi7Mg alloy after holding at 600°C for different time. Fig. 4(a) shows the microstructures of AlSi7Mg alloy after holding at 600°C for 1 min. Compared with the those held at 590°C and 595°C for 1 min, the microstructures of AlSi7Mg alloy held at 600°C for 1 min exhibited a larger grain size. The main dendrite became wider, and the new grains formed by the melting and breaking of the dendrite arms became larger and larger. At this time, liquid phases began to form in the grain. Fig. 4(b) shows the microstructures of AlSi7Mg alloy after holding at 600°C for 5 min. Dendrites were further broken and smaller grains were formed. The grain edges became round and spheroidized obviously. The short holding time results in a slow diffusion rate of atoms, so only a few grain boundaries appear burrs. Fig. 4(c) shows the microstructures of AlSi7Mg alloy after holding at 600°C for 10 min. The coalescence and growth of adjacent grains increased the grain size, and the rosette grains appeared. The recrystallization of grains was obvious, and the burr grains entered the liquid phases and gradually grew up. Fig. 4(d) shows the microstructures of AlSi7Mg alloy after holding at 600°C for 15 min. The size of the burrs generated in the previous stage increased and tended to coalesce into dendritic grains. The distance between large grains was far, and the rose grains began to break. Fig. 4(e) shows the microstructures of AlSi7Mg alloy after holding at 600°C for 20 min. With the extension of holding time, the old rosette broke and the new rosette was formed in a dynamic equilibrium. At this time, the grain size was generally large and irregular.

Materials Research Proceedings 44 (2024) 718-735 https://doi.org/10.21741/9781644903254-77

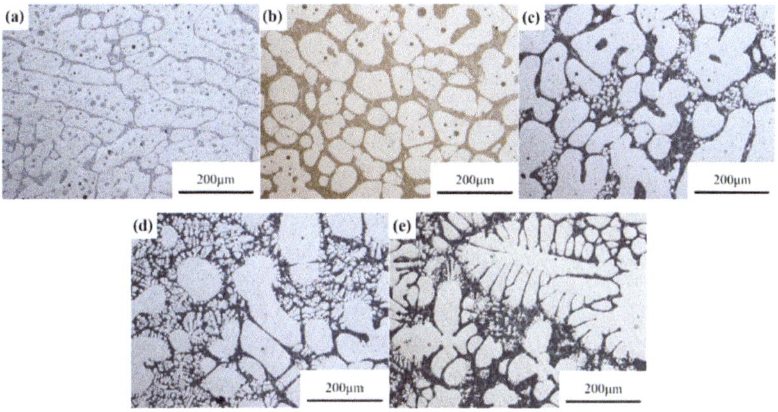

Fig. 4. Microstructure images of AlSi7Mg alloy held at 600°C for different time: (a) 1 min; (b) 5 min; (c) 10 min; (d) 15 min; (e) 20 min.

Fig. 5 shows the microstructures of AlSi7Mg alloy after holding at 605°C for different time. Fig. 5(a) shows the microstructures of AlSi7Mg alloy after holding at 605°C for 1 min. Due to the reasons of atomic diffusion and energy fluctuation, the composition of the alloy would be homogenized. This would lead to the gradual coarsening of dendrites, and the disappearance of the branching characteristics of dendrites, forming clumps or grains [26]. Dendrites were encapsulated in liquid phases and began to break up into smaller grains. Traces of primitive dendrites could be seen from grain orientation. Fig. 5(b) shows the microstructures of AlSi7Mg alloy after holding at 605°C for 5 min. The liquid phases began to appear in the grains, and the broken dendrite grains began to spheroidize. The burr grains began to break at the grain boundary. Fig. 5(c) shows the microstructures of AlSi7Mg alloy after holding at 605°C for 10 min. There were more rosette grains in the microstructures. The burrs in the intercrystalline liquid phases grew up gradually. Fig. 5(d) shows the microstructures of AlSi7Mg alloy after holding at 605°C for 15 min. The burrs disappeared, and the rosette grains were broken into several radial grains with larger grain size. The grains free in the liquid region were smaller and had higher roundness. Fig. 5(e) shows the microstructures of AlSi7Mg alloy after holding at 605°C for 20 min. The grain arrangement in the region had a certain directivity. The phenomenon of adjacent grains merging and growing was not obvious. The original grains were constantly flattening and elongating.

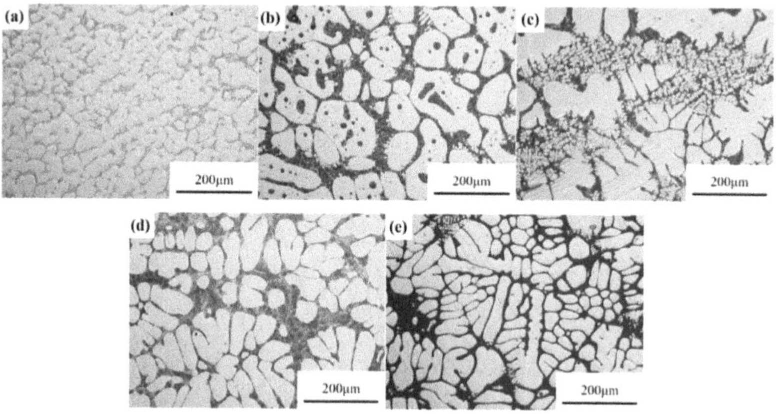

Fig. 5. Microstructure images of AlSi7Mg alloy held at 605℃ for different time: (a) 1 min; (b) 5 min; (c) 10 min; (d) 15 min; (e) 20 min.

Fig. 6 shows the microstructures of AlSi7Mg alloy after holding at 610°C for different time. Fig. 6(a) shows the microstructures of AlSi7Mg alloy after holding at 610°C for 1 min. Grain boundaries interpenetrated to form a network of structures. Fig. 6(b) shows the microstructures of AlSi7Mg alloy after holding at 610°C for 5 min. A small number of spherical particles had been suspended in the liquid phases. At this time, there were both fine grains in the microstructure and large rosette grains that were about to break. Fig. 6(c) shows the microstructures of AlSi7Mg alloy after holding at 610°C for 10 min. It could be seen that the rose-like grains without complete ripening. There were no obvious burr grains, and the structure was a large number of rose-like grains and petal grains formed after the fragmentation of rose-like grains. Fig. 6(d) shows the microstructures of AlSi7Mg alloy after holding at 610°C for 15 min. The long grains were gradually melting off. A large number of recrystallized grains were formed at the grain boundary, and the large grains were far apart. Fig. 6(e) shows the microstructures of AlSi7Mg alloy after holding at 610°C for 20 min. The microstructure morphology at this time was similar to that at 10 min. There were more spherical grains, but there were also many strip-like grains and fewer rose-like grains. It could be seen that the traces of the rose-like grains were broken off. But there were burr grains at the grain boundary edge.

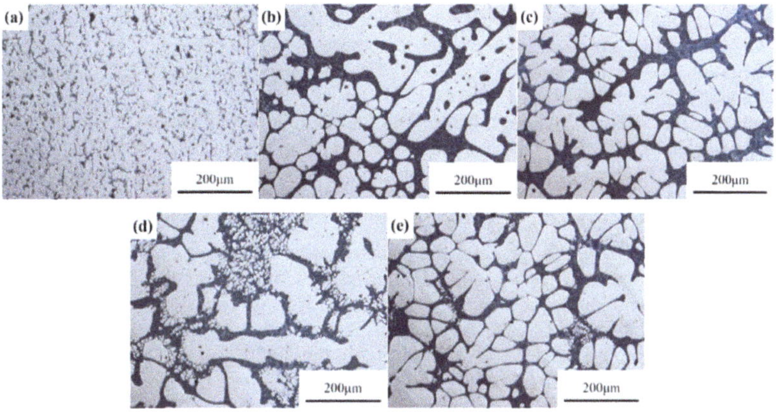

Fig. 6. Microstructure images of AlSi7Mg alloy held at 610°C for different time: (a) 1 min; (b) 5 min; (c) 10 min; (d) 15 min; (e) 20 min.

By comparing the microstructure under the same holding time and different temperature parameters, it can be noted that the increase of isothermal temperature is helpful to accelerate the evolution of semi-solid microstructures [27]. Under the same holding time, with the increase of isothermal temperature, the solid volume fraction of semi-solid billet gradually decreases, and the grains grow and spheroidize [28]. Compared with the original structure, the sample is heated at 590°C for 5 min, and the microstructures are clumped or massive grains formed by dendrite fragmentation because of the lower isothermal temperature and shorter holding time. When the isothermal temperature is increased to 600°C, the homogenization of composition begins to occur in the microstructures, and the eutectic microstructures existing in the intergranular begin to diffuse to the α-Al grains. The α-Al grains partially merge with each other, and the grain size increases. When the temperature rises to 610°C, the microstructure evolution process accelerates. The small grains in the liquid phases begin to spheroidize and grow further. The grain boundaries become smoother and more rounded. At the same time, large clumped grains also begin to break into small particles. It can be noticed that under the same holding time, the higher isothermal time is, the faster the transformation process is. The increase of isothermal temperature can significantly improve the speed of semi-solid microstructure evolution process. This is embodied in the advance of eutectic phase melting time and the increase of the emergence speed of spheroidized particles. However, due to the increase of coarsening power of grain growth, the grains also become larger [29].

It indicates that the evolution rate of semi-solid microstructure is affected by temperature. When the isothermal temperature is low, the driving force of eutectic phases to liquid phases is small. The evolution process of semi-solid microstructures is slow, which also delays the spheroidization rate of α-Al phase with low alloying elements. With the increase of isothermal temperature, the evolution rate of semi-solid microstructures accelerates obviously. When the liquid phases are generated, under the action of interfacial energy and interfacial curvature, small grains will begin to be absorbed by the liquid phases, and large grains will continue to grow and spheroidize, with smooth edges. Thus, the overall solid-liquid boundary area decreases, the system is stable, and the energy decreases [30, 31]. The coarsening and spheroidization of primary grains are based on the remote migration and diffusion of Al and Si atoms. In the absence of convection exerted by an external field in the melt, the migration and diffusion rate of atoms is quite low, resulting in a very low growth rate of α-Al phase. In the process of semi-solid isothermal treatment, as the particles grow up, the probability of solid particles bumping against each other's edges increases, thus the combining growth of grains is promoted and the grains are more uneven. When two grains with different grain sizes grow into new grains by merging, the newly formed grains become round under the drive of interface energy. However, when two or more grains of similar size merge and grow into new particles, the new particles will not occur spheroidization, but eventually grow into "8" shape or pellet shape. When the melt treatment temperature is too high, the proportion of solid phases is too small, the proportion of liquid phases is too high. This made the forming process will bring too high fluidity and close to casting, which has little significance for the prevention of casting defects.

In the semi-solid isothermal process, the primary dendrites also melt off at the root and the number of grains increases greatly. The eutectic phases with lower melting point tend to be enriched in the dendrite interspaces of the clumps. However, when the temperature of the melt is higher than the solidus temperature, the eutectic phases with lower melting point will become liquid [32], and the clump dendrites will dissociate in the liquid phases. The degree of convexity and concavity of dendrite grain boundaries in contact with the melt liquid phases is different, that is, the radius of curvature is different from each other. According to the influence of curvature radius on melting point, it can be seen that the convex part of grain boundary has small curvature radius, and its melting point is lower [33], so the concave part of grain boundary is further melted and fills in the convex part, and finally generates the spherical grains [34]. Driven by interfacial energy, the small particles suspended in the liquid will tend to have the same degree of concave and convex, and eventually spheroidize. In the process, with the increase of holding time, the grain distribution will be more uniform. This growth mechanism is called Ostwald Ripening mechanism.

The mechanism of grain growth also involves the other mechanism, coalescence of grains, which is that the adjacent primary α-Al phase grains with similar lattice orientation begin to grow together during the holding process. The α-Al grains continuously merge and grow. Due to the inclusion of a small amount of liquid phases during grain consolidation, there are liquid phases in the larger grains after the merger. There is a high liquid rate in the structure, with good fluidity, and the small solid phases in the liquid phases will gradually melt into the liquid phases, so that the liquid fraction will gradually increase. The grains have good roundness, and the primary α-Al grains grow uniformly in all directions and gradually become rounder due to the liquid-solid interfacial tension. There are two reasons for this phenomenon. First, due to the presence of liquid phases, under the action of interface curvature and interface energy, in order to make the solid-liquid interface of the whole system lower and reduce the energy of the system, the small primary α phase grains gradually melt, and the large primary α phase grains grow up and become more rounded. Second, When the liquid fraction is high, it is assumed that the solid-liquid phases are in equilibrium, but the small grains correspond to lower equilibrium melting point, and the large grains correspond to higher equilibrium melting point. Therefore, at the same temperature, the small grains correspond to the liquid phases with lower Si content, while the large grains correspond to the liquid phases with higher Si content, resulting in the diffusion of Si in the liquid phases to the liquid phases on the side of the small grains. At the same time, Al in the liquid phases diffuse to the side of the large grains, causing the continuous melting of the small grains and the continuous growth of the large grains.

Quantative analysis of semi-solid billet fabricated by SSITWA. Fig. 7 shows the point plot of average grain size and average shape factor of semi-solid billet fabricated by SSITWA at different isothermal temperatures and holding times. As shown in Fig. 7(a), when the specimen was heated at different isothermal temperatures, the average grain size generally changes in the same way with the extension of holding time. When the holding time is 590°C, with the increase of holding time, the size of primary α-Al phase increases and then decreases obviously. The average grain diameter increases gradually from 15 μm at 1 min to 30 μm at 5 min, and further increases to 55 μm at 10 min. Then it decreases to 30 μm at 15 min, followed by a slight decrease to 28 μm at 20 min. This is because in the early stage of semi-solid isothermal treatment, the solid fraction is high, adhesion and merger phenomenon between adjacent grains, like "8" word grains or petal-like grains appear. With the further extension of holding time, the eutectic phases with low melting point between grains melt, and the grain boundary becomes obvious. It is difficult to merge between grains, and the grains begin to spheroidize, which further reduces the free energy and accords with the Ostwald growth mechanism [35]. The average grain size of AlSi7Mg aluminum alloy semi-solid billet fabricated by SSITWA varies in a range of from 15 μm to 65 μm when the isothermal temperature and holding time change.

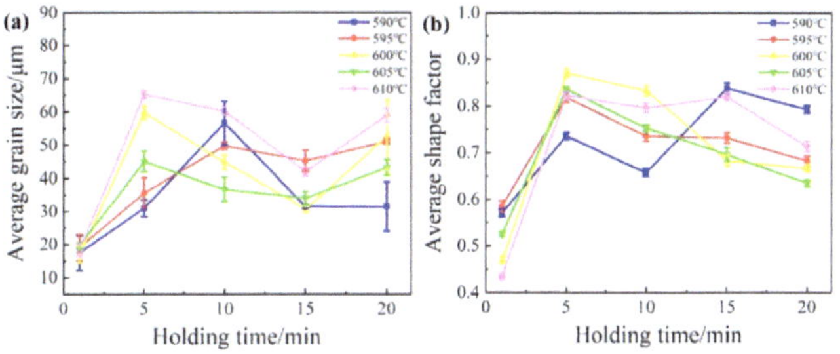

Fig. 7. Point plot of average grain size and average shape factor: (a) Average grain size; (b) Average shape factor.

In the other four temperature parameters, the grain size increases to different degrees when holding time is 20 min compared with holding time of 15 min. This is because when the isothermal temperature is lower, the evolution process of semi-solid microstructure is slower than that when the isothermal temperature is higher. When the holding time is 20 min, it is mainly rosette grains broken into massive or spherical small grains. With the increase of isothermal temperature, the process of microstructures evolution is accelerated. When the holding time is 20 min, the microstructures begin to have different degrees of grain growth or combined growth.

It can be seen from Fig. 7(b) that with the extension of holding time, the average shape factor generally tends to increase first and then decrease. When the holding time is 1 min, the grain sizes of the original dendrites and the grains produced by fragmentation are large and small. The grain edges are irregular, their growth rates are different, and the average shape factor is small. When the holding time reaches 5 min, the grains are gradually rounded and the average shape factor becomes larger. With the further extension of holding time, some grains begin to merge and grow, Some "8" shaped grains or petal-like grains appear with irregular shape, resulting in the decrease of the average shape factor. However, with the extension of holding time to 10 min and 15 min, the liquid phases increase, the grains are isolated, which are not easy to merge and grow, and begin to transform into spherical grains. The shape factor increases at 590°C and 610°C. The average shape factor decreases at all temperature parameters during the transition from 15 min to 20 min. The average shape factor varies in a range of from 0.65 to 0.85 when the holding time changes from 5 min to 20 min. It illustrates that good spheroidization effect is obtained. As reported by Kiuchi and Kopp [36], a shape factor greater than 0.5 indicates a good spheroidization of semi-solid billets. Compared with other methods in Table 3, including electromagnetic stirring (ES) method [37], cooling slope (CS) method [38], ultrasonic treatment (UT) method [39], strain induced melt activation (SIMA) method [40], rheoforged and semi-solid holding process(RFSSH) method [40]and equal-channel angular pressing (ECAP) method [41], the results show that the SSITWA method has a good effect on the preparation of AlSi7Mg semi-solid billet.

Table 3. The average grain size and average shape factor from different works.

Aluminum alloy	Method	Average grain size/μm	Average shape factor/μm
AlSi7Mg	SSITWA	15~65	0.65~0.85
A356	ES	113.62~127.25	3.9~5.9
A356	CS	74~105	0.41~0.54
A356	UT	50	0.7
A356	SIMA	70	0.85
A356	RFSSH	100	0.82
A356	ECAP	21~35	-

Energy spectrum analysis of AlSi7Mg aluminum alloy semi-solid billet prepared by SSITWA. Fig. 8 shows SEM point scanning analysis diagram of AlSi7Mg alloy after holding at 600°C for 5 min. Analysis equipment was Zeiss field emission scanning electron microscope (SUPRA55). Three test points were selected from the area in Fig. 8(a) for element analysis. Spot1 and Spot3 are intragranular, and Spot2 is intergranular. As shown in Fig. 8, the content of Al element in Spot3 is obviously higher than that in Spot1 and Spot2. It is because that the phase in Spot3 is α-Al phase and the phase in Spot1 and Spot2 is eutectic phase consisting of Al and Si elements.

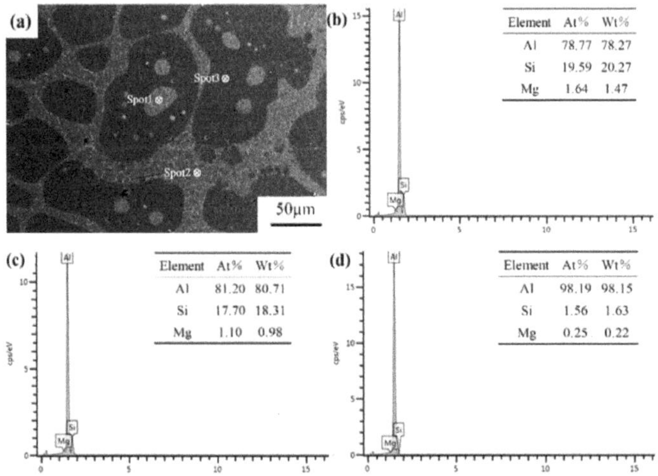

Fig. 8. SEM point scanning analysis diagram of AlSi7Mg alloy held at 600°C for 5 min: (a) The area being scanned; (b) Spot1 element analysis; (c) Spot2 element analysis; (d) Spot3 element analysis.

Fig. 9 shows the SEM line scanning analysis diagram of the AlSi7Mg alloy after holding at 600°C for 5 min. Fig. 9(b) lists the variation curves of the contents of main elements Al and Si in and between grains. It can be seen that Si is mainly enriched in the liquid phase region between grains and ingrains, while Al is relatively evenly distributed, and its content in the non-liquid phase region in grains is slightly higher than that in the liquid phase region between grains and ingrains.

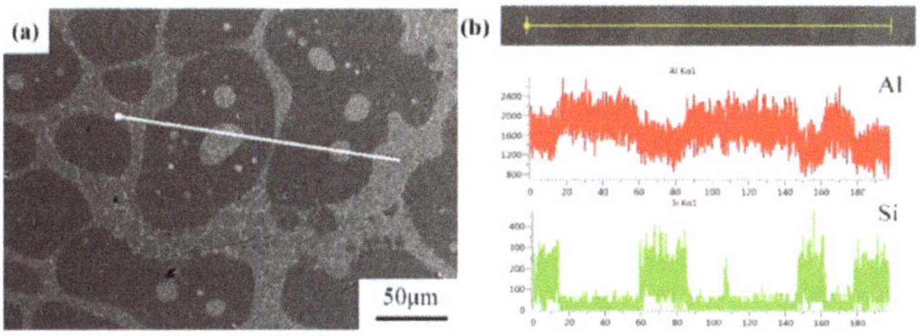

Fig. 9. SEM line scanning analysis diagram of AlSi7Mg alloy held at 600°C for 5 min.

Fig. 10 shows the SEM map scanning analysis diagram of AlSi7Mg alloy after holding at 600°C for 5 min. Fig. 10(a)-(f) shows the distribution of Al, Si, Mg, Ti and Fe elements on the selected region. It can be seen from the map scanning results that the eutectic phases first melt and form ingrain droplet and intergranular liquid phases due to its low melting point. In the whole isothermal process, the composition of Al and Si elements is different at different positions. It can be seen from Fig. 10(b) that Al element is distributed inhomogeneously in the whole, especially in the interior of the grains, and the content of Al element in the liquid phase region between grains and inside grains is slightly lower. Si element is obviously enriched in the liquid phase region between grains and inside grains, due to eutectic silicon mainly exists in these two regions. The distribution of Mg element is uniform as a whole, and there is little difference in the distribution within and between grains. The contents of Ti and Fe are small, but they are evenly distributed on the regional surface. The contents of other trace elements and impurity elements are relatively small, which can be ignored under the results of map scanning.

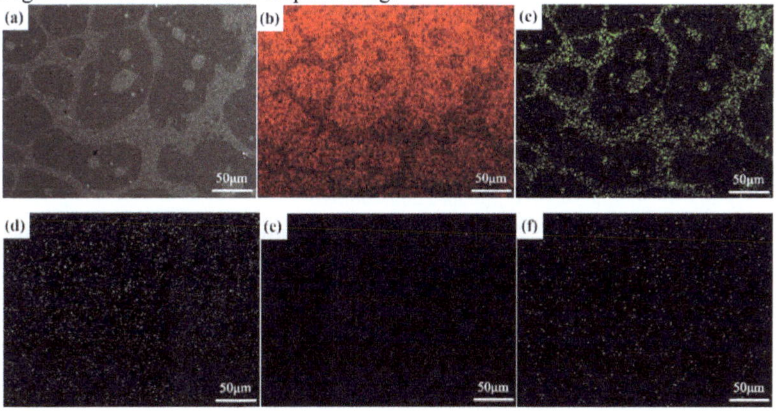

Fig. 10. SEM surface scanning analysis of AlSi7Mg alloy held at 600°C for 5 min: (a) The area being scanned; (b) Al element; (c) Si element; (d) Mg element; (e) Ti elements; (f) Fe element.

Conclusions

In this paper, semi-solid billet of AlSi7Mg aluminum alloy is fabricated by isothermal treatment in semi-solid temperature range after hot upsetting at 350°C for 50% deformation degree. The microstructure evolution during semi-solid isothermal treatment of wrought alloy was investigated. The research conclusions are as follows:

(1) With the increase of holding time in semi-solid isothermal process, the evolution process of primary phase morphology is as follows: dendrite growth → massive or granular → rosette → fragmentary massive or granular → spheroidization → grain growth. Under the same holding time, the higher isothermal temperature can increase the primary grain size and shorten the time of dendrite morphology transformation to granular form.

(2) During the semi-solid isothermal process, the average grain size increases first, then decreases and then increases. The trend of average shape factor increases first, then decreases, then increases and then decreases. The average grain size of AlSi7Mg aluminum alloy semi-solid billet fabricated by SSITWA varies in a range of from 15 μm to 65 μm when the isothermal temperature and holding time change. The average shape factor varies in a range of from 0.65 to 0.85 when the holding time changes from 5 min to 20 min. It illustrates that good spheroidization effect is obtained.

(3) The results of point, line and map scanning show that the content of Al element in α-Al is obviously higher than that in liquid eutectic phase due to different phase. Si element is mainly distributed in the eutectic organization of intergranular, while Mg element has no significant difference in the distribution in and out of grains.

Acknowledgement
This work is supported by the National Natural Science Foundation of China (NSFC) under Grant Nos.U2241232, U2341253 and 52375317 and the National Key R&D Program of China (No.2022YFB3404204).

Data Availability
The raw/processed data required to reproduce these findings cannot be shared at this time due to technical or time limitations.

Conflict of Interest
On behalf of all authors, the corresponding authors state that there is no conflict of interest.

References

[1] H. Yang, W.T. Tian, Metallurgical Structure of A356 Alloy Solidified by Mechanical Stirring, Solid State Phenom. 285 (2019) 183-188.

[2] Y.E. Albarbary, R. Afify, E.H. Mansour, T.S. Mahmoud, M. Khedr, The effect of pre-drilling on the characteristics of friction drilled A356 cast aluminum alloy, J. Manuf. Process. 82 (2022) 646-656. http://dx.doi.org/10.1016/j.jmapro.2022.08.040

[3] N.Z. Liu, B. Jiang, Y.Wang, Z.S.Ji, M.L.Hu, H.Y. Xu, Influence of trace amount chromium on microstructure and corrosion behavior of A356-5vol.%TiB2 alloy, Mater. Lett. 314 (2022) 131798.

[4] S.K. Thandalam, J. Nampoothiri, S. Shalini, T. Thankachan, Microstructure and Wear Characteristics of Nano Y_2O_3 Particles Reinforced A356 Alloy Composites Synthesized Through Novel Ultrasonic Assisted Stir Casting Technique, Trans. Indian Inst. Met. 75 (2022) 417-426. http://dx.doi.org/10.1007/s12666-021-02424-1

[5] P. Nelaturu, S. Jana, R.S. Mishra, G. Grant, B.E. Carlson, Effect of temperature on the fatigue cracking mechanisms in A356 Al alloy, Mater. Sci. Eng. A 780 (2020) 139175. https://doi.org/10.1016/j.msea.2020.139175

[6] V.H. Carneiro, J. Grilo, D. Soares, I. Duarte, H. Puga, The Influence of Precipitation Hardening on the Damping Capacity in Al-Si-Mg Cast Components at Different Strain Amplitudes, Metals 12 (2022) 804. http://dx.doi.org/10.3390/met12050804

[7] S. Zhang, L. Bo, L. Wang, Y.B. Hou, D. Zhao, Effects of thermal-rate treatment and modification of Ce on the microstructure and properties of A356 alloys, J. Phys.: Conf. Ser. 1885 (2021) 022064. http://dx.doi.org/10.1088/1742-6596/1885/2/022064

[8] M. Varmazyar, S. Yousefzadeh, M.M. Sheikhi, Effect of Ni on Microstructure and Creep Behavior of A356 Aluminum Alloy, Met. Mater. Int. 28 (2022) 579–588. http://dx.doi.org/10.1007/s12540-020-00892-6

[9] G.F. Xiao, J.F. Jiang, Y.Z. Liu, Y. Wang, B.Y. Guo, Recrystallization and microstructure evolution of hot extruded 7075 aluminum alloy during semi-solid isothermal treatment, Mater. Charact. 156 (2019) 109874.

[10] J. Liu, Y.S. Cheng, S.W.N. Chan, D. Sung, Microstructure and mechanical properties of 7075 aluminum alloy during complex thixoextrusion, Trans. Nonferrous Met. Soc. China 30 (2020) 3173-3182. http://dx.doi.org/10.1016/S1003-6326(20)65452-8

[11] Y.F. Wang, S.D. Zhao, Y. Guo, K.X. Liu, S.Q. Zheng, Deformation Characteristics and Constitutive Equations for the Semi-Solid Isothermal Compression of Cold Radial Forged 6063 Aluminium Alloy, Materials 14 (2021) 194. https://doi.org/10.3390/ma14010194

[12] H. Negini, A. Rezaei, M. Hajisafari, H. Momeni, Effects of semi-solid thermomechanical processing on microstructure and mechanical properties of Al 2024, Trans. Indian Inst. Met. 72 (2019) 1201-1209. http://dx.doi.org/10.1007/s12666-019-01608-0

[13] Q. Gao, B. Yang, G.S. Gan, Y.J. Zhong, L. Sun, W.Y. Zhai, W. Qiang, S.Q. Wang, Y.X. Lu, Microstructure and Wear Resistance of TiB2/7075 Composites Produced via Rheocasting, Metals 10 (2020) 1068. https://doi.org/10.3390/met10081068

[14] Y.Z. Liu, J.F. Jiang, Y. Zhang, M.J. Huang, Y. Wang, Semi-solid compression of 2A14 alloy with high solid fraction: rheology, constitutive equation and microstructure, J. Mater. Sci. 57 (2022) 16507–16527. http://dx.doi.org/10.1007/s10853-022-07656-0

[15] Y. Liu, M.Q. Gao, Y. Fu, W.R . Li, P. Yang, R.G. Guan, Microstructure Evolution and Solidification Behavior of a Novel Semi-Solid Alloy Slurry Prepared by Vibrating Contraction Inclined Plate, Metals 11 (2021) 1810. https://doi.org/10.3390/met11111810

[16] Y.F. Wang, Y. Guo, S.D. Zhao, Effects of Process Parameters on the Microstructure and Hardness of Semi-Solid AlSi9Mg Aluminum Alloy Prepared by RAP Process, Mater. Trans. 61 (2020) 1731-1739. http://dx.doi.org/10.2320/matertrans.MT-M2020088

[17] M. Jahanbakhshi, S. Nourouzi, R. Naseri, K. Esfandiari, Investigation of Simultaneous Effects of Cooling Slope Casting and Mold Vibration on Mechanical and Microstructural Properties of A356 Aluminum Alloy, Met. Mater. Int. 28 (2022) 1508-1516. https://ui.adsabs.harvard.edu/link_gateway/2022MMI....28.1508J/doi:10.1007/s12540-021-01056-w

[18] L.J. Cao, G.R. Ma, C.X. Wang, Z.J. Chen, J.H. Zhang, Effect of compression ratio on microstructure evolution of Mg–10%Al–1%Zn–1%Si alloy prepared by SIMA process, Trans. Nonferrous Met. Soc. China 31 (2021) 2597-2605. https://doi.org/10.1016/S1003-6326(21)65678-9

[19] J.F. Jiang, Y.Z. Liu, G.F. Xiao, Y. Wang, Thixoforming of Semisolid Slurry with High Fraction Solid Fabricated by Partial Melting of Commerical Wrought Aluminum Alloys, Solid State Phenom. 285 (2019) 210-218.

[20] T.Y. Guan, Z.F. Zhang, M. He, Y.L. Bai, P. Wang, Effects of Annular Electromagnetic Stirring Melt Treatment on Microstructure and Mechanical Properties of 7050 Rheo-Casting, Solid State Phenom. 285 (2019) 219-223.

[21] W.C. Liu, Y. Sun, C. Deng, L.X. Hu, S.J. Yuan, J.Y. Shen, F. Gao, M.Y. Ba, Cu strengthened Al-Si-Cu semi-solid billet fabricated by liquid phase reaction sintering, Mater. Charact. 188 (2022) 11925.

[22] Z.X. Zhang, T.J. Chen, H. Xue, G.L. Bi, Y. Ma, R.G. Guan, Microstructure Evolution of an Al-Si Alloy During Shear-vibration Coupling Sub-rapid Solidification and the Effects of Processing Parameters, Int. J. Metalcast. 17 (2022). http://dx.doi.org/10.1007/s40962-022-00809-6

[23] L.J. Yu, H. Yan, L.X. Xu, L.J. Zhang, Z.B. Liu, Rheological Research of the Semisolid ADC12 Slurry Prepared with High-Energy Ultrasound and Pr/Ce Addition, Trans. Indian Inst. Met. 75 (2022) 495-502.

[24] J.F. Jiang, Y.H. Zhang, Y.Z. Liu, Y. Wang, G.F. Xiao, Y. Zhang, Research on AlSi7Mg Alloy Semi-solid Billet Fabricated By RAP, Acta Metall. Sin. 57 (2021) 703-716. http://doi.org/10.11900/0412.1961.2020.00254

[25] P.K. Seo, C.G. Kang, The effect of raw material fabrication process on microstructural characteristics in reheating process for semi-solid forming, J. Mater. Process. Technol. 162-163 (2005) 402-409. https://doi.org/10.1016/j.jmatprotec.2005.02.012

[26] Z. Liu, K. Cao, H.B. Xu, M.Y. Huang, Evolution Characteristics of Primary Phase in A356 Al Alloy during Isothermal Holding in Solid-Liquid Phase Region, Adv. Mater. Res. 421 (2012) 39-42. http://dx.doi.org/10.4028/www.scientific.net/AMR.421.39

[27] S.B. Hassas Irani, A. Zarei-Hanzaki, B. Bazaz, A.A. Roostaei, Microstructure evolution and semi-solid deformation behavior of an A356 aluminum alloy processed by strain induced melt activated method, Mater. Des. 46 (2013) 579-587. http://dx.doi.org/10.1016/j.matdes.2012.10.041

[28] W. Khalifa, S. El-Hadad, Y. Tsunekawa, Microstructure characteristics and tensile property of ultrasonic treated-thixocast A356 alloy, Trans. Nonferrous Met. Soc. China 25 (2015) 3173-3180. https://doi.org/10.1016/S1003-6326(15)63949-8

[29] A. Mahdavi, M. Bigdeli, M. Hajian Heidary, F. Khomamizadeh, Study of Microstructural Evolution and Phase's Morphology after Partial Remelting in A356 Alloy, Solid State Phenom. 141-143 (2008) 367-372. http://dx.doi.org/10.4028/www.scientific.net/SSP.141-143.367

[30] H. Arami, R. Khalifehzadeh, H. Keyvan, F. Khomamizadeh, Effect of predeformation and heat treatment conditions in the SIMA process on microstructural and mechanical properties of A319 aluminum alloy, J. Alloy. Compd. 468 (2009) 130-135. http://dx.doi.org/10.1016/j.jallcom.2008.01.020

[31] K. Wang, Z.M. Zhang, H. Wen, D. Xia, W.J. Sun, Microstructural evolution of a fine-grained 7075Al alloy processed by friction stir process during partial remelting, Mater. Charact. 121 (2016) 1-8. https://doi.org/10.1016/j.matchar.2016.09.029

[32] S. Ashouri, M. Nili-Ahmadabadi, M. Moradi, M. Iranpour, Semi-solid microstructure evolution during reheating of aluminum A356 alloy deformed severely by ECAP, J. Alloy. Compd. 466 (2008) 67-72. https://doi.org/10.1016/j.jallcom.2007.11.010

[33] Y. Hu, F. Liu, L.Z. Zhao, Y.C. Tang, H.T. Jiao, Microstructural Evolution and Mechanical Properties of Semi-Solid $Al_{15}Mg_{45}Li_{39}Ca_{0.5}Si_{0.5}$ Light-Weight High Entropy Alloys Fabricated by Isothermal Heat Treatment, Met. Mater. Int. 29 (2023) 1489-1497. https://ui.adsabs.harvard.edu/link_gateway/2023MMI....29.1489H/doi:10.1007/s12540-022-01303-8

[34] C.P. Wang, Z.J. Tang, H.S. Mei, L. Wang, R.Q. Li, D.F. Li, Formation of spheroidal microstructure in semi-solid state and thixoforming of 7075 high strength aluminum alloy, Rare Met. 34 (2015) 710–716. http://dx.doi.org/10.1007/s12598-013-0123-0

[35] Y. Hu, S.Q. Fu, L.Z. Zhao, D.H. Wang, F. Liu, Microstructure evolution of semi-solid Mg2Si/A356 composites during remelting process, China Foundry 17 (2020) 384-388. http://dx.doi.org/10.1007/s41230-020-9158-7

[36] M. Kiuchi, R. Kopp, Mushy/Semi-Solid Metal Forming Technology – Present and Future, CIRP Ann. Manuf. Technol. 51 (2002) 653-670. https://doi.org/10.1016/S0007-8506(07)61705-3

[37] J.S. Roh, M. Heo, C.K. Jin, J.H. Park, C.G. Kang, Effect of Current Input Method on A356 Microstructure in Electromagnetically Stirred Process, Metals 10 (2020) 460. https://doi.org/10.3390/met10040460

[38] A. Kolandooz, S.A. Dehlzordi, Effects of important parameters in the production of Al-A356 alloy by semi-solid forming process, J. Mater. Res. Technol. 8 (2018) 189-198. http://dx.doi.org/10.1016/j.jmrt.2017.11.005

[39] W. Khalifa, S. El-Hadad, Y. Tsunekawa, Microstructure characteristics and tensile property of ultrasonic treated-thixocast A356 alloy, Trans. Nonferrous Met. Soc. China 25 (2015) 3173-3180. http://dx.doi.org/10.1016/S1003-6326(15)63949-8

[40] A. Dodangeh, M. Kazeminezhad, H. Aashuri, Effects of cold severe plastic deformation and heating on dendritic and non-dendritic structures: A356 alloy, Int. J. Cast Met. Res. 27 (2014) 312-320. http://dx.doi.org/10.1179/1743133614Y.0000000116

[41] M. Moradi, M. Nili-Ahmadabadi, B. Poorganji, B. Heidarian, T. Furuhara, EBSD and DTA Characterization of A356 Alloy Deformed by ECAP During Reheating and Partial Re-melting, Metall. Mater. Trans. A 45 (2014) 1540-1551. http://dx.doi.org/10.1007/s11661-013-2093-0

The continuous rotary extrusion (CRE) process of the AZ31 magnesium alloy with various deformation zone geometry

MITKA Monika[1,a *], MISIOLEK Wojciech Z.[2,b], LECH-GREGA Marzena[1,c], LIMANÓWKA Kamila[1,d] and SZYMAŃSKI Wojciech[1,e]

[1]Łukasiewicz Research Network - Institute of Non-Ferrous Metals, Skawina Division, Piłsudskiego 19 Street, 32-050 Skawina, Poland

[2]Loewy Institute, Lehigh University, Bethlehem, PA 18015, USA

[a]monika.mitka@imn.lukasiewicz.gov.pl, [b]wzm2@lehigh.edu, [c]mlgkrk@gmail.com, [d]kamila.limanowka@imn.lukasiewicz.gov.pl, [e]wojciech.szymanski@imn.lukasiewicz.gov.pl

Keywords: AZ31 Magnesium Alloy, CRE, Conform®, Δ Parameter, Mechanical Properties, EBSD

Abstract. The presented studies are focused on the continuous rotary extrusion (CRE) process, also known in the literature under the Conform® name. The CRE process is commonly used for aluminum and copper alloys but is new for magnesium alloys. A broader understanding of the application of the CRE process and the input material, the AZ31 magnesium alloy, enables the combination of a new, economical method of plastic deformation of the material with the desired final properties. The main aim of the presented work was to analyze how different deformation zone geometry influences the quality of the extruded AZ31 magnesium rods. Additionally, the properties of the CRE extruded rods were compared to the traditionally directly extruded (DE) rods. The research was conducted to analyze the impact of the CRE process parameters on the mechanical properties of the obtained rods, as well as on changes occurring in grain size, texture, and the level of recrystallization.

Introduction

Due to its attractive specific mechanical properties, the AZ31 magnesium alloy is a promising candidate for replacing certain aluminum alloys and steels in engineering applications, especially in the aviation and automotive industries [1].

Because of the hexagonal crystal structure with a limited number of slip systems, AZ31 alloy is difficult to deform at room temperature, and the deformation parameters must be carefully selected. The alloy formability increases as the deformation temperature increases. Magnesium and its alloys can be deformed in metal-forming processes at temperatures above 200-225°C when subsequent slip systems are activated, significantly improving magnesium's formability and its alloys [2]. However, the plastic processing temperature of such difficult-to-deform alloys should be controlled due to its increase related to the work of deformation and friction and the possibility of forming low-melting eutectics at specific temperatures on the other hand [2]. Additionally, the mechanical properties of the processed alloy could be improved and controlled by proper selection of the plastic deformation process parameters. Therefore, new methods of plastic deformation are being sought. A new, unconventional method of plastic processing for magnesium alloy is continuous rotary extrusion (CRE).

The CRE process is commonly used for well-deformable metals and their alloys due to its advantages, such as constant deformation conditions related to the continuous process, high efficiency, and uniform mechanical properties of the products [3].

In the CRE process, friction between the material and the tooling causes the material to move towards the deformation chamber, which gradually heats the material through its contact with the tooling. This way, the appropriate extrusion temperature is achieved [4, 5]. The same phenomenon occurs in the continuous rotary extrusion (CRE) of the AZ series of magnesium alloys [6].

Material and experimental methods

Rods of 10 mm in diameter made from commercial AZ31 magnesium alloy (chemical composition shown in Table 1) were obtained in direct extrusion (~400°C) and used as feedstock material for the CRE process using the MC-260 Conform-type device. During the CRE process, the speed of the rotary wheel was controlled, and the die temperature was measured when the process was stabilized.

Table 1. The chemical composition of the AZ31 alloy used in the presented work.

Element content, weight %									
Al	Zn	Mn	Cu	Ni	Fe	Si	Be	Ca	Mg
2,99	0,84	0,33	0,001	0,001	0,003	0,015	0,001	0,001	balance

During the CRE process, the process parameters, including different speeds of the rotating wheel and three variations of the deformations zone geometry, were used. A 9 mm diameter die with a rounded edge on the entrance to the die-bearing land was used for every variation. Fig. 1 shows the scheme of the CRE process and a description of the deformation zone. The deformation zone was defined by the feed plate in front of the additional deformation chamber on top of the extrusion die. The feed plate was added for metal flow control in front of the extrusion die.

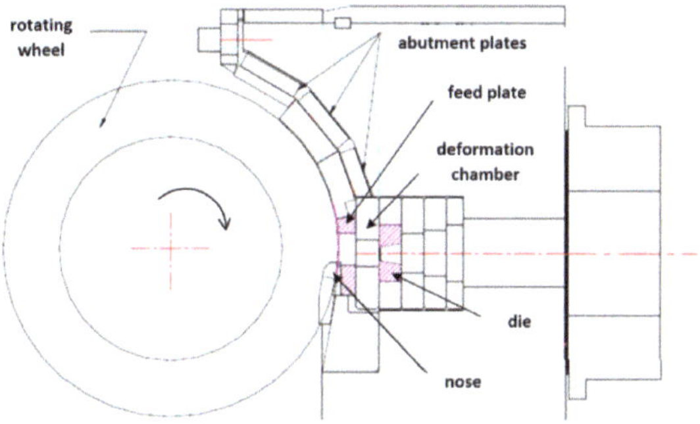

Fig. 1. CRE scheme with a description of the tooling in the deformation zone.

Providing some adjustments to the deformation zone geometry through tooling modifications it is critical to be able to describe these changes quantitatively. For each variation of the deformation zone geometry, the parameter Δ was calculated to allow better comparison of the deformation conditions. Parameter Δ (Eq. 1) is defined as the value of the average height of the deformation zone perpendicular to the deformation direction (H_{av}) divided by the total length ($L_{tot.}$) of the deformation zone measured in the direction parallel to the deformation direction [7]. Its value tells us how optimal the deformation zone geometry is and what the contributions are associated with the energy needed to overcome the friction work and related to the redundant work.

$$\Delta = H_{av.}/L_{tot.} \tag{1}$$

In the case of the continuous rotary extrusion process, determining the Δ parameter is more complicated due to the use of a deformation chamber in front of the extrusion die. Using a feed plate (pre-chamber) also complicates the geometry of the die set. As a result of these tooling changes, it is more complicated to define the parameters of the deformation zone, including the calculation of the Δ parameter. Therefore, the Δ parameter calculations considered the contribution of individual tools and their heights (feed plate, deformation chamber, and extrusion die) in the die set. The weighted average of the heights H was used. The weighted average is based on the percentage contribution (percentage share of the H total) of the height of a given tool (H_1, H_2, H_3) into the entire extrusion tooling (Fig. 2). Detailed data for Δ parameters calculations are presented in Table 2.

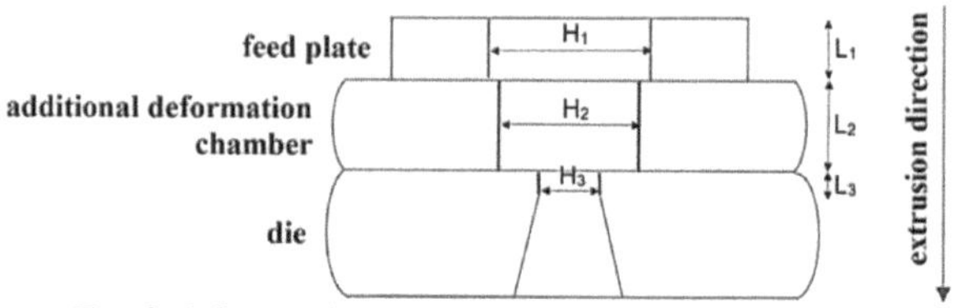

Fig. 2. Scheme of a die set used in the CRE process.

Table 2. Δ parameters calculations for each variation of deformations zone geometry.

Variation Number	Detailed data						Δ= $H_{av.}/L_{tot.}$
I	$H_{1,2,3}$ height [mm]	H_1	H_2	H_3	$H_{tot.}$	H_{av}	**1,840**
		32	0	9	41	26,951	
	$H_{1,2,3}$ weight [%]	78,049	0,000	21,951	100		
	$L_{1,2,3}$ lengths [mm]	L_1	L_2	L_3	$L_{tot.}$		
		12,9	0	1,75	14,650		
II	$H_{1,2,3}$ height [mm]	H_1	H_2	H_3	$H_{tot.}$	H_{av}	**0,815**
		32	30	9	71	28,239	
	$H_{1,2,3}$ weight [%]	45,070	42,254	12,676	100		
	$L_{1,2,3}$ lengths [mm]	L_1	L_2	L_3	$L_{tot.}$		
		12,9	20,0	1,75	34,65		
III	$H_{1,2,3}$ height [mm]	H_1	H_2	H_3	$H_{tot.}$	H_{av}	**1,355**
		32	40	9	81	33,395	
	$H_{1,2,3}$ weight [%]	39,506	49,383	11,111	100		
	$L_{1,2,3}$ lengths [mm]	L_1	L_2	L_3	$L_{tot.}$		
		12,9	10,0	1,75	24,65		

Table 3 presents the variations of used tooling geometries in the CRE process for which AZ31 alloy rods were obtained and the Δ parameters calculated.

Table 3. The deformations zone geometries, CRE process parameters, and the Δ parameters.

Variation Number	Additional deformation chamber	Process speed [rpm]	Process temperature [°C]	Δ
I	not used	3	400	1,840
		4	415	
		4,5	420	
		5	425	
II	H_2 = 30 mm L_2 = 20 mm	2,5	370	0,815
		3	380	
		3,5	400	
III	H_2 = 40 mm L_2 = 10 mm	3	390	1,355
		4	395	
		5	400	
		6	415	

The rods extruded in direct extrusion, and the CRE processes were evaluated in the static tensile test, where the hardness measurements were performed. The rods were also analyzed for grain size, texture, and level of recrystallization using the EBSD method. The level of recrystallization was calculated based on the Grain Orientation Spread (GOS) index. For AZ31 alloy, it was assumed that grains with the GOS orientation dispersion of less than or equal to 1° are dynamically recrystallized grains, while for GOS>1°, grains are deformed [8].

EBSD investigations were performed on cross-sections of extruded rods in the middle of the cross-section and at the edge (approx. 300 μm from the surface of the rod).

Experimental Results

The die temperature generated at a given process speed was monitored during the CRE process. The process speed was experimentally adjusted according to the temperature target. Fig. 3 shows the dependence of the achieved process temperatures on the process speed for each variation of deformations zone geometry used. Variant III showed the lowest process temperatures; however, the strongest relationship between the process temperature and speed.

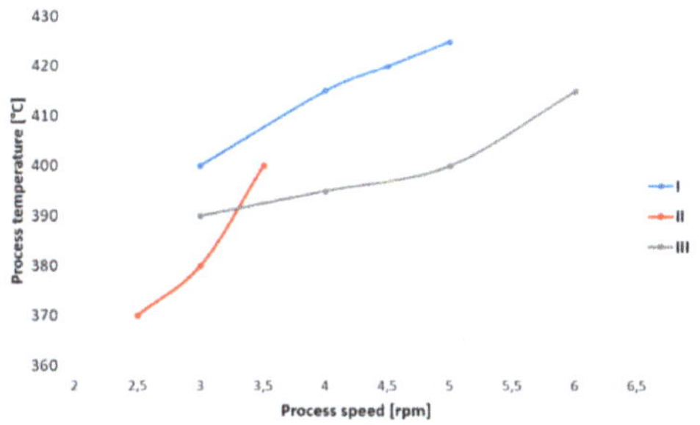

Fig. 3. CRE process temperature vs process speed relationship for different die designs marked as I, II, and III.

Hardness was measured at three points near the center of the rod's cross sections, and the average hardness was calculated. Fig. 4 shows the hardness of CRE rods as a function of the process temperature. Hardness measured for the direct extruded rod is marked with a black line.

A static tensile test determined the tensile strength, yield strength, and elongation. Three samples were tested, and the average values were calculated. Fig. 5 shows the tensile strength, yield strength, and elongation depending on the process temperature. Values measured for the directly extruded rods are marked by a black line.

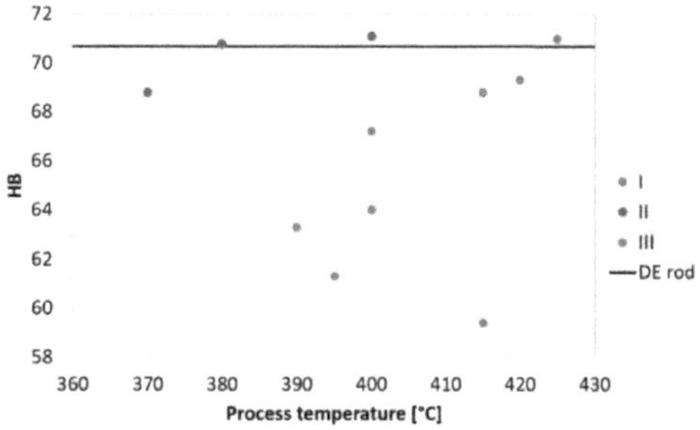

Fig. 4. Hardness of the CRE and DE rods vs process temperature.

For the CRE rods extruded at a speed of 5 rpm and at a temperature of 425°C with no additional deformation chamber, no tensile test was performed due to the deep surface tearing on the rods' surface.

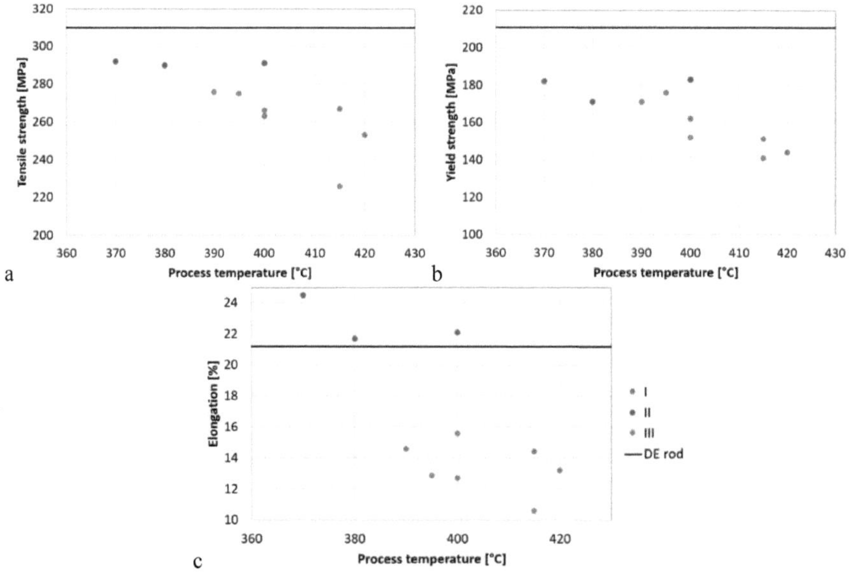

Fig. 5. Tensile strength (a), yield strength (b), and elongation (c) of the CRE and DE rods vs process temperature.

Materials Research Proceedings 44 (2024) 736-743 https://doi.org/10.21741/9781644903254-78

Grain size, texture level, and the share of recrystallized fraction were measured based on the Inverse Pole Figure (IPF) maps obtained by EBSD. The dependence of the grain size, texture level, and recrystallized fraction on the temperature of the CRE process for the center and edges of the rods is shown in Fig. 6. The black lines indicate the grain size of the DE rod.

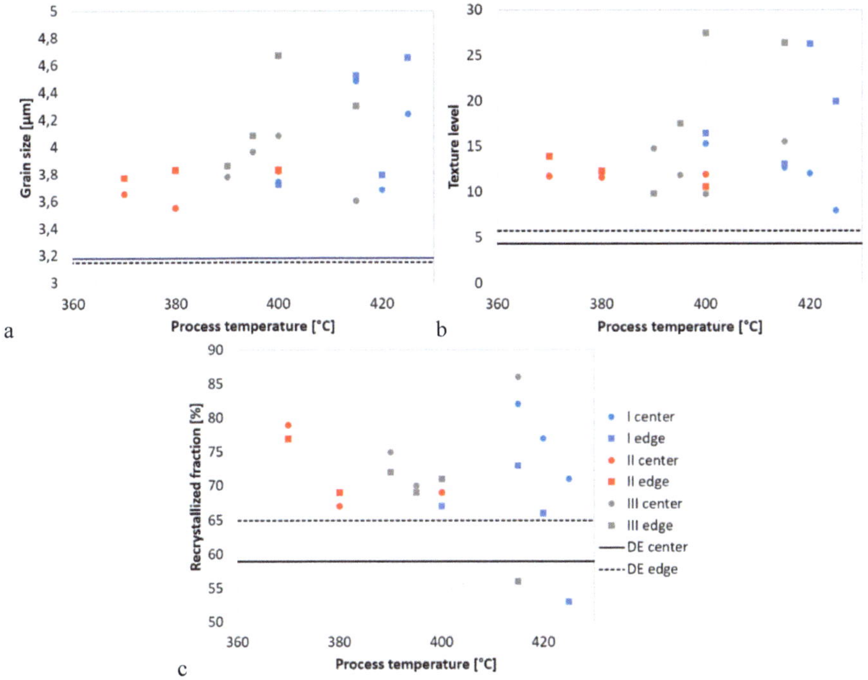

Fig. 6. Grain size (a), texture level (b), and the share of recrystallized fraction (c) for the CRE and DE rods vs process temperature.

Discussion of the Experimental Results

Each change in the geometry of the deformation zone resulted in a change in the deformation history and, consequently, a change in the parameters of the CRE process. It was found that by using variations of deformations zone geometries, it was possible to obtain different process outputs for the AZ31 alloy, measured by the process speed (Table 2).

During the CRE process, the process speed (rotary wheel speed) was correlated to the temperature obtained. For all variations of the deformation zone, an increase in process temperature was observed with an increase in speed. This was caused by the more intense friction phenomenon at higher speeds and the limited cooling of the extruded rod after the CRE process.

The use of various types of deformation zones affects the heat generated in the CRE process and, therefore, increases its temperature. Depending on the variation of the deformation zone, different process temperatures were obtained for the same process speeds.

It was also observed that the AZ31 alloy achieves lower process temperatures for the same speeds using both additional deformation chambers than CRE extrusion without additional tooling.

This is mainly related to different friction forces in the deformation zone when using different deformation chambers or the pre-chamber.

No rods with better mechanical properties than directly extruded rods were obtained for deformation zone variations labeled as I and III in the CRE process (Fig. 4 and 5).

Hardness measurements and static tensile tests of CRE rods indicate that better properties are obtained for extruded rods using an additional deformation chamber with dimensions H = 30, mm L = 20 mm (variation number 2). For this variation, higher values of hardness and elongation were obtained. Higher elongation values for variation II also indicate a better surface quality of the CRE rods extruded in this variant. This is the deformations zone geometry variant for which the Δ parameter is the highest, so the local deformation conditions in the deformation zone should be the most uniform. Better mechanical properties of CRE rods from variant II are also correlated with obtaining a smaller grain diameter in comparison to rods from variants I and III (Fig. 6a). The smaller grain size for variant II is the result of the lower temperature of the CRE process, which also affects the lower degree of recrystallization (Fig. 6c).

For variations I and III, higher process temperatures were achieved, which caused the material to tear at the surface. The presence of the surface cracks is responsible for lower values of both elongation and tensile strength (Fig. 5a, Fig. 5c).

Analyzing the results of mechanical properties, there is no clear dependence on the temperature of the CRE process. This means that the material in the deformation zone behaves differently for each variant of the deformation zone for different CRE process speeds. Significant fluctuations of mechanical properties for different speeds/temperatures of the CRE process are visible, especially for rods extruded in variant III, for which the Δ parameter is not the highest among those used.

The results of the grain size measurement, texture level, and degree of recrystallization also indicate non-uniform deformation conditions in the near-surface and central areas of rods extruded in the CRE process (Fig. 6). The Δ parameter lower than 1 in the extrusion process should indicate better uniformity of parameters in the deformation area. In the case of the extrusion variant II (Δ = 0.815) more uniform properties are visible when comparing the areas of the center and the edge of the rods, and the influence of lower Δ is visible.

However, for the deformation zone variations I and III, for which the Δ parameter is higher than 1, the differences in properties between the areas of the edge and the center of the rod are inadequate for the Δ parameter's value. For variation I, the Δ parameter is 1.840, which is higher than for variation III, where Δ is 1.355. Therefore, it can be expected that the differences for both rod areas would be more significant for variation I, while the results indicate greater differences for variation III.

It should also be noted that CRE rods with a process temperature similar to the direct extrusion process temperature (~400°C) were characterized by a higher degree of recrystallization than DE rods. This may indicate local changes in deformation conditions due to using a more complicated deformation path than in the traditional DE process.

Summary

Based on the experiments performed using the continuous rotary extrusion process, this technological process can extrude AZ31 magnesium alloy. However, it is more complicated than the traditional direct extrusion process, as it may be influenced by more process parameters.

The increased process temperature (measured in the area of the die-bearing land) was recorded for higher wheel speeds, and different properties of extruded rods were obtained for every variation of the extrusion tooling geometry. Based on the CRE trials, it was found that various levels of material flow heterogeneity took place in the deformation zone when additional deformation chambers were used. Results of mechanical properties and EBSD analysis indicate a more complicated deformation path in the CRE process, for which deformation conditions within the extruded metal are not uniform. This uniformity partially depends on the value of the Δ parameter.

 Materials Research Forum LLC
https://doi.org/10.21741/9781644903254-78

However, the Δ parameter traditionally used in the direct extrusion process can be used in the CRE process as an approximate indicator due to less symmetric metal flow.

It was also found that the CRE process performed in our experiments at room temperature does not lead to material strengthening compared to the direct extrusion process.

A study of continuous rotary extrusion confirmed the ability of AZ31 magnesium alloy to be extruded. However, it showed the need for further research to better understand and optimize this material's deformation conditions.

References

[1] E. Friedrich, B. L. Mordike, Magnesium Technology. Metallurgy, Design Data, Applications. Springer Science & Business Media, Berlin Heidelberg, 2006.

[2] E. Hadasik, D. Kuc, Obróbka plastyczna stopów magnezu, Obróbka plastyczna metali 24 (2013) 131-146.

[3] M. Mitka, M. Gawlik, M. Bigaj, W. Szymański, The Effect of Extrusion Parameters on Possibility of Extrusion of Magnesium Alloys in Continuous Rotary Extrusion Process, Arch. Metall. Mater. 60 (2015) 2981-2984. https://doi.org/10.1515/amm-2015-0476

[4] J. Zasadziński, T. Knych, E. Dziedzic, Badanie procesu wyciskania metodą Conform w produkcji wyrobów i recyklingu aluminium i jego stopów, Rudy i Metale Nieżelazne R-52 (2007) 757-767.

[5] Y. Kim, J. Cho, H. Jeong, K. Kim, S. Yoon, A study on optimal design for CONFORM process, J. Mater. Process. Tech. 80-81 (1998) 671-675. https://doi.org/10.1016/S0924-0136(98)00173-3

[6] M. Mitka, M. Gawlik, M. Bigaj, W. Szymański, W.Z. Misiołek, Continuous rotary extrusion of AZ81 magnesium alloy, Procedia Manuf. 15 (2018) 257-263. https://doi.org/10.1016/j.promfg.2018.07.217

[7] W.F. Hosford, R.M. Caddell, Metal Forming. Mechanics and Metallurgy, Cambridge University Press, 2011.

[8] M. Nienaber, G. Kurz, D. Letzig, K.U. Kainer, J. Bohlen, Effect of Process Temperature on the Texture Evolution and Mechanical Properties of Rolled and Extruded AZ31 Flat Products, Crystals 12 (2022) 1307. https://doi.org/10.3390/cryst12091307

Metal Forming 2024 | Materials Research Forum LLC
Materials Research Proceedings 44 (2024) 744-751 | https://doi.org/10.21741/9781644903254-79

Achievement of martensite strengthening in titanium alloy thin-walled components via non-equilibrium hot stamping

CHANG Shupeng[1,2], HUANG Shuaijun[3], WANG Kehuan[1,2,3,a *], LI Zhe[1,2], WEN Zehua[3] and LIU Gang[1,2]

[1]National Key Laboratory for Precision Hot Processing of Metals, Harbin Institute of Technology, Harbin, 150001, China

[2]School of Materials Science and Engineering, Harbin Institute of Technology, Harbin, 150001, China

[3]National Key Laboratory of Air-based Information Perception and Fusion, Luoyang, 471099, China

[a]wangkehuan@hit.edu.cn

Keywords: Rapid Heating, Hot Stamping, Non-Equilibrium Microstructure, Martensite, Mechanical Properties

Abstract. The hot stamping using cold die technology demonstrates great advantages in improving the forming efficiency of titanium alloy thin-walled components. Martensite has been widely employed to strengthen steels, yet few applications were reported in titanium alloys. The main reason is that the martensite microstructure embrittles titanium alloys easily. In this paper, the non-equilibrium hot stamping technology is proposed for titanium alloys to solve the strength-ductility trade-off caused by martensite microstructure. Rapid heating is used to control phase transformation and grain growth during the short-time heating and obtain non-equilibrium microstructure. Non-equilibrium hot stamping experiments of the Ti-6Al-4V alloy were carried out to validate the feasibility. Results show that the rapid heating in the single β-phase region could avoid overgrowth of β grains and lead to the formation of fully fine martensite after water quenching. An Ω-shaped component with fully martensite microstructure were successfully formed by non-equilibrium hot stamping technology. The formed component has a maximum tensile strength of 1153.9 MPa, with a total elongation of 8.0% at room temperature, and the tensile strength is 13.0% higher than that of the as-received sheet.

Introduction

Thin-walled titanium alloy components are widely used in aviation, aerospace and other fields. However, due to high deformation resistance, poor plasticity and serious springback, it is challenging to form thin-walled titanium alloy components [1]. The isothermal forming process has been used for the forming of titanium alloy components, yet it results in low efficiency and high cost [2,3]. Hot stamping using cold die technology has been widely used in the forming of high-strength steel thin-walled components, which not only has a high forming efficiency, but also could improve the post-form strength [4,5].

In recent years, studies have also been carried out on hot stamping of titanium alloys. However, the martensitic transformation temperatures (M_s and M_f) of titanium alloy are much higher than those of high-strength steels [6,7]. As a result, the martensitic transformation of titanium alloy may occur in the process of hot stamping using cold die, thereby increasing the deformation resistance and reducing plasticity, resulting in forming failure [7]. In order to improve the formability of titanium alloy during hot stamping using cold die, Mu et al. proposed the using of triple-layer sheet to avoid the temperature dropping [8]. Hamedon et al. [9] and Maeno et al. [10] employed resistance heating to avoid the temperature dropping by cancelling the transfer process and improve the formability. Recently, hot stamping with rapid heating technology was put forward for titanium alloy by Wang et al. [11], in which rapid heating was used to control the microstructure evolution during the heating process, and obtain the non-equilibrium microstructure. Because rapid heating provides an insufficient time for phase transformation, grain growth and oxidation, it can improve the forming limit of titanium alloy during the hot stamping using cold die [11].

It is well-known that high-strength steels could obtain martensite microstructure to strengthen the components after hot stamping. However, few applications of martensite strengthening were reported in the forming of titanium alloy components. This is because the fully martensite microstructure obtained by solution and water-quenching in the single β-phase region normally results in a much lower ductility. In this paper, the non-equilibrium hot stamping technology is proposed for titanium alloys to solve the strength-ductility trade-off caused by martensite microstructure. A Ω-shaped Ti-6Al-4V alloy component with fully martensite microstructure was successfully formed.

Experiments

The material used in this study is a Ti-6Al-4V alloy sheet with a thickness of 2 mm. The initial microstructure of the as-received alloy sheet consists of equiaxed α phase and fine β phase. The yield strength, tensile strength and elongation of the Ti-6Al-4V alloy are 975.2 MPa, 1020.9 MPa and 13.6%, respectively. Rapid heating was performed on flat dog-bone-shaped specimens with a Gleeble-3800 thermomechanical simulator. The fully martensite microstructure can be achieved through heating to temperatures in the single β-phase region and water-quenching. Therefore, specimens were heated to 1000°C and then water-quenched, with heating rates of $2°C·s^{-1}$ / $15°C·s^{-1}$ / $50°C·s^{-1}$ / $100°C·s^{-1}$. An additional specimen was tested with $100°C·s^{-1}$ heating and 120 s soaking. Microstructure samples and tensile specimens were cut from the uniform temperature zone of the water-quenched specimens.

An Ω-shaped component as shown in Fig. 1(a) was used as the demonstrator to validate the non-equilibrium hot stamping process. Rectangular-shaped initial sheets with a size of 200 mm × 120 mm were machined from the as-received Ti-6Al-4V alloy sheet, with its rolling direction along the longitudinal direction. The principle of the non-equilibrium hot stamping process is schematically illustrated in Fig. 1(b). The process routine includes rapid heating, transfer, forming and in-die quenching. The rapid heating was achieved through the contact heating method. In the contact heating, titanium alloy sheets were heated between the high-temperature heating plates within 60 s. The sheet was quickly transferred into the forming dies after heating within 1~2 s, and then formed with in-die quenching simultaneously.

Microstructure samples and tensile specimens were cut from the Ω-shaped components near different measuring points, as shown in Fig. 2. Microstructure observations were performed through a Zeiss Scanning electron microscopy (SEM). The room temperature tensile tests with a strain rate of $0.001\ s^{-1}$ were performed on a Shimadzu AGS-X electronic universal testing machine.

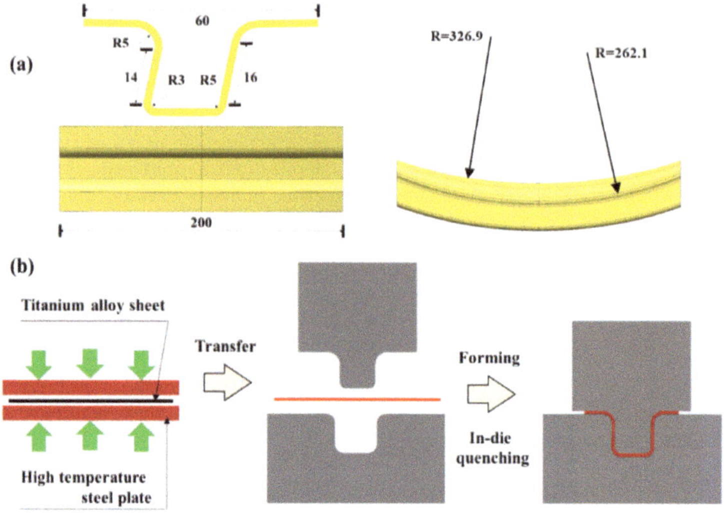

Fig. 1. (a) The dimensional information of the Ω-shaped component and (b) the principle of non-equilibrium hot stamping process (All dimensions are given in mm).

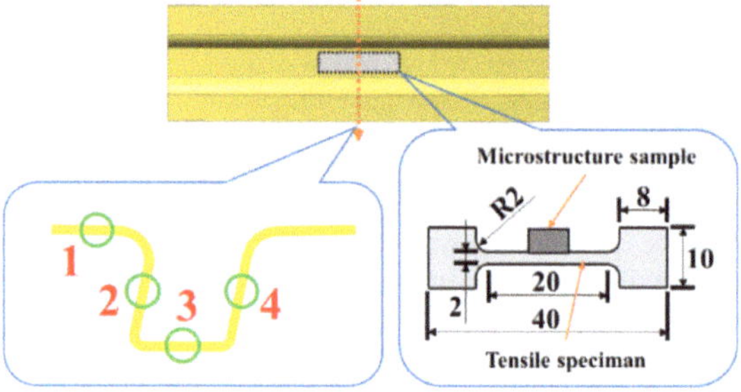

Fig. 2. Diagram of measuring points on the Ω-shaped component, and the dimensional information of tensile specimen and microstructure sample cut from the component around the measuring point (All dimensions are given in mm).

Results and discussion

Fig. 3 shows the fully martensite microstructure after heated to 1000°C through different heating rates then water-quenched. It can be seen that the average size of prior β grains decreases as the heating rate rises, as well as the average width of lamellar α. With soaking for 120 s after rapid heating, the average size of prior β grains and the average width of lamellar α rise apparently. The average size of prior β grains and the average lamellar α width in different microstructures are shown in Fig. 4(a). When the heating rate rises from 2°C to 100°C/s at temperature of 1000°C, the average size of prior β grains decreases from 58.3 μm to 19.5 μm, and the average width of lamellar α decreases from 679.1 nm to 352.0 nm. With soaking for 120 s after heating, the average size of prior β grains rises from 19.5 μm to 94.9 μm in the microstructure heated at 100 °C/s, and the average width of lamellar α increases from 352.0 nm to 719.7 nm.

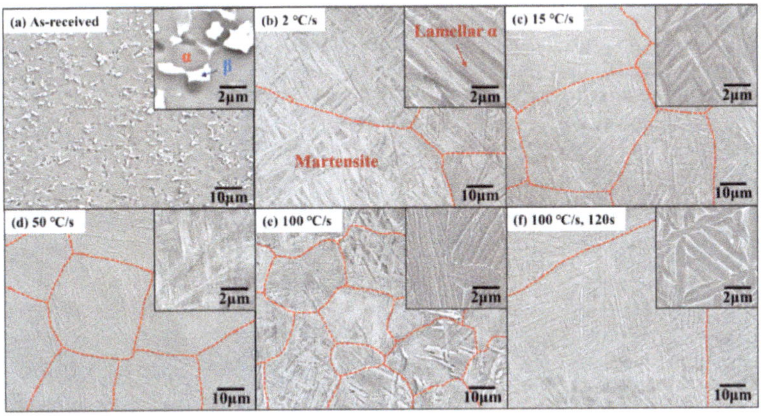

Fig. 3. Microstructures of (a) the as-received Ti-6Al-4V alloy and the Ti-6Al-4V alloy heated to 1000°C with a heating rate of (b) 2°C/s, (c) 15°C/s, (d) 50°C/s, (e) 100°C/s and (f) 100°C/s with 120 s soaking.

The strength and elongation of water-quenched Ti-6Al-4V alloy tensile specimens after heating to 1000°C with different heating rates are shown in Fig. 4(b). The yield strength of the alloy increases from 922.4 MPa to 1032.0 MPa when the heating rate rises from 2°C/s to 50°C/s, then decreases to 991.1 MPa when the rate reaches 100°C/s, compared to 982.7 MPa after 120s soaking. The tensile strength increases from 1116.6 MPa to 1263.0 MPa when the rate rises from 2°C/s to 50°C/s, then decreases to 1209.4 MPa when the rate reaches 100°C/s, compared to 1165.9 MPa after 120 s soaking. The yield strength of the specimen heated at 1000°C is higher than that of the as-received alloy when the heating rate reaches 15°C/s or higher, and the tensile strength is higher than that of the as-received alloy. The elongation increases from 5.2 % to 13.0 % when the heating rate rise from 2°C/s to 100°C/s, and decreases to 4.5 % after 120 s soaking. The water-quenched Ti-6Al-4V alloy heated to 1000°C has the highest yield strength and tensile strength at a heating rate of 50°C/s, with an elongation of 9.6%. Results above indicate that rapid-heated Ti-6Al-4V alloy can achieve better strength and ductility with fully martensite microstructure after heating to a temperature in the single β-phase region, due to finer prior β grains when the alloy is heated at a higher rate.

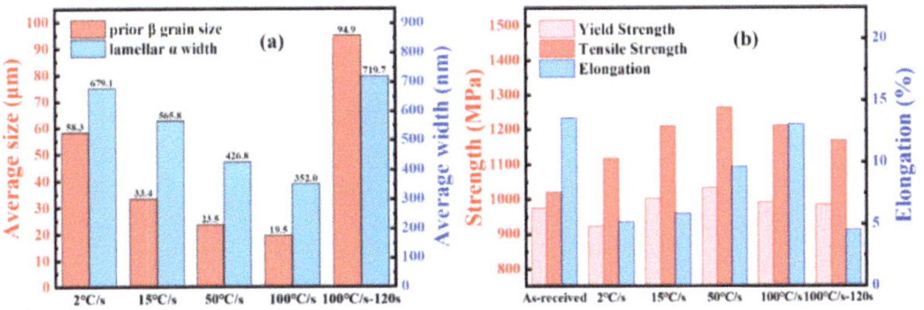

Fig. 4. Effects of heating rate on (a) the average prior β grain size and the average lamellar α width, along with (b) the strength and elongation of the Ti-6Al-4V alloy with fully martensite microstructure.

Considering both post-form property and heating device requirements, heating rates ranging from 10 to 50°C/s are recommended for the non-equilibrium hot stamping. Fig. 5 shows a qualified result of the Ti-6Al-4V alloy Ω-shaped component formed by the novel process. The sheet was heated to 1050°C within 60s, with an average heating rate of 17.5°C/s. It can be seen that no cracks appear on the component surface and the geometry features were also accurately formed. This proves that titanium alloys thin-walled components can be successfully formed by hot stamping using rapid heating to the temperature in the single β-phase region.

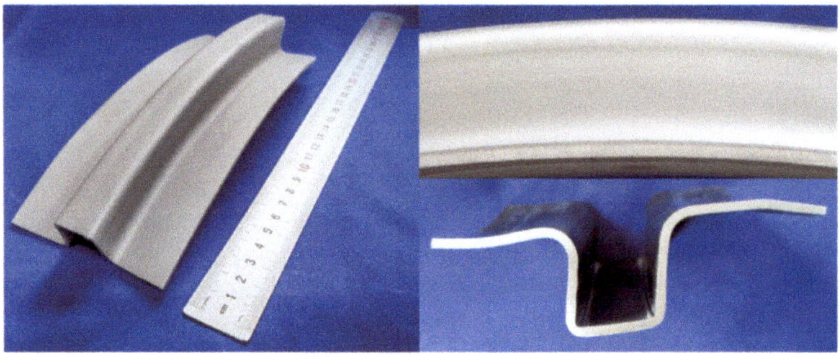

Fig. 5. A qualified result of Ω-shaped component formed at a sheet temperature of 1050°C.

Microstructures of the Ω-shaped component with a sheet temperature of 1050°C are shown in Fig. 6. It can be seen that the Ω-shaped component can achieve fully martensite microstructure after hot stamping with a sheet temperature of 1050 °C. The average sizes of prior β grains in the fully martensite microstructure near the measuring point 1 to 4 are 76.5μm / 65.0μm / 81.3μm / 60.4μm, respectively. Prior β grains in the microstructures near the measuring point 2 and 4 are finer than those in the microstructures near the measuring point 1 and 3.

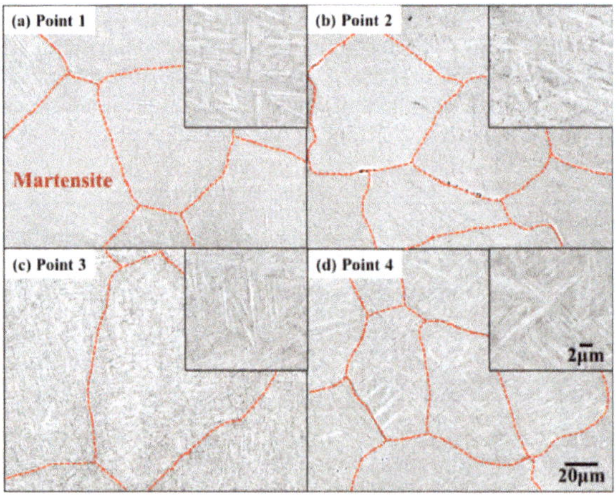

Fig. 6. The fully martensite microstructure in the Ω-shaped component near the (a) measuring point 1, (b) measuring point 2, (c) measuring point 3 and (d) measuring point 4.

The mechanical properties of the Ω-shaped component near different measuring points are shown in Fig. 7. It can be seen that the yield strength and the tensile strength of the component are higher than those of the as-received sheet. The yield strength of the component remains at a range of 1011.2 MPa ~ 1060.6 MPa, 3.7% ~ 8.8% higher than that of the as-received sheet. The tensile strength remains at a range of 1106.3 MPa ~ 1153.9 MPa, 8.4% ~ 13.0% higher than that of the as-received sheet. Furthermore, the elongation of the component exceeds a range of 7.7% ~ 8.5%. The distribution of the overall mechanical properties is relatively uniform.

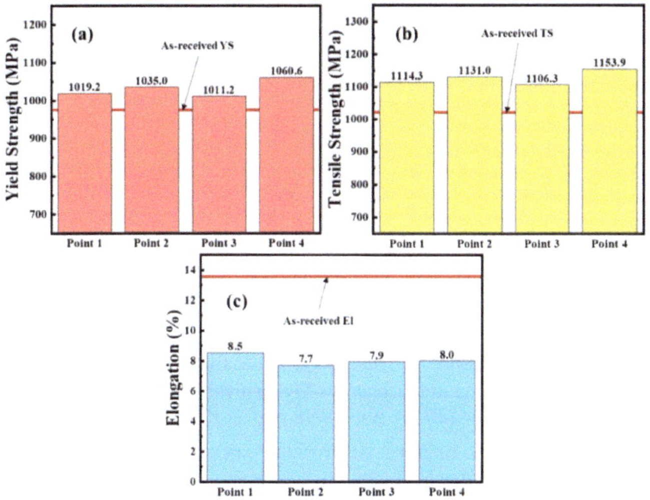

Fig. 7. The (a) yield strength, (b) tensile strength and (c) elongation of the Ti-6Al-4V alloy near different measuring points in the Ω-shaped component.

The above results demonstrates that one could use martensite microstructure to strengthen titanium alloy component and meanwhile avoid the crack failure by non-equilibrium hot stamping process. This is because when the heating temperature is in the single β-phase region, a higher heating rate leads to finer prior β grains, along with finer internal lamellar α after water-quenching [12]. The refined martensite microstructure could strengthen titanium alloys and reserve most plasticity at the same time [13].

Conclusion

In this study, the effects of rapid heating in the single β-phase region on the Ti-6Al-4V alloy microstructure evolution and mechanical properties was investigated. The Ti-6Al-4V alloy Ω-shaped component with fully martensite microstructure was formed via non-equilibrium hot stamping process. The distribution of microstructure and mechanical properties was summarized. The main conclusions are as follows:

(1) The martensite microstructure after rapid heating affects the mechanical properties of Ti-6Al-4V alloy significantly. The average size of prior β grains and the average width of internal lamellar α decrease as heating rate rises. A higher heating rate in the single β-phase region tends to increase both strength and elongation. The tensile strength of Ti-6Al-4V alloy increases from 1116.6 to 1263.0 MPa and the elongation increases from 5.2 % to 9.6 %, when the heating rate rises from 2°C/s to 50°C/s.

(2) The titanium alloy thin-walled components could be strengthened by fully martensite microstructure, which could be obtained by non-equilibrium hot stamping with rapid heating process. With heating temperature of 1050°C and average heating rate of 17.5°C/s, a Ti-6Al-4V alloy thin-walled component with fully martensite microstructure was successfully formed. A maximum tensile strength of the formed component could reach up to 1153.9 MPa, which is 13.0% higher than that of the as-received sheet. And a total elongation of 8.0% was remained at room temperature.

Financial Assistance
This research was funded by the Program of National Natural Science Foundation of China (No. 51905124), Aeronautical Science Foundation of China (No. 20220001077002) and China Postdoctoral Science Foundation（2021T140153）

References

[1] S. Yuan, Fundamentals and Processes of Fluid Pressure Forming Technology for Complex Thin-Walled Components, Engineering 7 (2021) 358-366. https://doi.org/10.1016/j.eng.2020.08.014

[2] K.H. Wang, L.L. Wang, K.L. Zheng, Z.B. He, D.J. Politis, G. Liu, S.J. Yuan, High-efficiency forming processes for complex thin-walled titanium alloys components: state-of-the-art and perspectives, Int. J. Extreme Manuf. 2 (2020) 20-43. https://doi.org/10.1088/2631-7990/ab949b

[3] E.-L. Odenburger, R. Pederson, M. Oldenburg, Finite Element Modeling and Validation of Springback and Stress Relaxation in the Thermo-Mechanical Forming of Thin Ti-6Al-4V Sheets, Int. J. Adv. Manuf. Technol. 104 (2019). https://doi.org/10.1007/s00170-019-04071-9

[4] K. Mori, P.F. Bariani, B.-A. Behrens, A. Brosius, S. Bruschi, T. Maeno, M. Merklein, J. Yanagimoto, Hot stamping of ultra-high strength steel parts, CIRP Ann.-Manuf. Techn. 66 (2017) 755-777. https://doi.org/10.1016/j.cirp.2017.05.007

[5] Y. Nakagawa, K. Mori, T. Maeno, Y. Nakao, Reduction in holding time at bottom dead centre in hot stamping by water and die quenching, Procedia Manuf. 15 (2018) 1111-1118. https://doi.org/10.1016/j.promfg.2018.07.380

[6] Y.F. Li, Y. Chen, S.H. Li, Phase transformation testing and modeling for hot stamping of boron steel considering the effect of the prior austenite deformation, Mater. Sci. Eng. A-Struct. Mater. Prop. Microstruct. Process. 821 (2021). https://doi.org/10.1016/j.msea.2021.141447

[7] M. Kopec, K.H. Wang, D.J. Politis, Y.Q. Wang, L.L. Wang, J.G. Lin, Formability and microstructure evolution mechanisms of Ti-6A-l4V alloy during a novel hot stamping process, Mater. Sci. Eng. A-Struct. Mater. Prop. Microstruct. Process. 719 (2018) 72-81. https://doi.org/10.1016/j.msea.2018.02.038

[8] Y.H. Mu, X.M. Yang, B.Y. Wang, H.T. Qu, F.L. Chen, A novel triple-layer hot stamping process of titanium alloy TC4 sheet for enhancing formability and its application in a plug socket part, Int. J. Lightweight Mater. Manuf. 6 (2023) 189-197. https://doi.org/10.1016/j.ijlmm.2022.10.003

[9] Z. Hamedon, K. Mori, T. Maeno, Y. Yamashita, Hot stamping of titanium alloy sheet using resistance heating, NMSTU 5 (2013).

[10] T. Maeno, Y.Yamashita, K. Mori, Hot stamping of titanium alloy sheets into U shape with concave bottom and joggle using resistance heating, Key Eng. Mater. 716 (2016) 915–922. https://doi.org/10.4028/www.scientific.net/KEM.716.915

Materials Research Forum LLC
https://doi.org/10.21741/9781644903254-79

[11] K.H. Wang, M. Kopec, S.P. Chang, B. Qu, J. Liu, D.J. Politis, L.L. Wang, G. Liu, Enhanced formability and forming efficiency for two-phase titanium alloys by Fast light Alloys Stamping Technology (FAST), Mater. Des. 194 (2020). https://doi.org/10.1016/j.matdes.2020.108948

[12] S.P. Chang, K.H. Wang, B. Wang, M. Kopec, Z. Li, L.L. Wang, G. Liu, Effects of rapid heating on non-equilibrium microstructure evolution and strengthening mechanisms of titanium alloy, Mater. Sci. Eng. A-Struct. Mater. Prop. Microstruct. Process. 880 (2023). https://doi.org/10.1016/j.msea.2023.145337

[13] Y. Chong, T. Bhattacharjee, J. Yi, A. Shibata, N. Tsuji, Mechanical properties of fully martensite microstructure in Ti-6Al-4V alloy transformed from refined beta grains obtained by rapid heat treatment (RHT), Scripta Mater. 138 (2017) 66-70. http://dx.doi.org/10.1016/j.scriptamat.2017.05.038

Study of die punching process for hot stamped high strength steel and its performance evaluation

CHEN Jiafeng[1], WEI Jinchan[1], HAN Xianhong[1,a*],
QU Shaofei[2] and SHI Yutong[2]

[1]School of Materials Science and Engineering, Shanghai Jiao Tong University, 1954 Huashan Road, 200030, Shanghai, China

[2]Gestamp China R&D Center, 56 Antuo Road, 201805, Shanghai, China

[a]hanxh@sjtu.edu.cn

Keywords: Hot Stamped Ultra-High Strength Steel, Die Punching, Process Study, Performance Evaluation

Abstract. Using the hot stamping process to produce ultra-high strength steel parts has been regarded as a good option for lightweight of automotive parts. Hot stamped workpieces usually require post-stamping processes like hole machining before they can be utilized. Considering industrial production, die punching process is more attractive than other hole machining methods such as laser cutting due to its high efficiency and low cost. In this study, the die punching process for hot stamped steel Usibor 1500P was investigated, and the performance of the punched parts was also explored. Firstly, the influence of process parameters including punching velocity, die clearance, punch corner radius, punch diameter, etc. was studied through a specifically designed trimming tool. Corner radius at the punch edge significantly improves the sheared edge quality, and tapered punch and conical punch evidently reduce and raise blanking force respectively. Secondly, several finite element fracture models for die punching process of Usibor 1500P were constructed and validated, where the Oyane and MMC damage models are considered suitable for punching simulation. Thirdly, the service performances of hot stamped ultra-high strength steel after die punching were evaluated through some typical experiments, including tensile tests, bending tests, and hydrogen embrittlement tests.

Introduction

Hot stamping steel has been widely used in the automobile industry because of its high properties. Meanwhile, the high strength and hardness bring challenges to trimming and punching processes, thus, the laser cutting is normally adopted as the subsequent treatment for hot stamped parts. In contrast, die punching method has the advantages of high efficiency and low cost, and its applicability in ultra-high strength steel products is worthy of study [1].

Gustafsson et al. [2] studied the effects of die clearance and edge pressure on the blanking force of different strength steels. Victor Hugo Cabrera et al. [3] discussed the effects of different convex die shapes on shear force through experiments and simulation. Li et al. [4] used finite element simulation to study the cross-sectional quality of high-strength steel plates in punching forming. Mori et al. [5] measured the hardness of 1.5GPa hot stamped ultra-high strength steel plate during cold stamping and laser cutting and found that the hardness of the cut edges was significantly improved. Sheng et al. [6] investigated the effects of several cutting processes on hydrogen embrittlement sensitivity and found that the punching specimen has the highest sensitivity to hydrogen embrittlement, which agrees with the results of Mori et al. [5].

This paper focuses on the die punching process of hot formed Usibor 1500P steel, studying the influence of process parameters on punching force and surface quality [7, 8]. Meanwhile, the mechanical properties and hydrogen embrittlement performance of the part after punching are also discussed.

Influence of process parameters on punching force and section quality
The chemical composition of Usibor 1500P in weight percent is 0.25% C, 1.40% Mn, 0.35% Si, 0.30% Cr, and 0.005% B, with Fe balance. After heat treatment consistent with hot stamping process, the yield strength, ultimate tensile strength, and elongation are respectively 1120 MPa, 1526 MPa, and 6.90%. The thickness of the blank is 1.6mm and the punch diameter is 12mm.

The punching experiment was designed with a combination of various process parameters. Where, the punch velocity changes from 8.25 to 82.5 mm/s, the die clearance changes from 0.12 to 0.26mm, the punch corner radius changes from 0 to 0.2mm, and the punch shape includes conventional flat one, conical one, and tapered one with flat bottom on the top.

From the experimental results, the blanking force is linearly related to the punch velocity, die clearance, and punch corner radius, but the amplitude of the change is relatively small. The punch shape has a great influence on the blanking force. Comparing to the conventional flat shape, the tapered punch can reduce the blanking force, while the conical punch increases the blanking force, as shown in Fig.1.

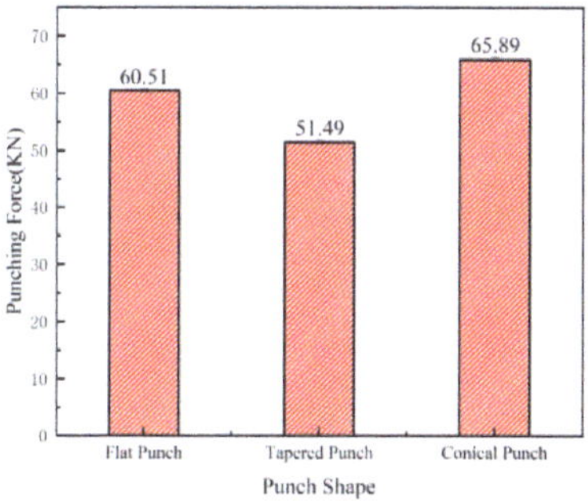

Fig.1. Effect of punch shapes on punching force.

To study the influence of process parameters on punching section, the distributions of different zones are observed. Higher quality of punched section needs small tear zone, burr length, and rollover zone, along with larger burnish zone. From the experimental results, it is concluded that among the process parameters, the punch corner radius has the most effect on the improvement of section quality as shown in Fig. 2. It should be noted that a smaller length of burr is important for the product quality, which reduced with the punch corner radius clearly.

Materials Research Forum LLC
https://doi.org/10.21741/9781644903254-80

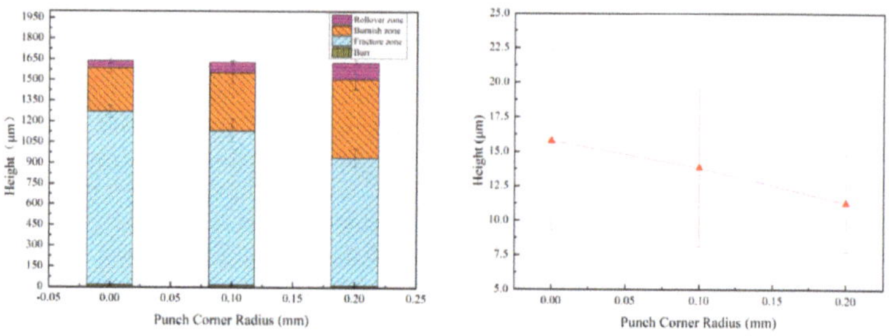

Fig.2 Sectional morphology characteristic zone and burr length

Fig. 3 shows the curves of punching force obtained by simulating different fracture criteria, it is observed that the curves for damage initiation, crack initiation, and crack propagation vary depend on the fracture criterion used. The curves corresponding to the Cockcroft-Latham, Rice-Tracey, and Freudental criteria exhibit a short damage initiation stage, and the punching force drops noticeably at small displacements, while the Oyane criterion and MMC criterion have better results by comprising with experiment.

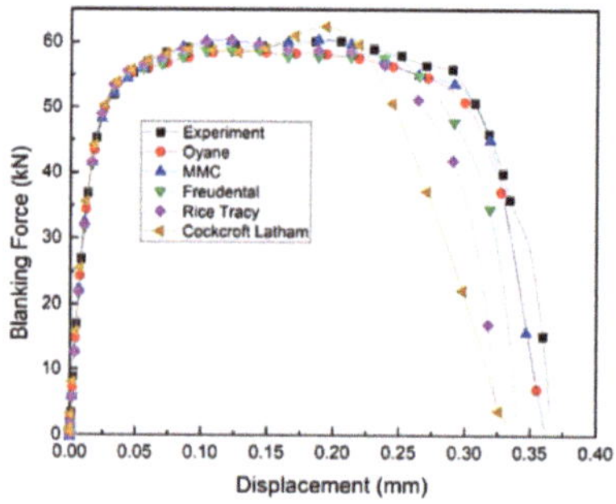

Fig. 3 Simulation results of punching force vs. displacement curve.

Performance evaluation of hot stamped boron steel after die punching

Fig. 4 shows the tensile load-displacement curves of samples corresponding to different shearing processes. Compared with the laser cutting and wire cut electrical discharge machining (WEDM) processes, perforated specimens processed by die punching exhibit reduced tensile strength and ductility after fracture during the stretching process. It is speculated that this difference may stem from the thermal processing involved in laser cutting and WEDM cutting acting as a heat treatment on the specimens, resulting in a larger but softer shear affected zone.

Meanwhile, it also can be found that different punching process has different effects on tensile properties. The die clearance has a relatively smaller impact than that of punch corner radius and punch shape. Especially for the tapered punch case, it has only half maximum tensile displacement comparing to others.

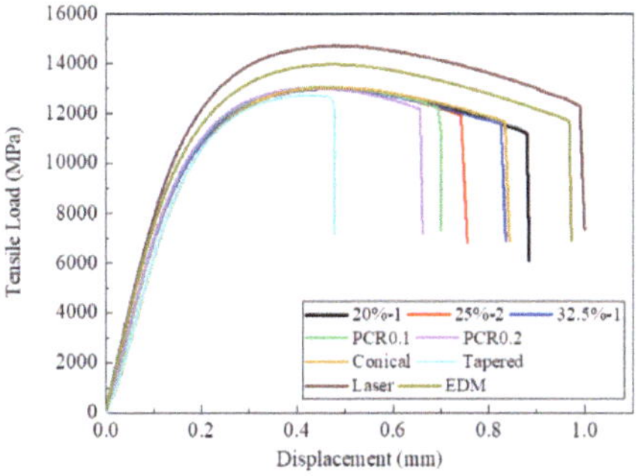

Fig. 4. Tensile load-displacement curves.

A series of three-point bending tests were also carried out through the bending test machine, and the results are shown in Fig. 5. According to Fig. 5(a), the ultimate cold bending angles of different samples are all above 50°, which meets the performance index of this steel. Fig. 5(b) shows the maximum bending loads, it can be found that the influence of the parameters on bending load is small. Meanwhile, the specimens processed by WEDM cutting and laser cutting show a relatively higher bending load compared to those processed by die punching.

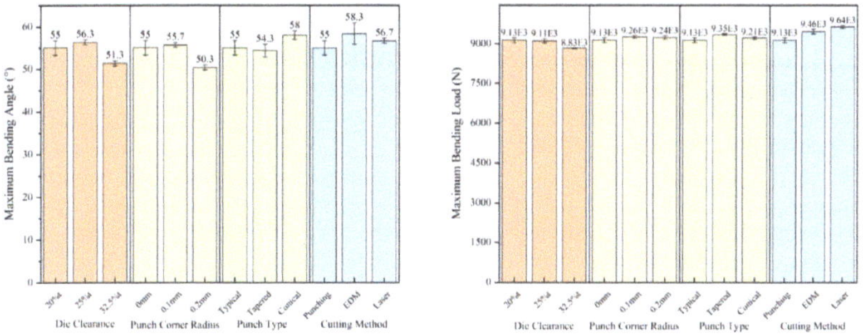

Fig. 5. Maximum bending angles and bending loads.

Through slow tensile tests with a constant strain rate of 10^{-4}/s, the influence of different punching processes on hydrogen embrittlement sensitivity was examined, before which hydrogen was firstly charged into the samples by the electrochemical hydrogen charging method with a current density of 5mA/cm^2 for 0.5h. As shown in Fig. 6, it can be found that the machinal properties decrease for all samples after hydrogen charging, especially for maximum displacement that reflects ductility. By comparing with different cutting methods, both strength and ductility of die cutting sample are worse than those of laser cutting and WEDM samples.

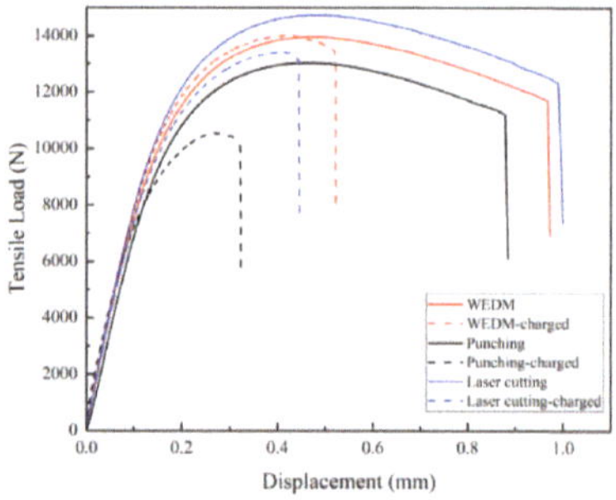

Fig. 6. Tensile load-displacement curves before and after hydrogen charging.

Summary

This paper studied the die punching process for hot stamped high strength steel, and evaluated the mechanical and hydrogen embrittlement performance after punching. The main conclusions are as follows:

(1) The punching force is linearly associated with the punching velocity, die clearance, and punch corner radius, but the variation is relatively small, while the punch shape has a larger impact. For the quality of the punching section, the main process parameter is punch corner radius, where larger radius leads to better surface quality. The finite element simulation with Oyane model can predict the punching process well.

(2) The mechanical properties under different parameters are mainly reflected in the maximum tensile deformation. The tensile performance of the conical punch sample is similar to that of the regular one, while that of the tapered punch sample shows a steep drop. As for the bending test, the maximum bending angle and bending load under all the parameters show a small relation to the change of parameters. Similarly, the HE susceptibility is also less dependent on the punching parameters.

(3) Compared with the die punching samples, the losses of tensile mechanical properties of laser cutting and WEDM cutting samples after hydrogen charging are relatively small, that is, the sensitivities of hydrogen fragility of these two samples are lower.

Acknowledgments

Support by the National Key Research and Development Programs of China (Grant No. 2022YFE0196600 and No. 2023YFB2504604) and the National Natural Science Foundation of China (Grant No. 52175349) are acknowledged.

References

[1] Y. Abe, R. Yonekawa, K. Sedoguchi, K. Mori, Shearing of ultra-high strength steel sheets with step punch, Proc. Manuf. 15 (2018) 597-604. https://doi.org/10.1016/j.promfg.2018.07.283

[2] E. Gustafsson, M. Oldenburg, A. Jansson, Experimental study on the effects of clearance and clamping in steel sheet metal shearing, J. Mater. Process. Technol. 229 (2016) 172-180. https://doi.org/10.1016/j.jmatprotec.2015.09.004

[3] V. H. Cabrera, W. Quillupangui, L. Juiña, M. Valdez, Cutting parameters in dies for sheet metal through inclined, concave, and convex shapes in round punches, Mater. Today Proc. 49, Part 1 (2022) 28-34. https://doi.org/10.1016/j.matpr.2021.07.437

[4] C. Li, Analysis and Application of the Instability of Thin-walled (2019), Beijing Institute of Technology.

https://kns.cnki.net/kcms2/article/abstract?v=Fhes7GDiHN1EPAoBTe3XyObMGwT3_I2JpCimxXCTCcz-rNOUfOyUkNAGN4O7cHR_nQ-Mld4gPmYvvHYE3ackiHTh6lyDbvnsgoV2PbdgeMayRQaa91GFS8Aj4pJfwVRqIneWa7PkEsMIMRoF9Q3LyQ==&uniplatform=NZKPT&language=CHS

[5] K. Mori, N. Nakamura, Y. Abe, Y. Uehara, Generation mechanism of residual stress at press-blanked and laser-blanking edges of 1.5 GPa ultra-high strength steel sheet, J. Manuf. Process. 68, Part A (2021) 435-444. https://doi.org/10.1016/j.jmapro.2021.05.047

[6] Z. Sheng, C. Altenbach, U. Prahl, D. Zander, W. Bleck, Effect of cutting method on hydrogen embrittlement of high-Mn TWIP steel, Mater. Sci. Eng. A 744 (2019) 10-20. https://doi.org/10.1016/j.msea.2018.11.128

[7] X. Han, K. Yang, Y. Ding, S. Tan, J. Chen, Numerical and experimental investigations on mechanical trimming process for hot stamped ultra-high strength parts, J. Mater. Process. Technol. 234 (2016) 158-168. https://doi.org/10.1016/j.jmatprotec.2016.03.025

[8] C. Yang, J. Wei, Z. Chen, S. Qu, X. Han, Study of fracture damage criteria and influence of process parameters on punching of hot stamped ultra-high strength steel, Int. J. Adv. Manuf. Technol. 128 (2023) 1493–1504. https://doi.org/10.1007/s00170-023-11980-3

Integration of automated roll pass design and simulation for the development of shape rolling technology

BIBA Nikolay[1,a *], STEBUNOV Sergey[2,b], MALTSEV Pavel[2,c], BELUGIN Vitaliy[3,d] and ALIMOV Igor[2,e]

[1]Micas Simulations Ltd., 107 Oxford Road, Oxford, UK

[2]QForm Group FZ LLC, Fujairah-Creative City, P. O. Box 4422 Fujairah, UAE

[3]FormingSoft LLP, Seifullin pr. 498, office 24, Almaty, A05A6G2, Kazakhstan

[a]nick@qform3d.com, [b]serg@qform3d.com, [c]maltsevpa@gmail.com, [d]belugin@qform3d.com, [e]m9190266236@gmail.com

Keywords: Shape Rolling, Simulation, Finite Element Method, Roll Pass Design, Industrial Implementation

Abstract. The shape rolling technology is a productive manufacturing process. Meanwhile, its efficiency is highly dependent on a proper roll pass design and accurate adjusting of a rolling mill, which is especially critical for new products and complicated profile shapes. Traditionally, the roll pass design methods have been developed relying on extensive experience and knowledge accumulation, resulting in empirical rules expressed by analytical formulae. Nowadays, some design methods are converted into computer programs to speed up calculations. Nonetheless, the approach based on empirical rules has limitations, especially when implemented to develop the rolling technology for new materials and complex shape profiles. The effectiveness of the roll pass design can be significantly enhanced with the help of finite element (FE) simulation of the material deformation that happens in the stands of the rolling mill. The material flow prediction provided by FE simulation is more accurate than empiric formulae, leading to more accurate load, torque, and energy estimations. The presented paper is focused on the integration of the FE simulation of the rolling process with the roll pass design to make it effective and automated for a wide variety of profile shapes.

Introduction

Roll pass design (RPD) methods have evolved significantly over the years, leveraging both analytical techniques and empirical knowledge [1,2]. Often integrated into software solutions, such methods were implemented for billet, bar, and rod mills [3]. For example, in the Freiberg Mining Academy was developed PyRolL, an open-source project that encapsulates the collective expertise of students, engineers, and scientists in RPD [4]. Concurrently, Finite Element Method (FEM) codes have proven effective in simulating shape rolling, though they necessitate detailed input for each stand, a task that demands substantial preparatory and analytical effort.

Several years ago, the authors developed a specialised module for shape rolling simulation within the QForm UK metal forming simulation software [5] that is now integrated with the newly developed program QKaliber for automated roll pass design. This integration enables the roll pass design software to generate and transfer detailed process parameters and groove geometries directly to the simulation program and get feedback from it. This integration, embodying Industry 4.0 ideals. It is applicable to a wide range of products in any rolling mill type. While focusing on product shapes, loads, and torques, the software also examines heat balances and is capable of simulating heat treatment processes and microstructure prediction, setting the stage for future discussions.

Materials Research Forum LLC
https://doi.org/10.21741/9781644903254-81

Materials and Methods

This study enhances our FEM simulation software for metal forming, specifically targeting the unique challenges of shape rolling simulations. Unlike forging, shape rolling involves localised, small deformation zones, with the rest of the product acting largely as a rigid body. Simultaneous rolling across multiple stands can influence adjacent stands via tension or support in continuous mills. Our solution employs a automatically densified FE mesh at contact nodes between the product and rolls for accurate deformation simulation (Fig. 1). Using Euler's method enables roll motion simulation through calculated nodal velocity vectors, keeping the FE mesh in rolls fixed in the space.

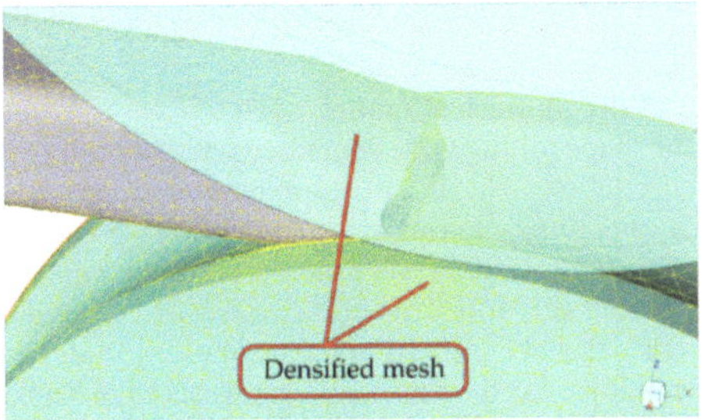

Fig.1. Quasi-rotational rolls for simulation of rolling in QForm UK with densified mesh in the contact area.

Moreover, it's crucial to refine the Finite Element (FE) mesh for the product within the deformation and contact zones. In contrast, a coarser mesh can be used for the rest of the product due to the absence of deformation. However, enlarging FE elements in rigid zones can compromise the precision of the product geometry representation. To overcome this issue, a dual mesh technique has been introduced (Fig. 2). This approach consists of a "Mechanical" mesh, densified in deformation areas and used for solving the material deformation task, and a "Geometry" mesh, dedicated to storing the product's shape and solving the thermal task. Data transfer between the "Mechanical" and "Geometry" meshes allows for the exchange of temperature and strain data to solve the general coupled thermo-mechanical task of the shape rolling process. Advanced algorithms have been developed to enhance the calculation of node trajectories across the roll's curved surface, ensuring continuous contact between the material being rolled and the rolls.

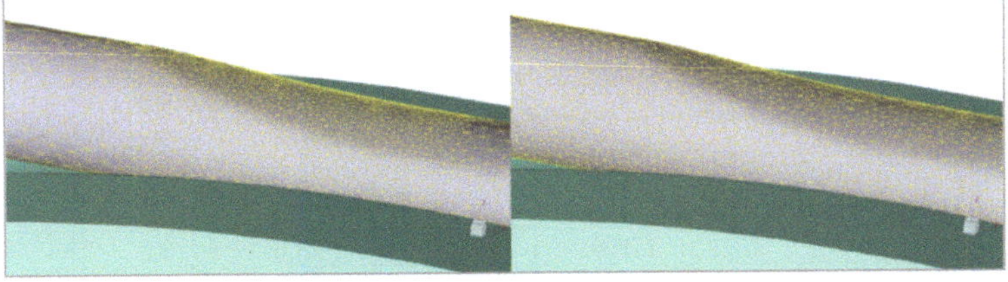

(a) (b)

Fig. 2. Dual mesh method for shape rolling simulation in QForm UK: the Mechanical mesh (a) and the Geometry mesh (b).

Contemporary shape rolling technology facilitates the manufacture of diverse profiles, necessitating the capability of software to simulate processes across various rolling mill types and configurations. These include reverse stands with multiple grooves, continuous rolling mills, and rolling lines that combine a sequence of duo stands and universal rolling stands. The influence of a pusher, the automatic positioning of the workpiece before rolling, and the influence of entry and exit guides are integral to the rolling simulation. To provide the input of roll geometry, several alternative methods have been developed, i.e. importing 3D solid models directly from CAD systems, importing 2D roll contours and then revolving these into 3D models, and direct transfer of roll pass designs from CAD system QKaliber.

Additional parameters essential for accurate rolling simulations encompass the material properties of both the billet and rolls, rolling speed, friction, initial temperature of the billet, and more. Once all necessary process parameters are established, the simulation software can automatically execute the entire sequence of passes in reverse or continuous mills to predict outcomes with high precision.

Industrial verification of the numerical model

The simulation results for shape rolling were compared with laboratory and industrial experiments, displaying a notable correlation. The industrial experiments served not only as a means for validation but also as an opportunity to refine the simulation's boundary conditions. During these industrial tests, technological parameters were measured in the continuous mill "450" during the production runs of the I-Beam IPE 160. This mill consists of 16 stands divided into roughing and finishing groups. The pass sequence utilised in the technology under the test is detailed in Fig. 3. The technological information used for simulation is presented in Table 1.

Fig. 3. The sequence of the passes in the mill "450" that is used for rolling the I-Beam IPE 160.

Table 1. Technological parameters of the I-Beam IPE 160 hot rolling in the continuous mill "450".

Stand №	Roll horizontal gap, [mm]	Roll vertical gap, [mm]	Roll diameter, [mm]	RPM drives	Distance to next stand, [m]	Gear ratio
1	36	-	620	156,0	4,2	9,68
2	6	-	620	152,0	4,2	8,19
3	12,0	-	605	207,0	15,0	8,25
4	14,0	-	685	205,0	5,0	6,69
5	29,6	-	603	255,0	5,0	5,55
6	10,0	-	680	209,0	25,0	4,85
7	16,0	166,4	1070	224,0	10,0	5,55
9	11,9	-	650	286,0	62,0	3,64
12	9,3	170,0	865	328,0	6,0	4,36
13	7,1	172,5	885	354,0	6,0	3,32
14	6,5	-	559	271,0	6,0	1,56
15	5,6	165,3	865	360,0	6,0	2,75
16	6,5	-	477	314,0	6,0	1,11

Materials Research Proceedings 44 (2024) 758-767 https://doi.org/10.21741/9781644903254-81

During the experiment with the I-Beam IPE 160, the procedure involved optimising the mill settings to minimise support and tension between the stands, achieving the profile shape within the standards, documenting the rolls' diameters, their rotational speeds, the real-time loads on the drives, and the temperature across passes. Extracted cross-cut templates from the product's end after every pass for geometric assessment are shown in Figure 4 and analysed in Figure 5.

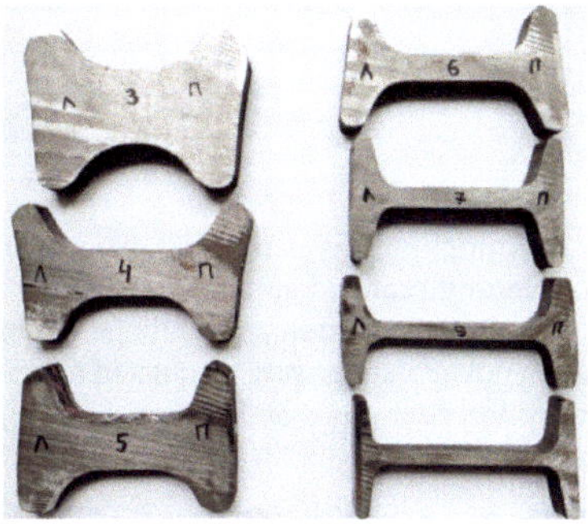

(a) (b)

Fig. 4. Photos of the templates of the I-Beam IPE 160 cut from the end of the product while rolling in the mill 450: (a) templates from 3rd to 5th stand; (b) templates from 6th to 9th and 11th stands.

The simulation of the continuous rolling process was executed using the preliminary data furnished by the manufacturing facility. The simulation segmented the rolling sequence into stand groups, mirroring actual mill operations. Given the notable impact of the inter-stand group distance on metal temperature, the cooling was simulated after each group. This approach enabled the analysis of how the workpiece's cooling duration between stand groups affects the overall process.

This paper presents only a subset of the findings from these analyses. A comparison of the pass templates' geometry is depicted in Fig. 5, showcasing an impressive accuracy of the shape prediction exceeding 96.3%.

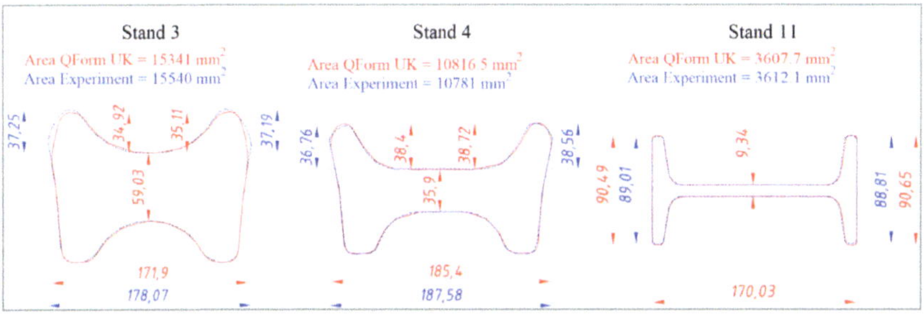

(a) (b) (c)

Fig. 5. The experimental templets and simulated shapes superposed over each other: (a) after the 3rd stand, (b) after the 4th stand, (c) after the 11th stand.

The largest discrepancy between simulation and actual measurements was observed after the 11th pass as 9.1%, while the smallest was after the 4th pass as 1%. Temperature comparisons between stationary pyrometer readings and simulations after the 3rd, 9th, and 16th stands are detailed in Table 2, with an average temperature prediction error about 4%.

Table 2. Comparison of the temperature during the rolling of I-Beam IPE 160.

Stand	Simulation	Experiment	Error
3	1096°C	1114°C	2%
9	1024°C	1012°C	1%
16	1045°C	958°C	9.1%
Mean error	4%		

A comparison of power consumption between actual experiments and simulations for the mill drives was conducted by converting real and simulated data into the drive power. Experimental rolling power was derived from average current and voltage readings and the rotational speed of main drives (RPM). Simulation power values were calculated using forces, torques, and roll RPMs, factoring in gearbox efficiency and gear ratios and compared in Fig. 6.

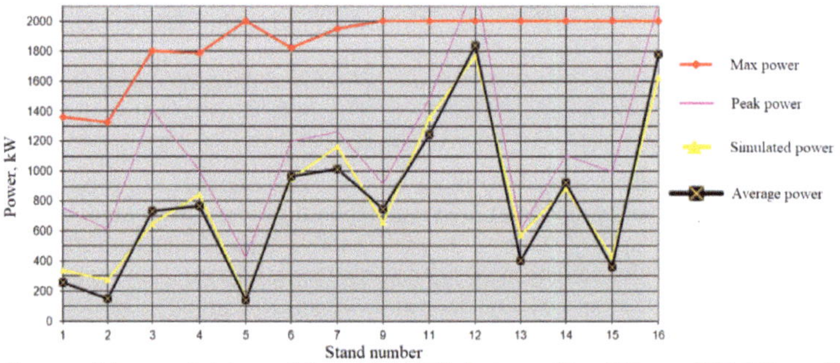

Fig. 6. Power of the stand drives of the mill "450" during rolling I-Beam IPE 160 profile.

The simulated power (yellow graph) and measured power (black coloured) have shown a total correspondence of 94.43%.

Automated roll pass design system QKaliber

QKaliber is our innovative software for automated roll pass design, offering a comprehensive selection of functionalities such as creating groove configurations for shape rolling, preparation of technical documentation, and initial data generation for simulation in QForm UK. It incorporates databases for rolling mill specifications and roll inventories, groove shapes design with rolling process parameters like temperature, speed, loads, and torques. It also conducts rolling process analysis within mill limitations. The software enhances understanding of metal deformation and bite conditions through visualizations, maintains an archive of groove variations, and conducts roll stress analysis. Additionally, it automates the generation of essential technical documentation and data preparation for simulations, streamlining the roll pass design process. The program's adaptability allows for the design of shaped profiles in two directions, i.e. from the workpiece to the finished profile and in the opposite direction. The typical breakdown passes are shown in Fig. 7.

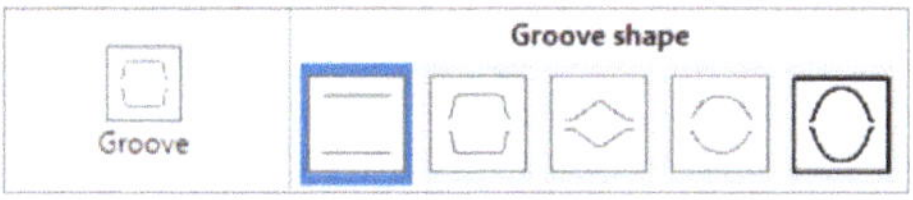

Fig. 7. The breakdown passes of the CAD QKaliber.

In QKaliber, breakdown grooves such as round, diamond, square and oval are defined in parametric form. The system enables the design of grooves from billet to final shape by specifying groove dimensions and calculating the spread and elongation of metal cross-cut sections along with rolling parameters for all the passes. An alternative design approach from the final section to the workpiece allows to work with complex shapes like I-beams, T-bars, rails, channels and others. Here, profiles undergo segmentation into parametric components, and their sizes are automatically determined through a built-in algorithm that considers the dimensional aspects of adjacent roll passes. The user can easily adjust this algorithm to achieve a well-balanced reduction pattern. Elements of the profile are quadrilateral with straight, rounded, or inclined sides. The example of the I-Beam profile divided into elements is shown in Fig. 8.

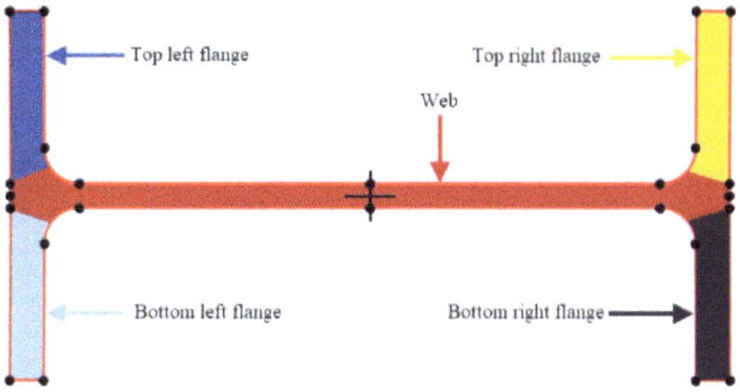

Fig. 8. The I-beam profile consists of five elements: four flanges and a web.

The strain in the passes is calculated automatically according to the total elongation between the final and leader profiles. It can be controlled separately to achieve the needed dimensions on the preliminary passes based on the chosen strategy. QKaliber helps the roll pass designer calculate all deformation coefficients in each pass in alignment with element dimensions. Fig. 9, using the example of a flange element, illustrates the labelling of element parameters that the software employs to calculate deformation coefficients.

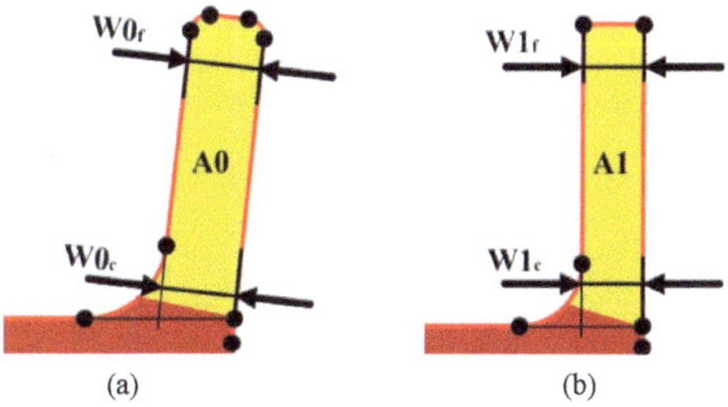

Fig. 9. Flange dimensions and areas of pre-finish (a) and finish (b) passes.

The reduction coefficients (Eq. 1), relative reductions (Eq. 2) and elongation factor (Eq. 3) are calculated as follows:

$$\mu_f = \frac{W0_f}{W1_f},\ \mu_c = \frac{W0_c}{W1_c}, \tag{1}$$

$$\mu_f, \% = \frac{W0_f - W1_f}{W0_f} \times 100\%,\ \ \mu_c, \% = \frac{W0_f - W1_f}{W0_f} \times 100\%, \tag{2}$$

$$E = \frac{A0}{A1}, \tag{3}$$

All the dimensions of elements are gathered into one table of passes, where the ratio of heights (coefficient *kH*) along with the absolute reduction (*dL*) are presented (Fig. 10). The obtained coefficients can be displayed as a graph along passes.

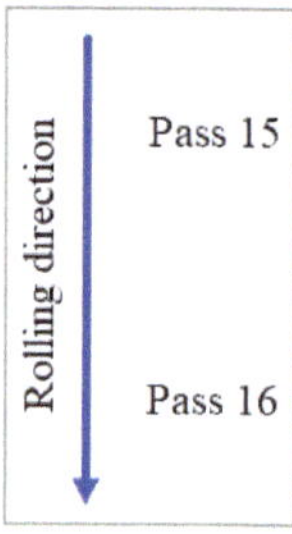

Pass 15	web	dL	11.6	kH	1.284			web	Length	172.8				
	Head_top	dL	-2.7	kHc	1.154	kHf	1.109	Head_top	Height	23.3	Width C	54.8	Width F	39.855
	Head_bottor	dL	-2.7	kHc	1.154	kHf	1.109	Head_bottor	Height	23.3	Width C	54.8	Width F	39.855
	Foot_top	dL	-3.6	kHc	1.15	kHf	1.371	Foot_top	Height	62.4	Width C	33	Width F	17
	Foot_bottom	dL	-3.6	kHc	1.15	kHf	1.371	Foot_bottom	Height	62.4	Width C	33	Width F	17
Pass 16	web	dL	-1.5	kH	1			web	Length	161.2				
	Head_top	dL	-0.4	kHc	0.965	kHf	1.027	Head_top	Height	26	Width C	47.5	Width F	35.938
	Head_bottor	dL	-0.4	kHc	0.965	kHf	1.027	Head_bottor	Height	26	Width C	47.5	Width F	35.938
	Foot_top	dL	5	kHc	0.973	kHf	0.886	Foot_top	Height	66	Width C	28.7	Width F	12.4
	Foot_bottom	dL	5	kHc	0.973	kHf	0.886	Foot_bottom	Height	66	Width C	28.7	Width F	12.4

Fig.10. Dimensions of profile elements are calculated according to the previous pass opposite to the rolling direction.

Let us consider the following case study. The roll pass design of the Rail 54E1 profile according to EN 13674-1 standard is created based on the modern rail rolling mill and verified by the simulation. The roll pass design for the Rail was made using both approaches, i.e. from the final shape to the preliminary profile (leader profile) and from the rectangular billet to the leader. This article provides a detailed analysis of the creation of RPD in the direction opposite to the rolling, utilising new features in the QKaliber program. The results are presented partly for the first half of the RPD, which was prepared following the rolling direction.

The first step is to select the pass sequence and determine the number of passes according to the rail type and size. We selected the typical rolling scheme, where the leader rail section is produced in two roughing stands by ten passes and then fed into the group of universal and edger stands with seven passes. Then, the rail shape is obtained in the semi-universal stand, which has only one vertical roll.

The next step was to create the Rail 54E1 profile according to EN 13674-1 standard in its hot state, excluding radii and inclines. The profile is constructed from five parametric elements, i.e. Top and Bottom Tips of the Rail, Top and Bottom Roots of the Rail, and the Web. Based on the selected mill layout, we proceeded with the getting of the leader profile in the roughing group and roughly estimated its dimensions. Subsequently, we employed automatic calculation of reduction coefficients opposite to the rolling direction from the finished profile to the leader profile (Fig. 11). During the calculation, all intermediate shapes are composed of the same elements but with varying dimensions, except for shapes that are rolled on edger stands and which were not considered.

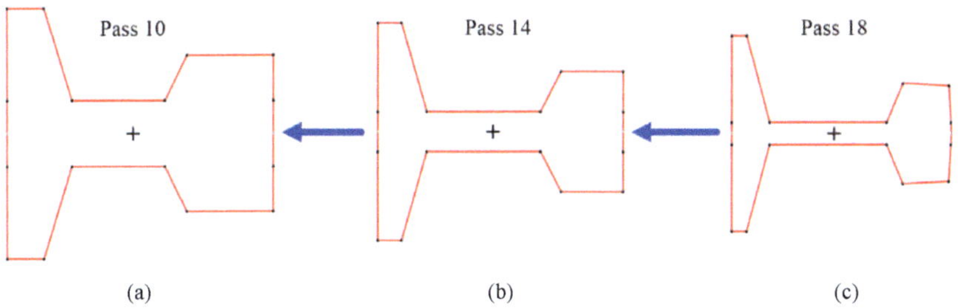

Fig. 11. The sequence of the roll passes created in the inverse direction from the 18th to the 10th pass: (a) leader shape, (b) one of the intermediate shapes, (c) final shape of the rail.

The analysis of reduction coefficients across passes (Fig. 12) indicates that the thickness on the Rail's Tip free side (F) exceeds that on the contact side (C) with Tip (F) being higher than Tip (C), which is also true for the Rail's Root. This distribution is consistent with the fundamental guidelines for rail RPD [7]. However, the reduction of the rail's Root is greater than of the Tip, which is incorrect. The next step in the RPD process should be to adjust the distribution ensuring the Tip experiences greater reduction than the Root.

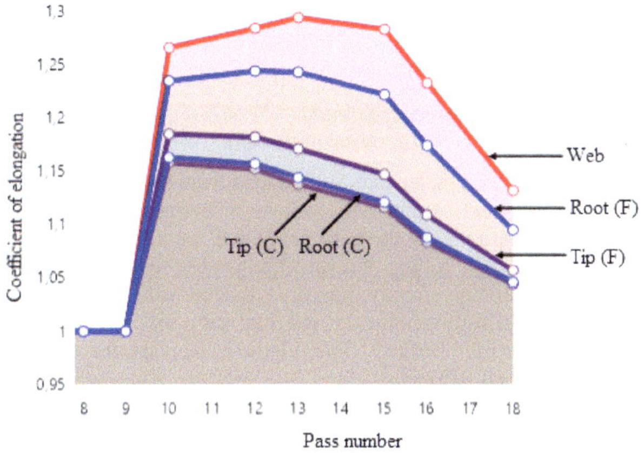

Fig. 12. The coefficients of reduction as a graph along passes after the automatic calculation.

In the next phase, we adjusted the width of the profile according to the configuration of the tandem group, refined the leader shape and adjusted the Tip and Root inclines in line with the roll configuration. By modifying the dimensions of the elements, we achieved the desired distribution of reduction coefficients across all the passes (Fig. 13). Fig. 14 shows the intermediate shapes of the rail after adjusting the reduction coefficients.

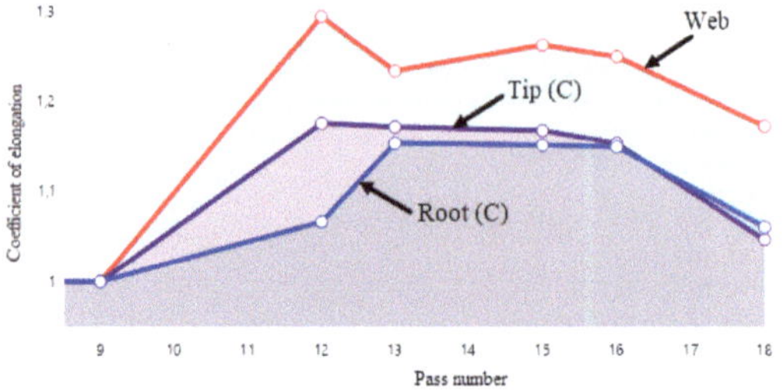

Fig.13. The coefficients of reduction as a graph along passes after the RPD correction.

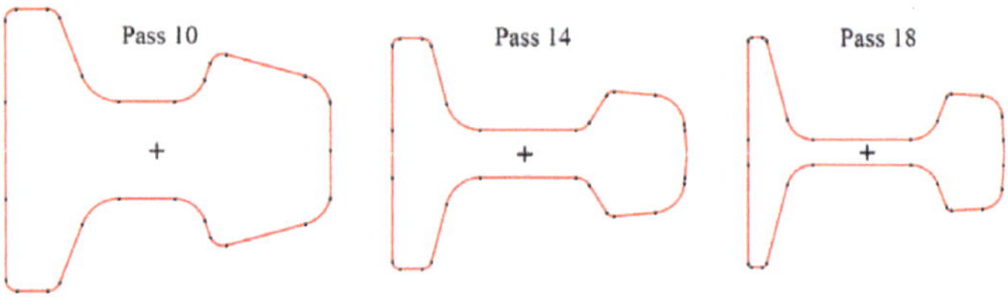

(a) (b) (c)

Fig. 14. The corrected RPD shapes: (a) leader, (b) one of the intermediate shapes, (c) final shape of the rail.

After refining the RPD, drive parameters were added from the rolling mill database, as well as all the necessary technological parameters, such as temperature, roll rotation speed, and the distance between stands. Integrating these technological parameters with the developed RPD makes it possible to simulate the rolling of a specified profile shape. Utilizing the API allows for the initiation of rolling simulation directly from the QKaliber software. The initial simulation was conducted solely to verify the rolling of the desired final rail profile from the selected leader profile shape, which served as a billet. Through finite element analysis, the simulation confirmed the effectiveness of the developed RPD (Fig. 15).

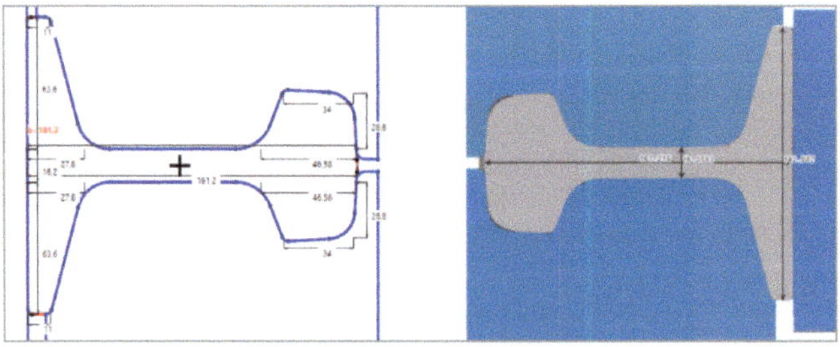

(a) (b)

Fig. 15. The finish pass design of the Rail 54E1 profile with dimensions (a) and results of its simulation (b).

A rectangular bloom with hot dimensions of 363x300 mm was selected as the workpiece, for which the first part of the RPD was prepared following the rolling direction on roughing stands to obtain the leader shape of the rail (Fig. 16). This was followed by verification using FE simulation, and finally, the entire project was simulated in full from the billet to the finished rail shape to prove the developed RPD.

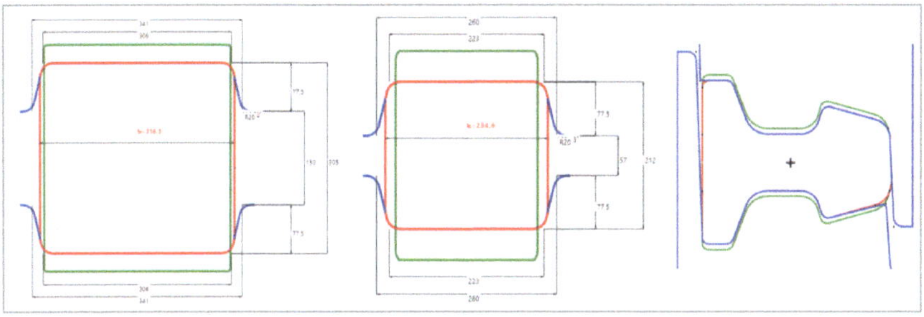

(a) (b) (c)

Fig. 16. Passes from the workpiece to the leader profile: (a) box groove in the first pass, (b) box groove in pass 5, (c) leader shape in pass 10.

Summary
This article presents an innovative automated roll pass design approach, seamlessly integrated with comprehensive FEM simulations. The system enables efficient management and creation of roll inventories and sequences. It offers significant enhancements, notably the streamlined design process for new profiles, optimisation of load and torque for existing designs within the limits of equipment capabilities, and the enhancing of new and established shape rolling technologies.

References

[1] B.P. Bahtinov, M.M. Shternov, Roll pass design; Metallurgizdat: Moscow, Soviet Union, 1953.

[2] Z. Wusatowski, GrundlagendesWalzens; VEBDeutscherVerlagfürGrundstoffindustrie: Leipzig, Germany, 1963.

[3] D. Matteo, Wicon EVO – a preview of Danieli's easy-to-use rolling process simulation software for wire rod and barmills. In Proceedings of the Rolling-12, Trieste, Italy, 25-28 October 2022.

[4] Weiner M.; Renzing, C; Stirl, M.; Schmidtchen, M., PyRolL. 2023. https://github.com/pyroll-project.

[5] S. Stebunov, N. Biba, A. Vlasov, A. Maximov, Thermally and Mechanically Coupled Simulation of Metal Forming Processes. In Proceedings of the 10th International Conference on Technology of Plasticity, Aachen, Germany, 25-30.09.2011.

[6] O.C. Zienkiewicz, Flow formulation for the numerical solution of metal forming processes. Numerical analyses of forming processes, London, UK, 1984.

[7] N.K. Gupta, Steel Rolling: Principle, Process & Application, London, UK, 2021. https://doi.org/10.1201/9781003182399

Metal Forming 2024
Materials Research Proceedings 44 (2024) 768-776

Materials Research Forum LLC
https://doi.org/10.21741/9781644903254-82

Advancing weathering steel through thermomechanical processing and simulation techniques

REFAIY Hoda[1,a *], EISSA Mamdouh[1,b], MATTAR Taha[1,2,c] and EL-SHENAWY Eman[1,d]

[1]Central Metallurgical R&D Institute (CMRDI), Egypt

[2]Galala University, Egypt

[a]Eng_huda_refaiy@yahoo.com, [b]mamdouh.eissa@gmail.com, [c]tahamattar@yahoo.com, [d]dr.emanelshenawy@yahoo.com

Keywords: Thermomechanical Processing (TMP), Weathering Steel, Numerical Simulation, Physical Simulation, Corrosion Resistance Steel

Abstract. Thermomechanical processing (TMP) techniques are pivotal in fabricating and advancing weathering steel, a high-strength, low-alloy steel admired for its exceptional corrosion resistance and aesthetic appeal. Combined with numerical and physical simulation methods, these techniques optimize mechanical properties, microstructure, and overall weathering steel performance. Selecting an appropriate steel composition is crucial before applying TMP techniques. Alloying elements like copper, chromium, nickel, and phosphorus are carefully incorporated to enhance corrosion resistance. Once the steel composition is determined, numerical simulations are employed to model and predict the behavior of weathering steel during processing. Hot rolling is the first stage of thermomechanical processing in which weathering steel is heated and deformed in rolling mills. Additionally, numerical simulations can be employed to study other heat treatment steps, normalization and stress relieving, to optimize their parameters and impact on weathering steel's microstructure and mechanical properties. In conjunction with thermomechanical processing techniques, physical simulations, accelerated corrosion testing, and environmental exposure, help assess weathering steel's long-term performance and durability. These techniques provide valuable data on the formation and behaviour of protective patina, contributing to the steel's corrosion resistance. In summary, combining thermomechanical processing techniques and numerical and physical simulations enables the precise control and optimization of weathering steel's mechanical properties, microstructure, and corrosion resistance. These advancements contribute to the development and application of weathering steel in various industries, where its unique combination of mechanical performance, aesthetic appeal, and long-term durability is highly valued.

Introduction

Weathering steel is a class of low-carbon steel alloys that may create stable protective rust coatings when exposed to outdoor elements such as rain, wind, ice, fog, and so on for several years. Rust layers form in the outer layer of the steel as a protective shield that prevents future corrosion. This beneficial corrosion process gives the material a generally appealing appearance, with a range of colours ranging from lighter browns to darker browns with probable purple tints, and helps to reduce maintenance requirements [1]. The main applications for weathering steels include civil structures such as bridges and other load-bearing structures, road installations, electricity poles, transmission towers, guide rails, ornamental sculptures, facades, and roofing [1][2][3].

Weathering steel alloys have a carbon concentration of less than 0.2% wt, with other alloying elements such as copper (Cu), chromium (Cr), nickel (Ni), phosphorus (P), silicon (Si), and manganese (Mn) accounting for a small percentage of total weathering steel alloys [1][4]. Patina refers to the protective rust layer produced on the outer surface of weathering steel [5]. The patina forming process is aided by frequent wetting and drying cycles on the surface of the weathering steel. It typically takes two to six years for the patina to fully develop. The rate of patina production and the ultimate hue of weathering can be affected by climatic conditions and surrounding air quality, such as moisture and pollution levels [1].

Recently, improving the properties of weathering steel and other grades is important in order to participate in achieving the sustainability requirements and solving environmental problems. Improving the mechanical properties of weathering steel, such as strength and stiffness, allows for the design of lighter and more efficient structures. This can lead to reduced material consumption, and increased energy efficiency during construction. By enhancing the mechanical properties of weathering steel, structures can be designed to withstand higher loads and stresses. This means that less steel is needed to achieve the required strength and performance, resulting in reduced resource consumption and lower environmental impact. Additionally, Stronger steel with enhanced mechanical properties improves the safety and resilience of structures, especially in critical applications such as bridges, high-rise buildings, and infrastructure exposed to extreme conditions. This ensures that structures can withstand natural disasters, reducing the need for reconstruction and the associated environmental impacts.

Simulation and Thermomechanical controlled processing of weathering steel

Thermomechanical properties play a crucial role in enhancing the properties of weathering steel. Thermomechanical Controlled Processing (TMCP) refers to the physical and thermal treatments applied to steel during manufacture. It precisely controls the temperature, strain, and strain rate during hot rolling of steel. Following controlled rolling, the steel is fully chilled using air cooling or accelerated cooling methods. The cooling pace and temperature profile have a considerable impact on the steel's ultimate microstructure and properties [6]. By subjecting steel to controlled heating and cooling processes, its microstructure and mechanical properties can be modified, resulting in improved performance [7][8]. Thermomechanical processing techniques, such as hot rolling, controlled cooling, and quenching and tempering, can increase the strength and toughness of steel. These processes refine the microstructure, promote the formation of fine and homogeneous grains, and induce the precipitation of strengthening phases, which leading to improved mechanical properties [9][10]. Proper thermomechanical processing can enhance the ductility and formability of steel. Techniques like annealing, hot working, and controlled rolling can refine the grain structure, reduce residual stresses, and improve the steel's ability to deform without fracture [11]. The manipulation of thermomechanical properties allows for the customization and optimization of steel's mechanical behavior, resulting in improved strength, toughness, formability, weldability, and resistance to fatigue and creep. These enhanced properties contribute to the development of more efficient and sustainable steel products for various applications across industries [3][12]. For industrial sectors, Sheet and plate manufacture uses TMCP with micro alloys to balance strength, toughness, and weldability through grain refinement. [12].

Any adjustment in process parameters or chemistry for improving weathering steel properties necessitates a detailed understanding of material behaviour. Furthermore, if done on a large scale, it is a time-consuming and expensive process. Sometimes it is not even feasible or economical. Furthermore, reduced study flexibility leads to fewer trials being conducted. This causes slow or constrained development, resulting in a lengthy product development cycle that is unprofitable or technologically infeasible. Hence, Numerical and physical simulation of thermomechanical processing plays a crucial role in the steel industry by providing valuable insights and benefits [6]. Numerical simulation software for thermomechanical processing offers significant advantages, including process optimization, predictive capabilities, resource and time savings, troubleshooting, material design, environmental impact assessment, and educational benefits. By leveraging these tools, the steel industry can enhance its efficiency, productivity, and sustainability while accelerating innovation and product development. Numerical Simulation software's allow to optimize thermomechanical processes by analyzing various parameters, such as temperature, strain rate, cooling rate, and deformation. It enables the exploration of different process conditions and helps identify the optimal combination of parameters to achieve desired material properties and performance. This optimization can lead to improved efficiency, reduced costs, and enhanced product quality. Moreover, they help in understanding the effects of thermal and mechanical treatments on the microstructure, phase transformations, and mechanical properties of steel. They provide predictive capabilities, allowing the assessment of the material's behavior under different processing conditions. This predictive capability aids in making informed decisions regarding process design and material selection. By using simulation software, the need for physical experimentation and trial-and-error approaches are reduced. Instead, they can virtually test different process scenarios and evaluate their outcomes, saving resources and time. This accelerates the development and optimization of thermomechanical processes, enabling faster time-to-market for new products. Additionally, by using numerical and physical simulations of TMCP in weathering production, Manufacturers can optimize the manufacturing processes and attain the required control to achieve weathering steel with predictable and dependable performance. This is a significant benefit that encourages the broad usage and application of weathering steel in a variety of industries.

The finite element technique (FEM) was used to simulate stress-strain curves of a duplex weathering steel. This article discussed effective aspects in duplex weathering steel, including the volume fraction of martensite and the yield stress ratio of martensite to ferrite. The simulated and measured findings are nearly identical, indicating the accuracy of the FEM developed in this study. The volume proportion of martensite clearly affected the plastic part of the stress-strain curve, rather than the elastic part. As the volume fraction of martensite decreases, the strain increases significantly. Furthermore, the yield stress ratio of duplex weathering steel clearly affected the stress-strain curve, rather than just the elastic component. The volume fraction of martensite has a greater impact on the stress-strain curve of duplex weathering steel than the volume fraction of martensite itself [13].

Physical simulation of TMCP reduces the cost and duration of studies by simulating real-world processes in the lab. It can also provide useful data for the creation of new goods, the optimization of process parameters, and so on [6]. The Gleeble machine is an outstanding integrated digital closed-loop control system that is used for physical simulation of TMCP. It can heat specimens at up to 10,000 C/s, maintain steady-state equilibrium temperatures, and cool at high rates (10,000 C/s). The Gleeble machine uses an infrared pyrometer to provide accurate feedback on specimen temperatures. Gleeble systems can perform thermal testing 3-10 times faster than traditional furnace-equipped devices. Simulation of the Gleeble machine results in complex microstructural changes caused by thermal, mechanical, and metallurgical impacts. The Gleeble machine provides researchers with insights into many stages of deformation, including dynamic recovery, dynamic recrystallization, and static recrystallization, by measuring mean flow stress versus deformation temperature [6][8].

Recently, 420 MPa grade weathering bridge steels were developed from Si Bearing steel alloys at various percentages (0.15-0.77 wt.%)[14] by applying TMCP which was simulated using a thermo-mechanical simulator. The TMCP included two stages of controlled rolling and an accelerated cooling process, for four experimental steels with varied Si concentrations (see Fig. 1). The study found that increasing the Si concentration in TMCP steel improved both yield and tensile strength.

Materials Research Forum LLC
https://doi.org/10.21741/9781644903254-82

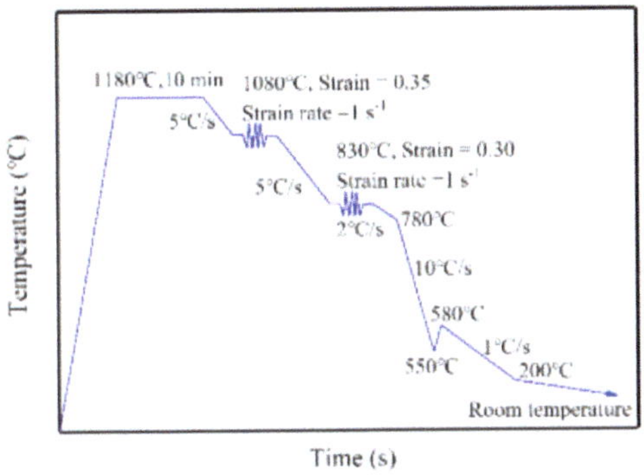

Fig.1. TMCP cycle of studied Si bearing weathering steel [14].

A novel Cu-P-Cr-Ni-Mo weathering steel was compressed using Gleeble 3500 to determine the mechanism of ferrite grain refining in the deformation temperature range from 750°C to 925°C [15]. The microstructure of the studied specimens revealed that the ferrite grain size decreased while the ferrite volume fraction rose with increasing compression temperature. Electron backscatter diffraction patterns indicated that some low-angle boundaries shifted to the high-angle boundary, resulting in fine ferrite grains surrounded by high-angle boundaries. At 750°C, several low-angle boundaries were found inside ferrite grains, indicating the presence of pre-eutectoid ferrite. Fine equiaxed ferrite grains with size ranging from 1.77 μm to 2.69 μm were produced in the $(\alpha + \gamma)$ dual phase region. As shown in Fig. 2, at temperatures ranging from 750°C to 775°C, ferrite grain refining may be caused by dynamic recrystallization, while at 800°C and 850°C, deformation-induced ferrite transformation is the primary mechanism. The flow curves produced at 875°C, 900°C, and 925°C demonstrated work hardening. Flow stress increased constantly as strain increased, and work hardening had a larger apparent peak stress than dynamic softening. Work hardening ability also decreased as the deformation temperature increased. This phenomenon could be related to the following factors: high distorted internal energy and less stable austenite, as well as the numerous nucleation sites offered by defects.

The Gleeble 3500 thermo-mechanical simulator was used for studying hot ductility behavior of 800 MPa weathering steel through tensile testing at different high temperatures. The reduction in area (RA) and tensile strength (TS) were acquired to draw hot ductility curve and hot strength curve. The results show that the third brittle zone of test steel is between 700°C and 800°C and which is mainly related to the formation of film-like ferrite along the prior austenite grain boundary and the precipitation of second phase. Moreover, the drop of hot ductility at 900°C is rooted in the reduction of grain boundary strength owing to the precipitation of sulfides. Therefore, the findings concluded that the straightening temperature of test steel should be kept over 900°C [16].

Materials Research Forum LLC
https://doi.org/10.21741/9781644903254-82

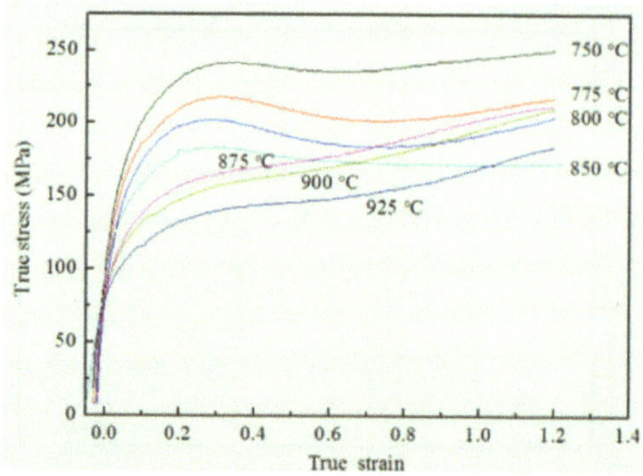

Fig. 2. Flow stress strain curves for different temperatures at strain rate of 0.1 s^{-1} [15].

Another research explored the microstructure evolution of Cu-P-Cr-Ni-Mo weathering steel after many passes of deformation (see Fig. 3) via hot compression simulation. At multi-pass deformation conditions, ultrafine ferrite grains of 1.8 µm were seen in the ($\alpha + \gamma$) area. The second phase of bainite and martensite was dispersed as a band on the ferrite matrix, becoming thinner with further passes. Microstructural analysis revealed that strain-enhanced ferrite transformation is the primary mechanism for structure refinement during initial pass deformation, while continuously dynamic recrystallization is the primary mechanism for ferrite grain refinement during later passes [17].

Evolution of weathering steel microstructure and characteristics

Because of their chemical composition, weathering steels are considered a popular alternative for applications requiring excellent atmospheric corrosion resistance. However, the decreased formability of weathering steels remains an issue that needs to be studied. Thus, Dual Phase (DP) weathering steel has been proposed as a steel grade that combines the better formability of DP steels and the strong atmospheric corrosion resistance of weathering steels [18]. The inter-critical annealed DP weathering steel has not only excellent corrosion resistance, but also a low yielding stress, a high elongation value, and a smooth flow-stress curve with a high strain-hardening coefficient. However, the intercritical annealing method has certain drawbacks, including a significant investment in equipment, high energy consumption, and low manufacturing efficiency and quality. Based on these parameters, development of directly hot-rolled DP weathering steel is required [19].

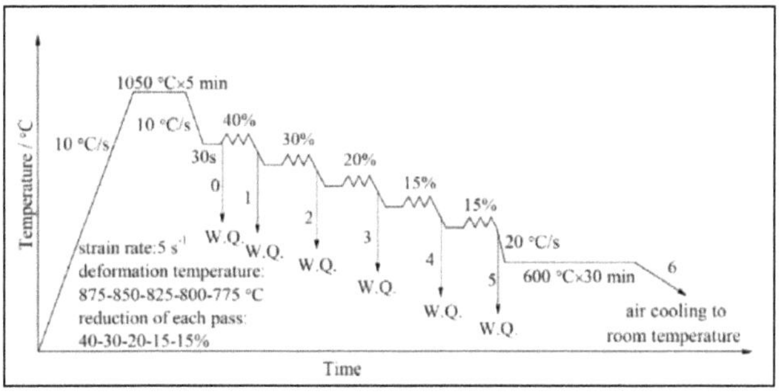

Fig. 3. Multi-pass deformation of the investigated weathering steel [17].

A prior study looked into developing novel Cu-P-Cr-Ni-Mo hot-rolled dual-phase weathering steels with a microstructure composed of ferrite-martensite phases. The Gleeble-3500 hot simulator was used to recreate the hot rolling process which shown in Fig. 4. The created hot-rolled dual-phase weathering steel containing 0.41% Mo with ferrite and martensite in its microstructure demonstrated superior comprehensive mechanical characteristics and formability compared to commercial weathering steel 09CuPCrNi[20].

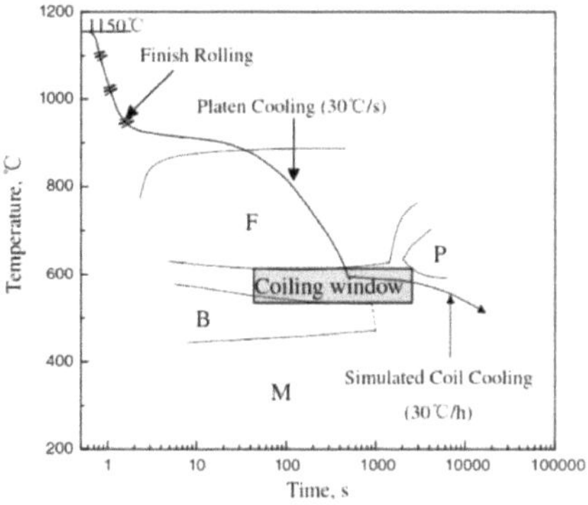

Fig. 4. Simulation of hot rolling processing cycles for the studied DP weathering steel.

Furthermore, electrochemical potentiodynamic experiments and simulated salt spray tests were used to investigate how dual-phase treatment affects corrosion resistance of the previously investigated 09CuPCrNi steels. The corrosion behaviour of DP780 and the as-received weathering steel is similar in 5% sodium hydroxide, 3.5% sodium chloride, 5% sulphuric acid, and different concentrations of hydrochloric acid solutions. However, dual-phase steels exhibit lower corrosion currents than the as-received steel. The rust layers of as-received steel and DP780 are similar, consisting primarily of three iron oxides (goethite, akaganeite, and lepidocrocite) and magnetite. The number and form of phase constituents impact the corrosion behaviour of DP steels [21].

The impact of microstructure evolution on the mechanical properties of a 500 MPa-grade weathering steel produced by an identical thermo-mechanical control process and different tempering treatments at 450-650 C was thoroughly investigated (See Fig.5)[22]. It is noticed from the results that the as-rolled steel has granular bainitic ferrite (GBF) and martensite/austenite (M/A) components, resulting in unsatisfactory mechanical characteristics. As tempering

Materials Research Proceedings 44 (2024) 768-776
https://doi.org/10.21741/9781644903254-82

temperature increased from 450°C to 650°C, the massive twin-type M/A constituents dissolved preferentially into carbides, with stabilization of only a minor amount of fine lath-type M/A constituents. Merging the bainitic ferrite laths resulted in improved matrix recovery, reduced dislocation density, and enhanced high-angle grain boundary (HAGB) proportion. The simultaneous precipitation of nanoscale particles (Ti, Nb) C and ε-Cu resulted in increased yield strength. M/A ingredient breakdown reduces strain hardening capacity and tensile strength. Tempering at 450-600°C reduced twin-type M/A components and increased HAGBs, whereas tempering at 650°C caused necklace-like M23C6 carbides to precipitate at grain boundaries, reducing impact toughness. Additionally, tempering at 550-600°C resulted in an optimal mechanical property combination [22].

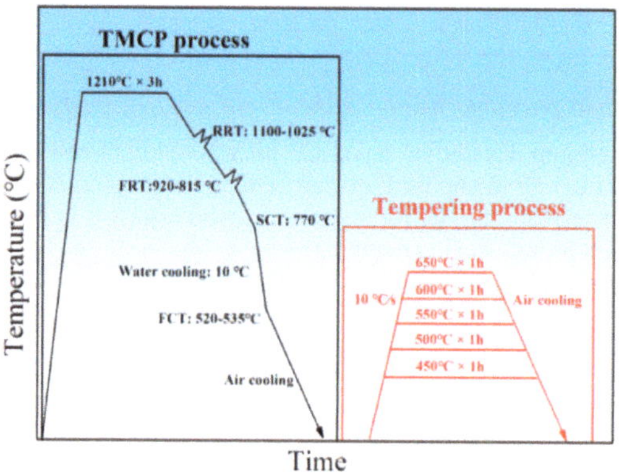

Fig. 5. The simulated TMCP of martensite/austenite 500 MPa weathering steel. RRT: rough rolling temperature, FRT: finish rolling temperature, SCT: start cooling temperature and FCT: finish cooling temperature [22].

Conclusion

- Weathering steel is a low alloy steel that combines simple carbon steel with minor amounts of Cu, Ni, Cr, and other alloy metals. This material offers superior toughness, ductility, and corrosion resistance compared to standard carbon steel. Weathering steel is commonly utilized in structures such as rails, automobiles, bridges, and towers that require long-term exposure.
- Thermomechanical processing are critical for improving the properties of weathering steel. Steel's microstructure and mechanical properties can be adjusted through controlled heating and cooling procedures, leading to increased performance. [6][7]. Hot rolling, controlled cooling, and quenching and tempering are examples of thermomechanical processing processes that can increase steel's strength and toughness. These processes enhance the microstructure, stimulate the development of fine and uniform grains, and induce the precipitation of strengthening phases, resulting in increased mechanical characteristics.
- Numerical simulation software for thermomechanical processing of weathering steel provides numerous benefits which promote optimization and analyzing processing parameters like temperature, strain rate, cooling rate, and deformation. It allows to find the best combination of parameters to obtain the desired material qualities and performance.
- Physical simulation of TMCP minimizes study expense and duration by imitating real-world processes in the lab. It can also provide useful data for new product development, process

parameter optimization, and other purposes. Different investigations utilized the thermo-mechanical simulator such as Gleeble machine to different grades of weathering steel via simulating thermomechanical process. Through these studies, they were able to develop different grades of steel with a maximum tensile strength value exceeded 500 MPa as well as determining effect of different TMCP on the microstructure and mechanical properties of weathering steel at low cost and time.

- However, the good corrosion resistance and tensile properties of weathering steels with Ferrite-Pearlite in its microstructure, they are characterized by lack of formability. Consequently, different researches focused on evolving dual phase Ferrite- Martensite weathering steel and via simulation of TMCP. They managed to enhance the tensile properties and formability of weathering steel without affecting its corrosion resistance.

References

[1] N.T.V.D.Viaducn, KTH Royal Institute of Technology, 2022.

[2] M. Morcillo, I. Díaz, B. Chico, H. Cano, D. de la Fuente, Weathering steels: From empirical development to scientific design. A review, Corros. Sci. 83 (2014) 6–31. https://doi.org/10.1016/j.corsci.2014.03.006

[3] S. Zeng, D. Xu, J. Chang, F. He, C. Sun, L. Zhang, Development of hot rolled weathering angle steel for transmission towers, Eng. Res. Express 1 (2019) 025002. https://doi.org/10.1088/2631-8695/ab4787

[4] European Committee for Standardization. EN 10155, 2004.

[5] M. Morcillo, B. Chico, I. Díaz, H. Cano, D. de la Fuente, Atmospheric corrosion data of weathering steels. A review, Corros. Sci. 77 (2013) 6–24. https://doi.org/10.1016/j.corsci.2013.08.021

[6] E.H. El-Shenawy, Physical simulation technology for thermo-mechanical processing of metallic alloys using Gleeble system, Materials Today: Proceedings 28 (2020) 998-1004. https://doi.org/10.1016/j.matpr.2019.12.339

[7] S.N. Prasad, D.S. Sarma, Influence of thermomechanical treatment on microstructure and mechanical properties of Nb bearing weather resistant steel, Mater. Sci. Eng. A 408 (2005) 53–63. https://doi.org/10.1016/j.msea.2005.06.018

[8] T. El-Bitar, E. El-Shenawy, M. El-Meligy, Physical Simulation of Thermo-Mechanical Processing of Ferritic-Bainitic Dual Phase (FBDP) Steel, Mater. Sci. Forum 879 (2017) 495–501. https://doi.org/10.4028/www.scientific.net/MSF.879.495

[9] E. El-Shenawy, R. Reda, Optimization of TMCP strategy for microstructure refinement and flow-productivity characteristics enhancement of low carbon steel, J. Mater. Res. Tech. 8 (2019) 2819–2831. https://doi.org/10.1016/j.jmrt.2019.04.021

[10] L. ying Song, X. hua Gao, Q. he Xue, J. bo Li, H. yan Wu, Effect of Cooling Rate and Coiling Temperature on Microstructure and Precipitation Behavior of a 700 MPa Weathering Steel, J. Mater. Eng. Perform. 31 (2022) 10225–10236. https://doi.org/10.1007/s11665-022-07034-6

[11] B. Buchmayr, Thermomechanical Treatment of Steels – A Real Disruptive Technology Since Decades, Steel Res. Int. 88 (2017) 1–14. https://doi.org/10.1002/srin.201700182

[12] K. Nishioka, K. Ichikawa, Progress in thermomechanical control of steel plates and their commercialization, Sci. Tech. Adv. Mater. 13 (2012). https://doi.org/10.1088/1468-6996/13/2/023001

[13] B. Liao, C.L. Zhang, J. Wu, D. Cai, C. Zhao, X. Ren, Q. Yang, Numerical simulation of the stress–strain curve of duplex weathering steel, Mater. Design 29 (2008) 562–567. https://doi.org/10.1016/j.matdes.2006.12.021

[14] Z. Chen, G. Shi, J. Qi, L. Zhao, Y. Liu, Q. Wang, G. Wang, Influence of Si Content on the Microstructure and Tensile Properties of Weathering Bridge Steel Produced via Thermal Mechanical Control Process, Metals 12 (2022). https://doi.org/10.3390/met12111901

[15] C. Zhang, M. Zhang, T. Guo, J. Yang, Y. Kong, D. Cai, Q. Li, Ferrite grain refinement in low carbon Cu–P–Cr–Ni–Mo weathering steel at various temperatures in the ($\alpha + \gamma$) region, Mater. Charact. 113 (2016) 10–16. https://doi.org/10.1016/j.matchar.2015.12.033

[16] C. Bin Huang, Y. Zhu, R. Wu, Q. Li, M. Zhang, Y. Zhong, L. Li, Research on Hot Ductility Behavior of 800 MPa Grade Ultra High Strength Weathering Steel, Adv. Mater. Res. 815 (2013) 205–211. https://doi.org/10.4028/www.scientific.net/AMR.815.205

[17] C. Zhang, Y. Kong, M. Zhang, J. Yang, D. Cai, Microstructure Evolution of a New Dual-Phase Steel via Multi-Pass Compression Deformation in the ($\alpha + \gamma$) Region, Steel Res. Int. 87 (2016) 157–164. https://doi.org/10.1002/srin.201400526

[18] T. Allam, M. Abbas, Mechanical Properties, Formability, and Corrosion Behavior of Dual Phase Weathering Steels Developed by an Inter-Critical Annealing Treatment, Steel Res. Int. 86 (2015) 231–240. https://doi.org/10.1002/srin.201400033

[19] C. Zhang, D. Cai, B. Liao, Y. Fan, Using direct hot-rolling approach to obtain dual-phase weathering steel Cu–P–Cr–Ni–Mo, J. Mater. Sci. 45 (2010) 490–495. https://doi.org/10.1007/s10853-009-3966-x

[20] C. Zhang, B. Liao, D. Cai, T. Zhao, Y. Fan, Development of hot-rolled dual-phase weathering steel Cu–P–Cr–Ni–Mo, Mater. Design 28 (2007) 1760–1766. https://doi.org/10.1016/j.matdes.2006.05.011

[21] C.L. Zhang, D.Y. Cai, K.Q. Zhang, B. Liao, Effect of Dual-Phase Treatment on the Corrosion Behavior of Weathering Steel 09CuPCrNi, Adv. Mater. Res. 740 (2013) 586–590. https://doi.org/10.4028/www.scientific.net/AMR.740.586

[22] L. Zhao, Q. Wang, G. Shi, B. Hu, S. Wang, M. Qiao, Q. Wang, R. Liu, The impacts of M/A constituents decomposition and complex precipitation on mechanical properties of high-strength weathering steel subjected to tempering treatment, J. Mater. Res. Tech. 23 (2023) 2504–2526. https://doi.org/10.1016/j.jmrt.2023.01.136

Keyword Index

f